ENVIRONMENTAL ENGINEERING

Gerard Kiely

The McGraw-Hill Companies

London · New York · St Louis · San Francisco · Auckland
Bogotá · Caracas · Lisbon · Madrid · Mexico · Milan
Montreal · New Delhi · Panama · Paris · San Juan
São Paulo · Singapore · Sydney · Tokyo · Toronto

Published by
McGraw-Hill Publishing Company
Shoppenhangers Road, Maidenhead, Berkshire, SL6 2QL, England
Telephone 01628 23432
Facsimile 01628 770224

British Library Cataloguing in Publication Data
Kiely, Gerard
  Environmental Engineering
  I. Title
  628
  ISBN 0-07-709127-2

Library of Congress Cataloging-in-Publication Data
Kiely, Ger.
    Environmental engineering / Gerard Kiely.
      p.   cm.
    ISBN 0-07-709127-2 (pbk. : alk. paper)
    1. Environmental engineering. I. Title.
  TD146.K54   1996
  628—dc20   95-30426 CIP

McGraw-Hill

A Division of The McGraw·Hill Companies

345 QC 9987

Printed and bound in the United States of America

CONTENTS

PREFACE   xiii

ACKNOWLEDGEMENTS   xix

LIST OF CONTRIBUTORS   xx

PART ONE   ESSENTIAL BACKGROUND TO ENVIRONMENTAL ENGINEERING   1

CHAPTER 1   HISTORY AND LEGAL FRAMEWORK   3
    1.1   Historical introduction to the water and wastewater environment   3
    1.2   Historical introduction to the air environment   4
    1.3   Engineering, ethics and the environment   7
    1.4   EU and US environmental law   8
    1.5   Evolution of EU environmental legislation   9
    1.6   Some relevant international environmental agreements   11
    1.7   EU environmental legislation   11
    1.8   US environmental legislation   19
    1.9   Comparison of EU and US environmental legislation   26
    1.10 Problems   27
    References and further reading   28

CHAPTER 2   ECOLOGICAL CONCEPTS AND NATURAL RESOURCES   29
    Paul S. Giller, Alan A. Myers and John O'Halloran
    2.1   Introduction to the ecological perspective   29
    2.2   The value of the environment   30
    2.3   Levels of organization in the biotic component of the environment   34
    2.4   Ecosystem processes   36
    2.5   The human dimension   43
    2.6   Environmental gradients, tolerance and adaptation   44
    2.7   Environmental changes and threats to the environment   48
    2.8   Problems   50
    References and further reading   51

* Unless otherwise stated, chapters are by Gerard Kiely.

v

CHAPTER 3 INTRODUCTION TO CHEMISTRY AND MICROBIOLOGY IN ENVIRONMENTAL ENGINEERING   52
  3.1 Introduction   52
  3.2 Physical and chemical properties of water   57
  3.3 Atmospheric chemistry   95
  3.4 Soil chemistry   102
  3.5 Microbiology   107
  3.6 Chemical and biochemical reactions   124
  3.7 Material balances and reactor configurations   129
  3.8 Problems   142
  References and further reading   144

CHAPTER 4 CONCEPTS IN HYDROLOGY   146
  4.1 Introduction   146
  4.2 Hydrological cycle   147
  4.3 Water balance   150
  4.4 Energy budget   152
  4.5 Precipitation   154
  4.6 Infiltration   163
  4.7 Evaporation and evapotranspiration   174
  4.8 Rainfall–runoff relationships   181
  4.9 Hydrologic instrumentation   189
  4.10 Flood flows   191
  4.11 Low flows   195
  4.12 Urban hydrology   198
  4.13 Groundwater   200
  4.14 Groundwater chemistry, contamination and pollution prevention 214
  4.15 Problems   221
  References and further reading   224

PART TWO POLLUTION ENVIRONMENTS   229

CHAPTER 5 ECOLOGICAL SYSTEMS, DISTURBANCES AND POLLUTION   231
  Paul S. Giller, Alan A. Myers and John O'Halloran
  5.1 Introduction   231
  5.2 The freshwater environment   233
  5.3 Marine systems   245
  5.4 Terrestrial ecosystems   252
  5.5 Ecological systems and pollution   255
  5.6 Problems   260
  References and further reading   260

CHAPTER 6 WATER POLLUTION: ECOLOGICAL PERSPECTIVES   263
  Paul S. Giller, Alan A. Myers and John O'Halloran
  6.1 Introduction   263
  6.2 Water quality standards and parameters   265
  6.3 Assessment of water quality   267
  6.4 Aquatic pollutants   276
  6.5 Freshwater pollution   276

6.6	Estuarine water quality	288
6.7	Marine pollution	292
6.8	Problems	299
	References and further reading	300

CHAPTER 7 WATER QUALITY IN RIVERS AND LAKES: PHYSICAL PROCESSES 303

7.1	Introduction	303
7.2	Parameters of organic content of water quality	304
7.3	Dissolved oxygen and biochemical oxygen demand in strcams	309
7.4	Tranformation processes in water bodies	311
7.5	Transport processes in water bodies	318
7.6	Oxygen transfer by interphase transfer in water bodies	319
7.7	Turbulent mixing in rivers	322
7.8	Water quality in lakes and reservoirs	324
7.9	Groundwater quality	328
7.10	Problems	330
	References and further reading	332

CHAPTER 8 AIR POLLUTION 334

8.1	Introduction	334
8.2	Air pollution system	335
8.3	Air pollutants	337
8.4	Criteria pollutants	340
8.5	Acid deposition	358
8.6	Global climate change—greenhouse gases	358
8.7	Non-criteria pollutants	362
8.8	Emission standards from industrial sources	362
8.9	Air pollution meteorology	366
8.10	Atmospheric dispersion	374
8.11	Problems	387
	References and further reading	388

CHAPTER 9 NOISE POLLUTION 390
Donncha O'Cinnéide

9.1	Introduction	390
9.2	Physical properties of sound	392
9.3	Noise and people	398
9.4	Noise criteria	401
9.5	Noise standards	404
9.6	Noise measurement	404
9.7	Outdoor propagation of sound	409
9.8	Noise contours	415
9.9	Noise section of an environmental impact assessment	416
9.10	Noise control	417
9.11	Problems	418
	References and further reading	419

CHAPTER 10 AGRICULTURAL POLLUTION   420
    Bill Magette and Owen Carton
    10.1 Introduction   420
    10.2 Nutrient cycles in agricultural systems   421
    10.3 Soil physical and chemical properties   424
    10.4 Waste production on farms   427
    10.5 Pollution potential of farm wastes   428
    10.6 Nutrient losses   429
    10.7 Other wastes and potential pollutants   431
    10.8 Legislation (EU)   432
    10.9 Summary   433
    10.10 Problems   433
    References and further reading   434

PART THREE   ENVIRONMENTAL ENGINEERING TECHNOLOGIES   435

CHAPTER 11 WATER TREATMENT   437
    11.1 Introduction   437
    11.2 Amount of water required   438
    11.3 Water quality standards   439
    11.4 Water sources and their water quality   443
    11.5 Water treatment processes   446
    11.6 Pre-treatment of water   449
    11.7 Sedimentation, coagulation and flocculation   451
    11.8 Filtration   465
    11.9 Disinfection   471
    11.10 Fluoridation   478
    11.11 Advanced water treatment processes   479
    11.12 US primary drinking water standards   487
    11.13 Problems   491
    References and further reading   492

CHAPTER 12 WASTEWATER TREATMENT   493
    12.1 Introduction   493
    12.2 Wastewater flow rates and characteristics   495
    12.3 Design of wastewater network   506
    12.4 Wastewater treatment processes   508
    12.5 Wastewater pre-treatment   511
    12.6 Primary treatment   519
    12.7 Secondary treatment   524
    12.8 Activated sludge systems   532
    12.9 Attached growth systems   540
    12.10 Nutrient removal   545
    12.11 Secondary clarification   550
    12.12 Advanced treatment processes   552
    12.13 Wastewater disinfection   555
    12.14 Diffusers for wastewater   556
    12.15 Problems   559
    References and further reading   561

CHAPTER 13 ANAEROBIC DIGESTION AND SLUDGE TREATMENT   563
     13.1   Introduction to anaerobic digestion   563
     13.2   Microbiology of anaerobic digestion   564
     13.3   Reactor configurations   565
     13.4   Methane production   570
     13.5   Applications of anaerobic digestion   573
     13.6   International regulations for biosolids   574
     13.7   Biosolids characteristics   576
     13.8   Processing routes for biosolids   584
     13.9   First stage treatment of sludge   584
     13.10 Second stage treatment of sludge   595
     13.11 Sludge disposal   606
     13.12 Integrated sewage sludge management   617
     13.13 Problems   617
     References and further reading   619

CHAPTER 14 SOLID WASTE TREATMENT   623
     Ejvind Mortensen and Gerard Kiely
     14.1   Introduction   623
     14.2   Sources, classification and composition of MSW   628
     14.3   Properties of MSW   635
     14.4   Separation   643
     14.5   Storage and transport of MSW   645
     14.6   MSW management   652
     14.7   Waste minimization of MSW   652
     14.8   Reuse and recycling of MSW fractions   652
     14.9   Biological MSW treatment   657
     14.10 Thermal treatment—combustion/incineration   665
     14.11 MSW landfill   673
     14.12 Integrated waste management   688
     14.13 Problems   688
     References and further reading   690

CHAPTER 15 HAZARDOUS WASTE TREATMENT   693
     Per Riemann
     15.1   Introduction   693
     15.2   Definition of hazardous waste   696
     15.3   Hazardous waste generation   699
     15.4   Medical hazardous waste   703
     15.5   Household hazardous waste   705
     15.6   Transportation of hazardous waste   705
     15.7   Hazardous waste treatment facility   712
     15.8   Planning a hazardous waste incinerator   716
     15.9   Planning an inorganic waste treatment plant   716
     15.10 Treatment systems for hazardous waste   719
     15.11 Handling of treatment plant residues   738
     15.12 Contaminated sites   742
     15.13 EU Hazardous Waste Directive (91/689/EEC) Annexes I, II, III   745
     15.14 Problems   748
     References and further reading   749

CHAPTER 16 INDUSTRIAL AIR EMISSIONS CONTROL 750
Sean Bowler
16.1 Introduction 750
16.2 Characterizing the air stream 751
16.3 Equipment selection 752
16.4 Equipment design 754
16.5 Special topics 773
16.6 Problems 778
References and further reading 779

CHAPTER 17 AGRICULTURAL POLLUTION CONTROL 781
Bill Magette and Owen Carton
17.1 Introduction 781
17.2 Obstacles to agricultural pollution control 781
17.3 Agricultural water pollution control principles 782
17.4 Point source controls 783
17.5 Non-point source (NPS) controls 787
17.6 Land application of wastes 787
17.7 Codes of practice for land application of animal and other wastes 794
17.8 Agricultural air pollution control 795
17.9 Problems 796
References and further reading 796

PART FOUR   ENVIRONMENTAL MANAGEMENT 799

CHAPTER 18 WASTE MINIMIZATION 801
Dermot Cunningham and Noel Duffy
18.1 Introduction 801
18.2 Life cycle assessment 802
18.3 Elements of a waste minimization strategy 809
18.4 Benefits of waste minimization 812
18.5 Elements of a waste minimization programme 815
18.6 Waste reduction techniques 820
18.7 Conclusion 826
18.8 Case study—Paint Industry (USEPA 1990) 826
18.9 Problems 828
References and further reading 830

CHAPTER 19 ENVIRONMENTAL IMPACT ASSESSMENT 831
Michael O'Sullivan
19.1 Introduction 831
19.2 Origins of EIA 832
19.3 EIA procedure 832
19.4 Project screening for EIA 835
19.5 Scope studies for EIS 838
19.6 Preparation of an EIS 838
19.7 Review of EIS 841
19.8 Multidisciplinary team management 841
19.9 Examples of project EIS 844

19.10 Case study   851
19.11 Problems   853
References and further reading   853
Appendix 19.1: project screening   854

CHAPTER 20 ENVIRONMENTAL IMPACT OF TRANSPORTATION   857
Donncha O'Cinnéide
20.1   Introduction   857
20.2   Transportation and development   858
20.3   Transportation planning   858
20.4   Matrix of environmental impact and transportation system stages   859
20.5   The environmental effects of roads and traffic   860
20.6   Vehicular impacts   861
20.7   Safety and capacity impacts   870
20.8   Roadway impacts   871
20.9   Construction impacts   874
20.10 The traffic generated by proposed developments   874
20.11 The environmental impact assessment of proposed road developments   876
20.12 Problems   876
References and further reading   877

CHAPTER 21 ENVIRONMENTAL MODELLING   878
21.1   Introduction   878
21.2   Mechanism of pollutant fate in the environment   880
21.3   Mathematics of mass transport: diffusion–advection   886
21.4   Population models and models of physical systems   893
21.5   Hydrodynamic modelling of rivers   903
21.6   Water quality modelling in riverine systems   909
21.7   Watershed modelling   918
21.8   Modelling water quality in estuaries   921
21.9   Modelling water quality in lakes and reservoirs   925
21.10 Groundwater modelling   927
21.11 Modelling of wastewater treatment: activated sludge   932
21.12 Fugacity modelling   932
21.13 Air quality modelling   936
21.14 Problems   943
References and further reading   945

GLOSSARY   947

APPENDICES A–D   **965**

INDEX   **969**

* Unless otherwise stated, chapters are by Gerard Kiely.

Humanity is now being forced to investigate the environmental consequences of its development actions, on a local, national and global scale. In the short time span since the industrial revolution, the face of this planet has been changed in many areas, sadly in some, irreversibly. Change was called progress, but now this generation, who are the beneficiaries of past progress, are also the inheritors of past environmental mistakes. The gains of the past will be retained and future progress will be attained, not based on the narrow forces of economics or engineering but on sustainable development. Somewhat an overused phrase, sustainable development is defined as 'the ability to meet the needs of the present without compromising the ability of future generations to meet their own needs'. The latter calls for a balanced use of resources. The evolution of the age of sustainable development will require radical changes for many professional disciplines as they are now known, but most particularly for engineering. Engineering now requires an ecological appreciation and a responsiveness to a public well educated in environmental conservation.

The engineering profession must include environmental protection in its brief if it is to retain public credibility. No longer can engineers design and construct projects without assessing their environmental impact on the environment. The onus of 'duty of care' now legislates for the developer or producer to be accountable for materials, waste or otherwise, from 'cradle to grave'. Engineers are now regularly called upon publicly to defend their proposals, sometimes losing, because of inadequate sensitivity of their proposals to impact on humans, flora and fauna. The democratization of the planning process with requirements of Environmental Impact Assessment necessitates that engineers not only be well versed in their own discipline but also be acquainted with and sensitive to the environment in a holistic way.

Engineers now work in multidisciplinary teams alongside ecologists, economists, sociologists, planners, environmentalists, lawyers and chemists. Environmental engineering and environmental science are modern disciplines (post Second World War) and have only found their way into general usage in the past few decades. Environmental engineering is defined by Peavy *et al.* (1985) as 'that branch of engineering that is concerned with protecting the environment from the potentially deleterious effects of human activity, protecting human populations from the effects of adverse environmental factors and improving environmental quality for human health and well being'. Figure P1.1 is a schematic of the author's present concept of the requirements of the academic education of an environmental engineer. Environmental engineering is closely associated with other branches of engineering, e.g. civil and chemical, and with the sciences of chemistry, physics and biology. In addition, environmental engineering is associated with subsets of the above, e.g. hydrology, meteorology and atmospheric science as subsets of physical science; water, air and soil chemistry as a subset of chemical science; microbiology and ecology as a subset of biological science. As such, environmental engineering may seem like 'all things to all men'. This, of course, identifies to some extent the fact that environmental engineering is not yet a mature engineering discipline with well-defined boundaries. It is an evolving branch of engineering and this contributes to its excitement as a profession. There is room in environmental engineering for students and practitioners with backgrounds differing from the traditional engineer.

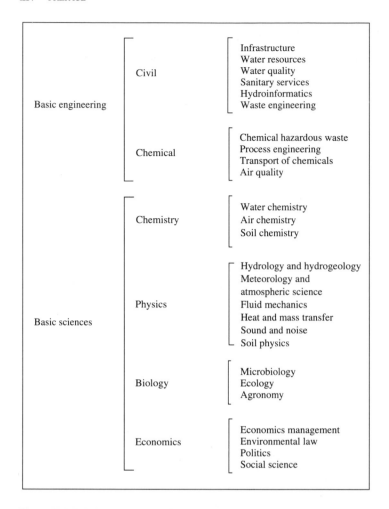

Figure Pl.1 Building blocks of environmental engineering.

Environmental engineering today addresses problems in the water, air and soil environments. Some consider that environmental engineering evolved out of sanitary engineering. The latter is an established subset of civil engineering. Civil engineering students have been taught sanitary (or public health) engineering courses for over 100 years. Traditional sanitary engineering addressed problems of water quality (drinking water and wastewater) and municipal solid waste disposal. This branch of engineering was sometimes called public health engineering or municipal engineering as it dealt with the services (water supply, wastewater and solid waste) that public municipalities (local authorities in Britain and Ireland) managed. In addition to the above, environmental engineering has evolved to address problems in the air and soil environment.

Civil engineers traditionally covered the areas of hydrology and water quality, water treatment and wastewater treatment. Chemical engineers, on the other hand, identified with chemical processes, industrial wastewater, hazardous waste and air pollution. Environmental engineering is considered in this text as having a wider brief and, as well as the core areas of water and air quality and water treatment, the following subjects are now considered essential in the education of an environmental engineer: ecology, microbiology, groundwater, solid waste, farm waste, noise pollution, environmental impact assessment, environmental legislation and environmental modelling and management.

Environmental engineers when trained in the above subjects will have a basis to adapt to the changing needs in the employment world. For instance, environmental engineers will be attracted to the area of waste management and reduction. This may mean wider employment opportunities not readily available to the traditional engineering disciplines. The urgency for solutions to waste treatment and disposal, be it in the solid, semi-solid, liquid or gaseous areas, will drive environmental engineers and environmental scientists to evolve innovative treatment and disposal technologies. As such, this field will not only create new employment opportunities, but will also stimulate many research opportunities.

Environmental engineering (within a multidisciplinary environment) as a recent development is exciting and the wealth of opportunities for environmental engineers in the wider arena, be it as project leaders or team players, is very challenging. It is now necessary that the environmental engineering/science undergraduate student be exposed to the non-engineering environmental disciplines (e.g. social science/law) during their academic years. They then would be not only capable of negotiating with non-engineering professionals but would be sensitive, in their own right, to other key players (e.g. the public) in the environmental debate.

Who This Book Is For

This book is designed to be used as an undergraduate textbook in an introductory course on environmental engineering. It is also broad enough to serve as a text in an elective subject of environmental engineering to allied engineering fields at undergraduate level. These may include: civil, chemical, mechanical, agricultural and food engineering. Undergraduate students in environmental science will find the wide range of material covered and its quantitative nature of benefit to them. The quantitative approach with a mathematical foundation and numerous examples and homework problems is essential to engineering and science education. Graduate students, new to the subject, will find the essentials for transition to environmental engineering within this book.

What This Book Contains

Part One of the book covers the legislative and scientific background, with chapters on environmental legislation, ecology, chemistry and microbiology, and hydrology. Chapter 1 opens with the background to environmental engineering and outlines the legislative details with which engineers are obliged to comply. Some of the landmarks in environmental history are presented. An overview of these requirements is presented for the different environments. The driving force for innovative technologies is sometimes environmental legislation. A brief history of the evolution of environmental legislation is included, with comparisons between environmental legislation in the European Union and the United States. The current status and detail of legislation is enumerated, indicating the international homogeneity in legislative compliance requirements in the 1990s.

Chapter 2, on ecosystems as a resource, is designed to introduce the engineering student to the need for sensitivity towards the ecological environment. Traditional engineering textbooks lacked ecology studies. The environmental engineer needs to be sensitive to the ecologist's view of the environment. Chapter 3 is an introduction to chemistry and microbiology. This chapter deals with the water, air and soil environment, including material balances, biological and chemical kinetics, and reaction engineering. The latter half of this chapter focuses on some microbiological aspects of water and wastewater treatment. Chapter 4 introduces concepts in hydrology. This includes the hydrological cycle and its components as well as the radiative energy balance. Rainfall/runoff relationships are introduced and include discussions of low flows and high flows in riverine systems. Groundwater, as a resource, is also treated.

Part Two contains six chapters on environmental pollution not always associated with civil or chemical engineering. These are ecological aspects of water pollution, physicochemical aspects of water pollution, agricultural pollution, noise and air pollution. Chapter 5 looks at ecological disturbances in the different environments. Chapter 6 provides an ecological perspective to water pollution, using biological parameters to assess pollution levels. The types of pollution are described and the response of the systems to the various types of pollution are discussed (freshwater and marine). Chapter 7 examines water quality from an engineering perspective, investigating BOD, DO and other parameters in rivers, lakes and estuaries using numerical tools to quantify water quality and pollution. Chapter 8 has in its introduction a brief quantitative outline of the environmental legislation requirements for the air environment. Criteria pollutants and their sources are discussed. Relevant aspects of meteorology and climatology are introduced and plume dispersion, including atmospheric dispersion, is examined. Chapter 9 introduces the topic of noise, its sources, in the outdoor and indoor environment. The impact that noise has on people and the calculation of noise magnitudes are covered. Traffic noise is discussed. Chapter 10 covers the topic of agricultural pollution. In the developed world, farm waste is arguably the most serious source of pollution yet to be addressed. The negative impact of farming activities is seen in the cultural eutrophication of freshwater systems, high levels of nutrients in groundwater systems and marine systems. The residues of agrichemicals are also being identified internationally in the water bodies. This chapter covers the areas of soil physics, farm practices and natural and artificial fertilizers.

Part Three is the core of the text, and traditional in its content. It contains eight chapters on water and waste treatment and disposal. These include water and wastewater treatment, sludge and solid waste treatment, hazardous waste treatment, air pollution control technologies and agricultural pollution control technologies. Chapter 11 is a traditional engineering style chapter on potable water treatment for municipal and industrial requirements. This chapter serves as an introduction to the topic covering unit operations and processes. The design aspects of municipal water treatment plant is also covered. Advanced water treatment processes for algae removal, iron and manganese removal are also addressed. Chapter 12 is on wastewater and sewage treatment for municipal, industrial and agricultural applications. It covers the unit processes, including physical, chemical and biological operations. Secondary treatment with nutrient removal is covered along with advanced wastewater treatment. Chapter 13 examines anaerobic digestion and sludge from wastewater treatment processes. The sludge composition, treatment and disposal are discussed with a view to ecologically sound disposal routes. A negative consequence of better wastewater treatment and better standards of liquid effluent is the generation of greater volumes of sludge. With the elimination of the sea disposal route and the tightening up of land spreading as a disposal route, innovative treatment and disposal methods are urgently required. Sludge and its fate are topical issues. By definition, 'clean sludge' and its use as a soil conditioner lead on to alternative uses, i.e. forest fertilizers, land reclamation, sacrificial land application and landscaping. This chapter examines such possibilities. Chapter 14 covers municipal solid waste in aspects from composition and quantity to landfilling, incineration, composting and anaerobic digestion. Recycling and waste reduction at source is addressed. Chapter 15 is an introduction to the treatment of hazardous waste, its definition, composition, quantification, transport, treatment and ultimate disposal. Aspects of the central hazardous waste treatment facility at Kommueikemi in Denmark are also discussed. An introduction to rehabilitation of contaminated sites is included. Chapter 16 examines industrial air pollution control technologies including those for particulate and gaseous emissions. Chapter 17 discusses, in a short chapter, methods of reducing agricultural pollution which are essentially good farmyard and agricultural practices.

Part Four introduces the areas of environmental management. These four chapters include waste minimization, impact assessment, impact of transportation and environmental modelling. Chapter 18 examines the area of waste minimization, life-cycle assessment and clean technology. A priority of environmental planning is waste minimization, particularly in relation to industry and its waste products, cither liquid, solid or gaseous. Advanced environmental policy has moved upstream from 'end of pipe' technologies to waste reduction at 'source'. Chapter 19 looks briefly at the historical development of the EIA process. The scoping of a project, the detailing of a study and, finally, the assessment are defined. Chapter 20 examines the impact of transportation systems on the environment and how to evaluate that impact. Chapter 21 introduces environmental modelling with aspects of kinetics, river and lake water quality and estuarine systems. Applied aspects of numerical modelling of the air environment is introduced. Specific areas of wastewater are also introduced. Software available (typically in the public domain) is discussed with views to applications.

How to Use This Book

Because of the extensive amount of material here, this book may be covered in a one-year or two semester course. Ideally, material in Part One (essential background) and Part Two (pollution environments—water, air, soil, etc.) would be covered in the first semester course. Parts One and Two serve as a solid introduction to the more engineering technical parts, Three and Four. Part Three (environmental engineering technologies) and Part Four (environmental management) could be covered in the second semester.

Reference

Peavy, H. S., D. R. Rowe and G. Tchobanoglous (1985) *Environmental Engineering*, McGraw-Hill, New York.

Gerard Kiely

ACKNOWLEDGEMENTS

This book has grown out of courses in environmental engineering and environmental science that I have been lucky enough to have been involved in at University College Cork, (UCC). These courses have been, *Introduction to Environmental Engineering* and *Advanced Wastewater and Solid Waste Management*, both at undergraduate level. At graduate level, I have also been priveleged to have coordinated a graduate Diploma in Environmental Engineering for the past four years. At UCC, I have had excellent cooperation from all my colleagues within the Civil and Environmental Engineering Department but particularly Dr. Eamon McKeogh, Dr. Donncha O'Cinnéide, Dr. Tony Lewis and Professor Philip O'Kane. Special thanks for their encouragement to Professor David Orr and Professor Eamonn Dillon. I have also drawn over the years from colleagues in other departments within UCC, but particularly; in the Zoology Department, Professor Paul S. Giller, Dr. John O'Halloran, Professor Alan A. Myers, Mr. Gerard Morgan and Professor Maire Mulcahy; in the Mathematical Physics Department, both Dr. Garrett Thomas and Dr. Jim Grannell; in the Food Microbiology Department, both Professor Kevin Collins and Dr. Alan Dobson. I would also like to thank my colleagues at the Cork Regional Technical College, Mr. Noel Duffy and Mr. Dermot Cunningham. From the consulting industry and government bodies within Ireland, there are several professionals who have contributed to the above mentioned coursework, namely: Mr. Tony Moloney of Malachy Walsh and Partners, Cork; Mr. Donal Daly of the Geologic Survey Ireland; Mr. John Aherne, Mr. Fred Willis, Mr. Brian Russell, Cork County Council; Mr. Richard Bryne, Kruger Engineering Ltd, Kilkenny, Ireland; Mr. Tadg O'Flaherty, Forbairt, Dublin; Mr. Gerard O'Sullivan and Ms Denise Barnett of E. G. Pettit's, Consultants, Cork; Dr. Gabriel Dennison of Dublin County Council; Mr. Michael O'Sullivan of RPS, Ireland; Mr. Sean Bowler of Cara Partners, Cork; Dr. Owen Carton of Teagasc, Wexford, Ireland; Mr. Donal Cronin of Weston FTA Consultants, Cork.

I would also like to thank: Dr. Robert Matthews of Bradford University; Mr. Liam Cashman of the European Commission in Brussels; Mr. Per Riemann of Chemcontrol A/S, Denmark; Mr. Ejvind Mortensen of Reno Sam Consultants, Denmark; Professor Bill Magette at the University of Maryland, College Park, MD.; Professor Ed Schroeder; Professor George Tchobanoglous and Dr. Masoud Kayhanian of the Civil and Environmental Engineering Department at the University of California at Davis. A very special thank you to Professor Mel Ramey and Professor Marc Parlange for allowing me to work at UC Davis as a Fulbright Fellow during the academic year 1993–94. And thanks to my friend Dr. John Albertson at UC Davis. Other Davis colleagues I would like to thank are Professor Ken Tanji and Mr. Tony Cahill. Finally, I would like to thank Mr. Charles Dolan for proofreading the manuscript and Ms. Marcia D'Alton for the glossary and also for proofreading. Thanks are also due to the staff at McGraw-Hill UK, for their patience and unfailing co-operation, particularly David Crowther, Camilla Myers, Rosalind Comer and Matthew Flynn.

I would like to acknowledge the following people who contributed chapters to this book:

• Professor Paul S. Giller, Professor Alan A. Myers, and Dr. John O'Halloran of the Zoology Department, University College Cork, Ireland for Chapters 2, 5 and 6.

• Dr. Donncha O'Cinnéide of the Civil and Environmental Engineering Department, University College Cork, Ireland for Chapters 9 and 20.

• Dr. Owen Carton of Teagasc, Johnstown Castle, Wexford, Ireland and Professor Bill Magette of Natural Resources and Environmental Engineering Department, University of Maryland, College Park, MD, USA, for Chapters 10 and 17.

• Mr. Ejvind Mortensen of Reno Sam, Association of Joint Municipal Waste Handling Companies, Roskilde, Denmark for Chapter 14.

• Mr. Per Reimann of Chemcontrol A/S, Nyborg, Denmark for Chapter 15.

• Mr. Sean Bowler of Cara Partners, Cork, Ireland for Chapter 16.

• Mr. Noel Duffy and Mr. Dermot Cunningham of the Clean Technology Centre, Cork Regional Technical College, Cork, Ireland for Chapter 18.

• Mr. Michael O'Sullivan of RPS Group Ltd, Cork, Ireland for Chapter 19.

Gerard Kiely

ONE

ESSENTIAL BACKGROUND TO ENVIRONMENTAL ENGINEERING

HISTORY AND LEGAL FRAMEWORK

1.1 HISTORICAL INTRODUCTION TO THE WATER AND WASTEWATER ENVIRONMENT

Edwin Chadwick, Secretary of the Poor Law Commission in the United Kingdom, penned the 'sanitary idea' as a means of promoting better health among the masses, in the 1842 report 'An Inquiry into the Sanitary Conditions of the Labouring Population of Great Britain'. Prior to this period, household wastes, liquid and solid, were simply dumped on the public streets and left to rot and blow away. Chadwick called for street and house cleaning by means of supplies of water and improved sewage collection and specifically stated that 'aid be sought from the science of the Civil Engineer, not from the Physician'. Chadwick and his physician collaborators identified the condition that solutions for environmental medical problems would come from engineering and not from the medical community (Petulla, 1987). Chadwick's engineering solutions included:

- Supplying each dwelling with clean drinking water
- Removing wastewater from dwellings and collecting it in a network of pipes
- Applying the collected sewage to agricultural land (away from the towns)

From that time onward, there was recognition that improvements to health could be brought about by improvements to the sanitary idea, with regards to drinking water, sewage and household wastes. Also in 1876 in the United Kingdom, the River Pollution Act forbade the discharge of sewage to streams and rivers, but not to the estuaries and seas. The land application of sewage was in many cases bypassed and ended up in the rivers, when engineers could show that there was 'adequate dilution' available in the river. The Royal Commission Report of 1912 permitted sewage discharge to rivers if it had a BOD (biochemical oxygen demand) of 20 mg/L and an SS (suspended solids) of 30 mg/L.

In England in 1914, Arden and Lockett discovered that when organic sewage was aerated in 'settling' tanks, after some time (days) the effluent from the tank underwent a treatment resulting in reduced oxygen demand of the effluent. This process, called activated sludge (see Chapter 12), used micro-organisms in suspension, in an aerobic environment, to break down the organic wastes. This was a breakthrough in technology and it should have led to widespread treatment of organic wastes by biological activated sludge. However, the process was patented and did not find itself in widespread use until much later. This technology has stood the test of time, and activated sludge is still by far the most common means of treating not only municipal sewage but also industrial organic wastewaters.

Around the turn of the century, similar developments were taking place in the United States. It was recognized that drinking water and sewage discharge water were two separate issues. Raw water for drinking purposes was most commonly surface water abstracted from rivers, lakes and man-made reservoirs. Minimalist treatment was not much more than settling of the visible solids. On the other hand, sewage continued to be discharged to rivers, thereby potentially polluting downstream water abstraction points with organic and microbiological contaminants. By 1900, many American cities were making progress in supplying piped drinking water to households and returning the sewage wastewater via an underground sewerage piped network, to rivers. There were still concerns with health and many cities had recurring typhoid epidemics. In 1910, when the city of Pittsburg, Pennsylvania, population 500 000 plus a suburban population of 500 000, required a State permit to extend its sewage network, the State Health Director requested a comprehensive sewage plan for the collection and disposal of all municipal sewage. Hazen and Whipple, Consulting Engineers, prepared a report and concluded that a sewage treatment plant would offer no increased protection to the downstream residents, as suburbia would still be discharging its sewage downstream. This study was a landmark and, regretfully, brought the American engineering community with it for decades to come, to preach the acceptability of discharging untreated sewage to rivers. It was not until 1959 that Pittsburg eventually got a wastewater treatment plant (Petulla, 1987).

The purification of water for drinking dates back to sand filters in Roman times. In the twentieth century, urban areas began to treat water using slow sand filters (detailed in Chapter 11). The quality of purification was excellent from these slow sand filters, but because of the slow rates of throughput, they required large land areas. In the 1890s and 1900s, rapid gravity filters, with their much increased throughput, began to replace the slow sand filter. Chlorination was added as a later treatment step for microbiological purification. In the 1950s, fluoridation of drinking water to reduce tooth decay was introduced and was popularized throughout the world for a few decades after its American introduction. Fluoridation of drinking water supplies is no longer thought to be desirable in several western developed countries as the availability of fluoride tablets and fluoride toothpaste and much improved diet and dental hygiene (since the 1950s) is now commonplace.

The preoccupation with human sewage treatment/disposal in the nineteenth century, resulted in ignoring the possibility of pollution from industrial wastewater discharges. It was thought, at the time, that the acids in industrial discharges would help to kill the human sewage microbes that caused human diseases and so be a benefit to river water quality and its downstream abstraction for drinking purposes. No attention was given to the incidents of fish kills, river water discoloration from industrial wastes or from abattoir wastes. Industrial wastewater treatment is a modern (post Second World War) development. In the United Kingdom, traditionally many industries discharged their liquid wastes into public sewers, which were subsequently treated at the sewage treatment plant. This practice is still dominant in the United Kingdom, while countries like Ireland (that were late to industrialize) insisted that their industries (especially chemical and pharmaceutical) treat their own industrial wastewaters separate to municipal sewage. In the short period of a few decades in Ireland, industry has installed wastewater treatment plants, many of which are superior in their technology to those of municipal plants. Table 1.1 is a chronological list of the key events of the nineteenth and twentieth centuries regarding public health and water/sewage (after Petulla, 1987).

1.2 HISTORICAL INTRODUCTION TO THE AIR ENVIRONMENT

Cities in the United Kingdom and the United States in the nineteenth century were well into the industrial revolution which predominantly used soft coal for energy and heating. An insignificant amount of gas began to be used from the 1800s on. Smoke from coal was then the bane of urban life. Many large cities world-wide, like London and New York, were sited on coastal areas or on inland waterways like Leeds or

Table 1.1 The Industrial Revolution and public health

Date	Science and technology	Citizen groups and professional opinion	Government activity
1842			Chadwick's monumental public health report calls for engineering solutions to environmental miasma. Griscom's New York study in 1845 reaches similar conclusion
1847		American Medical Association founded with intent to conduct sanitation surveys	
1848–9	Englishman William Budd links typhoid fever to water polluted by sewage. New cholera epidemic		England's National Public Health Act passes
1850s	Commuter railroads' smoke and noise begin to appear in American suburbs	'The sanitary question' gains national prominence in England. Miasma theory guides public opinion	For two generations, Royal Commissioners recommend purifying sewage by spreading it on the land
1860s	Pasteur experiments with microbes and vaccines		England's River Pollution Act of 1876 makes it an offence to discharge sewage into streams
1880s	United States boasts 598 waterworks systems. Most sewage systems dump untreated waste into rivers, streams, lakes, harbours	Engineering journals debate single-pipe versus separate sewer system designed by Col. Waring	
1884		Ladies Protective Association founded	
1890s	Water closets present engineers with overflow, collection, disposal problems. Series of typhoid epidemics occur	Medical community pitted against engineering societies over who should decide issues of public health	British Commissions give up land spreading of sewage as impractical, begin to advocate treatment and dilution methods
	Chlorine introduced to purify drinking water supplies	Engineers prefer dilution and filtration/purification methods for economic reasons. Physicians argue for wastewater treatment	Engineer Col. Waring appointed New York's first street-cleaning Commissioner
1900	3.5 million horses living in American cities, representing air/water pollution problems		
1905–7		International Association for the Prevention of Smoke established	Pennsylvania passes law that forbids cities to dump untreated sewage
1910	Hazan and Whipple Report concluded that a new sewage treatment plant for Pittsburg served no advantage from an economic and health standpoint		Resulted in widespread dumping of sewage to rivers and Pittsburg did not get a sewage plant until 1959
1914	Arden and Lockett discover 'activated sludge'.		

Adapted from Petulla, 1987. Reprinted by permission

Table 1.2 Air pollution episodes

Location	Date	Pollutants	Effects
Meuse Valley, Belgium	Dec. 1930	SO_2	63 excess deaths
Donora PA, USA	Oct. 1948	SO_2	20 excess deaths
Poza Rica, Mexico	Nov. 1950	H_2S	22 excess deaths
London	Dec. 1952	SO_2	4000 excess deaths
New York	Nov. 1966	SO_2	168 excess deaths

Adapted from Seinfield, 1986. © 1986, John Wiley & Sons, Inc. Reprinted by permission

Berlin. Such locations were susceptible to atmospheric inversions which, particularly in winter time, caused the particulate matter of black smoke to darken the airshed, resulting in the air environment being highly contaminated. The air pollutants affected human health, vegetation and materials. Incidents of death due to air pollution in the early part of the twentieth century are chronicled in Seinfield (1986) and repeated here in Table 1.2. The principal culprit was SO_2 from coal.

Health officers and Medical Associations identified the black lungs of dead city residents as being evidence of deadly smoke pollution causing consumption and bronchial disorders. Smoke Abatement Leagues and Ladies Health Clubs were organized around the United States and newspapers and journals attacked judges for leniency on industrial air polluters. Eventually, politicians listened and engineers became involved in technical innovation (Petulla, 1987). Thus turn-of-the-century reformers had begun to address smoke as a nuisance.

The catalyst for legislative change was the London smog of 1952, which lasted for several days and caused 4000 deaths. The Clean Air Act in the United Kingdom followed in 1956, which was preceded by the Air Pollution Control Act, in the United States in 1955. The emphasis of both was to reduce urban air pollution generated by the use of coal and oil, producing particulates (black smoke), carbon dioxide, carbon monoxide and unburnt hydrocarbons. Legislative improvements grew and by the 1980s very detailed legislative pieces were in place in all developed countries, with some reference even to international and transnational air pollutants.

With the expansion of industry and the proliferation of the chemical and pharmaceutical industries and nuclear power since the 1940s, air pollutants other than coal off-products were being detected in the air environment. The evolution of environmental air quality legislation took cognisance of these changes, spelling out the permissible values of a wide range of organic contaminants (see Chapter 8). With the spread of industry internationally, accidents occurred. The 1976 dioxin release at a chemical plant in Seveso, Italy, was the largest release of dioxins known and the dioxin cloud is still considered to have contaminated areas occupied by up to 37 000 people. As dioxin is a confirmed chemical carcinogen, this accident caused much international concern, and the long-term studies following Seveso have confirmed the widespread outbreaks of chloracne and many years later, cancer. This accident was the catalyst in Europe for the EEC directive on the handling, transport and safe preserving of hazardous chemicals, also called the Seveso directive (see Chapter 15). Further serious accidents, like the nuclear power plant disaster at Three Mile Island in the United States (1979), Chernobyl in 1986 and the explosion in a chemical plant at Bhopal, India, in 1984, moved the legislative protective policies forward to demand more accountability by industry and, particularly, those industries perceived by the public as higher risk, e.g. the chemical industry.

In addition to the disasters like Seveso, Chernobyl and Bhopal, the external environment was deteriorating in a more insidious way. Rachel Carson in Silent Spring (1962) noted the demise of plant and animal life in the rural countrysides of the United States which were previously havens for all life from insects and birds to rabbits. She viewed the widespread use of pesticides and other chemicals in agriculture as destroying non-target animals and their habitats, and noted that this was an international phenomenon that began with the industrialization of agriculture. Other problem areas addressed in Chapters 2 and 5 are the modern phenomenon of acid rain or the transnational destruction of forest lands.

We now know that air pollutants are damaging buildings and architectural works to the tune of an estimated $10 billion dollars annually in the United States. Other materials affected are clothing due to soiling, automobile tyres and windshield wipers due to ozone, and masonry and electrical contacts due to acid or alkaline pollutants containing sulphur. Crop losses in the United States due to pollution are estimated at $10 billion annually (Cunningham and Saigo, 1990). SO_2, PAN and other air pollutants are known to cause damage to vegetation, but others, including chlorine, hydrogen chloride, ammonia and mercury, are implicated in vegetation distress. Our concern for the air environment has therefore shifted emphasis from the products of soft coal in earlier days to the widespread organics and hydrocarbons produced by modern industry.

1.3 ENGINEERING, ETHICS AND THE ENVIRONMENT

Traditionally, engineers were able to practise their profession without having to address environmental ethics to the same depth that is now required. Traditionally, it was acceptable to design a road or change the course of a river, based on technical and economic conditions. Questions on the rights or wrongs of a river diversion were questions that engineers did not spend too much time on. These questions were left to others, usually the client, the relevant government agency or maybe an architect. Environmental engineering practice now requires us to address such questions. Legislation requires us to produce an environmental report (EIS) prior to the design stage of a project. It is at this stage that we must address the environmental ethic question. *An environmental ethic concerns itself with the attitude of people towards other living things and towards the natural environment* (Vesilind *et al.*, 1994). Engineers are not yet trained to address and understand the environmental ethic. However, it is fast becoming a requirement and some of the seniors in our profession are addressing the issue—because, on reflection, they are not totally happy with how engineering has moulded our modern environment. We are, of course, proud of the profession's achievements in the general area of public health, of water supply, sewage treatment, waste treatment, infrastructure of roads, housing, etc. What some are unhappy about are the negative long-term impacts of engineering projects on ecology as well as on humans. We relish the engineering achievements of Mulholland in bringing fresh water 300 km from the Sierra Nevada to Los Angeles, but engineers then (1910) did not question the long-term impact and did not forecast the decay of the fertile Owens Valley (Eastern California) from where the water was diverted. Similarly, we take engineering pride in the great water dams in the Tennessee Valley, but now question the ecological implications. In the development of the O'Shaughnessy Reservoir in Eastern California for a water supply for San Francisco, why were the conservationist views of the naturalist, John Muir, almost anathema to the engineering profession at the time? Today we thank John Muir and President Roosevelt for preserving one of the greatest natural parks in the world, i.e. Yosemite in California, though the engineers in the early part of this century, hungry for progress, wanted major road development rather than preservation. Recently, we have seen the opening of the largest series of dams in the world in Eastern Turkey, for energy and irrigation. The loss of water to adjacent nations is an issue of transnational water rights. The dilemmas for engineers on the design of such projects are many. Leaving aside the ecological issue, the ethical issue is a serious question. It is not only a political question. It is also an environmental ethical question. No longer can we engineers and scientists hide behind technology and economics; our profession must share responsibility for the ethical dilemmas or face the longer-term consequences of such issues returning to haunt us. The ethical question requests that we also put aside nationalist views for the long-term greater benefit of the global population and ecology.

There are precedents for issues returning to haunt us. For instance, in many areas of the world, a technological fix in the 1960s was to house people in high-rise suburban apartments. Less than thirty years later, these developments are considered failures. What was an engineering technological success turned out to be an environmental and social failure. A question that should have been addressed by the design engineers at the time was—would they themselves live in that high-rise, unserviced environment? Of course, none would have said 'yes'. This answer was a declaration for failure. As environmental engineers, we need to 'believe' in our projects as ecologically and humanely desirable. If I do not 'believe' in a project, say the establishment of a waste incineration plant, can I promote it and be technically involved? If the answer is negative, this does not necessarily give the project the 'thumbs down', but it does force us to be 'creative' and seek out innovative technical solutions, which are also environmentally suitable. As engineers, traditionally we have a 'fixed' solution for many technical problems. Today, we must be more adaptable and flexible and collaborate with groups (environmental and community groups) for input at the inception of projects. The engineering brief today is very different to that of a generation ago. Granted, we are more high-tech with solutions and more technically competent, but the added brief of public participation makes our task more demanding. Traditionally, an engineer could live out a successful technical career without public participation as a professional. Now we need to have the technical competence of a backroom technician with the ability to interact with groups, other than engineers, at a public level.

Environmental ethics for engineers is virtually non-existent in current engineering courses. Students do need exposure to this new and sensitive subject and readers are referred to the journal *Environmental Ethics*, published quarterly by The University of North Texas, and works by Vesiland *et al.* (1994).

1.4 EU AND US ENVIRONMENTAL LAW

Today's environmental problems pay no heed to geographic boundaries, as was well illustrated by the Chernobyl nuclear accident in 1986. On the day of the nuclear explosion, the wind was blowing northward, carrying the bulk of the air contaminants out of the Ukraine. The problems faced by the Ukraine people are related to groundwater, surface water and soil contamination, although airborne radiation is also a problem. The wind patterns at the time caused contamination to reach Northern Poland and Scandinavia the day after. Ultimately, the radioactive cloud spread throughout Europe and the United Kingdom and Ireland. In the highlands of the latter countries sheep grazing the grasses were not allowed on the food market for several years afterwards.

Environmental law is evolving to cover a wider catchment and even continental areas with the same regulations. This is the case in the United States where federal environmental regulations apply to all of the States. However, there are still some US States, e.g. California, that often adopt stricter environmental regulations than those set nationally. In the European Union, environmental legislation is following the same path as the United States, with the Commission of the European Union in Brussels setting Eurowide regulations. However, some countries such as Germany and Denmark in several cases adopt more stringent regulations than the European Union. It is not permitted in either the United States or the European Union for a state or country to adopt lesser stringencies. On an international scale, there are also regulations for those countries that have signed the Stockholm Agreement (1972) on the banning of ocean dumping of wastes/sludges, the Montreal Protocol (1987/1990) to eliminate CFCs and other ozone-depleting chemicals by the year 2000 and the Rio Summit (1992) with its objectives on sustainable development. However, not all countries, especially the less developing nations, are party to these agreements as they see no monetary gain for them. They, of course, contribute little to the problems of ocean dumping or greenhouse gases and they do not accept that they should adopt expensive technologies for environmental protection, especially as they contributed little to global environmental decay. The hierarchy in modern environmental legislation is shown in Fig. 1.1. At the international level, the 'honour' system is used, where penalties are difficult to impose on an offender. On the other hand, at the local level, the 'polluter pays principle' is enforceable and jail terms, while not common, have been enforced for offenders in some jurisdictions.

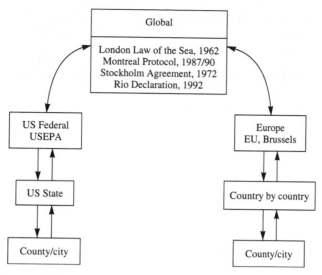

Figure 1.1 Levels of environmental legislation in the European Union and the United States.

The levels of legislation that industry and municipalities must abide by are shown in Fig. 1.1 for the United States and the European Union. Other countries such as Australia, New Zealand, South Africa, Japan, Canada and the Scandinavian countries are similarly structured. However, the larger part of Asia, Africa, Central and South America and Eastern Europe have insignificant environmental legislation. Today many development projects funded (by the World Bank and others) in the less developed countries are only done so on the basis of compliance with as near as possible to world environmental standards. The requisite standards are sometimes based in WHO (World Health Organization) guidelines or United Nations guidelines.

1.5 EVOLUTION OF EU ENVIRONMENTAL LEGISLATION

The original EEC treaty of 1958 made no mention of environmental protection. The first Community directive that can be considered 'environmental' was the 1967 directive on 'classification, packaging and labelling of dangerous substances'. In 1970, the EEC legislated on motor vehicle air and noise emissions. It is considered that the impetus for such legislation was as much economic as environmental. In 1972, the Community participated in a United Nations Conference on the 'Human Environment' in Stockholm. Out of this grew the United Nations Environment Programme (UNEP) and in 1973 the Community adopted its first environmental action programme (1973 to 1976). Since then, environmental action programmes have become a central feature of the EU environmental policy. To 1995, there have been five. The current programme is entitled 'Towards Sustainability' and was launched in March 1992. The 'Single European Act' brought into force in 1987, inserted an environment chapter into the Treaty (1958) and made explicit reference to the environment in a provision aimed at the realization of a frontier-free internal market. The period since 1987 has seen several significant developments. A new accent has been placed on the enforcement of EU environmental legislation, with the Commission processing a large number of infringement proceedings against non-complying Member States (countries). Some of these proceedings have arisen out of environmental complaints from the public, the number rising from 9 in 1982 to 480 in 1990. There has been a growing recognition of the role of market-based and financial instruments that seek to reflect environmental costs in the price of goods and services and legal instruments that provide public finance for environmental objectives. Global and regional issues have figured prominently with the

Community participating in several important international conventions. In 1992, the Maastricht Treaty was signed which strengthens the Community's environmental role, (Cashman, 1993).

1.5.1 Evolution of Environmental Legislation in Leading EU Countries

Northern Europe, particularly Germany, Denmark and Holland, played a leading role in establishing environmental legislation. In fact, today many of the legislation pieces in place in the European Union are derived from national legislation of some Northern European countries. Western and Southern Europe have lagged behind their northern European counterparts in development and implementation of environmental legislation. Where European Union legislation is not complete, as in the 'air environment', practitioners make use of standards such as the German TA Luft air quality standards or the WHO air quality guidelines. Similarly, advanced standards for solid waste treatment and disposal with regard to landfill liners, leachate collection, landfill closure, landfill and stability, adapted by the European Union, are similar to those practices in place in northern European countries. The United Kingdom has had environmental standards for water quality going back to the beginning of the century. A Royal Commission Standard in 1916 set wastewater discharge standards for public and industrial discharges at 20/30 (20 mg/L BOD, 30 mg/L SS). These same standards are still deemed adequate 80 years later. After the London smog of 1952, the United Kingdom introduced in 1956 the British Clean Air Act. This has since been amended several times. The United Kingdom wrote into law the UK EPA (Environmental Protection Agency) in 1991, taking a lead in several environmental legislative areas, with Ireland following suit in 1993, 20 years after the Danish and US EPAs.

1.5.2 Evolution of US Environmental Legislation

From 1948 onwards, the US Congress responded to increasing pollution from industrialization by expanding the federal role in both air and water pollution. Between 1948 and 1972 the Federal Water Pollution Control Act (FWPCA) was amended five times, as was the Air Pollution Control Act (APCA) between 1955 and 1972. The FWPCA of 1948 provided grants to state agencies. The 1955 APCA also provided grants to state agencies for feasibility studies and research, but not grants for contract work. Both the water and air acts were upgraded by Congress in tandem. The 1955 Federal Pollution Control Act (FPCA) replaced the 1948 FWPCA and developed a system of grants for municipal wastewater treatment plants. On the air side, the US Clean Air Act of 1955 provided federal assistance to states on air pollution issues. The 1960 Motor Vehicle Act required research into automobile exhaust gases and the revised 1963 Clean Air Act set about creating standards for air pollution control. In the area of solid waste the 1965 Solid Waste Disposal (Act) addressed both municipal and hazardous wastes disposal and set up management guidelines for collection, separation, transportation, recycling and disposal.

Solid waste laws have come a long way in the United States since 1965 and the most recent RCRA (Resource Conservation and Recovery Act) 1994 specifies that 50 per cent of all municipal waste collected be diverted from landfill by the year 2000. The hierarchy regarding waste is now, firstly, waste minimization at source, then recycling and reuse, then conversion to energy and, as a last resort, landfilling. The Clean Air Act has been updated in 1970, 1977, 1990 and 1992. The more general criteria pollutants were identified and standards set for them. The National Emission Standards for air toxics was elaborated in the 1990 amendment under the National Emission Standards for Harzardous Air Pollutants (NESHAP), where 189 toxic air pollutants were listed with standards. Goals to reduce SO_2 emissions by 10 million tons below the 1980 levels were identified in 1990.

While the Solid Waste Disposal Act was introduced in 1965, the follow-through on municipal solid waste (MSW) was delayed due to the momentum that was developed by the requirements of hazardous waste disposal. The episode of the 'Love Canal' tragedy accelerated the formation of the Comprehensive Environmental Response, Compensation and Liability Act (CERCLA) or Superfund in 1980. This was a five-year programme which hoped to clean up all the contaminated sites in the country. The volume of sites was grossly underestimated and even by 1994 there were only in excess of 50 sites fully cleaned up, with about 1700 in need of extensive remediation. The 1980 Superfund was reauthorized under the Superfund Amendments and Reauthorization Act (SARA) in 1986 and massive budgets were allocated to the work. However, most of these budgets have been consumed by legal consultancy and research fees, with the result that few sites have been cleaned up. The EPA is insisting on full remediation and this of course is both time and money consuming.

The Federal Water Pollution Acts have been amended several times and the most recent in 1992 identified 65 000 chemicals that are regulated in wastewater discharge. Many of these are either toxic to humans or fish. The Safe Drinking Water Act of 1974 has been upgraded several times and the 1986 Act specifies that the EPA publish a list of contaminants found in drinking water. This list was of 85 contaminants by 1992. Environmental legislation is very well developed in the United States and in many areas it is the most developed of all nations.

1.6 SOME RELEVANT INTERNATIONAL ENVIRONMENTAL AGREEMENTS

What is known as the 'Law of the Sea' was first signed in London in 1954. This was an international convention for the prevention of pollution of the sea by oil. Several annexes and protocols to the convention followed. The 1974 protocol was again signed in London and this was an international convention on the high seas in cases of pollution by substances other than oil. The 1978 protocol was signed as an international convention relating to the prevention of pollution from ships. The 1992 convention was on the prevention of marine pollution by dumping of wastes and other matter.

The Stockholm Declaration on the Human Environment was signed in 1972 and this included 109 recommendations for environmental action. It was very wide-ranging on the protection of the environment and included topics as disparate as protection of architectural heritage, rainforest protection and water quality in rivers, lakes and seas.

The USEPA banned the use of CFCs in non-essential aerosol propellants in 1978. The Montreal Protocol signed in 1987 and revised in 1990 called for the elimination of CFCs and other ozone-depleting gases by the year 2000. However, there is much debate still on the economic viability of such a major change. Many developing countries resist this agreement as they will be burdened by the high cost of CFC replacement even though it was the developed world that has been the dominant user of CFCs.

The most recent significant agreement was the Rio Declaration in 1992, twenty years after the Stockholm (or Helsinki) declaration. This aimed to reach global agreement on a number of key environmental issues: climate change, biodiversity, tropical forests and sustainable development. The conference produced two important conventions: one on climate change and the other on biodiversity, as well as the Rio Declaration which sets out basic principles to guide environment and development policy and Agenda 21, a long-term plan to integrate environment and development. Each of these helps to define the wider international setting in which not only future European Union and United States measures will take place, but all countries of the world.

1.7 EU ENVIRONMENTAL LEGISLATION

By 1992, the European Community had adopted more than 200 pieces of legislation covering pollution of the atmosphere, water and soil, waste management, safeguards in relation to chemicals and biotechnology, product standards, environmental impact assessment and marine protection (Johnson and Corcelle, 1989). The extent and scope of this legislation, its binding character, the dynamic nature of Community Environmental Policy and the general growth in supranational and global environmental regulation, all make the Community's environmental role of considerable relevance to the engineering and scientific professions. This is so, not only in the specific field of environmental engineering (including water and wastewater treatment, solid waste management, control of emissions and land use priorities) but also in more general fields (e.g. in relation to project development and product design and production).

Action by the Community is through a range of legal acts provided for in the Treaties (from the Treaty of Rome in 1958 to the present day).

1. *Regulations*   These are the most far reaching and lay down the same law for all the Community and apply in full throughout, conferring rights and imposing duties directly on citizens in the same way as domestic law.
2. *Directives*   These are addressed to Member States, binding them to achieve certain results (standards) but allowing them to choose the form and methods for doing so. The Directive is by far the most commonly used legal act in the field of the environment and is published in the *Official Journal of the European Communities*. Technical annexes are commonly found at the end of the Directive. It is generally necessary for Member States to adopt the contents of the Directive in some form of domestic legislation.
3. *Decisions*   These are legal acts binding on the persons to whom they are addressed including Member States and legal persons. Their main use in the area of the environment has been to authorize the Community to become a party to international conventions.
4. *Recommendations*   These are non-binding legal acts (rarely used in the environmental area).
5. *Opinions*   These are explanations of the other legal acts and are frequently issued after draft directives or regulations have been made public.

In 1990, the Commission endorsed 'A Community Strategy for Waste Management' which set out priorities as:

- Waste prevention
- Reuse and recycling
- Waste to energy
- Safe final disposal

In addition to environmental legislation, the Community adopts strategies on environmental policy. Similarly, the Fifth Environmental Action Programme (1992–1996) addresses several critical environmental themes: climate change, acidification and air quality, protection of nature and biodiversity, management of water resources, the urban environment, coastal zones and waste management.

1.7.1 Some of the Significant EU Environmental Directives

Table 1.3 lists the significant Directives. The list is meant to give the reader an idea of the range of directives and is not meant to be comprehensive. The reader is referred to Duggan (1992) from which Table 1.3 was extracted. The significant 'environments' are:

- Water
- Air
- Land (soil)

In addition, other 'areas' of significance are listed in Table 1.4. These areas are:

- Waste—non-hazardous and hazardous
- Chemical substances
- Noise
- Habitats

Table 1.3 Significant EU environmental directives in water, air and land environments

Environment	Directive name	Directive number and date
Water	Surface Water for Drinking	75/440/EEC
	Sampling Surface Water for Drinking	79/869/EEC
	Drinking Water Quality	80/778/EEC
	Quality of Freshwater Supporting Fish	78/659/EEC
	Shellfish Waters	79/923/EEC
	Bathing Waters	76/160/EEC
	Dangerous Substances in Water	76/464/EEC
	Groundwater	80/68/EEC
	Urban Wastewater	91/271/EEC
	Nitrates from Agricultural Sources	91/676/EEC
Air	Smoke in Air	80/779/EEC
	Sulphur Dioxide in Air	89/427/EEC
	Lead in Air	82/884/EEC
	Industrial Plants	84/360/EEC
	Large Combustion Plants	88/609/EEC
	Existing Municipal Incineration Plants	89/429/EEC
	New Municipal Incineration Plants	89/369/EEC
	Asbestos in Air	87/217/EEC
	Sulphur Content of Gas Oil	75/716/EEC
		82/219/EEC
	Lead in Petrol	85/210/EEC
	Emissions from Petrol Engines	70/220/EEC
	Air Quality Standards for NO_2	85/203/EEC
	Emissions from Diesel Engines	88/77/EEC
Land	Protection of Soil When Sludge is Applied	86/278/EEC

The environmental quality objectives are requirements—usually concentrations of substances—that must be fulfilled at a given time, now or in the future in a given environmental medium, water, air or soil. For instance, the 1976 Bathing Water Directive fixes microbiological and physicochemical standards for freshwaters and seawaters where bathing is explicitly authorized. This directive has had and still has major implications for wastewater discharges, both municipal and industrial. In addition to this directive, the more recent (1991) Urban Wastewater Directive ensures at least secondary treatment plants on all urban coastal locations where traditionally no plants existed. In addition to raising standards of wastewater discharges, the European Union is funding the setting up of these plants, particularly in the less developed regions of the Community.

1.7.2 European Union Quality Standards in the Water Environment

The water environment can be considered to consist of:

- Potable waters
- Raw water sources from freshwaters including groundwater
- Waters for fish habitats and bathing
- Wastewaters—municipal and industrial

Table 1.4 Significant EU environmental directives in other 'areas'

Environment	Directive name	Directive number and date
Waste	Incineration of Hazardous Waste	Proposal Com(2) Mar '92
	Hazardous Waste	91/689/EEC
	Packaging Waste	Proposal 92/C263/01
	Waste Framework Directive	75/442/EEC
	Toxic and Dangerous Waste	78/319/EEC
	Disposal of Oils	75/439/EEC
		87/101/EEC
	Transfrontier Shipment of Hazardous Waste	84/631/EEC
		85/469/EEC
		86/279/EEC
	Landfill of Waste	Proposal Com 91. May '91
	Disposal of PCBs and PCTs	74/403/EEC
	Prevention of Air Pollution from Municipal Incineration Plants	89/369/EEC
	Reduction of Air Pollution from Municipal Incineration Plants	89/429/EEC
Noise	Noise in the Workplace	82/188/EEC
	Construction Plant Noise	79/113/EEC
		81/1051/EEC
Chemical substances	Major Accident Hazards	82/501/EEC
	'Seveso' Directive	87/216/EEC
		88/610/EEC
Habitats	Conservation of Wild Birds	79/409/EEC
		85/411/EEC
General	Environmental Impact Assessment	85/337/EEC
	Packaging and Packaging Waste	92/C263/01
	Eco-audit Scheme (Regulation)	Proposal 92/C76/02

Potable waters   Table 11.3 in Chapter 11, lists the European Union drinking water standards. Briefly they are divided among:

- Organoleptic parameters—include colour, turbidity, odour and taste
- Physicochemical parameters—include temperature, pH, conductivity, Ca, Mg, K, Na
- Parameters concerning substances undesirable in excessive amounts—include nitrates, nitrites, ammonium, HCs, phenols, organochlorines, etc.
- Parameters concerning toxic substances—include pesticides, PAHs, heavy metals
- Microbiological parameters—include total and faecal coliforms and faecal streptococci
- Minimum requirements for softened water—include total hardness and alkalinity.

While all six parametric sets (55 parameters) are significant, a typical water treatment plant is required to test for the different parameters at different frequencies. This is discussed in Chapter 11 and enumerated in Annexe II of Directive 80/778/EEC. For instance, some of the parameters of immediate relevance to the design of a water treatment plant include colour, turbidity, odour and taste. These are the four organoleptic parameters and are required to be sampled according to C2 frequency. This is 3 times per year for a 10 000 population to 120 times per year for a population of one million. The reader is encouraged to become familiar with this technical Annexe II to the Directive. The four headings relating to the general monitoring frequency are: minimum monitoring (C1), current monitoring (C2), periodic monitoring (C3) and occasional monitoring in special situations or in case of accidents (C4).

Raw water sources from freshwaters including groundwater Sources of surface water to be used (and treated) as drinking water are classified according to three categories in the 'Surface Water Drinking Directive', 75/440/EEC. Category A1 requires only physical treatment (filtration) and disinfection. Category A2 requires normal physical and chemical (coagulation) treatment and disinfection. Category A3 requires intensive physical and chemical treatment and disinfection. Raw water is categorized by its values of 46 parameters (similar to those in the Drinking Water Directive, above) which are listed in Annexe II of the Directive. Only in exceptional circumstances can water of a lower quality than category A3 be used (possibly for blending with better water). Further comments on the application of this Directive are included in Chapter 12.

Groundwater As the extent of potable water sourced from groundwater varies from 25 to 75 per cent in the EU Community, the Groundwater Directive 80/68/EEC sets standards regarding the tipping of waste and activities on or in the ground involving List I (more dangerous) or List II substances. It is noted that List I and List II in this Directive are not identical to List I and List II of the directive on dangerous substances discharged into the aquatic environment, 76/464/EEC.

Waters for fish habitats and bathing The freshwater fish directive is designed to retain freshwater quality suitable for supporting either salmonoid or coarse fish. The more stringent standards associated with salmonoid rivers set out in an annex to Directive 78/659/EEC, 14 physical and chemical parameters which are to be attained (imperative values). Other higher standards called guide values are to be aimed for. For example, the values set for total ammonia are 1 mg/L (imperative) and 0.04 mg/L (guide). Values for coarse fish are also set with 14 physical and chemical parameters, but the values are less stringent. By comparison with the total ammonia for salmonoid rivers, the coarse fish river values are 1 mg/L (imperative) and 0.2 mg/L (guide). The other significant directive is related to bathing water quality, 76/160/EEC, which fixes imperative and guide values for microbiological, physicochemical and other substances. Imperative values for total and faecal coliforms are set at 10 000 and 2000 per 100 mL, while the guide values are set at 500 and 100 respectively. The reader is recommended to study this directive as it pertains not only to water quality but also to wastes which impact on water quality. The other directive relating to water quality is the Shellfish Directive 79/923/EEC. Twelve physicochemical and microbiological parameters are listed with imperative and guide values for brackish or coastal waters.

Wastewaters Several directives are addressed to industrial and municipal wastewaters. The Urban Wastewater Directive 91/271/EEC specifies that 'there is a general need for secondary treatment of urban wastewater' for industrial and municipal discharges. While values for BOD (25 mg/L), COD (125 mg/L) and total suspended solids (35 mg/L) are fixed, the more recent requirements are for nutrient removal (2 mg/L of P and 10–15 mg/L of total N) for 'sensitive areas'. The issue of sensitive areas is still being assessed. Some of the more conservationist countries like Germany and Denmark are likely to classify substantial areas of their land/water space as (environmentally) sensitive. The earlier directive on dangerous substances discharged into the aquatic environment, 76/464/EEC, contains an annex of List I and List II substances for which reduction programmes were required. List I includes substances selected on the basis of toxicity, bioaccumulation and persistence, e.g. carcinogenic substances, organochlorine compounds and mercury and cadmium. List II includes the less dangerous substances such as the heavy metals of copper, lead, zinc, ammonia and cyanide.

1.7.3 EU Quality Standards in the Air Environment

EU air quality regulations are not as well developed as those for water/wastewater. In the European Union there is still much reliance on German standards (TA Luft), USEPA standards and the WHO Air Quality Guidelines. The air quality regulations relevant to the European Union are described in Chapter 8 and are written into directives relative to emissions of gaseous and particulate matter from vehicles, industrial and municipal plants, and incinerators. Industrial and municipal plants include not only manufacturing, chemical, pharmaceutical, automotive plants, etc., but also waste plants as in industrial and wastewater treatment plants. In recent years, the air emissions from wastewater treatment plants have been given much attention, particularly in relation to odours, hydrogen sulphide, methane, ammonia gases, volatile organic compounds and hydrocarbons. In the European Union the TA Luft emission limits have been used in the absence of EU directives. Also, in the past, the ambient air quality standards as defined by WHO (1987) have been used in the absence of EU directives. Table 8.1 of Chapter 8 itemizes the most significant EU directives, resolutions and decisions. Table 8.4 lists the air pollutants defined as criteria pollutants and their associated ambient limit values. The averaging time for pollutant measurement is important, as clearly the maximum value tolerated for a single hour would be typically much higher than the annual average value. In the air environment, two types of limits are of interest:

- Ambient limits (those set by a regulatory authority pertaining to upper acceptable levels on a spatial urban environment).
- Emission limits (levels set by a regulatory authority pertaining to point emissions from industrial sources).

Again, emission limits set on an industrial air emission of, say, CO would be much higher than the ambient limit in the location of discharge. The difference is to allow for dilution and dispersion of the industrial discharge into the ambient environment. Very generally, the emission limit of some of the air pollutants is about 30 times that of the ambient limit. This of course does not always hold true. The criteria air pollutants (ambient air standards) and some limits and associated directives are listed in Table 1.5.

It is noted from Table 1.5 that not all criteria pollutants have EU ambient limit values. Where they are lacking, the USEPA value has been inserted. The WHO guidelines are very detailed with respect to both

Table 1.5 EU criteria pollutant standards in the ambient air environment

Pollutant	EU Limit		EU Directives
CO	30 mg/m^3:	1 h	EU Guideline: no EEC ambient limit
NO_2	$200 \text{ } \mu g/m^3$:	1 h	85/203/EEC
O_3	$235 \text{ } \mu g/m^3$:	1 h	USEPA: no EEC ambient limit
SO_2	$250–350 \text{ } \mu g/m^3$: $80–120 \text{ } \mu g/m^3$:	24 h Annual	80/779/EEC 89/427/EEC
PM_{10}	$250 \text{ } \mu g/m^3$: $80 \text{ } \mu g/m^3$:	24 h Annual	80/799/EEC 89/427/EEC
$SO_2 + PM_{10}$	$100–150 \text{ } \mu g/m^3$: $40–60 \text{ } \mu g/m^3$:	24 h Annual	80/779/EEC 89/427/EEC
Pb	$2 \text{ } \mu g/m^3$:	Annual	82/884/EEC
Total suspended particulates (TSP)	$260 \text{ } \mu g/m^3$:	24 h	USEPA: no ambient limits for EEC
HC	$160 \text{ } \mu g/m^3$:	3 h	USEPA: no ambient limits for EEC

criteria pollutants and non-criteria pollutants. The reader is referred to Chapter 8 for further details and references.

1.7.4 EU Quality Standards for Waste

The first EU directive on waste was in 1975 (75/442/EEC) and this placed the duty on each Member State to ensure that waste was disposed of without harming human health or the environment. It contained five mandatory requirements including the need to appoint competent authorities, prepare waste disposal plans and practise safe disposal of waste. Waste recycling was encouraged and waste hauliers and disposal sites were to be permitted. While this directive was in essence very general, it did establish a framework throughout Europe for the 'technical' management of waste. Countries like Denmark and others were well advanced at this stage while others, like Ireland, Spain and Greece, were not. Two directives were issued in 1975 (75/439/EEC) and 1987 (87/101/EEC) on the disposal of waste oils. This was first introduced because in several Member States large percentages of waste oil were dumped. The directive sought to redress this environmental problem while encouraging oil recycling. The proposed directive (92/C263/01) on packing waste seeks to cut down significantly the volume of this waste and its toxicity. Within 10 years of implementation, 90 per cent by weight must be removed from the waste stream for the purpose of recovery and 60 per cent by weight of each material of the packaging waste output must be removed from the waste stream for the purpose of recycling. The proposed landfill directive (Com 91 May 91) imposes strict criteria on the design, construction, maintenance, management, closure and post-closure of landfills. This proposal reiterates that the waste management philosophy be: firstly, waste reduction at source; secondly, recycling and reuse; thirdly, waste to energy, including incineration; and as a last choice, landfill.

Landfills are classified as monolandfill or multidisposal. There are also three classifications based on waste type: hazardous waste; inert waste; and municipal non-hazardous and other compatible waste. Certain wastes are not permitted to be landfilled, oxidizing or explosive waste or flammable waste, infectious wastes and some liquid wastes, if they are incompatible with the specific landfill. There is much detail to this proposed directive and it serves to a degree as a brief on the proper and acceptable design and management criteria by which a landfill should operate. The expensive element as regards many countries who traditionally depended on elementary dump sites are the stringent requirements set for groundwater protection by either natural or synthetic liners (see Chapter 14).

EU directives on hazardous waste are not nearly as well developed as those in the United States. The first EU directive on hazardous waste was in 1978 (78/319/EEC) on toxic and dangerous waste. This directive laid down a broad framework of control of both household and toxic wastes involving the establishment of competent authorities responsible for producing hazardous waste management plans and authorizing installations handling waste. Toxic waste was defined. The competent authorities responsible for implementation were also defined. A mechanism for plans, permits, recording, inspection, separation, packaging, transporting and disposal was identified. As regards payment for disposal, the principle of 'the polluter pays' was adopted. The annex to this directive listed 27 toxic or dangerous substances. Earlier regulations on the disposal of PCBs (polycyclic chlorinated biphenyls) were identified in Directive 74/403/EEC. A more recent directive on hazardous waste, 91/689/EEC, defined in greater detail what was hazardous waste and was done so in some cases by identifying the waste (e.g. wood preservatives) and in others by identifying the industry producing the waste (e.g. pharmaceuticals, medicines and veterinary compounds). Annexe IA lists 18 property types and Annexe IB an additional 22 constituent types. Annexe II lists 51 constituents of wastes that render them harmful (e.g. phenols, mercury, etc.) and Annexe III lists properties (14 sets) of wastes that render them hazardous (e.g. corrosive, carcinogenic, etc.). This is a significant directive in that there are definite guidelines in defining what is a hazardous waste. This directive identifies what is hazardous in relation to a waste constituent and also in relation to the impact such a constituent or waste has on the environment or humans. Three directives on the transfrontier shipment of waste were issued and revised since 1984 (84/631/EEC, 85/469/EEC, 86/279/EEC). The purpose of these directives was to control movement within the Community and into and out of the Community by ensuring that competent authorities provided a permit for such movement. The proposed directive (March 92) on incineration of hazardous waste is a technically detailed instrument focusing on emission quality from these incinerators. The hazardous emissions are specified with proposed limits (for international comparisons, see Chapter 15) for the compounds of total dust, total organic carbon, inorganic chlorine compounds, inorganic fluorine compounds, sulphuroxides, carbon monoxide, cadmium, thallium, mercury, other heavy metals, and dioxins and furans. The technologies and their operating conditions are detailed as well as costs of establishing an actual incinerator. It is interesting that this is one of the proposed directives that is more like a technical manual. It appears that there is some trend, at least in some of the directives, to contain more than volumes of legal comments and to contain engineering/scientific data more akin to an engineering code or standard. Two other directives relating to incineration are for prevention and reduction of air pollution from municipal incineration plants (89/369/EEC, 89/429/EEC). As in the case of hazardous incineration plants, emission limits are set for total dust, heavy metals (listed), HCL, HF and SO_2. The emission standards decrease as the plant production output (in tonnes/h) increase.

Another major directive that is related to hazards is that of the Major Accident Hazards of Certain Industrial Activities, 82/501/EEC. This is often referred to as the Seveso Directive, from Seveso in Northern Italy where an accidental industrial release of hazardous chemicals, including dioxins, occurred in 1980. This directive was updated in 1987 and 1988 (87/216/EEC and 88/610/EEC). The purpose of these directives was to establish measures that would address the risk of exceptional hazards such as fires, explosions and massive emissions of dangerous substances. These measures include preparing hazard surveys and emergency plans and notifying the competent authorities of dangerous substances retained on an industrial site (see hazardous waste treatment in Chapter 15).

1.7.5 EU Quality Standards in Other Areas

Other areas addressed in directives that are of relevance to environmental engineering/science include:

- Noise
- Habitats
- General

The directives on noise relate to construction plant noise (79/113/EEC and 81/1051/EEC) and noise in the workplace (86/188/EEC). The framework sets out an approval, verification and certification procedure for noise levels for specific plant types (e.g. generators, bulldozers, jackhammers, etc.). Methods of measuring noise and standards are also defined.

There are two directives on the conservation of wild birds, 79/409/EEC and 85/411/EEC. The purpose of these directives is to cover the protection, management and control of all species of naturally occurring birds. It also includes migratory birds that use Europe or any Community State as a breeding ground or even in a flyover capacity. Their habitats, nestings, eggs and, of course, the birds themselves, are included. General obligations are placed on the Member States to preserve and maintain and to re-establish biodiversity and also areas for bird habitats. There are also specific measures to be complied with and Annexe I lists the especially vulnerable species that require special conservation measures. Appropriate steps are to be taken by the Member States to avoid pollution and deterioration of habitats. Annexe II lists the species that can be hunted in Member States. Those species listed in Annexe I are prohibited from being sold or moved outside of their current habitat. This directive is not only important to environmental scientists from the perspective of conservation but it is also important to engineers, who should understand that there are areas or habitats which at first investigation may be amenable for development but on detailed investigation may be found to contain sensitive species habitats. Before too much effort and finance is expended on the engineering side of projects (e.g. roads), it is prudent to invest also in eco-investigation so that projects are not rejected at a later planning stage. It is recommended that readers study in more detail and compare the US Endangered Species Act with the EU Conservation of Wild Birds Directive.

Other directives of relevance include the Environmental Impact Assessment (EIA) Directive (85/337/EEC) and the proposed directive on Eco-auditing (92/C76/02). The EIA Directive provides for the systematic assessment of proposed projects in the public and private arenas. Items to be accounted for include the protection of human health and the environment, the quality of life and the need to maintain the diversity of the species and the continuity of the ecosystems. Annexe I of this directive includes project descriptions that must undergo EIA. These include motorways, hazardous waste disposal installations, crude oil refineries, power stations, airports, etc. A more comprehensive list is in Annexe II where Member States consider such projects in need of an EIA. The direct and indirect effect of a project is to be described on:

- Material and cultural heritage
- Human beings, fauna and flora
- Soil, water, air, climate and landscape

The proposed eco-audit directive has the objective of evaluation and improvement of the environmental performance of industrial activities and the provision of relevant information to the public (see Chapter 19 for further details on EIS and eco-audit).

1.8 US ENVIRONMENTAL LEGISLATION

By 1996, the Federal US government, principally through the offices of the USEPA, had substantial environmental legislation implemented in the areas of:

- Water pollution (surface water and groundwater)
- Drinking water
- Air environment
- Municipal solid waste
- Hazardous wastes
- Others (including endangered species and occupational safety and health)

The mechanism of putting an environmental law on to the statutes is a many-faceted process. Initially the USEPA develops the regulations by assessing the available technical and scientific information, preparing draft regulations and then submitting this draft for intra-agency review. For instance, in the case of setting water quality standards for contaminants under the SDWA (Safe Drinking Water Act), the USEPA is required to assess each contaminant with respect to human exposure, health effects and toxicity, monitoring and analysis, treatment technologies and costs. The reason for the intra-agency review process is to ensure the proposed regulations are: scientifically and technically defensible; are meeting legislative requirements; and are not at odds with other agency activities. Once peer review is obtained, the proposed regulations are submitted for public comment. The 'public' means the scientific community, those potentially impacted by the new regulations and the interested public. Public hearings and briefings are encouraged and prior to the proposed rule making an advance notice is made public, so that public participation can be made at this late stage. After an adequate public comment period of 30 to 120 days, the received public comments are reviewed and used in the final preparation of the proposed rule. Even after a CFR (Congressional Federal Register) has been published with a title number, interested parties can still use the courts for litigation against the rule. Each environmental rule is published as a CFR. For instance, in the case of the SDWA, the CFR Title 51, Part 63, 2 April, 1986, page 11396 states the maximum concentration limit (MCL) for fluoride.

1.8.1 Some Significant United States Federal Environmental Legislation

Tables 1.6, 1.7, 1.8 and 1.9 list the chronological development of United States Federal environmental legislation in the areas of:

- Environmental policy
- Water and wastewater
- Air
- Municipal solid waste
- Hazardous waste
- Others

It is seen that legislation began around 1955 and proceeded slowly up to the founding of the USEPA in 1970 by President Nixon. From 1970 to 1995, legislation has been continuous with Congress updating the Water Acts, the Air Acts and the Waste Acts typically every 4 to 6 years. As standards are now at an all-time restrictive set of values, it is likely that the next decade will not bring the same voluminous set of environmental laws. It is possible that as the depth of scientific knowledge increases, standards for some

Table 1.6 Significant US federal environmental legislation—policy and water quality

Name of Act	Abbreviation	Date	Comments
Policy			
Earth Day		1970	Signalled height of environmental awareness in US government
National Environmental Policy Act	NEPA	1969	Required EIS for Federal Projects
Environmental Pollution Agency	EPA	1970	Combines the Federal Water Quality Administration Agency and US Public Health Service
Water			
River and Harbors Act		1889	Protects waters from pollution
Federal Water Pollution Control Act	FWPCA	1948	Grants for feasibility studies
Amendments to FWPCA	FWPCA	1956	Grants for feasibility studies
Water Quality Act	WQA	1965	Enacted water quality standards
Amendments to FWPCA—now called the Clean Water Act	CWA	1972	Massive grants for wastewater treatment and collection
Safe Drinking Water Act	SDWA	1974	EPA sets standards for drinking water quality
Fishery Conservation and Management Act	FCMA	1976	
Soil and Water Resources Conservation Act	SWCRA	1977	
EPA Wetlands Regulations		1989	
Amendments to Safe Drinking Water Act	SDWA	1986	EPA publishes lists of contaminants found in public water supplies (85 by 1992)
Amendments to CWA	CWA	1987	Regulation of toxics and sewage sludge in wastewaters and non-point source (NPS) pollution control grants
Surface Water Treatment Rules	SWTR	1989	
National Pollution Discharge Elimination System	NPDES	1990	Permits for industrial and municipal stormwater discharges
Amendments to CWA	CWA	1992	65 000 chemicals regulated
Standards for the Use and Disposal of Sewage Sludge	49CFR503	1993	Sewage sludge standards
Amendments to CWA	CWA HR961	1995	Incentives for watershed management, a new wetlands permitting programme based on ecological values, etc.

Table 1.7 Significant US federal environmental legislation—air quality

Name of Act	Abbreviation	Date	Comments
Clean Air Act	CAA	1955	Federal and technical assistance to states
Motor Vehicle Act		1960	Required basic research on automobile exhausts
Amendments to Clear Air Act	CAA	1963	Department of HEW to publish criteria for national air quality standards
Air Quality Act—Amendments	AQA	1967	Established a framework to improve nation's air and guarantee air standards by region
Clean Air Act—Amendments	CAA	1970	EPA to establish minimum air quality and regulatory goals that state and local governments were to achieve
Clean Air Act—Amendments	CAA	1977	EPA imposed strict requirements on areas not meeting standards for criteria and toxic air pollutants (see Chapter 6)
Clear Air Act—Amendments	CAA	1990	Air toxics—National Emission Standards for Hazardous Air Pollutants (NESHAP)—goal to reduce by 50% the nationwide emission of 189 toxic air pollutants including asbestos, beryllium, vinyl chloride, benzene, radionuclides, arsenic, radon 222, etc. Also in the chemical industry to lower by 90% the fugitive emissions of 148 VOCs (volatile organic carbon)
Clean Air Act—Amendments	CAA	1990	EPA to ensure reduction of SO_2 emissions by 10 million tons from 1980 levels. Also NO_x reductions
Clearn Air Act—Amendments	CAA	1992	New Chemical Safety Board to police the above

Table 1.8 Significant US federal environmental legislation—solid wastes

Name of Act	Abbreviation	Date	Comments
Solid Waste Act	SWDA	1965	Policed by the US Public Health Service to promote solid waste management and resource recovery including guidelines for the collection, transportation, separation, recovery and disposal
National Environmental Policy Act	NEPA	1969	Requires EIS on solid waste projects; e.g. new landfills
Resource Recovery Act	RRA	1970	Shifted emphasis of national solid waste programme from disposal to recycling and reuse, and to conversion of wastes to energy
Resource Conservation and Recovery Act	RCRA	1976	Provided legal basis for implementation guidelines and standards for solid waste storage, treatment and disposal
		1978	EPA published guidelines on MSW management and revised them in 1984 and 1988
Comprehensive Environmental Response, Compensation and Liability	CERCLA	1980	Mechanism for responding to uncontrolled landfills, i.e. to sites with mixed municipal and hazardous wastes
Public Utility Regulation and Policy Act	PURPA	1981	Directs public and private utilities to purchase power from waste to energy sources
RCRA Amendments—Hazardous and Solid Waste Disposal Amendments	RCRA (HSWDA)	1984	Includes banning of landfilling for RCRA hazardous wastes
California Assembly Bill	AB	1993	Recycling goals—50% of all MSW to be directed from landfill by 2000: 25% by 1995
Amendments to RCRA	RCRA	1994	Includes recycling goals enumerated in line with those of California and Florida

Table 1.9 Significant US federal environmental legislation—hazardous waste

Name of Act	Abbreviation	Date	Comments
Solid Waste Act	SWDA	1965	Policed by the US Public Health Service; established a federal research and grant programme including hazardous and non-hazardous waste
Federal Insecticide, Fungicide and Rodenticide Act	FIFRA	1972	Manufacturers of pesticides to supply data to EPA of unreasonable effects on the environment
Amendments to FIFRA	FIFRA	1988	EPA mandated registration of pesticides by 1991: 23 000 pesticide products registered
Toxic Substance Control Act	TSCA	1976	Mandated testing prior to commercial manufacture of any new chemical and disclosure of information on toxicity with the objective of preventing distribution of toxic materials
Resource Conservation and Recovery Act	RCRA	1976	Listed 'acutely hazardous' and 'toxic' wastes separately, of chemicals to be disposed as hazardous wastes
Amendments to RCRA—Hazardous and Solid Waste. Disposal Amendments	RCRA (HSWDA)	1984	EPA bans all landfilling of hazardous wastes. Regulates small producers, prohibits non-containerized hazardous waste in landfills, provides minimum technical requirements for landfills, regulates underground storage tanks for wastes and medical wastes
Amendments to RCRA	RCRA	1992	Increased emphasis on recycling and toxic use reduction
Comprehensive Environmental Response, Compensation and Liability Act	CERCLA	1980	Designed to resolve all issues with abandoned, uncontrolled, inactive hazardous waste disposal sites. Identified 32 000 sites, 1200 put on priority list, 52 cleaned up to 1992
	CERCLA	1988	It set up: 1. A superfund to pay for clean-up (1200 sites) 2. A National Priority List—NPL 3. National Contingency Plan
Superfund Amendments and Reauthorization Act	SARA	1986	Revisions to Superfund including: new cleanup standards that require use of permanent remedies: sets a minimum number of superfund cleanup sites and specifies liability for improper hazardous waste disposal
Title III: The Emergency Planning and Community Right to Know Act	SARA	1986	Includes: emergency planning and notification. Community right to know reporting on chemicals, emissions, inventories. Strategy of first eliminating acute health threats and permanent remedies
Hazardous Materials Transportation Act	HMTA	1975	Governs safety aspects of hazardous materials transport—packing, handling, labelling, marking, placarding, transporting, unpacking
Hazardous Materials Transportation Uniform Safety	HMTWA	1990	Uniformity of regulations from state to federal. Governed by DOT, who register and charge a fee for hazardous materials. Train personnel, inspect paperwork, etc.
Pollution Prevention Act	PPA	1990	This Act requires the EPA to establish a source reduction programme including setting standards and determining effects of hazardous waste source reduction

parameters will be made more strict, while others will be relaxed. One area still to be expanded is that concerning solid waste, i.e. municipal solid waste. Many State programmes, particularly California and Florida, are ahead of the Federal Regulations with respect to waste minimization and recycling. The last half of the 1990s holds much debate for direction with respect to solid waste reduction and recycling.

Other areas that are addressed by United States environmental law include the following:

- Endangered Species Act (1993)
- Fish and Wildlife Coordination Act
- Wetlands Action Plan (1989)
- Wild and Scenic Rivers Act
- Wilderness Act
- Surface Mining Control and Reclamation Act
- Coastal Zone Management Act
- Ground Water Protection Strategy (1984)
- Indoor Air Quality
- National Estuary Program
- Pesticides in Groundwater
- Occupational Safety and Health Act
- Marine Protection, Research and Sanctuaries Act
- Acid Precipitation Act (1980)

1.8.2 US Quality Standards in the Water Environment

The water environment is here considered as:

- Drinking water
- Wastewater

Drinking water (including groundwater)   The key document on drinking water is the Safe Drinking Water Act (1974) and amendments. Tables 11.4 and 11.15 of Chapter 11 list the primary drinking water standards from the SDWA. Primary drinking water standards are those that are enforceable by law, while secondary drinking water standards are non-enforceable by law and are, as such, recommendations. The mandatory parameters are those that the states must comply with. However, many states have more stringent requirements. Briefly, the standards are subdivided as follows:

- *Primary standards*
  - Clarity (turbidity)
  - Microbiological (coliform bacteria)
  - Organic chemicals (list of 23)
  - Inorganic chemicals (list of 11)
  - Radioactivity (gross alpha and gross beta activity)
- *Secondary standards include*
  - Colour, odour, chloride, copper, foaming agents, iron, manganese sulphate, zinc, specific conductance, total dissolved solids, pH, hardness, sodium, calcium, potassium, magnesium, boron and nitrite

It is interesting to note that many of the secondary standards for use in the United States are 'mandatory' in the European Union. Also, many of the organic chemical standards that are mandatory in the United States are not mentioned in the EU regulations. It would seem that the US treatment facilities have achieved the standards on the basic physicochemical–organoleptic areas and have advanced to focus on the more insidious organic chemicals. See American Water Works Association (1990) for an excellent review of US water quality standards.

Wastewater   The principal document on wastewater standards is the CWA (1972) and its amendments. The objective of this Act was to restore and maintain the chemical, physical and biological integrity of US receiving waters. Progress was made in receiving waters quality by the introduction of limits on discharges for industry and municipalities. Under the National Pollution Discharge Elimination System (NPDES) Program, permits were issued. The minimum national standards for secondary treatment were set as 45 mg BOD_5/L and 45 mg SS/L for the average 7-day concentration. It was also set that the pH of the effluent at all times be in the range 6–9. The Water Quality Act of 1987 was a revision of the Clean Water Act and introduced more stringent standards and severe penalties for permit violations. The National Environmental Policy Act (NEPA) (1969) and its amendments require that an environmental impact statement (EIS) be prepared for all wastewater treatment plants. The impacts to be evaluated include social, economic, ecological, legal and political considerations. The development of an EIS for wastewater treatment projects is controlled by the Council on Environmental Quality Regulations as part of the NEPA. In 1993, major national regulations were introduced for the setting of standards for sludge disposal. This is 40CFR part 503 and has very detailed standards for land application, surface disposal, pathogen and vector attraction and incineration. Further details are included in Chapter 13.

Regarding surface water quality, the Clean Water Act of 1987 introduced a shift of emphasis. Rather than looking at categorizing surface waters, the future effort was to be spent in making sure that discharges to these waters were within standard. The Surface Water Treatment Rules (SWTR) of 1989 addressed those water bodies needing remedial action.

1.8.3 US Quality Standards in the Air Environment

The Clean Air Act was first introduced in 1955 and has been amended eight times up to 1992. The first Act was simple and provided federal assistance to states that had serious problems. Under the 1963 and 1967 amendments, the department responsible, the Health, Education and Welfare (HEW) Department, published criteria for national air quality standards and set up a legal, financial and technical framework to improve the nation's air and guarantee air standards by region. In 1970, the new EPA, in revising the CAA, established minimum air quality and regulatory goals that state and local governments were to achieve. Criteria pollutant standards were set for CO, NO_2, O_3, SO_2, PM_{10}, Pb, TSP (total suspended particulates) and HC. The CAA amendments of 1977 imposed strict requirements on areas not meeting criteria pollutant standards and toxic air pollutants. The 1990 CAA amendments, under the National Emission Standards for Hazardous Air Pollutants (NESHAP), set goals to reduce by 50 per cent the levels of 189 toxic air pollutants and by 90 per cent fugitive emissions from industry. In line with international policies on greenhouse gases, the 1990 CAA amendments set the goal of reducing SO_2 by 10×10^6 tons below that of 1980 levels. The CAA amendments of 1992 established a new Chemicals Safety Board to police the NESHAP standards.

1.8.4 US Quality Standards for Solid Waste

Waste is broadly subdivided into:

• Non-hazardous (most municipal waste)
• Hazardous

The Solid Waste Act of 1965 promoted a management system for waste that included resource recovery so as to improve the quality of air, land and water. Technical and financial assistance was provided to the states via the US Public Health Service. A national database program was initiated with the intent of improving collection, separation, recovery, recycling and disposal. Occupational training was initiated for operations in the waste business and guidelines and standards for the several processes from collection to disposal were initiated. New landfills and other significant projects related to waste management became the subject of required EISs under the National Environmental Policy Act (NEPA) of 1969. With responsibilities going to the EPA for administration, the Resource Recovery Act (1970), the Resource Conservation and Recovery Act (1976), set guidelines for solid waste management to the public. The RCRA established a legal framework for the implementation of the guidelines and the standards for solid waste handling, storage treatment and ultimate disposal. The Superfund or CERCLA (1980) established the mechanism of identifying and funding landfill sites that were the object of previous uncontrolled dumping, including solid waste and hazardous waste. RCRA was amended in 1984 and 1994 and essentially 50 per cent of all waste collected is to be diverted from landfills by the year 2000. The EPA philosophy for waste management is firstly waste minimization, then recycling and reuse, then energy recovery and, as a last resort, landfilling.

The 1965 Solid Waste Act included management guidelines for hazardous waste as well as non-hazardous waste. The Federal Insecticide, Fungicide and Rodenticide Act (FIFRA) (1972) required pesticide manufacturers to inform the EPA of their products with special reference to the unreasonable effects on the environment. FIFRA was updated in 1988 and at that time the national register of pesticides contained 23 000 entries. In 1976, the Toxic Substance Control Act (TSCA) mandated testing prior to the commercial manufacture of any new chemical and the public disclosure of data on toxicity with the goal of preventing the distribution of toxic materials. The Resource Conservation and Recovery Act (1976) listed 'acutely hazardous' and 'toxic' wastes separately, with chemicals to be disposed as hazardous wastes. RCRA has been updated in 1984 and 1994, and today's emphasis is on hazardous waste reduction at source (through the Pollution Prevention Act of 1990) and proper disposal of end wastes. Because of the public disquiet of the 1970s regarding landfill and other sites that were up to then badly managed, the Comprehensive Environmental Response, Compensation and Liability (CERCLA) was introduced in 1980. Also known as Superfund, this law set out to identify the badly managed sites and to clean them up. The plan was that the task would take five years. By 1985, CERCLA was bogged down in the legalities of ownership, responsibility, and engineering consulting and legal fees. It was reauthorized as the Superfund Amendments and Reauthorization Act (SARA) in 1986. To date, about 1200 sites have been identified as badly contaminated and needing to be cleaned up, but less than 100 have undergone permanent remedial treatments. The National Priority List is continuously updated as more data become available. Title III of SARA of 1986 instituted emergency planning and a community right to know reporting on chemicals and emission inventories.

Other Acts of relevance are the Hazardous Materials Transportation Act (1975) and its amendments, the Hazardous Materials Transportation Uniform Safety Act (1990). The latter instituted uniformity of regulations from state to state and is policed by the Department of Transport (DOT). Regulations for hazardous materials transport are in place and standardized for their packing, handling, labelling, marking, placarding, transporting and unpacking.

1.8.5 US Quality Standards in Other Areas

There are many areas besides the air, water and waste environments which are relevant to the environmental engineer and scientist. Included among these are:

- Conservation
- Public lands

There are several major legislative Acts under each of the above.

Conservation   The significant Acts in chronological order are:

1. *The Multiple-Use Sustained-Yield Act of 1960*   Congress deemed that the national forests are established and administered for outdoor recreation, range, timber, watershed, and wildlife and fish purposes. The Secretary of Agriculture was authorized and directed to develop and administer the renewable surface resources of the national forests for multiple use and sustained yield of the several products and services from these lands. The establishment of areas of wilderness are consistent within the provisions of this Act.
2. *The Coastal Zone Management Act of 1972*   Congress deemed that there was a national interest in the effective management, beneficial use, protection and development of the coastal zone as it was rich in a variety of natural, commercial, recreational, ecological, industrial and aesthetic resources, many of these fragile. It is national policy to encourage and assist the states to effectively protect, develop and enhance these resources, including floodplains, wetlands, estuaries, beaches, dunes, barrier islands, coral reefs, and fish and wildlife habitat. Policy also included assisting in the redevelopment of deteriorating urban waterfronts and ports and sensitive restoration of historical sites.
3. *The Endangered Species Act of 1973*   Congress declared that various species of fish, wildlife and plants in the United States have been rendered extinct as a consequence of economic growth and that other species are so depleted that their continuity is threatened. In keeping with international agreements, federal assistance to the States as a system of incentives is encouraged so as to develop and maintain conservation programmes that also maintain the ecosystems for these species. The federal policy is such that federal departments and agencies are to conserve endangered species and to use their authority to do so, and also to co-operate with State and local agencies to resolve water resource issues in concert with conservation of endangered species.
4. *Forest and Rangeland Renewable Resources Planning Act of 1974*   With amendments in 1978 and 1980, these Acts promote a sound technical and ecological base for the effective management for the nation's renewable resources. Because most of the forest lands are in private, State and local government ownership, Congress planned that the Federal Government is to be a catalyst to encourage and assist the owners in the efficient long-term use and improvement of these lands and their renewable resources, consistent with the policy of sustained yield and multiple use.

Public lands   The significant Act is the Federal Land Policy and Management Act of 1976. Congress declared that the US policy was that public lands be retained in federal ownership unless a parcel for disposal was deemed in the national interest. That the national interest is best served if the resources are periodically and systematically inventoried and their future use planned for. The public are to be informed of any proposed change of use or change of ownership. The lands are to be maintained in a manner that protects the quality of scientific, scenic, historical, ecological, environmental, air and water resources, and archaeological values. This also means the provision of food and habitat for fish, wildlife and domestic animals so as to provide outdoor recreation, and human occupancy and use. The management of such lands would be on the basis of multiple use and sustained yield.

1.9 COMPARISON OF EU AND US ENVIRONMENTAL LEGISLATION

Environmentalism is one of the most significant 'movements' of the second half of the twentieth century. The movement was essentially led by the United States and some northern European countries. New Zealand and Canada also showed a healthy regard for the environment. The beginning of the legislative era started with the Clean Air Act in 1955 in the United States and progressed slowly for a decade to 1965 with the Solid Waste Act, and then in 1970 with the establishment of the USEPA and a flourish of legislation during the 1970s. Superfund was set up in 1980 and SARA was reauthorized in 1986. The latter half of the 1980s and the 1990s also saw a proliferation of legislation in the hazardous and solid waste area. Amendments were instituted to the Clean Air Act and also the Clean Water Act.

European environmental legislation developed much more slowly. While some countries in northern Europe were making legislative progress, the European Economic Community, established in 1958, had no direct mandate to legislate for environmental matters until the Single European Act was signed in 1986. However, using the criteria of reducing barriers to trade, the EEC was able to establish legislation on environmental matters since 1970. Since then, significant legislative pieces have been introduced in the areas of air, water, soil, waste and other environments. The European Economic Community (EEC) has had name changes to the European Community (EC) (1986) and then to the European Union (EU) (1993). The volume of United States environmental legislation surpasses that of the European Union.

In the area of water treatment (potable), groundwater, surface waters, wastewater treatment and wastewater discharges, both the United States and European Union have reasonably equivalent standards and guidelines. There are differences, for instance, where the United States is now more focused on the objective of eliminating synthetic organics in waters and less so on the physical–chemical parameters of water quality (as this is conventional technology). The future changes in the water environment are likely to be associated with the synthetic organics, which are numerous, and their reactivity in the different water/soil/sediment environments is not fully understood.

The US legislation on waste, including hazardous waste, is more developed than that of the European Union. There is still a need in the European Union for elaboration of solid waste disposal, particularly in the areas of landfilling and alternatives to landfilling. For instance, the United States Sewage Sludge Regulations of 1993 are excellently detailed and far more elaborate than their EU counterparts. On hazardous waste, the European Union is still in the area of feasibility and identifying the hazardous sites, while the United States Superfund (1980) has provided finances for the remediation of hazardous sites. Also, it appears that the legislation on air quality is more advanced in the United States, having started in 1955 and with several amendments since then. Competent authorities in parts of the European Union still make use of the German regulations on air emissions (TA Luft) in the absence of an integrated European Union standard.

An invaluable study resource for environmental legislation is USEPA via the Internet. Less detailed access is also available for the EU and many other countries on their legislation.

1.10 PROBLEMS

1.1 Write a review of EU environmental legislation as it pertains to drinking water (use other references).

1.2 Write a review of US environmental legislation as it pertains to drinking water (use other references).

1.3 'Engineers have contributed more to the public health of mankind than medical doctors'. Write a two-page essay supporting the above.

1.4 Write a two-page essay against the statement of Problem 1.3.

1.5 Locate the relevant EU directives on municipal solid waste and write a two-page review of the state of the legislation.

1.6 Repeat Problem 1.5 for the US.

1.7 Compare the state of environmental legislation in Japan and Australia with regard to wastewater discharges to the river environment.

1.8 'Environmentalism is a spiritual movement'. Critique the above statement with specific reference to the state of the environmental legislation in New Zealand and the input to it of the Maori culture.

1.9 Identify the key pieces of international environmental legislation that would be used against me in a court of law, should I as an industrialist, dump a hazardous waste into an upstream reach of the River Danube.

1.10 'The US is over-legislated with regard to industrial liquid waste discharges'. With regard to the above, identify the key legislative pieces and suggest those that you would 'water down' to support the above statement.

REFERENCES AND FURTHER READING

American Water Works Association (1990) *Water Quality and Treatment. A Handbook of Community Water Supplies.* 4th edn, McGraw-Hill, New York.

Arbuckle, J. G. *et al.* (1989) *Environmental Law Handbook*, 10th edn, Government Publications Inc., Maryland.

Arbuckle, J. G., *et al.* (1993) *Environmental Law Handbook*, 12th edn, Government Institutes, Rockville, Maryland.

Arden, E. and W. T. Lockett (1914). 'Experiments on the oxidation of sewage without the aid of filters'. *J. of Soc. Chem. Ind.* Vol. 33, pp. 523–1122.

Bockrath, J. (1977) *Environmental Law for Engineers, Scientists and Managers*, McGraw-Hill, New York.

Carson, R. (1962). *Silent Spring*, Haughton-Mifflin, Boston.

Cashman, L. (1993). Lecture on environmental law to the Environmental Engineering Dept., at University College, Cork.

Camp Dresser & McKee (April 1993). *A Guide to EPAs New Sludge Regulations.*

Commission of the European Communities (March 1992) *The State of the Environment in the European Community. Com(92).*

Cunningham, W. P. and W. Saigo (1990) *Environmental Science*, W.C. Brown, New York.

Denney, R., M. Monahan and D. Black (1993) *California Environmental Law Handbook*, Government Institutes, Rockville, Maryland.

Department of Health, New Zealand (1992) *Public Health Guidelines for the Safe Use of Sewage Effluent and Sewage Sludge on Land.*

Duggan, F. (1992) *EC Environmental Legislation—A Handbook for Irish Local Authorities*, Environmental Research Unit, Dublin, Ireland.

Environmental Law Society (1992) *Environs. Environmental Law and Policy Journal*, University of California, Davis.

Federal Register (1 July 1988) *Secondary Treatment Regulations*, 40CFR Part 133.

Findley, R. W. and D. A. Farber (1992) *Environmental Law in a Nutshell*, West Publishing Co., Minnesota.

Freedman, M. and B. Jaggi (1993) *Air and Water Pollution Regulation. Accomplishments and Economic Consequences*, Quorum Books, New York.

Henrichs, R. (1988) '*Law, literature review,*' *J. WPCF*, **60** (6).

Johnson S. P. and G. Corcelle (1989) *The Environmental Policy of the European Community.*, Graham and Trolman Publishers, London.

Mackenthun, K. M. and J. I. Gregman (1992) *Environmental Regulations: Handbook*, Lewis Publishers.

Middlekauff, R. D. (1975) *Water Quality Control Legislation*, Practising Law Institute, New York.

Noyes, R. (1991) *Handbook of Pollution Control Processes*, Noyes Publications, Park Ridge, New Jersey.

O'Brien and Gere Engineers Inc. (1988) *Hazardous Waste Site Remediation. The Engineer's Perspective.* Van Nostrand Rheinhold, New York.

Osmanczy, E. (1990) *Encyclopedia of the United Nations and International Agreements.* Taylor and Francis Publications, New York.

Peavy, H. S., D. R. Rowe and G. Tchobanoglous (1985) *Environmental Engineering*, McGraw-Hill, New York.

Petulla, J. M. (1987) *Environmental Protection in the United States. Industry, Agencies, Environmentalists*, San Francisco Study Center, pp. 34–35.

Robinson, Snr. M. (1976) *Environmental Legislation*, Praeger Publications, New York.

Scannell E. *The Law and Practice Relating to Pollution Control in Ireland.*

Seinfeld, J. H. (1986) *Atmospheric Chemistry and Physics of Air Pollution*, John Wiley & Sons, Inc., New York, Table 2.2, p. 53.

Selected Environmental Law Statutes, (1993–4) Educational Edition, St Paul, West Publishing Co., St. Paul, Minnesota.

Stolaff, N. (1993) *Environmental Law Dictionary*, Oceana Publications Inc., New York.

Sutherland, T. E. (1993) *Environmental Law*. In *Greener Buildings, Environmental Impact of Property*. S. Johnston (ed.), Macmillan, Basingstoke.

Vesilind, P. A., J. J. Peirce and R. F. Weiner *Environmental Engineering*, 3rd edn, Butterworth-Heinneman, Oxford.

Water Environment Federation (1991) *Manual of Practice No. 8* (ASCE Manual and Report on Engineering Practice No. 76), *Design of Municipal Wastewater Treatment Plants*, Vol. 1.

Water Environment Federation (1993), Alexandria, VA, USA. *Standards for the Use and Disposal of Sewage Sludge*, 40 CFR Parts 257, 403, 503.

ECOLOGICAL CONCEPTS AND NATURAL RESOURCES

2.1 INTRODUCTION TO THE ECOLOGICAL PERSPECTIVE

The multifaceted role of present-day environmental engineers demands a greater understanding of the functioning of living systems and of their interaction with the environment on which the work of the engineer is based. This is the role of this chapter as a scene setter—the essential background of ecological concepts and natural resources—with a qualitative emphasis. See the references on page 51 for the quantitative side. As a starting point, we need to understand the currency we are dealing with—what is the environment? The global environment can be divided into two major components as depicted in Fig. 2.1.

Environmental engineers use and manipulate physical resources such as natural energy (waves, wind, hydroelectric) and water (for domestic supply and waste transport). They alter the topography of terrestrial and aquatic systems through road building and through structures for flood alleviation, protection from erosion, etc., creating new physical settings in which living systems have to exist and function. However, as shown in Fig. 2.1, the physical and chemical (abiotic) components are only one part of the natural environment and, as many would argue nowadays, are not as important as the biotic component of living organisms to the well-being of the human species and the earth as a whole.

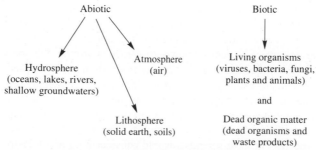

Figure 2.1 The major components and subcomponents of the natural environment.

Natural living systems supply humanity with an array of indispensable and irreplaceable services that support our life on earth (Erlich, 1991). These include direct resources such as building products (wood), food, medicines, clothing materials, etc. Living systems also provide functional services such as maintenance of the appropriate mix of atmospheric gases, generation and preservation of soils, disposal of wastes, restoration of systems following disturbance, control of pests, cycling of nutrients and pollination of crops. Thus, not only is humanity totally dependent on the living environment but the integrity of the planet is itself dependent on the maintenance of the natural environment and on the interactions between the living organisms and the physical/chemical components of the earth.

2.2 THE VALUE OF THE ENVIRONMENT

Natural resources may be renewable, non-renewable or abstract. Non-renewable resources include fossil fuels, minerals, clear-felled tropical hardwoods that are not replaced and rare animals or plants that are hunted or collected in an uncontrolled way. Renewable resources include energy from the sun and the biological and biogeochemical cycles (such as the water and energy hydrological and carbon cycles). At a more immediate level, renewable resources include forests that have been selectively felled and replanted, animal and plant populations that have been properly managed through controlled hunting, fishing and collecting, and waters with controlled inputs that can be readily recycled and reused. Abstract resources include animals, plants and the natural landscape as part of 'the countryside' used for recreation and tourism activities such as bird watching, fishing, hiking, sight-seeing, etc. Non-renewable resources are of course finite, while the other two categories are effectively infinite provided they are not overutilized or damaged. Our descendants will not thank us for exhausting finite resources, nor for destroying the renewable ones.

The biotic diversity of living systems must therefore be viewed as a common property resource for all mankind. It is subject, however, to 'the tragedy of the commons' (Hardin, 1968). The commons (in British terms) were communal areas set aside for villagers to graze their livestock. The tragedy was that individuals were free to increase their exploitation by grazing more and more animals without paying the additional costs of keeping the commons in a productive state. The costs were shared by all the villagers although the benefits accrued only to the individual who owned the livestock. The result was overgrazing and deterioration of the resource. Common property resources in today's world include scenic landscapes, wilderness areas, migratory animals including whales, biodiversity and clean air and water. The tragedy of the commons equally applies to these.

2.2.1 Environmental Auditing

Decisions on whether or not various engineering or other developments go ahead in an area are often based on economic considerations. However aesthetically pleasing an environment is, the economic considerations require that we also put a specific value on it. So how can we do this? The first step should be an *environmental audit*, a listing of all the resources of the area. This is fairly straightforward for non-biological resources such as fossil fuels or minerals but is not so straightforward for many biological resources. The number of species presently known to science is around one and a half million, but how many species are there yet to be discovered? Various methods have been used to try to estimate the number of species that actually exist (global biodiversity), e.g. by extrapolating from areas with well-known faunas and floras (such as temperate forests) to those less well known (such as the tropical rainforests) or by looking at the number of new species of each group of plants and animals added per unit of time searching or area searched in, say, the rainforest. Estimates of global biodiversity range from 2 million to 50 million species but most scientists would consider a figure of 20 to 30 million to be about right.

Table 2.1 Some components, functions and attributes of the environment viewed in relation to their value to mankind

Characteristics	Direct use	Indirect use	Non-use
Components			
Wild fauna resources	+		
Wild flora resources	+		
Agricultural resources	+		
Water supply	+		
Functions			
Sediment retention		+	
Nutrient retention		+	
Shoreline stabilization		+	
Flood control		+	
Ground water recharge		+	
Water transport	+		
Tourism and recreation		+	
Attributes			
Biological diversity	+	+	+
Heritage			+

Modified from Aylward and Barbier, 1992

Having arrived at an audit of biodiversity, we would need to examine the known value to mankind of the existing species, then estimate the, as yet, unknown value of those new species and finally extrapolate to the total value of the particular environmental area under study, which includes all the still to be discovered species and their yet to be discovered value. The uses of the living environment may be of direct or of indirect value (Table 2.1), but first let us consider some of the direct uses, which include foodstuffs, industrial and commercial products, tourism, recreation and medicines.

Food All our food ultimately comes from other organisms, most of which are cultivated plants or domesticated animals. Most plant food comes from about twenty major crop species, while the bulk of meat supplies originate from an even smaller number of domesticated animal species. Seafood is an exception, aquaculture currently supplying only about 10 per cent of the 70 million tons harvested annually, but this will change as wild stocks continue to decline. This dependence on a very small number of domesticated animal and plant species for our food supplies is a result of historical chance rather than design. Many wild plants and animals have great potential as human food sources. Indonesians make use of some 4000 native wild plants in their diet, but few of them have been explored for potential domestication and culture, while less than one-quarter of the 250 different fruits eaten by the peoples of Papua New Guinea have been cultivated.

The inhabitants of the Brazilian rainforest also utilize a wide range of food resources from the forest, among which is a palm tree called 'Milpesos' (*Jessenia bataua*). An amino acid analysis of this plant has revealed that the protein is equivalent to that of prime beef and 40 per cent higher than the biological value of soya bean protein (Balick and Gershoff, 1981). Such high-yielding plants of small size with large fruit clusters could be selected as an ideal crop for the lowland tropics (Balick, 1985), where high protein sources are especially desirable.

Those few crops that we do currently cultivate are derived from a very small number of plants, sometimes a single plant. They therefore have a very reduced genetic base. This may result in low disease resistance, inability to grow in a wide range of climatic conditions and so on. The wild populations from which they were originally collected have a large gene pool and hence much greater variability. The

Table 2.2 Estimated performance of cows and capybaras in the conditions of the Pantanal, Brazil

Species	Individuals for each 3 hectares	Age (years) at withdrawal	Weight at withdrawal	Median weight gained in each hectare (g/day)
Cow	0.1	4.5	490	283
Capybara	18	1.5	35	1134

Data from Dourojeanni, 1985

potato, for example, is subject to diseases such as blight and to infestations by eel worms. Genes conferring resistance to such diseases and pests may well exist in wild populations and these could be introduced into cultivated forms to improve them in many ways as long as adequate populations exist in the wild. Extinction is not the issue here but maintenance of sufficiently large wild populations to supply a repertoire of genes for selection by plant breeders.

The situation is no different where animals are concerned, with domestic stocks often derived from a small population of animals. The cow is domesticated almost world-wide as a source of animal protein, despite the fact that it is not well adapted to many habitats, such as the tropics. In Venezuela attempts have been made to domesticate the capybara, a large native rodent, as an alternative to the cow. The yields have been shown to be superior to that from cows under similar conditions (Table 2.2).

Industrial and commercial products   Industrial and commercial products provided by the environment include minerals and fossil fuels from the abiotic component; wools, hides, feathers and bone meals from animals; cotton, sisal, jute, rubber, waxes, resins, oils, gums and tannins from plants, to name just a few. Add to this thousands of as yet unrecognized products. The most commercially valuable of all products that we obtain from plants, however, is wood. Trees are cultivated in plantations, but mainly for whitewood. Hardwoods are such slow growers that we depend upon harvesting from natural forests and over 85 per cent of hardwoods come from virgin tropical forests. Most tropical forests are cut down completely with no selective logging and no replanting schemes.

Medicines   Wild plants and animals are a very important source of drugs, analgesics, antibiotics, anticoagulants and antiparasitics. Over half of all prescription drugs contain some natural products. A major success story is the Madagascar periwinkle (*Catharanthus roseus*). Alkaloids from this plant inhibit cancer cells and have contributed massively to the reduction in mortality from childhood leukaemia, as well as in improvements in survival from Hodgkin's disease. This plant is now widely cultivated, although Madagascar profits little from the exploitation of its flora. There are millions of possibly useful substances in living organisms, because life has had billions of years to evolve a repertoire of antipredator, antiparasite and antidisease chemistry. At the moment there is increasing interest in coral reef organisms as a source of antibiotics. Coral reefs, like rainforests, are highly diverse, so that competition between organisms for living space has resulted in the development of an array of chemicals for killing or dissuading competitors and reducing predation. Rainforests may contain many cures but they are also the source for very serious diseases like the HIV and Ebola viruses, which are most likely sourced from animals (monkeys) living in the rainforest.

Tourism and recreation   In many cases, land is more valuable as a wildlife preserve than it would be if converted to crops. The economic yield from tourists who come to see a lion in Amboseli National Park, Kenya, is equal to the income from a herd of 3000 cows. Hunting of a lion, including permits, vehicle hire, assistants, food, etc., would bring in only about 5 per cent of the revenue generated from tourism. Activities such as bird watching and fishing are enjoyed by millions of people world-wide. In the United

States there are 8 million bird watchers and 30 million anglers. The total amount spent each year in America on these activities is several billion dollars (Cunningham and Saigo, 1990).

2.2.2 Biodiversity

With regard to indirect uses of the environment, it is the biological processes themselves that provide the value (Table 2.1). Indirect uses are particularly difficult to put an economic value on, but are dependent to a large extent on the value of biodiversity. Pearce (1990) calculated a global damage cost for climate change of US$13 per tonne of carbon emitted into the atmosphere. He further estimated the indirect carbon credit due to a single hectare of conserved diverse tropical rainforest to be US$1300. A correct valuation of national and international indirect uses of managed and unmanaged natural systems might alter significantly perceptions of the worth of such systems (Aylward and Barbier, 1992).

Biodiversity does not imply just a collection of species. In the same way that our bodies are more than simply the sum of the parts, so the natural environment is an ordered complex in which the species are harmoniously co-adapted. Just as our bodies are made up of cells combined into tissues, which in turn make organs and finally a whole body, as we will see later, the environment consists of populations of species which together with other species form communities. Many communities combine to form landscapes, and finally culminate in global biodiversity. Biodiversity is a basic resource which acts as a human life-support system. Soil formation, waste degradation, air and water purification, nutrient cycling, solar energy absorption and maintenance of biogeochemical and hydrological cycles all depend upon plants and animals. Natural biological systems are the culmination of billions of years of evolution and they maintain ecological processes at no material costs to us.

Not all species play an equally important role; some species are 'key' species while others play a supporting role. When rabbits are removed from open grassland, shrubs freed from grazing pressure may take over and the whole appearance of the habitat changes from open grassland to impenetrable thickets. Limpets play a similar role on rocky shores, controlling the growth of seaweeds. Remove the limpets and open rocks with barnacles soon become clothed in dense growths of seaweed. Some other organisms, by contrast, can be removed without any obvious effect and their place is quickly taken by existing species or by newcomers. Nevertheless, ecosystems are highly complex, dynamic and little understood. To continue the analogy with the human body, damage to an organ can cause serious repercussions to health. So also the removal of a species can lead to knock-on effects so that a whole system becomes greatly impoverished or collapses. Recent experimental work in the laboratory-based 'ecotron' has shown that reduced biodiversity does impair the services that ecological systems provide, such as overall productivity, stability against disturbance and nutrient recycling (Naeem et al., 1994).

We can recognize three kinds of loss of biological resources; firstly the depletion of a once-common species, secondly local or global extinction and thirdly ecosystem disruption.

Depletion of a once-common species   This loss has occurred to many species. For example, vast herds of buffalo which once roamed the prairies of America have been replaced by cows. These herds, like other previously abundant species, could be restored if humanity so wished, because much of the habitat still exists and the species still survives in small numbers in wildlife refuges. The reduction in the numbers of a species, however, can reduce the gene pool of that species and hence the loss of some characteristics, so restored populations may suffer lowered ability to survive in a changing environment.

Local or global species extinction   Extinction is forever, so when a species is lost, we lose not only all the characteristics of the species at the time of extinction but also all of the potential adaptations that might have appeared in future offspring. Extinction is a natural process and the final end of all species. It has been occurring regularly since life evolved 3 billion years ago. However, it generally occurs at a slow background rate which allows natural ecological systems to adjust, as new species arise or existing species expand to replace the extinct species. There have been several mass extinctions in the past, the last of which occurred some 60 million years ago. It is also true that biodiversity rebounded after this mass extinction, but that took millions of years and continued until very recently when humanity originated the current mass extinction. The present rate of species loss from deforestation alone is about 10 000 times greater than the rate of natural extinction prior to the appearance of humans on this planet (Silver and DeFries, 1990). It is probable that the earth could sustain a mass extinction of life and bounce back again as it has done in the past, but are we prepared to wait the millions of years necessary for the recovery to occur and will the human race still be around to see it happen?

Ecosystem disruption   Beyond a certain point, ecological systems can become so impoverished that major ecological processes are disrupted and a catastrophic decline sets in. This is the most serious of all losses. Humans are directly responsible for habitat destruction and environmental degradation and this promises to escalate as the human population rises in the coming decades with an estimated 6 billion people by the year 2000 (Kim, 1993). Accelerated destruction of world-wide biodiversity is threatened and with it ultimately the capacity of the earth to sustain human life.

2.2.3 Cost–Benefit Analysis

Traditionally in economics, the state of the environment has not been used as a criterion to influence calculations of economic optima. Environmental services such as pollution absorption by biotic systems are not marketable and many environmental factors such as ecosystem stability, scenic and recreational values, historic importance and so on are not readily quantifiable. Such elements were always excluded from the analyses. It is even more difficult to put a monetary value on such things as beauty or health. With the publication of the World Conservation Strategy (IUCN, 1980), the idea of sustainability entered the equation and environmental parameters needed to be incorporated into calculations.

Sustainability involves the management of all assets and natural and human resources to increase long-term wealth and well-being for all. Sustainable development rejects policies which deplete the productive base and leave future generations with poorer prospects and/or greater risks than our own. Technologies that contribute to sustainable development include pollution control, renewable energy production, recycling and resource recovery, resource management and scientific research. Working principles include the 'safe minimum standards' (SMS) criterion (Bishop, 1978; Bishop and Ready, 1991), which attempts to provide a decision-making process for problems involving long periods of time, large uncertainties (such as extinction probabilities and irreversibilities) and the 'minimax rule' (minimize maximum losses), a necessarily pessimistic decision rule biased towards conservation. The SMS approach makes conservation the preferred option unless it can be demonstrated that the social cost of conservation (i.e. the foregoing of benefits) is unacceptably large (Pearce and Turner, 1989).

2.2.4 Environmental Ethics

Environmental ethics pose questions about the morality of the relationships between humans and the rest of nature. Do humans have obligations, duties and responsibilities to the natural environment? If so, how do we weigh these obligations and responsibilities against human values and interests? We must eradicate our feelings of superiority over the rest of the natural world and develop, as the theologian Martin Buber put it, an 'I–thou relationship' rather than an 'I–it relationship' with the environment (Cunningham and Saigo, 1990).

2.3 LEVELS OF ORGANIZATION IN THE BIOTIC COMPONENT OF THE ENVIRONMENT

One of the major axioms in ecology is that everything in the global environment is connected to everything else, so that changes in one component can affect many others over both space and time. This is clearly seen when we consider the nested levels in the organizational hierarchy of ecological systems on earth. At the lowest level, cells are the basic structural and functional units of life while organisms are the active processors of matter and energy. Six major levels of ecological organization are recognized.

2.3.1 Individual

These have physiological functions and respond to environmental conditions. Individual organisms belong to a species that includes all the individuals potentially able to interbreed with one another and produce fertile offspring. An example of a species is the salmon *Salmo salar*.

2.3.2 Population

This consists of a group of individuals of the same species living in a particular area at the same time. Each population is genetically distinct to some degree from other separate populations of the same species. They have a size and a birth, death and hence population growth rate.

2.3.3 Community

Populations of different species live together, many interacting with each other, forming a community, e.g. in a pond—a natural community of plants, animals and microbes forming a distinctive living system. These interactions lead to the formation of food webs, a hierarchy of who eats who. Communities occur in *habitats*, which refers to the kind of physical environment or place determined by the topography, vegetation structure, geology and surrounding medium (air or water), e.g. forest habitat, seashore habitat, etc. Some species live in just one habitat, e.g. fish in a lake. Others may use several different habitats, e.g. bird species like the crow use fields, seashore, hedgerows, etc.

Within each habitat one can describe for a species its place within the community—a combination of what it does and where it lives; this is termed the species *niche*. Competition occurs between similar species or species with similar niches for resources that are in short supply.

2.3.4 Ecosystem

This encompasses both the living (biotic) and non-living (abiotic) components of an area—a combination of the community and physical and chemical components of the local environment. The major feature of this ecological level is the strong interaction between the biotic and abiotic components as illustrated in Fig. 2.2. Major processes like nutrient cycling and energy flow occur at this ecological level (see below).

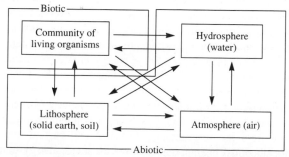

Figure 2.2 The dynamic nature of the ecosystem due to the interactions between and interdependence of the various components.

2.3.5 Biomes

Where environmental conditions (e.g. climate) are similar in different parts of the country, or on a larger scale in different parts of the world, the habitats (e.g. in terms of vegetation type) and communities are also often similar. We can thus discern a higher level of organization, the biome, e.g. tropical rainforest (high temperature and rainfall), coniferous forest or Taiga (low winter temperatures), grassland (warm temperatures, fairly low rainfall), desert scrub (high temperatures, very low rainfall).

2.3.6 Biosphere

The highest organizational level is the biosphere—that part of the earth and atmosphere in which life exists. It includes the surface layer of land, the oceans and sediments at the bottom of water bodies, and part of the atmosphere occupied by life. Large-scale biogeochemical cycling of materials (both natural elements and pollutants) occurs at this ecological level. The ultimate aim of ecological study is to understand how the biosphere functions (as if it were one large ecosystem) in order to be able to predict what affects certain human activities will have in the future and how to ameliorate the problems already caused.

2.4 ECOSYSTEM PROCESSES

As seen above, the ecosystem consists of a series of interacting biotic and abiotic sub-compartments (Fig. 2.2) and two important ecological processes act as the major linking pathways between these compartments. These two processes, energy flow and nutrient cycling, are essential for the survival and maintenance of the biotic environment.

2.4.1 Energy Flow

One of the most important interactions between living organisms and their environment is in the provision of food. This involves not only the supply of energy to survive but also of raw material for construction of body tissues and gametes for reproduction of the species. On earth, the ultimate source of energy for life is solar radiation or light, and this is eventually re-radiated back to space as heat (see Chapter 4). It is the change from non-random energy (light) to random energy (heat) that allows work to be done, and this drives life on earth.

Energy sources   In ecosystems, a proportion of the light energy is converted to chemical energy, which is the energy currency of living systems. This is stored in either living or dead organic matter (carbon-based compounds). In living organisms, some organic matter is converted to a certain chemical complex (called ATP, adenosine triphosphate), which is itself broken down during metabolism to release the stored chemical energy and allow work to be done (e.g. locomotion, cell division, biochemical reactions). In the environment there are essentially two sources of energy: autotrophic and heterotrophic. Autotrophic production of energy rich organic matter is carried out within the ecosystem by green plants in the presence of light via photosynthesis (see below). Some energy is also produced in deep-sea ecosystems around hydrothermal vents by sulphur-oxidizing bacteria. Green plants and chemosynthetic bacteria are called autotrophs. In contrast, a heterotrophic energy source is one where the chemical energy is imported as organic matter which originated from primary production in some other ecosystem (Ricklefs, 1979). An example is in heavily shaded forest streams, which are dependent on organic matter, such as dead leaves, entering the stream from the surrounding catchment and carried down from upstream. This imported (often called allochthonous) organic matter forms the major energy source on which the stream community is built.

Photosynthesis   All green plants create their own food through a complex series of chemical reactions driven by solar radiation (using pigments called chlorophylls). They can synthesize energy-rich organic molecules such as glucose containing chemical energy from carbon dioxide and water with the release of oxygen:

$$12H_2O + 6CO_2 + \underset{\text{(from light)}}{709 \text{ kcal}} \xrightarrow{\text{chlorophyll+enzymes}} \underset{\text{(carbohydrate)}}{C_6H_{12}O_6} + \underset{\text{(to air)}}{6O_2} + 6H_2O \qquad (2.1)$$

Photosynthesis was not only a major step for life, it was also a major step for the environment (Ricklefs, 1979). Firstly, photosynthesis releases oxygen, creating an atmosphere whereby organisms requiring oxygen can survive, and, secondly, through the formation of oxygen complexes an ozone layer has resulted, which blocked the penetrating rays of ultraviolet light and thus permitted life on land.

Photosynthesis is carried out only in daylight in leaves and often in the stems of green plants. It is only light in the visible spectrum that can be used and much of this is lost before it reaches the plant through reflection, scattering and absorption in the atmosphere and by clouds and reflection from plant surfaces. Of the total amount of solar radiation available, only 1 to 5 per cent is used in photosynthesis. The glucose produced by the plants may simply be stored as an energy-rich substance in the form of starch or be combined with other sugar molecules to form carbohydrates like cellulose, used in plant cell or tissue construction. Plants also require inorganic substances like nitrogen, phosphorus, magnesium and iron. These are obtained from the soil or in the case of some plants (like clover) nitrogen can be obtained from the atmosphere with the help of bacteria living in the roots. These substances, together with glucose, are then structured via chemical reactions into complex substances like fats, proteins and nucleic acids, which in turn are used to form plant tissues, etc. The production of organic matter by plants is called *primary production*.

When any organism requires energy, essentially the reverse chemical reaction to photosynthesis occurs, called respiration. Respiration is where the glucose molecule is broken down in the presence of oxygen to yield carbon dioxide, water and energy:

$$C_6H_{12}O_6 + 6O_2 \xrightarrow{\text{metabolic enzymes}} CO_2 + H_2O + \text{energy for work and maintenance} \qquad (2.2)$$

In animals, more complex organic molecules like fats and proteins can also be respired to produce energy for work. A waste product of respiration and work is heat, which is ultimately re-radiated back into space.

Primary production   Because the rate of photosynthesis and primary production is so vital, the amount of primary production in different ecosystems is an important parameter to consider. *Gross primary production* is the total amount of chemical energy (or biomass) stored by plants per unit area per unit time. However, since plants require energy for synthesis of organic matter and functioning of the plant itself, some of the gross primary production is used in the process of respiration. The remaining production, *net primary production*, can then be used in plant growth and reproduction; thus:

$$\text{Net primary production} = \text{gross primary production} - \text{respiration} \qquad (2.3)$$

This is normally 80 to 90 per cent of gross primary production levels. The primary production is affected by a number of environmental factors, but the key ones are water, light, soil nutrients and temperature. Net primary production levels vary considerably over the globe but can be classified into four broad groups, each with a characteristic productivity range. It is only this net primary production which is available for harvest by mankind or other organisms and which is passed along various food chains leading to the flow of energy through ecosystems. The *normal range* (1000 to 2000 g/m² yr) is found in many forests, some grasslands and highly productive temperate crops.

Low range (0 to 250 g/m² yr) ecosystems include deserts, semi-desert, Arctic tundra and tropical ocean water. The *middle range* group (250 to 1000 g/m² yr) includes non-forest communities limited by drought, cold, nutrients, etc., e.g. shrublands, tropical mangroves, grassland and most cereal crops.

Finally, the *very high range* (2000 to 3000 g/m² yr) includes rainforest, marshes, temperate inshore ocean water and intensive tropical crops, e.g. sugar-cane, rice. These ecosystems are all provided with plenty of water, warm temperatures and continuous nutrient replenishment.

Food chains and food webs   Autotrophs can manufacture their own food, but two other groups of organisms, saprobes and animals, collectively known as heterotrophs, cannot. All heterotrophs, whether directly or indirectly, are dependent on producers as the primary source of food. Saprobes or decomposers feed by absorption of dead organic matter, e.g. certain bacteria, yeasts and all fungi. Animals feed by ingesting ready-made organic foodstuffs from living or dead organisms (carbohydrates, fats and proteins).

Chemical energy produced by primary producers and the nutrients used by plants to build plant tissues are passed up through a chain of consumers—the food chain—providing each link in the chain with energy and nutrients. Each consumer population uses the food energy consumed to live and respire and the remaining energy can then be used to help produce new biomass by growth and reproduction. This production of new biomass by the consumer population is called *secondary production*. Secondary production of one consumer population then becomes a potential food and energy source for another further up the food chain. Ecosystems consist of a myriad series of such food chains. Species populations at each link in the various chains can in turn be grouped into what are known as *trophic levels*. The primary producers are the first step in the so-called grazing food chain and are the first trophic level. The primary consumer forms the next trophic level (i.e. herbivores feeding on the plants) followed in turn by the secondary consumer which may be a predator or a parasite feeding on herbivores. A tertiary consumer is a predator population feeding on the preceding consumer level (predator eating predator) and so on (see Fig. 2.6).

Thus energy initially bound up by plants flows up through the trophic levels along a simple food chain. There appears to be a limit to the length of grazing food chains—three to four levels in terrestrial and freshwater ecosystems and up to six in marine ecosystems. There are a number of theories proposed to account for this general pattern, ranging from energy limitation to size ratios of predators and prey to dynamical constraints on the stability of long chains, but the explanation is not yet clear. In most ecosystems food chains interlink, producing food webs. Some consumer populations also feed at more than one trophic level (e.g. omnivores feeding on both plants and animals) and in more than one food chain.

Plants and animals also produce waste organic matter (e.g. leaves and faeces respectively) and many individuals die before being eaten and are not consumed by the next trophic level in the grazing food chain. This dead organic matter or *detritus* then becomes a food source for further groups of consumer organisms, the detritivores (such as earthworms, woodlice, millipedes, etc.) and the decomposers (bacteria and fungi) in the decomposer food chains. More detailed discussion of the role these types of consumers play in the recycling of nutrients will follow in Sec. 2.4.2. Suffice it to say at this point that the largest proportion of the energy flow in terrestrial ecosystems is probably through this pathway, as herbivores probably use on average much less than 30 per cent of primary production.

Secondary production   As we have seen, energy flows from one trophic level of a grazing food chain to the next, but at each transfer stage the laws of thermodynamics operate and energy is lost from the system. As mentioned above, not all net primary production is actually eaten: much escapes herbivores and much is lost from the grazing food chain when plants die. Of the food actually consumed by animals, some is egested undigested in the faeces, e.g. cellulose, by many terrestrial herbivores. As 40 to 80 per cent of energy intake can be lost by herbivores and 10 to 50 per cent by carnivores it plays no part in animal production (Ricklefs, 1979).

From the food digested and assimilated into the body, the bulk of the energy is spent in respiration for metabolism and activity. For example, insects respire 63 to 84 per cent of assimilated energy, fish 91 to

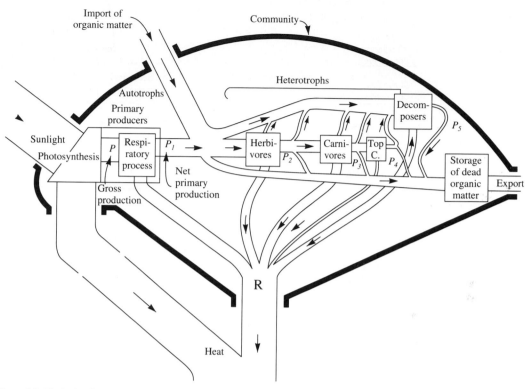

Figure 2.3 The hydraulic analogy of energy flow through an ecosystem. The energy is imagined as being channelled through pipes whose thickness is proportional to the rates of energy flow. Some hydraulic equivalent of a prism placed at the entrance deflects most of the sunlight from the community to represent that proportion of the incident light not used in photosynthesis. From then on the degradation of energy at each trophic level is shown by pipes running to the heat outlet (R = respiration; Top C. = top carnivore) (after Odum, 1956).

94 per cent and mammals 97 to 99 per cent. This energy is effectively lost as heat from the ecosystem. More energy is lost in excretion of the waste products of metabolism (as urine, for example). Therefore only a fraction of the original energy intake remains to be incorporated into new organic matter (either growth of the individual or reproduction of new animals) or secondary production and can be passed from one trophic level up to the next level in the food chain. Thus the flow of energy through food chains within an ecosystem is characterized by gradual reduction of energy available for each succeeding trophic level. This has been likened to energy being channelled through a series of pipes of decreasing diameter (Fig. 2.3). It is now generally accepted that there is about a 10 per cent energy transfer between levels; therefore up to 90 per cent of the potential energy available is lost between one trophic level and another. What one finds, therefore, is an energy pyramid within the ecosystem as one moves through the trophic levels (Fig. 2.4).

This energy pyramid is usually mirrored by a pyramid of numbers whereby there are more individuals in the primary producer trophic level than in the primary consumer level, more in the primary consumer level than in the secondary consumer level and so on (Fig. 2.4). The top trophic levels are characterized by having the lowest population size. We will see in Chapter 5 that this pyramid is also important for explaining how toxic compounds affect ecosystems.

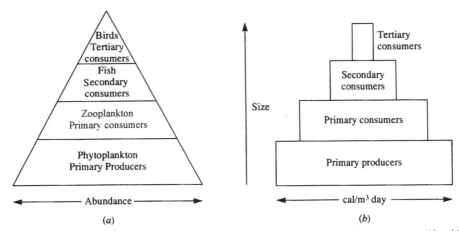

Figure 2.4 Pyramid of (a) numbers and trophic levels for an ecosystem and (b) the concept of the energy pyramid and individual size of a food chain.

2.4.2 Decomposition and Nutrient Recycling

Associated with the flow of energy through grazing food chains in ecosystems is the movement of nutrients: amino acids, minerals, sugars, salts and vitamins. These are passed from one organism to another during feeding. As organic molecules of food are broken down by respiration and metabolism, so chemical constituents are either incorporated into the body of the organism or released back into the environment as wastes and when the organism sheds body parts or dies (Begon *et al.*, 1990).

In living organisms, the chemical constituents released are carbon dioxide and water from respiration and excretory nitrogenous products of metabolism (e.g. urine in mammals, uric acid in birds and insects). In dead organisms and with faeces, the breakdown of organic matter or decomposition is carried out by the detritivores and ultimately by decomposing organisms mentioned above, although physical and chemical processes may also play a part.

Decomposer food chain On land most decomposition takes place in soil, while in aquatic ecosystems it occurs in sediments at the bottom of water bodies. Decomposer or detritus food chains (Fig. 2.5) are based on detritus (dead organisms, leaves, etc., undigested and partially digested faecal matter and excreted waste products from metabolism) as opposed to living plants in grazing food chains. The initial consumers of this food source, the detritivores, ingest and partially digest detritus, breaking down the organic matter to some degree, and after extracting some energy, egesting the remainder in faeces and excretory wastes as smaller particles. These wastes are then utilized by the next detritivore population in the chain, which repeats the process, breaking down the original organic matter further, and the wastes are again utilized by the next link in the chain, etc. The final breakdown of the organic matter to its original inorganic constituents of carbon, nitrogen, phosphorus, etc., is carried out by bacteria (true decomposers). At each stage, energy is extracted and the organisms respire; thus energy flows along these decomposer food chains with less energy being available to each succeeding stage (Fig. 2.5).

The importance of decomposition is that the complex organic molecules in the original detritus are gradually broken down to much simpler constituents and inorganic molecules (like nitrates and phosphates) as the material moves through the decomposer food chain. These are then incorporated into the soil or sediments or dissolved in water, where they become the nutrients available for reuse by plants. Thus there is a recycling of nutrients within the ecosystem.

Just as ecosystems are composed of a web of grazing food chains, so there is a network of decomposer food chains based on different detritus food sources. The detritus feeders have their own predators which creates a link between the grazing and detritus food chains. When the predators die, they

Decomposer food chain energy flow

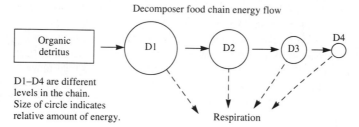

D1–D4 are different
levels in the chain.
Size of circle indicates
relative amount of energy.

Respiration

Figure 2.5 A schematic diagram of the detritus or decomposer food chains. Each link in the chain extracts nutrients and energy from the organic matter and loses energy through respiration before passing on the remaining organic matter to the next link in the food chain.

enter the decomposer food chains as well. So we see a cycle of nutrients in the ecosystem, from soil to plants where they are incorporated into organic matter, through a series of consumer populations in the grazing food chains. Waste and dead organisms from these pass to the decomposer food chains which gradually break the organic matter down to its original constituents. The end product of this food chain is the reintroduction of nutrients into the soil and atmosphere for reuse by plants, a vital ecosystem process to replenish nutrients for further primary productivity. Detailed examples of nitrogen and phosphorus nutrient cycles are presented in Chapter 10. Although energy flow and nutrient recycling are quite complex, they can be summarized in a simple diagram (Fig. 2.6) that demonstrates clearly the major differences in the two processes.

Biogeochemical cycles While ecosystems are often considered as black boxes for many of the processes that take place within them, ecosystem boundaries are in fact permeable to some degree or other; thus energy and nutrients can be transferred to and from one ecosystem to another via imports and exports. Movement of chemicals and elements therefore occurs on a global scale within the biosphere as if the biosphere was one large ecosystem. All parts of separate ecological systems on a local and indeed a global scale are ultimately linked in biogeochemical cycles.

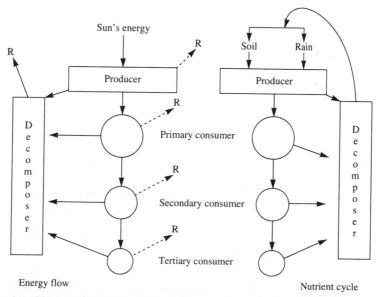

Figure 2.6 An overview of the two major ecosystem processes of energy flow and nutrient recycling (redrawn from Clapham, 1983).

A clear demonstration of this movement of materials was seen following the Chernobyl nuclear accident in Russia on 25 April 1986 (Apsimon and Wilson, 1986; Aoyama *et al.*, 1986). During the accident, 1 per cent of the reactor cores' radioactive ions rose high above the reactor site before dispersing downwind just like smoke out of a chimney. Most material rained down to earth in the vicinity of the plant, devastating an area 100 km across. However, the radiation cloud then moved around the globe in the atmosphere over the next few days, reaching the British Isles to the west, Scandinavia to the north, the Mediterranean to the south and Japan to the east. Deposition of radioactive material to the land occurred following rainfall. The interest in this example is that we can recognize the origin and movement of the radioactive material easily. However, nutrients and minerals can be moved in the same way in the biosphere within biogeochemical cycles, and by radioactively labelling nitrate or phosphate, for example, we can similarly trace their progress through the biosphere. From such studies, we know that biogeochemical cycles can have a number of phases and reservoirs:

1. *Organic phase*, where nutrients pass rapidly through biotic communities via food chains.
2. *Inorganic phases* are important to ecosystems, as the reservoirs for all nutrient elements are external to food chains:
   − *Sedimentary phase* involves interaction with the solid earth as rocks. This phase forms part of all element cycles and movement in and out of these sedimentary reservoirs is usually slow and naturally the result of geological activities such as volcanoes, weathering, etc.
   − *Atmospheric phase* forms part of some cycles, e.g. nitrogen and carbon, but not others, e.g. phosphorus.
   − There is also an *aquatic reservoir* for some elements including plant nutrients.

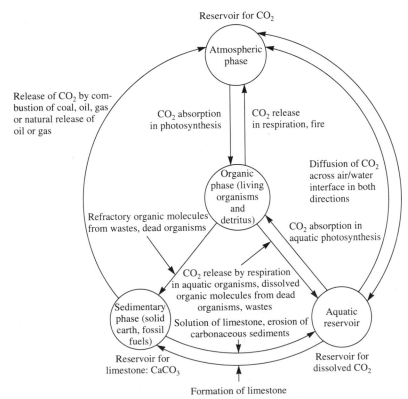

Figure 2.7 The carbon biogeochemical cycle (adapted from Clapham, 1983).

As an example, we have selected the carbon biogeochemical cycle which resembles the flow of energy through the ecosystem (Fig. 2.7). Almost all carbon enters food chains as carbon dioxide via plants. It then passes through food chains and is released back into the environment as carbon dioxide in respiration and from fires into the atmosphere or land, or from respiration into water in the sea. It is then reused by plants. This is the basic carbon cycle. Some storage of carbon occurs in sediments as carbonate/limestone, in soils as peat, coal or as oil and gas. These sedimentary phases can be released naturally from the reservoirs by geological activity such as erosion, volcanoes, etc.

As a result of man's activities through the combustion of fossil fuels, the rate of movement of carbon into the atmospheric phase has accelerated and is now much faster than natural recycling of sedimentary carbon. This has led to a disturbance in the balance of carbon throughout the biosphere, which in turn has led to the well-publicized increase of carbon dioxide in the atmosphere. The result is predicted to be a global rise in temperature of some 1 to 3 °C over the next 30 to 50 years through the so-called 'greenhouse effect'. Some predicted consequences of this effect include melting of the ice caps, increase in sea levels and flooding. A 1 °C increase in temperature at a location will be equivalent to effectively moving approximately 60 to 100 miles nearer to the equator and hence will result in a marked climatic shift (Houghton et al., 1993). Through range shift, animals and plants may be able to keep up with the movement of conditions over time as the climate changes providing the rate of change is not too fast, but it is likely that many species will not be able to track their preferred climate and habitat.

2.5 THE HUMAN DIMENSION

The 'greenhouse effect' is but one of the environmental problems that have resulted either directly or indirectly from the activities of man. The role of the human population on environmental change has been simply summarized by Erlich and Erlich (1990) in the simplified equation

$$I = PAT \tag{2.4}$$

where the impact I of the population on the environment results from the size of the population (P), the per capita affluence or consumption (A) and the damage caused by technologies (T) employed to supply each unit of consumption. As P increases, so too does T because supplies to additional people must be mined from deeper ores, pumped from deeper deposits, transported further. It is also suggested that the per capita consumption of commercial energy in a nation can be used as a surrogate for the AT part of the equation—a considerable proportion of the environmental damage involves use of commercial energy, from clearing tropical forests for agriculture to mining, manufacturing, road building and extraction of fossil fuels (Erlich, 1991).

The overall human population has more than doubled in the past 40 years although not evenly over the globe. Population growth rates are increasing exponentially in the less/underdeveloped countries while growth is slow or non-existent in most developed countries. Many resources are being depleted with little recycling, and waste products are being returned to the environment in a different form and at concentrations that are often toxic or otherwise damaging. Land use changes are taking place rapidly. The global human population lives on only about 2 per cent of the global land area, but a further 60 per cent is taken up growing crops, grazing livestock or being utilized for extraction of mineral resources and removal of forest. Much of the remaining land area is either desert or covered with ice or is too steep for use (Miller, 1990). Forests, grasslands and wetlands are disappearing rapidly and deserts are expanding due to soil erosion and a decline in underground water deposits and lowering of water tables. The effects of this level of stress on the environment is evident in the form of climate change (see the report of the Intergovernmental Panel on Climate Change in Houghton et al., 1993) and degradation in the quality of the environment from global warming and sea-level rise at the global scale to river pollution and urban smog at a local scale.

Scientists are concerned that climate change (specifically patterns of temperature and rainfall) may be occurring too rapidly for human societies and agricultural systems to adjust successfully (Schneider, 1989). Our technological capabilities and demands for natural resources have grown rapidly since the industrial revolution in the Western world and outstripped our understanding of the impact of these changes on the environment. It is only in the last 20 years or so that scientific research has started to provide some understanding of what is happening to our environment.

Human activity is therefore seen as a significant cause of environmental change, mainly as a result of the conflict between maintaining and using the environment; i.e. development and exploitation of physical resources, building and urbanization, changing land use and deposition of wastes, often at the expense of the integrity of the biotic component of the environment and biological resources. The fact that the biotic component has tended to be ignored and has suffered as a consequence of exploitation of the abiotic component has led to the emergence of extremist views by some parts of the environmental movement.

Ecological research has provided empirical data on the effects of environmental degradation, disturbance and pollution at both local and global scales and has provided methods of measuring the deterioration in the quality of the environment. This has led to a reappraisal of the balance between exploitation of physical and chemical components of the environment and the consequent changes to the biotic component. This, in turn, has led to the development of environmental and planning legislation in many countries world-wide (see Chapter 1). The application of various legal controls, such as the US Endangered Species Act of 1973, has led to the cessation or holding up of large-scale engineering and development projects due to attempts to protect single species such as the snail darter (in the Tellico Dam, Tennessee controversy of the 1970s; Cunningham and Saigo, 1990) or the spotted owl (halting forestry development in northern USA in the late 1980s and early 1990s). This may not be the best approach to use (as compared to protecting habitats), but does indicate the degree of change in approach to the planning process for various developments.

2.6 ENVIRONMENTAL GRADIENTS, TOLERANCE AND ADAPTATION

2.6.1 Environmental Gradients

One of the most obvious patterns in nature is the uneven distribution of organisms over the globe. Each species is not found in every type of habitat or in every part of the world. To discover why, we need to look at the environment as a place to live. Within the environment are a multitude of factors that can affect organisms. These include physical and chemical factors like light, temperature and pH, known as conditions; factors the organism actually uses like food, water, shelter, etc., known as resources; and the presence of other organisms like predators and competitors with which organisms interact.

The level of most environmental factors varies over some limited range, with a gradient between the extremes: high to low, large to small. For example, if we examine temperature; there is a global gradient from the equator towards the north or south, a regional gradient with altitude, a local gradient within vegetation (from top to bottom) and with aspect. Temporal gradients in temperature also occur over historical time with ice ages, over decades with sun spot activity, through the year with seasons and also on a daily cycle.

To live and thrive in a given situation, each species must have the essential resources and conditions necessary for growth and reproduction. A species population tends to inhabit those areas offering a particular combination of suitable conditions and resources. This is because each species only functions efficiently over a limited part of each environmental gradient—its tolerance range.

2.6.2 Tolerance

Consider just one environmental gradient. If we plot some measure of success of a species population (such as population density or number of individuals, survivorship or fitness—contribution to subsequent

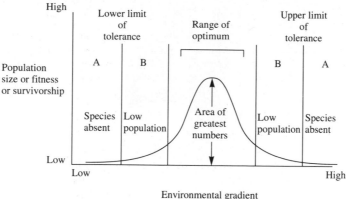

Figure 2.8 A schematic tolerance curve for a single species population existing on a single environmental gradient.

generations) against the gradient, a tolerance curve is generated. Figure 2.8 illustrates a 'universal' tolerance curve, which is usually depicted as a normally distributed, bell-shaped curve, but in reality may often be skewed as, for example, in relation to gradients of concentration of toxic elements, where the right side of the curve is often truncated.

Within the *range of optimum* of the curve, the majority of individuals of the species population can survive and reproduce; thus a large population size is maintained as the conditions along this part of the environmental gradient are ideal for the species. Beyond this range, towards the high and low ends of the environmental gradient, most individuals suffer increasing physiological stress—*the zone of physiological stress*—where conditions are less than optimal such that most individuals can stay alive but cannot function efficiently nor reproduce. Not all individuals are genetically identical and show exactly the same levels of tolerance to some environmental parameter. A few individuals with a slightly different genotype and tolerance level may in fact survive fairly well under conditions where the majority of individuals do badly. However, surrounding the areas along the gradient where the species population can exist successfully are areas where no individuals of the species can survive because physical conditions or lack of food resources are too extreme for that species to permit survival—*the zone of intolerance*. For example, boundaries of zones of lichen floras in England and Wales are clear and related to gradients in air pollution associated with patterns of urbanization (Fig. 2.9). Lichens are basically intolerant of air pollution and they have therefore been used as indicators of such pollution. The use of lichens and other biological species in this way is termed biomonitoring.

Thus the geographical range of a species is bordered by upper and lower tolerances to a factor that changes on a geographical scale. The same applies at a local scale where the distribution within a particular habitat is confined by boundaries set by conditions which change along environmental gradients in the habitat. Species populations, however, are affected by a range of environmental gradients at the same time, e.g. aquatic animals respond to temperature, food availability, oxygen level, flow rate, substrate type and many other factors (Chapter 5). Species populations have a range of optimum for all of these factors, and the species is most successful if it inhabits that part of the environment where these ranges of optima overlap to the greatest degree. Different species have more or less different ranges of optima for the same factors; thus as one moves along an environmental gradient, so the combination of species living at various points along the gradient—the community—will change. This is shown very clearly in Fig. 2.10 for vegetation types on mountains in Tennessee in relation to altitude, temperature and

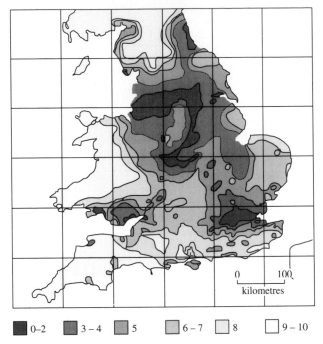

Figure 2.9 Approximate boundaries of zones of richness of lichen floras in England and Wales. Zones 9 and 10 are richest in flora and zones 0 to 2 poorest (after Holdgate, 1979; Hawksworth and Rose, 1970).

moisture. Different species are adapted to different conditions, with some overlap. Obviously if conditions change, e.g. become drier or colder, the whole pattern will change and some species will disappear while others will come to dominate.

2.6.3 Limiting Factors

The tolerances of species to different conditions or resource levels vary. Some species have wide tolerances to some environmental factors (*eurytopic*), with a wide tolerance curve and a broad range of optimum, and narrow tolerances to others (*stenotopic*). Also species growth and reproduction are often regulated by just one or a few conditions or resources in short supply. The resource in short supply or the condition over which the species has the smallest range of optimum will limit the species functions and is called the *limiting factor*. For example, plants need light, nutrients from soil and water. In arid climates, plant growth is strongly correlated with rain; there is sufficient light, but water is the limiting factor.

As discussed earlier, not only can too little of something be limiting but also too much. Many animals feed on dead organic matter (e.g. in freshwater or marine systems), but when this resource is present in too great an amount, (e.g. immediately downstream of sewage input) such organisms may not survive because the decomposition of the organic matter by bacteria affects levels of another resource— oxygen.

2.6.4 Normal Responses to a Changing Environment

The environment is changing all the time, with, for example, daily and seasonal temperature cycles and longer term climatic fluctuations over thousands and millions of years. Individuals, populations and

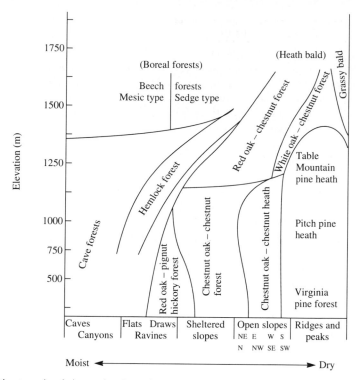

Figure 2.10 Vegetation types in relation to elevation and topography in the Great Smoky Mountains, Tennessee. The vertical axis incorporates gradients of temperatures and other factors related to elevation; the horizontal axis incorporates gradients of moisture relations and other factors from moist or mesic situations on the left to dry or xeric on the right, as affected by topographic position (after Whittaker, 1956).

species can make adjustments that can allow them to survive in conditions they might not otherwise be able to—this is provided that the changes are at a rate that can be matched by these adjustments.

A lizard species in a desert has a definite optimal range of temperature with upper and lower limits of tolerance. Individuals can maintain themselves within this range as temperatures fluctuate during the day by changing their behaviour (balancing time in the sun and shade at different times of the day to maintain their optimal body temperature). Animals can also avoid unsuitable conditions which effectively shift the environmental gradient at a location by undertaking a small-scale range shift (change in location) to remain within suitable environmental conditions along the environmental gradient. On a larger scale, this may involve seasonal migration as seen in many insects, birds and mammals that move to different areas offering better conditions for part of the year, returning to the initial area when conditions there improve. Some animals can create their own microclimate within an otherwise adverse set of conditions, such as digging burrows. They can then organize their daily activity outside to coincide with the more tolerable conditions during the daily temperature cycle, for example.

Organisms that are too big to burrow and/or cannot move, or cannot move fast enough to avoid changing and unsuitable conditions, have to remain and overcome the adverse conditions by changing their physiology. They are in effect temporarily moving their tolerance curve along the environmental gradient axis in one direction or another. For example, short-term responses to a decrease in temperature in mammals includes raising body heat production and constricting near-surface blood vessels. Slightly longer term responses may include increasing body fat layers, length of fur, etc., allowing the organism to survive under colder temperatures than previously. Some animals may alter their daily or seasonal activity times to suit the timing of favourable conditions. Thus many species hibernate, basically 'switching off' most physiological systems. Organisms can also adjust their physiology over time by acclimation (acclimatization to conditions). If a person goes to high altitude, they produce more red blood cells to cope with the lower partial pressures of oxygen.

In the face of more permanent changes in the environment, species can also adjust by acclimatization, a gradual and reversible alteration in physiology and morphology over a number of generations. On a longer term basis, species can undergo adaptation through natural selection (evolution), where the species actually changes genetically, as only the more tolerant individuals to the prevailing conditions can survive and reproduce well; thus the species as a whole shows improved survival and more successful reproduction. The tolerance curve of the species therefore moves permanently with respect to the environmental gradients. One of the best known examples is metal tolerance in the grass *Agrostis* in relation to pollution from mine wastes (Walley *et al.*, 1974). Metal-tolerant populations quickly arose and were able to colonize mine waste heaps and polluted soils. When the pollutant is newly introduced into the environment or is at very high concentrations, only a few individuals of a species that have a tolerant genotype (i.e. individuals that lie at the extremes of the species population tolerance curve) survive. Maintenance of genetic variability within a species is important in this context by providing a pool of genotypes among which some individuals may be capable of surviving a range of different environmental changes. Polluted areas such as those affected by mine wastes become species poor, as few species can survive. However, those tolerant species populations that do survive often grow to reach large population sizes.

The ability of organisms to respond to changes in the environment depends on the rate and extent of changes. If changes are too fast or too large, there may not be enough time for species to acclimate, to adapt or move to a new area with suitable conditions.

2.7 ENVIRONMENTAL CHANGES AND THREATS TO THE ENVIRONMENT

A major axiom in ecology is that the environment changes naturally over time and organisms, populations and communities have to respond. Depending on the time scale we are looking at, different responses are seen (Fig. 2.11).

Over very short time periods, individual organisms respond to changes in stimuli they receive from the environment, e.g. through nerve impulses, reflexes and behaviour. Over periods from days to years, depending on the life span of the species, individuals are born and die, but species populations remain and may adapt to small-scale environmental change. Over geological time, the temperature of earth has naturally changed, causing climatic changes into and out of ice ages, continental plates have moved around the surface of the globe and the topography of the land has changed. This has led to changes in the extent and distribution of various biomes on a global scale.

The climatic changes are usually slow enough to allow communities to change gradually over historical time at any one particular location, e.g. the beetle community at one particular site in Wales, sampled from fossil remains at various depths in the soil corresponding to various times in the past (Coope, 1987). Each community at a particular time for a particular depth in the soil, was adapted to the particular climatic regime that was prevalent at that time.

Thus gradual changes to the environment over long enough time periods do not seriously damage the environment or the ecosystem processes as species are able to track the environmental change, by either moving in space or changing in their degree of tolerance, keeping within their optimal type of environment. However, sudden dramatic changes in environmental conditions of an area over short time-scales can have serious and damaging effects on the living systems there. This is what we frequently see as a result of human activity.

We live in a world of increasing industrialization, development, use of resources, urbanization and intensification of agriculture. All cause rapid changes to the local or regional environment and frequently

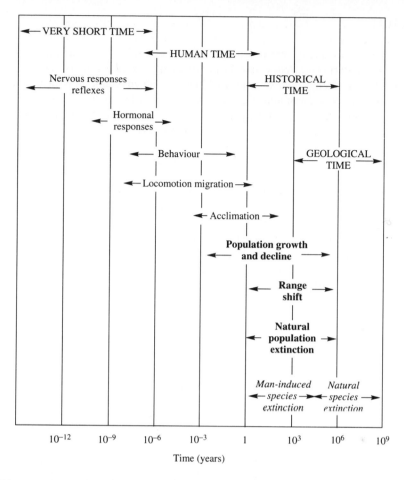

Figure 2.11 Diagrammatic representation of individual (plain text), population (bold text) and species (italic text) responses to environmental change over a range of time-scales.

at the level of the biosphere. These environmental changes are often too fast to allow organisms, populations, communities and ecosystem processes time to respond. It is the degree, timing and speed of habitat disturbances that causes most of the problems arising from human activity.

Basically, changes that would naturally occur over geological or historical time now happen over human time and cause large-scale disturbance to natural systems. One consequence is a decline in biological resources or environmental services, as discussed earlier. Many engineering processes and constructions can cause rapid environmental changes that can seriously disrupt environmental systems. Cultural eutrophication of rivers and lakes lead to a rapid increase of nutrients that normally takes hundreds to thousands of years and hence give rise to large-scale problems of water pollution.

Engineers are at the forefront of humanity's endeavours to change, harness and mould the environment and in some countries they are the decision makers as to whether or not new developments go ahead. It is therefore essential that engineers work hand in hand with environmental scientists; i.e. the biotic component of the environment should form an integral part of the equation for the assessment of any potential environmental manipulation, exploitation of resources or other development, and that environmental considerations should become part of an engineer's decision making and design processes.

2.8 PROBLEMS

2.1   Outline the pathway of energy through natural ecosystems and its role. In what ways can energy flow be altered or interrupted by the activities of humans?

2.2   Define secondary production and discuss how humans can maximize the yield from secondary producers, with minimum disruption to the environment.

2.3   Describe some of the outcomes resulting from the disruption of biogeochemical cycles.

2.4   Compare and contrast an autotrophic and heterotrophic ecosystem.

2.5   Why are decomposers so important in ecosystems?

2.6   How would you summarize the concerns of the environmental scientist over the interaction between humans and the environment?

2.7   Why do many anthropogenic disturbances have such dramatic and long-lasting effects on natural ecological systems?

2.8   What is meant by a 'key species'?

2.9   Why is it important to preserve wild populations of cultivated plants?

2.10 With regard to 'tolerance curves', do a brief literature search and describe such a curve for a real situation. Discuss why tolerance curves are more likely to be skewed rather than bell-shaped.

REFERENCES AND FURTHER READING

Aoyama, M., *et al.* (1986). 'High levels of radioactive nuclides in Japan in May', *Nature*, **321**, 819–820.

Apsimon, H. and J. Wilson (1986) 'Tracking the cloud from Chernobyl', *New Scientist*, **17**, July, 42–45.

Aylward, B. and E. B. Barbier (1992) 'Valuing environmental functions in developing countries', *Biodiversity and Conservation*, **1**, 34–50.

Balick, M. J. (1985) 'Useful plants of Amazonia: a resource of global importance', in *Key Environments—Amazonia*, G. T. Prance and T. E. Lovejoy (eds), Pergamon Press, Oxford, pp. 339–368.

Balick, M. J. and S. N. Gershoff (1981) 'Nutritional evaluation of *Jessenia bataua:* source of high quality protein and oil from tropical America', *Economic Botany*, **35**, 261–271.

Begon, M., J. Harper and C. Townsend (1990) *Ecology. Individuals, Populations and Communities*, Blackwell Scientific Publications, Oxford.

Bishop, R. C. (1978) 'Economics of a safe minimum standard', *American Journal of Agronomy and Economy*, **57**, 10–18.

Bishop, R. C. and R. C. Ready (1991) 'Endangered species and the safe minimum standard', *American Journal of Agronomy and Economy*, **73**, 309–312.

Clapham, W. B. (1983) *Natural Ecosystems*, 2nd edn, Macmillan Publishing Company, New York.

Coope, G. R. (1987) 'The response of late quarternary insect communities to sudden climate changes', In *Organisation of Communities: Past and Present*, J. H. Gee and P. S. Giller (eds), Blackwell Scientific Publications, Oxford, pp. 421–438.

Cunningham, W. P. and B. W. Saigo (1990) *Environmental Science*,. Wm. C. Brown, Dubuque, Iowa.

Dourojeanni, M. J. (1985) Over-exploited and under-used animals in the Amazon region', in: *Key environments—Amazonia*, G. T. Prance and T. E. Lovejoy (eds), Pergamon Press, Oxford, pp. 419–433.

Erlich, P. R. (1991) 'Forward: facing up to climate change', in *Global Climate Change and Life on Earth*, R. C. Wyman (ed.), Chapman and Hall, London.

Erlich, P. and A. Erlich (1990). *The Population Explosion*, Simon and Schuster, New York.

Hardin, G. (1968) 'The tragedy of the commons', *Science*, **162**, 1243–1248.

Hawksworth, D. and F. Rose (1970) 'Qualitative scale for estimating sulphur dioxide air pollution in England and Wales using epiphytic lichens', *Nature*, **227**, 145–148.

Holdgate, M. W. (1979) *A Perspective of Environmental Pollution*, Cambridge University Press.

Houghton, J. T., G. J. Jenkins and J. J. Ephraums (1993) *Climate change. The IPCC Scientific Assessment*, Cambridge University Press.

IUCN (1980) *World Conservation Strategy*, Gland, IUCN, UNEP, WWF.

Kim, K. C. (1993) 'Biodiversity, conservation and inventory: why insects matter', *Biodiversity and Conservation*, **2**, 191–214.

Miller, G. T. (1990) *Resource Conservation and Management*, Wadsworth, California.

Naeem, S., L. Thompson, S. Lawlor, J. Lawton and R. Woodfin (1994) 'Declining biodiversity can alter the performance of ecosystems', *Nature*, **368**, 734–736.

Odum, H. T. (1956) 'Efficiencies, size of organisms and community structure', *Ecology*, **37**, 592–597.

Pearce, D. W. (1990) 'An economic approach to saving the tropical forests', LEEC paper DP 90-06, International Institute for Environment and Development, London.

Pearce, D. W. and R. K. Turner (1989) *Economics of Natural Resources and the Environment,* The John Hopkins University Press, Baltimore.

Ricklefs, R. E. (1979) *Ecology,* 2nd edn, Nelson, Middlesex.

Schneider, S. (1989). *Global Warming,* Sierra Club Books, San Francisco.

Silver, C. S. and R. I. S. DeFries (1990) *One Earth One Future: Our Changing Global Environment,* National Academic Press, Washington, D. C.

Whalley, K., M. Khan and A. Bradshaw (1974) 'The potential for evolution of heavy metal tolerance in plants. I. Copper and zinc tolerance in *Agrostis tenuis*', *Heredity,* **32**, 309–319.

Whittaker, R. H. (1956) 'Vegetation of the Great Smokey mountains', *Ecological Monographs,* **26**, 1–80.

THREE

INTRODUCTION TO CHEMISTRY AND MICROBIOLOGY IN ENVIRONMENTAL ENGINEERING

3.1 INTRODUCTION

In addition to a solid foundation in physics and mathematics, the student of environmental engineering needs to be well versed in aspects of chemistry and microbiology as they relate to environmental engineering applications. The prerequisite for this chapter is a university level course in general chemistry. It is desirable but not a prerequisite to have taken a university level course in the biology of microorganisms or general bacteriology. However, after completing this course on the introduction to environmental engineering, the student wishing to progress in this field is recommended to take additional courses in chemistry, atmospheric chemistry, soil chemistry, general bacteriology, or microbiology of aerobic and anaerobic bacteria. What further courses the student takes will depend on the path of specialization chosen.

This chapter aims to introduce the student to the physical and chemical properties of water and also to some fundamentals in atmospheric and soil chemistry as they relate to environmental engineering. This chapter also introduces microbiology with outlines of the organization of the microbial world including bacteria and viruses. There are positive and negative aspects of microbiology in environmental engineering. For instance, bacteria are exploited in the purification of wastewater to reduce the organic contaminants, in a very commonly used process called activated sludge (Chapter 12). Also the presence of some bacteria or any viruses in drinking water or bathing water are potentially harmful and must be 'engineered' out of the water. This chapter also introduces the topic of chemical and biochemical reactions, stoichiometry and kinetics, so that the student can compute, for instance, how much of a specific chemical (for instance 'alum', also known as aluminium sulphate) is required to purify drinking water by removing very minute suspended or dissolved solid particles. The final topics in this chapter are on material balances and reactor configurations. Material balances are fundamental to 'accounting' of inputs and outputs in processes. In any process, be it chemical, biochemical or physical, the mass of materials input to the process must be equal to the mass of the products output, plus or minus any generation or loss within the 'process'. For instance, a trivial example is the case of an outfall pipe discharging a wastewater to a river. Consider that upstream of the outfall, a river has a low level of pollution, while downstream of the pipe, the level of pollution is higher and depends on the two inputs:

the upstream 'clean' river and the wastewater pipe discharge. The amount of 'pollution' discharged legitimately can be computed if the level of acceptable water quality downstream is known. A control volume of the reach of river of interest is, in this case, the 'reactor'. Many configurations of reactors exist in the natural and in the industrial world. One such simplified reactor is the 'complete mix', where the concentrated influent is rapidly mixed on entering the reactor and becomes diluted, so that when the influent mass is ready to be discharged as effluent (maybe a few days or weeks later) its concentration will be greatly reduced. Having completed this chapter, the student should have an adequate base with which to handle the chemistry and microbiological aspects as they relate to the later chapters on water, air, soil and mixed environments.

3.1.1 Methods of Expressing Concentration

The two methods of expressing the concentration of a constituent of a liquid or gas are:

1. *Mass/volume*   The mass of solute per unit volume of solution (in water chemistry). This is analagous to weight per unit volume; typically, mg/L = ppm (parts per million).
2. *Mass/mass or weight/weight*   The mass of a solute in a given mass of solution; typically mg/kg or ppm (parts per million).

If the density of a solution $= \rho = \dfrac{\text{mass of solution}}{\text{volume of solution}}$ (kg/L)

and concentration of a constituent in mg/L $= C_{A1} = \dfrac{\text{mass of constituent}}{\text{volume of solution}}$ (mg/L)

and concentration of a constituent in ppm $= C_{A2} = \dfrac{\text{mass of constituent}}{\text{mass of solution}}$ (mg/kg)

then rearranging,

$$\rho = \frac{C_{A1}}{C_{A2}}$$

$$\text{If } \rho = 1\,\text{kg/L, then } C_{A1} = C_{A2} \tag{3.1}$$

i.e. the concentration of a constituent in ppm mg/kg = concentration of a constituent in mg/L.

For most applications in water and wastewater environments, $\rho = 1\,\text{kg/L}$. For applications in the air environment, Eq. (3.1) does not hold. The use of mg/L is most common in water applications as the volume of the solution is usually determined as well as the mass of the solute. The unit ppm is typically used in sludges or sediments.

Example 3.1   Express the concentration of a 3 per cent by weight $CaSO_4$ solution in water in terms of mg/L and ppm.

Solution

$$3\% \text{ by weight} = \frac{3}{100} = \frac{30\,000}{1\,000\,000} = 30\,000\,\text{ppm}$$

Since the solution is water, from Eq. (3.1), then $C = 30\,000\,\text{mg/L}$.

Example 3.2   If a litre of solution contains 190 mg of NH_4^+ and 950 mg of NO_3^-, express these constituents in terms of nitrogen (N).

Solution

$$190 \text{ mg NH}_4^+/\text{L} = 190 \text{ mg NH}_4^+/\text{L} \times \frac{14 \text{ mg N}}{18 \text{ mg NH}_4^+} = 148 \text{ mg NH}_4^+\text{--N/L}$$

$$950 \text{ mg NO}_3^-/\text{L} = 950 \text{ mg NO}_3^-/\text{L} \times \frac{14 \text{ mg N}}{62 \text{ mg NO}_3^-} = 214 \text{ mg NO}_3^-\text{--N/L}$$

At the beginning of this section concentration in terms of mass or weight for a fixed weight or volume of solution is discussed, e.g. 1 L or 1 kg. Chemists sometimes prefer to use the concentration term *mole*, which is the mass of a constituent which is numerically equal to the molecular weight of the constituent. For instance:

$$1 \text{ gram mole of methane (CH}_4) = 18 \text{ g of methane}$$

where 1 mole is that amount of a constituent which contains the Avogadro number of molecules. Therefore the mole notation does not refer to a fixed weight but to a fixed number of particles. In the mole context there are four entities of concentration:

1. Molality (m), $\dfrac{\text{mole}}{\text{kg}} = \dfrac{\text{moles of solute}}{1 \text{ kg of solution}}$

2. Molarity (M), $\dfrac{\text{mole}}{\text{L}} = \dfrac{\text{moles of solute}}{1 \text{ L of solution}}$

3. Normality (N), $\dfrac{\text{eq}}{\text{L}} = \dfrac{\text{equivalent of solute}}{1 \text{ L of solution}}$

   where equivalent weigtht in $\dfrac{\text{g}}{\text{eq}} = \dfrac{\text{molecular weight (g)}}{\text{equivalence (}n)}$

   where n is the number of protons denoted in an acid–base reaction or is the total change in valence in an oxidation reduction reaction. If two different solutions have the same normality, they will react in equal proportions, i.e.

$$V_A N_A = V_B N_B \tag{3.2}$$

where V_A, V_B are the volumes of solutions A and B, and N_A, N_B are the respective normalities.

4. Mole fraction $X = \dfrac{\text{number of moles of solute}}{\text{total moles of solution}}$

Example 3.3   Using Example 3.1, i.e. a solution which is 3 per cent by weight of $CaSO_4$ in water, express the concentration in (a) molality, (b) molarity and (c) mole fraction.

Solution

Molecular weight of $CaSO_4 = 136$ g/mole

3 per cent by weight $= 30$ g/kg

(a) Molality (m), $\dfrac{\text{mole}}{\text{kg}} = \dfrac{30 \text{ g/kg}}{136 \text{ g/mole}} = 0.22 \text{ mole/kg}$

(b) Molarity (M), $\dfrac{\text{mole}}{\text{L}} = \dfrac{30 \text{g/L}}{136 \text{ g/mole}} = 0.22 \text{ mole/L} = 0.22 \; M$

(c) Mole fraction, $X_{CaSO_4} = \dfrac{30/136}{30/136 + 970/18} = 0.0041$

$$X_{H_2O} = \dfrac{970/18}{30/136 + 970/18} = 0.9959$$

Mass Concentrations as CaCO$_3$ A very common system for expressing hardness (calcium and magnesium) and alkalinity (HCO_3^-, CO_3^{2-} and OH^-) concentrations in water chemistry is the calcium carbonate system (see Sec. 3.2.5). This system of units can be thought of as normalizing concentrations to $CaCO_3$, a substance commonly used in water chemistry. In this system, the concentration of a substance as mg/L as $CaCO_3$ is determined by the equation:

$$\text{Number of equivalents of substance per litre} \times \frac{50 \times 10^3 \text{ mg } CaCO_3}{\text{equivalent of } CaCO_3}$$

For example, for hardness,

$$CaCO_3 \rightarrow Ca^{2+} + CO_3^{2-}$$

$$\underset{\text{1 mole}}{} \quad \underset{\substack{\text{1 mole} \\ \text{2 equivalents}}}{\phantom{Ca^{2+}}} \quad \underset{\substack{\text{1 mole} \\ \text{2 equivalents}}}{\phantom{CO_3^{2-}}}$$

For Ca^{2+} in precipitation or dissolution reactions,

$$\text{Equivalent weight of } CaCO_3 = \frac{100 \text{ g/mole}}{2 \text{ eq/mole}} = 50 \text{ g/eq}$$

The reader is referred to Snoeyink and Jenkins (1980) for a historical commentary on the origin of this unit system.

Example 3.4 Given the concentration of Ca^{2+} as 92 mg/L in a solution, express the concentration in eq/L and also in mg/L as $CaCO_3$.

Solution

The equivalent weight of Ca^{2+} in mg/meq $= \dfrac{\text{molecular weight}}{\text{charge}} = \dfrac{40}{2} = \dfrac{20 \text{ mg}}{\text{meq}}$

The normality (N) in eq/L $= \dfrac{\text{concentration in mg/L}}{\text{equivalence in mg/meq}}$

$$= \frac{92 \text{ mg/L}}{20 \text{ mg/meq}} = 4.6 \text{ meq/L}$$

However, the equivalent weight of Ca as $CaCO_3 = 50$ g/eq $= 50$ mg/meq

The concentration of Ca in mg/L as $CaCO_3 = 50 \dfrac{\text{mg}}{\text{meq}} \times 4.6 \dfrac{\text{meq}}{\text{L}}$

$$= 230 \text{ mg/L}$$

3.1.2 Stoichiometric Examples

If the gas methane is burned with oxygen to produce carbon dioxide and water, the reaction is

$$CH_4 + O_2 \rightarrow CO_2 + 2 H_2O \tag{3.3}$$

Oxygen (O) is an 'element' of atomic weight 16, hydrogen is also an element with atomic weight 1 and carbon is an element of atomic weight 12. An element is defined as 'a pure substance which cannot be split into any simpler pure substance'. They are usually classified into metals and non-metals. Methane (CH_4) is a 'compound' of molecular weight 16. Carbon dioxide is a compound of molecular weight 44 and water is a compound of molecular weight 18. A compound is defined as 'a pure substance composed of two or more elements, combined in fixed and definite proportions in a chemical reaction'. The 'molecular weight' is the sum of the atomic weights of all the constituent atoms. The molecular weight of methane is 16. A 'mole' has the Avogadro number (6.023×10^{23}) of molecules and is expressed as

$$\text{g mole} = \frac{\text{mass in g}}{\text{molecular weight}}$$

Reactions are expressed stoichiometrically as:

$$CH_4 \quad + \quad 2O_2 \quad \longrightarrow \quad CO_2 \quad + \quad H_2O$$

1 mole of methane	+	2 moles of oxygen	$\longrightarrow$	1 mole of carbon dioxide	+	1 mole of water

$$1 \times 16\,\text{g/mole} \quad + \quad 2 \times 32\,\text{g/mole} \quad \longrightarrow \quad 1 \times 44\,\text{g/mole} \quad + \quad 1 \times 36\,\text{g/mole}$$
$$16\,\text{g} \quad + \quad 64\,\text{g} \quad \longrightarrow \quad 44\,\text{g} \quad + \quad 36\,\text{g}$$

Mass balance: $80\,\text{g} = 80\,\text{g}$.

In stoichiometric examples, the left side of an equation computed in grams (or kg) must equal that of the right side of the equation. This is the most elementary of the concepts of mass balances that receive much attention throughout this book.

Example 3.5 In the treatment of potable water, an aluminium sulphate solution is used as a coagulant to produce an aluminium hydroxide (sludge) floc (see Chapter 11 for details). Compute the amount of sludge produced if 100 kg of alum coagulant is used daily. The stoichiometric analysis is as follows.

Solution

$$Al_2(SO_4)_3 \cdot 14H_2O \quad + \quad 3Ca(HCO_3)_2 \quad \longrightarrow \quad 2Al(OH)_3 \quad + \quad 3CaSO_4 \quad + \quad 14H_2O \quad + \quad 6CO_2$$

1 mole of alum	+	3 moles of ca bicarbonate	$\longrightarrow$	2 moles of alum hydroxide	+	3 moles of ca sulphate	+	14 moles of water	+	6 moles of CO_2

$$(3.4)$$

Mass Balance (molecular weights):

$$Al_2(SO_4)_3 \cdot 14\,H_2O = 27 \times 2 + (32 + 16 \times 4) \times 3 + 14(18) = 594\,\text{g}$$
$$3\,Ca(HCO_3)_2 = 3[40 + 2 \times (1 + 12 + 3 \times 16)] \qquad = 486\,\text{g}$$
$$2Al(OH)_3 = 2[27 + 3 \times (16 + 1)] \qquad = 156\,\text{g}$$
$$3CaSO_4 = 3(40 + 32 + 4 \times 16) \qquad = 408\,\text{g}$$
$$14H_2O = 14(2 \times 1 + 16) \qquad = 252\,\text{g}$$
$$6CO_2 = 6(12 + 2 \times 16) \qquad = 264\,\text{g}$$

Therefore, Eq.(3.4)becomes :

$$594\,\text{g} + 486\,\text{g} = 156\,\text{g} + 408\,\text{g} + 252\,\text{g} + 264\,\text{g}$$
$$1080\,\text{g} = 1080\,\text{g}$$

i.e.   594 g of alum produces 156 g of alum hydroxide sludge and so

i.e.   100 kg of alum used daily produces 26 kg of alum hydroxide sludge.

Example 3.6   If natural gas (98 per cent CH_4) is used to fuel a thermal power plant, compute the amount of oxygen required daily to produce 100 MW of power if the calorific value of gas is 50 MJ/kg.

Solution

$$\text{Power 100 MW} = 3600 \times 10^2 \text{ MJ/h}$$

$$\text{Gas required} = \frac{360 \times 10^3}{50 \times 10^3} = 7.2 \text{ T/h}$$

$$\text{Gas at 98\% } CH_4 \rightarrow CH_4 = 0.98 \times 7.2 = 7.06 \text{ T/h}$$

Stoichiometric equations:

$$CH_4 \quad + \quad 2O_2 \quad \rightarrow \quad 2H_2O \quad + \quad CO_2$$

| 1 mole of methane | + | 2 moles of oxygen | → | 2 moles of water | + | 1 mole of CO_2 |

$$1 \times 16 \text{ g} \quad + \quad 2 \times 32 \text{ g} \rightarrow 2 \times 18 \quad + \quad 1 \times 44$$

$$\text{Mass balance: } 80 \text{ g} = 80 \text{ g}$$

$$7.06 \text{ T/h of } CH_4 \text{ requires } \frac{64}{16} \times 7.06 = 28.2 \text{ T/h of } O_2$$

Example 3.7   The composition of air is given in % volume as: 78.1% N_2; 20.95% O_2; 0.05% Ar. Determine the average molecular weight of air and its composition by % weight.

Solution   The % volume is also the number of relative moles.

Component	% volume = moles	Molecular weight g/mole	grams	% weight
N_2	78.10	28	2186.8	75.5
O_2	20.95	32	670.4	23.12
Ar	0.95	40	38	1.3
Total	100.00	100	2895.2	100.0

The average molecular weight is ~ 2895.2 g/100 mole

$$\sim 28.952 \text{ g/mole}$$

3.2 PHYSICAL AND CHEMICAL PROPERTIES OF WATER

Water is never pure, except possibly in its vapour state. Water always contains impurities, which are constituents of natural origin. Frequently, water contains contaminants, which are constituents of anthropogenic origin. For instance, the presence of the chemical impurities of calcium and magnesium ions (Ca^{2+} and Mg^{2+}) in groundwater are usually of natural origin, being due to the dissolution of these minerals from the soil and underground rocks. However, the presence of the nitrogen compounds of ammonia nitrogen (NH_4), nitrite (NO_2^-) or nitrate (NO_3^-) in groundwater is possibly due to pollution from agricultural fertilizers, agricultural liquid wastes, sewage or industrial wastewaters. In environmental engineering, water is of central interest due to its many varied occurrences and uses, including:

- Surface freshwaters in rivers and lakes and groundwaters when used as drinking water
- Surface freshwaters as used in fish and other fauna habitats
- Surface freshwaters as used for anthropogenic liquid discharges
- Surface freshwaters and groundwaters as used for irrigation
- Surface waters as used for recreation
- Surface waters as used for navigation

The acceptability of a water for its defined use depends on its physical, chemical and biological properties, and sometimes on whether these properties can be modified to suit the defined use. The composition of water is the end result of many possible physical and/or chemical and/or biochemical processes.

3.2.1 Physical Properties of Water

The standard physical properties of water such as molecular weight, density, melting point, boiling point, temperature (at maximum density), etc., are reported in the Appendices. The variations of density, dynamic viscosity, etc., with temperature are also reported in Appendix A3. Other physical properties that show wide variation in magnitude include: colour, turbidity, odour, taste, temperature and solids content.

Colour   Colour in water is caused by dissolved minerals, dyes or humic acids from plants. The decomposition of lignin produces colour compounds of tannins and humic acids. The latter causes a brown-yellow to brown-black colour. Coloured wastes, including dyes or pulp and paper plants, also cause colour, as does the presence of iron, magnesium and plankton. Water colour caused by dissolved or colloidal substances that remain in the filtrate after filtration through a 0.45 mm filter is called 'true colour'. 'Apparent colour' is the term applied to coloured compounds in solution together with coloured suspended matter. Colour is measured in units of mg/L of platinum and in rivers ranges from 5 to 200 mg/L. As mentioned in Chapter 11, the European Union (EU) drinking water upper limit is 20 mg/L, with a guideline value of 1 mg/L.

Turbidity   Turbidity in water is a measure of the cloudiness. It is caused by the presence of suspended matter which scatters and absorbs light. In lakes, turbidity is due to colloidal or fine suspensions. Very clear lakes are notable in that light can penetrate to great depths, as determined by the depth at which a 200 mm diameter black and white plate (Seechi disc) when lowered into the water is still visible. The so-called Seechi depth was determined to be 35 m at Lake Tahoe, the Alpine lake on the California/Nevada border (Tchobanoglous and Schroeder, 1987). In summer eutrophic conditions, still water bodies may have a Seechi depth as low as 0.5 m. In rapidly flowing rivers, the particles in suspension are larger and turbid conditions occur during flood times. Turbidity can be correlated with suspended solids, but only for waters from the same source. In such cases, a simple turbidity measurement may replace the complex time-consuming suspended solids test. Turbidity is measured by a visual comparison test with standard turbidity suspensions in 1 L bottles, in units of mg/L of SiO_2. River values range from 2 to 200 mg/L of SiO_2. The EU drinking water upper limit is 10 mg/L of SiO_2 with a guideline value of 1 mg/L.

Odour   Clear (distilled) water is odourless. Many organic and some inorganic chemicals are odorous, including algae and other organisms. Hydrogen sulphide (H_2S), sometimes present in groundwater and in wastewaters, is malodorous. The threshold odour is determined by diluting a sample with odour-free water until the last perceptible odour is detectable. This result is expressed as a dilution ratio. The smallest odour threshold is 1. For drinking water the EU upper limit is a dilution ratio of 2 at 12 °C. In the United

States (Standard Methods 1992), the minimum detectable odour threshold concentration (MDOTC), is determined also on a dilution basis. This test is to some extent subjective and the MDOTC can vary from person to person.

Taste   Taste, like odour, may be due to decaying micro-organisms or algae. It may also be due to high concentrations of salts such as Ca^{2+} and Mg^{2+} and Cl^-. In the European Union, the drinking water test of dilution is set at an upper limit of 2 at 12 °C. Taste is usually only an issue in drinking water and rarely in waters not used for drinking.

Temperature   Temperature is perhaps the most significant parameter in lake waters with regard to lake stability. As density decreases from 1 kg/L at 4°C to 0.994 kg/L at 35°C, water at deeper depths is heavier and is lightest close to the surface. If large temperature gradients are reported, then stratification may occur between the upper warmer water body and the lower colder water body. With no mixing between both layers, the upper warmer layer may become susceptible to eutrophication. Similar problems may occur in saline waters. In wastewater treatment plants temperature may also be significant in that, above 36 °C, the aerobic micro-organism population tends to be less effective as wastewater purifiers. Temperature may also be important in the river environment, as increased water temperature reduces the amount of oxygen in the water, thereby making the river less desirable for fish, particularly the more sensitive salmonoids. The discharge of effluent wastewaters (if warm) will also elevate the river temperatures. For drinking water, the EU upper limit value is 25 °C.

Solids   The solids content of water is one of the most significant parameters. The amount, size and type of solids depend on the specific water. For instance, an untreated sewage wastewater may have organic particulate matter, including food scraps of size range in millimetres, while a purified drinking water may have particles in the size range 10^{-6} mm. A clear lake, like Lake Tahoe, may have solid particles of size range 10^{-3} mm while a eutrophic reservoir has particulate matter of size range in millimetres. There are several classifications of solids in water and wastewater as sketched in Figs 3.1 and 3.2.

Solids are annotated in the following ways:

- Total solids, TS
- Suspended solids, SS
- Total dissolved solids, TDS = TS − SS
- Total volatile solids, TVS
- Volatile suspended solids, VSS

The *total solids* (TS) of a water/wastewater sample is all of the residue remaining after evaporation at 105 °C. Which classification of Fig. 3.1 to use depends on the application. For instance, in investigations for the design of a settling tank for water or wastewater treatment, it would be desirable to know the *settleable* fraction. As such, the classification to use is: settleable/non-settleable. This is determined in a laboratory test using an Imhoff cone (Fig. 3.3) into which the water sample is poured and allowed to settle over a period of 1 h. The extent or degree of settlement is then expressed in mg/L.

In drinking water or river water samples it is sometimes desirable to know the fractions of solids that are *suspended* and also those that are *dissolved*. Sometimes the colloidal fraction is broadly included with the dissolved fraction. The suspended solids (SS) fraction is that which is retained on a membrane filter or a glass-fibre Whatman filter of pore size about 1.2 μm.

The *filterable* solids is the term given to the combination of colloidal and dissolved solids. *Colloidal* particles are of clay origin and of a size range of 10^{-3} to 1 μm. Colloids do not dissolve, but instead remain as a solid phase in suspension. Colloids usually remain suspended because their gravitational settling velocity is less than 0.1 mm/s. The process by which a colloidal suspension becomes unstable and undergoes gravitational settling is called coagulation (Chapter 11). The dissolved particles may be of

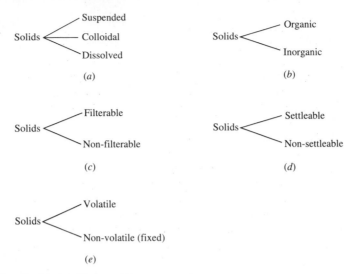

Figure 3.1 Classification of solids found in water and wastewater.

either organic or inorganic origin. In water or wastewater plants, these fractions are removed by either chemical coagulation (Chapter 11) or biological oxidation (Chapter 12).

In semi-solid sludges from wastewater treatment, it is sometimes relevant to further subdivide the suspended solids into *volatile* and *non-volatile* fractions. These correspond to the organic and inorganic fractions. The volatile fraction is gasified at a temperature of 550 °C and the remaining residue is the non-volatile or fixed suspended solids fraction.

3.2.2 Inorganic Chemical Properties of Water

Included among the chemical processes influencing the quality of water described by Dojlido and Best (1993) are the chemical processes of:

- Acid–base reactions
- Exchange processes between the atmosphere and water
- Precipitation and dissolution of substances
- Complex actions/reactions
- Oxidation–reduction reactions
- Adsorption–desorption processes

The chemical properties of water may be classified as either inorganic or organic. What properties to look for and what water analysis to carry out depends not only on the end use of the water but also on its origin and history. For instance, for a historically 'pure' groundwater with no known anthropogenic pollution, the key parameters to quantify may be the major ions of Ca^{2+} and Mg^{2+}. These are reported in concentrations of mg/L. An excess of these ions may be a water that is unpalatable and corrosive to plumbing fixtures. A surface water is more likely to contain organic chemicals from land runoff or anthropogenic pollution. The chemical properties of water are important to assess its quality as suitable for domestic or industrial use. The presence or absence of certain chemicals will define the suitability of water as being non-corrosive to metals or concrete. The assessment of a water quality may be either:

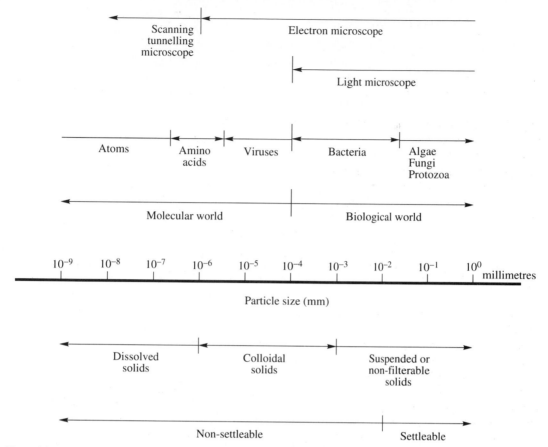

Figure 3.2 Particle size classification of solids in water and wastewater.

- Specific, for instance analysis for major ions (Ca^{2+}, Mg^{2+}) or heavy metals (Pb, Cu, Zn, Sn)
- General, for instance tests such as alkalinity, hardness, electrical conductivity, pH, etc.

Major ions The major ionic species in some natural waters are listed in Table 3.1. It is seen that all natural waters contain dissolved ionic constituents in varying amounts. The dominant ion in rainwater is chloride, as rainwater is largely derived from seawater. The predominant ionic species in either surface waters or groundwaters is that of bicarbonate and the dominant divalent ionic species are usually calcium and magnesium. In seawater, chlorides and to a lesser extent sodium predominate. Details of each element, occurrence, significance and method of determination are given in Dojlido *et al.* (1993).

Minor ions In addition to the major ionic species found in natural waters, there may also be minor ionic species. Table 3.2 lists these. They are classified as minor since their concentrations are in the order of ppb (parts per billion) or ppt (parts per trillion), while major ions are more typically in ppm concentrations.

Example 3.8 A groundwater was analysed and gave the following results. Use an anion–cation balance to check if the analysis is adequate.

Constituent	Concentration (mg/L)
Ca^{2+}	190
Mg^{2+}	84
Na^+	75
Fe^{2+}	0.1
Cd^{2+}	0.2
HCO_3^-	260
SO_3^{2-}	64
CO_3^{2-}	30
Cl^-	440
NO_3^-	35

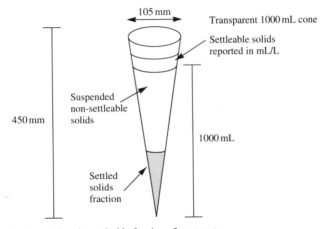

Figure 3.3 Imhoff cone for determining the settleable fraction of wastewater.

Table 3.1 Major ionic species in natural waters

		Concentrations in mg/L						
Ionic group	Constituent ion	Rainfall	World average river water†	Surface water, reservoir in California‡	Surface water, Niagara River§	Groundwater, Dayton Ohio§	Groundwater, Davis, California††	Sea-water§
Cations	Calcium, Ca^{2+}	0.09	1.5	4.0	36	92	34	400
	Magnesium, Mg^{2+}	0.27	4.1	1.1	8.1	34	66	1 350
	Sodium, Na^+	1.98	6.3	2.6	6.5	8.2	67	10 500
	Potassium, K^+	0.30	2.3	0.6	1.2	1.4	2.0	380
Anions	Bicarbonate, HCO_3^-	0.12	58.4	18.3	119	339	c244	142
	Sulphate, SO_2^-	0.58	11.2	16	22	84	57	2 700
	Chloride, Cl^-	3.79	7.8	2	13	9.6	39	19 000
	Nitrate, NO_3^-	—	1.0	0.41	0.1	13	13.9	
General characteristics	TDS	7.13	120	34	165	434	523	34 500
	Total hardness as $CaCO_3$	—	56	14.6	123	369	346	
	pH	5.7					7.4	

† Adapted from Montgomery (1985).
‡ Adapted from Tchobanoglous and Schroeder (1987).
§ Adapted from Snoeyink and Jenkins (1980).
†† Adapted from City of Davis, California, 1993 Annual Water Quality Report.

Table 3.2 Minor ionic species in natural waters

Cations	Anions
Aluminium, Al^{3+}	Bisulphate, HSO_4^-
Ammonium, NH_4^+	Bisulphite, HSO_3^-
Arsenic, As^+	Carbonate, CO_3^{2-}
Barium, Ba^{2+}	Fluoride, F^-
Borate, BO_4^{3-}	Hydroxide, OH^-
Copper, Cu^{2+}	Phosphate, $H_2PO_4^-$, HPO_4^{2-}, PO_4^{3-}
Iron, Fe^{2+}, Fe^{3+}	Sulphite, S^{2-}
Manganese, Mn^{2+}	Sulphate, SO_3^{2-}

Adapted from George Tchobanoglous and Edward D. Schroeder, *Water Quality* (p. 72), © 1987 by Addison-Wesley Publishing Company, Inc. Reprinted by permission of the publisher

Solution   If the analysis satisfies the following ion balance equation, it is then considered adequate:

$$\left| \sum \text{anions} - \sum \text{cations} \right| \leqslant 0.1065 + 0.0155 \sum \text{anions} \qquad (3.5)$$

	Cations					Anions			
Ion	Concentration (mg/L)	Atomic mass (g)	Equivalent mass (mg/meq)	Concentration (meq/L)	Ion	Concentration (mg/L)	Atomic mass (g)	Equivalent mass (mg/meq)	Concentration (meq/L)
Ca^{2+}	190	40.08	20	9.5	HCO_3^-	260	61	61	4.3
Mg^{2+}	84	24.3	12.2	6.9	SO_4^{2-}	64	96	48	1.33
Na^+	75	23.0	23	3.3	CO_3^{2-}	30	60	30	1.0
Fe^{2+}	0.1	55.85	27.9	0.004	Cl^-	440	35.5	35.5	12.4
Cd^{2+}	0.2	112.4	56.2	0.004	NO_3^-	35	62	62	0.6
Total				19.7					19.6

$$\left| \sum \text{anions} - \sum \text{cations} \right| = |19.6 - 19.7| = 0.1$$
$$0.1065 + 0.0155 \sum \text{anions} = 0.1065 + 0.3038 = 0.410$$

Therefore the analysis is adequate as per Eq. (3.5).

Silica, SiO_2   The presence of silica (a non-ionic mineral) along with calcium, magnesium, iron and aluminium can cause scaling in boilers. Most natural waters contain less than 5 mg/L of SiO_2, although higher values up to 100 mg/L have been reported. Silica can potentially be a limiting mineral in the process of surface water eutrophication. Silicon (Si) is a constituent of aquatic plants and animals in their skeletal structure. The concentrations in surface waters reduce in summer time due to its uptake by the accelerated growth of aquatic phytoplankton organisms in water, which is fuelled by sunlight and nutrients (phosphate). In Chow Valley Lake near Bristol, UK, the winter levels of SiO_2 were 6 mg/L, and in May during the spring plankton growth the values were reduced to 3.5 mg/L (Dojlido and Best, 1993).

Nutrients   The two nutrients of importance in water/wastewater are nitrogen and phosphorus. They are both essential nutrients for plant and organism growth, but in excess they can be undesirable, often leading to eutrophication.

Nitrogen   This is one of the basic components of proteins and in water it is used by the primary producers in cell production. Nitrogen exists in nine valence states. The largest amount of nitrogen is in the atmosphere, as 78 per cent by volume. In the nitrogen cycle, introduced in Chapter 2 (and again in Chapter 10), nitrogen cycles between its inorganic and organic forms. The inorganic forms of nitrogen of key interest are: N_2, NH_3 and NO_3^-. The organic forms of nitrogen of interest are: NH_3, NO_2^- and NO_3^-. Plants have the ability to 'fix' N_2 and convert it to nitrates. Animals cannot utilize inorganic nitrogen or nitrogen from the atmosphere, unless it is first converted into its organic form. The conversion of N_2 to NH_3 takes place when hydrogen combines with nitrogen. Ammonia is used to make nitrogen fertilizers and as ammonium nitrate, ammonium sulphate, urea and ammonium phosphate. In the water environment, nitrogen dissolved in water can be fixed by algae and bacteria. Nitrogen can also enter surface or groundwaters via sewage or industrial wastewaters resulting from the breakdown of proteins and other nitrogen compounds. A sewage effluent contains ammonia nitrogen if partial oxidation is used and contains nitrate if full oxidation is used (see Chapter 12). Large concentrations of organic nitrogen are indicative of organic pollution in a surface water, so typical limits are set at about 1 mg N_{org}/L for good quality rivers. Ammoniacal nitrogen exists in both the NH_4^+ ion and undissociated ammonia gas, NH_3. It is the free ammonia, NH_3, which is toxic to organisms and for salmonoid rivers upper limits are set at 1 mg NH_3-N/L. In the sludge treatment process of anaerobic digestion, values above 50 mg NH_3-N/L are toxic to the methanogenic bacteria (Chapter 13). Nitrite, NO_2^-, is a transitional compound in the nitrogen cycle and tends to be unstable. The EU drinking water directive sets upper limits of 0.1 mg NO_2-N/L and the salmonoid EU freshwaters fisheries directive sets an upper limit of 0.01 mg NO_2-N/L.

Nitrates in drinking water are harmful, and upper limit values of 40 mg NO_3^--N/L are typical for drinking water. For surface waters for salmonoids the upper limits are typically 1 mg NO_3-N/L. Dojlido and Best (1993) gives further details on nitrogen and analysis for nitrogenous compounds.

Phosphorus   This is an important nutrient in the aquatic environment and in freshwaters is most often the limiting nutrient of cultural eutrophication. Phosphorus was introduced to detergents in 1935 and is also a key crop fertilizer component. Phosphorus occurs in all living organisms and is important for cellular activity. Bones contain about 60 per cent $Ca_3(PO_4)_2$ and about 2 per cent of dry weight of protoplasm is phosphorous. About 80 per cent of phosphate production is in fertilizers. Other uses are chemicals, soaps, detergents, pesticides, alloys, animal feed supplements, catalysts, lubricant and corrosion inhibitors (Dojlido *et al.*, 1993). Phosphates are present in surface waters as a result of weathering and leaching of phosphorus-bearing rocks, from soil erosion, from municipal sewage, industrial wastewater effluent, agricultural runoff and atmospheric precipitation. Studies reported in Dojlido and Best (1993) indicate that from phosphorus-based fertilizer rates of about 30 kg P/ha applied to land, loss rates ranged from about 0.1 to 5 per cent, giving loss concentrations of about 0.03 to 1.5 kg P/ha. In sewage treatment plants with specific phosphorus removal technologies, about 75 per cent of phosphorus is removed. Typical phosphorus in sewage influent varies from 15 to 50 mg P/L. This initial figure is typical of sewage only, while the latter is more typical of plants treating a mix of municipal sewage and phosphorus-bearing industrial effluent. Very high phosphorus wastewaters are produced in the distillery industry (about 1000 mg P/L). Atmospheric precipitation on lands may account for between 0.01 and 1.43 mg P/L. Dojlido and Best (1993) report rainfall contributions of 0.6 kg P/ha. In freshwater and lakes, the input of phosphorus from municipal sewage plants and industrial wastewater plants is much reduced due to the installation of either biological phosphorus removal technologies or more typically technologies for chemical precipitation of phosphorus. Lakes in spring/summer may go eutrophic if the concentration of phosphorus exceeds about 30 µg/L of total phosphorus. EC guidelines for salmonoid water has an upper limit of 65 µg total P/L. If the ratio of nitrogen to phosphorus exceeds 14:1, the limiting nutrient is phosphorus and this is almost always the case for freshwater rivers and lakes.

The most commonly occurring compounds of phosphorus in water are:

Orthophosphates:	Na_3PO_4	Trisodium phosphate
	Na_2HPO_4	Disodium hydrogen phosphate
	$Na_2H_2PO_4$	Sodium dihydrogen phosphate
	$(NH_4)_2HPO_4$	Diammonium hydrogen phosphate

Polyphosphates:	$Na_3(PO_3)_6$	Sodium hexametaphosphate
	$Na_3P_3O_{10}$	Sodium tripolyphosphate
	$Na_4P_2O_7$	Sodium pyrophosphate

Reporting of phosphate is by the varying physical options:

- Dissolved total phosphate, determined after filtration through a 0.45 μm membrane
- Particulate phosphate, as in filterable solids
- Total phosphates, the summation of insoluble and soluble forms.

Figure 3.4, adapted from Dojlido and Best (1993), reports average concentrations of inorganic substances in surface waters (excluding groundwaters or wastewaters).

Gross chemical properties of water—inorganic   The gross chemical properties of water that are in widespread use when relating a water quality, be it drinking water, wastewater or river water, are:

- pH
- Alkalinity and acidity
- Hardness
- Conductivity

pH   pH is defined as the negative log (base 10) of the hydrogen ion concentration and is unitless, i.e.

$$pH = -\log[H^+] \tag{3.6}$$

Water dissociates slightly into hydrogen ions (H^+) and hydroxide ions (OH^-), often referred to as hydroxyl ions as per the following equation:

$$H_2O \rightleftharpoons H^+ + OH^-$$

$$K = \frac{[H^+][OH^-]}{[H_2O]} \tag{3.7}$$

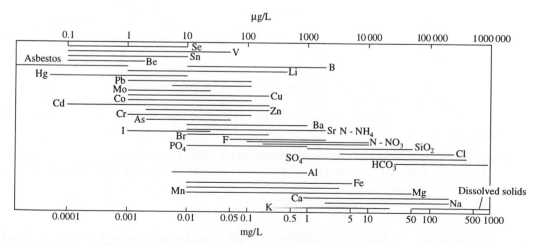

Figure 3.4 Average concentration ranges of inorganic substances in surface waters (adapted from Dojlido and Best, 1993. Reprinted by permission of Ellis Horwood/Prentice Hall).

where K is the equilibrium constant and [ ] is the concentration in mole/L. Introduce

$$K_w = [H^+][OH^-] \qquad (3.8)$$

where
$$K_w = \text{ion product or dissociation constant}$$
$$= 10^{-14} \text{ mol/L at } 25\,^\circ C$$

Taking the negative log of Eq. (3.8) gives

$$-\log K_w = -\log[H^+] - \log[OH^-] \qquad (3.9)$$

Introduce

$$pK_w = pH + pOH \qquad (3.10)$$

where

$$pH = -\log[H^+] \qquad (3.11)$$

and

$$pOH = -\log[OH^-] \qquad (3.12)$$

In the absence of foreign materials for (distilled) water $[H^-] = [OH^-]$, as required by electroneutrality ($\sum$ cations $= \sum$ anions). Hence the definition of 'neutrality' for water is at $pH = 7 = pOH$.

'Acidity 'implies
$$[H^+] > [OH^-]$$
$$[H^+] > 10^{-7} \text{ mole/L}$$
Therefore
$$pH < 7$$
'Basicity' implies
$$[H^+] < [OH^-]$$
$$[H^+] < 10^{-7} \text{ mole/L}$$
Therefore
$$pH > 7$$

Examples on the pH scale are reported in Fig. 3.5.

The pH of most mineral waters is 6 to 9. The pH remains reasonably constant unless the water quality changes due to natural or anthropogenic influences, adding acidity or basicity. As most ecological life forms are sensitive to pH changes, it is important that the anthropogenic impact (e.g. effluent discharges) be minimized. In Chapter 12 on wastewater treatment, it is seen that the pH control of wastewater influent to biological treatment systems is important to maintain within a specific range. An influent pH, too far to one side of the acceptable range (6 to 8) may kill off the active microbiological population, leading to untreated effluent discharges. Similarly, in Chapter 11 on water treatment, it is seen that the addition of alum as a coagulant depresses the pH, which may be required to be corrected by the addition of lime ($CaCO_3$). As such, the pH of water is of key parameter in numerous aspects of environmental engineering and is dependent on:

- the types of rock/soil from which acid/alkaline compounds can be eroded
- The carbonate system (Sec. 3.2.5) and the concentrations of carbonates and carbon dioxide; waters with low carbonate concentrations are usually acidic
- The exposure to wastewater or atmospheric pollutants

Alkalinity and acidity   'Alkalinity', the capacity of water to accept H^+ ions, is a measure of its acid neutralizing capacity (ANC) and is often described as the buffering capacity. Similarly, 'acidity' is a measure of the base neutralizing capacity (BNC). Alkalinity and acidity are capacity factors of a water.

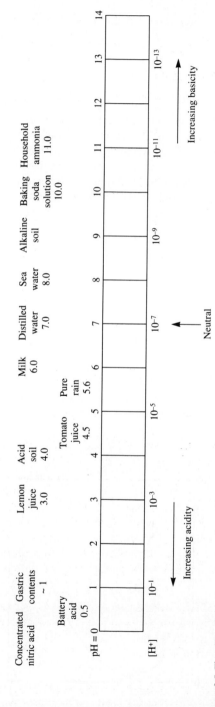

Figure 3.5 The pH scale.

From the carbonate system (Sec. 3.2.5), the following carbonate species contribute to alkalinity in relative amounts, as shown in Fig. 3.6:

- Hydroxide, OH^-
- Carbonate ion, CO_3^{2-}
- Bicarbonate ion, HCO_3^-
- Carbon dioxide, CO_2

Stumm and Morgan (1981) define alkalinity as

$$[Alk] = [OH^-] + 2[CO_3^{2-}] + [HCO_3^-] - [H^+] \qquad (3.13)$$

Besides the carbonate system species, other salts of weak acids such as borates, silicates and phosphates also contribute to alkalinity. In the 'anaerobic' environment the salts of weak acids, including acetic and propionic acids, contribute to alkalinity, as do ammonia and hydroxides in other environments. Table 3.3 gives typical alkalinity values for common environmental engineering applications. Alkalinity is measured volumetrically by titration with $N/50$ H_2SO_4 and is declared in mg $CaCO_3$/L. The amount of acid required to react with OH^-, CO_3^{2-} and HCO_3^- is called the 'total alkalinity'. The measured value may vary depending on the pH and end point of titration chosen (*Standard Methods*, 1992). The relative amounts of CO_2, HCO_3^- and CO_3^{2-} at various pH values are shown in Fig. 3.6.

Since alkalinity is made up of three components it is sometimes required to know the individual contributions. This can be determined if the pH and total alkalinity are known.

$$H_2O + H_2O \leftrightarrow H_3O^+ + OH^+$$
$$K_w = [H_3O^+][OH^-] = 10^{-14} \text{ mole}^2/L^2$$
$$HCO_3^- + H_2O \leftrightarrow H_3O^+ + CO_3^{2-}$$
$$K_2 = \frac{[H_3O^+][CO_3^{2-}]}{[HCO_3^-]} = 4.8 \times 10^{-11} \text{ mole/L}$$
$$[Alk] + [H_3O^+] = [OH^-] + 2[CO_3^{2-}] + [HCO_3^-]$$

Alkalinity is expressed in mg $CaCO_3$/L. The gram equivalent weight of $CaCO_3$ is 50. Therefore 1 mole/L is equivalent to 50 000 mg/L of alkalinity as $CaCO_3$. Solving the above equations, the individual alkalinity

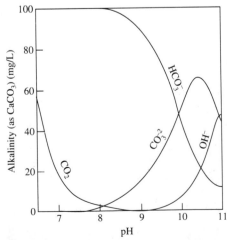

Figure 3.6 Relative amounts of CO_2, HCO_3^-, CO and OH_3^{2-} at various pH levels (adapted from Sawyer & McCarthy, 1989).

Table 3.3 Some typical alkalinity values

Application	Alkalinity (mg $CaCO_3$/L)
Upland (limestone) river	50–200
Lowland lake	10–30
Potable water	50–200
Domestic wastewater	200–400
Anaerobic sludge supernatant	2000–8000
Pig slurry	15 000–20 000
'Acidic' stream	10–20
'Non-acidic' stream	750
'Acidic' soil water	10–20

contributions (Sawyer and McCarty, 1989) are:

$$\text{Alkalinity in mg/L of } CaCO_3 = \frac{\text{total alkalinity} + 50\,000[H_3O^+] - 50\,000\,K_w/[H_3O^+]}{1 + [H_3O^+]/2K_2} \qquad (3.14)$$

$$\text{Bicarbonate alkalinity} = \text{carbonate alkalinity}\left[\frac{[H_3O^+]}{2K_2}\right] \qquad (3.15)$$

$$\text{Hydroxide alkalinity} = 50\,000\left[\frac{K_w}{[H_3O^+]}\right] \qquad (3.16)$$

Example 3.9 The supernatant from anaerobic co-digestion of the MSW food fraction and primary sewage sludge has an alkalinity of 4427 mg/L as $CaCO_3$. The pH is 7.27 at a temperature of 25 °C. Determine the individual alkalinity contributions.

Solution

$$\text{pH} = -\log[H_3O^+] = 7.27$$
$$[H_3O^+] = 10^{-7.27} = 5.37 \times 10^{-8} \text{ mole/L}$$

From Eq. (3.14):

$$\text{Carbonate alkalinity} = \frac{4427 + 50\,000 \times 5.37 \times 10^{-8} - 50\,000 \times 10^{-14}/(5.37 \times 10^{-3})}{1 + 5.37 \times 10^{-8}/(2 \times 4.8 \times 10^{-11})}$$
$$= 7.9 \text{ mg/L as } CaCO_3$$

Table 3.4 Ionic species responsible for hardness

Cations	Anions
Ca^{2+}	HCO_3^-
Mg^{2+}	SO_4^{2-}
Si^{2+}	Cl^-
Fe^{2+}	NO_3^-
Mn^{2+}	SiO_3^{2-}

Table 3.5 Relative hardness of waters

Degree of hardness	meq/L	mg/L as $CaCO_3$
Soft	< 1	0–75
Moderately hard	1–3	75–150
Hard	3–6	150–300
Very hard	> 6	> 300

From Eq. (3.15):

$$\text{Bicarbonate alkalinity} = 7.9 \times \frac{5.37 \times 10^{-8}}{2 \times 4.8 \times 10^{-11}}$$
$$= 4419.06 \text{ mg/L as } CaCO_3$$

From Eq. (3.16):

$$\text{Hydroxide alkalinity} = 50\,000 \times \frac{10^{-14}}{5.37} \times 10^{-8}$$
$$= 0.03 \text{ mg/L as } CaCO_3$$
$$\text{Total alkalinity} = 7.9 + 4419.06 + 0.03$$
$$= 4427 \text{ mg/L as } CaCO_3$$

Hardness 'Hardness' is expressed principally by the sum of the divalent metallic cations, Ca^{2+} and Mg^{2+}. These cations react with soap to form precipitate and with other ions present in water to form scale in boilers. The ions causing hardness have their origin in soil and geological formations. Table 3.4 lists the dominant ionic species, all responsible for hardness. Hardness is a water parameter used in potable water (not wastewater). Traditionally, hardness was calculated in mg/L as $CaCO_3$ (similar to alkalinity) or as meq/L. Table 3.5 is a qualitative listing of waters rated on hardness.
Hardness is made up of:

- Carbonate hardness or temporary hardness (TH) since this form is removed on prolonged boiling:

    Carbonate hardness $= \sum$ (bicarbonate + carbonate) alkalinity

    This is so, when alkalinity is < total hardness.
- Non-carbonate hardness (NCH)

Furthermore, since Ca^{2+} and Mg^{2+} are the dominant hardness-producing ions, it may be relevant to quantify their contributions, should softening be considered as a treatment process. Hardness is computed in mg/L as $CaCO_3$ as follows:

$$\text{Hardness in mg/L as } CaCO_3 = M^{2+} \text{ (in mg/L)} \times \frac{50}{\text{eq wt of } M^{2+}} \tag{3.17}$$

where M^{2+} is any divalent metallic ion (e.g. Ca^{2+}, Mg^{2+}).

$$\text{Hardness in mg/L as } CaCO_3 = M^{2+} \text{ (in meq/L)} \times 50 \tag{3.18}$$

Example 3.10   Determine the various hardnesses of the following water sample:

Constituent	Concentration	
	mg/L	meq/L
Ca^{2+}	60	3
Mg^{2+}	29.3	2.4
HCO_3^-	366	6

Recall:

$$\text{Concentration (meq/L)} = \frac{\text{concentration (mg/L)}}{\text{equivalent weight (mg/meq)}}$$

$$\text{Equivalent weight} = \frac{\text{Atomic weight}}{\text{Valence}}\text{(mg/meq)}$$

$$\text{Hardness (mg/L) as } CaCO_3 = \text{Concentration (mg/L)} \times \frac{50}{\text{equivalent weight}}$$

$$\text{Hardness (mg/L) as } CaCO_3 = \text{Concentration (meq/L)} \times 50$$

Solution   Using mg/L:

Ion	Concentration (mg/L)	Atomic weight	Valence	Equivalent weight	Hardness (mg/L as $CaCO_3$)
Ca^{2+}	60	40	2	20	$60 \times 50/20 = 150$
Mg^{2+}	29.3	24.31	2	12.2	$29.3 \times 50/12.2 = 120$
HCO_3^-	366	61	1	61	$366 \times 50/61 = 300$

$$\text{Total hardness (TH)} = Ca^{2+} + Mg^{2+} = 270 \text{ mg/L as } CaCO_3$$

$$\text{Carbonate hardness (CH)} = HCO_3^- = 300 \text{ mg/L as } CaCO_3$$

However, the CH cannot exceed the TH and the CH is therefore represented as 270 mg/L as $CaCO_3$.

Solution   Using meq/L:

Ion	Concentration (meq/L)	Hardness (mg/L as $CaCO_3$)
Ca^{2+}	3	$3 \times 50 = 150$
Mg^{2+}	2.4	$2.4 \times 50 = 120$
HCO_3^-	6	$6 \times 50 = 300$

Conductivity   Electrical conductivity or, as it is more commonly called, conductivity, is a measure of the ability of an aqueous solution to carry an electric current. The electric current is conducted in the solution by the movement of ions and so the higher the number of ions (i.e. the greater the concentration of dissolved salts) the higher the ionic mobility and so the higher the magnitude of conductivity. Chemically pure water does not conduct electricity since the only ions present are H^+ and OH^- and so the conductivity of very pure water is about 0.05 µS/cm (microsiemens/cm). On the other hand, a seawater with high salts has a conductivity of about 40 000 µS/cm. Typical values are reported in Table 3.6. The specific conductivity is the conductivity of 1 cm^3 of water across a 1 cm distance at 20 °C. Conductivity is

Table 3.6 Typical conductivity ranges for different waters

Water	Conductivity range (μS/cm)
Chemically pure	0.05
Distilled	0.1–4
Rainwater	20–100
Soft water	40–150
Hard water	200–500
Range of rivers	100–1 000
Groundwater	200–1 500
Estuarine water	200–2 000
Seawater	40 000

measured by placing a conductivity meter (made up of two platinum electrodes) in a water sample and recording the electrical resistance. In most waters, conductivity is due to the dissociation of inorganic compounds as organic compounds dissociate little. Therefore, a positive measure of conductivity is indicative of the concentration of dissolved inorganic salts. If the concentrations of ions (cations and anions) are known, then conductivity can be computed from

$$eC = \sum_{i=1}^{n} C_i f_i \qquad (3.19)$$

where
eC = electrical conductivity in μS/cm

C_i = concentration of ionic species i in solution in mg/L or meq/L

f_i = conductivity factor for the ionic species (see Table 3.7)

In water quality analysis, conductivity has been used to determine other parameters, since conductivity is easy to measure. For instance:

Salinity:
$$\text{in mg NaCl/L} = eC \times f_s$$
$$\text{where } f_s = \text{a conversion factor}$$
$$\cong 0.52\text{–}0.55$$

Total solids:
$$\text{in mg TS/L} = eC \times f_{ts}$$
$$\text{where } f_{ts} \cong 0.55\text{–}0.9, \text{ determined experimentally}$$
$$\text{for the particular water}$$

Total dissolved solids:
$$\text{in mg TDS/L} = eC \times f_{tds}$$
$$\text{where } f_{tds} \cong 0.55\text{–}0.7, \text{ determined experimentally}$$
$$\text{for the particular water}$$

Table 3.7 Conductivity factors for various ions

Cation	Conductivity factor f_i (μS/cm) per meq/L	per mg/L	Anion	Conductivity factor f_i (μS/cm) per meq/L	per mg/L
Ca^{2+}	52	2.6	HCO_3^-	43.6	0.72
Mg^{2+}	46.6	3.82	CO_3^{2-}	84.6	2.82
K^+	72	1.84	Cl^-	75.9	2.14
Na^+	48.9	2.13	NO_3^-	71.0	1.15
			SO_4^-	73.9	1.54

Adapted from Tchobanoglous and Schroeder, 1987

The various factors for salinity, TS and TDS may not be single valued but may approximate to a calibration-type curve, determined by experiment. For example, in physical model studies of salinity intrusion from estuarine to freshwater areas, conductivity is used as the measurement parameter to determine salinity. Studies of salinity intrusion in the $1\frac{1}{2}$ acre physical model of San Francisco Bay use conductivity probes at different depths to determine salinity profiles.

In addition, the ionic strength which is a measure of the intensity of the electric field can be approximated by the following equations (Snoeyink and Jenkins, 1980):

$$\mu = 2.5 \times 10^{-5} \times \text{TDS} \tag{3.20}$$

and
$$\mu = 1.6 \times 10^{-5} \times \text{eC}$$

$$\mu = \tfrac{1}{2} \sum_{i}^{n} (C_i Z_i^2)$$

where
$$\mu = \text{ionic strength}$$
$$Z_i = \text{charge of species } i$$

Although conductivity appears to have no health significance, the EU drinking water directive places an upper limit of 1500 µS/cm.

3.2.3 Organic Chemical Properties of Water

The main element of organic compounds is carbon, C. Organic substances may be natural or man-made. In fact, most are natural, and they are produced by plants and animals. Today in excess of 1.8 million synthetic organic compounds are produced, with about 250 000 new chemical compounds being synthesized annually and 300 to 500 going into production (Dojlido and Best, 1993). The presence of organic compounds in water is as a contaminant, whether it be naturally occurring or man-made. The objective in water/wastewater treatment is to minimize these compounds by biological, physical or chemical treatment processes. Figure 3.7 reports the average concentrations of some organic substances found in surface waters (Dojlido and Best, 1993).

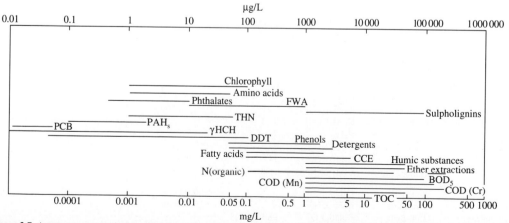

Figure 3.7 Average concentration ranges of organic substances in surface waters (adapted from Dojlido and Best, 1993. Reprinted by permission of Ellis Horwood/Prentice-Hall).

Organic compounds in water are typically divided into five groups, depending on their chemical structure:

1. *Hydrocarbons* These are organic compounds containing only carbon and hydrogen, e.g. ethane ($CH_3 CH_3$), ethylene ($CH_2 = CH_2$), benzene (C_6H_6), toluene ($C_6 H_5$—CH_3).
2. *Halogenated compounds* These are organic compounds in which a halogen is the principal atom (atoms include fluorine, chlorine, bromine and iodine), e.g. chloroform ($CHCl_3$), dichloromethane (CH_2Cl_2), carbon tetrachloride (CCl_4).
3. *Carboxylic acids and esters* These are organic compounds built around the carboxyl group (a carbon linked to oxygen with a double bond) and others with two functional groups attached to an oxygen atom, e.g. acetone (CH_3COCH_3), formaldehyde (CH_2O), ethyl ether ($CH_3CH_2COH_2CH_3$).
4. *Other organic compounds.*

Contaminants: naturally occurring organic substances The naturally occurring organic substances in water, wastewater and bottom sediments are:

1. *Proteins* These are made up of carbon, hydrogen, oxygen, sulphur and nitrogen with the basic building blocks of amino acids e.g. bacteria ($C_5H_7NO_2$) is mainly protein.
2. *Lipids* These comprise fats, waxes, oils and hydrocarbons, which are insoluble in water but soluble in some organic chemical solvents, and are slowly biodegradable.
3. *Carbohydrates* These contain carbon, hydrogen and oxygen. They include cellulose, hemicellulose, starch and lignin which are readily biodegradable (except lignin). An example is glucose, $C_6H_{12}O_6$.
4. *Plant pigments* These are made of chlorophyll, haemins and carotenes and include alcohols, ketones and carotenoids.

Synthetic organic chemicals (contaminants) The synthetic organic chemicals which are the products of the chemical, petroleum and agricultural industries include:

1. *Pesticides and agrichemicals* Table 3.8 reports on some of the pesticides, including chlorinated hydrocarbons and organophosphates, and also some herbicides. It is seen that some like DDT (now banned internationally) have a low solubility in water and are thus prone to vaporization. The 95 per cent degradation period for several of the pesticides is 1 to 25 years. The LCD_{50} which is the lethal concentration to kill 50 per cent of the population (fish, etc.) is detailed in Dojlido and Best (1993), but typically is < 50 mg/kg for the most toxic to > 5000 mg/kg for the least toxic.

Table 3.8 Synthetic substances in water environment

Classification	Compound	Formula	Solubility (mg/L)	Persistent half-life	Time for 95 per cent degradation (years)	WHO standards drinking water limit ($\mu g/L$)
Pesticides						
Chlorinated	DDT	$C_{14}H_9Cl_5$	0.0012	>6 months	4–30	1.0
hydrocarbons	Aldrin	$C_{12}H_6Cl_6$	0.01	>6 months	1–6	0.03
	Methoxychlor	$C_{16}H_{15}Cl_3O_2$	0.10	2–6 weeks		30
	Dieldrin	$C_{12}H_8Cl_6O$	0.18	>6 months	5–25	0.03
	Endrin	$C_{12}H_8Cl_6O$	0.23	—		
	Lindane	$C_6H_6Cl_6$	7.0	—		
Organic	Parathion	$C_{10}H_4NO_5PS$	24	<2 weeks		
phosphates	Melathion	$C_{10}H_4NO_5PS_2$	145	<2 weeks		
	Dimethoate	$C_5H_{12}NO_3PS_2$	2500	<6 months		
Herbicides	Simazine	$C_7H_2Cl N_5$	5	<6 months		
and	Propazine	$C_9H_{16}Cl N_5$	8	<6 months		
fungicides	2, 4, 5–T	$C_8H_5Cl_3O_3$	280	—		
	Diquat	$C_{12}H_{12}Br_2N_2$	70%			

2. *Surface active agents*   These are used for washing, emulsifying, wetting, foaming, etc., as they lower the surface tension of water when placed in it. The molecule of a surfactant (detergent) has two parts— a hydrophobic part, which is insoluble in water and soluble in non-polar liquids such as oils, and a hydrophilic part, which is soluble in water and insoluble in non-polar liquids. It is this dual property that makes them suitable for the earlier listed uses.

Surfactants are harmful to the aquatic environment as they may cause foaming and reduce the diffusion of atmospheric oxygen to the water. In sewage wastewaters, concentrations of surfactants may be up to 20 mg/L and in industrial wastewaters (those industries that use large quantities of solvents) concentrations may be up to 1000 mg/L. In rivers, typically values range from 0 to 1 mg/L in contaminated rivers. In recent years, biodegradable detergents have been introduced which has reduced the contamination level.

3. *Halogenated hydrocarbons*   These are the end products of the reaction of halogen with hydrocarbons, with the chlorinated hydrocarbons being of key interest. Low molecular weight chlorinated hydrocarbons are volatile and so measurable by gas chromatography. High molecular weight chlorinated hydrocarbons are of low volatility and difficult to measure. The most significant halogenated hydrocarbons are trihalomethanes (THMs) or haloforms. These compounds are represented by CHX_3, where X is the halogen, Cl, F, Br or I, and some are:

- Trichloromethane (chloroform), $CHCl_3$
- Tribromomethane (bromoform), $CHBr_3$
- Bromodichloromethane, $CHCl_2Br$
- Dibromochloromethane, $CHClBr_2$

THMs may be either discharged to aqueous environments in industrial (chemical) wastewaters or formed by the reaction of chlorine with organic compounds in the water environment. Of principal concern to environmental engineers/scientists is the reaction of chlorine (used extensively in water and wastewater treatment plants as a disinfectant) with organic compounds in the water. Chloroform is produced in the following reaction:

$$
\underset{\substack{\text{acetone} \\ \text{(precursor)}}}{CH_3COCH_3} \quad + \quad \underset{\substack{\text{sodium} \\ \text{hypochlorite}}}{3NaOCl} \quad \longrightarrow \quad \underset{\text{Trichloroacetone}}{CH_3COCCl_3} \quad + \quad \underset{\substack{\text{sodium} \\ \text{hydroxide}}}{3NaOH} \qquad (3.21)
$$

$$
\underset{\text{trichloroacetone}}{CH_3COCl3} \quad + \quad \underset{\substack{\text{sodium} \\ \text{hydroxide}}}{NaOH} \quad \longrightarrow \quad \underset{\substack{\text{sodium} \\ \text{acetate}}}{CH_3CO_2Na} \quad + \quad \underset{\text{trichloromethane}}{CHCl_3} \qquad (3.22)
$$

Acetone in Eq. (3.21) is called a trihalomethane precursor. Humic substances, from decaying vegetable matter, leaves, etc., are the main precursors of THMs, but petrochemical organic effluents may also be. Chlorophyll and algae may also be precursors (Dojlido and Best 1993).

Other synthetic organic chemicals include, chlorinated aromatic compounds (chlorinated benzenes, polychlorinated biphenols) dioxins (found in the water, air and soil environments and polynuclear aromatic hydrocarbons (PAHs). Many of these synthetic organics are present in the water environment in only trace amounts as low as 10^{-9} mg/L, but rarely as high as 1 mg/L. Figure 3.7, from Dojlido and Best (1993) reports some range values of some organics. Dojlido and Best (1993) is essential reading for further study in this area.

Determination of organic content of water   The determination of the organic content of water can be by:

1. Specific tests to measure the concentrations of specific compounds. The reader is referred to Standard Methods (1992) for details of specific tests.
2. Non-specific tests to measure the overall concentration of the organic content.

Tests for the overall concentrations include:

- BOD (a biochemical test that uses micro-organisms)
- COD (a chemical test)
- TOC (an instrumental test)
- TOD (an instrumental test infrequently used; see *Standard Methods*, 1992).

Before introducing these tests, the concept of dissolved oxygen is mentioned briefly. Oxygen from the atmosphere is transferred across the air/water interface by the principle of mass transfer. The amount of transfer depends on how much oxygen can be solubilized in water. Oxygen is regarded as weakly soluble in water with dissolved oxygen levels in water usually less than about 10 mg/L, with the concentration decreasing with increasing water temperature. This topic is covered in more detail in Sec. 3.2.4.

BOD—biochemical oxygen demand The BOD_5 is the amount of dissolved oxygen used up from the water sample by micro-organisms as they break down organic material at 20 °C over a 5-day period. It measures the readily biodegradable organic carbon. Clean waters have BOD_5 values of less than 1 mg/L. Rivers are considered polluted if the BOD_5 is greater than 5 mg/L. The EU freshwater fisheries directive sets an upper limit of 3 mg/L for salmonoid rivers and 6 mg/L for coarse fishwaters. The EU directive applicable to rivers for abstraction for drinking water is for 3 mg/L (for minimum treatment—chlorination only), 5 mg/L (for standard treatment—coagulation, flocculation, sedimentation, filtration; (see Chapter 11) and 7 mg/L (for special treatment—in addition to standard treatment, specific organic removal processes may be required; see Chapter 11). The BOD_5 is arbitrarily set at 5 days and this may not be long enough to determine the BOD ultimate, which is again arbitrarily set at 20 days. The BOD of municipal wastewaters ranges from about 150 to 1000 mg/L, while for industrial wastewaters (food industries) the value may be several thousands. Another term used is the BOD_u or the ultimate BOD, which may take 10 to 20 days for complete stabilization. Typically the $BOD_u \cong$ twice BOD_5. Further details on BOD in freshwater and saline waters are given in Chapter 7.

COD—chemical oxygen demand The COD test measures the total organic carbon, with the exception of some aromatics such as benzene which are not oxidized in the reaction. The test determines the amount of oxygen needed to chemically oxidize the organics in a water or wastewater. It is described in detail in *Standard Methods* (1992). A strong chemical oxidizing agent is used to oxidize the organics rather than using micro-organisms as in the BOD test. The oxidizing agent is potassium dichromate in an acid solution. COD is attractive as a test since it takes about 2 hours by comparison with 5 days for the BOD. A disadvantage is that it tells us nothing about the rates of biodegradation. In typical municipal wastewaters the BOD ultimate ($\approx BOD_{20}$) is approximately equal to the COD and

$$BOD_5 \approx 0.6 \text{ COD} \tag{3.23}$$

However, this factor (0.6) has to be established by calibration and does not hold for complex industrial wastewaters, particularly those with high contents of organic non-biodegradable fractions (e.g. ABS, alkyl benzene). Such a wastewater may register close to zero BOD while registering a high COD. The EU drinking water directive sets guideline values of 2 mg/L, with maximum admissible concentrations of 5 mg/L when using potassium permanganate (rather than potassium dichromate) as the oxidizing agent, for water to be treated and used as potable water. The EU urban wastewater directive sets an upper limit of 125 mg/L for treated wastewaters prior to discharge to rivers.

Example 3.11 If bacterial cells are represented by the chemical formula $C_5H_7O_2N$. Determine the potential carbonaceous BOD.

Solution As the cells require O_2 to stabilize them to end products we first stoichiometrically balance the equation

$$C_5H_7O_2N + 5O_2 \Rightarrow \underset{\substack{\text{stable end products} \\ \text{as a result of oxidation}}}{5CO_2 + 2H_2O + NH_3}$$

Therefore each mole of bacterial cells requires 5 moles of O_2 for oxidation and thus

$$COD = \frac{5 \text{ moles of } O_2}{1 \text{ mole of } C_5H_7O_2N} = \frac{5 \times 32}{1 \times 113} = 1.42$$

$$BOD_u = 0.92 \; COD$$

$$= 0.92 \times 1.42 = 1.31$$

If the bacterial cell concentration was, say, 1000 mg/L, then the potential $BOD_u = 1310$ mg/L.

TOC—total organic carbon   The TOC test measures all carbon as CO_2 in mg/L and therefore the inorganic carbon (HCO_3^-, CO_2, CO_3^{2-}, etc.) must be removed prior to the test. Acidifying and aerating the sample is the method used to remove inorganic carbon. This test can be carried out by oxidizing the organic carbon to carbon dioxide at a temperature of about 950 °C (evaporation) in the presence of a catalyst; the carbon dioxide is then determined spectrophotometrically by infrared absorption. The TOC test measures the mass of carbon per litre of sample, while the BOD and COD tests determine the amount of oxygen required for biochemical and chemical oxidation respectively. The TOC is a simple instrumental test and there are many TOC (instrument) analysers available commercially. TOC is now regularly specified in permits for industrial wastewater treated effluent. The theoretical relationship between COD and TOC is that the COD is 2.66 times greater, though practically closer to 2.5 times, as the COD rarely achieves full oxidation.

$$C_6H_{12}O_6 + 6O_2 \Rightarrow 6CO_2 + 6H_2O$$

$$COD = \frac{6 \times 32}{180} = 1.066$$

$$TOC = \frac{6 \times 12}{180} = 0.4$$

Therefore $$\frac{COD}{TOC} = \frac{1.066}{0.4} = 2.66$$

The COD test is still a more informative test as it quantifies the amount of oxygen that will be consumed by a wastewater, and it is knowledge of oxygen in, say, a river before and after effluent discharge that will describe the water quality states of that river. Figure 3.8 is a schematic of the BOD/COD relationship in a sample of wastewater.

In some situations, in wastewater analysis, the different components of carbon may be useful, and these are detailed in *Standard Methods* (1992). Total carbon is subdivided as per Fig. 3.9. Purgeable organic carbon is another term for volatile organic carbon.

Example 3.12   A wastewater is analysed and is shown to contain 100 mg/L of ethylene glycol ($C_2H_6O_2$) and 120 mg/L of phenol (C_6H_6O). Determine the COD and TOC.

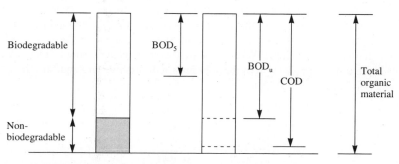

Figure 3.8 Schematic of BOD/COD/T.org matter relationship.

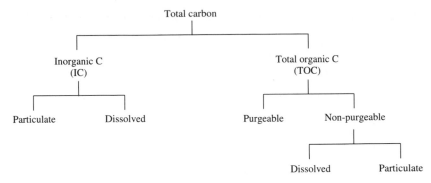

Figure 3.9 Schematic of subdivisions of carbon in water.

Solution

$$\text{Ethylene glycol, } C_2H_6O_2$$

$$\text{Atomic mass} = 62 \text{ g/mole}$$

$$C_2H_6O_2 + 2.5\,O_2 \Rightarrow 2CO_2 + 3H_2O$$

$$\text{COD} = \frac{2.5 \times 32}{62} \times 100 = 129.3 \text{ mg/L}$$

Ethylene glycol contains 2 atoms of carbon so,

$$\text{TOC} = \frac{2 \times 12}{62} \times 100 = 38.7 \text{ mg/L}$$

$$\text{Phenol, } C_6H_6O$$

$$\text{Atomic mass} = 94 \text{ g/mole}$$

$$C_6H_6O + 7\,O_2 \Rightarrow 6CO_2 + 3H_2O$$

$$\text{COD} = \frac{7 \times 32}{94} \times 120 = 286 \text{ mg/L}$$

Phenol contains 6 atoms of carbon so

$$\text{TOC} = \frac{6 \times 12}{94} \times 120 = 92.4 \text{ mg/L}$$

3.2.4 Solubility

Solids, gases and liquids may dissolve in water to form solutions. Water in this case is called the *solvent* and the substance, either solid, gas or other liquid, is called the *solute*. A measure of how soluble a substance is in water is determined by its *solubility*. For example, the solid compound NaCl is very soluble in water but the solid compound AgCl is almost insoluble. Similarly, ammonia gas (NH_3) is highly soluble in water while oxygen is weakly soluble. When a substance enters water, some of that substance will go into solution. At some time later, no more of the substance will dissolve and equilibrium is thus reached. The solubility reaction is generally written as:

$$A_aB_b \rightleftharpoons aA^+ + bB^-$$

For example,

$$Al(OH)_3 \rightleftharpoons Al^{3+} + 3\,OH^-$$

or

$$NH_3 + H_2O \rightleftharpoons NH_4^+ + OH^-$$

Introduce

$$K_s = [A]^a[B]^b$$

where $\qquad$ K_s = solubility constant, which never changes for a given substance

and $\qquad$ [ ] = molar concentration, mole/L ($\neq$ mg/L)

Introduce $\qquad$ $pK_s = -\log K_s$

For instance, for the solubility product of $Al(OH)_3$, K_s is 10^{-32} mole^2/L^2 or $pK_s = 32$. Similarly, the solubility product of NH_3 is 1.82×10^{-5} or $pK_s = 4.74$ mole^2/L^2.

Solubility of solids   When a solid is dissociating into its ionic components it is said to be undergoing 'dissolution'. When the ionic components of the solution are changing into solid, it is said to be undergoing 'precipitation'. Both processes of dissolution and precipitation are common in environmental engineering applications. For instance, ionic impurities washed off the land in flood times are usually dissolved in the recipient stream floodwaters. In wastewater treatment, phosphorus is removed from the treated wastewater via 'precipitation', more commonly chemical precipitation using ferric as:

$$Fe^{3+} + H_2PO_4^- \overset{H_2O}{\longleftrightarrow} \underset{\substack{\text{ferric} \\ \text{phosphate}}}{FePO_4} + 2H^+$$

ferric

Many 'solid' chemicals like lime, alum, activated carbon, etc., are introduced in solution form in water treatment processes for the purpose of water purification. For instance, $Al(OH)_3$ dissolves in water and dissociates as follows:

$$Al(OH)_3 \rightarrow Al^{3+} + 3OH^-$$

$$K_{sp} = [Al^{3+}][OH^-]^3 = 10^{-32}$$

where $\qquad$ K_{sp} = solubility product

The 'solubility product' is a measure of solubility, and is dependent on other parameters, including pH, temperature and pressure. The higher the K_{sp}, the more soluble the compound. Table 3.9 lists some common compounds in environmental engineering and their solubility products. For example, $Al(OH)_3$ is weakly soluble, whereas calcium sulphate ($CaSO_4$) is highly soluble.

Example 3.13   Determine the concentration of the aluminium ion (Al^{3+}) in pure water caused by the complete dissociation of $Al(OH)_3$. Remember that Al^{3+} was defined as a 'minor' ionic species, indicating its concentrations to be in the range ppt to ppb.

Solution

$$Al(OH)_3 \leftrightarrow Al^{3+} + 3OH^-$$

Table 3.9 Solubility products of some compounds

Application	Common name of compound	Equilibrium reaction	Solubility product K_{sp} at 25 °C
Coagulation	Aluminium hydroxide	$Al(OH)_3 \leftrightarrow Al^{3+} + 3OH^-$	1×10^{-32}
	Magnesium carbonate	$MgCO_3 \leftrightarrow Mg^{2+} + CO_3^{2-}$	4×10^{-5}
Hardness removal	Calcium carbonate (lime)	$CaCO_3 \leftrightarrow Ca^{2+} + CO_3^{2-}$	5×10^{-9}
Iron removal	Iron hydroxide	$Fe(OH)_3 \leftrightarrow Fe^{3+} + 3OH^-$	6×10^{-38}
Phosphate removal	Calcium phosphate	$Ca(PO_4)_2 \leftrightarrow 3Ca^{2+} + 2PO_4^{3-}$	1×10^{-27}
Fluoridation	Calcium fluoride	$CaF_2 \leftrightarrow Ca^{2+} + 2F^-$	3.9×10^{-11}
Metals removal	Copper hydroxide	$Cu(OH)_2 \leftrightarrow Cu^{2+} + 2OH^-$	1.6×10^{-19}
Flue gas desulphurization	Calcium sulphate	$CaSO_4 \leftrightarrow Ca^{2+} + SO_4^{2-}$	2.4×10^{-5}

From Table 3.9,

$$K_{sp} = 1 \times 10^{-32}$$

$$[Al^{3+}][OH^-]^3 = 1 \times 10^{-32}$$

From stoichiometry, it is seen that three times as many hydroxide ions (OH^-) dissolve as aluminium ions (Al^{3+}); i.e. the concentration of $[OH^-]$ is three times $[Al^{3+}]$. Therefore

$$[OH^-] = 3[Al^{3+}]$$

$$[Al^{3+}][OH^-]^3 = [Al^{3+}](3[Al^{3+}])^3 = 1 \times 10^{-32}$$

$$27[Al^{3+}]^4 = 1 \times 10^{-32}$$

$$[Al^{3+}] = 44 \times 10^{-10} \text{ mole/L}$$

Molecular weight of Al = 27 g/mole

$$[Al^{3+}] = 44 \times 10^{-10} \text{ mole/L} \times 27 \text{ g/mole} \times 10^3 \text{ mg/g}$$

$$= 120 \times 10^{-6} \text{ mg/L (120 ppt)}$$

Note that the concentration of Al^{3+} is ~120 ppt (parts per trillion), indicating that it is a 'minor ion' as defined in the minor ion species of Table 2.3.

Example 3.14 Determine the amount of ferric chloride required to precipitate phosphorus in a wastewater with a P concentration of 10 mg/L, if the flow rate is 36 400 m³/day. Assume that liquid ferric chloride is $FeCl_3 \cdot 6 H_2O$ and the equilibrium reaction is

$$\underset{\substack{\text{ferric} \\ \text{chloride}}}{FeCl_3 \cdot 6 H_2O} + H_2PO_4^- + 2 HCO_3^- \rightarrow \underset{\substack{\text{ferric} \\ \text{phosphate}}}{FePO_4} + 3 Cl^- + 2 CO_2 + 8 H_2O$$

Solution Molecular weights:

Liquid ferric chloride, $FeCl_3 \cdot 6 H_2O = 55.9 + 3 \times 35.5 + 6 \times 18 = 270.4$ g/mole

Ferric phosphate, $FePO_4 = 55.9 + 31 + 4 \times 16 = 150.9$ g/mole

The stoichiometric equation indicates that 1 mole of liquid ferric chloride produces 1 mole of ferric phosphate precipitate. This corresponds to a weight ratio of 270.4 : 150.9 or 1.8 : 1. Assume the density of liquid ferric chloride is 1.4 kg/L with a ferric strength of 50 per cent. Since 1 mole of Fe is required per mole of P, then the weight of Fe required per unit weight of P is:

$$1 \text{ kg} \times \frac{\text{mol wt of Fe}}{\text{mol wt of P}} = 1 \times \frac{55.9}{31} = 1.8 \text{ kg Fe/kg P}$$

The weight of liquid ferric chloride (ferric) per L is

$$1.4 \times 0.5 = 0.7 \text{ kg/L}$$

The weight of Fe per L of ferric is

$$0.7 \frac{\text{kg}}{\text{L}} \times \frac{\text{mol wt of Fe}}{\text{mol wt of ferric}} = 0.7 \times \frac{55.9}{270.4} = 0.145 \text{ kg/L}$$

The amount of ferric required per kg of P is

$$1.8 \frac{\text{kg Fe}}{\text{kg P}} \times \frac{\text{L of ferric}}{0.145 \text{ kg Fe}} = 12.4 \text{ L of ferric/kg P}$$

$$\text{Ferric daily requirement} = 36\,400 \frac{\text{m}^3}{\text{day}} \times 10 \frac{\text{mg P}}{\text{L}} \times 12.4 \frac{\text{L of ferric}}{\text{kg P}}$$

$$= 4513 \text{ L/day}$$

Solubility of gases   Some fraction of air constituents contacting water goes into solution in water. This means that of the numerous air constituents some will dissolve in water to different extents. The principal atmospheric gases that go into solution are oxygen, nitrogen and carbon dioxide and all waters exposed to the atmosphere will have some of these gases in solution. Other gases in solution in water are ammonia (NH_3), hydrogen sulphide (H_2S) and methane (CH_4), which are more likely to be associated with microbiological activity. For instance, methane in lakes may come from the anaerobic decay of sediment-laden organic matter. The amount of a particular gas dissolved in water depends on:

- Its solubility in water
- Its partial pressure at the air/water interface or sediment/water interface
- The water temperature
- The level of salts in the water

If water contains as much of a specific gas as it can hold in the presence of an abundant supply, then the water is said to be saturated. For instance, the saturated concentration of O_2 in water at 20 °C is 9.3 mg/L. If at 20 °C, a water contains, say, 7.5 mg/L of O_2, then this is equivalent to 80 per cent saturation. When oxygen dissolved in water is in equilibrium with the oxygen in the atmosphere, the water is saturated with oxygen to 100 per cent. When the O_2 content exceeds 100 per cent it is said to be supersaturated. Supersaturation may occur under the following cases:

1. Water flowing over a dam and plunging into the plunge pool at the bottom generates water in the plunge pool that is >100 per cent saturated. This is due to the excess air bubbles being trapped in the plunge pool water, which falls to the depths of the pool; on rising back to the surface, the pressure decreases rapidly and supersaturation occurs.
2. High photosynthetic activity (in summer) by aquatic plants and phytoplankton produce more oxygen than is expelled at the surface, thereby creating supersaturation.
3. When thermal plumes are discharged to rivers with temperatures (often 10 to 20 °C) greater than the recipient river water, the rapid rise in water temperatures causes oxygen supersaturation in the plume vicinity.

The solubility of gases in water is related to the partial pressure of the gas in the atmosphere above the water by Henry's law:

$$\text{Henry's law} \Rightarrow P_g = K_h x_g \tag{3.24}$$

where

P_g = partial pressure of gas, atm
K_h = Henry's law constant, atm (see Table 3.10)
x_g = equilibrium mole fraction of dissolved gas

Therefore

$$x_g = \frac{\text{mole gas } (n_g)}{\text{mole gas } (n_g) + \text{mole water } (n_w)} \tag{3.25}$$

The partial pressure P_g, of the gas in air is the volumetric concentration times the air pressure. As air contains ~ 21 per cent O_2, then the partial pressure for $O_2 \approx 0.21$ atm. Oxygen levels in rivers are an important parameter regarding the suitability of the river for particular fish life. For instance, salmonoid rivers (i.e. rivers capable of supporting salmon) require oxygen levels in excess of 6 mg/L. Coarse fish require O_2 levels in excess of 3 mg/L. Oxygen levels in rivers are reduced by oxygen consuming effluents, e.g. from a wastewater plant. Table 3.10 shows Henry's law constants for the most common gases that are slightly soluble in water.

Example 3.15   Determine the saturation concentration of O_2 in water at 10 °C and 20 °C at 1 atm.

Solution

$$O_2 \text{ is } 21\% \text{ of air } (v/v) \Rightarrow P_g = 0.21 \times 1 \text{ atm} = 0.21 \text{ atm}$$

Table 3.10 Henry's law constants for common gases soluble in H₂O

Temperature (°C)	$K_h \times 10^{-4}$, atm							
	Air	N_2	O_2	CO_2	CO	H_2	H_2S	CH_4
0	4.32	5.29	2.55	0.073	3.52	5.79	0.027	2.24
10	5.49	6.68	3.27	0.104	4.42	6.36	0.037	2.97
20	6.64	8.04	4.01	0.142	5.36	6.83	0.048	3.76
30	7.71	9.24	4.75	0.186	6.20	7.29	0.061	4.49
40	8.70	10.4	5.35	0.233	6.96	7.51	0.075	5.20

From Table 3.10, at 10 °C, Henry's law constraint is:

$$K_h = 3.27 \times 10^4 \text{ atm/mole}$$

From Eq. (3.24),

$$x_g = \frac{P_g}{K_h}$$

Therefore

$$x_g = \frac{0.21}{3.27 \times 10^4} = 6.42 \times 10^{-6}$$

Since 1 mole of water is 18 g then

$$[H_2O] = \frac{1000 \text{ g/L}}{18 \text{ g/mole}} = 55.6 \text{ mole/L}$$

From Eq. (3.25),

$$x_g = \frac{n_g}{n_g + n_w}$$

$$6.42 \times 10^{-6} = \frac{n_g}{n_g + 55.6}$$

Since $n_g \ll 55.6$, we can say:

$$n_g = 6.42 \times 10^{-6} \times 55.6 = 357 \times 10^{-6} \text{ mole/L}$$
$$= 3.57 \times 10^{-4} \text{ mole/L}$$

Saturation concentration (mg/L):

$$C_s = n_g M, \text{ where } M = \text{molecular weight of } O_2$$
$$= 3.57 \times 10^{-4} \text{ mole/L} \times 32 \text{ g/mole} \times 10^3 \text{ mg/g}$$
$$= 11.4 \text{ mg/L at } 10\,°C \text{ (at 1 atm pressure)}$$

Similarly,

$$C_s = 9.3 \text{ mg/L at } 20\,°C \text{ (at 1 atm pressure)}$$

As temperature increases, the saturation concentration decreases, so in warmer seasons river oxygen levels are more vulnerable to oxygen demand from effluents or algae, as discussed in Chapters 6 and 7. The saturation concentration of oxygen in water will also depend on the chloride concentration. For instance, at 20 °C, $C_s \sim 9.3$ mg/L for no chloride. At chloride levels of 20 000 mg/L, C_s for $O_2 \sim 7.4$ mg/L at 20 °C. Further details are given in Sawyer and McCarty (1989). As oxygen is only slightly soluble in H₂O, it is this factor primarily that limits the ability of freshwaters (and saline waters) to be used for dilution of organic wastes. As the dissolved oxygen (DO) reduces, the water body may become 'anaerobic' or without oxygen, during which time it can only support anaerobic fauna.

Example 3.16   Determine the saturation concentration of (a) nitrogen in water at 20 °C, at 1 atm and (b) carbon dioxide in water at 20 °C at 1 atm.

Solution

(a) Air is 79% N_2 (v/v) $\Rightarrow P_g = 0.79 \times 1$ atm $= 0.79$ atm

From Table 3.10 at 20 °C,

$$K_h = 8.04 \times 10^4 \text{ atm/mole}$$

$$x_g = \frac{0.79}{8.04 \times 10^4} = 9.8 \times 10^{-6}$$

$$n_g = \frac{n_g}{n_g + n_w}$$

$$9.8 \times 10^{-6} = \frac{n_g}{n_g + 55.6}$$

$$n_g = 5.46 \times 10^{-4} \text{ mole/L}$$

Saturation concentration, (molecular weight of $N_2 = 28$)

$$C_s = 5.46 \times 10^{-4} \times 28 \times 10^3$$

$$= 15.29 \text{ mg/L}$$

(b) Air is 0.033% CO_2 (v/v) $\Rightarrow P_g = 0.000\,33 \times 1$ atm $= 0.000\,33$ atm

From Table 3.10 at 20 °C,

$$K_h = 0.142 \times 10^4 \text{ atm/mole}$$

$$x_g = \frac{0.000\,33}{0.142 \times 10^4} = 0.233 \times 10^{-6}$$

$$n_g = \frac{n_g}{n_g + n_w}$$

$$0.233 \times 10^{-6} = \frac{n_g}{n_g + 55.6}$$

$$n_g = 0.13 \times 10^{-4} \text{ mole/L}$$

Saturation concentration,

$$C_s = 0.13 \times 10^{-4} \times 44 \times 10^3$$

$$= 0.57 \text{ mg/L at 20°C and 1 atm}$$

In summary, the saturation concentration at 20 °C and 1 atm is:

$$O_2 = 9.3 \text{ mg/L}$$

$$N_2 = 15.3 \text{ mg/L}$$

$$CO_2 = 0.57 \text{ mg/L}$$

The saturation concentration of N_2 in water is ~ 1.6 times that of O_2 in water. An environmental engineering application of the relevance of oxygen and nitrogen levels is seen in urban wastewater discharges. For instance, the typical untreated municipal wastewater has a chemical oxygen demand (COD) of about 600 mg/L and a total nitrogen concentration of about 40 mg/L. Traditional wastewater treatment plants reduced the COD level to about 50 mg/L, using the various unit processes described in Chapter 12. However, little attention was given to nitrogen reduction. An environmental problem resulting from excess nutrients is algae production. Algae can be represented by the chemical formula, $C_{106}H_{263}O_{110}N_{16}P$ (Randall *et al.*, 1992). Therefore, 1 kg of nitrogen could theoretically stimulate the production of 16 kg of algal biomass. This is equivalent to 20 kg of COD. Thus, 40 mg/L of nitrogen in an effluent discharged to a water body could result in the production of a COD equivalent to 800 mg/L. This is greater than the COD of the organics in the untreated urban wastewater. While the treated effluent (COD ~ 50 mg/L) should not impact on the river water quality, the fact that the nitrogen went untreated may have a major impact. Hence the more recent legislative requirement of nitrogen (and phosphorus) removal as well as organic removal from urban wastewaters prior to discharge.

Volatilization   Liquids and solids may vaporize into the atmosphere in a process known as volatilization. The mechanisms of volatilization are similar to those of evaporation of soil water. Figure 3.10 is a schematic of evaporization/volatilization.

The mechanisms of volatilization are:

1. Initially vapour escapes through the air/liquid or air/solid interface into the atmospheric boundary sublayer.
2. Then gas/vapour diffuses through the boundary sublayer by molecular and turbulent diffusion (see Chapter 21).
3. The gaseous compound is transported away from the site by advection and dispersion.

The phenomenon of volatilization occurs in many areas of environmental engineering, including:

1. VOCs are released to the atmosphere from aeration tanks, wastewater lagoons, constructed wetlands, freefall weirs, etc., and wastewater treatment plants.
2. Chlorine gas may be released accidentally at either wastewater or water treatment plants.
3. Methane gas and VOCs are released at landfill and other sites. See Gill (1995) on the emission of hazardous air pollutants (HAPs), such as benzene and vinyl chloride, from landfill sites.

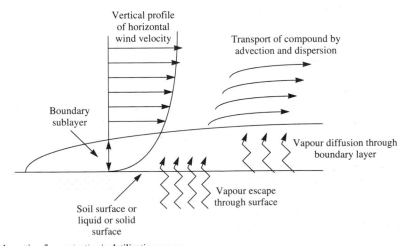

Figure 3.10 Schematic of vaporization/volatilization process.

In the European Union and the United States the emission of many VOCs are now regulated with guidelines and maximum concentration admissible values. Recent measures to reduce the amount of VOCs entering the atmosphere include covering aeration tanks in wastewater treatment plants and collecting all off-gas and treating it. For low concentrations of a VOC in water, the gas phase concentration at equilibrium above the water surface is proportional to the liquid phase concentration, with Henry's law constant the proportionality factor. As this constant increases, so does the relative rate of volatilization. Other physicochemical factors influencing the rate of volatilization include: molecular weight, diffusion coefficient in water and air, and the concentration of the VOC in both the gas and liquid phase. The simplest models used in determining VOC volatilization are based on the Lewis and Whitman (1924) two film theory model, described in Chapter 7.

The rate of vaporization (mass transfer) of a VOC is proportional to the difference between the saturation or equilibrium concentration in the water body and the existing concentration of the VOC in the water body. This is:

$$r = -k(C - C_s) \tag{3.26}$$

where
r = vaporization rate, g/m^2 h

k = mass transfer coefficient, m/h

C = existing VOC concentration in water body, g/m^3

C_s = saturation concentration of VOC in the atmosphere, g/m^3

If $C < C_s$, no vaporization occurs. If $C > C_s$, vaporization occurs. If a VOC is assumed to be vaporizing from a water body of finite area, say a wastewater tank of surface area A_s and depth h, then

$$\text{Outflow due to vaporization} = rA_s, \ g/h$$

$$\text{Loss of VOC to volume } A_s h \text{ is} = \frac{dC}{dt} A_s h$$

Assume vaporization is independent of surface temperature, wind speed, etc., then

$$\frac{dC}{dt} A_s h = rA_s$$

$$\frac{dC}{dt} = \frac{r}{h} = -\frac{k}{h}(C - C_s)$$

Integrating,

$$\frac{C - C_s}{C_0 - C_s} = \exp\left(\frac{-kt}{h}\right) \tag{3.27}$$

and so

$$C = C_s + (C_0 - C_s)\exp-\left(\frac{-kt}{h}\right) \tag{3.28}$$

where
C_0 = VOC concentration in air at time zero

and
C = VOC concentration in air at any time t

This is a simple model of VOC vaporization (Tchobanoglous and Schroeder, 1987), and assumes 'still' air conditions. Vaporization rates increase with wind speeds and air temperatures. As many VOCs are persistent in the air environment, an indication of their persistence is given by the duration of their half-life, i.e. the time it takes for the concentration to reduce to half its original.

Introduce $C_0/2$ as the VOC concentration equal to half the initial concentration C_0. Replacing C by $C_0/2$ in Eq. (3.27) gives

$$\frac{C_0/2 - C_s}{C_0 - C_s} = \exp\left(\frac{kt_{1/2}}{h}\right)$$

Typically, C_s in the atmosphere is low, so if $C_s \to 0$, then

$$t_{1/2} = 0.69 \frac{h}{k} \tag{3.29}$$

Example 3.17   Determine the time required for benzene (C_6H_6) and DDT ($C_{14}H_9Cl_5$) to vaporize to half their original concentrations from a wastewater treatment plant holding tank of depth 2 m. The mass transfer coefficient for benzene is 0.144 m/h and for DDT is 9.34×10^{-3} m/h. Values for transfer coefficients are given in Tchobanoglous and Schroeber (1987).

Solution
Equation (3.29) is

$$t_{1/2} = 0.69 \frac{h}{k}$$

for benzene, $k = 0.144$ m/h

and DDT, $k = 9.34 \times 10^{-3}$ m/h

Therefore for benzene,

$$t_{1/2} = \frac{0.69 \times 2}{0.144} = 9.5 \text{ h}$$

and for DDT,

$$t_{1/2} = \frac{0.69 \times 2}{9.34 \times 10^{-3}} = 147 \text{ h (6 days)}$$

Other VOCs, like dieldrin or lindane, have half-lifes as long as three years and one year respectively.

3.2.5 The Carbonate System

The carbonate system of acid–base reactions is ubiquitous in the environment, particularly so in the water medium. Inorganic chemical species originating in minerals (e.g. $CaCO_3$) and in the atmosphere (CO_2) can become dissolved in water. They have the effect of impacting on the pH, alkalinity and buffer capacity of waters. Regarding the various dissociation constants in the following description, the higher the dissociation constant, the more rapidly the chemical equation goes towards completion. For instance, the chemical reaction associated with the dissociation constant of 4.47×10^{-7} is more rapid than that associated with a dissociation constant of 4.8×10^{-7}. A simplified carbonate system is presented stepwise, beginning with atmospheric CO_2 being dissolved in water vapour.

Step 1   Gaseous $CO_2(g)$ in the atmosphere dissolves in water vapour (H_2O) to produce aqueous $CO_2(aq)$:

$$CO_2(g) \leftrightarrow CO_2 \tag{3.30}$$

With Henry's constant (see Table 3.10),

$$K_h = 0.164 \times 10^4 \text{ atm/mole at } 25\,°C$$

Using Henry's law,

$$P_{CO_2}(g) = K_h x_g \tag{3.31}$$

If the concentration of CO_2 in the atmosphere is 330 ppm (see Chapter 8), then $P_{CO_2}(g) = 0.000\,33$:

$$0.000\,33 \text{ atm} = (0.164 \times 10^4 \text{ atm})\,(x_g = \text{mole fraction})$$

Therefore

$$x_g = 2.01 \times 10^{-7}$$

so

$$x_g = \frac{n_g}{n_g + n_w} \tag{3.32}$$

where

$$n_w = 55.6 \text{ mole/L}$$

The molar concentration of

$$[CO_2(aq)] = n_g \approx n_w x_g$$
$$= 55.6 \times 2.01 \times 10^{-7}$$
$$[CO_2(aq)] = 1.12 \times 10^{-5} \text{ mole/L} \tag{3.33}$$

Saturation concentration in mg/L,

$$C_s = n_{CO_2} \times \text{mole weight}$$
$$= 1.12 \times 10^{-5} \text{ mole/L} \times 32 \text{ g/mole} \times 10^3$$
$$= 0.36 \text{ mg/L}$$

Step 2 Aqueous $CO_2(aq)$ reacts with H_2O in a water medium, e.g. a freshwater river, to form weak carbonic acid (H_2CO_3) as:

$$CO_2(aq) + H_2O \leftrightarrow H_2CO_3 \tag{3.34}$$

The equilibrium expression is

$$K = \frac{[H_2CO_3]}{[CO_2(aq)]} = 1.6 \times 10^{-3} \tag{3.35}$$

The molar concentration of

$$[H_2CO_3] = K[CO_2(aq)] \tag{3.36}$$
$$= 1.6 \times 10^{-3} \times 1.12 \times 10^{-5}$$
$$[H_2CO_3] = 1.79 \times 10^{-8} \text{ mole/L} \tag{3.37}$$

It is difficult to differentiate between $CO_2(aq)$ and H_2CO_3 in solution as $[CO_2(aq)] \approx 625[H_2CO_3]$. An 'effective carbonic acid' term is introduced and defined as

$$H_2CO_3^* = H_2CO_3 + CO_2(aq) \tag{3.38}$$

Molar concentration of

$$[H_2CO_3^*] = [H_2CO_3] + [CO_2(aq)] \tag{3.39}$$
$$= 1.79 \times 10^{-8} + 1.12 \times 10^{-5}$$
$$[H_2CO_3^*] \approx 1.12 \times 10^{-5} \text{ mole/L} \tag{3.40}$$

Step 3 In the water medium, carbonic acid $(H_2CO_3^*)$ is diprotic, i.e. it dissociates in two steps: firstly to the bicarbonate ion (HCO_3^-) and secondly to the carbonate ion (CO_3^{2-}). The dissociation to bicarbonate is:

$$H_2CO_3^* \leftrightarrow H^+ + HCO_3^- \tag{3.41}$$

The equilibrium expression is

$$K_1 = \frac{[H^+][HCO_3^-]}{[H_2CO_3^*]} \approx 4.47 \times 10^{-7} \text{ mole/L} \tag{3.42}$$

Secondly, the bicarbonate ion (HCO_3^-) dissociates to the carbonate ion (CO_3^{2-}) and the hydrogen ion (H^+) as

$$HCO_3^- \leftrightarrow H^+ + CO_3^{2-} \tag{3.43}$$

The equilibrium expression is

$$K_2 = \frac{[H^+][CO_3^{2-}]}{[HCO_3^-]} \approx 4.8 \times 10^{-11} \text{ mole/L} \tag{3.44}$$

Summary of major equilibria in the carbonate system

$$CO_2(g) \leftrightarrow CO_2(aq)$$
$$CO_2(aq) + H_2O \leftrightarrow H_2CO_3$$
$$H_2CO_3^* = CO_2(aq) + H_2CO_3$$
$$H_2CO_3 \leftrightarrow HCO_3^- + H^+$$
$$HCO_3^- \leftrightarrow CO_3^{2-} + H^+$$
$$H_2O \leftrightarrow H^+ + OH^-$$

Step 4   Limestone ($CaCO_3$) in solid form has the solubility reaction:

$$CaCO_3 \leftrightarrow Ca^{2+} + CO_3^{2-} \tag{3.45}$$

where the solubility product is

$$K_{sp} = [Ca^{2+}][CO_3^{2-}] = 5 \times 10^{-9} \text{ mole}^2/\text{L}^2 \tag{3.46}$$

Example 3.18   Determine the pH of rainwater if the concentration of CO_2 in the atmosphere is 330 ppm at 25 °C and 1 atm.

Solution

$$pH = -\log[H^+]$$

The mole concentration of $[H^+]$ is required. For electroneutrality, the hydrogen ion concentration is balanced by the negative ions of bicarbonate, carbonate and hydroxide, as inferred from Sec. 3.2.2:

$$[H^+] = \underset{\text{bicarbonate}}{[HCO_3^-]} + \underset{\text{carbonate}}{2[CO_3^{2-}]} + \underset{\text{hydroxide}}{[OH^-]} \tag{3.47}$$

The pH of rainwater is known to be <7; hence the values of the bases, $[CO_3^{2-}]$ and $[OH^-]$, are negligible. Thus:

$$[H^+] \approx [HCO_3^-] \tag{3.48}$$

The equilibrium expression from $H_2CO_3^*$ is

$$K_1 = \frac{[H^+][HCO_3^-]}{[H_2CO_3^*]} \approx 4.47 \times 10^{-7} \text{ mole/L} \tag{3.49}$$

Therefore

$$K_1 \approx \frac{[H^+][H^+]}{[H_2CO_3^*]} \tag{3.50}$$

However, from Eq. (3.40),

$$[H_2CO_3^*] = 1.12 \times 10^{-5} \text{ mole/L}$$

Therefore

$$[H^+]^2 = K_1[H_2CO_3*]$$
$$= 4.47 \times 10^{-7} \times 1.12 \times 10^{-5}$$
$$[H^+] = 2.24 \times 10^{-6}$$
$$pH = -\log[H^+]$$
$$= 5.65$$

Example 3.19 Determine (a) the pH and (b) the alkalinity of the following groundwater at 15 °C.

Constituent	Concentration (mg/L)
Ca^{2+}	190
Mg^{2+}	84
Na^+	75
Fe^{2+}	0.1
Cd^{2+}	0.2
HCO_3^-	260
SO_3^{2-}	64
CO_3^{2-}	30
Cl^-	440
NO_3^-	35

Solution

(a) pH

Consider the dissociation of HCO_3^-:

$$[HCO_3^-] \Rightarrow [CO_3^{2-}] + [H^+]$$

$$\text{Equilibrium constant } K_2 = \frac{[H^+][CO_3^{2-}]}{[HCO_3^-]}$$

At 15°C from Appendix A4, $K_2 = 3.72 \times 10^{-11}$ mole/L

Atomic mass of $HCO_3^- = 61$ g/mole

Therefore

$$[HCO_3^-] = \frac{260}{1000 \times 61} = 4.26 \times 10^{-3} \text{ mole/L}$$

Atomic mass of $CO_3^{2-} = 60$ g/mole

Therefore

$$[CO_3^{2-}] = \frac{30}{1000 \times 60} = 5.0 \times 10^{-4} \text{ mole/L}$$

From the above equation,

$$[H^+] = K_2 \frac{[HCO_3^{2-}]}{[CO_3^{2-}]} = 3.72 \times 10^{-11} \times \frac{4.26 \times 10^{-3}}{5.0 \times 10^{-4}}$$

$$= 3.17 \times 10^{-10}$$

$$pH = -\log[H^+] = 9.5$$

(b) Alkalinity

Ion	Concentration (mg/L)	Valence	Con: × valence mg/L
HCO_3^-	260	1	260
CO_3^{2-}	30	2	60

Alkalinity $= 260 + 60 = 320$ mg/L

Example 3.20 Given the following water quality analysis, determine the unknown values:

Ca^{2+}	$= 40$ mg/L
Mg^{2+}	$= ?$
Na^+	$= ?$
K^+	$= 39.1$ mg/L
HCO_3^-	$= ?$
SO_4^{2-}	$= 96$ mg/L
Cl^-	$= 35.5$ mg/L
Alkalinity	$= 3$ meq/L
Non-carbonate hardness	$= 1$ meq/L

Solution

(a) Use alkalinity to determine $[HCO_3^-]$.

Alkalinity is due to the presence of HCO_3^- only.

Atomic mass of $HCO_3^- = 61$ g

$$\text{Alkalinity} = 3 \text{ meq/L} = \frac{[HCO_3^-]}{61}$$

Therefore $[HCO_3^-] = 183$ mg/L

(b) Use hardness to determine $[Mg^{2+}]$:

$$\text{Total hardness} = \text{carbonate hardness} + \text{non-carbonate hardness}$$

$$\{Ca^{2+} + Mg^{2+}\} = \qquad 3 \text{ meq/L} \qquad + \qquad 1 \text{ meq/L}$$

$$\frac{[Ca^{2+}]}{\text{equivalent mass}} + \frac{[Mg^{2+}]}{\text{equivalent mass}} = 4 \text{ meq/L}$$

$$\frac{40}{20} + \frac{[Mg^{2+}]}{12.2} = 4$$

$$[Mg^{2+}] = 24.4 \text{ mg/L}$$

(c) Use the anion–cation balance to determine $[Na^+]$:

Cations	Concentration (mg/L)	Equivalent mass (mg/meq)	Concentration (meq/L)	Anions	Concentration (mg/L)	Equivalent mass (mg/meq)	Concentration (meq/L)
Ca^{2+}	40	20	2.0	HCO_3^-	183	61	3
Mg^{2+}	24.4	12.2	2.0	SO_4^{2-}	96	48	2
K^+	39.1	39.1	1.0	Cl^-	35.5	35.5	1
Na^+	x	23	$x/23$				
Total			$5 + x/23$				6

$$\text{Assume } \sum \text{anions} = \sum \text{cations}$$

$$6 = 5 + \frac{x}{33}$$

Therefore $\qquad\qquad [Na^+] = x = 33 \text{ mg/L}$

Buffers   Some rivers when subjected to acid rain suffer a reduction in pH and become acidified. Other rivers when subjected to acid rain suffer no serious pH reduction and do not become acidic. The reason for the difference in response is due to the presence or absence of 'buffering capacity'. A riverine system is said to have buffering capacity if it resists changes in pH. This 'resistance' is due to the presence of species of the carbonate system. Weak bases, such as HCO_3^-, CO_3^{2-} and OH^-, all contribute to resist pH changes if a strong acid is added. The 'acids' of acid rain are the strong acids of nitric acid (HNO_3) and sulphuric acid (H_2SO_4). Similarly, the weak acids, such as H_2CO_3 and H_3O^+, resist pH changes if a strong base (e.g. NaOH) is added. Most freshwater systems have a pH range of 6 to 9. As such, it is only the weak bases and weak acids that contribute to a riverine system's buffering capacity. Figure 3.11 shows the titration curve of the weak carbonic acid, H_2CO_3, with a strong base, NaOH. It is seen that in the pH range of interest, 6 to 9, the pH changes only slowly. This means that carbonic acid behaves as a buffer in this pH region. By comparison, it is also seen that the strong acid, H_2SO_4, has no buffering capacity in this pH region, when titrated with a strong base.

In the self-ionization of water, the ion product remains constant. The hydration of protons is generally assumed and so $[H_3O^+]$ is equivalent to $[H^+]$:

$$H_2O + H_2O \leftrightarrow H_3O^+ + OH^-$$

$$K_w = [H_3O^+][OH^-] = 10^{-14}$$

If the concentration of the weak bases $[OH^-]$ is increased by the addition of a strong base (e.g. NaOH) then the concentration of $[H_3O^+]$ decreases. In the major equilibrium equations of the carbonate system, as $[OH^-]$ increases the following equations move to the right as:

$$CO_2(aq) + H_2O \rightarrow H_2CO_3$$

$$H_2CO_3 + H_2O \rightarrow H_3O^+ + HCO_3^-$$

$$HCO_3^- + H_2O \rightarrow H_3O^+ + CO_3^{2-}$$

In moving to the right, more HCO_3^- is produced while reducing CO_2, thus reducing only slightly the pH. However, since there is only a finite amount of $CO_2(aq)$, eventually this will be all used up in the production of HCO_3^-. When this reservoir of HCO_3^- (or CO_2) is spent, the pH will decrease rapidly.

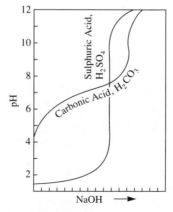

Figure 3.11 Titration curves of carbonic acid and sulphuric acid.

Buffering is of major interest in retaining water quality for aquatic life in riverine systems. Some anthropogenic activities (exotic forestry) whose acidifying effects (see Chapter 6) can be counteracted by natural buffering capacity can be permitted without having a negative impact on water quality. However, where the buffering capacity of the stream is poor, such activities may be harmful to aquatic life.

Buffering is also of importance in water and wastewater treatment. For instance, as many wastewater treatment processes are biological (e.g. activated sludge and anaerobic digestion), inadequate buffering capacity may cause the pH to go outside the optimum performance range for the micro-organisms. When this occurs, inhibition of micro-organisms occurs and plant operation may fail. In the same way, the nitrification process in wastewater treatment produces nitrous and nitric acids which may require the addition of lime or NaOH for neutralization. In potable water treatment, the addition of coagulants often causes a reduction in pH, and this too may need to be neutralized by the addition of lime. Liming is a common process in the management of exotic forestry where acidification of streams occurs (Howells and Dalziel, 1992). Lime is typically added from the air to eventually filter through the forest canopy, the soil and thence to the stream system, neutralizing the pH of the stream and permitting a habitat suitable for supporting fish life.

3.2.6 Oxidation–Reduction

In acid-base reactions, some compounds are proton (H^+) donors while others are proton acceptors, e.g.

$$\underset{\substack{\text{free}\\\text{ammonia}}}{NH_3} + H_2O \leftrightarrow \underset{\substack{\text{ammonium}\\\text{ion}}}{NH_4^+} + OH^-$$

i.e. the free ammonia accepts a proton and transforms to the ammonium ion. In an analogous way, oxidation–reduction reactions involve electron (e^-) donors (or reductants) and electron acceptors (oxidants). With no free electrons in water solutions, an oxidation reaction is always accompanied by a reduction reaction. Where both oxidation and reduction occur in a single equation, the cumulative is called a redox equation. The individual equations are sometimes termed half-reactions.

Reduction : $\qquad O_2 + 4\,H^+ + 4e^- = 2\,H_2O$

Oxidation : $\qquad 4\,Fe^{2+} = 4\,Fe^{3+} + 4e^-$

Redox reaction : $\qquad O_2 + 4\,Fe^{2+} + 4\,H^+ = 4\,Fe^{3+} + 2\,H_2O$

Oxidation–reduction phenomena are significant in environmental engineering aspects of water and wastewater. The reduction of oxygen by organic matter in freshwater is represented by

$$\underset{\text{oxidized}}{CH_2O} + \underset{\text{reduced}}{O_2} \rightarrow CO_2 + H_2O$$

Such oxygen depletion (reduction) is undesirable, particularly in freshwaters supporting fish life. In the nitrogen cycle, NH_4^+ is oxidized to NO_3^- in water and is represented by

$$\underset{\text{oxidized}}{NH_4^+} + \underset{\text{reduced}}{2\,O_2} \rightarrow NO_3^- + 2\,H^+ + H_2O$$

An 'oxidation number' is the charge which an atom of that element has or appears to have in a compound. The fundamentals of oxidation numbers or states are:

1. In free elements, each atom has a zero oxidation number.
2. In simple ions, the oxidation number is equal to the charge on the ion.
3. Halogens have an oxidation number of 1^-.
4. In most compounds, oxygen has an oxidation number of 2^-.
5. In most compounds, hydrogen has an oxidation number of 1^+. In hydrides it is 1^-.
6. In neutral molecules, the oxidation number of all atoms add up to zero. In complex ions, the oxidation number adds up to the charge of the ion.

Examples of oxidation numbers:

$$NH_4^+ : H = 1^+ : \Rightarrow N = 3^-$$
$$H_2S : H = 1^+ : \Rightarrow S = 2^-$$
$$CH_4 : H = 1^+ : \Rightarrow C = 4^-$$
$$N_2 : N_2 = 0 : \Rightarrow N = 0$$
$$NO_2^- : O = 2^- : \Rightarrow N = 3^+$$
$$NO_3^- : O = 2^- : \Rightarrow N = 5^+$$
$$SO_4^{2-} : O = 2^- : \Rightarrow S = 6$$
$$H_2CO_3^- : O = 2^- : H = 1^+ \Rightarrow C = 3^+$$

'Oxidation' is defined as the 'loss' of electrons from a substance. Oxidation is shown to occur if the oxidation state of the substance on the left side of the reaction equation 'increases' on going to the right side, e.g.

$$4\,Fe^{2+} \rightarrow 4\,Fe^{3+}$$

Typical oxidation agents include:

- Non-metallic elements, accepting electrons to form anions, e.g. Br_2, Cl_2, I_2, O_2
- Metal cations, accepting electrons to form neutral molecules as used in the electro-industries, e.g. Cu^{2+}, Al^{3+}, Ag^+, Au^+
- Ions with an element in a high oxidation number, e.g. MnO_4^-, $Cr_2O_7^{2-}$

'Reduction' is defined as the 'gain' of electrons by a substance. Reduction is shown to occur if the oxidation state of the substance on the left side of the reaction equation 'decreases' on going to the right side, e.g.

$$O_2 + 4\,H^+ + 4e \rightarrow 2\,H_2O$$

Examples of redox reactions:

Metals with acids :	$Mg^{2+} + H_2SO_4 \rightarrow MgSO_4 + H_2$
Metals with water :	$2\,Na^+ + 2\,H_2O \rightarrow 2\,NaOH + H_2$
Metals with non-metals :	$2\,Na^+ + Cl_2 \rightarrow 2\,NaCl$
Organic combustion :	$C_8H_{18} + 12.5\,O_2 \rightarrow 8\,CO_2 + 9\,H_2O$

Example 3.21 Balance the redox reaction of sodium dichromate ($Na_2Cr_2O_7$) with ethyl alcohol (C_2H_5OH) (Humenick, 1977) if the products of the reaction are Cr^{+3} and CO_2.

Solution

1. Balance the principal atom:

$$N_2Cr_2O_7 \rightarrow Cr^{3+}$$
$$C_2H_5OH \rightarrow 2\,CO_2$$

2. Balance the non-essential ions:

$$Na_2Cr_2O_7 \rightarrow 2\,Cr^{3+} + 2\,Na^+$$
$$C_2H_5OH \rightarrow 2\,CO_2$$

3. Balance oxygen with oxygen in water:

$$Na_2Cr_2O_7 \rightarrow 2\,Cr^{3+} + 2\,Na^+ + 7\,H_2O$$
$$3\,H_2O + C_2H_5OH \rightarrow 2\,CO_2$$

4. Balance hydrogen with hydrogen ions:

$$14\,H^+ + Na_2Cr_2O_7 \rightarrow 2\,Cr^{3+} + 2\,Na^+ + 7\,H_2O$$
$$3\,H_2O + C_2H_5OH \rightarrow 2\,CO_2 + 12\,H^+$$

5. Balance charges with the electrons:

$$6e^- + 14\,H^+ + Na_2Cr_2O_7 \rightarrow 2\,Cr^{3+} + 2\,Na^+ + 7\,H_2O$$
$$3\,H_2O + C_2H_5OH \rightarrow 2\,CO_2 + 12\,H^+ + 12e^-$$

6. Balance the number of electrons in each half reaction and add together:

$$6e^- + 14\,H^+ + Na_2Cr_2O_7 \rightarrow 2\,Cr^{3+} + 2\,Na^+ + 7\,H_2O, \text{ multiply by 2}$$
$$3\,H_2O + C_2H_5OH \rightarrow 2\,CO_2 + 12\,H^+ + 12e^-, \text{ multiply by 1}$$

Add:

$$\sum = 28\,H^+ + 2\,Na_2Cr_2O_7 + 3\,H_2O + C_2H_5OH \rightarrow 4\,Cr^{3+} + 4\,Na^+ + 2\,CO_2 + 14\,H_2O + 12\,H^+$$

7. Subtract common items from both sides of the equation:

$$16\,H + 2\,Na_2Cr_2O_7 + C_2H_5OH \rightarrow 4\,Cr^{3+} + 4\,Na^+ + 2\,CO_2 + 11\,H_2O$$

Application of oxidation–reduction reactions   The following are some of the more common redox reactions in environmental engineering:

- Solubilization and precipitation of iron and manganese, particularly in the treatment of groundwaters to be used by municipalities and pharmaceuticals/beverage industries
- The use of chlorine and ozone as oxidants in water and wastewater treatment for bacterial disinfection (see Chapter 11 and 12)
- In wastewater treatment, the removal of nitrogen by nitrification and dentrification either biologically or chemically assisted
- In anaerobic wastewater/sludge digesters producing methane
- In municipal and industrial wastewater treatment plants, the oxidation of organic substances, typified in BOD or COD reductions
- Corrosion of metals

Each reaction is characterized by a redox potential which is calculated from the Nerst equation:

$$E_h = E_0 + \frac{RT}{nF} \ln \frac{[\text{oxidized species}]}{[\text{reduced species}]} \tag{3.51}$$

where E_h = equilibrium redox potential, volts

E_0 = standard reduction potential at pH = 0 and at 25°C and at 1 atm

  = 1.23 V and $E_0 = E_h$ when [Ox] = [Red]

R = gas constant, 8.314 J/mole K°

T = absolute temperature, K°

n = number of electrons transferred

F = Faraday constant, 96 487 c/mole

The E_h potential is a measure of the potential, for a substance to be oxidized/reduced. The range of E_h in the natural environment is 0.6 V (well oxidized) to -0.8 (intensely reduced). Surface waters have a range of E_h from 0.2 to $+0.5$ V.

Consider the redox of water:

Oxidation :
$$2\,H_2O \leftrightarrow O_2 + 4\,H^+ + 4e^-$$

Reduction :
$$2\,H_2O + 4e^- \leftrightarrow H_2 + 2\,OH^-$$

(3.52)

According to Eq. (3.52), hydrogen ions are released when water is oxidized, thereby changing the hydrogen ion $[H^+]$ concentration of water, i.e. pH change. The typical range of pH in natural waters is 4 to 9. This is determined by the carbonate system components: CO_2, HCO_3^- and CO_3^{2-}. Values less than pH 4 are rare but may be due to oxidation of sulphide minerals. Similarly, pH > 9 is also rare and such alkaline environments are possibly due to sodium carbonate (O'Neill, 1991). The Merst equation for water is:

$$E_h = E_0 + \frac{RT}{4F} \ln\left(\frac{[O_2]\,[H^+]^4}{[H_2O]^2}\right)$$

$$= 1.23 + \left(\frac{8.314 \times 298}{4 \times 96\,487}\right) \ln\left([O_2]\,[H^+]^4\right)$$

$$- 1.23 + 0.0148 \log\left([O_2]\,[H^+]^4\right)$$

$$= 1.23 + 0.0148 \log\,[O_2] + 0.059 \log\,[H^+]$$

Recall that the atmosphere is made up of 21 per cent O_2, so the partial pressure of $O_2 = 0.21$ atm (at 25 °C). Therefore

$$E_h = 1.23 + 0.0148 \log\,[0.21] - 0.059\,\text{pH}$$

$$= 1.22 - 0.059\,\text{pH}$$

Figure 3.12 is a plot of the E_h/pH for some natural water environments. It is seen that typical rainwater, which has a pH of about ~ 5.7, has an $E_h \approx 0.5$ V. As 'rain' finds its way to streamwater, it becomes somewhat 'reduced' to an $E_h \sim 0.4$ V and a corresponding pH ~ 6.0. This assumes that in its transformation from rain to streamwater it does not encounter acidifying soil or vegetative influences (e.g. exotic forestry). If it does, and the rainwater remains in the soils and becomes waterlogged, it may become acidic with a pH ~ 4.5 and an $E_h \sim 0$. In other words, water remaining waterlogged becomes reduced.

3.3 ATMOSPHERIC CHEMISTRY

The following section is a brief introduction to atmospheric chemistry, necessary for the later chapters on air pollution and air pollution control. The region of the atmosphere of key interest to the environmental engineer is that closest to the earth's surface, where the chemical composition of the air is 78 per cent nitrogen and 21 per cent oxygen, which are essential for life in the biosphere. However, these elements react in the atmosphere and can produce undesirable features such as smog. The primary pollutants, which are emissions from anthropogenic sources are principally the oxides of sulphur, the oxides of nitrogen, the oxides of carbon, hydrocarbons, metals and particulates. The chemistry of these is briefly explained but the reader is referred to Seinfield (1986) for greater details, particularly with regard to secondary pollutants. The most significant secondary pollutants (resulting from the reactions of primary pollutants in the atmosphere) are those of ozone (O_3) and acid rain. Metals, including lead and mercury, are discussed in the chapter on air pollution (Chapter 8).

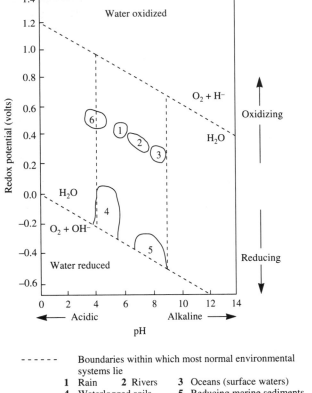

Figure 3.12 E_h/pH for some natural water environments (adapted from O'Neill, 1991. Reprinted by permission of Chapman and Hall).

3.3.1 Structures of the Earth's Atmosphere

The thin film of gas that surrounds the earth varies in structure as the distance increases outward from the surface. The earth's atmosphere is divided into regions based primarily on considerations of temperature gradients, as shown in Fig. 3.13. The temperature at the earth's surface varies from sub-zero °C in the polar regions and high mountainous areas to highs of about 70 °C in the arid desert regions. The corresponding air temperatures close to the earth's surface (within a few metres) are lows of sub-zero and highs of about 50 °C. In very warm areas, the air temperature is typically 10 to 20 °C cooler than the hot surface temperatures. Typically, at mid latitudes, the temperature falls with increasing altitude in the *troposphere*. This is known as a positive lapse rate. The increase continues to an altitude known as the *tropopause*, above which the temperature increases again in the region known as the *stratosphere*. The height of the troposphere is about 10 km above the earth's surface, while the stratosphere extends a further 20 to 30 km.

The lower 0 to 2 km of the troposphere can be further divided into several regions. This entire region (0 to 2 km) is called the *atmospheric boundary layer* (ABL). The ABL is that region where the wind velocity is affected by the shear resistance of the earth's surface. This ABL is shallowest over oceans or large inland waterways, where its height is about 500 m. The depth of the ABL may be up to 2 km in urban areas with many tall structures. In typical rural areas, the ABL depth is about 1 km. At the earth's surface, the wind velocity is lowest, and increases gradually (non-linearly) to the top of the ABL. Above

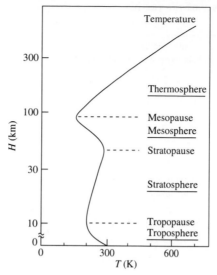

Figure 3.13 Vertical temperature profile of earth's atmospheric regions.

the ABL, the wind velocity is approximately constant, being unaffected by the shear resistance of the earth's surface. The region of most interest for atmospheric pollution is that within the ABL, though higher regions within the troposphere are of interest to large-scale air circulation behaviour and global climate circulation modelling. A region close to the earth's surface, called the sublayer of the ABL, is affected by local roughness and is characterized by high turbulence and strong mixing. This height depends on the fetch of the localized roughness, but may be as low as a few metres.

In addition to the atmospheric gases of Table 3.11, which are relatively constant, there are a number of variable constituents of natural origin. These are listed in Table 3.12.

Table 3.11 Average composition of the atmosphere

Gas	Composition by volume (ppm)
N_2	780 900
O_2	209 500
A	9 300
CO_2	300
Ne	18
He	5.2
CH_4	2.2
Kr	1
N_2O	1
H_2	0.5
Xe	0.08
O_3	0.02
NH_3	0.006
NO_2	0.001
NO	0.000 6
SO_2	0.000 2
H_2S	0.000 2

Table 3.12 Variable constituents of the earth's atmosphere of natural origin

Water as water vapour	HFL (of volcanic origin)
Airborne soil or sand	H_2S (Sulphide bacteria of volcanic origin)
NaCl	Ozone
Meteoric dust	Pollen
Volcanic dust and ash	Bacteria
SO_2	Spores
HCL (of volcanic origin)	Condensation nuclei

Adapted from Magill *et al.*, 1956

The atmosphere is computed to weigh 4.5×10^{15} tonnes. The density of the atmosphere at ground level is 1.29 g/L (1.29 kg/m^3) and decreases with altitude. It also decreases with increasing temperature. The above values are for standard temperature (273.16 K) and pressure (1013.25 mbar $= 101.3$ kPa). In terms of gas volume, it is known that 1 gram mole of an ideal gas, at STP (standard temperature and pressure) occupies 22.414 litres. An ideal gas is one that obeys the ideal gas law:

$$PV = nRT \tag{3.53}$$

where $P =$ pressure, Pa (N/m^2)

$V =$ volume, m^3

$n =$ number of moles

$R =$ gas constant $= 8.314$ J/K moles

$T =$ temperature, K

$R = R_d$ (in dry air) $= 287.04$ J/kg K

$R = R_w$ (for moist air) $= 461.5$ J/kg K

In addition to the existence of the atmosphere (the thin film of gas enveloping the planet earth), there is also a hydrosphere which contains all the water in the oceans, lakes and rivers. The hydrosphere is about 230 times heavier than the atmosphere. There is a constant flux of chemicals, primarily gaseous, between these two spheres, as well as a constant flux of heat and momentum. This area of science is called land (ocean)–atmosphere interaction.

3.3.2 Chemical Composition of the Earth's Atmosphere

Table 3.11 shows the concentration of gases in the earth's atmosphere (troposphere) close to ground level. This 'air' is about 78 per cent by volume of nitrogen (N_2), 21 per cent oxygen (O_2) and 1 per cent argon (Ar) with fractions of a per cent for CO_2, CH_4, H_2, CO, etc. This ABL layer is of most interest since it is the zone where most living matter exists. However, microbial life has been found in the stratosphere, but only in minute quantities.

3.3.3 Contaminant Gases in the Earth's Atmosphere

From Table 3.11 it is seen that many trace gases exist in the atmosphere at levels of parts per million (ppm) and parts per billion (ppb), or even as low as parts per trillion (ppt). For instance, SO_2 has background levels (in a 'clean' atmosphere) of about 200 ppt. However, in highly polluted air environments, the levels of SO_2 can be as high as 200 000 ppt or about 0.0002 per cent by volume. This is, in the overall picture, a small fraction; however, it is because of its toxicity to humans and other life forms that it is undesirable.

The *primary pollutants* are those that are emitted by identifiable man-made sources and are

- SO_2 and SO_x
- CO
- NO_x
- Metals
- Particulates
- Hydrocarbons
- Aerosols

The *secondary pollutants* are those formed in the atmosphere by chemical/photochemical reactions of the primary pollutants, and are

- O_3
- Photochemical oxidants including peroxyacetyl nitrate (PAN)
- Oxidized hydrocarbons
- Acid rain

There is also a regulatory (USEPA) list of pollutants which are defined and listed in Chapter 8.

Sulphur oxides   Pollution from sulphur oxides is primarily from two colourless gaseous compounds, sulphur dioxide, SO_2, and sulphur trioxide, SO_3. Together they are called SO_x. Sulphur dioxide has a pungent odour and does not burn in air. Sulphur trioxide is highly reactive. The simplified mechanism for the formation of SO_x is

$$S + O_2 \leftrightarrow SO_2 \tag{3.54}$$

$$2\,SO_2 + O_2 \leftrightarrow 2\,SO_3 \tag{3.55}$$

Typically, SO_3 is of the order of 1 to 10 per cent of the total SO_x. At low temperatures, eq. (3.55) proceeds so slowly that the equilibrium condition is rarely achieved and so, little SO_3 is produced. At high temperatures, the equilibrium mixtures contain lower SO_3 than at low temperatures. At high temperatures, the reaction rate is rapid, thus achieving equilibrium rapidly, but little SO_3 is present in the mixture. Also, SO_3 exists in the air as a gas only if the concentration of water vapour is very low. When sufficient water vapour is present, the SO_3 and water combine rapidly to form droplets of sulphuric acid, H_2SO_4, according to

$$SO_3 + H_2O \rightarrow H_2SO_4 \tag{3.56}$$

Thus, H_2SO_4 is the compound most commonly found in the atmosphere, rather than SO_3. In the atmosphere, SO_2 is in part converted to SO_3 and then to H_2SO_4 by photolytic and catalytic processes.

Carbon monoxide   Carbon monoxide, CO, is a colourless, odourless and tasteless gaseous compound. It is only slightly lighter than air ($\rho \sim 1.25$ g/L at STP) and is weakly soluble in water (see Table 3.10). It may be formed from:

1. Incomplete combustion of carbon or carbon compounds when there is an O_2 deficit for complete combustion.
2. By dissociation of CO_2 at high temperatures.
3. High-temperature reactions between CO_2 and carbon compounds.

The combustion of carbon fuels is:

$$2\,C + O_2 \rightarrow 2\,CO \tag{3.57}$$

$$2\,CO + O_2 \rightarrow 2\,CO_2 \tag{3.58}$$

The first reaction is about 10 times faster than the second; CO is thus an intermediary product and may show up as an end product if insufficient O_2 is available for the second reaction. The dissociation of CO_2 is

$$CO_2 \leftrightarrow CO + O \qquad (3.59)$$

and only occurs at very high temperatures (>1700 °C). High-temperature reactions of CO_2 and C are represented by

$$CO_2 + C \rightarrow 2\,CO \qquad (3.60)$$

which takes place only at very high temperatures. Automobiles are still the greatest contributors of CO to the atmosphere and soils provide a source for removing the CO.

Oxides of nitrogen   The basic chemistry of the formation of nitric oxide (NO) and nitrogen dioxide (NO_2) is

$$N_2 + O_2 \leftrightarrow 2\,NO \qquad (3.61)$$
$$2\,NO + O_2 \leftrightarrow 2\,NO_2 \qquad (3.62)$$

Nitric oxide is a colourless, odourless gas, while nitrogen dioxide is a red-brown gas with a pungent, choking odour. Although other oxides of nitrogen exist (NO_3, N_2O, N_2O_5), only NO and NO_2 are dominant. NO is emitted to the atmosphere in greater quantities than NO_2 (the reverse of CO and CO_2). As nitrogen makes up 78 per cent and oxygen 21 per cent of the typical 'air' volume, N_2 and O_2 tend to react with each other, but only at temperatures >1200 °C. This reaction thus occurs during high-temperature combustion processes and can be considered a side reaction of combustion (Stoker and Seager, 1972). Once in the atmosphere some NO is converted to NO_2 by photolysis (ultraviolet sunlight energy) and does not involve a reaction with O_2. Many of the serious effects of NO_x pollution result from their role in the formation of photochemical oxidants which are the harmful components of smog. The photolytic process is represented as

(a) $$NO_2 + hv \rightarrow NO + O^* \qquad 3.63$$

where hv represents the photon of solar radiation of energy
and O^* is a very reactive atomic oxygen

(b) $$O^* + O_2 \rightarrow O_3 \qquad 3.64$$
(c) $$O_3 + NO \rightarrow NO_2 + O_2 \qquad 3.65$$

This process is cyclic, as shown in Fig. 3.14. However, the process would be 'ideally' cyclic producing no net NO_2, just a rapid cycling of NO_2 (with O_3 and NO being produced and destroyed in equal quantities), were it not for the competing interests of hydrocarbons (HCs).

HCs interact in such a way that the cycle is unbalanced and NO is converted into NO_2 faster than NO_2 dissociates into NO and O. This causes a buildup of O_3. When NO_x, HCs and sunlight combine, they initiate a complex set of reactions that produce secondary pollutants called photochemical oxidants, ozone (O_3) being the most serious. This reaction can be represented in its simplest form by

$$HC + NO_x + sunlight \rightarrow photochemical\ smog$$

Nitric acid (HNO_3) is one of the acids of acid rain (another being sulphuric acid, H_2SO_4). One mechanism for the formation of HNO_3 is

$$O_3 + NO_2 \rightarrow NO_3 + O_2 \qquad (3.66)$$
$$NO_3 + NO_2 \rightarrow N_2O_5 \qquad (3.67)$$
$$N_2O_5 + H_2O \rightarrow 2\,HNO_3 \qquad (3.68)$$

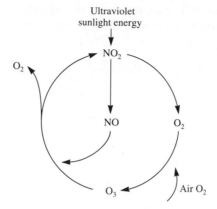

Figure 3.14 Photolytic NO_2 cycle (adapted from Stoker and Seager, 1972).

Another is

$$2\,NO_2 + H_2O \rightarrow HNO_3 + HNO_2 \qquad (3.69)$$

These topics are further discussed in Chapter 8.

Particulates   While most air pollutants in the previous sections are gases, there are also many particulate pollutants, which in some cases are by-products of gaseous pollutants, e.g. $N_2O_5 + H_2O \rightarrow HNO_3$. Table 3.13 shows a wide range of particulate matter components from coal combustion, which may be discharged into the atmosphere if adequate scrubbing facilities are not in place. Many industries still emit particulates, although environmental legislation has been successful in reducing many particulate emissions.

Particulate matter ranges in size from as small as 10^{-6} mm to as large as 1 mm. Because of their very small size, they remain in the atmosphere for long periods and can travel great distances. They will, on their own or on agglomeration with water droplets, settle to the earth's surface. Particulates of natural origin may be dust storms, volcanic ash and forest fires. An oddity of particulate natural pollution in recent years has been the migration of Saharan dust to northern Europe. Particulate matter, either of natural or anthropogenic form, is undesirable as it impedes lung efficiency in humans and animals.

Table 3.13 Fly ash composition of coal combustion

Component	Fly ash (%)
Aluminium (as Al_2O_3)	9.81–58.4
Silicon (as SiO_2)	17.3–63.6
Carbon	0.37–36.2
Iron (as Fe_2O_3 or Fe_3O_4)	2.0–26.8
Magnesium (as MgO)	0.06–4.77
Calcium (as CaO)	0.12–14.73
Sulphur (as SO_2)	0.12–24.33
Titanium (as TiO_2)	0–2.8
Carbonate (as CO_3^{2-})	0–2.6
Phosphorus (as P_2O_5)	0.07–47.2
Potassium (as K_2O)	2.8–3.0
Sodium (as Na_2O)	0.2–0.9

Adapted from Stoker and Seager, 1972

Particulate matter also interferes with plant growth when deposited on their leaves, it impedes photosynthesis by shielding sunlight from the plant and it interferes with the balance of CO_2 between the plant and the atmosphere. Some particulates are toxic and these are discussed further in Chapter 8.

Hydrocarbons By definition, hydrocarbon compounds contain only the elements of hydrogen and carbon. There are, however, many thousands of such compounds in all three physical states. Of interest are those HCs that are gases at normal temperature or are volatile, and they account for about 200 HC compounds in urban air. They are sometimes classified into three types, based on the molecular structure.

1. *Aliphatic or acyclic HCs,* where all carbon atoms are arranged in chains (no rings) and without branching, e.g.

Propane, C_3H_8

2. *Aromatic HCs* contain six carbon rings (benzene) in their molecules and each carbon atom has an H or C attached to it, e.g.

Benzene, C_6H_6

Toluene, C_7H_8

3. *Alicyclic HCs* are those containing a ring structure (six Cs) other than benzene.

Cyclohexane, C_6H_{12}

   In the urban air, HCs contain values typically in the level <ppm. For instance, $CH_4 \sim 3$ ppm, $C_6H_6 \sim 0.03$ ppm and C_7H_8 (toluene) ~ 0.05 ppm, were measured in Los Angeles urban air. Methane is by far the highest, but most of its concentration is from natural biological processes. It is added to by landfill sites, forest fires and automobiles. In general, automobiles account for most of the HCs in the urban air. HCs are undesirable in the atmosphere because of their toxicity, particularly the more common benzene compounds, e.g. C_6H_6 and C_7H_8.

3.4 SOIL CHEMISTRY

Soils are porous media formed at the earth's surface by the process of weathering over long periods, contributed to by biological, geological and hydrologic phenomena. Soils differ from rocks in that their buildup over time shows layers of different soil types on top of each other, with definite vertical stratification. Soils are considered as multicomponent, open, biogeochemical systems containing solids, liquids and gases. Being open systems, they are subject to fluxes of mass and energy with the atmosphere, biosphere and hydrosphere, and their composition is spatially highly variable and also changes with time (Sposito, 1989). Soils are made up of three phases of solids, liquids and gases (including air and water vapour). The composition of each phase depends on the climate, moisture content, closeness to the surface and a host of other factors. Soils may also be organic or inorganic, but usually a combination of both. Organic soils may contain a vastness of microbial activity. Ten grams of soil may contain a microbial population equal to that of the earth's human population. One kilogram may contain as much as 500 billion bacteria, 10 million actinomycetes and 1 billion fungi, with a length of root system in the first metre for a single plant of up to 600 km. Therefore, because of the different phases in a soil sample, the microbial population, the vast number of elements and of minerals and heterogeneity of structure, soils are a dynamic domain. It is because of these and other phenomena of transport of soils from area to area that the physics and chemistry of soils is undoubtedly most complex, much more so than that of air or water.

3.4.1 Chemical Composition of Soils

Table 3.14 indicates the main elements present in soil and in the rock crust. It is seen that in soils the 10 most abundant elements are:

In soils : $O > Si > Al > Fe > C > Ca > K > Na > Mg > Ti$

In crustal rocks : $O > Si > Al > Fe > Ca > Mg = Na > K > Ti > P$

The elements present in soils may be in either solid, liquid or gaseous form. Many soils have a porosity of 10 to 50 per cent depending on soil type and composition, so some of the elements listed in Table 3.14 may be in more than one phase. Soils in arid climates are typically 90 per cent inorganic matter. The Yolo clay soil in northern California is about 98 per cent inorganic. In humid temperate climates, the 'top soils' (usually 0.1 to 0.3 m depth) will be predominantly organic, giving way to inorganic soils at depths below the top-soil horizon. Soils are broadly non-homogeneous spatially. The structure of soils changes with time through weathering, biological activities and water movement. As such, the stoichiometry of soils is not straightforward. From Table 3.14, oxygen accounts for about 490 g/kg and silicon for about 310 g/kg. Together, they account for 80 per cent of the soil make-up. They exist not as singular elements, O_2 or Si, but combine to form minerals, some of which are listed in Table 3.15 (after Sposito, 1989). The dominant structural unit in these is the Si—O bond, which is much stronger (and less prone to weathering) than the typical metal–oxygen bond. The primary minerals are quartz, feldspar, mica, amphibole, pyroxene and olivine, which have their origin in the parent rock. The secondary minerals are those that result from weathering over time of the primary minerals (or silicates).

The non-solid fraction of soils is made up of soil air and soil water, ranging from 30 to 60 per cent of the total volume. While the soil air composition is similar to atmospheric air, there can be wide ranges in some compounds. Atmospheric air contains 209 500 ppm of O_2 while that of soil air can be as low as 20 000 ppm (near plant roots) and as high as 200 000 ppm. Similarly, CO_2 in the atmosphere is about 300 ppm but can be much higher in soil air, typically 3000 to 30 000 ppm near the earth's surface and as high as 100 000 ppm at greater depths. Other gases also vary, and this is due to the intense microbiological activity in soils.

Soil water is more commonly referred to as soil solution, as it is far from being 'pure'. It contains dissolved minerals and colloids and suspensions. Some dissolved solids dissociate into ions, and these in turn may attach to other ions or other solid particles. Section 3.4.2 discusses exchangeable ions.

Soil water may exist in small amounts as water vapour (<3 per cent by volume of soil). Other gases exist in the dry pore space or are dissolved from atmospheric air into the soil solution. The amount of a gas in the soil solution is dependent on the amount in the atmosphere and on Henry's law gas constant. If

Table 3.14 Mean elemental content in mg/kg in soil and crustal rock

Element	Soil	Crust	EF	Element	Soil	Crust	EF
Li	24	20	1.2	Zn	60	75	0.80
Be	0.92	2.6	0.35	Ga	17	18	0.94
B	33	10	3.3	Ge	1.2	1.8	0.67
C	25 000	480	52	As	7.2	1.5	4.8
N	2 000	25	80	Se	0.39	0.05	7.8
O	490 000	474 000	1.0	Br	0.85	0.37	2.3
F	950	430	2.2	Rb	67	90	0.74
Na	12 000	23 000	0.52	Sr	240	370	0.65
Mg	9 000	23 000	0.39	Y	25	30	0.83
Al	72 000	82 000	0.88	Zr	230	190	1.2
Si	310 000	277 000	1.1	Nb	11	20	0.55
P	430	1 000	0.43	Mo	0.97	1.5	0.65
S	1 600	260	6.2	Ag	0.05	0.07	0.71
Cl	100	130	0.77	Cd	0.35	0.11	3.2
K	15 000	21 000	0.71	Sn	1.3	2.2	0.59
Ca	24 000	41 000	0.59	Sb	0.66	0.20	3.3
Sc	8.9	16	0.56	I	1.2	0.14	8.6
Ti	2 900	5 600	0.52	Cs	4.0	3.0	1.3
V	80	160	0.50	Ba	580	500	1.2
Cr	54	100	0.54	La	37	32	1.2
Mn	550	950	0.58	Hg	0.09	0.05	1.8
Fe	26 000	41 000	0.63	Pb	19	14	1.4
Co	9.1	20	0.46	Nd	46	38	1.2
Ni	19	80	0.24	Th	9.4	12	0.78
Cu	25	50	0.50	U	2.7	2.4	1.1

Adapted from Sposito, 1989

the partial pressure of the gas is known and also the gas constant, then the amount of gas in the soil solution at equilibrium can be determined.

Example 3.22 Determine the amount of CH_4 in a soil solution for equilibrium conditions.

Solution

Partial pressure of CH_4 in atmosphere = 2.2 ppm $\rightarrow$ 2.2 × 10^{-6} atm

Henry's law gas constant for CH_4 at 20 °C = 2.97 × 10^4 atm/mole
(see Table 3.10)

$$\text{Mole fraction} = \frac{2.2 \times 10^{-6}}{2.97 \times 10^4} = 0.741 \times 10^{-10} \text{ mole}$$

$$\text{Moles of } CH_4 = 0.741 \times 10^{-10} \times 55.6$$

$$= 41.19 \times 10^{-10} \text{ mole/L}$$

$$\text{Concentration of } CH_4, \text{mg/L} = 41.19 \times 10^{-10} \times 16$$

$$= 0.066 \times 10^{-3} \text{ mg/L}$$

$$= 4.12 \times 10^{-3} \text{ mmole/m}^3$$

3.4.2 Exchangeable Ions

The interaction between the solid and liquid phase of a soil can be discussed in terms of the exchange of cations/anions. Particularly for those soils that are of clay content, they may be largely colloidal (particle

Table 3.15 The most common soil minerals

Name	Chemical formula	Importance
Quartz	SiO_2	Abundant in sand and silt
Feldspar	$(NaK)AlO_2[SiO_2]_3$	Abundant in soil that is not leached
	$CaAl_2O_4[SiO_2]_2$	extensively
	$K_2Al_2O_5[Si_2O_5]_3Al_4(OH)_4$	
Mica	$K_2Al_2O_5[Si_2O_5]_3(Mg,Fe)_6(OH)_4$	Source of K in most temperate-zone soils
	$(Ca,Na,K)_{2,3}(Mg,Fe,Al)_5(OH)_2$	
Amphibole	$[(Si,Al)_4O_{11}]_2$	Easily weathered to clay minerals and oxides
Pyroxene	$(Ca,Mg,Fe,Ti,AlSi,Al)O_3$	Easily weathered
Olivine	$(Mg,Fe)_2SiO_4$	Easily weathered
Epidote	$Ca_2(Al,Fe)_3(OH)Si_3O_{12}$	Highly resistant to chemical
Tourmaline	$NaMg_3Al_6B_3Si_6O_{27}(OH,F)_4$	weathering used as index mineral
Zircon	$ZrSiO_4$	in pedologic studies
Rutile	TiO_2	
Kaolinite	$Si_4Al_4O_{10}(OH)_8$	Abundant in clay as products of
Smectite	$M_x(Si,Al)_8(Al,Fe,Mg)_4$	weathering; source of exchangeable
Vermiculite	$O_{20}(OH)_4$, where	cations in soils
Chlorite	M = interlayer cation	
Allophane	$Si_3Al_4O_{12} \cdot nH_2O$	Abundant in soils derived from
Imogolite	$Si_2Al_4O_{10} \cdot 5\,H_2O$	volcanic ash deposits
Gibbsite	$Al(OH)_3$	Abundant in leached soils
Goethite	$FeO(OH)$	Most abundant: Fe oxide
Hematite	Fe_2O_3	Abundant in warm regions
Ferrihydrite	$Fe_{10}O_{15} \cdot 9\,H_2O$	Abundant in organic horizons
Birnessite	$(Na,Ca)Mn_7O_{14} \cdot 2.8\,H_2O$	Most abundant Mn oxide
Calcite	$CaCO_3$	Most abundant carbonate
Gypsum	$CaSO_4 \cdot 2\,H_2O$	Abundant in arid regions

Adapted from Sposito, 1989

size 10^{-3} m to 1 μm) and when in soil solution are in suspension. Thus, a fraction of soil water and much of these ions in soil solution are affected by the electrical charges. Soil colloids have a predominantly negative charge (−ve) and retain cations (+ve) in the water film on the colloid surface. This retention or bonding reduces the loss of cations such as Ca^{2+}, Mg^{2+}, K^+, Na^+ by processes such as leaching, while retaining those cations for plant uptake. The fraction of cations available for plant uptake is mostly in soil solution near colloidal surfaces. These ions can exchange for other ions either in natural or artificial processes of irrigation, liming or fertilization. For instance, an ammonium sulphate solution passed through a specific soil column may result in a calcium sulphate leachate. The dominant cation (NH_4^+) changes to the cation Ca^{2+} in the soil solution, because of the exchange within the soil column. Thus, soils can be identified in terms of their cation exchange capacities (CEC), i.e. their affinity for various cations. The CEC is also influenced by the pH of the soil/soil solution. Table 3.16 shows CEC values and major exchangeable cations for selected soils. The CEC values are low for coarse soils (~ 10 mmole/kg) and high for fine texture soils (~ 600 mmoles/kg).

Thus, an ion exchange reaction may be broadly defined as a reaction involving the replacement of one ionic species in a solid compound by another ionic species taken from an aqueous solution in contact with the solid (Sposito, 1989). For instance, the exchange of Mg^{2+} for Ca^{2+} is represented as

$$CaCO_3(s) + Mg^{2+}(aq) \leftrightarrow MgCO_2(s) + Ca^{2+}(aq) \qquad (3.70)$$

This equation can also be described in terms of its equilibrium constant, such as

$$k = \frac{[MgCO_3][Ca^{2+}]}{[CaCO_3][Mg^{2+}]} \qquad (3.71)$$

The higher the equilibrium constant, the more readily the reaction is likely to go towards equilibrium.

Table 3.16 CEC values and major exchangeable cations of selected soils

Soils	pH	CEC (mmole/kg)	Exchangeable cations (% of total)				
			Ca^{2+}	Mg^{2+}	K^+	Na^+	H^+Al^{3+}
Average of agricultural soils (Netherlands)	7.0	383	79.0	13.0	2.0	6.0	—
Average of agricultural soils (California)	7.0	203	65.6	26.3	5.5	2.6	—
Chernozem or Mollisoll (Russia)	7.0	561	84.3	11.0	1.6	3.0	—
Sodie Merced soil (California)	10.0	189	0.0	0.0	5.0	95.0	0.0
Lanna soil, unlimed (Sweden)	4.6	173	48.0	15.7	1.8	0.9	33.6
Lanna soil, limed (Sweden)	5.9	200	69.6	11.1	1.5	0.5	17.3

Adapted from Bohn *et al.*, 1985. Copyright © 1985. Reprinted by permission of John Wiley & Sons, Inc.

The higher the equilibrium constant, the more readily the reaction is likely to go towards equilibrium. Further details on the CEC are discussed in Chapter 10, and the reader is referred to Sposito (1989) and Bohn *et al.* (1985) for further details.

3.4.3 Soil Salinity

Salinity of soils is usually not an issue in temperate climates with organic soils. However, it is a major issue in the arid and semi-arid areas of the world. A soil solution is considered saline if the electrical conductivity (EC) is >4000 μs/cm. In the inland fertile areas of California, values in excess of 1000 μs/cm are regularly found. This is due mainly to the fact that, in these regions, evaporation exceeds precipitation. Salts are not leached from the soil and therefore accumulate. The main sources of soil salinity are: mineral weathering, atmospheric precipitation (near coasts), fossil salts and anthropogenic sources, such as irrigation.

Chemical and physical weathering of exposed rocks and minerals is the main source of salts (from soils). In humid climates, salts are leached down through the soil profile by infiltrating rainwater. When in the water table, they are further transported to rivers and seas and they enter the hydrological cycle. However, in arid areas, there is less leaching (downward movement of rainwater and salts) and so the salts accumulate. This is the case where high evaporation occurs and it is compounded by additional transpiration rates if these arid areas (valleys) are highly cultivated with irrigation water. The salt (chloride) concentration of rainwaters near the coast may be up to 200 mg/L and this can add salts to coastal areas by precipitation. Inland, this mechanism of salt accumulation diminishes as the inland concentration of rainwater rapidly reduces to ∼1 mg/L. In areas where reservoirs have been built on saline sediments, the latter have been transmitted to the downstream waters. Also when areas are irrigated with water, the water table may tend to rise and the resulting saline groundwaters, when eventually discharged, will increase the salinity of downstream areas. Some areas of the globe have naturally high water tables, within a couple of metres of the surface. The water is drawn to the surface by capillarity and also in water vapour, where it evaporates and leaves behind the salt. Some of these areas are called saltpans. For further details on salinity, the student is referred to Tanji (1990).

3.5 MICROBIOLOGY

Microbiology is the study of micro-organisms, which are distinct from all other living matter by their small size, in the range 10^{-5} to 10° mm. Micro-organisms are important in the water, air and soil environments, not only because of their ubiquity but also because of their activity, beneficial or otherwise, in that environment. For instance, 1 gram of rich organic soil may contain as many as 2.5 billion bacteria, half a million fungi, 50 000 algae and 30 000 protozoa. The water environment, as in rivers and lakes, may contain undesirable micro-organisms such as algae, viruses, worms, biological slimes, etc. On the other hand, bacteria are exploited in wastewater treatment processes to biodegrade the organic wastes. In the air environment, many undesirable organisms such as mould spores, bacteria, yeasts, etc., are found. It is therefore the task of the environmental engineer and scientist to understand the role of micro-organisms in the particular environment, so as to beneficially transform that environment. For instance, some understanding of microbiology is required by the water specialist when determining how much chlorine to use when disinfecting potable water supplies, when designing a process for purifying wastewater, when bioremediating a contaminated soil site or when designing an air circulation system for a contaminated air environment.

3.5.1 Organization of the Microbial World

A simplistic classification of the microbial world in decreasing order of size and level of cell evolution is presented in Table 3.17.

Micro-organisms can also be classified by their cell make-up. For instance, the lowest form, the virus, does not have a cell structure and is composed of a double strand of genetic material and a protein coat. Above the virus are the lower protista, including the bacteria, which are single-cell structures. In the case of bacteria, these single cells are of a simple primitive kind, called procaryotes, in which the cell nucleus is not encompassed by a membrane, but is free to move within the cytoplasm. Above the procaryotes (bacteria) are the higher protista of plants and animals which have a well-developed cell structure surrounded by a membrane. The organisms are either of single-cell or multicell composition, and are classed as eucaryotic. Those organisms that are made up of cell structures, either single or multicellular,

Table 3.17 Classification of microbial world

Microbial kingdom	Some microbes	Cell structure
Animals	Worms Helminths	Unicellular or multicellular
Plants	Aquatic plants Macrophytes Seed plants Ferns Mosses	Well-evolved cell structure —eucaryotic
Higher protista	Fungi Algae Protozoa Rotifers Crustacea	
Lower protista	Bacteria Blue-green algae Cyanobacteria	Primitive cell structure— procaryotic
Viruses	Many	No cell structure

procaryotic or eucaryotic, can survive on their own, if the nutrient and the food supply is adequate. The virus, on the other hand, is a non-living obligate parasite and needs other living cells for its reproduction. The development and continuity of micro-organism life requires cell synthesis, and then cell maintenance. To sustain cell synthesis and cell maintenance, food in the form of carbon, energy and nutrients (N, P, K, S, Ca, Mg, etc.) is required. Also some micro-organisms require oxygen for survival and are called *aerobes*, while others cannot survive in the oxygen environment and are known as *anaerobes*. Those organisms that can survive in either environment are called *facultative anaerobes*. Those micro-organisms that use CO_2 as the food carbon source are called *autotrophic* while those that utilize organic carbon are called *heterotrophic*. In a similar way, those micro-organisms that use light as their energy source are called *phototrophic*, while those that use energy from an inorganic chemical source are called *chemotrophs*. Thus, those organisms that use light as an energy source and CO_2 as a food carbon source are called *photoautotrophic* (e.g. algae). Those organisms that use light as an energy source and organic material as the food carbon source are called *photoheterotrophic* (e.g. photosynthetic bacteria). Many bacteria that use inorganic matter as the energy source and CO_2 as the food carbon are called *chemoautotrophs*. Those bacteria and protozoa that use inorganic chemicals as the energy source and organic matter as the food carbon source are called *chemoheterotrophs*.

The microbial nomenclature is binary, i.e. it consists of two words indicating:

- Genus (genera = plural), e.g. *Vibrio*
- Species, e.g. *cholera*

The genera, written with a capital letter, are defined in terms of

- Physiological characteristics, e.g. *Vibrio*, *Nitrosomonas*
- Pigmentation, e.g. *Chromobacterium*
- Diseases, e.g. *Pneumococcus*
- Nutrition, e.g. *Amylobacter*

3.5.2 Animals, Plants, Fungi, Algae, Protozoa and Viruses

This listing includes most organisms (mostly micro-organisms) of interest in the water, air and soil environments, except bacteria (which are treated in Sec. 3.5.3). The dominant environment for microbial proliferation is the soil, where usually there is an abundant supply of carbon, energy, moisture and nutrients. Soil is also suitable for microbial development since it protects them from the sun's radiation. The greatest density of microflora is in the upper layers, sometimes called the organic top soil. The soil surface tends to be free of micro-organisms, as is the case at greater depths. In the same way the water environment usually contains sufficient nutrition to maintain micro-organism development, which usually increases in density during the summer periods when river waters increase in temperature and decrease their volume and velocities. Rainwater contains few micro-organisms due to its 'purity'. Generally the air environment is not suitable for micro-organism growth and development. However, organisms that are found in the air have their origin in the soil (or water environment).

Animals (worms)   The animals of interest to water quality are those of worms, which are of the mm size and affect human health. The presence of flatworms (several species, including tapeworms) is hazardous to health if found in water supplies. For instance, different species of nematodes (eel worms, roundworms and threadworms) are responsible for diseases such as roundworm, hookworm and filarisis. Nematodes serve a beneficial function in wastewater treatment (percolating filters) where the nematodes break loose some of the biofilm attached to the media and thus prevent excessive growth on the media and clogging. Nematodes are insignificant contributors to activated sludge wastewater treatment since the residence time is too short for their reproduction period.

Plants Plant growth in the river environment is generally regarded as undesirable. This is particularly the case with *heterotrophic slimes* (sewage fungus) and *phototrophic organisms* (plants and some bacteria). The latter include planktonic algae of microscopic plants which drift freely in the water, benthic algae which grow on the river bed and solid objects such as logs, which are known as *periphyton* and macrophytes or larger plants which are often rooted. Undesirable biological growths in rivers/lakes are associated with elevated concentrations of organic matter (sewage fungus) and nutrients (plants). Heterotrophic growths respond to elevated concentrations of simple sugars (measured by BOD). Phototrophs require inorganic forms of nitrogen (ammonium and nitrate) and phosphorus (dissolved reactive phosphorus). Undesirable sewage fungus growths do not occur when the BOD of the river/lake is less than 1 mg/L. To prevent benthic algal growth, levels of dissolved inorganic nitrogen should be less than 0.04 to 0.1 mg/L and levels of dissolved reactive phosphorus should be less than 0.15 to 0.03 mg/L in the rivers/lakes.

Fungi Fungi are non-photosynthetic, chemo-organotrophic, aerobic, multicellular organisms. They are primarily soil inhabitants, but can also be found in sea and freshwater. Fungi are mainly employed in the degradation and composting of dead organic material, a behaviour described by the term *saprophytic*. They are important members of the food chain as they recycle essential plant nutrients. Besides that, they are also parasitic fungi of plants, animals and humans. Reproduction occurs by a variety of means, either asexually by fission (genetic division by mitosis) involving either budding or spore formation or sexually by fusion (genetic division by meiosis) of the nuclei of two parent cells. Fungi are classified on the basis of their sexual reproduction cycle into four divisions as listed in Table 3.18.

Moulds are filamentous fungi. They grow by extending long thread-like structures called hyphae, which form a mass called mycelium. The vegetative mycelium penetrates into the substrate to absorb dissolved nutrients, while the reproductive mycelium forms reproductive structures such as spores, spore sacs, etc.

Yeasts are unicellular, non-filamentous organisms. The cell is considerably larger than bacteria (1 to 5 μm in width and 5 to 30 μm in length) and is generally egg-shaped, spherical or ellipsoidal. Yeasts are widely distributed in nature and reproduce asexually by binary fission or by budding. In contrast to moulds, they can grow both aerobically and anaerobically. Yeasts have been employed by man for many hundreds of years for the production of wine, beer, cheeses, etc., and are also used nowadays in large-scale fermentations to produce antibiotics (penicillin) and other biochemicals.

Mushrooms are highly differentiated forms of fungi. The mycelium is in the soil, and in the right conditions the macroscopic fruiting body, termed basidia, forms above the ground as the structure called a mushroom.

Because of their saprophytic nature, fungi play a significant role in biological wastewater treatment and the composting of municipal refuse. On the other hand, they cause damage through deterioration and rotting of products made from natural materials. Of the approximately 100 000 species of fungi, only about 100 are pathogenic to humans and animals. They mainly cause infections of hair, nails, skin as well as serious infections of internal organs such as the lung. One of the worst toxins, the aflatoxin, is produced by a fungus called *Aspergillus flavus*. Improperly dried foods like peanuts, grain, etc., can be contaminated by aflatoxin, thus causing fatty degeneration of the liver and liver cancer. The white rot

Table 3.18 Classification of soil and aquatic fungi

Type	Division	Characteristics/examples
Moulds (filamentous)	Phycomycetes	Sexual or asexual spores: *Mucor, Rhicupus*
	Fungi imperfecti	No sexual stage: *Penicillium, Aspergillus*
Yeasts (non-filamentous)	Ascomycetes	Sexual spores in sacs: *Neu-raspora, Candida*
Mushrooms (macroscopic)	Basidiomycetes	Sexual stage on basidia: common mushroom

Adapted from Mitchell, 1974

fungus is now known to be capable of degrading hazardous organic compounds and is used *in situ* to clean up contaminated soil environment.

Algae   Algae are essentially plant like. Most are aquatic organisms and they may inhabit fresh or saline waters. The terrestrial species normally grow in soil or on the barks of trees and some have established a symbiotic relationship with fungi to form lichens. The size of algae ranges from the microscopic unicellular phytoplankton to the large multicellular seaweeds. The shapes of unicellular algae can be spherical, cylindrical, club-like or spiral. Multicellular colonies can grow in filaments or long tubes or simple masses of single cells that cling together.

Regardless of their size or complexity, all algal cells contain photosynthetic pigments and are thus capable of photosynthesis. Because of the vast abundance of oceans on earth, it is probable that the algae fix more carbon dioxide, and thus release more oxygen, than all the land plants combined. The pigments are found in distinct bodies called *plastids*, *chloroplasts* or *chromatophores*.

The classification of algae is based on their cellular properties, the nature of the cell wall, photosynthetic pigments and the arrangement of the flagella in motile cells. Seven general groups are shown in Table 3.19. Groups I, II, IV and VII are of interest in the environmental field because of their appearance in both clean and polluted water.

Algae can be a problem in water supplies, since they contribute to tastes and odours, clog water intakes, shorten filter runs in water treatment sand filters and cause high chlorine demand in disinfection processes. Another environmental problem is caused by the massive growth of some marine species. The *dinoflagellates*, unicellular, flagellated marine organisms, comprise the 'red tides' sometimes seen in large areas of the sea. Shellfish which consume the algae cannot be eaten during these events as the toxin produced by some species of dinoflagellates is toxic to humans. Excessive growth of algae in rivers or lakes is often an indicator of *eutrophication*. Due to excessive nutrient input of especially nitrogen and phosphorus compounds—most often caused by human activities such as the use of agricultural fertilizers, detergents, untreated waste water, etc.—algal growth may get out of balance. Dying algal masses usually cause a high BOD and lead to anaerobic conditions with the development of methane and hydrogen sulphide.

Protozoa   Protozoa are the most highly specialized unicellular organisms. Most are non-photosynthetic, reproduce asexually by binary fission and lack true cell walls, the latter being a distinguishing feature of algae and fungi. Table 3.20 shows a classification scheme for common aquatic and soil protozoa.

The size and shape of protozoa vary greatly; some are as small as 1 μm in diameter while others are as large as 2000 μm. They are primarily aquatic organisms and widespread in nature. They survive

Table 3.19 Classification of algae

Division		Colour	Environment/cell arrangement/comments
I	Chlorphyta	Grass-green	Freshwater; mainly clean-water algae except *Chlorella*, *Scenedezmus*, mostly colonial, filamentous
II	Chrystohyta	yellow-green	Clean, cold water; mainly cellular, some colonial; diatoms have silica in cell walls
III	Pyrrophyta	yellow-brown	Mostly marine; 90% unicellular, two flagella
IV	Euglemophyta	Green	Freshwater; requires organic nitrogen; will grow as a protozoa in absence of light; unicellular motulity by flagellum
V	Rhodophyta	Red	Mostly marine; very clean, warm water, colonial; sheets are common
VI	Phyophyta	Brown	Marine; cool water; colonial, large. Example: *Macrocystis*, giant kelp
VII	Cyanophyta	Blue-green[†]	Freshwater, warm, often polluted; unicellular, gelatinous clumps; no chloroplasts or true nucleus; nitrogen fixers, often responsible for algal blooms

[†] Blue-green algae are now generally referred to as blue-green bacteria or cyano-bacteria.
Adapted from G. W. Heinke, 'Microbiology and epidemiology', in *Environmental Science and Engineering*, J. G. Henry and W. W. Heinke (eds), © 1989, p. 256. Reprinted by permission of Prentice-Hall, Inc., Englewood Cliffs, N.J.

Table 3.20 Classification scheme for common aquatic and soil protozoa

Pseudopods (Sarcodina)
  Motile by pseudopods, flowing amoeboid motion: *Amoeba, Entamoeba*

Flagellates (Mastigophora)
  Motile by flagella; many photosynthetic: *Euglena, Volvox, Giardia*

Ciliates (Ciliophora)
  Free-swimming; motile by many cilia that move in unison: *Paramecium*
  Attached; fixed by stalk to a surface: *Vorticella*

Parasitic protozoa
(Suctoria)
  Free-swimming ciliates early in life cycle, then tentacles in later adult stalked stage
(Sporozoa)
  Usually non-motile; rarely free-living; parasitic: *Plasmodium*

Adapted from Mitchell, 1974

adverse conditions by forming cysts with thick cell walls. Their nutrition type may be saprophytic, but mainly they act as efficient predators on bacteria, and are thus found in particular where huge amounts of bacteria are prevalent. Of the 32 000 species 10 000 are parasitic, some causing serious diseases such as malaria and sleeping sickness. The size and shape and the type of locomotory system serves as a criterion for classification.

The *sarcodina* have a cell membrane that continually changes shape. They move by extending their cytoplasm in search of food. These extensions are called *pseudopodia*, or false feet, and are typical of amoebae. Sarcodina are saprophytic. *Entamoeba histolytica* is a common pathogen causing amoebic dysentery in humans. Figure 3.15 is a sketch of *Amoeba*.

The *mastigophora* have flagella, and some species such as *Euglena* are photosynthetic, exhibiting some of the characteristics of both protozoa and algae. The *ciliophora* are characterized by having fine hairs or cilia. In addition to providing motility, cilia aid in the capture of food. *Paramecium* is a typical ciliate. Figure 3.16 shows a sketch of *Paramecium*. The parasitic protozoa include *suctoria* (free-swimming) and *sporozoa* (non-motile). Four species of *Plasmodium*, the cause of malaria, are members of the latter group. The vector (carrier) that conveys these parasites to a human host is the female Anopheles mosquito.

The waterborne diseases causing protozoa, include:

- *Giardia lamblia*, reservoired in beavers, sheep, dogs and cats. Causes early spring diarrhea, found particularly in cold mountain water.
- *Cryptosporidum*, reservoired in many species of wild and domestic animals. Causes acute gastrointestinal illnesses, including diarrhea, nausea and stomach cramps.
- *Naegleria fowleri*, found in warm waters, pools and lakes. Causes primary amebic meningoencephalitis (PAM).

As protozoa act as predators for bacteria, they are found in many wastewater treatment facilities such as activated sludge processes, percolating filters, etc. The considerable removal of bacteria (>90 per cent for *Escherichia coli*) in such treatment processes is believed to be caused mainly by protozoan activities, as shown in Table 3.21.

In a typical activated sludge plant, the mixed liquor will contain approx 5×10^4 protozoa per ml. These consist mostly of ciliates, but also significant numbers of amoebae and flagellates may develop under certain conditions. The dominant ciliates include *Opercularia, Vorticella, Aspidisca, Carchesium* and *Chilodonella*. The majority of these are attached to or crawl over the surface of the floc and feed on the bacteria. By scavenging free-swimming cells, protozoa efficiently reduce effluent turbidity and soluble BOD.

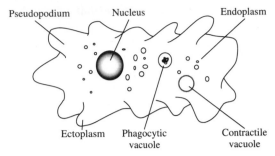

Figure 3.15 Structure of the *amoeba* protist (adapted from Lansing M. Prescott *et al.*, *Microbiology*, 2nd edition. Copyright © 1993. Wm. C. Brown Communications, Inc., Dubuque, Iowa. All Rights Reserved. Reprinted by permission).

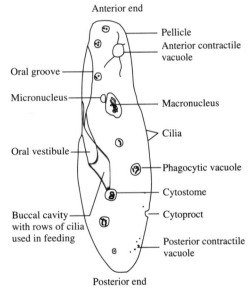

Figure 3.16 Structure of the *Paramecium* protist (adapted from Lansing M. Prescott *et al.*, *Microbiology*, 2nd edition. Copyright © 1993. Wm. C. Brown Communications, Inc., Dubuque, Iowa. All Rights Reserved. Reprinted by permission.)

Table 3.21 The effect of ciliated protozoa on effluent quality parameters in laboratory scale activated sludge systems

Effluent parameter	Ciliates absent	Ciliates present
Total BOD (mg/L)	53–70	7–24
Soluble BOD (mg/L)	30–35	3–9
Suspended solids (mg/L)	86–118	26–34
Viable count (10^6 mL^{-1})	160	1–9

Adapted from Curds, 1982

Viruses   Viruses are a unique group of infectious agents. They are different from protista as they do not have a cellular structure. The simplest possible form of a virus is merely an RNA molecule. As they cannot reproduce themselves, many biologists do not consider viruses to be *obligately parasitic*. The virus RNA (or DNA) injected into the host cell 'reprograms' the cell's reproduction mechanism to produce exclusively viral material. Due to this interference with the normal function of the cell, viruses are responsible for diseases in bacteria, plants, animals and humans. Viruses are extremely small, about 5 to 10 nm in diameter and up to 800 nm in length for some of the long thin ones. Due to their smallness, it was not possible to detect them prior to the age of the electron microscope in 1931. Their morphology is simple by comparison with algae, fungi or protozoa. Figure 3.17 shows the size and morphology of some selected viruses. In general, viruses are composed of a nucleic acid core (either DNA or RNA, single or double stranded). This is surrounded by a protein covering called a 'capsid'. The capsid unit is made up of smaller units called 'capsomores'. A complete virus is called a virion. Viruses can exist in two phases: extracellular and intracellular. Virions, in the extracellular phase, pass few if any enzymes and cannot reproduce independently of living cells. In the intracellular phase, viruses exist primarily as replicating nucleic acids and induce host metabolism to synthesize virion components. Eventually, complete virus particles or virions are released (Prescott *et al.*, 1993).

Viruses can be distinguished on the basis of structural arrangement of the capsid. There are four general morphological types of capsids and virion structure:

1. Icosahedral is a regular polyhedron with 20 equilateral triangular faces and 12 vertices as shown in Fig. 3.17(*h*), (*j*), (*k*), (*l*) and Fig. 3.18(*a*)

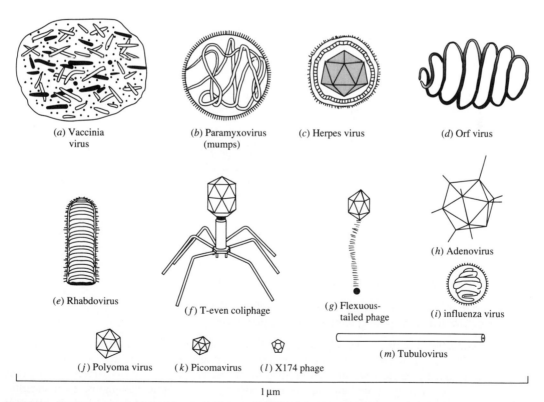

(*a*) Vaccinia virus   (*b*) Paramyxovirus (mumps)   (*c*) Herpes virus   (*d*) Orf virus

(*e*) Rhabdovirus   (*f*) T-even coliphage   (*g*) Flexuous-tailed phage   (*h*) Adenovirus   (*i*) influenza virus

(*j*) Polyoma virus   (*k*) Picomavirus   (*l*) X174 phage   (*m*) Tubulovirus

1 μm

Figure 3.17 The size and morphology of selected viruses (adapted from Lansing M. Prescott *et al.*, *Microbiology*, 2nd edition. Copyright © 1993. Wm. C. Brown Communications, Inc., Dubuque, Iowa. All Rights Reserved. Reprinted by permission).

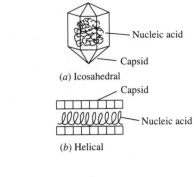

(a) Icosahedral

(b) Helical

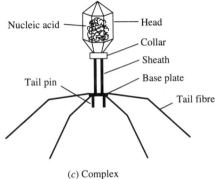

(c) Complex

Figure 3.18 Major structural arrangements in viruses (adapted from Sterritt and Lester, 1988. Reprinted by permission of E&FN Spon Ltd).

2. 'Helical' is shaped like hollow protein cylinders, either rigid or flexible, as shown in Fig. 3.17 (*m*) and Fig. 3.18 (*b*).
3. 'Envelope' is a roughly spherical outer membrane layer surrounding the nucleocapsid. The latter is either icosahedral or helical. 'Envelopes' are shown in Fig. 3.17(*b*), (*c*), (*i*).
4. 'Complex' is capsid symmetry that is neither icosahedral nor helical. They may possess tails or other structures as shown in Fig. 3.17(*a*), (*d*), (*f*), (*g*), (*e*) and Fig. 3.18(*c*) (Prescott *et al.*, 1993).

The life cycle of viruses consist of three stages:

- Entry to the cell
- Reproduction
- Release

The life cycle has two important variants:

- 'Lytic' cycle, which is followed by virulent types often killing the host cell
- 'Lysogeny', which is the state entered by the avirulent types

The classification of viruses is largely based on structural and chemical characteristics. The properties involved are the type of nucleic acid, capsid geometry, capsid size and the presence or absence of an envelope. Many other viruses, however, are classified simply on the basis of their observed effects on the host. Viruses cause infections in humans, such as poliomyelitis, aseptic meningitis, gastroenteritis, chicken pox, ebola, HIV, etc. Their removal from potable water, sewage waters and bathing waters is therefore of great importance. Removal of viruses in sand filters is highly variable and depends on filter design and operation. Coagulation prior to filtration seems to be more successful. In disinfection processes, viruses generally need higher doses of the disinfectant and longer disinfection times for complete inactivation than those needed for pathogenic bacteria. Viruses are small enough to pass through most sand filters and have been known to resist chlorination. Many viruses survive outside the host for varying periods of time. This latter point has recently been given attention (Farzadegan, 1991) with respect to HIV. Virus-kill in relation to effluent wastewaters is performed in the United States using chlorine and, more recently, ultraviolet light, while disinfection of wastewaters is not a common West European practice (except in France). Table 3.22 is a listing of common human diseases caused by viruses (Stanier *et al.*, 1986).

3.5.3 Bacteria

Bacteria are the dominant organisms in biological wastewater treatment systems, and in many ecological systems, including running and still waters, in soils and also in the air. In some instances, bacteria and their abilities are exploited, for instance in wastewater treatment, or in biodegradation of groundwater contamination. Bacteria are classified in different methods but in wastewater treatment bacteria are often described as being aerobic (needing oxygen) or anaerobic (not requiring oxygen). About 80 per cent of the bacterial cell consists of water and so the growth of bacteria is closely linked to the available water supply. Water also serves as a food source, providing the necessary substances in a dissolved stage, and are thus able to penetrate the cell membrane. Furthermore, physical conditions such as temperature, pH, salt concentrations, pressure, the gaseous environment, etc., are all growth determining factors.

Bacteria (procaryotes) are distinguished from eucaryotes by several features. Eucaryotes have a true nucleus which is surrounded by a membrane, whereas the procaryotes nucleus is arranged in a loop and lies free in the cytoplasm, as shown in Fig. 3.19. Bacteria exist in three principal shapes: spherical (cocci),

Table 3.22 Common human diseases caused by viruses

Viral group and type	Disease
Herpes viruses	
Cytomeglovirus	Respiratory infections
Epstein–Barr virus	Mononucleosis
Herpes simplex viruses	Oral and genital cold sores
Varicella virus	Chickenpox, shingles
Hepatitus B virus	Serum hepatitis
Influenza viruses	Viral influenza and viral pneumonia
Polio viruses	Poliomyelitis
Pox viruses	
Orf virus	Contagious pustular dermatitis
Variola virus	Smallpox
Picorna viruses	
Coxsackie viruses	Herpangina
Hepatitis A virus	Infectious hepatitis
Poliomyelitis virus	Poliomyelitis
Rhino viruses	Most colds
Parainfluenza viruses	Measles, mumps, rubella
Rhabdo viruses	Rabies
Reo viruses	Diarrhoeal diseases
Retro viruses	
Human T-cell leukaemia virus	T-cell leukaemia
Human immunodeficiency viruses	Acquired immunodeficiency syndrome (AIDs)

Adapted from Stanier *et al.*, *Microbial World*, 5th edn, © 1986, p. 650. Reprinted by permission of Prentice-Hall, Inc., Englewood Cliffs, N.J.

rod-like or cylindrical rods, and spiral shapes. An essential unique feature of bacteria is that they are single-cell organisms ranging in size from 0.5 to 5 μm long and 0.3 to 1.5 μm in width. They are arranged singly, in pairs, in clusters or in chains, as shown in Fig. 3.20.

To describe a bacterium properly, several specifications can be made, including:

- Morphological characteristics

    General shape   —coccus, rod, spirillum

    Arrangement of cells—chains, e.g. streptococcus, filamentous

                               —cluster e.g. micrococcus

                               —pairs or packages

    Locomotion features—existence or absence of flagella

    Formation of spores

    Capsules and slime envelopes

    Gram coloration in positive or negative

- Physiological characteristics

    Oxygen compatibility—aerobe, i.e. grows in the presence of oxygen

                                 —anaerobe, i.e. grows in the absence of oxygen

                               —facultative aerobe/anaerobe, i.e. grows under both conditions

                               —obligate aerobe/anaerobe, i.e. grow under one of two conditions

    Energy gain from   —respiration

                          —fermentation

                        —photosynthesis

    Pigmentation of cells

    Pathogenity

For further details on this topic, the student is referred to Prescott *et al.* (1993).

The most commonly encountered bacteria are either of the coccus or bacillus (rod) shape. Coccus are roughly spherical in shape. They can exist as individual cells. Diplococci arise when cocci divide and remain together in pairs. Long chains of cocci are formed when cells adhere after repeated division. This is the pattern in the genera *Streptococcus*, *Enterococcus* and *Lactococcus*. *Staphylococcus* divide in random planes to produce irregular clumps.

The rod (bacillus) bacteria differ in their length to width ratio. After division, the *Bacillus megaterium* is a long-chain rod-like bacteria. A few rod-shaped bacteria, the *Vibrios*, are curved to form distinct

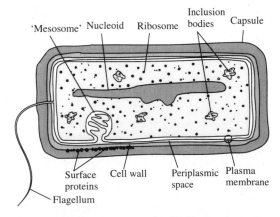

Figure 3.19 Schematic of bacterial cell morphology (adapted from Lansing M. Prescott *et al.*, *Microbiology*, 2nd edition. Copyright © 1993. Wm. C. Brown Communications, Inc., Dubuque, Iowa. All Rights Reserved. Reprinted by permission).

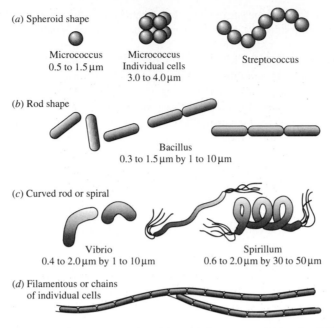

(a) Spheroid shape

Micrococcus
0.5 to 1.5 μm

Micrococcus
Individual cells
3.0 to 4.0 μm

Streptococcus

(b) Rod shape

Bacillus
0.3 to 1.5 μm by 1 to 10 μm

(c) Curved rod or spiral

Vibrio
0.4 to 2.0 μm by 1 to 10 μm

Spirillum
0.6 to 2.0 μm by 30 to 50 μm

(d) Filamentous or chains
of individual cells

Figure 3.20 Examples of bacterial cell morphology (adapted from Tchobanoglous and Schroeder, 1987).

commas or incomplete spirals. The smallest bacteria (of the genus *Mycoplasma*) are 0.1 to 0.2 μm in diameter, while the mid-size is about 1.1 to 1.5 μm in diameter (*Escherichia coli*) with the longest up to 500 μm.

The bacterial cell   The basic elements of the bacterial cell are shown in Fig. 3.19 and include:

1. The 'cell wall' of bacteria contains structures and substances (e.g. murein) not found in animals or plants. It is the Achilles heel of bacteria and modern drugs (e.g. penicillin) attack and interferes with the cell wall synthesis, without harming the host (e.g. human). The reaction of the cell wall to a staining technique, the 'Gram stain', distinguishes two important types of bacteria. Gram-positive bacteria react positively to stain and become blue while those that are negative stain pink. The cell walls of Gram-positive bacteria contain up to 95 per cent peptidoglycan existing as a single layer. The Gram-positive cell wall also contains polysaccharides, techoic and techuronic acids. The Gram-negatives contain typically ~ 50 per cent proteins, ~ 25 per cent lipopolysachoride and ~ 25 per cent phospholipid. The Gram-positives contain two cell walls while the Gram-negatives contain three distinct cell walls.
2. The 'cell membrane' or 'cytoplasmic membrane' serves as an osmotic barrier controlling and regulating material transfer. It contains about 10 per cent of the total cell dry weight and is composed of ~ 60 per cent protein and ~ 25 per cent lipids.
3. The 'cytoplasm' is the liquid content inside the cell where granular food reserves of glycogen, sulphur, etc., are stored.
4. The 'nucleus' contains the genetic information on one DNA filament arranged in a loop of fibrillar appearance. The double-stranded DNA forming the chromosome is about 0.3 nm in length.
5. 'Ribosomes' are producers of proteins and contain ~ 80 per cent of the bacterial RNA. The RNA contains the genetic code for the synthesis of proteins, mostly enzymes which serve as specific catalysts in biochemical reactions.

6. The 'capsule' is the slime envelope surrounding the bacterial cell. It increases the resistance of the cell against phagocytose and results in a higher virulence if the bacterium is pathogenic. 'Sewage fungus', *Spaerotilus natans*, has such a slime envelope.
7. 'Flagella', either singly or in tufts, are the means of mobility of bacteria. The flagella shape and arrangement are used in bacterium differentiation. The speed of *Vibrio cholera* is ~ 12 mm/min.
8. 'Spores' are formed by some groups of bacteria, e.g. the genera *Bacillus* and *Clostridium*. Due to the impermeable cell wall and dehydration of the cell content, spores serve as a survival technique resisting even adverse conditions, like heat, dryness, radiation or chemicals. While all living bacteria are easily killed by heating at 80 °C for a few minutes, spores can survive for hours at this temperature. This can be an issue in using wastes for compost production. In this stage of dormant existence, bacteria resist unsuitable conditions for long periods. In water disinfection procedures, spores can also cause problems as they need higher concentrations of the disinfectant, usually chlorine.

Composition and characterization of bacterial cells   Typical elemental composition of bacterial cells, based on dry weight is: 50 per cent carbon, 20 per cent oxygen, 15 per cent nitrogen, 8 per cent hydrogen, 3 per cent phosphorus and less than 1 per cent each of sulphur, potassium, sodium, calcium, iron and magnesium. Approximately 80 per cent of the bacterial cell is made up of water and it is through water that the cells receive their food in dissolved form. Bacterial cells are represented chemically by empirical forms such as:

$$C_5H_7NO_2 \text{ (most common)}$$

and $\qquad\qquad C_{42}H_{100}N_{11}O_{13}P$

However, the above formulae are not universal and are environment dependent and so should be used only cautiously. For instance, the theoretical oxygen demand (TOD) of 1 gram of $C_5H_7NO_2$ is 1.42 g, while that of $C_{42}H_{100}N_{11}O_{13}P$ is 1.71 g.

Bacteria categorized according to nutrition are grouped according to Table 3.23 in broad metabolic groups.

Examples

1. Green plants and *Cyanobacteria* are *photolithotrophs*, i.e. they use sunlight and the inorganic carbon (e.g. CO_2) as the electron donor to produce their organic food.
2. Nitrifying bacteria (discussed in Chapter 12 on wastewater treatment) such as *Nitrobacter* and *Nitrosomonas* are *chemolithotrophs*, meaning that they use the oxidation reactions

$$NO_2^- \leftrightarrow NO_3^- + 2\,e^- \qquad \text{and} \qquad NH_4^+ \leftrightarrow NO_2^- + 6\,e^-$$

respectively, as energy sources and the inorganic compounds nitrate and ammonia as electron donors.
3. Animals and most micro-organisms are *chemo-organotrophs*, meaning that they use an oxidation reaction as the energy source and organic compounds as electron donors.

Bacteria may be categorized according to the physical parameters of temperature, pressure, salinity and pH.

Table 3.23 Broad metabolic groups of bacteria

Carbon source	Inorganic (CO_2)	Autotrophs
	Organic	Heterotrophs
Energy source	Sunlight	Phototrophs
	Oxidation reaction	Chemotrophs
Electron donor	Inorganic	Lithotrophs
	Organic	Organotrophs

Temperature   The fact that micro-organisms are found in the permafrost soils of the Antarctic as well as in hot springs is indicative of their ubiquity. Table 3.24 shows the different temperature ranges and optima and the microbiological terms used to describe the particular types, and some examples of bacteria living under these conditions.

Pressure   Micro-organisms living at great depths in the earth or in the oceans experience high-pressure conditions. The term used to describe this characteristic is called *barophil*.

Salt concentrations   Salt water normally kills most micro-organisms. However, all marine organisms and some special types adapted to very high salt concentrations are found in places like the Dead Sea and the Great Salt Lake in Utah. They need these conditions for their growth. They are called *halophiles*.

pH conditions   The optimum pH conditions for bacteria to grow range between 6.5 and 7.5, with maximum limits between 4.0 and 10.0. Nevertheless, there are some who live in a strongly acidic environment (*acidophiles*). *Thiobacillus thiooxidans*, for example, oxidizes S^{2-} to SO_4^{2-}, thus producing sulphuric acid with pH values of about 1. The type is found in sewage systems and causes serious damage to materials such as concrete in pipes, culverts and holding tanks.

Growth and death of bacterial cells   Bacteria reproduce themselves by cell division, a process called *binary fission*. Under a given set of conditions, the rate for this division process is characteristic for each organism and is called *generation time*. The time to double the number of cells in the original population ranges from 15 minutes for *Enterobacteria* to 5 to 10 hours for *Nitrosomonas* and *Nitrobacter*. Table 3.25 shows the doubling time and optimum temperature for a range of micro-organisms. *Escherichia coli*, for instance, has a doubling time of ~ 0.35 hours at a temperature of $40\,°C$. The algae, protozoa and fungi tend to have much longer doubling times than bacteria.

The rate of growth of the bacterial population is directly proportional to the number of bacteria present and the growth rate constant:

$$dN/dt = kN \tag{3.72}$$

where

$dN/dt =$ rate of growth

$N =$ number of bacteria at time t

$k =$ first-order growth rate constant

Integration yields

$$\ln(N/N_0) = kt \tag{3.73}$$

where

$N_o =$ initial population concentration

If G is the generation (doubling) time, then $N = 2N_0$. Equation (3.73) can now be rewritten as

$$kG = \ln 2$$

$$k = \frac{\ln 2}{G} \tag{3.74}$$

Table 3.24 Micro-organism types related to temperature

Temperature range (°C)	Temperature optimum (°C)	Types	Examples
0–30		Psychrotrophs	Marine bacteria
	< 20	Psychrophiles	*Gallionella*
30–40	37	Mesophiles	*E. coli*
40–70	> 50	Thermophiles	*Bacillus stearthermophilus*
70–80	> 65	Extreme thermophiles	*Bacillus, Clostridia*

Table 3.25 Generation times for some micro-organisms

Micro-organism	Temperature (°C)	Generation time (hours)
Bacteria		
Beneckea nutriegens	37	0.16
Escherichia coli	40	0.35
Bacillus subtilis	40	0.43
Clostridium botulinum	37	0.58
Mycobacterium tuberculosis	37	≈ 12
Anacystis nidulans	41	2.0
Anabaena cylindrica	25	10.6
Rhodospirillum rubrum	25	4.6–5.3
Algae		
Chlorella pyrenoidosa	25	7.75
Scenedesmus quadricauda	25	5.9
Asterionella formosa	20	9.6
Skeletonema costatum	18	13.1
Ceratium tripos	20	82.8
Euglena gracilis	25	10.9
Protozoa		
Acanthamoeba castellanii	30	11–12
Paramecium caudatum	26	10.4
Tetrahymena geleii	24	2.2–4.2
Leishmania donovani	26	10–12
Giardia lamblia	37	18
Fungi		
Saccharomyces cerevisiae	30	2
Monilinia fructicola	25	30

Adapted from Lansing M. Prescott *et al.*, *Microbiology*, 2nd edn.
Copyright © 1993 Wm C. Brown Communications, Inc., Dubuque, Iowa.
All rights reserved. Reprinted by permission

By substituting this value of k into Eq. (3.73), the bacterial population N as a function of time t is expressed as

$$N = N_0 2^{t/G} \qquad (3.75)$$

Taking the logarithm of this equation,

$$\log N = \log N_0 + (t/G)\log 2 \qquad (3.76)$$

A plot of N against t on semilog paper produces a straight line with a slope of $0.3/G$ ($\log 2 = 0.3$) and a y intercept of N_0.

Example 3.23 If a bacterial cell count increases from 10^3 to 10^9 in 10 hours, determine the generation (doubling) time, G.

Solution
From Eq. (3.76),

$$\log N = \log N_0 + \frac{t}{G}\log 2$$

Then
$$N = 10^9; \qquad N_0 = 10^3; \qquad t = 10 \text{ hours}; \qquad G = ?$$

$$\frac{t}{G} \log 2 = \log N - \log N_0$$

$$\frac{10}{G} \log 2 = \log(10^9) - \log(10^3)$$

$$\frac{10}{G} \times 0.301 = 9 - 3$$

$$G = 0.5 \text{ h}$$

In reality, the *growth curve* of bacteria described by these equations represents just one phase among many. Bacteria brought into a nutrient grow until one factor comes to a minimum. If there is no continuous addition of nutrient and no elimination of metabolic products, the growth under the given conditions is called a static or *batch culture*, and is comparable to that of multicell organisms with genetic limited growth.

Figure 3.21 represents several phases of growth. During the initial period, called the *lag phase*, the cells adjust to their new environment, often connected with a synthesis of new enzymes. This is followed by the *exponential* or *log phase* and is characterized by a constant maximum division rate, which is a bacterial characteristic for a given set of conditions. The bacterial population is the most uniform during the log phase in terms of chemical composition, metabolic rates and other physiological characteristics. This is followed by the *stationary phase*. Food limitation, population density, and the accumulation of toxic by-products of cell metabolism lead to a decreasing growth rate, thus introducing the stationary phase. The *death* or *declining* or *endogenous* rate is reached when the death rate starts to exceed the growth rate.

Figure 3.21 is a schematic of the growth/death and time history of a single-cell batch culture. In realistic situations, like biological wastewater treatment, the environment is made up of many bacteria and the time history of the multiorganism (mixed culture) is not necessarily represented by Fig. 3.21. It may be that while one species of bacteria are in the log growth phase, a neighbouring bacterial mass may be in its endogenous phase.

Most bacterial processes found in engineering applications are either aerobic, anoxic or anaerobic. An aerobic process is a respiration process in which free molecular oxygen (O_2) serves as the terminal electron acceptor. An anoxic respiration process is that in which the organic compounds serve as the terminal electron acceptor. An anaerobic process is that which does not require free molecular oxygen. The general model of the aerobic process is represented by

$$\text{Organic} + O_2 \xrightarrow{\text{aerobes}} \underset{\text{biomass}}{\text{new}} + CO_2 + H_2O + NH_4 + NO_3{}^- + NO_2{}^- \qquad (3.77)$$

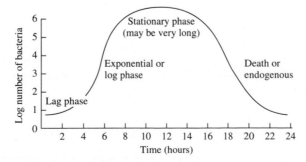

Figure 3.21 Schematic of the growth of a bacterial batch culture.

The corresponding model of the anaerobic process is represented by:

$$\text{Organic matter} + H_2O + \text{nutrients} \xrightarrow{\text{anaerobes}} \text{new biomass} + CH_4 + CO_2 + NH_3 + H_2S + \text{heat} \qquad (3.78)$$

Further details of these processes are given in Chapters 12 and 13.

Aspects of bacteria of special interest to environmental engineering   It is seen from Tables 3.26 and 3.27 that bacteria play a major role in all subject areas associated with the environment, e.g. soils, water, air, etc. Many waterborne diseases are caused by *pathogenic bacteria* and it is one of the major tasks of water treatment (Chapter 11) to reduce their numbers and render them harmless. Bacteria like *E. coli*, which normally live in the intestines of warm-blooded animals and are excreted with faeces, are used as *indicator bacteria*. If present in a water sample, they indicate that a contamination of the water has taken place and thus the potential presence of pathogens exists. On the other hand, many natural *environmental processes* would not work without the beneficial bacterial activities such as nitrification and nitrogen-fixing (see Table 3.27). The significance of bacteria to ecology and the agricultural environment are discussed in Chapters 2, 5 and 10 respectively.

The treatment processes for potable water are essentially physical and chemical (see Chapter 11) (unless, of course, we regard some of the mechanisms of sand filtration as biological). As such, microbiological unit processes are not dominant. Thus the presence of bacteria or other micro-organisms is seen as undesirable. The task is therefore to eliminate harmful bacteria and this is most traditionally accomplished by chlorine disinfection. However, the identification of harmful bacteria is expensive and complex. Indicator bacteria such as *E. coli*, if found present in potable water, are indicative of faecal pollution. Other indicator organisms are faecal coliforms, total coliforms, faecal streptococci and *Clostridium perfringens*. The most common potable water test is that for 'total coliforms', which is

Table 3.26 Pathogenic bacteria relevant to drinking water

Genus	Species	Host	Disease	Route
Salmonella	*S. typhi* *S. enteritides* *S. typhimurium*	Human and animal gut; Polluted water; contaminated food	Typhoid fever	Water and food processed with contaminated water
Shigella	*S. sonni* *S. flexneri* *S. bodyii* *S. dysentericae*	Man	Shingellosis	As above, plus person to person contact
Myobacterium	*M. tuberculosis* *M. balnei* *M. bovis*	Man Cattle	Tuberculosis	Airborne or sewage contaminated water
Vibrio	*V. cholera*	Man	Cholera	Contaminated water, person to person
Leptospira	*L. pomona* *L. autumnalis* *L. australis*	Infected humans; excreted in urine	Leptospirosis	Transmitted to bloodstream from animal carriers; polluted water
Entero pathogenic *E. coli*	Variety	Warm-blooded animals	Urinary infections	Sewage contaminated water or food

Adapted from Montgomery, 1985

Table 3.27 Pathogenic bacteria of special interest in environmental engineering

Group of bacteria	Genus	Environmental significance
Pathogenic bacteria	*Salmonella*	Cause typhoid fever
	Shigella	Cause dysentery
	Mycobacterium	Cause tuberculosis
Indicator bacteria	*Escherichia*	Faecal pollution
	Enterobacter	Faecal pollution
	Streptococcus	Faecal pollution
	Clostridium	Faecal pollution
Decay bacteria	*Pseudomonas*	Degrade organics
	Flavobacterium	Degrade proteins
	Zooglea	Floc-forming organism in activated sludge plants
	Clostridium	Produce fatty acids from organics in anaerobic digester
	Micrococcus	Produce fatty acids from organics in anaerobic digester
	Methanobacterium	Produce methane gas in anaerobic digester from fatty acids
	Methanococcus	Produce methane gas in anaerobic digester from fatty acids
	Methanosarcina	Produce methane gas in anaerobic digester from fatty acids
Nitrifying bacteria	*Nitrobacter*	Oxidize inorganic nitrogenous compounds
	Nitrosomonas	Oxidize inorganic nitrogenous compounds
Dentrifying bacteria	*Bacillus*	Reduce nitrate and nitrite to nitrogen gas or nitrous oxide
	Pseudomonas	Reduce nitrate and nitrite to nitrogen gas or nitrous oxide
Nitrogen-fixing bacteria	*Azotobacter*	Capable of fixing atmospheric nitrogen to NH_3
	Beijerinckia	Capable of fixing atmospheric nitrogen to NH_3
Sulphur bacteria	*Thiobacillus*	Oxidize sulphur and iron
	Desulfovibrio	Involved in corrosion of iron pipes
Photosynthetic bacteria	*Chlorobium*	Reduce sulphides to elemental sulphur
	Chromatium	Reduce sulphides to elemental sulphur
Phosphorus bacteria	*Acinetobacter*	Responsible for phosphorus removal in wastewater
Iron bacteria		
Filamentous	*Sphaerotilius*	Responsible for sludge bulking in activated sludge plants
Iron oxidizing	*Leptothrix*	Oxidize ferrous iron

Adapted from G. W. Heinke, 'Microbiology and epidemiology', in *Environmental Science and Engineering*, J. G. Henry and G. W. Heinke (eds), © 1989, p. 264. Reprinted by permission of Prentice-Hall, Inc., Englewood Cliffs, N.J.

defined as 'all aerobic and facultative anaerobic, gram negative, non spore forming, rod shaped bacteria that ferment lactose with gas formation within 48 hours at 35 °C', (*Standard Methods* by Greenberg *et al.*, 1992). The total coliform group is made up of *Escherichia coli, Enterobacter aerogenes, Citrobacter fruendii* and others. Differentiation between total and faecal coliforms is their ability or inability to grow at 45 °C. The dominant intestinal coliform in temperate climates is *E. coli*, but not so in tropical areas. In the latter, the total coliform test is used. If animal pollution is suspected, a measure of its significance is in the ratio of the faecal coliforms to that of faecal streptococci. If this ratio exceeds 4.0, the contamination is considered human. If it is < 0.7 it is considered to be from animal wastes, since faecal streptococci are more common in animals, (Henry and Heinke, 1989).

For potable water bacterial counts, the most probable number method (MPN) or the membrane filter method (MF) are used. These are detailed in *Standard Methods* (1992). Some relevant drinking water standards set for the European Community are shown in Table 3.28.

Table 3.28 Coliform standards for potable water

Parameter	MNP for EC
Total coliforms	<1/100 ml
Faecal coliforms	<1/100 ml
Faecal streptococci	≤1/100 ml
Sulphide-reducing clostridia	<1/20 ml

[†] No positive coliform results in more than 5% of samples collected each month.

3.6 CHEMICAL AND BIOCHEMICAL REACTIONS

Many processes in environmental engineering take place in reactors or in natural systems which can be called pseudoreactors. For instance, a volume of wastewater in an activated sludge tank undergoes changes to the contents over time. Initially, the contents may be a high COD waste, but the presence of a suitable population of micro-organisms in the tank will over time degrade the organic waste, producing ultimately a low COD effluent. In this example, the tank is the reactor and the process of change is called reaction kinetics or, more specifically, biological or biochemical reaction kinetics. It is important in the design of the 'process' to know what are the rates at which various components (e.g. organic material) are removed from the wastewater and the rate at which the biomass sludge is produced in the tank reactor. Knowledge of the reaction rates determines the size of the reactors required for a specific degree of treatment. A process may be biological, biochemical or chemical. Examples of such environmental processes include:

- Biological growth/decay of biomass/organic material in activated sludge, in anaerobic digestion, in wastewater lagoons, in trickling filters, and in rotating biological contactors, etc.
- Chemical processes of disinfection of potable water by chlorination or chlorine dioxide
- Gas–water transfer, e.g. removal of H_2S from groundwater
- Diffusion of effluents in rivers and estuaries
- Chemical reactions of contaminants in the air environment
- Biochemical production of methane in a landfill site

Many reactions go to completion only after extended time periods. As such, some reactions are introduced as going to, say, 80 per cent completion and this affects the mass or materials balance of the process (see Sec. 3.7). Knowledge of the extent of the reaction is often required to size and cost the reactor. The most important biochemical wastewater treatment process is activated sludge and is an extremely slow reaction, requiring large reactor volumes and long times of retention before the reaction goes towards completion. On the other hand, many industrial chemical processes are rapid, requiring small reactors. Anaerobic digestion of organic wastes may take typically 20 days 'retention time'.

3.6.1 Kinetics

For the example of using the process of activated sludge (described in Chapter 12) in wastewater treatment, rate expressions are required to describe the removal of organic material, the growth of microbial population and the utilization of oxygen. When reactions are described on a kinetic basis, different reaction orders occur for a variety of organisms, substrates and environmental conditions (see Fig. 3.22). In general, the relationship between the rate of reaction, the concentration of the reactant and the reaction order, n, (0, 1, 2), is given by:

$$r = C^n \tag{3.79}$$

$$\log r = n \log C \tag{3.80}$$

where
$$r = \text{rate of reaction}$$
$$n = \text{order of reaction}$$
$$C = \text{concentration of element}$$

Zero order is defined where the rate of reaction is independent of the concentration. First order is defined where the rate is directly proportional to the concentration. Second order is defined where the rate is proportional to the square of the concentration.

Zero-order reactions Consider the following zero-order reaction (Fig. 3.23), where the single reactant A is converted to a single product P:

$$A \rightarrow P$$

The rate of conversion of reactant A, according to zero-order kinetics, is

$$-\frac{d[A]}{dt} = k_0 \tag{3.81}$$

where the minus sign indicates that A is reducing with time. If C represents the concentration of A at any time t, and k_0 is the reaction rate constant then

$$-\frac{dC}{dt} = k_0 \tag{3.82}$$

Integrating:

$$C = -k_0 t + \text{constant}$$

If
$$C = C_0 \text{ at time } t = 0, \text{ then}$$

$$C - C_0 = -k_0 t \tag{3.83}$$

A useful measure of performance is knowing the time it takes the reaction to proceed to 50 per cent completion or half its initial concentration, i.e.

$$\frac{C_0}{2} - C_0 = -k_0 t_{1/2} \tag{3.84}$$

$$t_{1/2} = \frac{C_0}{2k_0} \tag{3.85}$$

This is sometimes referred to as the (half) saturation constant.

First-order reaction For the conversion of a single reactant A to a single product P, the first-order

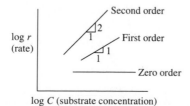

Figure 3.22 Orders of reaction.

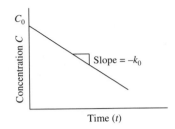

Figure 3.23 Zero-order reaction.

reaction behaviour (Fig. 3.24) is

$$A \rightarrow P$$

Then

$$-\frac{dC}{dt} = k_1 C \tag{3.86}$$

where k_1 is the first-order reaction rate constant and C the concentration at any time t. Integrating:

$$\ln\left(\frac{C_0}{C}\right) = k_1 t \tag{3.87}$$

or

$$\log\left(\frac{C_0}{C}\right) = \frac{k_1 t}{2.3} \tag{3.88}$$

The half-saturation constant is,

$$\ln\left(\frac{C_0}{C_0/2}\right) = k_1 t_{1/2} \tag{3.89}$$

Therefore

$$t_{1/2} = \frac{\ln(2)}{k_1} = \frac{0.69}{k_1} \tag{3.90}$$

Second-order reaction   The rate of reduction of a reactant A for a second-order reaction (Fig. 3.25) is described by

$$-\frac{dC}{dt} = k_2 C^2 \tag{3.91}$$

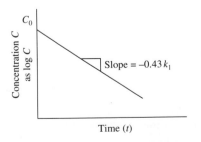

Figure 3.24 First-order reaction.

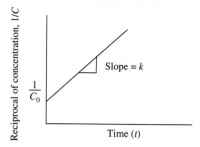

Figure 3.25 Second-order reaction.

where k_2 is the second-order reaction rate constant. Integrating:

$$\frac{1}{C} - \frac{1}{C_0} = k_2 t \tag{3.92}$$

The half-saturation constant is

$$\frac{1}{C_0/2} - \frac{1}{C_0} = k_2 t_{1/2} \tag{3.93}$$

$$t_{1/2} = \frac{1}{k_2 C_0} \tag{3.94}$$

Example 3.24 A laboratory test was used to determine the reaction order of a batch of microorganisms to degrade a food waste. The parameter used was COD (mg/L) over time. From the following data on concentration, C, determine the most appropriate reaction rate.

C (mg/L)	$1/C$ (L/mg)	Time (min)
400	2.5	0
320	3.15	5
280	3.57	10
240	4.17	20
180	5.6	30
110	9.1	40
50	20	50
40	25	60

Solution Plot the three figures associated with Fig. 3.26. For zero-order, simply plot the concentration C against time. For the first order, plot log C against time, and for the second order plot $1/C$ against time. From inspection of Fig. 3.26, the best fit is the second-order rate reaction so the general equation for this substrate utilization is

$$\frac{1}{C} - \frac{1}{C_0} = k_2 t$$

Bacterial rate processes In wastewater treatment the rate expressions of significance are:

- Rate of removal of organic matter
- Rate of growth of microbial population
- Rate of oxygen use

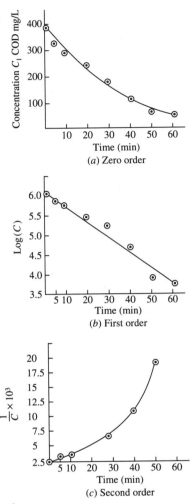

Figure 3.26 Trial reaction orders for a batch process.

These are described by the Monod model (see Chapter 12) and only briefly introduced here. Figure 3.27 is a schematic of the relationship of these parameters. At time $t = t_0$, the concentration of substrate (food) is $S = S_0$; and as the process time increases, the concentration of S decreases (in a batch process). At the same time, the biomass starts from a concentration of $X = X_0$, and increases over time, as the organic matter which is being removed is converted to cell mass. Also, at the same time, as the cell viable mass increases, initially somewhat exponentially, much O_2 is required, but this requirement stabilizes as the biomass concentration becomes constant.

The various growth/death phases of the microbial process as shown in Fig. 3.21 can be represented quantitatively. The exponential or log growth phase is represented by

$$\frac{dX}{dt} = \mu X \tag{3.95}$$

where $\qquad$ $X = $ microbial population as total or volatile solids, mg/L

$\qquad\qquad\quad$ $\mu = $ specific growth rate, d^{-1}

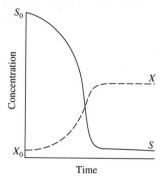

Figure 3.27 Change in substrate concentration during bacterial growth.

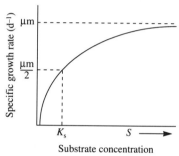

Figure 3.28 Monod model—effect of substrate concentration on growth rate.

The most commonly used model relating microbial growth to substrate (total organic carbon matter) utilization is the Monod model (1949). Monod identified that the growth rate, dX/dt, was not only a function of microbial concentration but also of some limiting substrate(S) or nutrient concentration. He described the relationship as in Fig. 3.28:

$$\mu = \mu_{max} \frac{S}{K_s + S}$$ (3.96)

where

μ = specific growth rate, d^{-1}

μ_{max} = maximum specific growth rate, d^{-1}

S = substrate concentration, mg/L

K_s = saturation (half) constant, mg/L

Therefore, Eq. (3.95) can be replaced by

$$\frac{dX}{dt} = \mu_{max} \frac{S}{K_s + S} X$$ (3.97)

which is the Monod model. For further details, refer to Chapter 12 and Gray (1990).

3.7 MATERIAL BALANCES AND REACTOR CONFIGURATIONS

Many quantities in environmental engineering are determined using material balances. All applications where material is input to a process, be it chemical, biological or even hydrochemical, can be quantified, in terms of its output and accumulation within the reactor process using the technique of material

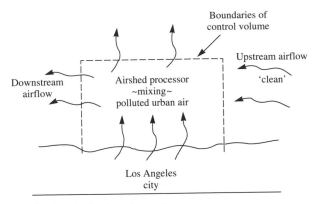

Figure 3.29 Schematic of control volume in the air environment.

balances. This tool is used extensively in the areas of fluid mechanics and hydraulics, in chemical engineering and food engineering. Simply stated, we identify a 'control volume', which maps the boundaries of the process or process reactor. For example (see Fig. 3.29), if the airshed over Los Angeles is taken as the control volume or box volume, then into that box will flow (vertically) air contaminants at the lower level from urban activity. Through the side wall (upstream) will flow the local air circulation, bringing clean or polluted air. Out the top of the box and the downstream side goes 'processed' air. This may be more contaminated or purified than the air entering. This depends on the process in the box. If the process in the box is highly turbulent with high mixing, then the contaminated inflows will be diluted by the clean inflows. If there is no mixing, then the outflows will be undiluted and behave like a plug of air being transported through the control volume, unchanged. This process is schematized in Fig. 3.29. Quantities of air can be input in this model. So also can qualities of air, and the processes, such as mixing. From hydrodynamics and chemistry, it is therefore possible to develop a model of the behaviour of the airshed process and thus the air quality.

Many similar applications can be visualized. For instance, the flow of groundwater through a 'purifying' soil can be studied using material balances, as can most of the process activities of water and wastewater treatment.

The alteration of a material usually takes place in a reactor. The reactor may be a defined 'box' like a rectangular, reinforced concrete activated sludge tank or a reach of a river or an airshed, as already mentioned, or a specific industrial chemical engineering reactor. Reactors are simplified into two types as being either 'completely mixed' or plug flow, and these are detailed later in this section.

3.7.1 Material Balances

The laws of conservation of mass and energy arise in engineering frequently. Of special interest to environmental engineering is the law of conservation of mass which underpins material balances. It states: 'The sum of the weights (masses) of substances entering into a reaction is equal to the sum of the weights (masses) of the products of the reaction.'

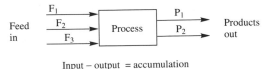

If there is no accumulation within the process, input = output. Therefore

$$\sum_{i=1}^{n} F_i = \sum_{j=1}^{k} P_j \qquad (3.98)$$

Example 3.25 A wastewater treatment plant with an output of 38 400 m³/day discharges the liquid effluent with a BOD of 20 mg/L into a river. If the BOD of the river upstream of the discharge point is 0.2 mg/L, at a minimum flow of 20 m³/s, compute the BOD of the river downstream of the discharge, assuming complete mixing.

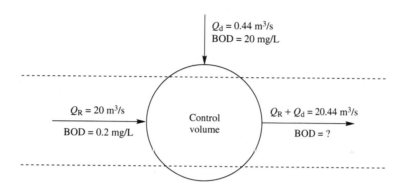

Solution

Let the basis be 1 second time interval. The component material (flow) balance, since it is assumed there is no accumulation within the control volume, is

$$\text{Input} = \text{output}$$

$$Q_R + Q_d = (Q_R + Q_d)$$

The polluting load balance is

$$P_R + P_d = P_{\text{downstream}}$$

where

$$P_R = \text{mg of polluting load of upstream/s}$$
$$= 20 \times 10^3 \times 0.2 = 4000 \text{ mg/s}$$
$$P_d = 0.44 \times 10^3 \times 20 = 8800 \text{ mg/s}$$
$$P_{\text{downstream}} = (Q_R + Q_d) \times 10^3 \times \text{BOD}$$
$$= 20.44 \times 10^3 \times \text{BOD}$$

Therefore

$$4000 + 8800 = 20.44 \times 10^3 \times \text{BOD}$$

$$\text{BOD}_{\text{downstream}} = 0.63 \text{ mg/L}$$

A 'process' is one or a series of reactions or operations or treatments that result in an end product. Examples are:

- Chemical manufacture
- Alum coagulation
- Fluoridation
- Chlorination

- Anaerobic reactors—sludge digestion
- Activated sludge processes
- Fluid transport
- Gas absorption
- NO_x and SO_x generation in the air environment

Material balances may be:

- Single unit process without chemical reaction or biological reaction
- Single unit process with chemical reaction or biological reaction
- Multiple unit processes with/without chemical reaction or biological reaction

A 'process' or a 'system' has a boundary, maybe arbitrarily defined, across which there may or may not be activity (e.g. flow). The two types of systems are:

1. 'Open' system, where material is transferred across the boundary, i.e. it enters or leaves the system.
2. 'Closed' system, where no material crosses the boundary. This is sometimes termed a 'batch' system, as no material passes the boundary until the process is complete.

A comprehensive formulation of the principle of material balances is as follows:

$$\begin{array}{lll} \text{Accumulation} & \text{input} & \text{output} & \text{generation} & \text{consumption} \\ \text{within the} & = \text{throughout the} & - \text{through the} & + \text{within the} & - \text{within the} \\ \text{system} & \text{system} & \text{system} & \text{system} & \text{system} \\ & \text{boundaries} & \text{boundaries} \end{array} \quad (3.99)$$

The material balance can refer to:

- Total mass or total moles
- Mass or moles of a chemical compound
- Mass or moles of an 'atomic' species
- The species may be chemical or biological
- Mass flow rates

In the case of total mass, generation is considered equal to consumption (within the boundaries), even if there is or is not a chemical/biological reaction and so:

$$\text{Accumulation} = \text{input} - \text{output} \quad (3.100)$$

Example 3.26 Primary (PSS) and secondary (SSS) sewage sludge is thickened together in a picket fence thickener. If the PSS is produced at 100 kg/h at 1 per cent dry solids (DS) and the SSS at 150 kg/h at 3 per cent DS, determine the per cent DS of the end product.

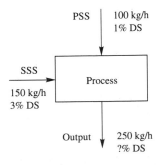

PSS 100 kg/h 1% DS

SSS 150 kg/h 3% DS

Process

Output 250 kg/h ?% DS

Solution

$$\text{Input} = 100 \text{ kg/h at } 1\% \text{ DS}$$
$$+ 150 \text{ kg/h at } 3\% \text{ DS}$$
$$\text{Output} = 250 \text{ kg/h at } x\% \text{ DS}$$

% Dry solids:

$$\text{Mass flow input} = 100 \times 0.01 + 150 \times 0.03 = 5.5 \text{ kg/h}$$
$$\text{Mass flow output} = 250x, \text{ kg/h}$$
$$250x = 5.5 \text{ kg/h}$$

Therefore

$$\text{Output } x = 2.2\% \text{ DS}$$

Note: It is recommended that the students always sketch the material balance arrrangement, as here.

Methodology of material balances   The following 'strategy' for analysing material balances is adapted in part from Himmelblau (1989):

1. Sketch a flow chart or figure defining the boundary of the process.
2. Label the flow of each stream and their compositions with symbols.
3. Show all known flows and compositions on the figure. Calculate additional compositions from the data where possible.
4. Select the basis for calculations, e.g. 1 h, 1 day, 1 kg, etc.
5. Write the material balances which includes the total balance and component balances. There must be x independent equations if there are x unknowns.
6. Solve the equations and check the solutions.

To explain the methodology of material balances, let us do some examples.

Example 3.27   A slurry containing 20 per cent by weight of limestone ($CaCO_3$) is processed to separate pure dry limestone from water. If the feed rate is 2000 kg/h, how much $CaCO_3$ is produced per hour?

Solution

Step 1. Organize the data in flowchart format:

There is one input stream (1) and two output streams (2) and (3).

Stream number	(1)		(2)		(3)	
Stream name	Slurry feed S		Limestone L		Water W	
Component	Fraction	kg/h	Fraction	kg/h	Fraction	kg/h
$CaCO_3$	0.2		1.0		0	0
H_2O	0.8		0.0	0	1.0	
Total	1.0	2000	1.0		1.0	

Step 2. Material balance equations:

(1) Total: $2000 = L + W$
(2) $CaCO_3$: $2000 \times 0.2 = L \times 1.0 + W \times 0.0$
(3) H_2O: $2000 \times 0.8 = L \times 0.0 + W \times 1.0$

Solution

$$\text{Limestone production } L = 400 \text{ kg/h}$$
$$\text{Water produced } W = 1600 \text{ kg/h}$$

Example 3.28 As a fuel source 20 kg of ethylene (C_2H_4) is burned with 400 kg of air. Determine the composition of the resulting mixture. What is the percentage of excess air, assuming complete conversion.

Solution

$$C_2H_4 + 3O_2 \rightarrow 2CO_2 + 2H_2O$$

Air: $\qquad\qquad N = 79\%; \qquad O_2 = 21\%$

Molecular masses: $\qquad C_2H_4 = 28; \qquad O_2 = 32; \qquad N_2 = 28$

$$CO_2 = 44; \qquad H_2O = 18; \qquad \text{air} = 28.84$$

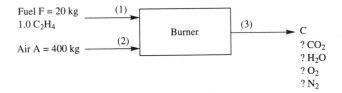

Note:

$$1 \text{ mole of } C_2H_4 = 28 \text{ g}$$
$$1 \text{ kmole of } C_2H_4 = 28 \text{ kg, therefore 20 kg of } C_2H_4 = \frac{20}{28} = 0.71 \text{ kmole.}$$

Stream number	(1)		(2)		(3)	
Stream name	Fuel F		Air A		Combined C	
Component	Fraction	kmole	Fraction	kmole	Fraction	kmole
C_2H_4	1.0	$\frac{20}{28} = 0.71$	—	—	?	0.71
O_2	—	—	0.21	$\frac{0.21 \times 400}{28.84} = 2.91$	?	2.91
N_2	—	—	0.79	$\frac{0.79 \times 400}{28.84} = 10.96$	?	10.96
CO_2	—	—	—	—		
H_2O	—	—	—	—		
Total	1.0	0.71	1.0	13.87	1.0	14.58

Input = output table in kmole

Component	Input	Generated	Consumed	Output
C_2H_4	0.71	—	0.71	0
O_2	2.91	—	2.13	0.78
N_2	10.96	—	—	10.96
CO_2	—	1.42	—	1.42
H_2O	—	1.42	—	1.42
Total				14.58

The input–output table is produced from the following. No C_2H_4 is generated and all is consumed. There are 2.91 kmole of O_2 in the feed and this reacts with C_2H_4 according to:

$$C_2H_4 \quad + \quad 3\,O_2 \quad \rightarrow \quad 2\,CO_2 \quad + \quad 2\,H_2O$$
$$0.71 \qquad 3 \times 0.71 \qquad 2 \times 0.71 \qquad 2 \times 0.71$$
$$0.71 \qquad 2.13 \qquad\quad 1.42 \qquad\quad 1.42$$

Of the 2.91 kmole of O_2 in the feed, 2.13 kmole are consumed leaving $2.91 - 2.13 = 0.78$ kmole in the output. N_2 is inert and passes through to the output unchanged; 1.42 kmole of CO_2 are generated, as are 1.42 kmole of H_2O. From the table, 14.58 kmole are in the output stream.

Example 3.29 Each day 3780 m^3 of wastewater is treated at a municipal wastewater treatment plant. The influent contains 220 mg/L of suspended solids. The 'clarified' water has a suspended, solids concentration of 5 mg/L. Determine the mass of sludge produced daily from the clarifier.

Solution

$$\text{Input sludge to clarifier} = 3780 \times 10^3 \times 220 \times 10^{-6} = 832 \text{ kg/day}$$
$$\text{Effluent from clarifier} = 3780 \times 10^3 \times 5 \times 10^{-3} = 19 \text{ kg/day}$$

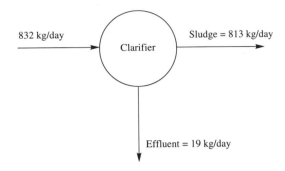

Write a mass balance around the clarifier:

$$\text{Dry solids sludge in} = \text{dry solid sludge out}$$
$$\text{Dry solids sludge out} = \text{effluent} + \text{sludge}$$

Therefore
$$\text{Dry solids removed as sludge} = 832 - 19 = 813 \text{ kg/day}$$

3.7.2 Reactor Configurations

In environmental engineering, particularly in wastewater and water treatment, reactors are essentially of three types:

1. 'Batch reactors' (BR), where the reactants are input to the reactor at the desired conditions and the reaction takes place over a specified period of time. The contents are then discharged. The longer the reaction time, the more complete the conversion. Many chemical processes are batch produced. For example, the BOD test is a batch test. A BR is shown in Fig. 3.30.
2. 'Continuously stirred tank reactor' (CSTR), where the reactants are 'continuously' (maybe once per day, per hour, etc.) fed to the reactor and the products (including unused reactants) are continuously discharged from a well-mixed vessel. Being well mixed, the contents are assumed uniform in concentration throughout with no concentration gradients and therefore equal to the effluent concentration. Increasing 'residence' time in the tank increases the extent of conversion. This reactor is common in wastewater treatment and anaerobic processes. It is shown in Fig. 3.31.
3. 'Plug flow reactors' (PFR), where the input is fed in at one end of a long reactor and products are discharged at the other end after spending a minimum retention time in the system. As the distance of travel along the length of the reactor is a function of time, the extent of conversion depends on the length. As such, the longer the reactor, the greater the conversion. As the 'plug' of reactants moves forward, it is well mixed within itself, but not with the remainder of the reactor contents. This reactor is

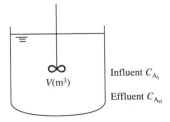

Figure 3.30 Batch reactor (BR).

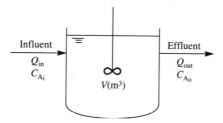

Figure 3.31 Continuously stirred reactor (CSTR).

most common in activated wastewater treatment. Figure 3.32 shows the configurations of the PFR. This type of reactor is often used in a variety of simulations, for instance, the mixing of a pollutant in a river flow (see Chapters 7 and 21). There is a gradient of concentration from input end to output end.

Analysis of performance of reactor types   The basis of all reactor analysis is the material balance equation:

$$\text{Input} - \text{output} + \text{generation} = \text{accumulation}$$

In reactor analysis the following may occur:

1. Accumulation may be positive or negative.
2. Input may be through the system boundaries or by generation within the reactor due to reaction.
3. Output may be flow through the system boundaries or by consumption due to reaction.

The material balance for a material A may be written as

$$QC_{A_i} - QC_{A_o} + r_A V = V\frac{dC_A}{dt} \qquad (3.101)$$

$$\text{Inflow} - \text{outflow} + \text{generation} = \text{accumulation}$$

where

$$Q = \text{the flow rate, m}^3/\text{s}$$
$$C_A = \text{concentration of material A, mg/L}$$
$$C_{A_i} = \text{influent concentration, mg/L}$$
$$C_{A_o} = \text{effluent concentration, mg/L}$$
$$V = \text{volume of fluid in the reactor, m}^3$$
$$r_A = \text{rate of reaction of material A, mg/L s}$$

Note that the batch process and plug flow process are considered 'similar' for analysis.
Batch process
Batch process $\Rightarrow Q = 0$ so $V = \text{constant}$
Equation (3.101) becomes

$$r_A = \frac{dC_A}{dt}$$

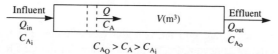

Figure 3.32 Plug flow reactor (PFR).

for the zero-order equation (see Sec. 3.6.1):

$$r_A = -k_0$$

where k_0 is the zero order, reaction rate constant (mg/Ls). Therefore

$$-k_0 = \frac{dC_A}{dt}$$

and

$$-k_0 \int_0^t dt = \int_{C_{A_i}}^{C_{A_0}} dC_A$$

$$C_{A_i} - C_{A_0} = k_0 t \qquad (3.102)$$

For this first-order equation:

$$r_A = -k_1 C_A$$

$$-k_1 C_A = \frac{dC_A}{dt}$$

$$-k_1 \int_0^t dt = \int_{C_{A_i}}^{C_{A_0}} \frac{dC_A}{C_A}$$

$$-k_1 t = \ln \frac{C_{A_0}}{C_{A_i}} \qquad (3.103)$$

Completely Stirred Tank Reactor (CSTR) process   In the steady state case, accumulation $= 0$. Therefore

$$V \frac{dC_A}{dt} = 0$$

Equation (3.101) becomes

$$QC_{A_i} - QC_{A_0} + r_A V = 0$$

$$-r_A = \frac{Q}{V}(C_{A_i} - C_{A_0})$$

For the zero-order equations:

$$r_A = -k_0$$

$$C_{A_i} - C_{A_0} = k_0 \frac{V}{Q}$$

Introduce the term 'hydraulic retention time' (HRT) and define as

$$\emptyset = \frac{V}{Q} \qquad \text{(see Chapter 12)}$$

Therefore

$$C_{A_i} - C_{A_0} = k_0 \emptyset \qquad (3.104)$$

For the first-order equation,

$$r_A = -k_1 C_{A_0}$$

$$\frac{C_{A_i} - C_{A_0}}{C_{A_0}} = k_1 \emptyset \qquad (3.105)$$

The first-order reaction may occur late into the reaction (see Chapter 12) when the concentration is approaching C_{A_0} (i.e. the effluent concentration). Further details on the CSTR as regards tracers in the riverine environment is in Section 21.4.1.

Reactions of variable order—effluent treatment The removal of organic material by microbial action is normally carried out in a continuous mode. The rate governing equation is

$$r_A = \frac{-k[A]}{K_s + [A]} \tag{3.106}$$

$$k = \frac{k_{max} C}{K_s + C} \tag{3.107}$$

where
k = specific substrate uptake rate,
units of mass of contaminant (organic) removal
per unit mass of micro-organisms present per
time, (g/g day)

$C(= S)$ = substrate concentration (BOD or COD), g/L

K_s = constant, i.e. the half saturation concentration of
substrate at $k = k_{max}/2$

k_{max} = maximum specific uptake rate

At low substrate concentrations, the first-order equation is

$$k = \frac{k_{max} C}{K_s} \tag{3.108}$$

At high substrate concentrations, the zero-order equation is

$$k = k_{max}$$

Example 3.30 Consider two effluent treatment systems:

Activated sludge—CSTR
Trickling filter—PFR

and two effluent types:

Domestic—with a BOD concentration of 200 mg/L
Industrial—with a BOD concentration of 4000 mg/L

The operating conditions are given as:

$$k_{max} = 0.05\,h$$
$$K_s = 100\ \text{mg/L}$$
$$\varnothing_{AS} = 5\,h$$
$$\varnothing_{TF} = 0.1\,h\ \text{(domestic)}$$
$$\varnothing_{TF} = 0.2\,h\ \text{(industrial)}$$
$$M_{AS} = 5000\ \text{mg/L (MLSS or micro-organism concentration)}$$
$$M_{TF} = 50\,000\ \text{mg/L}$$

Determine the output from each reactor of effluent concentrations using both the first-order model and actual kinetic models.

(1) Activated sludge—CSTR—domestic effluent

(a) Consider the first-order model, determine k:

$$k_1 = \frac{k_{\max} C}{K_s + C} \approx \frac{k_{\max} C}{K_s}$$

(3.109)

$$k_1 M = \frac{k_{\max} M C}{K_s}$$

where M = mass concentration mg/L

$$k_1 M = \frac{0.05 \times 5000 \times C}{100} = 2.5C$$

Therefore

$$k_1 = 2.5$$

The first-order equation for the CSTR is Eq. (3.105)

$$k_1 \varnothing = \frac{C_1 - C}{C}$$

(3.110)

where C_1 = influent concentration and C = effluent concentration

$$2.5 \varnothing = \frac{200 - C}{C}$$

For $\varnothing = 5$ h, the effluent concentration is $C = 14.9$ mg/L

(b) Consider actual kinetics:

$$\text{Since } dC/dt \Rightarrow 0$$

(3.111)

$$QC_1 - QC + rV = 0$$

$$QC_1 - QC - V \frac{M k_{\max} C}{K_s + C} = 0$$

$$\frac{Q}{V}(C_1 - C) = \frac{M k_{\max} C}{K_s + C}$$

$$= \frac{5000 \times 0.05 \times C}{100 + C}$$

$$\frac{C_1 - C}{\varnothing} = \frac{250C}{100 + C}$$

$$\frac{200 - C}{5} = \frac{250C}{100 + C}$$

Therefore

$$C = 17.1 \text{ mg/L}$$

So the first-order model and actual kinetics are compatible for domestic effluent, activated sludge process.

(2) Activated sludge—industrial effluent

(a) Consider the first-order model:

$$2.5\varnothing = \frac{C_1 - C}{C}$$

$$2.5 \times 5 = \frac{4000 - C}{C}$$

$$C = 296 \text{ mg/L}$$

(b) Consider actual kinetics:

$$\frac{C_1 - C}{\varnothing} = \frac{250C}{100 + C}$$

$$C = 2489 \text{ mg/L}$$

(inserting $C = 296$ mg/L $\Rightarrow \varnothing = 19.8$ h)

The first-order model and actual kinetics are not compatible in this case.

(3) Trickling filter—domestic waste

(a) Consider the first-order model (batch or plug flow):

$$-k_1 t = \ln \frac{C}{C_1} \tag{3.112}$$

As before,

$$k_1 M = \frac{M k_{max} C}{K_s} = \frac{50\,000 \times 0.05 \times C}{100} = 25C$$

Substituting into Eq. (3.112)

$$-25 \times 0.1 = \ln \frac{C}{200}$$

Therefore

$$C = 16.4 \text{ mg/L}$$

(b) Consider actual kinetics:

In a batch process, $Q=0$, therefore $V=$constant, as the material balance equation (3.101) becomes $r = dC/dt$.

Therefore

$$-\frac{dC}{dt} = \frac{M k_{max} C}{K_C + C} = \frac{50\,000 \times 0.05 \times C}{100 + C} = \frac{2500C}{100 + C}$$

$$-2500 \int_0^t dt = \int_{C_{A_i}}^{C_{A_o}} \frac{100 + C}{C} dC$$

$$-2500t = 100 \int_{C_{A_i}}^{C_{A_o}} \frac{dC}{C} + \int_{C_{A_i}}^{C_{A_o}} dC$$

$$2500t = 100 \ln \frac{C_{A_i}}{C_{A_o}} + (C_{A_i} - C_{A_o})$$

for the trickling filter, $t = \varnothing = 0.1 \text{ h}$

and $C_{A_i} = 200 \text{ mg/L}$

Solving $C_{A_o} = 64 \text{ mg/L}$

$$(\text{for } C_{A_o} = 16.4 \text{ mg/L}, \ \varnothing = 0.174 \text{ h})$$

(4) Trickling filter—industrial effluent

    (a) First-order model, with hydraulic retention time of 0.2 h,

$$-25 \times 0.2 = \ln \frac{C_{A_o}}{4000}$$

$$C_{A_o} = 27 \text{ mg/L}$$

    (b) Actual kinetics:

$$4000t = 100 \ln \frac{4000}{C_{A_o}} + (4000 - C_{A_o})$$

For $t = 0.2 \text{ h}$

$$C_{A_o} = 3220 \text{ mg/L}$$

$$(\text{for } C_{A_o} = 27 \text{ mg/L} \Rightarrow t = 1.12 \text{ h})$$

Summary of results

Treatment system	Waste type	Influent BOD mg/L	Model	Effluent BOD mg/L
Activated sludge	Domestic	200	First-order	14.9
			actual kinetics	17.1
	Industrial	4000	First-order	296
			actual kinetics	2488
Trickling filter	Domestic	200	First-order	16.4
			actual kinetics	64
	Industrial	4000	First-order	27
			actual kinetics	3200

It is seen from the above that several anomalies exist:

1. For domestic effluents a first-order model is sufficient for CSTR, since the reactant concentration is everywhere equal to the outlet concentration.
2. The first-order model is erroneous for domestic effluents using TF, since this reaction type had a concentration profile. The concentration is highest at the input end and lowest at the outfall end.
3. For high strength effluents, the first-order assumption is grossly unsatisfactory.

See Tchobanoglous and Schroeder (1987) and Himmelblau (1989) for many additional appropriate examples on material balances.

3.8 PROBLEMS

3.1 Explain why hydrogen bonding is responsible for the high heat capacity of water.

3.2 Explain how electrical conductivity can be used to determine the flow rates in a stream.

3.3 Balance the following equations:

$$Fe(OH)_2 + H_2O + O_2 \rightarrow Fe(OH)_3$$
$$Cl_2 + KOH \rightarrow KCl + KClO_3 + H_2O$$
$$FeS + HCl \rightarrow FeCl_2 + H_2S$$
$$MnO_2 + NaCl + H_2SO_4 \rightarrow MnSO_4 + H_2O + Cl_2 + Na_2SO_4$$

3.4 If a waste is characterized by its COD and 192 mg of COD is equivalent to 1 mmole of glucose, determine the glucose equivalent of a waste of 5500 mg/L COD.

3.5 How many moles of H_2SO_4 are required to produce 100 kg of $CaSO_4$ from $CaCO_3$?

3.6 If biomass (microbial cells) is represented by $C_5 H_7 NO_2$, determine the theoretical oxygen demand of 1 kg of biomass.

3.7 Determine the mass fractions of nitrogen, oxygen and argon in air if the respective mole fractions are 0.781, 0.21 and 0.009.

3.8 Calculate the pH of a solution if, before dissociation, the solution contains:
(a) 20 mg/L of hydrochloric acid,
(b) 15 mg/L of acetic acid,
(c) 50 mg/L of hypochlorous acid.

3.9 Determine the concentration of Fe^{3+} in pure water caused by the complete dissociation of $Fe(OH)_3$.

3.10 A water analysis is given below:

Cation	Concentration
Ca^{2+}	95 mg/L
Mg^{2+}	42 mg/L

Determine the total hardness.

3.11 Analysis of a wastewater sample gave the following:

Total alkalinity	88 mg/L as $CaCO_3$
Temperature	27 °C
pH	8.8

Determine the carbonate, bicarbonate and hydroxide alkalinities.

3.12 An industry discharges its treated effluent with a flow rate of $1\,m^3/s$ into a river of flow rate $250\,m^3/s$. If the BOD of the river background is 1.5 mg/L, determine the maximum BOD of the effluent discharge if the river BOD should not be greater than 1.7 mg/L.

3.13 A fuel source of 25 kg of ethylene is burned with 250 kg of air. Determine the composition of the resulting mixture. What is the percentage of excess air, assuming complete conversion.

3.14 In an anaerobic digestion reactor, the specific growth rate of the methanogenic bacteria is given by:

$$\mu = \mu_{max} \left(\frac{1}{1 + (K/SUB) + (AH/K_i)} \right)$$

where
K = saturation constant of methanogens
SUB = substrate used by methanogens
AH = acetic acid (unionized)
K_i = inhibition constant of methanogens.

Sketch with time the relationship of μ, K, SUB, AH and K_i, explaining the significance of each term.

3.15 In a CSTR the chemical reaction rate is given as $r_A = -0.1[A]$. If the reaction goes to 80 per cent completion, determine the volume required for a volumetric flow rate of 100 L/s if $[A]_o$ is 0.15 mol/L.

3.16 A wastewater CSTR of 50 m^3 volume operates on an irreversible first-order reaction basis,

$r_A = -k[A]$, where $k = 0.2$ day^{-1}. The efficiency is 95 per cent. Determine the maximum flow rate through the reactor. If the efficiency acceptable is 90 per cent, what is the optimum flowrate?

3.17 A plug flow reactor in an activated sludge wastewater system has a length of 100 m by a width of 10 m by a depth of 3 m. If the retention time is 2.5 h, compute the flow rate.

3.18 Write a brief review of 'HIV survivability in wastewater' by P. Gupta, in *Survival of HIV in Environmental Waters*, H. Farzadegan (ed.), 1991.

3.19 If a bacterial culture of 1000 cells doubles every hour, compute the number of bacteria after 24 h, assuming the same growth pattern. Is your answer logical? Comment with respect to the death rate.

3.20 Discuss briefly anaerobes and aerobes and their utility in wastewater treatment.

3.21 Search the literature and/or the Internet and write a note on 'The health risks (if any) of bacterial spore survival in compost made from wastewater sludges, and its subsequent use in mushroom production'.

3.22 Consider a primary settling tank in the treatment of wastewater. Assume an inflow of 2000 m^3/day, with a BOD$_5$ of 250 mg/L and a SS of 350 mg/L. If the SS reduction is 60 per cent once the BOD$_5$ is half of that, compute the flows in the two effluent streams, assuming the sludge stream contains 1 per cent DS. What is the BOD$_5$ in the sludge stream?

3.23 The growth of micro-organisms in an activated sludge (wastewater treatment) plant is assumed to be represented by the saturation growth rate model, i.e. $k = k_o S/(k_s + S)$ where k_o is the maximum growth rate, (constant), S is the substrate and k_s is the half saturation constant. A laboratory study shows that:

S, mg/L	7	9	15	25	40	75	100	150
k, day^{-1}	0.29	0.37	0.48	0.65	0.80	0.97	0.99	1.07

Determine the values of k_o and K_s.

3.24 Consider a completed mixed reactor (CSTR) with an inflow Q of concentration Ci and a tank and outflow concentration of C.
(1) Show that the mass balance for the reactor is: $V(dC/dt) = Q(C_i - C)$
(2) Solve the above equation analytically if $C = Co$ at $t = o$.
(3) If $Ci = 150$ mg/L, $Q = 7200$ m^3/day, $V = 200$ m^3 and $Co = 30$ mg/L, plot c versus time.

3.25 If bacterial cells are represented by $C_5H_7NO_2$, determine the potential carbonaceous BOD of 1 gram of cells.

REFERENCES AND FURTHER READING

Barnes, D., P. J. Bliss, B. W. Gould and H. R. Valentine (1986) *Water and Wastewater Engineering Systems*, Longman Scientific and Technical, London.

Bohn, H., B. McNeal, and G. O'Connor (1985) *Soil Chemistry*, John Wiley, New York.

Curds, C. R. (1982) 'The ecology and role of activated sludge', *Annual Review, Microbiology*, **36**, 27–46.

Dojlido, J. R. and J. A. Best (1993) *Chemistry of Water and Water Pollution*, Ellis Horwood/Prentice-Hall, Chichester/Englewood Cliffs, New Jersey.

Dolan, C. (1993) 'Anaerobic co-digestion of MSW and primary sewage sludge, MEngSc thesis, University College, Cork.

Farzadegan, M. (December 1991) 'Survival of HIV in environmental waters, Proceedings of a symposium at John Hopkins University.

Fogg, G. E., W. D. P. Stewart, P. Fay and A. E. Walsby (1973) *The Blue-Green Algae*, Academic Press, New York.

Gaudy, A. F. and E. T. Gaudy (1980) *Microbiology for Environmental Scientists and Engineers*. McGraw-Hill, New York.

Gill, D. L. (1995) *Report on Hazardous Air Pollutants from Landfill Sites*. University of Maryland School of Medicine.

Gray, N. (1990) *Activated Sludge: Theory and Practices*. Oxford Science Publishers, UK.

Harrison, R. M., S. J. de Mara, S. Rapsomanikis and W. R. Johnston (1991) *Introductory Chemistry for the Environmental Sciences*,

Cambridge University Press.

Henry, J. G. and G. W. Heinke (eds) (1989) *Environmental Science and Engineering*, Prentice-Hall, Englewood Cliffs, New Jersey.

Himmelblau, D. M. (1989) *Basic Principles and Calculations in Chemical Engineering.* 5th edn, Preintice-Hall, Englewood Cliffs, New Jersey.

Howells, G. and T. R. K. Dalziel (1992) *Restoring Acid Waters—Loch Fleet 1984–1990*, Elsevier, Amsterdam.

Humenick, M. J. (1977) *Water and Wastewater Treatment. Calculations for Chemical and Physical Processes.* Marcel Dekker Inc. USA.

Magill, P. L., F. R. Holden and C. Ackley (1956) *Air Pollution Handbook*, McGraw-Hill, New York.

Mahan, B. M. and R. J. Myers (1987) *University Chemistry*, 4th edn, Benjamin/Cummings Publishers.

Manahan, S. E. (1991) *Environmental Chemistry*, 5th edn, Lewis Publishers.

Mitchell, R. (1974) *Introduction to Environmental Microbiology*, Prentice-Hall, Englewood Cliffs, New Jersey.

Monod, J. (1949) 'The Growth of Bacterial Cultures'. *Annual Review of Microbiology*, vol. 3.

Monteith, J. L. and M. H. Unsworth (1990) *Principles of Environmental Physics*, Edward Arnold, London.

Montgomery, J. M. (1985) *Water Treatment, Principles and Design*, John Wiley, New York.

Mudrack, K. and S. Kunst (1986) *Biology of Sewage Treatment and Water Pollution Control*, Ellis Horwood, Chichester.

National Research Council (1993) *Soil and Water Quality—An Agenda for Agriculture*, National Academy Press, Washington.

O'Neill, P. (1991) *Environmental Chemistry*, Chapman and Hall, London.

Prescott, L. M., J. P. Harley and D. A. Klein (1993) *Microbiology*, 2nd edn, Wm. C. Brown, Dubuque, Iowa.

Randall, C. W., J. L. Barnard and H. D. Stensel (1992) *Design and Retrofit of Wastewater Treatment Plants for Biological Nutrient Removal.* Technomic Publishing, Lancaster, PA. USA.

Raiswell, R. W., P. Brimblecombe, D. L. Dent and P. S. Liss (1992) *Environmental Chemistry*, Edward Arnold, London.

Ryle, S. D. (1988) 'Optimization of the rotating biological contactor process for nitrification/dentrification of wastewater, MEngSc thesis, University College, Cork.

Sawyer, C. N. and P. L. McCarty (1989) *Chemistry for Environmental Engineering*, McGraw-Hill, New York.

Segel, I. H. (1993) *Enzyme Kinetics—Behaviour and Analysis of Rapid Equilibrium and Steady State Enzyme Systems*, John Wiley, New York.

Seinfeld, J. H. (1986) *Atmospheric Chemistry and Physics of Air Pollution*, John Wiley, New York.

Senior, E, (1990) *Microbiology of Landfill Sites*, CRC Press, Boca Raton, Florida.

Skinner, F. A. and J. M. Shewan (1977) *Aquatic Microbiology*, Academic Press, New York.

Snoeyink, V. L. and D. Jenkins (1980) *Water Chemistry*, John Wiley, New York.

Sposito, G. (1989). *The Chemistry of Soils.* Oxford University Press, UK.

Standard Methods for the Examination of Water and Wastewater, (1992). American Water Works Association (AWWA). eds Greenberg, A. E., L. S. Clesceil and A. D. Eaton.

Stanier, R. Y., J. L. Ingraham, M. L. Sheelis and P. R. Painter (1986) *Microbiol World*, 5th edn, Prentice-Hall, Englewood Cliffs, New Jersey.

Stern, A. C. (1976) *Air Pollution*, Vols I–V, Academic Press, New York.

Sterritt, R. M. and J. N. Lester (1988) *Microbiology for Environmental and Public Health Engineers*, E&FN Spon, London.

Stoker, H. S. and S. L. Seager (1972) *Environmental Chemistry: Air and Water Pollution*, Scott, Foresimon and Co., London.

Stumm, W. (1972) *Chemistry of the Solid–Water Interface*, John Wiley, New York.

Stumm, W. and J. J. Morgan (1981) *Aquatic Chemistry. An Introduction Emphasizing Chemical Equilibria in Natural Waters*, 2nd edn, John Wiley, New York.

Tanji, K. (1990) *Agricultureal Salinity Asessment and Management*, ASCE Manual and Report on Engineering Practices No. 71, New York.

Tchobanoglous, G. and E. Schroeder (1987) *Water Quality*, Addision-Wesley, Reading, Massachusetts.

Water Pollution Control Federation (WPCF) (1990) *Wastewater Biology. The Microlife*, WPCF, Alexandria, Virginia.

FOUR

CONCEPTS IN HYDROLOGY

4.1 INTRODUCTION

Hydrology is the study of water and its movement along its various pathways within the hydrological cycle; in the atmosphere; in the rivers and oceans; in the soil and in water containing rocks. Hydraulics is the engineering of water flow in pipes, conduits, lakes or rivers. Water resources engineering is the art, science and engineering of surface and groundwaters for human use. Hydrology is applied by engineers who use hydrological principles to compute, for instance, river flows from rainfall, water movement in soils from knowledge of soil characteristics including hydraulic conductivity, evaporation rates from water balance or energy balance techniques. Applied hydrology uses many engineering assumptions in attempting to quantify soil or river responses to rainfall events. It is easy, for instance, to quantify a rainfall event and to quantify streamflow after this event, if field instrumentation has recorded the event. However, it is still almost impossible to predict or model with accuracy what happens to rainfall once it has fallen on land. Does 100 per cent of that rainfall go as surface runoff to the nearest stream or does 100 per cent infiltrate to the soil and show up in the streams, days or weeks later, with impact not only on the streamwater volumes but also on the streamwater quality? In reality, either situation can occur but more likely some precipitation goes as surface run-off, some as infiltration and some is returned to the hydrological cycle via evaporation. What is the role of evaporation? Does precipitation exceed evaporation or vice versa? Can we compute with any accuracy the regional scale evaporation if all our evaporation studies are at the point scale? When can we expect remote sensing to deliver the answers? The influence of surface vegetation, soil type, soil moisture status and topography is significant on water and energy fluxes and the response from one site or watershed to another may be orders of magnitude different. Rain intensity, duration and spatial distribution also play a significant role in the fate of land-fallen precipitation. As such, art, science and engineering are all used to understand the pathways of water in the hydrological cycle. There are still many, unquantified issues, particularly as we take hydrology into meteorology and into climate studies.

This chapter discusses the hydrological cycle and its components as well as the energy cycle. It explains the differences in infiltration and surface runoff. It defines evaporation and shows how to quantify it at a point in space. This chapter also examines why our lack of understanding of evaporation, particularly at the regional scale, is the missing link in closing the hydrological cycle water balance. It explains water balance for catchments, looking at hyetographs and hydrographs. It examines rainfall–runoff relationships for flood flows and low flows. It examines the influence of urbanization on hydrological responses. It briefly introduces the student to the physics of the energy cycle. The final section of this chapter looks at both physical and chemical concepts of groundwater. By the end of this chapter it is hoped that the student will have an introductory qualitative and quantitative understanding of the physics of water in the hydrological cycle and also in the sun's energy cycle.

4.2 HYDROLOGICAL CYCLE

The hydrological cycle is central to hydrology. It is a continuous process with no starts or finishes. It is shown schematically in Fig. 4.1. Water *evaporates* from the earth's oceans and other water bodies, and to a lesser extent from the land surfaces. Approximately seven times more evaporation occurs from the oceans than from the earth's land surface. Remember the ocean surface area of the earth is 2.5 times the land surface area. The evaporated water or water vapour rises into the atmosphere until the lower temperatures aloft cause it to condense and then *precipitate* in the form most globally as rain but sometimes as snow. The latter occurs at the more alpine elevations or in cold seasons. The global annual average water balance relative to 100 units of land precipitation is enumerated in the water balance diagram of Fig. 4.2. A schematic of a modeller's flow chart of the hydrological cycle is shown in Fig. 4.3.

Figure 4.2 is a simplistic way of showing the hydrological cycle but the objective is quantitative, whereas Fig. 4.1 is qualitative. The material balance of Fig. 4.2 is based on the conservation of matter. The four subregions of Fig. 4.2 are numerically in equilibrium on their own or taken as the totality of the four. For instance, the equilibrium of the 'oceans balance' is satisfied by two inputs of precipitation plus inflow, being equal to one output of evaporation, i.e.

For landmass :   Input ± change in storage = output

For oceans :   Precipitation + inflow = evaporation

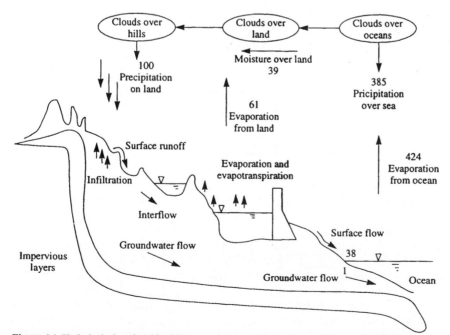

Figure 4.1 Hydrological cycle with global annual average water balance given in units relative to a value of 100 for the rate of precipitation on land (adapted from Chow *et al.*, 1988).

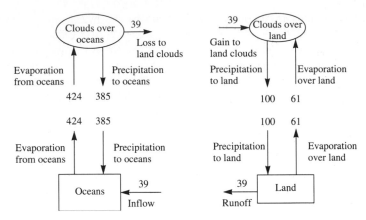

Figure 4.2 Material balance on aspects of the hydrological cycle.

Tables 4.1 and 4.2 show the distribution of the earth's water resources. The oceans contain 96.5 per cent of all water while the rivers occupy only 0.0002 per cent. The great store of usable freshwater is held in groundwater at 30.1 per cent, while soil moisture stores are 0.05 per cent, or 250 times that of rivers.

Understanding of the hydrologic cycle as it relates to precipitation on land is required by many different professionals—be it an engineer designing a water supply, an agriculturist designing an irrigation scheme, a freshwater biologist investigating adequacy of river flows for fisheries habitats, an industrialist abstracting water or discharging liquid effluent or a meteorologist forecasting weather patterns. What is of

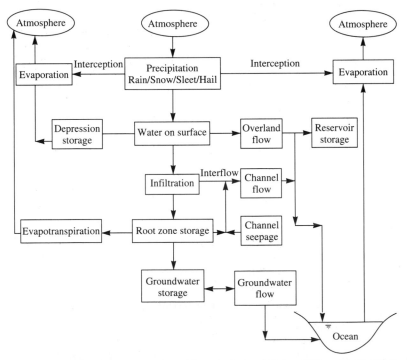

Figure 4.3 Components of the hydrological cycle (adapted from Bedient and Huber, 1988, p. 55, © 1988 by Addison-Wesley Publishing Company, Inc. Reprinted by permission of the publisher).

Table 4.1 Estimated world water quantities

Item	Area (10^6 km^2)	Volume (km^3)	Total water %	Fresh water %	Rates of exchange
Oceans	36.31	1 338 000 000	96.5		3000–30 000 yrs
Groundwater					
Fresh	134.8	10 530 000	0.76	30.1	Days to 1000 yr
Saline	134.8	12 870 000	0.93		
Soil moisture	82.0	16 500	0.001 2	0.05	2–52 weeks
Polar ice	16.0	24 023 500	1.7	68.6	1–16 000 years
Other ice and snow	0.3	340 600	0.025	1.0	
Lakes					
Fresh	1.2	91 000	0.007	0.26	1–100 years
Saline	0.8	85 400	0.006		10–1000 years
Marshes	2.7	11 470	0.000 8	0.03	
Rivers	148.8	2 120	0.000 2	0.006	10–30 days
Biological water	510.0	1 120	0.000 1	0.003	7 days
Atmospheric water	510.0	12 900	0.001	0.04	8–10 days
Total water	510.0	1 385 984 610	100.0		2800 years
Fresh water	148.8	35 029 210	2.5	100.0	

Adapted from UNESCO, 1978

most pragmatic interest, then, is what happens to the land-fallen precipitation on a mesoscale of a catchment or region, rather than the global annual water balance. The meteorologist has interests in the hydrological cycle on a larger scale, sometimes global. Precipitation may be intercepted by vegetation, i.e. grass, crops or trees. *Interception* is the evaporation of water from the outer surface of leaves during and after rainfall. *Transpiration* is evaporation of water through foliage. Some may lodge on the soil surface and be retained there in depressions. This is called depression storage or ponding. Some water may become overland flow and eventually reach a stream or river and be discharged as surface runoff. It may infiltrate into the soil and flow horizontally as interflow. It may percolate through the deeper soil layer into the groundwater zone and recharge the waters in the aquifers. A significant volume of precipitation may be returned to the atmosphere through *evaporation* from water bodies and *evapotranspiration* from vegetated surfaces. The extent of the latter depends on many factors, including climate, type of surface vegetation, amount of rainfall and rain intensity. In general, of 100 units of rain that falls on grassland in temperate zones, 10 to 20 units will go to groundwater, 20 to 40 units will evapotranspire and 40 to 70 units will become stream runoff. In arid and semi-arid areas, with little precipitation, not all of the above phenomena may be experienced, as high evaporation tends to dominate the hydrologic cycle.

Table 4.2 Global annual water balance

		Ocean	Land
Area (km^2)		361 300 000	148 800 000
Precipitation	(km^3/yr)	458 000	119 000
	(mm/yr)	1 270	800
Evaporation	(km^3/yr)	505 000	72 000
	(mm/yr)	1 400	484
Runoff to ocean			
Rivers	(km^3/yr)		44 700
Groundwater	(km^3/yr)		2 200
Total runoff	(km^3/yr)		47 000
	(mm/yr)		316

Adapted from UNESCO, 1978

The way precipitation becomes spatially distributed depends on climate, soils, geology, topography and land use. For instance, if a soil vegetation matrix is saturated with water from a previous rainstorm, a new precipitation event may become distributed solely to streamflow (via overland flow) with no contribution to evaporation, infiltration or percolation. Alternatively, if a soil matrix is very dry with a low water table, a precipitation event may be distributed to infiltration followed by percolation to groundwater, with no quantity to streamflow. Therefore, to be able to quantify the distribution of precipitation, knowledge of the soil and the response of soil to water is required.

4.3 WATER BALANCE

The water balance or water budget is the accounting of water for a particular catchment, region or even for the earth as a whole. As seen in the preceding sections, the hydrologic cycle considers all the phenomena of water phases in a qualitative description. The water balance is the quantitative account of the hydrologic cycle. The input to the cycle is precipitation, either as rainfall, snow or sleet. The precipitation is distributed as surface runoff, evaporation, infiltration to the unsaturated zone, changing its storage, and deep percolation to the saturated zones.

The equation for the water balance, which is the conservation of mass in a lumped or averaged hydrological system on a regional or catchment scale is

$$P = R + E \pm \Delta S \pm \Delta G \tag{4.1}$$

where

P = precipitation, mm/day

R = stream runoff

E = evaporation

ΔS = change in soil moisture status

ΔG = change in groundwater status

Equation (4.1) assumes that there is no 'flow' across catchments. While this is correct for surface water, it is not always possible to verify that there is zero flow in the subsoil regions across catchment boundaries, i.e. no interflow. If Eq. (4.1) is averaged over the hydrologic year (in northern temperate climates this is typically 1 October to 30 September), there may be no significant change in ΔS or ΔG. Thus

$$P = R + E \tag{4.2}$$

and so

$$E = P - R \tag{4.3}$$

Equation (4.3) is often used to determine evaporation from the 'annual' water balance of closed systems.

Water balance data are required for a myriad of uses. If water is to be abstracted from surface waters for irrigation, hydropower, cooling water or industrial requirements, it is necessary to understand not only the absolute values of precipitation, evaporation and streamflow but also the trends over time. If a land use change is proposed for a catchment, it may alter the water balance. For instance, it is most likely that a grassland catchment in the temperate zone, if converted to forestry, would see an increased evaporation on maturation of the plantations. This is due to the increased transpiration rates of trees over grassland. This would leave less water for streamflow and its human and ecological uses may be impacted. Table 4.3 lists water balance results for many catchments throughout the world for different land uses (mainly forest). Evaporation losses (defined as evaporation/precipitation) vary from 15 per cent for upland moorland catchments in the United Kingdom to about 70 per cent for fully forested catchments. Table 4.4 shows the water balance of the continents. It is seen that the water loss due to evaporation varies significantly with about 60 per cent for South America and 93 per cent for Australia.

Table 4.3 Water balance for different land uses

Author	Location	Land use	Annual rainfall P(mm)	Runoff Q(mm)	Total evaporation losses (mm)	Losses (%)
Law (1956) 1955–6	Stocks Reservoir (UK)	Coniferous Forest 100%	984	273	711	72
Institute of Hydrology 1967–70	Stocks Reservoir (450 m²)	100% forest	1496	555	953	64
Institute of Hydrology 1956–70	Stocks Reservoir (37.5 km²)	22% forest	1662	1204	454	27
Institute of Hydrology 1956–70	Stocks Reservoir (10.6 km²)	70% forest	1544	1049	495	32
Law (1956)	Stocks Reservoir (UK)	Grassland, moorland	1135	717	421	37
Law (1956) and Calder et al. (1982)	Stocks Reservoir	Grass, lysimeter irrigated	1702		467 (PET)	28
Law (1956) and Calder et al. (1982)	Stocks Reservoir	Heather, lysimeter irrigated	1702		520 (PET)	31
Caspary (1990)	Black Forest	Norway spruce, 100% forest	Dormant season 950	484	466	49
			Growing season (1975) 600	200	400	67
			Growing season (1985) 600	350	250	41
Mulholland et al. (1991)	Walker branch, Tennessee	100% deciduous	1400	728	672	48
Farrell (1991)	Ballyhooley, N. Cork, Ireland	100% forest	1022	(throughfall) 576	446	44
Bishop (1991)	Loch Fleet, Scotland	Grassland, moorland	2200	1740	460	21
Cooper (1980)	Thetford, East Anglia	100% forest	640	—	430	67
Shuttleworth (1988)	Amazonia	100% rainforest	2593	—	1393	53
Kirby et al. (1991)	Wye, Plynlimon, Severn	Grassland, moorland	2394	2041	353	15
Kirby et al. (1991)	Plynlimon, Wales (1977)	68% forest	2620	1820	770	30
FRI New Zealand (1980) McDonnell (1990)	Mamai, New Zealand	100% beech forest	2600	1500	1000	39
Pearce et al. (1976)	Big Bush, New Zealand	100% beech forest	1500	600	800	54
Fahey, Watson	New Zealand	Tussock grassland, pine forest	1150	620	530	46
			1150	500	650	57

Table 4.4 Water balance of continents[1]

Continent	Area (10^6 km^2)	P (mm/yr)	Evapotranspiration (mm/yr)	Runoff (mm/yr)
Europe	10.0	657	375	282
Asia	44.1	696	420	276
Africa	29.8	695	582	114
Australia	7.6	447	420	27
North America	24.1	645	403	242
South America	17.9	1564	946	618
Antarctica	14.1	169	28	141
Total land	148.9	746	480	266

[1] Data from Baumgartner and Reichel (1975)

4.4 ENERGY BUDGET

The energy received at the earth's surface is essentially all solar (shortwave) radiation. Some of this energy is reflected back from the earth's surface to the atmosphere, and some penetrates the earth. The earth also re-radiates some of the solar energy. Like the water budget, the energy balance is the accounting of the distribution of the incoming shortwave solar radiation from space, through the atmosphere and onto the earth's surface of land and ocean, (see Fig. 4.4). The energy balance also accounts for the outgoing longwave terrestrial radiation from the earth's surface. This distributes to evaporation flux, sensible heat flux and net radiant emission by the surface. What is of most interest to hydrology is the net incoming radiation at the earth's surface and the subsequent partitioning of this energy (measured in watts/m^2) to

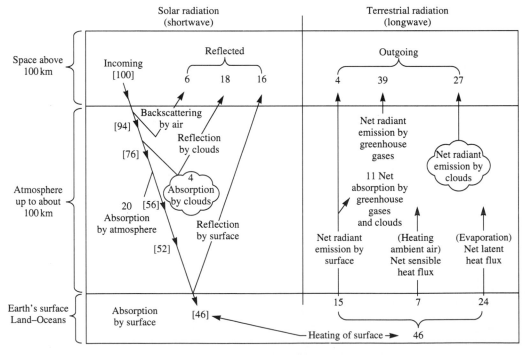

Figure 4.4 Average global energy balance of the earth–atmosphere system. Numbers indicate relative energy fluxes; 100 units equals the solar constant 1367 W/m^2. Modified from Shuttleworth (1991) and Dingman (1994).

evaporation, sensible heat and heat absorbed by the soil. The quantity of radiant energy remaining at the earth's surface is known as the net radiation, R_n, typically in units of watts/m^2, and is measured by a simple instrument called a net radiometer. For a simple lumped system, the energy budget is expressed as

$$R_n = LE + H + G + PS + M \tag{4.4}$$

where
R_n = specific flux of net incoming radiation, W/m^2
L = latent heat of vaporization
E = evaporation
H = specific flux of sensible heat into the atmosphere (i.e. the energy in watts/m^2 used to heat the ambient air)
G = specific flux of heat into or out of the soil
PS = photosynthetic energy fixed by plants
M = energy for respiration and heat storage in a crop canopy

Simplifying, by neglecting PS and M, then

$$R_n = LE + H + G \tag{4.5}$$

Like Eqs (4.1) to (4.3), Eqs (4.4) and (4.5) can be applied to either a single plant or a cropped field, a catchment, a region or the global scale. In any given system, the connecting link between the water budget and the energy budget equations is evaporation. Most of the net incoming solar radiation is absorbed near the surface of the earth and converted to internal energy, either as longwave back radiation, evaporation, downward conduction of heat into the soil, upward thermal conduction and convection of sensible heat (Brutsaert, 1982).

A useful parameter derived from Eq. (4.5) is the Bowen ratio

$$Bo = \frac{H}{LE} \tag{4.6}$$

which is a ratio of the sensible heat flux and the evaporation rate. For instance, in arid regions, Bo is >1, since evaporation is limited by limited water supplies. In moist, warm, tropical areas, Bo < 1. Further details are to be found in Brutsaert (1982) and Rosenberg *et al.* (1983). Estimates of the mean global energy budget from Brutsaert (1982) are depicted in Table 4.5. Over the global land surface, it is seen that, of the 50 kcal/cm^2 yr of net radiation, approximately 25 units go to evaporation and 25 go to sensible heat flux. This complies with the values of evaporation for specific catchments in Table 4.4. In the case of oceanic energy balance, approximately 85 per cent of the net radiation energy is distributed to evaporation. It is therefore seen from these tables that evaporation plays a major role in the water and energy balance of the globe and, more significantly, for the hydrological budget of the earth's land surface.

Table 4.5 Estimates of mean global energy budget at earth's surface in kcal/cm^2 yr

	Land			Ocean			Global		
Reference	R_n	LE	H	R_n	LE	H	R_n	LE	H
Budyko (1974)	49	25	24	82	74	8	72	60	12
Baumgartner and Reichel (1975)	50	28	22	81	69	12	72	57	15
Korzun *et al.* (1978)	49	27	22	91	82	9	79	67	12

Adapted from Brutsaert, 1982. Reprinted by permission of Kluwer Academic Publishers

4.5 PRECIPITATION

Precipitation in the form of rain, hail or snow is one input to the hydrologic cycle. If we are interested in predicting or assessing a hydrologic response we need to be able to determine the amount, rate and duration of precipitation on a spatial and temporal basis. In Sec. 4.14 we discuss the water quality aspects of rainfall.

Precipitation occurs when air rises, expands (on cooling) and cools sufficiently for the water vapour in the air to reach condensation point. The atmosphere is rich in nuclei, mainly soil/clay particles, hydrocarbon waste products, sea salts, etc., with a size requirement greater than about 0.1 μm. Additionally, for precipitation to occur, there must also be:

1. The presence of condensation nuclei on which condensation can start. In their absence, the air can become supersaturated.
2. These condensed droplets should not evaporate when passing through drier air and should be of sufficient size to free-fall under gravity to the earth's surface. If the droplets are too small, they may have an inadequate 'settling' or falling velocity to reach the ground.

Rain droplets increase in size either by coalescence (liquid to liquid) producing rain or when solid aggregates with solid as with snow. An intermediate phase of aggregation of solid with liquid produces hail, Bras (1990) identified the forms of rainfall precipitation, as in Table 4.6.

Precipitation in the form of rainfall has a large spatial variability for local thunderstorms covering an area as small as 5 km^2 to a synoptic storm covering up to 250 000 km^2. Table 4.7 outlines the spatial characteristics of general storms. In general, we have cellular thunderstorms during the warmer periods (but not exclusively so). Details of the physics of rainfall can be found in many books, including Bras (1990) and Eagleson (1970).

4.5.1 Measurement of Precipitation

The three means of determining the magnitude of rainfall, spatially and temporally, are:

- Precipitation gauges
- Radar
- Satellite remote sensing

The traditional means of measurement was to use a network of raingauges which were read manually on a daily basis, and this gave daily rainfall at a single point in space. Today, raingauges are predominantly continuous rainfall recorders, with attached electronic data recorders. Typically, these will record rainfall at a point for a particular magnitude of rainfall, e.g. in increments of 0.2 mm. The mechanism may be a tipping bucket, of capacity 0.2 mm and each time 0.2 mm falls, the start and finish time are recorded. Analysis of the record can then be for hourly, daily, weekly, rainfalls. If an area has a sufficient number of raingauges, then the temporal and spatial distributions of rainfall may be determined. Raingauges connected to a telemetry system are now being used for real-time runoff forecasting. Raingauge networks

Table 4.6 Forms of precipitation

Name	Description	Size
Drizzle	<1 mm/h	0.1–0.5 mm
Rain	Light <2.5 mm/h	>0.5 mm
	Moderate 2.5–7.5 mm/h	
	Heavy >7.5 mm/h	

Adapted from Bras, 1990

Table 4.7 Characteristics of general storms

Name	Size (km^2)	Intensity (mm/h)	Duration
Synoptic	25 000–250 000	0.2–2	Few days
Large mesoscale	2300–4600	1–3	<12 h
Small mesoscale	100–400	2–5	<3 h
Cellular	<10	>5	Minutes

Adapted from Bras, 1990

have been used in determining rainfalls that should be used in flood analysis, low flow in stream analysis, groundwater recharge analysis, water balance studies of catchments and, to a lesser extent, water quality analysis of rainfall. It is important to be aware that there are serious limitations to using raingauge data from an insufficient network of gauges. Essentially a raingauge is a point measurement of rainfall and rainfall will vary widely (spatially and temporally) depending on the type of rainfall. For instance, two raingauges, 2 km apart, may record significant differences during a thunderstorm, but most likely similar falls during a mesoscale storm.

Errors in the absolute magnitudes from raingauges can occur from poor siting (too close to buildings or trees), overgrowth of ground cover, winds and other types of shielding. Figure 4.5 is a typical bar chart (hyetograph) of a heavy rainstorm. It is seen that the record is not continuous. The storm duration is 24 h with a total rain of 91.8 mm. The peak intensity is 14.2 mm/h with an average intensity of 3.8 mm/h. This rainstorm would be considered an infrequent event in a temperate wet climate like Ireland.

Ground-based radar is used to estimate the areal distribution of the instantaneous precipitation rates in clouds. As such, it should be a more sophisticated and reliable method of rainfall determination than raingauges. In theory, it should be able to provide a continuous description of rainfall over the cone of influence of the radar. The radar image needs to be calibrated with on-ground raingauges or raindrop size measurements. Because of many factors, including evaporation of falling rain and distortion of the precipitation field by winds at elevations lower than clouds, a precise image of precipitation cannot be obtained. While the precise magnitude of rainfall estimates from radar can be in error by factors of 0.5 to 2.0, radar does give a good picture of the areal extent of precipitation. The student is referred to Bras (1990) and to Collinge and Kirby (1987) for further details on rain-radar.

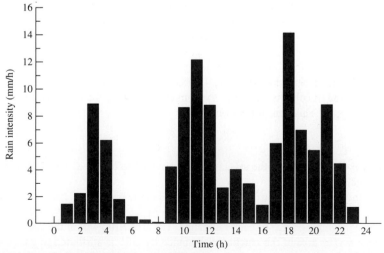

Figure 4.5 Typical hyetograph of a severe 24-hour winter rainstorm.

Satellite observations can provide information on the areal distribution of precipitation working on the principle that the atmosphere selectively transmits radiation at various wavelengths, and more particularly in the visible and thermal infra-red wavelengths. The visible wavelengths are of the order of 0.77 to 0.91 μm (Bras, 1990) and give information on the distribution of clouds, and therefore possible areal locations of rainfall. The infra-red wavelengths, 8 to 9.2 μm and 17.0 to 22.0 μm (see the electromagnetic spectrum in Chapter 8), can be used to locate high clouds and their associated convective precipitation cells. In the United States the polar orbiting satellites provide two visible and one infra-red pass per day and from geostationery satellites providing images at intervals of a half hour (Dingman, 1994). Very obvious benefits of satellite imagery is for areas of low inhabitation, where raingauges or radar are not available and particularly remote island locations, e.g. the South Pacific.

4.5.2 Precipitation Analysis

Analysis of precipitation results include the determination of:

- Areal precipitation
- Depth–area–duration analysis
- Precipitation frequency
- Intensity–duration–frequency analysis
- Extreme values of precipitation

Determination of areal precipitation from point measurements   The mean areal precipitation of a storm event is

$$P_1 = \frac{1}{A} \int_A p(x) \mathrm{d}x \tag{4.7}$$

and the time-averaged mean areal precipitation is

$$P_2 = \frac{1}{T} \frac{1}{A} \sum_{i=1}^{m} \int_A p(x, t_i) \mathrm{d}x \tag{4.8}$$

where $p(x)$ is a function describing the total accumulation of precipitation at all points x_i in the catchment and $p(x, t_i)$ describes the total precipitation at x and time t_i. A is the catchment area and T is the total storm period. Several methods are used to determine the areal average of a storm's precipitation, including the arithmetic mean, the Thiesson polygon, the Isohyetal method, the hypsometric method and the multiquadratic method (Shaw, 1994). The use of the above methods is best illustrated by means of examples.

Example 4.1   The catchment shown in Fig. 4.6 has six raingauges which recorded the intensities of a storm event as illustrated in column 2. Two of the gauges are outside the watershed line. Determine the areal precipitation using the Thiessen polygon method.

Solution

Step 1. Join with broken lines each of the six gauges as shown, 1 to 6, 1 to 3, 1 to 2, 1 to 5, 6 to 3, 3 to 4, 4 to 5, 5 to 2, 3 to 2, etc.

Step 2. Draw the orthogonal bisectors of these lines, i.e. AB is the bisector of 1 to 6, AC the bisector of 3 to 6, etc.

Step 3. Identify the contributing areas to each raingauge. The area BAC within the catchment is attributed to raingauge 6. The area of the catchment bounded by GEF is attributed to gauge 4. These areas are divided by the the total watershed area and reported as the Thiessen weights in column 3.

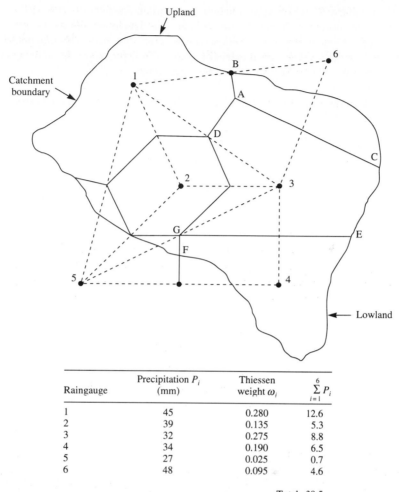

Figure 4.6 Areal average precipitation determination by Thiessen polygon.

Raingauge	Precipitation P_i (mm)	Thiessen weight ω_i	$\sum\limits_{i=1}^{6} P_i$
1	45	0.280	12.6
2	39	0.135	5.3
3	32	0.275	8.8
4	34	0.190	6.5
5	27	0.025	0.7
6	48	0.095	4.6

Total 38.5

Step 4. The total areal precipitation is then computed from weighted contributions of each gauge as in column 4. Therefore,

$$P = 38.5 \text{ mm}$$

Another common method for areal precipitation determination is the isohyetal method. The isohyetal map of a catchment shows the contours of precipitation. These could be composed from Example 4.1. The contours of precipitation as determined from the raingauges and the contour map are drawn in fine increments. The weights attributed to a contour interval are assigned w_i, similar to the way weights of area were assigned in the Thiessen polygon method. Depending on the range of rainfall, the contour increments may or may not exceed the number of raingauges. For instance, in Example 4.1 with a range of 27 to 48 mm, i.e. 21 mm, a contour precipitation interval range might be, say, increments of 3 mm each. The reader is referred to Shaw (1994) for further details.

Depth–area–duration analysis As the area of a catchment increases, typically the depth of precipitation decreases and this is accounted for in the UK *Flood Studies Report* (NERC, 1975) which uses an areal reduction factor (ARF) for precipitation. For short duration storms, the ARF is significant, as short duration storms also tend to spread over smaller land areas than longer duration storms. As the depth and duration increase, so does the areal average precipitation. For many environmental engineering applications, it is relevant to know the areal extent of a particular depth of precipitation and to know how the depth varies with area. This is best illustrated by an example.

Example 4.2   Determine the precipitation depth–area curve for the hypothetical storm given in Fig. 4.7(a).

Solution   Table 4.8 is computed as follows:

Step 1. Identify the isohyets, as shown in Fig. 4.7a, as being 100, 90, 80, 70, 60 and 50 mm, total precipitation. Associate with each interval its contributing catchment area and enter as column 2 of Table 4.8.
Step 2. Identify the area between isohyets as column 3 and the average rain between isohyets as column 4.
Step 3. The volume of rainfall between isohyets is the product of columns 3 and 4 and is entered in column 4. Column 6 is the cumulative rain.
Step 4. Column 7 is the areal rain which is column 6 divided by column 2.
Step 5. Figure 4.7(b) is now drawn with column 7 as ordinate and column 2 as abscissa. This is usually drawn on a log scale for the x axis if the range covers several orders of magnitude. It is seen that as area increases, rainfall depth decreases.

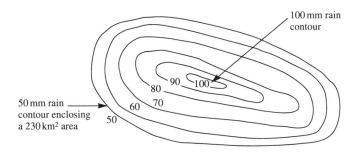

Figure 4.7(a) Schematic of isohyets of a single storm cell (in mm) (Example 4.2).

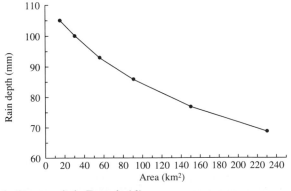

Figure 4.7(b) Precipitation depth–area analysis (Example 4.2).

Table 4.8 Computation of depth–area precipitation

Isohyet (mm)	Total area enclosed (km^2)	Area between isohyets (km^2)	Average rain between isohyets (mm)	Volume of rain between isohyets (10^{-12} mm^3)	Cumulative volume (10^{-12} mm^3)	Areal rain (mm)
100	15	15	105	1 575	1 575	105
90	30	15	95	1 425	3 000	100
80	55	25	85	2 125	5 125	93
70	90	35	75	2 625	7 750	86
60	150	60	65	3 900	11 650	77
50	230	80	55	4 400	16 050	69

Depth–area analysis of single storm events can be extended to depth–area–duration analysis where typically the durations are of the order of hours. This analysis is more detailed and the reader is referred to Shaw (1994, p. 216) for a detailed example. Figure 4.8 represents a typical depth–area–duration set of curves for a single storm. It is seen that as duration increases, so does the rainfall depth. Figure 4.8 can be developed for any locality with rainfall records.

Precipitation frequency   Most hydrologic parameters, including precipitation, streamflow, evaporation, etc., are characteristically, time series i.e. their magnitude varies with time. They may be continuous like a streamflow record (hydrograph) or possibly discrete like a rainfall record over a period of time with actual magnitudes followed by zero readings (see Fig. 4.5). The determination of the frequency of a rain event of a particular magnitude is of particular relevance to environmental engineering. For instance, we need to be able to know if a rainstorm magnitude of 30 mm/h has a frequency of once per year or once per 50 years. In other words, we need to be able to ascertain the return period or frequency. This determination will be more accurate if the length of known records is long. For instance, if we know the annual rainfall for 100 years, we can, using statistics, determine and forecast many properties from this data set. For instance, Fig. 4.9 (from Shaw, 1994) shows the different statistical distributions of the daily rainfall record for the United Kingdom. The daily record (*a*) is shown to follow the J-type (or exponential decay) distribution, while the monthly rainfall (*b*) follows the log normal distribution and the annual rainfalls (*c*) follow the

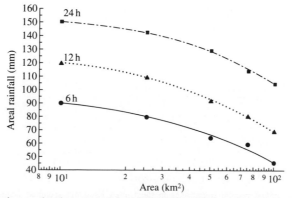

Figure 4.8 Precipitation depth–area–duration curves (adapted from Shaw, 1994).

normal distribution. Different areas on the globe and even different locations within the United Kingdom may follow different distributions to those of Fig. 4.9.

Intensity–duration–frequency analysis Many textbooks in hydrology detail intensity–duration–frequency (IDF) analysis. A conceptual IDF curve is shown in Fig. 4.10. A different set of curves pertain to different locations. As rainfall intensity increases, its duration decreases, i.e. $I \propto t^{-1}$. This may be represented by

$$I = \frac{c}{t^n} \tag{4.9}$$

where
I = rain intensity, mm/h

t = storm duration, (min or hours)

c, n = locality constants

Dillon (1954) derived the following equation for Cork, Ireland, from data for 35 years:

$$I = 152.4 \; \frac{T_{\mathrm{p}}^{1/5}}{t^{3/5}} \tag{4.10}$$

where T_{p} is the recurrence interval in years, e.g. 4, 10, 20, 30, etc.

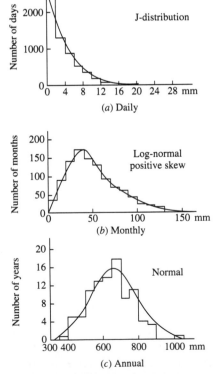

(a) Daily

(b) Monthly

(c) Annual

Figure 4.9 Daily, monthly and annual rainfall frequencies in the United Kingdom (adapted from Shaw, 1994. Reprinted by permission of Chapman and Hall).

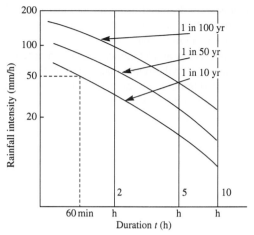

Figure 4.10 Schematic of rainfall frequency–intensity–duration relationships.

The historical equation for IDF in the United Kingdom is that of Bilham (1936) and re-issued by the UK Meteorological Society in 1962:

$$R = 25.4 \left(\frac{t}{48N}\right)^{0.282} - 2.54 \tag{4.11}$$

where

R = rainfall depth, mm

t = duration of rain, min

N = number of occurrences in 10 years

A revised version of Bilham's equation is by Holland (1967):

$$R = 25.4 \left(\frac{t}{60N}\right)^{0.318} \tag{4.12}$$

The above equations have been used in engineering design for many years for the design of sewers. However, the basic concept of IDF is to obtain a single-valued rain intensity for a particular storm duration and frequency. For instance, with respect to the schematic of Fig. 4.10, it is seen that the 60 minute storm with a return period of 10 years has an intensity of approximately 50 mm/h. This assumes that the storm has a rainfall of uniform intensity. This is most unlikely to occur in reality. The origin of IDF was to determine a uniform rainfall intensity to be applied to the rational formula in determining runoff

$$Q = \text{CIA} \tag{4.13}$$

where

Q = runoff, m^3/s

C = locality coefficient

I = intensity of rainfall, mm/h

A = catchment area, km^2

Therefore, the IDF rainfall intensities do not represent actual time histories of rainfall. Also, neither is the duration in IDF curves the actual length of the storm; rather it is merely a 60 minute period, say, within a longer storm of any duration during which the average intensity happened to be the specified value (e.g. 70 mm/h). In fact, the IDF curves are smoothed contours and unless a data point actually falls on the curve, it is hypothetical.

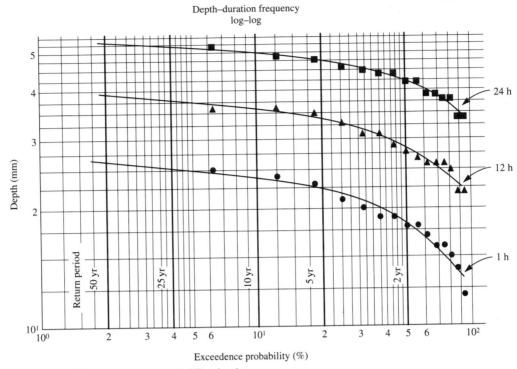

Figure 4.11 Rainfall depth–exceedance probability duration.

fraction of precipitation that infiltrates on a global scale is about 76 per cent. However, on a regional or local scale it has a large seasonal and spatial variation, even within a few hectares of catchment area.

4.6.1 Elemental Properties of Soils

The elemental properties of soil in relation to infiltration are:

- Bulk density
- Particle density
- Porosity
- Volumetric water content
- Degree of saturation

Bulk density   Bulk density ρ_b or the dry density of a soil is

$$\rho_b = \frac{M_d}{V_t} \tag{4.14}$$

where M_d = dry mass of a soil volume (dried at 105 °C for > 16 h)

and V_t = total volume (original undried)

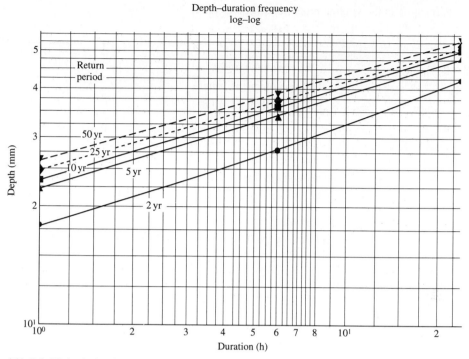

Figure 4.12 Rainfall depth–duration return period.

Typical values of ρ_b are 0.7 kg/m³ for peats to 1.7 kg/m³ for sands or loams. Clays are typically about 1.1 kg/m³.

Particle density   Partical density ρ_m is

$$\rho_m = \frac{M_d}{V_d} \qquad (4.15)$$

where V_d = dry volume (no air, no water)

typical values for ρ_m are 2.65 kg/m³ for most soils

Table 4.11 Depth–duration return period

Return period (years)	Duration (hours)		
	1	6	24
2	18	28	42
5	22.5	34	48
10	23.5	36	50
25	25	37.5	51
50	26.5	39	53

Porosity   Porosity ϕ is the volume proportion of pore space,

$$\phi = \frac{V_a + V_w}{V_s} = 1 - \frac{\rho_b}{\rho_m} \tag{4.16}$$

where
$$V_a = \text{volume of air}$$
$$V_w = \text{volume of water}$$
$$V_s = \text{volume of solids}$$

Values of porosity range from about 35 to 45 per cent for fine sands to 50 to 55 per cent for clays, with peats at about 80 per cent.

Volumetric water content   The water content θ is

$$\theta = \frac{V_w}{V_s} = \frac{M_{wet} - M_{dry}}{\rho_w V_s} \tag{4.17}$$

This is an important soil property and varies from 0 (when dry) to saturation (about 40 per cent for sands) and as we will see it varies over space and time. The most successful methods for determining field soil moisture are the neutron probe, the soil moisture capacitance probe or time domain reflectometry. Details of some of the techniques are found in Shaw (1994).

Degree of saturation   The degree of saturation s is the proporation of water containing pores and is a measure of the 'wetness':

$$s = \frac{V_w}{V_a + V_w} = \frac{\theta}{\phi} \tag{4.18}$$

4.6.2 Soil Horizons

From Tables 4.1 and 4.2, it is seen that groundwater contributes only 0.5 per cent of what rivers contribute to the oceans. Also, these tables show that groundwater contains 30.1 per cent of the earth's freshwater supply whereas rivers and lakes contain only 0.266 per cent. Groundwater, while huge in volume, is also almost static with very slow movement in the horizontal direction. The level of the water table, however, rises and falls vertically, depending on climate and soil type. Soils play a major role in what happens to precipitation, as large volumes of water can be held in the soil matrix or none at all, depending on soil texture, porosity, structure, hydraulic conductivity and existing soil moisture. From Table 4.1, it is seen that the freshwater held as soil moisture is almost ten times greater than the freshwater held in rivers.

Figure 4.13 shows an idealized vertical profile through a series of soil layers. The top layer is usually a vegetation of grass, crop or trees, but it may be bare soil. Below this is the litter layer, more easily identifiable in forest areas as composed of dead leaves, bark and other decomposed growth. Below this is the soil proper and is described in *horizons* or layers. The upper or A horizon in mineral soils is generally friable and rich in humus. This layer corresponds to the surface soil (sometimes called the topsoil). It is that part of the soil in which living matter is most abundant and in which the organic matter is most plentiful. Being closest to the surface, it becomes more leached by rainfall than the lower layers. The middle level or B horizon, often called the subsoil, is mainly composed of well-weathered parent material interwoven with roots and micro-organisms. Lying between the A and C horizons, it has some of the properties of both, with fewer living organisms than A but more than C. By comparison with the A horizon, the B horizon has a high content of iron and aluminium oxides, humus or clay that have been partly leached from the A horizon. The lower C horizon is unconsolidated rock material composed of a wide range of stones of different sizes. Below the C horizon is the parent consolidated rock. The depth of each layer varies from millimetres to metres.

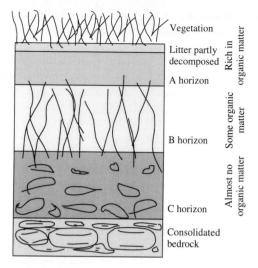

Figure 4.13 Idealized soil section (adapted from Hillel, 1980).

In environmental hydrology, there are two distinct zones above bedrock which may hold and transmit water. They are: the upper unsaturated zone and the lower saturated zone. They are shown in Fig. 4.14. Water movement in the unsaturated zone is more complex than that of the saturated zone. In the latter, the key parameter is hydraulic conductivity or the rate of movement of water. This is readily measured and tends to be reasonably constant. However, in the unsaturated zone the hydraulic conductivity can vary by orders of magnitude within a field, depending primarily on the degree of saturation and the current state of soil suction.

4.6.3 Soil Water Content

Soil moisture is a complex phenomenon well described but poorly quantified. All soils will have a maximum soil moisture magnitude when they are saturated. Similarly, if they are in extreme moisture

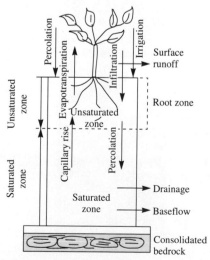

Figure 4.14 Unsaturated/saturated zone (adapted from Bras, 1990).

deficit, their soil moisture will be lowest (not zero). At any point in time, their soil moisture status will vary from close to zero to maximum. It is therefore a dynamic action and responds to the antecedent soil moisture conditions and the current rainfall and solar heat pattern. It is easy enough to quantify, in a vertical profile of a soil, the different levels of soil moisture (i.e. the percentage of moisture content). However, because of continuous activity beneath the surface and above the surface, the fluxes of moisture from one horizon to another are not constant. At times of rainfall, the movement of water in the soil column will be downwards under gravity or upwards towards the water table through capillarity. At times of drought, the direction of water movement will be upward towards the soil surface by capillarity from the groundwater. The fate of rainfall depends largely on:

- Climate zone
- Soil characteristics and
- Soil antecedent moisture conditions

Figure 4.15 is a schematic of water in soil. Within the soil column there are three zones: *aeration*, *capillary* and *groundwater*. The groundwater zone exists below the water table. The capillary zone is the zone through which water will rise through the soil pores by capillary action. The upper zone is the aeration zone where the pores are occupied by air. After rainfall events, the air may be expelled from the pores by hydrostatic pressure to allow the infiltration water to occupy the pores. The soil column is sometimes subdivided into two zones, the unsaturated upper zone and the lower saturated zone. The unsaturated zone is the subject of intense research by hydrologists and studies on hillslope hydrology help to elucidate the physics of unsaturated zone flow.

With respect to soil water, it occupies three different phases in a soil matrix. These are:

- Pore water
- Hygroscopic or adsorbed water
- Absorbed water

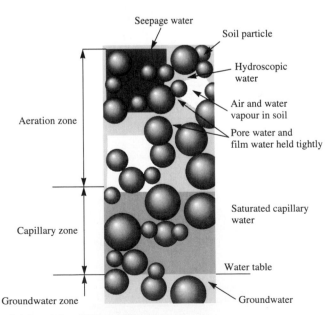

Figure 4.15 Water in the soil (adapted from Weisner, 1970).

The pore water is by far the greatest volume of soil water and is the easiest to expel. The hygroscopic water is adsorbed on to the surface of the grain particle and is held there by surface tension forces. The absorbed water (internal to each grain) requires the removal of the pore water and hygroscopic water before it can be dried out. Sandy soils have large pores and thus can be dried out easily. However, clay particles have small pores (but higher porosity than sand) and small particles with intense hygroscopic activity and require high suction forces to break the hygroscopic surface tension forces.

The phenomenon of soil suction is illustrated by placing a drop of water on to a dry soil particle. The water is quickly drawn into the soil until it is saturated and then a thin film attaches to the perimeter of the soil grains. This hygroscopic film is held with intense surface tension forces. These forces are expressed in bars, i.e. 1 bar is the pressure equivalent to 10.23 m height of water column.

Field capacity and *wilting point* are further soil moisture parameters most often used in agriculture soil studies. After the soil has been saturated and the excess water drained away, the soil is then at field capacity. Vegetation extracts moisture from soil until it cannot do so any more. At this point wilting occurs and the moisture content is known as the wilting point. Figure 4.16 shows a general relationship between soil moisture and soil texture.

4.6.4 Movement of Water in Soil and Hydraulic Conductivity

Water movement occurs in soils under three distinct conditions:

- Saturated flow
- Unsaturated flow and
- Vapour phase flow

All water movement beneath the water table is of the saturated flow type. However, a soil may be temporarily saturated above the water table and this occurs if *all* the pores are filled with water. From a two-dimensional perspective, the movement of water may be vertically downwards or laterally as interflow. The rate of movement depends on hydraulic conductivity of the soil. Unsaturated flow takes place in response to gravity or moisture gradient. Once field capacity exists, capillary action draws the water upwards to the roots and vegetation. After wetting of soils, water flows downwards due to gravity. The mechanism of water movement in unsaturated flow is from pore to pore. Water may exist in the vapour phase in the pores of a soil and be drawn upward to evaporate. The rate of movement depends on the temperature gradient, relative humidity, pore size and pore continuity and the amount of available water. So it is important to conceive of evaporation also from the depths of a soil column.

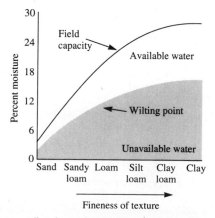

Figure 4.16 General relationship between soil moisture characteristics and soil texture.

Water moves in rivers due to a slope or gradient in its water surface. The steeper the gradient the faster the flow. As lake surfaces have little gradient, water flows slowly. In soils and aquifers, water also moves if it has a gradient, although at several orders of magnitude slower than river flow. This gradient is called the hydraulic gradient. In rivers, water always flows practically horizontally (assuming one-dimensional flow). However, beneath the soil surface, water may flow in either the x, y or z direction. The way water flows in soils is dependent on soil type and its current moisture status. For instance, in summer time, a sandy soil matrix may be dried out, and if rain falls, this rainfall will move vertically down through the soil to help fill the soil pores with water. However, if the soil moisture status is close to field capacity, then the principal direction of movement of water may be in the horizontal direction. This direction is usually along the gradient of the water surface line, which may follow the topographic slope. The rate at which water moves is called its *hydraulic conductivity*. It is easy to evaluate flow behaviour in a saturated porous medium. This is usually the case in aquifers. However, there are times when the soil status is also unsaturated. There may still be water movement in the soil, but it may be restricted due to excessive soil suction.

Darcy's law states:

$$q = -Ki = K\frac{dh}{dz} \qquad (4.19)$$

where
$$q = \text{the Darcy flux, m}^3/\text{m}^2.\text{s}$$
$$i = \text{the hydraulic gradient, } = dh/dz, \text{m/m}$$
$$K = \text{the hydraulic conductivity, m/s}$$

Usually h is the height relative to a datum, but for unsaturated flow the total head is

$$h = \Psi + z \qquad (4.20)$$

where
$$\Psi = \text{the suction head}$$

The suction head, responsible for holding water to the surface of soil particles in unsaturated flow, becomes significant as soil moisture decreases. The variation in the soil column of hydraulic conductivity and soil suction head is shown for different moisture contents in Fig. 4.17. Soil suction or soil tension is measured using tensiometers in the field.

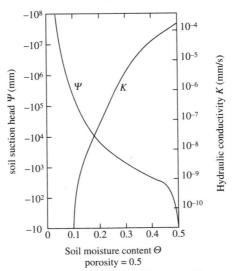

Figure 4.17 Variation of soil suction head ψ and hydraulic conductivity K with moisture content Θ for Yolo light clay (adapted from Raudkivi, 1979).

4.6.5 Soil Moisture Deficit

Soil moisture deficit (SMD) is a term commonly used in agricultural engineering. When the soil moisture is below field capacity, it is said to have a soil moisture deficit. When the soil is saturated it does not have a soil moisture deficit. SMD is a quantifiable parameter and is related to rainfall magnitude, degree of moisture in soil and evapotranspiration. A catchment loses water at rates greater or less than PE (potential evaporation), depending on whether the soil moisture is above or below field capacity. ET (actual evapotranspiration) is less than PE when the vegetation cannot abstract water from the soil. After a rainfall event (if the soil is saturated), it will hold no more water, thus producing runoff. The soil in this case will continue to 'give up' water to the vegetation until a temporary equilibrium stage is arrived at, when ET = PE, i.e. field capacity. At this stage SMD = 0. As the soil dries out, SMD increases and ET decreases. The magnitude of SMD and ET varies. As SMD increases further, ET becomes less and at wilting stage SMD is greatest and ET negligible. It is important to note that SMD is a cumulative number, depending on the previous months' SMD.

Figure 4.18 is an idealized and simplified schematic of the time sequence of soil moisture related to rainfall and PE for an annual cycle in the northern temperate zone. Three vegetation types are sketched: grass, shrubs and trees. Each has a different root depth, depicted as three distinct horizontal layers.

In spring when PE > P (precipitation), the soil enters an SMD, first at the surface layers. As the spring goes into summer SMD penetrates deeper until all the root zones (trees included) are in SMD. In the autumn, P > PE and the upper soil layers become replenished with water, while the lower layers are

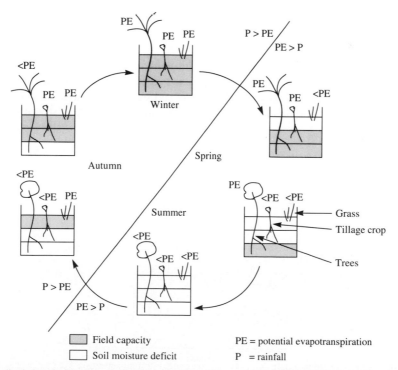

Figure 4.18 Idealized annual moisture cycle for three vegetational types. (Adapted from Bedient and Huber, 1988.)

still in SMD. At this time, the movement of water is vertically downward in the soil column. As autumn moves into winter, the depth of water replenishment deepens until all soil layers are filled with water and there is no SMD at any depth. It can be visualized that in springtime the direction of movement of soil water is downward while the reverse holds in the autumn.

Knowing the soil moisture deficit is important for agriculture and hydrology. During times of high soil moisture deficits catchments tend to be less susceptible to producing flood events. A parameter used in the United Kingdom and Ireland from the *Flood Studies Report* (*FSR*) (NERC, 1975) is the effective mean soil moisture deficit in mm. For instance, some areas in the south-west of Ireland have an EMSMD of 2 mm by comparison with values of 16 mm in East Anglia. The former is susceptible to flooding while the latter is not.

4.6.6 Simple Infiltration Models

Infiltration is the mechanism of water movement into soil under gravity and capillary forces. It was suggested by Horton (1933) that the rate of infiltration of rainfall into soil decreases exponentially with time during a rain event. Some hours into a rain event, the infiltration rate may be close to zero as the soil becomes saturated. The concept of infiltration, as seen by Horton, is shown schematically in Fig. 4.19.

Where $i > f$ at all times, Horton's empirical equation is

$$f = f_c + (f_0 - f_c)e^{-kt} \qquad (4.21)$$

where
$f_0 = $ initial infiltration rate
$f = $ infiltration rate at any time, mm/h
$f_c = $ final infiltration rate
$k = $ empirical constant
$i = $ rainfall rate, mm/h

Often f_c is referred to as infiltration potential. In Horton's equation k is a function of surface texture, where k decreases with increasing vegetation. Also, f_c and f_0 are functions of soil type and cover. Figure 4.20 indicates the variation of f with soil cover, rain intensity and topographic slope. Low rain intensity will have a higher proportion of its rainfall infiltrate than a high-intensity event, as shown in Fig. 4.20(*b*).

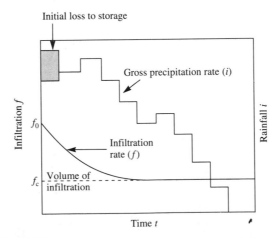

Figure 4.19 Horton's infiltration concept.

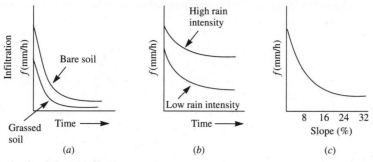

Figure 4.20 Schematic of variation of infiltration capacity.

Example 4.5   Given $f_0 = 100$ mm/h, $k = 0.35/h$ and $f_c = 10$ mm/h, find f at $t = 1$, 2 and 6 h and also F_{total} (cumulative infiltration)

Solution   From the Horton equation,

$$f = f_c + (f_0 - f_c)e^{-kt}$$
$$f = 10 + 90e^{-0.35t}$$
$$\text{at } 1 \text{ h} \rightarrow f = 73.4 \text{ mm/h}$$
$$2 \text{ h} \rightarrow f = 54.6 \text{ mm/h}$$
$$6 \text{ h} \rightarrow f = 21.0 \text{ mm/h}$$

$$\text{Cumulative infiltration after 6 h} = F_{total} = \int f \, dt = 285 \text{ mm}$$

The ϕ *index method* of infiltration is sometimes used. This is the simplest method and is measured by finding the loss difference between total precipitation and surface runoff (measured on the stream hydrograph). The infiltration is assumed uniform over the full duration of the rainfall event. It is shown schematically in Fig. 4.21. When considering rainfall events of, say, less than a day, the computation of gross rainfall and effective rainfall will usually ignore evapotranspiration (ET). Longer term events of greater than about two weeks will take ET into account.

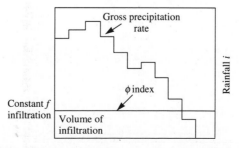

Figure 4.21. The ϕ index infiltration concept.

The reader is referred to Dingman (1994) and Bras (1990) for a more serious mathematical treatment of infiltration.

4.7 EVAPORATION AND EVAPOTRANSPIRATION

Evaporation is the process by which water is returned to the atmosphere, from the liquid or solid state into the vapour state. Transpiration into the atmosphere also occurs through the leafy parts of plants and trees. Because these processes are so interlinked, the 'all in' term used is *evapotranspiration*. In temperate climates, forest land has about twice the evapotranspiration rates of grassland (typically 40 to 70 per cent of total annual rainfall, by comparison with 20 to 40 per cent for grassland, as shown in some British research). This means, of course, that less water infiltrates the soil or becomes part of runoff. About 70 per cent of the mean annual rainfall in the United States is returned to the atmosphere via evapotranspiration, as shown in Table 4.12. In areas of scarce water supplies forest development with higher evapotranspiration losses can reduce the water yield to rivers and lakes.

The global (land + oceans) annual average precipitation of about 1 m is of course equal to the evaporation. As the land surface of the earth evaporates approximately 70 per cent of its rainfall, allowing the remaining 30 per cent to become runoff, then it is clear that on the ocean surface of the earth there is more evaporation than precipitation (Brutsaert, 1982). Figure 4.22 shows the latitudinal distribution of global precipitation and evapotranspiration. Figure 4.23 shows the relationship between evaporation, precipitation and interception for the Amazonian rainforest, after Shuttleworth (1988).

Three types of evaporation/evapotranspiration are:

- Evaporation from a lake surface, E_0
- Actual evapotranspiration, ET
- Potential evapotranspiration, PE

E_0 is the evaporation from a lake or open water body surface. ET is very complex as it includes the evaporation and transpiration from a land surface, vegetated or otherwise. This means that ET for any one surface type will vary depending on its present soil moisture status and is therefore a dynamic parameter. ET will be greater for a soil which is saturated than if it were unsaturated. In an effort to simplify ET the term PE was introduced, which is potential evapotranspiration. This is the evapotranspiration from a soil matrix when the soil moisture is held constant at field capacity. This is achieved by spraying regularly. Meteorological data will normally give values of E_0 and PE but not ET. The latter is usually only determined in catchment research projects, when measurements of the radiation and heat budgets are taken.

Table 4.12 Precipitation–evapotranspiration from continents

Continent	Precipitation (mm/yr)	Evapotranspiration (mm/yr)	Runoff (mm/yr)
Europe	657	375	282
Asia	696	420	276
Africa	695	582	114
Australia	447	420	27
North America	645	403	242
South America	1564	946	618
Antarctica	169	28	141
Total land	746	480	266

Data from Baumgartner and Reichel, 1975

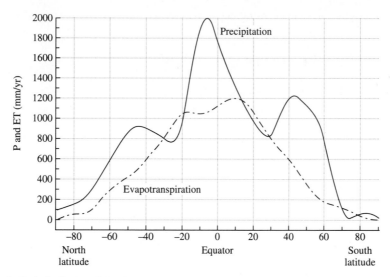

Figure 4.22 Latitudinal distribution of global precipitation and evapotranspiration.

Figure 4.24 shows a comparison of rainfall and potential evaporation at a number of sites in Ireland, averaged over the period 1961 to 1990. The data for this figure is shown in Table 4.13. This is a typical rainfall/potential evaporation plot for the temperate climate. PE exceeds rainfall in summer and as such the soils require artificial watering. In winter, rainfall exceeds PE and this can lead to high runoff, with potential for streamwater pollution from agricultural activities such as slurry spreading.

Two of the factors causing evaporation from any surface are:

- The availability of a supply of heat energy to provide the latent heat of vaporization
- The availability of a transport process to move the water vapour away from the surface, i.e. wind

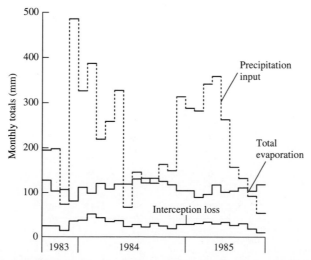

Figure 4.23 Monthly values for precipitation, total evaporation and the interception component for the 25-month period in the Amazonian rainforest (adapted from Shuttleworth, 1988. Reprinted by permission of The Royal Society).

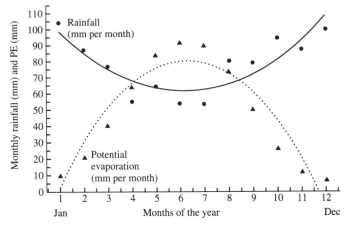

Figure 4.24 Monthly rainfall and potential evaporation trends, Ireland (prepared from Table 4.13).

Solar radiation provides the heat source while wind, along with a vertical humidity gradient, supplies the transport source. These are shown schematically in Fig. 4.25.

Evaporation from a lake surface depends on:

- Available energy as heat
- Solar radiation and more specifically net radiation
- The temperature of the air and water surface
- The wind speed
- Saturation vapour deficit $(e_0 - e_z)$

Table 4.13 Rainfall and potential evaporation at four sites in Ireland from 1961 to 1990

Month	South Coast* Rain	South Coast* PE	East Coast† Rain	East Coast† PE	Midland‡ Rain	Midland‡ PE	West§ Rain	West§ PE
January	104	10	69	9	93	2	121	3
February	87	21	50	21	66	14	83	13
March	77	40	54	39	72	30	96	28
April	55	64	51	61	59	53	62	49
May	64	84	55	83	72	74	78	69
June	54	92	56	94	66	82	71	75
July	53	90	50	91	62	78	64	68
August	80	74	71	73	81	61	97	54
September	79	50	67	50	86	39	104	33
October	95	26	70	25	94	16	124	14
November	88	12	65	10	88	2	118	2
December	100	7	76	5	94	1	124	1
Year average	935	570	732	561	934	446	1143	408

* Cork—Roches Point on South Coast.
† Dublin—Dublin Airport, 3 km from East Coast.
‡ Mullingar—Central Ireland.
§ Claremorris—West of Ireland.
Data from Irish Meteorological Office, 1993

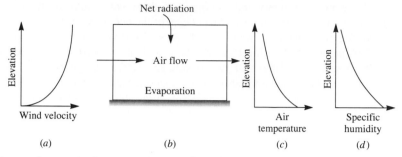

Figure 4.25 Concept of evaporation from an open water surface (adapted from Chow *et al.*, 1988).

Figure 4.25 shows the inputs and outputs to a control volume of 'evaporating air' and those natural processes—radiation, temperature and wind speed—that effect evaporation. The reader is referred to Chow *et al.* (1988), Bras (1990) and Brutsaert (1982) for further details.

At the earth's surface, evaporation is the connecting link between the water budget and the energy budget (Brutsaert, 1982). The most simplified energy budget is represented by

$$R_n = LE + H + G \qquad (4.5)$$

where
R_n = specific flux of incoming radiation, k cal/m^2 y or W/m^2

L = latent heat of evaporation, J/m^3

E = rate of evaporation, m/y

H = specific flux of sensible heat into the atmosphere, k cal/m^2 y or

W/m^2 (i.e. the energy used in heating the 'air')

G = specific flux of heat into the earth, W/m^2

Figure 4.26 shows the diurnal variation of the energy budget over an irrigated wet bare soil in Davis, California, in August 1993. The peak net radiation (post-midday) is about 630 W/m^2. The energy used up in evaporation, *LE*, peaks at about 400 W/m^2. The sensible heat, *H*, is then only less than 100 W/m^2. This is to be expected for clear skies over a wet soil. For instance, over dry desert conditions, we might expect about 10 to 30 W/m^2 for *LE* and 300 to 400 W/m^2 for *H*. It is important to realize that evaporation

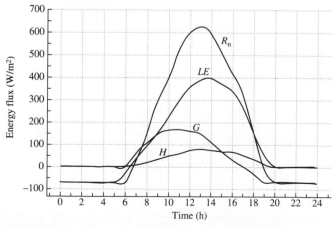

Figure 4.26 Energy balance 23 August 1993, Davis, California, on a wet irrigated bare soil.

can only occur if there is moisture to evaporate. In Fig. 4.26 there was significant water available in the top layers of the soil to produce evaporation with 70 per cent of the net radiation going to evaporation use.

On a global scale, on an annual basis where $G \to 0$, $R_n \approx 70\text{--}80 \text{ kcal/cm}^2$ y. Over land, $LE \approx H$ and over the oceans $LE \approx 90$ per cent of R_n. This suggests that on a global scale $LE \approx 80$ per cent of R_n. While significant spatial and temporal variations of the constituents of the energy budget occur, the above figures emphasize the overwhelming importance of the evapotranspiration process to the overall heat budget and also to the water budget (Brutsaert, 1982). Typically, the annual cyclic behaviour of evaporation parallels that of the cycle of solar radiation and daily air temperatures for land surfaces and shallow water bodies. However, deep waters show peaks in the fall of the year by comparison with peaks in the summer for shallow lakes. Also deep water bodies show minimum evaporation in the spring and while shallow water bodies show minimum in winter (like land). The daily cycle of evaporation follows the cycle of temperature over land and also over water.

4.7.1 Mass Transfer Method of Determining E_0

E_0 is the evaporation from an open water body as distinct from a moist soil surface. This method, sometimes called the vapour flux method, calculates the upward flux of water vapour from the evaporating surface. The equation conceived by Dalton, an English chemist, in the nineteenth century, was

$$E_0 = f(u)(e_s - e_a) \tag{4.22}$$

where
$E_0 = $ the evaporation from the water body

$e_a = $ the vapour pressure in the air

$e_s = $ saturation vapour pressure at the temperature of the water surface

$f(u) = $ a function of wind speed

$f(u) = a(b + u)$ for Europe

or
$f(u) = Nu$ for the United States and Australia

Modifications to this equation are

$$E_0 = (A + B\bar{u})(e_s - e_a) \tag{4.23}$$

where
$A = $ an empirical constant of 0.0702

and
$B = $ an empirical constant of 0.003 19 (not the Bowen ratio)

and
$\bar{u} = $ wind speed at 7.5 m above ground

Still another equation is

$$E_0 = N\bar{u}_2 (e_s - e_a) \tag{4.24}$$

where
$N = 0.11$

and
$\bar{u}_2 = $ wind speed at 2m above ground

A variation on N is where

$$N = 0.291 \, A^{-0.05}$$

where
$A = $ area, m^2

Example 4.6   Compute the evaporation as an annual water loss from a lake which has an area $\sim 20 \text{ km}^2$ with u_2 at 3 m/s. Assume e_s and e_a are 15 and 10 mm Hg respectively.

Solution

$$E_0 = N\bar{u}_2(e_s - e_a)$$

$$N = 0.11$$

$$\bar{u}_2 = 3 \, \text{m/s}$$

$$e_s = 15 \, \text{mm Hg} = \frac{15}{760} \, \text{bar} = 19.7 \, \text{mbar}$$

$$e_a = 10 \, \text{mm Hg} = \frac{10}{760} \, \text{bar} = 13.2 \, \text{mbar}$$

Therefore

$$E_0 = 0.11 \times 3 \times (19.7 - 13.2) = 2.2 \, \text{mm/day}$$

4.7.2 Energy Budget Method to Determine E_0

The simplified energy budget equation is

$$R_n = LE + H + G \tag{4.5}$$

$$LE = R_n - H - G \tag{4.25}$$

This equation assumes no water advected input of energy and no change in storage of energy. It also assumes a finite time period:

$$LE = \rho_w \lambda_v E \tag{4.26}$$

where
LE = latent heat flux

ρ_w = water density

λ_v = latent heat of vaporization $(2.47 \times 10^6 \, \text{J/kg})$

$\quad = 597 - 0.564 \, T, \;\; \text{cal/g with } T \text{ in } °C$

$$E = \frac{R_n - H - G}{L} = \frac{R_n - H - G}{\rho_w \lambda_v} \tag{4.27}$$

Introducing the Bowen ratio gives a ratio of the sensible heat flux to evaporating flux as

$$B = \frac{H}{LE}$$

Therefore

$$H = B(LE) = B\rho_w \lambda_v E$$

and

$$E = \frac{R_n - G}{\rho_w \lambda_v (1 + B)} \tag{4.28}$$

Over land surfaces, $B \cong 1$, with sensible heat about similar to evaporative flux. Over the ocean surfaces, $B \cong 0.1$ as evaporation is much more significant. Further details are found in Dingman (1994), Bras (1990) and Brutsaert (1982). Also refer back to Table 4.5.

Example 4.7 Determine the evaporation from a lake with the following data:

$$R_n - G = 70 \, \text{W/m}^2 = LE + H$$

$$B = 0.4$$

Lake temperature
$$T \cong 20 \, °C$$

Solution

$$E = \frac{R_n - G}{\rho_w \lambda_v (1 + B)}$$

$$\lambda_v = 597 - 0.564 \times 20 = 586 \text{ cal/g}$$

$$\begin{aligned}
\rho_w \lambda_v &= 10^6 \text{ g/m}^3 \times 586 \text{ cal/g} \\
&= 586 \times 10^6 \text{ cal/m}^3 \\
&= 586 \times 10^6 \times 4.2 \text{ J/m}^3 \\
&= 586 \times 10^6 \times 4.2 \text{ W.s/m}^3 \\
&= 28.5 \times 10^3 \text{ W.d/m}^3
\end{aligned}$$

$$\begin{aligned}
E &= \frac{70 \text{ W/m}^2}{28.5 \times 10^3 \text{ W.d/m}^3} \\
&= 1.76 \text{ mm/day}
\end{aligned}$$

This method can also be used for evaporation from land surfaces since $B = 0.4$, this implies that the evaporation flux $\gg$ sensible heat flux.

4.7.3 Water Balance Method of Determining ET or E_0

Lysimeters are used at field scale to determine the 'point' measurement of evaporation. A lysimeter is a 'pan', typically a metre to several metres in diameter and up to a metre in depth. It contains as near as possible 'undisturbed' soil from the local site. It sits into the ground with its top surface flush with the adjacent ground surfaces. It sits on time-calibrated weighing scales and responds to an increase or decrease of the moisture content of the soil in the pan. The loss in 'weight' represents the loss to evaporation and so evaporation is determined. This is a reliable method of determining field evaporation. On a regional scale, E is synonymous with ET and the water balance equation is used to determine E, if the other parameters are known:

$$E = P - R - \Delta G - \Delta S$$

where
$$P = \text{precipitation}$$
$$R = \text{stream runoff}$$
$$\Delta G = \text{change in groundwater storage}$$
$$\Delta S = \text{change in soil water storage.}$$

Over long periods (e.g. a year), we have

$$E = P - R$$

so if P and R are measured, estimates of E are attained. The reader is referred back to Tables 4.3 and 4.4 for values of E from different continents and different land use surfaces respectively

4.7.4 Determination of Potential Evapotranspiration, PE

Evaporation (from a water body $= E_0$), potential evapotranspiration (PE) and actual evapotranspiration (E) have already been defined. The previous sections dealt only with the determination of E_0 and it is seen how complex a phenomenon evaporation from a water body is. Actual evapotranspiration is even more complex as the effects of vegetation, and associated soil physics, are to be accounted for. This complexity is reduced somewhat by simplification to potential evapotranspiration. This considers that the soil matrix is continuously moist (at field capacity) and the evaporation from the vegetated surface is close to maximum. PE from a grassland surface is approximately equal to evaporation from a large water body. Thus methods used to determine E_0 are also used to calculate PE. Actual evapotranspiration is a dynamic parameter varying with the season, but is especially dependent on the soil moisture status. ET may exceed PE in vegetation with a high surface leaf area and high ambient temperatures. It was also mentioned that the ET of coniferous forests in the United Kingdom were approximately twice that of grassland or moorland at similar elevations. Penman (1948) inferred that:

$$PE = f E_0 \tag{4.29}$$

where f is an empirical constant from British data which varied with season for vegetated land surface as follows:

$$\text{November, December, January, February} \rightarrow f \simeq 0.6$$
$$\text{March, April, September, October} \rightarrow f \simeq 0.7$$
$$\text{May, June, July, August} \rightarrow f \simeq 0.8$$

So typically, PE is about 70 per cent of lake evaporation. Since actual evaporation is $\leqslant$ PE, we can generalize and say that actual evaporation $\leqslant$ ET $\leqslant 0.7 E_0$.

4.8 RAINFALL–RUNOFF RELATIONSHIPS

When rainfall occurs on the land surface it may follow different routes depending on topography and soil conditions and soil moisture. If there are surface depressions they are apt to be be filled up early in a storm event. Whether the rainfall converts to surface runoff or infiltration depends on principally two factors:

- Land slope
- Infiltration capacity

On steeply sloping sites, surface runoff is more likely to occur, while infiltration lags behind. On sites more remote from rivers and streams and where the land gradient is not steep, infiltration may be the primary mechanism and surface runoff lags behind. Generally, infiltration is seen as the controlling factor in the availability of rain for runoff. Surface runoff is also called overland flow. Many hydraulic equations exist to help quantify this runoff and some include the 'all-in' friction coefficient of Mannings, n. For instance, understanding the behaviour of overland flow is fundamental to predicting the volume of water arriving at outlets for storm sewer design. The parameter sometimes used for time is that of the time of concentration (T_c), i.e. the time for water falling on the remotest part of the catchment to arrive at the stream outlet. If the storm persists after T_c, it is assumed that all of the catchment is contributing runoff to the stream or sewer system. This concept is used in the design of sewage networks and small drainage networks (see example in Chapter 12). It is not considered comprehensive for large rural catchments, with a variety of land uses. Other concepts used include the classic concept of Hortonian overland flow and also the concept of subsurface interflow responding to infiltration status.

Figure 4.27 illustrates an idealized and simplified response of a catchment to different levels of infiltration. The storm is assumed to have uniform rainfall intensity i_r and duration t_r. The minimum infiltration rate for the soil is A_i and the time it takes to saturate will vary depending on the intensity. There is no runoff (R = 0) if:

1. The storm duration t_r is less than that required to saturate the soil surface.
2. The storm intensity i_r is less than the minimum infiltration rate A_i.

Case 1, shown in Fig. 4.27(a), is when the storm duration t_r is less than t_0, the minimum time required to saturate the soil for the rainfall rate, i_r. There is no runoff and the soil moisture deficit is decreased. The rainfall rate, i_r, is greater than the rate A_i. Case 2, shown in Fig. 4.27(b), is when storm intensity i_r is less

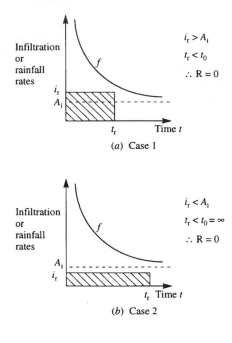

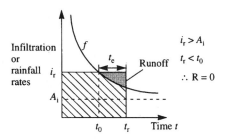

Figure 4.27 Storm characteristics versus infiltration capacity (adapted from Bras, 1990).

than the minimum intensity A_i used to saturate the soil. There is no surface runoff. Case 3, shown in Figure 4.27(c), is when the storm intensity and storm duration are greater than A_i and t_0 respectively. There is then surface runoff i.e. $R > 0$. At the beginning of this storm, there is no surface runoff, as all rainfall initially goes to soil infiltration and the soil moisture is continuously increasing until eventually there is no moisture deficit. Surface runoff occurs after the storm intensity i_r equals the infiltration rate f, and this occurs at t_0. This time period has been defined as the time it takes to saturate depression storage or ponding. Ponding time is dependent on soil type but more specifically on antecedent soil moisture status. The reader is referred to Bras (1990) for further details.

Two types of surface runoff mechanisms are the Horton mechanism and the Dunne mechanism, best explained in Bras (1990). The classical Horton mechanism is described with reference to Fig. 4.28(a) as follows. Prior to a rainfall event, the vertical soil moisture profile is indicated by the curve t^0. Assume the rainfall event has a precipitation rate (P) greater than the saturated hydraulic conductivity (Ksat). As the rain event proceeds, the vertical soil moisture profile goes from t^0, to t^1, to t^2, to t^3. At t^3, the surface is saturated. At this time, the infiltration rate drops below the rainfall rate, and overland flow begins. This is known as the ponding time. Necessary conditions for the Horton mechanism (Freeze, 1980) are:

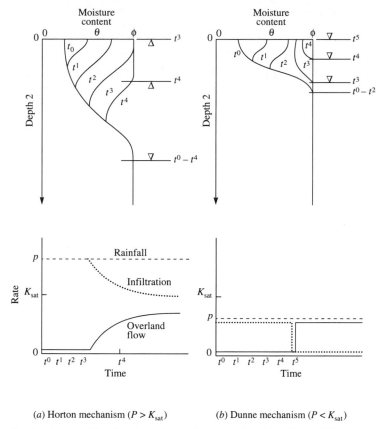

Figure 4.28 Vertical soil moisture profiles changing during a rain event (adapted from Bras, 1990).

- A rainfall rate greater than the saturated hydraulic conductivity
- A rainfall duration longer than the ponding time

The Dunne mechanism is explained with reference to Fig. 4.28(b). It is assumed that the rainfall rate (P) is less than the saturated hydraulic conductivity (Ksat). The vertical soil moisture profile prior to the rain event is indicated as t^0. As the rain event proceeds, the vertical soil moisture profile goes from t^0, to t^1, to t^2, to t^3, to t^4, to t^5. At t^5, the water table has risen to the water surface, causing surface saturation and ponding, followed by overland flow (Bras, 1990; Freeze, 1980).

The response of a catchment to rainfall will depend on the catchment topography and the distance away from streams. Generally it is found that the Horton flow mechanism occurs on upslope areas remote from the stream, while the Dunne mechanism is more likely adjacent to the streams. It has also been found that it is most unlikely that 100 per cent of a catchment will contribute surface flow during a storm event. What does occur, however, is that partial areas of a catchment contribute and this varies with soil and topographic factors. Where surface hydraulic conductivity is lowest, Horton overland flow is generated from partial areas of upslope lands. Where the water table levels are closest to the surface, Dunne overland flow is generated from partial areas of upslope lands.

Many studies have concluded that only a fraction of a catchment area contributes to runoff. Catchment studies by Betson (1964) concluded that, on average, about 22 per cent of the land surface contributes to runoff. The reality of course is that catchment areas will contribute to runoff to varying extents. The concept of variable source runoff holds that runoff is generated directly from precipitation on to areas that are saturated with a rising water table.

Runoff thus produced has two components:

1. Precipitation that cannot penetrate the soil surface becomes direct runoff.
2. Subsurface water, on rising to the surface, becomes surface runoff.

The latter component is sometimes called return flow and is one of the mechanisms by which subsurface water becomes streamflow. Subsurface water can also follow the subsurface route and eventually becomes streamflow, but this mechanism is very much delayed as the subsurface water velocity (hydraulic conductivity) is maybe five to ten orders of magnitude slower than surface overland flow. This subject is treated in detail by Bras (1990).

Bishop (1991) conceptualized streamflow as composed of surface runoff and subsurface runoff. He treated surface runoff as 'new water', e.g. from a current storm. He treated subsurface runoff as 'old water' or pre-event water (e.g. from a storm some weeks before). His interest was in the correlation between stream discharge and stream acidity. He concluded from studies in Scotland and Sweden that 'new water' was much less acidic than 'pre-event' water and that 'pre-event' water contributes little to the streamflow hydrograph.

Water from the unsaturated zone may, in certain catchments, be the principal source of baseflow. It has also been identified in studies that there is a strip along the sides of the stream channels whose width varies with rainfall and which is perennially saturated. This strip produces subsurface flow to form the flood peak of the flow hydrograph.

4.8.1 Concepts of Rainfall versus Runoff

Figure 4.29 by Bishop (1991) shows eight sketches of varying concepts in rainfall versus runoff. Sketch (*a*) is the traditional hydrological approach where rainfall is considered to split only into two parts—overland flow and infiltration. Overland flow is assumed uniform over the catchment. The infiltration is also assumed uniform. This concept is called the *infiltration excess overland flow model*. More recently, instead of considering overland flow as being uniform, some studies have indicated that only parts of the catchment contribute to overland flow. This concept is called the *partial contributing area model* and is shown in sketch (*b*).

Another concept, shown in sketch (c), is that of *saturation excess overland flow*, where overland flow occurs only after an area has saturated subsoil conditions and exfiltrates, causing overland flow. A still more recent concept of *macropore flow* is where the subsurface runoff occurs in the unsaturated zone through subsurface macropores with or without surface flows. The concept of *subsurface streamflow* is shown in sketch (*e*), where both the subsoil of the unsaturated zone and that below the water table contribute interflow to the stream. *Saturated wedge flow* is shown in sketch (*f*) and *groundwater ridging* in sketch (*g*). The recent concept of *transmissivity feedback* is sketched in (*h*). Here the bulk of the runoff is kept within the soil matrix by the postulated feedback relationship between a rising water table and hillslope transmissivity, which breaks the rate of water table rise.

4.8.2 The Hydrograph

The hydrograph is a plot of stream discharge versus time. Hydrographs from different storm types are shown in Fig. 4.30. The more classical shape hydrograph is shown in Fig. 4.30(*d*). Baseflow is that component of the flow which is supplied by groundwater. Interflow is that supplied by subsurface flow for the unsaturated soil matrix. The shape of the hydrograph depends on overland flow, subsurface flow, groundwater flow, land slope, stream slope, land and channel roughness and rainfall pattern, intensity and duration. For Fig. 4.30 it is assumed that the flow recorder is at the downstream end of the catchment in all cases. Figure 4.30(*a*) is a hydrograph that may result from a rainstorm at the upper end of a watershed, producing increased baseflow and some surface runoff. Figure 4.30(*b*) is the response to a storm at the

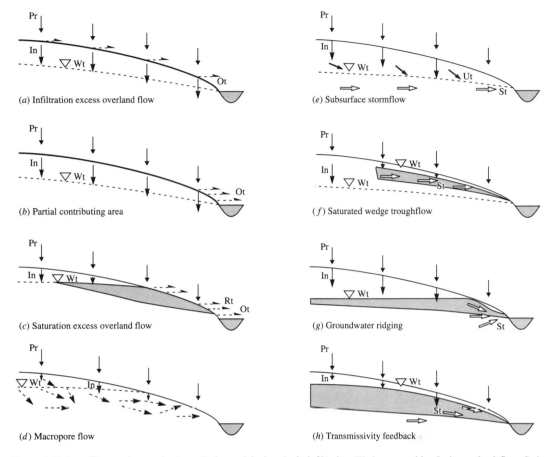

Figure 4.29 Runoff generation mechanisms: Pr is precipitation, In is infiltration, Wt is water table, Ot is overland flow, St is saturated lateral flow through the soil matrix and Ut is unsaturated lateral flow. The shading indicates areas saturated transiently during runoff events where pre-event water may play an active role in runoff generation (after Bishop, 1991. Reprinted by permission).

downstream end of the watershed where the river response is quick. Figure 4.30(*c*) is a case of a surface runoff hydrograph with rain throughout the catchment. Elements of interest for the hydrograph include the peak magnitude, the time to peak, the hydrograph duration and the separation of baseflow, interflow, and surface runoff.

In Fig. 4.30(*a*), the catchment shape focuses to a point at the stream outfall and the rainstorm covers the upland half of the catchment. The hydrograph shape shows a delayed response, by comparison with Fig. 4.30(*b*). Comparing (*c*) and (*d*), it is seen that the more wide-bodied upland catchment (*c*) generates an attenuated and delayed hydrograph peak. See Shaw (1994) for details of hydrograph construction and methods of separating direct flow from baseflow.

4.8.3 The Unit Hydrograph

The unit hydrograph is defined as the basin outflow resulting from 10 mm of direct runoff generated uniformly over the drainage area at a uniform rainfall rate for a unit of time', e.g. 1 h or 1 day. The T hour

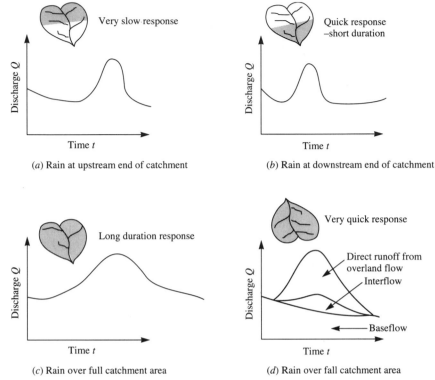

Figure 4.30 Interaction between basin shape and storm coverage in producing the hydrograph (adapted from Veissman *et al.*, 1977).

unit hydrograph is defined as resulting from a unit depth of effective rainfall falling in T hours over the full catchment. The three assumptions in this theory are:

1. The river flow rate Q is proportional to effective rainfall. In Fig. 4.31(*a*) (both rain events of equal duration) it is seen that increasing the rainfall intensity from i to ni produces a hydrograph with twice the peak flow magnitude.
2. The effective rainfall/runoff relationship does not vary with time. In assumption 1 above it is seen from Fig. 4.31 that both hydrographs (Fig. 4.31(*a*)) have the same duration, because both rainfall events were of the same duration.
3. The principle of superposition applies. From Fig. 4.31(*b*), the first rainfall event produces the hydrograph number 1. A second consequent rainfall event produces the hydrograph numbered 2 and a third rainfall event produces hydrograph numbered 3. The cumulative effect of the three rain events is to produce a hydrograph which is the addition of the three individual hydrographs (i.e. 1, 2, and 3).

Dooge (1973) summarized that the unit hydrograph theory modelled hydrologic systems as linear and time-invariant. The assumptions of the unit hydrograph theory simplify hydrologic understanding, even though none of the assumptions are likely to be strictly correct. For engineering applications, these assumptions are widely used. For example, if we have a catchment and a record of the hydrographs due to ten different rainstorms, we can for each rainstorm produce ten unit hydrographs corresponding to unitized rainstorms (i.e. normalize all hydrographs as if all storms were unit storms of, say, 10 mm of rain). We can then take an average of those ten unit hydrographs and call this the unit hydrograph for the catchment, due to a 10 mm rainstorm. So when there is another storm, of, say, 50 mm, we can compute the new hydrograph from the unit hydrograph.

4.8.4 The Rational Method

Mulvaney (1851), Kuichling (1889) and Lloyd-Davis (1906) are credited with outlining this empirical method of computing runoff, according to:

$$Q_p = 0.278 \, CIA \qquad \text{m}^3/\text{s} \tag{4.30}$$

$$Q_p = \text{peak river flow, m}^3/\text{s}$$

where
C = a runoff or impermeability coefficient varying
from 0.05 for flat sandy soils to 0.95 for impervious urban typography

I = uniform rainfall intensity in mm/h over the time of concentration T_c

A = the catchment or contributing area, km^2

This is a simplistic method and should not be used other than to obtain a preliminary rough estimate for Q_p. The runoff coefficient C is dependent on the full range of catchment response parameters including antecedent conditions, soil moisture status, soil type, soil porosity, land slope, land use, depth of water table, etc., and also on the storm parameters including duration, spatial distribution, etc. This formula may be used with care in small catchments for low-return period storms. For return periods greater than about five years and catchment areas greater than 10 km^2 the rational method estimates Q_p higher than the unit hydrograph method or the FSR (Flood Studies Report 1975) catchment characteristics method (see Sec. 4.10.1).

The rational method has been adopted in urban sewer design by Hydraulics Research, Wallingford (1983). The Wallingford method uses the modified rational method as follows:

$$Q_p = 0.278 \, C_v C_R IA \tag{4.31}$$

where
C_v = a volumetric runoff coefficient varying from 0.6 for rapidly draining
soils to 0.9 for heavy soils and urban areas as being approximately 0.75

C_R = a routing coefficient, usually 1.3

I = rainfall intensity, mm/h

A = catchment area km^2

Urban runoff is treated in Sec. 4.12 where the modified rational method is still used. It is also still usable for foul/storm sewer design as given in an example in Chapter 12.

4.8.5 Catchment Modelling

Catchment modelling varies from the rational method type model to the fully integrated component, deterministic, physically based, continuous models of SHE (Abbott *et al.*, 1986) or HSPF (1984). One of the more detailed models is HSPF (USEPA, 1984) which has been in development since the 1960s. It had its origin in Harvard, Stanford and the US Corps of Engineers. This is a one-dimensional integrated hydrological–hydrochemical model, ideal for identifying best-management practices for water quality management of river basins. HSPF is a physically based component model, i.e. each component of the hydrological cycle is described mathematically by means of appropriate empirical or exact governing equations. All of the components are then linked. The components include the different soil layers as described by the parameters: upper zone storage, lower zone storage, lower zone evapotranspiration, etc. This model and the SHE model are complex and the misinterpretation of a single parameter is sufficient to make results meaningless. Both models are also *continuous*, i.e. they require data as time series input (e.g. 15 minute rainfall) increments and translate into time series output (e.g. full hydrographs, rather than just peak flows). The flow sequence in the SHE model is shown schematically in Fig. 4.32.

Vieira *et al.* (1994) discuss the model of MIKESHE. MIKE is a hydrodynamics model for rivers, estuaries, channels, etc., and SHE is the catchment model. Both MIKE and SHE are linked for optimum

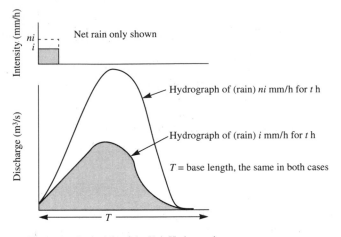

Proportional principle of the Unit Hydrograph

(*a*)

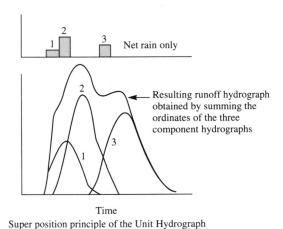

Super position principle of the Unit Hydrograph

(*b*)

Figure 4.31 Unit hydrograph assumptions.

catchment modelling and planning. Vieira *et al.* (1994) itemize that it can be used for river basin planning, water supply, irrigation and drainage planning, management and real-time control, contaminant assessment from waste disposal sites, assessment of impacts of changes to land uses and farming practices, soil and water management, studies of effects of climate change and ecological evaluations.

More simple models in use are the *lumped* parameter models. These transform actual rainfall into runoff output by conceptualizing that all the catchment processes occur at one spatial point. Not all physical parameters may be included in such a scheme. The unit hydrograph approach, which can use a single rainstorm magnitude or a hyetograph, combines with singular catchment parameters, like catchment area, stream slope, antecedent rainfall magnitude and soil type, to produce an outflow hydrograph.

In such lumped models, the end parameter of interest is that of peak flow, Q_p. The hydrograph shape or duration in such models are not usually required. As such, the lumped models give an estimate of peak flows in, say, flood examples.

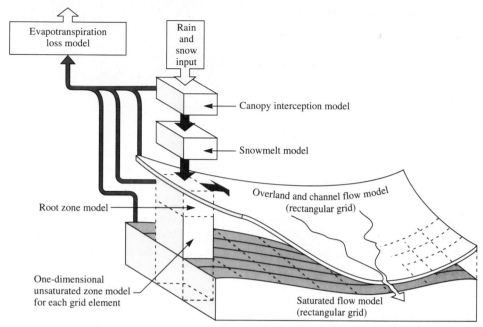

Figure 4.32 Schematic representation of the structure of the European hydrology model (SHE), the hydrological system (adapted from Abbot *et al.*, 1986. Reprinted by permission of Elsevier Science).

The relationship of water to soil is most important in catchment modelling. Some of the significant parameters include upper zone storage (ponding), lower zone storage (unsaturated zone) and lower zone evapotranspiration. Minor changes in any of these three parameters may cause significant changes to the output hydrodynamics. For instance, summer water levels in lakes and rivers may be very dependent on groundwater outflow. If this is not understood, lake levels in summer periods may not be correctly modelled. Additionally, because of the strong correlation of water quality with discharge, if a coupled hydrodynamic and water quality model like HSPF or SHE is used, then it is important that the hydrodynamics are modelled as accurately as possible. Sediment levels and nutrient loads are significantly dependent on accurate outflow hydrographs. More details of modelling are described in Chapter 21.

4.9 HYDROLOGIC INSTRUMENTATION

This chapter on hydrological concepts cannot detail to any extent (because of space limitations) the wide range of instrumentation being used in hydrology. Readers are referred to current manufacturers and other textbooks on hydrology, e.g. Rosenberg *et al.* (1983), Shaw (1994), Bras (1990), Dingman (1994), Bediènt and Huber (1988), and Chow *et al.*, (1988). Table 4.14 summarizes some pieces of instrumentation.

Traditionally, instrumentation for determining hydrological parameters, (e.g. rainfall, streamflow, evaporation etc.) was *in-situ*. Typically, raingauges (or sheets of canvas/plastic on forest floors) determined rainfall, an evaporation pot was used to estimate evaporation and water level recorders were used to determine streamflow. Also the meteorology/atmosphere parameters (e.g. air temperature, energy, humidity, etc.) were also ground-based or *in-situ*. This tradition has been used not only over land areas but also over seas and oceans.

Table 4.14 Hydrological instrumentation[1]

Purpose	Instrumentation	Parameters
Precipitation	Raingauge	—rainfall intensity
	Radar	—rainfall duration
Snowfall	Snowgauge	—snow depth
Streamflow	V notches, weirs	—small streams
	Water level recorders	—others
Energy/meteorology/atmosphere	Sonic anemometers	—wind speed
	Infra-red thermisters	—soil surface temperature
	Cup anemometers	—wind speed
	Wind vanes	—wind direction
	Hygrometers	—humidity
	Thermocouples	—air temperature
	Dew point hygrometers	—temperature/humidity
	Soil heat flux plate	—radiant energy to soil
Evaporation	Lysimeter	—evaporation of soil surface
	Evaporation pan surface	—evaporation from water surface
Soil moisture	Neutron probe	—soil moisture over a volume
	Capacitance probe	—soil moisture over a volume
	Time domain reflectometry	—soil moisture
Hydraulic conductivity	Time domain reflectometry	—hydraulic conductivity
Groundwater	Wells	—water level rise/fall

[1]Satellite and remote sensing are now used to determine each and every parameter in this table.

4.9.1 Remote Sensing in Hydrology

However as we enter the twenty-first century, most if not all of the hydrological parameters will be determined remotely by satellite or radar. Remote sensing is that field of hydrology/meteorology/ atmospheric science/earth science etc., which determines the requisite parameters, from measurements, not of the parameter itself, but by the way it changes the electromagnetic spectrum from its known state. Photography in the visible wavelengths is probably the easiest form of remote sensing. Today techniques are capable of making measurements over the entire electromagnetic spectrum. Different sensors provide unique information about the properties below the earth's surface, at the earth's surface (ground temperature, air temperature, humidity, rainfall, etc.) and in the several regions of the atmosphere (e.g. clouds). Measurements of the reflected solar radiation gives information on albedo, thermal sensors measure surface temperature, (land or sea) and microwave sensors measure the dielectric properties of surface soil or snow. The challenge for the remote sensing specialist and water resources scientist is to interpret these remotely sensed properties in a way that can be used for effective management and monitoring (Engman and Gurney, 1991).

The unique aspect of remote sensing in hydrology is primarily its ability to measure spatial information as opposed to point data from which most of the hydrologic concepts and models have been developed. Additionally, the ability to measure state variables (e.g., soil moisture, surface temperature, etc.) over a catchment area is 'almost' possible with remote sensing. Another fact of remote sensing is limited to satellite sensors and this is the potential ability to assemble long term temporal data sets (e.g., over decades), Engman and Gurney, (1991).

4.10 FLOOD FLOWS

Determination of flood flows in river and lake systems after rainfall events has always been of interest to hydrologists. Additionally, hydrologists would like to be able to predict river flows for rainfall events that have not yet occurred. The term *return period* or *recurrence interval* is used in flood planning or even the hydraulic design of a bridge to identify the significance of a storm or flood event. A flood with a return period of five years has 98 per cent probability of occurring once within that five years or a 20 per cent probability of occurring in only one year. It is important to note that the return period has nothing to do with the time sequence of an event. It merely states that, say, the five year flood has a 98 per cent probability of occurring within a five year time span. It may occur *any time* during that five years. The five year flood does not occur like clockwork every five years. We may in fact get the 100 year flood within the next five years. Designs of sewer networks are often based on a return period of 2 to 5 years. However, if this is in an urban renewal project with basements, the return period may be 50 to 100 years. Typically, river structures are designed for return periods of 30 to 50 years. The higher the return period, the higher the flood intensity. Higher return periods are reserved for structures potentially liable to cause catastrophe. Dams might be designed for between 1000 and 100 000 years. In such cases, the PMF, or probable maximum flood, is of interest. A typical relationship of flow and return period is shown in Fig. 4.33. It is seen that the relationship is approaching linearity below five years. Beyond this, it is non-linear and the difference between the 500 and 1000 year flows are approximately 15 per cent. For Ireland, the computation used for determining Fig. 4.33 (from the *Flood Studies Report, UK*, 1975) is

$$\frac{Q_t}{\bar{Q}} = -3.33 + 4.2e^{-0.05y} \tag{4.32}$$

where

$$y = \ln[-\ln(1 - 1/t)] \tag{4.33}$$

and

$\bar{Q} =$ the mean annual flood

$Q_t =$ the flood magnitude with a return period t years.

Equations (4.32) and (4.33) are empirical equations from the *FSR* (NERC, 1975). The latter document is a compilation of all available rainfall and river flow data in Ireland and the United Kingdom up to 1970

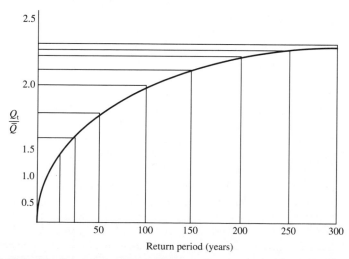

Figure 4.33 Typical relationship of river flow and return period.

(some rainfall and river flow records date from the early 1800s) which was analysed and empirical regression equations produced for flood flow determination. The *FSR* document produced what is widely known as the method of catchment characteristics for flood flow prediction. The methods are not applicable outside the UK/Ireland.

The determination of return periods for storms and runoff is usually based on time series historical data. For instance, to determine the return period for flows, the maximum flows in each year for the history of the record are assembled. These are then ranked from highest to lowest. Statistical techniques, including least squares estimates, moments estimates or maximum likely estimates, are used then to determine the return period and associated flows.

A sample of different methods used to determine flood flows are itemized in Fig. 4.34 for the ungauged and gauged catchments. A gauged catchment in one case has at least one water level recorded at one location and has an associated stage (height) versus discharge for this location.

4.10.1 Catchment Characteristics for Ungauged Catchments

For the ungauged catchment, the *FSR* (NERC, 1975) produced several equations for flood flow calculation based on the characteristics of catchments, i.e. area, stream slope, etc. The more detailed the equation the more robust is the prediction. The calculation is based on $\bar{Q}$, i.e. the mean annual flood, and this can be modified for any return period based on equations similar to Eqs (4.32) and (4.33). The mean annual flood does not have a return period of one year. In the United Kingdom/Ireland combination from the *FSR* (NERC, 1975), $\bar{Q}$ has a return period of 2.4 years. The peak flow is taken for each of the number of years of data available and is then 'averaged'. Thus $\bar{Q}$ may be greater or less than the annual flood (of any year) and is of course greater than the mean of the average annual flood. The empirical equations apply to the United Kingdom and Ireland. The six-variable equation is

$$\bar{Q} = C \times \text{AREA}^{0.94} \times \text{STMFRQ}^{0.27} \times S_{1085}^{0.16} \times \text{SOIL}^{1.23} \times \text{RSMD}^{1.03}(1 + \text{LAKE})^{-0.85} \qquad (4.34)$$

where
$\quad$ AREA = catchment area, km^2

$\quad$ STMFRQ = stream frequency, number of junctions/km^2

$\quad$ SOIL = an index based on five soil types, S1 to S5, where S1 is indicative of a low runoff soil and S5 is a high runoff (rock outcrops)

$\quad S_{1085}$ = stream slope, between the 10 and 85 per cent locations, m/km

$\quad$ RSMD = net 1 day rainfall with a 5 year return period

$\quad$ LAKE = the fraction of catchment area occupied by lakes

$\quad$ C = regional multiplier, i.e. ~ 0.018 for Ireland, ~ 0.020 for Scotland

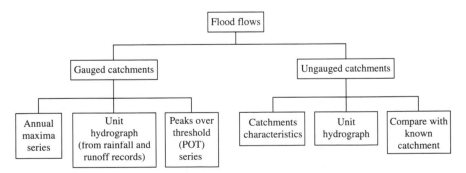

Figure 4.34 Methods of flood flow determination for gauged and ungauged catchments.

Details of how to use the above equation and an explanation of the parameters are found in Vol. 1 of the *FSR* (NERC, 1975).

The five-variable equation is:

$$\bar{Q} = C \times \text{AREA}^{0.87} \times \text{STMFRQ}^{0.31} \times \text{SOIL}^{1.23} \times \text{RSMD}^{1.17} (1 + \text{LAKE})^{0.97} \qquad (4.35)$$

where
$$C \sim 0.0183 \text{ for Ireland}$$
$$\sim 0.0224 \text{ to } 0.0362 \text{ for Scotland/United Kingdom}$$

The two-variable equations are

$$\bar{Q} = 0.0236 \times \text{AREA}^{1.19} \times S_{1085}^{0.84} \qquad (4.36)$$

and

$$\bar{Q} = 2.242 \times 10^{-7} \times \text{AREA}^{0.84} \times \text{SAAR}^{2.09} \qquad (4.37)$$

The single-variable equation is

$$\bar{Q} = 0.667 \times \text{AREA}^{0.77} \qquad (4.38)$$

Example 4.8 Determine the mean annual flood using the six-, five-, two- and one-variable equations for the following catchment in Ireland:

$$\text{Catchment area} = 1762 \text{ km}^2$$
$$\text{Average annual rainfall} = 1100 \text{ mm}$$
$$\text{Stream slope} = 1.35\text{m/km}$$
$$\text{Stream length} = 109 \text{ km}$$
$$\text{Stream frequency, STMFRQ} = 0.77$$

Soil parameters:

$$\text{Class S2} = 41\%$$
$$\text{Class S4} = 31\%$$
$$\text{Class S5} = 28\%$$
$$\text{Lake fraction} = 0$$

The soil parameter as used in the equation is

$$\text{SOIL} = 0.15 \text{ S1} + 0.3 \text{ S2} + 0.4 \text{ S3} + 0.45 \text{ S4} + 0.5 \text{ S5} = 0.403$$

The rainfall parameter as used in the equation is approximated by

$$\text{RSMD} = 2.48\sqrt{\text{SAAR}} - 40 = 42.4 \text{ mm}$$

Where SAAR is the standard annual average rainfall. A more exact method is shown in the *FSR* (NERC, 1975).

The results are summarized in Table 4.15 and the variation at ± 25 per cent of the mean value of $\bar{Q}$. It is prudent to stay on the high side and so adopt $\bar{Q} \sim 300 \text{ m}^3/\text{s}$. Methods pertinent to other countries, e.g. the US, Australia, *etc.*, are detailed in Chow *et al.* (1988).

4.10.2 Analysis of Peak Flows for Gauged Catchments

As mentioned in Sec. 4.10, if the catchment being analysed is gauged for river flow, then the records are integrated to produce flood flows in correspondence to return periods. The following discussion is

Table 4.15 $\bar{Q}$ values for different equations

Catchment characteristics method	$\bar{Q}$ (m³/s)
Six-variable equation	304
Five-variable equation	289
Two-variable equation 1	221
Two-variable equation 2	272
One-variable equation	211

independent of country and is based purely on statistics. Several different interpretations of the same data set can be examined. With reference to Fig. 4.35, the following is noted:

1. The *annual maximum series* is made up of P_1, P_2, P_3.
2. By defining a threshold at p, it is seen that the five peaks over this threshold P_1, p_1, P_2, P_3, p_3 form the peaks over threshold (POT) series. This is sometimes called the *partial duration series*.
3. A threshold may be set that there are N peaks over N years where not every year yields a peak. This is known as the *annual excedence series*.

It can be expected, then, that different methods yield different magnitudes of floods for the same return period or different methods ascribe the same flood magnitude to different return periods. The values will tend to be closer together when the period of record is longer. The following is an example of the annual maximum series method.

Example 4.9 Given the series of annual maximum floods shown below from 1975 to 1989 for the Bandon River, Ireland, compute the return periods and flood magnitudes from 1 to 100 years:

Year	1975	1976	1977	1978	1979	1980	1981	1982	1983	1984	1985	1986	1987	1988	1989
Flood (m³/s)	298.4	85.7	144.7	211.9	121.4	113.7	131.7	223.5	108.3	102.9	95.6	239.8	132.6	261.2	145.2

Prepare Table 4.16 as follows

1. Rank the 15 floods from top to bottom as in column 2.
2. In column 3 give each a rank number m from 1 to 15 ($n = 15$).

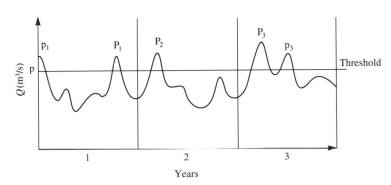

Figure 4.35 Flood peak data series.

3. Use the Weibull formula to attach a return period $T(x)$ in column 4 to each ranked flood as per:

$$T(x) = \frac{(n+1)}{m} \qquad \text{Weibull equation for return period} \qquad (4.39)$$

where

$$P(x) = \frac{1}{T(s)}$$

or

$$P(x) = \frac{m}{n+1} \qquad \text{Weibull equation for probability}$$

The Gringorton equation $P(x) = (m - 0.44)/(n + 0.12)$ is often used instead of the Weibull equation.

4. In column 5 attach the associated $P(x)$ value, where $P(x)$ is the probability of an annual maximum, equalling or exceeding a flood value x m^3/s in any given year, i.e. $P(x) = 1/T(x)$.
5. In column 6 attach the associated $F(x)$ value, where $F(x)$ is the probability of an annual maximum being less than x in any given year, i.e. $F(x) = 1 - P(x)$.
6. In column 7 place the associated value of y, as defined by $y = -\ln\{-\ln[1 - 1/T(x)]\}$.
7. These can now be plotted on extreme value gumbel probability paper, as seen in Fig. 4.36.
8. Draw the straight line of 'best fit' or fit a line using least squares or other method and extrapolate for higher return periods. However, care needs to be used in extrapolating data beyond the number of years for which data are available.

The line fitted to Fig. 4.36 is the least squares best-fit line. It is seen from the line of best fit that there is a shortage of data available for flows between 150 and 250 m^3/s. From this figure, values for extended return periods are taken off by extrapolation and entered on Table 4.17. Because the records extend only for 15 years, care is required in interpreting flood return values for periods much longer than 25 years.

4.11 LOW FLOWS

Low flows are also significant parameters in hydrology. Traditionally, hydrologists were preoccupied with flood alleviation and so analysis for high flows is more commonplace than that of low flow analysis.

Table 4.16 Determination of flood flows versus return periods for Example 4.9

Flood year	Flood flow x	Rank m	Return period $T(x)$ (yr)	Probability $P(x) \geqslant Q$	Probability $F(x) < Q$	Reduced variate y
1975	298.4	1	16	0.0625	0.9375	2.74
1988	261.2	2	8	0.125	0.875	2.01
1986	239.8	3	5.33	0.1876	0.8124	1.57
1982	223.5	4	4	0.25	0.75	1.24
1978	211.9	5	3.2	0.3125	0.6875	0.981
1989	145.2	6	2.66	0.375	0.625	0.755
1977	144.7	7	2.286	0.437	0.563	0.554
1987	132.6	8	2.0	0.50	0.50	0.366
1981	131.7	9	1.77	0.565	0.435	0.183
1979	121.4	10	1.6	0.625	0.375	0.019
1980	113.7	11	1.45	0.69	0.31	-0.158
1983	108.9	12	1.3	0.75	0.25	-0.327
1984	102.9	13	1.23	0.813	0.187	-0.517
1985	95.9	14	1.143	0.874	0.126	-0.728
1976	85.7	15	1.066	0.9375	0.0625	-1.0199
	$\bar{x} = 161.2$					

$n = 15$

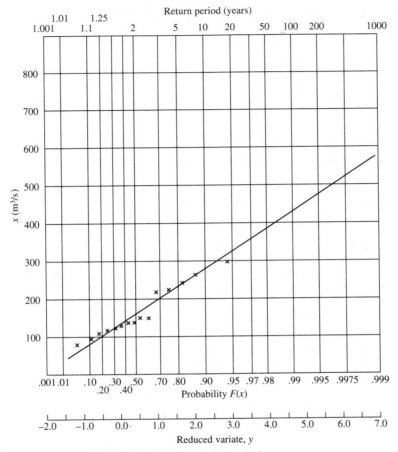

Figure 4.36 Flood frequency (Gumbel plot) flood flows for Bandon, Ireland.

However, analyses of low flows are of significant interest, particularly in relation to water abstractions for water supply or hydroelectricity and most particularly for water quality. Whether effluent discharge permits will be written or not depends on low flow magnitudes and the sustenance of fisheries or aquatic habitats. There are many terms used in low flow analysis, some of which are defined in the following section.

4.11.1 Parameters of Low Flow

A flow duration curve is a plot, as Fig. 4.37, with ordinates of flow (m³/s) and abscissa of percentage exceedence. For example, the flow magnitude of 7 m³/s is exceeded 50 per cent of the time.

Percentage exceedence is the percentage of time that a given discharge is exceeded. The 95 per cent exceedence is flow that is equalled or exceeded 95 per cent of the time. In Fig. 4.37 this value is about 1.5 m³/s.

Table 4.17 Return period versus flood flows for Example 4.9

Return period (yr)	2	5	10	25	50	100	200
Flow (m³/s) (Weibull)	160	230	275	330	380	430	470

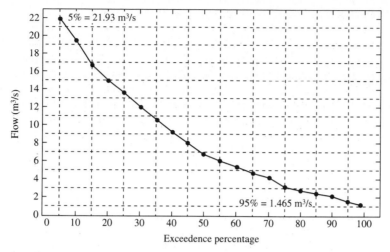

Figure 4.37 Flow duration curve.

Daily mean flow is the mean of the instantaneous discharge throughout a 24-hour period.

Annual daily mean flow is the mean over a year of the daily mean flows.

Dry weather flow is the annual minimum daily mean flow with a selected return period. This term or multiples of it are often used in foul and storm sewer design.

Baseflow is the contribution to streamflow from groundwater.

D-Day is a term used for the duration in days, e.g. 10-day is of ten days' duration.

D-day flow is the average flow over D-consecutive days.

Sustained low flow is defined as the lowest mean flow that is not exceeded for a given duration.

D-Day sustained low flow (SLF) is the largest daily mean flow in the D-day period, e.g. the 7-day SLF is found by getting the driest week's average flow and then the largest daily mean flow in that seven-day period is the SLF for the year. Thus, the 1-day SLF is the minimum DMF (daily mean flow) for the year.

Minimum flow is the minimum flow observed in the period.

Exceedence percentage D-day flow is best defined by an example:

> Q95, 10 is the 95 percentage exceedence flow, *averaged* over ten days
> (Q95 per cent, 10-day).

Authorities in different countries specify different low flow parameters. For instance, within Ireland, the 7-day SLF is often used in planning permits for effluent discharges. Other parameters used are the Q95, 10.

4.11.2 Frequency of Low Flows

It is of significance to establish frequencies of low flows as it was also significant to establish frequencies of flood flows. It is of interest to establish return periods to magnitudes of low flows. The methods used are several but that of the extreme value Gumbel distribution can be used as it was in Sec. 4.10.2 for flood analyis.

4.12 URBAN HYDROLOGY

Urbanization affects the response of a catchment to rainfall in many ways, depending on the location of the urbanization with respect to the upstream or downstream end of a river network. Complete urbanization may reduce the hydrograph rise time by up to 70 per cent (Fig. 4.38) and increases the mean annual flow by between 200 and 600 per cent, depending on the responsiveness of the catchment before urbanization (NERC, 1979). The introduction of impervious surfaces and an efficient drainage system increases the volume of runoff (reducing the amount of infiltration) and reduces flow travel time, yielding a hydrograph that is faster to peak, faster to recede and of increased peak discharge. Correspondingly, the flood frequency distribution is affected and floods of all return periods are, in general, increased. The magnitude of the increase depends on the extent of urbanization and the relationship of the urban response to the original rural response.

The following are listed as significant factors in the hydrograph response to urbanization in the *FSR* No. 5 (NERC, 1979):

1. Catchments characterized by low percentage runoff and slow response are more affected by urbanization than catchments already characterized by high percentage runoff and rapid response.
2. Urbanization has a greater effect on the response to small storms which previously yielded low percentage runoff and little overbank flow than on the response to severe storms. Thus the mean annual flood will be increased by a greater proportion than rarer floods.
3. Since urban catchments respond faster and yield runoff from smaller events, the *T*-year flood after urbanization tends to be caused by a shorter duration storm of smaller rainfall depth but higher intensity. Therefore, the effect of urbanization on the *T*-year flood depends on local rainfall characteristics and on the relationship between rainfall intensities for short- and long-duration storms.
4. The effect of urbanization depends on the location of urban development within the catchment. Urbanization in upstream areas may result in a rapid urban response which coincides with and reinforces the slower rural response from downstream. Urbanization in downstream areas may cause the urban response to pass before the slow rural response from upstream areas arrives.

4.12.1 FSR Assessment of Urban Runoff

The *FSR* (NERC, 1975) estimates the effect of urbanization from

$$\frac{\bar{Q}_u}{\bar{Q}_r} = \left[(1 + \text{URBAN})^{1.5} \left(1 + 0.3 \, \text{URBAN} \times \frac{70}{\text{PR}_r} \right) \right] - 1 \qquad (4.40)$$

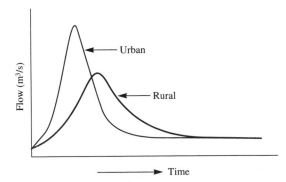

Figure 4.38 Conceptual of hydrograph response due to urbanization.

where $\quad\bar{Q}_u$ = mean annual urban flood

$\bar{Q}_r$ = mean annual flood for rural catchment, computed from Eq. (4.35)

URBAN = fraction of catchment urbanized

PR_r = percentage runoff from rural catchment

$\quad$ = 102.4 × SOIL + 0.28 (CWI − 12.5)

$\quad$ (for SOIL see Sec. 4.10.1)

CWI = catchment wetness index (ranges 125 to 180)

$\quad\sim$ 125 for dry catchment

$\quad\sim$ 180 for supersaturated catchment

For instance, with a 50 per cent urbanized catchment and $PR_r = 0.5$, the mean annual flood of the urbanized catchment is 23 per cent greater than the rural catchment. Further details are in the *FSR* Supplementary Report No. 5 (NERC, 1979).

4.12.2 Urban Runoff from the Rational Method

As described in Sec. 4.8.4, the rational method is a simple method sometimes used in determining urban runoff. The flow rate Q is represented by

$$Q_p = 0.278 \; CIA \qquad \text{m}^3/\text{s}$$

The runoff coefficient for typical applications is shown in Table 4.18. The other parameters were defined in Sec. 4.8.4.

4.12.3 Modelling of Urban Runoff

To comply with environmental regulations of surface runoff water quality, more sophisticated techniques than that of the rational design method are now required. Several systems exist including those that accommodate both flooding and pollution analysis in sewers. Surface runoff models compute the inflow hydrographs to the sewer network using time series historical rainstorms (if they are available). The runoff

Table 4.18 Runoff coefficient for different surfaces

Description of area	Runoff coefficient
Streets	0.7–0.9
Driveways	0.75–0.85
Grassed lawns (sandy)	
<2% slope	0.05–0.1
2–7%	0.1–0.15
>7%	0.15–0.2
Grassed lawns (clay)	
<2% slope	0.13–0.17
2–7%	0.18–0.22
>7%	0.23–0.35
Roofs	0.78–0.95
Light industrial area	0.5–0.8
Heavy industrial area	0.6–0.9
Business areas	0.7–0.9
Apartment dwelling area	0.5–0.7
Individual residential areas	0.25–0.4

is routed through the pipe network of pipes, manholes and hydraulic structures in an unsteady analysis. Such analysis allows examination of the hydraulic performance of complex looped sewer systems including storm overflows, detention basins and pumping stations. The extent and return period for floods can be identified and bottlenecks and weaknesses in the hydraulic system can be modified so as to reduce extreme flood events to acceptable levels. The third aspect is assessment of the pollution load so that the design of the downstream wastewater treatment plant can be more valid. Also, it is important that exact pollution loads be identified for storm overflows. For instance, in sensitive receiving waters, knowledge of the input of the eutrophication producing nutrients of nitrogen and phosphorus is very important, not only in their quantities but also in the time-varying inputs.

Modelling software systems accommodating these aspects of flooding analysis, hydraulic design and pollutant assessment are available from many international organizations, including the Danish Hydraulic Institute (MOUSE), the Hydraulic Research Station Wallingford and the USEPA. The MOUSE system is also used in real-time control (RTC) for combined sewer overflow systems.

4.13 GROUNDWATER

As the focus of this book is on environmental engineering, a brief introduction follows into the physics of groundwater. In Sec. 4.14, aspects of groundwater chemistry, contamination and aquifer protection schemes are discussed.

Earlier parts of this chapter examined surface water and water movement in the unsaturated zone. Groundwater is here defined as that water below the water table, in other words in the geologic strata where the pore space is ~ 100 per cent occupied by water. 'Hydrogeology' is another word given to the study of groundwater. Groundwater is a significant natural water resource. In Ireland, groundwater is used for approximately 25 per cent of the raw water requirements. In parts of Europe the corresponding percentage is about 80.

4.13.1 Aquifers

Figure 4.39 is a schematic of different aquifer types. An 'aquifer' is defined as a water-bearing rock formation that contains water in sufficient amounts to be exploited and brought to the surface by wells. An aquifer may be either 'confined' or 'unconfined'. In Fig. 4.39, the upper aquifer is unconfined, i.e. it has a natural 'water table line' free to move up or down. A confined aquifer is restrained by an upper impermeable strata called an 'aquaclude', inhibiting the upward movement of water. When a series of wells is drilled into the confined aquifer, the water will rise and attain its own 'water table line'. This line is the 'piezometric' line or that due to the hydrostatic pressure of the confined aquifer (see also Figs 4.45 and 4.46).

The piezometric line is a theoretical level if there are no wells. In aquifers, it is assumed that all the pore space is occupied by water. As such, the 'porosity' is a significant parameter. It is defined as

$$\text{Porosity } n = \frac{V_v}{V_t} \tag{4.41}$$

where
$$V_v = \text{volume of voids}$$
$$V_t = \text{total volume}$$
$$\text{Void ratio } e = \frac{V_v}{V_s}$$

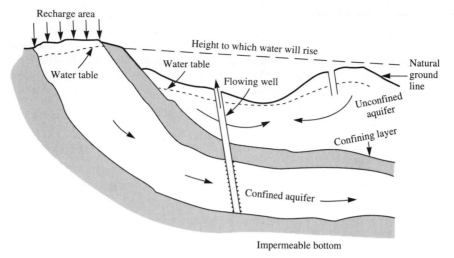

Figure 4.39 Schematic of aquifer types.

where
$$V_s = \text{solid volume } (V_t - V_v)$$

Therefore

$$n = \frac{e}{1+e} \quad \text{and} \quad e = \frac{n}{1-n} \tag{4.42}$$

Figure 4.40 shows the porosity variation with rock type. Table 4.19 shows the typical values of porosity, and other parameters for different material types. Effective porosity is defined as the percentage of interconnected pore space.

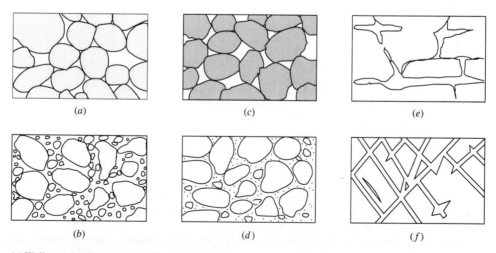

(*a*) Well-sorted sedimentary deposit with high porosity
(*b*) Poorly-sorted sedimentary deposit with low porosity
(*c*) Well-sorted sedimentary deposit of porous pebbles, giving high total porosity
(*d*) Well-sorted sedimentary deposit where mineral matter has deposited in interstices
(*e*) Rock rendered porous by solution (e.g. limestone area)
(*f*) Fractured porous rock

Figure 4.40 Relationship between texture and porosity of rocks (adapted from Meinzer, 1923; Domenico and Schwartz, 1990).

Table 4.19 Porosity values by material

Material	Porosity range (%)	Specific yield (%)	Range of hydraulic conductivity[1] K (m/s)
Crushed stone	> 30	> 20	10^{-4} to 10
Coarse gravel	24–36	~22	10^{-4} to 10^{-2}
Fine gravel	25–38	~22	—
Coarse sand	31–46	~25	10^{-7} to 10^{-3}
Fine sand	26–53	~10	10^{-7} to 10^{-4}
Silt	34–61	~8	10^{-9} to 10^{-5}
Clay	36–60	~3	10^{-11} to 10^{-9}
Sandstone	5–30	~5	10^{-10} to 10^{-6}
Limestone	5–50	~2	10^{-6} to 10^{-2}
Shale	0–10	~3	10^{-13} to 10^{-9}
Basalt	3–35	—	10^{-10} to 10^{-5}
Till	~32	~16	10^{-12} to 10^{-6}
Peat	~92	~44	—

[1]Saturated hydraulic conductivity also noted as $K h_{sat}$.
Adapted from Johnson, 1962, 1967

4.13.2 Water Yield

The water yielding capacity of an aquifer is a significant physical parameter in hydrogeology and this depends on and varies with several parameters, including:

1. Specific yield: the amount of aquifer water expressed as a percentage that is free to drain under the influence of gravity. By definition it is less than the porosity since some water is not free to drain due to attractive and bonding forces such as surface tension. Some typical values are given in Table 4.19.
2. Storage coefficient, S: somewhat similar to specific yield, this parameter expresses the volume of water that an aquifer releases (or accumulates) per unit of surface area per unit length change in piezometric head. For confined aquifers $10^{-5} < S < 10^{-3}$. For unconfined aquifers $10^{-2} < S < 0.35$ (Davis and Cornwall, 1991). The units are m^3 of water/m^3 of aquifer.
3. Hydraulic gradient, dh/dx: the slope of the piezometric surface line in m/m. The magnitude of the 'head' determines the pressure on the groundwater to move and at what velocity.
4. Hydraulic conductivity, K: in the unsaturated zone, discussed in earlier sections of this chapter, hydraulic conductivity was defined as a measure of the ability of a medium (soil) to allow the passage of water in units of m/s. With respect to aquifers, the medium is not soil but usually rock. K is a property of both the medium and the fluid and is dynamic, and varies with moisture content. In this book, we use the terms hydraulic conductivity and permeability interchangeably. The values can range over 12 orders of magnitude, with the highest values for gravels and limestones (about 10^{-2} to 10^{-4}) and lowest values for unifractional igneous and metamorphic rocks and also clays (10^{-8} to 10^{-14}). Some typical values are listed in Table 4.19.
5. Transmissivity, T: the rate of flow per unit width of the aquifer under a unit hydraulic gradient, $10^{-4} < T < 10^{-1}$:

$$T = Kb \qquad \text{m}^2/\text{s} \qquad\qquad (4.43)$$

where $\qquad\qquad b =$ thickness (height) of the aquifer

we will refer to Table 4.19 again when discussing clay liners for landfill (Chapter 14) which requires almost negligible permeability at values lower than 10^{-9} m/s.

4.13.3 Groundwater Flow

As with flow in the unsaturated zone, flow in aquifers is to some degree three dimensional. If the hydraulic gradient is predominantly undirectional, however, the flow will be almost one-dimensional. If the conductivity $K = K_x = K_y = K_z$, then the aquifer is regarded as 'isotropic' and this is the simplest flow case to analyse. If K is independent of location within the aquifer, it is said to be 'homogeneous'. Usually in analysis, the flow is assumed to be isotropic and homogeneous. Darcy's law for groundwater flow is given as

$$Q = -KA\frac{dh}{dx}$$
$$= KA\frac{h_2 - h_1}{l_2 - l_1} \qquad (4.44)$$

where
Q = flow (horizontal) through the aquifer, m^3/s
K = hydraulic conductivity, m/s
A = cross-sectional area, m^2
$h_2 - h_1$ = head drop, m
$l_2 - l_1$ = length difference (along horizontal x direction between h_2 and h_1), m

also,
$$q = \frac{Q}{A} = K\frac{dh}{dl}$$
$$q_x = K_x\frac{\partial h}{\partial x}; \qquad q_y = K_y\frac{\partial h}{\partial y}; \qquad q_z = K_z\frac{\partial h}{\partial z} \qquad (4.45)$$

where
q = specific discharge or flow per unit area

Example 4.10   Determine the daily flow capacity and the transmissivity of a sandstone aquifer if:

The aquifer height is 15 m
The aquifer width is 800 m
The aquifer length is 2 km
The head change over the length of 2 km is 3 m

Solution   Assume the hydraulic conductivity

$$K = 6 \times 10^{-7} \text{ m/s} = 5.2 \times 10^{-2} \text{ m/day}$$
$$Q = KA\frac{dh}{dl}$$
$$Q = 5.2 \times 10^{-2} \times 800 \times 15 \times \frac{3}{2000} = 0.94 \text{ m}^3/\text{day}$$

Transmissivity, $T = Kb$
$$= 5.2 \times 10^{-2} \times 15 = 0.78 \text{ m}^2/\text{day}$$

The general flow equations for groundwater flow are derived in several standard hydrology and hydrogeology textbooks (Bras, 1990; Domenico and Schwartz, 1990). They are briefly presented here.

Flow in a saturated medium   Figure 4.41 represents a unit control volume of a saturated medium. The concept of mass balance is applied to determine the flow equation. The net mass balance in the x

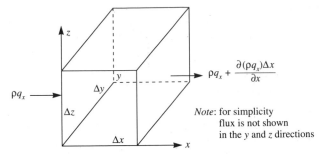

Figure 4.41 Element of saturated rock with no free surface.

direction is

$$\begin{array}{ccccc} \text{Mass flow rate} & - & \text{mass flow rate} & = & \text{change in mass} \\ \text{into the unit} & & \text{out of the unit} & & \text{storage with time} \end{array}$$

$$\rho q_x \Delta y \Delta z \quad - \quad \left[\rho q_x + \frac{\partial(\rho q_x)\Delta x}{\partial x} \right] \Delta y \Delta z \quad = \quad \frac{\partial(\rho n)}{\partial t} \Delta x \Delta y \Delta z \qquad (4.46)$$

where

n is an effective porosity (units of length) and pn is the mass per unit volume

and

q_x = flow per unit area orthogonal to face $\Delta y \Delta z$

Equation 4.46 then reduces to

$$-\frac{\partial q_x}{\partial x} = \frac{\partial n_x}{\partial t}$$

In three dimensions

$$-\left(\frac{\partial q_x}{\partial x} + \frac{\partial q_y}{\partial y} + \frac{\partial q_z}{\partial z} \right) = \frac{\partial n}{\partial t} \qquad (4.47)$$

Introduce the term 'specific storage', S_0, as the volume of water released from storage per unit volume of aquifer per unit change in the pressure head (Bras, 1990), i.e. dimensions of inverse length.

From hydrostatics:

$$P = \rho g(h - Z_0)$$

where

$$h = \text{piezometric head } (Z_0 = \text{datum})$$

$$-\left(\frac{\partial q_x}{\partial x} + \frac{\partial q_y}{\partial y} + \frac{\partial q_z}{\partial z} \right) = -S_0 \frac{\partial h}{\partial t} \qquad (4.48)$$

For steady state conditions,

$$-\left(\frac{\partial q_x}{\partial x} + \frac{\partial q_y}{\partial y} + \frac{\partial q_z}{\partial z} \right) = 0 \qquad (4.49)$$

Introduce Darcy's Law, i.e. $q_x = K_x(\partial h/\partial x)$. Then

$$\frac{\partial}{\partial x}\left(K_x \frac{\partial h}{\partial x} \right) + \frac{\partial}{\partial y}\left(K_y \frac{\partial h}{\partial y} \right) + \frac{\partial}{\partial z}\left(K_z \frac{\partial h}{\partial z} \right) = 0 \qquad (4.50)$$

If the rock can be considered as isotropic, i.e. $K_x = K_y = K_z = K$, then

$$\frac{\partial^2 h}{\partial x^2} + \frac{\partial^2 h}{\partial y^2} + \frac{\partial^2 h}{\partial z^2} = 0 \qquad (4.51)$$

i.e. (the Laplace equation)

$$\nabla^2 h = 0 \qquad (4.52)$$

Different techniques for the solution of Eq. (4.52) include graphical methods, electrical analogues and numerical methods which are discussed in Shaw (1994), Bras (1990) and Wang and Anderson (1982). Analytical solutions exist based on the simplifying assumptions of Dupuit and Forcheimer for unconfined flow. These assumptions are:

1. The hydraulic gradient dh/dx approximates the slope of the water table and the slope of the free surface.
2. The water table and free surface are 'practically' horizontal.
3. The discharge is constant throughout the depth of flow being evaluated.

The Laplace equation (Eq. 4.52) is the basis of solving numerical problems of flow in porous media. This can be extended beyond the hydrodynamics to include the water chemistry and so solve problems of contaminant flow in groundwater situations. This topic is further discussed in the modelling section (Chapter 21).

Unconfined flow   Figure 4.42 shows the pattern of groundwater flow between two rivers whose water levels are different.

  If the flow is assumed to be one dimensional and steady state with a hydraulic conductivity K then the Laplace equation is:

$$\frac{d^2 h}{dx^2} = 0 \qquad (4.53)$$

Integrating:

$$h^2 = ax + b \qquad (4.54)$$

Boundary condition 1: $\qquad\qquad h = h_L$ at $x = 0$

Therefore

$$b = h_L^2 \qquad (4.55)$$

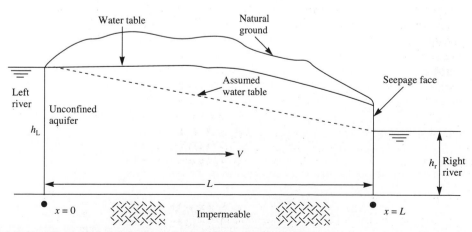

Figure 4.42 Unconfined aquifer flow between two water bodies without recharge.

Differentiating Eq. (4.54):

$$2h\frac{dh}{dx} = a$$

From Darcy's equation,

$$q = -Kh\frac{dh}{dx}$$

Equation (4.54) becomes

$$h^2 = 2h\frac{dh}{dx}x + h_L^2$$

$$= h_L^2 - \frac{2q}{K}x \tag{4.56}$$

Boundary condition 2:

$$h = h_R \text{ at } x = L$$

$$h_R^2 = h_L^2 - \frac{2q}{K}L$$

Therefore

$$q = \frac{K}{2L}(h_L^2 - h_R^2) \tag{4.57}$$

Equation (4.56) becomes

$$h^2 = h_L^2 - \frac{x}{L}(h_L^2 - h_R^2) \tag{4.58}$$

Equation (4.58) is known as the Dupuit parabola and Eq. (4.57) as the Dupuit equation. Equation (4.58) gives the variation in the height of the water table going across 'the island' from the left river towards the right river of Fig. 4.42.

The above equations hold only in the event of no recharge. Recharge is that proportion of rainfall that 'eventually' finds its way into the aquifer and raises the water level. If recharge is R, then

$$\frac{dq}{dx} = R \tag{4.59}$$

From Darcy's law,

$$q = -Kh\frac{dh}{dx}$$

Therefore

$$\frac{dq}{dx} = -\frac{K}{2}\frac{d^2h^2}{dx^2} = R \tag{4.60}$$

Integrating Eq. (4.60) twice:

$$h^2 = -\frac{Rx^2}{K} + ax + b$$

As in the non-recharge case, the boundary conditions are the same, giving

$$b = h_L^2$$

and

$$a = \frac{h_R^2 - h_L^2}{L} + \frac{RL}{K}$$

Substituting and rearranging, we arrive at

$$h^2 = h_L^2 - \frac{x}{L}(h_L^2 - h_R^2) + \frac{Rx}{K}(L - x) \tag{4.61}$$

This equation determines the shape of the water table line and is parabolic. The flow rate through the aquifer can now be determined from the following:
Differentiate Eq. (4.61):

$$2h\frac{dh}{dx} = \frac{h_R^2 - h_L^2}{L} + \frac{R}{K}(L - 2x) \tag{4.62}$$

Equation (4.62) becomes

$$\frac{-2q}{K} = \frac{h_R^2 - h_L^2}{L} + \frac{R}{K}(L - 2x)$$

$$q = \frac{K}{2L}(h_L^2 - h_R^2) - \frac{R}{2}(L - 2x) \tag{4.63}$$

Equation (4.63) is the Dupuit equation for flow with the effect of recharge.

Example 4.11 From Fig. 4.43, determine the height and location of the water table at the water divide if recharge is estimated at 0.05 mm/day, to a coarse sand aquifer.

Solution For a coarse sandy aquifer assume $K \cong 6.9 \times 10^{-4}$ m/s (Table 4.19), i.e. $K \cong 10$ m/day. At the water divide, the flow splits and that to the left goes left and that to the right goes right. Therefore, there is no flow where $h = h_{max}$ and so, from Eqn. (4.63),

$$q = 0 = \frac{K}{2L}(h_L^2 - h_R^2) - \frac{R}{2}(L - 2x)$$

Substituting for h_L, h_R, L, k and R, yields

$$x = 3.2 \text{ km}$$

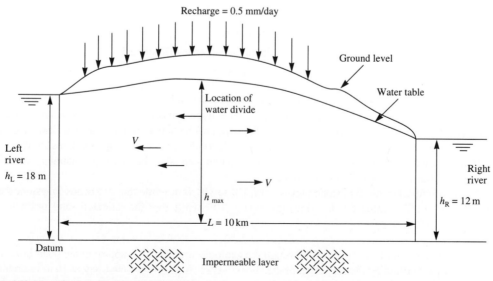

Figure 4.43 Unconfined aquifer island with recharge.

From Eqn. (4.61),

$$h^2 = h_L^2 - \frac{x}{L}(h_L^2 - h_R^2) + \frac{Rx}{K}(L - x)$$

Substitution yields

$$h = 19.4\,\text{m}$$

Note that the height is greater than the higher of the two rivers above datum.

4.13.4 Groundwater Investigations

Groundwater investigations are required for a variety of purposes including:

- Determination of volume/flows for a potential water supply
- Determination of water quality
- Determination of extent and magnitude of groundwater pollution
- Determination of drawdown levels on property boundaries
- The assessment of an aquifer for protection.
- Assessment of risks to pollution of groundwater

The locations and magnitudes of an aquifer can be determined using:

- Resistivity surveys
- Seismic survey
- Wells and boreholes

Resistivity tests   The composition of a geologic formation and the location of sites of groundwater can be determined by passing an electric current into the geology via two electrodes spaced a distance L metres apart, and measuring the resulting drop in voltage with a second set of electrodes:

$$r = \frac{ER \times A}{L} \tag{4.64}$$

where

r = resistivity

ER = electrical resistance

A = cross-sectional area over the distance L

L = length between electrodes

The measure varies with the composition of the geology and the water content. In porous strata, the resistivity depends on the water content. Fine clays have lesser resistivity values than coarse sands and gravels. The spacing of the electrodes determines the depth of the penetration of the electrical current and so relationships (calibrations) are developed between electrode spacing and apparent resistivity. Figure 4.44 is a schematic of a resistivity set-up. This method can also be used for contaminant plume identification (Domenico and Schwartz, 1990).

The resistivity of dense limestone ranges from 10^3 to 10^6 $\Omega\,\text{m}$ while that of porous limestone ranges from $10^{1.5}$ to 10^3. The value for dry sand/gravel is $\sim 10^4$, while that for saturated sand/gravel can be ~ 10 to 10^1.

Seismic surveys   Sound waves are sent through the geologic strata and the approximate (near horizontal) boundaries are identified by the refraction of the shock waves. A grid is selected, say at 10 m intervals (on plan), and sound wave recorders, called geophones, are set at these points. A seismic wave is set up by an

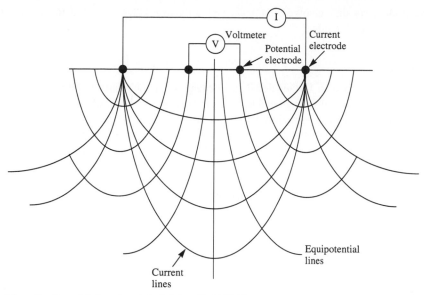

Figure 4.44 Schematic of a resistivity set-up (adapted from Todd, 1980).

explosive charge at a particular point. From the recorded wave arrival times, the sound velocities can be determined and hence the identification of the geologic strata type. It is an approximate method, and the water table line can also be identified.

Wells and boreholes   By far the most precise method of determining the geologic strata is to develop a series of boreholes and record the soil/rock composition with distance downwards from the ground surface. Typically core sizes may vary from 40 to 150 mm and continuous undisturbed cores yield detail data of the hydrogeologic formations. Spatial data on the extent of aquifers are determined by selected borehole sites and geostatistics is often applied to ascertain the optimum site locations. The identification and extent of an aquifer groundwater source is then followed by well pump tests to ascertain the aquifer flow rates and actual drawdown of the water table. See Domenico and Schwartz (1990) for further details.

4.13.5 Well Tests

Well tests are carried out *in situ* with the objectives including:

- Determination of hydrogeologic parameters of hydraulic conductivity, transmissivity and storage coefficient
- Determination of quantity/quality of water supply
- Determination of sustainability of maximum yield
- To evaluate the impact of drawdown on neighbouring wells
- To provide baseline data on well performance and characteristics (Daly, 1994)

The range of small-scale tests that result in little well discharge that are used to determine K include:

- Rising and falling head tests
- Slug and bail tests
- Constant head tests
- Packer tests
- Tracer tests

Those larger scale tests that involve significant well discharges that are used to determine, quantity, quality and performance include:

- Step drawdown tests
- Constant discharge tests

The tests are described in further detail in Domenico and Schwartz (1990) and in most texts on hydrogeology. In the rising and falling head test a fixed volume of water, sufficient to cause a measurable instantaneous rise in water level, is rapidly introduced to the well. The well water level is recorded as the water level falls with the water seeping into the surrounding aquifer. In the rising head test, a fixed volume is removed. When the head changes are plotted with time, the hydraulic conductivity can be determined (Daly, 1992). The other small-scale tests are variations of this. In the large-scale well tests, the well yield in quantity and quality and well performance may be determined. Water levels in the well and observation wells as well as those on nearby streams are recorded prior to a pumping test. The test typically exceeds 24 hours and is typically 72 hours, during which either a constant pumping rate is maintained or a step series of constant pumping rates, i.e. start, say, with 40 m^3/h for 6 h followed by 35 m^3/h for 6 h, etc. Drawdown measurements are plotted at intervals as small as 30 s in the first 30 min to about every 2 hours in the second and third day. Geologic institutions in different countries have their specific recommendations for time intervals and plotting regimes. From these results, information on the significant parameters of quantity, quality, etc., can be determined.

Example 4.12   Determine the approximate transmissivity (m^3/day) from the stepped pumping tests below (Daly, 1992):

Step	Q (m^3/day)	S_t (m)
1	115.6	0.235
2	146.2	0.355
4	200.7	0.5
6	242.2	0.705

Each step had a duration of 60 min S_t is the total (or cumulative) drawdown.

Solution   The specific capacity, SC, for each step is given as

$$SC_1 = 115.6/0.235 = 491.9 \text{ m}^3/\text{d.m}$$
$$SC_2 = 146.2/0.355 = 411.8 \text{ m}^3/\text{d.m}$$
$$SC_3 = 200.7/0.5 \quad = 401.4 \text{ m}^3/\text{d.m}$$
$$SC_4 = 242.2/0.705 = 343.5 \text{ m}^3/\text{d.m}$$

The specific capacity therefore decreases with increased pumping rate. Drawdown is composed of 'aquifer loss' plus 'well loss'. Aquifer loss is constant for the different discharge rates while the well loss increases with discharge:

$$S_t = BQ + CQ^2 \tag{4.65}$$

where B = aquifer loss coefficient

and C = well loss coefficient

The well loss coefficient C is determined from Jacob (1950) as follows:

$$C = \frac{\Delta S_i/\Delta Q_i - \Delta S_{i-1}/\Delta Q_{i-1}}{\Delta Q_{i-1} + \Delta Q_i} \tag{4.66}$$

For steps 1 and 2:

$$C = \frac{\Delta S_2/\Delta Q_2 - \Delta S_1/\Delta Q_1}{\Delta Q_1 + \Delta Q_2} = \frac{0.12/30.6 - 0.235/115.6}{146.2} = 12.9 \times 10^{-6} \text{m}/(\text{m}^3/\text{d})^2$$

Similarly, for steps 2 and 3:

$$C = \frac{\Delta S_3/\Delta Q_3 - \Delta S_2/\Delta Q_2}{\Delta Q_2 + \Delta Q_3} = \frac{0.145/54.5 - 0.12/30.6}{30.6 + 54.5}$$
$$= -14.8 \times 10^{-6} \text{ m}/(\text{m}^3/\text{d})^2$$

For steps 3 and 4:

$$C = \frac{\Delta S_4/\Delta Q_4 - \Delta S_3/\Delta Q_3}{\Delta Q_3 + \Delta Q_4} = \frac{0.205/41.5 - 0.145/54.5}{54.5 + 41.5}$$
$$= 23.7 \times 10^{-6} \text{ m}/(\text{m}^3/\text{d})^2$$

Taking the average as 7.2×10^{-6} m/(m^{3}d)^2. Therefore the well loss associated with each pumping rate can be determined from the relationship

$$S_w = CQ^2 \tag{4.67}$$

where
$$C = 7 \times 10^{-6} \text{ m}/(\text{m}^3/\text{d})^2$$

For each of the pumping rates used the well losses are:

$$S_{w1} = 7 \times 10^{-6} \times 115.6^2 = 0.094 \text{ m}$$
$$S_{w2} = 7 \times 10^{-6} \times 146.2^2 = 0.149 \text{ m}$$
$$S_{w3} = 7 \times 10^{-6} \times 200.7^2 = 0.282 \text{ m}$$
$$S_{w4} = 7 \times 10^{-6} \times 242.2^2 = 0.411 \text{ m}$$

The corresponding aquifer losses are:

$$S_{a1} = 0.235 - 0.094 = 0.141 \rightarrow S_{a1}/Q_1 = 1.22 \times 10^{-3}$$
$$S_{a2} = 0.355 - 0.149 = 0.206 \rightarrow S_{a2}/Q_2 = 1.41 \times 10^{-3}$$
$$S_{a3} = 0.5 \quad - 0.282 = 0.218 \rightarrow S_{a3}/Q_3 = 1.09 \times 10^{-3}$$
$$S_{a4} = 0.705 - 0.411 = 0.294 \rightarrow S_{a4}/Q_4 = 1.21 \times 10^{-3}$$
$$\text{Average} = 1.22 \times 10^{-3}$$

These can also be determined graphically by plotting S_t/Q against Q. The y intercept is the aquifer loss coefficient and the slope is the well loss coefficient.

A preliminary estimate of the transmissivity is as follows. At the highest pumping rate of 242.2 m^3/d a total drawdown of 0.705 m was registered, giving a specific capacity of $242.2/0.705 = 343.5$ m^3/d.m. Of the total drawdown, the well loss was 0.411 m. Therefore the theoretical specific capacity for a 100 per cent efficient well is $242.2/(0.705 - 0.411) = 823.8$ m^3/dm. Transmissivity as given by Daly (1994) is

$$T = \frac{1.22}{(S_a/Q)_{av}}$$
$$= \frac{1.22}{1.22 \times 10^{-3}}$$
$$= 1000 \text{ m}^3/\text{d} \tag{4.68}$$

This is approximate and obviously depends on the losses and the well diameter. For further details refer to Domenico and Schwartz (1990), and Daly (1992).

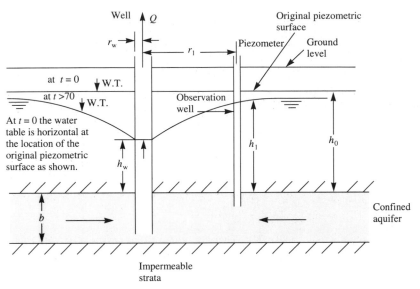

Figure 4.45 Steady flow in a confined aquifer.

4.13.6 Steady State Well Hydraulics

Figure 4.45 outlines radial flow to a well in a confined aquifer. It is of interest to determine the parameters of hydraulic conductivity K and transmissivity T. It is seen that the drawdown curve varies with distance from the well face. A pumping well is surrounded by two non-pumping observation wells. The flow is considered as two-dimensional and the aquifer is assumed homogeneous and isotropic. From Darcy's law,

$$Q = KA\frac{dh}{dx}$$
$$= K \times 2\pi rb\frac{dh}{dr} \tag{4.69}$$

where r = radial distance to an arbitrary point on the drawdown curve

                b = height (thickness of the confined aquifer)

i.e. volume $\pi r^2 b$ is that volume of aquifer available to yield water. For the boundary condition at $r = r_w$, $h = h_w$ (see Fig. 4.45). Integration after separation of variables gives

$$Q = 2\pi Kb\frac{h - h_w}{\ln(r/r_w)} \tag{4.70}$$

For the derivation of equations, refer to Bras (1990):

$$T = Kb = \frac{Q}{2\pi(h_2 - h_1)}\ln\left(\frac{r_2}{r_1}\right) \tag{4.71}$$

where T is the transmissivity of the aquifer.

    Where the observation wells, h_1 and h_2 are adjacent to each other at a radial distance r_1 and r_2 from the well centre-line (see Fig. 4.45) and $r_2 > r_1$.

    Figure 4.46 outlines the radial flow to a well in an unconfined aquifer. Darcy's equation is

$$Q = 2\pi r\,Kh\frac{dh}{dr}$$

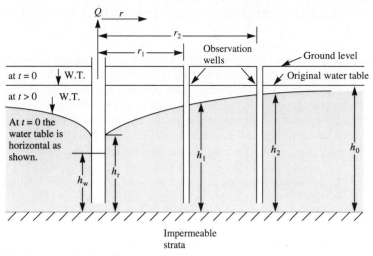

Figure 4.46 Steady flow in an unconfined aquifer.

Integrating:

$$Q = \pi K \frac{h_2^2 - h_1^2}{\ln(r_2/r_1)} \tag{4.72}$$

$$K = \frac{Q}{\pi(h_2^2 - h_1^2)} \ln\left(\frac{r_2}{r_1}\right)$$

where h_1 = height of the water table at observation well
                       1 at a radial distance r_1 and h_2 is the water table
                       height for the observation well 2 at a radial
                       distance $r_2(r_2 > r_1)$.

Example 4.13   Determine K for an unconfined aquifer 10 m thick if a well delivers $360\,\mathrm{m^3/d}$. Observation well 1 is sited at 20 m from the pumping well and records a drawdown of 6 m. Observation well 2 is sited at 600 m and the drawdown is 3 m. The original water table was recorded at 12 m.

Solution

$$K = \frac{Q}{\pi(h_2^2 - h_1^2)} \ln\left(\frac{r_2}{r_1}\right)$$

At observation well 1, the drawdown is 6 m. If h_0 is the original piezometric height then

$$h_1 = 12 - 6 = 6\text{ m, and } h_2 = 12 - 3 = 9\,\mathrm{m}$$
$$r_1 = 20\,\mathrm{m},\ r_2 = 600\,\mathrm{m}$$
$$K = \frac{360}{\pi(81 - 36)} \ln(600/20)$$
$$= 8.65\,\mathrm{m/day}$$

4.14 GROUNDWATER CHEMISTRY, CONTAMINATION AND POLLUTION PREVENTION

This section is adapted from Daly (1994b) with kind permission. The natural chemistry of groundwater varies depending on the nature of the subsoils and rocks that it passes through. For instance, in Ireland, limestone bedrock and limestone-dominated subsoils are common and consequently groundwater is often 'hard', containing high concentrations of calcium, magnesium and bicarbonate. However, in areas where volcanic rocks or sandstones are present, softer water is normal. The variations are shown in Table 4.20 for five sites in Ireland, where analyses of groundwater from different rock types are given. In considering the impact of human activities, it is necessary to first take the natural (or baseline) water quality into account. Groundwater is usually considered to be pure and safe to drink as it undergoes a filtering and cleansing process through a subsoil cover and rock medium that surface waters do not have. However, this does not guarantee groundwater purity. Problems can arise either due to the natural conditions in the ground or to pollution by human activities.

4.14.1 Natural Groundwater Quality Problems

The principal natural groundwater quality problems are caused by hardness, iron, manganese, hydrogen sulphide, sulphate and sodium chloride. With the exception of hardness and sodium chloride, they pose occasional problems in the minor and poor aquifers rather than in the major aquifers.

Hardness   Groundwater passing through limestone dissolves the calcium and magnesium compounds which cause hardness. Consequently, hard groundwaters are common in limestone areas, with total hardness concentrations varying from 200 to 400 mg/l. This may be beneficial to health and give a pleasant taste. However, very high levels can be a nuisance, resulting in scale formation in kettles, pipe systems and boilers.

Table 4.20 Water chemistry of some Irish groundwater

Parameter	Gorey Wexford	Knocktopher Kilkenny	Ballaghdereen Roscommon	Ballincurry Tipperary	Mortarstown Carlow
Total dissolved solids	92	280	360	283	500
Total hardness (as $CaCO_3$)	60	185	307	225	320
Total alkalinity (as $CaCO_3$)	18	184	295	237	304
Calcium	80	54	102	60	118
Magnesium	9.7	12.2	12.9	13.4	6.3
Sodium	24	16	8.8	18.8	12
Potassium	3.1	1.9	3	1.3	0.6
Chloride	35	20	26	17	22
Sulphate	45	nil	nil	2	2
Nitrate (as N)	0.5	3.0	0.9	1.4	2.8
Free and saline ammonia (N)	0.1	0.01	0.05	0.04	0.04
Albumincid ammonia (N)	—	0.01	0.08	—	0.01
Iron	0.1	nil	nil	nil	nil
Manganese	nil	nil	nil	0.01	nil
Aquifer	Ordovician volcanics	Devonian sandstones	Carboniferous limestone	Westphalian sandstone	Sand and gravel

Note: (1) The analyses were carried out by the Irish Geological Survey State Laboratory
     (2) All units in mg/l
Adapted from Daly, 1994b

Iron and manganese   Excess concentrations of iron do not usually cause health problems but are of concern for aesthetic and taste reasons. When drawn from a well or tap, the water may be colourless but, when it comes into contact with air, the iron precipitates to form a reddish-brown deposit resembling rust. This gives a metallic taste to the water and stains plumbing fixtures and clothes. Manganese causes a black discoloration of the water.

The source of iron can be iron minerals in the rocks or soils, pollution by organic wastes or occasionally the corrosion of iron fittings in the water system. Groundwater from certain rock types such as dark muddy limestones, shales and sandstones and from peat areas may contain very high iron concentrations. The breakdown of organic wastes from septic tank systems, farmyards and other sources can cause the formation of carbon dioxide and oxygen deficient conditions and can bring the iron into solution in the groundwater. Manganese is frequently associated with iron although it is less prevalent. It is also a good indicator of pollution by wastes with high BOD, such as agricultural silage effluent.

Hydrogen sulphide   Hydrogen sulphide is a gas that is recognizable by its 'rotten egg' smell. It is present only in deoxygenated water, from rocks such as black clay like limestones or shales that contain pyrite, or from evaporite beds. It is often associated with iron problems. See Chapters 12 and 13 for a discussion of hydrogen sulphide with regard to wastewater and anaerobic digestion.

Sulphate   Significant concentrations of sulphate may be present where lenses of evaporites in limestones have produced levels of sulphate of up to 800 mg/l. The problems caused by these constituents can be solved by using water treatment systems and, where applicable, by removing the pollution sources.

Sodium chloride   Saline intrusion into aquifers can cause high NaCl levels in groundwater. Problems occur in areas where the rocks are highly permeable and where there is a low hydraulic gradient. The problem may be exacerbated by groundwater abstraction wells near the coast.

4.14.2 Some Industrial Contributions to Groundwater Pollution

Pesticides   In considering pesticides, it is worth distinguishing between those that are used for agricultural and non-agricultural purposes, and between point and diffuse pollution sources. Non-agricultural uses are the control of weeds on road verges, playing fields, around housing and industrial estates and on railway tracks. Point sources are pesticide spraying of small areas, disposing of spent sheep dip in soakways, soakways for road runoff, spillages (often during loading, mixing and rinsing), leakages and careless disposal of containers and washings.

There are more than 32 000 different pesticide compounds being used, containing some 1800 active ingredients (Houzim *et al.*, 1986). They vary in their solubility, persistence, mobility and toxicity, and consequently in the risk they pose to groundwater. However, there is increasing concern about pesticides and so the EC MAC (maximum admissible concentration) for individual pesticides is low—0.1 μg/l—and for total pesticides 0.5 μg/l. According to Hallberg (1989), pesticides are occurring in groundwater in the United States far more commonly that would have been predicted in the early 1980s. Also, detailed studies in the United States (Hallberg, 1989) have shown that pesticides can move through the topsoil and subsoil along preferential flow paths much faster and to greater depths than would be predicted by using the Darcian concept of flow. The potential problem with pesticides is further compounded by the fact that the breakdown products—the metabolites or degradates—may also be toxic and mobile (Hallberg, 1989). Aldicarb, a highly toxic product used on sugar beet, has a very short half-life, but it degrades into compounds which are not only equally or more highly toxic, but are also more persistent and considerably more mobile (OECD, 1986).

The triazine herbicides, atrazine and simazine (pre-emergent weed killers) account for 40 per cent of consumption of non-agricultural herbicides in Britain. They are mobile in soil and rocks, and are detected more frequently in surface water than any other pesticides (ENDS Report, 1993). They are used for weed control along railway tracks and roadways. Research by the Water Research Centre in the Granta Catchment in Cambridgeshire found that herbicides applied to cereals entered surface streams, although concentrations varied with river flow (stage) and time of year. The main pesticide peak occurred within hours of heavy rainfall. The main risk from agricultural usage of pesticides is to surface water and not groundwater.

Studies and monitoring in Britain have highlighted the pollution risk from spent sheep dip. The main pollutants in sheep dip are organophosphates (diazinon and propetamhos) although solvents may also be present. While organophosphates have a high solubility in water, they are relatively immobile (Houzim *et al.*, 1986). Yet when disposed of in soakaways, they have been detected at distances up to 400 m from the soakaway (ENDS Report, 1993).

In the United States and Canada, a total of 39 pesticide compounds have been detected in wells in 34 States or provinces (Hallberg, 1989). Compounds used as soil fumigants and namaticides, such as aldicarb, EBD, 1,2-D, and DBCP, which by nature are mobile and/or volatile, have been widely detected. These pesticides are used on vegetables and speciality crops. The other commonly detected compounds are herbicides—alachlor, atrazine (and metabolites).

Industrial solvents   Groundwater supplies face a growing threat from a wide range of synthetic organic chemicals, as a result of their casual disposal, leakage or spillage (Lawrence, 1990). Of these, the chlorinated solvents which are widely used in industry appear to be the most commonly occurring contaminants. These solvents are used as degreasers for various processes in the metal, electronic, chemical, paper, textile and leather industries. Also, they are usually present in landfill leachate. The four chlorinated solvents most widely used in Britain today are perchlorethylene, trichlorethylene (TCE), methylene chloride and methyl chloroform (Lawrence, 1990).

Solvents and other synthetic organic chemicals are both a complicated and a significant hazard for groundwater:

- They are of environmental significance at very low concentrations—parts per billion (ppb or μg/l) quantities. A small spill of a few litres could potentially pollute many millions of litres of groundwater. For instance, one kilogram of pentachlorophenol entering groundwater would pollute 2×10^6 m^3 (440 million gallons).
- They are difficult to sample due to volatilization and analyses are expensive.
- They are resistant to degradation and so are persistent.
- Breakdown products do not always result in a harmless or less harmful product (MacKay *et al.*, 1985). Biotransformation of TCE can result in the formation of hazardous products such as vinyl chloride, which is a confirmed human carcinogen (Burmaster, 1982). Also, even if biotransformation products are less hazardous, they may be more mobile and create bigger problems.

Refined mineral oils   These include petrol, aviation fuel, diesel fuel and heating oils of various grades. They range in viscosity, but have densities less than that of water and a heterogeneous composition dominated by pure hydrocarbons (Ashley and Misstear, 1990). In groundwater they cause taste and odour problems, so the EC MAC is 10 μg/l. Pollution arises from petrol stations, oil storage depots, airfields and spillages during transport. Private and farm wells are occasionally polluted in Ireland by leakages from domestic oil storage tanks and by spillages that occur on farms when tractors are being filled with diesel. Petrol stations are the other main potential source of pollution, particularly from leaking underground storage tanks (LUSTS) and spillages. However, several of the petrol distribution companies are now checking their service stations.

Recently, a new threat from petrol has been realized. This is an additive used in unleaded petrol as an octane booster, called methyl butyl tertiary ether (MTBE). It forms up to 5 per cent of regular blends and as much as 15 per cent of super blends. It is ten times more soluble in water than other constituents in petrol, and it dissolves and spreads rapidly in groundwater. Contamination of groundwater by MTBE is now being recorded in Britain, although it is not particulary toxic (ENDS Report, 1993). MTBE has one advantage—it is a good indicator of petrol contamination because of its solubility and low taste threshold (10 μg/l) and so it is likely to enable easier and quicker identification of pollution of groundwater than the more toxic ingredients such as benzene.

4.14.3 Some Indicators of Groundwater Contamination

Introduction   There has been a tendency in analysing groundwater samples to test for a limited number of constituents. A 'full' or 'complete' analysis, which includes all the major anions and cations, is not generally recommended for routine monitoring and for assessing pollution incidents. This enables (*i*) a check on the reliability of the analysis (by doing an ionic balance), (*ii*) a proper assessment of the water chemistry and quality, and (*iii*) a possible indicator of the source of contamination. A listing of recommendations and optional parameters are given in Table 4.21. It is also important that the water samples taken for analysis have not been chlorinated. The following parameters are good contamination indicators: E. coli, nitrate, ammonia, potassium, chloride, iron, manganese and trace organics.

Faecal bacteria and viruses   E. coli is the parameter tested as an indicator of the presence of faecal bacteria and perhaps viruses; constituents which pose a significant risk to human health. The most common health problem arising from the presence of faecal bacteria in groundwater is diarrhoea, but typhoid fever, infectious hepatitis and gastrointestinal infections can also occur. Although E. coli bacteria are an excellent indicator of pollution, they can come from different sources, e.g. septic tank effluent, farmyard waste, landfill sites, birds. The faecal coliform : faecal streptococci ratio has been used as a tentative indicator to distinguish between animal and human waste sources (Henry *et al.*, 1987). However, care is needed in interpreting the results.

Viruses are a particular cause for concern as they survive longer in groundwater than indicator bacteria (Gerba and Bitton, 1984). They published data on elimination of bacteria and viruses in groundwater has been compiled by Pekdeger and Matthess (1983), who show that in different investigations, 99.9 per cent elimination of E. coli occurred after 10 to 15 days. The mean of the evaluated investigations was 25 days. They show that 99.9 per cent elimination of various viruses occurred after 16 to 120 days, with a mean of 35 days for Polio-, Hapatitis-, and Entero-viruses.

The natural environment, in particular the soils and subsoils, is effective in moving bacteria and viruses by predation, filtration and absorption. There are two high risk situations: (*i*) where permeable sands and gravels with a shallow water are present; and (*ii*) where fractured rock, particularly limestone, is present close to the ground surface. The presence of clay like gravels, tills, and peat will, in many instances, hinder the vertical migration of microbes, although preferential flow paths, such as cracks in the clay like materials, can allow rapid movement and bypassing of the subsoil.

Nitrate   Nitrate is one of the most common contaminants identified in groundwater and increasing concentrations have been recorded in many developed countries. The consumption of nitrate rich water by young children may give rise to a condition known as methaemoglobinaemia (blue baby syndrome). The formation of carcinogenic nitrosamines is also a possible health hazard and epidemiological studies have indicated a positive correlation between nitrate consumption in drinking water and the incidence of gastric cancer. However, the correlation is not proven according to some experts (Wild and Cameron, 1980). The EC MAC for drinking water is 50 mg/l.

The nitrate ion is not absorbed on clay or organic matter, (Kolenbrander, 1975). It is highly mobile and under wet conditions is easily leached out of the rooting zone and through soil and permeable subsoil. As the normal concentrations in uncontaminated groundwater is low (less than 5 mg/l), nitrate can be a good indicator of contamination by fertilizers and waste organic matter.

Table 4.21 Parameters used to assess groundwater contamination

Recommended Parameters

Appearance	Calcium (Ca)	Nitrate (NO$_3$)*
Sediment	Magnesium (Mg)	Ammonia (NH$_3$)*
pH (lab)	Sodium (Na)	Iron (Fe)*
Electrical conductivity (EC)*	Potassium (K)*	Manganese (Mn)*
Total hardness	Chloride (Cl)*	
General coliform	Sulphate (SO$_4$)*	
E. coli*	Alkalinity	

Optional Parameters (depending on local circumstances and reasons for sampling

Fluoride (F)	Fatty acids*	Zinc (Zn)
Orthophosphate	Trace organics*	Copper (Cu)
Nitrite (NO$_2$)*	TOC*	Lead (Pb)
BOD.*	Boron (B)*	Other metals
Dissolved oxygen*	Cadmium (Cd)	

* Good indicators of contamination
Adapted from Daly, 1994

In the past there has been a tendency to assume that the presence of high nitrates in well water indicated an impact by inorganic fertilizers. This assumption has frequently been wrong, as examination of other constituents in the water showed that organic wastes—usually farmyard waste, probably soiled water—were the source. The nitrate concentrations in wells with a low abstraction rate—domestic and farm wells—can readily be influenced by soiled water seeping underground in the vicinity of the farmyard or from the spraying of soiled water on adjoining land. Even septic tank effluent can raise the nitrate levels; if a septic tank system is in the zone of contribution of a well, a four-fold dilution of the nitrogen in the effluent is needed to bring the concentraton of nitrate below the EC MAC.

The EC Directive (91/676/EEC) on nitrates from agricultural sources allows for the designation of 'vulnerable zones', which are areas of land that drain into surface or groundwaters which are intended for the abstraction of drinking water and which could contain more than 50 mg/l nitrate if protective action is not taken. If areas are designated as vulnerable it will have repercussions for farmers in these areas as the application of livestock manure/slurry and inorganic fertilizers will be restricted. This is discussed further in Chapter 10.

Ammonia Ammonia has a low mobility in soil and subsoil and its presence in groundwater indicates a nearby waste source and/or vulnerable conditions.

Potassium Potassium (K) is relatively immobile in soil and subsoil. Consequently, the spreading of manure, slurry and inorganic fertilizers is unlikely to significantly increase the potassium concentrations in groundwater. In most areas in Ireland, the background potassium levels in groundwater are less than 3.0 mg/l. Higher concentrations are found occasionally where the rock contains potassium, e.g. certain granites and sandstones. The background potassium: sodium ratio in most Irish groundwaters is less than 0.4 and often 0.3. The K : Na ratio of soiled water and other wastes derived from plant organic matter is considerably greater than 0.4. Consequently, a K : Na ratio greater than 0.4 can be used to indicate contamination by plant organic matter—usually in farmyards, occasionally landfill sites (from the breakdown of paper). However, a K : Na ratio lower than 0.4 does not indicate that farmyard wastes are not the source of contamination, as K is less mobile than Na (phosphorus is increasingly a significant pollutant and cause of eutrophication in surface water. It is *not* a problem in groundwater as it usually is not mobile in soil and subsoil).

Chloride   The principal source of chloride in uncontaminated groundwater is rainfall and so in any region, depending on the distance from the sea and evapotranspiration, chloride levels in groundwater will be fairly constant. Chloride, like nitrate, is a mobile cation. Also, it is a constituent of organic wastes. Consequently, levels appreciably above background levels have been taken to indicate contamination by organic wastes. While this is probably broadly correct, chloride can also be derived from potassium fertilizers.

Iron and manganese   Although they are present under natural conditions in groundwater in some areas, they can also be good indicators of contamination by organic wastes. Effluent from the wastes cause deoxygenation in the ground which results in dissolution of iron (Fe) and manganese (Mn) from the soil, subsoil and bedrock into groundwater. With reoxygenation in the well or water supply system the Fe and Mn precipitate. High Mn concentrations can be a good indicator of pollution by silage effluent. However, it can also be caused by other high BOD wastes, such as milk, landfill leachate and perhaps soiled water and septic tank effluent.

Groundwater sampled immediately after a borehole is drilled often contains suspended solids which usually diminish after a few hours pumping. Analysis of samples taken with suspended solids can show high Fe and/or Mn concentrations which are due to the suspended solids and not the natural groundwater quality. Consequently, samples taken during pumping tests on new wells should be filtered in order to get representative Fe and Mn concentrations.

4.14.4 Groundwater Vulnerability to Pollution—A Hydrogeological Element of Risk

Vulnerability is a term used to represent the intrinsic geological and hydrogeological characteristics that determine the ease with which groundwater may be contaminated by human activities (Daly and Warren, 1994). In considering the location and/or control of a potentially polluting activity in an area, it is essential to appreciate that the vulnerability is a natural inherent (or fixed) characteristic of any area whereas the pollution loading can usually be controlled or modified.

In general, groundwater depends on the time of travel of groundwater (and contaminants), on the relative quantity of contaminants that can reach the groundwater and on the contaminant attenuation capacity of the geological materials. As all groundwater is hydraulically connected to the land surface, it is the effectiveness of this connection that determines the relative vulnerability to contamination. Groundwater that readily and quickly receives water (and contaminants) from the land surface is considered to be more vulnerable than groundwater that receives water (and contaminants) more slowly and in lower quantities. The travel time, attenuation capacity and quantity of contaminants are a function of the following natural attributes of any area:

1. the subsoils that overlie the groundwater;
2. the recharge type—whether point or diffuse;
3. in the case of sands/gravels, the thickness of the unsaturated zone;
4. in the case of diffuse pollution sources, the topsoil;
5. hydraulic conductivity.

In general, little attenuation of pollutants occurs in the bedrock because flow is almost wholly via fissures. Consequently, the subsoil—sands, gravels, glacial tills (or boulder clays), lake and alluvial silts and clays, peat—are the single most important natural features in influencing groundwater vulnerability and groundwater pollution prevention. Groundwater is most at risk where the subsoils are absent or thin and, in karstic areas, where surface streams sink underground at swallow holes. The influence of the various geological and hydrogeological factors is summarized below and in Table 4.22.

Table 4.22 Range of groundwater vulnerability

Low vulnerability (good protection)	High vulnerability (poor protection)
1. High clay or organic content	1. Low clay or organic content
2. Low permeability subsoil, e.g. clay	2. High permeability subsoil, e.g. gravel
3. Thick subsoil	3. Thin or absent subsoil
4. Thick unsaturated zone	4. Thin unsaturated zone
5. Intergranular flow	5. Fissure or karstic flow
6. Diffuse recharge	6. Point recharge

Adapted from Daly, 1994

Hydrogeological factors

Influence of subsoils   The subsoils act as a protecting layer over groundwater by both physical and chemical/biochemical means. Fine grained sediments such as clay like till, lacustrine clays and peats have a low permeability and consequently can act as a barrier or hindrance to the vertical movement of pollutants. In areas where these sediments are present, surface water is more at risk than groundwater as most, if not all the contaminants cannot migrate downwards and can only move laterally. Even if the permeability is sufficiently high to allow slow intergranular movement of pollutants, for instance in sandy tills or silts, the sediments can strain out and absorb bacteria and viruses. In contrast, high permeability deposits—sands and gravels—allow easy access of pollutants to the water table although they provide opportunities for dispersion of the pollutants among the pore spaces.

Sorption, ion-exchange and precipitation are vital chemical processes in attenuating pollutants. The cation–exchange capacity of subsoils depends on the clay and/or organic content and ranges from essentially zero for sands to about 50 meq/100 g for clay soils to over 100 meq/100 g for peat. Consequently, clays and peats can attenuate bacteria, viruses and chemical pollutants such as cadmium, mercury, lead, potassium and ammonium whereas clean sand and gravel has little effect. In general, the higher the clay content and the lower the permeability, the greater the protection of groundwater from pollution.

Influence of permeability type   Permeability, discussed earlier, is a measure of the capacity of a rock to transmit water, can be subdivided into two types. First, where the water moves between the grains of subsoils such as sand and gravel, it is called primary or intergranular permeability. Secondly, where the water moves through fractures or fissures or joints and along bedding planes, it is called secondary or fissure permeability.

Intergranular flow is slower than fissure flow in rocks under most conditions. Void or pore sizes are usually smaller and flow paths are more irregular. Also, the amount of water stored in granular rocks is generally greater than in fissured rocks. These factors have an important bearing on pollutant attenuation. In contrast to rocks in which fissure flow dominates, the slow flow rate in rocks with an intergranular permeability delays the entry of pollutants into groundwater and, particularly in the unsaturated zone, allows time and opportunities for interactions between pollutants and rock grains. Also, the relatively small pore sizes allow filtration and absorption of bacteria and viruses. The irregular flow paths within a porous matrix causes hydrodynamic dispersion which decreases pollutant concentration. For pollutants that reach the water table and enter groundwater, dilution is much greater in rocks with an intergranular permeability and thus the resultant pollutant concentrations are much less. The worst situation is in karst limestone area where flow rates are very high—over 100 m/h in some instances—due to widening of fissures by solution and there is little scope for attenuation other than by somewhat limited dilution. Consequently, there is generally far greater degradation and purification of pollutants in rocks with an intergranular permeability than in those with a fissure permeability.

Importance of unsaturated zone   In sands/gravels, a deep water table reduces the likelihood of pollution because the pollutants have to travel farther and are slower to reach the groundwater. This allows for the beneficial physical, chemical and biological processes, that occur in the unsaturated zone, to attenuate the pollutants (Daly and Wright, 1982).

Vulnerability and mapping

The geological and hydrogeological factors can be examined and mapped, thereby providing a groundwater vulnerability assessment for any area or site. Four groundwater vulnerability categories are used by the Geological Survey of Ireland (GSI)—extreme, high, moderate and low (see Table 4.22). Vulnerability assessments should be an essential element when considering the location of potentially polluting activities, such as landfill sites. The level of assessment required, and consequently the geological and hydrogeological data needs, will depend on the degree of hazard provided by the pollutant loading. In most instances a desk study and a simple, rapid site investigation using trial pits and existing wells will be adequate. However, a comprehensive site investigation is required for landfill sites.

Case study—groundwater vulnerability in Ireland On an international scale, Irish aquifers are relatively vulnerable to pollution:

1. Irish aquifers are mostly unconfined. (An aquifer is said to be confined when it is overlaid by impermeable materials so that the aquifer is completely filled with water and the water is under pressure.)
2. The subsoils overlying bedrock aquifers are often thin (< 10 m), are sometimes absent, and are often relatively permeable, consisting either of sands/gravels or sandy till. However, there are areas in virtually every county in Ireland, substantial areas in some counties, where the subsoils are thick (> 10 m).
3. Irish aquifers are generally shallow—most bored wells are less than 100 m deep.
4. Irish aquifers generally have a shallow water table, and, therefore only a thin unsaturated zone.
5. Irish aquifers exhibit only fissure flow; the sands/gravels are the only significant exception.

However, two factors help to mitigate the adverse consequences of such vulnerability:

1. Ireland has a high rainfall which provides substantial dilution for contaminants.
2. The generally high watertable and rapid through-put ensure that in many areas groundwater is discharged fairly rapidly to surface streams. The groundwater circulation therefore tends to be localized both in space and time. Thus pollution tends not to affect a wide area and is often of short duration (weeks and months rather than years).

Conclusion to vulnerability mapping In conclusion, the vulnerability concept and vulnerability maps are useful in the location of potentially polluting developments. Firstly, they indicate and are a measure of the likelihood of pollution. Secondly, they enable developments to be located in relatively low vulnerability and therefore in relatively low risk areas, from a groundwater point of view, and/or they enable suitable engineering preventative measures to be taken. See Domenico and Schwartz (1990) for a more detailed treatment of hydrogeology.

4.15 PROBLEMS

4.1 Prepare a flow chart for 100 units of precipitation falling on land and ultimately returning to the hydrologic cycle via runoff to the sea or evaporation. What are the respective percentages of the rainfall experienced at each process for a wet temperate climate (Ireland) by comparison with a tropical climate (Brazil).

4.2 Determine the volume of water lost through evaporation during one year from a 10 km^2 lake, if the annual rainfall is 1200 mm and the net increase in depth of the lake is 200 mm. State your assumptions.

4.3 Write a brief report on a significant hydraulic project in your area (dam, river diversion, etc.) which impacted on the natural waters. Describe the impacts, positive and negative.

4.4 Determine the infiltration rate based on the $\varnothing$ index method for the following rainstorm:

Time	Rainfall (mm/h)	Runoff equivalent (mm/h)
0–5 min	55	20
5–10 min	60	30
10–15 min	65	40
15–20 min	70	50
20–25 min	80	65
25–30 min	60	50
30–35 min	40	35
35–40 min	20	15
40–45 min	15	10

4.5 For Problem 4.4, fit a Horton-type equation to the infiltration and determine the initial infiltration f_o, the final infiltration f_c and the rate constant k. From this equation verify the infiltration rates at 30 min and 45 min and determine the total infiltration.

4.6 In your college library, find the section that contains precipitation and streamflow records. For a local watershed, record the monthly rainfall averages for a one year period. Determine the watershed area. Also record the monthly average streamflows. Now compute and plot the trends of rainfall and streamflow for each of the twelve months. Plot the rainfall and streamflow on similar units.

4.7 Explain why in Problem 4.6 the differences between rainfall and runoff are not constant throughout the year (i.e. not the same percentage of rainfall goes to runoff for all rain events).

4.8 For Problem 4.6, also record the monthly average for evaporation for the same location as the precipitation station. Now plot precipitation and evaporation. Discuss the trends. Also plot evaporation plus energy runoff and compare with precipitation. Discuss why these two lines are not similar.

4.9 In Raudkivi (1979) locate the section on evaporation with respect to evaporation from totally dry surfaces and water-only surfaces. With figures, explain the differences in the radiation budget parameters.

4.10 Compute the evaporation as an annual water loss from a lake of 100 km^2 if the mean wind speed at 2 m is 4 m/s. Assume values of $e_s^* \cong 15 \text{ mm Hg}$ and $E_a \cong 10 \text{ mm Hg}$ respectively. What effect would a higher wind speed have on evaporation?

4.11 Describe three field computation methods for the determination of evaporation.

4.12 With regard to Example 4.7, compute the series of evaporations for $R_n - G = 70 \text{ w/m}^2$, take temperature $= 20°C$, but vary B from 0.1 to 5 in increments of 0.1. Graph your results.

4.13 As with Problem 4.12, repeat the exercise if $R_n - G = 70 \text{ w/m}^2$ and $B = 0.1$ if T is 10, to $40°C$ in increments of $2°C$. Graph your results.

4.14 Prepare intensity–duration–frequency (IDF) curves for 5 years and 2 years for the following data of annual maximum intensity in mm/h:

Year	30 min duration	60 min duration	90 min duration	120 min duration
1980	48.2	33.1	27.9	26.2
1981	26.8	17.6	15.2	14.8
1982	35.4	29.8	17.6	23.2
1983	47.8	39.2	31.8	27.6
1984	44.3	34.8	32.1	25.8
1985	51.8	40.6	33.2	29.1
1986	46.3	41.4	36.4	30.3
1987	47.9	39.8	32.1	27.6
1988	44.8	34.2	29.8	24.8
1989	60.3	50.1	42.1	37.6
1990	18.2	16.8	15.1	12.8

(Refer to Shaw, 1994, or Bras, 1990, for methodology.)

4.15 Go to your library and compile the data for the period 1980–1995 compatible with rainfall records for your own area to that of Problem 4.14. Then compute the 2 year and 5 year IDF curves.

4.16 Review the paper 'Field studies of hillslope flow processes' by Dunne (1978).

4.17 Compute the 30 year flood from the following data: catchment area, 15 km^2, average annual rainfall, 1200 mm, stream slope, 2.9 m/km, stream length ~9.5 km, stream frequency, 1.76. Assume a lake occupies 15 per cent of the catchment and the subsoil classification is 100 per cent class S2. Use the British Flood Studies method of catchment characteristics to compute the flood magnitude.

4.18 You have been asked to size a culvert to carry a small stream beneath a new roadway. Outline the method(s) you would use to do this, assuming there is no hydrologic data.

4.19 Go to your library and compile for your area the highest annual floods for at least a 10 year period. A longer record is better. Use the method of Example 4.9 to compute the return period floods from 1 year to 100 years.

4.20 Repeat Problem 4.19, but now compile your data for the annual low flows. A suitable low flow may be the lowest 7 day sustained low flow. Other definitions also apply.

4.21 Two rivers are separated by a 15 km wide unconfined aquifer. The left river water level is 28 m above datum and the right river is 14 m above datum. Compute the height of the water divide if $K = 10^{-5}$ m/s and there is no recharge.

4.22 Repeat problem 4.21 for the effect of recharge.

4.23 For Problem 4.21, compute the flow rate and direction in the aquifer.

4.24 Use Table 4.19 for the parameter hydraulic conductivity (or a more detailed table from more specific texts) and compile a graph of K versus soil/rock type. Indicate the ranges for each soil/rock type. If the environmental requirement for the base of a landfill is to have $K < 10^{-9}$. State what are the most suitable soil/rock strata.

4.25 Compute the approximate aquifer transmissivity from the following stepped pumping tests:

Step 1	Q (m^3/day)	S_t (m)
1	52.1	0.17
2	68.2	0.32
4	83.8	0.46
6	99.4	0.69

4.26 Determine in your area the water sources used for potable water. Write a brief report to the City Engineer advising him of the current water demands, the current water sources (groundwater, freshwater) and future demands. Suggest how future demands may be met.

4.27 For Example 4.12, plot the cumulative drawdown divided by appropriate flow rate against flow rate. What are the aquifer loss coefficient and the well loss coefficient from this plot?

4.28 Consider a location you are familiar with that abstracts groundwater through wells for drinking purposes. Identify on a map (e.g. scale 1:20 000):

  (*a*) the wells;
  (*b*) the geology;
  (*c*) the land use activities (industrial and agricultural).

Prepare a map showing the areas that need protection, categorizing the areas at highest, modest and lowest risk from the land use activities.

REFERENCES AND FURTHER READING

Abbot, M. B., J. C. Bathurst, J. A. Cargo, P. E. O'Connell and J. Rosmussen (1986) 'An introduction to the European Hydrologic System—SHE', *J. Hydrology*, **87**, 61–77.

Ackers, P., *et al.* (1978) *Weirs and Flumes for Flow Measurement*, John Wiley, New York.

Ashley, R. P. and B. D. Misstear (1990) *Industrial development: the threat to groundwater quality.* Paper presented to the Institution of Water and Environmental Management, East Anglia Branch.

Baumgartner, A. and E. Reichel (1975) *The World Water Balance*, Elsevier Science Publishers, Amsterdam and New York, 179 pp., 31 maps.

Bedient, P. B. and W. C. Huber (1988) *Hydrology anbd Floodplain Analysis*, Addison-Wesley, Reading, Massachusetts.

Betson, R. S. (1964) 'What is watershed runoff?', *J. Geophy. Research*, **69**(8), 1541–1552.

Bilham, E. G. (1936) 'Classification of heavy (rain) falls in short periods'. *British Rainfall, 1935*, pp. 262–280.

Bishop, K. H. (1991) 'Episodic increases in stream acidity, catchment flow pathways and hydrograph scenario' PhD thesis, Jesus College, Cambridge, October 1991.

Bras, R. L. (1990) *Hydrology—An Introduction to Hydrological Science*, Addison-Wesley, Reading, Massachusetts.

Brutsaert, W. (1982) *Evaporation into the Atmosphere*, Kluwer Academic Publishers, Dordrecht, The Netherlands.

Budyko, M. I. (1974) *Climate and Life*, Academic Press, New York, 508 pp.

Burmaster, D. E. (1982) 'The new pollution.' *Environment.* **24**, 2, pp. 6–36.

Calder, I. R. (1990) *Evaporation in the Uplands*, John Wiley, New York.

Calder, I. R., M. D. Newson and P. D. Walsh (1982) 'The application of catchment lysimeter and hydrometeorological studies of coniferous afforestation in Britain to land use planning and water management'. *Proceedings of Symposium on Hydrological Research Basins*. pp. 853–863. Bern, Switzerland.

Campbell, J. B. (1987) *Introduction to Remote Sensing*, Guildford Press, USA.

Carton, O. T., M. Sherwood, V. Power and J. J. Lenehan (1991) *Soils as an Assimilation of Chemical Loadings from Fertilizers and Agricultural Wastes*, IEI Conference, Johnstown Castle, Wexford, Ireland, November 1991.

Caspany, H. J. (1990) 'An echohydrological framework for water yield changes of forested catchments due to forest decline and soil acidification'. *Water Resources Research*. Vol. 26, No. 6, pp. 1121–1131.

Chow Ven Te, D. R. Maidment and L. Mays (1988) *Applied Hydrology*, McGraw-Hill, New York.

Clark, T. P. and R. Piskin (1977) 'Chenical quality and indicator parameters for monitoring landfill leachate in Illinois' *J. Environmental Geology*, **1**, 329–339.

Collinge, V. and C. Kirby (1987) *Weather Radar and Flood Forecasting*. John Wiley and Sons, Chichester, UK.

Commission of the European Community (1991) 'Council Directive concerning the protection of waters against pollution caused by nitrates from agricultural sources' (91/676/EEC). *Official Journal of the European Communities*, No. L375/1-8.

Cunnane, C. (1978) 'Unbiased plotting positions—a review', *J. Hydrology*, **37**, 205–222.

Daly, D. and G. R. Wright (1982) *Waste Disposal sites. Geotechnical guidelines for their selection, design and managment.* Geological Survey of Ireland Information Circular 82.1 50 pp.

Daly, D. (1991) 'Groundwater protection schemes'. *Proceedings of Annual Spring Show Conference of Local Authority Engineers.* Department of the Environment, Dublin.

Daly, D. (1994a) 'General guidelines on aquifer definition'. *The GSI Groundwater Newsletter*, No. 25.

Daly, D. (1994b) *Lecture Series on Groundwater*, Civil and Environmental Engineering Department, University College Cork.

Daly, D. and W. P. Warren (1994) 'Mapping groundwater vulnerability to pollution: Geological Survey of Ireland guidelines'. *The GSI Groundwater Newsletter*, No. 25, pp. 10–15.

Daly, E. P. (1995) *Groundwater resources of the Nore River Basin*. Geological Survey of Ireland, Report Series.

Davis, M. L. and D. A. Cornwell (1991) *Introduction to Environmental Engineering*. McGraw-Hill, New York.

Dillon, E. C. (1954) 'Analysis of 35-year automatic recording of rainfall at Cork', *Trans. Inst. Civ. Engng. Ireland*, **80**, 191–283.

Dingman, S. L. (1994) *Physical Hydrology*, Macmillan, London.

Domenico, P. A. and F. W. Schwartz (1990) *Physical and Chemical Hydrology*, John Wiley, New York.

Dooge, J. C. I. (1973) 'Linear theory of hydrologic systems', US Department of Agriculture, Technical Bulletin 1468.

Dunne, T. (1978) 'Field studies of hillslope flow processes', in *Hillslope Hydrology*, M. J. Kirby (ed.), John Wiley, New York.

Eagleson, P. S. (1970) *Dynamic Hydrology*. McGraw-Hill, New York.

ENDS Report (1993) *Cutback on triazine herbicides gathers pace*. Environmental Data Services Ltd Publishers, No. 194, pp. 7.

ENDS Report (1993) *Research underlines pollution risks from sheep dip chemicals*. Environmental Data Services Ltd Publishers, No. 218, pp. 8–18.

ENDS Report (1993) *Unleaded Petrol*. Environmental Data Services Ltd Publishers, No. 225.

Engman, E. T. and R. J. Gurney (1991) *Remote Sensing in Hydrology*, Chapman and Hall, New York.

Fahey, B. D. and A. J. Watson (1991) 'Hydrological impacts on converting tussock grassland to pine plantation, Otago, New Zealand'. *N.E. Journal of Hydrology*, Vol. 30. No. 1.

Farrell, E. P. and G. M. Boyle (1991). *Monitoring of a forest ecosystem in a region of low level anthropogenic emissions—Ballyhooley Project*. Report No. 4, Forest Ecosystems Research Group, University College Dublin, Ireland.

Fitzpatrick, E. A. (1986) *An Introduction to Soil Science*, 2nd edn, Longman Scientific and Technical, London.

Forest Research Institute (1980). *What's new in Forest Research*. Report No. 92, Forest Research Institute, New Zealand.

Freeze, R. A. (1980) 'A stochastic conceptual analysis of rainfall–runoff processes on a hillslope, *Water Resources Res.* **16**(2), 391–408.

Gannon, J. (1993) 'An hydrological study of the Dripsey Catchment, Co. Cork, Ireland', MEngSc thesis, University College, Cork, Ireland.

Gardiner, M. J. and T. Radford (1980). *Soil Associations of Ireland and Their Land Use Potential*, National Soil Surveys of Ireland, An Foras Taluntais, Ireland.

Hall, M. J. (1986) *Urban Hydrology*, Elseiver Applied Science, Amsterdam.

Hallberg, G. R. (1989) 'Pesticide pollution of groundwater in the humid United States'. In: J. M. Sturrock and T. L. V. Ulbricht (eds), *Agriculture, Ecosystems and Environment*, Elsevier Sciences Publishers, pp. 209–367.

Harbeck, G. E. (1962) 'A practical field technique for measuring reservoir evaporation utilizing mass transfer theory', US Geological Survey, Professional Paper 272-E.

Harbeck, G. E. and J. S. Meyers (1970) 'Present day evaporation measurement techniques', *Proc. ASCE*, **HY7**, 1381–1389.

Henry, H., R. H. Thorn, E. M. Brady and M. Doyle (1987) 'Septic tanks and groundwater—some recent Irish research'. In: *Proceedings of International Association of Hydrogeologists* (Irish Group), Portlaoise.

Hillel, D. (1980) *Applications of Soil Physics*, Academic Press, New York.

Hoeksima, R. J. and P. K. Kitanidis (1985) 'Analysis of spatial structures of properties of selected aquifers', *Water Resource Research*, **21**, 563–572.

Holland, D. J. (1967) 'Rain intensity frequency relationships in Britain'. *British Rainfall 1961, part III*, pp. 43–51.

Horton, R. E. (1933) 'The role of infiltration in the hydrological cycle', *Trans. Am. Geophysics Union*, **14**, 443–460.

Houzim, V., J. Vavra, J. Fuksa, V. Pekny, J. Vrba and J. Stibral (1986) 'Impact of fertilizers and pesticides on groundwater quality'. In: J. Vrba and E. Romijn (eds), *Impact of Agricultural Activities on Groundwater*, International Association of Hydrogeologists, Vol. 4, pp. 89–132.

Hydraulics Research Ltd (1983) *Design and Analysis of Urban Storm Drainage—The Wallingford Procedure*, Hydraulics Research Ltd Wallingford.

Institute of Hydrology (1976) *Water Balance of the headwater in catchments of the Wye and Severn 1970–1975*. Report No. 33, Institute of Hydrology, UK.

Irish Meteorological Office (1993) Personal Communication regarding rainfall and evaporation data.

Jacob, C. E. (1950) *Flow of Groundwater Engineering Hydraulics*, H. Rouse (ed.), John Wiley, New York, pp. 321–386.

Johnson, A. I. (1962) 'Physical and hydrologic properties of water bearing deposits from core holes in the Los Banos–Kettleman City Area, California, Denver Co. USGS Open File Report.

Johnson, A. I. (1967) 'Specific yield compilation of specific yields for various materials', USGS Water Supply Paper 1662-D, 74 pp.

Kaimal, J. C. and J. J. Finnigan (1994) *Atmospheric Boundary Layer Flows. Their Structure and Measurement*, Oxford University Press.

Kirby, C., M. D. Newson and K. Gillman (1991). *Plynlimon Research—The first two decades*. Report No. 109, Institute of Hydrology, UK.

Kolenbrander, G. J. (1975) 'Nitrogen in organic matter and fertilizer as a source of pollution'. In: *Proceedings of IAWPR Conference "Nitrogen as a Water Pollutant"*, Copenhagen.

Korzun, v.I. *et al.* (eds) (1978) *World Water Balance and Water Resources of the Earth*, USSR National Committee for the International Hydrological Decade, UNESCO Press, Paris, 663 pp.

Kuichling, E. (1989) 'The relationship between the rainfall and the discharge of sewers in populous districts'. *Transactions ASCE*, **20**, pp. 1–56.

Law, F. (1956) 'The effect of afforestation upon the yield of water catchment areas', British Association for Advancement of Science, Sheffield, 5 September 1956.

Lawrence, A. (1990) 'Groundwater pollution threat from industrial solvents'. *NERC News*, No. 13, pp. 18–19.

Lazaro, T. R. (1990) *Urban Hydrology. A Multidisciplinary Perspective*, Technomic. Lancaster, Pennsylvania, USA.

Lindsley, R. K. and J. B. Franzini (1979) *Water Resources Engineering*, McGraw-Hill, New York.

Lloyd-Davies, D. E. (1906) 'The elimination of storm water from sewerage systems'. *Proceedings of Inst. of Civil Engineers U.K.* No. 164, pp. 41–67.

MacKay, D. M., P. V. Roberts and J. A. Cherry (1985) 'Transport of organic contaminants in groundwater'. *Environ. Sci. Technol.* Vol. 19, pp. 384–392.

Maidment, L. (1993) *Hydrology Handbook*, McGraw-Hill, New York.

Marani, A. and A. Rinaldo (1992) *Transport Processes in the Hydrological Cycle*, Instituto Veneto Di Scienze, Lettere ed Arti, Venice.

Morris, P. (1992) 'The hydrological effects of afforestation on upland catchments', MEngSc thesis, University College, Cork, Ireland.

Mulholland, P. J., G. V. Wilson and P. M. Jardin (1990) 'Hydrogeochemical response of a forest watershed to storms. Effect of preferential flow along shallow and deep pathways'. *Water Resources Research*. Vol. 26, No. 12.

Mulvaney, T. J. (1851) 'On the use of self registering rain and flood gauges in making observations of the relations of rainfall and flood discharges in a catchment'. *Transactions of the Institute of Civil Engineers of Ireland, part II.* 4. pp. 13–33.

Natural Environmental Research Council (NERC) (1975) *Flood Studies Report 1975*, Charing Cross Road, London, UK.

Natural Environmental Research Council (NERC) (1979) *Flood Studies Report No. 5*, Institute of Hydrology, UK.

Obsermet (1991) *Catalogue of Meteorological Instrumentation*, Obsermet, Holland.

O'Driscoll, L. (1992) 'Catchment modelling of the lee', MEngSc thesis, University College, Cork, Ireland.

OECD (1986) *Water pollution by fertilizers and pesticides*. Organisation for Economic Co-operation and Development, Paris, 145 pp.

Pearce, A. J., L. K. Rowe and J. B. Stewart (1980) 'Nightime wet canopy evaporation rates and the water balance of an evergreen mixed forest'. *Water Resources Research*. Vol. 16. No. 5, pp. 955–959.

Pekdeger, A. and G. Matthess (1983) 'Factors of bacteria and virus transport in groundwater'. *Environmental Geology*, Vol. 5, No. 2, pp. 49–52.

Penman, H. L. (1948) 'Natural evaporation from open water, bare soil and grass', *Proc. Roy. Soc. London*, **A193**, 120–146.

Perroux, K. M. and L. White (1988) 'Designs for disc permeameters', *Soil Sci. Soc. Am. J.*

Philip, J. R. (1986) 'Linearised unsteady and multidimensional infiltration', *Water Resources Res.*, **22**, 1717–1727.

Priestley, C. H. B. and R. J. Taylor (1972) 'On the assessment of surface heat flux and evaporation using large scale parameters', *Monthly Weather Review*, **100**, 81–92.

Raudkivi, A. J. (1979) *Hydrology*, Pergamon Press, Oxford.

Rosenberg, N. J., B. L. Blad and S. B. Verma (1983) *Microclimate—The Biological Environment*, John Wiley, New York.

Shaw, E. M. (1994) *Hydrology in Practice*, Chapman and Hall, London.

Shuttleworth, W. J. (1988) 'Evaporation from Amazonian rainforest', *Proc. Roy. Soc. Lond.*, **B233**, 321–346.

Shuttleworth, W. J. (1991) 'The Modelling Concept', *Reviews of Geophysics*, **29**: pp. 585–606.

Steffan, W. L. and O. T. Denmead (1988) *Flow and Transpoert in the Natural Environment: Advances and Applications*, Springer-Verlag, Berlin.

Teagasc (Agriculture and Food Development Authority) (1990) *Environment Impact of Landspreading of Wastes*, Conference Proceedings, Wexford, Ireland.

Thorn, R. H. and C. Coxon (1992) 'Nitrates, groundwater and the nitrate directive'. In: J. Feehan (ed.), *Proceedings of Environment and Development in Ireland Conference*, University College Dublin, pp. 483–486.

Todd, D. K. (1980) *Groundwater Hydrology*, Studies and Reports in Hydrology, No. 25, John Wiley, New York.

UNESCO (1978) *World Water Balance and Water Resources of the Earth*, UNESCO Press, Paris.

USEPA (1984) *Hydrological Simulation Program FORTRAN-HSPF*, Environmental Research Laboratory, Athens, Georgia.

USEPA (1984) *Hydrological Simulation Program-Fortran* by R. C. Johnson and John C. Imhoff. USEPA, Athens, Georgia, USA.

Veissman, Jr, W., J. W. Knapp, G. L. Lewis and T. E. Harbaugh (1977) *Introduction to Hydrology*, Harper and Row, New York.

Vieira, J. R., P. Lindgaard-Jorgensen and I. R. Warrent (1994) 'Management support systems for the aquatic environment—concepts and technologies', *J. IAHR, Hydroinformatics*, **32**.

Wang, H. F. and M. P. Anderson (1982) *Introduction to Groundwater Modelling: Finite Difference and Finite Element Methods*. W. H. Freeman and Company, New York.

Weisner, C. J. (1970) *Climate, Irrigation and Agriculture*, Angus and Robertson.

White, I. (1988) 'Measurement of soil physical properties in the field', in *Flow and Transport in the Natural Environment: Advances and Applications*, W. L. Steffan and O. T. Denmead (eds), Springer-Verlag, Berlin.

Whitehead, P. G. and I. R. Calden (1993) 'The Balquhidder Catchment and process studies', *J. Hydrology Special Issue*, **145** (3–4), 15 May 1993.

Wild, A. and K. C. Cameron (1980) *Nitrate leaching through soil and environmental considerations with special reference to recent work in the United Kingdom. Soil Nitrogen–Fertilizer or Pollutant.* IAEA Publishers, Vienna, pp. 289–306.

Wilson, E. M. (1990) *Engineering Hydrology,* Macmillan, London.

Wright, G. R. (1984) *Aquifer Map of Ireland,* Proceedings of International Association of Hydrogeologists (IAHR), Irish Group, Portlaoise.

POLLUTION ENVIRONMENTS

FIVE

ECOLOGICAL SYSTEMS, DISTURBANCES AND POLLUTION

5.1 INTRODUCTION

We introduced the concept of tolerance in Chapter 2. Each species has a particular range of optimum for physicochemical factors such as temperature, pH, light, nutrients and biological factors such as food, competitors and predators that vary in space and time within the environment. A species is most successful in that area or place where the ranges of optima for the different factors overlap to the greatest degree; this is, in effect, the species niche—the entire set of optimal conditions under which the species survives and reproduces itself indefinitely. Each species within a community and habitat has a different and unique niche, some more similar to each other than others. In crude terms we can think of the habitat as a box into which the niches of all the species living in the community fit just like balls. This habitat niche space itself fits into the entire niche space of the biome and in turn the biosphere.

This diversion into ecological theory is important to set the scene for the discussion on disturbance and pollution of ecological systems. Disturbance can be defined as a discrete, punctuated killing, displacement or damaging of one or more individuals or colonies that directly or indirectly creates an opportunity for new individuals to become established (Sousa, 1984) and so causes a temporary or permanent shift in the community. Pollution refers to any change in the natural quality of the environment brought about by chemical, physical or even biological factors, and normally refers to the activities of man.

Physical conditions can change naturally in the short term (e.g. through natural disasters like flood, fire, storms, landslides, etc.) or in a directed way in the longer term (e.g. gradual climatic change), or the habitat can be disturbed by man (building, drainage, forest clearance). Biological processes like predation or grazing, non-predatory effects like digging and man-induced events like tree felling, hunting, mowing, etc., can also cause disturbances. Chemical conditions can be altered through the elevation of concentrations of substances (like nutrients during eutrophication) or the addition of toxic substances (like PCBs) through pollution. Under all of these circumstances, the prevailing environmental conditions in a particular area have changed and may no longer encompass the ranges of optima—the niches—of a few, many or even all the original constituent species of the community living in that area. Organisms will then suffer physiological stress. In the longer term, if the adverse conditions persist, organisms may be forced to move, if that is feasible, or may simply die out. The end result is the same; conditions have now become unsuitable for the species in the original community, which changes, usually to a less diverse, simpler community. If the species has a very specialist set of requirements that are only found in a few places, adverse changes to the environment can lead to their extinction—they simply have nowhere else to go.

If conditions recover and the environment at that location reverts to its original configuration, the original species may return provided that a source pool or refuge exists in an undisturbed/unpolluted area from which species can recolonize. If the species is driven to extinction, this is of course impossible! Thus communities as a whole have the potential to recover from disturbances, but both the ability to recover and the rate of recovery are dependent on the regime of the disturbance (Sousa, 1984), which in turn depends on a number of factors: (a) the nature of the disturbance; (b) the size of the disturbed area; (c) the magnitude and duration of the event—the intensity or strength of the disturbing force; (d) the timing and frequency of the disturbance; (e) the predictability of the disturbance and (f) the turnover rate—the mean time required to disturb the entire area.

Communities and ecosystems show differential abilities to resist disturbances and hence show different degrees of stability in their composition and structure. Three community properties are important in this regard:

1. Stability, the ability of the community to recover and return to its original configuration after a disturbance. Some communities may be locally stable to small-scale disturbances but if perturbed beyond a certain critical point they are forced into a new configuration, with species ending up with different relative abundances or the community with a different composition, and effectively a new community will replace the original one. Such communities are globally unstable. If the community reverts back to its original configuration after both small- and large-scale disturbances it is said to be both locally and globally stable.

2. Resilience. This is the measure of the speed with which the community returns to its former state following a disturbance. A measure of resilience must be specific to the type of disturbance imposed, as communities will be more resilient to some disturbances than to others. Greater resilience will be conferred on a community where species can readily recolonize the area from an undisturbed area.

3. Resistance describes how much disturbance a community can absorb before it flips into a new configuration—its resistance to change. Again, communities will tend to vary in their resistance depending on the type of disturbance imposed and on the type of community itself.

The overall stability of any community depends on the environment in which it exists as well as the species composition of the community. Some communities are only stable within a narrow range of environmental conditions while others may be stable over a wide range of conditions. The former, so-called *dynamically fragile* (Begon *et al.*, 1990), communities are much more likely to be damaged by natural or man-made disturbances than the latter, *dynamically robust*, communities. Some communities are subject to naturally low levels of disturbance in a stable and predictable environment and have the tendency to reach and remain in a fairly stable state, but, in turn, normally show relatively low resistance and resilience to severe perturbations. On the other hand, communities living in a disturbed environment usually show high levels of resilience and are able to recover from individual disturbance events quite rapidly.

While disturbances by their very nature can lead to changes in an ecosystem, disturbances are not always a 'bad thing'. Some ecosystems that are disturbed actually have a higher biodiversity than if they had not been disturbed. In these cases, the disturbance removes or reduces the dominant species or superior competitor which in turn allows subordinate, weaker competitors the opportunity to increase in abundance or colonize a habitat they were prevented from living in previously. In other systems, disturbances may be so frequent and/or large that the environment becomes extremely harsh, where only a few tolerant species are capable of surviving; such systems will obviously have a low biodiversity. The most important factors in this context are the magnitude, extent and frequency of disturbances relative to the life span of the major species in the community (Giller and Gee, 1987). This will apply equally to both natural and man-made disturbances.

Before we are in a position to discuss the impact of pollution on ecological systems, it is necessary to explore the major factors governing freshwater, terrestrial and marine ecosystems and to briefly examine the effects of natural disturbances and engineering activities on these systems.

5.2 THE FRESHWATER ENVIRONMENT

We can conveniently divide freshwater systems into lotic—running waters—and lentic—still waters. The characteristic features, certainly of running waters, are determined by the physical setting, in which land–water linkages are important. Therefore it is important to consider freshwater systems in the context of an integrated system of the river catchment including both the aquatic habitat and the terrestrial drainage basin. What happens in the catchment has direct and indirect effects on the freshwater ecosystems.

5.2.1 Oxygen

Oxygen is essential for animal and plant life as the major requirement for respiration, as described in Chapter 2. Oxygen is 30 times less abundant in water (≈ 10 mg/L) than in air, and can therefore become a limiting factor. Oxygen concentration increases as water temperature decreases and as turbulence and mixing in the water body increase. Thus a fast, shallow, turbulent stream has higher dissolved oxygen levels than a slow-moving, deep river. Many species have narrow tolerance ranges to oxygen and can only survive where levels are very high (> 10 mg/L), such as in cool, upland streams with a fairly fast current. For example, among the fish, salmon and trout need high oxygen levels and are confined to relatively shallow, fast streams or cold, upland lakes. In contrast, coarse fish like roach and rudd have wider tolerance ranges and can survive under considerably lower levels of oxygen (> 6 mg L), such as in deeper, slower rivers and warmer, lowland lakes. The difference in tolerance to oxygen levels among invertebrate animals has been used in the development of biotic indices for identifying and monitoring organic water pollution and water quality (see Chapter 6). Water depth also influences oxygen levels and oxygen concentration–depth profiles are common in lakes (see Fig. 5.6). Water regulation activities, such as construction of dams, can directly affect oxygen levels in rivers.

Any kind of disturbance that reduces oxygen levels will have a dramatic effect on the functioning of freshwater communities and ecosystems. Freshwater systems are dominated by macroinvertebrates, including insect larvae and other invertebrates like freshwater shrimps and worms that play a major role in the ecosystem processes. Particularly relevant in the present context is their ability to break down organic matter that enters freshwaters. Some freshwater streams rely on allochthanous energy inputs in the form of detritus, which is gradually broken down through decomposer food chains (see Chapter 2). This process relies on oxygen. Organic matter from low levels of sewage discharge or agricultural runoff can naturally be dealt with in the same way. Excessive levels of organic matter, however, can result in drastic reductions in oxygen levels in the water body, with consequent dramatic disturbance of the natural communities (see Chapter 6).

5.2.2 Current

Current speed or velocity is a dominant physical factor in the functioning of running water ecosystems, affecting the type of substrate, eroding nature of the stream channel, oxygen levels and sediment loads and, in turn, therefore, the ecology. It is also a factor often influenced by engineering activities. The shear stress of the current on the substrate is proportional to velocity^2 ($\tau \propto v^2$), and influences the stability of the substrate and the ability of animals living on and in the substrate (benthic macroinvertebrates dominated by insect larvae) to retain their position on the stream bed. Fish and plants are also influenced by the current. Different types of organisms have differential tolerance to flow conditions and show strong preferences for a fairly narrow range of current speeds. High current speeds outside tolerance limits can dislodge organisms and carry them downstream. In addition, high flows can cause direct mortality through physical damage, while deposition and sedimentation of substrate can cause smothering of organisms during a flood.

Flooding disturbance Flooding of rivers and streams is a frequent feature of many landscapes and much attention is directed towards control and alleviation by engineers due to the impact on agriculture, industry and urban areas. The destructive force of floods can be immense and amounts to considerable economic loss. For example, floods in the United States resulting from Hurricane Agnes led to damage estimated at $3000 million (Ward, 1978). Floods also act as natural disturbance agents on the aquatic environment which may, in turn, result in economic loss through destruction of natural fisheries, related losses through tourism, reduction in the assimilation capacity of the system for organic and other contamination (largely through losses in invertebrate animals) and possibly indirect effects on water resources and quality. The disturbance effects of flooding on the biology of aquatic systems is due mainly to the rapid increase in shear stress on the substrate as a result of increased depth and velocity of water in the stream channel.

During floods, the stream bed of riffle areas (shallow, fast flowing, usually steep sloped sections of rivers) may be stirred up and detritus can be swept away. When stream depth reaches about 75 per cent bankfull in the channel (Hynes, 1970) or about twice mean annual flow (Sagar, 1986), medium-sized gravel (< 10 mm) begins to move. Severe flooding can cause the stream bed to be scoured by sand and gravel particles entrained in the high-velocity flow, removing plant communities growing on stones (periphyton, another invertebrate food source). It can also lead to large-scale displacement (10 to 20 m or more) and overturning of boulders and stones and removal of sediments to depths of between 20 cm and 2 m. In contrast, in stream sections with gentle slope (e.g. pools and glides), large amounts of sediments may be deposited. In many streams, discharge fills the channel about once every 1.5 years (Hynes, 1970). Climatic irregularities, storms and locally heavy rain can cause discharges much greater than bankfull, but these catastrophic events are rare (e.g. the 100 year flood). The timing of such natural disturbance events is put into context of other physical and biological temporal patterns in rivers and streams in Fig. 5.1.

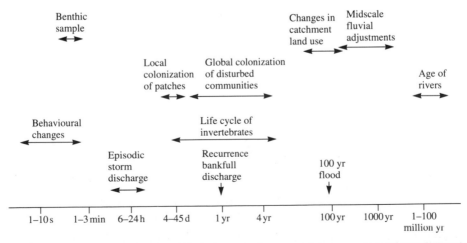

Figure 5.1 Temporal scale of some biological and physical processes in stream and river ecosystems. Length of arrows indicates the range of magnitude.

The impact on freshwater communities depends on the magnitude, timing, duration and aerial extent of the flood disturbance. Resilience to small floods is usually high with rapid recovery of invertebrate densities within several weeks to a few months. Similar recovery has been demonstrated under fluctuating water levels in a regulated stream subject to hydroelectric power (HEP) generation (Gilason, 1985) and flushing out of reservoirs (Bournard *et al.*, 1987). Because variation in stream discharge and flood/spate events are a natural part of the lotic environment, benthic invertebrates show adaptations to avoid relatively small increases in current speeds by retreating deeper into the substrate (the depth controlled by particle size, subsurface flow and oxygen tension gradients), and both invertebrates and fish can move into more benign locations such as pools and dead zones where current speeds are negligible. The success of this strategy is, however, dependent on the morphology of the stream channel. This kind of behaviour and rapid recolonization from upstream contribute to rapid recovery times of benthic communities in the face of small-scale disturbances.

Catastrophic disturbances caused by discharge levels equivalent to the 50 to 100 year flood have much longer-term consequences. Samples taken from the Yoshino River in Japan six years prior to the 1959 Ise-wan Typhoon showed an invertebrate biomass (wet weight) of $32 \, g/0.25 \, cm^2$. Four years after the disturbance, the value was only $0.5 \, g$ (Tsuda and Komatsu, 1964). Studies in Ireland have shown similar dramatic and long-lasting effects (Giller, 1990; Giller *et al.*, 1991). Multievent floods in August 1986 (associated with Hurricane Charlie) caused greater than 1.5 times bankfull discharge in a tributary of the River Araglin (a 1 in 50 year event), which led to a 95 per cent reduction in invertebrate density and a 30 per cent reduction in the number of different taxa (Fig. 5.2). Further flooding disturbances in June and October 1988 retarded recovery somewhat, although it had proportionally less significant impact on the already depleted fauna. Complete recovery to the pre-flood levels had still not occurred by 1989.

Fish can be affected by floods, through destruction of spawning areas, eggs or fry and loss of habitat or food resources. These effects can lead to the destruction of an entire year class, delay in growth and lower production. In some cases, recovery of fish communities can also be rapid, e.g. about 8 months following a 16 year record flood, (Matthews, 1986), although in other cases several years were required. Floods during the reproductive season may have more serious consequences for fish communities than even very major ones at other times.

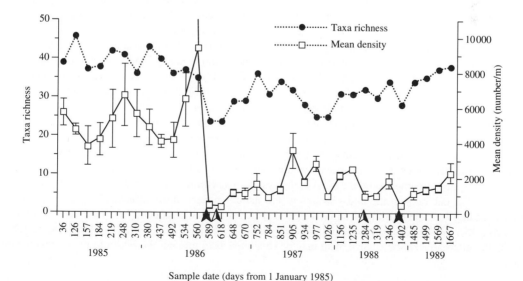

Figure 5.2 Changes in the mean density of macroinvertebrates (± 1 s.d.) and number of taxa before and after catastrophic flooding events at a site on a tributary of the river Araglin. Arrows show major (closed) and minor (open) disturbances (after Giller *et al.*, 1991).

Obviously, catastrophic floods causing severe scouring and removal of substrate will dramatically affect river and stream communities. The impact of less severe floods will, however, vary, as different types of streams are predicted to show different levels of resilience to flooding disturbance. Where floods are seasonal and regular (i.e. predictable), as in high-gradient, spatey streams influenced by winter and spring floods, the impact is less and recovery is relatively rapid. Unpredictable, aseasonal, irregular disturbances (e.g. summer floods in temperate streams), especially to systems infrequently disturbed, have severe impacts. Size of the disturbed area will also affect the rate and dynamics of recovery of the ecosystem (Minshall, 1988). Similarly, multievent disturbances and additional stochastic flooding events during the recovery period are also known to delay recovery of stream systems.

Engineering implications to the effects of floods on natural ecosystems   The way water interacts with the surrounding catchment area has important repercussions on the influence of heavy rainstorms on the running water system. Certain streams and rivers are naturally 'spatey'; the catchment areas surrounding them are such that there is rapid and considerable runoff and drainage of water into the system. Human influence on the landscape can increase this input from the catchment and hence increase the natural occurrence of catastrophic flooding events. Drainage of farmland causes water to run off the land more quickly, leading to sharp hydrograph peaks in streams during storm events. In temperate climates, pre-afforestation drainage (involved in preparation of the ground for planting) also changes the storm hydrograph, with higher hydrograph peaks and a reduction of as much as 50 per cent to the time of peak flow (Institute of Hydrology, 1984–87). The storage capacity of the soil is effectively reduced. This is similar to the comparison in Fig. 4.38. As the canopy closes with the growth of trees and with the infilling of drainage ditches, evapotranspiration by plants and reduced runoff (O'Halloran and Giller, 1993) can reduce the amount of water reaching streams by as much as 30 per cent under mild to moderate-sized storm events, but severe storm conditions produce similar hydrographs in afforested compared to open moorland streams. Clearfelling 40 to 50 years later again leads to increased runoff and 'spateyness'. Also, aperiodic releases of water from dams can cause similar 'unnatural' floods. Finally, dredging of rivers will not only reduce the preferred habitat for fish but can also greatly increase the flow rates and hence impact of spates on the system.

Much engineering work on streams and rivers is based on prevention of flooding of surrounding land, e.g. channel modification to enlarge the carrying capacity, reduce water levels and reduce the frequency of overbank flows; river discharge modification through dams, storing peak discharge and regulating flow levels downstream; systematic removal of snag boulders and debris dams to aid transportation; channelization to improve flow and flood control, etc. In the United States today, for example, few rivers remain totally unaffected (Dahn *et al.*, 1987). All these activities fundamentally alter the river systems and have been well reviewed by Hellawell (1986). Flood control procedures combined with human modification and degradation of catchment areas can lead to extreme flooding and sedimentation of abnormally long duration (Sousa, 1984). The spectacular flooding episodes seen in the Mississippi and Missouri catchments in the United States in summer 1993 and spring 1995 is evidence of this. Water regulation schemes (reservoirs, HEP), for example, usually have some legal stipulation as to minimum flow, but the contrasts between low and high flows may be very great and rapid (Moss, 1988), leading to catastrophic impacts on the biota of the system. The communities present prior to regulation are adapted to the natural regime, and changes in the natural pattern must influence these communities, usually leading to impoverishment and damage to fisheries and other aquatic life. Recovery may occur, e.g. a five- to sevenfold decrease in fauna following dredging recovered after one year (Crisp and Gledhill, 1970), but, alternatively, the environment may be so altered as to prevent the return to its original natural state.

Current, substrate and longitudinal changes   Current also controls the nature of the substrate. Fast currents lead to coarse gravels and cobbles, while slow currents lead to fine sediments, sands and mud. Different species and communities are associated with different substrates. The physical stability of the substrate and the propensity to be moved or scoured by spates and floods will also have an effect on the benthic communities (Winterbourn and Townsend, 1991). The less stable the substrate, the greater the disturbance effect of a spate or flood of a given magnitude. Pfankuch (1975) developed a rating system for relative channel stability which involved a number of important criteria. For stream/river banks this included the degree of vegetation cover, evidence of undercutting and the amount of newly deposited sand and gravel. Riparian vegetation zones act as nutrient filters, sediment traps, climatic regulators and wildlife refuges (Winterbourn and Townsend, 1991). Clearing of the natural riparian vegetation on river banks, either through grazing activity of domestic animals or as part of river regulation schemes, can have significant effects on sedimentation and bank erosion. For bed stability, Pfankuch's criteria included rock angularity, size and degree of packing of substrate, amount of scouring and deposition and the distribution of mosses and algae covering the substrate. Engineering activity such as dredging, bankside clearance and stream straightening and widening tend to have a negative effect on these kinds of parameters and lead to reduced substrate and bank stability. Often such work is carried out in such a crude way that recovery of the ecosystem is prevented (Maitland, 1990).

As one moves through a river system from headwaters to the lowland river, current speed tends to drop as the slope of catchment declines, water depth gradually increases as the river enlarges through the addition of tributaries, water temperature increases as one moves from cool uplands to warmer lowlands and there is therefore an overall trend for oxygen levels to decrease in the downstream direction. These changes along physical gradients lead to longitudinal changes in communities, as different species are adapted to survive only within a restricted range of conditions of current, temperature and oxygen.

5.2.3 Water Chemistry

Most of the dissolved and particulate organic and inorganic matter in freshwaters is carried in from the surrounding catchment via surface runoff, subsurface flow and from groundwaters, although there is some direct atmospheric input as well. As such, the chemical nature of the freshwater system strongly reflects the type of land being drained and the land use (Table 5.1). In limestone regions, streams and lakes have high alkalinity and pH, while in granite regions, and catchments draining bogs etc., the freshwaters are low in dissolved salts and are often acidic. In agricultural areas, nitrogen and phosphates from fertilizers enter the water system (see also Chapter 10).

Table 5.1 The effect of land use and catchment type on streamwater chemistry compared with rainfall in the same area. Note that the differences in streamwater quality are far greater than those for rainfall

Chemical parameters (mg/L)	New Hampshire, USA Igneous (insoluble) rocks, undisturbed forest		Norfolk, UK Chalk and glacial drift, lowland agriculture		Rift Valley, Kenya, Thorn bush and rangeland	
	Rainfall	Stream	Rainfall	Stream	Rainfall	Malewa River
Na^+	0.12	0.87	1.2	32.5*	0.54	9.0
K^+	0.07	0.23	0.74	3.1	0.31	4.3
Mg^{2+}	0.04	0.38	0.21	6.9	0.23	3.0
Ca^{2+}	0.16	1.65	3.7	100.0	0.19	8.0
Cl^-	0.47	0.55	< 1.0	47.0	0.41	4.3
HCO_3^-	0.006	0.92	0	288.0	1.2	70.0
SO_4^{2-}					0.72	6.2
pH	4.14	4.92	3.5	7.7		

From Burgis and Morris, 1990

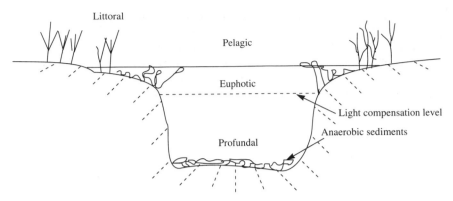

Figure 5.3 Lake zonation showing the major subdivisions of the lake ecosystem.

Changing land use can thus have dramatic effects on water chemistry. This was elegantly demonstrated by the famous Hubbard Brook (US) experiments where clearfelling of forests in an entire subcatchment led to dramatic changes in a whole range of chemical and hydrological parameters (Likens *et al.*, 1970). Different species of plants and animals are adapted to different chemical conditions, so one finds very different communities in acidic (soft) and alkaline (hard) waters for example. In general, acidic waters tend to be species poor and are not favoured by salmonid fish; alkaline waters in contrast are much richer in macroinvertebrates. Changing land use can therefore have knock-on effects to the freshwater ecosystems through changes in water chemistry.

5.2.4 Light and Lake Zonation

Light is an important factor in still water and slow, large rivers. Light penetration in water is poor and therefore light can be a major limiting factor to photosynthesis by aquatic plant life. Aquatic plants are restricted to fairly shallow depths, and are much dependent on the clarity of water. Because of this, there is a clear zonation of plants in lakes (Fig. 5.3).

The *littoral zone*, around the edges of lakes and ponds, extends down to the depth of the innermost population of rooted plants. Macrophytes dominate, either submerged (pond weeds), floating leafed (water lilies) or emergent (reeds). The open water *euphotic zone* extends down to the depth where average light intensity allows plant production to equal respiration—the *light compensation point*. These plants are phytoplankton, single cells or small colonies of algae, passively drifting or capable of only limited mobility. As such, they are easily entrained in horizontal and vertical water movements. Below the light compensation point, there is insufficient light for photosynthesis and plant survival. This point varies with season, cloud cover and water clarity. Mixing depth plays an important role in primary production of phytoplankton. Phytoplankton tends to sink, but turbulent mixing of surface waters can provide an external force to keep the plants in the lighted zones. If wind-driven mixing is too vigorous and at too great a depth, the plankton can receive insufficient light on average, as they spend too long below the light compensation point. This critical depth of mixing is usually 5 to 10 times the light compensation point, a mixing depth frequently occurring in winter. Plant growth thus ceases. Increasing mixing depth artificially offers a useful but costly management strategy for controlling algal production in drinking water reservoirs. For a fuller discussion of factors affecting phytoplankton production, see reviews by Reynolds (1987, 1994).

The *profundal or deeper zone* lacks plants but receives a heterotrophic energy source in the form of a rain of detritus from above (dead organisms, faeces, etc.). Decomposition of detritus by bacteria and fungi uses up oxygen; thus a complex interrelationship occurs between the amount of nutrients in the water, the level of plant production, the rate and amount of decomposition and oxygen levels. High levels of nutrients lead to high levels of plant production and a large rain of detritus to the profundal zone when the plants die. This, in turn, leads to high levels of decomposition, which consequently leads to low oxygen levels. Different communities exist in lakes and rivers with different primary production levels (see below), and the addition of large quantities of organic matter (from sewage outfalls or agricultural runoff) can severely disrupt those communities naturally adapted to low productivity systems with relatively high oxygen levels.

5.2.5 Lake Classification

Lakes and large rivers can be classified on the basis of the level of primary productivity. Low productivity systems are known as *oligotrophic*, characterized by high levels of oxygen and low nutrient concentrations. High productivity systems are *eutrophic*, usually with low oxygen levels and high nutrient concentrations. These are very different types of ecosystem with markedly different communities and general characteristics (Table 5.2). Oligotrophic lakes contain organisms with little tolerance for low oxygen whereas eutrophic lakes contain organisms more tolerant of low oxygen levels.

Because of the importance of phosphorus concentrations on primary productivity in lakes, the Organisation for Economic and Cultural Development (OECD) has a classification scheme based on concentrations of phosphorus and chlorophyll-a (a surrogate for phytoplankton densities), which are in turn related to concentration of phosphorus inflowing to the lake and water residence time of the lake (Moss, 1988) (Fig. 5.4).

Lakes are generally not permanent features of the landscape. They gradually become filled with sediment brought in from rivers and washed off the land and then eventually disappear completely, leaving highly organic soils. This is the natural process of eutrophication, with the lake gradually changing from oligotrophic to eutrophic as nutrient levels increase over time, usually several hundreds to thousands of years. This process can be dramatically speeded up through human activity, where excess nutrients added from fertilizers or sewage input leads to pollution. Similarly, damming a river leads to concentration of organic matter upstream of the dam and can lead to enriched, lake-like, conditions with increased plant production.

5.2.6 Water Density and Thermal Stratification

Another important property of water is that density is greatest at approximately $4\,^{\circ}C$—water above or below this temperature floats on water at $4\,^{\circ}C$. Also, warmer water floats on cooler water. This density differential per degree increases progressively with higher temperatures. During the year, as the water body warms and cools seasonally, there is a changing temperature profile with depth. This has direct and indirect effects on a number of ecosystem processes. As an example, consider a lake experiencing a maritime climate (Fig. 5.5). During winter, temperatures are relatively uniform throughout the lake and wind can mix lake water from top to bottom. Progressing into spring and early summer, the surface waters heat up, decrease in density and float on cooler water below. These density differences resist mixing of waters within the water body by wind. As temperature increases, this discontinuity becomes more marked until the water body is fully stratified into three parts. A warm *epilimnion* at the top, a cold *hypolimnium* at the bottom and a narrow region in between, called the *thermocline*, where temperature changes rapidly with depth. This effectively creates a barrier preventing the two water bodies from being mixed by the surface wind. This stratification may last through the summer and into autumn, when the surface waters start to cool again. When the water body as a whole reaches a similar temperature again, wind can mix the whole lake from top to bottom, causing the *overturn*.

This kind of lake is known as monomictic (one overturn per year) and examples of temperature depth–time profiles for a number of Irish lakes of varying depth are shown in Fig. 5.6. Dimictic lakes

Table 5.2 General characteristics of eutrophic and oligotrophic lakes

Character	Eutrophic	Oligotrophic
Basin shape	Broad and shallow	Narrow and deep
Lake substrate	Fine organic salt	Stones and inorganic silt
Lake shoreline	Weedy	Stony
Light penetration to 1% surface value (m)	−20	20–120
Water colour	Yellow or green	Green or blue
Net primary production (g/m^2 yr)	150–500	15–50
Chlorophyll-a concentration (μg/L)	−15+	0.3–2.5
Alkalinity range (year) (meq/L)	1+	Up to 0.59
Total P (ppb)	10–30	< 1–5
Total N (ppb)	300–650	< 1–200
Oxygen	High at surface, low under ice or thermocline	High
Macrophytes	Many species abundant in shallows	Few species, some in deep water
Phytoplankton	Few species, high numbers	Many species, low numbers
Zooplankton	Few species, high numbers	Many species, low numbers
Macroinvertebrates	Many species, high numbers	Moderate species, low numbers
Fish	Many species	Few species

Adapted from Maitland, 1990, and other sources

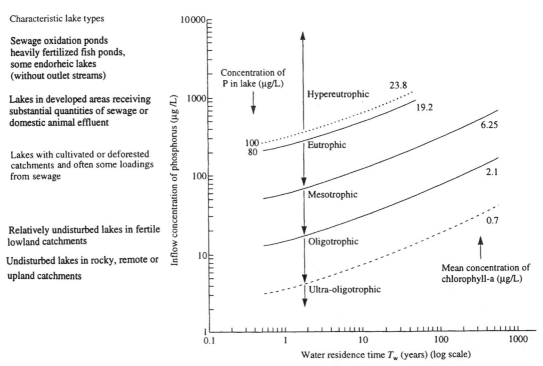

Characteristic lake types

Sewage oxidation ponds
heavily fertilized fish ponds,
some endorheic lakes
(without outlet streams)

Lakes in developed areas receiving
substantial quantities of sewage or
domestic animal effluent

Lakes with cultivated or deforested
catchments and often some loadings
from sewage

Relatively undisturbed lakes in fertile
lowland catchments

Undisturbed lakes in rocky, remote or
upland catchments

Figure 5.4 Lake classification based on relationships between inflow concentration of phosphorus and water residence time (T_w). The consequent most likely concentrations of phosphorus and chlorophyll-a in the lake are shown. Larger residence time leads to greater deposition of phosphorus in the sediment, so lower in-lake phosphorus and chlorophyll-a concentrations will be obtained for a given inflow concentration. Characteristic lake types in relation to inflow concentration of phosphorus are also shown (after Moss, 1988; Burgis and Morris, 1990; and Vollenweider and Kerekes, 1981).

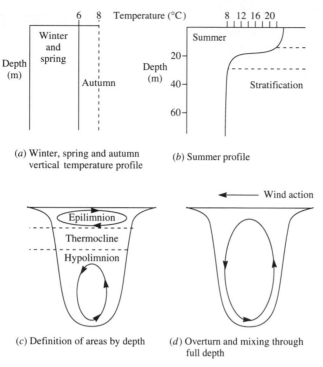

Figure 5.5 Diagrammatic representation of seasonal thermal stratification in a temperate, maritime lake, (a) showing no stratification and more or less constant temperatures throughout the lake. During this time, wind action can cause the lake to mix from top to bottom as in (d). Stratification occurs during the summer, with a marked discontinuity in the temperature–depth profile (b) dividing the lake into three strata (c).

(continental lakes in areas suffering cold winters) undergo a period of inverse stratification during the winter, when colder surface waters lay on top of a warmer hypolimnion. A spring overturn occurs as surface water temperatures rise. Shallow lakes like Lough Neagh in Northern Ireland (average depth only 12 m) will rarely demonstrate a stable thermal stratification except in the hottest of summers, while very deep lakes, like the African Rift Valley lakes are permanently stratified (Moss, 1988).

Thermal stratification has major effects on both oxygen concentration (see Fig. 5.5b) and nutrient supplies. When the lake is stratified, no mixing occurs between the top and bottom layers. The hypolimnion receives no oxygen that has diffused into the surface waters and becomes increasingly anoxic. The epilimnion, where the plants are, receives no dissolved nutrients from the bottom, where decomposition occurs, so primary productivity becomes nutrient limited and declines over the summer. When the overturn occurs, the hypolimnion is replenished with oxygen and the epilimnion with dissolved nutrients. An end result of such seasonal cycles are seasonal blooms of phytoplankton, due to the replenishment of nutrients in autumn and increasing temperatures and light levels in spring. Excessive deoxygenation of the hypolimnium in summer, which can arise as a result of strong eutrophication of the water body, can result in dramatic disturbances to the rest of the lake system on overturn and decreasing quality of the water resource.

5.2.7 Water Regulation

Although this topic has been introduced earlier, the effects of major water regulation schemes involving damming and impoundments deserve special attention. Dams are built to regulate river discharge

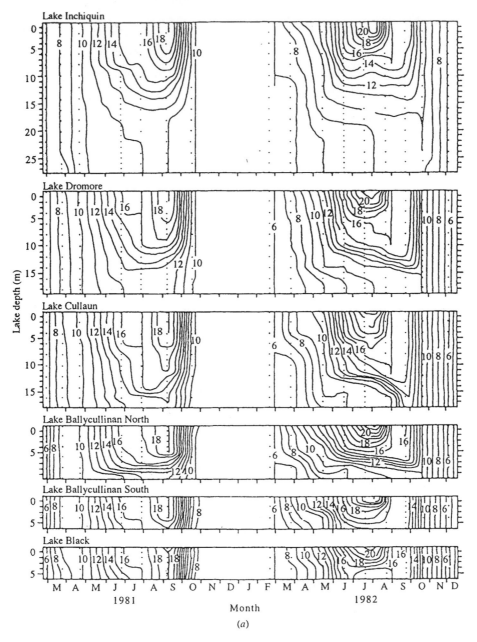

Figure 5.6 Temperature–depth (a) and oxygen–depth (b) profiles in six Irish lakes. Note that thermal stratification is unstable in shallower lakes and that parts of the hypolimnion can become anoxic during the summer (after Allott, 1986).

preventing flooding of downstream settled lands, to create reservoirs of water and often to produce hydroelectric power. When a barrier is put across a river, ecological conditions change dramatically upstream as the aquatic system is switched from lotic to lentic conditions—from a river to a lake. Filling time can vary depending on the size of the reservoir/man-made lake basin in relation to the size of the river. For example, when the basin is small relative to river flow, as in Lake Kainji in Nigeria, filling took

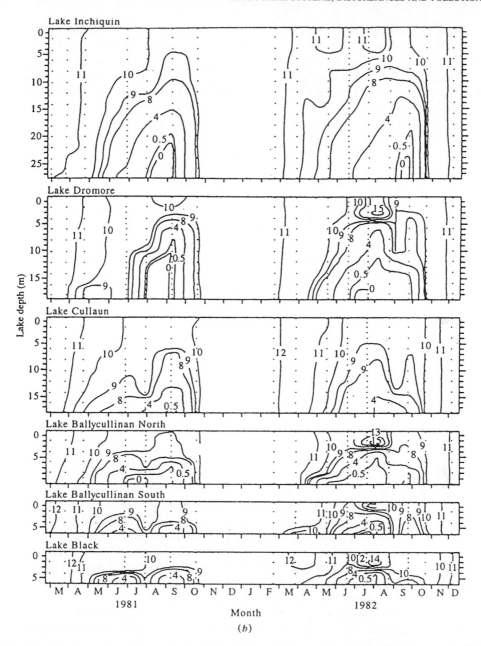

Figure 5.6 continued Temperature–depth (a) and oxygen–depth (b) profiles in six Irish lakes. Note that thermal stratification is unstable in shallower lakes and that parts of the hypolimnion can become anoxic during the summer (after Allott, 1986).

only three months, whereas Lake Nasser (capacity 130 000 million m^3), behind the Aswan High Dam on the Blue Nile in Egypt, took 10 years to fill (Burgis and Morris, 1990). There are dramatic changes in the flora and fauna post-impoundment that often take many years to stabilize. Provided there is not too much subsequent disturbance to the new lakes, the systems will eventually settle down and resemble the local natural lake ecosystems.

The most obvious effects of dams are local upstream changes in the aquatic ecosystem. The reduction in flow rate behind the dam will lead to deposition of fine sediments on to the original river substrate. Massive deposition of such sediments on the rocky bed of the Blue Nile near the Roseires Dam destroyed populations of the giant freshwater mussel for several kilometres (Hammerton, 1972). Drowned vegetation also starts to decompose, which releases nutrients but results in a decrease of oxygen levels and possibly deoxygenation of bottom waters due to the bacterial activity, which can be exacerbated by thermal stratification. Slowing of current speed, sedimentation, holding back of transported organic matter and nutrients and release of nutrients from increased decomposition all add up to produce quite dramatic increases in primary production in the newly created lakes. For example, following filling of the Sennar and Roseires Dams on the Blue Nile, there was a 10 to 200-fold increase in phytoplankton over a 650 km upstream stretch of the river (Hammerton, 1972). Such increased productivity can also lead to increased production of certain fish species.

There are clear economic advantages to construction of a dam. For example, the Aswan High Dam led to an increase in arable land from 7 to 9 million acres and HEP output in excess of $4500 million per year at 1970 prices (Hammerton, 1972). However, these gains must be weighed against a large number of immediate and long term disadvantages, in addition to the changed upstream water quality. Moss (1988) gives a lucid discussion of the pros and cons of man-made tropical lakes in particular. Many species of organism disappear, while other species appear, often in great abundance in the disturbed system. The introduced alien water fern is one example (*Salvinia molesta*). This appeared in Lake Kariba and rapidly spread to cover 21 per cent of the lake surface (Maitland, 1990). It severely impedes navigation, fishing and affects water quality, largely through indirect deoxygenation of water under the weed cover. As a floating plant, it shades out other plant life in the water so there is no oxygen release by photosynthesis and no wind mixing of surface waters (Burgis and Morris, 1990). Other pests, especially insects, are also favoured under the new environmental conditions and increase in abundance over very short periods of time. After construction of the Sennar and Aswan High Dams, blackfly larvae (Simuliidae) which carry river blindness (schistosomiasis) bred in the turbulent waters near sluice gates. Mosquitoes and midges flourished in the newly created shallow areas of the lake edges (Hammerton, 1972).

Dams also cut upstream–downstream linkages, which have serious consequences for fish migration in species that breed in the headwaters of rivers. This is particularly important for salmon, because, though the discharge may be sufficient to maintain the survival of the non-migratory species, it may be insufficient to enable movement of migratory species. However, the minimum acceptable flow for these requirements is not needed for the whole year (adult salmon ascend river systems in June/July to spawn in November/December) and a lower flow may be sufficient at other times for spawning requirements and survival of progeny. Creation of artificial spates has been shown to be useful to encourage fish to migrate up streams. The problem has also been overcome through the use of fish ladders and electrically powered lifts (Moss, 1988). As a last resort, it may be necessary for the fisheries authority to catch the salmon below the dam and physically transport the fish upstream. This practice, though it helps the overall fish production, is unsatisfactory and indeed costly. Another way to maintain a fishery in a regulated river is to undertake an introduction programme of reared fish which are released in large numbers upstream of the dam. Turbines can also cause problems to fish moving downstream when the smolts (young salmon) are returning to sea unless design or operating procedures are modified. The alternative is physical transportation downstream. Fluctuating water levels in the littoral areas of the artificially created lakes also cause problems to the flora and fauna (Smith *et al.*, 1987).

Downstream of dams, the river loses much of its dynamic nature. Flow patterns are mediated and become more regular but can also be more extreme. Downstream temperature regimes are altered (Winterbourn and Townsend, 1991) and oxygen levels can fall if oxygen-deficient waters from deeper layers in the lake are released. These waters may also contain suspended iron and manganese hydroxides and dissolved hydrogen sulphide (Moss, 1988). The hydrogen sulphide caused by the excessive decomposition can corrode the metal of turbines and the concrete of the dams. The regulation of water decreases or eliminates regular inundation and deposition of nutrient rich sediments on the flood plain and the downstream agricultural productivity is usually decreased quite dramatically. Sediments and nutrients are retained behind the dam, thus causing downstream phytoplankton productivity to be reduced; in the Nile, for example, it is only a fraction of the pre-dam density (Hammerton, 1972). Downstream biological systems will, without doubt, be altered as a result of such changes in food supply and the physicochemical environment.

Changes brought about by the construction of dams are not restricted solely to the local river system. Historically, the annual discharge of freshwater from the Nile to the Mediterranean was 62 km^3, but after closure of the Aswan High Dam the annual flow decreased to 10 per cent of this figure (Hargrave, 1991). There was also a shift in peak flow from the autumn rainy season to the winter months. This altered the physical and chemical conditions in the south-east part of the Mediterranean Sea near the Nile Delta. Sediment trapped behind the dam normally replenished sediments eroded from the Nile Delta by the sea, as well as carrying nutrients to the marine coastal ecosystems. The result of these changes was twofold (Hargrave, 1991): firstly, gradual erosion of the Nile Delta, with consequent shifts in navigation channels and, secondly, a decrease in productivity of coastal waters and consequent decline in important fisheries (Fig. 5.7). It was also speculated that the reduced freshwater input into the Mediterranean could result in the loss of the freshwater barrier across the northern end of the Suez Canal to the Red Sea and thus lead to movement of species between the two seas (Hammerton, 1972). The concomitant cascade effects on the flora and fauna of either one or the other sea depend on the main invasion direction of species. We thus see the importance of the link between freshwater and marine ecosystems.

5.3 MARINE SYSTEMS

The oceans cover a little over 70 per cent of the surface of the globe. On land, life exists as a thin veneer, extending only a relatively short distance down into the soil and up into the atmosphere. In the oceans, life exists from the surface down to the deepest parts of the oceans (about 11 000 metres). Over half of the globe lies beneath 4000 metres of sea, the largest ecosystem in the world, known as the abyss, where it is permanently dark, has a constant temperature of approximately 4 °C and experiences a constant pressure in excess of 400 atmospheres. The deep sea is, however, scarcely exploited commercially, apart from mining of the often-abundant manganese nodules. It is probable that as land resources become depleted,

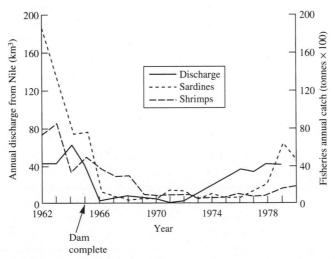

Figure 5.7 The effect of the Aswan High Dam on annual discharge from the River Nile to the Mediterranean Sea and the subsequent changes to sardine and shrimp fisheries (after Wadi, 1984).

the deep sea will attract more exploration and, when appropriate technologies have been developed, will become more exploited. Of course, the seas are of great commercial significance to human beings for food. It should be realized, however, that over 95 per cent of fish (wild and cultured) and shellfish are harvested from continental shelf areas, which make up only 8 per cent of the total sea area. The following sections outline some of the main physicochemical features of marine ecosystems and discusses the impact of various natural and anthropogenic disturbances.

5.3.1 Temperature

One of the most characteristic features of the oceans is its uniformity in physicochemical properties. Temperatures range from around 0 °C in high latitudes (as low as −2 °C in the Arctic deep water) to +30 °C in shallow tropical waters (up to 35 °C in the Persian Gulf), a range of only 30 to 37 °C. This compares with a range of about 145 °C in terrestrial ecosystems, which experience temperatures of −88 °C in Antarctica to +57 °C in the Libyan Desert.

5.3.2 Salinity

The total amount of inorganic material dissolved in seawater is termed the salinity and is usually about 3.5 per cent or 35 parts per thousand. In the open ocean, where rainfall is low and evaporation high, salinities may rise to 37 parts per thousand, e.g. in the Sargasso Sea. On the other hand, in the Arctic during the summer, melting ice may lower salinities to around 30 parts per thousand. Inshore waters are a little more variable, with salinities reaching 40 parts per thousand or more in the Eastern Mediterranean and Gulf of Arabia, due to high temperatures with consequent evaporation and little freshwater run-off, and as low as 5 parts per thousand in the Northern Baltic, where there is considerable input from large rivers coupled with relatively low temperatures.

Seawater is not simply 'salt water' but an extremely complex substance containing all naturally occurring cations and anions (Broecker, 1974). Solutes weathered from rocks enter the sea with freshwater drainage from the land. The chief cations are sodium, magnesium, calcium, potassium and strontium; the principle anions are chloride, sulphate, bromide and bicarbonate, which together make up 99.9 per cent of the dissolved matter. This complexity is far from irrelevant to the animals and plants living in the ocean. Many trace elements are necessary for the proper functioning of marine organisms. For example, Tunicates, sessile marine animals often known as sea squirts, require vanadium, which they extract from seawater to incorporate in the blood. Seawater also contains organic substances leached from the tissues of organisms, secretions and excretions, antibiotics and so on, which are necessary to the survival of many marine organisms. Thus seawater can be seen to be an extraordinarily complex substance, which should be borne in mind when we look later at pollution in marine systems in Chapter 6.

5.3.3 Stratification and Productivity

Temperature and salinity together influence the density of seawater and, to a much lesser extent, so does pressure, which increases with depth. The fact that the density of seawater varies with temperature has important consequences for primary productivity in the oceans. In low latitudes where the sea surface is heated by the sun more or less throughout the year, a thermal stratification similar to that in lakes is established, where the warmer surface water floats on the denser colder water with a thermocline in between. As plants and animals decompose, they sink and the consituent nutrients are not recycled into the upper layers, so that plant nutrients become limiting, despite the abundant light for photosynthesis. For this reason tropical waters are relatively unproductive (Koblentz-Mishke et al., 1970) and that is why they always look so clear and blue. In high latitudes, there is no stratification because the surface waters are cold and of much the same temperature as the deeper water. Instead there is continuous mixing which brings nutrients back up to the surface. In the long day-length summers of high latitudes, there is ample sunlight and ample nutrients so productivity is very high. It is the richness of cold Arctic and Antarctic waters that allows large numbers of seals, walruses and the huge whales to flourish. Temperate waters are intermediate between these two extremes, acting like the tropics in summer and like the high latitudes in winter (Table 5.3).

During the summer in temperate waters, thermal stratification is established and when the nutrients are used up from the surface waters they are not replenished, resulting in low productivity. In the autumn, the surface waters start to cool, allowing some mixing with nutrient-rich deeper waters, but at about the same time the amount of sunlight drops so at best there can be only a slight increase in productivity. In the winter low light levels are a limiting factor, but, come the spring, the light levels increase and the nutrients are available in abundance, leading to a spring phytoplankton bloom, which lasts until the thermocline sets in and the nutrients are exhausted. This pattern of productivity is similar to that in many temperate lakes.

5.3.4 pH

Unlike freshwater, the pH of seawater only varies between about pH 7.5 and 8.4, with the highest values occurring at the surface during periods of high productivity when carbon dioxide is withdrawn during photosynthesis. An increase in temperature or pressure causes a slight decrease in pH and under great pressure, below 6000 metres, calcium carbonate goes into solution. Calcareous deposits are thus conspicuously absent from the deep sea, and deep sea bivalves (shellfish) tend to have weakly calcified shells, while deep sea fish have weakly calcified skeletons. It is the buffering properties of seawater, resulting from the presence of strong bases and weak acids (H_2CO_3 and H_2BO_3), that maintains the pH.

5.3.5 Oxygen

The mixing properties of the oceans unlike those in lakes, operate on a global scale, and supply oxygen to all ocean depths, including the deepest trenches, so that oxygen is rarely a limiting factor. This is not to say that oxygen is uniformly distributed. There is, for example, an oxygen minimum layer at about 400 to 1000 metres. Enclosed seas such as the Mediterranean may experience deoxygenation at times, as may the Norwegian Fjords. The Black Sea, cut off from the Mediterranean by the Bosphorus, is permanently stagnant below 200 m, and therefore devoid of animal life, although anaerobic bacteria flourish.

5.3.6 Circulation

The oceans therefore are in general well oxygenated, with a fairly even temperature, chemical composition and pH. This benign nature is dependent upon the mixing properties of the oceans which are brought

Table 5.3 Seasonal changes in productivity in temperate waters.

Season	Light	Nutrients	Productivity
Spring	Increasing	High	High
Summer	High	**Low**	Low
Autumn	Decreasing	Recycling	Moderate
Winter	**Low**	High	Low

Limiting factors in bold

about by the current systems generated by the action of the winds on the surface waters and the differences in the density of seawater resulting from variations in salinity. The direction in which surface currents flow is affected principally by the Coriolis effect and the shape of the land masses.

As the earth rotates on its axis, points at different latitudes rotate at different velocities. The rotational velocity is proportional to its distance from the axis of rotation and ranges from 0 km/h at the poles to 1600 km/h at the equator. The result of this is that objects veer to the right of their intended path in the northern hemisphere and to the left in the southern hemisphere—this is the Coriolis effect. The prevailing wind systems initiate surface water motion by transferring energy to the surface layer through frictional stress. The trade winds blowing out of the north-east in the northern hemisphere and south-east in the southern hemisphere form the major ocean gyres (circular currents, Fig. 5.8). These gyres, because of the Coriolis effect, are clockwise in the northern hemisphere and anticlockwise in the southern hemisphere.

Surface currents have, of course, been known and charted for centuries, but only relatively recently have deep water currrents been identified. Surface currents transfer their energy layer by layer to deeper water and each layer moves with less velocity progressively to the right in the northern hemisphere and to the left in the southern hemisphere. This is known as an Ekman spiral after the Danish oceanographer who first described it. Between 100 and 200 m depths all the wind energy is used up. The net transport of water in an Ekman spiral is about 90° to the right (northern hemisphere) or to the left (southern hemisphere) and is called the Ekman transport.

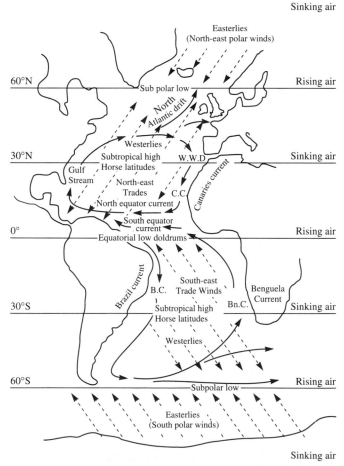

Figure 5.8 Wind systems (broken line) and surface currents (continuous line) of the Atlantic Ocean.

Surface currents move only the top 200 m or so of water, but deeper currents are initiated partly by differences in density and partly by upwelling. Highly saline and/or cold, and hence dense, water sinks, to be replaced by less saline and/or warm water. Deep water currents in the Atlantic, for example, are the result of cold water from Antarctica moving northwards along the bottom and Arctic basin water pouring over the Wyville–Thomson ridge and flowing south (Fig. 5.9). Upwelling occurs when winds blowing from the land move surface waters away from the land. Deeper water then upwells to replace the surface water.

Ocean currents are of great biological importance because in addition to mixing the waters and providing a physicochemically uniform habitat for marine life, they also bring plant nutrients to the surface waters where plant photosynthesis can take place, distribute larval stages of animals that are fixed to the bottom, distribute plankton and govern the migration of marine animals to spawning grounds. The speed of bottom currents also has a direct influence on the nature of bottom sediments and therefore on benthic (bottom-living) communities. Just as in rivers or streams, where currents are slow fine sediments accumulate, while where they are fast the bottom consists of shell gravels or other coarse substrates.

5.3.7 Waves

Another feature of oceans is the presence of waves. On the surface they are usually produced by winds. In a moving wave in deep water, water particles move in a nearly circular orbit, rising to meet each wave crest as it passes them. Waves move with the wind but the individual particles of water do not, although there is a slow mass transport of water in the general wind direction (Fig. 5.10(a)).

Wave size and speed are dependent upon wind speed, wind duration and fetch (the distance over which the wind blows). In theory, wave height cannot exceed 0.14 of its wave length. Breaking waves with white caps are evidence of this factor being exceeded. This also occurs when waves approach a beach (Meyer, 1972). As the waves enter shallow water, friction against the sea bed slows the waves down, causing the wave length to decrease. When this happens the wave height increases, causing the waves to collapse (Fig. 5.10(b)). A dramatic example of this phenomenon is the Tsunami (Murty, 1977). These are low amplitude (about 1 metre), long wave-length (about 24 km) waves (sometimes incorrectly called tidal waves), which travel at great speed (up to 750 km/h), usually as a result of an earthquake or volcanic eruption. Out at sea these Tsunami can go unnoticed, but as they reach shallow water at a coastline, friction reduces the wave length with a concomitant increase in wave height and they can rear up to 10 metres or more in height. When they reach low-lying islands such as in Japan or Micronesia, they cause devastation and much loss of life.

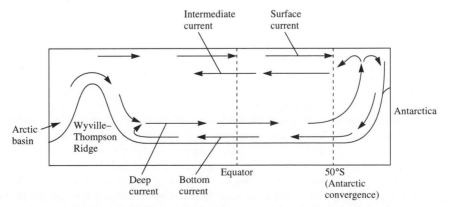

Figure 5.9 Longitudinal section through the Atlantic from pole to pole to show surface and deep water currents.

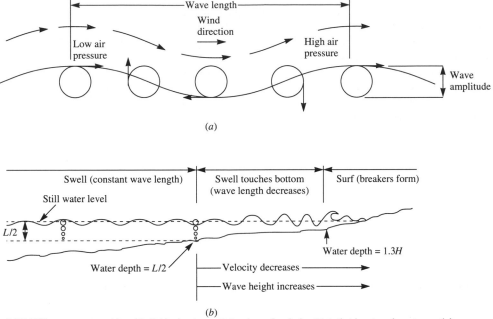

(a)

(b)

Figure 5.10(a) Wave parameter orbits of individual water particles shown by circles. Note that in a trough, water particles are moving backwards while at the peak they are moving forward. (b) Release of wave energy in the surf zone, leading to breaking of waves (after Thurman and Weber, 1984).

5.3.8 Natural Disturbances

Because of the immensity and homogeneity in conditions of the oceans, natural disturbances such as hurricanes, floods, droughts, etc., generally have little impact upon marine systems except on a very local scale, in inshore waters. In general, animals living in the intertidal and in shallow coastal waters of temperate regions have adapted to environmental variation since they experience regular differences in temperature and salinity and are exposed to the air for short periods as a result of the daily immersion from the tidal cycle. These factors also vary seasonally. The organisms are thus scarcely affected by the disturbances that can cause havoc on land. Organisms of tropical regions, on the other hand, do not experience significant seasonality and may be more damaged by natural disturbances. Shallow coral reefs, for example, may be killed by excessive rain during typhoons on the Great Barrier Reef of Australia, while corals in the northern Red Sea have been killed by exposure to hot sun during extremely low tides. Earthquakes and volcanic eruptions also damage coastal ecosystems.

Coastal areas exposed to the open ocean and with a long fetch may be periodically battered by storms. The damage is from the force of the waves rather than from the direct effects of winds. On temperate shores, winter storms often lead to the destruction and removal of large quantities of algae (seaweeds). This loss is, however, quickly made good by regrowth the following year, so that there is little long-term damage and thus the systems are resilient. Tropical shores are dominated by corals rather than algae and coral reefs have a definite structure, including heavily cemented buttress zones and surge channels to ameliorate the effects of heavy surf and strong waves. When hurricanes or cyclones hit a reef, massive destruction can nevertheless occur, coral being destroyed not only by the force of the waves but also by abrasion from sand and broken coral pieces. When such storms hit the leeward side (i.e. on the side away from the prevailing winds) of a coral atoll, which is less protected, damage can be particularly high. Badly affected reefs may take many years to recover.

Because of the relative constancy of the marine environments, marine organisms are relatively stenotopic in their physiology (have narrow tolerance limits); thus small changes in the environment can have correspondingly large effects. Many tropical areas of world seas have experienced massive die-off of corals. The corals are bleached white as a result of death of the symbiotic algae living within their tissues and this in turn leads to death of the coral. This so-called 'white death' is thought to be a result of raised temperatures caused by the El Nino Southern Oscillation (ENSO—cyclic changes in current strength shifting the current system and so affecting the global climate). While changes in the ENSO may be quite natural, there is also the possibility that they are an early warning of human-induced global warming and they have the strongest effect on tropical systems which are less resilient to such large-scale environmental change.

5.3.9 Anthropogenic Disturbances

Marine systems, because of their size, are buffered against disturbance and man has long regarded the oceans as useful dumps for a wide variety of rubbish. We are now becoming aware, however, that such a cavalier attitude to the seas is no longer acceptable. Recall the international uproar when Shell Oil intended to dump one of their defunct North Sea platforms in the North Atlantic. The problems of viral infection from swimming on beaches affected by sewage have brought home the real and immediate dangers of polluting coastal waters. The discovery of PCBs (polychlorinated biphenyls) in Antarctic penguins and lead residues in Arctic ice have demonstrated very clearly that marine pollution is a global problem requiring co-operation from all nations.

At a local scale, marine environments can be damaged by engineering works such as causeways, bridges and so on. A rich and diverse reef system in Florida was irrevocably changed into a low-diversity, algal-dominated muddy biotope by the construction of a causeway which altered the circulation regime. Dredging of sands, gravels, etc., for engineering works also seriously damages shallow-water habitats. Dredging and filling have, for example, caused more seagrass destruction than any other human activity, yet seagrass beds are of extreme importance as nursery grounds for young fish, including many commercial species, and a considerable quantity is transported offshore where it serves as food for a variety of surface feeding and benthic organisms (Zieman *et al.*, 1979).

Just as in terrestrial and freshwater systems, we have transported and introduced foreign species into many land areas. Shipping has brought alien species to many parts of the world (Chapman and Carlton, 1991), often from very distant regions, e.g., the New Zealand barnacle to Europe (Crisp, 1958). It has been estimated that the biomass of introduced animals to Southampton water in the United Kingdom is greater than that of native species. What the effect of this is on marine ecosystems has not been evaluated.

The opening of the Suez Canal put the Indian Ocean back into contact with the Mediterranean for the first time since the suturing of Africa with Europe over 50 million years ago. At first, the high salinity of the Great Bitter Lakes prevented immigrant marine organisms from making the trip between the Red Sea and the Mediterranean, but as the salt was leached away from the lakes, animals started to enter and colonize the Mediterranean from the Red Sea (Por, 1978; Vermeij, 1978), but interestingly scarcely any have made the reverse trip. The reason for the polarization is unclear, but it may be a result of failure in establishment in the new area rather than failure to arrive there. If the organisms from the Red Sea have a competitive advantage over those from the Mediterranean, then few Mediterranean species would be able to establish themselves in the Red Sea. This possibility worried scientists when the building of a sea-level canal in Panama was considered. The Caribbean species were thought to be competitively superior to those of the east Pacific. If the two seas were put into contact, and if the hypothesis about relative competitiveness was correct, the effect on the east Pacific fauna would have been disastrous.

5.4 TERRESTRIAL ECOSYSTEMS

Terrestrial ecosystems are characterized by their vegetation—the carpet of plant life covering the ground that confers a three-dimensional structure to the habitat. Vertical structure consists of the different layers or strata of the vegetation—a tropical forest has a very complex vertical structure compared to a simple pasture or tundra. There is also a horizontal component with different patches of different species combinations, each with its own degree of stability. There is also a temporal element to the nature of the vegetation at a particular place, as pioneer assemblages of ephemeral weeds on uncultivated or disturbed ground gradually change to climax forests with a complex array of life forms and layers.

Patches of vegetation growing under similar environmental conditions and with similar histories of environmental change often resemble each other in composition and structure (Miles, 1979); this leads to the idea of the biome described in Chapter 2. However, what environmental factors are the most important?

5.4.1 Temperature and Moisture

Land heats up and cools down much more quickly than water, thus terrestrial habitats have greater daily and seasonal temperature fluctuations. Thus from a global perspective, the distribution of vegetation can be associated with temperature (Krebs, 1985). However, at the level of the individual plant species, there is a less clear-cut effect of temperature. In cold climates, plants have usually evolved adaptations to cope with low temperatures, but they cannot anticipate unusual conditions, so late spring frosts, for example, can cause severe damage to temperate zone plants. The effect of temperature on animals does seem a little sharper, but again at a global rather than a local level. Temperature is more likely to affect activity patterns of animals than distribution *per se*.

Moisture, alone or in conjunction with temperature, is probably the most important physical factor affecting the ecology of terrestrial organisms. Living matter is entirely dependent on water. Terrestrial organisms lose water by evaporation and also in the excretion of waste products. The humidity of the air plays a major role in water loss through the skin or lungs of animals and from the leaves of plants. Water losses can be countered by reducing loss from evaporative surfaces (by behaviour, by altering activity periods, colour and body shape or through possession of an impermeable cuticle, etc.) and by excreting drier wastes (e.g. uric acid instead of urine). Alternatively, small animals can avoid the problem by remaining in damp environments. Water can be obtained by animals by drinking or from food and by plants from the soil. Water availability is critical and drought occurs when adequate amounts of water are not present or available (e.g. frozen in the soil). In fact plants differ markedly in their ability to tolerate drought and flooding.

Temperature and moisture levels interact strongly—as temperature increases so does the evaporation rate. In fact, the global distribution of the major biomes can be explained by the combined effect of temperature and mean annual precipitation (Begon *et al.*, 1990). Wind can also influence evaporative rates and the effect of the interaction of these three factors is clearly illustrated by the tree line so evident as one climbs with altitude. The temperature declines with increasing altitude, rainfall increases and wind velocity increases but soil moisture decreases due to freezing soils. Water desiccation or frost drought then ensues and is the primary determinant of the tree line on mountains. Figure 2.10 in Chapter 2 provides a good illustration of the roles of temperature and moisture on the distribution of plant communities.

5.4.2 Light, Nutrients and Soils

Light is vital for photosynthesis. Generally plants are either shade-tolerant or shade-intolerant, depending on their ability to function efficiently at low light levels. Under dense shade there is normally little ground vegetation. This is usually attributed to lack of light reaching the forest floor, but competition between plants for soil water and nutrients may also play a part. All plants require the same basic set of essential nutrients including P, N, Mg, Fe, S, K and Ca, but they do not use the nutrients in the same proportions, hence soil type can affect the distribution of plants. Soils are in turn affected by the plants that grow on them. While it is true that most plants have a fairly wide tolerance for soil types (Krebs, 1985), some species are restricted to specific soils. Bogs, for example, are very nitrogen-poor, but some plant species possess nodules in their roots that contain bacteria capable of fixing atmospheric nitrogen into usable nitrates for the plant. Soil pH can also exert a strong influence on the type of vegetation either directly through acidity levels or indirectly through the influence on the availability of nutrients and concentration of toxic metals. For example, below pH 4 to 4.5, mineral soils contain high concentrations of organic forms of aluminium, toxic to many plants.

5.4.3 The Influence of Humanity

In much of the world, the vegetation is not natural, but is semi-natural as a result of the activities of humanity. The change from tropical rainforest to poor agricultural grazing land in large tracts of Africa and South America is the most obvious example. However, what is not often realized is that very little of the so-called 'countryside' of Europe is actually the natural vegetation for the region. It is the result of hundreds of years of manipulation, tree felling, intensive grazing by cattle and sheep, application of fertilizers and other activities. Removing the agricultural pressures from most of Europe and North America would probably lead to a return to the natural forests that once covered these large tracts of land. Humanity's role in shaping terrestrial habitats should not be overlooked.

5.4.4 Natural Changes in Terrestrial Vegetation and Disturbance

Vegetation is naturally in a state of flux. Short-term, reversible, fluctuations occur about some notional mean state on a seasonal or perhaps year-to-year basis, with differences among the constituent species responding to fluctuations in the environmental conditions. Over a longer time frame, successional changes occur which markedly alter the appearance of the vegetation over time. These changes are directional from the initial state, involving the appearance of new species and loss of others. All vegetation appears to be subject to successional change although the rate may often be too slow to be detectable (Miles, 1979).

Succession can occur on pristine ground that has not previously been influenced by a community, e.g. on new sand dunes, on lava flows or on substrate exposed by a retreating glacier. Such succession is known as *primary succession*. An example would be the gradual colonization of sand dunes first by grasses, then shrubby plants, then trees and finally climaxing in a forest habitat. This kind of sequence is known as a sere and would take several hundred years to complete its course. The succession proceeds deterministically in this case as each successive vegetation type (or seral stage) establishes itself because the preceding type had modified the area in some way favourable to the successor (e.g. in dune succession, the grasses hold the sand/soil in place, allowing larger plants to root; they also increase the nutrient levels). This is known as the facilitation model of succession (Connell and Slatyer, 1977). Such a successional series would end in a climax stage that was stable and self-sustaining, providing that the conditions do not change markedly. Theoretically, if the climax is destroyed by a disturbance of some form, the process would be repeated and the same climax vegetation restored (Miles, 1979).

If vegetation is partially or wholly removed by a disturbance, subsequent succession is known as *secondary succession,* and may not operate as in the facilitation model due to the remaining seed banks in the soil. In this case, subsequent changes in the vegetation may result from differential growth rates, reproduction or survival of the species present. Weedy species may grow to full size first, with other species occurring only as seeds or seedlings. Annual weeds then die back and grasses dominate, with woody plants present, but only in a small or dormant form. As each successive stage drops out, a new group of species, present from the start, dominate, until eventually the climax state is reached, e.g. a forest. The gradual change is either (a) a result of different species having different strategies for exploiting resources, and later species being able to tolerate lower levels of resources, growing to maturity in the presence of earlier species and outcompeting them (the so-called tolerance model of succession), or (b) due to the fact that each species can resist invasion of competitors (such as by monopolizing resources like light), but later species gradually accumulate by replacing individuals of earlier species populations as they die (the inhibition model) (Connell and Slatyer, 1977). Examples of successions influenced by the processes outlined by the three models of Connell and Slatyer are given in Fig. 5.11.

Very often, vegetation is not allowed to reach its climax stage because of the impact of various catastrophic disturbance events preventing stability over a time-scale longer than the life-span of the primary species, e.g. as seen for the grasslands of the Great Plains and much of the natural forest of North America (Miles, 1979). These disturbances fall into three different types:

1. Physiographic or geomorphic processes that create areas for colonization by other species, including: coastal and other soil erosion by water, soil movement due to gravity (landslides, avalanches, rockfalls), meandering rivers, silting of lakes and estuaries, deposition of material by glaciers, soil erosion and deposition by wind, volcanic eruptions and flooding.
2. Climatic processes initiating vegetation change, including: drought, fires caused by lightning, wind-blown trees, winter cold and early autumn or late spring frosts, storms and long-term climatic change.

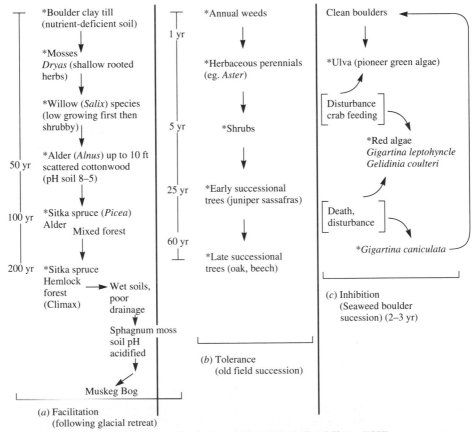

Figure 5.11 Examples of successions influenced by the three models of Connell and Slatyer (1977).

3. Biotic processes that lead to the death of plants and create openings for new species to colonize, including: the effect of other plants (competition), grazing by animals, plant diseases and epidemics.

Clearly, the frequency of these disturbances relative to the time for the full successional sequence to run will determine whether or not the community will ever reach a stable climax. Frequent disturbances may keep the ecosystem in a developmentally young state—this is frequently the case in so-called managed terrestrial systems associated with agriculture, recreation, etc. Whether or not this is a 'good thing' depends on the objectives of the management and the value of the resources held in the system. However, not all disturbances are 'natural' ones like those above. So-called 'unnatural' disturbances are the result of the activity of people. Clearly, engineering works can simulate physiographic and climatic processes, creating areas for colonization within a natural vegetation and initiating vegetation change. Biotic processes can be simulated directly by people through ploughing, use of herbicides, tree felling, etc., and indirectly through starting of fires, eutrophication of water bodies, pollution and influencing grazing pressures. Large-scale pollution can also stimulate climatic disturbances and change resulting from air pollution, as in the greenhouse effect. We conclude this chapter by concentrating on these indirect disturbances resulting from pollution.

5.5 ECOLOGICAL SYSTEMS AND POLLUTION

The movement of pollutants and toxic compounds through the environment is very similar to the movement of energy and nutrients within the ecosystem and on a larger scale through the biosphere. These chemicals (commonly known as pollutants) may cause harm to living organisms and the environment. The study of pollutant movement through the environment is known as ecotoxicology. However, before discussing the processes involved, let us first define a pollutant.

5.5.1 Definition and Classification of Pollutants

A pollutant is defined as 'a substance that occurs in the environment, at least in part, as a result of human activities, and which has a deleterious effect on the environment' (Moriarty, 1990). Pollutants are now unfortunately part of our environment as a result of industry and other activities. It is estimated that there are about 63 000 chemicals in use today (Maugh, 1978) and there are many hundreds being developed each year. The term 'pollutant' is a broad term and refers to a range of compounds, from a superabundance of nutrients giving rise to enrichment of ecosystems (see later) to toxic compounds which may be carcinogenic (cancer causing), mutagenic (cause damage to genes) or teratogenic (compounds that cause abnormalities in developing embryos). One of the most useful classifications of pollutants divides them into two major groups: (a) those that affect the physical environment and (b) those that are directly toxic to organisms, including humanity (Moriarty, 1990).

Pollutants that change the physical environment   Some pollutants do not have any obvious direct effect on living organisms but simply change the physical environment in such a way as to make conditions less suitable for life or unsuitable for the community present in the ecosystem at the time. The substances or conditions may have always been present, but now their concentrations or levels are altered. A classic pollutant of this type is the 'too much of a good thing' pollutant. There are two good examples of this phenomena. On a global scale, the atmospheric increase in levels of carbon dioxide, even though it is a gas that is essential for life through its role in photosynthesis, can give rise to significant global changes and may lead to global warming (Nisbet, 1991). In the same way, when normally limiting plant nutrients such as phosphate and nitrate become superabundant in waterways, there may be an increase in primary productivity. When this primary organic material dies and begins to decay, deoxygenation of water can ensue and a decline in environmental quality and species diversity follows. In these examples, when the natural balance in the distribution of these elements between the various phases of their biogeochemical cycle is disturbed, there are major consequences for the environment. In both cases, the rate of change of conditions is far in excess of natural changes; thus ecosystems are damaged—they become polluted.

Toxic pollutants   Some compounds, however, directly affect an organism's health and these are called toxic pollutants. Toxic pollutants include a range of compounds from heavy metals, polychlorinated biphenyls (PCBs) and dioxins to radioactive ions. Their toxicity depends on a number of factors.

Concentration   It is important to point out that although many elements may be required by organisms in trace amounts for normal physiological functioning they may also be toxic in large quantities; i.e. toxicity depends on concentration. A good example of this are the heavy metals, such as copper and zinc. Not all metals, however, have been shown to be essential in trace concentrations. Lead, aluminium and mercury have no known physiological role and are highly toxic to organisms. It is important to point out that all these compounds and metals have to be bioavailable, i.e. in a form that can be taken into the organism, before they become toxic. The bioavailability of compounds very much depends on their chemical form.

Chemical forms or species of compounds   Most heavy metals are only absorbed by the individual organism and distributed through ecosystems if they are in the methlyated form, where methyl group(s) (CH_3) are added to the element making it easier to enter organisms. In the same way, only certain species of metals with particular charges are toxic to organisms. For example, the particular form of aluminium which is toxic to fish in streams is a type called labile monomeric aluminium (Howells, 1990), and this form only occurs at certain stream pH levels. Therefore a complete examination of the form and 'species' of metal must be established in any pollutant before its toxicity can be determined.

Persistence   Some compounds disappear very quickly from the environment and are said to have a very short half-life, i.e. the time for 50 per cent of the compound to disappear or be broken down into a non-toxic form. Modern herbicides fall into this category. However, other compounds, such as organochlorines and polychlorinated biphenyls are highly persistent and linger in the environment for decades, and in some case generations. DDT is a well-known example of a persistent organochlorine insecticide that remains widely dispersed throughout ecosystems for prolonged periods of time, long after it was banned from international use (see also Table 3.8).

5.5.2 Bioaccumulation and Biomagnification

All organisms are made up of individual cells and in order for these cells to obtain nutrients and essential trace elements they selectively absorb and store a great variety of molecules. This is a natural process called bioaccumulation or bioconcentration. Because of the similarity between many toxic compounds, in particular heavy metals and essential trace elements, many toxins that are rather dilute in the environment can reach dangerous levels inside cells and tissues through bioaccumulation. Some of these compounds may even end up in the nucleus of cells. For example, O'Halloran and Duggan (1984) have shown that some of the spent lead shot ingested by swans in Ireland ends up in the nucleus of the kidney cells (Fig. 5.12). The important point to emphasize here is that the mechanisms of bioconcentration are the natural mechanisms whereby cells and indeed organisms obtain their basic nutrients, elements and vitamins and that the pathways are not unique to toxic compounds.

Biomagnification, on the other hand, occurs not at the cellular but at the ecosystem level. It generally occurs when the toxic compound concerned is not readily excreted from the organism but is stored and the toxic burden of a large number of organisms at a lower trophic level is accumulated and concentrated further by an organism (e.g. predatory bird or fish) at a higher trophic level as the material is passed up food chains. For example, if we consider phytoplankton and bacteria in an aquatic ecosystem, these organisms absorb toxic compounds from the water and sediment. Phytoplankton are preyed upon by

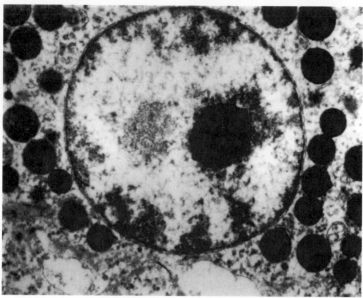

Figure 5.12 Electromicrograph of the kidney tubule cell showing a dense compact granular region, termed an inclusion body, made up of lead from a poisoned swan (× 20 000) (see the text for details) (from O'Halloran and Duggan, 1984).

zooplankton or small fish and these collect and retain the compound from the many prey organisms, building up higher concentrations of toxins. The top carnivores in the food chain, e.g. game fish (salmon or trout), fish-eating birds (kingfishers or heron) and indeed humans, can accumulate such high toxin levels that they suffer ill effects and even death. One of the classic examples of biomagnification was in the United States between 1949 and 1957 in Lake Clear, California (Mason, 1993). The lake was sprayed in 1949 with a chemical, closely related to DDT, called DDD, in order to reduce a public nuisance caused by non-biting midges. The spraying killed over 99 per cent of the insects, but by 1951 the population had almost fully recovered. In 1954, a second treatment was carried out using a higher concentration of pesticide. Shortly after the application of pesticide, a number of fish-eating grebes were found dead. The level of DDD found in the dead birds showed that the compound had biomagnified by 80 000 times the level found in the water. The result of this biomagnification was a significant decline in the grebe population from 3000 to 300 individuals by the end of the 1950s (Hunt and Bischoff, 1960). In Europe, the most widely cited example of biomagnification was the accumulation through food chains of DDT resulting in a decline in the peregrine falcon, a predatory bird (Newton, 1979).

5.5.3 Mixtures of Compounds or Pollutants

As mentioned in Chapter 2, organisms are not exposed to just a single environmental condition or factor but to many simultaneously. In the same way, organisms are rarely exposed to individual compounds in the environment. In fact, all environments consist of mixtures of natural organic and inorganic compounds and most also include a range of manufactured compounds. These mixtures may have significant effects on the toxicity of pollutants. If two or more compounds are present in a mixture they exert a combined effect on an organism which may be additive (Mason, 1993). The opposite may also be true, whereby certain mixtures of chemicals may reduce the toxicity of each chemical; these may be termed 'antagonistic'. However, in some situations and with certain combinations of environmental factors, a synergistic effect can occur in which the injury caused by exposure to two factors is more than the sum of the exposure to each factor individually. A classic example of this type of effect was described by

Guderian (1977) using spinach plants subjected to air pollution. These plants were exposed to (a) normal air, (b) air with a specified level of one of two pollutant compounds and (c) a combination of the two compounds. In this case a reduction in the photosynthetic rate of 18 per cent occurred when plants were exposed to air contaminated with hydrochloric acid (HCl) alone, an 11 per cent reduction in the photosynthetic rate when exposed to air contaminated with sulphur dioxide (SO_2) alone, while a reduction of 50 per cent occurred when both were combined. Clearly HCl and SO_2 act synergistically, giving a total reduction in photosynthesis which exceeds the sum due to the individual chemicals.

These complex interactions point out the unpredictability of the effects of pollutants in the environment. They also indicate that care must be taken when interpreting the effects of pollutants on organisms and ecosystems—effects that are normally predicted following single species toxicity tests which are carried out before chemicals are brought into industrial production.

5.5.4 Lethal and Sublethal Effects

Chemical compounds and mixtures of chemicals can have three types of effect on organisms. One is no effect at all, the second is a lethal effect and the third a sublethal effect. The first case is self-explanatory whereby a compound is effectively biologically inert to the organism (although it should be pointed out that some compounds may have no effect on some organisms but may prove lethal or sublethal to other types). In the second case a substance may be lethal to an organism at a given concentration. In this context there are specified protocols for defining the toxicity of substances and lethal doses based on LD_{50} and LC_{50} (i.e. the lethal dose or concentration of a toxic compound at which 50 per cent of organisms die when exposed to that concentration for a certain duration of time, usually 48 hours). However, it is important to remember again that these lethal doses are calculated using one or maybe two compounds in the laboratory and may not really reflect what is happening in the environment. For this reason, biologists and ecotoxicologists have begun to look at levels of compounds in wild animals in an attempt to better assess the effects of environmental pollutants and contaminants in the environment.

Sublethal effects, though they are often less dramatic than death of individual organisms, may have a much greater effect on the overall population. For example, the loss of one or two individuals of a population through lethal poisoning, though undesirable, may not be very significant at the population level, while a sublethal effect of compounds which damage genes (DNA) or affect developing embryos (such as in birds' eggs) may have very significant effects on the entire population. Sublethal effects can occur at the genetic, biochemical, physiological, behavioural or life cycle level. One of the difficulties identifying such effects is that biologists often do not not know the 'normal' range of biochemical, physiological, behavioural or life cycle patterns for organisms and hence are unable to clearly distinguish sublethal effects of pollutants. This lack of knowledge does not mean that such effects are not occurring, but may be undetectable at this point in time.

A modern approach to identify sublethal effects of environmental compounds involves the use of biochemical markers (termed biomarkers), recently reviewed by Peakall (1992). One of the most widely described examples of biomarkers is the reduction in an enzyme called aminolaevalinic acid dehydratase (ALDA) (Lansdown and Yule, 1986) and protoporphyrin IX (O'Halloran et al., 1988) in red blood cells during exposure to lead pollution. The levels of these biomolecules increase in the blood at low levels of exposure to lead, reflecting sublethal metabolic damage caused by pollutants. A number of constituents of plant biochemistry are also affected by pollutants, and some of the changes in the rates of photosynthesis, such as those mentioned in the previous section, may be explained by the sublethal effects of compounds.

5.5.5 Environmental Factors Affecting Toxicity

A large number of environmental factors influence the toxicity of chemical compounds. These environmental parameters not only affect the metabolism of organisms themselves but also influence the bioavailability of chemicals to organisms.

Chemicals that are released into the atmosphere, water or soil are often transformed from one form to another, e.g. elements may be transformed from an inorganic state to an organic state or vice versa. These chemical changes may be brought about by oxidation, methylation or other chemical processes in the soil, water or air. This often results in increased toxicity of compounds. For example, consider mercury (Hg), a compound released in large concentrations from crematoria (Mills, 1990) and formerly a fungicide widely used in agriculture. Inorganic mercury in itself is virtually unavailable to biological systems and hence has no toxicity. However, when the mercury is transformed into methylated mercury by bacteria and fungi in soil and water it becomes extremely toxic to biological systems (see Fig. 5.13 for a summary of the transformation).

In the same way, the hardness or softness of water has a direct influence on acidification, leading to hydrogen ion toxicity (H^+) in freshwater ecosystems. In aquatic systems whose catchments are under the influence of acid rain, the effects on the biology are many orders of magnitude greater if the alkalinity and total hardness of the water are low. In contrast, if the water has an alkalinity value of greater than 20 mg/L $CaCO_3$, then there should be sufficient buffering capacity present to neutralize the effect of the hydrogen ions (O'Halloran and Giller, 1993). In the same way, some heavy metals are only toxic at certain pH levels. For example, aluminium can exist in a number of different forms or 'aluminium species' in the environment, depending on the pH of the water. It is most toxic to aquatic organisms such as fish at pH values of 4.0 to 4.5 when the aluminium exists as labile monomeric aluminium.

There is a whole range of other environmental parameters that affect the toxicity of compounds, such as the level of organic compound present, environmental temperature, etc., but there are also particular attributes of biological systems that affect the toxicity of compounds. Temperature is one of the most obvious parameters affecting the toxicity of compounds which is related to the fact that the metabolism of an organism is closely related to environmental temperatures, reviewed by Cairns *et al.* (1975). Thus as the organisms increase their metabolism, they often increase the uptake of chemicals and pollutants. The increase in metabolism is often associated with increased absorption of compounds across the gut wall in animals (Mason, 1993). Natural processes of detoxification of compounds carried out by organisms are also important factors affecting toxicity. Some compounds are less easily absorbed and persistent in the body in their original form (e.g. the insecticide DDT) than when they have been metabolized (e.g. to DDE) by organisms. Thus in an attempt to detoxify compounds, the organism may produce an end product which is in fact more toxic.

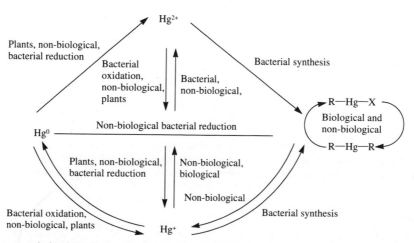

Figure 5.13 Elements and chemicals can be transformed by biological processes, making them more bioavailable. Here the transformation of inorganic mercury is illustrated as an ecosystem (after Mason, 1993).

This chapter has set the scene for this part on pollution. It started with an overview of the basic properties, processes and controlling factors of freshwater, marine and terrestrial ecosystems and how these ecosystems can be perturbed by disturbances and pollution. We then outlined the fundamental characteristics of pollution and pollutants and their general effects on ecological systems. The remaining chapters in this part deal with specific types of pollution covering water, air, noise and agricultural pollution.

5.6 PROBLEMS

5.1 In what ways do engineering processes and activities impinge on the functioning of freshwater systems?

5.2 How can changes in primary productivity of aquatic systems affect the use of aquatic resources?

5.3 What is eutrophication? What are the differences between eutrophic and oligotrophic systems?

5.4 Why are there seasonal cycles of productivity in temperate seas?

5.5 Explain the significance of ocean currents to marine systems.

5.6 Describe the properties of seawater that are of biological significance.

5.7 Compare and contrast the pathways of pollutants and nutrients in a named ecosystem.

5.8 Outline the physical and chemical characteristics of a waterbody that make them susceptible to pollution.

5.9 Why can the deposition of small quantities of some pollutants give rise to large-scale disruption of ecological systems? Evaluate the importance of sublethal toxicity of a pollutant on ecosystems.

5.10 What are the short-term and long-term effects on the environment that can arise from the construction of dams? Carry out a library search to identify examples other than those described in this chapter.

5.11 Sketch a concept of the comparative hydrograph shapers for the same catchment resulting from a storm for the following land, using conditions in a temperate climate.

    (a) original upland catchment

    (b) pre-afforest drainage catchment

    (c) mature forest catchment

    (d) clearfelled forest catchment.

REFERENCES AND FURTHER READING

Allott, N. (1986). 'Temperature, oxygen and heat budgets of six small western Irish lakes', *Freshwater Biology*, **16**, 145–154.

Begon, M., J. Harper and C. Townsend (1990) *Ecology.* Blackwell Scientific Publications, Oxford.

Bournard, M., H. Tachet and A. Roux (1987) 'The effects of seasonal and hydrological influences on the macroinvertebrates of the Rhone River, France. 2. Ecological aspects', *Archiv fur Hydrobiologie*/Supplement, **76**, 25–51.

Broecker, W. S. (1974) *Chemical Oceanography*, Harcourt, Brace, Jovanovich, New York.

Burgis, M and P. Morris (1990) *The Natural History of Lakes*, Cambridge University Press.

Cairns, J., A. G. Heath and B. C. Parker (1975) 'The effects of temperature upon the toxicity of chemicals to aquatic organisms', *Hydrobiologia*, **47**: 135–171.

Chapman, J. W. and J. T. Carlton 1991 'A test of criteria for introduced species: the global invasion by the isopod *Synidotea laevidorsalis* (Miers, 1881)', *Journal of Crustacean Biology*, **11**(3), 386–400.

Connell, J. and R. Slatyer (1977) 'Mechanisms of succession in natural communities and their role in community stability and organisation', *American Naturalist*, **111**, 1119–1144.

Crisp, D. J. (1958) 'The spread of *Elminius modestus* (Darwin) in North–West Europe', *Journal of the Marine Biological Association of the U.K.*, **37**, 483–520.

Crisp, D. and T. Gledhill (1970) A quantitative description of the recovery of the bottom fauna in a muddy reach in a mill stream in Southern England after draining and dredging', *Archiv fur Hydrobiologie*, **67**, 502–541.

Dahn, C., E. Trotter and J. Seddell (1987) 'Role of anaerobic zones and processes in stream ecosystem productivity', in *Chemical Quality of Water and Hydrologic Cycle*, R. Averett, and D. McKnight (eds), Lewis Publishing, Chelsea, Michigan, pp. 157–178.

Gilason, J. (1985) 'Aquatic insect abundance in a regulated stream under fluctuating and stable diel flow patterns', *North American Journal of Fisheries Management*, **5**, 39–46.

Giller, P. S (1990) 'After the deluge', *Technology Ireland*, **22**, 41–44.

Giller, P. S. and J. H. R. Gee (1987) 'The analysis of community organisation : the influence of equilibrium, scale and terminology', in *Organisation of Communities Past and Present*, J. Gee and P. Giller (eds), Blackwell Scientific Publications, Oxford, pp. 519–542.

Giller, P. S., N. Sangpradub and H. Twomey (1991) 'Catastrophic flooding and macroinvertebrate community structure', *Internationale Vereinigung fur Theoritische und Angewandte Limnologie*, **24**, 1724–1729.

Guderian, R. (1977) *Air Pollution. Phytotoxicity to Gases and Its Significance in Air Pollution Control*, Springer, Berlin.

Hammerton, D. (1972) 'The Nile—a case history', in *River Ecology and Man*, R. Oglesby, C. Carlson and J. McCann (eds), Academic Press, New York, pp. 171–214.

Hargrave, B. T. (1991) 'Impacts of man's activities on aquatic ecosystems', in *Fundamentals of Aquatic Ecosystems*, R. S. K. Barnes and K. H. Mann (eds), Blackwell Scientific Publications, Oxford, pp. 245–264.

Hellawell, J. M. (1986). *Biological Indicators of Freshwater Pollution and Environmental Management*, Pollution Monitoring Series, Elsevier Applied Science Publishers, London.

Howells, G. (1990) *Acid Rain and Acid Waters*, 1st edn, Ellis Horwood Ltd, Chichester.

Hunt, E. G. and A. I. Bischoff (1960) 'Inimical effects on wildlife of DDD application to Clear Lake', *Californian Fisheries and Game*, **46**: 91–106.

Hynes, H. B. N. (1970). *The Ecology of Running Waters*, University of Liverpool Press, Liverpool.

Institute of Hydrology (1984–87). *Research Report 1984–1987*, Natural Environment Research Council, UK.

Koblentz-Mishke, O. J., V. V. Volkovinsky and J. G. Kabanova (1970) 'Plankton primary production of the world ocean', in *Scientific Exploration of the South Pacific*, Standard Book No. 309 01755 6, National Academy of Science, Washington, pp. 183–193.

Krebs, C. J. (1985) *Ecology. The Experimental Analysis of Distributions and Abundance*, Harper and Row, New York.

Landsdown, R. and W. Yule (1986) *The Lead Debate: The Environment, Toxicology and Child Health*, Croom Helm, London.

Likens, G. E., *et al.*. (1970) 'Effects of forest cutting and herbicide treatment on nutrient budgets in the Hubbard Brook watershed ecosystem', *Ecological Monographs*, **40**, 23–47.

Maitland, P. (1990) *Biology of Freshwaters*. 2nd edn, Blackie, London.

Mason, C. F. (1993). *Biology of Freshwater Pollution*, 2nd edn, Longman Scientific and Technical, New York.

Matthews, W. (1986) 'Fish faunal structure in an Ozark stream: stability, persistance and a catastrophic flood', *Copeia*, 1986(2), 388–397.

Maugh, T. H. (1978) 'Chemicals: how many are there?', *Science (New York)*, **199**, 162.

Meyer, R. E. (ed.) (1972) *Waves on Beaches*, Academic Press, London and New York.

Miles, J. (1979). *Vegetation Dynamics*. Chapman and Hall, London.

Mills, A. (1990). 'Mercury and crematorium chimneys', *Nature*, **346**, 615.

Minshall, G. W. (1988) 'Stream ecosystem theory: a global perspective', *Journal of the North American Benthological Society*, **7**, 263–288.

Moriarty, F. (1990) *Ecotoxicology: A Study of Pollutants in Ecosystems*, 2nd edn, Academic Press, London.

Moss, B. (1988) *Ecology of Freshwaters. Man and Medium*, Blackwell Scientific Publications, Oxford.

Murty, T. S. (1977) 'Seismic sea waves: Tsunami', Fisheries Research Board of Canada, Bulletin 198, Ottawa.

Newton, I. (1979) *Population Ecology of Raptors*, Poyser, Berkhamstead.

Nisbet, E. G. (1991) *Leaving Eden: To Protect and Manage the Earth*, Cambridge University Press, Cambridge.

O'Halloran, J. and P. F. Duggan (1984) 'Lead levels in Mute swans in Co. Cork', *Irish Birds*, **4**, 501–514.

O'Halloran, J. and P. S. Giller (1993) 'Forestry and the ecology of streams and rivers: lessons from abroad', *Irish Forestry*, **50**, 35–52.

O'Halloran, J., A. A. Myers and P. F. Duggan (1988) 'Blood lead levels and free red blood cell protoporphyrin as a measure of lead exposure in Mute swans', *Environmental Pollution*, **52**, 19–38.

Peakall, D. B. (1992) *Animal Biomarkers as Pollution Indicators*, Chapman and Hall, London.

Pfankuch, D. J. (1975) *Stream Inventory and Channel Stability Evaluation*, United States Department of Agriculture, Forest Service, Region 1, Missoula, Montana.

Por, F. D. (1978) 'Lessepsian migration', in *The influx of Red Sea Biota into the Mediterranean by Way of the Suez Canal*, Springer-Verlag, Berlin.

Reynolds, C. R. (1987) 'Community organisation in the freshwater plankton', in *Organisation of Communities Past and Present*, J. Gee and P. Giller (eds), Blackwell Scientific Publications, Oxford, pp. 267–295.

Reynolds, C. R. (1994) 'The role of fluid motion in the dynamics of phytoplankton in lakes and rivers', in *Aquatic Ecology, Scale Pattern and Process*. P. S. Giller, A. G. Hildrew and D. Rafaelli (eds), Blackwell Scientific Publications, Oxford, pp. 141–187.

Sagar, P. (1986) 'The effects of floods on the invertebrate fauna of the streams of Rawalpindi and Wah', *Pakistan Journal of Forestry*, **19**, 227–234.

Smith, B. D., P. Maitland and S. M. Pitnnock (1987) 'A comparative study of water level regimes and littoral benthic communities in Scottish lochs', *Biological Conservation*, **39**, 291–316.

Sousa, W. P. (1984) 'The role of disturbance in natural communities', *Annual Review of Ecology and Systematics*, **15**, 353–391.

Thurman, H. and H. Weber (1984) *Marine Biology*, C. E. Merrill, Columbus, Ohio.

Tsuda, M. and T. Komatsu (1964) 'Aquatic insect communities of Yoshino River, four years after the Ise–Wan Typhoon', *Japanese Journal of Ecology*, **14**, 43–49.

Vermeij, G. J. (1978) *Biogeography and Adaptation; Patterns of Marine Life*, Harvard University Press, Cambridge, Mass.

Vollenweider, R. and J. Kerekes (1981) 'Background and summary results of the OECD cooperative programme on eutrophication', Appendix 1 in *The OECD Co–operative Programme on Eutrophication, Canadian Contribution*, compiled by L. L. Janus and R. A. Vollenveider, Environment Canada, Scientific Series 131.

Wadi, W. F. (1984) *Acta Adriat*, **25**, pp. 29–43.

Ward, R. (1978) *Floods: a Geographical Perspective*, Macmillan, New York.

Whittaker, R. H. (1975) *Communities and Ecosystems*. 2nd edn, Macmillan, New York.

Winterbourn, M. and C. Townsend (1991) 'Streams and rivers: one way flow systems', in *Fundamentals of Aquatic Ecology*, R. S. K. Barnes and K. H. Mann (eds), Blackwell Scientific Publications, Oxford, pp. 230–243.

Zieman, J. C., G. W. Thayer, M. B. Robblee and R. T. Zieman 1979 'Production and export of sea grasses from a tropical bay', in *Ecological Processes in Coastal and Marine Systems*, R. J. Livingston (ed.), Plenum Press, London, pp. 21–33.

WATER POLLUTION: ECOLOGICAL PERSPECTIVES

6.1 INTRODUCTION

Water is a renewable resource, which, as discussed in Chapter 4, is naturally recycled in the hydrological cycle. Surface waters tend to have a short residence time in the hydrological cycle while groundwaters have a very long residence time. This recycling renews water resources and potentially provides a continuous supply. With the advent of industrialization, intensification of agriculture and increasing populations, the demand for water has increased. The main uses of freshwater (consumer processes) can be divided into two broad catagories: abstraction and instream uses (Table 6.1).

The patterns of use vary from region to region and with different stages of development of the country. The marine environment and marine resources are also becoming increasingly important, not so much from direct use of seawater (except in desalination plants in arid areas) but more in terms of exploitation of biological (fisheries, seaweed) and geological resources (oil, gas, minerals), amenity and use in power generation (wave generators).

Despite the fact that water is renewable, freshwater resources are finite. Inputs of water (via rainfall) to a catchment are balanced by recharging of groundwaters and outputs from the catchment via river surface flow, evapotranspiration and abstraction. For example, most rainfall in the United Kingdom falls

Table 6.1 Main consumer processes of aquatic resources

Abstraction	Instream uses
Domestic supply	Biological exploitation
Irrigation	Power generation
Industry—manufacturing	Transportation/navigation
Industry—cooling	Recreation/amenity
Flushing of canals	Flood control
Diversion between catchments	Waste transportation
	Political boundaries

in the west and north uplands, and most excess rainfall (leading to surface runoff) occurs in the winter. Regions of high population and intensive industry are found in areas of low rainfall and demand for water is greatest in the summer (Mason, 1991). Arable farming also tends to be concentrated in drier parts of the country and requires irrigation. This supply–demand inbalance means that local water resources are not able to meet demand at all times. Similar problems are apparent throughout the world. For example, the desert lands of southern California and the metropolis of Los Angeles, in particular, get most of their water supply from northern California by means of a man-made aquaduct some 300 km long. Aside from such redistribution networks and intercatchment diversion of water supplies, this shortcoming can be overcome by the fact that water is also to a large extent a recyclable resource—once used, it can potentially be reused. However, each water use, including abstraction, leads to specific impacts on the water resources. Domestic, industrial and agricultural consumer processes produce large quantities of waste products for which the natural waterways offer cheap and readily available conduits for disposal. In rivers and lakes, therefore, the effluents of some water users may become the water supplies of others. The process of water treatment before and after use in consumer processes is therefore an important part of modern society and is discussed in detail in Chapters 11 and 12.

Indirect influences on water resources are also apparent, largely related to land–water linkages discussed in Chapter 5. Land use for urbanization, agriculture, afforestation and deforestation, and leaching of wastes from landfill sites all influence the nature of aquatic resources. A clear example is seen in relation to the intensification of arable farming through application of fertilizers to the land and the consequent change in nutrient levels in lowland rivers draining such catchments (Fig. 6.1). Because of the problem in the demand and supply ratios, much water abstraction occurs from these lowland reaches of rivers rather than from more pristine upland headwaters. Engineering-based interventions in the hydrological cycle, such as canalization, damming, diversion of water within and between catchments also have effects on aquatic resources (see Chapter 5).

All of the above effects obviously relate to the quality of the aquatic environment in terms of the physicochemical conditions and the state of the flora and fauna. However, in the present context, the concern is more with the effects of direct consumer processes and indirect human activities on the use of water as a resource. In this sense, we are interested in *water quality*.

For any of the above specific consumer processes, there are a set of requirements regarding the quality of water to be used (usually related to concentrations of various chemical parameters, suspended material and bacterial content). If water fulfills these requirements or standards, it is said to be of good quality for the particular consumer process. If it does not, it is deemed unacceptable and of poor quality.

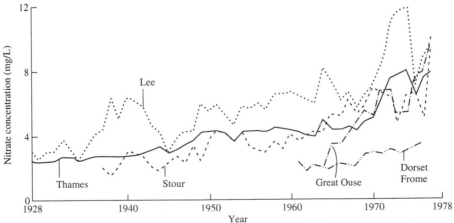

Figure 6.1 Changes in nitrate-nitrogen concentration in five British rivers associated with intensification of agriculture in the river catchments (after Moss, 1988).

Water quality is therefore a term that implies some value judgement of the water with respect to a particular use. A simple definition of water quality is therefore difficult due to the complexity of factors influencing it and the range of functions water resources are required to fulfill, often at the same time. Abstraction for domestic supply requires the most stringent standards and highest water quality, whereas use for navigation can be met with the poorest quality. The largest demands for quantity (e.g. navigation and industrial cooling) often require the least in terms of water quality, but to manage the freshwater ecosystem to allow usage by all consumer processes, water quality needs to be maintained at the highest quality demanded by the most stringent user.

Any change in the natural water quality implies pollution. Some natural events can lead to local deterioration in water quality, such as hurricanes, mud flows, torrential rainfall and unseasonal lake overturn. More serious, longer term and larger scale problems arise as a result of the activities of man. Pollution of the aquatic environment means the introduction by man, directly or indirectly, of substances or energy (heat) that result in deleterious effects, including harm to living (biological) resources, hazards to human health (pathogens), hindrance to aquatic activities including fishing and impairment of water quality with respect to a desired consumer process such as agriculture, industry, amenity or domestic supply (Chapman, 1992).

Temporal variability in water quality occurs on a number of scales:

- Minute–minute to day–day: from small-scale water mixing and fluctuations in inputs to the aquatic system
- Diurnal (24 h) variability: related to biological and light–dark cycles, such as oxygen, pH, and related to cycles in pollution inputs (e.g. domestic wastes)
- Days to months: related to climatic factors (hydrologic regimes, lake overturn) and to pollution sources (industrial wastes, agricultural activity in the catchment)
- Seasonal: related to climatically driven hydrological and biological seasonal cycles
- Year–year: related mostly to human influences in the catchment (e.g. changing land use, clearing of vegetation, building works)

The adverse pressures on the quality of the aquatic environment are particularly marked in technologically advanced countries, where the range of requirements for water resources is increasing together with greater demands for high-quality water. There is thus a conflict in use of aquatic resources for high-quality water for drinking and recreation, on the one hand, and for more water for improved sanitation, industrial uses, waste disposal, etc., on the other, all from the same source. Thus the amount of wastewater increases and water treatment (see Chapters 11 and 12) is required in an attempt to maintain sufficiently high quality of natural aquatic resources to satisfy the consumer demand. The conflict of uses and strong land–water linkages have contributed to establishing a range of freshwater quality issues and limitations to the use of water resources at the global scale, which are outlined in Table 6.2.

6.2 WATER QUALITY STANDARDS AND PARAMETERS

Pollution and water quality degradation thus interfere with vital and legitimate water uses at scales from local to regional to international levels (given the transboundary unidirectional nature of river systems and vastness of seas and oceans). Water quality criteria and standards are therefore necessary to ensure that the appropriate quality of resource is available to a particular consumer process. The related legislation is used as an administrative means to manage and maintain water quality for the maximum number of users of the water body.

Water quality and standards vary and may originate in a number of possible ways; there are international standards set by the WHO and EU (formerly EC), regional standards set by individual states or local standards set by individual local authorities. In most European countries, absolute standards for concentration of chemicals are set for all rivers. The standards and levels are set (on the basis of chemical

Table 6.2 Major freshwater quality issues on a global scale (a) and limitations to water uses due to degradation of water quality by various pollutants (b)

(a)

Issue	Water body			
	Rivers	Lakes	Reservoirs	Groundwaters
Pathogens	xxx	x†	x†	x
Suspended solids	xx	na	x	na
Decomposable organic matter‡	xxx	x	xx	x
Eutrophication§	x	xx	xxx	na
Nitrate as a pollutant	x	0	0	xxx
Salinization	x	0	x	xxx
Trace elements	xx	xx	xx	xx¶
Organic micropollutants	xxx	xx	xx	xxx¶
Acidification	x	xx	xx	0
Modification of hydrological regimes††	xx	x		x

xxx Severe or global deterioration found.	† Mostly in small and shallow water bodies.
xx Important deterioration.	‡ Other than resulting from aquatic primary production.
x Occasional or regional deterioration.	§ Algae and macrophytes.
0 Rare deterioration.	¶ From landfill, mine tailings.
na Not applicable.	†† Water diversion, damming, overpumping, etc.

(b)

Pollutant	Consumer process						
	Drinking water	Aquatic wildlife, fisheries	Recreation	Irrigation	Industrial uses	Power and cooling	Transport
Pathogens	xx	0	xx	x	xx†	na	na
Suspended solids	xx	xx	xx	x	x	x‡	xx§
Organic matter	xx	x	xx	+	xx¶	x††	x¶¶
Algae	x††,‡‡	x§§	xx	+	xx¶	x††	x¶¶
Nitrate	xx	x	na	+	xx†	na	na
Salts†††	xx	xx	na	xx	xx‡‡‡	na	na
Trace elements	xx	xx	x	x	x	na	na
Organic micropollutants	xx	xx	x	x	?	na	na
Acidification	x	xx	x	?	x	x	na

xx Marked impairment causing major treatment or excluding the desired use.	† Food industries.
x Minor impairment.	‡ Abrasion.
0 No impairment	§ Sediment settling in channels.
na Not applicable	¶ Electronic industries.
+ Degraded water quality may be beneficial for this specific use.	†† Filter clogging.
? Effects not yet fully realized.	‡‡ Odour, taste.
	§§ In fish ponds higher algal biomass can be accepted.
	¶¶ Development of water hyacinth (*Eichhomia crassipes*).
	††† Also includes boron, fluoride, etc.
After Chapman, 1992	‡‡‡ Ca, Fe, Mn in textile industries, etc.

and microbial parameters) by taking cognizance of the differing uses for which water quality must be maintained. The ultimate objective of the imposition of standards is the protection of the end user, be these humans, domestic animals or industrial plants. In this chapter we are concerned, however, with the safeguarding of public health and the protection of ecosystems (which by definition of the standards is often through the protection of fisheries). Both have very high quality requirements which complement each other to a great extent, on the basis that if a reservoir, lake or river is suitable to sustain fisheries it will most likely be suitable for all other users.

There are also guidelines for drinking water quality issued by the World Health Organization and the US National Academy of Sciences and the US Environmental Protection Agency. The relevant standards are enumerated in Chapter 11 for drinking water and in Chapter 12 for wastewater effluent discharges.

6.3 ASSESSMENT OF WATER QUALITY

The maintenance and assessment of water quality are important procedures in modern society. Earliest and simplest methods were purely subjective—does the water look clean, smell right, etc? Such assessment of water quality is sufficient for some consumer processes but, for most, the fact that water is such a good solvent and can contain all kinds of dissolved substances led to requirements for more precise assessment methods. These have been met through hydrochemical analytical techniques. Each chemical parameter has an associated standard, as discussed in Chapter 11, and water is chemically tested as a routine measure to ensure it fulfils the quality standards for the various consumer processes. However, there are thousands of chemical pollutants, but only a small number can be analysed in any one sample.

Scientists have also found that biological surveillance of aquatic systems can prove valuable in assessing water quality and detecting pollution. Aquatic organisms show a lasting response to the intermittent pollution episodes which may be missed by routine chemical monitoring which only samples a relatively tiny volume of water at a particular point in time. Aquatic organisms also provide an assessment of the average water quality that has pertained over a period of time. Organisms can also accumulate and magnify low levels of chemicals (see Chapter 2) which may be below the detection point of analytical chemical methods but which can be analysed from biological tissues. Biological methods also provide information on the impact of pollutants on the ecology of the system, which chemical methods, by themselves, cannot do. However, most biological techniques have the disadvantage of not being able to give an accurate measure of the exact quantity of pollutants and concentrations of chemicals and they may not pick up slight changes in water quality that do not impact significantly on the ecological system but which nevertheless may be of importance to some consumer processes.

Thus modern approaches to the description of water quality utilize three approaches:

- Quantitative measurements, such as of physicochemical parameters in water, in sediments or in biological tissues
- Biochemical/biological tests (including BOD estimation, toxicity testing, etc.)
- Semi-quantitative and qualitative descriptors involving biological indicators and species inventories (Chapman, 1992)

The actual process of water quality assessment is an evaluation of the physicochemical and biological nature of the water in relation to its natural quality, human effects and intended uses; i.e. basically to verify whether the observed water quality is suitable for its intended use (Chapman, 1992). It includes the use of *monitoring*. Monitoring involves the collection of information at set localities and at regular intervals to:

- Obtain information concerning substances entering the environment, the quantities, sources and distribution
- Evaluate the effects of these within the environment
- Provide a basis for detecting trends in concentrations and effects and to establish cause and effect relationships (e.g. acidification and eutrophication)
- Examine how far inputs, concentrations and trends can be modified, by what and at what cost (Chapman, 1992; Mason, 1991).

The water quality assessment processes have the same basic philosophies in marine and freshwater systems, although much more has been published concerning freshwater largely because of the greater importance to human societies. Chapman (1992), on behalf of UNESCO and WHO, presents a comprehensive outline of the design of water quality assessment procedures and protocols and the selection of water quality variables for rivers, lakes and groundwaters.

6.3.1 Assessment Methods

Chemical assessment techniques   Chemical assessment techniques are well known and involve regular sampling of water in the natural system and/or at some point in the abstraction and treatment processes and of most effluents before they are released back into the environment (see Chapters 11 and 12). The assessment involves rigorous testing for the presence and concentration of the major chemical parameters described above. Descriptions of these tests can be found in many texts and manuals, e.g. *Standard Methods* (1992).

Biological assessment techniques   Use of biological assessment is probably less well known to readers and therefore some details of the main approaches are given in this section. The effects of changes in the environment on organisms were reviewed in Chapters 2 and 5—the responses include death, migration, decrease in reproduction and lower populations. Some of the commonest effects of pollution on aquatic ecological communities are shown in Table 6.3. Once the responses of particular organisms to any given change in the environment have been identified (as, for example, in Fig. 6.5 for heavy organic pollution of rivers), these responses may be used as an analytical tool to determine the quality of water. There are several impressive texts and reviews on all methods for biological analyses of water quality, e.g. Washington (1984), Hellawell (1986), Chapman (1992) and Spellerberg (1991).

Ecological methods   Most of these methods have been devised to monitor and assess organic pollution. The simplest approach is to look for indicator species in a sample for the aquatic habitat, where one can infer from the presence of known intolerant or sensitive organisms (e.g. stoneflies and mayflies in rivers and streams) that the water is of sufficiently good quality to sustain normal aquatic life (e.g. high oxygen levels and hence low organic pollution). The absence of these and the presence of high numbers of known tolerant organisms (e.g. in rivers and streams sewage fungus and tubificid worms) would indicate polluted water (see Fig. 6.5). An example of changing numbers of plant species in response to a pollution outfall and subsequent recovery downstream is shown in Fig. 6.2. Similar sets of tolerant and intolerant species are also known from the marine environment (e.g. the worm *Capitella* is a well-known tolerant species). Microbial organisms can also be used in a similar way to animals and plants.

This idea can be extended to include the whole community through the calculation of a *biotic index*. Although some indices have been developed to monitor marine pollution (see Washington, 1984), the most comprehensive approaches have been derived for pollution of freshwaters. Samples are collected

Table 6.3 Common effects of pollutants on aquatic life in natural waters

Response	Cause
Changes in species composition	Death or migration of intolerant populations from the area and colonization by tolerant forms
Changes in dominant groups	Decline in populations of previously dominant forms, increase in other, tolerant, forms
Impoverishment of species	Loss of intolerant forms
High mortality of sensitive life stages, e.g. eggs	Due to toxicity, lack of oxygen
Changes in behaviour	Due to physiological and/or biochemical responses to pollutants
Changes in physiology, metabolism histological changes and morphological deformities	Due to cellular responses to sublethal, toxic pollutants

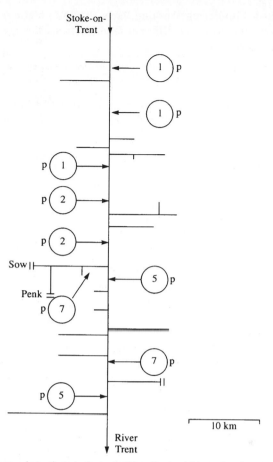

Figure 6.2 The number of species of macrophyte plants in the upper River Trent, which receives heavy pollution from Stoke-on-Trent. Tributary streams are shown (e.g. Sow, Penk). P indicates the presence of pollution-tolerant species *Potamogeton pectinatus*. Note how the macrophyte diversity is low just below Stoke-on-Trent but gradually increases downstream (adapted from Haslam, 1978).

from a site and the animal community is used to derive an index of water quality. There are three kinds of qualitative methods involved in this approach:

1. Each type of organism is given a score on some scale according to its tolerance to pollution (low score = tolerant, high score = intolerant clean water forms). The points are then totalled for the sample to give the biotic index, which can be compared to samples in the past or from other, unpolluted, sites. The best example of this approach is the British Monitoring Working Party method used by the National Water Council of the United Kingdom (Table 6.4). Scores of lightly polluted rivers may exceed 100, those for heavily polluted systems less than 10. This method has been used in a 5 yearly monitoring programme since the 1980s and the results for 1985 are shown in Table 6.5.
2. Other types of biotic index also take into account the abundance of various organisms but do not require a great degree of taxonomic expertise. The Trent Biotic Index (Woodiwiss, 1964) was devised for organic pollution in the River Trent in the United Kingdom but does respond to other forms of pollution (James and Evison, 1979). In its expanded form, the index varies from 0 (low quality) to 15 (highest quality) based on the presence of six key indicator taxa types of known tolerance weighted by a number of 'defined groups' present (Table 6.6). The value of the index is given by the score for the

Table 6.4 The British Monitoring Working Party (BMWP) biotic index scheme for rivers. Scores are summed for each species of the different families present in the sample

Families of freshwater macroinvertebrates		Score
(a)	Siphlonuridae, Heptageniidae, Leptophlebiidae, Ephemerellidae, Potamanthidae, Ephmeridae (mayflies)	
(b)	Taeniopterygidae, Leuctridae, Capniidae, Perlodidae, Perlidae, Chloroperlidae (stoneflies)	
(c)	Aphelocheiridae (bugs)	10
(d)	Phryganeidae, Molannidae, Beraeidae, Odontoceridae, Letpoceridae, Goeridae, Lepidostomatidae, Brachycentridae, Sericostomatidae (caddis-flies)	
(a)	Astacidae (crayfish)	
(b)	Lestidae, Agriidae, Gomphidae, Cordulegasteridae, Aeshnidae, Corduliidae, Libellulidae (dragonflies)	8
(c)	Psychomyiidae, Philopotamidae (net-spinning caddis-flies)	
(a)	Caenidae (mayflies)	
(b)	Nemouridae (stoneflies)	7
(c)	Rhyacophilidae, Polycentropodidae, Limnephilidae (caddis-flies)	
(a)	Neritidae, Viviparidae, Ancylidae (snails)	
(b)	Hydroptilidae (caddis-flies)	
(c)	Unionidae (bivalve molluscs)	6
(d)	Corophiidae, Gammaridae (crustacea)	
(e)	Platycnemididae, Coenagriidae (dragonflies)	
(a)	Mesoveliidae, Hydrometridae, Gerridae, Nepidae, Naucoridae, Notonectidae, Pleidae, Corixidae (bugs)	
(b)	Haliplidae, Hygrobiidae, Dytiscidae, Gyrinidae, Hydrophilidae, Clambidae, Helodidae, Dryopidae, Elminthidae, Crysomelidae, Curculionidae (beetles)	5
(c)	Hydropsychidae (caddis-flies)	
(d)	Tipulidae, Simuliidae (dipteran flies)	
(e)	Planariidae, Dendrocoelidae (triclads)	
(a)	Baetidae (mayflies)	
(b)	Sialidae (alderfly)	4
(c)	Piscicolidae (leeches)	
(a)	Valvatidae, Hydrobiidae, Lymnaeidae, Physidae, Planorbidae, Sphaeriidae (snails, bivalves)	3
(b)	Glossiphoniidae, Hirudidae, Erpobdellidae (leeches)	3
(c)	Asellidae (Crustacea)	
(a)	Chironomidae (Diptera)	2
(a)	Oligochaeta (whole class) (worms)	1

Adapted from Moss, 1988

most tolerant taxa type present in the sample, but it is dependent on the number of species (or groups) identified in that taxa type.

A simpler and more robust index has been devised by the Environmental Research Unit (ERU) in Ireland where the species are arranged into four basic groups depending on their known tolerance to organic pollution—sensitive, less sensitive, tolerant and most tolerant. From the presence and relative abundance of animals in each group one can derive a biotic index quality rating for the water body from 1 to 5 (Table 6.7). This approach has been used for successful monitoring of river quality over 20 years in Ireland (Table 6.8).

The most complex but one of the best known indices is the Chandler Score, which is often considered the most satisfactory. It divides the fauna into groups following the Trent Scheme but includes an abundance factor at five levels for different species that reflects their tolerance to organic pollution. Intolerant species receive high scores and scores increase further with increasing abundance

Table 6.5 Results of 1985 survey of Brisith rivers based on the BMWP scheme. The percentage of surveyed rivers which fall into each class of water quality (1A-4) are given (class 1A—high quality—to class 4—grossly polluted)

| Water Authority | Percentage of river length in each class | | | | | | Total river length surveyed (km) |
| | Good | | Fair | Poor | Bad | Poor and bad | |
	1A	1B	2	3	4	3+4	
Anglian	10	48	32	9	0.2	9	4 328
Northumbrian	62	25	10	2	0.8	3	2 784
North West	49	12	17	17	5	22	5 323
Severn Trent	18	36	33	12	1	14	5 150
Southern	31	44	22	2	0.2	3	1 992
South West	25	41	27	6	0.6	6	2 941
Thames	24	41	28	7	0.1	7	3 546
Welsh	53	30	11	6	0.6	6	4 600
Wessex	26	35	32	6	0.6	7	2 467
Yorkshire	40	37	12	8	3	12	5 767
England and Wales	34	34	22	9	2	10	38 896

Percentages are rounded to the nearest per cent or 1 significant figure and may not sum to 100 per cent.
After Chave, 1990

in the sample (Table 6.9). Tolerant species receive low scores and scores decrease with increasing abundance in the sample. Each species present is awarded a score and the sum of individual scores gives the overall biotic index for the site. There is no upper limit but sites with values < 300 are considered to be moderately polluted while, 300 to 3000 are mildly polluted to unpolluted. A modification is the Average Chandler Score, which corrects for variation in environmental conditions over and above pollution (such as seasonal change or floods), (P. S. Giller unpublished data).

Most of these indices have been derived for assessment of organic pollution. A new Index of Community Sensitivity has been devised for heavy metal pollution (Clements *et al.*, 1992). A measure of sensitivity of each of 13 dominant taxa to known concentrations of heavy metals in experimental streams is multiplied by their relative abundance and scores for each taxa are summated to give an overall score for the site. This approach is likely to be most useful on a national basis, (Clements *et al.*, 1992).

3. A third general approach is based on changes in the overall number of species and dominance levels. A mild yet important pollution problem may not eliminate all clean water indicator species but might reduce their numbers and increase numbers of more tolerant forms. Such changes may not be detected by biotic indices but can be from the calculation of *diversity indices*. The more diverse the community the better as far as water quality is concerned, but generally one would compare values over time or between unpolluted and suspect sites. A drop in diversity would normally indicate a lowered water quality and pollution or some other disturbance. The most commonly used index, particularly in the United States (Washington, 1984), is the Shannon–Wiener diversity index based on information theory where $H = \Sigma P_i \log_e P_i$, where P_i is the proportion of a taxa i to the total number of individuals of all taxa in the sample.

The advantages of diversity indices over biotic ones are that:

(a) In the simple diversity indices, species or organisms need only be distinguished and not necessarily identified

(b) No information on pollution tolerances is required.

Table 6.6 The expanded Trent Biotic Index scheme of Woodiwiss.* The index for a site is given by the score for the least tolerant animal form present in a sample with x (0–1 to 41–45) total number of 'groups' present, e.g. 1 Plecoptera species and 23 total 'groups' give an index of 10

		Total number of 'groups' present§								
	0–1	2–5	6–10	11–15	16–20	21–25	26–30	31–35	36–40	41–45
					Biotic indices					
Biogeographical region: Midlands, England										
Plecoptera nymphs present — More than one species	—	7	8	9	10	11	12	13	14	15
One species only	—	6	7	8	9	10	11	12	13	14
Ephemeroptera nymphs — More than one species†	—	6	7	8	9	10	11	12	13	14
One species only†	—	5	6	7	8	9	10	11	12	13
Trichoptera larvae present — More than one species‡	—	5	6	7	8	9	10	11	12	13
One species only‡	4	4	5	6	7	8	9	10	11	12
Gammarus present — All above species absent	3	4	5	6	7	8	9	10	11	12
Asellus present — All above species absent	2	3	4	5	6	7	8	9	10	11
Tubificid worms and/or Red Chironomid larvae present — All above species absent	1	2	3	4	5	6	7	8	9	10
All above types absent — Some organisms such as Eristalis tenax not requiring dissolved oxygen may be present	0	1	2	—	—	—	—	—	—	—

† Baetis rhodani excluded.
‡ Baetis rhodani (Ephem.) is counted in this section for the purpose of classification.
§ The term 'group' used for the purpose of the biotic index means any one of the species included in the following list of organisms or sets of organisms:

Each known species of Plathyhelminthes (flatworms)
Annelida (worms excluding genus Nais)
Genus Nais (worms)
Each known species of Hirudinea (leeches)
Each known species of Mollusca (snails)
Each known species of Crustacea (hog louse, shrimps)
Each known species of Plecoptera (stone-fly)
Each known genus of Ephemeroptera (mayfly, excluding Baetis rhodani)
Baetis rhodani (mayfly)

Each family of Trichoptera (caddis-fly)
Each species of Neuroptera larvae (alderfly)
Family Chironomidae (midge larvae except Chironomus Ch. thummi)
Chironomus Ch. thummi (blood worms)
Family Simulidae (black-fly larvae)
Each known species of other fly larvae
Each known species of Coleoptera (beetles and beetle larvae)
Each known species of Hydracarina (water mites)

*James and Evison, 1979

Table 6.7 The ERU biotic index scheme used in Ireland, based on five index scores or 'Q' values (Q5, good quality; Q4, fair quality; Q3, doubtful quality; Q2, poor quality; Q1, bad quality). The Q index is based on the relative abundance of five faunal indicator groups, A–E (ERU, 1992)

Relationship between biotic index (Q) and the five faunal groups

	Biotic index Q	Community diversity	Faunal groups				
			A	B	C	D	E
Eroding substrata	5	High	+ + + +	+ + +	+ +	+ −	+ −
(i.e. riffles)	4	Slightly reduced	+ +	+ + + +	+ + +	+ +	+ −
	3	Significantly reduced	−	+ −	+ + + +	+ + +	+ +
	2	Low	−	−	+ −	+ + + +	+ + +
	1	Very low	−	−	−	+ −	+ + + +
Depositing substrata	5	Relatively high	+ −	+ + + +	+ + +	+ +	+ −
(i.e. slow-flowing	4	Slightly reduced	−	+ +	+ + + +	+ +	+ −
areas)	3	Significantly reduced	−	+ −	+ +	+ + +	+ + +
	2	Low	−	−	+ −	+ + +	+ + +
	1	Very low	−	−	−	−	+ + + +

+ + + + = well represented or dominant   + + + = common   − = absent
+ + = present in small numbers   + − = sparse or absent

Indicator groups: key taxa

Group A Sensitive forms	Group B Less sensitive forms	Group C Relatively tolerant forms	Group D Tolerant forms	Group E Most tolerant forms
Plecoptera (excluding *Leuctra*, Nemouridae) Heptageniidae Siphlonuridae	*Leuctra* Nemouridae Baetidae (excluding *B. rhodani*) Leptophlebiidae Ephemerellidae Ephemeridae Cased Trichoptera (excluding Limnephilidae, Hydroptilidae Glossosomatidae) Odonata (excluding Coenagriidae) *Aphelocheirus* *Rheotanytarsus*	Tricladida Ancylidae Neritidae Astacidae Gammarus *Baetis rhodani* Caenidae Limnephilidae Hydroptilidae Glossosomatidae Uncased Trichoptera Coleoptera Coenagriidae Sialidae Tipulidae Simuliidae Hemiptera (excluding *Aphelocheirus*) Hydracarina	Hirudinea Mollusca (excluding Ancylidae, Niritidae) *Asellus* Chironomidae (excluding *Chironomus* and *Rheotanytarsus*)	Tubificidae *Chironomus*

Table 6.8 Results of a twenty year period of river pollution surveys in Ireland on a 2900 km baseline based on the ERU quality rating scheme. Figures represent the river lengths falling into the particular quality class (with percentages)

Q rating		1971	1981	1986	1990
Class A	Unpolluted	2400(83)	2250(78)	2000(69)	1900(65)
Class B	Slightly polluted	150(5)	324(11)	580(20)	570(20)
Class C	Moderately polluted	150(5)	206(7)	240(8)	380(13)
Class D	Seriously polluted	200(7)	120(4)	80(3)	50(2)

ERU, 1992

However, diversity indices cannot give any information on the type of pollutant. Also there is no real consensus as to what value indicates a polluted system. Wilm's survey of the United States (Washington, 1984) suggested Shannon–Wiener values of below 1.0 as substantially polluted, 1.0 to 3.0 as moderately polluted and 3.0 to 5.0 as unpolluted. However, values between 1.7 and 2.5 have often been found in unpolluted systems, thus comparisons over time at one site or between similar sites are needed to accurately detect pollution. All in all, despite its widespread use, the Shannon–Wiener index must be used with caution as far as its biological relevance is concerned (Washington, 1984).

Use of organisms in controlled environments   These mainly laboratory-based assessment techniques have been a traditional way to test for the effects of certain chemical concentrations, e.g. assessing the toxic effect of water samples or effluent on organisms in toxicity tests (bioassays; see Chapter 5). An alternative is to make use of effects of water quality change on the behaviour of certain organisms, e.g. fish. These tests can be conducted in the laboratory or on-site under controlled conditions in so-called *dynamic tests* or 'Toxalert systems' (Fig. 6.3) and can give an early warning of negative water quality changes.

Biological accumulation   Samples of various species can be examined from the natural environment for monitoring for the presence and relative levels of substances that can become bioaccumulated, such as heavy metals, organochlorines and PCB, e.g. O'Halloran *et al.* (1993) for birds, Mason and O'Sullivan (1992) for mammals and Philips and Segar (1986) for marine invertebrates. Alternatively, indicator

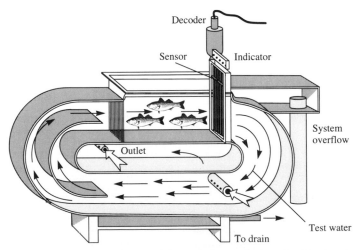

Figure 6.3 An example of a dynamic fish test which is used for continuous monitoring of toxicity in water directed from a water body through the test apparatus (after Chapman, 1992).

Table 6.9 The Chandler Score biotic index scheme. Each specified species is given a score corresponding to its relative abundance level in a five minute kick sample of a stream or river. The overall Chandler Score for the sample is the sum of all individual species scores

Groups present in sample		Abundance level				
		P	F	C	A	V
				Points scored		
Each species of	*Planaria alpina*					
	Taenopterygidae	90	94	98	99	100
	Perlidae, Perlodidae					
	Isoperlidae, Chloroperlidae					
Each species of	Leuctridae, Capniidae					
	Nemouridae (exd. Amphinemura)	84	89	94	97	98
Each species of	Ephemeroptera (exd. Baetis)	79	84	90	94	97
Each species of	Cased caddis, Megaloptera	75	80	86	91	94
Each species of	Ancylus	70	75	82	87	91
	Rhyacophila (Trichoptera)	65	70	77	83	88
Genera of	Dicranota, Limnophora	60	65	72	78	84
Genera of	Simulium	56	61	67	73	75
Genera of	Coleoptera, Nematoda	51	55	61	66	72
	Amphinemura (Plecoptera)	47	50	54	58	63
	Baetis (Ephemeroptera)	44	46	48	50	52
	Gammarus	40	40	40	40	40
Each species of	Uncased caddis (exd. Rhyacophila)	38	36	35	33	31
Each species of	Tricladida (exd. *P. alpina*)	35	33	31	29	25
Genera of	Hydracarina	32	30	28	25	21
Each species of	Mollusca (exd. Ancylus)	30	28	25	22	18
	Chironomids (exd. *C. riparius*)	28	25	21	18	15
Each species of	Glossiphonia	26	23	20	16	13
Each species of	Asellus	25	22	18	14	10
Each species of	Leech, exd. Glossiphonia, Haemopsis	24	20	16	12	8
	Haemopsis	23	19	15	10	7
	Tubifex sp.	22	18	13	12	9
	Chironomus riparius	21	17	12	7	4
	Nais	20	16	10	6	2
Each species of	Air-breathing species	19	15	9	5	1
	No animal life			0		

Levels of abundance in 'score' system

Level	Nos. per 5 min. sample	Remarks
P—present	1–2	May be drift from upstream
F—few	3–10	Probably indigenous, but rare
C—common	11–50	
A—abundant	51–100	
V—very abundant	100	

After James and Evison, 1979

Table 6.10 Aquatic pollutants classified according to their biodegradability

Degradable	Non-degradable
Sewage and farm manures	Inert particulates (clay, colliery waste, etc.)
Agricultural fertilizers and plant nutrients	Man-made plastics
Food-processing waste (including breweries)	Heavy metals (e.g. chromium, copper, lead)
Organic paper-mill wastes	Halogenated hydrocarbons (DDT, PCBs)†
Industrial wastes (especially petrochemical)	Radioactivity†
Oils	Acids and alkalis
Anions of sulphide and sulphite	Industrial emission gases
Detergents	Organophosphates†
Oil dispersants	

† Non-degradable in the short term but will degrade/decay over the longer term.

organisms can be placed in the environment in order to monitor water quality, e.g. marine mussels in the 'mussel watch' scheme (Bayne, 1978).

Pathological/morphological methods   Many organisms suffer from abnormal growth and morphological changes in body tissues as a result of long-term exposure to pollutants. These changes can be readily identified and often give a first indication of potential negative changes in water quality, e.g. fish in estuaries (Mulcahy *et al.*, 1987), shellfish in relation to TBT (Bryan *et al.*, 1986).

In the following sections the major aquatic pollutants will be described and some of the major effects and long-term changes that have significantly altered water quality in rivers, lakes and the sea over the past few decades are reviewed.

6.4 AQUATIC POLLUTANTS

The term pollutant was described in Chapter 5 and two main types were considered:

- Those pollutants that affect the physical environment
- Those that are directly toxic to organisms

Some 1500 substances have been listed as pollutants in aquatic systems and Mason (1991) has recently summarized these (Table 6.10). To this list can be added heat, which cannot be placed in these categories.

Some of the compounds may interact additively or antagonistically or synergistically to give different responses in aquatic systems. The influence of polluting substances in natural waters will vary according to the pollutant, the local conditions and the organisms concerned. Pollutants can act in at least three ways (Maitland, 1990):

- Settling out and smothering life, e.g. mining effluents, poorly treated sewage and sewage sludge
- Being acutely toxic and killing directly, e.g. some industrial effluents, heavy metals in relatively high concentrations
- Influencing organisms indirectly, e.g. through reduction in oxygen supply, addition of fertilizers (see eutrophication) or sublethal effects of compounds acting on growth, reproduction, etc.

In the following sections, freshwater, estuarine and marine pollution are dealt with separately. Although there are obvious overlaps in many of the processes and sources of pollutants, the uses and importance of the various aquatic resources are clearly different and deserve individual attention.

6.5 FRESHWATER POLLUTION

This section will deal specifically with organic pollution, the major form of pollution of freshwater systems, and then discusses two other subjects of serious concern, namely eutrophication and acidification, as illustrative case studies of freshwater pollution.

6.5.1 Organic Pollution

By far the greatest volume of discharge into freshwater systems is composed of organic material, municipal sewage, industrial wastewater and agricultural wastewaters. These liquid waste streams are rich in organic matter, and as they break down in the presence of oxygen under bacterial activity, the dissolved oxygen levels in the water are affected as well as liberating nutrients like nitrates and phosphates. These wastes are said to have a high *oxygen demand*. The reduction in the oxygen concentration in the water, brought about by the activities of the aerobic bacteria, is compensated for by diffusion of oxygen from the water surface and from surrounding areas of higher oxygen concentration. This replenishment process is, however, slow, and as the oxygen level drops, anaerobic bacteria, which can oxidize organic compounds without the presence of oxygen, start to thrive. The end products from the activities of these bacteria are hydrogen sulphide, methane and ammonia. These products are toxic to most higher organisms.

Consider the effects of organic pollution on a river community. When organic waste enters an aquatic system, in this case a river, there is a characteristic response of decreasing oxygen levels, through the processes described above, immediately below the source. The longitudinal profile of oxygen concentration is called an *oxygen sag curve*. The extent of the sag and the downstream length of river affected depends on the level of pollution (Fig.6.4). The shape of the curve will also depend on the flow and can change seasonally, e.g. under low flow conditions the minimum oxygen levels can prevail over longer distances. The degree of deoxygenation will depend on parameters such as temperature, dilution of the effluent, degree of river aeration, BOD of the discharge and of the receiving waters, and the amount of other organics in the river, etc. Recovery of oxygen levels results from dilution of the effluent and reduction in the effluent through decomposition. Reoxygenation in a downstream direction also elevates oxygen levels. This oxygen sag will have a profound effect on the biology of the system. In serious pollution incidents, it may lead to complete deoxygenation and an anoxic environment and hence complete elimination of the biota. However, in most cases, the level of pollution is not as serious as this. Under continuous outfall of heavy organic pollution, the effect on water quality and oxygen levels is such as to impose a clear downstream zonation of animal, plant and microbial populations (Hynes, 1960) (Fig. 6.5).

Figure 6.5(*a*) and (*b*) trace the chemical changes in a river downstream of an organic effluent outfall. The levels of suspended solids and BOD are high at the outfall and the level of oxygen quickly declines, which coincides with the high BOD. In the same way, the level of ammonia, nitrate and phosphate are high at the outfall but have different peaks of concentration in the receiving waters at various distances from the pollution source as the organic matter is decomposed. These changes can be seen to be influenced by micro-organisms and affect macro-organisms in the river in Fig. 6.5(*c*) and (*d*). The

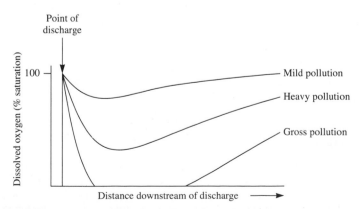

Figure 6.4 The effect of different levels of organic effluent discharge on the oxygen content of river water illustrating the oxygen sag curve (after Mason, 1991).

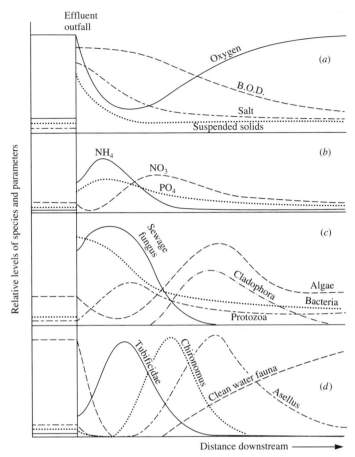

Figure 6.5 Diagrammatic presentation of the effects of an organic effluent on a river and the changes as one passes downstream from a sewage outfall: (*a*) and (*b*) are physical and chemical changes, (*c*) changes in micro-organisms, (*d*) changes in larger animals (after Hynes, 1960).

abundance of bacteria and sewage fungus is high below the outfall and has a significant impact on the level of oxygen. This reduction in oxygen leads to a decline in the diversity of clean water macroinvertebrates (Fig. 6.5(*d*)). In essence, the most tolerant species (e.g. tubificid worms) survive and dominate near the effluent outfall and cleaner water forms progressively reappear as water quality improves further downstream from the outfall. These changes in fauna are in fact used to help monitor pollution and to gauge water quality as discussed in the previous section. Further details of quantitative aspects of the oxygen sag curve are included in Chapter 7.

6.5.2 Eutrophication

Eutrophication can be defined as the enrichment of waters by inorganic plant nutrients. The nutrients are usually nitrogen and phosphorus and these result in an increase in primary productivity. In this discussion we are talking about artificial enrichment which has been termed 'cultural eutrophication'. This is an important distinction because eutrophication of waters is a natural process in the life history of freshwater lake systems which tend to gradually change from an oligotrophic to a eutrophic system as they age (see

Table 6.11 Modified version of OECD classification scheme based on values for annual maximum concentrations of chlorophyll-a. Indicators related to water quality and the probability of pollution are also shown

Lake trophic category (codes in parentheses)		Annual maximum chlorophyll-a (mg/m^3)	Algal growth	Degree of deoxygenation in hypolimnion	Probability of pollution	Impairment of multipurpose use of lake
Ultra-oligotrophic/ oligotrophic (0)		< 8	Low	Low	Very low	Probably none
Mesotrophic (M)		8–25	Moderate	Moderate	Low	Very little
Eutrophic	Moderately (m-E)	26–35	Substantial	May be high	Significant	May be appreciable
	Strongly (s-E)	36–55	High	High	Strong	Appreciable
	Highly (h-E)	56–75	High	Probably total	High	High
Hypereutrophic (H)		> 75	Very high	Probably total	Very high	Very high

Chapter 5). The general characteristics of oligotrophic and eutrophic systems were outlined in Chapter 5, Table 5.2. Eutrophication also occurs in marine systems (see Sec. 6.7.1).

Before going into further detail it is important to point out that a number of factors affect the occurrence of eutrophication: firstly, the nutrient or trophic status of the water body (see Fig. 5.6, Chapter 5), secondly, the characteristics of the water body (e.g. size, water residence time) and, thirdly, its susceptibility to temperature and oxygen stratification and whether it is a monomictic or dimictic lake (see Chapter 4). The degree of productivity can be classified according to the annual mean level of phosphate entering a system and the annual mean production of plant growth in the form of chlorophyll-a. For a variety of lake and water body types in Europe the OECD has the classification for water body status shown in Table 6.11.

Sources of nutrients   The sources of nutrients that cause the cultural eutrophication are essentially:

- Urban sewage effluent discharge which may be in the form of treated or untreated sewage or
- Agricultural activities, especially animal wastes and fertilizers (see Chapter 10 for further details), which are found particularly in the Western world and the United States.

The pollution enters as a point source outfall in the case of treated sewage or is dumped into lakes/coastal waters or is carried by rivers from diffuse sources in their catchments (associated with agriculture, animal husbandry practices, poultry and pigs, ploughing and forestry). Sediments, too, are an important source of nutrients in many water bodies and nutrient loads in many ways result from the past events in the catchment. This is especially true in shallow waters where sediments are more easily disturbed by wind action. Several studies have shown that significant fractions of nutrients from the sediments enter the water column under reducing conditions. The sources of nutrients have been a target for control of eutrophication (see below).

Ecological effects of cultural eutrophication   Whatever the source of nutrients, the effects are generally the same, and are summarized in Fig. 6.6. Plant population growth will increase, often leading to increased macrophytes in littoral zones of lakes. In extreme cases the excess input of plant nutrients (nitrogen and phosphorus) leads to massive algal blooms. Blue-green algae and diatoms like *Asteriorella*

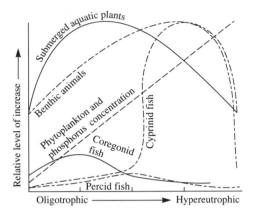

Figure 6.6 Changes in phytoplankton, macrophytes, benthic invertebrates and fish from oligotrophic to hypereutrophic lakes (after Moss, 1988).

and *Fragillaria* and Cyanobacteria like *Anabaena* dominate (examination of pollen and spores of plants from lake sediments can tell us when the eutrophication occurred in the past). Such blooms lead to a massive reduction in transparency to light caused by an increase in turbidity of the water, which can lead to death of macrophytes due to shading-out. Such blooms are aesthetically unpleasant, resulting in slimy masses and smells and flies. In flowing waters, eutrophication can lead to extreme macrophyte growth including the large filamentous algae *Cladophora*, especially in sewage-effluent rich rivers, or in the tropics to huge populations of the water hyacinth *Eichormia cassipes* (Mason, 1991).

Secondly, a marked increase in organic detritus caused by the massive algal populations which die off seasonally leads to increased decomposition in the lake bottom, and if thermal stratification occurs, then rapid deoxygenation of the hypolimnion (bottom water) may occur (see Chapter 5). Over and above the expected differences in communities between oligotrophic and eutrophic systems, this series of activities can lead to a massive reduction in invertebrate communities, particularly in diversity, and can change fish species from salmonids and coregonids (white fish) to cyprinids (carp and tench). Other environmental effects result from changes to the habitat structure from loss of macrophytes, loss of spawning grounds and loss of living habitats for fish and invertebrates. Effects on the whole food web can follow with reduction in fish food and fish-eating and plant-eating birds. Increased vulnerability of bank edges to erosion by waves can also ensue as macrophyte populations decline. Loss of fish through eutrophication can have knock-on effects on the whole ecology of some lakes through the so-called trophic cascade response (Carpenter and Kitchell, 1993). The zooplankton prey and consequently phytoplankton communities on which they feed are markedly altered, again often contributing to blooms of algae. In extreme cases of cultural eutrophication, primary production is also so high that subsequent decomposition of organic matter and respiration at night completely deoxygenate the water and fish and most invertebrates may be eliminated altogether.

Effects on man Eutrophication often occurs in large reservoirs which are used by people for water abstraction, recreation, fishing, etc. These changes will clearly damage this resource and makes the treatment of this water expensive for potable purposes (Table 6.12). Decreased amenity value occurs due to cloudy water, interference with fishing, sailing and swimming affected by surface scum during algal blooms, smelly decaying algae on the lake shores and often dense swarms of midges emerging from lakes, causing a nuisance. The famous study on Lake Washington illustrates the problem of cultural eutrophication. By 1926 the population of Seattle had increased to around 50 000 and raw sewage effluent was entering the lake such that, by 1954, 56 per cent of total phosphorus input was from sewage (Fig. 6.7). Dense blooms of algae and cyanobacterium occurred in 1955, decreasing the amenity value of the

Table 6.12 Some of the major effects of eutrophication on the receiving aquatic ecosystem and the problems that ensue

Effects on physical, chemical and biological parameters
1. Species diversity decreases and the dominant biota change
2. Plant and animal biomass increases
3. Turbidity increases
4. Rate of sedimentation increases, shortening the life span of the lake
5. Anoxic conditions may develop

Problems
1. Treatment of potable water may be difficult and the supply may have an unacceptable taste or odour
2. The water may be injurious to health
3. The amenity value of the water may decrease
4. Increased vegetation may impede water flow and navigation
5. Commercially important species [such as salmonids and coregonids (whitefish)] may disappear

After Mason, 1991

lake, (Mason, 1991). Water purification and supply problems arise if lakes are used for abstraction for drinking water. Problems are caused by the increase in phytoplankton blocking filtering processes. This results in reduced throughput of water at water treatment plants. Small algal cells can also get through to the consumer supply, decompose in the pipes, with growth of bacteria and fungi leading to unpleasant taste and odours and coloured water. Invertebrates can even sometimes occur in the domestic supply.

The blue-green algae themselves may also be toxic (Table 6.13). Reports of toxic strains of a particular algae type *Cyanophyta*, have come from various areas including Australia, the United States, South Africa and Israel. There have also been reports of poisoning incidences of birds or fish from 26 European countries (Table 6.14). Death of cattle and sheep in Norway was attributed to toxic blooms of *Microcystis* and also deaths of cattle in Cheshire, United Kingdom, have also been due to algal blooms.

Table 6.13 A comparison of the toxicities of biological toxins

Toxin	Source	Common name	Lethal dose† (LD_{50})
Botulinum toxin-a	*Clostridium botulinum*	Bacterium	0.000 03
Tetanus toxin	*Clostridium tetani*	Bacterium	0.000 1
Ricin	*Ricinus communis*	Castor bean plant	0.02
Diphtheria toxin	*Corynebacterium diphtheriae*	Bacterium	0.3
Kokoi toxin	*Phyllobates bicolor*	Poison arrow frog	2.7
Tetrodotoxin	*Sphaeroides rubripes*	Puffer fish	8
Saxitoxin	*Aphanizomenon flos-aquae*	Blue-green algae	9
Cobra toxin	*Naja naja*	Cobra	20
Nodularin*	*Nodularia spumigena*	Blue-green algae	30–50
Microcystin-LR*	*Microcystis aeruginosa*	Blue-green algae	50
Anatoxin-a*	*Anabaena flos-aquae*	Blue-green algae	200
Microcystin-RR*	*Microcystis acruginosa*	Blue-green algae	300–600
Curare	*Chrondodendron tomentosum*	Brazilian poison arrow plant	500
Strychnine	*Strychnos nux-vomica*	Plant	500
Amatoxin	*Amanita phalloides*	Fungus	600
Muscarin	*Amanita muscaria*	Fungus	1 100
Phallatoxin	*Amanita phalloides*	Fungus	1 800
Glenodin toxin	*Peridinium polonicum*	Dinoflagellate algae	2 500
Sodium cyanide			10 000

* Toxins produced by blue-green algae associated with eutrophication.
† The acute LD_{50} in μg per kg bodyweight by intra-peritoneal injection: some with mice, some with rats.
After National Rivers Authority, 1990

Table 6.14 Incidence of toxicity of blue-green algal blooms in European freshwaters up to 1989

Origin	Number of sites with blooms tested	Number of sites with toxic blooms	Incidence of bloom toxicity (%)
United Kingdom	24	18	75
Norway, Finland and Sweden (joint programme)	51	30	59
Sweden	27	15	56
Finland	103	45	44
Hungary	3	3	100
Greece	4	4	100
Italy	2	2	100

After National Rivers Authority data, 1990

People swimming in eutrophic waters containing blue-green algae have been reported to suffer from skin and eye irritations, gastroenteritis and vomiting. Similar problems can arise in marine systems.

High nitrogen levels in the water supply, causes a potential risk, especially to infants under six months. This is when methaemoglobin results in a decrease in the oxygen-carrying capacity of the blood as nitrate ions in the blood readily oxidize ferrous ions in the haemoglobin. The net result of these additions is increased cost and additional treatment processes.

The consequence of enrichment of coastal and marine waters is also eutrophication and many of the effects are similar to those seen in freshwater systems (Sec. 6.7).

Reduction and control of plant growth A number of methods have been used to control eutrophication in surface waters, including control of nutrient release into water bodies or nutrient removal, accelerating the outflow of nutrient material, sealing the bed or lake bottom and manipulating the hypolimnion by aeration. Control of nutrient release into water bodies is best achieved by limiting the phosphate loading, but where the water is potable and used for human consumption the nitrogen also needs to be reduced, e.g. using biological processes–nitrification and denitrification—with bacteria with further treatment may give up to 90 per cent removal. Organic pollution control can be achieved by treatment of waste to remove the nutrients and reduce the affects on water quality. In Lake Washington, 99 per cent of sewage effluent was diverted from the lake to the sea in 1967, which led to the rapid recovery of the lake, with phytoplankton levels falling to a fifth of the 1963 levels within three years (Fig. 6.7).

Similar phosphate removal has been partially successful in Lough Neagh, Northern Ireland, which has the largest lake area in Europe. In 1967 a bloom of green algae disrupted treatment processes in the local water treatment plant, commercial fishery and recreational use. At that time it was the most productive lake in the world. During the critical period of April to August, 80 per cent of the phosphate in the lough entered from sewage effluents fed in by six major rivers. Land drainage also introduced nutrients. Methods of removing phosphorus were introduced, reducing levels in the lough, and it has been estimated that 50 per cent of the phosphate could be reduced by removal of phosphate from sewage (Smith, 1977). See Chapter 12 for further details of phosphorous removal from municipal wastewaters.

Other possible methods of nutrient removal have also been employed, including the addition of chemicals to the water to minimize the availability of phosphate. Ferric sulphate solutions have been successfully used to reduce phosphate and reduce chlorophyll-a concentrations in Barton Broad, East Anglia (Mason, 1991). Removal of phosphate from household detergents has meant a significant improvement. The use of detergents in the dairying industry, i.e. phosphoric acid as a cleaning agent, has also ceased, therefore reducing the amount of phosphate entering surface waters.

Accelerating the outflow of nutrient material by increasing the flow of water means there is less time for phytoplankton growth and therefore the phosphorus is shifted out of the system faster. Removal of

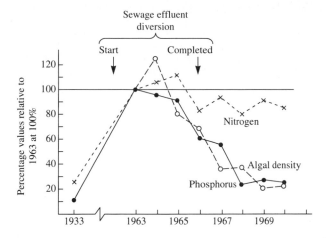

Figure 6.7 Recovery of Lake Washington from 1963 to 1970 after diversion of sewage effluent. Phorphorus dropped rapidly because sewage was the main source of phosphorus to the lake. Nitrogen dropped less because the surface waters feeding the lake are relatively rich in nitrogen. The amount of phytoplankton (measured by chlorophyll content of the water) dropped in parallel to the phosphorus (after Krebs, 1955).

sediment by dredging, etc., is also used, but this may release more nutrients into the system exacerbating conditions. Removal of vegetation will also help export the nutrients from the system.

Sealing the bed or bottom of the lake or reservoir in order to prevent the exchange of phosphorus between water and the sediment offers another method. Here membranes such as polythene are placed on the bed of the reservoir or lake and a layer of sand spread on it. The membrane must have holes in it to allow the release of anaerobic gases (CH_4, H_2S, etc.). Manipulating the hypolimnion by aeration with increasing surface circulation has been used in reservoirs in London to inhibit the algal blooms. This prevents existing algae from remaining for long periods of time in the euphotic zone. Modelling approaches to eutrophication control have been described in Jorgenson (1980) and Henderson-Sellars and Markland (1987).

6.5.3 Surface Water Acidification

Eutrophication is, for the most part, a problem of the populated lowlands. In the uplands and in more remote unpopulated regions, with poorly weathered rocks and thin soils on poorly buffered geologies, acidification of lakes and rivers is much more of a problem. Over the last two decades, acidification of surface waters has been a major focus for political and scientific activity largely because of the transboundary nature of the problem. Effects on the chemical, ecological and economic and aesthetic status of rivers and lakes have been detected and have serious implications. There are now many water bodies which today have low pH, low alkalinity and elevated metal concentration as a result of acidification, and in many cases changes have been induced by man through the disruption of biogeochemical cycles (see Chapter 2).

Acidification of rivers and lakes has been described from a variety of countries in western Europe, North America and Scandinavia (Harriman and Morrison, 1982; Harriman and Wells, 1985; Wellburn 1988, Edwards *et al.*, 1990). The extent of these changes can be far-reaching. For example, in Norway an area of 13 000 km^2 is devoid of fish populations, with lesser changes over 20 000 km^2. About 18 000 lakes in Sweden (20 per cent of those lakes > 1 ha) have a pH < 5.5 at some time of the year and fish stocks have been affected in 9000 lakes. Before discussing the effects in detail, the mechanism of acidification is considered.

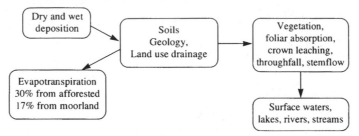

Figure 6.8 A simplified summary of the factors affecting surface water acidification (after Gee and Stoner, 1989).

Causes of freshwater acidification

Atmospheric deposition   Initially it was considered that acid rain resulting from atmospheric pollution was the exclusive cause of freshwater acidification. It is now clear that the process is more complex. The atmospheric deposition itself is only part of the problem and acidification of surface waters will only occur if the total acid deposition is not neutralized by the vegetation, soils and rocks with which it comes into contact prior to entering the stream, river or lake (see Fig. 6.8 for a summary of the pathway).

The primary atmospheric pollutants causing acid rain include oxides of sulphur and nitrogen emitted from fossil fuel power generating plants and nitrogen oxides and unburnt hydrocarbons emitted by the internal combustion engine of motor vehicles. Other potential sources of sulphur oxides include volcanic activity and oceanic plankton which generate dimethyl sulphoxide. The primary pollutants undergo chemical change in the atmosphere, leading to the production of secondary pollutants (HNO_3 and H_2SO_4) which cause water acidification (refer also to Chapter 8). Pollutants may reach the earth's surface as dry deposition or dissolved in atmospheric moisture, which may fall as rain or snow or be deposited as droplets of mist, termed 'occult deposition'.

Climatic influences ensure that both the atmospheric concentration and the rate of deposition of polluting gases is highly episodic and shows large spatial and temporal variation. The highly episodic nature is also significant in that all the rain may fall in a short period of time. In mid Wales, for example, 30 per cent of the total deposited acidity falls on < 5 per cent of wet days, while snow samples also collected in mid Wales in February 1986 had pH values as low as 3.5, resulting in a large acid pulse into the streams after snow melt (Gee and Stoner, 1989).

Scavenging by vegetation   Vegetation can scavenge 'dry' and 'occult' pollutants very effectively. As rainwater passes through a forest canopy, deposited materials are washed from the surfaces of the leaves.

Table 6.15 A comparison of mean values of three chemical parameters in the throughfall of bulk deposition (rainfall) under two age classes of the conifer, spruce, and under oak trees, indicating the scavenging effect of vegetation

	pH	SO_4 (μequiv/L)	Cl (μequiv/L)
Bulk deposition	4.6	54	145
(rainfall)	(3.4–7.1)	(21–240)	(128–873)
Oak	4.7	115	242
	(4.1–6.1)	(39–444)	(56–451)
Spruce			
12 year old	4.27	144	186
	(3.7–6.0)	(37–1181)	(85–958)
25 year old	4.32	296	268
	(3.6–5.9)	(51–1512)	(56–592)

After Gee and Stoner, 1989

Conifers are particularly good scavengers of atmospheric pollutants and sea salts, as illustrated in Table 6.15. These are often planted in upland areas in which headwaters of many streams and lakes rise. Coniferous trees are not the only vegetation types that scavenge pollutants. The levels of sulphate and chloride ions under oak are higher than in precipitation. However, there are sufficient base cations for neutralization to occur, and consequently throughfall in the soil under oak is generally less acidic than precipitation. Studies in Wales have also shown that evapotranspiration in conifer plantations can be as much as 30 per cent of the total water yield compared with 17 per cent for adjacent moorland catchments. Evapotranspiration can, therefore, concentrate the solutions of materials deposited on conifer forests, thus exacerbating acidification (Hornung and Newson, 1986).

Underlying geology and soils   Surface waters are only acidified if the underlying soil and geology is unable to buffer the effects of acid precipitation. The differences in effect are shown most clearly and consistently where acid and/or thin soils overlie massive, base poor bedrock: e.g. brown podzolic soils, stagnopodzols, podzols, stagnohumic gley soils, etc. The base poor rocks include igneous rocks, like granites, slates and mudstones. It has also recently been shown that for a given soil type, concentrations of aluminium in soil water are between two and three times higher under 25 year old conifers than in moorland (Gee and Stoner, 1989). Red sandstone and calcium-rich soils seem particularly good at buffering the effects of acidity.

The ability of a catchment to neutralize deposited acidity is also determined by the route by which the precipitation reaches the receiving stream. For example, in some catchments there is sufficient neutralizing capacity under low flow conditions, but under high flow events a large proportion of streamflow can enter via surface runoff and is not in contact with neutralizing bases. Drainage changes associated with land improvements or afforestation schemes exacerbate runoff rates, particularly under high flow conditions.

Land use   From the above it is clear that land use type will therefore have a significant influence on surface water quality. This interaction between surface waters and catchment use has been known for decades (Hynes, 1975) (see also Chapter 5). In temperate upland regions, many of the catchments from which streams and rivers drain are only suitable for sheep farming and/or afforestation, both of which are economically marginal and are thus frequently subsidized by government. Conifer afforestation is a natural vegetation type in northern temperate regions. In many other countries, afforestation with exotic conifer plantations has increased considerably in upland catchments in recent years. Clearly, based on the well-established link between forestry and acidification, such land use in the catchment will influence the stream's water quality (hydrochemistry) and ultimately its ecology. Some acid-sensitive lands, however, have been improved in the past for agricultural purposes and this has resulted in an improved neutralizing capacity and thus reduced surface water acidification.

Effects of surface water acidification

Surface water quality   The influence of acidification on surface water quality has received considerable attention, not only because of changes in potable water quality, but also due to changes in the ecology of systems caused by a reduction in pH. Because of the paucity and unreliability of historical data, it has been difficult until recently to describe spatial and temporal trends in acidification. In the United Kingdom, the Welsh Acid Waters Research Group looked at 57 major pH data sets for British sites and found only six sites with a marked downward trend in surface water pH (UKAWRG, 1986). Battarbee and his co-workers (1985) have looked at lake diatoms from sediment cores and constructed curves showing the changes in pH over the last 150 years. Examination of diatom diversity and composition, in particular the change towards acid-resistant species, has shown a pH decline by 0.5 to 1.2 pH units over 15 years in some lakes. Comparison of cores taken from forested and unafforested catchments in Wales have illustrated an acceleration in acidity under afforested catchments. More recently, the Welsh Water Authority has carried out extensive weekly sampling (150 sites) in acid-sensitive areas throughout 1984. Of the sites sampled 78 per cent had a minimum pH < 5.5 and 34 per cent had a minimum pH < 4.5.

There is now unequivocal evidence that afforestation in acid-sensitive areas depresses pH and elevates aluminium concentration.

Several studies have now shown that ecological damage is caused by a combination of acidity and elevated aluminium when calcium concentrations are low. The threshold beyond which ecological damage might be anticipated is pH < 5.5 with dissolved labile monomeric aluminium > 0.2 mg/L and hardness < 12 mg/L as C_aCO_3. Seasonal and episodic variations in water quality are also extremely important, since they have a dramatic ecological consequence. Under high flow conditions, particularly in winter, aluminium levels are generally at their highest, while pH and calcium concentrations are at their lowest. For example, in studies in Wales, a drop of 2 units of pH (6.0 to 4.0) and a 1 mg increase in aluminium concentration in an 11 hour period occurred during a storm event (Gee and Stoner, 1989). In an adjacent stream draining a moorland there was a pH drop of 0.9. The low pH of acidic waters is no different from the pH of many beverages and foods normally consumed by people and thus is not likely to pose any risk to us. However, the acidification of the potable water supply will increase its propensity to dissolve certain materials, in particular metals, from soil and from pipework (UKAWRG, 1989). The actual concentration of these substances in the raw water at the point of abstraction will depend on the form and the amounts in which they are present in the soils and rocks of the catchments from which the water is derived. For example, in 11 Scottish lochs, including those whose mean pH was less than 5.6, the concentration of 13 metals were low and below the maximum acceptable concentration of the EU water directive at the time (EC, 1980). There is some evidence of elevated aluminium, iron and manganese in more acidic waters, i.e. pH < 4.6. Water containing these metals and organic matter may deposit these substances in the domestic distribution system, which may result in temporary discoloration of water at the tap. However, since the majority of potable water (i.e. of public water supplies) is purified before consumption (including pH adjustment, see Chapter 11), the combined effect of acidification is likely to increase the cost of water treatment rather than pose any major health risk.

Biological effects of acidification Organisms can be affected by acidification either directly by physiological stress or indirectly by such changes as food supply, habitat provision and predation. As expected, the ecosystem response to acidification is very complex, indicating the complexity of both ecological and pollutant processes. Microbial activity seems to be reduced with reduced decomposition. For example, in the United States, studies have shown that both the number and activity of epilithic bacteria were reduced in acidified streams (UKAWRG, 1989). In lakes, diatoms usually decline at pH < 5.5 and filamentous green algae become dominant in littoral habitats. In extremely acid waters very few species are present. In studies in the United States and the United Kingdom some reductions in zooplankton species diversity have been found. For example, Fryer and Forshaw (1979) found more acidic waters (pH 5.14) had fewer species (mean 9.18) than less acid (pH 6.48) waters (mean species 11.26). In particular, daphnid and copepod species were missing from acid waters. It is felt that nutrients, e.g. calcium for their exoskeleton, and food availability were more important than acidity *per se*.

Stream organisms affected by acidification are the benthic macroinvertebrates which are important in the self-cleaning processes of freshwater systems. Relationships between acid–base status of streams and lakes and their macroinvertebrate faunas have been widely described and pronounced and consistent differences in invertebrate assemblages in several geographical areas have been related to acidity (Townsend *et al.*, 1983; Hildrew and Giller, 1994). Typically acid waters have significantly reduced diversity (Fig. 6.9). Two hypotheses have been proposed to explain the fewer species:

1. The direct physiological effects of acid-related factors, such as hydrogen ions and some heavy metals, exclude sensitive taxa.
2. Acid-related factors influence invertebrates indirectly by interactions through foodwebs, either top down (e.g. by release from fish predation) or bottom up (for instance by the alteration of food supplies).

Changes in the geographical distribution of fish due to acidification is now well described. Acid waters are characterized by reduced species diversity and in waters with pH < 4.5 eels may be the only

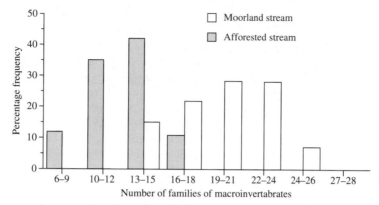

Figure 6.9 The percentage frequency of macroinvertebrate family richness (number of families) at 25 study sites on moorland and afforested catchments illustrating the effect of increased stream acidity on the biology of afforested streams in this region (after Omerod *et al.*, 1987).

fish present. The acidity affects the fish in many ways, principally through gill and blood physiology (impaired ion regulatory and acid–base status), reproductive physiology, developmental effects and demineralization, metal accumulation and changes in behaviour. These effects may be either the effect of pH or the combined effect of pH and aluminium.

The relative importance of hydrogen ions and aluminium toxicity varies substantially in natural waters. At one extreme, certain Canadian brown water rivers are characterized by very low proportions of the labile inorganic monomeric aluminium groups, even though total filterable aluminium may be high (Lacroix and Townsend, 1987). Complexing by organics effectively eliminates aluminium toxicity in such waters, and high mortality of salmon parr has been attributed to hydrogen ion toxicity at pH < 4.6. In contrast, Scandinavian studies have reported the dominance of labile aluminium toxicity in clearwater lakes (UKAWRG, 1989). In some studies hydrogen ion toxicity alone caused plasma electrolyte loss, but only at pH < 4.6. Other animals can also be affected but most other research has been directed at amphibians (Cummins, 1988) and birds, mostly dippers (O'Halloran *et al.*, 1990; Ormerod *et al.*, 1990).

Potential Solutions   The acidification of surface waters has resulted from deposition of acids, and this has been accelerated in some areas by upland conifer afforestation. Potential mitigation techniques may therefore be possible by policy changes in these areas either individually or in combination. Possibilities include reduction in emissions, liming of moorland catchments, short-term restriction in afforestation and direct liming of waters. Emissions, in particular of sulphur dioxide, are dominated by power stations and are likely to remain about the same over the foreseeable future. Although some attempts at control of emissions are being made, e.g. the United Kingdom proposes to reduce emissions by 14 per cent by 1997 by introducing flu gas desulphurization (FGD) at their plants (Gee and Stoner, 1989). Some workers, however, have suggested that with increased demands in electricity the reduction will be in the order of 5 per cent. In addition, recent studies have indicated that reduced sulphate emissions have also resulted in reduced depositions of other cations, which in the past have helped counteract the acidity to some extent. Thus the net effect is likely to be still further reduced (Hedin *et al.*, 1994). Atmospheric concentrations of nitrous oxides are dominated by emissions from two sources: power stations and vehicles. Power stations again are hoping to reduce emissions by 40 per cent, and with increased legislation there is likely to be a reduction in nitrous oxides from vehicles (Gee and Stoner, 1989). It is expected that nitrous oxide levels will drop by 6 to 12 per cent as a consequence.

However, despite these reductions, model predictions, calibrated in large catchments in Wales, suggest that reducing emissions by 50 per cent from 1984 levels would cause no significant increase in pH over the next 140 years, due to low weathering rates and base saturation in soils in those regions. The

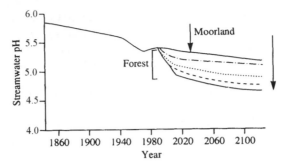

Figure 6.10 Simulation of the pH of streamwater from the Dargall Lane catchment (UK), comparing moorland response (——) assuming increased evaporation (–·–·–) with different levels of pollutant scavenging (·······) 20% additional sulphate, - - - - - 40% additional sulphate, ——— 60% additional sulphate (after Anon, 1987, Institute of Hydrology Research Report 1984–87, Wallingford, UK).

effects if the catchments were forested is also illustrated (Fig. 6.10). Therefore reductions in emissions will stop the decline in pH but may not help with recovery.

Liming is also a possible option and the liming of moorland catchments has been evaluated in Scotland (Howells, 1986). However, more specific targeting and avoidance of ecological damage are needed, e.g. preventing damage to acid-tolerant flora and fauna. Short-term restrictions to afforestation and mitigation techniques may also provide an opportunity to minimize acidification in those areas where plantation forests are planned to increase. No effective treatments have yet been proven, and the efficacy of various forestry practices, e.g. contour ploughing, buffer stripping, etc., is questionable at the present. Clearly the most appropriate future plan in such regions is to avoid planting on acid-sensitive catchments (e.g. Giller *et al.*, 1993). Such options are, of course, unavailable to regions where coniferous forests are the natural vegetation. One of the most direct methods of pH adjustment in freshwater systems is by direct liming of surface waters. However, liming is a costly operation which is only beneficial in the short term unless repeated on a regular basis. It can, however, be justified when the overriding consideration is the protection or restoration of a fishery.

6.6 ESTUARINE WATER QUALITY

Before discussing estuarine pollution in detail, we will identify some of the main characteristics of estuaries and illustrate how they differ from other aquatic systems.

6.6.1 Characteristics of Estuaries in Relation to Pollution

There are at least five important characteristics of estuaries which relate not only to the pollution itself but also how it is distributed and how it may affect estuarine ecosystems. The main points we will consider are tidal movement, salinity, fluctuations in temperature and oxygen, reduced species diversity and sediments and sedimentation.

Movement of the tide has a profound effect on the distribution, character and adaptations of the estuarine organisms, and not only on the organisms but also on the destination and the effects of pollution. The daily cycle of the tide means that organisms living on the upper parts of the intertidal zone generally must be able to withstand prolonged exposure to the air and short periods of inundation with brackish or saline waters. In addition to the twice daily changes in the levels, there are the monthly patterns of spring or neap tides. These tides will also determine the distribution of pollutants in an estuary. It is often common practice to discharge on a high tide, but if waste is discharged at the wrong state of the tide, the waste may be driven further above an outfall, rather than out to sea as desired. Superficially it appears that in estuaries there is a one-way flow of water from the river to the sea. This gives the false impression that anything poured into estuaries would simply flow out to sea, to be diluted by the vastness of the oceans. We now know differently. Estuaries are very complex, affected not only by tidal mixing but also by the Coriolis force (see Chapter 5), causing cyclonic circulation and increasing the residence time of effluents in estuaries, with greater opportunity for wastes to settle to bottom sediments.

Salinity is the amount of inorganic material dissolved in water expressed as a weight in grams per kilogram of water, i.e. parts per thousand (see Chapter 5). The salinity in an estuary will change over time. Saline water is heavier than freshwater, so it tends to sink below the inflowing freshwater as one progresses down an estuary, forming a salt wedge (Fig. 6.11). This salt wedge varies longitudinally within an estuary and during the course of a tidal cycle. Associated with the tidal movements, there are large changes in oxygen and temperature during the course of a tidal cycle. These changes, in addition to those of salinity, will also make it difficult for organisms to survive. Some workers, e.g. Wilson (1984), argue that it is sometimes difficult to distinguish the effects of pollutants on estuarine organisms from effects imposed by the physical variability in the environment. Oxygen consumption of the cockle *Cerastoderma edule*, for example, is more affected by a 50 per cent change in salinity than by a 10^3 fold increase in nickel concentration (Wilson, 1984).

The diversity and distribution of organisms is an important consideration in evaluating the effects of pollution in estuarine ecosystems. Estuaries have a naturally low diversity of species and higher biomass than other aquatic systems (Fig. 6.12). The majority of species present are euryhaline (tolerant of wide salinity changes) and occur in high abundance. The substrate will also influence the species present, in that at the top of the estuary it will tend to be muddy, whereas at the seaward end it would tend to be stony.

In addition to these changes in chemistry, there are also possibilities for stratification of the water body, the long residence time, the nature of mixing zones and sedimentation of material when the river enters the estuary to be taken into consideration. The change of pH and redox potentials as freshwater encounters the sea causes intense flocculation of clay and other particles, with increased adsorption of metals and other materials on the flocculates. The sedimentation in estuaries leads to extensive mudflats containing much organic material, metals and pesticides from the water column. No two estuaries are the same and detailed examination of the hydrological flow and the estuarine processes (mentioned above), etc., needs to be established to allow the prediction of the geochemical behaviour of each element and its possible effects on organisms as well as the role of estuaries in governing the mass balance of nutrients between rivers and oceans.

6.6.2 The Origin of Pollutants in Estuaries

Most rivers discharge into estuaries; therefore any pollutants or nutrients that the river may be carrying will end up in the estuary. The type of catchment activity that can influence the streamwater quality will also influence the surface water quality in estuaries. Ports have historically developed on estuaries which provided shelter and transport up the rivers into hinterlands. It is no coincidence that some of the major cities in the world are sited on estuaries. As these cities attracted industries around them, so more and more pollutants were discharged into the estuaries.

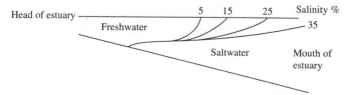

Figure 6.11 A generalized salinity salt wedge gradient in an estuary (redrawn after Prater, 1981).

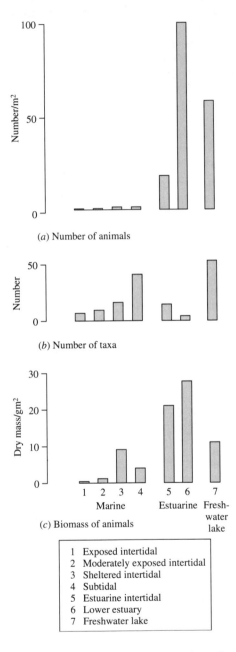

Figure 6.12 The number of animals, taxa and their biomass in freshwater, estuarine and marine aquatic habitats in the United Kingdom (data from McLuskey, 1981).

The main classes of pollutants in the estuarine and marine environment are organic waste, heavy metals, synthetic organic chemicals (e.g. organochlorines), thermal water and oil. There is overlap with marine pollutants but we have dealt with the two environments separately to highlight the major differences in cause and effect.

Organic waste   The bulk of municipal and licensed industrial waste into estuaries is organic matter. Most of the domestic organic waste is sewage (municipal wastewater). This organic waste will have a high biological oxygen demand (BOD). For example, unpolluted estuarine waters would be expected to have a BOD of 1 to 2 mg/L; in contrast, crude sewage has a BOD in the range 300 to 400 mg/L, while some industrial waste can have BODs exceeding 1000 mg/L. These BOD values may lead to deoxygenation and to the creation of anaerobic conditions with the production of hydrogen sulphide, as discussed earlier.

In addition to the high BOD, the waste may also contain high levels of nitrogen and phosphate, which may lead to eutrophication of estuarine systems. This effect, however, is often prevented by the natural turbidity of estuarine waters which reduces the amount of light reaching the plant life, especially algae fixed to the substrate in shallow water. In addition, there is often an increase in bacteria and viruses of faecal origin associated with sewage, many of which are pathogenic.

In many instances the sewage is simply discharged into the estuaries in the hope that the waste will be taken out on the tide. The complex nature of currents and mixing in estuaries, however, often leads to a high residence time of such effluents, further exacerbating the pollution problem, as has been seen in many Mediterranean lagoons, e.g. at Marseille and Venice.

Heavy metals   In addition to the organic contamination of the water, there is an added load of heavy metals from sewage and industrial waste. Metals such as copper and zinc in a sewage often become bound to the sediments of the estuary. These metals, unlike the organic waste, are not biodegradable and persist in the environment. The metals often precipitate out because of the displacement of the metal ions by the ions in the salt water. They are often concentrated in filter feeding molluscs such as the mussel, *Mytilus* (Bayne, 1978; Philips and Segar, 1986). This accumulation may have adverse effects on the aquatic flora and fauna and may constitute a public health problem where contaminated organisms are used for food.

It is worth mentioning that the behaviour of any heavy metal in an estuary will depend on its chemical state: whether it is in solution or particulate. The chemical behaviour of river-borne trace elements within an estuary is determined to a large extent by the chemical form in which it is transported by rivers: in solution, adsorbed on to surfaces, as solid organic particles, coating on detrital particles, lattice or crystalline form and precipitated (as pure phases on detrital particles). The behaviour of the metal or pollutant will very much depend on these properties, but the important point to bear in mind is that the pollutant must be bioavailable (i.e. accessible to the biota) before it will cause difficulty.

Synthetic organic chemicals, e.g. organochlorines   Synthetic organic compounds are also likely to enter the estuarine ecosystem. Some of these compounds such as pesticides and PCBs are harmful and may have a significant impact on the biota. Usually licences are not issued to discharge these wastes, and they seem to enter the estuaries via leachate (from, for example, landfill sites or dumps) or through illicit dumping.

Thermal waste   The origin of thermal water is from cooling water in industry and power stations. This will increase turbidity due to an increase in suspended solids in the waste and can also lead to greater oxygen demand by other waste effluents. The increased water temperatures can also support some unusual assemblages of organisms (often of semi-tropical origin in temperate estuaries) that are not usually found under the normal climatic conditions.

6.7 MARINE POLLUTION

The sea is a major sink or reservoir for all pollutants, covering as it does over 70 per cent of the surface of the globe. Pollutants entering the sea may be degradable or non-degradable and are similar to the range found in other aquatic environments (Table 6.10).

Pollutants enter the sea directly from outfalls and sometimes from coastal towns, but most frequently they enter from estuaries. Pollutants also reach the sea indirectly from rivers, which receive many pollutants from their drainage catchments, as discussed earlier. Particulate atmospheric pollutants may enter the sea directly as fall-out while non-particulate pollutants enter the sea through precipitation. Atmospheric pollution tends to be more regional or even global in scale than the previously mentioned inputs. The total global input of lead to the sea from natural and man-made sources is estimated to be about 400 000 t/yr. Of this about half is derived from vehicle exhausts containing lead petrol additives which reach the atmosphere and are then rained out (Clark, 1992). In addition to the above, there are also some offshore inputs of pollutants. These include chronic inputs from ships, resulting from emptying of bilge tanks, ballast water, etc., or acute inputs resulting from shipwrecks or losses of cargo during storms, etc.

We can recognize five very broad categories of pollutant which are dealt with separately below.

6.7.1 Sewage and Other Oxygen Demanding Wastes

Man has been aware of the need to control the 'nuisance' value of sewage for more than a century. In addition to the reduction in oxygen levels caused by the high BOD levels of sewage, as discussed earlier, sewage pollution also can lead to eutrophication, a process that has already been described for freshwaters and estuaries. In the marine environment, a frequent sign of eutrophication is the growth on the shore of the green algae *Enteromorpha* and *Ulva*, which turn parts of the beach bright green. An increase in the incidence of red tides (caused by blooms of red algae) has also been attributed to organic enrichment of coastal seas, although the precise factors initiating blooms of these toxic phytoplankton species are not known. These red tides can be economically destructive to fisheries by being toxic in themselves and also through the bans on shellfish sales during these periods. A number of conditions have been diagnosed in birds and other organisms known as paralytic shellfish poisoning which is brought about by red tide toxins and is fatal to organisms.

While sewage was traditionally discharged directly into the sea, most countries now insist on some form of physical/biological treatment (See Chapter 12). The end product after secondary treatment is an inert sludge and a clear liquid effluent. The sewage sludge has generally been dumped at sea. The sludge is not particularly toxic to marine life and the currents generally prevent deoxygenation of the seabed. Nevertheless, the buildup of fine material blankets the bottom and kills most of the natural organisms, so that only a low diversity fauna develops, characteristically dominated by species of worm of the genus *Capitella* (Reish, 1957). In the Firth of Clyde, sludge has been dumped for more than a century at a site at Garroch head. The dump site was relocated to a new site, about 5 km from the old site, in 1974. The new site received about one and a half million tons of sludge per year (Clark, 1992). Such sites are effectively the same as waste tips or dumping grounds on land. Examination of the old site has shown that the fauna does gradually return to normal, but the sediments remain heavily contaminated with metals. In many countries, sludge dumping at sea is now prohibited. By 1995 sea dumping of sludge had virtually stopped internationally.

One other problem associated with sewage is the presence of bacteria and viruses hazardous to human health. Faecal contamination of seawater is usually measured by counting the numbers of the bacterium *Escherischia coli*. This is called the *coliform count*. *E. coli* is always present in the human intestine and is not pathogenic (disease causing). It is merely a good indicator of the degree of faecal contamination of the seawater. In general people do not drink seawater, which therefore only becomes a hazard to bathers if it is heavily contaminated. The major risk from contamination is from contaminated seafood. Shellfish, which routinely filter large quantities of seawater in the feeding processes, may as a result harbour pathogenic bacteria and viruses (and also accumulate heavy metals). To counteract the hazard, public health authorities generally forbid the commercial gathering of mussels, clams, etc., from foreshores contaminated with sewage. All shellfish, even from less polluted areas, must be depurated (cleansed) before sale, usually by keeping them for a period of time in clean water, circulated over a weir irradiated by UV light. It is right that the public should be protected in this way, but somewhat unjust that fisherman should bear the cost of shellfish decontamination, when the root cause is inadequate sewage disposal by the same authorities that impose the regulation. Little is known about the fate of viruses in the sea, but these are in many ways more dangerous than bacteria. They are more difficult to depurate, and whereas bacteria need to occur in moderate numbers to cause disease, infection can result from a single virus. It has been noted that the highest incidence of lymphocystis (a cancer possibly induced by viruses) in fish, especially flatfish, is in estuaries (Mulcahy *et al.*, 1987) and it is these that tend to have the highest pollution load.

6.7.2 Oil

Crude oil is a complex mixture of hydrocarbons which must be refined before it is available for any of its multiple uses (lubricating, combustion, pharmaceuticals, etc.). Oil enters the sea from a variety of sources (Table 6.16).

Oil is a natural product which results from plant remains fossilized over millions of years, under marine conditions. It is not surprising, therefore, that all components of oil are readily biodegradable by bacteria, although different components degrade at different rates, tars being one of the slowest. When oil is spilled on the sea, being light, it spreads over the surface as a slick. The lightest components, which are also often the most toxic, either evaporate or dissolve in the water. Immiscible components become emulsified and dispersed in the water, while heavy residues form tar balls. The immiscible fraction forms a water-in-oil emulsion called 'chocolate mousse', which contains about 75 per cent water. This forms sticky brown masses when it comes ashore and causes major problems on tourist beaches. Tar balls float and are common in ocean waters, particularly along shipping routes.

Table 6.16 Estimated world input of petroleum hydrocarbons to the marine environment (m t/yr)

Source	Estimated input
Transportation	
Tanker accidents,	
bilge and fuel oil	0.5
Fixed installations	
Offshore production	
Terminal operations	0.2
Coastal refineries	
Other anthropogenic sources	
Municipal and industrial wastes	
Runoff (river and urban)	
Atmospheric fall-out	1.4
Ocean dumping	
Natural inputs	
Underwater seepage	0.2
Total input	2.3

Modified from Clark, 1992

While oil at sea was a cause for concern, oil pollution did not impinge much upon public awareness, and hence received little scientific attention until the first shipwrecking of tankers carrying crude oil. The earliest were the *Tampico Maru* in 1957 off Baja, California, and the *Torrey Canyon* in 1967 off Cornwall. The *Torrey Canyon* was wrecked in March, off Lands End, and large quantities of oil came ashore on the beaches of Cornwall. Many of the oil-contaminated beaches in Cornwall were tourist beaches and authorities were desperate to try to rehabilitate the beaches before the oncoming tourist season. Dispersants were used in an effort to remove the oil, and the resulting damage to the beaches was a result of the use of the dispersants rather than the direct effects of oil. The principal damage was to grazing animals such as limpets, which were eliminated. In both Cornwall and Baja, California, the elimination of herbivores allowed the rapid growth of green and brown algae which changed the appearance of the shore from one of open rocks with limpets and barnacles to one dominated by seaweeds. This change also occurs when the dominant grazers are killed by other causes such as red tides (Southgate *et al.*, 1984). The return of grazers occurs gradually but may take up to ten years (Southward and Southward, 1978).

The early dispersants were soon replaced by less toxic products. Modern dispersants are up to a thousand times less toxic than those used on the *Torrey Canyon* oil, although they cannot be said to be non-toxic. Dispersants act by breaking up oil into very fine particles, thus increasing the surface–volume ratio and allowing bacteria to degrade the oil much more rapidly.

Cleaning of beaches   Beaches can be cleaned in three ways.

Mechanical cleaning   Stranded oil in embayments can often be pumped ashore into a tanker. On shore, however, mechanical removal is usually only possible on sandy or pebble beaches. In these environments the use of dispersants is contraindicated because the dispersed mixture simply drains downwards to appear again at a later date. The use of heavy mechanical plant does of course damage the fauna and flora and the integrity of sandy beaches can be put at risk by the excessive use of bulldozers. Where there are large quantities of oiled seaweeds, they may need to be removed from beaches and require disposal. Small quantities can be ploughed into the land where bacterial activity will slowly break the oil down. Large quantities are generally simply tipped into waste dumps, an unsatisfactory but unavoidable solution.

Absorption   Large quantities of straw may be scattered on to a beach to absorb the oil. After a period of time this can be collected for removal. In theory this is the most environment-friendly approach, but it is very labour intensive and generally somewhat limited in its use (eg. access to site, complexity of terrain, etc.) and the problems of disposal of the oiled straw remain.

Dispersants   As already mentioned, these products have relatively low toxicity if used correctly (with large volumes of seawater). After use the shore should be doused with large quantities of seawater. Where possible, application can be carried out from tugs or barges standing off from the shore. There is some evidence that juvenile stages of some shore animals, e.g. barnacles, may be adversely affected by dispersants (Myers *et al.*, 1980).

The ecological impact of oil pollution   The most emotive impact of oil on marine life is the oiling of birds. Oiled, dying seabirds are the first and often only marine organisms to receive media attention. Seabirds are at risk because they come into contact with oil floating on the sea surface. The damage is caused more by the physical attributes of the oil than by its toxicity (when ingested) (Cox, 1993). It clogs the feathers, replaces the air normally trapped by the feathers between them and the skin and so reduces the ability of the birds to maintain their body temperature. Birds may drown if the plumage waterlogs or may die of hypothermia. Seal pups are occasionally reported as being badly oiled, but in general adult seals, sealions and whales do not appear to be at high risk from oil on the sea, since their body is insulated by thick layers of subcutaneous fat. Sea otters along the Pacific coast of the Americas insulate their body with thick fur in much the same way as birds do with feathers. These are at serious risk from floating oil and more than a thousand were killed by oil from the *Exxon Valdez* in 1989 (Miller, 1992).

In the intertidal zone, the effects of an oil spill are generally more marked on the high shore than on the middle and lower shores. This is doubtless because oil is pushed up the shore with each rising tide,

Table 6.17 Numbers of % cover/0.25 m^2 of three key organisms at three sites in Bantry Bay in July 1978 and July 1979

Organism	Site	July 1978	July 1979	Decrease (%)
Barnacles (number)	Cooskeen Cove	1872	925	51
Mussels (%)	Eagle Point	56	29	48
	Dereenacarrin	72	38	47
Limpets (number)	Dereenacarrin	287	83	71

After Myers *et al.*, 1980

causing an accumulation in the high tide region.

In January 1979, the French tanker *Betelgeuse* exploded in Bantry Bay, Ireland, pouring 2000 tons of crude oil into the bay (Cross *et al.*, 1979). A study of three key rocky shore species (limpets, mussels and barnacles) showed a marked drop in the numbers of these three species between July 1978 prior to the incident and July 1979, following the incident (Table 6.17).

These data might have been compulsive, if circumstantial, evidence for a significant effect of oil pollution. An analysis of data collected on a monthly basis between these two dates, however, revealed that the drop in numbers of mussels and barnacles could be attributed to the predatory effects of dog-whelks, and that the limpet numbers dropped prior to January 1979 when the oil pollution occurred (Myers *et al.*, 1980). Rarely does an oil pollution incident occur at a site coincident with an active intertidal marine research project. The above data reveal how careful one must be in assessing the results from a typical environmental impact survey using data collected 'before and after'. Natural changes occur all the time. Indeed, we should be more surprised if there were no change in the relative numbers of organisms in a community from season to season or from year to year. The magnitude of these changes may be so great that the background noise conceals the signal that we hope to perceive.

Commercial damage from oil pollution   Particularly at risk are man-made installations at the sea surface. This includes aquaculture installations such as fish cages. One of the problems of oil pollution is its unpredictability and hence it is difficult to guard against. Where possible, floating booms can be used to protect aquaculture installations and it is also sometimes possible to move cages to another and unaffected site, although when cages contain fish, they must be towed very slowly to avoid damage to the fish. An indirect effect of oil pollution is the tainting of the flesh of fish and shellfish. This may continue long after the pollution incident, from the leaching of oil residue from oiled equipment. Oily flavours are generally repulsive to human taste and while not necessarily hazardous to human health, avoidance of seafood by the public after a spill may seriously affect the market and hence the livelihood for both the fishing and aquaculture sectors.

Public health and oil pollution   Oil includes polycyclic aromatic hydrocarbons (PAH), some of which are known to be carcinogenic. It was at one time feared that these might accumulate in food chains and that consumption of animals high in the food chain, such as carnivorous fish, might pose a health hazard to humans. There now seems to be little evidence for the accumulation of PAH in marine organisms.

6.7.3 Heavy Metals

Metals are known to be essential for living organisms. They are used in respiratory pigments (iron, copper, vanadium) enzymes (zinc), vitamins (cobalt) and other metabolic processes. Only when normal concentrations are exceeded do heavy metals become potentially toxic (see the discussion in Chapter 5). Marine organisms tend to accumulate heavy metals from the environment and are adapted to handle

Table 6.18 Natural and anthropogenic inputs of heavy metals to the marine environment

Natural	Anthropogenic
Coastal supply	Direct processes
Rivers	Mining
Glaciers	Smelting
Wave action	Refining
Erosion	
Deep sea supply	Indirect processes
Volcanism	Electroplating
Tectonic activity	Catalysts
Chemical processes in sediments	Petrochemical Industry
Atmospheric	Atmospheric
Particles	Fossil fuel burning
Vapour (mercury)	

normal fluctuations in intake. In the sea, the concentrations of heavy metals are so low that they are readily increased to levels that marine organisms have never previously encountered. The natural and artificial sources of heavy metals are listed in Table 6.18.

The following subsections discuss several of the more important heavy metals in the marine environment.

Mercury   The first indication that mercury in the marine environment could be a hazard to human life came in the late 1950s when more than a hundred people in Minimata Bay, Japan, were killed or disabled through eating fish and shellfish contaminated with methyl mercury. While the toxicity of mercury was well known, this was and still is the only pollutant introduced into the marine environment known to be directly responsible for human deaths. The Minimata Bay mercury had been released into the environment from a factory using it as a catalyst to manufacture acetaldehyde from acetylene. Since 1965, mercury contamination has been recorded in several European countries, particularly in Swedish lakes and rivers, as well as in parts of Canada and the United States. Mercury is taken up by shellfish, especially bivalve molluscs, and by fish. Cod taken from the sound between Denmark and Sweden which is heavily contaminated with mercury contain up to 1.29 ppm mercury, while those from the North Sea 0.15–0.2 ppm (Clark, 1992). Large oceanic fish such as tuna, swordfish and marlin have high levels of mercury in their muscles. When first discovered it was thought that this indicated that mercury pollution had reached a global status. However, examination of mercury levels in the muscles of preserved specimens in the British Museum from the last century showed that high mercury levels in these fish was quite normal. These fish swim continuously and routinely pass large quantities of water through their open mouths. The result is a large uptake of metals which they cannot excrete and so concentrate in their tissues.

Both birds and sea mammals concentrate mercury—birds in their feathers and liver, mammals in the liver. Interestingly, marine mammals can accumulate quite large quantities without coming to harm. Selenium antagonizes the toxic effect of the mercury. Mercury selenide in the liver of dolphins appears to be a detoxification product of the methyl mercury taken in from their food.

Mercury is now recognized as a potential hazard on a regional scale and following a recommendation by the World Health Organization (WHO) of a maximum tolerable consumption in food of 0.3 mg of mercury, standards for the maximum permitted levels of mercury in seafood have been set in many countries. In the European Union a uniform standard of 0.3 μg/g has been adopted. Sweden initially introduced a limit of 0.5 μg/g but discovered that most of the fish from the Baltic and inland lakes already exceeded this, so were forced to increase the limit to 1.0 μg/g. They advised the public not to eat more than two fish meals per week (Clark, 1992).

Cadmium   Cadmium levels are quite high in some coastal waters such as the Severn Estuary in Britain and the Hardanger Fjord in Norway. In the former case the contamination is natural; in the latter it is due to smelter wastes. Nevertheless, no ecological effects have been reported. Molluscs are known to accumulate large concentrations of the metal and in Tasmania the consumption of oysters with high concentrations of cadmium led to nausea and vomiting in people consuming them. Cadmium can cause permanent damage to the kidney and can give rise to nephrii proteinum, where proteins are lost in the urine.

Lead   That lead levels are increasing globally in world seas is apparent from records from annual ice layers in Greenland. Lead aerosols are distributed world-wide in the atmosphere and enter the sea through rain. Local levels may be enhanced by sludge dumping or other special circumstances. However, lead does not appear to be particularly toxic to marine organisms and is known to be accumulated in some species without apparent harm. In the contaminated Sorfjord in Norway seaweeds and plants contain levels up to 3000 ppm. Lead is responsible for serious damage to the health of humans and birds (O'Halloran et al., 1988) but does not at the moment appear to be a problem in the marine environment. Interestingly lead nitrate actually enhances the growth of some diatoms.

6.7.4 Man-made Organic Poisons

These include an array of pesticides based on organochlorine or organophosphorus compounds, PCBs and synthetic herbicides.

Organochlorine pesticides   Unlike some early pesticides like nicotine, pyrethrum and rotenone that are obtained from plants, DDT and its relatives are organic compounds devised by man and synthesized from petrochemicals and chlorine. The commonest breakdown product reaching the sea is DDE, which is very persistent and fat soluble. Although technically a proscribed insecticide in developed countries, DDT is still in use illegally and in the Third World.

Minute amounts of DDT have been shown to reduce photosynthesis in phytoplankton, so rising levels in the sea could theoretically affect overall primary productivity. Dietry DDT levels of 2 to 5 μg/g body weight caused 30 to 100 per cent mortality in shrimp, crab and fish within 2 to 10 weeks in toxicity tests. Current recorded levels of DDT in some estuaries are around 0.01 to 0.2 μg/g body weight (Myers et al., 1980). Hardy animals concentrate the residues while they live and pass even more potent doses on to predators higher up the food chain. The evidence of the damaging effects of organochlorines on predatory birds was outlined in Chapter 2. In the marine environment smaller fish-eating birds appear to have been less affected, but a chemical factory near Rotterdam manufacturing Dieldrin and Endrin was thought to be responsible for the decline in the size of a Sandwich tern colony from 20 000 to 650 in 1965. When measures were taken to prevent the discharge of effluent from the factory, the numbers of terns rose to 5000.

Organophosphorus compounds   These biocides include Malathion, Parathion and Dipterex. Like DDT they are nerve poisons used to control insect pests on crops. They are very toxic to fish, although less so than DDT, but unlike DDT they are not retained but are slowly inactivated and excreted. Warning signals exist, however. In 1964, fish were found to be dying in an area around a marine outfall of a Danish factory manufacturing Parathion (Clark, 1992). Lobsters were affected over a much wider area. The factory effluent was found to be lethal to lobsters at a dilution of 1:50 000.

Herbicides   Early herbicides were based on common plant hormones (auxins), but now a variety of artificially synthesized compounds are used. Most of these are much less stable than organochlorines and have not yet been recorded as causing damage to the marine environment. It should be borne in mind, however, that algae, including phytoplankton, react to herbicides in a similar way to terrestrial plants.

Polychlorinated biphenyls (PCBs) PCBs have been used in industry since the 1930s and are complex mixtures of chlorobiphenyls, each characterized by the substitution of the biphenyl nucleus equally or unequally by one or more chlorine atoms. Unlike most other pollutants, PCBs were in the environment long before their presence or role as pollutants was recognized. PCBs are virtually insoluble in seawater and particularly stable at temperatures which would decompose almost all natural and many synthetic organic compounds. Their stability to chemical reagents is also high. Not surprisingly, therefore, once introduced into the environment, including the seas, these compounds are exceedingly persistent. Being fat soluble they move readily through the environment and within tissues or cells. Whatever the sampling locality, PCBs are always in greater concentrations than other chlorinated hydrocarbons, in a variety of organisms such as shrimps, plankton, pelagic and demersal fish and marine mammals (Safe, 1987). As second-order carnivores, seals have been shown to acquire the highest tissue levels of PCB of any marine mammals. PCB levels in the blubber of seals from the North Sea, Irish Sea and the Baltic are very much higher than those from the Arctic, Antarctic and Pacific Oceans (Nixon, 1991). There is no evidence that seals in a healthy condition are affected by these residues, but when fats are mobilized during periods of poor feeding the higher concentrations in the remaining fat could have a detrimental physiological effect, although there is currently no supporting evidence for this.

6.7.5 Radioactivity

Like heavy metals, radioactivity occurs naturally in seawater, principally from potassium-40, although it also contains other decay products. Radionucliides tend to accumulate in sediments and in some parts of the world reach quite high natural levels. At one popular bathing beach near Rio de Janeiro, visitors are exposed to a dose rate of $20\,\mu Gy/h$ (Clark, 1992). A gray (Gy) is the amount of radiation causing 1 kg of body tissue to absorb 1 J of energy. The principal anthropogenic inputs of radioactivity into the oceans are from nuclear weapons testing and liquid wastes from nuclear power stations and fuel reprocessing plants.

The ecological effects of radioactivity have been investigated, particularly in relation to the discharge from Sellafield, in the United Kingdom. Radionucliides can be biomagnified in food webs (see Chapter 2), with bottom-living fish receiving some of the highest loads. Experimental plaice tested up to two and a half years after release had required a mean dose of $3.5\,\mu Sv/h$, where Sv are sieverts, the dose equivalent. The minimum dosage which results in measurable physiological or metabolic disturbances to plaice in the laboratory is $100\,\mu Sv/h$.

The human population is exposed to natural and man-made radiation continuously. Exposure to radiation from the sea is from the consumption of seafood or from radioactive sediment on beaches. Unless massive doses of radiation are received, the chronic effects are not manifested for considerable periods of time. Furthermore, they may only manifest themselves in the offspring through genetic deformities. The effects of exposure to radioactivity are thus difficult to quantify.

6.7.6 Heat

Cooling waters from power generating stations, for example, are often discharged into the sea at high temperatures. The large volume of receiving water usually results in relatively modest increases in sea temperatures in the vicinity of the outfall, usually less than $2\,°C$. In temperate regions this has little effect on the communities, although breeding seasons may be extended. In enclosed areas such as docks, the survival of exotic organisms may be enhanced (Chapman and Carlton, 1991). In tropical waters, the effects of heating may be more severe, since many tropical organisms, such as corals, are already living near their thermal upper limit. A rise of sea temperature by as little as 2 °C may kill many tropical corals, sponges, crustaceans and molluscs. The bleaching of corals in the Galapagos and the Caribbean after modestly raised temperatures following an El Nino Southern Oscillation (ENSO) event is testimony to this (Glynn, 1993).

6.8 PROBLEMS

6.1 How would you define water quality? Why is the maintenance of water quality so important in water management?

6.2 On the basis of field observations and information from the local authority and water suppliers, identify the major consumer processes utilizing your local river system. What possible conflicts could arise and how might these conflicts be rationalized?

6.3 What is an 'oxygen sag curve' and outline its importance in understanding the effects of organic pollution?

6.4 Compare and contrast the causes and effects of acidification with cultural eutrophication of

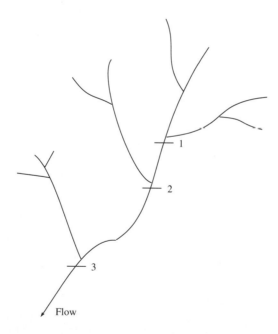

freshwater ecosystems.

6.5 How would you rate the water quality of the river from which the following invertebrate sample was taken? Is the same conclusion reached using the various biotic assessment indices?

6.6 The map above represents a stream system in an intensively farmed catchment. A serious fish kill occurs in summer between points 1 and 2 and fish populations do not recover to normal levels until point 3. Explain in detail how you would attempt to locate the source and identify the nature of the problem.

6.7 Compare and contrast the causes and consequences of pollution in estuaries and freshwater ecosystems.

6.8 Imagine you are a planner and have to site a new paper/pulp factory. Given the choice, where would you locate it—on an estuary, on a river or on land? Explain the possible problems in each case and justify your decision.

6.9 What procedures are generally used for cleaning up oil spills? Which are the most biologically friendly?

6.10 What are the ecological consequences of sewage input to the marine environment?

6.11 In what way do man-made organic poisons affect marine organisms?

REFERENCES AND FURTHER READING

Battarbee, R. W., R. J. Flower, A. C. Stevenson and B. Rippey (1985). 'Lake acidification in Galloway—a palaeoecological test of competing hypothesis', *Nature*, **314**, 350–352.

Bayne, B. L. (1978) 'Mussel watching'. *Nature*, **275**, 87–88.

Bryan, G. W., P. E. Gibbs, L. G. Hummerstone and G. R. Burt (1986) 'The decline of the gastropod *Nucella lappilus* around south-west England: evidence for the effect of tributyl tin from anti-fouling paints'. *Journal of the Marine Biological Association* **66**, 611–640.

Carpenter, S. and J. Kitchell (eds) (1993) *The Trophic Cascade in Lakes*, Cambridge Studies in Ecology, Cambridge University Press, Cambridge.

Chapman, D. (ed.) (1992) *Water quality assessments*, Chapman and Hall, London.

Chapman, J. W. and J. T. Carlton (1991) 'A test of criteria for introduced species: the global invasion by the isopod *Synidotea laevidorsalis* (Miers, 1881)'. *Journal of Crustacean Biology*, **11**, 386–400.

Chave, P. (1990) 'Waterways of the future: the role of biological assessment in protecting our aquatic environment'. *Biologist*, **37**, 129–131.

Clark, R. B. (1992) *Marine Pollution*, 3rd edn, Clarendon Press, Oxford.

Clements, W., D. Cherry and J. Van Hassel (1992) 'Assessment of the impact of heavy metals on benthic communities at the Clinch River (Virginia): evaluation of an index of community sensitivity', *Canadian Journal of Fisheries and Aquatic Sciences*, **49**, 1686–1694.

Cox, G. W. (1993) *Conservation Ecology*, Wm C. Brown Publishers, Dubuque, Iowa.

Cross, T. F., T. Southgate and A. A. Myers (1979) 'The initial pollution of shores in Bantry Bay, Ireland, by oil from the Tanker *Betelgeuse*. *Marine Pollution Bulletin*, **10**, 104–107.

Cummins, C. P. (1986). 'Effects of aluminium and low pH on growth and development in *Rana temporaria* tadpoles', *Oecologia*, **69**, 248–252.

Cummins, C. P. (1988) 'Effect of calcium on survival times in *Rana temporaria* embryos at low pH', *Functional Ecology* **2**, 297–302.

EC (1980). Council Directive relating to the quality of water intended for human consumption. 80/778/EEC.

Edwards, R. W., J. H. Stoner and A. S. Gee (eds) (1990) *Acid Waters in Wales*, Kluwer Academic Publishers, The Hague.

Environmental Research Unit (ERU) (1989) *Cork Harbour Water Quality*, Environmental Research Unit, Dublin, 113pp.

Environmental Research Unit (ERU) (1992) *Water Quality in Ireland 1987–1990*, Environmental Research Unit, Dublin.

Fryer, G. and O. Forshaw (1979) 'The freshwater crustacea of the island of Rhum (inner Hebrides): a faunistic and ecological survey', *Biological Journal of the Linnaean Society*, **11**, 333–367.

Gee, A. S. and J. H. Stoner (1989) 'A review of the causes and effects of acidification of surface waters in Wales and potential mitigation techniques', *Archives, Environmental Contamination and Toxicology*, **18**, 121–130.

Giller, P. S., J. O'Halloran, R. Hernan, N. Roche, C. Clenaghan, J. Evans, G. K. Kiely, N. Allott, M. Brennan, J. Reynolds, D. Cooke, M. Kelly–Quinn, J. Bracken, S. Coyle and E. Farrell, E. (1993) 'An integrated study of forested catchments in Ireland,' *Irish Forestry*, **50**, 70–88.

Glynn, P. W. (1993) 'Coral reef bleaching: ecological perspectives', *Coral Reefs*, **12**, 1–17.

Harriman, R. and B. R. S. Morrison (1982) 'Ecology of streams draining forested and non-forested catchments in an area of Central Scotland subject to acid precipitation', *Hydrobiologia*, **88**, 251–263.

Harriman, R. and D. E. Wells (1985) 'Cause and effect of surface water acidification in Scotland', *Journal of Water Pollution Control* **2**, 61–65

Haslam, S. M. (1978) *River Plants*, Cambridge University Press, Cambridge.

Hedin, L. O., L. Granat, G. E. Likens, T. A. Buishband, J. N. Galloway, T. J. Butler and H. Rodhe (1994) 'Steep declines in atmospheric base cations in regions of Europe and North America', *Nature*, **367**, 351–354.

Hellawell, J. M. (1986) *Biological Indicators of Freshwater Pollution and Environmental Management*, Applied Science Publishers, London.

Henderson-Sellars, B. and H. R. Markland (1987) *Decaying Lakes*, John Wiley, Chichester.

Hildrew, A. G. and P. S. Giller (1994) 'Patchiness, species interactions and disturbance in the stream benthos', in *Aquatic Ecology, Scale, Pattern and Process*, P. Giller, A. Hildrew and D. Rafaelli (eds), Blackwell Scientific Publications, Oxford, pp 21–61.

Hornung, M. and M. D. Newson (1986). 'Upland afforestation influences on stream hydrology and chemistry, *Soil Use Management*, **2**, 61–65.

Howells, G. D. (1986) *The Loch Fleet Project—Report of the Prevention Phase 1984–1986*, Central Electricity Generating Board, Leatherhead, 74pp.

Hynes, H. B. (1960) *Biology of Polluted Waters*, Liverpool University Press, Liverpool.

Hynes, H. B. N. (1975) 'The stream and its valley', *Verhandlung der International Vereinigung fur Limnologie*, **19**, 1–15.

Institute of Hydrology Research Report (1987). 1984–87, Great Yarmouth, UK.

James, A. and L. Evison (1979) *Biological Indicators of Water Quality*, John Wiley, New York.

Jorgensen, S. E. (1980). *Lake Management*, Pergamon, Oxford.

Krebs, C. J. (1985) *Ecology.* 3rd edn, Harper and Row, New York.

Lacroix, G. L. and D. Townsend (1987) 'Response of juvenile Atlantic salmon stocks to episodic increases of some rivers in Nova Scotoa, Canada', in *Ecophysiology of Acid Stress in Aquatic Organisms*, H. Witters and O. Vanderborght (eds), Annals of Royal Society of Zoology, Belgium, Vol. 117, Supplement 1, pp. 197–307.

Lansdown, R and W. Yule (1986) *The Lead Debate: The Environment, Toxicology and Child Health*, Croom Helm, London.

Likens, G. E., F. H. Bormann, R. S. Pierce, J. S. Eaton and N. M. Johnson, (1977) *Biogeochemistry of a Forested Ecosystem*, Springer-Verlag, New York.

McLuskey, D. S. (1981). *The Estuarine Ecosystem*, Blackie and Son, Glasgow.

Maitland, P. S. (1990). *Biology of Fresh Waters*, 2nd edn, Blackie and Son, Glasgow.

Mason, C. F. (1991). *Biology of Freshwater Pollution*, 2nd edn, Longman, Harlow.

Mason, C. F. and W. O'Sullivan (1992) 'Organochlorine pesticide residues and PCBs in otters (*Lutra lutra*) from Ireland', *Bulletin of Environmental Contamination and Toxicology*, **48**, 387–393.

Miller, G. T. (1992) *Living in the Environment*, 7th edn, Wadsworth, Belmont, California.

Moss, B. (1988) *Ecology of Freshwaters. Man and Medium*, 2nd edn, Blackwell Scientific Publications, Oxford.

Mulcahy, M. F., E. Twomey, A. Petersen and C. T. Maye (1987) 'Pathobiology of estuarine fish and shellfish in relation to pollution', in: *Biological Indicators of Pollution*, D. H. S. Richardson (ed.), Royal Irish Academy, Dublin, p. 210.

Myers, A. A., T. Southgate and T. F. Cross (1980) 'Distinguishing the effects of oil pollution from natural cyclical phenomena on the biota of Bantry Bay, Ireland', *Marine Pollution Bulletin*, **11**, 204–207.

National Rivers Authority (NRA) (1990) *Toxic Blue-Green Algae*, Water Quality Series 2, National Rivers Authority, London, 125pp.

Nixon, E. (1991) 'PCBs in marine mammals from Irish coastal waters', *Irish Chemical News*, **7**, 31–38.

O'Halloran, J., A. A. Myers and P. F. Duggan (1988) 'Lead poisoning in Mute Swans *Cygnus olor* in Ireland: a review', *Wildfowl Supplement*, **1**, 389–395.

O'Halloran, J., S. D. Gribbin, S. J. Tyler and S. J. Ormerod (1990) 'The Ecology of dippers *Cinclus cinclus* in relation to stream in upland Wales: time activity budgets and energy expenditure, *Oecologia*, **85**, 271–280.

O'Halloran, J., S. J. Ormerod, P. Smiddy and B. O'Mahony (1993) 'Organochlorines and mercury content of dipper eggs in south west Ireland', *Biology and Environment, Proceedings of the Royal Irish Academy*, **93**, 25–31.

Ormerod, S. J., N. S. Wetherley and A. S. Gee (1990) 'Modelling the ecological changing acidity of Welsh streams', in *Acid Waters in Wales*, R. N. Edwards, J. H. Stoner and A. S. Gee (eds), Kluwer Academic Publishers, The Hague, pp. 279–298.

Prater, A. J. (1981) *Estuary Birds*. T & A. D. Poyser, Calton.

Phillips, D. J. H. and D. A. Segar (1986) 'Use of bio–indicators in monitorining conservative contaminants', *Journal of Marine Biological Association*, **17**, 10–17.

Reish, D. J. (1957) 'The relationship of the polychaetous annelid *Capitella capitata* (Fabricius) to waste discharges of biological origin', in *US Public Health Service Biological Problems in Water Pollution*, Cincinnati, pp. 195–200.

Safe, S. (1987) *Polychlorinated Biphenyls (PCBs): Mammalian and Environmental Toxicology*, Springer-Verlag, Berlin.

Smith, R. V. (1977) 'Domestic and agricultural contributions to the inputs of phosphorus and nitrogen to Lough Neagh', *Water Research*, **11**, 453–459.

Southgate, T., K. Wilson, T. F. Cross and A. A. Myers (1984) 'Recolonisation of a rocky shore in S.W. Ireland following a toxic bloom of the dinoflagellate *Gyrodinium aureolum*', *Journal of the Marine Biological Association of the United Kingdom*, **64**, 485–492.

Southward, A. J. and E. C. Southward (1978) 'Recolonisation of rocky shores in Cornwall after use of toxid dispersants to clean up the *Torrey Canyon* spill', *Journal of the Fisheries Research Board of Canada*, **35**, 682–706.

Spellerberg, I. F. (1991). *Monitoring Ecological Change*, Cambridge University Press, Cambridge.

Standard Methods for the Examination of Water and Wastewater. (1992) American Public Health Association, American Water Works Association and Water Pollution Control Federation. L. Clesceri, A. Greenberg and R. Trussell (Eds).

Stoner, J. H. and A. J. Gee (1985) 'Effects of forestry on water quality and fish in Welsh rivers and lakes', *Journal of Institute of Water Engineering Science*, **39**, 27–45.

Townsend, C. R., A. G. Hildrew and J. Francis (1983) 'Community structure in some southern English streams: the influence of physicochemical factors', *Freshwater Biology*, **13**, 521–544.

United Kingdom Acid Waters Review Group (UKAWRG) (1986). 'Acidity in United Kingdom freshwater's, Interim report, Department of the Environment, London, 46pp.

United Kingdom Acid Waters Review Group (UKAWRG) (1989) Second Report to the Department of the Environment, HMSO, London.

Washington, H. G. (1984). 'Diversity, biotic and similarity indices. A review with special reference to aquatic ecosystems', *Water Research*, **18**, 653–694.

Wellburn, A. (1988). *Air Pollution and Acid Rain: The Biological Impact*, Longman. Harlow.

Welsh Water Authority (1987). 'Llynn Brianne Acid Waters Project', First technical summary report, Welsh Water Authority, Llanelli, 96pp.

Wilson, J. G. (1984) 'Assessment of the effects of short–term salinity changes on the oxygen consumption of *Cerastoderma edule*, *Macoma balthica* and *Tellina tenuis* from Dublin Bay, Ireland, *Journal of Life Sciences, Royal Dublin Society*, **5**, 57–63.

Woodiwiss, F. S. (1964) 'The biological system of stream classification used by the Trent River Board', *Chemistry and Industry*, 443–447.

WATER QUALITY IN RIVERS AND LAKES: PHYSICAL PROCESSES

7.1 INTRODUCTION

Chapter 11 discusses the water quality parameters with respect to drinking water. Chapter 12 similarly discusses water quality parameters of treated wastewater being discharged to either fresh or saline water bodies. Water bodies as in streams, lakes and estuaries support a variety of fish life, whose ability to survive in their natural habitat may be inhibited if the water quality is unsatisfactory, as discussed in Chapters 5 and 6. The quality of water to sustain aquatic life differs from species to species. For instance, coarse fish can survive in freshwaters with dissolved oxygen levels greater than about 3 mg/L. However, pelagic fish may need twice this amount. Other water uses such as water abstraction for human consumption set a further series of water quality standards on a lake or river. Recreational use may set further standards, e.g. low coliform counts for bathing waters. As such, there are a host of parameters relevant to water quality at different levels (see Table 11.3). However, probably the most significant parameter relating to the sustainability of fish life is dissolved oxygen (DO). A parameter intimately tied up with DO is biochemical oxygen demand (BOD), described in Chapter 3 with the test detailed in Standard Methods (1992); BOD is generally applied to a wastewater quality. It is a measure of the potential of the wastewater to reduce the oxygen levels of the receiving water body. Of course, the greater the dilution of the receiving waters, the less the negative impact of a discharge. As discussed in Chapter 12, wastewater is most often characterized by its effluent BOD strength. This BOD strength is diluted by the receiving water and if the resulting receiving water body BOD is impaired then the BOD effluent strength has been too high. The most relevant lotic water quality parameters are:

- Dissolved oxygen (DO)
- Suspended solids (SS)
- Biochemical oxygen demand (BOD)
- Temperature
- pH
- Nutrients, especially N and P
- VOCs

- Metals, e.g. Hg, Pb, Cd, etc.
- Pesticides

As mentioned, the DO level is significant to aquatic life, as is the level of suspended solids. The latter increase turbidity and over time settle out on the bottom, increasing the nutrient, metal and toxic levels of the settled sediments. It also can cause a sediment oxygen demand (SOD). Higher temperatures cause lower densities and also cause lower saturated dissolved oxygen levels. Nutrients, especially P and N, can lead to eutrophication of freshwaters and saline waters respectively. Volatile organic compounds (VOCs) can be toxic to either fish or humans as can excess metals or pesticides, all discussed qualitively in Chapter 6. The purpose of this chapter is to examine quantitively the physics of the more significant water quality parameters.

7.2 PARAMETERS OF ORGANIC CONTENT OF WATER QUALITY

The organic content of a wastewater effluent or a streamwater is determined using either of the following tests:

- BOD_5—biochemical oxygen demand
- COD—chemical oxygen demand
- TOC—total organic carbon

When an organic waste is discharged to a stream, the organic content of the effluent undergoes a biochemical reaction, i.e. assisted by micro-organisms, as follows:

$$\frac{\text{Organic}}{\text{matter}} + O_2 + \text{nutrients} \xrightarrow{\text{micro-organisms}} \frac{\text{new}}{\text{biomass}} + CO_2 + H_2O + \frac{\text{stable}}{\text{products}} \tag{7.1}$$

This is an oxidation reaction and consumes O_2 from the water body. If the oxygen demand of the waste (BOD) is high enough, it may deplete the O_2 and the worst-case scenario is an anaerobic water body. BOD is defined as the amount of oxygen required by living organisms in the stabilization of the organic matter of a water/wastewater. As the saturation concentration of O_2 in water at $20\,^{\circ}C$ is $9.2\,mg/L$ in the BOD test, dilution of a wastewater sample with BOD-free, oxygen-saturated water is necessary if the BOD of the wastewater is $> 3\,mg/L$. A sample of wastewater is diluted with a seeded water and is placed in an airtight bottle, measuring the concentration of DO at day 0 and again at day 5. The difference in DO is the BOD_5. The standard test uses a $300\,ml$ BOD bottle and incubation is performed at $20\,^{\circ}C$ over 5 days in a light-excluded environment. The various dilutions of 1, 5, 10, 20, 50, 100, etc., are now readily available as given in Standard Methods (1992). The BOD test is fully described in Standard Methods (1992) and readers are recommended to read it.

Example 7.1 The results from a BOD test, diluted by 100, are given in columns 1 and 2 in the table below. Compute the BOD.

$$BOD_5 = p(DO_I - DO_f) \tag{7.2}$$

where
$$p = \text{dilution of the sample in the BOD bottle, (e.g. 100)}$$

$$DO_I, DO_f = \text{the initial and final DO concentrations}$$

Solution   Column 3 contains the BOD volumes for the seeded sample.

Time (days)	Diluted sample DO (mg/L)	BOD (calculated) (mg/L)
0	7.95	—
1	3.75	420
2	3.45	450
3	2.75	520
4	2.15	580
5	1.80	615

$$BOD_5 = p(DO_1 - DO_5)$$
$$= 100(7.95 - 1.8) = 615 \text{ mg/L}$$

In addition to the oxygen demand by the wastewater sample, the seeded dilution water is likely to exert some oxygen demand of its own, however small. Therefore, if the above method were used, uncorrected, the BOD_5 would be too high. This is accounted for in the corrected BOD_5 defined as follows:

$$BOD_5 = p[(DO_1 - DO_f) - (B_1 - B_f)f] \tag{7.3}$$

where p = dilution factor

B_1, B_f = initial and final DO concentrations of the seeded diluted water (blanks)

f = ratio of seed DO in sample to seed in blanks

$$= \frac{\% \text{ seed in } DO_1}{\% \text{ seed in } B_1}$$

Example 7.2   Use Example 7.1 with corresponding DO values for the blank to determine the corrected BOD_5.

Solution   Column 3 contains the given values for the DO of the blank and column 5 the corrected BOD values.

Time (days)	Diluted sample DO (mg/L)	Blank seeded sample DO (mg/L)	BOD (Example 7.1) (mg/L)	BOD corrrected (mg/L)
0	7.95	8.15	—	—
1	3.75	8.1	420	415
2	3.45	8.05	450	440
3	2.75	8.00	520	505
4	2.15	7.95	580	560
5	1.80	7.9	615	590

$$f = \frac{\% \text{ seed in } DO_1}{\% \text{ seed in } B_1} = \frac{99\%}{100\%} = 0.99$$
$$BOD_5 = p[(DO_1 - DO_5) - (B_1 - B_5)f]$$
$$= 100[(7.95 - 1.80) - (8.15 - 7.90) \times 0.99]$$
$$= 590 \text{ mg/L}$$

In discussing water quality, we consider the rate of decomposition of organic matter to be proportional to the amount of organic matter available. This relationship is formulated as a continuous first-order reaction:

$$\frac{dL_t}{dt} = -K_1 L_t \tag{7.4}$$

where
$$L_t = \text{BOD, i.e. BOD remaining, mg/L}$$
$$K_1 = \text{BOD deoxygenation rate coefficient, day}^{-1}$$

Integrating, therefore, between L_o and L_t, we get

$$L_t = L_o e^{-K_1 t} \tag{7.5}$$

where
$$L_o = \text{BOD}_u, \text{ i.e. ultimate BOD (or ultimate carbonaceous BOD)}$$
$$\text{(or initial BOD of effluent at point of discharge to a stream)}$$

With reference to Fig. 7.1, there is

$$\text{BOD}_t = \text{BOD}_u - \text{BOD}_r \tag{7.6}$$

when
$$\text{BOD}_t = \text{BOD exerted at time } t \text{ (or oxygen demand)}$$
$$\text{BOD}_r = \text{BOD remaining at time } t$$
$$\text{BOD}_t = L_o - L_t$$
$$= L_o - L_o e^{-K_1 t}$$

Therefore

$$\text{BOD}_t = L_o (1 - e^{-K_1 t}) \tag{7.7}$$

It is important to note that the BOD_5 is not the ultimate BOD_u and BOD_5 is always less than BOD_u.

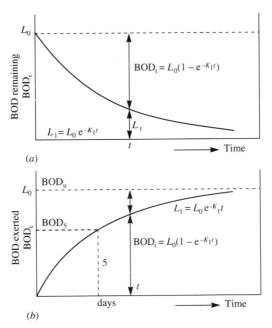

Figure 7.1 (a) Idealized BOD_r, (b) idealized BOD_t.
(a) Idealized BOD_r. The organic sample has an initial BOD $= L_o$, and when exposed to an oxygen source, the BOD decreases exponentially.
(b) Idealized BOD_t. The oxygen demand by microbes in the process of stabilization or organic content of water or wastewater sample.

Table 7.1 Typical K_1 values

Environment	K_1 (to base e)
Untreated wastewater	0.35–0.7
Treated wastewater	0.10–0.25
Polluted river	0.10–0.25

Example 7.3   If Example 7.1 has a deoxygenation rate coefficient of 0.15 day^{-1} determine the ultimate CBOD (carbonaceous BOD) i.e. BOD_u or L_o.

Solution

$$BOD_t = L_0(1 - e^{-K_1 t})$$

For $t = 5$ days,
$$BOD_5 = L_0(1 - e^{-0.15 \times 5})$$

Therefore

$$L_o = \frac{BOD_5}{1 - e^{-0.15 \times 5}}$$

$$= \frac{615}{0.527}$$

and so
$$L_o = 1165 \text{ mg/L}$$

Table 7.1 gives values for a range of different BOD constants. The effect of higher rate constants is to achieve higher BOD_5 values, but note that the BOD_u is independent of the rate (see Fig. 7.2).

Temperature has an effect on K_1 and is related by

$$K_1 = K_{20}\theta^{(T-20)} \tag{7.8}$$

where
$$K_T = \text{the rate at temperature } T(^\circ C)$$
$$K_{20} = \text{the rate at } 20\,^\circ C \text{ (known)}$$
$$\theta = \text{coefficient} \approx \begin{cases} 1.047 & \text{for } 20\,^\circ C < T < 30\,^\circ C \\ 1.35 & \text{for } 4\,^\circ C < T < 20\,^\circ C \end{cases}$$

When a carbonaceous waste biodegrades, it exerts an oxygen demand. Note that BOD is often synonomous with CBOD (i.e. the carbonaceous biochemical oxygen demand). In addition there can also be an oxygen demand exerted by the oxidation of nitrogenous compounds. The nitrogen cycle is

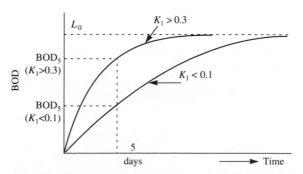

Figure 7.2 Influence of K_1 values on BOD.

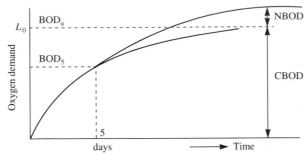

Figure 7.3 Relationship of CBOD and NBOD.

discussed in Chapters 3 and 10. Suffice to say here that ammonia is oxidized to nitrite and nitrite is oxidized to nitrate in the following microbial assisted reactions:

$$2NH_3 + 3O_2 \xrightarrow{\text{Nitrosomonas}} 2NO_2{}^- + 2H^+ + 2H_2O \tag{7.9}$$

$$2NO_2{}^- + O_2 \xrightarrow{\text{Nitrobacter}} 2NO_3{}^- \tag{7.10}$$

These two reactions are termed nitrification. The oxygen demand associated with the oxidation of ammonia to nitrate is called the nitrogenous biochemical oxygen demand (NBOD). Typically the NBOD occurs later than 5 days and does not show up in the CBOD test. This is sketched in Fig. 7.3.

Example 7.4 From the data of Example 7.1, determine the total oxygen demand if the sample contains 25 mg/L of nitrogen.

Solution

From Example 7.1: $BOD_5 = 615$ mg/L

From Example 7.3: $BOD_u = 1165$ mg/L

Nitrification :

$$\underset{\substack{1\ \text{mole} \\ 17\text{g/mole}}}{NH_3} + \underset{\substack{2\ \text{moles} \\ 16\text{g/mole}}}{2O_2} \longrightarrow \underset{1\ \text{mole}}{NO_3{}^-} + \underset{1\ \text{mole}}{H^+} + \underset{1\ \text{mole}}{H_2O}$$

Therefore

1 mole of NH_3 requires 2 moles of O_2

and

$$NBOD = 25 \times \frac{2 \times 32}{14} = 114 \text{ mg/L}$$

Thus

$$\text{Total oxygen demand} = CBOD + NBOD$$
$$= 1165 + 114 = 1279 \text{ mg/L}$$

If the concentration of nitrogen as organic nitrogen and ammonia nitrogen is known then it is possible to directly determine the NBOD. It is

$$NBOD = TKN \times \frac{2 \times 32}{14} \tag{7.11}$$

Therefore

$$NBOD = TKN \times 4.57$$

where

$$TKN = \text{total Kjeldahl nitrogen}$$

i.e.

Each 10 mg N/L exerts a 45.7 mg O_2/L oxygen demand

Hence the importance to perform nitrification of wastewaters as discussed in Chapter 12.

Chemical oxygen demand (COD) is a much used parameter, particularly as it can be completed in about 2 h. (using the titrimetric or colorimetric method) and also because it takes into account those organics that are not readily biodegradable. COD is a measure of the total organic carbon with the exception of certain aromatics such as benzene which are not completely oxidized in the reaction. Some organic materials such as cellulose, phenols, etc., resist biodegradation along with pesticides and PCBs which are toxic to micro-organisms and these materials are oxidized in the COD test. A chemical oxidizing agent is used instead of micro-organisms as is the case in the BOD test. The result is a COD that is always higher than the associated BOD. In municipal wastewaters,

$$COD \approx 1.6 \ BOD_5$$

As such, when wastewater characteristics are consistent over time, they can be calibrated; thus, if the COD is determined we can generally infer the BOD.

7.3 DISSOLVED OXYGEN AND BIOCHEMICAL OXYGEN DEMAND IN STREAMS

The Streeter and Phelps (1925) model of the relationship of BOD and DO is still valid. They considered that when biodegradable waste was discharged to a stream or river it consumed oxygen, and the latter was only renewed by atmospheric reaeration. The process is

$$\frac{dDO}{dt} = K_1 L_1 - K_2 DO = K_1 L_o e^{-K_1 t} - K_2 DO \tag{7.12}$$

where
$\quad$ DO = saturation DO deficit, mg/L
$\quad$ or the difference between saturation
$\quad$ (maximum DO) and the actual DO, in mg/L.

$\quad L_t =$ CBOD, mg/L

$\quad K_1 =$ deoxygenation rate, day^{-1}

$\quad K_2 =$ reaeration rate, day^{-1}

The reader is referred to Chapter 21 for a fuller mathematical description. The solution of Eq. (7.12) is

$$DO(t) = \frac{K_1 L_o}{K_2 - K_1} (e^{-K_1 t} - e^{-K_2 t}) + DO_o \ e^{-K_2 t} \tag{7.13}$$

where
$\quad L_o =$ oxygen demand at $t = t_o$, (or the BOD)

$\quad DO_o =$ dissolved oxygen deficit at $t = t_o$

$\quad DO(t) =$ dissolved oxygen saturation deficit at any time t

Equation (7.13) is the Streeter–Phelps oxygen sag formula and is sketched in Fig. 7.4.

From Chapter 21, the time (and also the distance downstream) of the occurrence of the minimum DO or the maximum dissolved oxygen deficit DO_c is obtained by differentiating Eq. 7.13 with respect to t and setting it equal to zero. Then

$$t_c = \frac{1}{K_2 - K_1} \ln \left\{ \frac{K_2}{K_1} \left[1 - \frac{DO_o(K_2 - K_1)}{K_1 L_o} \right] \right\} \tag{7.14}$$

and

$$DO_c = \frac{K_1}{K_2} L_o e^{-K_1 t} \tag{7.15}$$

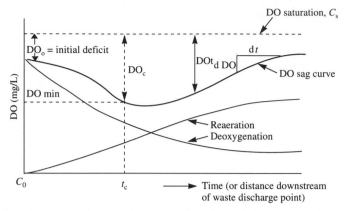

Figure 7.4 Dissolved oxygen sag curve in a river downstream of an effluent discharge point located at *x*.

Equation (7.15) is obtained by setting Eq. (7.12) equal to zero. The maximum oxygen deficit is also obtained by setting the value obtained from Eq. (7.14) into Eq. (7.13).

Example 7.5 A wastewater with a BOD of 25 mg/L is discharged through an outfall to a freshwater stream of mean velocity 0.1 m/s. The DO upstream of the outfall is 8.5 mg/L. Assuming the deoxygenation rate $K_1 = 0.25$ d^{-1} and the reaeration rate $K_2 = 0.4$ d^{-1}, determine (a) the time and distance downstream (x_c) where DO is a minimum and (b) the minimum DO.

Solution

Saturation DO = 9.2 mg/L

Initial DO = 8.5 mg/L

Initial deficit DO$_o$ = 9.2 − 8.5 = 0.7 mg/L

(a) The time at which DO$_o$ is a minimum from Eq. (7.15) is

$$t_c = \frac{1}{K_2 - K_1} \ln\left\{\frac{K_2}{K_1}\left[1 - \frac{DO_o(K_2 - K_1)}{K_1 L_o}\right]\right\}$$

$$= \frac{1}{0.4 - 0.25} \ln\left\{\frac{0.4}{0.25}\left[1 - \frac{0.7(0.4 - 0.25)}{0.25 \times 25}\right]\right\}$$

Therefore

$$t_c = 3.02 \text{ days}$$

$$x_c = \bar{U}t_c = 0.1 \times 3.02 \times 24 \times 3600 \times 10^{-3} = 26.1 \text{ km}$$

(b) The maximum deficit DO$_c$ from Eq. (7.15) is

$$DO_c = \frac{K_1}{K_2} L_o e^{-K_1 t}$$

$$= \frac{0.25}{0.4} \times 25 \times e^{-0.25 \times 3.02}$$

Therefore

$$DO_c = 7.3 \text{ mg/L}$$

Thus the minimum value of DO is

$$DO_{min} = DO_{sat} - DO_c = 9.2 - 7.3 = 1.9 \, mg/L$$

(which is too low to support even coarse fish life and would be considered seriously polluting).

7.4 TRANSFORMATION PROCESSES IN WATER BODIES

Constituents in water bodies are subjected to a range of transformation processes. Those pertaining to a specific constituent will depend on whether the constituent is conservative (which means the constituent does not react in water, e.g. metals) or non-conservative. The constituents of key interest in water quality are:

- DO and BOD (mg/L)
- Temperature (°C)
- Salinity (mg/L of Cl)
- Algae as chlorophyll-a
- Nitrogen as organic N
- Nitrogen as ammonia N
- Nitrogen as nitrate N
- Nitrogen as nitrite N
- Phosphorus as organic and dissolved
- Coliforms
- Dissolved solids or salts
- Metals
- Organics

The transformation processes in water bodies include:

- Influent 'clean' flows improve DO
- Influent 'waste' flows disimprove DO by adding BOD
- Biological oxidation of carbonaceous and nitrogenous organic matter
- Reaeration of surface layers—oxygen source
- Reduction of BOD by solids settling (sedimentation)
- Photosynthesis—oxygen source
- Respiration—oxygen sink
- Oxygen diffusion into benthic zone—SOD (sediment oxygen demand)—oxygen sink
- Increase in BOD of resuspended benthic sediments
- Bacterial cell mass decay
- Volatilization (VOCs)
- Chemical oxygen demands (organic chemicals)
- Adsorption (chemical constituents)

In addition to the transformation processes, the transportation processes also impact on water quality constituents. The key transportation processes are:

- Advection
- Diffusion
- Buoyancy

The general one-dimensional conservation of mass equation as described in Chapter 21 for a water body is

$$\frac{\partial C}{\partial t} = \underbrace{\frac{\partial}{\partial x}\left(D_L \frac{\partial C}{\partial x}\right)}_{\text{dispersion}} - \underbrace{\frac{\partial (UC)}{\partial x}}_{\text{advection}} \pm \underbrace{\sum S}_{\text{source/sink}} \tag{7.16}$$

where
C = concentration of a constituent (mg/L)

D_L = longitudinal dispersion coefficient (m^2/s)

U = mean of longitudinal velocity (m/s)

S = source or sink terms

To evaluate the fate of a constituent, information is required about the specific transformation and transport processes that it is subject to. Probably the most commonly studied constituent is DO as it is this constituent that determines the quality of water body and its ability or inability to support fish life. The following sections examine some of the transformation and transportation processes. The longitudinal dispersion coefficient is discussed in detail in Fischer *et al.* (1979).

7.4.1 Influent 'Clean' and 'Waste' Flows

The quantity of oxygen in an incoming flow is typically specified as an initial condition in Eq. (7.16) as it was in Example 7.5 ($DO_I = 8.5$ mg/L). The waste flow BOD is also typically specified as an initial condition, again as in Example 7.5 ($BOD_I = 25$ mg/L). For instance, in lake water quality it may be that the incoming river flows have a higher DO_I and contribute to raising the level of DO in the lake. Water quality in lakes is discussed briefly in a later section and in Chapter 21.

7.4.2 Biological Oxidation of Carbonaceous and Nitrogenous Organic Matter

Sometimes this transformation is simply called deoxygenation. The rate of deoxygenation is assumed to be a first-order reaction process; i.e. the rate of deoxygenation is proportional to the remaining BOD:

$$r_o = K_1 L_t \tag{7.17}$$

where
r_o = deoxygenation rate constant, d^{-1}

L_t = BOD remaining after time t, mg/L

Also
$$L_t = L_o e^{-K_1 t}$$

L_o = BOD of initial mixture of wastewater and streamwater

Therefore
$$r_o = K_1 L_o e^{-K_1 t} \tag{7.18}$$

If a pollutant of flow rate Q_w and BOD strength L_w is discharged to a river which has a flow rate Q_s and a BOD strength L_s, then the initial BOD strength L_o of the mixture is determined from mass balance.

$$L_o = \frac{Q_s L_s + Q_w L_w}{Q_s + Q_w}$$

where
$$Q_s = \text{flow in stream, m}^3/\text{s}$$
$$Q_w = \text{flow of wastewater discharge, m}^3/\text{s}$$
$$L_s = \text{BOD of stream, upstream of discharge, mg/L}$$
$$L_w = \text{BOD of wastewater discharge, mg/L}$$

There is no distinction made in the above equations between the rates for CBOD and NBOD. The rate K_1 is a 'lumped' parameter, deemed adequate in many cases. Values for K_1 range from 0.1–$0.25\,\text{d}^{-1}$ for polluted river to 0.35–$0.7\,\text{d}^{-1}$ for untreated wastewater (see Table 7.1).

7.4.3 Reaeration of Surface Layers of Water Bodies

As in deoxygenation, the rate of reaeration is assumed to be a first-order reaction, dependent on the current DO deficit, D, as

$$r_R = K_2 D \tag{7.19}$$

where
$$D = \text{DO deficit} = \text{DO}_s - \text{DO, mg/L}$$
$$\text{DO}_s = \text{DO saturation value, mg/L (9.2 mg/L at 20\,°C)}$$
$$\text{DO} = \text{actual DO, mg/L}$$
$$K_2 = \text{reaeration constant, d}^{-1}$$

The rate K_2 may vary over one order of magnitude, typically slow moving lakes with K_2 as low as 0.1 and riffle areas of fast moving rivers with K_2 values ~ 1.0. Typical values of K_2 are reproduced in Table 7.2.

O'Connor and Dobbins (1958) suggested a generalization for K_2 of

$$K_2 = \frac{3.9\bar{U}^{1/2}}{H^{3/2}} \tag{7.20}$$

where
$$\bar{U} = \text{mean longitudinal velocity, m/s}$$
$$H = \text{water depth, m}$$

Readers are referred to O'Connor and Dobbins (1958) for further details and also Tchobanoglous and Schroeder (1987).

Example 7.6   If a stream flows at 30 m wide × 2.0 m deep with a velocity of 0.1 m/s, compute its reaeration rate constant. If the river now goes into flood, again at 30 m wide but now at 3 m deep and an average velocity of 2.0 m/s, what is its new reaeration rate constant?

Table 7.2 Typical reaeration constants, K_2

Environment	Range of K_2 at 20°C (d^{-1})
Small ponds and backwaters	0.1–0.23
Sluggish streams, large lakes	0.23–0.35
Low velocity rivers	0.35–0.46
Average velocity rivers	0.46–0.69
Fast rivers	0.69–1.15
River rapids	> 1.15

Adapted from Tchobanoglous and Schroeder, 1987

Solution   Using

$$K_2 = \frac{3.9 \bar{U}^{1/2}}{H^{3/2}}$$

(a)
$$K_2 = \frac{3.9 \times 0.1^{1/2}}{2.0^{3/2}} = 0.436 \text{ d}^{-1}$$

(b)
$$K_2 = \frac{3.9 \times 2.0^{1/2}}{3.0^{3/2}} = 1.06 \text{ d}^{-1}$$

7.4.4 Reduction of BOD by Solids Settling (Sedimentation)

Suspended solids discharged in wastewater to streams may eventually settle to the bottom of the river or estuary. In doing so they reduce the contaminant load in the receiving water. The rate of settlement is directly proportional to the concentration of suspended solids and to the settling velocity:

$$r_s = \frac{v}{H} C_{ss} \tag{7.21}$$

where
r_s = rate of settling by suspended solids, mg/L day

v = settling velocity, m/day

C_{ss} = concentration of suspended solids, mg/L

H = depth of flow, m

As settling velocities are very low, laboratory tests are often used and values are of the order of 0.1 to 10 m/day (Metcalf and Eddy, 1991). The contribution of sedimentation to the mass balance equation (7.16) is not often included.

7.4.5 Photosynthesis and Respiration

If phytoplankton algae are present in a water body (typically lakes but sometimes slow moving rivers or estuaries) they produce oxygen (photosynthesis) during daylight hours and use up oxygen (respiration) continuously. This is shown schematically in Fig. 7.5.

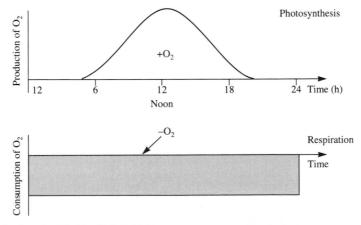

Figure 7.5 Photosynthesis and respiration of phytoplankton.

The growth of phytoplankton is dependent on the nutrient input, and for low nutrient supply the photosynthesis production of O_2 can be in balance with the respiration consumption of O_2. However, higher levels of nutrients lead to excess phytoplankton growth and a net consumption of oxygen. The photosynthesis–respiration requirement for O_2 has been described by Orlob (1981) as

$$O_2 = K_b(\mu_p P)\gamma V - K_b(r_p P)V$$
$$\underbrace{}_{\text{photosynthesis}} \quad \underbrace{}_{\text{respiration}}$$

(7.22)

where
V = volume (of lake or water body), m^3

O_2 = oxygen concentration, mg/L

K_b = biota activity coefficient, dimensionless

P = phytoplankton concentration, mg/L

μ_p = growth coefficient, d^{-1}

r_p = respiration coefficient, d^{-1}

γ = stoichiometric oxygenation factor for algae growth, mg/mg

Some of these parameters are quantified in manuals of computer programs, e.g. QUAL2 (as described in Chapter 21). If phytoplankton photosynthesize oxygen during the day, in a water environment which does not have a significant oxygen deficit, then the oxygen produced may achieve saturation levels and the excess is lost to the atmosphere. It is in this situation that a net loss of oxygen may occur, although typically photosynthesis produces about 1.5 times the oxygen required for respiration. O'Connor (1960) suggested the following expression for oxygen release during photosynthesis:

$$P(t) = \frac{P_m \sin t}{p\pi}, \qquad 0 \le t \le P$$

(7.23)

where
$P(t)$ = rate of oxygen release

P_m = maximum rate of release over the period p

Under bloom conditions Nemerow (1985) reported that from 0.5 to 0.96 g of O_2 could be produced by photosynthetic action per square metre per day.

7.4.6 Oxygen Diffusion into the Benthic Zone—SOD

This is often referred to as the sediment oxygen demand (SOD) and occurs when wastewater organics settle on the river bottom and biodegrade, thereby representing another dissolved oxygen sink. The zero-order rate model of SOD is

$$r_{SOD} = \frac{K_{SOD}}{H}$$

(7.24)

where
r_{SOD} = rate of O_2 consumption, mg/L day

K_{SOD} = oxygen uptake in sediment, mg/m^2 day

H = depth, m

In areas of significant sediments, the SOD may be a significant sink in the oxygen mass balance. Table 7.3 reports typical K_{SOD} values.

7.4.7 Modified Streeter–Phelps Oxygen Sag Model

Section 7.3 discussed the original oxygen sag curve from Streeter and Phelps (1925) based on the two processes of BOD oxidation and surface reaeration. This section extends that development to include

Table 7.3 Typical K_{SOD} values

Description of sediment	K_{SOD}†(mg/m² day)
Municipal sludge near outfall	− 4
Aged municipal sludge downstream of outfall	− 1.5
Estuarine mud	− 1.5
Sandy bed	− 0.5
Mineral bed	− 0.07

† Note the different units of K_{SOD} and K_1 and K_2.
Adapted from Tchobanoglous and Schroeder, 1987

for SOD and other parameters of significance. The Streeter and Phelps model assumes a constant pollution load discharged at a single point in a river with constant flow rate and uniform cross-section. This of course is not very real as typical wastewater BOD loads vary with time (diurnal) and a river may be subject to a series of discharge points with loads of different magnitude. Also, the flow rates of rivers vary with time and cross-sections are rarely uniform or constant. In the Streeter and Phelps model, the lateral and vertical DO and BOD concentrations are assumed uniform throughout each section, i.e. essentially a plug flow assumption. Also the deoxygenation and reaeration rates are assumed first order with constant rates. This may be too simplistic for delicate water bodies. The many additional potential processes are listed in Sec. 7.4 (besides deoxygenation and reaeration), some of which are included in more detailed one-dimensional models (such as QUAL2) or in two-dimensional models. Recall that Eq. (7.12) for dissolved oxygen is

$$\frac{dDO}{dt} = \underset{\text{deoxygenation}}{K_1 L} \underset{\text{reaeration}}{- K_2 DO} \tag{7.12}$$

Considering that some of the BOD may be removed by sedimentation, Orlob (1983) writes for BOD,

$$\frac{dL}{dt} = -(K_1 + K_s)L \tag{7.25}$$

where K_s = rate constant for BOD removal by sedimentation, d^{-1}

Thus

$$L_t = L_0 e^{-(K_1 + K_s)t} \tag{7.26}$$

and $$DO(t) = \frac{K_1 L_0}{K_2 - (K_1 + K_s)}[e^{-(K_1 + K_s)t} - e^{-K_2 t}] + DO_0 e^{-K_2 t} \tag{7.27}$$

In the above two equations, which are modifications to Eqs. (7.12) and (7.13), it is assumed that no DO is drawn from the sediment and that there is no photosynthesis (Orlob, 1983).

Further modifications by Camp (1963) for the BOD and DO profile equations include

$$\frac{dL}{dt} = -(K_1 + K_s)L + \underset{\text{BOD from sediment}}{B} \tag{7.28}$$

$$\frac{dDO}{dt} = \underset{\text{deoxygenation}}{K_1 L} \underset{\text{reaeration}}{- K_2 DO} \underset{\text{photosynthesis}}{- P} \tag{7.29}$$

where B = the addition of BOD to the water body from the benthic deposits, mg/L day

The analytic solution to Eqs (7.28) and (7.29) is

$$L(t) = \left(L_0 - \frac{B}{K_1 + K_s}\right)e^{-(K_1 + K_s)t} + \frac{B}{K_1 + K_s} \tag{7.30}$$

and

$$DO(t) = \frac{K_1}{K_2 - (K_1 + K_s)}\left(L_0 - \frac{B}{K_1 + K_s}\right)[e^{-(K_1 + K_s)t} - e^{-K_2 t}]$$

$$+ \frac{K_1}{K_2}\left(\frac{B}{K_1 + K_s} - \frac{P}{K_1}\right)(1 - e^{-K_2 t}) + DO_0 e^{-K_2 t} \tag{7.31}$$

If the BOD is exerted in the water body by the benthic deposits but there is no BOD reduction by settling, (i.e. $K_s = 0$), Eqs (7.30) and (7.31) reduce to

$$L(t) = \left(L_0 - \frac{B}{K_1}\right)e^{-K_1 t} + \frac{B}{K_1} \tag{7.32}$$

$$DO(t) = \frac{K_1}{K_2 - K_1}\left(L_0 - \frac{B}{K_1}\right)(e^{-K_1 t} - e^{-K_2 t})$$

$$+ \frac{B - P}{K_2}(1 - e^{-K_2 t}) + DO_0 e^{-K_2 t} \tag{7.33}$$

An almost endless series of modifications can be accounted for, even including the mixing process. The reader is referred to the concise descriptions by Orlob (1983).

Example 7.7   Consider the case in Example 7.5. If in addition to deoxygenation and reaeration, assume the process of BOD removal by sedimentation is occurring with $K_s = -0.05\,d^{-1}$ determine:

(a) the equation for the time to attain DO_{min},
(b) the actual time to attain DO_{min},
(c) the actual distance to attain DO_{min},
(d) the maximum DO deficit.

Solution

(a) $DO(t) = \dfrac{K_1 L_0}{K_2 - (K_1 + K_s)}(e^{-(K_1 + K_s)t} - e^{-K_2 t}) + DO_0 e^{-K_2 t}$

For convenience set $K_1 + K_s = K_1^*$. Then

$$\frac{dDO}{dt} = \frac{K_1 L_0}{K_2 - K_1^*}(-K_1^* e^{-K_1^* t} + K_2 e^{-K_2 t}) - K_2 DO_0 e^{-K_2 t}$$

$$\frac{dDO}{dt} = 0$$

(i.e. where DO is a minimum, $t = t_c$

$$\frac{K_1 L_0}{K_2 - K_1^*}(-K_1^* e^{-K_1^* t} + K_2 e^{-K_2 t}) = K_2 DO_0 e^{-K_2 t}$$

$$\frac{K_1 L_0}{K_2 - K_1^*}\left(-\frac{K_1^*}{K_2}e^{(K_2 - K_1^*)t} + 1\right) = DO_0$$

$$e^{(K_2 - K_1^*)t} = -\frac{DO_0(K_2 - K_1^*)}{K_1 L_0} \frac{K_2}{K_1^*} + \frac{K_2}{K_1^*}$$

$$e^{(K_2 - K_1^*)t} = \frac{K_2}{K_1^*}\left[1 - \frac{DO_0}{L_0}\left(\frac{K_2 - K_1^*}{K_1}\right)\right]$$

Taking the natural log (ln) of both sides:

$$(K_2 - K_1^*)t = \ln \frac{K_2}{K_1^*}\left[1 - \frac{D_0}{L_0}\left(\frac{K_2 - K_1^*}{K_1}\right)\right]$$

Therefore

$$t_c = \frac{1}{K_2 - K_1^*} \ln \frac{K_2}{K_1^*}\left[1 - \frac{D_0}{L_0}\left(\frac{K_2 - K_1^*}{K_1}\right)\right]$$

replacing K_1^* by $K_1 + K_s$. Therefore

$$t_c = \frac{1}{K_2 - (K_1 + K_s)} \ln \frac{K_2}{K_1 + K_s}\left[1 - \frac{D}{L_0}\left(\frac{K_2 - K_1 + K_s}{K_1}\right)\right] \tag{7.34}$$

(b) $t_c = \dfrac{1}{0.4 - (0.25 - 0.05)} \ln \dfrac{0.4}{0.25 - 0.05} \left\{1 - \dfrac{0.7}{25}\left[\dfrac{0.4 - (0.25 - 0.05)}{0.25}\right]\right\}$

$= 6.82$ days (> 3.02 days of Example 7.5)

(c) The distance to location of maximum deficit, assuming $\bar{U} = 0.1$ m/s is

$$X_c = \bar{U}t_c$$
$$= 0.1 \times 6.82 \times 3600 \times 24 \times 10^{-3} = 58.9 \text{ km}$$

(d) $DO_c(t_c = 6.82) = \dfrac{K_1 L_0}{K_2 - (K_1 + K_s)}\{e^{-(K_1 + K_s)t} - e^{-K_2 t}\} + DO_0 e^{-K_2 t}$

$$= \frac{0.25 \times 25}{0.4 - (0.25 - 0.05)}(e^{-0.3 \times 6.82} - e^{-0.4 \times 6.82}) + 0.7e^{-0.4 \times 6.82}$$

Therefore $DO_c = 4.04$ mg/L

i.e. the maximum deficit is 4.04 mg/L and saturation O_2 is 9.2 mg/L, so there is still a dissolved oxygen level of 5.16 mg/L, enough to support coarse fish life but not enough to support salmonoids, so allowing BOD removal by sedimentation makes the water quality viable.

7.5 TRANSPORT PROCESSES IN WATER BODIES

The key transport processes (of a solute) in water bodies, be it a river, a lake or an estuary are:

- Advection (by the mean velocity of the water body) and
- Diffusion (molecular and turbulent)

The general one-dimensional conservation of mass equation as described in Chapter 21 for a water body is

$$\frac{\partial C}{\partial t} = \underbrace{\frac{\partial}{\partial x}\left(D_L \frac{\partial C}{\partial x}\right)}_{\text{dispersion}} - \underbrace{\frac{\partial}{\partial x}(UC)}_{\text{advection}} \pm \sum S$$

The previous equations for DO/BOD assumed rates of reaction only with no transport. If longitudinal mixing is also considered then the DO/BOD profiles are more complex. Camp (1963) described the profiles for a constant BOD source as:

BOD profile:

$$D_L \frac{d^2 L}{dx^2} - U \frac{dL}{dx} - (K_1 + K_s)L + B = 0 \qquad (7.35)$$

DO profile:

$$D_L \frac{d^2 DO}{dx^2} - U \frac{dDO}{dx} + K_2(DO_s - DO) - K_1 L + P + 0 \qquad (7.36)$$

where
D_L = dispersion or longitudinal turbulent mixing coefficient, m^2/s
U = average stream velocity, m/s
DO = dissolved oxygen, mg/L
DO_s = dissolved oxygen saturation, mg/L

These equations were further simplified by O'Connor (1960) who considered longitudinal mixing as insignificant, lateral mixing as significant to the dilution process. Further details are developed in the modelling in Chapter 21.

7.6 OXYGEN TRANSFER BY INTERPHASE TRANSFER IN WATER BODIES

While O_2 is the gas of key interest in water quality, other gases including nitrogen, methane, hydrogen sulphide and VOCs etc., are of interest also. The ease or difficulty by which a gas becomes absorbed by water depends principally on the solubility of the gas in water. Ammonia, which is highly soluble in water, readily becomes absorbed by the water, while oxygen or carbon dioxide, which are weakly soluble in water, are less readily absorbed by the water (refer to Chapter 3 for solubility coefficients). There are also gases that are somewhere between being strongly soluble and weakly soluble in water; determination of the mass transfer for these gas types is more complex than that of the other two.

Where a gas interacts with a liquid, there is an assumed gas/liquid interface. On the gas side of the interface there is a thin 'gas film', within which there is no convective mixing. On the bulk gas side of this thin film, rapid convective mixing is assumed. This is shown schematically in Fig. 7.6. In considering a specific gas, there is a partial pressure gradient across the thin gas film, represented as $P_g - P_i$, where P_i is the partial pressure of the gas at the interface, and P_g is the partial pressure of the 'bulk gas'. In the same way, on the liquid side of the interface, there is a thin 'water film', within which there is no convective mixing. In the bulk water phase outside this water film, there is rapid convective mixing once the concentration of, say, O_2 in water is constant. Across the water film, there is a gradient of concentration of, say, O_2, represented by $C_i - C_L$, where C_i is the concentration (of O_2) at the interface and C_L the concentration (of O_2) in the water body ($C_i > C_L$).

The basis of this theory of gas absorption is the two-film model as detailed by Lewis and Whitman (1924). The assumptions are essentially:

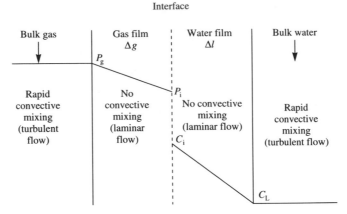

Figure 7.6 Schematic of air–water interface.

1. Steady state conditions exist.
2. Laminar flow exists in the gas and water films.
3. Instantaneous equilibrium is established at the interface.

The mechanism of gas transfer is assumed to be purely by diffusion (no turbulence). The rate of gas transfer is thus dependent on the resistance of either the gas film or the water film to diffusion. Diffusion through the gas film (from the bulk gas towards the interface) goes at a rate proportional to the solute (O_2 in air) concentrations in the bulk gas and in the gas film. Diffusion through the liquid water film is controlled by the difference in concentration between C_i and C_L. In the gas film, there are much fewer molecules of gas by comparison with the tightly packed water molecules in the water film. The diffusional resistance encountered in the liquid film is greater than that in the gas film.

According to Lewis and Whitman (1924), the amount of solute (O_2 in air) absorbed per unit time by diffusion through the two films is

$$\frac{1}{A}\frac{dW}{dt} = k_g(P_g - P_i) = k_L(C_i - C_L) \tag{7.37}$$

where
$\quad W$ = weight of solute, mg = CV

$\quad A$ = interfacial area, m^2

$\quad k_g$ = gas diffusion coefficient (or mass transfer coefficient)

$\quad k_L$ = water diffusion coefficient

$\quad P_g, P_i$ = partial pressure of solute in bulk gas and at the interface, atm

$\quad C_L, C_i$ = solute concentration in bulk water and at the interface, mg/L

Therefore

$$kg(P_g - P_i) = k_L(C_i - C_L) \tag{7.38}$$

From Henry's law, the interface concentration C_i is at equilibrium with the gas-phase partial pressure P_i:

$$C_i = HP_i \tag{7.39}$$

where
$\quad H$ = Henry's law constant, mg/L atm

For highly soluble gases in water (e.g. ammonia)

$$\frac{1}{A}\frac{dW}{dt} = k_g P_g \tag{7.40}$$

since the concentration in the gas phase at the interface is very small as the gradient across the gas film is very large. As such, once the gas (ammonia) reaches the interface, it is rapidly sucked through the liquid film. The gas film is thus the controlling limitation.

For weakly soluble gases in water, e.g. oxygen,

$$\frac{1}{A}\frac{dW}{dt} = k_L(C_g - C_L) \tag{7.41}$$

where C_i of Eq. (7.37) is replaced by C_g. This is so because there is very little partial pressure gradient across the gas film because $P_g \cong P_i$. Therefore the value of C_i is the same as that of a liquid saturated with oxygen (or other weakly soluble gas) at P_g and may be expressed as C_g and in this case $k_L = K_L$ as defined by Eq. (7.44).

For intermediately soluble gases,

$$\frac{1}{A}\frac{dW}{dt} = K_g(P_g - P_i) = K_L(C_g - C_i) \tag{7.42}$$

where K_g, K_L are overall coefficients, $K_L \leq k_L$, $K_g \leq k_g$:

$$K_g = \frac{Hk_2 k_g}{Hk_2 + k_g} \tag{7.43}$$

$$K_L = \frac{K_g}{H} \tag{7.44}$$

The time rate of solution of a gas in water or the flux of gas into water is then (O_2, weakly soluble)

$$\frac{dC}{dt} = K_L a(C_g - C_t) = K_2(C_g - C_t) \tag{7.45}$$

Intergrating, we get $\qquad \ln\dfrac{C_g - C_o}{C_g - C_t} = K_L$ at $= K_2 t \tag{7.46}$

$$a = \frac{A_i}{V}$$

where
$$C_g = \text{saturation concentration, mg/L}$$
$$C_o = \text{initial concentration, mg/L}$$
$$C_t = \text{concentration after time } t, \text{mg/L}$$
$$A_i = \text{interface area, m}^2$$
$$V = \text{volume of liquid, m}^3$$
$$K_L = \text{liquid film mass transfer coefficient, d}^{-1}$$

In relation to reaeration of water bodies, the reaeration coefficient K_2 (often quoted as Ka) is

$$Ka = K_2 = K_L a = K_L \frac{A_i}{V} = \frac{K_L}{H} \tag{7.47}$$

where
$$H = \text{depth of water body}$$

A wide range of estimates for Ka are to be found in Orlob (1983). Typical values were quoted in Table 7.1. Equation (7.46) becomes

$$\frac{C_g - C_o}{C_g - C_t} = e^{-K_2 t} \tag{7.48}$$

Therefore

$$C_t = C_g(1 - e^{-K_2 t}) + C_o e^{-K^2 t} \tag{7.49}$$

Example 7.8   If the saturation concentration of O_2 in water is 9.2 mg/L and the initial concentration C_o is 2.1 mg/L, determine the time it takes for the concentration C_t to become 7.5 mg/L if K_2 is $0.25/day^{-1}$. What is the time required for C_t to be 5.0, 6.0, 7.0, 8.0 and 9.0 mg/L?

Solution

$$\ln\frac{C_g - C_o}{C_g - C_t} = K_2 t$$

$$t = \frac{1}{K_2}\ln\frac{C_g - C_o}{C_g - C_t}$$

For $C_t = 7.5$ mg/L :

$$t = \frac{1}{0.25}\ln\frac{9.2 - 2.1}{9.2 - 7.5} = 5.7 \text{ days}$$

Similarly,

C_t (mg/L)	Time t (days)
5	2.1
6	3.2
7	4.8
7.5	5.7
8	7.1
8.5	9.3
9	14.3

Therefore, if left to molecular diffusion alone, it is seen that several days are required for reaeration or reoxygenation to occur. However, in most riverine systems, there is water flow and so some element of turbulent diffusion occurs in addition to molecular diffusion. So the realistic reaeration time is significantly less than the above would indicate.

7.7 TURBULENT MIXING IN RIVERS

If the flow in a river or water body is turbulent, it will actualize much more aeration through turbulent mixing than it will through Fickian diffusion. However, diffusion may be the prime 'aerator' of quiescent water bodies. In rivers, mixing can be imagined as occurring in the vertical, lateral (horizontal) and longitudinal axes. For instance, Fischer *et al.* (1979), suggested the following mixing coefficients:

Vertical:

$$\varepsilon_v = 0.067 \, du^* \tag{7.50}$$

Transverse:

$$\varepsilon_t \cong 0.15 \, du^* \tag{7.51}$$

$$\cong (0.01 - 0.2)\, du^* \text{ for straight reaches}$$
$$\cong (0.4 - 0.8)\, du^* \text{ for real channels}$$

Longitudinal:

$$\varepsilon_{\mathrm{L}} > 0.15\, du^* \qquad\qquad (7.52)$$

where

$$d = \text{depth of water body}$$

and

$$u^* = \text{shear velocity} \approx \sqrt{gds} \qquad\qquad (7.53)$$

where

$$s = \text{channel bed slope}$$
$$g = \text{gravity}$$

The above coefficients vary, depending on the channel type or estuary. Fischer *et al.* (1979) also indicated that for spreading plumes the concentration at the centre-line can be approximated by

$$C_{\max} = \frac{QC_0}{Ud} \frac{1}{\sqrt{4\pi\varepsilon_{\mathrm{t}} x/\overline{U}}} \qquad\qquad (7.54)$$

where

$$Q = \text{flow rate, m}^3/\text{s}$$
$$C_{\mathrm{o}} = \text{initial concentration, ppm}$$
$$x = \text{distance downstream}$$
$$\overline{U} = \text{mean velocity, m/s}$$

Fischer *et al.* (1979) also give the downstream length for complete mixing for a discharge at the centre-line of a river estuary as

$$L \cong 0.1\, UW^2/\varepsilon_{\mathrm{t}} \qquad\qquad (7.55)$$

where

$$W = \text{the width of the river}$$

For complete mixing, for a discharge from the sidewall of a stream, the length L is twice that given in Eq. (7.55).

Example 7.9   Determine the shear velocity, the transverse mixing coefficient and the downstream length of complete mixing for a sidewall discharge of wastewater, of $0.5\,\mathrm{m}^3/\mathrm{s}$ with an initial BOD concentration of $30\,\mathrm{mg/L}$, if the bed slope is 0.001 and the depth $\cong 5\,\mathrm{m}$. Also determine the maximum concentration $1.5\,\mathrm{km}$ downstream if the river is meandering and has a width of $100\,\mathrm{m}$ and a mean velocity of $0.5\,\mathrm{m/s}$.

Solution

$$\text{Shear velocity } u^* = \sqrt{gds}$$
$$= \sqrt{9.81 \times 5 \times 0.001} = 0.22 \text{ m/s}$$

$$\text{Transverse mixing coefficient } \varepsilon_{\mathrm{t}} \cong 0.6\, du^*$$
$$= 0.6 \times 5 \times 0.22 = 0.66 \text{ m}^2/\text{s}$$

$$\text{Downstream length for complete mixing } L \cong 0.2\bar{U}W^2/\varepsilon_{\mathrm{t}}$$
$$\cong 0.2 \times 0.5 \times 100^2/0.66$$
$$\cong 1.5 \text{ km}$$

$$\text{Maximum concentration at 1.5 km downstream } C_{\max} = \frac{QC_0}{Ud} \frac{1}{\sqrt{4\pi\varepsilon_t x/\overline{U}}}$$

$$= \frac{0.5 \times 30}{0.5 \times 5} \frac{1}{\sqrt{4\pi \times 0.66 \times 1500/0.5}}$$

$$C = 0.04 \text{ mg/L}$$

The reader is encouraged to refer to Fischer *et al.* (1979) for an excellent treatise on mixing in inland and coastal waters.

7.8 WATER QUALITY IN LAKES AND RESERVOIRS

Water quality in lakes and reservoirs was discussed in Chapter 5 from an ecological perspective. This section introduces some of the numerical concepts and further modelling concepts are dealt with in Chapter 21. Water quality problems in reservoirs are often due to cultural eutrophication which is caused by:

- Municipal sewage discharges
- Industrial wastewater discharges
- Urban runoff
- Agricultural runoff with natural or artificial fertilizers causing high nutrient loading
- Biocides from aquaculture

In most developed countries the first three items are generally insignificant due to the wealth of environmental legislation that encouraged the upgrading of treatment facilities. However, in the 1990s, nutrient-rich agricultural runoff is by far the most common problem left to deal with. It is a difficult problem as the sources are diffuse and solution techniques, by way of changes to land use practices, are slow and complex to implement. *In situ* lake solutions via mechanical aeration are sometimes used in small lakes. Lakes and reservoirs are characterized by long residence times of maybe 3 months to tens of years. In the latter case, the inflows have the insignificant flushing effect on water quality. The inflow if they are rich in nutrients is undesirable. The key physical parameters affecting water quality in lakes are:

- Wind movements
- Temperature changes
- Inflows/outflows

Examining the main lake parameters over the annual cycle typically in the northern hemisphere we have (see Fig. 7.7):

- Reduced inflows in summer months with corresponding reduced water depths
- Increased solar radiation in the summer months with corresponding elevated water temperatures, particularly closer to the surface
- Reduced dissolved oxygen values in the summer months, due to increased temperature and reduced depths.

Three distinct zones can be set up in a lake and these are sketched in Fig. 7.8. The summer temperature profile is most profound; in the spring and autumn, the vertical temperature profile will vary between both the winter and summer extremes. Water at great depths is usually $4\,^\circ\text{C}$ where the density is $1000\,\text{kg/m}^3$. The deeper depths are called the hypolimnion, where the temperature and density do not change throughout the year. The surface volume, known as the epilimnion, is very dependent on solar radiation. In the winter, the temperatures may go to freezing, as shown in Fig. 7.8. In summer, the epilimnion

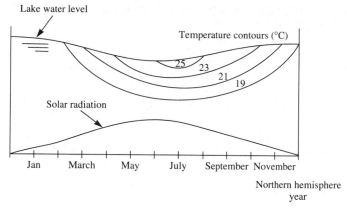

Figure 7.7 Schematic of trends in water level, temperature and solar radiation over the year in a moderately quiescent lake.

temperatures may go up to 20°C. Between the epilimnion and hypolimnion there is a depth of water where a significant temperature gradient occurs, particularly in summer.

Lakes are defined with respect to stratification by their densiometric Froude number as

$$\text{Froude number } F = \frac{V}{\sqrt{Dg}} = \frac{\text{inertial force}}{\text{gravitational force}} \qquad (7.56)$$

$$\text{Densimetric } F = F_D = \frac{V}{\sqrt{\dfrac{\Delta\rho}{\rho_o}D_g}} \qquad (7.57)$$

where

$$\rho_o = \text{reference density}$$
$$\Delta\rho = \text{density change over depth } D$$
$$(\text{if } \Delta\rho > 0.01 \text{ g/m}^3 \Rightarrow \text{strong stratification})$$
$$\text{If } F_D > 0.32 \Rightarrow \text{no stratification}$$
$$0.01 < F_D < 0.32 \Rightarrow \text{moderately stratified}$$
$$F_D < 0.01 \Rightarrow \text{strongly stratified}$$

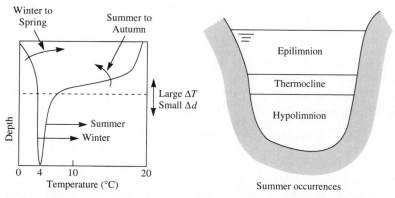

Figure 7.8 Temperature profiles in stratified lake.

Lakes are also defined with respect to stability by their Richardson number:

$$N_R = \frac{PE}{KE}$$
$$= \frac{-\frac{1}{2}g(\Delta\rho/\Delta Z)\Delta Z}{\frac{1}{2}\rho(\Delta u^2/\Delta Z)}$$
$$= \frac{-g(\Delta\rho/\Delta Z)}{\rho(\Delta u^2/\Delta Z^2)} \qquad (7.58)$$

where $\qquad$ PE = potential energy

and $\qquad$ KE = kinetic energy

If $\dfrac{\Delta\rho}{\Delta Z} = 0 \Rightarrow$ neutrally stable (or meta stable)

If $\dfrac{\Delta\rho}{\Delta Z} < 0 \Rightarrow$ stable

If $\dfrac{\Delta\rho}{\Delta Z > 0} \Rightarrow$ unstable

where $\qquad$ ΔZ is a change in height and u is a mean velocity

Example 7.10 Determine the stratification category for a lake if its length by width by depth is $10\,km \times 2\,km \times 25\,m$. The summer discharge is $10\,m^3/s$. The surface temperature in the summer is 25 °C.

Solution

$$V = \frac{Q}{BD} = \frac{10}{2000 \times 25} = 2 \times 10^{-4}\ m/s$$

$$\rho_{surface} = 997\ kg/m^3$$

$$\rho_0 = 1000\ kg/m^3$$

Therefore $\quad F_D = \dfrac{V}{\sqrt{(\Delta\rho/\rho_0)Dg}} = \dfrac{2 \times 10^{-4}}{\sqrt{[(1000-997)/1000]25 \times 9.81}} = 2.3 \times 10^{-4} \ll 0.01$

Therefore this lake is strongly stratified (i.e. $F_D < 0.01$).

The water quality in lakes has a trophic state depending on the value of several parameters, as indicated in Table 7.4. Chlorophyll-a, one of the green pigments involved in photosynthesis, is related to the total phosphorus concentration by

$$\log(\text{chlorophyll-a}) = -1.09 + 1.46P$$

where chlorophyll-a is concentration in mg/L and P is the total phosphorous concentration in mg/L. It can be seen from Table 7.4 that there is not a wide variation in several of the parameters between

Table 7.4 Trophic lake quality

Parameter	Oligotrophic	Mesotrophic	Eutrophic
Total P, µg/L	< 10	10–20	> 20
Chlorophyll-a, µg/L	< 4	4–10	> 10
Seechi depth, m	> 4	2–4	< 2
Hypolimnion O_2 % saturation	> 80	10–80	< 10

Table 7.5 Diffusion coefficients in lakes

Diffusion type	Diffusion coefficient (cm^2/s)
Eddy diffusion	10^{-2}–10^6
Molecular diffusion	10^{-5}–10^{-4}
Thermal diffusion	10^{-8}–10^{-6}

oligotrophic and eutrophic. Chlorophyll-a and, and total P are parameters of significant interest. Models of lake dynamics usually incorporate total P (or other P components) and chlorophyll-a.

The diffusion in lakes varies significantly and the magnitudes can be identified as in Table 7.5.

7.8.1 Simple Phosphorus Balance in a Lake

As phosphorus is one of the most common limiting nutrients for lake eutrophication, many research papers have looked at enumerating its significance (Vollenwerder, 1975; Fischer *et al.*, 1979; Imberger, 1982; and Havis and Ostendorf, 1989). A simple mass balance of phosphorus is sketched in Fig. 7.9 and is given below:

mass rate in − mass rate out − mass of P settling in lake + mass generation = rate of accumulation

$$Q_{in}C_{pin} \quad - \quad Q_{out}C_p - V_sA_sC_p \quad + \quad 0 \quad = \quad \frac{dM}{dt}$$

$$(7.59)$$

In this case, assume there is no generation of phosphorus within the lake. Also assuming steady state conditions $\frac{dM}{dt} = 0$. The phosphorus concentration leaving the lake is assumed to be equal to the phosphorus concentration in the lake (well mixed) and the outflow rate is assumed to be equal to the inflow rate; therefore equation (7.59) becomes

$$QC_{p_{in}} = QC_p + V_sA_sC_p$$
$$= C_p(Q + V_sA_s)$$

Lake P (concentration) $C_p = \dfrac{QC_{p_{in}}}{Q + V_sA_s}$ $\quad(7.60)$

Example 7.11 A $20\,km^2$ lake (surface area) has an inflow from streams of $10\,m^3/s$ with $C_{p_{in}} = 0.01$ mg/L. A municipal wastewater plant discharges its treated effluent with a flow rate of $0.05\,m^3/s$ and a $C_{p_{in}}$ of 10 mg/L. Determine the steady state concentration of phosphorus in the lake. Assume the settling rate $V_s \sim 20\,m/year$ ($0.6 \times 10^{-6}\,m/s$). Determine the permissible wastewater influent concentration if the maximum C_p allowable in the lake is 0.01 mg/L.

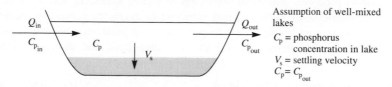

Figure 7.9 Mass balance of lake phosphorus.

Solution

$$\text{Equation (7.60) } C_p = \frac{QC_{P_{in}}}{Q + V_s A_s}$$

$$Q = 10 + 0.05 = 10.05 \text{ m}^3/\text{s}$$

$$QC_{P_{in}} = 10 \times 0.01 + 0.05 \times 10 = 0.6 \text{ g/s}$$

$$V_s A_s = 0.6 \times 10^{-6} \times 20 \times 10^6 = 12 \text{ m}^3/\text{s}$$

Therefore

$$C_p = \frac{0.6}{10.05 + 12} = 0.027 \text{ g/m}^3$$

$$= 0.027 \text{ mg/L} > 0.01 \text{ mg/L}$$

The wastewater influent concentration needs to be reduced in P concentration:

$$QC_{P_{in}} = C_p(Q + V_s A_s)$$

$$= 0.01(10.05 + 12) = 0.2205 \text{ g/s}$$

$$0.2205 = 10 \times 0.01 + 0.05 \times C_{P_{in}}$$

Therefore

$$C_{P_{in}} = 2.41 \text{ mg/L}$$

Thus the acceptable wastewater discharge concentration of P should be less than 2.41 mg/L.

7.9 GROUNDWATER QUALITY

As discussed in Chapter 4, groundwater is a major resource in terms of freshwater. Groundwater typically moves slowly, its rate of movement being defined by the hydraulic conductivity K which was indicated in Table 4.19. Coarse gravel has a K value of $\sim 6.4 \times 10^{-3}$ m/s while that of shale is $\sim 1.2 \times 10^{-12}$ m/s. Thus groundwater moves at rates substantially different for different bed material. The rate of movement is also complicated by the fact that K is three dimensional, K_x, K_y, K_z, and these may vary as material is rarely homogeneous. Further complications arise from the fact that heterogeneity is ubiquitous and K varies from one field to the next. The determination of groundwater movement and the associated movement of contaminants can be studied by simplistic models or extremely complex models and these are briefly introduced in Chapter 21.

Groundwater may be contaminated by either point sources or diffuse sources. Typically in rural farmyards where there are water wells, there is always the possibility of wastewater runoff and other agricultural contaminants entering the groundwater. The spreading of fertilizers and pesticides on land are potential diffuse sources of pollution. Accidental spillages of oil or leaking oil tanks in petroleum depots is a potential source of point pollution. When the pollution has occurred it is the task of the engineer and the scientist to quantify the pollutant and track its progress. Modelling the future plume development may then lead to remedial measures.

The mass transport processes determine the extent of plume spreading and the geometry of the distribution of concentration. Some pollutants may attenuate or aggregate depending on the biological, chemical or nuclear processes occurring. The transport process is essentially advection, with diffusion and/or hydrodynamic dispersion being insignificant. The magnitude and direction of transport are governed by:

- The three-dimensional hydraulic conductivity, K
- The water table and its gradient

- The existence of sources (underground streams) or sinks (limestone caverns)
- The shape of the flow domain

The most simple model of decay of a contaminant plume is to assume the decay to be a first-order reaction, i.e. $r = -kC$. Thus,

$$C_t = C_o e^{-kt} \tag{7.61}$$

where
    $C_t =$ concentration at time t, mg/L

    $C_o =$ concentration at time zero, mg/L

    $t =$ time, d

    $k =$ first-order decay coefficient, usually determined from site investigation

Example 7.12   A confined sandstone aquifer of 5 per cent porosity, 15 m thick, has two monitoring wells in it, spaced at 1 km. If the hydraulic conductivity is 5.8×10^{-7} m/s and the gradient between the two wells is 10 m, determine the time it takes a contaminant plume to travel from the upstream well to the downstream well.

Solution

$$\text{Gradient} \frac{dh}{dL} = \frac{10}{1000} = 0.01$$

$$\text{Flow rate } q = Kh \frac{dh}{dL}$$

$$= 5.8 \times 10^{-7} \times 15 \times 10^{-2}$$

$$q = 87 \times 10^{-9} \text{ m}^2/\text{s}$$

$$\text{Darcy superficial velocity } v = K \frac{dh}{dL}$$

$$= 5.8 \times 10^{-7} \times 10^{-2}$$

$$= 5.8 \times 10^{-9} \text{ m/s}$$

$$\text{Average linear or pore velocity } v_L = \frac{5.8 \times 10^{-9}}{0.05} = 116 \times 10^{-9} \text{ m/s}$$

$$\text{Travel time } t = \frac{s}{v_L} = \frac{1000}{0.116} \times 10^{+6} \text{ s}$$

Therefore
$$t = 274 \text{ years}$$

As such there is not too much concern about contaminant movement in this highly impermeable strata.

Example 7.13   For Example 7.12, determine the concentration of a contaminant plume at the downstream well if the upstream concentration is 80 mg/L. Assume a decay coefficient $k \sim 10^{-4} \text{ d}^{-1}$

Solution

$$C_t = C_o e^{-kt}$$

$$kt = 10^{-4} \times 274 \times 365 = 10.0$$

Therefore
$$C_t = 80 \, e^{-10.0}$$

$$= 0.004 \text{ mg/L}$$

The one-dimensional dispersion advection equation as described in Chapter 21, for a contaminant plume is

$$\frac{\partial C}{\partial t} = D_x \frac{\partial^2 C}{\partial x^2} - v_x \frac{\partial C}{\partial x}$$

The Ogata-Banks (1961) solution for the one-dimensional case, presented in more detail in Chapter 21, is

$$\frac{C_t}{C_o} = \frac{1}{2}\left[\text{erfc}\left(\frac{x - \bar{v}t}{\sqrt{4Dt}}\right) + \exp\left(\frac{\bar{v}x}{D}\right)\text{erfc}\left(\frac{x + \bar{v}t}{\sqrt{4Dt}}\right)\right] \tag{7.62}$$

where

C_t = concentration at time t, mg/L

C_o = concentration at time zero, mg/L

x = distance, m

T = time, days

D = hydrodynamic dispersion, m/s.

The reduced form is

$$\frac{C_t}{C_o} = \frac{1}{2}\text{erfc}\left(\frac{x - \bar{v}t}{\sqrt{4Dt}}\right) \tag{7.63}$$

Example 7.14   A conservative tracer is discharged through a vertical soil layer, 100 cm in depth, with a velocity of 10^{-2} cm/s. If the ratio C_t/C_o is 0.33 after 1 hour, determine the hydrodynamic dispersion, D.

Solution

$$\frac{C_t}{C_o} = \frac{1}{2}\text{erfc}\left(\frac{x - \bar{v}t}{\sqrt{4Dt}}\right)$$

$$0.33 = \frac{1}{2}\text{erfc}\left(\frac{100 - 10^{-2} \times 3600}{\sqrt{4D \times 3600}}\right)$$

$$0.66 = \text{erfc}\left(\frac{0.53}{\sqrt{D}}\right)$$

$$\text{From erfc tables} \Rightarrow \frac{0.53}{\sqrt{D}} = 0.31$$

Therefore

$$D = 2.96 \text{ m/s}$$

The reader is referred to Domenico and Schwartz (1990) and Bear and Verruijt (1992) for further reading.

7.10 PROBLEMS

7.1 Define the water quality parameters BOD, COD, DO and NBOD, using figures where necessary to explain your answer.

7.2 The results of a BOD test, diluted by 10, are given below. Determine the BOD_5.

Time (days)	DO of diluted sample (mg/L)
0	8.9
1	4.7
2	4.4
3	3.7
4	3.1
5	2.8

7.3 For Problem 7.2 determine the corrected BOD_5 if the corresponding blank seeded sample DO is given as follows:

Time (days)	DO blank seeded sample (mg/L)
0	9.05
1	9.0
2	8.95
3	8.9
4	8.85
5	8.8

7.4 Explain with the aid of a sketch the relationship between BOD_5 and BOD_u.

7.5 If the sample of Problem 7.2 contains 15 mg/L of nitrogen, determine the total oxygen demand.

7.6 Sketch and explain the dissolved oxygen sag curve.

7.7 A wastewater with a BOD of 10 mg/L is discharged through an outfall into a freshwater stream. The mean velocity of the stream (at low flow) is 0.2 m/s. The upstream DO is 8.3 mg/L. Assuming a deoxygenation rate of $k_1 = 0.20$ d^{-1} and a reaeration rate of 0.35 d^{-1}, determine:
  (a) the time and distance downstream when the DO is a minimum,
  (b) the maximum DO.

7.8 Describe and quantify how you would engineer the outfall to have minimum impact on the stream dissolved oxygen values.

7.9 Write out the conservation of mass equation in one-dimension as it pertains to a pollutant discharged into a river environment. Explain the terms *diffusion*, *dispersion* and *advection*.

7.10 Review the paper 'Mechanisms of reaeration in natural streams', by O'Connor and Dobbins (1958).

7.11 Review the paper 'A study of the pollution and natural purification of the Ohio River', by Streeter and Phelps (1925).

7.12 For Problem 7.7, if in addition to the deoxygenation and reaeration processes there is also a BOD sedimentation rate of $K_s = -0.07$ y^{-1}, compute:
  (a) the time to attain the minimum DO,
  (b) the distance to minimum DO,
  (c) the value of the maximum DO deficit.

7.13 Determine the vapour density of water in air at $25\,^\circ$C.

7.14 Explain the physical chemical phenomenon of using a water tower to absorb ammonia into water from an ammonia pollution air source. Write the mass balance equation for a structure with influent and effluent gas and influent and effluent water. The influent gas and effluent water are at the bottom of the tank while the others are at the top.

7.15 Review the paper 'Principles of gas absorption' by Lewis and Whitman (1924).

7.16 Explain the difference between molecular diffusion and turbulent diffusion. Give examples of when each process is true.

7.17 Determine the maximum concentration 2.5 km downstream of a sidewall discharge of concentration 25 mg/L and flow rate $0.02\,\mathrm{m^3/s}$. Assume the river is straight with a width of 50 m and a low flow velocity of 0.25 m/s.

7.18 Explain with sketches the terms *epilimnion*, *thermocline* and *hypolimnion*.

7.19 Consider a shallow lake to be a continuous flow stirred reactor. Determine the mass balance equation for a pollutant of concentration C. Solve this equation.

7.20 For Problem 7.17, determine the time required for a lake to reduce the concentration of an influent pollutant to 10 per cent of an initial concentration C_o of 250 mg/L. Assume the lake volume is $5 \times 10^5\,\mathrm{m}$ and an influent stream has a flow rate of $200\,\mathrm{m^3/s}$. Assume first-order decay with $k = 0.01\,\mathrm{day^{-1}}$.

7.21 Review the paper 'Approximate dynamic lake phosphorous budget models' by Havis and Ostendorf (1989).

7.22 Explain with sketches, where necessary, the physical, chemical and biological processes involved in the movement of a pollutant through the vadoze zone.

7.23 If the water wells spaced at 5 km in a 20 m thick confined aquifer are used to monitor groundwater quality and the upstream well detects a pollutant of concentration 100 mg/L. Assuming a decay coefficient of $k \sim 10^{-4}\,\mathrm{d^{-1}}$ and a hydraulic conductivity of $10^{-3}\mathrm{m/s}$, determine the concentration in the downstream well.

REFERENCES AND FURTHER READING

American Public Health Association and American Water Works Association (1990) *Standard Methods for the Examination of Waste and Wastewater*, 17th edn.

Bear, J. and A. Verruijt (1992) *Modelling of Groundwater Flow and Pollution. Theory and Applications of Transport in Porous Media*, D. Reidal Publishing Co., Dordrecht, Holland.

Bingham, D. R. and T. H. Feng (1980) 'Mathematical modelling of recovery of a eutrophic lake', Report Env.E.65-80-1, Department of Civil Engineering, University of Massachusetts, Amherst, Mass.

Biswas, A. K. (1981) *Models for Water Quality Management*, McGraw-Hill, New York.

Camp, T. R. (1963) *Water and Its Impurities*, Chapman and Hall, London.

Camp, T. R. and R. L. Meserve (1963) *Water and Its Impurities*, Dowden, Hutchinson and Ross Inc.

Camp, T. R. and R. L. Meserve (1974) *Water and Its Impurities*, Dowden, Hutchinson and Ross Inc.

Casamitjana, X. and G. Schladow (1993) 'Vertical distribution of particles in a stratified lake', *ASCE Journal of Environmental Engineering*, **119**(3), May/June.

De Pinto, J. V., W. Lick and J. F. Paul (1994) *Transport and Transformation of Contaminants near the Sediment Water Interface*, Lewis Publishers.

Domenico, P. A. and F. W. Schwartz (1990) *Physical and Chemical Hydrogeology*, John Wiley, New York.

Eckenfelder, W. (1970) *Water Quality Engineering for Practicing Engineers*, Barnes and Noble Inc., New York.

Eckenfelder, W. (1989) *Industrial Water Pollution Control*, McGraw-Hill, New York.

Fischer H. B. (1976) 'Mixing and dispersion in estuaries', *Annual Review of Fluid Mechanics*, **8**.

Fischer, H. B., E. J. List, R. C. Y. Koh, J. Imberger and N. H. Brooks (1979) *Mixing in Inland and Coastal Waters*, Academic Press, New York.

Grady, W. G. (1986) 'Physics-based modelling of lakes, reservoirs and impoundments', ASCE Report.

Harbold, H. S. (1979) *Sanitary Engineering Problems and Calculations for the Professional Engineer*. Ann Arbor Science, Illinois.

Havis, R. N. and D. W. Ostendorf (1989) 'Approximate dynamic lake phosphorous budget models', *ASCE Journal of Environmental Engineering*, **115**(4), August.

Havis, R. N., *et al.* (1983) 'A mathematic model of phosphorus in completely mixed lakes with special application to Lake Warner, Mass', Report Env.Eng. 78-83-9, Department of Civil Engineering, University of Massachusetts, Amherst, Mass.

Hocking, G. C. and J. C. Patterson (1991) 'Quasi-two dimensional reservoir simulation model', *ASCE Journal of Environmental Engineering*, **117**(5), September/October.

Imberger, J. (1982) 'Reservoir dynamic modelling' in *Prediction in Water Quality*, E. M. O'Loughlin and P. Cullen (eds), Australia Academy of Science, Canberra, pp. 223–248.

Kay, D. (1992) *Recreational Water Quality Management*, Ellis Horwood, Chichester.

King, I. P. (1990) 'Modelling of flow in estuaries using combinations of one and two dimensional finite elements', *Hydrosoft*, **3**(3).

Kirby, M. J. (1978) *Hillslope Hydrology*, John Wiley, New York.

Lamb, J. C. (1985) *Water Quality and Its Control*, John Wiley, New York.

Lewis, W. K. and W. C. Whitman (1924) 'Principles of gas absorption', *J. Ind. Engng. Chem.*, **16**.

Logan, B. E. and G. A. Wagensoller (1993) 'The HBOD test: a new method for determining biochemical oxygen demand', *Water Environment Research*, **65**(7).

McGauhey, P. W. (1968) *Engineering Management of Water Quality*, McGraw-Hill, New York.

Mason, C. F. (1991) *Biology of Freshwater Pollution*, Longman Scientific and Technical, London.

Masters, G. M. (1991) *Introduction to Environmental Engineering and Science*, Prentice-Hall, Englewood Cliffs, New Jersey.

Metcalf and Eddy Inc., (1991) *Wastewater Engineering—Treatment, Disposal and Reuse*, G. Tchobanoglous and F. Burton (principal authors), McGraw-Hill, New York.

Meybeck, M., D. Chapman and R. Helmar (1990) *Global Freshwater Quality—A First Assessment*, WHO and UNEP, Blackwell, Oxford.

Miazaki, T. (1993) *Water Flow in Soils*, Marcel Dekker Inc., New York.

Ministry for the Environment (New Zealand) (1992) *Water Quality Guidelines*, No. 1.

Nemerow, N. L. (1985) *Stream, Lake, Estuary and Ocean Pollution*, Van Nostrand Reinhold, New York.

O'Connor, D. J. (1960) *Oxygen Balance of an Estuary*. J. Son. Eng. Dio ASCE, Vol 86. SA3. pp 35.

O'Connor, D. J. and W. E. Dobbins (1958) 'Mechanisms of reaeration in natural streams', *Transactions of the ASCE*, **123**, 641–666.

Ogata, A. and R. B. Banks (1961) 'A solution of the differential equation of longitudinal dispersion in porous media', USGS Professional Paper 411-a, Washington D.C.

O'Kane, P. (1980) *Estuarine Water Quality Management with Moving Element Models and Optimization Techniques*, Pitman Advanced Publishing, London.

Orlob, G. (1981) 'Models for stratified impoundments', in *Models for Water Quality Management*, A. K. Biswas (ed.), McGraw-Hill, New York.

Orlob, G. (1983) *Mathematical Modelling of Water Quality: Streams, Lakes and Reservoirs*, John Wiley, New York.

Parisod, J. P. and E. D. Schroeder (1978) 'Biochemical oxygen demand progression in mixed substrates', *Journal of Water Pollution Control Federation*, July.

Peavy, H. S., D. R. Rowe and G. T. Tchobanaglous (1985) *Environmental Engineering*, McGraw-Hill, New York.

Pedersen F. B. (1986) *Environmental Hydraulics: Stratified Flows*, Lecture Notes on Coastal and Estuarine Studies, Springer-Verlag, Berlin.

Schroeder E. (1977) *Water and Wastewater Treatment*. McGraw-Hill, New York.

Snoeyink V. L. and D. Jenkins (1980) *Water Chemistry*, John Wiley, New York.

Streeter H. W. and E. B. Phelps (1925) 'A study of the pollution and natural purification of the Ohio River', US Public Health Bulletin 146.

Tchobanoglous G. and E. Schroeder (1987) *Water Quality*, Addison Wesley, Reading, Mass.

Thanh N. C. and A. K. Biswas (1990) *Environmentally Sound Water Management*, Oxford University Press, Delhi.

Thibodeaux L. J. (1979) *Chemodynamics: Environmental Movement of Chemicals in Air, Water and Soil*, John Wiley, New York.

Thoman R. V. and J. A. Mueller (1987) *Principles of Surface Water Quality Modelling and Control*, Harper and Row, New York.

Vesilind P. A., J. J. Peirce and R. Weiner (1988) *Environmental Engineering*, Butterworths, Oxford.

Vollenwerder R. A. (1975) 'Input–output models with special reference to the phosphorous limiting concept in limnology', *Schweitz Z. Hydrol.*, **37**, 53–83.

Vreugdenhill C. B. (1989) *Computational Hydraulics. An Introduction*, Springer-Verlag, Berlin.

World Resources Institute, UNEP and UNDP (1992–3) *World Resources*, Oxford University Press.

Wrobel L. C. and C. A. Brebbia (1991) *Water Pollution, Modelling, Measuring and Prediction*, Elsevier, Amsterdam.

Wroebel L. C., T. R. Buge and J. H. Prodanoff (1989) 'A study of river pollution using the QUICKEST finite difference algorithm', *Hydrosoft*, **2**(4).

CHAPTER EIGHT

AIR POLLUTION

8.1 INTRODUCTION

The reported deaths of 4000 people from the London 'smog' in 1952 was the catalyst for the introduction of the UK Clean Air Act in 1956. In the United States, the Air Pollution Control Act was introduced in 1955. In the decades prior to the 1950s, air pollution was a problem in highly industrialized urban areas where coal was burned as an industrial and domestic fuel. The pollutants from coal include particulates (black smoke), carbon dioxide and unburnt hydrocarbons.

In 1986 the EC stated that 'several of the traditional causes of air pollution, such as smoke and particulates, were now under control in the European Community'(EC, 1987). This improvement in local urban air quality is due to the replacement of coal in industry and households by natural gas, oil and nuclear power. While great advances have been made in the reduction of 'coal'-type pollutants in the European Community and the United States, little progress has been made in Eastern Europe or other lower income countries. While coal is still a significant energy source, the pollution potential of large coal burning power stations has been reduced by the addition of wet/dry scrubbers to the exhaust gases prior to emission. Significant air pollutants still exist in the urban air environment due to transportation. Approximately 70 per cent of the total carbon monoxide pollution is due to transportation as is ~ 10 per cent of unburnt hydrocarbons and 10 per cent of nitrous oxides.

The improvement in air quality may be regarded as a great environmental success as noted by the fact that 80 per cent of the United States (urban areas) now meet the National Ambient Air Quality Standards (NAAQS). However, the South Coast Air Quality Management District (in Los Angeles) in 1990 introduced 160 rules so as to further clean up the air in the Los Angeles Basin. Of course Los Angeles is unfortunate in its basin-type topography. Continuing major problems occur and are growing with the growth of the super cities where, by the year 2000, many cities will have populations in excess of 25 million. These include Mexico City, Cairo and Beijing. According to the EC (1992a), some progress has been made towards reducing emissions of sulphur dioxide, suspended particulates, lead and CFCs (chlorofluorocarbons) at European Community level, but serious problems persist. These are particularly with greenhouse gases such as carbon dioxide, oxides of nitrogen, atmospheric ozone and methane. The concentration and combined (synergistic) effect of pollutants gives cause for concern in most towns and cities due to the increasing emissions from motor vehicles.

Broadly there are two sets of air quality standards:

- Those for ambient air quality
- Those for industrial emissions

A general rule to be used with caution is that an emission limit for criteria pollutants would be of the order of 30 times the ambient air standard. This attempts to take into account the potential of an emission to be diluted in the air environment. Clearly the ability of the ambient air to dilute an emission will depend on many factors, including ambient air quality and density of emissions. The more polluted the ambient air environment, the less capable it is of diluting an emission. The emission flow rate is also important and the higher the emission rate, the lower the emission standard.

The key references for both ambient and emission standards are:

- WHO (1987)—*Air Quality Guidelines for Europe*
- TA Luft (1987)—*Technical Instructions on Quality Control, Germany*
- EC Directives (see Table 8.1)
- USEPA (1990)—*National Air Quality Standards*

EC air pollution standards began in 1970 with emission standards for petrol vehicles. Prior to 1970, individual EC countries had their own standards. The United Kingdom had its Clean Air Act in 1956. Ireland had 'some' air quality standards since 1906 under the Alkali Act, now repealed and replaced by the Air Pollution Act of 1987. In the United States, the Air Pollution Control Act was introduced in 1955. Although it provided funding for research and not control, it was a key legislative instrument. This was followed in 1963 by the substantial legislative Clean Air Act. In 1971 the Canadian Clean Air Act was passed. In the 1990s all of the original documents have been repealed, superseded or updated as new data unfolds and new and more stringent air quality standards and emission limits are set.

Table 8.1 lists the significant Directives/Acts on air quality standards and emissions in both the European Community and the United States. For further details, the reader is referred to Chapter 2 and the References.

Table 8.2 lists the trace species concentrations in clean and polluted air. The units are given in parts per billion.

Table 8.3 lists broadly the sources of major air pollutants. It is seen that traffic and power stations are key contributors to air pollution. The trends in air pollution and the sources are covered in Sec. 8.4.

8.2 AIR POLLUTION SYSTEM

An atmospheric condition in which substances exist at concentrations higher than normal background or ambient levels is said to be polluted if it has measurable effects on humans, animals, flora or materials (e.g. acid rain on buildings). A block diagram from Seinfeld (1986) depicts an air pollution system in Fig. 8.1.

An air pollution source might be a coal burning power station. The source control might be flue gas scrubbers prior to emission. The detectors might be continuous monitors for SO_2. The receptors might be a biological indicator, i.e. lichens. When the detector or receptor shows unacceptable emissions, an automatic control response might result, i.e. reduce the emission rate. An unacceptable emission may be pursued by legislation, forcing modifications to the emission or source control, i.e. installation of activated carbon filters for the flue gas.

Table 8.1 Some EC and US regulatory items on ambient air and emission standards

EC Directives—decisions and resolutions	United States regulatory items
70/220/EEC on air pollution from petrol vehicles	1955—Air Pollution Act
72/306/EEC on air pollution from diesel vehicles	1963—Clean Air Act
75/716/EEC on sulphur content of certain liquid fuels	1965—Motor Vehicle Air Pollution Control Act
76/611/EEC and 85/210/EEC on lead in petrol	1967—Air Quality Act
Resolution—30 May 1978—on fluorocarbons in the environment	
	1970—Clean Air Act Amendments
Resolution—15 July 1980—on transboundary pollution by SO_2 and PM_{10}	
	1974—Energy Supply and Environmental Coordination Act
80/779/EEC on ambient quality and guidelines on SO_2 and PM_{10}	1977—Clean Air Act Amendments
Decision—80/372/EEC and 82/795/EEC on chlorofluorocarbon in air	USEPA regulations on implementation of the Clean Air Act include:
Decision—81/462/EEC on long-range transboundary air pollution	40 CFR Part 50—National Primary and Secondary Ambient Air Quality Standards
Decision—82/459/EEC on reciprocity of data on air pollution between Member States	40 CFR Part 53—Ambient Air Monitoring Methods
82/884/EEC on lead in air	40 CFR Part 60—New Source Performance Standards
84/360/EEC on air pollution from industrial plants	
85/203/EEC on air quality standards for NO_2	40 CFR Part 61—National Emission Standards for Hazardous Air Pollutants
Regulation—3528/86 on forest protection against air pollution	
88/77/EEC on gaseous emissions from diesel vehicles	
88/609/EEC on air emissions from large combustion plants	1990—Clear Air Act Amendments Air Toxics—National Emissions Standards for Hazardous Air Pollutants
89/369/EEC on air pollution from new municipal waste incinerators	1990—Clean Air Act Amendments
89/429/EEC on air pollution from existing municipal waste incinerators	
Proposal—19 March 1992—on incineration of hazardous waste	1992—Clean Air Act Amendments

Table 8.2 Trace species concentrations in clean and polluted air

Parameter	Concentration (ppb)		Approximate residence time
	Clean air	Polluted air	
Particulates		$>100 \, g/m^3$	
CO	120	1 000–10 000	65 days
CO_2	320 000†	400 000	15 years
SO_2	0.2–10	20–200	40 days
NO	0.01–0.6	50–750	1 day
NO_2	0.1–1	50–250	1 day
HNO_2	0.001	1–8	
HNO_3	0.02–0.3	3–50	1 day
O_3	20–80	100–500	
NH_3	1–6	10–25	20 days
CH_4	1500	2500	8–10 years
N_2O	300		10–150 years
H_2S	0.2		
Pb	$5 \times 10^{-3} \, \mu g/m^3$	$0.5–3 \, \mu g/m^3$	

† Carbon dioxide is not a pollutant in the strict sense, as its damage is to the upper stratosphere and not to the air quality at ground level.
Adapted in part from Seinfeld, 1986. Copyright © 1986. Reprinted by permission of John Wiley & Sons, Inc.

AIR POLLUTION   **337**

Table 8.3 Major air pollutants and their sources

Pollutant	Power stations	Traffic	Domestic heating	Oil refining	Quarrying, mining	Chemical, pharmaceutical	Manufacturing metals, etc.	Waste incineration	Agriculture
Particulates	✓	✓	✓		✓		✓	✓	✓
CO	✓	✓	✓					✓	
CO_2	✓	✓	✓					✓	
SO_x	✓	✓	✓	✓				✓	
NO_x	✓	✓	✓	✓				✓	
VOCs	✓	✓	✓	✓				✓	
O_3		✓							
HC	✓	✓	✓	✓		✓		✓	
Heavy metals									
Pb		✓			✓		✓	✓	
Hg	✓		✓		✓	✓	✓	✓	
Cu					✓		✓	✓	
Cd			✓		✓	✓	✓	✓	✓
Zn					✓		✓	✓	
Radionuclides†								✓	
CFCs							✓	✓	

† For example Sellafield reprocessing plant (UK).

8.3 AIR POLLUTANTS

'Primary pollutants' are those emitted by an identifiable source. The more significant of these are:

- SO_2
- CO
- NO_x
- SO_x
- Particulates
- Hydrocarbons
- Metals

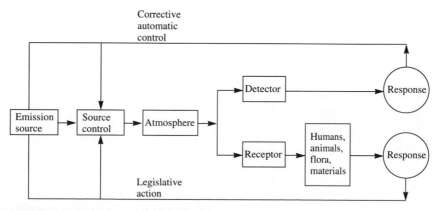

Figure 8.1 Air pollution system (adapted from Seinfeld, 1986).

'Secondary pollutants' are those formed in the atmosphere by chemical reactions and include:

- O_3
- Other photochemical oxidants—peroxyacetyl nitrate
- Oxidized hydrocarbons

The 'criteria pollutants' as defined by the USA, EC and WHO include:

- CO
- NO_2
- O_3
- SO_2
- PM-10 (particulate matter of diameter $< 10 \, \mu$m)
- Lead

Other pollutants such as some of the hydrocarbons have been dropped from the American list. All these standards undergo periodic review and the lists and the standards change. For instance, black smoke (particulate matter) was traditionally a winter problem in Dublin due to domestic coal fires. However, due to the introduction of 'natural gas' and 'smokeless coals', black smoke is no longer an air pollution problem. It is probable that in the near future there may be no need for a 'black smoke' standard in the EC. However, it is a persistent problem in Eastern Europe.

8.3.1 Ambient Air Quality Standards for Criteria Pollutants

Table 8.4 shows the Ambient Air Quality Standards for the United States, California, the European Community, WHO and Germany. Where no standard in the EC exists for a particular pollutant it is common for countries to adopt the TA Luft or WHO standard. Table 8.5 shows the properties and pollution significance of the criteria pollutants.

The criteria pollutants may be considered as likely to exist in all urban areas. Their concentrations will vary depending on the level of industrial and traffic activity and the degree of sophistication of control. Besides the six criteria pollutants, hundreds of other pollutants may also be emitted to the atmosphere. These latter tend to be industry specific and not as ubiquitous as the criteria pollutants. These 'non criteria' pollutants are discussed in Sec. 8.7.

8.3.2   Units of Concentration

The different ways of expressing concentrations of air pollutants are:

- ppm (v/v)
- ppb (v/v)
- mg/m^3
- mg/Nm^3 (Nm^3 = normal dry m^3, i.e. at STP, temp $= 0\,°C$, pressure $= 1013$ mb)

The 'normal cubic metre' is important if the emission temperature is greater than ambient as one mole of the emission gas does not occupy 22.4 litres. At non-standard temperatures and pressures, corrections have to be made.

At STP ($0\,°C$ and $101.3 \, kP_a$), 1 mole of an ideal gas occupies 22.4 L

At NSTP (non standard), 1 mole $= 22.4 \times \dfrac{T}{273\text{K}} \times \dfrac{101 \text{ kPa}}{P}$

where T is the gas temperature in degrees kelvin and P is the gas pressure in kPa

Table 8.4 International Ambient Air Quality Standards for criteria pollutants

Pollutant	Averge time	USA Federal USEPA ($\mu g/m^3$)	California ($\mu g/m^3$)	EC Directives ($\mu g/m^3$)	WHO Air quality guidelines ($\mu g/m^3$)	TA Luft, Germany ($\mu g/m^3$)
CO (carbon monoxide)	15 min				100 000	
	30 min				60 000	
	1 h	40 000	23 000		30 000	30 000
	8 h	10 000	10 000		10 000	10 000
NO_2 (nitrogen dioxide)	1 h		470	200	400	200
	24 h				150	80
	Annual	100				
O_3 (ozone)	1 h	235	180		200	
	8 h					
	24 h				65	
	100 day				60	
SO_2 (sulphur dioxide)	10 min				500	
	1 h		655		350	
	3 h	1300				400
	24 h	365	105	250–350	125	140
	Annual	80		80–120	40–60	60
Particulates (PM-10)	24 h	150	50	250	125	150–300
	Annual	50	30	80	50	
SO_2 + PM-10	24 h			100–150		
	Annual			40–60	60–90	
Pb (lead)	1 month		1.5			
	3 month	1.5				
	Annual			2	0.5–1	2
Total suspended particulates (TSP)	24 h	260				
	Annual	75				
HC	3 h	160 (non-CH_4)				

Table 8.5 Properties and pollution significance of criteria pollutants

Pollutant	Properties	Pollution significance
Carbon monoxide	Colourless, odourless gas	Formed during incomplete combustion of hydrocarbons. Causes greenhouse effects and climatic changes
Nitrogen dioxide	Brown-orange gas	Significant component of photochemical smog and acid deposition
Ozone	Highly reactive	A secondary pollutant, produced during formation of photochemical smog. Damages flora and materials
Sulphur dioxide	Colourless, choking gas, soluble in H_2O to produce sulphurous acid, H_2SO_3	Principal component of acid deposition. Damages humans, flora, fauna and materials
PM-10	Particulate matter $< 10 \, \mu m$ in diameter—black smoke	Coal burning power station, traffic, domestic coals, quarrying, incineration. Can cause respiratory problems
Lead	Heavy metal, bioaccumulative	Principal source leaded petrol. Also from lead pipes, quarrying, incineration. Damages humans and fauna when in excess

Adapted from WHO, 1987

If the gas temperature is 25°C, then 1 mole $= \dfrac{22.4 \times 298}{273} = 24.5\,\text{L}$

If the gas temperature is 1000°C, then 1 mole $= \dfrac{22.4 \times 1273}{273} = 104.5\,\text{L}$

It is therefore important to be aware that the standards are written for STP conditions.

Example 8.1 From Table 8.4, it is seen that the 1 h standards according to WHO (1987) for CO, NO_2 and SO_2 are 30 mg/m³, 400 µg/m³ and 350 µg/m³ respectively. Compute these concentrations in ppm at STP.

Solution One mole of an ideal gas at standard temperature (0°C) and pressure (101.325 kPa) occupies 22.4 litres.

$$[\text{conc}]\ \text{ppmv}\,\dfrac{m^3}{m^3} = \dfrac{L}{L} = \dfrac{[\text{conc}]g/m^3}{(\text{mol wt})\,g/\text{mol}} \times (V_{\text{ideal}} = 22.4)\dfrac{L}{\text{mole}} \times 10^{-3}$$

where   ppmv = parts per million by volume

and   ppmm = parts per million by mass $= [\text{conc}]g/m^3$

(a) mol wt of CO $= 12 + 16 = 28\,\text{g/mol}$

  Therefore

  $$V_{\text{CO}} = \dfrac{30 \times 10^{-3}g/m^3}{28\ g/\text{mol}} \times 22.4 \times 10^{-3}m^3/\text{mol}$$
  $$= 24 \times 10^{-6}g/g = 24\ \text{ppm}$$

(b) mol wt of $NO_2 = 14 + 2 \times 16 = 46\,\text{g/mol}$

  Therefore

  $$V_{\text{NO}_2} = \dfrac{400 \times 10^{-6}\ g/m^3}{46\ g/\text{mol}} \times 22.4 \times 10^{-3}m^3/\text{mol}$$
  $$= 195 \times 10^{-9}g/g = 195\ \text{ppb} = 0.195\ \text{ppm}$$

(c) mol wt of $SO_2 = 32 + 2 \times 16 = 64\,\text{g/mol}$

  Therefore

  $$V_{\text{SO}_2} = \dfrac{350 \times 10^{-6}g/m^3}{64\ g/\text{mol}} \times 22.4 \times 10^{-3}m^3/\text{mol}$$

  $$= 0.125\ \text{ppm} = 125\ \text{ppb}$$

8.4 CRITERIA POLLUTANTS

8.4.1 Carbon Monoxide—CO

Carbon monoxide is a colourless, odourless, tasteless gas that is the most abundant of the criteria pollutants with a per capita per annum emission in excess of 100 kg. It is a product of incomplete combustion of carbonaceous fuels, giving CO instead of CO_2. About 70 per cent of all CO comes from mobile sources (see Tables 8.6 and 8.7), with practically all of that from motor vehicles. It has adverse effects on human health, replacing oxygen in the bloodstream and forming carboxyhemoglobin (COHb). If the percentage of COHb exceeds about 2 per cent, health is temporarily impaired, and this level occurs in people engaged in heavy physical activity if the ambient CO level is greater than about 30 ppm. It has been shown by Petersen and Allen (1982) that subjects in moving motor vehicles were exposed to far higher levels of CO than fixed site monitors of ambient air CO indicate. Table 8.6 shows the total CO emissions on a per capita basis. This indicates that EC levels are at approximately 100 kg per capita, while those in the United States are approximately 300 kg per capita. Table 8.7 shows the total annual CO emitted from mobile and stationary sources. In the European Community and the United States, mobile sources still account for about 70 per cent of all CO. These tables also show that while the United States, Germany and the Netherlands reduced their output by approximately half from 1970 to 1980, both the United Kingdom and Ireland have increased their output. Figure 8.2 shows the trend of five countries from 1970 to 1989. In 1984 the United States still had almost three times the per capita load of the EU countries. It is interesting to note that levels since 1980 have not shown any improvement.

Natural CO production is about 25 times that of the anthropogenic sources. The major source of natural CO is the oxidation of methane in the troposphere. Thus the cycles of CO and CH_4 are interdependent.

8.4.2 Nitrogen Oxides—NO_x

The oxides of gaseous nitrogen include:

- NO—nitric oxide
- NO_2—nitrogen dioxide
- NO_3—nitrogen trioxide
- N_2O— nitrous oxide
- N_2O_5—nitrogen pentoxide

The acids of nitrogen include:

- HNO_2—nitrous acid
- HNO_3—nitric acid

Table 8.6 Total carbon monoxide emissions per capita

| Year | Carbon monoxide (kg CO/capita/year) | | | | |
	USA	West Germany	Netherlands	UK	Ireland
1970	495	240	148	87	—
1975	389	226	140	83	122
1980	350	195	100	86	146
1981	337	175	91	86	145
1982	311	162	87	89	142
1983	318	151	85	89	135
1984	303	152	84	90	131
1985	250	146	80	94	131
1986	268	148	78	98	—
1987	262	143	76	103	129
1988	260	141	76	106	—
1989	244	133	78	114	—

Data from EC, 1992b with permission

Table 8.7 Total carbon monoxide emissions—mobile (M) and stationary (ST) sources

						Carbon monoxide (1000 t CO)					
	USA		West Germany		Netherlands		UK		Ireland	
Year	M	ST	M	ST	M	ST	M	ST	M	ST
1970	74 400	27 000	8 920	5620	1490	438	3097	1747	—	—
1975	65 000	19 100	10 152	3835	1495	423	3508	1157	331	57
1980	56 100	23 500	8 813	3193	1043	369	3896	933	420	77
1981	55 400	22 100	7 768	3001	938	360	3938	893	421	79
1982	52 900	19 400	7 355	2620	919	331	4109	884	408	85
1983	52 400	22 100	6 900	2394	896	321	4161	854	384	88
1984	50 600	21 200	6 746	2577	859	346	4335	750	366	98
1985	47 900	21 800	6 314	2580	806	357	4431	887	355	107
1986	44 600	19 400	6 599	2416	780	353	4658	877	—	—
1987	43 300	20 900	6 539	2238	765	352	5074	818	388	119
1988	41 200	23 800	6 477	2194	768	358	5355	785	—	—
1989	40 000	20 900	6 100	2172	795	357	5792	730	—	—

Data from EC, 1992b with permission

NO_x are produced during the combustion of fossil fuels i.e. oil, coal, timber and gas, via two processes. 'Fuel NO_x' are produced in the oxidation of nitrogen containing compounds in the fuel. Negligible amounts of fuel nitrogen exist in natural gas but up to 3 per cent by weight of nitrogen compounds may exist in coals and oils. In liquid fuels, fuel NO_x contributes 80 to 90 per cent of the total NO_x for fuel nitrogen contents of 1 to 2 per cent (Martin *et al.*, 1979). 'Thermal NO_x' are produced in the oxidation of atmospheric molecular N_2 at high temperatures of combustion in the presence of oxygen (Seinfeld, 1986). Most of the NO_x emissions are in the form of NO which rapidly oxidizes to NO_2 in the presence of O_2 or O_3 according to:

$$2NO + O_2 \leftrightharpoons 2NO_2 \tag{8.1}$$

$$NO + O_3 \rightarrow NO_2 + O_2 \tag{8.2}$$

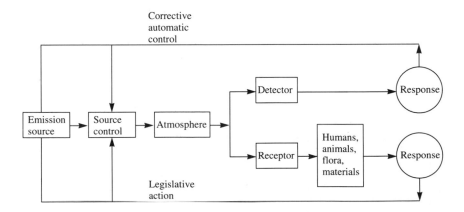

Figure 8.2 Total carbon monoxide emissions per capita, 1970–1989.

NO_2 is heavier than air and is soluble in water. NO_2 may in turn dissociate to NO or be further oxidized to HNO_3 or HNO_2 according to:

$$2NO_2 + H_2O \rightarrow HNO_3 + HNO_2 \tag{8.3}$$
$$3NO_2 + H_2O \rightarrow 2HNO_2 + NO + O_2 \tag{8.4}$$

Furthermore, NO_2 may react with organic compounds to produce peroxyacetyl nitrates (PAN) or with hydrocarbons in the presence of sunlight to produce smog:

$$HC + NO_x + sunlight \rightarrow photochemical\ smog \tag{8.5}$$

Significant health and environmental effects are caused by NO_x. NO_2 can cause respiratory problems. NO and NO_2 can cause smog, which can cause pulmonary and bronchial diseases. PAN during smog can cause eye irritation while ozone can produce respiratory effects. Ozone is acutely damaging to crops, inhibiting crop yield. Table 8.8 shows the NO_2 emissions per capita per annum. It is seen that values in the European Community are about 40 kg (per capita, per year) by comparison with 80 kg in the United States. Table 8.9 shows that the NO_2 emissions are related to mobile and stationary sources. Here it is seen that about 40 to 70 per cent comes from mobile sources and the values for all countries has been relatively stable since about 1970. Figure 8.3 is a plot of the per capita emissions, which have changed little since 1970.

Example 8.2 Compute the annual production of NO_x from the 50 000 vehicles in Cork City, if the NO_x emission rate is 2.0 g/km per vehicle.

Solution Assume the annual travel is 20 000 km per vehicle. Then

$$Each\ vehicle\ produces\ 20\,000 \times 2\,g = 40\,kg\ NO_x$$
$$50\,000\ vehicles\ produce\ 40 \times 50\,000\,kg = 2000\ tonnes\ NO_x$$

Note With 1 million vehicles in Ireland travelling an average of 20 000 km per year, this accounts for an annual mobile NO_x production of 40 000 tonnes, which is of the order of the 1987 figure for Ireland as shown in Table 8.9.

Table 8.8 Total nitrogen dioxides emissions per capita

		Carbon monoxide (kg CO/capita/year)			
Year	USA	West Germany	Netherlands	UK	Ireland
1970	89	39	35	45	—
1975	89	42	34	43	19
1980	90	48	39	43	20
1981	89	47	39	42	19
1982	84	46	38	41	18
1983	81	47	38	41	18
1984	83	48	38	41	17
1985	83	48	38	42	19
1986	83	49	39	44	—
1987	78	48	39	45	32
1988	80	47	40	46	—
1989	—	44	37	47	—

Data from EC, 1992b with permission

Table 8.9 Nitrogen dioxides emissions—mobile (M) and stationary (ST) sources

| | Nitrogen dioxide (1000 t NO$_2$) | | | | | | | | | |
| | USA | | West Germany | | Netherlands | | UK | | Ireland | |
Year	M	ST	M	ST	M	ST	M	ST	M	ST
1970	7700	10 600	1059	1322	211	244	943	1567	—	—
1975	9000	10 200	1308	1263	258	206	997	1430	16	44
1980	9300	11 100	1604	1376	340	218	1056	1386	20	47
1981	9400	11 000	1570	1326	336	216	1034	1325	21	43
1982	9000	10 600	1593	1271	333	211	1050	1272	20	44
1983	8500	10 500	1626	1277	333	207	1079	1251	19	44
1984	8600	11 100	1687	1278	340	214	1141	1152	18	43
1985	8800	11 000	1730	1229	335	212	1160	1242	19	49
1986	8500	10 800	1818	1190	345	220	1199	1276	—	—
1987	8400	11 100	1830	1097	350	228	1289	1289	54	61
1988	8100	11 700	1849	1010	365	220	1378	1264	—	—
1989	—	—	1837	870	346	204	1460	1230	—	—

Data from EC, 1992b with permission

8.4.3 Sulphur Oxides—SO$_x$

Sulphur oxides are the product of fossil fuel combustion, usually oil and coal. The dominant gaseous emission of sulphur is as sulphur dioxide with small amounts of sulphur trioxide. Fuels contain significant quantities of sulphur (< 1 per cent) as inorganic sulphides or organic sulphur and, on burning, SO$_2$ and SO$_3$ are released. The following reactions show how sulphuric acid is produced on release of SO$_2$:

$$SO_2 + OH^- \rightarrow HOSO_2^- \qquad (8.6)$$

$$HOSO_2^- + O_2 \rightarrow SO_3^- + HO_2^- \qquad (8.7)$$

$$SO_3^- + H_2O \rightarrow H_2SO_4^- \qquad (8.8)$$

Sulphate particles (SO$_4^{2-}$) are found either as dry deposition or wet deposition. In wet deposition water vapour combines with H$_2$SO$_4$ to produce acid rain droplets. The pH of rainfall is normally about 5.7 (see Chapter 3). By definition rainfall is acidic if the pH is < 5.5. Acid rain is common in the central and more particularly the eastern part of the United States. It is also common in Europe, particularly mainland Europe, with countries such as Germany and further east being worst affected. The negative impact of

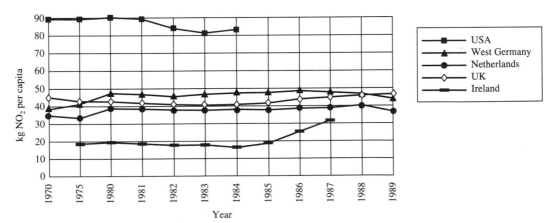

Figure 8.3 Total nitrogen dioxide emissions per capita, 1970–1989.

SO_2 levels is on humans and plants. Impaired bronchial functioning is noted at ambient levels of approximately 25 mg/m^3 for 10 minute exposures. Forest growth is inhibited at levels as low as 50 μg/m^3, (see Table 8.4 for ambient air standards).

Table 8.10 and Fig. 8.4 show the annual levels in kg of SO_2 per capita per year. It is seen that mainland European countries produce about 30 kg per person per year while the United Kingdom and the United States produce about 60 and 90 kg respectively. Table 8.11 shows the SO_2 emissions for mobile and stationary sources. Clearly SO_2 is a stationary source problem, with 66 per cent of total SO_2 coming from the power generating industries using coal and oil, while 25 per cent comes from other industries producing their own power using fossil fuels. A further 7 per cent comes from oil refineries and only 3 per cent from transport sources. Emissions from domestic sources are now considered insignificant (except in eastern Europe still), but were primarily responsible for the London smog of 1952.

8.4.4 Particulate Matter—PM-10

Airborne particulate matter represents a complex mixture of organic and inorganic substances typically divided into two groups as shown in Table 8.12. Terms used to describe particulate matter include: suspended particulate matter, total suspended particulates, black smoke, inhalable thoracic particles (which deposit on the lower respiratory tract, below the larynx), PM-10 (the term used by the USEPA indicates particulate matter of aerodynamic diameter less than 10 μm). Analytical methods include 'black smoke' measurements, represented by the darkness of staining on white filter paper through which air has been drawn. Total suspended particulates are measured by gravimetric methods with concentrations for this method about 2 to 3 times that of the black smoke method. This is so because gravimetric methods also measure larger particles of 2 to 10 μm, which are not seen in the smoke method.

Particulate matter is emitted in urban areas from power plants, industrial processes, vehicular traffic, domestic coal burning and industrial incinerators. Table 8.13 shows values for rural and urban areas. Table 8.14 and Fig. 8.5 show that PM has decreased substantially since 1970 and is now at about 5 kg per capita per year in the European Union. Table 8.15 shows that stationary sources emit about three times that of mobile sources. It is also noted that in Ireland stationary sources emit about ten times more PM than mobile sources. This is due to the presence of coal, peat and oil fired power plants which up to 1995 did not have state of the art flue gas scrubbing systems.

Table 8.10 Total sulphur dioxide emissions per capita

Year	USA	West Germany	Netherlands	UK	Ireland
		Sulphur dioxide (kg SO_2/capita/year)			
1970	139	62	62	115	—
1975	120	54	31	96	59
1980	103	52	35	87	64
1981	98	49	33	79	55
1982	92	47	28	75	45
1983	89	44	22	69	40
1984	91	43	21	66	37
1985	88	39	19	66	39
1986	86	37	19	69	—
1987	82	32	18	68	49
1988	83	20	18	67	—
1989	83	16	15	65	—

Data from EC, 1992b with permission

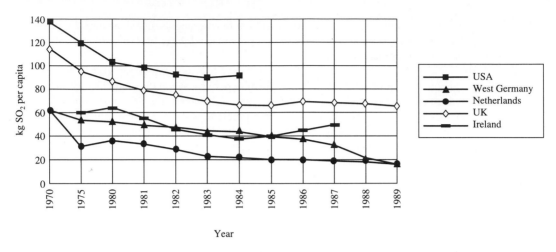

Figure 8.4 Total sulphur dioxide emissions per capita, 1970–1989.

Table 8.11 Sulphur dioxide emissions—mobile (M) and stationary (ST) sources

| | Sulphur dioxide (1000 t SO$_2$) | | | | | | | | | |
| | USA | | West Germany | | Netherlands | | UK | | Ireland | |
Year	M	ST	M	ST	M	ST	M	ST	M	ST
1970	607	27 800	155	3588	47	760	199	6224	—	—
1975	650	25 200	133	3201	41	388	153	5217	3	183
1980	889	22 500	107	3087	38	453	117	4777	5	212
1981	884	21 700	107	2932	33	435	117	4316	5	184
1982	824	20 600	103	2765	34	360	116	4092	4	151
1983	784	20 000	101	2589	33	286	101	3760	5	135
1984	825	20 700	94	2509	32	270	106	3613	4	125
1985	864	20 200	87	2309	32	237	102	3617	5	133
1986	869	19 800	98	2165	34	239	103	3792	—	—
1987	884	19 500	90	1843	36	231	97	3801	7	167
1988	938	19 800	73	1164	37	222	105	3707	—	—
1989	952	—	74	927	32	186	121	3578	—	—

Data from EC, 1992b with permission

Table 8.12 Particulate matter size

| | | Particle size | |
Group description	Composition	WHO	USEPA (PM-10)
Coarse	Dust, earth, crust matter	> 2.5 μm	$\geq$ 10 μm
Fine	Aerosols, combustion particles, recondensed organic and metal vapours (primary and secondary pollutants)	< 2.5 μm	$\leq$ 10 μm

Table 8.13 Typical values of black smoke and PM concentrations

	Annual concentrations	
Location	Black smoke ($\mu g/m^3$)	Suspended particles by gravimetry ($\mu g/m^3$)
Rural	0–10	0–50
Urban	10–40	50–150
Maxima	100–150	200–400

Table 8.14 Total particulates per capita

	Total particulates (kg PM/capita)				
Year	USA	West Germany	Netherlands	UK	Ireland
1970	90	19	14	19	—
1975	49	10	11	12	24
1980	37	8	12	10	28
1981	35	8	11	10	28
1982	31	7	10	10	28
1983	30	7	9	9	29
1984	31	7	8	9	32
1985	30	7	7	10	33
1986	29	6	7	10	—
1987	30	6	7	9	30
1988	28	5	7	9	—
1989	28	5	5	9	—

Data from EC, 1992b with permission

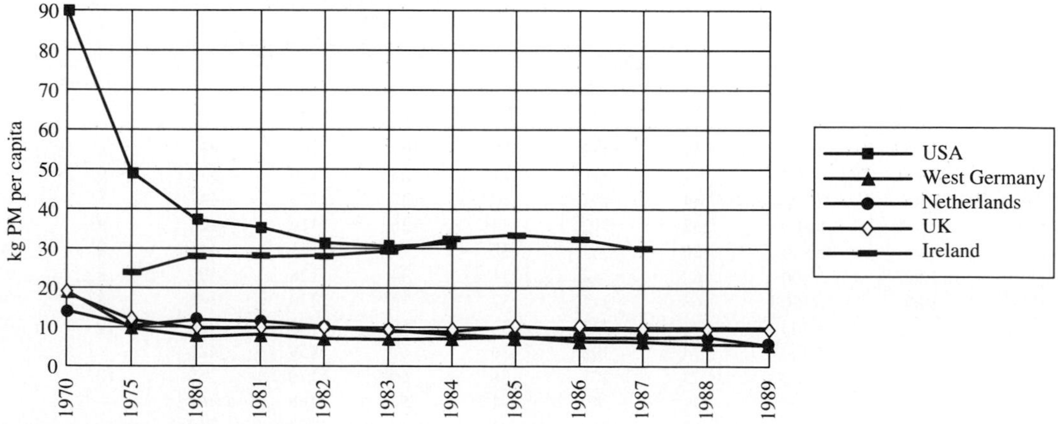

Figure 8.5 Total particulates per capita, 1970–1989.

Excessive concentrations of SO_2, black smoke and total suspended particulates are associated with increased mortality, morbidity and pulmonary difficulties. It has been noted that 24 hour values in excess of 500 $\mu g/m^3$ for combined SO_2 and smoke result in increased morbidity. The EU ambient limit standard for the 24 hour period is 100 to 150 $\mu g/m^3$ for combined SO_2 and smoke and is 40 to 60 $\mu g/m^3$ for the annual period, as shown in Table 8.4.

8.4.5 Volatile Organic Compounds—VOCs

Organic air pollutants comprise hydrocarbons and other substances (about 50 per cent hydrocarbons). Many are reactive (excluding CH_4) in the air environment and have considerable environmental and health implications. The most abundant HC is methane with ambient concentrations of 1 to 6 ppm. The less abundant but more reactive volatile organic compounds include: ethylene oxide, formaldehyde, phenol, phosgene, benzene, carbon tetrachloride, CFCs and PCBs. These are almost all manufactured and are known or suspected carcinogens. Many are precursors for photochemical oxidants and react with NO_x and O_2 with sunlight to produce smog and aerosol pollution. They may irritate the eye, throat and lungs and inhibit plant growth. VOC emissions come from a wide range of sources as shown in Table 8.16 and Fig. 8.6. This table is incomplete as some countries include methane while others do not. Typically VOCs are sourced, 27 per cent from road transport, 17 per cent from the solvents industry, 15 per cent from coal mining, 17 per cent from landfilling (CH_4), 10 per cent from gas distribution, 12 per cent from natural sources (forests, etc.) and 2 per cent others. The total VOC emissions for 1985 were 30 to 90 kg VOC per capita. Ambient air quality standards do not impose limits to VOCs. This is so because there is no safe limit for most of these compounds. For instance, benzene is a known carcinogen, with no known safe threshold limit. WHO (1987) states that an air concentration of benzene of 1 $\mu g/m^3$ imposes a lifetime risk of leukaemia of 4×10^{-6}. However, regulatory authorities set emission limits for specific industries as discussed in Sec. 8.8.

8.4.6 Hydrocarbons—HC

Hydrocarbons are one species of VOC emissions. They are organic compounds containing only carbon and hydrogen. Typically these are petroleum products and are classified as shown in Table 8.17.

Table 8.15 Particulates—mobile (M) and stationary (ST) sources

Year	USA M	USA ST	West Germany M	West Germany ST	Netherlands M	Netherlands ST	UK M	UK ST	Ireland M	Ireland ST
					Particulate matter (1000 t PM)					
1970	1200	17 300	84	1084	18	165	104	945	—	—
1975	1300	9 300	61	581	21	128	114	574	6	69
1980	1300	7 200	64	453	30	133	123	447	8	86
1981	1300	6 800	65	412	30	126	116	424	9	88
1982	1300	5 800	66	365	30	115	121	417	8	90
1983	1300	5 800	67	342	31	94	128	392	8	92
1984	1300	6 100	68	344	32	78	139	342	8	104
1985	1400	5 700	70	327	34	68	145	410	10	107
1986	1400	5 400	74	308	36	62	159	426	—	—
1987	1400	5 600	73	271	38	60	170	368	10	97
1988	1400	5 500	71	249	41	57	188	345	—	—
1989	—	—	72	214	21	54	202	310	—	—

Data from EC, 1992b with permission

Table 8.16 VOC emissions per capita for 1985

Source	Total VOC emissions (kg VOC/capita)				
	EC12	West Germany	Netherlands	UK	Ireland
Road transport	16	19	15	14	∼7
Solvent evaporation	10	18	11	12	∼7
Solid fossil fuel					
Mining	8	17	—	16	
Landfilling	10	29	—	12	
Distribution of gas	5	4	—	7	∼8
Production process	<2	<0.1	<1	4	
Combustion industry	<0.5	<0.1	<1	<1	<1
Oil refineries	<0.5	<0.1	<1	<1	<1
Heat for commerce, residential and institutional	2.0	<0.5	<0.5	<1	
Nature					
Miscellaneous	7	4–3	1	1.5	7
Total	∼60	∼95	∼29	∼69	∼31

Data from EC, 1992b with permission

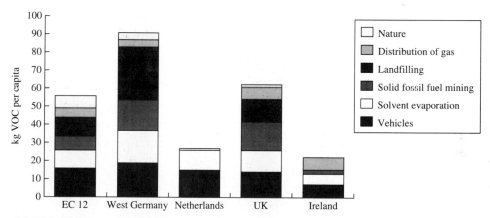

Figure 8.6 VOC emission per capita for 1985.

Table 8.17 Hydrocarbon classification

Group	Sub-group	Reactivity	Example	
Aliphatic	Alkanes	Inert	Methane	
	Alkenes (olefins)	Highly reactive	Ethylene $+ NO_2 \rightarrow$ PAN, O_3	
	Alkynes	Reactive	Rare	
Aromatic (related to benzene)	Benzene	Not very reactive	PAHs	
			Benzo(a)pyrene	
			Benz(e)acephenanthrylene	
			Benzo(j)fluroanthene	} Carcinogenic
			Benzo(z)pyrene	
			Benz(a)anthracene	

The most common hydrocarbon is methane, as was mentioned in the previous section. Because it is inert it is not a serious tropospheric pollutant. However, it is now deemed significant as one of the greenhouse gases and is produced by animals, forestry, boglands, landfill sites, vehicles, etc. Its damaging influence to the ozone layer is only now being recognized. The alkenes (or olefins) are highly reactive; when combined with NO_x ethylene produces peroxyacetyl nitrate (PAN) and ozone. The intimidating hydrocarbons belong to the aromatic or benzene group. While not very reactive, several of the benzene-related compounds are carcinogenic or suspected carcinogenic. These include polynuclear hydrocarbons and the benzene compounds.

The major sources of man-made hydrocarbons include traffic, organic chemical production, transport and processing of crude oil, and distribution of natural gas. Table 8.18 and Fig. 8.7 show that the HC production varies from 30 to 90 kg HC per capita per year. Table 8.19 shows that in the EC the HC production is about 50 per cent from mobile sources and 50 per cent from stationary sources, while in the United States, the HC from stationary sources is about twice that of mobile sources; HC production is only gradually decreasing since 1980. Gas leakages were a significant source of HC production in both Ireland and the United Kingdom up to about 1985.

Example 8.3 Compute the HC discharged from a population centre of 1 million if:

- 300 000 vehicles travel 12 000 km per annum emitting 1 g/km each
- The per capita consumption of oil-based paint is 2 litres per annum with a HC content of 1 kg per litre
- The per capita HC from dry cleaning solvents is 1 kg per annum

Solution

$$\text{Vehicles} : 300\,000 \times 12\,000 \times \frac{1}{10^6} = 3600 \text{ tonnes}$$

$$\text{Paints} : 10^6 \times 2\,\text{kg} = 2000 \text{ tonnes}$$

$$\text{HC solvents} : 10^6 \times 1\,\text{kg} = 1000 \text{ tonnes}$$

$$\text{Total} = 6600 \text{ tonnes}$$

$$= 6.6\,\text{kg per capita per year}$$

Table 8.18 Total hydrocarbon emissions per capita

Year	\multicolumn{5}{c}{Hydrocarbons (kg HC/capita)}
	USA	West Germany	Netherlands	UK	Ireland
1970	128	48	41	31	—
1975	102	45	41	31	15
1980	98	45	35	33	18
1981	91	43	33	34	18
1982	84	43	32	34	18
1983	87	43	31	34	18
1984	91	43	30	34	18
1985	83	43	29	34	18
1986	79	44	28	34	—
1987	78	43	27	35	30
1988	74	42	27	35	—
1989	—	42	27	36	—

Data from EC, 1992b with permission

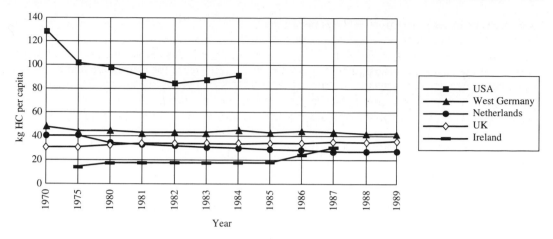

Figure 8.7 Total hydrocarbon emissions per capita, 1970–1989.

8.4.7 Ozone—O_3

Ozone is considered a 'criteria pollutant' because of its association with urban smog. However, it is a secondary pollutant. Nitrogen oxides and various hydrocarbons in the presence of sunlight initiate a complex set of reactions that produce secondary pollutants or photochemical oxidants. The most abundant oxidant is ozone (O_3). The formation of smog is simplified as

$$\text{Hydrocarbons} + NO_x + \text{sunlight} \rightarrow \text{photochemical smog}$$

The source of HC and NO_x in urban areas is primarily from vehicles. The irradiation of air containing hydrocarbons and oxides of nitrogen leads to:

- Oxidation of NO to NO_2
- Oxidation of HCs
- Formation of O_3

Table 8.19 Hydrocarbon emissions—mobile (M) and stationary (ST) sources

	Hydrocarbons (1000 t HC)									
	USA		West Germany		Netherlands		UK		Ireland	
Year	M	ST	M	ST	M	ST	M	ST	M	ST
1970	11 100	15 100	1030	1851	289	251	403	1347	—	—
1975	9 200	12 800	1210	1598	289	266	447	1286	29	19
1980	7 400	14 900	1310	1444	236	266	606	1281	37	25
1981	7 200	13 700	1242	1419	218	256	619	1274	38	25
1982	6 800	12 800	1258	1379	217	246	639	1273	36	27
1983	6 700	13 800	1266	1366	215	232	632	1271	34	28
1984	6 800	14 700	1287	1364	212	217	648	1259	32	32
1985	6 400	13 600	1269	1355	203	213	646	1280	32	32
1986	6 200	13 200	1322	1339	202	202	664	1293	—	—
1987	6 000	13 600	1329	1304	202	196	697	1287	65	43
1988	6 100	12 400	1334	1269	207	189	729	1284	—	—
1989	—	—	1273	1263	213	186	788	1287	—	—

Data from EC, 1992b with permission

Table 8.20 Lead emissions from vehicles per capita

Year	Lead (kg Pb/capita)			
	USA†	Netherlands	UK	Ireland
1970	0.64	0.12	—	—
1975	0.46	0.18	—	—
1980	0.22	0.09	0.13	0.295
1981	0.18	0.09	0.12	0.252
1982	0.18	0.09	0.12	0.245
1983	0.15	0.09	0.12	0.144
1984	0.12	0.09	0.13	0.137
1985	0.05	0.08	0.11	0.130
1986	0.02	0.05	0.05	0.048
1987	—	0.023	0.05	0.048
1988	—	0.023	0.05	0.048
1989	—	0.02	0.04	0.048

† USEPA (1988a).
Data from EC, 1992b with permission

Clean tropospheric background levels of O_3 are 20 to 80 ppb, while urban polluted areas may reach concentrations of up to 500 ppb. High levels of ozone are associated with health effects of chest constriction and irritation of the mucous membrane. It is also associated with deterioration of rubber products (tyres) and damage to vegetation. USEPA ambient standards are set at $235 \, \mu g/m^3$ for the 1 hour limit. WHO limits are recommended for Europe at 150 to $200 \, \mu g/m^3$ for 1 hour.

8.4.8 Lead—Pb

The heavy metals of cadmium, lead and mercury are significant air pollutants. Lead is a bluish-grey soft metal with a melting point of $327.5\,°C$ and a boiling point of $1740\,°C$. Organic lead compounds such as tetraethyl and tetramethyl lead are extensive fuel additives. They are colourless liquids and are less volatile than most petrol components. As such, they tend to become concentrated when petrol evaporates. Prior to 1986 80 to 90 per cent of lead in ambient air was from the combustion of leaded petrol. The mining and smelting of lead is also a source but is site specific. Secondary lead smelters and refining and manufacture of compounds containing lead as well as waste incinerators also produce lead emissions. Background levels of lead in air are about $5 \times 10^{-5} \, \mu g/m^3$. Environments with high traffic densities have polluted concentrations of 0.5 to $3 \, \mu g/m^3$. The annual ambient air quality standards are set internationally at $2 \, \mu g/m^3$.

Lead in ambient air is of particulate matter of size $< 3 \, \mu m$. Lead is also found in water and food. Lead is bioaccumulative and 30 to 50 per cent of inhaled lead lodges in the respiratory system with the remainder absorbed into the body. Elevated levels in blood lead to haematological problems, particularly at blood levels in excess of $0.2 \, \mu g/ml$. Since the introduction of unleaded petrol, Table 8.20 and Fig. 8.8 show that the Pb emissions reduced from 640 to 20 g/capita in the United States from 1970 to 1986. This is one of the great improvements in air quality. Typically Pb emissions into urban air are 20 to 60 g/capita. Table 8.21 shows that in 1991, of petrol used in West Germany and Denmark, about 60 per cent is unleaded, while the corresponding figure for Spain and Italy is less than 5 per cent. A substantial improvement on the 1991 figures for lead in petrol had occurred by 1995.

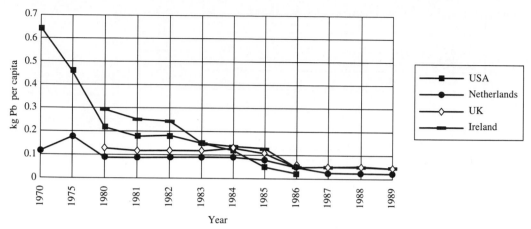

Figure 8.8 Lead emissions from vehicles per capita, 1970–1989.

Table 8.21 Percentage of automotive gasoline (petrol) deliveries (unleaded) in 1990

Country	%
USA	—
West Germany	68
Netherlands	42
UK	34
Ireland	19
France	15
Spain	1
Italy	5
Denmark	57

8.4.9 Summary of Criteria Pollutants

Figure 8.9 shows a plot of the summation of CO, NO_2, SO_2, PM-10 and HC on a per capita basis for the United States, West Germany and Ireland. In 1984 the US emissions were 600 kg in comparison with West Germany's 300 kg and Ireland's 250 kg. A dramatic improvement is seen in US levels from 1970 to 1984. A gradual and continuing improvement is seen in West German levels for 1970 through to 1989. Almost no improvement in levels is seen in Ireland since 1975. Ireland has undergone extensive industrialization since 1975, particularly with the chemical, pharmaceutical, computer, agriculture and power industries. As such, gains in environmental quality have been offset by increased industrialization. However, Ireland is particularly polluting with regard to particulates, which is due in part to new fossil fuel power installations.

Example 8.4   Consider a thermoelectric power plant of 915 MW total capacity with a load factor (or annual capacity) of 72.5 per cent and an efficiency of 40 per cent. Determine the amount produced of particulates, CO_2 and SO_2 if coal is used. The ultimate analysis and calorific value of coal is as follows:

Moisture	Ash	Carbon	Hydrogen	Nitrogen	Sulphur	Oxygen	Calorific value
8%	7.7%	77.0%	3.0%	1.25%	1.0%	2.05%	29.7 MJ/kg

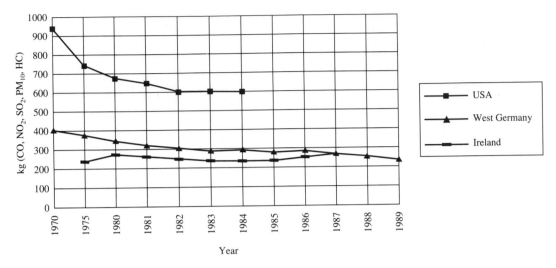

Figure 8.9 Sum of CO, NO_2, PM_{10} and HC per capita, 1970–1989.

See Chapter 14 for an explanation of ultimate analysis.

Solution:

$$\text{Power } 915\,\text{MW} = 915 \times 10^3\,\text{kW}$$
$$= 915 \times 10^3 \times 10^3\,\text{J/s}$$
$$= 3294 \times 10^3\,\text{MJ/h}$$
$$\text{Coal required} = \frac{3294 \times 10^3 \times 0.725}{0.4 \times 29.7} = 201.0\,\text{t/h}$$

Assume 80 per cent of ash is as fly ash or particulates:

$$\text{Particulates} = 0.8 \times 0.077 \times 201 = 12.4\,\text{t/h}$$
$$\text{Sulphur dioxide} = SO_2 = S + O_2$$
$$\text{Atomic mass} = 32 + 32 \text{ (equal parts of S and } O_2)$$
$$\text{Sulphur at } 1\% = 0.01 \times 201 = 2.01\,\text{t/h}$$
$$\text{Sulphur dioxide} = S + O_2 = 2.01 + 2.01 = 4.02\,\text{t/h}$$
$$\text{Carbon dioxide} = CO_2 = C + O_2$$
$$\text{Atomic mass} = 12 + 32$$
$$\text{Carbon at } 77.0\% = 0.77 \times 201 = 155\,\text{t/h}$$
$$\text{Carbon dioxide} = C + O_2$$
$$= 155.06 + \frac{32}{12} \times 155.06\,\text{t/h} = 568\,\text{t/h}$$

The annual production is as follows:

$$\text{Particulates} = 108 \times 10^3\,\text{t}$$
$$SO_2 = 35 \times 10^3\,\text{t}$$
$$CO_2 = 5.0 \times 10^6\,\text{t}$$

While SO_2 and CO_2 are emitted into the atmosphere, the particulates are captured by flue gas cleaning devices, generally by electrostatic precipitators (ESP). If an ESP removes 99.5 per cent of particulates, compute the amount of particulate emissions:

$$\text{Particulates} = \frac{0.5}{100} \times 12.9 = 0.065\,\text{t/h}$$

$$= 565\,\text{t/annum}$$

Note   Other emissions of VOC, HC, NO_x and others may be released in lesser quantities, but their magnitudes depend on power plant technology. Modern thermal power plant technologies can achieve almost insignificant NO_x emissions.

Example 8.5   Consider the thermoelectric power plant of Example 8.4, of total capacity 915 MW with a load factor of 72.5 per cent and an efficiency of 40 per cent. Determine the amount of particulates, CO_2 and SO_2 if oil is the fuel source. The ultimate analysis and calorific value is as follows:

Moisture	Ash	Carbon	Hydrogen	Nitrogen +oxygen	Sulphur	Calorific value
0.3%	0.04%	85.2%	11.3%	0.36%	2.8%	40.5 MJ/kg

Solution:

$$\text{Power 915 MW} = 3294 \times 10^3\,\text{MJ/h}$$

$$\text{Oil required} = \frac{3294 \times 10^3 \times 0.725}{4 \times 40.5} = 147\,\text{t/h}$$

Assume 80 per cent of ash is as fly ash or particulates:

$$\text{Particulates} = 0.8 \times \frac{0.04}{100} + 147 = 0.047\,\text{t/h}$$

Additional particulates are generated from the oil during the combustion process and this may double the emission rate. However, the actual emission rate will depend on the boiler technology:

$$\text{Sulphur dioxide} = S + O_2$$
$$\text{Atomic mass} = 32 + 32$$
$$\text{Sulphur at 2.8\%} = \frac{2.8}{100} \times 147 = 4.1\,\text{t/h}$$
$$\text{Sulphur dioxide} = 4.1 + 4.1 = 8.2\,\text{t/h}$$
$$\text{Carbon dioxide} = C + O_2$$
$$\text{Atomic mass} = 12 + 32$$
$$\text{Carbon at 85.2\%} = \frac{85.2}{100} \times 147 = 125.2\,\text{t/h}$$
$$\text{Carbon dioxide} = 125.2 + \frac{32}{12} \times 125.2 = 459\,\text{t/h}$$

The total annual production is as follows:

$$\text{Particulates} = 412\,\text{t}$$
$$SO_2 = 72 \times 10^3\,\text{t}$$
$$CO_2 = 4.02 \times 10^6\,\text{t}$$

Example 8.6 Consider the thermoelectric power plant of Example 8.4, of total capacity 915 MW with a load factor of 72.5 per cent and an efficiency of 33 per cent. Determine the amount of particulates, CO_2 and SO_2 if milled peat is the fuel source. The ultimate analysis for milled peat is as follows:

Moisture	Ash	Carbon	Hydrogen	Nitrogen +oxygen	Sulphur	Oxygen	Calorific value
50%	2.5%	27.3%	2.6%	0.7%	0.3%	16.6%	14.5 MJ/kg

Solution:

$$\text{Power } 915 \text{ MW} = 3294 \times 10^3 \text{ MJ/h}$$

$$\text{Milled peat required} = \frac{3294 \times 10^3 \times 0.725}{0.33 \times 14.5} = 500 \text{ t/h}$$

Assume 80 per cent of ash is as fly ash or particulates:

$$\text{Particulates} = 0.8 \times \frac{2.5}{100} \times 500 = 10 \text{ t/h}$$

$$\text{Sulphur dioxide} = S + O_2$$
$$\text{Atomic mass} = 32 + 32$$

$$\text{Sulphur at } 0.3\% = \frac{0.3}{100} \times 500 = 1.5 \text{ t/h}$$

$$\text{Sulphur dioxide} = 1.5 + 1.5 = 3 \text{ t/h}$$

$$\text{Carbon dioxide} = C + O_2$$
$$\text{Atomic mass} = 12 + 32$$

$$\text{Carbon at } 27.3\% = \frac{27.3}{100} \times 500 = 136.5 \text{ t/h}$$

$$\text{Carbon dioxide} = 136.5 + \frac{32}{12} \times 136.5 = 500.5 \text{ t/h}$$

Using an ESP with an efficiency of 99.5 per cent the particulate emissions reduce to

$$\text{Particulates} = \frac{0.5}{100} \times 10 = 0.05 \text{ t/h}$$
$$= 439 \text{ t/annum}$$

The total annual production is as follows:

$$\text{Particulates} = 87.6 \times 10^3 \text{ t}$$
$$SO_2 = 26.3 \times 10^3 \text{ t}$$
$$CO_2 = 4.4 \times 10^6 \text{ t}$$

If particulates are reduced by flue gas filters with an efficiency of 99 per cent, compute the emission rate of particulates:

$$\text{Particulates} = \frac{1}{100} \times 87.6 \times 10^3 = 876 \text{ t/annum}$$

Example 8.7   The thermoelectric power station of 915 MW capacity and a load factor of 72.5 per cent uses natural gas as its fuel source and a plant efficiency of 40 per cent. Determine the annual production of CO_2, water vapour, and NO_x. The natural gas parameters are as follows. Composition is per cent by volume. There are no particulates or SO_2 emissions.

CO_2	N_2	CH_4	C_2H_8	C_3H_8	C_4H_{10}	Density (kg/m^3)	Calorific value
0.1%	0.6%	98%	1.0%	0.2%	0.1%	0.72	40 MJ/kg

Solution

$$\text{Power } 915\,\text{MW} = 3294 \times 10^3\,\text{MJ/h}$$

$$\text{Natural gas calorific value} = \frac{40}{0.72} = 55.5\,\text{MJ/m}^3$$

$$\text{Gas required} = \frac{3294 \times 10^3 \times 0.725}{0.4 \times 40} = 149.3\,\text{t/h}$$

$$CH_4 \text{ at } 98\% = 0.98 \times 149.3 = 146.3\,\text{t/h}$$

$$\text{Stoichiometrically: } CH_4 + 2O_2 \rightarrow 2H_2O + CO_2$$

$$\text{Atomic mass: } \quad 16 \qquad 64 \qquad 36 \qquad 44$$

$$\text{Carbon dioxide} = \frac{44}{16} \times 146.3 = 402.3\,\text{t/h}$$

$$\text{Water vapour } (H_2O) = \frac{36}{16} \times 146.3 = 329.2\,\text{t/h}$$

$$\text{Nitrogen } N_2 \text{ at } 0.60\% = \frac{0.60}{100} \times 149.3 = 0.9\,\text{t/h}$$

$$\text{Stoichiometrically: } N_2 + 2O_2 = 2NO_2$$

$$\text{Atomic mass: } 28 \qquad 64 \qquad 92$$

$$\text{Nitrogen dioxide} = \frac{92}{28} \times 0.9 = 2.96\,\text{t/h}$$

The total annual production is as follows:

$$\text{Particulates} \rightarrow 0$$

$$CO_2 = 3.5 \times 10^6\,\text{t}$$

$$SO_2 \rightarrow 0$$

$$H_2O \text{ (water vapour)} = 2.9 \times 10^6\,\text{t}$$

$$NO_2 = 7900\,\text{t}$$

$$CO \rightarrow 0$$

It is relevant to note that very significant amounts of water vapour (H_2O) are produced, in fact almost as much as CO_2. Water vapour is also a greenhouse gas, of greater numerical significance than CO_2. But because it is in the atmosphere naturally from evaporation, we tend not to notice it (see Example 8.8).

8.5 ACID DEPOSITION

Figure 8.10 shows the mechanism of acid deposition. Emissions of sulphur oxides, nitrogen oxides and hydrocarbons from industry, transport, households and power production are transformed in the atmosphere into sulphate or nitrate particles. When combined with sunlight and water vapour, a complex chemical reaction results in mild sulphuric acid or nitric acid. These acids in turn return to the earth as dew, drizzle, fog, sleet, snow or rain. The deposition of acids is either as dry deposition, which may be in particulate or gaseous form, or a wet deposition in rain or snow. Normal 'clean' rainwater has a pH ~ 5.7.

In areas of northern Europe (Scandinavia) and the eastern parts of the United States, the pH in rainfall has been monitored regularly below 5.0, sometimes 4.0 and on rare occasions at 3.0. Acid deposition has a serious negative effect on forest land, on aquatic life and on some natural stone building materials.

Much of the acid deposition is transnational. For instance, 77 per cent of sulphur deposited in the Netherlands was from other countries, as was 64 per cent in Denmark (EC, 1987). Similarly, much of Canada's acid deposition is from the United States and 50 per cent in central Ontario is from the United States midwest (Henry and Heinke, 1989). It is obviously desirable to reduce NO_x and SO_x emissions. However, while there was significant reduction in NO_x and SO_x from 1970 to 1985 (see Figs 8.3 and 8.4), there has been almost no improvement in the past decade.

8.6 GLOBAL CLIMATE CHANGE—GREENHOUSE GASES

The major greenhouse gases in order of greenhouse contribution are CO_2, CFCs, CH_4 and N_2O and O_3. These gases have the effect of absorbing incoming short-wave solar energy (at wavelengths $< 4\,\mu m$). They also have the ability (when in the atmosphere) of absorbing some of the outgoing earth's radiated energy of long wavelength ($> 4\,\mu m$). Each 'greenhouse gas' has its own ideal wavelength bands at which it absorbs solar or earth's radiation energy best. Figure 8.11 shows these wavelength bands. For instance,

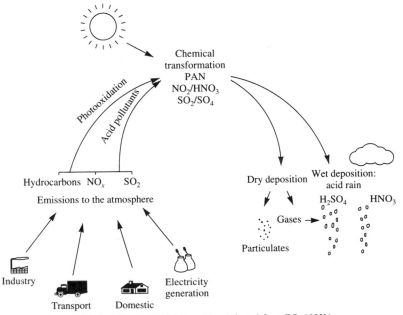

Figure 8.10 The formation of atmospheric acidity and acid deposition (adapted from EC, 1992b).

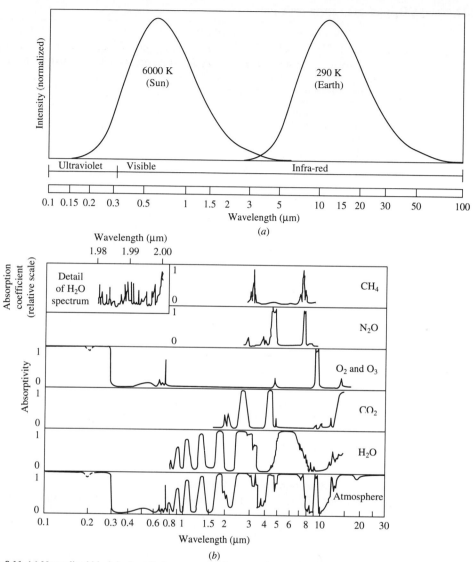

Figure 8.11 (*a*) Normalized black body radiation curves for the sun and earth. (*b*) Atmospheric absorption on a clear day (adapted from Fleagle and Businger, 1963. Reprinted by permission of Academic Press).

CO_2 absorbs thermal radiation in at least three bands, 2.3 to 3.1, 4.1 to 4.5 and 13 to 18 μm. CH_4 absorbs radiation in at least two narrow bands, centred at 3.2 and 8 μm. Water vapour (H_2O) absorbs in bands centred at 0.95, 1.1, 0.4, 1.9, 2.7, 6.2 and above 20 μm. As such, several of the greenhouse gases absorb radiation at the same wavelengths. While some of the above-mentioned gases absorb radiation both above and below 4 μm, the term greenhouse gases is reserved for those gases that absorb radiation at wavelengths > 4 μm, i.e. gases that absorb the long wavelength earth radiation. They therefore trap much of the outgoing earth's radiative energy, therefore heating the atmosphere and subsequently radiating this energy back to earth and out to space. These greenhouse gases have the effect of acting like a thermal blanket around the globe, raising its temperature.

Table 8.22 Total CO_2 and per capita emissions

	USA		West Germany		Netherlands		UK		Ireland	
Year	Total $(10^6 t)$	Per capita (t)	Total $(10^6 t)$	Per capita (t)	Total $(10^6 t)$	Per capita (t)	Total $(10^6 t)$	Per capita (t)	Total $(10^6 t)$	Per capita (t)
1960	791	4.38	531	8.7	75	5	619	10.5	10.5	3
1965			624	10.3	99	6.6	663	11.6	13.5	3.9
1970			741	12.1	131	8.7	675	11.8	19.1	5.5
1975			709	11.6	134	8.9	599	10.5	20.6	5.9
1980			809	13.3	154	10.2	583	10.2	25.2	7.2
1985			718	11.8	145	9.7	561	9.8	17.2	7.8
1986			719	11.8	156	10	567	9.9	28.3	8.1
1987	1224	5.03	709	11.7	155	10.3	583	10.2	30.1	8.6
1988			706	11.7	153	10.2	577	10.1	30.3	8.7
1989			689	11.3	155	10.3	579	10.1	30.9	8.8

Data from EC, 1992a with permission

8.6.1 Carbon Dioxide

Carbon dioxide is not a pollutant in the conventional sense. It is a normal component of the atmosphere (0.033 per cent) and is essential for plant growth. Fossil fuel combustion, including coal fired thermal power stations and forest fires, have increased the background levels of CO_2 from ~ 315 ppm in 1960 to 350 ppm in 1990. It is now believed that man-made CO_2 is the most significant of the greenhouse gases. Table 8.22 and Fig. 8.12 show the total emissions of CO_2 from five countries and the per capita emissions from 1960 to 1989. The production of CO_2 in the developed world is approximately 5 to 10 t/capita per year. In 1987, 5.6×10^9 t were emitted globally into the atmosphere with the US contribution being 1.2×10^9 t. Forest clearing is another source of CO_2, as the exposed soil degrades and unlocks and emits the CO_2 into the atmosphere. Estimates of the biomass CO_2 contribution vary from 0 to 1×10^9 t.

8.6.2 Chlorofluorocarbons—CFCs

CFCs are man-made molecules that contain chlorine, fluorine and carbon. They absorb radiation in the atmospheric radiative window range of 7 to 12 μm. In the atmosphere they have long residence times and

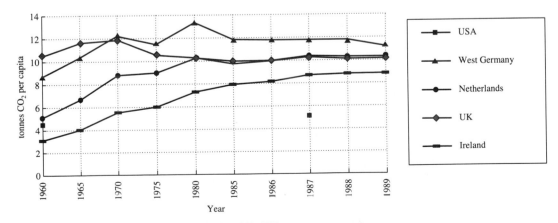

Figure 8.12 Total CO_2 emissions, tonnes per capita 1960–1989.

are inert and non-water soluble. Chlorine is freed from CFCs by the process of photolysis by short-wave radiation and drifts upwards to damage the ozone layer. The significant CFCs are CFC-11 (trichloro-fluoromethane) and CFC-12 (dichlorofluoromethane). The EC 1990 production of CFC-11 and CFC-12 was 210×10^3 t, down from a high of 376×10^3 t in 1987. The use of CFCs in the European Community for 1988 was as aerosols (~ 45 per cent), foam plastics (~ 40 per cent), refrigerants (~ 10 per cent) and solvents (~ 5 per cent). Other CFCs are CFC-113, CFC-114 and CFC-115, which had an EC production of 75×10^3 t in 1990. CFC-113 is a solvent used in the electronics industry. CFCs are to be phased out as per the Montreal Protocol of 1987. Because of their long residence times, the 50 per cent cut still assumes that the ppb level of CFC will increase almost linearly with time, going from 0.4 ppb in 1986 to 1.0 ppb by 2010.

8.6.3 Methane—CH_4

Methane is a naturally occurring gas that is produced under anaerobic conditions. This occurs in swamps, rice fields, cattle stocks and in the production and consumption of fossil fuels. Like the CFCs, CH_4 has a long residence time of about 10 years, after which it may be oxidized with OH radicals. It is estimated that concentration in the atmosphere has increased in the past 200 years, in correspondence with a population increase. It absorbs thermal radiation at narrow-band wavelengths around 3.2 and $7.6\,\mu m$.

8.6.4 Nitrous Oxide—N_2O

Nitrous oxide (N_2O) absorbs thermal radiation at the same wavelength as methane $\sim 7.6\,\mu m$. It is produced in the nitrogen cycle via nitrification, i.e. $NH_4 \rightarrow N_2 \rightarrow N_2O$. It has a residence time of about 150 years and is about 200 times as potent a greenhouse gas as CO_2. The quantities of N_2O produced are insignificant by comparison with CO_2. Emissions arise from wastewater treatment plants, industrial sources and gas combustion.

8.6.5 Water Vapour—H_2O

Water vapour (H_2O) is often ignored as a greenhouse gas. The examples on power plant emissions (Sec. 8.5) show that the amounts of H_2O produced are similar to the amounts of CO_2. Also the long wavelength energy radiated by the earth is absorbed not only by CO_2, N_2O and CH_4 but also by H_2O. Figure 8.11 shows that water vapour absorbs energy over several bands within the range of 1 to $8\,\mu m$. However, although the anthropogenic quantities of H_2O emitted are comparable to those of CO_2, they are miniscule in comparison with the natural quantities from evaporation, as shown in Example 8.8.

Example 8.8 Compute the annual global natural emission of water vapour by evaporation from the earth's land and ocean surface. Compare this to the amount of water vapour emitted by industrial power plants. State the assumptions made.

Solution From Brutsaert (1982, p. 3):

$$\text{Evaporation from land (global average)} = 0.45\,\text{m/yr}$$

$$\text{Land area} = 1.49 \times 10^8\,\text{km}^2$$

$$\text{Therefore water vapour for land evaporation} = 67 \times 10^{12}\,\text{m}^3/\text{yr}$$

$$\text{Evaporation from oceans} \cong 1.3\,\text{m/yr}$$

$$\text{Ocean area} = 3.6 \times 10^8\,\text{km}^2$$

$$\text{Therefore water vapour evaporated from oceans} \cong 468 \times 10^{12} \, m^3/yr$$

$$\text{and total evaporated water vapour from earth surface} \cong 535 \times 10^{12} \, m^3/yr$$

$$\text{Density of water vapour} \cong 0.8 \, kg/m^3$$

$$\text{Therefore total evaporated water vapour from earth's surface} \cong 420 \times 10^9 \, t/yr$$

In Sec. 8.6.1, it was mentioned that $\sim 6 \times 10^9$ t/yr of CO_2 were emitted anthropogenically and that the emission of water vapour from man-made sources is about this order also. Therefore:

$$\text{Anthropogenic water vapour emission} \cong 6 \times 10^9 \, t/yr$$

This is approximately 1.4 per cent of the natural emission values.

Note This is an order of magnitude analysis only.

8.7 NON-CRITERIA POLLUTANTS

The criteria pollutants of CO, NO_2, SO_2, PM-10, VOCs, HC and Pb are listed in Table 8.4 along with the international standards for each and its associated exposure time. The criteria pollutants are regularly found in urban environments, and the standards are set to uphold the quality of air for urban dwellers, the flora, fauna and materials of that environment. In addition, there are many other air pollutants for which emission limits from industry are set. The so-called non-criteria pollutants listed by WHO (1987) are set out in Tables 8.23 and 8.24. The ambient air quality standards as set by WHO are set for both organic and inorganic substances. Many of these are carcinogenic, mutagenic and damaging to the central nervous system (CNS) as well as having a host of other negative health effects. Some of the air pollutants are man-made synthetic chemicals such as 1,2-dichlorethane, while others such as radon are naturally occurring in the background from the earth's crustal geology. From those pollutants listed in Tables 8.23 and 8.24, most, however, are the products of industry and more specifically the chemical, pharmaceutical and petroleum products industries. The heavy metals of Cd, Cr and Hg are ubiquitous in the manufacturing and chemical industries, particularly electroplating, plastics, paints, chlorine and chemicals.

8.8 EMISSION STANDARDS FROM INDUSTRIAL SOURCES

Generally emissions are from point sources, though line and areal sources may also occur. Fugitive emissions are another significant industrial source, the control of which is legislated for in the United States. Emissions standards are usually set by the local licensing authority whose task it is to consider the impact of new and existing emissions. In Europe, the German TA Luft standards are referred to widely. Irish industrial air emission licenses frequently refer and use these standards. The procedure is as follows:

1. Identify the emission substances by name and, from Annexe E (in TA Luft), determine whether they are Class I, II or III. There are 145 substances listed alphabetically in this Annexe and classed. Class I substances are those with the most severe emission standard and Class III the least severe.
2. The Class I substances should not exceed a concentration of $0.1 \, mg/m^3$ at a mass flow rate of $0.5 \, g/h$ or more.
3. The Class II substances should not exceed a concentration of $1 \, mg/m^3$ at a mass flow rate of $5 \, g/h$ or more.
4. The Class III substances should not exceed $5 \, mg/m^3$ at a mass flow rate of $25 \, g/h$ or more.
5. If substances of Classes I and II exist, then the concentration should not exceed $1 \, mg/m^3$. If Classes I and III or II and III exist, the mass concentration should not exceed $5 \, mg/m^3$.

Table 8.23 Some non-criteria pollutants

Pollutant	Description and source	Health effects	WHO ambient guidelines
Acrylonitrile (AN)	Volatile, flammable, colourless liquid, soluble in water. Man-made, used in acrylic fibre and resins	Carcinogen	No safe level
Benzene (C_6H_6)	Colourless clear liquid, slightly soluble in water. Component of petrol and petroleum products	Carcinogen	No safe level
Carbon disulphide (CS_2)	Colourless, volatile, inflammable liquid. Used in viscose rayon production, about 20 g CS_2 to 1 kg viscose	Brain damage, muscle atrophy	$100\,\mu m/m^3$, 24 h
1,2-Dichlorethane ($C_2H_4C_{12}$)	Flammable colourless liquid, soluble in water. Man-made, used in synthesis of other chemicals	Mutagen, liver, lung, kidney damage	$700\,\mu g/m^3$, 24 h
Dichloromethane (CH_2C_{12})	Non-flammable, clear liquid, highly volatile. Paint remover, solvent, polyurethane foam blowing agent	Carcinogen (animals)	$3000\,\mu g/m^3$, 24 h
Formaldehyde (HCHO)	Common aldehyde at room temperature is a gas. Intermediary in CH_4 cycle. Insulating material	Carcinogen (animals)	$100\,\mu g/m^3$, 30 min
PAH	Polynuclear aromatic hydrocarbons group of synthetic chemicals from incomplete combustion of organic materials	Carcinogen	No safe level
Styrene (C_6H_5CH)	Volatile, colourless liquid used in manufacture of polymers, reinforced plastics and polystyrene	Suspected mutagen	$70\,\mu g/m^3$, 30 min
Tetrachloroethylene (C_2C_{14})	Non-flammable compound, insoluble in H_2O. Solvent in dry cleaning and metal cleaning, etc.	Toxic to CNS and liver	$5\,mg/m^3$, 24 h
Toluene	Non-corrosive volatile liquid. Sourced at petroleum refinery and styrene production, etc. In paint thinners, inks and adhesives, some cosmetics	Toxic to CNS	$7.5\,mg/m^3$, 24 h
Trichloroethylene (C_2HC_{13})	Man-made from ethane or dichloroethane. Degreasing fabricated metals, dry cleaning, printing, paint production, adhesives, carpet cleaners, etc.	Neurobehavioural, liver and kidney effects	$1\,mg/m^3$, 24 h
Vinyl chloride (VC)	Colourless gas from VC production, PVC facilities, landfills	Carcinogen	No safe level

Data in part from WHO, 1987 with permission

The 'mass flow rate of x g/h' is defined as the total emission occurring for 1 h of operation of a facility under operating conditions which are most unfavourable to the maintentace of air quality.

The 'mass concentration' of air pollutants in the waste gas is defined such that:

1. All daily means shall not exceed the emission standard mass concentration.
2. Of all half-hourly means 97 per cent shall not exceed six-fifths of the established mass concentration.
3. All half-hourly means shall not exceed the established mass concentration more than twice.

Table 8.25 lists the carcinogenic substances and their classes and some non-carcinogens and their classes.

8.8.1 Emission Standards from Waste Incinerators

The most up to date EU emission standards for waste incinerators are set down in the EU directive on incinerators of hazardous waste. Table 8.26 shows a comparison of EU limits and those for Germany and the Netherlands. In comparing these 1992 emission limits with the TA Luft limits (1986) it is seen that new emission limits are many times more stringent than the 'old' limits. For instance, the 'new' limit for HCl is 5 times lower than its previous value. The 'new' limit for CO is $50\,mg/m^3$, which is only 5 times that of the WHO (1987) ambient standard.

Table 8.24 Some non-criteria pollutants—inorganic substances

Pollutant	Description and source	Health effects	WHO ambient guidelines
Arsenic (As)	Ubiquitous in nature—metallic and non-metallic volcanic activity, smelting of metals, fuel combustion pesticides	Carcinogen	No safe level
Asbestos	Group of naturally occurring fibrous serpentine or amphibole minerals. Used in building industry, heat insulation. Fibres $< 3\ \mu m$ inhalable	Carcinogen	No safe level
Cadmium (Cd)	Soft silver white metal, by product of zinc production. Metal electroplating, plastics, etc.	Animal carcinogen	$< 20\ mg/m^3$
Chromium (Cr)	Grey hard metal. Cr^{3+}, Cr^{6+}. Ubiquitous in nature and soils. Used in tanning industry	Cr^{6+} carcinogen	No safe level
Hydrogen sulphide (H_2S)	Colourless gas, soluble in water and alcohol. Formed from organic matter in absence of O^2. In viscous industry, wastewater treatment, oil refining, tanning, pulp industry.	Intoxicant—eye irritant	$150\ \mu g/m^3$, 24 h
Manganese (Mn)	Earth crust's fifth most abundant metal. Used in metallurgy processes, alloy constituent, fertilizer, leather, textile, glass industry	Toxic at high levels to CNS and lungs	$1\ \mu g/m^3$, annual average
Mercury (Hg)	Metallic, mercurous or mercuric (-3) states. Inorganic mercury $\rightarrow$ methyl mercury by microbes. Mining, chloralkali plants, paint preservative, batteries, medical equipment, etc.	Bioaccumulative—CNS damage, kidneys	$1\ \mu g/m^3$, indoor annual average
Nickel (Ni)	Silver-white hard metal found in earth's crust. Used in steel production, electroplating, coinage, etc.	Carcinogen	No safe level
Radon (Rn)	Radioactive noble gas in several isotropic forms. Background level is about $3\ Bq/m^3$. Uranium mining, from soils and rocks, in groundwater and air	Risk of lung cancer	$100\ Bq/m^3$ in buildings
Vanadium (Va)	Ubiquitous bright white metal. Used in metallurgy, coal combustion, fuel combustion	Bronchitis, pneumonitis, upper respiratory tract effects	$1\ \mu g/m^3$, 24 h

Data in part from WHO, 1987 with permission

Table 8.26 includes emission limits for dioxins and furans of $0.1\ mg/m^3$. Dioxins represent a family of chemicals referred to as TCDD or 2,3,7,8-tetrachlordibenzopara-dioxin. This chemical occurs as an impurity in the manufacture of many chemicals and pesticides having a trichlorophenol base. When plastics are burned, small amounts are released to the atmosphere, as from pulp and paper mills, pesticide plants, vehicles, forest fires and cigarette smoke. Corresponding chloro-substituted isomers of the parent molecule dibenzofuran (PCDF) also occur and are about one-tenth as toxic as TCDD. Dioxins are formed optimally in the temperature range 180 to 400 °C as by-products during the synthesis of certain herbicides, PCBs and naphthalenes (Heffron, 1993). Degradation is not yet possible biologically but molecular destruction does occur in liquid injection incinerators operating at temperatures in excess of 1200 °C. Toxic contamination from TCDD or PCDF is exhibited by chloracne and related dermal lesions. It is a suspected carcinogen, but not a genotoxic agent.

Table 8.27 shows the acceptable level of daily intake of 2,3,7,8-TCDD by different agencies. These limits are set because 2,3,7,8-TCDD is found in the air, in food (butter, meat, fish, eggs, etc.), in cigarette smoke, in vehicle exhaust fumes, etc. The limits are set in pg (picogram $= 10^{-12}$ g) per kg of body weight per day. USEPA values on dioxins were dramatically reduced in 1995. Evidence from Seveso, Italy, is now indicating that dioxins are more hazardous than originally considered.

Table 8.25 TA Luft emission classes

Class	Carcinogenic substances	Some non-carcinogenic substances†
I	Asbestos	Acetaldehyde
I	Benzo(a)pyrene	Chloromethane
I	Beryllium	Formaldehyde
I	Dibenz(a,h)anthracen	Nitrobenzene
I	2-Napthalen	Phenol, etc.
II	Arsenic	Chlorobenzene
II	Chromium (VI)	Acetic acid
II	Cobalt	Carbon disulphide
II	3,3-Dichlorobenzidine	Naphthaline
II	Dimethyl sulphate	Propionic acid
II	Ethyleneimine	Vinyl acetate
II	Nickel	Xylenes
III	Acrylonitril	Acetone
III	Benzene	Chloroethane
III	1,3-Butadiene	Ethyl acetate
III	1-Chloro-2,3-epoxipropane	Pinenes
III	1,2-Dibromemethane	Paraffin HCs
III	1,2-Epoxipropane	Methyl benzoate
III	Ethylen oxide	Dichloromethane
III	Hydrazine	Trichlorofluoromethane
III	Vinyl chloride	

† For detailed list refer to Annexe E of TA Luft (1987).
Reproduced with permission

Table 8.26 Emission limits from waste incinerators

Pollutant	Averaging Time (hours)	Germany (mg/m^3)	Netherlands (mg/m^3)	EC Directive 1994 (mg/m^3)
Total dust (PM-10)	24	10	5	10
Total organic carbon (TOC)	24	10	10	10
Inorganic chlorine Compounds HCl	24	10	10	10
Inorganic fluorine Compounds HF	24	1	1	1
Sulphur oxides (SO$_2$)	24	50	40	50
Carbon monoxide (CO)	24	50	50	50
Cadmium (Cd) ⎱	30 min–8 h	0.05	0.05	−0.1
Thallium (Ti) ⎰	30 min–8 h			
Mercury		0.05	0.05	−0.1
Total other heavy metals (Sb + As + Pb + Cr + Co + C$_4$ + Mn + Ni + V + Sn)	0.5–4	0.5	0.5	−0.1
Dioxins + furons	6–16 h	0.1 ng/m^3	0.1 ng/m^3	0.1 ng/n^3

Data from EC, 1992 and EC 1994, with permission

Table 8.27 Acceptable daily intakes of 2,3,7,8-TCDD by different agencies

Agency	Acceptable daily intake (pg/kg/day)
Canada	10
Netherlands	4
USEPA	0.006
WHO Europe (1991)	10
Germany	1–10

8.9 AIR POLLUTION METEOROLOGY

When a gaseous or particulate emission, be it from a vehicle exhaust, an industrial stack or other source, is released into the atmosphere its fate is almost impossible to predict. This is so because of the complex factors that influence its subsequent pathways. The influencing factors are primarily:

- Meteorological
- Source
- Process

The meteorological factors of interest are:

- Wind speed and direction
- Temperature and humidity
- Turbulence
- Atmospheric stability
- Topographic effects on meteorology

Air pollution emissions are of interest at three scales:

- Microscale—of the order of 1 km (e.g. chimney plumes)
- Mesoscale—of the order of 100 kms (e.g. mountain-valley winds)
- Macroscale— ~ thousands of km (e.g. highs/lows over oceans or continents)

These scales are also time related and since wind speeds are ~ 5.0 m/s, the microscale meteorological effects occur at durations of minutes to hours, the mesoscale from hours to days and the macroscale at days to weeks.

For instance, after the Chernobyl nuclear explosion of 1986, the plume on a microscale very seriously affected the local region within hours of the emission, while the highlands of Wales were affected with wet deposition approximately four days later. For air pollution transport with sources such as power station plumes, industrial plumes or plumes from accidents, most transport interest is within the atmospheric boundary layer (ABL). The ABL is the lower (500 to 1000 m) layer of the earth's atmosphere, which is influenced by the earth's surface shearing and heating effects. Dispersion of pollutants within the ABL is controlled by turbulence which varies strongly with stratification. In the unstable or convective boundary layer (CBL), turbulence is characterized by large coherent eddies with scale of the order of the CBL or ~ 1 to 2 km. In the stable boundary layer (SBL), turbulence is much weaker with eddy size of the order of tens of metres or less. These differences lead to widely different dispersion rates. For example, in the CBL, a tall stack plume can be brought down to the earth's surface within a few kilometres of the source. In the SBL, elevated plumes may remain aloft with negligible surface impact for tens of kilometres (Weil, 1988). In the troposphere (i.e. the layer closest to the ground extending to an altitude of

10 to 15 km), the temperature decreases with height at a rate of approximately 9.8 °C per km. This vertical temperature gradient sets up convective currents or eddies where warm air rises and cool air falls.

Wind speed at the earth's surface is zero due to friction of the surface roughness. Remote from this surface, the wind speed increases with distance due to the earth's motion. In the near surface atmospheric boundary layer, random fluctuations of wind speed (and direction) are set up due to the surface roughness. In this zone, an instantaneous velocity measurement will have a mean and a fluctuating component. (Of course, wind speed is three dimensional, but the vector wind direction will for a specific location have a dominant direction.) This is represented by

$$U = u + u'$$

where
$$U = \text{instantaneous velocity}$$
$$u = \text{mean velocity component}$$
and
$$u' = \text{fluctuating component}$$

The mean component of velocity is due to the earth's motion, while the magnitude and sign of the fluctuating component is due to the presence and complexity of surface roughness (i.e. city buildings, a forest, etc.). The fluctuating velocity component superimposes eddy-type fluid structures on the mean velocity. This fluctuating velocity phenomena with eddies is called turbulence and because of surface roughness is called mechanical turbulence. Mechanical turbulence is partly responsible for air pollution dispersion.

Turbulence is also produced by the vertical temperature gradient. The heated surface of the earth causes the hot air or 'thermals' to rise, producing currents of air with hot air rising and cool air falling. This turbulent motion is called thermal turbulence or buoyancy. The opposite to buoyance may also occur on cold clear nights, when the ground radiates its heat away from the earth. The cold ground in turn cools the air above it, leading to sinking density currents. Figure 8.13 shows typical vertical profiles of temperature and wind speed. Both vary significantly in the ABL from day to night.

8.9.1 Ambient and Adiabatic Lapse Rates

In the lower troposphere, the temperature of the ambient air usually decreases with altitude. The rate of temperature change or gradient is known as the 'lapse rate'. A balloon equipped with a thermometer, when released, will move upwards through the atmosphere and record the temperature of the ambient air. This temperature gradient is known as the 'ambient lapse rate', and it varies from day to day, day to night, season to season. Most of the time the ambient lapse rate decreases with altitude, but there can be occasions when the reverse occurs.

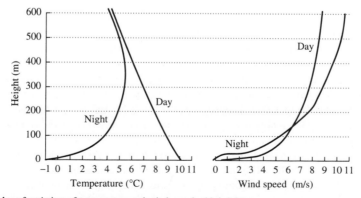

Figure 8.13 Examples of variation of temperature and wind speed with height.

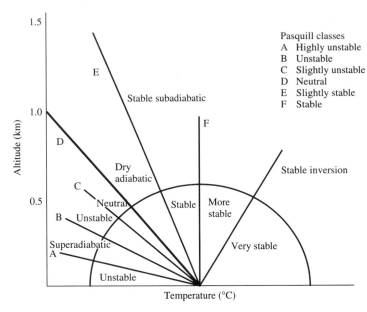

Figure 8.14 Pasquill–Gifford stability classes—vertical temperature profile.

The reference lapse rate by which the 'ambient lapse rate' is compared is the 'dry adiabatic lapse rate' (DALR). Under adiabatic (i.e. without the addition or loss of heat) conditions, a rising warm parcel of air behaves like a rising balloon. The air within the parcel expands on meeting air of lower density. It will expand until its own density is equal to the density of the surrounding air. Dry air expanding adiabatically cools at a rate of $9.8\,°C/km$ or $\sim 1\,°C/100\,m$. Figure 8.14 shows the relationship of the variety of possible ambient lapse rates to the dry adiabatic lapse rate as corresponding to the Pasquill stability classes.

8.9.2 Atmospheric Stability

A 'neutrally stable' atmosphere occurs when the ambient lapse rate (ALR) is equal to the dry adiabatic lapse rate (DALR); i.e. the rate of cooling with altitude is $\sim 1\,°C/100\,m$. In this case, if a parcel of air is moved upwards or downwards, its temperature adjusts to that of its surroundings. In any new position, it experiences no forces that encourage it to further adjust its position. It is stable in its old position and stable in its new position.

An 'unstable' atmosphere occurs when the ambient lapse rate exceeds the dry adiabatic lapse rate, i.e. the rate of cooling with altitude is $> 1\,°C/100\,m$. This steeper temperature gradient encourages greater thermal turbulence. In this case, if a parcel of air is moved upwards (say by mechanical turbulence or updraughts near a building complex), within itself it cools by $\sim 1\,°C/100\,m$, so it is warmer than its new surroundings. Due to buoyancy it will continue to climb. Similarly, if a parcel of air is moved downwards (say by a downdraught due to topography), it is cooler and denser than its surroundings and will thus continue to sink. This ambient condition is said to be 'unstable' with a 'superadiabatic' lapse rate.

A 'stable' atmosphere occurs when the ambient lapse rate is less than the dry adiabatic lapse rate, i.e. the rate of cooling is $< 1\,°C/100\,m$. The temperature gradient is less steep and thus responsible for less turbulent activity. A variant of this condition is the isothermal class where there is no variation of temperature with height. If a parcel of air is moved upwards, it will within itself cool at $\sim 1\,°C/100\,m$. It

Table 8.28 Percentage of time in each stability class (Ireland)

Stability class		Occurrence–hourly observations (%)	
		Cork 1980–87	Dublin 1963–88
A	Highly unstable	0.1	0.3
B	Unstable	1.7	2.2
C	Slightly unstable	5.0	5.7
D	Neutral	79,1	75.2
E	Slightly stable	6.8	8.6
F	Stable	7.3	7.9

will find itself cooler than the surroundings and therefore be forced to sink. Similarly, if a parcel of air is moved downwards it will heat at $\sim 1\,°C/100\,m$. It will find itself warmer than its surroundings and due to buoyancy will be forced back upwards. As such, the parcel of air does not want to move up or down from its 'stable' position. The ambient condition is said to be 'stable' with a 'subadiabatic' lapse rate.

A 'stable inversion' condition is a variant of the stable atmosphere. Here, the temperature increases with altitude. If a parcel of air is moved upwards it will within itself cool at $\sim 1\,°C/100\,m$. It will find itself much cooler than its surroundings, and so will be forced to sink. If a parcel of air is moved downwards, it will heat at $1\,°C/100\,m$. It will find itself warmer than its surroundings and due to buoyancy will be forced back upwards. An inversion temperature condition is a very stable condition, forcing air pollutants to remain trapped in the atmosphere for long periods.

The occurrence of the different stability classes depends on meteorological conditions. Table 8.28 lists the percentage of time in each stability class at two locations in Ireland (1980 to 1987). Typically in Ireland, neutral conditions may apply for about 80 per cent of the time. Neutral conditions are likely to occur with overcast conditions. Dispersion in this case is aided by mechanical turbulence rather than by thermal turbulence as in an unstable condition.

Table 8.29 is a relationship of stability class and wind speed. These classes were described earlier as the Pasquill stability classes. They are also dependent on the strength of the incoming solar radiation and the extent of cloud cover. If Ireland has an extensive cloud cover with slight to moderate incoming solar radiation, it is seen that the dominant neutral condition occurs with wind speeds ranging from 3 to 6 m/s.

8.9.3 Variation of Wind Speed with Altitude

Wind speed is recorded at a standard height of 10 m and is called U_{10}. Wind speed varies with height, from a low of zero at the ground surface to a maximum at some height above the influence of buildings and topography. This is called U_∞. The height at which U_∞ occurs depends on the terrain. In level rural country U_∞ occurs at a height of about 250 m, while the corresponding height is $> 500\,m$ in urban

Table 8.29 Pasquil stability classes

Wind speed U_{10} (m/g)	Solar radiation			Night time	
				cloud cover fraction	
	Strong	Moderate	Weak	$\geq \frac{4}{8}$	$\leq \frac{3}{8}$
< 2	A	A–B	B	$\geq$	$\leq$
2–3	A–B	B	C	E	F
3–5	B	B–C	C	D	E
5–6	C	C–D	D	D	D
> 6	C	D	D	D	D

complexes. To determine the magnitude of U_z, i.e. at any height, a power law relationship is sometimes used:

$$U_z = U_{10}\left(\frac{Z}{Z_{10}}\right)^p \tag{8.9}$$

where

U_z = wind (horizontal) speed at height z (required)

U_{10} = wind speed at 10 m

Z = height Z (stack tip height)

Z_{10} = 10 m

p = exponent

The exponent p varies with terrain and also with stability class, from about 0.1 to 0.4. Table 8.30 shows some typical urban/rural values. Z_∞ is the height of the boundary layer, typically ~ 250 m in rural terrain and 600 m in high urban areas. Other relationships for wind speed and height are used and are covered in Seinfeld (1986), Stern (1976) and Hanna *et al.* (1982).

Example 8.9   Use the power low velocity profile equation to determine U_{20}, U_{50}, U_{100}, U_{200} if $U_{10} = 5$ m/s and the terrain is on the bounds of rural/urban.

Solution   Assume stability class D—neutral. From Table 8.30, assume $p = 0.2$:

$$U_z = U_{10}\left(\frac{Z}{Z_{10}}\right)^{0.2}$$

$$U_{20} = 5\left(\frac{20}{10}\right)^{0.2} = 5.75 \text{ m/s}$$

$$U_{50} = 5\left(\frac{50}{10}\right)^{0.2} = 6.9 \text{ m/s}$$

$$U_{100} = 5\left(\frac{100}{10}\right)^{0.2} = 7.9 \text{ m/s}$$

$$U_{299} = 5\left(\frac{200}{10}\right)^{0.2} = 9.1 \text{ m/s}$$

It is relevant to note that while U_z increases with altitude, it does so only by 15 to 80 per cent in this example, over a height change of 20 to 200 m.

8.9.4 Variation of Wind Direction with Altitude

A time series plot of the 'along' wind direction of a plume would show that the 'mean' along wind direction is longitudinal. However, superimposed on this is a fluctuating lateral wind direction. This

Table 8.30 Exponent p power low velocity profit

Stability class	Smooth rural	Forest $Z_\infty = 400$ m	Urban Liverpool	Urban $Z_\infty = 500$ m
A–B–C	0.07–0.10		0.15–0.2	
D	0.14–0.16	0.28	0.21	0.4
E–F	0.2–0.33		0.21–0.33	

Adapted in part from Stern, 1976

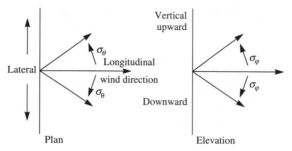

Figure 8.15 Definition of σ_θ and σ_ϕ standard deviations of wind direction.

fluctuation is termed 'standard deviation of lateral wind direction'. This is defined in Fig. 8.15. It varies with atmospheric stability class and to a lesser extent with height. Obviously, the more unstable the condition, the greater the standard deviation of the lateral wind direction, $\sigma\theta$. This is shown in Fig. 8.16 to vary from ~ 0 (stable) to $25\,^\circ$C (unstable). In the same way, the along wind direction defined in Fig. 8.15 has a vertical standard deviation, $\sigma\varphi$. This is shown in Fig. 8.17 to vary from 0 (stable) to 15–$25\,^\circ$C (unstable). Again, $\sigma\varphi$ varies with height. The impact of the magnitude of $\sigma\theta$ and $\sigma\varphi$ on the ability of the atmosphere to disperse a pollutant is significant, as explained in the following sections.

A very unstable atmosphere with $\sigma\theta \sim 25\,^\circ$C will cause greatest dispersion in the lateral direction. It will also cause dispersion in the vertical wind direction. Lateral dispersion is usually desirable. Vertical dispersion upwards is also desirable. However, vertical dispersion downwards may not be desirable if the contaminant plume contacts the ground surface. Buoyancy or thermal turbulence does not change with altitude for neutral or stable conditions, but increases significantly, with altitude, for unstable conditions. Mechanical turbulence impacts most at very stable conditions at low heights. Mechanical turbulence expires at heights above building or topographical complexes, above which buoyancy is the mechanism of turbulence.

8.9.5 Lapse Rates and Dispersion

If the ambient temperature profile is known (i.e. ambient lapse rate) and is compared to the adiabatic lapse rate, it is possible to estimate the fate of an air pollutant emission. Stack emissions produce plumes which vary with atmospheric stability class. Figure 8.18 shows the types of plumes developed for unstable, neutral and stable conditions. In Fig. 8.18(a), the unstable or superadiabatic ambient lapse rate produces a

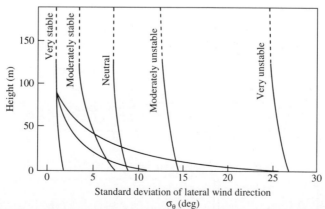

Figure 8.16 Vertical variation of the standard deviation of the lateral wind direction σ_θ (adapted from Seinfield, 1986. Copyright © 1986. Reprinted by permission of John Wiley & Sons, Inc.).

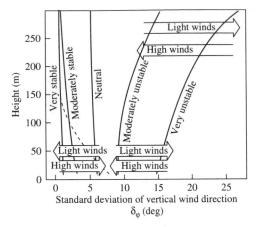

Figure 8.17 Vertical variation of the standard deviation of the vertical wind direction σ_ϕ (adapted from Seinfield, 1986. Copyright © 1986. Reprinted by permission of John Wiley & Sons, Inc.).

plume that is characterized by a 'looping' longitudinal elevation. From Fig. 8.17, it was seen that $\sigma\varphi \sim 20\,°C$ for the unstable condition. However, $\sigma\varphi$ might be either vertically upwards or downwards at a particular instant so the plume will be forced upwards and downwards by thermal turbulence. In Fig. 8.18(b), the neutral ambient lapse rate produces a plume with a characteristic 'coning' longitudinal elevation (and plan). In terms of vertical dispersion, $\sigma\varphi \sim 5\,°C$, as seen in Fig. 8.17. This causes a reduced vertical dispersion, in both the vertically upwards and downwards directions. The plume is unlikely to impact on ground in the near stack environment, but will at some distance downwind, depending on stack height and topography. In Fig. 8.18(c), the stable ambient lapse rate produces a plume with a characteristic 'fanning' longitudinal elevation. Vertical dispersion is almost nil since $\sigma\varphi \sim 0\,°C$, from Fig. 8.17.

Discontinuities in atmospheric stability   Figure 8.19 shows three possible discontinuities in the vertical temperature profile. Close to ground level, on clear winter nights, the earth radiates its heat back to space, causing the air close to the ground to be cooler than the air above it. Nocturnal inversions begin at dusk close to the ground and extend upwards as the night progresses. The inversion clears during the day.

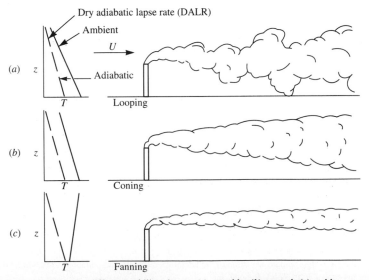

Figure 8.18 Chimney stack plumes for different stability classes: (a) unstable, (b) neutral, (c) stable.

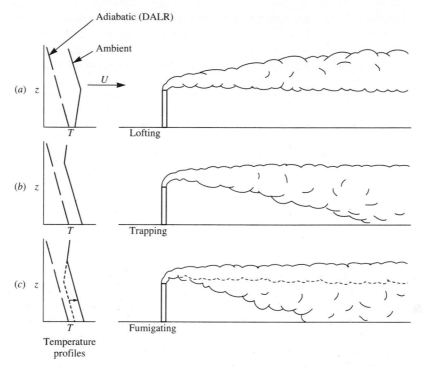

Figure 8.19 Chimney stack plumes for discontinuous stability classes: (*a*) inversion below, neutral aloft, (*b*) inversion aloft, neutral below, (*c*) inversion aloft, neutral below.

Figure 8.19(*a*) shows the profile of a very stable inversion below the stack tip and neutral conditions aloft. In this case there is mild mixing in the vertical, above the discontinuity. Below the discontinuity, there is no mixing and the plume remains aloft. Such plumes are called 'lofting'. Figure 8.19(*b*) shows a profile with near neutral conditions below the stack tip and a very stable inversion aloft. The latter inhibits any mixing above the stack height, while the near neutral condition below the stack tip encourages mixing in the vertical downwards direction. As such, the plume impacts the ground at some distance downwind. Such plumes are called 'trapping'. Figure 8.19(*c*) shows the profile of a plume resulting from neutral conditions below the stack tip and an inversion aloft. The discontinuity is closer to the ground than in Fig. 8.19(*b*). The result is that the plume impacts the ground much closer to the stack. Such a plume is called 'fumigating'.

8.9.6 Terrain Effects on Dispersion

It may be required to investigate the dispersion of a pollutant over a variety of terrain of topographical features including:

- Heat islands
- Land/sea interfaces
- Valleys or hillslopes

A heat island might be natural or man made. In an urban complex for instance, heat is absorbed and re-radiated at higher rates than in rural areas. Vertical convection currents are set up which are superimposed on the prevailing meteorological condition. As such, there is less atmospheric stability over such a feature and urban chimney plumes are more likely to impact on ground more rapidly than rural plumes.

At the interface of land and sea, convection currents are set up with opposite rotation from day to night. At night time, the land mass cools quicker than the sea mass, so cool air from the land at ground level flows towards the warmer area over the sea, which due to buoyancy rises and sets up a circulation. The reverse occurs during the day period. As such, at the interface area a nocturnal inversion is produced, creating a 'fanning' plume.

Valleys produce their own microclimate. Urban valleys produce unique dispersive plume characteristics. The sides of valleys are warmed by solar radiation. They in turn radiate heat, and particularly at night warm the 'cool' air in the valley thus setting up convective currents. In the absence of wind flowing through the valley, flushing of plumes may not occur as frequently as it does in open terrain. The convective valley currents then may be responsible for trapping plumes and increasing ground level impacts.

8.10 ATMOSPHERIC DISPERSION

Plume dispersion from a chimney stack is a convenient phenomena to discuss atmospheric dispersion, although many other areas are also of interest, e.g. fugitive emissions from pipes and vents, vehicle exhaust emissions, plumes from fires or explosions and emissions from dump sites. In an effort to predict concentrations of pollutants remote from the source, many techniques of air quality modelling have evolved. These include:

- Gaussian
- Numerical
- Statistical
- Empirical
- Physical

Traditionally, 'physical modelling' utilizing wind tunnels was used. However, only a few centres in the world had adequately sized wind tunnels to investigate urban pollutant plumes. To overcome scaling effects, models of urban areas required to be of adequate size so as to properly represent the variation in vertical velocity and temperature profile. Physical modelling is desirable in multisource emissions on a complex terrain.

'Statistical or empirical' techniques are used if inadequate information on the physicochemical processes exists to satisfy input requirements of numerical or Gaussian modelling. Such modelling will tend to produce information to assist with understanding of the fundamentals of the air quality problem. Such an evaluation may require the data monitoring to assist with understanding of the problem.

'Numerical modelling' requires the solution of the equations for conservation of mass, energy and momentum in three dimensions. Such modelling capability is most desirable but is as yet not readily available. This is due to the complexity of fluid and mass transport and particularly the turbulent terms in these equations which are yet not fully amenable to solution.

'Gaussian' modelling is the more widely used technique for estimating the impact of non-reactive pollutants (USEPA, 1986). Gaussian modelling is far from being exact, as some of the model assumptions compromise accuracy. These assumptions include:

1. There is no variation in wind speed and direction between the source and the receptor.

2. All effluent remains in the atmosphere and no provision is made for wet or dry deposition or chemical conversion. Any plume impacting on the ground is totally reflected.

3. Dispersion does not occur in the downwind direction. It only occurs in the vertical and cross wind directions. The dispersion is stochastic and is described exactly by the Gaussian distribution.

4. Emission rates are assumed constant and continuous.

8.10.1 Chimney Plume Characteristics

Figure 8.20 is a schematic elevation of a plume concentration profile from a chimney stack. The stack height is H_s. The pollutant is emitted with a flow rate Q and an exit velocity W. The plume, due to exit velocity and buoyancy, will gain in vertical height, before being turned into the downwind direction by the prevailing atmospheric wind speed. The gain in plume height is called plume rise ΔH. The effective plume height is then

$$H_e = H_s + \Delta H$$

As the plume progresses in the downwind direction, it is assumed to diffuse and the concentration profile takes on a Gaussian distribution. If the atmospheric condition is neutral, then a 'coning' plume will develop. The concentration at the centre-line of the plume will be greatest close to the stack and will decrease in a downwind direction. As the downwind direction increases, the extremities of the plume may impact on the ground as shown in Fig. 8.20. As the plume development is three dimensional, the dispersion prediction model must be able to account for changes in the x, y, z positions. The concentration of a pollution at any point is such that

$$C(x, y, z) \propto \frac{1}{U} Q G$$

$$\frac{1}{U} \qquad (U \text{ is wind speed}) \tag{8.10}$$

$$Q \qquad (Q \text{ is the emission flow rate}) \tag{8.11}$$

$$G \qquad (G \text{ is normalized Gaussian curve in the } yz \text{ plane}) \tag{8.12}$$

8.10.2 Gaussian Distribution

Figure 8.21(a) shows the normal (Gaussian) curve, used to describe an occurrence governed by probability. The height of the central ordinate is called the 'mean' and the width of the curve is described

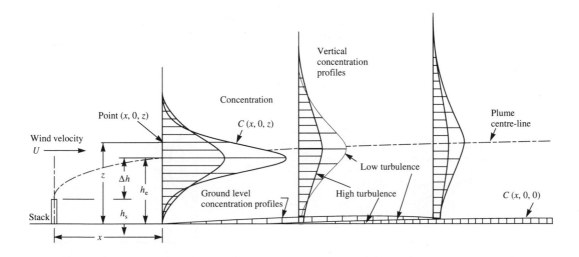

Figure 8.20 Schematic elevation of plume concentration profile.

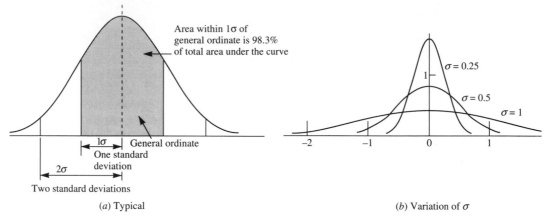

(a) Typical (b) Variation of σ

Figure 8.21 Normal (Gaussian) distribution curve.

in 'standard deviations'. For instance, 68.3 per cent of the total area under the curve resides within 'one' standard deviation of the centre, 95.5 per cent corresponds to 'two' standard deviations and 99.75 per cent corresponds to 'three' standard deviations. The Gaussian distribution in the lateral y direction is described by

$$G_y = \frac{1}{\sqrt{2\pi}\sigma_y} \, \exp\left[-\frac{1}{2}\left(\frac{y}{\sigma_y}\right)^2\right]$$

Similarly in the z direction

$$G_z = \frac{1}{\sqrt{2\pi}\sigma_z} \, \exp\left[-\frac{1}{2}\left(\frac{z}{\sigma_z}\right)^2\right]$$

As the distance downwind increases, the peak centre-line concentration decreases as the plume spreads out wider in the y and z directions. Figure 8.21(b) indicates that σ increases as the distance downwind increases. Figure 8.22(a) shows a plan of the plume being tracked at different times. At any instant, the plume shows a meandering boundary. At time 10 min or 1 h, the boundaries extend and with it, the peak centre-line concentrations decrease, as shown schematically in Fig. 8.22(b).

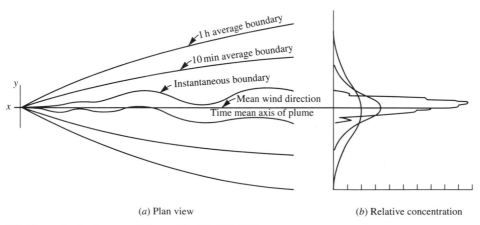

(a) Plan view (b) Relative concentration

Figure 8.22 Time series development of plume (after Seinfield, 1986).

8.10.3 Diffusion Equation

The concentration of a pollutant plume at any x, y, z location can be computed from the diffusion equation below.

$$C(x, y, z) = \frac{Q}{2\pi\sigma_y\sigma_z U}\exp\left[-\frac{1}{2}\left(\frac{y}{\sigma_y}\right)^2\right]\exp\left[-\frac{1}{2}\left(\frac{z}{\sigma_z}\right)^2\right] \tag{8.13}$$

For the coordinate system : $x = 0$ at stack

$\qquad\qquad\qquad\qquad y = 0$ at plume centre-line

$\qquad\qquad\qquad\qquad z = 0$ at ground level

If the effective stack height is H, then the z coordinate of the plume centre-line is $(z - H)$, where z is measured upwards from ground level:

$$C(x, y, z) = \frac{Q}{2\pi\sigma_y\sigma_z U}\exp\left[-\frac{1}{2}\left(\frac{y}{\sigma_y}\right)^2\right]\exp\left[-\frac{1}{2}\left(\frac{z-H}{\sigma_z}\right)^2\right] \tag{8.14}$$

In the conservation of mass assumption, that all plume contact with the ground is totally reflected, a second term must be added to account for this. This is shown schematically in Fig. 8.23. Then Eq. (8.14) becomes:

$$C(x, y, z) = \frac{Q}{2\pi\sigma_y\sigma_z U}\exp\left[-\frac{1}{2}\left(\frac{y}{\sigma_y}\right)^2\right]\left\{\exp\left[-\frac{1}{2}\left(\frac{z-H}{\sigma_z}\right)^2\right] + \exp\left[-\frac{1}{2}\left(\frac{z+H}{\sigma_z}\right)^2\right]\right\} \tag{8.15}$$

The appropriate units are:

$\qquad\qquad Q =$ any property per unit time, e.g. kg/s, m³/s

$\qquad\qquad C =$ that property per unit volume, e.g. kg/m³, m³/m³ or ppb

$\qquad\sigma_y, \sigma_z =$ diffusion coefficients, in m, as functions of downwind distance x

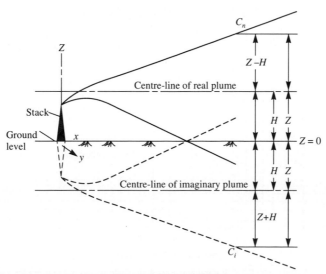

Figure 8.23 Coordinate system and ground reflection of plume development.

Figures 8.24 and 8.25 are the Pasquill–Gifford curves for σ_y and σ_z. Knowing the downwind distance x and the stability criteria, σ_y and σ_z can be read from the figures.

Most interest is in ground level concentrations ($z = 0$), so Eq. (8.15) reduces to

$$C(x, y, o) = \frac{Q}{\pi \sigma_y \sigma_z U} \exp\left[-\frac{1}{2}\left(\frac{y}{\sigma_y}\right)^2\right] \exp\left[-\frac{1}{2}\left(\frac{H}{\sigma_z}\right)^2\right] \tag{8.16}$$

The maximum ground level concentration occurs at the plume centre-line, at $y = 0$, so

$$C(x, o, o) = \frac{Q}{\pi \sigma_y \sigma_z U} \exp\left[-\frac{1}{2}\left(\frac{H}{\sigma_z}\right)^2\right] \tag{8.17}$$

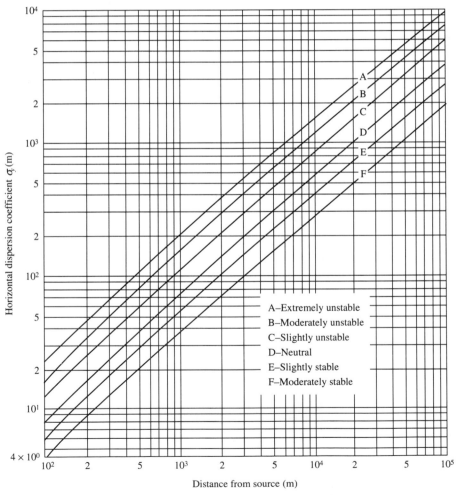

Figure 8.24 Correlations for σ_y based on the Pasquill stability classes A–F (Gifford, 1961). These are the so-called Pasquill–Gifford curves.

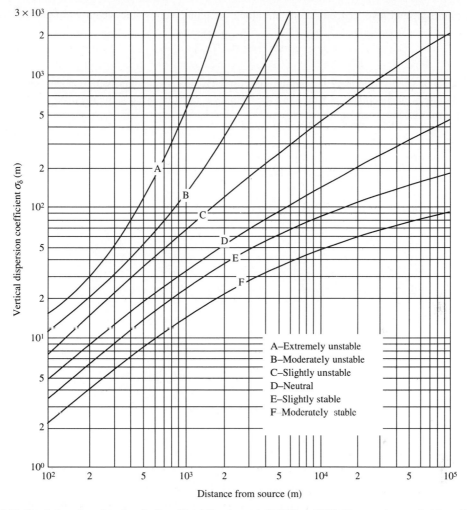

Figure 8.25 Correlations for σ_z based on the Pasquill stability classes A–F (Gifford, 1961). These are the so-called Pasquill–Gifford curves.

8.10.4 Ground Level Emission

Equation (8.15) for ground level emissions, such as fires or explosions or fugitive gases or smouldering landfill sites, becomes

$$C(x, y, z) = \frac{Q}{\pi \sigma_y \sigma_z U} \exp\left[-\frac{1}{2}\left(\frac{y}{\sigma_y}\right)^2 \right] \exp\left[-\frac{1}{2}\left(\frac{z}{\sigma_z}\right)^2 \right] \tag{8.18}$$

For ground level concentrations, $z = 0$:

$$C(x, y, o) = \frac{Q}{\pi \sigma_y \sigma_z U} \exp\left[-\frac{1}{2}\left(\frac{y}{\sigma_y}\right)^2 \right] \tag{8.19}$$

The maximum ground level concentrations are along the plume centre-line where $y = 0$:

$$C(x, o, o) = \frac{Q}{\pi \sigma_y \sigma_z U} \qquad (8.20)$$

Example 8.10 A dumpsite fire emits 3 g/s of NO_x. Determine the NO_x concentration at 2 km downwind if the windspeed $U_{10} = 5$ m/s and the stability is class D. What is the maximum concentration at ground level and also at 50 m above ground?

Solution

$$C(x, o, o) = \frac{Q}{\pi \sigma_y \sigma_z U} \qquad (8.20)$$

From Figs. 8.24 and 8.25, for $x = 2$ km,

$$\sigma_y = 150 \text{ m}$$
$$\sigma_z = 50 \text{ m}$$

Therefore $C(x = 2 \text{ km}, y = 0, z = 0) = \dfrac{3 \times 10^6}{\pi \times 150 \times 50 \times 5} = 25.5 \ \mu g \ NO_x/m^3$

$$C(x, y = 0, z = 50 \text{ m}) = \frac{Q}{\pi \sigma_y \sigma_z U} \exp\left[-\frac{1}{2}\left(\frac{z}{\sigma_z}\right)^2\right]$$

$$= \frac{3 \times 10^6}{\pi \times 150 \times 50 \times 5} \exp\left[-\frac{1}{2}\left(\frac{z}{\sigma_z}\right)^2\right]$$

$$= 15.5 \ \mu g \ NO_x/m^3$$

8.10.5 Plume Rise

When a chimney stack emits a pollutant plume, it does so with an exit velocity. This exit velocity (momentum) and buoyancy forces the emission upwards into the atmosphere before adapting to the direction of the prevailing wind direction.

The height of rise is called plume rise, ΔH. Briggs (1969) reviewed the phenomenon of plume rise and the many formulae and methods used to determine ΔH. Plume rise can have an effect on the ultimate ground level concentrations, reducing them significantly. Plume rise increases the effective stack height by 10 to 200 per cent. Early work on plume rise produced the Holland or Oak Ridge formula (US Weather Bureau, 1953):

$$\Delta H = \frac{2V_s r_s}{U}\left[1.5 + 2.68 \times 10^{-2} P\left(\frac{T_s - T_a}{T_s}\right) 2r_s\right] \qquad (8.21)$$

where

V_s = stack exit velocity, m/s

r_s = stack tip radius, m

U = wind speed, m/s

P = atmospheric pressure, kPa

T_s = stack temperature, K

T_a = air temperature, K

This equation includes a term for momentum and a term for buoyancy, where the latter was based on photographs of plume behaviour at Oak Ridge. Since then, many other formulae have come into existence and the recommendations of Briggs (1969) are most comprehensive.

For 'neutral' or 'unstable' conditions (A–B–C or D stabilities):
For $x < x_f$:

$$\Delta H = \frac{1.6 F^{1/3} x_f^{2/3}}{U} \tag{8.22}$$

where
x_f = downwind distance to final plume rise, m
U = wind speed at stack tip, m/s
F = buoyancy flux parameter

Then

$$F = g V_s r_s^2 \frac{(T_s - T_a)}{T_s} \qquad m^4/s^3 \tag{8.23}$$

and

$$x_f = \begin{cases} 2.16 F^{0.4} H_s^{0.6} & \text{for } H_s < 305 \text{ m} \tag{8.24} \\ 674^{0.4} & \text{for } H_s > 305 \text{ m} \tag{8.25} \end{cases}$$

For $x > x_f$:

$$\Delta H = \frac{1.6 F^{1/3} x_f^{2/3}}{U} \left[0.4 + 0.64 \frac{x}{x_f} + 2.2 \left(\frac{x}{x_f} \right)^2 \right] \left(1 + 0.8 \frac{x}{x_f} \right)^{-2} \tag{8.26}$$

For fossil fuel power plants with heat emissions > 20 MW (Briggs, 1969),

$$\Delta H = \begin{cases} \dfrac{1.6 F^{1/3} x_f^{2/3}}{U} & \text{for } x < 10 H_s \tag{8.27} \\[3mm] \dfrac{1.6 F^{1/3} (10 H_s)^{2/3}}{U} & \text{for } x > 10 H_s \tag{8.28} \end{cases}$$

For 'stable' conditions (E–F):

$$\Delta H = 2.4 \left(\frac{F}{US} \right)^{1/3} \tag{8.29}$$

where S is a stability parameter:

$$S = \frac{g}{T_a} = \left(\frac{\Delta T_a}{\Delta Z} + 0.01 \,°C/m \right) \tag{8.30}$$

Example 8.11   For the 915 MW power plant of earlier examples, compute the effective stack height for neutral or unstable conditions, using (a) the Holland equation, and (b) the Brigg's equation. If the 'stable' condition is such that $\dfrac{\Delta T_a}{\Delta z} = 2\,°C/km$, then determine (c) the plume rise using the appropriate Brigg's equation.

Stack tip radius = 4 m
Stack height = 250 m
Ambient temperature = 20 °C (293K)
Exit gas velocity = 15 m/s

$$\text{Exit gas temperature} = 140\,°C\ (413K)$$
$$\text{Atmospheric pressure} = 100\,kPa\ (1000\ mb)$$
$$\text{Stack tip wind speed} = 5\,m/s$$

Solution

(a) Holland's equation:

$$\Delta H = \frac{2 V_s r_s}{U}\left[1.5 + 2.6 \times 10^{-2} P\left(\frac{T_s - T_a}{T_s}\right) 2r_s\right]$$

$$= \frac{2 \times 15 \times 4}{5}\left[1.5 + 2.6 \times 10^{-2}\left(\frac{413 - 293}{413}\right)8\right] = 181\,m$$

(b) Brigg's equation for neutral or unstable conditions:

$$\Delta H = \frac{1.6 F^{1/3} x_f^{2/3}}{U}$$

$$F = g V_s r_s^2\left(\frac{T_s - T_a}{T_s}\right)$$

$$= 9.81 \times 15 \times 4^2\left(\frac{413 - 293}{413}\right) = 684\,m^4/s^3$$

$$x_f = 2.16 F^{0.4} H_s^{0.6}$$

$$= 2.16 \times 684^{0.4} \times 250^{0.6} = 807\,m$$

$$\Delta H = \frac{1.6 \times 684^{1/3} \times 807^{2/3}}{5} = 244\,m$$

Therefore for neutral or unstable conditions, Brigg's equations give $\Delta H = 244\,m$

(c) For stable conditions (E or F) we have Eq. (8.29)

$$\Delta H = 2.4\left(\frac{F}{US}\right)^{1/3}$$

$$S = \frac{g}{T_a}\left(\frac{\Delta T_a}{\Delta Z} + 0.01\right)$$

$$= \frac{9.81}{293}(0.002 + 0.01) = 4 \times 10^{-4}$$

$$\Delta H = 2.4\left(\frac{684}{5 \times 4 \times 10^{-4}}\right)^{1/3} = 168\,m$$

It is noted that the plume rise is about the same order of height as the stack height.

Example 8.12 A 915 MW power plant with a load factor of 72.5 per cent and an efficiency of 40 per cent uses coal as a fuel source. The coal has a 1 per cent sulphur content and a calorific value of 30 MJ/kg. The stack tip is 200 m high with a diameter of 7 m. If neutral conditions prevail, determine the maximum ground level concentrations of SO_2 at 1, 10 and 100 km from the plant. $U_{10} = 4\,m/s$, $T_s = 150\,°C$, $T_a = 20\,°C$ and $V_s = 15\,m/s$.

Solution:

$$\text{Power } 915 \text{ MW} = 3294 \times 10^3 \text{ MJ/h}$$

$$\text{Coal required} = \frac{3294 \times 10^3 \times 0.725}{0.4 \times 30} = 199 \text{ t/h}$$

$$\text{Sulphur at } 1\% = \frac{1}{100} \times 199 = 1.99 \text{ t/h}$$

$$\text{Sulphur dioxide SO}_2 = S + O_2$$

$$\text{Atomic mass} \quad 32 \quad 32$$

$$\text{Sulphur dioxide} = 1.99 \text{ t/h} + 1.99 \text{ t/h} = 3.98 \text{ t/h}$$

$$\text{Emission rate of SO}_2 = 1.1 \text{ kg/s}$$

$$\text{Wind speed at stack tip } U_s = U_{10}\left(\frac{z}{10}\right)^p$$

$$\text{For rural terrain } p \sim 0.16$$

$$U_s = 4\left(\frac{200}{10}\right)^{0.16} = 6.5 \text{ m/s}$$

Plume rise by Brigg's formula:

$$\text{Buoyancy flux } F = gV_s r_s^2 \left(\frac{T_s - T_a}{T_s}\right)$$

$$= 9.8 \times 15 \times 3.5^2 \left(\frac{423 - 293}{423}\right) = 553 \text{ m}^4/\text{s}^3$$

$$\text{Distance to full plume rise } x_f = 2.16 F^{0.4} H_s^{0.6}$$

$$= 2.16 \times 553^{0.4} \times 200^{0.6} = 648 \text{ m}$$

For $x = 1$ km:
$$\Delta H = \frac{1.6 F^{1/3} x_f^{2/3}}{U_s}$$

$$= \frac{1.6 \times 553^{1/3} \times 648^{2/3}}{6.5} = 150.5 \text{ m}$$

For $x = 10$ km $(>10H_s)$:
$$\Delta H = \frac{1.6 F^{1/3} \times (10 H_s)^{2/3}}{U}$$

$$= \frac{1.6 \times 553^{1/3} \times 2000^{2/3}}{6.5} = 320 \text{ m}$$

For $x = 100$ km:
$$\Delta H = 320 \text{ m}$$

Maximum ground level concentrations at 1, 10 and 100 km:

$$C(x, y = 0, z = 0) = \frac{Q}{\pi \sigma_y \sigma_z U} \exp\left[-\frac{1}{2}\left(\frac{H}{\sigma_z}\right)^2\right]$$

At 1 km:
$$H = H_s + \Delta H = 200 + 150 = 350 \text{ m}$$

From Figs 8.24 and 8.25:

$$\sigma_y = 75 \text{ m}$$

$$\sigma_z = 33 \text{ m}$$

Therefore

$$C(x = 1\,\text{km}) = \frac{1.1}{\pi \times 75 \times 33 \times 6.5} \exp\left[-\frac{1}{2}\left(\frac{350}{33}\right)^2\right]$$

$$= 8.2 \times 10^{-21}\,\mu\text{g/m}^3$$

At 10 km: $\qquad\qquad\qquad H = 200 + 320 = 520\,\text{m}$

From Figs. 8.24 and 8.25:

$$\sigma_y = 550\,\text{m}$$
$$\sigma_z = 140\,\text{m}$$

Therefore

$$C(x = 10\,\text{km}) = \frac{1.1}{\pi \times 550 \times 140 \times 6.5} \exp\left[-\frac{1}{2}\left(\frac{520}{140}\right)^2\right]$$

$$= 0.71\,\mu\text{g/m}^3$$

At 100 km: $\qquad\qquad\qquad H = 511\,\text{m}$

From Figs 8.24 and 8.25:

$$\sigma_y = 4000\,\text{m}$$
$$\sigma_z = 450\,\text{m}$$

Therefore

$$C(x = 100\,\text{km}) = \frac{1.1}{\pi \times 4000 \times 450 \times 6.5} \exp\left[-\frac{1}{2}\left(\frac{520}{450}\right)^2\right]$$

$$= 15.3\,\mu\text{g/m}^3$$

Recall that the WHO SO_2 standards for ambient air are $\sim 40\,\mu\text{g/m}^3$ for the annual exposure to $\sim 500\,\mu\text{g/m}^3$ for the 10 min exposure. In the above example, if the plume behaved strictly under neutral conditions, the local landscape around the plant would have negligible negative impact. Different atmospheric conditions plus rainfall would produce a very different result. As this example shows, the maximum concentration at ground level increases with distance from the stack. A 200 m stack height is considered very tall and ensures maximum atmospheric dispersion before ground impact. However, there may be local or source effects that negatively impact on such idealistic dispersion. These source effects are covered in the following section. Readers are referred to Turner (1970) for more detailed worked examples. The reader is encouraged to plot the above results and carry out further computations to identify at what distance downward the maximum ground level concentrations will occur.

Example 8.13 Determine the stack height for an industry source emitting 150 kg/day of 1,2-dichloroethane ($C_4H_4Cl_2$) if a residential complex is sited 1.5 km downwind and the ambient limit should not exceed $700\,\mu\text{g/m}^3$. The neutral conditions (D) occur 85 per cent of the time and this is to be the design atmospheric condition. The characteristics are:

$$\text{Gas exit velocity} = 15\,\text{m/s}$$
$$\text{Gas exit temperature} = 150\,°\text{C}$$
$$\text{Stack tip diameter} = 3\,\text{m}$$
$$\text{Ambient temperature} = 20\,°\text{C}$$

$$U_{10} = 4\,\text{m/s (assume } U_s \sim 6\,\text{m/s)}$$
$$Q = 150\,\text{kg/day} = 1.7\,\text{g/s}$$

Solution From Figs. 8.24 and 8.25, at 1.5 km, $\sigma_y = 100\,\text{m}$ and $\sigma_z = 33\,\text{m}$. Therefore

$$C(x = 1.5\,\text{km}, y = 0, z = 0) = \frac{Q}{\pi \sigma_y \sigma_z U} \exp\left[-\frac{1}{2}\left(\frac{H}{\sigma_z}\right)^2\right]$$

$$700 \times 10^{-6} = \frac{1.7}{\pi \times 100 \times 33 \times 6} \exp\left[-\frac{1}{2}\left(\frac{H}{\sigma_z}\right)^2\right]$$

$$\exp\left[-\frac{1}{2}\left(\frac{H}{\sigma_z}\right)^2\right] = 25.6$$

$$\exp\left[-\frac{1}{2}\left(\frac{H}{33}\right)^2\right] = 25.6$$

$$H^2 = 2178 \ln 25.6$$

Therefore

$$H = 84\,\text{m}$$

Effective plume $\qquad\qquad\qquad H = 84\,\text{m} = H_s + \Delta H$

From this, $\qquad\qquad\qquad\qquad H_s \sim 30\text{--}60\,\text{m}$

To determine H_s, iterate the following Brigg's equations:

$$\Delta H = \frac{1.6 F^{1/3} \times (10 H_s)^{2/3}}{U}$$

$$F = g V_s r_s^2 \frac{T_s - T_a}{T_s}$$

$$= 9.8 \times 15 \times 1.5^2 \frac{423 - 293}{423}$$

$$= 102$$

If $H_s = 50\,\text{m}$: $\qquad\qquad \Delta H = \dfrac{1.6 \times 102^{1/3} \times 500^{2/3}}{3} = 156\,\text{m (too high)}$

If $H_s = 30\,\text{m}$: $\qquad\qquad \Delta H = \dfrac{1.6 \times 102^{1/3} \times 300^{2/3}}{3} = 111\,\text{m (too high)}$

If $H_s = 10\,\text{m}$: $\qquad\qquad \Delta H = \dfrac{1.6 \times 102^{1/3} \times 100^{2/3}}{3} = 52\,\text{m}$

i.e. $\qquad\qquad\qquad\qquad H = 10 + 52 = 62\,\text{m} < 84\,\text{m (thus too low)}$

If $H_s = 15\,\text{m}$: $\qquad\qquad \Delta H = 69\,\text{m}$

$$H = 15 + 69 = 84\,\text{m}$$

Therefore

Minimum stack height required $H_s = 15\,\text{m}$. These computations are elementary. See Seinfeld (1986), Stern (1976) for more detailed references.

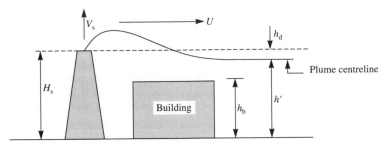

Figure 8.26 Downwash phenomenon—chimney stack downwind of buildings.

8.10.6 Source Effects on Plume Behaviour

Power plant stacks and incinerator chimneys are usually tall, as are many industrial gaseous venting structures. Similarly, cooling towers for power plant water vapour (H_2O) are also tall. In these cases, the venting structure is well above ancillary buildings to eliminate or reduce source effects. However, many industries use short vents on building roofs and low-level boiler chimneys. In these situations, as well as on residential and commercial structures, short exhausts are susceptible to interference by adjacent buildings. 'Downwash' is the term used to describe the phenomenon of a plume being drawn downwards after emission. It occurs if there is low pressure in the wake of a chimney stack. It is shown schematically in Fig. 8.26. If $V_s/U > 1.5$ (where V_s = stack emission velocity and U = horizontal wind speed), downwash does not normally occur.

Figure 8.27 shows the flow patterns around a bluff body. As the flow impinges on the upstream face, it creates a stagnation point, above which flow moves upwards and below which flow moves downwards.

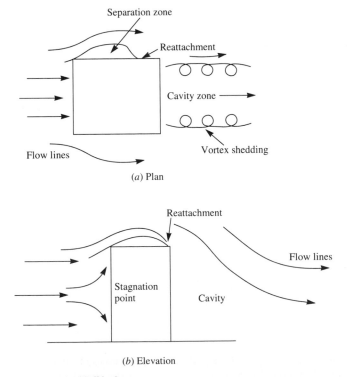

Figure 8.27 Schematic of flow around a bluff body.

On the roof and side walls, the flow separates from the building at the corners and reattaches on the roof (and side walls) further downstream. Between the separation point and reattachment points is a separation or cavity zone, with little exchange of air flow. Similarly, on the downwind face, the flow separates at the corners, creating a cavity zone for a short distance downwind. Again, in this cavity zone, little exchange of air occurs. There are therefore zones around buildings where air pollutants may gather and concentrate. Emission vents should account for such potential hot spots and be properly located and given adequate height to minimize the impacts. In locating emission sources upwind of buildings, it has been found (Hanna *et al.*, 1982) that if the source is $> 2H$ upstream and if $H_s > \frac{2}{3} H_b$, the plume will tend to rise over the building face. Parts of the plume at heights $< \frac{2}{3} H_b$ will be caught in the downwash in the frontal eddy for the lower part of the building. If the building length (L) is short, reattachment will not occur on the roof or walls and the plume will be deflected above the wake cavity. However, if the building is long, reattachment occurs and high concentrations are found at the roof locations and some of the plume becomes partially trapped in the building wake. If this occurs, recirculation at ground level of the pollutant occurs. A rule of thumb often applied is to keep the emission source at 2.5 times the building height. In many cases this is relaxed. Briggs (1973) developed methodologies to ascertain building effects on plumes.

8.11 PROBLEMS

8.1 An industry emits SO_2 at 24 h concentrations of 10, 60, 48, 57, and 11 ppm from Monday to Friday, 2–6 May 1994. On what days is the standard of 140 $\mu g/m^3$ exceeded?

8.2 If methane is found in the atmosphere at 1.5 ppm by volume, determine the concentration in ppm by mass.

8.3 If the total mass of particulates in the atmosphere from all sources is 10^6 tonnes and if it is assumed to be distributed entirely within the boundary layer 1 km high, compute the average concentration in ppm.

8.4 Prepare a table for the ambient air quality standards for the criteria pollutants for your particular legislative area (state or country) and compare to WHO (1987) recommendations.

8.5 Sketch a map of your part of the world and locate on it the power stations (excluding hydro). Identify whether the power is from nuclear, gas, coal or other sources.

8.6 From Table 8.6, estimate the amount of CO that is emitted per vehicle kilometre.

8.7 If Los Angeles has 10 million cars that are driven 20 000 km per year, compute the amount of CO and NO_x that they discharge into the Los Angeles Basin.

8.8 For a significant coal burning power station in your area, compute the annual output of particulates, SO_2 and CO_2. What technology does this plant use for particulate scrubbing.

8.9 If the power station of Problem 8.6 was to be replaced by a natural gas plant, for gas quality 98 per cent CH_4, determine the annual output of particulates, SO_2 and CO_2.

8.10 From your map of the power plants in your area (country) and knowing the power output, compute the amount of CO_2 emitted annually. How would you as a legislator plan to reduce this carbon load by 50 per cent by 2020?

8.11 Compute the annual evaporation from the land area assumed in Problem 8.8. Compare the natural and anthropogenic emissions of water vapour.

8.12 Use the power law to compute and draw the vertical velocity profile for an urban area and a flat rural area if $U_{10} = 8.5$ m/s in both cases. Draw the profile up to 100 m in height. Comment on the differences.

8.13 With respect to Table 8.28, determine for your area the percentage occurrences of each of the stability classes. What is the dominant stability class for dry desert conditions?

8.14 If a particular day was cloud free at midday, with a wind speed of 1.8 m/s at 10 m above the ground, what is the stability class?

8.15 For Problem 8.14, determine the wind speed at 180 m above the ground.

8.16 Derive the diffusion equation if it is assumed that there is no ground reflection.

8.17 An industrial plant has an emission rate of 0.11 kg/s of SO^2 from a stack of height 40 m. The tip exit velocity is 10 m/s. The gas exit temperature is 100 °C and the ambient temperature is 20 °C. Compute the ground level concentration at distances of 0.1, 0.2, 0.5, 1.0, 1.5, 2 and 2.5 km downwind of the stack. Draw a longitudinal profile of the concentrations.

8.18 For Problem 8.14, sketch the shape of the concentration profile for the atmospheric stability classes of highly unstable and highly stable. Explain your answers.

8.19 A source located in a rural area emits 0.5 kg of SO_2 per s from a 30 m high stack. The average plume rise is assumed to be 20 m and the wind speed at 10 m is 4.5 m/s. Determine the average maximum ground level concentration in $\mu g/m^3$.

8.20 For Problem 8.17, determine the maximum concentration in ppm. How far downwind does the maximum concentration occur.

REFERENCES AND FURTHER READING

American Conference of Government Industrial Hygienists (1990) *Advances in Air Sampling*, Lewis Publishers.

An Foras Forbartha (1987) *EEC Environmental Legislation. A Handbook for Irish Local Authorities*, AFF, Dublin.

Benarie, M. M. (1980) *Urban Air Pollution Modelling*, Mcmillan Press Ltd, New York.

Briggs, G. A. (1969) *Plume Rise*, US Atomic Energy Commission.

Briggs, G. A. (1973) *Diffusion Estimates for Small Emissions*, ATDL Contribution File 79, US Atomic Energy Commission.

Brookings, D. G. (1990) *The Indoor Radon Problem*, Columbia University Press, New York.

Broome, J. (1992) *Counting the Cost of Global Warming*, The White Horse Press, Cambridge, UK.

Brutsaert, W. (1982). *Evaporation into the Atmosphere*, Kluwer Academic Publishers, Dordrecht, Holland.

Carras, J. N. (1989) 'Some implications of measurements of plume characteristics for Gaussian models', *Journal of Clean Air*. May.

Carras, J. N. and D. J. Williams (1984) *Measurements of plume dispersion coefficients during convective conditions at various sites around Australia*. Proceedings of 8th International Clean Air Conference, Australia.

Corbitt R. A. (1989) *Standard Handbook on Environmental Engineering*, McGraw-Hill, New York.

Davis, M. and D. Cornwell (1991) *Introduction to Environmental Engineering*, McGraw-Hill, New York.

Duffy, N, and D. Cunningham (1992) Lecture Notes to Postgraduate Diploma in Environmental Engineering, University College Cork, Ireland.

EC (1987) *The State of the Environment in the European Community, 1986*, EC Official Publication, Brussels, March.

EC (1992a) *Proposal for a Council Directive on the Incineration of Hazardous Wastes*, Com (92), Brussels, 19 March.

EC (1992b) *The State of the Environment in the European Community*, Com (92), 23, Vol. III, Overview, Brussels, 27 March.

EC (1994) Council Directive, 94/67/EC *On the Incineration of Hazardous Waste*. OJ.L. 365. December.

Fleagle, R. G. and J. A. Businger (1963) *An Introduction to Atmospheric Physics*, Academic Press, New York, 346pp.

Gifford, F. A. (1961) 'Use of routine meteorological observations for estimating atmospheric dispersion, *Nuclear Safety*, **2**, 47–51.

Godish, T. (1988) *Air Quality*, Lewis Publishers.

Hanna, S. R., G. A. Briggs and R. P. Hosker (1982) *Handbook on Atmospheric Diffusion*, US Department of Energy.

Harrison, R. (1990) *Pollution Causes, Effects and Control*, Royal Society of Chemistry, London.

Heffron, J. J. (1993) A review of the toxicity of dioxins and furons, Proceedings of 1st Irish Atmospheric Conference, University College Cork, Ireland, February.

Henry, J. G. and G. W. Heinke, (1989) *Environmental Science and Engineering*, Prentice-Hall, N.J., USA.

Heskith, H. E. and F. L. Cross (1989) *Odor Control Including Hazardous/Toxic Odors*, Technomic, Lancaster, PA, USA.

HMSO (1989) *Dioxins in the Environment*, Pollution Paper 27, HMSO, London.

HMSO (1992) *Dioxins in Food*, Food Surveillance Paper 31, HMSO, London.

Institution of Engineers Ireland (1989) 'Industry and the environment', Proceedings of Conference, Cork, February.

Jackson, M. H., G. P. Morris, P. G. Smith and J. F. Crawford (1990) *Environmental Health Reference Book*, Reed International/Butterworth, Heinneman, London.

Jakeman, A. J., Jun Bai and J. A. Taylor (1988) 'On the variability of the wind speed exponent in urban air pollution models', *Journal of Atmospheric Environment*, **22**(9).

Maguill, P, F. Holden and C. Ackley (1956) *Air Pollution Handbook*, McGraw-Hill, New York.

Mangan, H. C. (1993) 'An overview of the atmospheric policy of the European Community, with specific reference to the Fifth EC Environmental Action Programme', *Local Authority News Ireland*, January.

Martin, G. B, R. E. Hall and J. S. Biomin (1979) 'Nitrogen oxides control technology for stationary area and point sources and related implementation posts', Technical Symposium on *Implications of a low NO$_x$ Vehicle Emission Standard*, USEPA, Reston, Virginia, May.

Masters, G. M. (1991) *Introduction to Environmental Engineering and Science*, Prentice-Hall, Englewood Cliffs, New Jersey.

Moloney, D. (1993) *Proceedings of the First Irish Atmospheric Environment Conference*, REMU, University College Cork, Ireland, February.

Monteith, J. L. and M. H. Unsworth (1990) *Principles of Environmental Physics*, Edward Arnold Publishers, London.

O'Neill, P. (1991) *Environmental Chemistry*, Chapman and Hall, London.

Osborn, P. D. (1989) *The Engineers Clean Air Handbook*, Butterworths, Oxford.

Painter, D. E. (1974) *Air Pollution Technology*, Reston Publishing Co., Virginia.

Peavy, H., D. Rowe and G. Tchobanoglous (1985) *Environmental Engineering*. McGraw-Hill, New York.

Petersen, W. B. and R. Allen. (1982) 'Carbon monoxide exposures to Los Angeles area commuters', *J. Air Pollution Control Assoc.*, **32**, 826–833.

Seinfeld, J. (1986) *Atmospheric Chemistry and Physics of Air Pollution*, John Wiley, New York.

Steffan, W. L. and O. T. Denmead (1988) *Flow and Transport in the Natural Environment: Advances and Applications*, Springer-Verlag, Berlin.

Stern, A. C. (1976) *Air Pollution*, Vols I, II, III, IV and V, Academic Press, New York.

Szepesi, D. J. (1989) *Compendium of Regulatory Air Quality Simulation Models*, Akademiai Kiado, Budapest.

Luft, T. A. (1987) *Technical Instructions on Air Quality Control*, Germany, February.

Technica International Ltd. (1988) *Whazan—User Guide*, Technical International Ltd, London.

Turner, D. B. (1964) 'A diffusion model for an urban area', *Journal of Applied Meteorology*, February.

Turner, D. B. (1970) *Workbook on Atmospheric Dispersion Estimates*. US Department of HEW, May.

US Congress, Office of Technology Assessment (1992) *Changing by Degrees. Steps to Produce Greenhouse Gases*, Cutter Corp., Arlington, Mass.

USEPA (1986) 'Guidelines on air quality models', EPA-450/2-78-027R, July.

USEPA (1987) 'Industrial source complex (ISC) dispersion model—users guide', EPA-450/4-88-002a, December.

USEPA (1988a) *National Air Pollution Emission–Estimates (1940–1986)*, US Environmental Protection Agency, Washington, D.C.

USEPA (1988b) 'Meteorological processor for regulatory models. Users guide', July.

USEPA (1986c) 'Screening procedures for estimating the air quality impact of stationary sources', PB 89-159396, August.

USEPA (1990) *National Air Quality Standards*, US Environmental Protection Agency, Washington, USA.

USEPA PTPLU (1982) 'A single source Gaussian dispersion model', EPA-600/8-82-014, August.

US Weather Bureau (1953) 'A meteorological survey of the oak ridge area: final report covering the period 1948–1952', USAEC Report ORO-99, Holland.

Vigneswaram, S., T. Mino and C. Polprasert (1989) *Selected Topics on Clean Technology*, Asian Institute of Technology, Bangkok, Thailand.

Weil, J. C. (1988) *Atmospheric Dispersion— Observation and Models in Flow and Transport in the Natural Environment. Advances and Applications*, Springer-Verlag, Berlin.

Wellburn, A. (1991) *Air Pollution and Acid Rain. The Biological Impact*, Longman Scientific and Technical, London.

WHO (1987) *Air Quality Guidelines for Europe*, European Series 23, World Health Authority Regional Publications. Copenhagen, Denmark.

WHO (1991) *Consultation on Tolerable Daily Intake from Food of PCDDs and PCDFs*, World Health Authority Regional Office for Europe, Copenhagen.

NOISE POLLUTION

9.1 INTRODUCTION

It is mainly in the highly industrialized countries, characterized by high volumes of traffic, that noise is perceived as a serious environmental nuisance. As with all pollutants, it reduces the 'quality of life' and causes a significant health hazard. For example, people living adjacent to busy roads tend to have higher blood pressure. High noise levels at work or at home can exacerbate existing health conditions. In addition to the risks to human health, noise pollution has negative ecological impacts on species sensitive to noise.

Noise is defined as unwanted sound, consequently it can be considered as the wrong sound in the wrong place at the wrong time. The degree of 'unwantedness' is frequently a psychological matter since the effects of noise can range from moderate annoyance to permanent hearing loss, and may be rated differently by different observers. Therefore it is often difficult to determine the benefits of reducing a specific noise. Noise does affect the inhabitants, humans, fauna, etc., in the natural environment. Although the impact of a particular noise source is limited to a specific area, noise is so pervasive that it is almost impossible to escape from it. Community social surveys almost always rate noise among the most annoying environmental nuisances. Common noise sources include traffic, industry and neighbours, the latter being often the most annoying although industrial noise is usually the source of most noise complaints.

High noise levels of sufficient duration can result in temporary or permanent hearing loss. This is generally associated with those working in industrial plants or operating machinery but can also occur at discotheques or close to aircraft on the ground if the exposure period is long enough. However, measurable hearing damage from most industrial sounds requires daily exposure for several years. On the other hand, environmental noise intrusions such as traffic noise can interfere with speech communication, disturb sleep and relaxation and interfere with the ability to perform complex tasks.

9.1.1 Sources of Noise Pollution

Noise can be emitted from a point source (electric fan) an areal source (discotheque) or a line source (moving train). Noise diffuses rapidly from its source and at an adequate distance from the source the noise is not felt. Noise pollution comes from a wide variety of sources, including:

- Traffic—the main source
- Industrial equipment
- Construction activities
- Sporting and crowd activities
- Low-flying aircraft

9.1.2 Levels of Noise in the Environment

The following statement is taken in total from the EC document on *The State of the Environment in the European Community* (EC, 1987) with permission:

> Because of the strong subjective component in the perception of noise, physical measurements of noise pollution provide only a partial picture of the real problem. One of the most reliable indicators of the problem is therefore public opinion; the opinion poll on the environment conducted by the European Commission in 1982, for example, showed that noise was seen as the second most important (environmental) problem at the local level. Similarly, the incidence of official complaints by the public about noise nuisances give a useful measure of the extent of the problem. According to this barometer, the problem is getting worse [as Fig. 9.1 illustrates]: numbers of complaints in England and Wales have doubled between 1975 and 1985. In France, in 1986, approximately 25% of the population were exposed to an average daily noise level of more than 65 dB(A); in Germany the figure is 15–20%.
>
> Among the different sources of noise, traffic is undoubtedly the most important. Aggregated data on traffic, however, do not provide a particularly sensitive or reliable index of noise generation, because different components of the traffic have different emission levels. Cars, for example, produce relatively little noise, and in any case have increased relatively slowly in numbers in recent years. Conversely, lorries are a much more important source of noise, and these have increased much more rapidly, with a rise of 20 per cent in total freight volumes between 1973 and 1981. [Figure 9.2 shows the range of noise levels for different vehicle types.]
>
> In addition to noise from road traffic, public complaints indicate concern with noise from several other sources, including industrial plants, building sites, aircraft landing and taking off, low-flying military aircraft and sporting activities. Undoubtedly, air passenger traffic has expanded considerably since 1970, though improvements in engine design have meant that noise emissions may not have risen proportionately. In general, however, there are too few data on these other sources of noise to make valid assessments of the situation in the European Union.

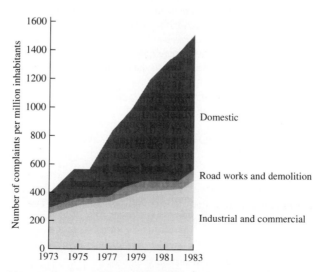

Figure 9.1 Official complaints about noise pollution in the United Kingdom 1973–83 (adapted from EC, 1987). Reprinted by permission of the Commission of the European Communities).

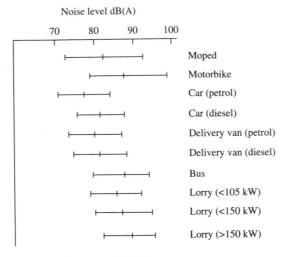

Figure 9.2 Noise emissions from vehicles in Germany (adapted from EC, 1987). Reprinted by permission of the Commission of the European Communities).

On the basis of the available evidence, it is stated that about 25 per cent of the total population complain about noise pollution. About 15 to 25 per cent of the population are probably exposed to a noise level that is sufficient to pose a significant health threat. It is not possible with the available data to determine whether the measures already taken are reducing the scale of the problem in the European Union, but as far as road traffic is concerned it seems that the problem is in fact getting worse. The situation seems to be deteriorating, particularly in urban areas, despite the more widespread construction of noise barriers along roads.

In the future, significant reductions in noise are likely to come mainly from the reduction of vehicle noise by more rigorous controls and by the introduction of quieter vehicles. Certainly this is possible, for the technology already exists to produce lorries with noise levels 10 dB(A) less than those currently on the road. On this basis, 20 of the new lorries would produce no more noise than one existing vehicle of similar size and capacity. The same is true of motor cycles. The adoption of such technology would inevitably increase the price of the vehicles, but this must be set against the costs of the passive measures of protection that would otherwise be needed (or the damage costs if no preventive action is taken). At the same time, improvements in the situation are also likely to be achieved by the application of environmental impact assessments. By requiring environment effects of infrastructural developments to be taken into account at the planning stage, these will encourage better design against noise from major projects such as road and industrial developments.

9.2 PHYSICAL PROPERTIES OF SOUND

9.2.1 Sound Waves

Sound is defined as any pressure variation that the human ear can detect. The simplest sound pressure variation (caused by a pure tone) produces the sinusoidal wave pattern shown in Fig. 9.3. Some of the elemental noise terms are:

- Amplitude (A)—the maximum or minimum pressure
- Wavelength (λ)—the distance between successive troughs or crests
- Period (P)—the time between successive peaks or troughs

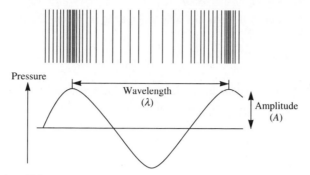

Figure 9.3 Parameters of a sinusoidal wave pattern.

- Frequency (f)—the number of complete pressure variations or cycles per second
- Speed of sound (c)

Period and frequency are related by

$$P = \frac{1}{f} \tag{9.1}$$

Wavelength and frequency are related by

$$\lambda = \frac{c}{f} \tag{9.2}$$

The speed of sound in air at sea level at 20 °C is approximately 340 m/s. Sound waves result from the vibration of solid objects or the separation of fluids as they pass over, around or through holes in solid objects. Vibrations are characterized by their frequency, amplitude and phase. Since noise generally consists of a large number of frequencies combined in random phase, the phase characteristics are generally not important and can be ignored. Only some mechanical vibrations can be perceived by the hearing mechanism of the human ear. They must be of a certain amplitude to be audible and the frequency must be within certain limits. Such audible vibrations occur in the hearing range which varies from person to person and is also dependent on age, extent of hearing loss and even physiological condition. The hearing range extends from a frequency of about 20 to 20 000 Hz.

Most sounds are not the purely sinusoidal vibrations as shown on Fig. 9.3. They vary both in frequency and magnitude over time. To quantify their magnitude over a measurement time T, the root mean square (r.m.s.) sound pressure (p_{rms}) is defined as follows:

$$p_{rms} = \left(\overline{p^2}\right)^{1/2} = \left[\frac{1}{T}\int_0^T p^2(t)\,\mathrm{d}t\right]^{1/2} \tag{9.3}$$

The r.m.s. sound pressure is therefore obtained by squaring the value of the pressure (amplitude) at each instant in time, summing the squared values over the measurement time T, dividing by T and taking the square root of the total.

9.2.2 Sound Power and Intensity

The rate at which energy is transmitted by sound waves is called the sound power (W) measured in watts. The average sound power per unit area normal to the direction of propagation of a sound wave is termed the acoustic or sound intensity (I). For example, if a small noise source, such as a vibrating or pulsating sphere, radiates a sound power of W watts in a spherical manner in a non-dissipative medium, the acoustic intensity at a distance r is

$$I = \frac{W}{4\pi r^2} \qquad \text{watts/m}^2 \qquad (9.4)$$

Also, at a sufficient distance away from the noise source, it can be shown that the intensity is proportional to the square of the sound pressure. The exact relationship is as follows:

$$I = \frac{p^2}{\rho c} \qquad (9.5)$$

where

I = acoustic intensity, W/m^2

p = sound pressure (r.m.s. value in pascals, Pa)

ρ = density of the medium, kg/m^3

(air = 1.185 kg/m^3 at 20 °C at standard pressure)

c = speed of sound in the medium, m/s

$$\left(\text{Note}: \ 1\,\text{watt} = 1\,\text{joule/sec} = \frac{1\text{N} - \text{m}}{\text{sec}} \right)$$

9.2.3 The Decibel

An enormous range of sound pressures are perceived by the human ear. The ratio of the weakest sound power to the largest perceived without pain is approximately one to one million. Also, the hearing mechanism responds to changes in sound pressures in a relative rather than in an absolute manner. Consequently, a scale based on ten times the logarithm of the ratios of the measured quantities to specified reference quantities is used for noise measurement purposes. The reference power level is 10^{-12} watts and the sound power level in decibels (dB) is defined as follows:

$$L_{\text{w}} = 10 \ \log_{10} \frac{W}{10^{-12}} \qquad (9.6)$$

where

L_{w} = the sound power level in dB for 10^{-12}W

W = the sound power of the noise source, watts (W)

Since sound power is proportional to the square of the sound pressure, the sound pressure level in decibels is defined as follows:

$$L_{\text{p}} = 10 \log_{10} \frac{p^2}{p_0^2} = 20 \log_{10} \frac{p}{p_0} \qquad (9.7)$$

where

L_{p} = the sound pressure level in decibels, dB

p = the measured pressure (r.m.s. value in Pa)

p_0 = the reference pressure (20 μPa)

The reference pressure is taken as the threshold of hearing, i.e. the weakest sound that the ear can detect. Such a sound would have a sound pressure level of zero decibels. However, sound pressure levels less than 25 decibels are not normally encountered except in specially constructed rooms such as broadcasting studios. Sound level meters measure the sound level in decibels and the lowest sound that can be measured by conventional equipment is about 38 dB. Typical sound pressure levels are shown on Fig. 9.4.

It should be noted that the term level denotes a relative measurement. The given quantity has a certain level above a specified reference value. For most practical purposes the sound power level is independent of the environment in which the noise source is located. The sound pressure level depends on both the power output of the noise source and the environment of the measurement position.

9.2.4 Combining Sound Pressure Levels

Decibel values cannot be added directly because they are logarithms.

Example 9.1   If a sound source has a pressure of 2000 μPa at 10 m distance, compute:

(a) the sound pressure level in dB,
(b) the sound intensity in W/m^2,
(c) the sound power in W.

Solution

(a) From Eq. (9.10) sound pressure $L_p = 20 \log_{10} \dfrac{p}{p_0}$

$$= 20 \log_{10} \left(\frac{2000 \times 10^{-6}}{20 \times 10^{-6}} \right)$$

$$= 40 \, \text{dB}$$

(b) From Eq. (9.5) sound intensity $I = \dfrac{p^2}{\rho c}$

$$= \frac{(2000 \times 10^{-6})^2}{1.185 \times 340}$$

$$= 9.9 \times 10^{-9} \text{W/m}^2$$

(c) From Eq. (9.4) sound power $W = 4\pi r^2 I$

$$= 4\pi \times 10^2 \times 9.9 \times 10^{-9}$$

$$= 12.5 \times 10^{-6} \text{W}$$

If two similar sound sources are added the power and intensity are doubled but not the pressure (dB). However, since the received sound pressure is proportional to the square root of the intensity, the new sound pressure would equal the original pressure multiplied by $\sqrt{2}$.

Example 9.2   If two sound sources have equal pressures of 2000 μPa, determine the sound pressure in dB.

Solution

$$\text{From Eq. (9.10) sound pressure } L_p = 20 \log_{10} \frac{p}{p_0}$$

$$= 20 \log \sqrt{2} \left(\frac{2000 \times 10^{-6}}{20 \times 10^{-6}} \right)$$

$$= 43 \, \text{dB}$$

Therefore doubling the existing sound source results in an increase in the sound pressure level of 3 dB. Similarly if ten equal sound sources were present instead of one, the sound pressure level would increase by 10 dB. Figure 9.4 shows a comparative scaling of sound pressure levels and decibels.

For noise pollution work, sound levels should be rounded to the nearest whole number. To simplify calculations the approximate addition rules (accurate to about 1 dB) shown in Table 9.1 may be used. If there are a number of noise sources to be combined, they should be added two at a time, starting at the two lowest values and adding on the next highest value to the result.

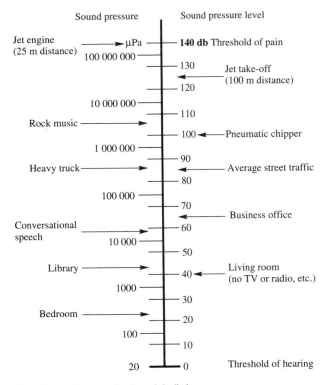

Figure 9.4 Comparative scaling of sound pressure levels and decibels.

Example 9.3 Determine the sound pressure level from combining the following four levels: 58, 62, 65 and 68 dB.

Solution Select the two lowest levels, 58 and 62 dB. Their difference is 4 dB and means adding 1 (or 1.5) to the higher, i.e.

$$58\,\text{dB} + 62\,\text{dB} = 63\,\text{dB}$$
$$63\,\text{dB} + 65\,\text{dB} = 67\,\text{dB}$$
$$67\,\text{dB} + 68\,\text{dB} = 71\,\text{dB}$$

A graph for solving decibel addition is shown in Fig. 9.5

9.2.5 Frequencies

Sounds of a single frequency, called pure tones, rarely exist except under artificial conditions. Most environmental sounds consist of a large number of frequencies. Audible sound frequencies range from

Table 9.1 Decibel addition (approximate)

Difference in sound pressure levels (dB)	Add to higher level
0 or 1	3
2 or 3	2
4 to 9	1
10 or more	0

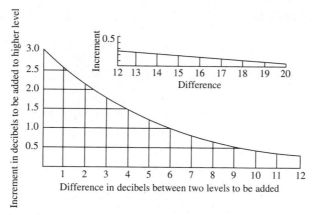

Figure 9.5 Graph for solving decibel addition problems (adapted from Davis and Cornwell, 1991. Reprinted by permission of McGraw-Hill, Inc.).

0.015 to 15 kHz. At frequencies less than 0.015 kHz the sound is not generally audible, although if sufficiently strong it can be felt as a vibration (infrasonic frequencies). Sound of frequencies greater than 15 kHz often cannot be heard by older listeners (ultrasonic frequencies). Between these two extremes, vibrations can be heard if they are of sufficient magnitude. The human voice contains frequency components between 0.08 and 8 kHz but is mainly concentrated in the range from 0.5 to 2 kHz (the piano contains frequencies from 0.0275 to 4.168 kHz). In practice sounds with a frequency greater than 8 kHz are not often commonly encountered and are not likely to give rise to complaints. Consequently sounds of frequencies above 8 kHz can usually be ignored in environmental noise monitoring.

For noise control purposes it is often necessary to identify the frequency components or spectra of sounds. The wide range of audible sound frequencies is divided into octave bands. An octave band is the frequency interval between a given frequency and twice that frequency. Thus the frequency intervals 0.05–0.1 kHz, 0.1–0.2 kHz and 1–2 kHz are all octave bands. For noise analysis a number of fixed octave bands are internationally recognized. These are centred on 0.0315 kHz as shown in Table 9.2.

Octave band analysis requires an octave filter set which may be incorporated with the sound level meter. All sound frequencies outside the selected band are rejected by electronic filters. When it proves difficult to identify an offending noise source, octave band analysis can be used to compare the measured noise frequency spectrum with those of different types of machinery. For remedial analysis of machinery a more refined analysis involving one-third octave or narrow-band frequency analysis is required.

Table 9.2 Octave bands

Octave band *centre* frequency (kHz)	Octave band frequency limits (kHz)	Human sensitizing
0.0315	0.022–0.044	Infrasonic
0.063	0.044–0.088	
0.125	0.088–0.176	Audible
0.250	0.176–0.353	
0.500	0.353–0.707	
1.000	0.707–1.414	
2.000	1.414–2.825	
4.000	2.825–5.650	
8.000	5.650–11.300	Ultrasonic
16.000	11.300–22.500	

9.2.6 Classification of Sounds

Sounds are classified as:

- Continuous
- Intermittent
- Impulsive

A continuous sound is an uninterrupted sound level that varies less than 5 dB during the period of observation. A typical example would be a household fan. Intermittent sound is a continuous sound that lasts for more than one second but then is interrupted for more than one second, e.g., a dentist's drill. If a sound is of short duration, less than one second, it is classified as an impulsive sound. Typewriter or hammering noises are typical impulsive sounds. A more rigorous classification of impulse sound would require a change of sound pressure of 40 dB or more within 0.5 second with a duration of less than one second. Gunfire would be an example of the latter, where the noise level may go from background of, say, 50 dB to 100 dB in a fraction of a second. It should be noted that special sound level meters are required to measure impulsive sounds, as their duration is only fractions of a second.

9.3 NOISE AND PEOPLE

The perception of sound by the human ear is a very complicated process and is not completely understood. The ear is divided anatomically into three sections, the outer, middle and inner ear. The outer and middle ear convert the sound pressure variations to vibrations and also protect the inner ear. The perception of sound is by nerve fibres in the inner ear. The hearing process consists of a number of separate processes. It should be noted that no simple and unique relationship exists between the physical measurement of sound and the human perception of the same sound. For further details, the reader is referred to Anderson and Anderson (1993).

9.3.1 Loudness

Loudness is a person's perception of the strength of a sound and to some extent is subjective. It varies with both the magnitude (sound pressure level) and pitch (frequency). Based on a large series of laboratory psychoacoustical experiments, a set of internationally agreed equal loudness level contours has been produced. These contours show how the loudness level of pure tones, with constant sound pressure, vary with frequency. However, since it is unlikely that this type of analysis is ever required in environmental noise abatement it is not considered further here.

As well as varying with frequency, loudness varies with the sound pressure level in a non-linear manner. If the physical intensity of a sound is increased so that the sound appears twice as loud, the increased sound pressure level is about 10 dB. Consequently, an increase of 10 dB approximates to a doubling of subjective loudness. Similarly a 10 dB decrease is considered as halving the noise. A 1 dB difference is the smallest discernible difference between two sounds identical in frequency and character but a difference of 2 or 3 dB would be required before people would notice differences. A difference of 5 dB is definitely perceptible.

Two sounds with the same sound pressure level in decibels but at different frequencies will be heard as different loudness levels. As previously noted, the range of audible frequencies is from about 20 Hz to about 16 kHz. However, young people and women have an upper limit of about 20 kHz. As one gets older, from the age of about 20, the top end of the range is reduced. The ear is most sensitive to frequencies in the range 1 to 5 kHz. Therefore a sound at this frequency would be rated much louder than one at the same sound pressure level at a frequency of, for example, 50 Hz or 10 kHz. To compensate for the frequency-dependent sensitivity of the ear, sound level meters incorporate electronic weighting networks which

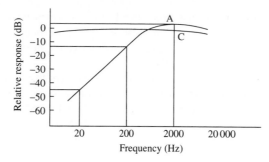

Figure 9.6 Frequency responses of A and C weighting networks.

correspond to the ear's response. Two principal weighting networks have been established, A and C, as shown in Fig. 9.6. The most important is the A network since there is general agreement that traffic, machinery, industrial and neighbourhood noise can be appropriately measured using this network. Sounds at higher frequencies (1 to 5 kHz) are given considerably more weight on the A network. Sound pressure levels measured on the A scale are referred to as dB(A) or dBA and are usually called sound levels. It is useful to note that the C network incorporates little modification by frequency, i.e. almost no filtering. Consequently if a measured sound level is much higher on the C scale than on the A scale, much of the noise is likely to be at low frequencies. Table 9.3 lists the internationally agreed weightings for the A and C networks.

9.3.2 Hearing Damage

The hearing damage potential of a given noise source depends not only on its level but also on its duration. It is generally accepted that a sound environment below 75 dB is not harmful (although much lower levels can cause annoyance and disturb sleep), while a single sound above 140 dB may produce permanent hearing damage. Between these two levels, the amount of hearing damage varies with the sound level, the length of exposure and the individual's susceptibility to noise. Other contributing factors are the number and length of quiet periods between exposures, the type of sound (continuous, intermittent or impulsive) and its frequency distribution. Sounds with most of their energy in the speech frequencies are more damaging. Hearing loss may be either temporary or permanent. Exposure to high noise levels for a short period of time can result in a temporary loss of hearing (temporary threshold shift) which may last for several hours depending on the duration and noise level. A ringing in the ears (tinnitus) may also occur. Repeated exposure to high noise levels may result in permanent hearing damage (permanent threshold shift). Permanent hearing damage can occur before the individual becomes aware of difficulties in communication. However, sounds that do not result in temporary hearing loss after two to eight hours of exposure tend not to produce permanent hearing loss if continued longer.

Hearing tests are carried out with an *audiometer* which produces a frequency related graph called an *audiogram*. The audiogram compares the individual's hearing against a reference standard. To determine the effects of exposure to a specific noise, a baseline audiogram, taken before the exposure, is essential in

Table 9.3 A and C network weighting values

Weighting network	Octave band centre frequency (Hz)								
	31.5	63	125	250	500	1k	2k	4k	8k
A (dB)	− 39.4	− 26.2	− 16.1	− 8.6	− 3.2	0	1.2	1.0	− 1.1
C (dB)	− 3.0	− 0.8	− 0.2	0	0	0	− 0.2	− 0.8	− 3.0

order to distinguish the impact of the noise from other hearing impairments such as ageing. Consequently, industrial workers and others exposed to high noise levels require hearing checks at regular intervals. However, environmental noises are seldom high enough to cause hearing damage and such tests are unlikely to be necessary for the general public.

The EC Directive (86/188/EEC) on the protection of workers from the risks related to exposure to noise at work is incorporated into the laws of EC member states (Commission of the European Communities, 1986). It specifies that certain actions must be taken where the daily personal exposure (eight hour equivalent) of a worker to noise is likely to exceed 85 dBA or where the maximum value of the unweighed instantaneous sound pressure is likely to be greater than 200 Pa, equivalent to 140 dB.

9.3.3 Speech Interference

The quality of speech communication depends on the noise level and the distance. It may also vary with the individuals involved. For normal conversation at about one metre distance, the background noise should not exceed 70 dBA. Shouted conversations at the same distance are possible up to about 85 dBA. To permit normal conversation at distances of about five metres would require a background noise level below 50 dBA. Satisfactory telephone conversations need background levels less than about 80 dBA.

9.3.4 Work Interference

When work does not involve spoken communication it is difficult to determine the effects of noise levels on performance. High noise levels may reduce the accuracy of the work being undertaken rather than the quantity. Steady noises appear to have little effect on work performance unless the A-weighted noise level exceeds about 90 dB (Davis and Cornwell, 1991). However, irregular noises, such as bangs or clicks, may interfere with performance at lower noise levels. Consequently, it is desirable to remove such features from the background noise.

9.3.5 Annoyance

The annoyance caused by noise varies greatly between people. For example, what is considered music by one person may be noise to another. The extent of annoyance from a given sound depends not only on the sound level and its duration but also on the listener and on the activity being undertaken at that time. The type of sound (continuous, intermittent or impulsive) and the time of day are also significant. Sounds at night are often considered twice as loud as the same sound during the day. To determine the overall community annoyance at a particular sound involves the sociological, political and demographic characteristics of the community in addition to the characteristics of the sound. It is not possible to state noise levels below which no one will be annoyed and above which everybody will be annoyed. According to a WHO task group, daytime noise levels of less than 50 dBA outdoors cause little or no serious annoyance in the community (OECD, 1986). However, lower complaint thresholds have been reported by Herbert et al. (1989) and a significant proportion of the community may be annoyed at levels below 50 dBA.

Sleep interference by noise causes great annoyance to many people. Intermittent or impulsive noises are particularly disturbing. Because of differences between people and locations, it is difficult to determine the noise level below which sleep interference will not occur.

9.3.6 Pattern of Noise Exposure Over the Day

Figure 9.7 shows broad bands to the levels of generalized individual noise exposure over 24 hours. It is seen that at night time, the levels go from about 35 dBA in suburban areas to levels of about 52 in urban areas. The commuting periods in busy roads causes exposure of about 80 dBA. The type of work or school activity determines the exposure during daytime hours and this can range from lows of 55 dBA in quiet offices to 90 dBA in noisy factories.

9.4 NOISE CRITERIA

As previously explained, the loudness of a sound is determined by its sound pressure level and its frequency. There is general agreement that environmental or community noise can be appropriately measured in dBA units which closely replicate the loudness perceived by the ear. However, noise levels often vary over time and criteria are required to quantify such varying levels. The severity of a noise problem can then be determined by the amount that the noise level exceeds a threshold or standard value of the specified criterion. The principal requirement of a noise criterion is that its values should correlate reasonably well with the perceived community annoyance at different noises. It should also be easy to measure and predict. Ideally a noise criterion should be simple to comprehend and applicable to any noise

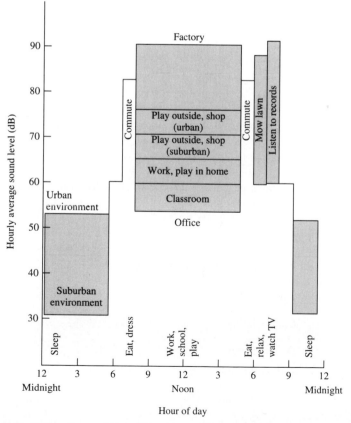

Figure 9.7 General individual noise exposure patterns (adapted from Corbett, 1989. Reprinted by permission of McGraw-Hill, Inc.).

source. Because of the wide differences in peoples' responses to noises it is unlikely that any single criterion would fulfil all the above requirements. A number of different criteria have been proposed (including those in ISO 1996) (1971) and the most commonly used criteria are:

- L_{Aeq}
- L_{AE}
- L_{AN}

9.4.1 L_{Aeq}, The Equivalent Continuous Level

The equivalent continuous level, L_{Aeq}, can be applied to a fluctuating noise source. It is the constant noise level over a given time period that produces the same amount of A weighted energy as the fluctuating level over the same time frame.

$$L_{Aeq} = 10 \log_{10}\left[\frac{1}{T}\int_0^T \frac{p(t)^2}{p_0^2}\, dt\right] \qquad (9.8)$$

$$= 10 \log_{10}\left(\frac{1}{T}\int_0^T 10^{0.1L_i}\, dt\right) \qquad (9.9)$$

where
$$T = \text{time period over which } L_{Aeq} \text{ is determined}$$
$$p(t) = \text{the instantaneous A-weighted sound pressure}$$
$$p_0 = \text{the reference sound pressure (20 } \mu\text{Pa)}$$

If the sampling methodology is discrete then L_{Aeq} becomes

$$L_{Aeq} = 10 \log_{10}\left(\frac{1}{T}\sum_{i=1}^{n} 10^{0.1L_i} t_i\right) \qquad (9.10)$$

where
$$n = \text{number of samples}$$
$$L_i = \text{noise level in the } i\text{th sample}$$
$$t_i = \text{the fraction of total time}$$

Example 9.4   An air conditioner generates a noise level of 75 dB for five minutes every hour. If the background noise level is 55 dB, compute the L_{Aeq}.

Solution

$$L_{Aeq} = 10 \log_{10}\left(\frac{1}{T}\int_0^T 10^{0.1L_i(t)} dt\right)$$

where
$$T = 1 \text{ hour}$$

$$L_{Aeq} = 10 \log_{10}\left[\frac{1}{T}(10^{0.1L_1}\Delta T_1 + 10^{0.1L_2}\Delta T_2)\right]$$

where
$$\Delta T_1 = 5 \text{ min} \quad \text{and} \quad L_1 = 75 \text{ dBA}$$
and
$$\Delta T_2 = 55 \text{ min} \quad \text{and} \quad L_2 = 55 \text{ dBA}$$

$$L_{Aeq} = 10 \lg_{10}\left[\frac{1}{60}(10^{7.5} \times 5 + 10^{5.5} \times 55)\right]$$

$$= 64.3 \text{ dBA}$$

9.4.2 L_{AE}, The Sound Exposure Level

The sound exposure level (SEL) is the constant level in dBA lasting for one second which has the same amount of A weighted energy as a transient noise. It may be used to express the energy of isolated noise events such as aircraft flyovers. It can be measured using some sound level meters.

$$L_{AE} = 10 \log_{10} \left[\frac{1}{t_0} \int_0^T \frac{p(t)^2}{p_0^2} \, dt \right] \tag{9.11}$$

where t_0 = the reference duration (1 second)

Further examples can be found in BRE and CIRIA (1993).

9.4.3 L_{AN}, The Sound Level Exceeded for $N\%$ of the time in dBA

This parameter is a statistical measure that indicates how frequently a particular sound level is exceeded. Time-varying noise may also be quantified in terms of the levels exceeded for different percentages of the measurement duration. The required percentile levels are derived from a statistical analysis of the noise. Measurements should be made using the fast response time setting of noise meters. Percentile levels reveal maximum and minimum noise levels and are often used in baseline studies taken prior to the introduction of new industrial or highway noise sources. Commonly used percentile levels include the L_{A10}, the sound level exceeded for 10 per cent of the time, which is sometimes used to represent maximum noise levels. The L_{A50} is the level exceeded for 50 per cent of the time. The L_{A90} is the level exceeded for 90 per cent of the time, and is often used to represent the background noise level. Alternatively, maximum levels may be represented by the L_{A1} or L_{A5} and background levels by the L_{A95}.

Percentile levels may be used as a complement to the L_{Aeq} as they provide information on the noise variation range. However, there is no direct relationship between the L_{Aeq} and the L_{AN}. An empirical relationship between the L_{Aeq} and the L_{A10} has been established for traffic noise where the traffic volume is in excess of 100 vehicles per hour. Over the 12 hour time period, 08.00 to 20.00 hours, the L_{A10} is approximately 2 dBA higher than the L_{Aeq} and 3 dBA higher for the 18 hour period 06.00 to 24.00 hours.

In addition to the three basic quantities defined in ISO 1996 (1971), other noise rating systems used in some countries include the day–night level, L_{dn}, a long-term measure of the L_{Aeq} with a 10 dB penalty applied to night time levels from 22.00 to 07.00 hours and the noise pollution level, L_{NP}, which is defined as follows:

$$L_{NP} = L_{Aeq} + K\sigma \tag{9.12}$$

where K is a constant usually taken as 2.56 and σ is the standard deviation. For traffic noise evaluation the traffic noise index (TNI), which correlates well with subjective annoyance, is sometimes used:

$$\text{TNI} = 4(L_{A10} - L_{A50}) + L_{A90} - 30 \tag{9.13}$$

Apart from the L_{Aeq}, the L_{AE} or selected values of L_{AN}, a number of different rating systems are used for aircraft noise monitoring These take the frequency spectrum associated with aircraft into account. For example, the perceived noise decibel, PNdB, is a complex rating requiring one-third octave band analysis. However, for many purposes it is sufficiently accurate to measure directly in dBA and add 13.

There is an international trend towards the use of the L_{Aeq}, supplemented by various L_{AN} levels, for the evaluation of community noise nuisance. However, impulsive sounds are particularly annoying and sound level meters equipped with an 'I' time weighting are required to determine the maximum levels of such sounds. Because of the very short rise time (less than $50\,\mu s$) sound level meters will not accurately determine the peak levels from small firearms and special equipment is required to determine the hearing hazards involved with noise from such weapons.

9.5 NOISE STANDARDS

Noise standards or thresholds are commonly specified as part of the planning permission (consent) for proposed developments. The specified values vary with the existing land use, the background noise level in the area and the type of development. There are often significant differences between the noise standards specified by different planning authorities. Specified external noise standards at nearby residences or at site boundaries can vary from an L_{Aeq} of 40 to 70 dBA by day and from 35 to 60 dBA by night. Typical values for residential areas would be at the lower end of this range.

9.5.1 EU Noise Directive

The EU Directive (86/188/EEC) is on the protection of workers from the risks related to exposure to noise at work. The objective of the directive is to reduce the level of noise experienced at work by taking action at the noise source. Two exposure levels are used:

- Daily personal noise exposure of a worker, $L_{\text{EP,d}}$:

$$L_{\text{EP,d}} = L_{\text{Aeq},T_e} + 10\log_{10}\frac{T_e}{T_0}$$

where

$$L_{\text{Aeq},T_e} = 10\log_{10}\left\{\frac{1}{T_e}\int_0^{T_e}\left[\frac{p_A(t)}{p(0)}\right]^2 dt\right\}$$

T_e = daily duration of a worker's exposure to noise

$T_0 = 8\text{h}$

p_A = A-weighted instantaneous sound pressure in Pa

- Weekly average of the daily values, $L_{\text{EP,w}}$:

$$L_{\text{EP,w}} = 10\log_{10}\left\{\frac{1}{5}\sum_{k=1}^{m}10^{0.1(L_{\text{EP,d}})_k}\right\}$$

where $(L_{\text{EP,d}})_k$ = the values of $L_{\text{EP,d}}$ for each of the m working days in the week being considered

The EU directive specifies that when the daily exposure level exceeds 85 dBA, the worker is to be advised of the risks and trained to use ear protectors. If the daily exposure level exceeds 90 dBA, a progamme to reduce levels should be put in place.

9.5.2 US Department of Labour Permissible Noise Exposure Levels

The US Department of Labour (1971 to 1979) defined the maximum permissible duration of exposure to noise levels which are reproduced in Table 9.4.

9.6 NOISE MEASUREMENT

9.6.1 Sound Level Meters

Although sound power cannot be measured directly, sound intensity can be measured with modern instrumentation (ISO 9614). Sound level meters are used to measure the sound pressure level. Noise meters are classified as follows:

Table 9.4 US Department of Labour permissible occupational noise levels

Sound level (dBA)	Duration (h/day)
90	8
92	6
95	4
97	3
100	2
102	1.5
105	1
110	0.5
115	< 0.25

Type 0   For laboratory reference situations
Type 1   Precision grade, used for accurate field measurements
Type 2   Industrial grade, for non-critical survey work
Type 3   Survey grade are low cost sound level indicators

Construction tolerances for various functions of the instrument system are specified in International Electrotechnical Committee (IEC) publications IEC 651 (conventional sound meters) and IEC 804 (integrating sound meters) and also in similar national standards such as BS 5969 and BS 6698, ANSI S1.4, etc. Each part of the instrument is controlled and tests specified to ensure compliance. The overall accuracy at specified reference conditions are 0.7 dB for type 1, 1.5 dB for type 2 and 2.5 dB for type 3. It is recommended that type 1 instruments be used for industrial measurements and for environmental measurements involving legislation.

The microphone is one of the most important elements of the sound level meter system and it typically determines the type of instrument. It must be protected from mechanical damage, moisture and low-frequency turbulence from wind. A special windshield and rain cover fitted over the microphone should be used whenever possible (even indoors) to shield the microphone from dust and dirt as well as from wind noise.

Sound level meters usually incorporate A and C weighting networks and provide both 'fast' and 'slow' settings. The slow setting is used for manually recorded observations only. Other features which may be provided include digital displays, impulse and peak level measurement, automatic calculation of the L_{Aeq}, L_{AN}, etc. Most types of 1 and 2 sound level meters include a standard RS-232 connection for data transfer to computers or printers. Sophisticated computer-based analysis programs which permit the determination and presentation of a wide range of data are common. Octave or one-third octave filtering and outputs for graphic level recorders or tape recorders are also available. It should be noted that tape recorders must match the frequency of the sound level meter used. However, tape recording is seldom necessary unless a specific analysis of the noise measurements that is not available with the measuring instrumentation is required (e.g. octave band analysis) or where it would be useful to demonstrate annoying sounds in legal court proceedings.

Digital audio tape (DAT) systems are available which can be used to record sounds directly in the field (instead of using sound level meters). All analyses and calculations of noise parameters are carried out by subsequent computer analysis. For example, narrow-band frequency analysis can be carried out. Full compliance with type 1 measurement specifications is provided by such systems. Available computer programs can give simultaneous audio and graphical outputs which can be useful for those involved in judging noise nuisances.

An individual worker's exposure to noise can be measured by a personal noise dose meter. Since the microphone is attached to the worker's clothes or headgear, a noise dose meter gives sound levels that are about 3 dB higher than the level that would be recorded in the absence of the worker, because of reflection effects.

9.6.2 Calibration

Sound level meters must be calibrated by a pistonphone or a sound level calibrator both before and after use. These provide a known acoustic signal at one or more frequencies. A thorough annual calibration either by the instrument manufacturer or by an acoustical test laboratory qualified to perform calibration is also necessary. A record of each annual calibration and of all calibration adjustments should be kept for each noise meter.

9.6.3 Measurement Procedure

The recommended procedure varies with the purposes of the noise measurement. For example, measurement of workplace sound pressure should preferably be made in the undisturbed sound field in the workplace (the person concerned being absent), with the microphone located at the position normally occupied by the ear exposed to the highest value of exposure (EC, 1986). Similarly, environmental noise measurements are normally taken at locations where the maximum noise nuisances are likely to occur. This is usually assumed to be at the nearest adjacent residences.

The microphone is often placed at a height of 1.2 m above the ground surface. The presence of adjacent sound reflecting surfaces must be taken into account. When noise intrusions are being measured, it is recommended that the background noise level should be at least 10 dB lower than the intrusive noise level. For the purposes of the Noise Insulation Regulations for housing in the United Kingdom, traffic noise is assessed at a reception point located 1 m in front of the most exposed part of an external window or door (EU Directive 86/188/EEC). However, where there are significant non-traffic noise sources present a different microphone position is specified.

9.6.4 Physical Conditions for Measurement

The wind speed and direction relative to the microphone must be noted. Valid measurements can be taken in wind speeds up to 5 m/s. At higher wind speeds turbulent noise caused by the wind may mask the noise source being measured. However, measurements may be acceptable with wind speeds up to 10 m/s. In general the peaks of wind noise should be at least 10 dB below the noise source being measured. Windshields should always be used for outdoor measurements.

Humidity levels up to about 90 per cent and pressure variations of ± 10 per cent have neglible effects on noise measurements. Unless the temperature is below $-10°C$ or above $50°C$ measurements are usually not affected. However, sudden changes in temperature may lead to condensation on the microphone. Sound level meters are relatively insensitive to vibration while magnetic or electrostatic fields have negligible effects. Wet road surfaces give increased noise levels; consequently traffic noise measurements are normally taken when the road surface is dry.

9.6.5 The Noise Measurement Report

Environmental noise data are often used in legal proceedings and it is essential that the conditions of the measurement be carefully documented in a formal measurement report. According to Bruel and Kjaer (1992), the following items should be included in a noise report:

1. The model type and serial number of the sound level meter used.
2. The date of the last laboratory calibration.
3. A statement of on-site calibration before and after the measurements (including the calibration equipment used).
4. The weighting networks and meter responses (fast or slow) used.
5. A description of the area and of the sound sources, including the type of sound (continuous, intermittent, impulsive, pure tones).
6. A sketch of the measurement site showing the locations of the microphone and the sound sources.
7. The time and date of the measurement.
8. A description of the meteorological conditions.
9. The background noise level if measuring noise intrusions.
10. The names of the people involved.
11. A general description of the measurements, including a summary of the levels of the various noise criteria for the relevant time periods. Special note should be made of the presence of pure tones (e.g. whistles) or impulsive sounds such as bangs, clanks, etc., which would have a startling effect.

In addition to the above items, environmental noise measurement reports typically include tabular values of the noise criteria levels for each hour of the measurement period and graphs showing the variation of these criteria throughout the measurement period. An example of an environmental noise measurement report is shown in the following section.

9.6.6 Case Study: Report on Noise Measurements at the ABC Sand and Gravel Plant

Plant noise source description   The principal noise sources are the compressor plant and the sand and gravel screening plant. Other occasional noise sources are from the ripping of rock and the loading of lorries with stone and gravel by overhead tipping. The plant location in a former quarry ensures complete protection from noise emissions for all adjacent residences on the south side of the river. There are no residences directly across the river which are not shielded by ground contours except that of Mr Newbury who operates a farm unit adjacent to his residence. However, the noise originating from this farm unit (about 64 dBA) completely masks any noise from the ABC plant.

Overall noise description   The maximum noise level within the site during full operation is about 70 dBA adjacent to the sand and gravel screening plant. At the inland boundary of the site the noise reduces to 50 dBA or lower due to the screening effect of the rising ground. No pure tones were noted but occasional impulsive noises resulted from the tipping of stones into trucks from overhead bins.

Noise measurements   Two sets of noise measurement were taken:

- From 17.00 h on Wednesday, 24 February to 17.00 h on Thursday, 25 February
- From 17.00 h on Thursday, 25 February to 17.00 h on Friday, 26 February

The microphone was positioned 30 m from the office building at a height of 1.2 m above the ground. On day 1, the wind was fresh to moderate. On day 2, the wind was moderate. Equipment used was an integrating sound level meter, Cirrus model type CRL 702, an outdoor microphone type MK 425 and a sound level calibrator type 5.11D. The instrument was set to sample the A-weighted sound levels at 16 times per second.

Results The results for day 1 are shown in Table 9.5 and Fig. 9.8. Analysis of Table 9.5 gives:

	24 h	$L_{eq} = 53.1$ dBA
Daytime:	17 h–20 h	$L_{eq} = 56.0$ dBA
Intermediate time:	6 h–7 h and 20 h–22 h	$L_{eq} = 40.4$ dBA
Night-time:	22 h–6 h	$L_{eq} = 38.3$ dBA

The lower level of detection of the meter was 35 dBA. During the night-time hours from 22 h to 6 h, it is seen from Fig. 9.8 that wind gusts raise the noise level from 38 up to 50 dBA even though the L_{eq} night-time reading was 38.3 dBA. Wind can raise the background level of noise appreciably.

The results for day 2 are shown in Table 9.6 and Fig. 9.9. Analysis of Table 9.6 gives:

	24 h	$L_{eq} = 55.5$ dBA
Daytime:	17 h–20 h	$L_{eq} = 58.3$ dBA
Intermediate time:	6 h–7 h and 20 h–22 h	$L_{eq} = 35.9$ dBA
Night-time:	22 h–6 h	$L_{eq} = 35.6$ dBA

As the night-time had only moderate winds, the L_{eq} night-time reading was 35.6 dBA. This compared with 38.3 dBA for the previous night. However, the L_{eq} daytime reading on day 2 was 58.3 dBA by comparison with 56.0 dBA on the previous day. Generally, both days gave similar results on the long averaging times. However, it is visible from the traces in both figures that instantaneous values exceeded 70 dBA, but this was as a result of impulsive noises of the loading of trucks.

Table 9.5 Noise measurements for day 1, at sand and gravel quarry site, all dBA

Period (h. min)	L_{eq}	S (standard deviation)	L_1 maximum levels	L_{10} ~ subjective annoyance	L_{95} ~ background noise
17.27	55.2	10.5	61.5	51.5	40.5
18.27	35.6	0.1	35.7	35.6	35.6
19.27	41.8	5.8	50.3	45.3	35.6
20.27	40.6	5.0	50.3	42.0	35.6
21.27	35.6	0.2	36.2	35.6	35.6
22.27	36.1	1.0	41.0	36.1	35.6
23.27	36.5	1.4	42.2	37.4	35.6
0.27	37.6	2.3	44.9	40.2	35.6
1.27	38.8	3.0	48.4	40.8	35.6
2.27	41.9	3.8	49.2	45.1	35.7
3.27	37.0	1.9	43.7	39.0	35.6
4.27	39.0	3.2	47.3	41.9	35.6
5.27	36.1	1.1	40.8	36.0	35.6
6.27	39.2	3.9	50.9	37.5	35.6
7.27	52.8	10.6	63.3	56.0	35.6
8.27	55.2	3.0	64.3	56.6	51.0
9.27	55.9	2.5	62.9	58.0	52.6
10.27	57.1	3.1	65.0	60.0	52.8
11.27	57.8	2.8	64.9	60.2	53.8
12.27	59.2	2.3	64.7	61.3	55.9
13.27	58.9	5.2	70.2	60.5	51.8
14.27	55.9	3.1	65.0	57.7	51.8

Overall L_{eq}: 53.1 dBA

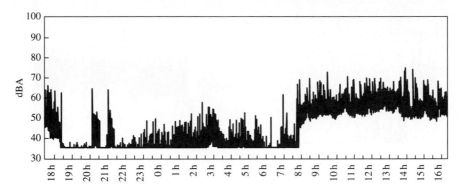

Figure 9.8 Noise measurement trace for day 1 at sand and gravel quarry.

9.7 OUTDOOR PROPAGATION OF SOUND

9.7.1 Geometrical Spreading

The sound pressure level received by a listener at a point due to a noise source with sound power W watts must frequently be estimated. Consider sound that spreads spherically from a source that may be regarded as a point. This can be assumed to occur when there are no nearby solid surfaces or fluid boundaries and

Table 9.6 Noise measurements for day 2, at sand and gravel quarry site, all dBA

Period (h. min)	L_{eq}	S (standard deviation)	L_1 maximum levels	L_{10} ~ subjective annoyance	L_{95} ~ background noise
16.29	58.0	2.3	63.4	60.1	54.8
17.29	56.3	6.2	63.0	59.1	47.3
18.29	40.1	4.5	52.2	37.8	35.6
19.29	35.9	0.8	40.1	35.7	35.6
20.29	35.7	0.4	36.9	35.6	35.6
21.29	35.6	0.1	35.7	35.6	35.6
22.29	35.6	0.0	35.6	35.6	35.6
23.29	35.6	0.0	35.6	35.6	35.6
0.29	35.6	0.0	35.6	35.6	35.6
1.29	35.6	0.1	35.6	35.6	35.6
2.29	35.6	0.1	35.6	35.6	35.6
3.29	35.6	0.1	35.6	35.6	35.6
4.29	35.6	0.1	35.6	35.6	35.6
5.29	35.6	0.1	35.6	35.6	35.6
6.29	65.1	2.1	45.8	37.2	35.6
7.29	55.4	4.0	62.0	58.0	49.9
8.29	58.0	2.5	64.8	60.1	54.5
9.29	58.7	2.5	65.1	60.9	55.1
10.29	60.5	4.4	72.8	61.9	54.1
11.29	60.2	2.5	65.3	62.7	56.5
12.29	59.5	2.7	65.5	62.0	55.2
13.29	58.7	3.2	67.1	61.0	54.1
14.29	61.2	2.4	66.2	63.6	57.4
15.29	58.0	2.4	64.1	60.2	54.6

Overall L_{eq}: 55.5 dBA

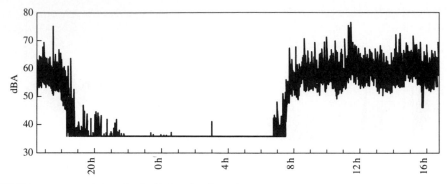

Figure 9.9 Noise measurement trace for day 2 at sand and gravel quarry.

when the dimensions of the sound source are small relative to the distance to the receiving point. As shown previously, the acoustic intensity (I) at a distance r metres is

$$I = \frac{W}{4\pi r^2} \qquad \text{watts/m}^2 \tag{9.4}$$

Consequently, the acoustic intensity is inversely proportional to the square of the distance. This is the *inverse square law*. The sound pressure level in decibels, L_p, as previously defined, is

$$L_p = 10 \log_{10} \frac{p^2}{p_0^2}$$

Since the acoustic intensity (I) is proportional to the square of the sound pressure, the sound pressure level received at a distance r metres from the point source of W watts is

$$L_p = 10 \log_{10} \frac{W/4\pi r^2}{W_0} \tag{9.14}$$

$$= 10 \log_{10} \frac{W}{W_0} - 10 \log_{10} 4\pi r^2 \tag{9.15}$$

$$= L_W - 20 \log r - 11 \tag{9.16}$$

where $\qquad W_0 =$ the reference power level, 10^{-12} watts

$L_W =$ the sound power level in dB for 10^{-12} watts

It should be noted that the sound power level is independent of the environment (it depends entirely on the noise source). The sound pressure level depends on the power output of the source, on the characteristics of the transmission path and on the environment of the measurement position.

Using Eq. (9.14) it can be shown that, if the *sound power* from the source is doubled, the resulting increase in the sound pressure level at distance r would be approximately 3 dB. Similarly, if there are ten identical sound sources instead of one, the resulting increase would be 10 dB. If the *distance* from a point source is doubled, the resulting decrease in the sound pressure level as determined from the above equation is approximately 6 dB.

9.7.2 Directivity

Most sound sources do not radiate sound uniformly in all directions. This occurs because of directional characteristics of the sound source (which may be frequency dependent) or because of external constraints by nearby surfaces. Directivity data for industrial sounds are seldom available and may have to be measured or estimated for each individual source. The directivity index (dB) in a specific direction is the difference between the measured sound power level and the value based on the assumption of uniform radiation in all directions.

The directivity effects due to surface constraints may be estimated by examining the position of the source. Examples of the latter include sound sources located close to the ground or those located remote from the ground close to a wall. Instead of spherical sound radiation, a hemispherical radiation pattern occurs. Consequently, the surface area through which the sound spreads is halved and the sound intensity is twice as high as for spherical radiation. The $4\pi r^2$ term in the above equation is replaced by $2\pi r^2$ and the sound pressure level at distance r becomes

$$L_p = L_w - 20 \log r - 8 \qquad (9.17)$$

The sound pressure level is therefore 3 dB higher than for the unconstrained situation.

9.7.3 Non-point Sound Sources

Noise sources cannot be assumed to be point sources unless the distance to the receiving point is large compared with the dimensions of the source. However, many environmental or community noise sources may be taken as point sources, e.g. a small machine or a car at distances greater than 20 to 30 m. An example of a non-point source is an infinitely long source which radiates sound waves that are cylindrical in shape. The surface area through which the sound radiates at a distance r from the source is $2\pi r$ per unit length. Therefore the sound pressure level at distance r is

$$L_p = L_{w/m} - 10 \log 2\pi r \qquad (9.18)$$
$$= L_{w/m} - 10 \log r - 8 \qquad (9.19)$$

where $\qquad\qquad L_{w/m} =$ sound power level per metre

From the above equation the noise decrease due to a doubling of the distance from the source is 3 dB (compared with 6 dB for a point source). The noise from a motorway or a long train approximates to such a source, giving a reduction of about 3 dB for doubling of distance.

9.7.4 Acoustic Near Field

The *near field* in noise means distances less than twice the maximum linear dimension of the noise source. Outside this distance, the position is considered as the *far field*. In the near field, the sound pressure level fluctuates with distance, and does not obey the inverse square law. Generally environmental noise measurements are not usually concerned with the near field.

9.7.5 Attenuation

The sound pressure level received from a non-directive source at a distance greater than the near field is attenuated by geometrical spreading and also by the environmental conditions between the source and the receiving position.

Attenuation due to distance   As previously described, the extent of geometrical spreading depends on the type of source and on the presence of nearby boundaries. Doubling of distance gives a 3 dB reduction for a line source or 6 dB for a point source.

Atmospheric attenuation Energy is always dissipated by transmission through a fluid due to heat conduction, viscosity, etc. The rate of sound absorption with distance depends on the frequency of the noise source. It also varies with the temperature and humidity, but pressure variations have little effect. As the temperature increases to about 20 °C (depending on the sound frequency) absorption increases, but above 25 °C the rate of absorption decreases. For a frequency of 1 kHz and a temperature of 10 °C, the reduction is approximately 3 dB per km. The rate of loss is larger for higher frequencies and for lower relative humidity values. Atmospheric attenuation is often relatively small compared with that from other factors except for the higher frequencies. Table 9.7 gives atmospheric attenuation values for the centre frequencies of the lowest third octave of each band. Frequency values in brackets are the octave band centre frequencies. When making calculations in octave bands, the one-third octave data quoted above may be used.

Attenuation due to meteorological conditions Relatively little information is available on the attenuation due to rain, mist, fog or snow. A value of 0.5 dB/km in fog has been quoted (Davis and Cornwell, 1991). These effects are frequently ignored in noise level predictions. However, variable temperature or wind speed gradients can result in large variations in noise levels at distances greater than 100 m from a noise source. When upwind of the source or when the temperature decreases with height, the sound waves are refracted away from the ground, resulting in decreased sound levels. The opposite occurs when downwind or when there is a temperature inversion. The effects of temperature inversions are negligible for short distances but may exceed 10 dB at distances over 800 m (Davis and Cornwell, 1991). Typically, meteorological attenuations range over ± 6 dB up to a frequency of 0.5 kHz and ± 10 dB above 0.5 kHz (Herbert et al., 1989). Environmental noise evaluations frequently omit wind and temperature effects because of their variability.

Ground surface effects Acoustically soft ground surfaces like grass, cultivated land or gravel absorb sound energy and reduce the received sound levels. Acoustically hard surfaces, such as concrete or water, reflect sound waves and absorb little sound energy. The extent of the sound attenuation from acoustically soft surfaces varies with frequency and with the heights of the noise source and receiver. Both should be less than about 10 m above ground and the grazing angle of the ground reflected ray should be less than about 3° for significant effects and less than about 0.5° for the maximum benefit (Herbert et al., 1989). Because of significant frequency effects, it is difficult to give any general absorption data for acoustically soft surfaces; attenuations range from 0 to more than 20 dB per 100 m. However, grass gives high attenuation values at low frequencies (0.3 to 1 kHz) where noise control is usually more difficult. Further information is available elsewhere (Herbert et al., 1989; Piercy et al., 1977, and Attenborough, 1982).

Table 9.7 Atmospheric absorption values, dB/km at 10 °C

Frequency (Hz)	Relative humidity (%)				
	60	70	80	90	100
50 (63)*	0.1	0.1	0.1	0.0	0.0
100 (125)	0.3	0.2	0.2	0.2	0.2
200 (250)	0.8	0.7	0.7	0.6	0.6
400 (500)	1.7	1.7	1.7	1.7	1.6
800 (1000)	3.0	3.1	3.1	3.1	3.2
1600 (2000)	7.2	6.2	6.2	6.0	5.9
3200 (4000)	23.1	17.8	17.8	16.3	15.2
6300 (8000)	82.3	62.4	62.4	55.9	50.9

*Those numbers in brackets are the octave band centre frequencies
Adapted from Herbert et al., 1989)

Sound attenuation by trees   Contrary to popular opinion, little sound reduction is provided by thin belts of trees. A wide (more than 50 m) and dense planting with foliage down to ground level is required for significant sound absorption. A reduction of about 0.1 dB/m thickness may be achieved.

Effect of ground topography   This varies with the closeness of the sound waves to the ground surface; all ground attenuation may be lost across a valley. No generalized information into the effects of topography is available and site measurements are normally required.

Reflecting surfaces and noise barriers   The sound level near a hard smooth vertical surface such as a building facade is the result of both direct and reflected sound waves. Immediately at the surface the combined effects give an increased sound level of 6 dB, reducing to about 3 dB within about 1 m of the surface. The effect of the vertical surface decreases as the measuring point is moved away and become neglible at distances greater than about 10 m.

A sound shadow is created when the line of sight from a noise source to a receiver is cut by a barrier. However, sound shadows are not clearly defined like light shadows because the wavelengths of sound are comparable with the dimensions of practical noise barriers. Consequently, sound waves tend to bend around the top and the ends of a barrier. The extent varies with frequency; the lower the frequency the greater the diffraction and the smaller the resulting noise attenuation. To be effective barriers must be as close as possible either to the noise source or to the receiver. They must also be sufficiently high and long to ensure that noise cannot get around the ends. Sound transmission through a barrier must also be minimized. Typically the barrier material requires a minimum mass per unit area of about $10 \, \text{kg/m}^2$. This permits the use of lightweight materials such as asbestos sheeting or chipboard. However, earth mounds are often preferable on appearance or convenience grounds. It is important to ensure that no cracks or gaps are present in a noise barrier or the expected attenuation may not be achieved.

The effects of barriers are complex functions of the difference between the direct and deflected noise path lengths and of the wavelengths of the sound. The estimation of the sound attenuation from finite length barriers is outside the scope of this text. The approximate attenuation from a point source by an infinitely long thin barrier of sufficient mass to ignore direct transmission can be determined as follows (Attenborough, 1991):

$$A = 10 \log_{10}(3 + 20N)$$

where
$$N = \pm \frac{\text{path difference}}{\text{wavelength}}$$

and
$$A = \text{attenuation in dB}$$

More detailed information is available elsewhere (HMSO, 1988; Herbert *et al.*, 1989). It should be noted that the attenuation due to ground effects are largely lost when a barrier is inserted because of the raised level of the new effective noise source (top of the barrier). Practical attenuations from noise barriers seldom exceed 10 to 15 dB.

9.7.6 Outdoor Noise Level Prediction

The sound pressure level received at a point depends on the source sound power level, including its directivity and location relative to nearby surfaces, and on the attenuation along the transmission path. The presence of surfaces close to the reception point may also affect the sound pressure level. The sound pressure level received from a point source can be estimated by the following equation:

$$L_p = L_w + D - 20 \log r - 11 - A_1 - A_2 - A_3 - A_4 + R \tag{9.20}$$

where

L_p = sound pressure level, dB

L_w = sound power level of source, dB for 10^{-12} watts

D = the directivity index, dB; for a source located near a hard flat surface $+3$ dB must be added

r = direct distance from source to receiving point, m

A_1 = atmospheric attenuation, dB

A_2 = attenuation by meteorological conditions, dB

A_3 = attenuation by ground, dB

A_4 = attenuation by barriers, dB

R = increase due to sound reflection at receiving point, dB

Where the source sound power level (in dB for 10^{-12} watts) is available for each octave band centre frequency, the attenuation should be calculated separately for each octave band. The resulting sound pressure level should then be calculated by combining the different dB levels as previously explained.

Example 9.5 A house is separated from a proposed new industrial building, first by a garden 10 m in length with a 2.0 m high hedge and then by 20 m of a level surfaced parking area. The noise from a fan located at a height of 1.8 m on the nearest wall of the proposed new building is given below. Estimate whether this additional noise source will cause an increased noise level at the house. The existing ambient noise level is 55 dBA.

Octave band centre frequency (Hz)	63	125	250	500	1k	2k	4k	8k
Sound power level (dB)	96	104	103	98	91	86	84	79

Solution (Using Eq. (9.20)):

$$L_p = L_w + D - 20 \log r - 11 - A_1 - A_2 - A_3 - A_4 + R$$

The atmospheric absorption will be insignificant for the distance involved and the variable effects of the meteorological conditions are ignored. Since the grassed area is short, little attenuation will be available, while no attenuation will be provided by the hard surfaced parking area. The hedge will not provide any noise barrier effect. Reflection effects are initially ignored.

Given $r = 30$ m and $D = 3$ dB for a hemispherical radiation pattern (no specific directivity data is provided), Eq. (9.20) reduces to,

$$L_p = L_w + 3 - 20 \log 30 - 11$$
$$= L_w - 37.5$$

Octave band centre frequency (Hz)	63	125	250	500	1k	2k	4k	8k
L_w (dB)	96	104	103	98	91	86	84	79
$L_p = L_w - 37.5$	58.5	66.5	65.5	60.5	53.5	48.5	46.5	41.5
A-weighting (dB)*	−26.2	−16.1	−8.6	−3.2	0	1.2	1.0	−1.1
A-weighted L_p	32.3	50.4	56.9	57.3	53.5	49.7	47.5	40.4

* Use Table 9.3 or Fig. 9.6.

The resulting sound pressure level is next calculated by combining the different A-weighted L_p levels (added two at a time, starting with the two lowest values) using Table 9.1 to give a level of 63 dBA. Combined with the existing ambient noise level of 55 dBA, the expected noise level is 64 dBA (63 dBA + 55 dBA = 64 dBA). Reflection effects close to the house facade could increase this to 67 dBA or more. Therefore there is a significant increase in the noise level at the house (55 dBA → 67 dBA).

9.8 NOISE CONTOURS

Many proposed developments require knowledge on what level of noise might be expected in the surrounding area. Figure 9.10 shows typical noise contours at an airport runway. Table 9.8 shows the land use guidance (LUG) zones for large airports. It is seen from Fig. 9.10 that the noise level at the airport boundary is approximately 75 dBA. As per Table 9.8, this noise level is between normally unacceptable and clearly unacceptable. At this range, positive noise abatement is essential. The noise levels as shown in Fig. 9.10 are defined as being the day–night average, L_{dn}. Here, daytime is defined as 7 a.m. to 10 p.m. and night-time is the remainder of the 24-hour period. The partial L_{dn} values are calculated for each significant noise intrusion. They are then summed according to the following equation to obtain the total L_{dn} due to all aircraft operations.

$$L_{dn} = 10 \log_{10} \sum_i \sum_j 10^{0.1 L_{dn}(i,\,j)} \tag{9.21}$$

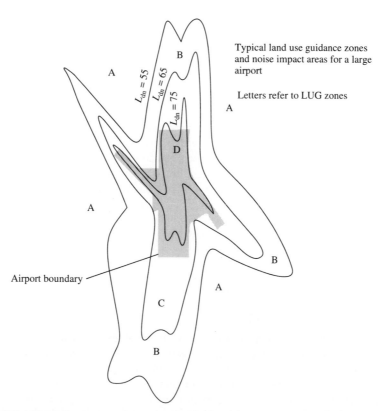

Figure 9.10 Typical noise contour map (adapted from Airport-Land Use Compatibility Planning FAA Advisory Circular 150/5050-6, 1977).

Table 9.8 Land use guidance chart for various levels of airport noise

Land use guidance zones	Noise exposure class	L_{dn} Day–night average sound level	Noise assessment guidelines	Suggested noise controls
A	Minimal exposure	0–55	'Clearly acceptable'	No special consideration
B	Moderate exposure	55–65	'Normally acceptable'	Land use controls
C	Significant exposure	65–75	'Normally unacceptable'	Controls recommended
D	Severe exposure	≥ 75	'Clearly unacceptable'	Controls recommended

Adapted from FAA advisory circular 150/5050-6, 1977)

Computer programs for noise prediction developed by the FAA are available in the public domain, including the integrated noise model (INM). Noise contours are important for proposed developments around existing 'noisy' sources so as to limit the amount of development, i.e. new houses near motorways.

9.9 NOISE SECTION OF AN ENVIRONMENTAL IMPACT ASSESSMENT

A formal environmental impact assessment statement (EIS) is required for larger proposed developments under EU legislation (see Chapter 19). Such assessment should also form part of the design process for other proposed developments. The noise section of an EIS should include the following.

9.9.1 Baseline Noise Survey

The objective of this survey is to record the noise climate in the potential impact area prior to the construction of the proposed development. It would normally include the measurement of the L_{Aeq} to indicate the overall noise level and selected L_{AN} values to represent the maximum and background noise levels—typically, either the L_{A1} or the L_{A5} to represent maximum levels and the L_{A95} for background levels. For road proposals the L_{A10} is frequently used to represent maximum traffic noise levels. The minimum acceptable measurement period would be at least 24 hours on a weekday. However, where there are significant daily noise variations, a longer period would be required. Overall, daytime and night-time (and perhaps intermediate-time) values of the selected noise criteria are normally determined. Hourly values of each criteria are also produced. The number of positions at which noise measurements are required depends on the locations of adjacent residences or other noise-sensitive areas relative to the likely noise sources.

9.9.2 Probable Noise Emission Levels

The potential noise sources, their sound power levels and their exact location within the proposed development are identified and the likely external emission levels at the site boundary, at nearby residences and at other noise-sensitive locations are calculated. The presence of any noise-reducing features either within the proposed development site or along the transmission path to the noise-sensitive locations would be included as part of the noise prediction process. Any impulsive noises or pure tones should be clearly noted. Potential noise emission levels are predicted for at least daytime and night periods. Consequently, the hours of operation of the various noise sources must be taken into account. The probable hours of operation of the proposed development, including any weekend working, should be clearly stated.

9.9.3 Statement of Probable Impact

The likely noise impact of the proposed development on residents and other users of the surrounding area is summarized. The predicted noise levels are compared with the existing levels and with acceptable values of the noise criteria used.

9.9.4 Proposed Remedial Measures

Where the predicted noise levels at noise-sensitive areas are considered excessive, recommendations on methods for reducing the noise levels should be made.

9.10 NOISE CONTROL

When the noise level received is excessive, the solution may involve attention to one or all of the three elements involved:

$$\text{Source} \longrightarrow \text{transmission path} \longrightarrow \text{receiver}$$

1. *Source*   The source could be modified by the acoustic treatment to machine surfaces, design changes, etc. This is a specialized area which is outside the scope of this chapter. However, an offending noise source could be stopped or its operation limited to certain times of the day.
2. *Transmission path*   The transmission path could be modified by containing the source inside a sound insulating enclosure, by the construction of a noise barrier or by the provision of absorbing materials along the path.
3. *Receiver*   The protection of the receiver by altering the work schedule or by the provision of ear protection is mainly applicable to those working with noisy machinery and is not considered here.

9.10.1 Sound Insulation Provided by Buildings

Environmental noise levels are typically specified at building facades due to large differences in the sound insulation provided by individual buildings. External sound will enter a building through the weakest transmission path, which is often through the windows. Consequently, the amount of insulation provided by the different building elements is of interest and is briefly considered in the following paragraphs.

When sound pressure waves meet a wall or another building surface they exert a fluctuating pressure, causing it to vibrate. Sound is then radiated by the vibrating surface into the space on both sides of the surface. Part of the sound energy is reflected, part is absorbed by the surface and the remainder is transmitted. All building elements, walls, ceilings, windows, etc., will vibrate to some extent, but the vibrations will be larger at certain frequencies, the lowest of these being called the natural frequency of the element. The natural frequency depends on the weight, suface area and rigidity of the building element. Since the natural frequencies of floors and walls tend to be low, it is much more difficult to insulate against noise with predominantly low frequencies.

The amount of sound stopped by a building element such as a door is known either as the sound reduction index or as the transmission loss. The amount of sound reduction is dependent on the frequency of the sound (more for higher frequencies) but depends primarily on the weight and rigidity of the

Table 9.9 Sound reduction indices of building elements

Element	Sound reduction index (dB)
Walls	
Concrete blocks (hollow, unpainted)	37
Brick (228 mm, plastered both sides)	50
Reinforced concrete (200 mm)	50
Stud partition (plasterboard + plaster on both sides)	35
Doors	
Panel (hollow core)	14
Hardwood	26
Acoustic	44
Floors	
Timber (with plasterboard ceiling + plaster skim coat underneath)	37
Reinforced concrete	
100 mm	45
200 mm	50
300 mm	52
Windows	
Single open	5–10
Single closed	15–20
Double glazing	
50 mm cavity	30
200 mm cavity	40

building element. The sound reduction index is measured in 16 one-third octaves from 100 to 3150 Hz and is quoted in one-third or octave bands in dB. It is often quoted as a single figure, e.g. 26 dB hardwood door. This is the arithmetic average of the 16 one-third octave values. Some typical sound reduction indices are shown in Table 9.9. However, it must be borne in mind that the actual sound reduction achieved may be less, due to cracks, ventilators or other leakage around the edges, etc.

9.11 PROBLEMS

9.1 If a sound source has a pressure of 3000 μPa at 10 m distance, compute:
  (a) the sound pressure level in dB
  (b) the sound intensity in W/m^2
  (c) the sound power in W.

9.2 Repeat Problem 9.1 for a sound pressure source of 3000 μPa at a distance of 20 m.

9.3 If two sound sources have equal pressures of 3000 μPa, determine the sound pressure in dB.

9.4 Determine the sound pressure level from combining the following four levels: 56, 68, 71 and 48 dB.

9.5 If an industrial fan generates a noise level of 65 dB for 10 minutes out of every hour compute the equivalent continuous level (L_{Aeq}) if the background level is 55 dB.

9.6 Repeat Problem 9.5 for a fan that runs for 20 minutes of every hour and 30 minutes of every hour. Plot your results and comment on the relationship.

9.7 Locate a reference in your library detailing the physiology of hearing noise and write a brief report with sketches of the mechanism of hearing.

9.8 An industrial complex operates at 80 dB for 5 hours per day and at 65 dB for 3 hours per day. Compute the $L_{EP,d}$.

9.9 A fan unit of an industry is located at 40 m from a house. Adjacent to the fan is a 20 m length of concrete paved area. At 20 m, there is a 2 m high, 0.3 m thick concrete wall. The 20 metres adjacent to the house is a grassed garden. The noise from the fan is given as:

Octave band centre frequency (Hz)	63	125	250	500	1k	2k	4k	8k
Sound power level (dB)	89	97	98	92	88	81	78	77

Determine the increased noise level at the house if the existing ambient noise level is 56 dB.

9.10 Go to your local planning authority and collect data on a noise pollution problem that was created by new development, either residential, industrial or traffic. Describe briefly the problem and the solution if it was identified. How would you have solved the noise problem?

REFERENCES AND FURTHER READING

Anderson, J. S. and M. B. Anderson (1993) *Noise, Its Measurement, Analysis Rating and Control*, Avebury Technical.

Ashford N. and P. H. Wright (1992) *Airport Engineering*, 3rd edn, John Wiley, New York.

Attenborough, K. (1982) 'Predicted ground effect for highway noise', *Journal of Sound and Vibration*, **81** 1982.

Attenborough, K. (1991) 'Noise pollution', chapter 9 in *Highway Pollution* R. S. Hamilton and R. M. Harrison (eds), Elsevier, Amsterdam.

BRE and CIRIA (1993) *Sound Control for Homes*, Building Research Establishment, Watford, UK.

Bruel and Kjaer. *Environmental Noise Measurement*, 2850 Naerum, Denmark.

Commission of the European Communities (1986) 'Council Directive on the protection of workers from the risks related to exposure to noise at work', 86/188/EEC.

Corbett R. (1989) *Standard Handbook on Environmental Engineering*, McGraw-Hill, New York.

Davis M. L. and D. A. Cornwell (1991) *Introduction to Environmental Engineering*. McGraw-Hill, New York.

EC (1986). *The State of the Environment in the European Community*, Publications of the European Communities, Luxembourg.

Faulkner L. L. (1976) *Handbook of Industrial Noise Control*. Industrial Press Inc.

Foreman, J. E. K. (1990) *Sound Analysis and Noise Control*, Van Nostrand Reinhold, New York.

Herbert, A. G. *et al.* (1989) 'Sound and vibration analysis and control', *Kempe's Engineering Yearbook*.

HMSO (1988) *Calculation of Road Traffic Noise*, HMSO, London.

INM (1982) *Integrated Noise Model, Version 3, Users Guide*, Prepared by CACI Inc. Federal Aviation Administration, FAP-EE-81-17, October.

INM (1989) *Integrated Noise Model, PC Version, Contour Plotting Program, Release 2*, Prepared by Unisys. Federal Aviation Administration, FAA-EE-90-02, November 1989.

ISO 1996 (1971), *Assessment of Noise with Respet to Community Response*, International Standards Organisation, Geneva, Switzerland.

ISO 9614 (1993), *Instrumentation for Noise Measurement*. International Standards Organisation, Geneva, Switzerland.

Magrab, E. B. (1975) *Environmental Noise Control*, John Wiley, New York.

OECD (1986). *Environmental Effects of Automotive Transport*, The OECD Compass Project, OECD, Paris.

Piercy, J. E., *et al.* (1977) 'Review of noise propagation in the atmosphere', *Journal of Sound and Vibration*, **61**(6).

Thumann, A. and R. K. Miller (1976) *Secrets of Noise Control*, The Fairmont Press, Atlanta, Georgia.

US Department of Labour (1971–79) *Occupational Noise Exposure. Code of Federal Regulations*, Title 29, Part 1926, US Government Printing Office, Washington, D.C.

AGRICULTURAL POLLUTION

10.1 INTRODUCTION

Since the introduction of the USEPA in 1972, great progress has been achieved in identifying pollutants and their associated environments. With identification came the application of technology to remedy the pollution, methods which are detailed in the chapters of Part Three of this text. Pollution from industry and municipalities, while still not eliminated, is being addressed with greater and greater success as we march towards the twenty-first century. However, pollution from agriculture into the air, water and soil environments has not been addressed with the same level of success as that from industry. Pollution from agriculture is to a large extent non-point source (NPS) pollution and this makes the task of identification and characterization difficult. Additionally, politicians internationally have not had the will to legislate against agriculturally sourced pollution. However, the impact of Rachel Carson and such organizations as the Sierra Club have fostered a community environmental movement that has also put the agricultural industry on notice re its pollution activities. Thus aspects of environmental legislation in both the European Union and the United States now address agricultural pollution.

The objectives of this chapter are to examine:

- The biogeochemical cycles relevant to agriculture
- Aspects of soil physics and soil chemistry
- Farmyard wastes
- Nutrient losses
- Chemical wastes
- Environmental legislation of relevance

The production of food and fibre by an industry called 'agriculture' is an essential and strategic component of any society. Throughout history, civilizations unable to feed their people have vanished. Conversely, societies that have advanced and developed have done so only by first achieving high food and fibre production efficiencies. In developed countries, only a small percentage of the population is involved in agricultural production, freeing most of the populace for other pursuits, such as commerce, science, the arts and manufacturing. Intensification, using external inputs (energy, crop protection chemicals, fertilizers, etc.), has been the critical factor in agriculture achieving such success in producing food and fibre. However, undesirable side effects on environmental quality have been correlated with the development of modern agricultural systems. The impact of modern agriculture on the environment can be minimized. However, agricultural pollution control does require techniques and strategies that are quite different from those used in other industries. This chapter introduces key concepts about agricultural practice and pollution. Chapter 17 is an introduction to agricultural pollution control techniques and strategies.

10.2 NUTRIENT CYCLES IN AGRICULTURAL SYSTEMS

Agriculture produces food and fibre. Plants are the basic output from agriculture, whether produced for direct consumption by humans, used as a food source for animals or processed into fibres and other organic products. The soil is the basic medium for plant production. It is itself a non-homogeneous system having widely varying physical, chemical and biological properties (Sec. 10.3).

A major facet of any agricultural production system involves managing nutrients, primarily nitrogen (N) and phosphorus (P). N and P are essential for all living systems. These also are the two nutrients most often associated with agriculture as water pollutants. An important objective of land-based agricultural systems is to achieve a balance between nutrient inputs (e.g. purchased feeds, fertilizers and organic wastes) and nutrient outputs (e.g. milk, meat and wool) while minimizing nutrient 'leaks' to the environment and meeting production targets. Achieving this objective at farm level is difficult, however, as agriculture operates in an 'open' production environment, and the system must accommodate uncontrolled weather events and soils having variable characteristics (see Chapter 17).

When animals are involved in the production system, nutrient management is even more difficult. Animals retain only approximately 15 per cent of the nutrients contained in feedstuffs by converting them into animal product; the remaining 85 per cent are excreted. Therefore, animal wastes contain significant quantities of nutrients that must be managed in an environmentally acceptable way. Animal wastes typically are managed by recycling them to the land that produced the feedstuffs consumed by the animals. However, doing so in the open environment (as distinct from a closed factory environment) creates certain pollution risks, which are amplified by practical management and technological problems associated with land spreading of wastes. On intensive pig and poultry farms, nutrient management difficulties are magnified further because nutrient inputs (feeds) are produced on other farms and purchased by the animal producers. This creates a large nutrient surplus on pig/poultry farms which requires significant land areas for spreading wastes to avoid overapplication of nutrients, particularly P.

Management of N and P cannot be accomplished without cognizance of the transformations of the nutrients that occur in nature, represented conveniently as N and P 'cycles' (Figs 10.1 and 10.2 respectively). In agricultural systems, these transformations largely occur in the soil and are a function of complex interactions between the atmosphere, soil particles, soil bacteria, plant and animal life, and soil water.

10.2.1 Nitrogen Cycle

Inorganic nitrogen is an ubiquitous element, having nine different chemical oxidation states. Soil bacteria are responsible for most nitrogen transformations in the soil; plant uptake also plays a role. Therefore, microbial activity and plant growth govern the rates at which N transformations occur, which, in turn, are determined by a number of environmental variables. These variables include soil moisture content, temperature and oxygen concentrations, all of which depend upon the weather.

Globally, most N exists as elemental nitrogen gas, N_2, in the atmosphere, and it is to this stable form that N from various compounds always tends to return. In agricultural systems, elemental N can be transformed into organic forms by leguminous plants and certain bacteria and algae.

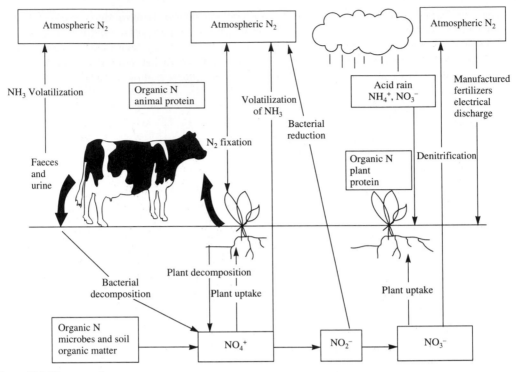

Figure 10.1 Nitrogen cycle.

Organic N (in plant remains and soil microbial biomass) is the largest 'pool' of N in the soil. A typical mineral soil in temperate climates contains 3000 to 5000 kg N/ha in the upper 0.3 to 0.5 m. Plants cannot utilize organic N. However, this N is mineralized, or changed into inorganic forms that are plant available, by soil bacteria at an approximate rate of 2 to 3 per cent annually. Mineralization is an oxidative

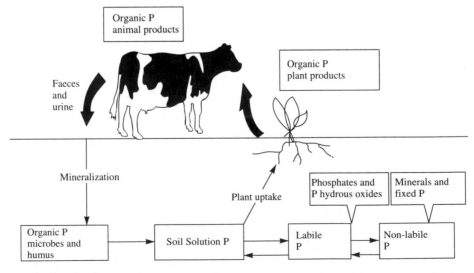

Figure 10.2 Phosphorus cycle.

process; therefore, in well-aerated soils there is a tendency for N (whether from fertilizers, wastes or soil organic matter) to be transformed to nitrate N (NO_3^-–N) by soil micro-organisms. Nitrate N remains in the soil solution and, if not taken up by plants, can leach downwards through the soil with drainage water, ultimately reaching groundwater. Nitrate leaching is especially likely from free draining soils (e.g. sands or gravels). This form of N loss from agriculture is both an economic and environmental concern; nitrate in groundwater used for human consumption can cause health risks when present in high concentrations (> 50 mg/L). Immobilization, the reverse process of mineralization (changing inorganic N to an organic form), is also accomplished by soil micro-organisms and plants as they incorporate inorganic N into microbial and plant tissue respectively.

Elemental N from the atmosphere can be transformed into inorganic forms naturally by lightning or artificially by energy-intensive manufacturing processes, the latter being far more important for agriculture. Inorganic N (NO_3^-–N, NH_4^+–N), either converted from organic forms by mineralization or added as supplemental fertilizer, is available to plants for uptake and for biological and chemical transformations in the soil. In very wet soils, where oxygen in the soil atmosphere is limited, heterotrophic bacteria can transform NO_3^-–N into gaseous N through a reduction process called denitrification. The products of denitrification are gases, nitrous oxide (N_2O) and elemental N (N_2), which are released to the atmosphere. Gaseous losses of N as ammonia (NH_3–N) can also occur through a process called volatilization when the ammonium ion (NH_4^+–N) is added to the soil either as chemical fertilizers, animal wastes or as the result of other transformations. Denitrification and volatilization provide 'short cuts' for nitrogen in agricultural systems to return to the atmosphere.

Animals consume organic N in plants (as protein), incorporate a portion of the nutrient into tissue and bone, and excrete the remainder as waste products (faeces and urine) in both organic and inorganic forms. Mineralization and chemical processes (hydrolysis of urea in urine) convert organic N in animal wastes into inorganic forms usable by plants (NH_4^+–N and NO_3^-–N). Urea N in urine is converted almost immediately to NH_4^+–N. Conversely, as a general rule, approximately 50 per cent of the organic N in animal wastes is converted to NH_4^+–N in a period of 12 months, but the actual transformation rate is dependent on environmental conditions. Subsequently, 50 per cent of the remaining organic N will be converted per year in a typical decay series.

10.2.2 Phosphorus Cycle

Like N, the behaviour of P in agricultural systems is complex. Phosphorus exists in the soil either in dissolved or solid form, but the latter is dominant. Dissolved P is typically less than 0.1 per cent of total soil P, usually existing as *ortho*-phosphates, inorganic polyphosphates and organic P. Phosphorus in solid form (particulate P) can be classified as :

- Adsorbed P (attached to soil particles)
- Organic P (in dead and living plant material and organisms)
- Precipitate P (P that has reacted with calcium, aluminium and iron in the soil) and
- Mineral P (in soil minerals)

In a mineral soil as little as 33 per cent to as much as 90 per cent of the total P is in the inorganic form. Both organic and inorganic P are involved in transformations that release water-soluble P from solid forms (and vice versa). Only soluble P is used by plants.

Organic P compounds undergo mineralization and immobilization (analogous to N transformations) with the aid of soil bacteria and growing plants. Transformations of inorganic P are related to the ease with which various forms become soluble, soil pH, and the presence and amounts of soluble aluminum, iron and calcium.

The direction (i.e. whether mineralization or immobilization) and magnitude of P transformations determine the physical and chemical status of P in the soil and, in turn, the potential of the soil system to supply P to plants or to contribute to phosphorus pollution. A key difference between N and P in soil is the fact that P attaches strongly to soil particles, particularly clay-sized particles, whereas N (especially $NO_3^- - N$) does not. For this reason, P does not leach through the soil profile (except from organic soils such as peats or very sandy soils). Phosphorus is lost from agricultural systems in runoff either in soluble or adsorbed forms. Where soil erosion occurs, P losses generally are associated with eroded soil particles as adsorbed P. Where soil erosion is not a problem, P is transported by runoff to rivers and streams in soluble form.

More complete discussions of the N and P cycles in soil are given elsewhere (Alexander, 1977).

10.3 SOIL PHYSICAL AND CHEMICAL PROPERTIES

Transformations of N and P in agricultural systems are very much soil-related, as indeed are waste degradation processes (mineralization of organic matter). Soil physical, and to a lesser extent chemical, properties also control the movement of air and water through soil. The importance of soil physics and soil chemistry for nutrient movement in soils cannot be overstated. However, as with other topics in this chapter, complete texts and comprehensive courses are available on soil physics and soil chemistry (Jury *et al.* 1991, Hillel, 1980). The treatment of these topics in this chapter only provides a cursory explanation of basic concepts. Further discussion is given in Chapter 17.

It is important to distinguish between *soil* and the *soil system*. The latter is a living, non-homogeneous, biological entity consisting of soil, soil fauna (microbes and macroinvertebrates), and soil flora. In sanitary and environmental engineering parlance, the soil system is, in fact, a biological reactor.

As a component of this complex system, soil is a mixture of inorganic minerals that have developed from bedrock, organic matter, air and water. The term 'soil' is applied to the top 1 to 2 m of the regolith (Fig. 10.3). The distinguishing characteristic of soil is the relatively high concentration of organic matter, compared to that of the lower regolith. The organic matter derives from decomposed plant, animal and microbial life that inhabit the soil. Soils are broadly classified as mineral (having generally less than 10 per cent by weight organic matter) and organic (having more than 10 per cent organic matter by dry

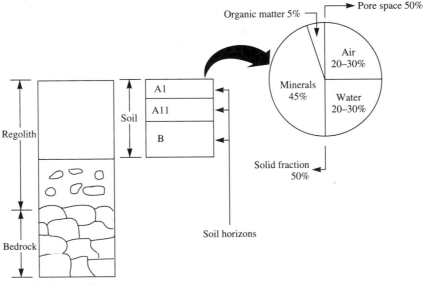

Figure 10.3 Soil and soil composition.

weight). In mineral soils, the concentration of organic matter decreases rapidly as depth below the surface increases.

10.3.1 Soil Physical Properties

In any given volume of a mineral soil, there is an approximately equal proportion of solids and pore space (Fig. 10.3). The former typically consists of 45 per cent minerals and 5 per cent organic matter and the latter consists of water and air. For waste treatment, as well as for crop production, the proportion of air and water should be approximately equal so as to create favourable conditions for biological activity (both microbial and plant).

Air and moisture movement through the soil are important factors in the decomposition of organic matter and crop growth. The relative ease with which water moves into and through the soil also influences the extent to which leaching (N losses) and runoff (P losses) occur, and thereby the potential for ground and surface water pollution respectively. Soil *texture* and soil *structure* are key determinants affecting the movement of both air and water through the soil profile.

Soil texture refers to the proportion of minerals of varying sizes that comprise the solid fraction of the soil. Well-defined classification schemes exist for categorizing soil particles according to size and, as there is no world-wide standard, result in some overlap among different systems in the size ranges that encompass the various particles. In the United Kingdom, sand includes particles with mean diameters between 60 and 2000 μm, clay particles have mean diameters less than 2 μm and silt particles are those with mean diameters between 2 and 60 μm. Textural classification schemes generally categorize soils based on the mixture of sand, silt and clay particles (primary particles) through use of a 'textural triangle' (Fig. 10.4). See also Chapter 4.

Soil texture is related to both hydraulic conductivity and water retention, two key properties that influence leaching and runoff. Coarse textured soils, having high sand contents, possess higher hydraulic conductivities and lower water retention capacities than fine textured soils, which have high silt and clay contents. Hydraulic conductivity is the constant in Darcy's law of liquid movement through porous media and expresses the readiness of the media (in this case, soil) to allow flow at any particular hydraulic

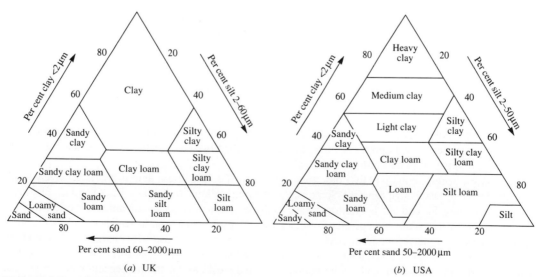

Figure 10.4 Textural classification triangles: (*a*) UK, (*b*) US.

gradient. Thus, coarse textured soils with high hydraulic gradients allow water to move freely (rapidly) through the profile (see Fig. 4.16).

Soil structure refers to the way in which individual primary soil particles (i.e. sand, silt and clay) are arranged and held together as more or less distinct, recognizable units. At one end of the structural spectrum are single grains (such as sand particles) and at the other extreme are groups of particles that have been packed tightly together into a seemingly continuous mass (massive structure). In the former, air and water move freely; in the latter, air and water movement is severely restricted. Within this spectrum are soils with more or less well-defined segregations of soil particles and pore spaces that afford varying degrees of air and water movement characteristics.

The stability of these soil structural units (i.e. their resistance to destruction by climatic influences, cultivation or animal traffic) determines the extent to which soils retain their abilities to transmit air and water and, consequently, their abilities to degrade wastes or facilitate the movement of pollutants. This is because the stability of pores that separate the units, the sizes of these pores and the degree to which the pores are interconnected influence hydraulic conductivity and water retention, which, in turn, influence leaching and runoff potential. Infiltration capacity, which controls the rate of entry of water into the soil profile, also influences leaching and runoff potential. Infiltration capacity is controlled by soil texture and structure, as well as by the presence and type of vegetation, degree of slope and soil moisture status.

10.3.2 Soil Chemical Properties

In addition to imparting physical characteristics to soil, the mineral and organic fractions also determine chemical properties of soil. It is these fractions that hold most of the soil's nutrients, in forms that are not plant available. Nutrients that *are not* plant available generally do not pose a threat to environmental quality (unless soil particles or organic matter are eroded and transported to receiving waters). Nutrients that *are* plant available are also available to cause water pollution. Soil chemical properties, especially *cation* and *anion exchange capacities*, tend to mitigate this potential, however.

Cation exchange capacity (CEC) defines the ability of a soil to retain positively charged ions, or cations. CEC is measured in terms of milliequivalents per 100 gram of soil (meq/100 g) and ranges from 2 to 3 meq/100 g for sand and silt to 100 to 200 meq/100 g for organic matter. Clay-sized inorganic particles and humus (colloidal organic matter) are responsible for most of the chemical properties of soil. Both clay and humus have a net negative electrical charge that can attract, and to varying extents adsorb, positively charged cations from the soil solution. The negative charges of the colloidal fraction of soil derives from:

- Replacement of trivalent by divalent cations within clay minerals
- From unsatisfied electrical charges at the edges of broken clay particles and
- From dissociation of protons (hydrogen) from organic acids and hydrous oxide surfaces

The latter is pH dependent; consequently the capacity of some soils to 'adsorb' and retain cations increases as pH increases. Important cations that are adsorbed by electrostatic attraction are Na, K, Mg and Ca, approximately in that order.

Anion exchange capacity (AEC) describes the ability of soil to retain anions and is relatively small compared to cation exchange capacity. Some clay and humic colloids (especially Fe, Al and Ca clays and humus containing Fe and Al hydroxides) have considerable ability to retain anions, particularly the phosphate ion. Anion adsorption results from either a chemical interaction, in which the anion becomes co-ordinated with a metal ion, or an electrostatic attraction between a positively charged colloid and the negatively charged ion. Chemical interaction is specific and strong, as for the adsorption of the phosphate anion. Electrostatic attraction tends to be non-specific and weak, as for the nitrate and chloride anions. Selectivity for anion exchange is generally in the order phosphate, molybdate, sulphate and nitrate.

Soil reaction (acidity and buffering capacity) is an inherent chemical property influencing crop uptake of nutrients and the mobility of some potential pollutants (such as metallic ions found in some wastewater sludges). Soil reaction is expressed by pH and is controlled by the amount of exchangeable calcium and magnesium and amount of free calcium carbonate present in the soil profile. Some soils are naturally rich in calcium and magnesium because of their parent materials and require little or no pH adjustment; lime is added to other soils to achieve appropriate pH levels.

The optimum pH at which soils should be maintained depends on the crop to be produced and whether the soil is mineral or organic. Soil pH affects the concentration of ions in soil solution that are available for uptake. To optimize nutrient uptake, and thereby minimize nutrient pollution risks, soil reaction must be managed carefully. Most agricultural advisory services publish recommended soil pH levels and liming rates. Soil pH affects the chemical form and therefore the mobility of heavy metals, which may be added to the soil in land-applied wastewater sludges. Generally, heavy metal mobility increases as soil pH decreases. Maintaining soil pH at 6.0 to 6.5 or greater minimizes potential leaching of heavy metals from soils used as receptors for wastewater sludges.

10.4 WASTE PRODUCTION ON FARMS

While the agricultural industry includes all facets of agribusiness from the production of basic materials to the processing of final products, this discussion of waste generation is confined to that which occurs at farm level. Pollutants from farms can be classified as physical (e.g. eroded soil, gaseous emissions), chemical (e.g. nutrients) or biological (e.g. bacteria). The sources of these pollutants are varied: animal wastes, silage effluent, contaminated runoff from farmyard areas, dairy washings, pesticides and fuel oil. Animal manures are by far the most significant of these, especially in Western Europe. Within the European Union approximately 8 tonnes of animal wastes are produced per hectare of utilizable agricultural land annually, with a range of from 2.6 t/ha in Greece to 42 t/ha in the Netherlands (Lee and Coulter, 1990). An estimated 87 million tonnes of manure are produced annually in Ireland alone (Table 10.1); of this, almost 30 million tonnes are produced indoors and require management. Animal wastes that require management include all pig and poultry manure and the manure produced by grazing animals during the indoor winter feeding period. If not utilized properly, these wastes have significant potential to cause both air (in terms of ammonia volatilization and nuisance odours) and water pollution.

Excreta (faeces and urine—excreta from animals on a forage diet will consist of approximately 65 per cent faeces and 35 per cent urine on a volume basis) consist of the partially digested remains of the animal's feed intake diluted with varying quantities of water. Excreta include a range of materials from undigested food remains to the primary components of plant tissues, carbon dioxide, minerals and water combined with microbes from the animal's digestive tract. The microbial contents of animal wastes make them biologically active materials, one of their important characteristics. Microbial activity is responsible for the mineralization of nutrients in the organic matter.

Table 10.1 Estimated annual manure production from livestock and the quantity requiring management in Ireland

Animal type	Production (kilotonne/annum)	Requiring management (kilotonne/annum)
Bovine	76 000	28 327
Ovines	8 258	454
Pigs	1 804	1 804
Poultry	227	227
Total	86 919	30 812

10.4.1 Mineral/Nutrient Composition of Farm Slurries

The nutrient composition or fertilizer value of slurry is influenced by the type of animal, the animal's diet, waste storage conditions and the extent of the dilution with either water, bedding or litter. An indication of the mean nutrient composition of various slurries is given in Table 10.2. The nutrients include significant quantities of N, P, potassium (K) and sulphur (S). They also contain, in smaller quantities, calcium, magnesium and trace elements. Cattle slurry compared with pig slurry is high in K and low in P. This reflects the differences in nutrient concentrations between the grass and the cereal diets of cattle and pigs respectively. Extra mineral P is added to pig diets to satisfy growth requirements because of their inability to absorb sufficient P from the diet's cereal component. The apparently higher nutrient concentration in poultry compared with either cattle or pig slurry reflects the higher dry matter content of the former. Due to the high variability of nutrient concentrations in animal wastes, individual analyses are required on a given farm for an accurate determination of the nutrient value of animal wastes.

10.5 POLLUTION POTENTIAL OF FARM WASTES

Animal wastes are high-strength wastes and have potential to cause serious water pollution problems. Table 10.3 compares various animal wastes (at 4 per cent dry matter) to the composition of typical raw (i.e. untreated) domestic sewage.

As evidenced by the high concentrations of both BOD_5 (biochemical oxygen demand) and COD (chemical oxygen demand), animal wastes have considerable amounts of organic matter and reactive inorganic species (e.g. ammonium) that will exert excessive oxygen demands on surface waters. These high oxygen demands also preclude treating animal wastes by conventional processes, as is done with domestic sewage and other industrial wastes. Consequently, land application is an economically viable and environmentally sustainable method of 'treating' animal wastes. The application of animal wastes to land will not result in soil pollution when applied at agronomic rates and at the correct times of the year (Chapter 17). Application to land of some high-strength wastes at very high rates may cause the soil to become temporarily anaerobic due to the oxygen demand of the wastes.

Repeated heavy applications of animal wastes to soil also can cause a buildup of soil P levels, although instances where phytotoxic P levels have been reached are very rare. Increased soil P levels are more often associated with increased potential for release of water-soluble P and the resulting increase in potential for water pollution. In freshwater systems, minute concentrations of soluble P (0.01 mg/L) are sufficient to cause algal blooms if other environmental conditions are satisfactory. These blooms, in turn, reduce oxygen levels in water to levels that result in fish kills, increase water treatment costs (if the water

Table 10.2 Dry matter and nutrient composition of animal manures

Manure type	Dry matter (g/kg)	Composition of farm slurry (kg/10 t)		
		Nitrogen	Phosphorus	Potassium
Cattle slurry†	69	36	6	43
Pig slurry†	32	46	9	26
Dungstead manure‡	17	35	9	40
Farmyard manure‡	20	45	10	60
Poultry‡				
Deep litter	70	260	90	120
Layers	24	140	50	60

† O'Bric *et al.* (1992).
‡ Tunney and Molloy (1975).

Table 10.3 Composition of various animals' manures (4 per cent dry matter) and raw domestic sewage

Component	Water source (mg/L)				
	Dairy cattle	Beef cattle	Swine	Poultry	Humans
Total solids (TS)	40 000	40 000	40 000	40 000	500
Votatile solids (VS)	29 700	31 000	31 600	31 100	350
BOD_5	6 000	6 700	12 800	9 800	200
COD	36 200	35 600	32 800	36 000	450
Nitrogen as N	1 600	1 900	2 500	2 900	30
Phosphorus as P	300	400	950	1 100	10
Potassium as K	860	1 100	1 400	1 100	10

US Department of Agriculture, 1975

is used for public drinking supplies) and are aesthetically unacceptable. In saltwater systems, small concentrations of N can stimulate algae growth.

Odours are the most recognizable air 'pollutant' arising from animal wastes; usually these emissions are associated with the land application process. While less noticeable to laypersons, the volatilization of NH_3 from animal wastes has been identified as a serious air pollutant in some regions (e.g. the Netherlands). Strict legislation has been passed in these areas to control gaseous N losses. In confined buildings (i.e. animal housing) where animal wastes are stored below the buildings, the release of both ammonia and hydrogen sulphide can be health-threatening air pollutants for both the animals and humans. Proper building and manure store design features can reduce NH_3 emissions to acceptable levels; H_2S levels generally are not a problem until the animal wastes are removed from the buildings. At these times, animals must be removed from the buildings and workers must take special care and use protective breathing devices to avoid health risks.

10.6 NUTRIENT LOSSES

Nutrients from fertilizers or wastes not utilized by plants may 'leak' from agricultural systems to either ground or surface water. These losses are economically and environmentally undesirable. Nitrogen and phosphorus are the two nutrients of major agricultural importance that have the greatest potential to create water pollution. Both nutrients can be either:

- Taken up by growing plants
- Move to surface water in runoff (P)
- Move to groundwater in leaching (N)
- Be immobilized in the soil/organic matter pool

In addition, N from fertilizers or wastes can be lost as a gas to the atmosphere. Figure 10.5 illustrates the pathways for nutrient losses from agricultural systems.

10.6.1 Nitrate Leaching

Leaching of nitrate N is the major pathway by which N leaks from agricultural systems. Although N is the preferred form of nitrogen by plants, it is only weakly adsorbed by the soil and remains in the soil solution, making it a highly mobile anion that can readily move downwards (leach) through the soil profile. Whether nitrate reaches groundwater is influenced by many factors: abundance of the ion, carbon–nitrogen ratios, soil type, depth to groundwater, type of plant, time of year and climatic variables.

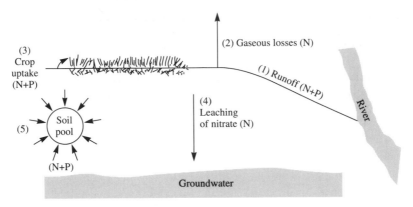

Figure 10.5 Potential pathways from land-based agricultural systems.

In soil, there is a tendency for N from all forms to be converted to nitrate N by soil microbes. When wastes are applied to soil, the rate at which this occurs is dependent on waste characteristics, especially the ratio of carbon to nitrogen (C:N ratio) or the C:N ratio of the soil–waste mixture if the waste is incorporated. When C:N ratios are 20:1 or greater, soil micro-organisms will use the relatively abundant carbon supply as an energy source to support rapid growth and multiplication. As a result, most of the N will be immobilized as it is incorporated into microbial biomass. This minimizes the potential for nitrate leaching, at least in the short term. Nitrogen applied as inorganic fertilizers (urea, ammonium or nitrate) or in wastes with low C:N ratios will tend to be converted rapidly to nitrate, providing a supply of nitrate for crop uptake as well as leaching.

Soil type (texture and structure) influences nitrate leaching by controlling the rate and amount of water moving downwards (leaching) through the soil profile as well as the aeration status of the profile. Freely draining sandy, gravelly and karst soils transmit large quantities of water rapidly, together with dissolved substances such as nitrate. In addition, these soils tend to be well aerated, providing conditions favourable for the conversion of N forms to nitrate. Conversely, clay soils, with smaller pores, tend to transmit water less readily and have lower oxygen concentrations in the pores. Nitrate leaching from these soils is usually not a problem. In fact, anaerobic soil conditions can easily occur in soils with high clay contents, resulting in a loss of N as a gas to the atmosphere through denitrification. Soil water retention capacities for various soils are on the order of clay > loam > sand (Chapter 17). For a given amount of precipitation, nitrate will move to greater depths in a sandy soil compared to a clayey soil. Similarly, the risk of nitrate leaching to groundwater is greater for soils with shallow water tables, all other factors being equal, than for soils with deep water tables.

Plants exert an influence on the extent of nitrate leaching through their N uptake patterns. Crops with long growing seasons, such as grasses, have a greater opportunity for N uptake than do crops with short growing seasons, such as spring-sown cereals. Growing season length is especially important considering the continual mineralization of inorganic N from soil organic matter, which supplies N for plant uptake or leaching irrespective of N added by either fertilizers or wastes. Spring cereals, for example, cease N uptake in June, while mineralization of N in the soil continues to some extent throughout the year. Consequently, nitrate tends to leach to groundwater from free draining soils where a large amount of cereal production occurs. Application of fertilizers or wastes to soils on which spring cereals are grown exclusively tends to exacerbate nitrate leaching problems.

Nitrogen uptake patterns of plants strongly influence the availability of nitrate for leaching. Therefore, applying fertilizer or wastes at times that make N available when plants need it is critical for minimizing nitrate leaching potential. In Ireland, grass begins to incorporate N actively in mid to late February and continues through late August, providing large N uptake potential and minimal nitrate leaching risk. Each crop has a characteristic N uptake pattern, which should be matched by N applications (timing and amount) with due consideration for N supplied by the soil. Agricultural advisory services in most countries provide fertilizer recommendations for agronomic and horticultural crops and grassland.

Because water is the transporting agent for nitrate, the occurrence of soil water draining through the profile and the availability of nitrate in the soil profile determine the extent to which nitrate will leach. In general, the net movement of water downwards through a soil profile occurs when precipitation exceeds evapotranspiration plus soil moisture storage. This condition is usually reached when plant activity is low (late fall, early spring) or non-existent (winter).

10.6.2 Phosphorus Runoff

In contrast to N, P is lost from agricultural systems in runoff. Runoff is the amount of precipitation in excess of infiltration, interception and depression storage. Losses of P in runoff tend to be higher from 'heavy' soils (with high clay contents) than from 'light' soils (with high sand contents) because the former have lower infiltration capacities and thus are more likely to generate runoff. Leaching of P from mineral soils generally does not occur, although P can be lost via leaching from highly organic (especially peat) soils and from very sandy soils.

Phosphorus in runoff can be either in soluble (dissolved) or 'attached' (adsorbed to soil particles) form. The concentration of P in the soil solution within the upper 1 to 3 mm of soil can be diluted by precipitation, permitting more P to be released into solution from the labile (slowly available) pool of soil P. For this reason, soils with high concentrations of soil P pose a greater risk, in general, of P losses in runoff than do soils with lower P contents. Highly erodible soils also are susceptible to P losses in runoff and must be managed so that soil particles carrying attached P do not leave the site and reach surface waters.

Losses of P in runoff can be high where organic wastes have been applied to soils at the application site that are prone to generate runoff. Such sites would have low infiltration capacities, high soil moisture contents, high water tables or other restrictions to absorption of precipitation. In these situations, the timing of runoff relative to the application of wastes is an important determinant in the extent to which P losses will occur. In general, as the time between runoff and waste application increases, losses of P (and BOD_5) decline exponentially. Where erosion is controlled, losses of P in runoff occur mainly in the soluble (*ortho*-phosphate) form. While not all P lost from a site in runoff actually reaches receiving waters (being assimilated along the way in intervening land uses), minimizing P losses is crucial given the minute concentration (0.01 mg/L) of soluble P required to stimulate algae growth in surface waters.

10.7 OTHER WASTES AND POTENTIAL POLLUTANTS

A variety of wastes other than those from animals are applied to agricultural land. In the best circumstances, these wastes are applied at rates and using techniques that minimize the risk of pollution and maximize the uptake of nutrients by plants. In the worst circumstances, wastes are applied according to a 'disposal' strategy that ignores soil and agronomic principles, resulting in elevated pollution potential.

The variety of wastes that are applied to agricultural land is too large for an exhaustive listing and characterization. Sludges from industrial, domestic and combined wastewater treatment facilities (as well as treated wastewater) and wastes from food processing factories (blood, paunce, whey, residuals from wastewater treatment) are dominant on a volume basis. In most cases, these wastes contain only organic matter and inorganic nutrients, though the relative amounts of each vary widely for different wastes. Pollution potential is mitigated by the same soil physical, chemical and biological processes described above for strictly agricultural wastes.

In the case of sludges, particularly those of industrial origin, heavy metals and other potential micropollutants (dioxins, PCBs) may be an environmental concern. Sludges from secondary sewage treatment facilities are generally 3 to 7 per cent dry matter. The types and concentrations of heavy metals depends on the nature of the industries producing the sludge. Careful chemical analyses are essential for characterizing any sludge prior to applying it to land. In addition, when agricultural land is utilized for sludge application, agronomic principles must be respected to schedule sludge applications into other farm operations. It is prudent to keep accurate records of the amounts of characterization of all sludge applications to agricultural land. Guidelines for sludge utilization on agricultural land typically are available from agricultural research and advisory agencies, universities and regulatory authorities.

10.7.1 Pesticides

Crop protection chemicals (pesticides) are an integral part of intensive arable agriculture, responsible in part for helping produce abundant food supplies that keep consumer food costs relatively low. With these benefits are associated certain environmental risks, since by formulation most pesticides are toxic materials. Most pesticides are synthetic organic compounds, the physical and chemical properties of which vary widely, resulting in very different behaviours of pesticides in the environment. Soil properties (especially texture and organic matter content), application techniques (foliar applications versus surface application or soil incorporation) and environmental conditions (soil moisture, temperature and aeration) also influence pesticide behaviour.

Three pesticide characteristics are especially critical in influencing the potential of a pesticide to be 'lost' from an application site in runoff or by leaching. The *soil sorption index* of a pesticide measures the tendency of the chemical to attach to soil particles and organic matter. Soil sorption indices are defined by a chemical relationship, K_{oc}, which measures the relative degree to which pesticides are 'sorbed' by soil and organic matter from an aqueous solution. Solubility in water is an inherent property of the active ingredient of a pesticide, and defines the amount of the pesticide that can be dissolved in water. A pesticide's *half-life in soil* is the time in days required for its original concentration in soil to be degraded, or reduced, by one-half. Unlike solubility, pesticide half-lives are not a constant value, being influenced by soil temperature and soil moisture (see also Chapter 3).

Pesticides most likely to leave an application site in runoff are those that are foliar or surface applied without incorporation, have high sorption indices (K_{oc} greater than 1000), low water solubilities and long half-lives. Pesticides likely to leach from an application site are those that are incorporated in the soil and have low sorption indices, high water solubilities and long soil half-lives. Chemicals with short half-lives tend to be degraded before they can contaminate surface or ground waters, but even these chemicals can move from the application site if heavy precipitation occurs shortly after application.

10.8 LEGISLATION (EU)

Public interest in environmental protection is increasing all over the world, but especially in developed countries that have the social, financial and technological resources to address pollution issues. In addition, demographic changes in these countries are increasing the size of the non-farm population, while the farm population is continually declining. One result of these two important social changes is that agriculture in developed countries has come under scrutiny for contributing to environmental pollution.

The European Union has issued several directives to protect environmental quality. The legislation sets a minimum standard for environmental protection that must be adopted by all member countries. One of the most important of these is the 'Drinking Water Directive' (80/778/EEC), which sets maximum concentrations of various contaminants that are allowed in public drinking water. Similar legislation (Directive 78/659/EEC) sets water quality limits for the protection of fish habitats. Together, these directives afford a general means by which to control water pollution from a variety of sources, including agriculture.

Legislation designed specifically to protect groundwater from agricultural sources of nitrates is the 'Nitrates in Ground Water Directive' (91/676/EEC). The directive reaffirms a maximum allowable concentration (MAC) of nitrate in groundwater at 50 mg/L which was established by Directive 80/778/EEC. In addition, the legislation directs that codes of good agricultural practice will be established by member countries, to be implemented on a voluntary basis by farmers.

Agricultural land is becoming a favoured receptor for wastewater sludges, prompting the passage of regulations to control the use of sewage sludges so as to prevent harmful effects to soil, vegetation, animals and humans. Directive 86/278/EEC establishes maximum sludge application rates and stipulates that sewage sludge should be applied to land in accordance with codes of good practice to avoid pollution of waters and emissions of nuisance levels of odours.

Odours and other atmospheric contaminants (e.g. ammonia) generally have not been addressed by EU legislation. However, several EU member countries have passed regulations to do so. For example, in Ireland, the Air Pollution Act 1987 (No. 6 of 1987) provides the statutory framework for controlling air quality. By this Act, it is illegal to cause or permit an emission in a quantity or in such a manner as to be a 'nuisance'. Determining whether or not a nuisance exists is the responsibility of courts of law. A 'good defense clause' affords potential offenders legal protection if they have used the best practicable means to prevent or control emissions.

Similarly, ammonia emissions from agriculture have been addressed by specific legislation in some EU countries. In the Netherlands, for example, ammonia emissions from agriculture must be reduced by 30 and 70 per cent of 1980 levels by the years 1994 and 2000 respectively, under the National Environmental Policy Plan.

In the United States, laws passed by the Congress have the same effect in the various states as do EU Directives in member countries, by establishing minimum standards for environmental protection. Control of water pollution by agriculture (as well as by other pollutant sources) is mandated by the Water Pollution Control Act Amendments of 1972 (PL 92-500) and its successor, the Clean Water Act (PL 95-217). Generally, only large confined animal feeding facilities are specifically included in the legislation; however, the Acts establish minimum water quality standards that are applicable to all potential pollution sources. Agriculture and other diffuse sources of pollution are coming under more rigid controls under PL 95-217, which directed the implementation of 'nonpoint source' pollution programmes. The 1990 Farm Bill (the major US agricultural legislation) required specific measures for environmental protection on farms that intended to participate in federal agricultural assistance programmes. As in the European Union, individual US states and localities can, and do, pass environmental legislation that is more restrictive than federal law (see Chapter 13 for details of rule 503, 1995).

10.9 SUMMARY

Land-based agricultural production utilizes large land areas and occurs in the presence of uncontrolled and unpredictable weather events. Consequently, agriculture has the potential to cause both water and air pollution. Soil pollution typically is not an issue due to the nature of agricultural pollutants, which are organic matter and nutrients (and eroded soil in some instances). Nitrogen is usually lost from agricultural systems by leaching, volatilization and denitrification. Phosphorus is typically lost by runoff, as is organic matter and eroded soil. Environmental legislation addressing agricultural sources of pollution exists mainly to protect water resources rather than air quality.

10.10 PROBLEMS

10.1 Contrast the pathways by which nitrogen and phosphorus are lost from agricultural systems. At what times of the year would one expect each pathway to be most important?

10.2 Texturally, compare a clay loam with a sandy loam. What hydraulic differences would be expected between the two soils?

10.3 Why are animal wastes so different in character from human wastes? What implications do the differences in waste characteristics have for animal waste management?

10.4 Soil behaves as a fixed-film biological reactor. Describe key differences between the soil and (a) trickling filters and (b) activated sludge basins in terms of wastewater treatment.

10.5 A freshwater lake suffers eutrophication from excess phosphorus runoff. Consider a catchment with a series of streams, leading to a river which outfalls from the catchment via a lake. Describe how you would go about doing a mass balance on the nutrient phosphorus.

10.6 Write a brief two page report on the possibilities, advantages and disadvantages of using anaerobic digestion to alleviate farm pollution. Refer to Chapter 13.

10.7 Explain why farmyard wastes have a much higher COD value than BOD value.

REFERENCES AND FURTHER READING

Alexander, M. (1977) *Introduction to Soil Microbiology*, 2nd edn. John Wiley, New York.

Archer, J. (1988) *Crop Nutrition and Fertiliser Use*, 2nd edn, Farming Press, Ipswich.

ASAE (1990) *Standards, Engineering Practices and Data*, 37th edn, American Society of Agricultural Engineers, St Joseph, Michigan.

Barth, C. L. (1985) 'The rational design standard for anaerobic livestock lagoons', in *Agricultural Waste Utilization and Management: Proceedings of the 5th International Symposium on Agricultural Wastes*, American Society of Agricultural Engineers, St Joseph, Michigan.

Department of Agriculture, Ireland (1985) *Guidelines and Recommendations on Control of Pollution from Farmyard Wastes* (revised), Department of Agriculture and Food, Dublin, Ireland.

Grundy, K. (1980) *Tackling Farm Waste*, Farming Press, Ipswich.

Halley, R. J. and R. J. Soffe (1988) *The Agricultural Notebook*, 18th edn, Blackwell Scientific Publications, Oxford.

Hillel, D. (1980) *Fundamentals of Soil Physics*, Academic Press, New York.

Hudson, N. (1981) *Soil Conservation*, 2nd edn, Cornell University Press, Ithaca, New York.

Jury, W. J., W. R. Gardner and W. H. Gardner (1991) *Soil Physics*, 5th edn, John Wiley, New York.

Lee, J. and B. Coulter (1990) 'A macro view of animal manure production in the European Community and implications for environment', in *Manure and Environment Seminar—VIV Europe*, Utrecht, the Netherlands, 14 November.

McCuen, R. H. (1989) *Hydrologic Analysis and Design*, Prentice-Hall, Englewood Cliffs, New Jersey.

MAFF (1991) *Code of Good Agricultural Practice for the Protection of Water*, Ministry of Agriculture, Fisheries and Food, London.

MAFF (1992) *Code of Good Agricultural Practice for the Protection of Air*, Ministry of Agriculture, Fisheries and Food, London.

Merkel, J. A. (1981) *Managing Livestock Wastes*, AVI Publishing, Westport, Connecticut.

Midwest Plan Service (1985) *Livestock Waste Facilities Handbook*, 2nd edn (MWPS-18), Midwest Plan Service, Iowa State University, Ames, Iowa.

Novotny, V. and G. Chesters (1981) *Handbook of Nonpoint Pollution*, Van Nostrand Reinhold, New York.

O'Bric, C., O. T. Carton, P. O'Toole and A. Cuddihy (1992) 'Nutrient values of cattle and pig slurries on Irish farms and the implications for slurry application rates, *Irish Journal of Agricultural Research*, **31**(1), 89–90.

Schwab, G. O., D. D. Fangmeier, W. J. Elliot and R. K. Frevert (1993) *Soil and Water Conservation Engineering*, 4th edn, J. Wiley, Somerset, New York.

Shaw, E. M. (1988) *Hydrology in Practice*, 2nd edn, Chapman and Hall, London.

Teagasc (1989) *Farmyard Wastes and Pollution*, Agriculture and Food Development Authority, Dublin, Ireland.

Teagasc (1992) Miscellaneous data (unpublished), Johnstown Castle Research and Development Centre, Wexford, Ireland.

Tunney, H. and S. M. Molloy (1975) 'Variations between forms of N, P, K, Mg and dry matter composition of cattle, pig and poultry manures, *Irish Journal of Agricultural Research*, **14**, 71–79.

US Department of Agriculture (1975) *Agricultural Waste Management Field Manual*, Soil Conservation Service, US Department of Agriculture, Washington, D.C.

USEPA (1975) *Land Treatment of Municipal Wastewater Effluents: Design Factors II*, US Environmental Protection Agency, Washington, D.C.

Wesseling, J., W. R. van Wijk, M. Fireman, B. D. van't Woudt and R. M. Hagan (1957) 'Land drainage in relation to soils and crops', in *Drainage of Agricultural Lands*, J. N. Luthin (ed.), American Society of Agronomy, Madison, Wisconsin.

ENVIRONMENTAL ENGINEERING TECHNOLOGIES

ELEVEN

WATER TREATMENT

11.1 INTRODUCTION

The objectives of this chapter on the purification and treatment of raw water to bring it to drinking water standards are:

- To understand the differences in standards of raw water and purified water
- To examine the various physical–chemical treatment processes involved

Natural waters are rarely of satisfactory quality for human consumption or industrial use and nearly always need to be treated. The level of treatment required will depend on how acceptable or 'pure' the natural water is.

Raw freshwater is abstracted from rivers, lakes or underground sources and treated to standards acceptable for human consumption or industrial requirements. In the United States and the United Kingdom, by far the most common sources of raw freshwater are rivers and lakes, though in recent decades more sources of groundwater are being utilized. On the European mainland groundwater is used extensively. Some groundwater sources are so pure that no treatment is necessary, although when used for public supplies, local water authorities (public and private) tend to apply a disinfection process, but this is primarily for disinfection purposes of the distribution network. Some upland river or lake sources may also be relatively pure and again need little treatment. At the other end of the scale, when downstream reaches of rivers are used for abstraction, extensive treatment may be needed, particularly if the abstraction is downstream of urban, industrial or agricultural developments. In practice, all public water supplies undergo some form of treatment, with the degree of that treatment being dependent on the quality of the raw water supply. The quality of treated water is now almost standardized in the developed world, with treatment facilities having to satisfy many water quality parameters on a frequent monitoring basis.

The objectives of water treatment are to produce:

- Water that is safe for human consumption
- Water that is appealing aesthetically to the consumer and
- Water at a reasonable cost

These objectives are readily met in the developed world and the technology of treatment is similar world-wide. Water treatment technologies are, however, constantly undergoing research, not only to improve the end water product but also to find ways of treating water that was once deemed unsuitable as a raw source. Advanced water treatment processes are often required by industry, e.g. the beverage or pharmaceutical industries. Such industries may have higher standards than those for potable supplies. Advanced methods are also used by public waterworks to remove contaminants such as organics.

11.2 AMOUNT OF WATER REQUIRED

Public water supplies normally service the requirements of:

- Domestic households
- Fire fighting
- Industrial
- Commercial

The demand for water varies with the end user and also with the country. For instance, the average daily per capita (ADPC) water consumption varies in the United States from a low of 130 to a high of 2000 litres. The European average is approximately 225 litres, with some countries in northern Europe (e.g. Denmark and Germany) consuming less than 200 litres. In the design of a new water treatment plant or upgrading of an existing one, surveys and metering of supply pipelines determine the values of per capita consumption. Fire services always require a minimum water volume to be on hand and to be available at an adequate pressure. This demand in urban areas can be serviced by reservoirs that also service the domestic requirements. In small urban areas, special water reservoirs may be required to meet the fire service demands. Commercial and industrial volume requirements are industry specific, but the availability of an adequate water supply infrastructure is a priority for attracting new industry to a locality.

Water is also consumed by 'leakage', a major problem in urban areas with old distribution networks. Water quality also reduces if sent through old distribution networks, particularly lead and iron pipes. Tables 11.1 and 11.2 show examples of water use and consumption rates for the United States.

Further details are given in McGhee (1991) and Cunningham and Saigo (1992). It is clear that as a public, we abuse water use, e.g. car washing and garden sprinklers with expensively treated water. Northern Europe, particularly Denmark, is making great strides in water reduction. In some areas, this is being achieved by charging consumers the real cost of water. In parts of Denmark the 1995 domestic water rate is approximately ECU 175 per household per annum. However, the above figure is only for domestic water supply. On top of this, domestic consumers pay an additional ECU 350 for wastewater treatment. The connection fee for water is approximately ECU 1200 and the connection fee for

Table 11.1 Examples of water use

Category	Description	Consumption (litres)
Home	Bath	100–150
	Shower per 5 min	100
	Clothes wash	75–100
	Cooking	30
	Toilet flush	10–15
Industry	US automobile	400 000
	1 kg steel	250
	Newspaper	500–1000

Table 11.2 US projected water consumption by year 2000

Use	Consumption (litres/capita/day)
Domestic	300
Public	60
Leaks	50
Industry	160
Commercial	100
Total	670

wastewater is approximately ECU 3000 (Mortensen, 1993). Hence conservation of water is encouraged by requiring households not only to pay for (metered) water but also to pay for wastewater.

11.3 WATER QUALITY STANDARDS

In discussing water quality, two sets of standards exist:

- One for the quality of raw water and
- The second for the quality of the treated potable water

The quality of raw water is governed in the European Union by two directives. Firstly, for raw surface water, Directive 75/440/EEC is known as the 'surface water directive intended for abstraction of drinking water'. Secondly, for raw ground water, Directive 80/68/EEC is known as the 'groundwater protection directive against pollution by dangerous substances'. The drinking water directive (80/778/EEC) relates to the quality of treated water intended for human consumption. The above three directives address water quality at source and at the point of delivery to the public. Throughout the developed world, similar legislation exists, driven to some extent by the World Health Organization. *Guidelines for Drinking Water Quality* was published by WHO in two volumes in 1984. In the United States, the EPA published a series of reports in 1980 on 'quality criteria for water'. The USEPA primary drinking and secondary drinking water regulations set standards for the United States. Also applicable are the USEPA advisories on direct and indirect additives as well as the USEPA drinking water health advisories. The Safe Drinking Water Act (1974) and the Surface Water Treatment Rules (1989) set the standards at national level in the United States. As well as the federal guidelines in the United States, there are a myriad legislative pieces concerning drinking water quality at state level.

The EU water quality parameters are discussed in detail by Flanagan (1992) with regard to methods of analysis, occurrence or origin of parameter, the health or sanitary significance, background, information, comments and criteria for recommended or mandatory limits. A summary of the parameters is shown in Table 11.3, with the maximum admissible concentration values listed. In some cases, the guide value is listed. Raw water quality in the European Union is defined in one way by the degree of treatment that is needed. For instance, category A1 water only requires simple physical (filtration) and disinfection treatment. Category A2 water requires normal physical and chemical and disinfection treatment. Poor quality water is category A3 and requires intense physical and chemical disinfection treatments. These categories are defined in Directive 75/440/EEC.

The testing of water for the water quality parameters are not detailed in this chapter. The reader is referred to Standard Methods in Greenberg *et al.* (1992). The parameters to be monitored in the EU drinking water directive are described in the following paragraphs and listed in Table 11.3. All the tests mentioned are detailed in Standard Methods (Greenberg *et al.*, 1992).

11.3.1 Organoleptic Parameters

Raw water will usually have impurities quantified in colour, turbidity, odour and taste. These are called organoleptic parameters, i.e. as sensed by the human organs of eye, nose and throat. Colour in water means that the water will absorb light in the visible spectral range (400 to 700 nm). Clear water is colourless. Colour in water is caused by dissolved minerals, dyes or humic acid from plants. The latter causes a brown-yellow unsightly colour. Traditionally, it was thought that this colour was harmless, but recently some correlation has been identified between colour and the formation of haloforms in the chlorination of drinking waters (Dojlido and Best, 1993). Colour is measured in units of mg/L on the platinum cobalt (Pt/Co) scale. Raw water is of very good colour quality at values less than 10 mg/L, is acceptable at 100 mg/L and is unacceptable at levels greater than 200 mg/L. In the same way, disinfectants are most effective when used on low turbidity waters.

Turbidity is due to the presence of particulate matter and is a measure of the ability of water to scatter light. It is caused by the presence of very fine suspended or clay particles. Turbidity is measured in nephelopmetric trubidity units. Water is of very good turbidity quality if turbidity is less than 0.1 NTU. It is of acceptable turbidity if levels are less than 1 NTU and considered unacceptable if the value is greater than 5 NTU. Turbidity is sometimes described as the cloudiness of water. Colour in water is best measured when there is no turbidity as the latter will mask colour.

Odour and taste in water are caused by the presence of the by-products of plant and animal micro-organisms, especially hydrogen sulphide. Basic water treatment processes address the elimination of colour, turbidity, odour and taste and in the 'process' improve the quality of other parameters (e.g. microbiological).

11.3.2 The Physicochemical Parameters

Depending on its source, raw water may need particular treatment to satisfy eleven parameters listed in Table 11.3. The maximum admissible concentrations of these parameters for drinking water are specified. Sometimes the raw supply needs pH correction. Peaty upland water tends to have a pH ~ 4.5. The presence of photosynthetic algae sometimes raises the pH to ~ 10. The presence of chlorides is indicated by a salty taste and may suggest a sewage-polluted water source. High calcium values increase hardness, which may be beneficial for health (Pocock *et al.*, 1981). Magnesium also contributes to hardness. Hardness is a measure of the potential scaling effect on boilers, etc., and is identified by the presence of cations, i.e. Ca^{2+}, Mg^{2+}. High levels of potassium may suggest artificial fertilizer pollution. Excessive levels of aluminium may be associated with Alzheimer's disease (Craig and Craig, 1989).

11.3.3 Parameters Concerning Substances Undesirable in Excessive Amounts

Nitrogen with its ten oxidation states has MAC values attached to four states. Of principal health concern is nitrate. The infant disease methanoglobinaemia (blue baby syndrome) is due to excessive levels of nitrate. Other nitrogen states, particularly ammonia nitrogen, may indicate organic pollution. The presence of hydrogen sulphide indicates decomposed organic matter. The presence of phenols, which are toxic in extremely low doses, is attributed to road and roadworks runoff and some industrial effluents. The presence of zinc, copper, iron, manganese, barium and silver are likely to be due to background geology or industrial effluents. Excessive levels of iron and manganese are common, particularly in groundwater, although while not specifically a health hazard they lead to brown staining of sanitary ware. Fluoride is not often found in raw water, but, if so, it can in excess amounts lead to mottling of teeth (see Sec. 11.10 on fluoridation).

11.3.4 Parameters Concerning Toxic Substances

Eleven parameters are mentioned in this category (see Table 11.3) including arsenic, cadmium, cyanide, chromium, lead, mercury, nickel, antimony and selenium. Their presence may be from background geology, but if found in excessive amounts are most likely due to industrial discharges. Pesticides and related products are undesirable as some of them are categorized as synthetic carcinogens. Polycyclic aromatic hydrocarbons or PAHs are synthetic carcinogens and are the products and by-products of soot, tar, car exhausts and benzenes.

11.3.5 Microbiological Parameters

The presence of undesirable pathogens (bacteria, viruses, etc.) is due to human and animal excreta. In water treatment, raw water is not routinely analysed for bacteria, viruses, etc., because of the great expense and huge variety of these pathogens. The usual analytical process is to use indicator organisms which will confirm the presence of pathogens if they exist. This is a simple test process and while six parameters are listed in the EU Directive, most of the time it is satisfactory to look at only two: total coliforms and faecal coliforms. As discussed in Chapter 3, most waterborne pathogens are introduced to water through faecal contamination, and the strain of bacteria known as *Escherichia coli* is an ideal indicator organism in that it has a long survival time in a water environment. Faecal coliform organisms in themselves are not pathogenic. Statistical methods are used to determine the most probable number (MPN) of coliform bacteria in 100 m/L of the water sample. The maximum admissible concentrations are given in Table 11.3. See Tchobanoglous and Schroeder (1987) for further details.

11.3.6 Minimum Required Concentration for Softened Water

Total hardness is set at a minimum concentration of 60 mg/L as $CaCO_3$. Excess hardness scales water boilers. Levels of the order of 200 are considered desirable for health purposes. Hard water has been positively correlated with reduced heart attacks. The bicarbonate, sulphates and chlorides of calcium and magnesium cause hardness. It is a measure of the presence of the cations, Ca^{2+} and Mg^{2+}. Water for industry is usually further softened if hardness levels exceed 100 mg/L, due to the scaling of boilers and plumbing equipment and unacceptable taste. Further details on hardness are given in Chapter 3. Alkalinity is defined as a measure of the water's ability to neutralize acids. It is not a measure of its acidity, as in pH. It is computed from the presence of carbonate species anions, HCO_3^-, CO_3^{2-} and OH^-. Further details are given in Chapter 3.

11.3.7 Monitoring Frequency

The 55 water quality parameters mentioned in the previous section are required to be monitored in different frequencies depending on the source and quality of the raw water. The monitoring proposals dictate that the sampling and analysis increase with the population being served. The three categories are: minimum monitoring (C1), current monitoring (C2), periodic monitoring (C3) and occasional monitoring in special situations or accidents. The minimum frequencies are set out in Annexe II of the Directive, but the Member States have latitude as to the setting of the frequencies. More specifically the monitoring frequencies are related to population size and volume of water produced per day. For instance, for a population of 150 000 the frequency of C1 is 180, C2 is 18 and C3 is 3 samples per year. The undesirable parameters, nitrates, nitrites and ammonia are in the C2 monitoring group. The reader is referred to Annexe II and Table B of Directive 80/778/EEC.

Table 11.3 EU drinking water directive parameters

Group	Group description	Parameter	Maximum admissible concentration (MAC)	Comments†
A	Organoleptic parameters	Colour	20 mg/L Pt/Co scale	
		Turbidity	10 mg/L SiO_2	
		Odour	Dilution of 2 at 12 °C	
		Taste	Dilution of 2 at 12 °C	
B	Physicochemical parameters	Temperature	25 °C	
		pH	$6.5 < pH < 8.5$	GV
		Conductivity	400 μS/cm	GV
		Chlorides	250 mg/L Cl	GV
		Sulphates	250 mg/L SO_4	
		Calcium	100 mg/L Ca	GV
		Magnesium	50 mg/L Mg	
		Sodium	150 mg/L Na	
		Potassium	12 mg/L K	
		Aluminium	0.2 mg/L Al	
		Total dry residues	1500 mg/L	
C	Parameters concerning substances undesirable in excessive amounts	Nitrates	50 mg/L NO_3	
		Nitrites	0.1 mg/L NO_2	
		Ammonium	0.5 mg/L NH_4	0.05 mg/L GV
		Kjeldahl N	1 mg/L N	
		Oxidizability	5 mg/L O_2	
		Hydrogen sulphide	Undetectable μg/L	
		Substances extractable in chloroform	No increase above background	0.1 mg/L dry residue in GV
		Hydrocarbons	10 μg/L	
		Phenols	0.5 μg/L C_6H_5OH	
		Boron	1000 μg/L B	GV
		Surfactants	200 μg/L (lacryl sulphate)	
		Organochlorines	1 μg/L	GV
		Iron	200 μg/L Fe	
		Manganese	50 μg/L Mn	
		Copper	100 μg/L Cu	GV
		Zinc	100 μg/L Zn	GV
		Phosphorus	5000 μg/L P_2O_5/L	
		Fluoride	1000 μg/L F	
		Suspended solids	0	
		Barium	100 μg/L Ba	GV
D	Parameters concerning toxic substances	Arsenic	50 μg/L As	
		Cadmium	5 μg/L Cd	
		Cyanides	50 μg/L Cn	
		Chromium	50 μg/L Cr	(Total)
		Mercury	1 μg/L Hg	
		Nickel	50 μg/L Ni	
		Lead	50 μg/L Pb	
		Antimony	10 mg/L Sb	
		Selenium	10 mg/L Se	
		Pesticides	0.1 μg/L	
		PAHs	0.2 μg/L	
E	Microbiological parameters	Total coliforms	MPN $\leq$ 1/100 mL	
		Faecal coliforms	MPN $\leq$ 1/100 mL	
		Faecal streptococci	MPN $\leq$ 1/100 mL	
		Sulphite reducing clostridia	MPN $\leq$ 1/20 mL	
F	Minimum required for softened water	Total hardness	> 60 mg/L as $CaCO_3$	
		Alkalinity	> 30 mg/L HCO_3	

† GV = guide value.

Example 11.1   Calculate the hardness in mg/L $CaCO_3$ of the following water sample.

Cation	Concentration (mg/L)	Equivalent weight
Na^+	35	23
Mg^{2+}	9	12.2
Ca^{2+}	48	20
K^+	1	39

Hardness is computed on the presence of Mg^{2+} and Ca^{2+}:

$$\text{Hardness, mg/L } CaCO_3 = \frac{Mg^{2+}(\text{mg/L}) \times 50 \text{ mg/meq}}{\text{eq wt } Mg^{2+}}$$

$$+ \frac{Ca^{2+}(\text{mg/L}) \times 50 \text{ mg/meq}}{\text{eq wt } Ca^{2+}}$$

$$= \frac{9 \times 50}{12.2} + \frac{48 \times 50}{20}$$

$$\text{Hardness} = 156.9 \text{ mg/L } CaCO_3$$

Further details on hardness are found in Chapter 3.

11.3.8 US Primary Drinking Water Standards

Some of the US primary drinking water standards are shown in Table 11.4. The parameters shown are divided into organic and inorganic chemicals, radionuclides, microbiological and other substances. The contaminants (parameters), the health effects, the parameter sources and the maximum concentration level (MCL) are included in Table 11.4 and in more detail in Table 11.15 of Sec. 11.12. This is somewhat comparable to the EU drinking water directive with variations on some parameters. The MCL is a USEPA enforceable regulation. Another term used is MCLG, which is the maximum concentration level goal and is a non-enforceable health goal.

11.3.9 Forms of Water Impurities

Raw water may contain impurities in several forms including:

- Particulate (size $> 10^{-1}$ mm) $\sim$ dust
- Suspended (10^{-3} mm $<$ size $< 10^{-1}$ mm) $\sim$ turbidity
- Colloidal (10^{-6} mm $<$ size $< 10^{-3}$ mm) $\sim$ clay minerals
- Dissolved (size $< 10^{-6}$ mm) $\sim$ humic/tannic acid, colour

It is the objective of the water treatment industry to reduce these impurities to acceptable levels. The forms of the impurities will define the type and level of treatment used, as detailed in Sec. 11.4. Figure 11.1 shows the breakdown of these impurities.

11.4 WATER SOURCES AND THEIR WATER QUALITY

Water for treatment and subsequent public consumption is normally sourced from:

- Rivers: upland and lowland
- Lakes and reservoirs
- Groundwater aquifers

Table 11.4 Excerpts from US primary drinking water standards

Contaminants	Health effects	MCL (mg/L)
Inorganic chemicals		
Cadmium	Kidney	0.005
Chromium	Liver/kidney, skin and digestive system	0.1
Copper	Stomach and intestinal distress; Wilson's disease	TT†
Fluoride	Skeletal damage	4
Lead	Central and peripheral nervous system damage; kidney; highly toxic to infants and pregnant women	TT
Mercury	Kidney, nervous system	0.002
Nitrate	Methanoglobinaemia, 'blue-baby syndrome'	10
Nitrite	Methanoglobinaemia, 'blue-baby syndrome'	1
Total (nitrate and nitrate)	Not applicable	10
Microbiological		
Giardia lamblia	Stomach cramps, intestinal distress (Giardiasis)	TT
Legionella	Legionnaires' disease (pneumonia), Pontiac fever	TT
Total coliforms	Not necessarily disease-causing themselves, coliforms can be indicators of organisms that can cause gastroenteric infections, dysentery, hepatitis, typhoid fever, cholera and other diseases. Coliforms also interfere with disinfection	
Turbidity	Interferes with disinfection	0.5–1.0 NTU (nephelometric turbidity unit)
Viruses	Gastroenteritis (intestinal distress)	TT
Other substances		
Sodium	Possible increase in blood pressure in susceptible individuals	None (20 mg/L reporting level)

† TT = treatment technology required.

The selection of the source is governed by many factors, including proximity to the consumer, economics, long-term adequacy of supply and raw water quality. The first three factors tend to be site specific and are not discussed further in this section. For further details refer to Twort *et al.* (1990) and Linsley and Franzini (1979). A typical required basic analysis of a raw water is listed as follows:

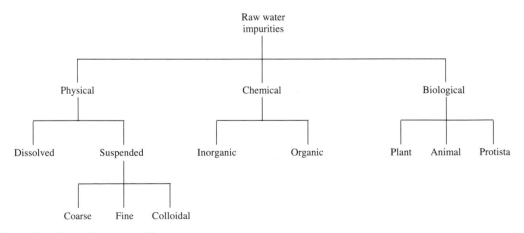

Figure 11.1 Forms of water impurities.

- Physical/Chemical
  - pH
  - Acidity
  - Alkalinity
  - Suspended solids
  - Colour
  - Turbidity
  - Dissolved oxygen
- Biological
  - Total coliforms (37°C, 24 h)
  - *E. coli* (37°C, 48 h)
- Aesthetic
  - Colour
  - Taste
  - Odour

Raw water quality varies with the source and if the source is surface water, the quality will vary seasonally, particularly with flooding. Tables 11.5 and 11.6 show a typical analysis of raw water quality from different sources. Table 11.7 highlights some differences between surface water and groundwater quality.

It is seen from Table 11.7 that the selection of a raw water source based on quality requires the investigation of many parameters. Traditionally, water sources were primarily surface water and not much groundwater was used. However, in recent decades, groundwater is becoming the favoured source not least because treatment costs are much less. Groundwater is most likely to be of better quality than surface water. Fears of surface water pollution and public disquiet about man-made reservoirs and dams have led to recent extensive developments of groundwater sources. In the United States, groundwater is becoming more important since:

Table 11.5 Typical raw water analysis

Parameter	Deep well water	Moorland water	River water	Arid zone water	Brackish well	Seawater
Colour	Clear	Slightly yellow	Turbid	Turbid		
Conductivity, μS/cm	580	150	915	1000–7000	2250	51 000
pH	7.3–7.9	6.5–7.2	7–8	7.5–8.5	7.45	7.9
TDS, ppm	410	105	640	700–5000	1500	36 200
Cations mg/L as $CaCO_3$						
Ca^{2+}	250	30	200	250–1500	60	350
Mg^{2+}	75	15	75	150–500	73	1330
Na^+	25	35	200	150–2000	257	10 300
K^+					15.4	350
Mn^+					0.002	
Fe^+					0.06	
Anions mg/L as $CaCO_3$						
Cl^-	40	30	125	< 2000	502	20 500
SO_2^-	500	15	175	< 1500	162	2850
F^-					2.05	
Si^-		6	10	10–20	24.5	20
HCO_3^-	250	30	125			170
NO_3^-	10	5	50			

Adapted from LORCH, 1987

Table 11.6 Typical analyses of raw water quality

Parameter	Upland catchment	Lowland river	Chalk aquifer	Sandstone/gravel/clay aquifer, Davis, Calif.*
pH	6.0	7.5	7.2	7.8
Total solids, mg/L (suspended and dissolved)	50	400	300	523
Alkalinity, mg/L	20	175	110	—
Hardness, mg/L $CaCO_3$	10	200	200	346
Colour, mg/L SiO_2	70	40	< 5	< 3
Turbidity, NTU	5	50	< 5	0.1
Coliforms, MPN/100 mL	20	20×10^3	5	Detectable in 1.2% of 1993 samples

*1994 Annual Report from Davis City Council

Table 11.7 Comparison of surface water and groundwater quality

Parameter	Surface water	Groundwater
Temperature	Varies with season	Relatively constant
Turbidity and suspended solids	Varies and is sometimes high	Usually low or nil
Mineral content	Varies with soil, rainfall, effluents, etc.	Relatively constant at high value
Divalent iron and manganese in solution	Some	Always high
Aggressive carbon dioxide	None	Always some
Dissolved oxygen	Often near saturation, except when polluted	Usually low, requires aeration
Ammonia	Only in polluted water	Levels found to be increasing
Hydrogen sulphide	None	Usually some
Silica	Moderate levels	
Nitrate	Generally none	Levels found to be increasing due to agricultural pollution
Living organisms	May have high levels	Usually none

- 50 per cent of the US population use groundwater for drinking water
- Groundwater makes up 95 per cent of US freshwater resources
- 75 per cent of US cities use groundwater in some way
- 95 per cent of US rural dwellers use groundwater

However, groundwater is not immune to pollution, as was described in Chapters 3, 4 and 9, and therefore aquifer protection schemes (Chapter 4) are essential. A major concern in the adoption of groundwater is the leaching of nitrates from agricultural activities. Excessive agricultural contamination of groundwater has led to the abandonment of some groundwater sources in northern Europe, as has industrial and military pollution in the United States, Europe and the former USSR.

11.5 WATER TREATMENT PROCESSES

There are four classes of water treatment outlined in Table 11.8.

Many single family rural dwellings and small rural conurbations, taking their water supply from boreholes, have no treatment. This is generally considered acceptable when there is a very short line of supply from source to user.

In class B, where borehole water or occasional upland water is used for public supplies, disinfection is practical to maintain purity along the supply pipeline. However, much debate is ongoing about the

Table 11.8 Classes of water treatment

Class	Description	Source
A	No treatment	Some borehole water
		Occasional upland water
B	Disinfection only	Some borehole water
		Occasional upland water
C	Standard water treatment	Lowland rivers and reservoirs
D	Special water treatment	Some rural supplies (Fe and Mn)
		Colour removal
		Trace element removal
		Industrial water
		Electronics industry requirement
		Algae removal
		Organics removal

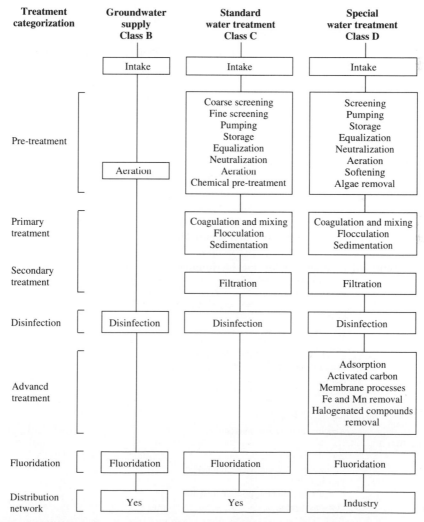

Figure 11.2   Flow chart outline of unit process in different raw water classes.

widespread use of chlorine disinfection for such purposes and alternatives to chlorine are now being adopted. In northern Europe, there is the tendency not to disinfect borehole water but to maintain intensive analytical vigilance over the microbiological quality as delivered to the consumer. This is because of the chemical reaction of chlorine with organics producing trihalomethanes (THMs). However, not disinfecting must be balanced with the potential waterborne diseases that may become pipe borne. In water of class B, aeration is sometimes used to eliminate hydrogen sulphide odours and tastes and increase the oxygen levels in the water.

Class C is what is internationally known as standard water treatment and is applied to water sources from lowland areas and reservoirs. A typical chart of the processes in this form of treatment is shown in column 3 of Fig. 11.2. Each of these processes is detailed in later sections. In general, the processes include pre-treatment, standard treatment (sedimentation and filtration), disinfection and possibly fluoridation.

Class D is special water treatment and is used when the source is downstream of urban developments or when industries (e.g. pharmaceutical) require high-quality water. Additional processes include membrane technology, iron and manganese removal, chemical oxidation, carbon adsorption, etc.

11.5.1 Selection of Treatment Processes

The selection of the set of treatment processes is preceded by detailed raw water quality analysis. The analyses should run over a period of a minimum of one year and, where possible, longer. It should sample the raw water at periods of low, medium and high flows from a surface water source. The parameters to be looked at should be all those listed in the EU drinking water directive (or equivalent legislative document). The report on the raw water quality analysis should then be evaluated in conjunction with other engineering reports on site regarding suitability, availability and continuity of water supply, proximity to the consumers and available land and its suitability for structures.

The treatment processes selected will depend on the water quality report. Table 11.9 is a brief outline of the general processes and their suitability for removing particular impurities. For instance, turbidity, which is a measure of the very fine suspended or colloidal impurities, is treated by the processes of coagulation and sedimentation and also filtration. Pathogens are normally removed by the processes of pre- or post-chlorination. However, other processes, including disinfection by chloramines, ozone or ultraviolet irradiation, are gaining popularity over chlorination.

Table 11.9 Recommended treatment for specific impurities

Parameter	Treatment process
Floating matter	Coarse screens, fine screens
Suspended matter	Microscreens
Algae	Microscreens, pre-chlorination, carbon adsorption, rapid filtration
Turbidity	Coagulation, sedimentation, post-chlorination
Colour	Flocculation, coagulation, filtration
Taste and odour	Activated carbon
Hardness	Coagulation, filtration, lime softening
Iron and manganese	
$> 1\,mg/L$	Pre-chlorination
$< 1\,mg/L$	Aeration, coagulation, filtration
Pathogens, MPN/100 ml	
< 20	Post-chlorination
20–100	Coagulation/filtration/post-chlorination
> 100	Pre-chlorination
	Coagulation/filtration/post-chlorination
Free ammonia	Post-chlorination
	Adsorption

11.6 PRE-TREATMENT OF WATER

If the raw water is of adequate quality, it may be pumped directly to the standard treatment processes of flocculation/coagulation and sedimentation. However, generally there are some steps that are required prior to this. They may include:

- Screening: equalization and neutralization
- Storage: equalization and neutralization
- Aeration
- Chemical pre-treatment: softening, algae removal and pre-chlorination

11.6.1 Screening

Coarse screens, typically inclined bars of 25 mm diameter and 100 mm spacing prevent large floating material from entering the treatment plant. Raking is facilitated by the inclination of the bars. Velocities are usually limited to about 0.5 m/s through the screens, which may be manually or automatically raked down.

If storage is not provided, fine screens are fitted after the coarse screens. If there is storage then fine screens are placed at the outlet of the storage tanks. Fine screens are typically mesh with openings about 6 mm diameter or square. Proprietary forms of fine screens are now conventionally used and these are usually automatically cleaned. They tend to be either of a circular drum type or a travelling belt (as in a vertical escalator). Screens introduce a head loss across the screen and this must be accounted for in hydraulic calculations.

A third type of screening used in water treatment is microscreening, where the mesh openings range from 20 to 40 μm. Such screens are used only as the main (physical) treatment process for relatively uncontaminated waters and moderately coloured waters. They have also been used upstream of slow sand filters to allow filter runs (2 to 6 months) and flow rates to be increased. For further details on design and proprietary screens, refer to Pankratz (1988).

11.6.2 Storage, Equalization and Neutralization

If raw water is abstracted from a river it is prudent to provide raw water storage tanks. They serve as a safety line in the event of the river becoming polluted. They also serve as reservoirs in times of low flow. Storage may be an open impoundment on a fast stream or a smaller reservoir to balance the flows going to the treatment plant. In the event of low river flows, it might not be possible to provide an 'equal' (or consistent) flow to the plant. The storage tanks alleviate this problem by always having a minimum volume for supply to the plant and the design selection (and operation) of water pumps is then more consistent. Storage tanks also assist in the treatment process as water allowed to sit in tanks will permit some of its suspended settleable matter to begin to settle. As such, storage tanks may act as initial settling tanks (type I settling, see later sections). Storage should be equivalent to 7 to 10 days of the average water demand. This period may be adequate to reduce most pathogens by exposure to daylight. The period should not be long, so as not to encourage the growth of other organisms, including undesirable algae. Water allowed to settle for some days in a storage tank will be cheaper to treat at the plant. However, the cost of building storage tanks is expensive and their maintenance (bed load silt removal) is also expensive. Therefore, in the design of an overall treatment plant, detailed cost analyses are required to optimize the benefit that storage tanks will provide. Another type of storage system is when tanks provide a storage time of about 12 h. This is commonly used to reduce pumping costs and of using pumping at more economical night-time electricity rates. In this situation, no settling benefits occur. An example of how to compute the size of an equalization basin is given in Chapter 12.

11.6.3 Aeration

Aeration is the supply of oxygen from the atmosphere to water to effect beneficial changes in the quality of the water. It is a common treatment process for groundwater and less common for surface waters. Aeration is used:

1. To release excess H_2S gas which may cause undesirable tastes and odours.
2. To release excess CO_2 which may have corrosive tendencies on concrete materials.
3. To increase the O_2 content of water in the presence of undesirable tastes due to photosynthetic algae (fishy smell), which release volatile oils on decomposition.
4. To increase the O_2 content of water which may have negative taste, colour and stain properties due to the presence of iron and manganese in solution. The addition of oxygen assists the precipitation of iron and manganese.

Aeration can be a simple mechanical process of spraying water into the air and allowing it to fall over a series of cascades (waterfalls), while absorbing or desorbing (stripping) oxygen in its journey. For further details on oxygen transfer to water and in particular other gas transfers (including chlorine or ozone for pathogen removal), the reader is referred to Montgomery (1989) and Reynolds (1982).

11.6.4 Chemical Pre-treatment

Chemical pre-treatment to remove undesirable properties of water (algae or excess colour) is a more expensive process than chemical post-treatment. In pre-treatment, greater amounts of chemicals are required to effect the same result as some of the chemical is masked and absorbed by turbidity in the water. For instance, pre-chlorination may be at doses five times greater than post-chlorination. Only two chemical pre-treatment processes are discussed here:

- Pre-chlorination
- Activated carbon

Pre-chlorination is used on low turbidity water with a high coliform count. The chlorine is injected into the water stream and over the period that it stays in the settling tanks, it oxidizes and precipitates iron and manganese. It also causes pathogenic kill and reduces colour. Doses as much as 5 mg/L are used, (distinct from 1 mg/L in post chlorination) and this is expensive. Water authorities tend to use pre-chlorination at times of the year when the surface water supply is likely to be polluted from agricultural or industrial sources or when excess organic matter is transported during flooding. Pre-chlorination is also beneficial in the reduction of ammonia in both surface and groundwater supplies. Chlorine may also be added by the dissolution of chlorine gas in water by the process of gas absorption, but this technique is preferable for post-chlorination. Chlorination is discussed in Sec. 11.9.5.

The addition of activated carbon as an adsorbent is used for many purposes including:

- Removal of photosynthetic algae
- Improvement of colour and odour
- Removal of selective organic compounds.

Activated carbon can be used either as powdered activated carbon (PAC) or granular activate carbon (GAC). PAC was the traditional choice in water treatment but increasingly GAC is required where tastes and odours in water have an industrial base. PAC has a lower capital cost but also a lower efficiency than GAC. PAC, in slurry form, is usually added prior to coagulation or just before the sand filters. Doses may vary from 3 to 20 mg/L. The mechanism is that the PAC is deposited in the sand filters and the water impurities causing the undesirable tastes and odours are adsorbed to the PAC during filtration. The longer the filter runs (time between cleaning of filters—backwashing) the more efficient the PAC. Typically runs in excess of 4 h are satisfactory. However, if the filter runs are shorter, it is necessary that the PAC be added prior to the coagulation process. PAC is only normally used for intermittent control of taste and odour problems. Where the problem is persistent GAC is used. GAC consists of a specifically designed GAC filter bed which is used in either the upflow or downflow modes. However, the use of GAC beds are more applicable to advanced water treatment processes in the production of very high quality water and are discussed further in Sec. 11.11.3.

11.7 SEDIMENTATION, COAGULATION AND FLOCCULATION

Standard treatment is the set of unit processes that reduce colour, turbidity and particulate impurities to acceptable levels. In doing so, additional benefits occur, such as iron and manganese reductions, algae reductions, pathogenic reductions, etc. Standard treatment can be considered to consist of the following unit processes:

- Sedimentation
- Coagulation and flocculation
- Sedimentation of flocculent particles
- Filtration

11.7.1 Sedimentation—General

Sedimentation by definition is 'the solid–liquid separation using gravity settling to remove suspended solids' (Reynolds, 1982). In water treatment, sedimentation processes used are:

Type I. To settle out discrete non-flocculent particles in a dilute suspension. This may arise due to the plain settling of surface waters prior to treatment by sand filtration.

Type II. To settle out flocculent particles in a dilute suspension. This may arise after chemical coagulation and flocculation where the non-discrete particles are chemically assisted to coagulate.

Other types of settling are combinations of I and II.

Sedimentation of discrete particles—type I Settling tanks are of two types: rectangular and circular. A rectangular settling tank is shown in Fig. 11.3. They tend to have a length–width ratio of about 2 and a depth of the order of 1.5 to 6 m. A sludge draw-off well is located at the upstream base, and the sludge is drawn to this by a travelling scraper board.

Figure 11.4 shows a circular settling tank. Dimensions typically are 10 to 50 m in diameter and 2.5 to 6 m in depth. Water enters to the central well either at the top or up through a central pipe. As the influent water settles, it spreads out and a sludge scraper moves the sludge towards a central sludge withdrawal hopper at bed level. The clarified water exits over a weir along the perimeter of the tank at surface level.

The key parameters and typical values in the design of settling tanks are:

- Surface overflow rate $\sim$ 20–35 $m^3/day/m^2$
- Detention times $\sim$ 2–8 h
- Weir overflow rate $\sim$ 150–300 $m^3/day/m^2$

The above values vary depending on whether the water treated is raw water for potable treatment or coagulated raw water. Similar settling tanks are used in wastewater treatment. Refer to Reynolds (1982) or Metcalf and Eddy (1991) for further details.

In type I settling, the particles settle out individually and it is assumed that there is no flocculation or coagulation between the particles. Such sedimentation may typically be the first physical process of surface waters where discrete organic and grit particles are given adequate time to settle. This type of settling is somewhat similar to grit channel settling (for wastewater) described in Chapter 12. The design

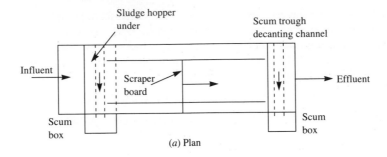

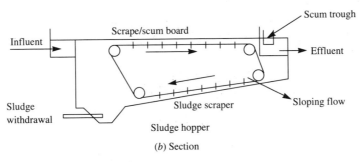

Figure 11.3 Schematic of rectangular settling tank.

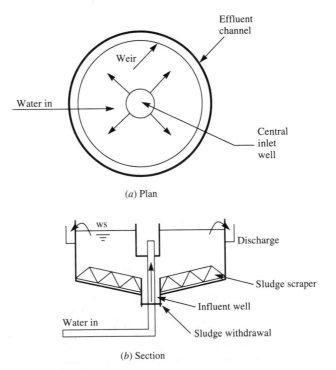

Figure 11.4 Schematic of circular settling tank.

of such tanks is based on a knowledge of settling velocity as described in the following text. In type I settling, a particle will accelerate vertically downwards until the drag force F_D equals the impelling force F_I and thereafter the particle settles at a constant velocity known as Stokes' velocity V_s:

The impelling force.

$$F_I = (\gamma_s - \gamma_w) V_{ol} \tag{11.1}$$

where $\qquad\qquad\qquad \gamma_s$ = weight density of solid particles $\rho_s g$

and $\qquad\qquad\qquad \gamma_w$ = weight density of water $\rho_w g$

$\qquad\qquad\qquad\qquad V_{ol}$ = volume of particle

The drag force

$$F_D = C_D A_s \rho_w \left(\frac{V_s^2}{2}\right) \tag{11.2}$$

where $\qquad\qquad\qquad C_D$ = drag coefficient ≈ 0.4 for spheres

$$C_D = \frac{24\nu}{V_s d} \qquad \text{for laminar flow for } Re < 100 \tag{11.3}$$

$\qquad\qquad\qquad A_s$ = sphere section area orthogonal to velocity vector

$\qquad\qquad\qquad V_s$ = settling or Stokes' settling velocity

$\qquad\qquad\qquad \nu$ = kinematic viscosity $- \mu/\rho$

Equating

$$F_I = F_D \tag{11.4}$$

$$(\gamma_s - \gamma_w)\frac{\pi}{6}d^3 = \frac{24\nu}{V_s d}\frac{\pi}{4}d^2 \rho_w \left(\frac{V_s^2}{2}\right) \tag{11.5}$$

Solving

$$V_s = \frac{g}{18\mu}(S_p - 1)d^2 \tag{11.6}$$

where $\qquad\qquad\qquad S_p$ = specific gravity of particles

or

$$V_s = \frac{g}{18\mu}(\rho_s - \rho_w)d^2 \tag{11.7}$$

This is known as Stokes' law for settling velocity of discrete particles and applies for $Re < 0.5$. A tank for preliminary settling of raw waters is assumed to behave as follows:

1. Type I settling applies.
2. The flow entering and leaving the tank is uniform.
3. There are three zones within the tank (see Fig. 11.5):
    (a) an inlet zone,
    (b) an outlet zone,
    (c) a sludge zone.
4. The particle distribution throughout is uniform.
5. The particles, on entering the sludge zone, stay there until scraped off the bottom.

Such an arrangement is shown in Fig. 11.5.
    For a rectangular tank:

$$\text{Retention time } t = \frac{H}{V_s} = \frac{L}{V} \tag{11.8}$$

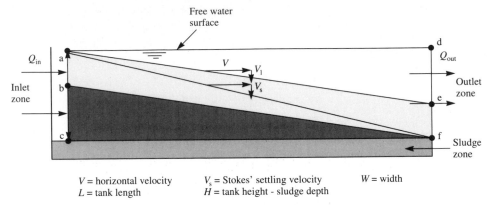

V = horizontal velocity V_s = Stokes' settling velocity W = width
L = tank length H = tank height - sludge depth

Figure 11.5   Elevation of type 1 settling tank.

Note that $L \geq 2W$ and $L \gg H$.

$$\text{Horizontal velocity } V = \frac{Q}{WH} \tag{11.9}$$

$$t = \frac{WHL}{Q} = \frac{V_{ol}}{Q} \tag{11.10}$$

Therefore

$$\frac{V_{ol}}{Q} = \frac{H}{V_s} \tag{11.11}$$

$$V_s = \frac{Q}{LW} = \frac{Q}{A_p} \tag{11.12}$$

However,

$$\frac{Q}{A_p} = \text{surface overflow rate} \tag{11.13}$$

where $\qquad A_p = \text{plan area}$

Therefore, the settling velocity, V_s, is equal to the surface overflow rate for a rectangular tank. The same is true for a circular tank.

The question remains as to what percentage of discrete particles are removed. From Fig. 11.5 it is seen that a particle entering the tank at point a, if it settles at V_1, then leaves the tank at point e. Similarly, a particle entering at point b, settling at V_1, leaves the tank at point f. It is also seen (in a vector sense) that $V_1 < V_s$. The percentage of particles (settling with V_1) removed is then:

$$X_1 = \frac{b-c}{a-c} = \frac{(V_1/V)L}{(V_s/V)L} = \frac{V_1}{V_s} \tag{11.14}$$

Note that a particle entering at point a (larger than the particle of the previous paragraph), if settling at V_s, will exit the tank at point f. So all particles with a settling velocity greater than V_s will settle out. Therefore, if all the particles were of one size (all entering at point a) and settling at V_s then theoretically 100 per cent settling could be removed. However, a water usually has a range of particle sizes and proper design of settling tanks requires a particle size distribution analysis or a settling column test (described in the next section).

Typically, a cumulative distribution of particle settling velocity is computed for a water sample. A schematic of such a curve is shown in Fig. 11.6. In any settling tank, all particles with a settling velocity $> V_s$ will settle, plus an additional as yet unknown fraction of those smaller particles with a settling velocity $< V_s$. The total fraction that settles is then

$$X_r = (1 - X_s) + \int_0^{X_s} \frac{V}{V_s} dx \qquad (11.15)$$

where $(1 - X_s)$ is the fraction of particles with settling velocity greater than V_s and

$$\int_0^{X_s} \frac{V}{V_s} dx = \text{fraction of particles removed with velocity less than } V_s$$

The discrete settling equation (11.15) becomes

$$X_r = (1 - X_s) + \frac{1}{V_s} \Sigma V \Delta X \qquad (11.16)$$

Example 11.2   Size a square type I settling tank to treat $36\,400\,\text{m}^3/\text{day}$ of raw water, with a surface overflow rate of $12\,\text{m}^3/\text{day}/\text{m}^2$ and a detention time of 6 h. If the particle size distribution is given below, determine the overall removal when the specific gravity is 1.15.

Particle size mm	0.1	0.08	0.07	0.06	0.04	0.02	0.01
Weight fraction %	10	15	35	65	90	98	100
V_s, mm/s	0.81	0.52	0.40	0.30	0.13	0.03	0.008
Re	0.08	0.042	0.028	0.018	0.005	0.0006	0.00008

Solution

$$\text{Surface area required} = \frac{Q}{\text{overflow rate}}$$

$$A_p = \frac{36\,400}{12} = 3033\,\text{m}^2$$

$$L = W = 55\,\text{m, say } 60 \times 60$$

$$\text{Depth } H = V_s t$$

However, the settling rate equals the surface overflow rate (SOR):

$$\text{Actual SOR} = \frac{36\,400}{60 \times 60} = 10.1$$

Therefore

$$H = 10.1 \times \frac{6}{24} = 2.57\,\text{m}$$

Say

$$H = 2.6\,\text{m}$$

Check the weir overflow rate (WOR):

$$\text{WOR} = \frac{Q}{W}$$

$$= \frac{36\,400}{60}$$

$$= 606\,\text{m}^3/\text{day}/\text{m}$$

To stay within a WOR of $< 300 \, \text{m}^3/\text{day/m}$ the width should be close to 120 m. Recall $Re = V_s d/V$ where $V = $ kinematic viscosity.

Particle size, mm	0.1	0.08	0.07	0.06	0.04	0.02	0.01
Weight fraction greater than, %	10	15	35	65	90	98	100

The settling velocity from Stoke's law, with particles of specific gravity of 1.15 is:

$$V_s = \frac{g}{18\mu}(\rho_s - \rho_w)d^2$$

$$= \frac{9.81}{18 \times 1.002 \times 10^{-3}}(1.15 - 1.0)d^2$$

$$= 81.6d^2$$

As $Re \ll 0.5$, Stokes' law applies. All particles will be removed that have a settling velocity greater than the actual surface overflow rate (SOR) (plus another fraction):

$$\text{SOR} = 10.1 \, \text{m}^3/\text{day/m}^3$$

$$= 10.1 \, \text{m/day}$$

$$= 0.12 \, \text{mm/s}$$

From the above table it is seen that slightly more than 90 per cent of the particles (by weight fraction) or greater than about 0.04 mm in size are removed. To determine the precise number, a cumulative particle settling curve can be prepared (Fig. 11.6).

Sedimentation of flocculent particles—type II As defined in Sec. 11.7.1, type II settling is settling of flocculent groups of particles. Flocculent particles are those particles that are chemically assisted to come together and produce large particles, and thus settle. Coagulation is the first process of adding the coagulated chemical which changes the particles' electric charge and is then amenable to aggregation. Flocculation is the second process of getting the 'coagulated mix' to form larger flocs. While the particles are settling, they are also flocculating and so increase in size and mass during the settling process. This phenomenon occurs in settling of chemically coagulated potable waters and wastewaters. Primary settling of wastewaters (Chapter 12) is also type II settling. Because the mass/size increases with depth, the process does not lend itself to direct analyses as was the case for type I settling. Instead, to determine the settling rates, laboratory tests are carried out using a batch settling column shown in Fig. 11.7. This

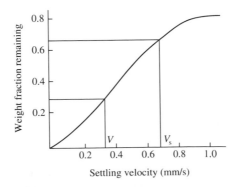

Figure 11.6 Typical discrete particle settling curve.

column has openings at different points, where samples are withdrawn at regular time intervals and the suspended solids concentration is determined. The column is typically 100 to 200 mm in diameter with a height equal to the proposed settling tank height ($\cong 1$ to 3 m). The column is filled initially with a well-mixed sample of known total solids concentration, such that initially the same solids concentration exists for the full depth. For the samples taken during the settling process, the percentage removal (of solids) is computed and plotted as shown in Fig. 11.7, with equi-percentage removal curves, R_A, R_B, etc. The overflow rates (equal to the settling rates) are

$$V_s = \frac{H}{t_x} \times \text{scale-up factors} \qquad (11.17)$$

The reader is referred to Reynolds (1982) for further details.

11.7.2 Coagulation

Raw water, after screening, has impurities in suspension and in solution. Particulate matter in suspension has a particle size range of 10^{-7} to 10^{-1} mm. Inorganic clay colloids range in size from 10^{-6} to 10^{-3} mm and form the dominant component of particulates in suspension. The minor component is that of organic colloids or micro-organisms. Because of the very small size, the suspended matter has negligible settling velocity, whether it is organic or inorganic. One of the objectives of water treatment is to promote the settlement of suspended particulate matter. Settlement of particles occurs when their settling velocity is adequate to cause settling in a short (economic) time period.

The coagulation process utilizes what is known as a chemical coagulant (aluminium or iron salt) to promote particle agglomeration. Before identifying the ideal coagulant, specific properties of the suspended particles (impurities) are identified. These are: its classification and electric charge. Classification determines whether a particulate in suspension has an affinity for water adsorption or not. Particles with an affinity for water adsorption are hydrophilic and those that do not adsorb water are termed hydrophobic. Most suspended particles are of a negative electrostatic charge. This means that they repulse each other and thus stay in suspension. Particles that remain in suspension are said to be stable. If their electrostatic charge can be changed, they would become destabilized, attract each

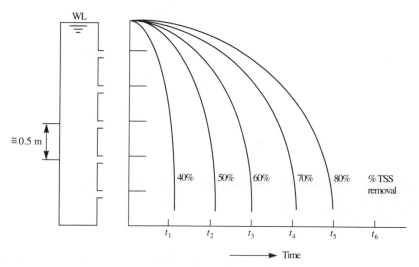

Figure 11.7   Schematic batch settling column of settling diagram for type II.

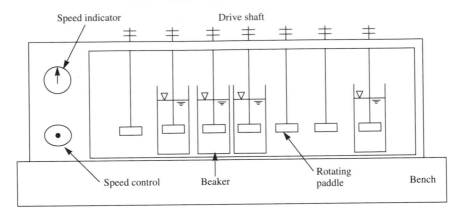

Figure 11.8   Jar test instrument.

other, agglomerate and settle. Further details on particle agglomeration are described in Chapter 13 on sludge treatment.

Chemical coagulants (e.g. aluminium sulphate or ferric sulphate) are added to the raw water and for a brief period (20 to 60 s) rapid mixing is carried out. This is done in tanks of a variety of designs, with the objective of producing a microfloc. The latter is produced as a result of the destabilization of the initially stable suspended impurities, followed by the interparticle bridging.

Having produced the microfloc (itself not yet very settleable), the objective is then to produce a floc of adequate size that will settle under gravity. The next process is to subject the microfloc 'solution' to a slow flocculation procedure. This is carried out in tanks over a time period of 20 to 60 min at very slow agitation rates. If the flocculation mixing rotor is too fast, there is the risk of breaking up the initial microfloc and thus negating the process. Depending on the raw water quality, microfloc promotion may be inadequate by chemical coagulants alone and may require the addition of coagulant aids, also known as polyelectrolytes. These are added, after the coagulant in a small rapid mixing tank and before flocculant mixing. Coagulation and flocculation processes may be accomplished in a variety of geometrical tank configurations, which are discussed in the following section.

The amount of chemical coagulant and/or polyelectrolyte required for a particular raw water quality can be determined using the laboratory jar test, although more sophisticated bench scale/pilot studies are more desirable. The jar test instrument is shown in Fig. 11.8. The jar test is a simple test made up of several one litre beakers with samples of the raw water. To each is added a different and increasing amount of coagulant and rapid mixing is followed for 20 to 60 s. The samples are allowed to settle and that sample with the best settlement characteristic is selected as the coagulant. If the settlement or microfloc generation is inadequate, the coagulant may be assisted with a polyelectrolyte (at varying doses). After a series of tests, it is possible to ascertain the best-dose combination of coagulant and polyelectrolyte that promotes the optimum floc size. Further details of the test are described in Barnes *et al.* (1986).

Coagulants   The three most common coagulants are:

- Aluminium sulphate (alum)
- Ferrous sulphate (ferric)
- Ferric chloride

As aluminium sulphate is corrosive on its own, it is packaged with H_2O. When alum is added to raw water, the following reactions take place:

$$Al_2(SO_4)_3 \cdot 14H_2O + 3Ca(HCO_3)_2 \longrightarrow 2Al(OH)_3 + 3Ca(SO_4) + 14H_2O + 6CO_2 \qquad (11.18)$$

$\qquad$ alum $\qquad\qquad\qquad$ calcium $\qquad\qquad$ aluminium $\qquad$ calcium
$\qquad\qquad\qquad\qquad\qquad$ bicarbonate $\qquad\qquad$ hydroxide $\qquad\quad$ sulphate
$\qquad\qquad\qquad\qquad\qquad\qquad\qquad\qquad\qquad\qquad$ floc

The object is the production of a floc, and in the above case this is an aluminium hydroxide floc. However, if there is insufficient alkalinity in the water, alkalinity is added by means of lime (calcium hydroxide) addition. This reaction is as follows:

$$Al_2(SO_4)_3 \cdot 14H_2O + 3Ca(OH)_2 \longrightarrow 2Al(OH)_3 + 3Ca(SO_4) + 14H_2O \qquad (11.19)$$

$\qquad\qquad$ alum $\qquad\qquad\qquad$ calcium $\qquad\qquad$ aluminium $\qquad$ calcium
$\qquad\qquad\qquad\qquad\qquad$ hydroxide $\qquad\qquad$ hydroxide $\qquad\quad$ sulphate
$\qquad\qquad\qquad\qquad\qquad\qquad\qquad\qquad\qquad\qquad$ floc

The optimum pH range for aluminium hydroxide floc formation is 4.8 to 7.8, as within this range this floc is insoluble. Different water treatment plants use different coagulants and combinations of coagulants and coagulant aids. For more details, refer to Montgomery (1989). The constant production of a satisfactory floc at treatment plants is a process of continuous review, as the raw water quality may vary seasonally. Different coagulants/coagulant aids or different doses may be required during flood periods when the raw water may be high in turbidity and high in colour.

$\qquad$ Coagulants are sometimes assisted with further chemicals, known as coagulant aids. They essentially are:

- Polyelectrolytes
- Lime alkalinity addition
- pH correction: lime, sulphuric acid

Polyelectrolytes are long-chain synthetic organic chemicals (SOC) which are used to optimize coagulation. They can be cationic, anionic or polyampholites and the huge range and strengths available allow almost every conceivable coagulant requirement to be satisfied. The dose range of polyelectrolytes is 0.05 to 0.5 mg/L, by comparison with a dose range of 5 to 100 mg/L for alum. Polyelectrolytes are very expensive by comparison with alum. The optimum dose of coagulant plus coagulant aid (including the pH correction dose) can be determined using the jar test.

$\qquad$ Table 11.10 indicates the initial selection process of coagulant/coagulant aid for a raw water based on its level of turbidity and alkalinity. High turbidity and high alkalinity water are easy to treat, as flocs form readily. High turbidity and low alkalinity water may need lime to improve the alkalinity and optimize coagulation. Low turbidity and low alkaline water is difficult to treat, requiring alkalinity correction with lime and high doses of high molecular weight polyelectrolyte. The addition of alum causes a reduction in

Table 11.10 Coagulant and polyelectrolyte uses for turbidity treatment

Water class	water description	Alum	Ferric	Polyelectrolyte
A	High turbidity > 5 NTU High alkalinity > 250 mg/L HCO_3 (easy to treat)	Effective if pH is 5–7	Effective if pH is 5–7	Not required
B	High turbidity Low alkalinity < 50 mg/L HCO_3	Effective if pH is 5–7 + lime	Effective if pH is 5–7 + lime	Not required
C	Low turbidity	Polyelectrolyte	Polyelectrolyte	
	High alkalinity	aid essential	aid essential	Essential
D	Low turbidity < 1 NTU Low alkalinity < 50 mg/L HCO_3 (difficult to treat)	Only possible with lime and polyelectrolyte	Only possible with lime and polyelectrolyte	Essential

the pH of the water. In lowland waters, this pH reduction needs to be counteracted, which is done by the addition of soda ash.

Example 11.3 Determine the daily requirement of alum, lime and polyelectrolyte to coagulate a flow of 200 L/s, if the jar test indicates that optimum coagulation occurs when 1 litre of water is dosed with 3 mL of 10 g/L alum solution, 1.8 ml of 5 g/L suspension of lime and 0.2 mg/L of polyelectrolyte.

Solution:

$$\text{Daily flow rate} = 200 \times 60 \times 60 \times 24 = 17.28 \times 10^6 \text{ L}$$
$$\text{Alum requirement 3 mL of 10 g/L} = 30 \text{ mg/L} \times 17.28 \times 10^6 \text{ L} = 518.4 \text{ kg/day}$$
$$\text{Lime requirement 1.8 mL of 5 g/L} = 9 \text{ mg/L} \times 17.28 \times 10^6 \text{ L} = 155.5 \text{ kg/day}$$
$$\text{Polyelectrolyte} = 0.2 \text{ mg/L} \times 17.28 \times 10^6 \text{ L} = 3.46 \text{ kg/day}$$

Example 11.4 How much alkalinity will be destroyed if 110 mg/L of bulk ferric sulphate is applied to the water at a water treatment plant? Assume that bulk ferric sulphate is 20 per cent by weight Fe.

Solution

$$Fe^{3+} + 3HCO_3^- \longrightarrow Fe(OH)_3^{2+} + 3CO_2$$
$$\text{1 mole} \qquad \text{3 moles}$$

Molecular weight: 55.8 g/mol   61 g/mol

The equivalent weight of $CaCO_3$ is 50 mg/L, so $3 \times 50 = 150$ mg of alkalinity react with 55.8 mg of Fe^{3+}.

$$\text{The amount of iron applied} = \frac{20}{100} \times 110 = 22 \text{ mg/L}$$

$$\text{The alkalinity reacted} = 22 \times \frac{150}{55.8} = 59.1 \text{ mg/L}$$

Coagulation and flocculation infrastructure The water treatment plant infrastructure required for coagulation/flocculation typically consists of:

- Coagulant dosing and rapid mixing unit
- Polyelectrolyte preparation tank and dosing unit
- Flocculation basin

Coagulant rapid mixing unit Several different types of tank or pipe configuration are used to inject the coagulant and produce rapid mixing over the short time period of 20 to 60 s. Very high velocity gradients of 700 to 1000 m/s/m achieve the rapid mixing. This high shear environment can be: a hydraulic jump, jet injection, propeller mixer or paddle mixers or combinations. Details of configurations and their design are given in Reynolds (1982).

Polyelectrolyte mixing unit The polyelectrolyte is first prepared in a mixing basin to achieve the right concentration. It is then added to the treatment process a short distance downstream of the coagulant rapid mixing unit.

Flocculation basin Before arrival at this stage, the water has been coagulated, and so microflocs have been produced. The object now is to encourage the microflocs (pin flocs) to agglomerate and produce larger flocs. As such, detention times of 20 to 60 min are now required and therefore the flocculation basin is about 50 times larger than the rapid mixing unit. Gentle agitation is required in this unit to promote thorough mixing. However, mixing should not be so strong as to cause break up of the weak influent microfloc. Flocculation basins are categorized as being either axial flow type (hydraulic) or cross flow type (mechanical), as shown in Fig. 11.9.

Example 11.5   Determine the basin dimensions of a uniform depth axial flow flocculation basin to treat 8 MGD ($36\,400\,\text{m}^3/\text{day}$). The detention time is 50 min. Assume the basin width is 25 m consisting of five equal width units separated by perforated concrete walls.

Solution

$$\text{Basin volume } V = \frac{36\,400}{24} \times \frac{50}{60} = 1264\,\text{m}^3$$

$$\text{Long section area } A = \frac{1264}{25} = 50.5\,\text{m}^2$$

$$\text{Length} \times \text{depth} = 50.5\,\text{m}^2$$

If the cross-section of each unit is square, i.e. 5 m × 5 m, i.e. depth = 5 m, then

$$\text{Length} = \frac{50.5}{5} = 10.1\,\text{m}$$

Therefore   Basin dimensions = length × width × depth = $10.1\,\text{m} \times 5\,\text{m} \times 5\,\text{m}$

Example 11.6   A water treatment plant consists of the following unit processes: coagulation, flocculation, sedimentation, filtration and disinfection. The suspended solids concentration of the raw water is 500 mg/L and the plant treats $36\,400\,\text{m}^3/\text{d}$ (8 MGD). Alum [$Al_2(SO_4)_3 \cdot 14H_2O$] is used as a

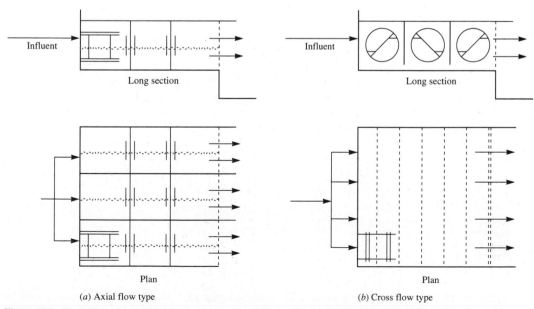

(a) Axial flow type   (b) Cross flow type

Figure 11.9   Flocculation basins (adapted from Reynolds, 1982. Reprinted by permission of PWS Publishing Company).

coagulant with a dose rate of 50 mg/L. Compute the sludge solids produced daily if complete reaction of alum to aluminium hydroxide occurs and 98 per cent total solids are removed by sedimentation/filtration.

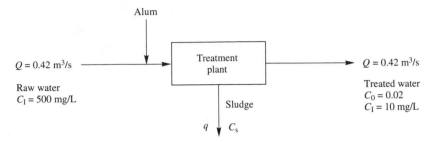

Solution   Materials balance for suspended solids:

$$\begin{array}{l} \text{Accumulation} = \text{Input} - \text{output} \;+\; \text{generation} \;-\; \text{consumption} \\ \text{within system} \qquad\qquad\qquad\; \text{within system} \quad\;\; \text{within system} \end{array}$$

$$0 = \text{input} - \text{output} + 0 \quad -0$$

Therefore

$$\text{Input} = \text{output}$$
$$QC_I = QC_0 + qC_s$$
$$qC_s = Q(C_I - C_0) = 0.42 \times 490 \times 10^3 \,\text{mg/s} = 206 \,\text{g/s}$$

Stoichiometric materials balance for aluminium hydroxide:

$$Al_2(SO_3)\cdot14H_2O + ? \longrightarrow 2Al(OH)_3 + ? + ?$$

$$\text{molecular weights} \; 594 \,\text{g/mol} + ? \longrightarrow 156 \,\text{g/mol} + ? + ?$$

i.e. 594 g of alum produces 156 g of alum hydroxide (sludge)

1 g of alum produces 0.26 g of sludge

$$\begin{aligned} \text{Using } 50 \,\text{mg/L alum} &= 50 \times 10^3 \,\text{mg/m}^3 \\ &= 50 \times 10^3 \times 0.42 \,\text{mg/s} = 21 \,\text{g/s} \\ 21 \,\text{g/s of alum} &\rightarrow 21 \times 0.26 \,\text{g/s of sludge} \\ &= 5.46 \,\text{g/s} \end{aligned}$$

Therefore

$$\begin{aligned} \text{Total solids} &= \text{suspended solids removed} + \text{alum hydroxide sludge} \\ &= 206 \,\text{g/s} + 5.46 \,\text{g/s} \\ &= 211.5 \,\text{g/s} \\ &= 18\ 274 \,\text{kg/day} \\ \text{Total solids} &= 18.3 \,\text{t/day} \end{aligned}$$

Example 11.7   For the water treatment plant of Example 11.5, design the flash mixing unit for flocculation.

Solution   Assume a detention period in the flash mixer of 40 s.

$$\text{Tank size required} = \text{Vol} = Qt$$

$$= \frac{36\,400}{24 \times 60 \times 60} \times 40$$

$$\text{Vol} = 16.85\,\text{m}^3$$

$$\text{Tank size} \sim 2.5\,\text{m} \times 2.5\,\text{m} \times 2.5\,\text{m}$$

Example 11.8   For example 11.6, design the flocculation basin.

Solution   Assume a detention time of 40 minutes.

$$\text{Tank size required} = \text{Vol} = Qt$$

$$= \frac{36\,400}{24 \times 60} \times 40$$

$$\text{Vol} = 1011\,\text{m}^3$$

For a 4 m basin, the surface area (SA) required is

$$\text{SA} = 252.7\,\text{m}^2$$

Choose 1 tank with dimensions 16 m × 16 m.

11.7.3 Sedimentation Infrastructure for Flocculent Particles

Sedimentation of flocculent particles (settling or clarification) allows adequate time in a basin for flocs to settle and be ultimately drawn off as sludge. There are two principal methodologies, as shown in Fig. 11.10.

The traditional US practice is to have separate flocculation and settling tanks, while the UK practice has been to promote flocculation and settling in one basin. In the case of separate flocculation/settling, the flocculation tank normally runs into the settling tank, and so they are usually of the same width (for rectangular tanks only).

11.7.4 Upflow Sludge Blanket Clarifiers

As mentioned in Sec. 11.7.3, it is more common in the United Kingdom to use a combined flocculation/settling tank, called a sludge blanket clarifier. There are two types: either flat-bottomed or hopper bottomed. A schematic of a flat-bottomed unit is shown in Fig. 11.11.

Water enters the tank at the bottom and is distributed through a series of perforated pipes laid close to the floor. The necessary agitation to provide effective flocculation is provided by turbulence caused by discharging the inlet downwards, assisted in the case of the flat bottomed tank by a device designed to create uneven flow conditions. The upflow flocculated water flows through a zone of suspended sludge affording the opportunity for further contact and aggregation of floc particles. The upper level of this zone is controlled by providing a quiescent zone which creates a density gradient, causing the sludge to flow in the direction of the central sludge withdrawal cone. Sludge is bled (withdrawn) regularly to ensure that the sludge blanket is maintained at the right density and the sludge blanket produced is at the optimum thickness, thereby reducing the amount of wastewater. The flat-bottomed unit provides true upflow while the hopper unit does not, as it sets up undesirable weak circulation patterns. As such, the flat-bottomed unit can be operated at twice the upflow rate of the hopper unit. When sludge blanket clarifiers are

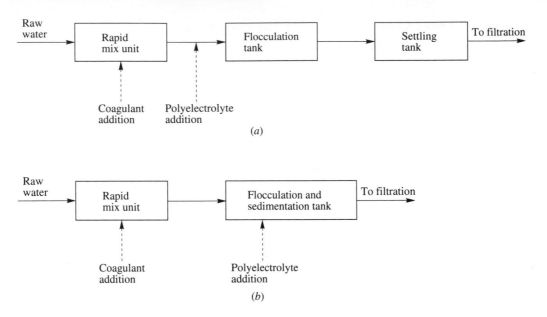

Figure 11.10 Primary water treatment: (*a*) US practice, (*b*) UK practice.

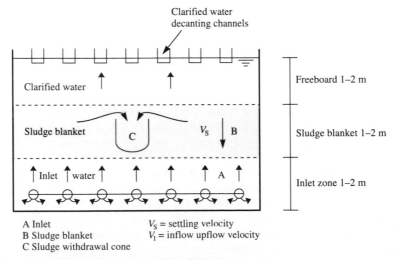

Figure 11.11 Schematic of upflow sludge blanket clarifier—flat bottom.

operated to capacity or in excess, the sludge blanket top level is liable to rise and some water with floc may be decanted with the clean water. This is extremely undesirable as the floc contains aluminium hydroxide. Overloaded plants sometimes have this problem.

$$\text{The particle settling velocity } V_s = \frac{g}{18\mu}(\rho_p - \rho_w)d^2$$

$$\text{The inflow upflow velocity } V_I = \frac{Q}{A_p}$$

where	$Q =$ the inflow rate
and	$A_p =$ the plan area
In the limit	$V_s = V_I$

$$\text{and so} \qquad V_s = \frac{Q}{A_p}$$

Particles will be removed only if the settling velocity exceeds the upflow velocity.

In the case of upflow sludge blanket clarifiers the principle of sedimentation is type III. This type is when settling occurs, but is hindered, in this case by upflow velocity. Additionally, settling occurs in zones, in this case the top of the blanket is most dense while the bottom is weakly defined. The settling velocities are highest at the top and lowest at the bottom. This is shown schematically in Fig. 11.12. Refer to Reynolds (1982) for further details of type III settling.

11.8 FILTRATION

Filtration is the process of passing water through a porous medium with the expectation that the filtrate has a better quality than the influent. The medium is usually sand. Filtration has been in use since the nineteenth century when the process of slow sand filtration was generally the only water treatment process. Slow sand filtration is credited with improving the aesthetic quality of water and also with the removal of pathogens. The latter benefit was not known in the nineteenth century. A classical example of the effectiveness of slow sand filtration is recorded when in 1892 in Hamburg 8500 people died in a cholera epidemic. Hamburg used untreated water from the River Elbe. Its 'suburb', Altara, treated its water by sedimentation and slow sand filtration and did not experience the cholera epidemic.

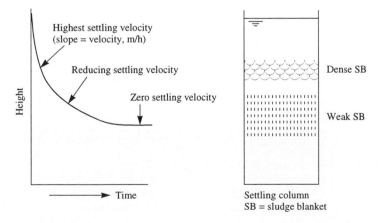

Figure 11.12 Concept of upflow sludge blanket settling.

There are many classifications of filtration systems and they include:

- Gravity or pressure
- Rapid, slow or variable filtration rates
- Cake or depth filtration

Gravity filtration is the process where the water goes through the filter unassisted except by gravity. Pressure filters are usually contained in vessels and the water is forced through the filter media under pressure. Pressure type applications are usually for industrial use rather than municipal use. Slow sand filtration operates with filtration rates ranging from 0.1 to 0.2 m/h, while rapid filters have rates from about 5 to 20 m/h. Cake filtration is the process of filtration in slow sand filters where a filter cake builds up on the filter surface (sand/air interface) and filtration through this surface is by both physical and biological mechanisms. Depth filtration is where most of the depth of the filter medium is active in the filtration process and filtrate quality improves with depth, as in rapid sand filters.

The most common type is the rapid gravity depth filter. This took over almost exclusively from the slow sand filter (gravity cake) in the 1930s. However, the slow sand filter (SSF) has widespread application in small rural communities. The exception is London, where still in the 1990s all the surface-derived water is treated and slow sand filtered. This type of filter is undergoing a renaissance, as it has recently been shown to have excellent abilities in improving the microbiological water quality. Slow sand filters have removal rates of up to 99.9 per cent for *Giardia* and *Cryptospiridium* cysts. Slow sand filtration may become popular again as *Giardia* cysts are known to be somewhat resistant to chlorination.

11.8.1 Slow Sand Filtration

A schematic of a slow sand filter is shown in Fig. 11.13. Typically it is a rectangular concrete open box structure containing:

- Supernatant layer of raw water
- Bed of fine sand, supported on a thin layer of gravel
- System of under drains
- Inlet and outlet structures

The inlet structure allows the water to flow on to the schmutzdecke layer without damaging it. This layer is at the top of the sand bed and is composed of living and dead micro-organisms. The inlet structure also facilitates the drainage of the supernatant water during the cleaning process. The supernatant water provides a head of water sufficient to drive the water through the sand filter while creating a detention period of several hours for the raw water. The filter bed is usually of fine sand of size 0.15 to 0.3 mm with

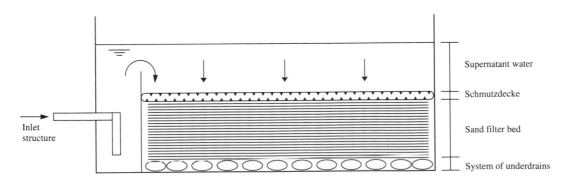

Figure 11.13   Slow sand filter.

a uniformity coefficient no greater than 2. Some design criteria as used in the United States and United Kingdom are shown in Table 11.11.

Filtration Mechanisms and Performance   The removal of impurities is thought to be primarily in the schmutzdecke layer. The mechanisms are thought to be both physical and biological, the latter contributing to the fact that many microbiological parameters improve significantly on flowing through the slow sand filter. The effective sand size is usually of the order of about 0.2 mm and this effectively contains all particles greater than about 0.02 mm. *Giardia* cysts are about this size and thus would be filtered out in the schmutzdecke. Some conflicting evidence exists as to whether the schmutzdecke is essential in removing *Giardia* as some research has shown their elimination still occurs in slow sand filters without a fully developed schmutzdecke. A review of slow sand filtration in the United States by Logsdon and Fox (1988) indicated significant improvement in other water quality parameters. Total coliform bacteria removal was 99.4 per cent or better. This removal rate depended on the bed depth and decreased as the bed depth decreased. Cyst particles of sizes of the order of 7 to 12 μm had removal rates of 96.8 per cent or better. Total particulate removal for particle sizes of 1 to 60 μm was 98.1 per cent or better. Colour removal is not significant in slow sand filters, with removal rates only of the order of 25 per cent. Turbidity removal parallels removal of other particulates with high removal rates and values being reduced to 0.5 NTU and sometimes as low as 0.1 NTU. However, water with very fine clay particles from mountain runoff is found to have turbidity removal rates of only about 25 per cent. Slow sand filters have filter runs varying from 2 to 6 months. Cleaning means scraping off the top surface or schmutzdecke and starting again about two days later.

Example 11.9   Design a slow sand filter (SSF) to treat a flow of 800 m³/day.

Solution   Assume a filtration rate of 0.15 m/h. Therefore

$$\text{Required tank area} = \frac{800}{24} \times \frac{1}{0.15}$$

$$= 222 \, \text{m}^2$$

Table 11.11 Design criteria for slow sand filters

Parameter	UK recommended level	US range and (average)
Design life	10–15 yr	
Period of operation	24 h/day	
Filtration rate	0.1–0.2 m/h	0.04–4 m/h(0.1)
Filter bed area	5–200 m²/filter (minimum of 2 filters)	
Height of filter bed		
Initial	0.8–0.9 m	0.46–1.5 m (0.9)
Minimum	0.5–0.6 m	
Sand specification effective size	0.15–0.3 mm	0.15–0.4 mm (0.3)
Uniformity coefficient	< 3	1.5–3.6 (2)
Height of under drains including gravel layer	0.3–0.5 m	0.15–0.9 m (0.5)
Height of supernatant water	1 m	0.7–4 m (1.2)

Adapted from Visscher *et al.*, 1987 with permission

Choose a tank 23 m long × 10 m wide. From Table 11.11, the height of tank required is:

(a) System underdrain ~ 0.5 m
(b) Filter bed ~ 0.9 m
(c) Supernatant water ~ 1 m

Therefore

$$\text{Total height} \sim 2.5 \text{ m}$$

and

$$\text{Tank size} = 2.5 \text{ m high} \times 23 \text{ m long} \times 10 \text{ m wide.}$$

11.8.2 Rapid Gravity Filters

Since the 1930s rapid gravity filters have taken over almost exclusively from slow sand filters, except in the case of small rural conurbations. In the latter case, slow sand filtration is commonly used without any prior coagulation. The popularity of rapid gravity filtration is due to its enhanced filtration rate of 5 to 20 m/h by comparison with 0.1 to 0.2 m/h for slow sand filters (a factor of about 50).

Rapid gravity filtration (RGF) is used to filter chemically coagulated water and so produce a high quality drinking water. Filtration removes suspended particles (turbidity) by the simple physical method of filtration. Some biological activity may occur in the breakdown of ammonia to nitrate as in nitrification:

$$\underset{\text{ammonia}}{NH_4^+ + 2O_2} \xrightarrow[\substack{Nitrobacter \\ bacteria}]{\substack{Nitrosomonas \\ \text{and}}} \underset{\text{nitrate}}{NO_3^- + 2H^+ + H_2O} \tag{11.20}$$

The solids removal mechanisms in filtration are a combination of settlement, straining, adhesion and attraction. Thus particles, much smaller than the interstitial spaces between sand grains, are removed. Both RGF and SSF operate in a mode where the water filters vertically down through the media under gravity. As mentioned in Sec. 11.8.1, the filter runs on slow sand filters are two to six months. However, because of the increased filtration rate, filter runs on rapid filters are 20 to 60 hours. Cleaning is achieved by agitating the bed either mechanically or with compressed air and washing water upwards through the bed to the surface, from where it is decanted as wastewater. This 'backwash' water is then wasted or returned to the beginning of the plant for treatment.

Rapid gravity filters may be of three possible media types:

- Single medium, usually sand or anthracite
- Dual medium, usually sand and anthracite
- Multimedia, usually garnet, sand and anthracite

Table 11.12 shows the media characteristics in the single, dual and multimedia filters. In the latter two, the top media is the coarse material of anthracite, followed by sand and the final material of garnet.

Figure 11.14 is a schematic of a rapid gravity filter in filtration (downflow) and backwash (upflow) modes. Filtration operates with a head of about 1 m of water. As filtration proceeds, the turbidity particulates lodge on the filter media and reduce efficiency over time. To maintain the same filtration rate the head of water would have to be increased, which is not possible. In filtration mode, once the filtration rate drops below a predetermined value (or time) the system stops and performs a backwash. This is where water or compressed air from the bottom is sent up through the media, dislodging the particulates and sending them into a suspension of contaminated water, which is overflowed to the central backwash channel as waste (see Fig. 11.14b). Usually this water is recycled through the plant. Filtration operates on hydraulic principles and the Darcy–Weisbach and Rose equations are used in the design of filter units.

Table 11.12 Rapid gravity filter media characteristics

Media type	Medium	Depth (m)	Effective size (mm)	Uniformity coefficient	Filtration rate (m/h)
Single	Sand	0.7	0.6	<2	10
	or anthracite	0.7	0.7	<2	10
Dual	Anthracite and	0.6	1.0	<2	
	sand	0.15	0.5	<2	
					12
Multi	Anthracite,	0.5	1.0	<2	
	sand and	0.2	0.5	<2	
	garnet	0.1	0.2	<2	
					15

The backwash velocity is usually $> 0.3\,\text{m/min}$ and less than $10D_{60}$ for sand and less than $4.7D_{60}$ for anthracite, where D_{60} is the 60 per cent size in mm. The velocity just to fluidize the bed from McGhee (1991) is

$$V_f = V_t \times f^{4.5} \tag{11.21}$$

where

V_f = minimum fluidization velocity

V_t = terminal velocity to wash medium from bed

f = porosity

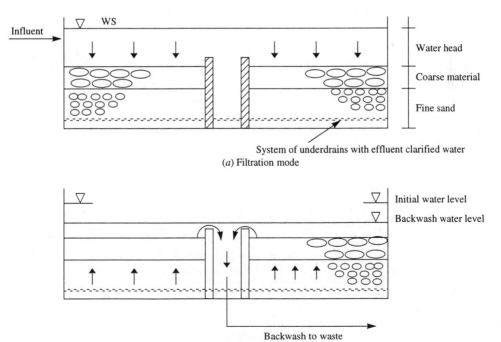

Figure 11.14   Schematic of rapid gravity dual media filter: (*a*) downflow filtration mode, (*b*) upflow backwash mode.

Sand has a specific gravity of 2.65 and anthracite 1.55. The velocity to (fluidize and) backwash is

$$V_b(T) = V_{b(20)} \times \mu_T^{-1/3} \tag{11.22}$$

where
$$\mu_T = \text{the viscosity}$$
$$V_b(20) = 0.1\ V_t \text{ for sand}$$
$$= 0.47\ D_{60} \text{ for anthracite}$$

Example 11.10   Design a rapid gravity filter to treat $36\,400\,\text{m}^3/\text{day}$.

Solution   Assume a filtration rate of 12 m/h (from Table 11.12).

$$\text{Surface area} = \frac{36\,400}{24} \times \frac{1}{12} = 126\,\text{m}^2$$

Choose 1 tank 10 m wide × 13 m long. From Table 11.12, the depth should be:

(a) System underdrain $\sim 0.5\,\text{m}$
(b) Filter bed $\sim 0.7\,\text{m}$
(c) Water head $\sim 1.0\,\text{m}$

Therefore

$$\text{Tank size} = 2.2\,\text{m high} \times 13\,\text{m long} \times 10\,\text{m wide}.$$

This compares to 77 tanks of the same size for the slow sand filter. The filtration rate of the RGF is 80 times that of the SSF.

Example 11.11   Determine the terminal velocity and the minimum fluidization velocity of a sand filter of effective size 0.60 mm, with a uniformity coefficient 1.6, a specific gravity 2.65 and a porosity of 0.44. Determine also the backwash rate at 10 and 20 °C.

Solution

$$\text{Terminal velocity } V_t = 10\,D_{60}$$
$$= 10 \times 1.6 \times 0.6$$
$$= 9.6\,\text{m/min}$$
$$\text{Fluidization velocity } V_f = V_t \times f^{4.5}$$
$$= 9.6 \times 0.44^{4.5}$$
$$= 0.24\,\text{m/min}$$
$$\text{Backwash velocity } V_{b(T)} = V_{b(20)} \times \mu_T^{-1/3}$$
$$\mu_{10\,°C} = 1.31$$

Therefore

$$V_{b(10)} = 0.96 \times 1.31^{-1/3}$$
$$= 0.87\,\text{m/min}$$

11.9 DISINFECTION

As practised in water treatment, disinfection refers to operations aimed at killing or rendering harmless, pathogenic micro-organisms. Sterilization, the complete destruction of all living matter, is not the objective of disinfection. Figure 11.15 shows the reduction of waterborne diseases at the beginning of the twentieth century by water treatment, e.g. typhoid eradication.

The other treatment procedures like coagulation and filtration should remove > 90 per cent of bacteria and viruses. Also the lime softening process is an effective disinfectant due to the high pH involved. However, to meet the standards given by water standard directives, such as the European Union (EU) the US Environmental Protection Agency (EPA) or the World Health Organization (WHO), and to provide protection against regrowth, additional disinfection has often been practised. The requirements of a good disinfectant are that it should:

- Be toxic to micro-organisms at concentrations well below the toxic threshold to humans and higher animals
- Have a fast rate of kill
- Be persistent enough to prevent regrowth of organisms in the distribution systems

The rate of destruction of micro-organisms is often postulated as a first-order chemical reaction (Chick's law):

$$\frac{dN}{dt} = -kN_t \tag{11.23}$$

$$\ln\frac{N_t}{N_o} = -kt$$

$$N_t = N_o\, e^{-kt} \tag{11.24}$$

where $\qquad N_t$ = number of organisms at time t

N_o = number of organisms at time zero

and $\qquad k$ = rate constant characteristic of the type of disinfectant,
micro-organism, and water quality aspects of system

The rate determining step is the diffusion of the disinfectant into the cell of the micro-organism. This must be completed before the water reaches the consumer. Complete disinfection cannot be accomplished

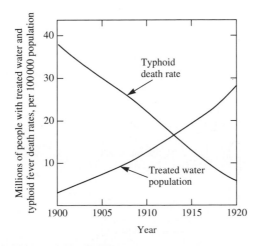

Figure 11.15 Typhoid reduction with use of treated water.

because N_t, the number of organisms remaining at time t, will approach zero asymptotically as time increases, as shown schematically in Fig. 11.16. However, since N_o should be small, 99.9 per cent kill can be affected in a reasonable time.

The model type of a disinfection curve is shown in Fig. 11.16. It has three 'zones'. The initial lag phase is followed by the exponential die-off rate. The final phase is the slow phase where little extra benefit is achieved.

The following factors can result in low efficiency of disinfection:

- Turbidity
- Resistant organisms (*Giardia*)
- High amount of organic material
- Deposits of iron and manganese
- Oxidizable compounds

Viruses are more resistant to disinfectants than bacteria and need additional exposure time and higher concentrations. Turbidity producing colloids, and iron and manganese deposits can shield organisms and use up the disinfectant. The most commonly used industrial scale disinfectants are:

- Chlorine dioxide
- Chloramines
- Ozone
- UV radiation
- Chlorination

Chlorine is by far the most commonly used disinfectant. Because its application involves a series of disadvantages, (not least the production of THMs) the use of other disinfectants should be considered. However, it must be stated that no disinfectant is perfect—all have advantages and disadvantages for a given water. The distinct advantage of a reliable disinfection of pathogenic micro-organisms must be weighed very carefully against all the possible disadvantages. Cryptospiridium outbreaks in the western world (particularly the US since 1990) have occurred in public water supplies, many of which use chlorination as the disinfectant. Recent reports indicate that the cryptospiridium cysts are not always removed in filtration systems, although some evidence suggest that rapid gravity filters are successful. This success is not necessarily due to the filter design or media, but more to the rate of water throughput. Chlorine, chlorine dioxide and monochloramine have all been found to be ineffective. Ozone is however known to effectively reduce the cyst numbers to non infective levels.

11.9.1 Chlorine Dioxide

With regard to disinfection, chlorine dioxide possesses theoretically 25 times greater oxidizing power than chlorine. At normal conditions it is a yellowish or yellow-green or reddish gas that liquefies at

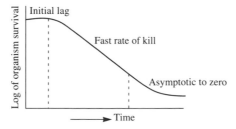

Figure 11.16 Model type disinfection curve.

approximately $10\,^{\circ}$C. The solution in water is not stable and degrades, especially when exposed to light. At higher temperatures it is explosive. Therefore chlorine dioxide must be produced on site before application. The properties of chlorine dioxide in comparison with chlorine are:

1. Its application does not cause any deterioration of taste and odour.
2. Its disinfection efficiency is largely pH dependent.
3. The formation of trihalogen methanes (THM) can be neglected (although other chlorinated compounds can also be formed).
4. It does not react with ammonia.

In contrast to other disinfectants in use, chlorine dioxide forms toxic inorganic compounds. It has been established that 50 per cent of the used ClO_2 is transformed into chlorite (ClO_2^-) and chlorate (ClO_3^-). These substances may provoke methanoglobinemia in babies (like nitrates NO_3^-) if concentrations in drinking water exceed the value of $0.1\,mg/L$. Studies of mutagenic activities indicate that at concentrations of $< 1\,mg$ chlorine dioxide per litre, no increase in the mutagenic activity in drinking water occurs, whereas there is a significant increase in mutagenic activity at higher doses.

11.9.2 Chloramines

When chlorine and ammonia (NH_3) are both present in water, they react to form products collectively known as chloramines. As opposed to 'free chlorine', the chloramines are referred to as 'combined chlorine'. The inorganic chloramines consists of three species:

$$NH_{3(aq)} + HOCl \rightleftharpoons NH_2Cl + H_2O \qquad \text{(monochloramine)} \qquad (11.25)$$

$$NH_2Cl + HOCl \rightleftharpoons NHCl_2 + H_2O \qquad \text{(dichloramine)} \qquad (11.26)$$

$$NHCl_2 + HOCl \rightleftharpoons NCl_3 + H_2O \qquad \text{(trichloramine)} \qquad (11.27)$$

The species formed as a result of the combination of chlorine and ammonia depend on a number of factors, including the ratio of chlorine to ammonia-nitrogen, chlorine dose, temperature, pH and alkalinity. As higher ratios of chlorine to ammonia-nitrogen are reached, the ammonia is eventually oxidised to nitrogen gas, a small amount of nitrate (NO_3^-) or a variety of nitrogen containing inorganic oxidation products. If ammonia is present, either as a natural constituent of the raw water or as a chemical deliberately added to produce combined chlorine rather than free chlorine, a hump-shaped breakpoint curve is produced (see breakpoint chlorination, Fig. 11.18). Compared to chlorine dioxide or chlorine, chloramines:

- Are less effective disinfectants
- Have an algicidous effect
- Have a detrimental effect on taste and colour
- Show an efficiency that is heavily dependent on the pH value, a higher efficiency having been observed at lower pH values
- Do not react with organic material or phenols
- Are persistent and provide continued protection against regrowth in the distribution system

11.9.3 Ozone

Ozone is a bluish gas with an unpleasant smell. It is one of the most powerful oxidizing agents that is practical for water treatment. It can be produced in a high-strength electrical field from oxygen in pure form or from ionization of clean dry air:

$$3O_2 + \text{energy} \rightarrow 2O_3$$

Because ozone is chemically unstable it must be produced on site and used immediately. Substantial amounts of energy are required to split the stable oxygen bond to form ozone. Energy consumptions of 10 to 20 kWh per kg ozone are necessary to produce the typical dosages ranging from 1 to $5\,g/m^3$. Therefore, costs of ozonation are 2 to 3 times higher than the costs for chlorination. Ozone is able to break down large molecules of organic material such as humic or fulvic acids. It degrades such health impairing compounds as polycyclic aromatic hydrocarbons (PAH), phenols and chlorophenols. However, not all organic compounds can be reduced or even mineralized. Some of the properties of ozone as a disinfectant are:

- Especially effective in killing viruses
- Improvement of odour and taste
- Transformation of almost non-degradable substances into easily degradable ones
- Microflocculation effect
- Largely pH independent
- Regrowth of micro-organisms within the water supply system due to the production of more easily degradable substances
- Formation of a number of toxic compounds
- No remaining residuals (of disinfectant)

Studies on the mutagenic activities of ozonized drinking water gave conflicting results. Because of the regrowth of micro-organisms, the treatment of water with ozone as the (last) disinfection step does not occur very often. Since no residual remains, it is necessary to use small amounts of chlorine after ozonization to provide continuous protection against regrowth in the distribution system.

11.9.4 UV Radiation

Irradiation with UV light is a promising method of disinfection. Although it leaves no residuals, this method is effective in disabling both bacteria and viruses. UV light spans the wavelength of 200 to 390 nm. The most effective band for disinfection is in the shorter range of 250 to 280 nm. This is the range where the UV light is absorbed by the DNA of the micro-organisms which then lead to a change in the genetic material so that they are no longer able to multiply. Light of this wavelength range can be generated with low-pressure mercury vapour lamps which emit peak light radiation at a wavelength of 254 nm. The properties of UV radiation as a disinfectant include:

- Necessity of having clear water (turbidity free) and thin sheets of water
- No residual
- Photo-oxidation of compounds may occur
- No odour, or taste problems
- No chemicals added

11.9.5 Chlorination

The application of chlorine for disinfection of drinking water goes as far back as the nineteenth century. By the year 1800 sewage had already been treated with chlorinated lime in England and France. The large-scale technical application of chlorination to drinking water was first performed in the United States at the beginning of this century. It replaced the slow sand filtration in use at that time. During the past 50 years, it has developed into the most widely used procedure for the treatment of surface water. In many countries it is still performed as a 'breakpoint' chlorination following sedimentation and filtration (see Fig. 11.18). However, the application of high doses of chlorine run the risk of developing large amounts of potentially carcinogenic and/or mutagenic by-products.

Under natural conditions, chlorine (Cl_2) is a yellow-greenish gas showing high toxicity to humans and animals. Due to its high reactivity, it does not occur naturally as Cl_2, but forms many compounds found on earth such as the well-known NaCl (common salt). Most chlorine is produced on a large-scale industrial process known as the chlor-alkali electrolysis. Chlorine can be liquefied at room temperature. Therefore, it can be stored and transported. It is a very efficient oxidizing, bleaching and disinfecting agent. In water, chlorine reacts as follows:

$$Cl_2 + H_2O \rightleftharpoons H^+ + Cl^- + HOCl \tag{11.28}$$

HOCl is hypochloric acid. Depending on the pH value of the water, HOCl can dissociate to hypochlorite:

$$HOCl \rightleftharpoons H^+ + OCl^- \tag{11.29}$$

Hypochloric acid and hypochlorite ions represent the 'free chlorine residual' which is the primary disinfectant employed. Experiments have shown that the undissociated HOCl molecule is the most effective compound for the disinfecting process. Its efficiency is considered to be 80 times as high as that of the hypochlorite ion (OCl^-). Figure 11.17 shows the dissociation of HOCl as a function of the pH value. At a pH < 5, the reaction equilibrium is shifted to the left, i.e. the lower the pH value the greater the quantity of Cl_2 occurring in the water. At a pH > 5, the concentration of OCl^- increases until it reaches 100 per cent at a pH of 10. The optimum pH range for disinfectant application lies in the range 6 to 8.

The mechanism of pathogen kill is considered to be:

- A penetration into the cell of micro-organisms with subsequent blocking of an essential enzyme
- Destruction of cell walls

Factors affecting the process are:

- Chemical form of chlorine
- pH
- Concentration
- Contact time
- Type of organisms
- Suspended solids
- Temperature

Most of the chlorine feed systems in use are gas-to-solution systems, implemented only for indirect chlorination. Indirect chlorination means the preparation of a chlorine solution from Cl_2 gas and water on site, which then serves as the disinfectant. Instead of adding Cl_2 gas to the water it is also possible to use chlorine in an ionized state, e.g. as the compounds $Ca(ClO)_2$ or $NaClO_2$. This procedure is chosen particularly when small amounts of water are to be treated with a relatively small expenditure

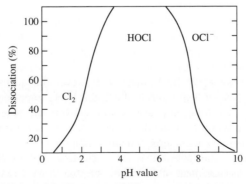

Figure 11.17 Dissociation of HOCl as a function of pH value.

on technical equipment. Another consideration for choosing hypochlorites is for safety reasons. Chlorine gas is very toxic and its handling requires extreme care. In spite of the relatively safe injections systems, chlorine gas cylinders have to be transported, stored and changed when empty. Since chlorine gas is heavier than air, it spreads slowly at ground level. Therefore, in highly populated areas, the use of hypochlorites may be advantageous. Hypochlorite solutions in water are alkaline, which is a disadvantage as the disinfection efficiency is less at higher pH values, as seen in Fig. 11.17. There is some difficulty involved in dissolving $Ca(ClO)_2$. This compound contains approximately 70 to 80 per cent available chlorine. It is often used on a short-term basis in the form of 'Chlorine' tablets at special locations, i.e. water storage containers or after some minor works on the distribution system. $NaClO_2$ solutions are more commonly used. The solution contains only 15 to 17 per cent chlorine and is not very stable. It degrades with time, particularly when exposed to light. The application of both hypochlorites is much more expensive than the application of chlorine gas.

The accurate measurement of chlorine is a very important control tool for efficient but careful dosage of chlorine. In the German Drinking Water Directive, a residual of at least 0.1 mg/L must be detectable after the disinfection step. The maximum level must not exceed 0.3 mg/L (except for extraordinarily occurring events: 0.6 mg/L are allowed over a short period). These relatively low levels take into account the disadvantages associated with chlorination and seek to minimize these. In the case of some groundwater supplies, the water is naturally free from pathogens. Either there is no chlorination at all or the addition of chlorine is performed on a very low level (about 0.02 mg/L detectable). This procedure is not a disinfection, only a 'conservation precaution' to ensure the microbiological quality of the water during the distribution step.

A method used for the minimization of the chlorine application is the measurement of the redox potential. The target is to use the smallest amount possible to fully disinfect, while avoiding side reactions, i.e. associated with unpleasant taste and odour of chlorine and the formation of undesirable compounds. The redox potential is determined by the Nernst equation:

$$E = E_o + RT/nF \ln (Ox)/(Red) \tag{11.30}$$

where

E = redox potential

E_o = potential against normal-hydrogen electrode

R = gas constant

F = Faraday constant

n = number of electrons transferred

Ox = concentration of oxidized compounds

Red = concentration of reduced compound

T = absolute temperature, K

The equation (11.30) shows the dependence of the redox potential on the concentrations of red/ox pairs. This means that if strong oxidants like oxygen, chlorine, etc., are present, the redox potential is high and positive. Groundwater containing no oxygen would have a negative value. The advantage of determining the redox potential lies in the fact that it gives information about the disinfective ability of water. There may be waters with a high concentration of the disinfectant. However, this quantity might not be available for the disinfection process, because of the presence of chlorine using impurities. All the chlorine would be used for oxidation rather than disinfection. Due to the fact that the redox potential measures the relation of both oxidizing and reducing compounds it gives a reliable statement about the still available disinfection power within a water (see also redox potential, Chapter 3).

Especially at low levels, the redox potential is more sensitive than the chlorine analysis. Positive experience was found with the measurement of the redox potential in the Bremen Water Works (Baxter, 1992). During the 'conservational' chlorination mode, it was sometimes not even possible to detect the chlorine by analysis. However, studies of the continuously measured redox potential gave information on the microbiological quality at any time and helped the setting of the chlorine dose. Later, when no chlorination was performed at all, the level of the redox potential remained greater than 600 mV. This assured the good microbiologial quality of the water. It was empirically derived that at a redox potential value of 600 mV bacteria cannot survive. The pH value and temperature influences must be considered.

Due to its reactivity, chlorine reacts with a multitude of inorganic and organic materials present in the water. In some cases, it is used as an oxidizing agent first to break down these materials. This type of chlorination is called 'breakpoint' chlorination. Due to the high dosages of chlorine normally employed, this procedure is no longer performed in Germany, but is still widely used in the United States and other countries that predominantly use surface water as the potable water source.

Figure 11.18 shows a generalized curve of chlorine added versus chlorine residual, obtained during breakpoint chlorination. The oxidizable materials (even small amounts of BOD) consume chlorine before it has a chance to react as a disinfectant. At the breakpoint, these reactions are complete and it is only then that disinfection can take place. Continued addition of chlorine past the breakpoint will result in a directly proportional increase in the free available chlorine (unreacted hypochlorite).

The by-products of organics oxidized by chlorine are often undesirable. Minute quantities of phenolic compounds react with chlorine to form severe taste and odour problems, due to the formation of chlorophenols ('medicinal' smell and taste). A concentration as low as 1:20 000 000 can still be detected by the human nose. Materials (e.g. glue) used during work in wells or pipes can sometimes contain phenolic compounds. To avoid customer complaints, it is recommended that the materials used in pipeline repairs be tested in advance or that the repair areas be thoroughly cleaned afterwards by rinsing with unpolluted water.

Another important reaction is the formation of halogenated hydrocarbons, including the trihalogen methanes (THM). The most important classes of by-products are:

- Trihalogen methanes
- Chlorinated phenols
- Halogenated methanes, ethanes and ethenes
- Halogenated polynuclear aromatic hydrocarbons
- Chlorinated aldehydes and ketones

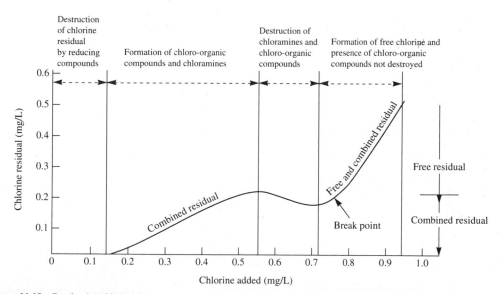

Figure 11.18 Breakpoint chlorination.

All these products carry a carcinogenic and/or mutagenic risk to human health and their reduction is therefore highly undesirable.

Example 11.12 To obtain a free residual of chlorine of 0.5 mg/L, a dose of 10 mg/L is added to a 36 400 m^3/day water treatment plant. Calculate the amount of Cl_2 and NH_3 required to achieve this.

Solution

$$Cl_2 \text{ required} = 36\,400 \times 10 \times 10^3 \times 10^{-6} = 364\,\text{kg/day} \tag{11.31}$$

Stoichiometrically : $$NH_3 + HOCl \rightarrow NH_2Cl + H_2O$$

1 mole of NH_3 reacts with 1 mole of HOCl to produce NH_2Cl (monochloramine). However, if this dose is applied 1:1, then the reaction also proceeds to form dichloramines (NH_2Cl) as in

$$NH_2Cl + HOCl \rightarrow NHCl_2 + H_2O \tag{11.32}$$

Dichloramines are undesirable in taste and odour. It is more common to apply Cl_2/NH_3 dose ratios of about 3:1 or 4:1 but not 1:1 (Montgomery, 1985) to prevent the formation of dichloramines. Therefore

$$NH_3 \text{ required} \cong 0.5\,\text{mg/L} \times 0.25 = 0.125\,\text{mg/L}$$
$$= 36\,400 \times 0.125 \times 10^{-3} = 4.6\,\text{kg/day}$$

Table 11.13 indicates some chlorine residual values and associated water pH and contact times.

The method of application of chlorine to a water system is significant in relation to efficiency of micro-organism kill. Typically the chlorine gas is injected into a tank or pipe and undergoes rapid mixing at high turbulence ($Re > 10^5$) for a few seconds. This is somewhat similar to coagulation using rapid mixers followed by slow mixing in a downstream tank. The injection of chlorine to a system is typically achieved by a diffuser inserted into the path of the water flow or directly to a rapid mixing propeller for instantaneous diffusion. Once injected, the rapid mixing can be achieved via several different hydraulic methods, including hydraulic jumps in open channels, venturi flumes, pipelines, pumps, static mixers, etc. Once the initial rapid mixing has taken place, so as to achieve full diffusion, the mixed chlorine solution is allowed to sit in a chlorine contact tank for periods of 10 min to 1 h. The contact time is a function of pH, required free chlorine residual and microbiological water quality. See Montgomery (1985) for further details.

11.10 FLUORIDATION

Fluoride is a natural trace element, found in small but widely varying amounts in waters. Groundwaters are more likely to have a higher fluoride content. However, most surface waters have negligible amounts.

Table 11.13 Chlorine residuals, pH and contact time

pH	Contact time (min)	Free residual (mg/L)	Combined residual
6–8	10	0.2	
8–9	10	0.4	
9–10	10	0.8	
10 +	10	> 1.0	
6–7	60		1.0
7–8	60		1.5
8–9	60		1.8
9 +	60		> 1.8

The presence of fluoride in water was identified in the United States in the 1950s to improve the dental health of growing children. The relationship between dental caries (tooth decay) and dental fluorosis (discolouration or mottling of teeth) and fluoride levels is shown in Fig. 11.19. This figure shows the results of research in the 1960s.

At the optimum concentrations in potable water, fluoride reduces dental caries from 20 per cent to 40 per cent among children who ingest the water from birth. Evidence of the effectiveness of fluoridation is world-wide (from 1950 to 1980). However, in the past decade, the relative beneficial impact of fluoridation is masked as other sources of fluoride have appeared, e.g. fluoride toothpaste, fluoride tablets, etc. Fluoride has been added to water in different forms:

- Sodium fluoride
- Sodium silicofluoride
- Hydrofluosilicic acid

The procedure in countries that have fluoride public water supplies is to add it in doses of 1 mg/L, and this is now mostly in the form of hydrofluosilic acid. In Eire, all public water supplies are fluoridated (by law). In the United States, several states do not fluoridate. In the United Kingdom, the level of fluoridation is decreasing and local water authorities decide regionally whether to fluoridate or not. In northern Europe, almost no areas practice fluoridation.

11.11 ADVANCED WATER TREATMENT PROCESSES

The purposes of advanced water treatment processes are:

- To take a water treated by standard processes and to improve it to an exceptionally high quality as often required by particular industries, e.g. beverage, pharmaceutical
- To treat a water containing specific chemical or microbiological 'contaminants' to an acceptable standard, e.g. the removal of iron and manganese, the removal of blue-green algae, the removal of specific organics

There is a wide range of physical, chemical and microbiological techniques that can be utilized to achieve the above objectives. Some of these include:

- Iron and manganese removal
- Ion exchange and inorganic absorption

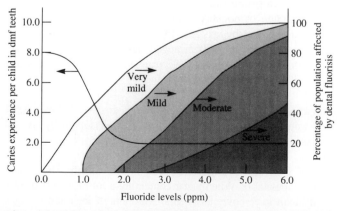

Figure 11.19 Dental caries and dental fluorosis in relation to fluoride levels in public water supplies (adapted from Dunning, 1962).

- Adsorption of organics
- Membrane processes including reverse osmosis
- Oxidation including chemical oxidation

11.11.1 Iron and Manganese Removal

Traces of iron and manganese are found in many waters. Quantities can range from $\mu g/L$ to mg/L. Both metals are often found to occur together, but there are also waters with only iron and more rarely just manganese. The natural occurrence of iron on earth is large: 4.7 per cent of the earth crust consists of iron, making it the fourth most common element. Manganese occurs as commonly as carbon, taking a share of 0.08 per cent of the earth's crustal solids. Therefore in water the concentration of iron is usually higher than that of manganese. Both elements are essential trace elements for all living creatures. Iron, for example, plays an important role in the formation of the red blood colour haem (together with protein it is called haemoglobin). Manganese in animals is important for growth and healthy functioning of the nervous system. In spite of this fact, excessive quantities of both elements must be removed from water, as otherwise disturbance and damage within the distribution system would occur.

To understand how Fe and Mn manage to get into the water the occurrence of micro-organisms in the ground is examined. Bacteria use organic material as food, converting it ultimately to CO_2 and H_2O. For that purpose, they need oxygen. The easiest way to obtain oxygen is to take it out of water. Therefore, a raindrop, for example, gradually loses its oxygen on its way through the soil towards the groundwater table. After having used all oxygen in the water, the micro-organisms look for another source of oxygen. This is obtained by scavenging the oxygen from commonly occurring metal oxides, like Fe_2O_3 and MnO_2. Iron and manganese are therefore reduced to ferrous and manganous (Fe^{2+} and Mn^{2+}) compounds, which dissolve in the water. Groundwaters that do not contain any oxygen are called 'reduced' groundwaters. The properties of this kind of water are:

- Clear, colourless
- Metallic smell and/or taste
- Free of oxygen
- Sometimes a smell of H_2S

When a 'reduced' water comes in contact with air, it is aerated easily and the Fe is oxidized after a short time. This is seen clearly in the formation of brown flocs which then precipitate. Manganese is not oxidized purely by aeration alone, in the normally occurring pH range.

Forms of iron and manganese in groundwater When reduced to their $+2$ state, Fe and Mn can form ionized compounds like hydrogen carbonates (bicarbonates), chlorides, sulphates, etc. These are the forms which are easily oxidized (Fe) by oxygen. In the presence of humic acids, Fe can also form complexes. Complexes are stable, geometrically arranged, large molecules with a positive charged centre atom, here Fe^{2+} or Fe^{3+}, which is surrounded by electronegative ligands. The stability of complexes is often very high.

Forms of iron and manganese in surface water For reasons discussed above, Fe^{2+} compounds are not found in natural surface waters. The iron (Fe^{3+}) there can be dissolved by the formation of colloids consisting of Fe^{3+} hydroxide molecule groups.

Undesirable properties of iron and manganese Iron is not harmful, but undesirable on aesthetic grounds, making the water unpalatable. When water containing a lot of iron is used for laundry it causes brown stains which are difficult to remove. Even small amounts of iron and manganese can lead to the accumulation of large deposits in a distribution system. As well as being unacceptable to the customer, such deposits can give rise to iron bacteria, which in turn cause further deterioration in the quality of the water by producing slimes with objectionable odours. Also, other types of bacteria use these deposits to adsorb and start breeding, which then leads to a regrowth in the distribution system. The EU Directive recommends a guide level of 0.05 mg/L Fe and a maximum concentration of 0.2 mg/L. It should be noted, however, that derogations from the EU Directive can be granted to take into account the nature of the ground from which the supply emanates. This takes into account the good quality groundwaters which are acceptable in all other respects but can contain more than 0.2 mg/L Fe. For manganese the guide level is 0.02 mg/L and the maximum concentration should not exceed 0.05 mg/L.

The principal method of removing iron from these waters is to oxidize the ferrous ion to ferric iron, which can be done by aeration, and to provide a suitable filter or other method to remove the precipitated iron compound.

It has been shown that oxidation of Fe^{2+} by oxygen is slow and is therefore the rate determining step. The whole process consists of several reactions but can be satisfactorily described by the following equation:

$$2Fe^{2+} + 0.5O_2 + (x + 2)H_2O \rightarrow Fe_2O_3 \cdot xH_2O + 4H^+ \qquad (11.33)$$

Removal procedures for low concentrations of iron   If the amount of iron in a water is fairly low (< 1 mg/L), the removal of iron is often achieved in a pressure filter containing the usual bed of sand, but sometimes incorporating a layer of proprietary material such as 'Polarite' or an iron core which acts as a catalyst. As the amount of oxygen required is small, it is provided merely by introducing air into a space of the filter shell. Much faster rates can be used for this type of filtration than are possible for waters where chemical coagulation precedes sand filters: rates > 50 m/h have been successfully used. As an alternative to sand filtration, it is possible to use semi-calcined dolomitic limestone ('Akdolit') as the filter medium (contains $Ca/MgCO_3$ and oxides). This material can be effective for the removal of low concentrations of soluble iron by precipitating ferric hydroxide at its alkaline surface and the material can help render the water less aggressive. Furthermore, it can increase the amount of magnesium in the water which is believed to have beneficial effects on health.

Removal procedures for high concentrations of iron   Where the amount of iron is large (> 10 mg/L), the traditional practice was to treat the water as a surface water (where the iron is dissolved as a colloid) providing a coagulation step before rapid sand filtration. That is, however, not always necessary and should be avoided because of the high demand in equipment and personnel. It is now possible to decrease filtration rates to approximately 5 m/h or use a sedimentation tank to allow a large proportion of the ferric hydroxide to settle out. Multimedia filtration, involving the use of a filter containing several layers of different materials, has proven to be highly effective. Raw water with about 15 mg/L Fe and high concentrations of humic acids can be treated at filtration rates of 15 m/h.

Poorly buffered waters   Waters which are described as being difficult in terms of removal of iron are often poorly buffered; i.e. the amount of hydrogen carbonate is low. As seen from Eq. (11.33), there is the formation of protons during the oxidation process. In well-buffered waters these H^+ do not cause problems, but poorly buffered waters may suffer acidification. During aeration of water, a lot of carbon dioxide is removed, the pH rises and iron begins to be oxidized. The H^+ produced reacts with HCO_3 to give more CO_2. In poorly buffered waters the pH will finally reach a low point where the reaction stops (self-inhibition). Because there is still a large amount of Fe^{2+} left, it was thought necessary to complete the reaction by adding oxidizing agents. This was later identified to be wrong, because after that it was no longer possible to destabilize the Fe^{3+} during the following filtration. The reason lies in the formation of stable colloidal solutions in which destablilization is just possible when there is still enough Fe^{2+} left. Coagulation takes place only in a definite pH range which becomes narrower as the amount of buffer present decreases. Above and below this range very stable colloidal solutions exist formed by equally positive or negative charges. The stability is often so high that destabilization by filtration succeeds only in a very thin layer on the filtration material.

Removal of manganese   Manganese occurs in natural waters as Mn^{2+}. There is no way to oxidize it by aeration in the normal pH range even if the amount is 10 mg/L or higher. The pH has to be 9 or 10 before oxidation by oxygen takes place. This can be achieved only with strongly alkaline filter materials. The processes going on during the oxidation are even more complicated than for iron, as Mn can occur in different oxidation states (II, III, IV occur naturally, V, VI, VII can be produced artificially). For the removal of Mn^{2+} in practice it has proven favourable to add a $KMnO_4$ solution as a conditioning agent prior to the filtration step without changing the pH. The stoichiometrically evaluated value of 1.9 kg $KMnO_4$ per kg Mn is often found to be too high (or too low), depending on simultaneously occurring reduction and oxidation processes. The addition of $KMnO_4$ has to be carried out for several days (3 days to 1 month) before the process works on its own.
Figure 11.20 is a schematic of a groundwater treatment plant in Germany for the removal of iron and manganese. Initial levels of iron are 8000 μg/L and manganese 250 μg/L. The iron is removed by multimedia pressure filters. The manganese is removed by multimedia gravity sand filters. The final levels of iron and manganese are $< 10\,\mu$g/L (Eberhardt, 1980).

11.11.2 Water Softening by Chemical Precipitation

As mentioned at the beginning of this chapter, hardness in water is caused by the presence of any polyvalent metallic cations but principally Ca^{2+}, Mg^{2+} and, less so, Fe^{2+} and Mn^{2+}. Total hardness is usually computed based on the concentration of Ca^{2+} and Mg^{2+} and is expressed in mg/L as $CaCO_3$. Hardness is more commonly associated with groundwater than with surface waters. Table 11.14 classifies the scale of hardness.
    The characteristics of the water source will predetermine the softening process. Four processes are listed by AWWA (1990):

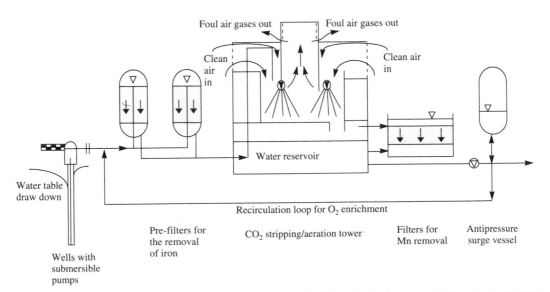

Figure 11.20   Schematic of groundwater treatment plant for the removal of iron and manganese (adapted from Eberhardt, 1980).

Table 11.14 Classification of hardness

Hardness (mg/L as $CaCO_3$)	Description
0–75	Soft
75–150	Moderately hard
150–300	Hard
> 300	Very hard

1. The single-stage lime process is used when the raw water has high Ca^{2+} hardness and low Mg^{2+} hardness (< 40 mg/L as $CaCO_3$) and no non-carbonate hardness.
2. The excess lime process is used when the raw water has high Ca^{2+} hardness and high Mg^{2+} hardness and no non-carbonate hardness.
3. The single-stage lime-soda ash process is used when the raw water has high Ca^{2+} hardness and low Mg^{2+} hardness and some calcium non-carbonate hardness.
3. The excess lime-soda ash process is used when the raw water has high Ca^{2+} hardness and high Mg^{2+} hardness and some non-carbonate hardness.

Example 11.13 A groundwater analysis indicated the following quality. Determine the lime dosage if straight lime softening is used.

$$pH = 7.5$$
$$Ca^{2+} = 150 \, mg/L \text{ as } CaCO_3$$
$$Mg^{2+} = 20 \, mg/L \text{ as } CaCO_3$$
$$Alkalinity = 200 \, mg/L \text{ as } CaCO_3$$

Solution

$$\text{Lime dose} = \text{carbonic acid concentration} + \text{calcium carbonate hardness}$$

Bicarbonate concentration:

$$[HCO_3^-] = 200 \times \frac{61}{50} \times 10^{-3} \times \frac{1}{61}$$
$$= 4.0 \times 10^{-3} \, mol/L$$

The dissociation constants for carbonic acid at $10\,^{\circ}C$ (Eq. (3.42)) are

$$K_1 = \frac{[H^+][HCO_3{}^-]}{[H_2CO_3{}^*]}$$
$$= 4.47 \times 10^{-7} \, mol/L \text{ at STP}$$

$$K_2 = \frac{[H^+][CO_3{}^{2-}]}{[H_2CO_3{}^*]}$$
$$= 4.8 \times 10^{-11} \, mol/L \text{ at STP (Eq. (3.44))}$$

The total carbonic species concentration is

$$C_T = [H_2CO_3{}^*] + [HCO_3{}^-] + [CO_3{}^{2-}]$$
$$C_T = \frac{[HCO_3{}^-]}{\alpha_1}$$

where
$$\alpha_1 = \frac{1}{1 + [H^+]/K_1 + K_2/[H^+]}$$

$$= \frac{1}{1 + 10^{-7.5}/(4.47 \times 10^{-7}) + (4.8 \times 10^{-11})/10^{-7.5}} = 0.93$$

$$C_T = \frac{4 \times 10^{-3}}{0.93} = 4.29 \times 10^{-3}\,\text{mol/L}$$

$$[H_2CO_3^{*}] = C_T - [HCO_3^-] - [CO_3^{2-}]$$

$$= 4.29 \times 10^{-3} - 4.0 \times 10^{-3} = 0.29 \times 10^{-3}\,\text{mol/L}$$

$$= 29\,\text{mg/L as CaCO}_3$$

Therefore

$$\text{Lime dose} = 29 + 150 = 179\,\text{mg/L as CaCO}_3$$

$$= 179 \times \frac{37}{50} = 133\,\text{mg/L as Ca(OH)}_2$$

Further and more detailed examples are given in AWWA (1990, Chapter 10).

11.11.3 Ion Exchange

Hard water (> 50 mg/L CaCO$_3$) contains an excess of calcium and magnesium cations. The process of water softening, i.e. reduction of hardness, but not its elimination, can be carried out by exchanging the undesirable calcium and magnesium cations with sodium. Ion exchange is this process. If Na$_2$R is an exchange sodium resin (R is the complex base) then water softening is represented by

$$Mg^{2+} + Na_2R \rightleftharpoons MgR + 2Na^+ \tag{11.34}$$

$$Ca^{2+} + Na_2R \rightleftharpoons CaR + 2Na^+ \tag{11.35}$$

Ion exchange processes are reversible and the direction of the reaction depends on the concentrations and the level of saturation of the sodium resin. A water softening unit consists of a bed of the medium about 0.5 to 2 m high with a 'filtration' rate of about 4 L/s m^2. The process of water softening does not remove all the hardness, as minimum values are required for health purposes. The softening capacity of exchange resins varies from 100 to 1500 eq/m^3 (equivalents).

Example 11.14   A synthetic zeolite with a capacity of 400 eq/m^3 and a 'filtration' rate of 4 L/s m^2 is used to soften water with a flow rate of 10 L/s and a hardness of 4 meq/L (250 mg/L as CaCO$_3$). The bed depth is 1.5 m and an 85 per cent exchange utility rate is achieved before 'breakpoint'. Determine the medium diameter and the volume of water passed through before regeneration is required. Also determine the regeneration time required.

Solution

$$\text{Area} = \frac{Q}{V} = \frac{10}{4} = 2.5\,\text{m}^2$$

$$\text{Vessel diameter } \varnothing = 1.78\,\text{m}$$

$$\text{Bed volume} = 1.5\,\text{m} \times 2.5\,\text{m} = 3.75\,\text{m}^3$$

$$\text{Total energy capacity} = 3.75 \times 400 = 1500\,\text{equivalents}$$

$$\text{Required exchange capacity} = 85\%\text{ of }1500 = 1275\,\text{equivalents}$$

$$\text{Volume of water passed before regeneration} = \frac{1275}{4} \times 10^3 = 318\,750 \text{ L}$$

$$\text{On-line time between regenerations} = \frac{318\,750}{4} = 79\,687\,\text{s} = 22.1\text{h}$$

Ion exchange processes are also used for the removal of other undesirable cations including barium, strontium and radium—and undesirable anions including fluoride, nitrate, humates, silicates, chromates, etc.

11.11.4 Adsorption

Some undesirable contaminants can be adsorbed on to solid adsorbents. Adsorption is both the physical and chemical process of accumulating a substance at an interface between the liquid and solid phases. Adsorbents used in the water treatment industry include:

- Activated carbon—PAC or GAC
- Activated alumina
- Clay colloids
- Hydroxides
- Adsorbent resins

Activated carbon is commonly used to adsorb organics that cause taste, odour, colour and microbiological problems. It is used to adsorb algae which can be the cause of undesirable tastes, colours and odours. PAC can be fed (in slurry form) to the water stream either at coagulation stage or just prior to filtration. Dose rates range from 1 to $100\,\text{g/m}^3$. In this application, the PAC produces additional sludge and is not available for regeneration.

GAC is sometimes used as a final filter bed after sand filtration. It produces a very high quality water and can reduce chlorine levels. Some industries receiving municipal water will treat it with GAC prior to use, e.g. pharmaceutical industry.

For PAC the significant properties are bulk density and filterability, the latter meaning that they will eventually be filtered out in the sand filter backwash. Bulk density means that since mass is proportional to adsorption capacity, the higher bulk densities have higher adsorption capacities. For GAC the significant properties are hardness and particle size. The harder the particle, the less is lost by attrition. Also, the smaller the particle, the greater the availability of macropore space and also less head is required.

11.11.5 Chemical Oxidation

Chemical oxidation is the resultant reaction when two or more chemical species are added with the purpose of increasing the oxidation state of one. The processes of oxidation and reduction occur in the same reaction. For one chemical species to increase its oxidation state (lose electrons) the other chemical(s) in the reaction must be reduced (gain electrons). The following equation is where Fe^{2+} is oxidized by HOCl:

$$2Fe^{2+} + HOCl + 5H_2O \rightarrow 2Fe(OH)_3 + Cl^- + 5H^+ \qquad (11.36)$$

The ferrous ion increased in oxidation state from $+2$ to $+3$, i.e. it lost one electron. Chlorine was reduced from Cl^{1+} to Cl^{1-}, i.e. it gained two electrons. For compatibility two iron atoms are oxidized for each hypochloric acid atom reduced.

In water treatment, natural oxidation occurs in several areas. Chemical oxidation is also carried out in a number of areas. Natural oxidation can occur in open water bodies, such as lakes, storage basins and settling tanks. The oxidation process is essentially microbiological but may also be photochemical assisted. Iron and manganese can be oxidized as well as natural organics.

Chemical oxidation is a common practice within water treatment plants. Traditionally chlorine was used but in recent years attention is focused on chlorine alternatives, because of the production of trihalomethanes (THM) when chlorine reacts with natural organics. Oxidants are used for the following purposes:

- Oxidation of iron
- Oxidation of manganese
- Colour removal
- Improvement of taste
- Improvement of odour
- Flocculent aid

The alternatives to chlorine as an oxidant are:

- Chloramines
- Ozone
- Potassium permanganate
- Chlorine dioxide

Further reading on chemical oxidants is in AWWA (1990).

11.11.6 Membrane Processes Including Reverse Osmosis

Membrane separation techniques include:

- Microfiltration (MF)
- Ultrafiltration (UF)
- Reverse osmosis (RO)
- Electrodialysis (ED)

These are broadly sophisticated filtration techniques used to 'filter' out minute impurities. MF and UF separate molecules according to their size and molecular mass. Figure 11.21 is a schematic of the particle size and associated separation technique.

Microfiltration allows macromolecules (10^{-4} to 10^{-3} mm) to flow through the membrane. Bacteria that are larger than 10^{-4} mm are prevented from passing through the membrane wall. The wall pores in these units are 10^{-3} to 10^{-2} mm. Generally, as in conventional filtration techniques, the retained particle size is approximately one order of magnitude smaller than the particle/pore size of the filter. For ultrafiltration the cut-off threshold lies in the range of 10^{-6} to 10^{-4} mm pore size.

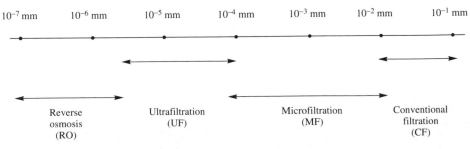

Figure 11.21 Separation processes and associated particle sizes.

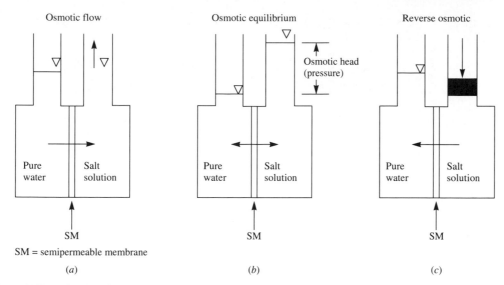

Figure 11.22   Schematic of reverse osmosis (Courtesy of E. I. du Pont de Nemours and Co., 1980).

Reverse osmosis is very different to MF or UF. This is a solubilization diffusion technique that makes use of a semi-permeable membrane which acts as a barrier to dissolved salts and inorganic molecules. It also confines organics with molecular weights greater than 100. RO membranes do not have identifiable pores as in MF or UF (Bilstad, 1992). RO has been used in desalination. The process is shown schematically in Fig. 11.22. In Fig. 11.22 (c), the pressure that is applied exceeds the osmotic pressure of the saline solution against a semi-permeable membrane, thus forcing pure water through and leaving only salts behind, i.e. all the ions are retained on the right side.

Electrodialysis (ED) is an electrically charged membrane process where the ions are transferred through a membrane from a less concentrated to a more concentrated solution. The flow of pure water is tangential to the membrane while the flow of ions is perpendicular (Bilstad, 1992).

Membrane processes that were traditionally used in desalination had limited uses. However, today, because of the vast array of impurities in waters, membrane processes are undergoing a revived interest. Some of the more recent uses include liquid or liquid effluents from industry where conventional treatment is deemed inadequate to satisfy standards with respect to complex organics and inorganics. Applications to solving contamination problems include heavy metals, oily waters, chlorinated hydrocarbons and also sludges.

11.12 US PRIMARY DRINKING WATER STANDARDS

The US primary drinking water standards are listed in detail in Table 11.15 and most recently updated in USEPA (1995).

Table 11.15 US primary drinking water standards

Contaminants	Health effects	MCL (mg/L)	Sources
Organic chemicals			
Acrylamide	Probable cancer, nervous system	TT†	Flocculents in sewage/wastewater treatment
Alachlor	Probable cancer	0.002	Herbicide on corn and soybeans; under review for cancellation
Aldicarb	Nervous system	0.003	Insecticide on cotton, potatoes, restricted in many areas due to groundwater contamination
Aldicarb sulfone	Nervous system	0.002	Degraded from aldicarb by plants
Altrazine	Reproductive and cardiac	0.003	Widely used herbicides on corn and on non-crop land
Benzene	Cancer	0.005	Fuel (leaking tanks); solvent commonly used in manufacture of industrial chemicals, pharmaceuticals, pesticides, paints and plastics
Carbofuran	Nervous system and reproductive system	0.04	Soil fumigant/insecticide on corn/cotton; restricted in some areas
Carbon tetrachloride	Possible cancer	0.005	Commonly used in cleaning agents, industrial wastes from manufacture of coolants
Chlordane	Probable cancer	0.002	Soil insecticide for termite control, corn; potatoes; most uses cancelled in 1980
2,4-D (current MCL = 0.1)	Liver, kidney, nervous system	0.07	Herbicide for wheat, corn, rangelands
Dibromochloropropane (DBCP)	Probable cancer	0.0002	Soil fumigant on soybeans, cotton; cancelled 1977
Dichlorobenzene *p*-	Possible cancer	0.075	Used in insecticides, mothballs, air deodorizers
Dichlorobenzene *o*-	Nervous system, lung, liver, kidney	0.6	Industrial solvent; chemical manufacturing
Dichloroethane (1,2-)	Possible cancer	0.005	Used in manufacture of insecticides, gasoline
Dichloroethylene (1,1-)	Liver/kidney effects	0.7	Used in manufacture of plastics, dyes perfumes, paints, SOCs (synthetic organic chemicals)
Dichloroethylene (*cis*-1,2-)	Nervous system, liver, circulatory	0.07	Industrial extraction solvent
Dichloroethylene (*trans*-1,2-)	Nervous system, liver, circulatory	0.01	Industrial extraction solvent
Dichloropropane (1,2-)	Probable cancer, liver, lungs, kidney	0.005	Soil fumigant, industrial solvent
Endrin	Nervous system/kidney effects	0.0002	Insecticides used on cotton, small grains, orchards (cancelled)
Epichlorohydrin	Probable cancer, liver, kidneys, lungs	TT	Epoxy resins and coatings, flocculents used in treatment
Ethylbenzene	Kidney, liver, nervous system	0.7	Present in gasoline and insecticides; chemical manufacturing
Ethylene dibromide (EDB)	Probable cancer	0.00005	Gasoline additive; soil fumigant, solvent cancelled in 1984; limited uses continue
Heptachlor	Probable cancer	0.0004	Insecticide on corn; cancelled in 1983 for all but termite control
Heptachlor epoxide	Probable cancer	0.0002	Soil and water organisms convert heptachlor to the epoxide
Lindane (current MCL = 0.004)	Nervous system, liver, kidney	0.0002	Insecticide for seed/lumber/livestock pest control; most uses restricted in 1983
Methoxychlor (current MCL = 0.1)	Nervous system, liver kidney	0.04	Insecticide on alfalfa, livestock
Monochlorobenzene	Kidney, liver, nervous system	0.1	Pesticide manufacturing; metal cleaner; industrial solvent
Pentachlorophenol	Probable cancer, liver kidney	0.001	Wood preservative and herbicide; non-wood uses banned in 1987

Table 11.15 US primary drinking water standards—(continued)

Contaminants	Health effects	MCL (mg/L)	Sources
Polychlorinated byphenyls (PCBs)	Probable cancer	0.005	Electrical transformers, plasticizers; banned in 1979
Styrene	Liver, nervous system	0.1	Plastic manufacturing; resins used in water treatment equipment
Tetrachlorethylene	Probable cancer	0.005	Dry-cleaning/industrial solvent
Toluene	Kidney, nervous system, lung	1	Chemical manufacturing; gasoline additive; industrial solvent
Toxaphene (current MCL = 0.005)	Probable cancer	0.003	Insecticide/herbicide for cotton, soybeans; cancelled in 1982
2-4-5-TP (Silvex) (current MCL = 0.01)	Nervous system, liver, kidney	0.05	Herbicide on rangelands, sugar cane, golf courses; cancelled in 1983
Total trihalomethanes (TTMH) (chloroform, bromoform, bromodichloromethane, dibromochloromethane)	Cancer risk	0.1	Primarily formed when surface water containing organic matter is treated with clorine
Trichloroethane (1,1,1)	Nervous system problems	0.2	Used in manufacture of food wrappings, synthetic fibres
Trichloroethylene (TCE)	Possible cancer	0.005	Waste from disposal of dry-cleaning materials and manufacturing of pesticides, paints, waxes and varnishes, paint stripper, metal degreaser
Vinyl chloride	Cancer risk	0.002	Polyvinyl chloride pipes and solvents used to joint them; industrial waste from manufacture of plastics and synthetic rubber
Xylenes	Liver, kidney, nervous system	10	Paint/ink solvent; gasoline refining by-product, component of detergents
Inorganic chemicals			
Arsenic	Dermal and nervous system toxicity effects	0.05	Geological, pesticide residues, industrial waste and smelter operations
Asbestos	Benign tumours	7 MFL	Natural mineral deposits; also in asbestos/cement pipe
Barium	Circulatory system	2	Natural mineral deposits; oil/gas drilling operations; paint and other industrial uses
Cadmium	Kidney	0.005	Natural mineral deposits; metal finishing; corrosion product plumbing
Chromium	Liver/kidney, skin and digestive system	0.1	Natural mineral deposits; metal finishing, textile, tanning and leather industries
Copper	Stomach and intestinal distress; Wilson's disease	TT	Corrosion of interior household and building pipes
Fluoride	Skeletal damage	4	Geological; additive to drinking water; toothpaste; foods processed with fluorinated water
Lead	Central and peripheral nervous system damage; kidney; highly toxic to infants and pregnant women	TT	Corrosion of lead solder and brass faucets and fixtures; corrosion of lead service lines
Mercury	Kidney, nervous system	0.002	Industrial/chemical manufacturing; fungicide; natural mineral deposits
Nitrate	Methomoglobinemia, 'blue-baby syndrome'	10	Fertilizers, feedlots, sewage, naturally in soil, mineral deposits
Nitrite	Methomoglobinemia, 'blue-baby syndrome'	1	Unstable, rapidly converted to nitrate; prohibited in working metal fluids

Table 11.15 US primary drinking water standards—(*continued*)

Contaminants	Health effects	MCL (mg/L)	Sources
Total (nitrate and nitrate)	Not applicable	10	Not applicable
Selenium	Nervous system	0.05	Natural mineral deposits; by-product of copper mining/smelting
Radionuclides			
Beta particle and photon activity	Cancer	4 mrem/yr	Radioactive waste, uranium deposits, nuclear facilities
Gross alpha particle activity	Cancer	15 pCI/L	Radioactive waste, uranium deposits, geological/natural
Radium 226/228	Bone cancer	5 pCI/L	Radioactive waste, geological/natural
Microbiological			
Giardia Lamblia	Stomach cramps, intestinal distress (Giardiasis)	TT	Human and animal faecal matter
Legionella	Legionnaires' disease (pneumonia), Pontiac fever	TT	Water aerosols such as vegetable misters
Total coliforms	Not necessarily disease-causing themselves, coliforms can be indicators of organisms than can cause gastroenteric infections, dysentery, hepatitis, typhoid fever, cholera, and other. Also coliforms interfere with disinfection	TT	Human and animal faecal matter
Turbidity	Interferes with disinfection	0.5–1.0 NTU‡	Erosion, runoff, discharges
Viruses	Gastroenteritis (intestinal distress)	TT	Human and animal faecal matter
Other substances			
Sodium	Possible increase in blood pressure in susceptible individuals	None (20 mg/L reporting level)	Geological, road salting

† treatment technique requirement in effect.
‡ NTU = nephelometric turbidity unit.

11.13 PROBLEMS

11.1 Identify the source of raw water used as potable water in your area. Determine the raw water quality parameters (similar to, say, Table 11.5). If there is more than one source, compare the water quality parameters.

11.2 With the data you have collected for Problem 11.1, determine the hardness (Ca^{2+}, Mg^{2+}) as mg/L as $CaCO_3$. Is the level of hardness soft, moderate, hard or very hard?

11.3 If the water in your area is hard (> 150 mg/L as $CaCO_3$), describe the process used to reduce the hardness to acceptable levels. After treatment, what level of hardness is still in the water?

11.4 A groundwater supply is odorous from the presence of H_2S. Describe how you would arrange for this to be eliminated.

11.5 The settling velocity V_s is shown in Section 11.7.1 to be equal to the overflow rate. Show that the same is true for a circular sedimentation tank.

11.6 For Example 11.2 in the text, compute the dimensions of a circular sedimentation tank for discrete particle settling.

11.7 For Example 11.2 in the text, prepare a circulative particle settling curve to determine the precise (> 90 per cent) weight fraction of solids removal.

11.8 Determine the daily requirement of alum, lime and polyelectrolyte to coagulate a flow of $36\,400\,m^3$/day, if the results of the jar test indicated optimum coagulation when 1 litre of water was dosed with 5 ml of a 10 mg/L alum solution, 2.0 ml of a 10 g/L suspension of lime and 0.3 mg/L of a commercial polyelectrolyte.

11.9 For Problem 11.8, compute the amount of chemical sludge due to the use of alum, lime and polyelectrolyte only.

11.10 An alternative coagulant to alum is ferric sulphate. Explain stoichiometrically the impact of adding ferric on pH and alkalinity. If a raw water has an alkalinity of 80 mg/L as $CaCO_3$, determine the final alkalinity if 50 mg/L of ferric sulphate is added as a coagulant. Assume that bulk ferric sulphate is 20 per cent by weight Fe.

11.11 Design a slow sand filter to treat a water supply rate of $20\,000\,m^3$/day. Also size a rapid gravity filter for the same supply. If the filter is required as an extension to a land-locked urban treatment plant, which filter do you recommend?

11.12 Explain the process of backwashing in a rapid gravity filter. How frequently is this carried out?

11.13 Discuss the advantages and disadvantages of using chlorine as a disinfectant.

11.14 Explain in detail the process of UV disinfection.

11.15 Calculate the amount of Cl_2 and NH_3 required to disinfect a supply of $20\,000\,m^3$/day with a free residual of 1 mg/L if the chlorine dose is 10 mg/L.

11.16 Explain the process of 'breakpoint' chlorination.

11.17 Fluoridation of drinking water supplies is no longer utilized in many areas around the world. Explain why this is occurring, using references to support your comments.

11.18 A raw water contains 220 mg/L Ca, 65 mg/L Mg, 160 mg/L HCO_3 and 180 mg/L CO_2, all expressed as $CaCO_3$. Determine the lime softening dosage required without removing magnesium. What will the ultimate hardness be?

11.19 If the water of Problem 11.18 is to be softened to a final 100 mg/L as $CaCO_3$ and not to contain more than 35 mg/L Mg as $CaCO_3$, find the required lime dosage.

11.20 Review the paper 'Chemical products and toxicological effects of disinfection' by Benjamin (1986).

11.21 Review the paper 'Surface water supplies and health' by Craun (1988).

11.22 Explain the process of reverse osmosis using figures to aid your explanation.

11.23 A water softening unit is made of a synthetic zeolite with a capacity of 300 eq/m^3 and a filtration rate of 5 L/s m^2, to soften water with a flow rate of 20 L/s and a hardness of 4 meq/L. The bed depth is 1.8 m and a 75 per cent exchange rate is achieved before breakthrough. Determine the medium diameter and water volume throughput before regeneration is required.

REFERENCES AND FURTHER READING

American Society of Civil Engineers (ASCE) (1969) *Water Treatment Plant Design*, American Water Works Association.

American Water Works Association (AWWA) (1990) *Water Quality and Treatment*, McGraw-Hill, New York.

Barnes, D., P. J. Bliss, B. W. Gould and H. R. Valentine (1986) *Water and Wastewater Engineering Systems*, Longman Scientific and Technical, London.

Baxter, K. L. (1992) Lecture Notes on Water Treatment, Civil and Environmental Engineering Department, University College Cork, Eire.

Benjamin, W., *et al.* (1986) 'Chemical products and toxicological effects of disinfection,' *Journal of the American Water Works Association*, **78** (11), 66.

Bilstad, T. (1992) *Membrane Processes*, Industrial Water Pollution Technology Course in London, Technomic Publishers, London.

Byrne, R. (1988) Water Treatment Lecture Notes, University College Cork, Eire.

Craig, F. and P. Craig (1989) *Britain's Poisoned Water*, Penguin Books, Harmondsworth.

Craun, G. F. (1988). 'Surface water supplies and health', *Journal of the American Water Works Association*, **80** (2), 40.

Cunningham, W. P. and B. W. Saigo (1992) *Environmental Science—A Global Concern*, Wm C. Brown Publishers, Dubuque, Iowa.

Degremont (1991) *Water Treatment Handbook*, Vols 1 and 2, Lavoisier Publishing, France.

Dojlido, J. R. and J. A. Best (1993), *Chemistry of Water and Water Pollution*. Ellis Horwood/Prentice-Hall, Englewood Cliffs, New Jersey, New York.

du Pont Company (1982) *Permasep Engineering Manual* Tech. Bulletin 502. Wilmington, Delaware, USA.

Dunning, J. M. (1962) *Principles of Dental Public Health*, Harvard University Press, Cambridge Mass.

Eberhardt, M. (1980) DVG Schriftenreiche Wasser No. 206. *Enteisung und Entmanganung. Removal of Iron and Manganese.* 17-1.

Feacham, R. G., D. Bradley, H. Garelick and D. Mara (1980) *Appropriate Technology for Water Supply and Sanitation*, World Bank, December.

Flanagan, P. J. (1990) *A Handbook on Implementation for Sanitary Authorities*, Environmental Research Unit, Dublin.

Flanagan, P. J. (1992) *Parameters of Water Quality*, Environmental Research Unit, Dublin.

Graham, N. J. D. (1988) *Slow Sand Fltration—Recent Developments in Water Treatment Technology*, Ellis Horwood, Chichester.

Greenberg, A. E., L. S. Clescori and A. D. Eaton (1992) *Standard Methods for the Examination of Water and Wastewater*, American Public Health Association.

Humenick, M. J. (1977) *Water and Wastewater Treatment. Calculations for Chemical and Physical Processes*, Marcel Dekker, New York.

Jackson, M. H., G. P. Morris, P. G. Smith and J. F. Crawford (1989) *Environmental Health Reference Book*, Butterworth-Heinemann, Oxford.

Linsley, R. K. and J. B. Franzini (1979) *Water Resources Engineering*, McGraw-Hill, New York.

Logsdon, G. and K. Fox (1988) Slow sand filtration in the US, in: *Slow Sand Filtration*, N. J. H. Graham (ed.), Ellis Horwood, Chichester.

Lorch, W. (1987) *Handbook of Water Purification*, John Wiley, New York.

McGhee, T. J. (1991) *Water Supply and Sewerage*, McGraw-Hill, New York.

Mackereth, F. J. M., J. Heron and J. F. Talling (1989) *Water Analysis*, Freshwater Biological Association. Ambleside, Cumbria, UK.

Matthew, R. (1993) Water Treatment Lecture Notes, University College Cork, Eire.

Metcalf and Eddy, Inc., G. Tchobanoglous and F. L. Burton (1991). *Wastewater Engineering Treatment, Disposal and Reuse*, 3rd edn. McGraw-Hill, New York.

Montgomery, J. Consulting Engineers (1985) *Water Treatment. Principles and Design*, John Wiley, New York.

Mortensen, E. (1993) Private Communication, Reno Sam, Denmark.

O'Neill, P. (1991) *Environmental Chemistry*, Chapman and Hall, London.

Pankratz, T. M. (1988) *Screening Equipment Handbook—for Industrial and Municipal Water and Wastewater Treatment*, Technomic Publishers, London.

Peavey, H. S., D. R. Rowe and G. Tcholbanoglous (1985) *Environmental Engineering*, McGraw-Hill, New York.

Pocock, S., A. G. Shaper and R. F. Packham (1981) Studies of water quality and cardiovascular disease in the UK, *Sci. Tot. Env.*, **18** 25–34.

Reynolds, T. D. (1982) *Unit Operations and Processes in Environmental Engineering*, PWS-Kent Publishing Company, Boston, Mass.

Smethurst, G. (1990) *Basic Water Treatment for Application World Wide*, Thomas Telford, London, UK.

Tebbutt, T. H. Y. (1992) *Principles of Water Quality Control*, Pergamon Press, Oxford.

Thomas, G. and R. King (1991) *Advances in Water Treatment and Environmental Management*, Elsevier Applied Science, Oxford.

Twort, A. G., F. M. Law and F. W. Crowley (1990) *Water Supply*, Arnold Publishers, Sevenoaks, Kent, UK.

US Safe Drinking Water Act (1986) PL-99-339, 19 June.

USEPA (1995) *National Primary Drinking Water Standards*, USEPA, New York.

Visscher, J. T., R. Paramasiaam, A. Raman and H. A. Hijnen (1987) 'Slow sand filtration for community water supply planning, design, construction and maintenance', Technical Paper 24, IRC, UK.

White, G. C. (1992) *Handbook of Chlorination and Alternative Disinfectants*, Van Nostrand, New York.

WASTEWATER TREATMENT

12.1 INTRODUCTION

This chapter is meant to be an introduction to wastewater engineering. The objectives in studying this chapter on wastewater treatment are:

- To understand the effluent quality standards required for both domestic and industrial wastewaters
- To be able to quantify the hydraulic loads that arrive at a treatment plant
- To be able to determine the composition and concentrations of the influent to a treatment plant
- To be able to design a primary treatment (sedimentation) facility and quantify its performance
- To understand the secondary biological treatment process and to be able to design a secondary treatment and final clarifier facility
- To understand what processes are required in tertiary and advanced treatment

Domestic wastewater is sewage only and does not include rain runoff. Urban (or municipal—the traditional term) wastewater is defined as domestic wastewater or a combination of domestic and industrial wastewater, with or without rain runoff. The words 'urban' and 'municipal' are synonymous in this text. The EU Directive 91/271/EEC, 'Concerning Urban Wastewater Treatment' sets down the effluent standards to be achieved in member countries for the treatment of wastewater. The significant parameters are shown in Table 12.1. Wastewater is first treated in settling tanks where the solids which settle are removed. The partially treated wastewater is then processed in a biological treatment plant, where micro-organism's degrade the organic water to biomass (sludge) and water (plus gases). Further settling follows. This biological treatment is by far the most common treatment process for municipal and industrial wastewaters.

It is significant that the two key effluent parameters of BOD_5 and TSS at concentrations of 25 and 35 mg/L respectively are almost identical to the standards set out in the UK Royal Commission Standards (8th Report, 1912) as 20 mg/L BOD_5 and 30 mg/L TSS. The 20/30 standard, as it was known, was adopted world-wide. However, the new EU standards of 25/35 are to be complied with in all cases, but the competent authority i.e. local water authorities, may adopt more stringent standards where the effluent discharges into sensitive waters. This is commonly the case in parts of northern Europe including Denmark and Germany. It is not uncommon for standards of 10/10 to be set by the local water authorities.

Table 12.1 EU requirements for urban wastewater

Parameter	Concentration	Minimum reduction (%)	Comments
BOD† (at 20 °C without nitrification)	25 mg/L	70–90	All plants
COD	125 mg/L	75	
TSS	35 mg/L		
	(p.e. > 10 000)‡	90	
	60 mg/L		
	(p.e. < 10 000)	70	
P (total phosphorus)	2 mg/L P		
	(p.e. 10 000–100 000)	80	Sensitive areas only*
	1 mg/L P		
	(p.e. > 100 000)		
N (total nitrogen)	15 mg/L N		
	(p.e. 10 000–100 000)	70–80	Sensitive areas only
	10 mg/L N		
	(p.e. > 100 000)		

* Sensitive waters ecologically.
† BOD = BOD_5.
‡ p.e. = population equivalent, defined as contributing 0.06 kg BOD_5 per person per day.
Adapted from EC Directive 91/271/EEC

To achieve these standards, advanced wastewater treatment processes or long detention times are used in treatment processes.

The BOD_5, or biochemical oxygen demand (discussed in Chapters 3 and 7), is a measure of biodegradable organic carbon. The COD, or chemical oxygen demand, is a measure of the total organic carbon in a wastewater. For each wastewater treatment plant receiving consistent waste loads, there will be an approximate relationship between BOD_5 and COD. Typically for urban wastewater, COD is about 1.5 BOD_5. This of course depends on whether the load is purely domestic or domestic with rain runoff or domestic plus industrial. The TSS, or total suspended solids, is a measure of the suspended material (non-settleable), mostly of organic content, though colloidal clay minerals may also form part. It is one of the indicators of pollution potential of wastewater and its percentage reduction is a measure of the efficiency of the wastewater treatment plant.

When effluents are discharged into sensitive areas, they must also comply with nutrient standards. Sensitive areas are broadly defined as those water bodies that may intermittently suffer eutrophication. Two additional important parameters are total phosphorus (P) and total nitrogen (N). The standards for P are set at 1–2 mg/L and for N are set at 10–15 mg/L. Again some countries in Europe abide by more stringent nutrient standards. Nutrient removal from wastewater only became a major issue in the past decade, though wastewater processes were already in operation in South Africa in the 1960s for nutrient removal from wastewater.

Urban wastewaters typically have raw influent BOD_5 values ranging from 150 to 400 mg/L. In the United States, typical BOD_5 influent values are closer to 150 mg/L and this is due to the large dilutions as domestic consumption of water is typically up to 1000 L/person/day by comparison with 250 L/person/day in the European Union. Raw influent values for TSS range from 150 to 400 mg/L. It is important to note that these are the raw unsettled values. The raw settled values are about 50 per cent of the raw unsettled values (i.e. equivalent to primary treatment). Typical raw untreated wastewater values for P range from 5 to 15 mg/L P and for N range from 40 to 80 mg/L N.

The objective in wastewater treatment is to protect the quality of the receiving waters and this is achieved by wastewater plants designed (among others) to:

- Reduce the BOD_5
- Reduce the TSS

- Reduce N and P
- Reduce faecal coliforms

There are other objectives regarding the quality of the effluent and these depend on the type of water body to which the effluent is discharged. For instance, in the European Union, quality compliance may be required to the EU Bathing Water Directive, which sets particular standards on microbiological quality should an effluent be discharged to the vicinity of a bathing area. Some of the EU Directives which may pose additional requirements on wastewater treatment are listed in Table 12.2.

In the United States, the compliance standards are the 'Effluent National Pollutant Discharge Elimination System' (NPDES). They are intended to protect and preserve the beneficial uses of the receiving water body, based on water quality criteria or technology-based limits, or both. The receiving water quality criteria are based on the 7 day (consecutive), 10 year, low-flow regime. These criteria define the permissible release of conservative and non-conservative pollutants. The national minima are termed secondary treatment equivalent and for municipal effluent discharge from publicly owned treatment works (POTWs) are shown in Table 12.3.

The whole effluent toxicity is determined by bioassay that measures both the acute and chronic toxicity of the effluent to the aquatic specimens, e.g. fathead minnows or daphnia (ASCE and WEF, 1991). Comparing Tables 12.1 and 12.3, it is seen that EU standards are very similar to US standards.

12.2 WASTEWATER FLOW RATES AND CHARACTERISTICS

Depending on the country, two distinct wastewater treatment policies have evolved. One is where municipal wastewater is treated independently of industrial wastewater, as is the case in Ireland. The other is where municipal wastewater is treated in combination with industrial wastewater, as is more often the case in the United Kingdom. In many countries a combination of both policies is adopted.

12.2.1 Domestic Wastewater Flow Rates

Sewer networks (the underground piping collection system) are of three types:

- Separate sewers—transporting only wastewater
- Combined sewers—transporting wastewater plus stormwater
- Partially combined sewers

The objective in the design of new systems is to install separate sewer systems as this leads to the minimum amount of water that will be treated in the wastewater treatment plant. However, existing networks are usually combined or partially combined. It is a design task to 'offload' the excess stormwater

Table 12.2 EC water quality directives relevant to effluents from wastewater treatment plants

Directive number	Title
91/271/EEC	Concerning Urban Wastewater Treatment
76/160/EEC	Bathing Water Directive
76/464/EEC	Dangerous Substance Directive
78/659/EEC	Freshwater Fish Directive and Salmonoid Waters
79/223/EEC	Shellfish Directive
80/68/EEC	Groundwater Directive
75/440/EEC	Surface Water Directive
80/778/EEC	Drinking Water Directive

Table 12.3 US minimum national performance standards for POTWs

	7 day average	
Parameter	Secondary treatment activated sludge process	Trickling filters and stabilization ponds
BOD$_5$	40–45 mg/L	60–65 mg/L
SS	45 mg/L	65 mg/L
pH	6–9	6–9
WET (whole effluent toxicity)	Site specific	Site specific
Faecal coliforms (MPN/100 ml)	400	400

in a storm event, via storm overflows to rivers, so that the treatment plant does not become overloaded with 'clean' stormwater.

The components of domestic wastewater flow are:

- Wastewater from homes and commercial premises
- Infiltration wastewater that enters the pipe sewers from groundwater and underground sources (e.g. streams)
- Water from rain runoff

Wastewater flow rates from homes and commercial premises   Wastewater flow rates are used in the hydraulic design and also in the process design of the treatment plant. However, it is easy to overdesign a wastewater plant on a hydraulic basis and so have no hydraulic overload problems. However, this same hydraulic overdesign can lead to inefficient process operation, which may not be realistic. Plant breakdown may then occur due to insufficient food substrate for the micro-organisms in the secondary treatment biological plant. It is imperative that the process design be realistic so that the micro-organisms are operating at optimum conditions producing optimum degradation of the organics in the wastewater. Traditionally, the hydraulic design of the plant based on wastewater flow rates was given more attention than the process design. The process design is now considered inseparable from hydraulic design and designers plan systems based on the dynamic nature of the process.

Figure 12.1 is a schematic of the possible variation of hourly flow over the 24 hours into a typical municipal plant (no rain runoff, no industrial flow). The flow varies from lows in the night-time (22.00 to 06.00 h) to highs in the daytime. The peak hourly flow rate is about twice the minimum hourly flow rate and about 1.5 times the 24 hour mean value. The shape of this curve depends on the community being serviced. If there is no commercial, institutional or industrial activity, then peaks will occur around 9 a.m. and 6 p.m. with slightly lower values around mid-day. In the latter case, the peaks tend to be higher (up to twice the daily mean) than those of a single peak shown in Fig. 12.1. Ideally, in designing a plant, an early objective is to collect the local data and compose such a curve for the specific community being serviced. Figure 12.1 is schematic and each urban catchment has its own individual response. Figure 12.1 shows the typical flows (m^3/s) and also the waste load in kg/BOD$_5$/s. The relative ratio of peak to mean is higher for waste load than for flow rates.

In Europe, hydraulic flow rates are defined, using the term dry weather flow (DWF), i.e. the flow rate in sewers without rain runoff. DWF is about 225 L/head/day. Again, this rate is general and specific surveys of existing sewer networks will vary. The above rate is taken to include infiltration/exfiltration. In Denmark DWF is less than 200 and in parts of the United States the figure exceeds 1000.

Infiltration, exfiltration flow rates   The values of flows in sewers due to underground sources is site specific. In the design of new plants, utilizing existing sewer networks, it is possible by surveying a

1. Maximum hourly flow ~ 0.06 average daily flow
2. Minimum hourly flow ~ 0.03 average daily flow

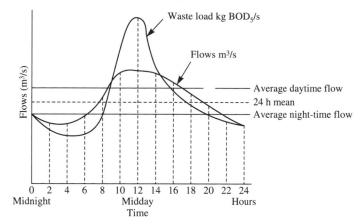

Figure 12.1 Schematic of hourly variation in flow and waste load rates.

number of sewer lines, at varying flow rates, to determine the infiltration and exfiltration rates. Suggested values for infiltration in new sewer lines are about $10\,\mathrm{m^3/ha/day}$ for sewered areas less than 50 ha (Metcalf and Eddy, 1991). These rates can be reduced if additional care during construction of sewer lines is taken to adequately seal the joints.

Rain runoff flow rates   Rain runoff in urban areas was examined briefly in Chapter 4, where the elementary rational method was used to determine urban flow rates. This method is a static analysis and should be used only as an initial guide. It is now desirable to use dynamic analysis for efficient and thorough design of sewer networks and the hydraulics of wastewater treatment plants. Traditionally, rain runoff flow rates were related to DWF. Typically storms might produce runoff events equivalent to some multiplier of DWF, e.g. 6 DWF, 12 DWF, 30 DWF, etc., increasing with storm return period. The combined sewers transported the full stormflow towards the treatment plant. Along its way, excess flow was discharged to rivers via storm overflows. It was usual that treatment plants would be designed to treat 3 DWF, with holding tanks for a further 3 DWF, and those flows in excess of 6 DWF were overflowed, untreated, to the receiving water. If discharge was to sensitive areas (e.g. a beach) maybe the treatment plant capacity was designed for 6 DWF and holding tank facility for a further 6 DWF. Again, discharge in excess of 12 DWF was overflowed. In this methodology, DWF was equivalent to 225 L/head/day. Once the decision was taken to treat, say, 3 DWF, then overflow structures were designed to carry in excess of 3 DWF. Different design conditions may impose different multipliers of DWF than those quoted above.

  Dry weather flow (DWF) is simply a definition, and in the United Kingdom it is sometimes defined as:

$$DWF = LP + I + E \tag{12.1}$$

where
$\quad\quad\quad\quad L$ = per capita consumption

$\quad\quad\quad\quad P$ = population

$\quad\quad\quad\quad I$ = infiltration (rain runoff)

$\quad\quad\quad\quad E$ = industrial effluent contribution

When using the term DWF, the reader should know which definition is being used.

  A further question is, for what return period should a sewer network be designed? Typically a flow commensurate with a specific return period was selected, though if urban areas contained buildings with basements the return period could be 50 years or for new separate sewer networks in housing developments a 2 year return period was deemed adequate. This type of analysis was static; i.e. a single value of flow, e.g. $Q_5\,m^3/s$, was adopted and each pipeline within the network was designed (for gradient and pipe size) to accommodate this.

A more detailed flood routeing analysis is required today, based on detailed rainfall and runoff characteristics. Most often the critical flows in sewer networks result from the short intensive summer rainstorm and not the protracted winter rains. As such, knowledge of the rainstorm profile is essential, and in particular its degree of peakedness. If there are substantial records of rainstorms giving flow hydrographs, the latter are routed through the network in a dynamic analysis, identifying where flow attenuation and backwater effects occur. Such analysis is possible with the computer program MOUSE, from the Danish Hydraulics Institute (1990), or MICRODRAINAGE, from Hydraulics Research, Wallingford (1985), or with SWMM from the United States (EPA, 1988).

As mentioned earlier, it is not really possible in wastewater treatment plant design to separate the hydraulic analysis from the process analysis. An attraction of the computer packages now available is that hydraulic and process analysis can be carried out together. In other words, it is possible with MOUSE, for instance, to view a longitudinal section of a pipeline with a water surface profile of a storm event together with a profile of pollutant loads in terms of BOD_5, COD and SS. Other rain runoff contaminants including hydrocarbons from roads can also be analysed as to quantity and concentrations.

It is important in the 1990s to reduce, where possible, the volume and pollutant load going to a treatment plant. Areas that afford consideration in conservation plans are:

- Regulation of domestic consumption
- Reduction of leaks/infiltrations
- Provision of porous pavements for rain runoff and so reduce the rain runoff component to the sewers

12.2.2 Domestic Wastewater Characteristics

The characteristics of wastewater are physical, chemical and microbiological, as shown in Table 12.4. Domestic wastewaters are not usually as complex as industrial wastewaters, where specific toxic and hazardous compounds may exist, i.e. phenols and toxic organics. Table 12.5 shows a typical major pollutant composition of a raw domestic wastewater.

The non-specific parameters of SS, BOD_5, etc., are of limited use. To be able to design the required wastewater treatment processes, more detailed knowledge is required. The solids can be further broken down as shown in Fig. 12.2, using typical values of 225 L/h/day and 800 mg/L total solids concentration. Figure 12.3 defines particle sizes in relation to solids. Organic materials in wastewater form the bulk of wastewater constituents. As such, the dominant treatment process is biological which degrades the organics over time. A further classification was detailed by Levine *et al.* (1985) for organics and is shown in Table 12.6.

Suspended solids make up about 40 per cent of the total solids or a concentration of about 350 mg/L. Of this, 200 mg/L is normally settleable and removed by physical processes; i.e. primary settling removes about 60 per cent of suspended solids if given adequate settling time (one to two hours). The remaining 100 mg/L is non-settleable and requires either chemical or biological processes for its removal. The most common secondary treatment process is biological. In a physical/biological sequence of processes, most of the settleable suspended solids (organic and inorganic) are removed, as is the organic fraction of the non-settleable solids. What is left after secondary treatment is a small fraction of inorganics of the non-settleable solids. Thus the treatment process reduces the suspended solids from about 300 mg/L to about 25 mg/L.

There is also a substantial component of solids called the dissolved fraction which is about 450 mg/L of the total 800 mg/L solids. As this fraction is dissolved or in solution, it is not conventionally regarded as 'true solids'. Its existence and treatment is taken as being 'part' of BOD_5 or COD. This means that the

Table 12.4 Classification of some wastewater parameters

Class	Parameter
Physical	Total solids
	Total suspended solids
	Temperature
	pH
	Colour
	Odour
Chemical	Carbohydrates ⎫
	Proteins ⎪
	Lipids ⎬ Organic
	Fats, oils, grease ⎪
	BOD_5, COD, TOC, TOD ⎭
	Alkalinity ⎫
	Grit ⎪
	Heavy metals ⎪
	Nutrients N, P ⎬ Inorganic
	Chlorides ⎪
	Sulphur ⎪
	Hydrogen sulphide ⎪
	Gases ⎭
Microbiological	Bacteria
	Algae
	Protozoa
	Viruses
	Coliforms

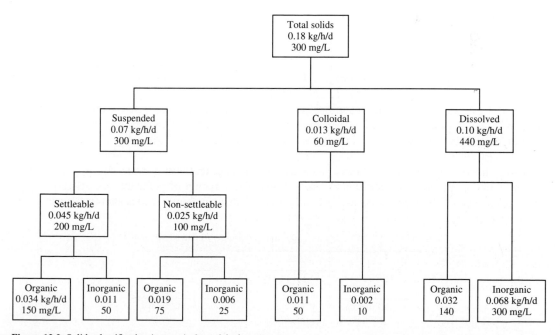

Figure 12.2 Solids classification in a typical municipal wastewater.

Table 12.5 Typical major pollutant characteristics of a raw domestic wastewater

Parameter type	Parameter	Concentration by phase (225 L/h/d)			Total (mg/L)
		Total load (kg/h/d)	Soluble* (mg/L)	Particulate* (mg/L)	
Physical	Suspended solids				
	Volatile	~80%			240
	Inert	~20%			60
	Total	~0.07 kg/h/d			300
	Dissolved solids				
	Volatile	~40%			175
	Inert	~60%			265
	Total	~0.10 kg/h/d			440
	Temperature				10–20 °C
	Colour				Fresh—grey
					Old—black
Chemical	BOD_5	~0.06 kg/h/d	(30%) 65	(70%) 135	250
	COD	~0.11 kg/h/d	130	370	500
	TOC				160
	Total nitrogen	0.01 kg/h/d	25	15	40
	Organic N				15
	Free ammonia				25
	Nitrites				0
	Nitrates				0
	Total phosphorus	0.002 kg/h/d	5	4	9
	Organic				4
	Inorganic				6
	Alkalinity				100
	FOGs†				100
Microbiological	Total coliforms				100–1000 million MPN/L
	Faecal coliforms				10–100 million MPN/L
	Total viruses				1000–10 000 infectious units/L

† FOGs = fats, oils and grease.

physical/chemical/biological treatment processes that are designed to reduce BOD_5/COD are in fact also reducing the dissolved solids fraction. The dissolved solids are made up of, in large part, carbohydrates, as seen in Table 12.6. The latter also shows that it is the soluble fraction of total solids (i.e. carbohydrates) that is most amenable to biochemical oxidation. Additionally, part of the dissolved solids fraction is actually solids in concentration, similar to the background concentrations in rivers. As such, parameters like calcium, magnesium and nutrients are part of this. The variation of pollutant load (BOD) over 24 hours parallels the flow rate curve shown in Fig. 12.1. The maximum hourly load is about 8 per cent of the

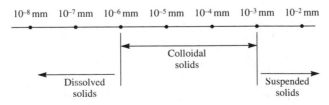

Figure 12.3 Solids particle size classification.

Table 12.6 Composition of municipal wastewater organics

Parameter	Classification by Size			
	Soluble $< 10^{-6}$ mm	Colloidal 10^{-4} to 10^{-1} mm		Settleable $> 10^{-1}$ mm
COD (% of total)	25	15	26	34
Organic composition (% of total solids)				
Grease	12	51	24	19
Protein	4	25	45	25
Carbohydrates	58	7	11	4
Biochemical oxidation rate $k\,\mathrm{d}^{-1}$	0.39	0.22	0.09	0.08

Adapted in part from Levine *et al.*, 1985

average daily load. The minimum hourly load is about 2 per cent of the average daily load. As such, the pollutant load varies from low to high by a fraction of 4, by comparison with a factor of 2 for the flow rates (see Fig. 12.1).

12.2.3 Industrial Wastewater Flow Rates and Characteristics

More precise flow rates can be determined from industry using continuous processes than from industry using batch processes. Many chemical and pharmaceutical industries operate batch processes for periods as short as one week. In this event, not only the flow rates change on moving to the next batch product but also the pollutant loads. Each industry is individual, and a waste survey is obligatory to determine flow rates and pollution load.

It is desirable to develop a flow duration curve and also a pollution load duration curve (like Fig. 12.1) at specific points along the waste streams. The latter may be defined by SS or BOD_5 or COD or some other chemical parameter. Figure 12.4 is a schematic of a flow and waste load duration curve. In Fig. 12.4, it is seen that the flow is less than $180\,\mathrm{m}^3/\mathrm{day}$ for 99 per cent of the time and is only less than $70\,\mathrm{m}^3/\mathrm{day}$ for 50 per cent of the time. Similarly, the BOD_5 is less than $600\,\mathrm{mg/L}$ for 99 per cent of the time and is only less than $120\,\mathrm{mg/L}$ for 50 per cent of the time. The selection of the chemical parameter depends on the product and waste type being generated. The value duration curves take time to complete and may be required on each of a number of different waste streams before they enter the wastewater treatment plant. It may be adequate to establish the flows, once per day for a period of one year, and then compute the flow duration curve. However, with automatic continuous flow recorders, the sample intervals can be set much smaller than one day, in fact as low as one minute. However, the task of producing a pollution load–duration curve is expensive in time and labour. Chemical sampling equipment can be placed on line or expensive autosampling–autoanalytical equipment can be used to produce satisfactory pollution load–duration curves.

In industrial complexes, as with municipal wastewater, rain runoff and infiltration may also be required to be accommodated. Many industries operate on a 5-day week and equalization basins of raw influent are used to even out 5-day flows over seven days. It is important that gaps in flow data and pollution loads be part of waste survey analysis.

Industrial wastewater characteristics It is not possible to enumerate wastes from all industry types as many wastes are specific to particular industries. Eckenfelder (1989) cites the following among undesirable waste constituents and their negative effect:

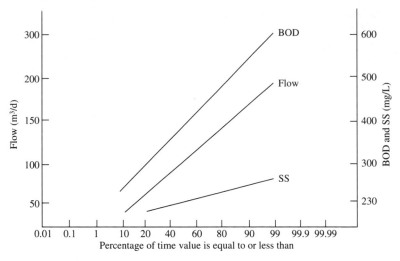

Figure 12.4 Schematic of flow and pollution load–duration curves.

• Soluble organics	Deplete DO
• Suspended solids	Ultimately deplete DO and release undesirable gases
• Trace organics	Affect taste, odours and toxicity
• Heavy metals	Are toxic
• Colour and turbidity	Affect aesthetics
• Nutrients (N and P)	Cause eutrophication
• Refractory substances resistant to biodegradation	Toxic to aquatic life
• Oil and floating substances	Unsightly
• Volatile substances	H_2S and other VOC cause air pollution

The levels of the above parameters that are acceptable in influents to streams or rivers are governed in the European Union by Directives and in the United States by EPA regulations. Table 12.7, from O'Sullivan (1991), itemizes some industries, mainly food processing, their dominant pollutant and their range of BOD_5.

Mass balance for industrial wastewater   Because of the uniqueness of most industries, waste flow diagrams or mass balances of waste flows and characteristics are carried out prior to the design of a wastewater treatment plant. The survey involves:

• Identifying the distinct processes from start to finish
• Identifying the liquid waste streams
• Computing flows of all waste streams
• Determining pollutant loads of all waste streams
• Analysing the pollutant loads in terms of the most suitable parameters for the type of waste stream, i.e. BOD_5, COD, SS, VSS, etc.

The reader is referred to AWWA (1992) for further details on mass balances for municipal plants and to Eckenfelder (1989) for industrial plants.

Table 12.7 BOD$_5$ range for some typical industries

Industry	Principal pollutants	BOD$_5$ range (mg/L)
Dairy, milk processing	C,F,P†	1000–2500
Meat processing	SS,P	200–250
Poultry processing	SS,P	100–2400
Bacon processing	SS,P	900–1800
Sugar refining	SS,C	200–1700
Breweries	C,P	500–1300
Canning, fruit, etc.	SS,C	500–1200
Tanning	SS,P, sulphide	250–1700
Electroplating	Heavy metals	Minimal
Laundry	SS,C, soaps, oils	800–1200
Chemical plant	SS, extremes of acidity and alkalinity	250–1500

† C, carbohydrate; F, fat; P, protein.
Adapted from O'Sullivan, 1991

Table 12.8 lists the approximate wastewater flow rates and per capita BOD$_5$ values for a range of water users. As the figures are based on US uses, they are likely to be higher than comparable European figures. These figures should be of use as guidelines and not for design. Precise figures for use in design should be obtained from specific surveys, as the range of values below can be wide.

Example 12.1 Determine the average daily wastewater flow rate and the BOD$_5$ concentration for an urban area composed of:

(a) Population = 150 000
(b) Hospital (1000 bed)
(c) 40 restaurants serving 40 customers per day (each)
(d) One 15 000 student college with cafeteria
(e) An equivalent 30 000 student high school without cafeteria

Table 12.8 Approximate wastewater flow rates and BOD$_5$ (per day)

Utility	Unit	Range (L/day)	Average (L/day)	BOD$_5$ (kg/day)
Domestic				
US	Per capita	250–1100	630	0.1
EU	Per capita	—	225	0.1
Schools				
Boarding	Per student	180–370	280	0.1
Day schools with cafeteria	Per student	40–80	60	0.03
Day schools without cafeteria	Per student	20–60	40	0.02
Restaurants	Per patron	20–40	30	0.03
Hotel	Per guest	160–240	200	0.1
Hospitals	Per patient	300–1000	600	0.14
Offices	Per employee	30–80	60	0.02
Department store	Per employee	30–50	40	0.02

Adapted in part from Benefield and Randall, 1980

Solution   Using the rates from Table 12.8 we get:

Utility	Flow rate	Flow rate (m^3/day)	BOD_5	BOD_5 (kg/day)
Population	150 000 × 0.225	33 750	150 000 × 0.10	15 000
Hospital	1 000 × 0.6	600	1 000 × 0.14	40
Restaurants	1 600 × 0.03	48	1 600 × 0.03	48
College	15 000 × 0.06	900	15 000 × 0.03	450
Schools	30 000 × 0.04	1 200	30 000 × 0.02	600
		$\Sigma = 36\,498\,m^3/day$		$\Sigma = 16\,238\,kg/day$

Example 12.2   Determine the average BOD_5 concentration of a municipal wastewater if the results at the influent for 12 consecutive days are as listed below.

Also, what are the standard deviation, the 90 percentile and 50 percentile concentration?

Day	BOD_5 (mg/L)
1	525
2	350
3	475
4	200
5	250
6	300
7	300
8	375
9	425
10	525
11	475
12	400

Solution   Rearrange the values in ascending order. Determine the plotting position of each of the following values from plotting position $(m - 0.5)/n \times 100$, (see hydrology examples, Chapter 4),

where m = rank of value

and n = number of observations

Rank	BOD_5	Plotting position
1	200	4.17
2	250	12.5
3	300	20.8
4	300	29.2
5	350	37.5
6	375	45.8
7	400	54.2
8	425	62.5
9	475	70.8
10	475	79.2
11	525	87.5
12	525	95.8

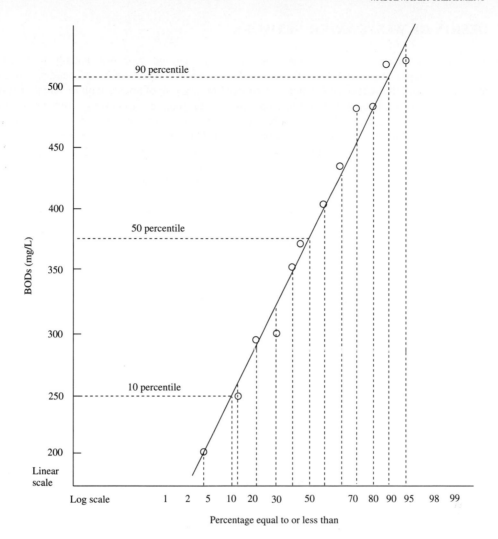

Then plot the points on logarithmic probability paper and trace the line of best fit.

$$\text{Average} = 383\,\text{mg/L}$$
$$\text{The 90 percentile value} = 510\,\text{mg/L}$$
$$\text{The 50 percentile value} = 380\,\text{mg/L}$$
$$\text{The standard deviation} = \tfrac{2}{3}\,(90\text{ percentile} - 10\text{ percentile})$$
$$= \tfrac{2}{3}\,(510 - 255)$$
$$\text{s.d.} = 170\,\text{mg/L}$$

Note: This is somewhat similar to examples of flood flow determination used in Chapter 4.

12.3 DESIGN OF WASTEWATER NETWORK

Wastewater in urban areas is collected via a series of underground pipelines, usually of concrete construction. The pipes are usually laid about 1 to 3 m below the road surface (depending on their gradient). Pipelines usually start out collecting wastewater from a group of houses with an initial pipe size of 0.15 m diameter, as this is the smallest practicable size. As the number of houses (or industries) grow, so do the pipeline sizes. On the perimeter of urban areas, pipe sizes start out at about 0.15 m but by the journey's end, having collected the wastewater from, say, 100 000 houses and many industries, plus road runoff, pipeline sizes increase to several metres in diameter. For instance, the proposed outfall pipe for the Boston wastewater plant into the sea is 8.3 m in diameter. It is not unusual to find sewer pipelines in urban areas that are 1 or 2 or 3 m in diameter. The pipes are laid at depths below road surfaces, depending on the geology and road surface loading.

The design of the pipe network to carry sewage and other flows is best understood by means of an example. Today, such design is via routine computer programs and uses real time rain data but Example 12.3 is prepared manually for the benefit of understanding the methodology. The example is adapted from Brassil (1978). The student is referred to the Wallingford procedure for Hydraulics Research (1985).

In the design of foul sewers, the following are usually adopted:

1. Minimum pipe velocity, for flushing > 0.75 m/s.
2. Maximum pipe velocity to prevent scouring and separation of liquids from solids < 3.5 m/s.
3. In the design of foul sewers only, use a flow rate

$$Q_F = 3\text{DWF (at 700 L/person/day)}. \tag{12.2}$$

to account for peak flows and infiltration to the pipeline from rainwater/groundwater.

4. For combined sewers use a flow rate equal to

$$Q_C = 3\text{DWF} + I + E \tag{12.3}$$

where
$$I = \text{storm runoff}$$
$$E = \text{industrial flow}$$

5. For industrial contributions, unless the flows and pollutant loads are known specifically, flows may be estimated from the following:

Light industry	2 L/s per ha
Medium industry	4 L/s per ha
Heavy industry	8 L/s per ha

These figures assume no recycling and no water conservation. Good practice, as is the case in Germany, has comparable figures of 0.5 L/s for light, 1.5 L/s for medium and 2 L/s for heavy industry.

6. For domestic flows, assume a habitation factor of 4, i.e. 4 persons per house, so

$$Q \text{ per house} = 4 \times 0.225 = 0.9 \text{ m}^3/\text{day/house}$$

The values given above are guides only and specific local authorities or consultants will use their own guideline values. Readers in Europe should refer to the Wallingford method (UK Hydraulics Research at Wallingford) for a simplified treatment or the D.H.I (Danish Hydraulics Institute) for computer software.

Example 12.3 Consider a foul sewer network layout as shown in Fig. 12.5. This example is reproduced from Brassill (1981). There are seven blocks of developments, housing mixed with industry. In block 1 there are 200 houses and 0.3 ha of light industry. The sewer pipe is a single line of 120 m length at a gradient of 1/180. The other six development blocks are downstream of block 1. The task is to design each pipeline to satisfy the requirements of Sec. 12.3. Use the Colebrook–White equation to compute the velocity in each pipe:

$$V = -2\sqrt{2gDS_f}\,\log_{10}\left(\frac{k_s}{3.7D} + \frac{2.51v}{D\sqrt{2gDS_f}}\right) \tag{12.4}$$

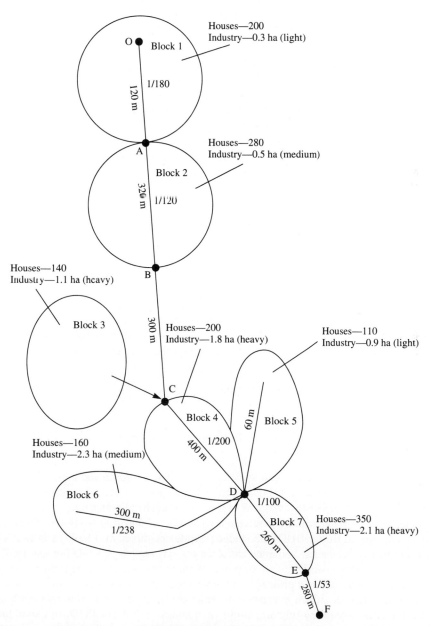

Figure 12.5 Sewer layout for Example 12.3 (adapted from Brassil, 1981, by permission).

where $\quad$ $D =$ pipe diameter, m

$\quad\quad\quad\quad\quad\quad$ $S_f =$ hydraulic gradient, m/m

$\quad\quad\quad\quad\quad\quad$ $k_s =$ roughness height, m

$\quad\quad\quad\quad\quad\quad$ $v =$ kinematic viscosity $= 1.003 \times 10^{-6}\,\text{m}^2/\text{s}$ at $20\,^{\circ}\text{C}$

Use a roughness height of 0.3 mm for concrete pipes.

$\quad$ *Solution* $\quad$ Prepare a table into 17 columns, as shown in Table 12.9:

Column (1)	Identifies the (new) area of contribution
Column (2)	Identifies previous area of contribution
Column (3)	Identifies the total area contributing
Column (4)	Identifies flows from upstream
Columns (5–10)	Identify the number of houses and type of industry and flows from each
Column (11)	Identifies the total flow from housing and industry
Column (12)	Is the total flow from (11) plus upstream flows
Column (13)	Is the pipe gradient
Column (14)	Is a guess of size of pipe diameter
Columns (15 and 16)	Are capacity and flows from either hydraulic tables or Colebrook–White equation. If the first size is inadequate, the next size up is tried
Column (17)	Makes remarks

The flow rates of column (12) are compared to the actual capacity of column (15) and the capacity then deemed adequate or not. If not, the pipe is increased in size, etc.

12.4 WASTEWATER TREATMENT PROCESSES

Municipal wastewater is primarily organic in content and a significant number of industries including chemical, pharmaceutical and food industries have high organic waste loads. This means that the main treatment processes are geared towards organics removal. In a typical treatment plant, the wastewater is directed through a series of physical, chemical and biological processes each with a specific waste load reduction task. The tasks are typically:

- Pre-treatment $\quad\quad\quad$ Physical and/or chemical
- Primary treatment $\quad\quad$ Physical
- Secondary treatment $\quad$ Biological
- Advanced treatment $\quad\;$ Physical and/or chemical and/or biological.

Table 12.10 is a flow chart of possible unit operations (unit processes) for three wastewater types: domestic, chemical and food. In some countries like the United Kingdom and parts of the United States, industrial wastewater is treated to a standard acceptable at publicly owned treatment works (POTWs) where it is further treated before discharge as effluent to a receiving water body. Municipal wastewater treatment processes treat water with influent values (typically) of 300/400 for BOD_5/SS and produce an effluent with values of 20/30 for BOD_5/SS. Industrial influent values may be high in BOD_5 and low in SS and high in metals or organics. Some industrial wastes may have high COD but low BOD values. As such, pre-treatment, both physical (e.g. air stripping of ammonia) and chemical (oxidation/reduction for heavy metals), is normally required prior to biological treatment. The most common biological treatment process is 'activated sludge', which is capable of removing BOD_5/SS of a wide range of particle sizes. Dairy processing usually requires roughing biofilters to reduce its influent BOD_5 of about 2000 mg/L to values of 300 to 400 mg/L before being treated in activated sludge systems.

Table 12.9 Design of foul sewer for Example 12.3

Area number contribution (1)	Previous area contribution (2)	Total area contribution (3)	Previous flow (L/s) (4)	Number of houses (5)	Flow (L/s) (6)	Industrial area (ha) (7)	Industrial type (8)	Unit flow (9)	Industrial flow (L/s) (10)	Total flow from area (L/S) (11)	Total flow contribution at point (L/s) (12)	Gradient (13)	Pipe size (mm) (14)	Capacity (L/s) (15)	Velocity (m/s) (16)	Remarks (17)
1	—	1	—	200	12.5†	0.3	Light	2L/s/ha	0.6‡	13.1	13.1	1/180	150	14.9	0.81	Use 225 mm pipe to minimize risk of blockage
													225	43.6	1.06	
2	1	1+2	13.1	280	17.5	0.5	Medium	4L/s/ha	2.0	19.5	32.6	1/220	225	39.3	0.96	Adequate
	1+2	1+2	32.6								32.6	1/150	225	47.8	1.16	Adequate
3	1+2	1+2+3	32.6	140	8.75	1.1	Heavy	8L/s/ha	8.8	17.6	50.2	1/200	225	41.3	1.01	225 mm size inadequate, use 300 mm diameter
													300	88.2	1.21	
4	Areas 1–3	Areas 1–4	50.2	200	12.5	1.8	Heavy	8L/s/ha	14.4	26.9	77.1					
5	—	5	—	110	6.975	0.9	Light	2L/s/ha	1.8	8.8	8.8	1/250	150	12.6	0.7	Design decision to use 225-future extension
													225	36.8	0.9	
6	—	6	—	160	10.0	2.3	Medium	4L/s/ha	9.2	19.2	19.2	1/238	225	37.8	0.92	Design of outfall sewer from area 6
1–6	—	Areas 1–6	105.1								105.1	1/100	300	125.0	1.72	Adequate
7	Areas 1–6	Areas 1–7	105.1	350	21.9	2.1	Heavy	4L/s/ha	16.8	38.7	38.7	1/53	300	174.0	2.38	Adequate

Flow breakdown

† For domestic flows 4 persons/house 225L/person/day. Peak = 6 DWF therefore Q = 0.0625 L/s/house.
‡ Industrial flows peak flow = 6 average DWF
Adapted from Brassill, 1981

Table 12.10 Flow chart outline of unit processes in domestic and typical industrial wastewater treatment

Treatment categorization	Municipal wastewater	Chemical industry wastewater	Milk processing wastewater
Physical pre-treatment	Equalization Coarse screening Fine screening Grit removal	Equalization Air stripping Oxidation/reduction Flotation	Coarse screening Fine screening Grit removal Flotation
Chemical pre-treatment		Neutralization	Neutralization
Primary treatment	Primary treatment	Primary clarification	
Secondary treatment	Biological treatment Activated sludge or Trickling filters or Aerated lagoons or RBC Secondary clarification	Biological Activated sludge	Biotowers Activated sludge
Nutrient removal	Biological (N) Chemical (P) Biological (P)		
Tertiary treatment	Sand filters	Sand filters Adsorption Chemical oxidation Ozonation	Sand filters
Advance wastewater treatment		Ion exchange	
Sludge treatment† and disposal	Yes	Yes	Yes

† See Chapter 13.

Figure 12.6 is a more detailed flow chart of the processes in a municipal wastewater treatment system. There are two end products: sludge solids and liquid effluent. The processes are identified and the approximate extent of treatment in terms of BOD_5/SS reduction is found. The influent characteristics are about 300 mg/L BOD_5 and 400 mg/L SS. The values after primary clarification are about 200/200. The primary clarification (typically) removes about 30 per cent of the BOD_5 and about 60 per cent of the SS. The secondary treatment process of activated sludge and secondary clarification (sedimentation) further reduces the BOD_5/SS to 20/30. Therefore it is possible in well-managed plants to produce an effluent after secondary clarification good enough to meet international standards for discharge to most watercourses. Sand filtration may be added as an advanced treatment step if the effluent is being discharged to a sensitive area and so reducing values to 10/10. Nutrient removal (N and P) is achieved by a combination of biological and chemical processes, although purely biological processes are also possible.

A liquid–solid sludge is produced from several of the unit processes. The 'sludge' from screens and grit channels may be separated and the grit stone recycled in part with the excess going to landfill. The primary sludge (which is very different to secondary sludge in composition) is sometimes anaerobically digested (as it is still high in organic content) or mixed with secondary sludge and waste activated sludge. It is then conditioned, thickened and dewatered for further disposal. The sludge effluent or supernatant is usually returned to the primary clarifier in a further cycle. Chapter 13 contains details of sludge and its disposal.

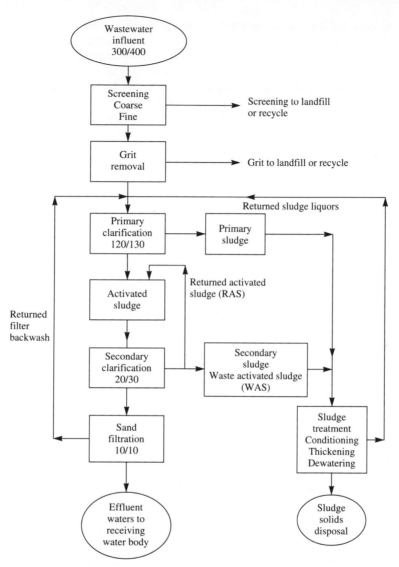

Figure 12.6 Unit processes in municipal wastewater treatment *300/400 means 300 mg/L of BOD and 400 mg/L of SS.

12.5 WASTEWATER PRE-TREATMENT

By definition, pre-treatment is the process or processes that prepare a wastewater to a condition that it can be further treated in conventional secondary treatment biological processes. In municipal wastewater it means the removal of floating debris and grit and the removal of oily scums. These pollutants would inhibit the biological process and possibly damage mechanical equipment. Ideal influent parameters for municipal activated sludge, the principal biological treatment processes are in the range 100 to 400 mg/L for BOD_5 and SS. There may be occasions when municipal wastewater (if also taking industrial effluents) may have a pH either too acidic or too alkaline for optimum biological degradation and may thus need pH correction. This may be achieved by the addition of sulphuric acid (H_2SO_4) or lime. There may also be requirements when the flow rate is inconsistent (e.g. five day week industrial effluent) that flow balancing in a storage tank be provided. This balancing or equalization tank may also be used to balance the organic loading if that varies substantially. If a wastewater is deficient in nutrients, essential for biological treatment, then nutrients may be added in the pre-treatment stage. Pre-treatment for municipal wastewaters is normally only physical, i.e. flow balancing, screenings removal and grit or oily scum removal. Industrial influents may additionally require chemical pre-treatment in the form of air stripping (ammonia removal), oxidation, reduction (heavy metal precipitation) and air flotation (oil removal). Figure 12.7 shows some pre-treatment processes used for industrial effluents. Figure 12.8 shows some of the pre-treatment processes required for municipal effluents. If industrial effluents are further treated in a municipal plant, they would usually first undergo the processes shown in Fig. 12.7. Details on industrial wastewater pre-treatment are well covered in Eckenfelder (1989) and Eckenfelder *et al.* (1992).

12.5.1 Screenings

The objective of screens is to remove large floating material (e.g. rags, plastic bottles, etc.) and so protect downstream mechanical equipment (pumps). There are four types of screens in normal use:

1. Coarse screens with openings greater than 6 mm that remove large material.
2. Fine screens with openings in the range 1.5 to 6 mm, which are sometimes used as a substitute for primary clarification (e.g. when activated sludge is used).
3. Very fine screens with openings in the range 0.2 to 1.5 mm, which reduce the SS to primary clarification levels.
4. Microscreens with openings in the range 0.001 to 0.3 mm, which are used for effluent polishing as a final treatment step. These are not used in pre-treatment except as a single one step treatment process for predominantly inorganic wastewater, e.g. quarry washings.

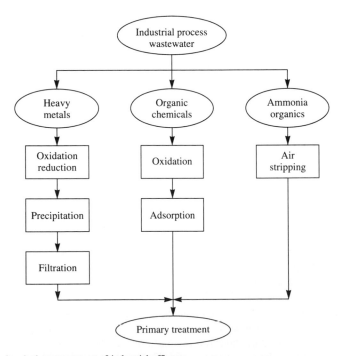

Figure 12.7 Typical chemical pre-treatment of industrial effluents.

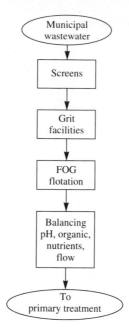

Figure 12.8 Typical physical pre-treatment of municipal and combined municipal/industrial wastewater.

Comminutors are a traditional method of screening and shredding the retained material and then allowing it back into the flow. They are no longer recommended since items like plastic pieces can find their way to the biological plant creating inhibition conditions for the microbial population. Screens are designed to accommodate through flow velocities greater than 0.5 m/s and less than 1.2 m/s with maximum head losses of about 0.7 m.

12.5.2 Grit Channels

Grit is inorganic sand or gravel particles of size about 1 mm which are washed into sewer collection systems from roads and pavements. Grit usually does not exist in industrial process wastewater but is part of municipal systems where the collection systems combine foul water and stormwater. Grit is removed (after screenings) because its inclusion within the system can abrade mechanical equipment and also because it can settle out in the biological treatment plant, reducing its space efficiency.

The two common types of grit collection devices are:

- The helical flow aerated grit chamber
- The horizontal flow grit channel

Figure 12.9 shows a schematic of the helical flow aerated grit chamber. Air is introduced along one side of the channel near the bottom and this causes a spiral motion perpendicular to the main flow direction. The heavier grit particles settle while the lighter organic matter remains in suspension and is carried on to primary clarification. The design philosophy is based on an adequate retention time in the basin of about 3 min. Aerated grit chambers have proved more efficient than the horizontal flow type and the grit tends to be 'cleaner'.

The more traditional grit chamber is of the horizontal flow type. The design philosophy is based on a minimum through-flow velocity of about 0.3 m/s and a detention time of about 1 min. They are typically

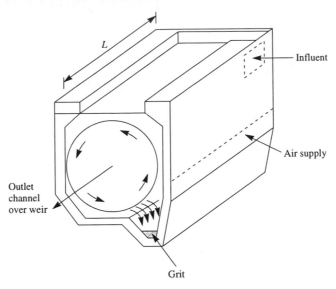

Figure 12.9 Helical flow pattern in aerated grit chamber.

designed to remove grit particles greater than 0.15 mm with associated settling velocities of about 0.01 m/s. The reader is referred to sedimentation in Chapter 11 for details of settling processes.

Example 12.4   Design a horizontal flow grit chamber to remove grit of size greater than 0.2 mm if the through flow is $10\,000\,\text{m}^3/\text{d}$. The specific gravity of the particles is 1.9.

Solution   From Stokes' law,

$$\text{Settling velocity } V_s = \frac{g(\rho_p - \rho_w)d^2}{18\mu} \tag{12.5}$$

$$= \frac{9.81(1.9 - 1.0)d^2}{18 \times 1.002 \times 10^{-3}}$$

$$= 490d^2$$

$$= 490 \times 0.2^2 = 19.6\,\text{mm/s}$$

$$\text{say} \quad V_s = 0.02\,\text{m/s}$$

Assume the depth is 1.5 times the width and the through-flow velocity $= 0.3\,\text{m/s}$. Thus

$$\text{Cross-sectional area } A = W \times D = 1.5\,W^2$$

$$= \frac{Q}{V_H}$$

$$= \frac{10\,000}{3600 \times 24 \times 0.3} = 0.39\,\text{m}^2$$

$$\text{Therefore } W = 0.51\,\text{m} \quad \text{and} \quad D = 0.76\,\text{m}$$

$$\text{Chamber length } L = V_H \times t_d$$

$$\text{Detention time } t_\text{d} = \frac{D}{V_\text{s}}$$

$$= \frac{0.76}{0.02} = 38\,\text{s}$$

$$\text{Therefore} \quad L = V_\text{H} \times t_\text{d} = 0.3 \times 38$$

$$= 11.4\,\text{m}$$

and dimensions are $11.4\,\text{m} \times 0.51\,\text{m} \times 0.76\,\text{m}$

12.5.3 Flotation

Sedimentation is the gravity unit process of separating solids from liquids. Flotation is the buoyancy unit process of separating 'solid' particles from a liquid phase. In municipal works, the solids are typically fats, oils and greases (FOGs), although in many municipal plants their amount is insignificant and so flotation is not an essential unit process. In industrial plants the 'solids' may be waste oil products. The milk producing industry usually has fats and grease among its liquid wastes. The process of separation involves introducing air bubbles at the bottom of a flotation tank. These air bubbles attach themselves to the particulate matter and their combined buoyancy encourages the particles to rise to the surface where they can be removed by skimming. Flotation is used when suspended particles have a settling velocity so low that they are not settleable in sedimentation tanks. The particulates may settle if chemically assisted by a coagulant, as is the case in the removal of suspended particles in potable water treatment. Flotation systems include:

- Gravity flotation
- Vacuum flotation
- Electroflotation
- Dissolved air flotation (DAF)
- Air flotation

Gravity flotation is accomplished by what is known as a 'grease trap' or a series of them. They are common in very small industries and automobile garages. The waste liquid flows through a series of chambers and in the process the grease or oil particles, being lighter than water, rise to the surface and are mechanically removed. A hydraulic retention time of about 30 min and a flow-through velocity of 4 to 6 m/h are required for successful operation. They are not suitable for urban wastewater treatment plants because of the large tank size required.

Vacuum flotation consists of saturating the wastewater with air in an aeration basin and then applying a partial vacuum to a covered tank. Minute bubbles are released from the liquid and become attached to the suspended particles which then migrate to the surface where they are removed. The flotation procedure is common with the US fruit and vegetable processing industry.

Electroflotation is the process where electrodes placed towards the floor of a tank produce microbubbles when the liquid in the tank is electrolysed by means of a direct current. The bubbles of oxygen produced at the anode end rise and attach themselves to the suspended particulates, producing a surface scum which can be removed. This process has a high cost of electrode replacement.

The most successful method is dissolved air flotation. Usually (but not always) part of the effluent is recycled from a point downstream of the DAF unit. Recycled flow is retained in a pressure vessel for a few minutes where mixing and saturation of the flow with air occurs. This recycled effluent is added to the DAF unit where it is mixed with the incoming raw effluent. As the pressure returns to atmospheric, the dissolved air comes out of solution, forming fine bubbles, which rise to the surface bringing grease matter

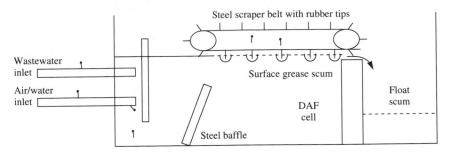

Figure 12.10 Dissolved air flotation (DAF) unit.

with them, where they are removed. The design upward flow velocity is in the range of 4 to 6 m/h and the air to solids ratio by weight is in the range of 1 to 5 per cent. Figure 12.10 is a schematic of a DAF unit. Air flotation is a variation of DAF where air is directly introduced to the flotation tank by means of an impeller.

12.5.4 Equalization

In order that a wastewater treatment plant receives an effluent that it is capable of handling without distress, equalization (balancing) may be required. This may include one or more of the following:

- Flow equalization
- Organic equalization
- Nutrient balancing
- pH balancing (neutralization or pH correction)

Flow balancing is common in industries that operate a 5 day week. Here the flow is balanced or spread out equally over 7 days so that the flow arriving into the plant is the same for each of the 7 days. Similarly, in the case of organic or pollutant load balancing, industry may at different times during the week have a high COD effluent, lasting only a few hours. If this were sent directly through the treatment plant it may cause a shock load with consequent problems. It is therefore usual to balance the high load such that a more even load is sent to the plant for treatment. This is done by retaining the pollutant load in a balance or equalization tank, prior to treatment. Nutrient balancing is where nutrients may be added to the influent wastewater should the wastewater be deficient in nutrients. pH balancing (neutralization) may be required should an influent to a wastewater treatment plant be too high or too low in pH for optimum secondary treatment of that waste. It is desirable that the pH be in the range of 6.5 to 8.5 for activated sludge treatment systems. Many industrial wastewaters are not within this range and therefore need to be 'corrected' or balanced. The objectives of balancing may be summarized as:

- Equalization of flows to minimize flow surges
- Equalization of organic loads to dampen fluctuations
- Neutralization of pH variations to bring it to the range 6.5 to 8.5
- Provision of continuous influent to the plant
- Provision of continuous effluent from the plant to the receiving water body
- Control of high toxicity loads

Equalization and neutralization are achieved by the provision of a tank, usually after screenings and grit removal and prior to primary sedimentation. Two layouts are common: the in-line system shown in Fig. 12.11 and the side-line system of Fig. 12.12. With in-line equalization, all the flow passes through the

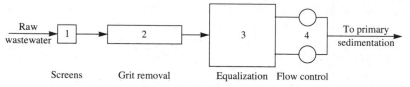

Figure 12.11 In-line equalization (adapted from ASCE and WEF, 1991).

equalization basin, resulting in a significant flow and load (BOD$_5$) dampening. With side-line equalization, only that flow greater than, say, 1 DWF goes to the equalization basin. Less effective dampening occurs but pumping costs are much less than for the in-line system.

The type of tank equalization facility used may be of two kinds:

- A tank with minimum mixing to inhibit septicity
- A tank with sufficient mixing to act as a pre-aeration unit

Both types of facility are beneficial to consequent primary sedimentation and secondary biological treatment. However, the provision of aeration (either mechanical or diffused air) has extra benefits to the later treatment processes. Using pre-aeration results in pre-flocculation of some suspended solids and improves the settling characteristics. Equalization benefits biological treatment by providing an almost steady waste flow and load, thereby allowing the biological system to operate at near steady state conditions. As such, optimum operation occurs. In-line aeration equalization may provide up to a 20 per cent BOD$_5$ load reduction to the biological system.

Equalization may be used to upgrade existing plants or can be provided in new plants. The design decision to be made is whether it is more economical to include equalization basins or to increase slightly the size of downstream unit processes (primary clarification and biological secondary treatment). According to the AWWA (1992), equalization facilities may not be economical for municipal plants if the peaking factors (peak flow/average flow) are less than 2.

Equalization facilities generally may be of three types:

- Constant flow, variable waste strength
- Constant waste strength, variable flow
- Constant flow, constant waste strength

With some industries, flow rates are small but waste strengths vary significantly. In this case it may be desirable to produce a constant waste strength by equalization and permit the small flow to remain

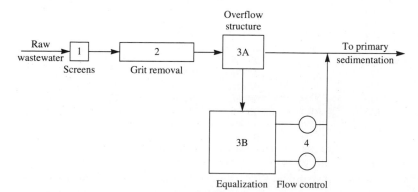

Figure 12.12 Side-line equalization (adapted from ASCE and WEF, 1991).

variable. However, in providing an equalization facility to produce a constant flow, some benefits of dampening the waste strength also occur.

Example 12.5 Design the size of an equalization tank to balance flow rates from a municipal wastewater as given in columns (1) and (2) of Table 12.11.

Solution Columns (1) and (2) are the given hourly increments and associated flow rates. Column (3) shows the associated hourly flow rates. Column (4) shows the cumulative flows in m^3 from midnight to midnight. The average flow rate is computed from the cumulative flow as $12\,000\,m^3/24\,h \times 3600\,s = 0.1388\,m^3/s$. If the averaged flow is computed over an hour, this equals $500\,m^3$. As such, if $500\,m^3$ is sent through the system every hour then after 24 hours, a total flow of $12\,000\,m^3$ is processed. Column (7) shows the difference between the cumulative flow and the cumulative equalized flow. The equalized flow exceeds the cumulative flow by a maximum of $1260\,m^3$. This then is the size of the equalization basin required to process the equalized flow of $500\,m^3/h$. It is of course prudent to design for a space capacity of, say, 20 per cent, giving the design volume size of the equalization basin as 1260×1.2 at $1500\,m^3$.

These calculations are shown graphically on a plot of time versus cumulative flow. The inflow cumulative mass diagram is a curve. The equalized cumulative flow is a straight line. The differences between the two lines shows when an equalization basin is required. This is shown schematically in Fig. 12.13.

Table 12.11 Municipal flow rates to assess equalization basin size

Time (h) (1)	Q_{av} (m³/s) (2)	Volume (m³/h) (3)	Flow cumulative (m³) (4)	$Q_{equalized}$ (m³/s) (5)	Equalized flow cumulative (m³) (6)	Column (6) − (4) (m³) (7)
0–1	0.13	468	468	0.1388	500	32
1–2	0.12	432	900	0.1388	1 000	100
2–3	0.11	396	1 296	0.1388	1 500	204
3–4	0.10	360	1 656	0.1388	2 000	344
4–5	0.08	288	1 944	0.1388	2 500	556
5–6	0.06	216	2 160	0.1388	3 000	840
6–7	0.08	288	2 448	0.1388	3 500	1052
7–8	0.10	360	2 808	0.1388	4 000	1192
8–9	0.12	432	3 240	0.1388	4 500	1260
9–10	0.14	504	3 744	0.1388	5 000	1256
10–11	0.16	576	4 320	0.1388	5 500	1180
11–12	0.18	648	4 968	0.1388	6 000	1032
12–13	0.20	720	5 688	0.1388	6 500	812
13–14	0.19	684	6 372	0.1388	7 000	678
14–15	0.18	648	7 020	0.1388	7 500	480
15–16	0.17	612	7 632	0.1388	8 000	368
16–17	0.16	576	8 208	0.1388	8 500	292
17–18	0.15	540	8 748	0.1388	9 000	252
18–19	0.16	576	9 324	0.1388	9 500	176
19–20	0.17	612	9 936	0.1388	10 000	− 64
20–21	0.18	648	10 584	0.1388	10 500	− 84
21–22	0.16	576	11 160	0.1388	11 000	− 160
22–23	0.127	456	11 700	0.1388	11 500	− 200
23–24	0.107	384	12 000	0.1388	12 000	0
	Average 0.1388					

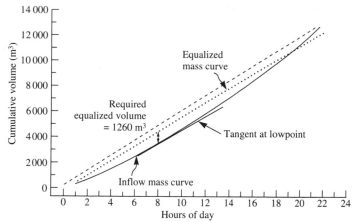

Figure 12.13 Mass diagram for equalization basin.

12.6 PRIMARY TREATMENT

Primary treatment is often called clarification, sedimentation or settling. This is the unit process where the wastewater is allowed to settle for a period (≈ 2 h) in a settling tank and so produce a somewhat clarified liquid effluent in one stream and a liquid–solid sludge (called primary sludge) in a second stream. The objective is to produce a liquid effluent of suitably improved quality for the next treatment stage (i.e. secondary biological treatment) and to achieve a solids separation resulting in a primary sludge that can be conveniently treated and disposed of. The benefits of primary treatment include:

- Reduction in suspended solids
- Reduction in BOD_5
- Reduction in the amount of waste activated sludge (WAS) in the activated sludge plant
- Removal of floating material
- Partial equalization of flow rates and organic load

Primary treatment is quiescent sedimentation with surface skimming of floating matter and grease, and bed level collection and removal of settled sludge. Sedimentation is carried out in a variety of tank configurations including:

- Circular—most common
- Rectangular
- Square

They may be flat bottomed or hopper bottomed. Wastewater enters the tank, usually at the centre, through a well or diffusion box. Figure 12.14 shows a cross-section of a typical circular primary clarification tank. The tank is sized so that retention time is about 2 h (range 20 mins to 3 h). In this quiescent period, the suspended particles settle to the bottom as sludge and are raked towards a central hopper from where the sludge is withdrawn. The clarified water is discharged over a perimeter weir at the surface of the tank, at a rate known as the basin overflow rate or surface overflow rate (SOR). The units of SOR are $m^3/day/m^2$. The last m^2 is the plan area of the tank. The SOR is different to the weir overflow rate (WOR). The units of the WOR are $m^3/day/m$. The last m is the length (perimeter) of the weir.

Primary sedimentation is among the oldest of wastewater treatment processes. According to the AWWA (1992), money spent on primary treatment often provides the greatest return on the investment in terms of dollars per kg of pollutant removed. Traditionally, the design criteria were:

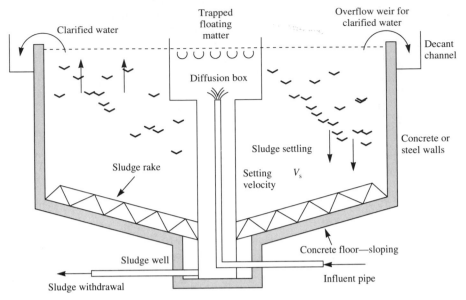

Figure 12.14 Typical circular primary settling tank.

- Basin overflow rate (surface loading $m^3/day/m^2$)
- Depth
- Surface geometry
- Hydraulic retention time
- Weir rate ($m^3/day/m$)

These criteria are physical and while they may be adequate for the design of the tank they say nothing about the performance and operation of the sedimentation process. Therefore, additional parameters, called performance criteria, were established to monitor and improve the day-to-day performance. These criteria include:

- Influent flow rates and their variation (daily variation)
- Influent waste strength rates and its variation
- Recycle influent streams:
  - From activated sludge or septage
  - Supernatants from sludge dewatering
  - Washings from tertiary filter processes

The above may vary from hour to hour or from day to day. The flow rates may have peaks several times the daily average, and waste strengths may vary accordingly. Recycle streams can come from several sources and in hugely varying waste strengths. Septage, for instance, may have a BOD_5 value 30 times greater than municipal raw wastewater. Supernatants from anaerobic digestion processes or filtrate backwashings may also be very high in waste strength. As such, the performance of a primary clarification is not solely dependent on influent flow variations. For instance, plants that may have been overdesigned for flow may find that the retention time in the tank is not the 2 hours of the original design but several times that. Excessive retention time leads to septicity as there is no mixing in primary sedimentation. It is also possible that where the operation and maintenance of primary tanks is poor (i.e. long retention times and infrequent sludge withdrawals) the quality of clarified water is no improvement on the influent wastewater. However, with good performance management, removal rates of 50 to 70 per cent for suspended solids and 25 to 40 per cent for BOD_5 can be achieved. Figure 12.15 shows the percentage removal rates versus surface overflow rates. Surface overflow rates range from 32 to 48 m³/day/m² at average flow rates. Based on peak flow rates, the corresponding values are 80 to 120 m³/day/m². When waste activated sludge from activated sludge treatment is recycled through primary settling tanks the overflow rates are approximately 75 per cent of the above. The depth of tanks varies from about 2.5 to 5 m (refer to Chapter 11 for further details on sedimentation tanks).

Example 12.6   Design a primary settling tank to remove 60 per cent of the SS if the average flow is 5000 m³/day with a peak factor of 2.5. What is the corresponding BOD_5 reduction?

Solution   From Fig. 12.15, we see that to achieve a 60 per cent SS reduction a surface overflow rate (SOR) of 35 m³/day/m² is required. This also affords a 32 per cent BOD_5 reduction.

$$\text{Surface area required} = \frac{Q}{\text{SOR}} = \frac{5000 \text{ m}^3/\text{d}}{35 \text{ m}^3/\text{m}^2/\text{d}} = 143 \text{ m}^2$$

Using a circular tank,

$$\text{Required diameter} = 13.5 \text{ m}$$

Assuming a side wall depth of 3 m,

$$\text{Volume} = 143 \times 3 = 429 \text{ m}^3$$

$$\text{Detection time} = \frac{V_{ol}}{Q} = \frac{429 \text{ m}^3 \times 24}{5000 \text{ m}^3/\text{m}^2/\text{day}} = 2.06 \text{ h}$$

At peak flows,

$$\text{SOR} = \frac{2.5 \times 5000}{143} = 87 \text{ m}^3/\text{m}^2/\text{d}$$

This gives a detention time of 50 min and an SS removal rate of 38 per cent and a corresponding BOD_5 reduction of 20 per cent.

For further details on the principle of sedimentation the reader is referred to Chapter 11.

12.6.1 Chemically Enhanced Primary Sedimentation

The addition of coagulant chemicals (iron salts, lime, alum) before sedimentation promotes flocculation of fine suspended matter into more readily settleable flocs. This increases the efficiency very substantially

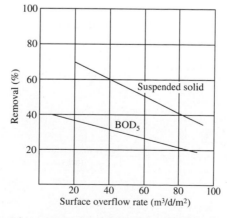

Figure 12.15 Surface overflow rate and percentage removal (adapted from McGhee, 1991).

Table 12.12 Comparison of pollutant removal efficiencies for primary sedimentation with and without coagulation

	Removal efficiencies of primary settling	
Parameter	With coagulants (%)	Without coagulants (%)
TSS	60–90	40–70
BOD_5	40–70	25–40
COD	30–60	20–30
TP	70–90	5–10
Bacteria	80–90	50–60

Adapted from Harleman, 1991

of SS and BOD_5 removal rates. Table 12.12 shows a comparison of removal efficiencies using primary sedimentation with and without chemical coagulation.

Chemical enhancement sustains the high removal efficiency over a wide range of removal rates. In conventional primary sedimentation tanks, as the surface overflow rate increases, the removal efficiency decreases. With chemical coagulants, the removal efficiencies are almost constant over an SOR range of 20 to $80\,m^3/m^2/day$ (Heinke and Tay, 1980). A disadvantage of chemical coagulants is an increase in primary sludge which is a chemical-type sludge quite different to the biological sludge, from primary sedimentation. The chemical sludge is difficult to dewater as is known from water treatment chemical type sludges (Chapter 13).

The mechanism of chemically enhanced primary sedimentation is to use an aeration tank prior to the settling tank. The chemicals are added to the aeration tank. One such set up is shown in Fig. 12.16 from Canada, where the chemical coagulant and coagulant aids are added continuously to a pre-aeration tank.

In several plants in Scandinavia, the total treatment process is pre-aeration with coagulant addition, followed by sedimentation. No secondary biological treatment follows. It is possible to satisfy most international water quality standards using this process. However, this process is not popular in Western Europe or the United States. This process deserves consideration particularly for upgrading of existing plants, where space and cost may be limiting conditions. For further details refer to Harleman (1991).

12.6.2 Sludge Quantities from Primary Settling

The amount of sludge produced during primary settling will depend on the throughflow, the total suspended solids and the efficiency of solids removal. This quantity can be readily determined from:

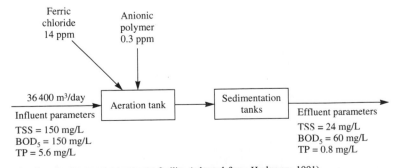

Figure 12.16 Ontario, Canada, wastewater treatment facility (adapted from Harleman, 1991).

$$S_m = Q \times \text{TSS} \times E \qquad \text{kg/day} \tag{12.6}$$

where
$$Q = \text{throughflow, m}^3/\text{day}$$
$$\text{TSS} = \text{total suspended solids, mg/L}$$
$$E = \text{removal efficiency}$$
and
$$S_m = \text{sludge quantity, kg/day}$$

Example 12.7

(a) Determine the amount (in kg/day and m^3/day) of primary sludge solids generated in a domestic treatment plant for a p.e. of 20 000, with an efficiency of (TSS) removal of 60 per cent.
(b) What is the density of primary sewage sludge if it is 2 per cent dry solids (i.e. 98 per cent water)?

Solution

(a) Assume per capita flow $= 225$ L/day

Therefore
$$Q = 20\,000 \times 225 = 4500 \, \text{m}^3/\text{day}$$

Assume TSS concentration $= 300$ mg/L

Therefore

$$S_m = 4500 \, \text{m}^3/\text{day} \times 300 \, \text{mg/L} \times 0.6 \times \frac{10^3}{10^6}$$

$$= 810 \, \text{kg/day}$$

or
$$= \frac{810}{4500} \, \text{kg/m}^3/\text{day} = 0.18 \, \text{kg/m}^3/\text{day}$$

Assuming the density of sludge (ρ_s) to be $1000 \, \text{kg/m}^3$ ($\rho_w = 1000 \, \text{kg/m}^3$), then the volume of sludge generated daily is

$$V = \frac{M}{\rho_s} = \frac{810 \, \text{kg/day}}{1000 \, \text{kg/m}^3} = 0.81 \, \text{m}^3/\text{day}$$

Chemically enhanced primary settling may increase the above values by 50 to 100 per cent. Typically, unaided treatment produces about $0.15 \, \text{kg/m}^3$ while chemically assisted with lime, alum or ferric produce produces, 0.4, 0.25 and $0.25 \, \text{kg/m}^3$ respectively.

(b) Density of water ρ_w $\qquad = 1000 \, \text{kg/m}^3$
(Assume) particle density of sludge $\rho_p = 1700 \, \text{kg/m}^3$

Particle density is defined by $\rho_p \qquad = \dfrac{M_p}{V_p}$

where
$$M_p = \text{mass of particles (solid)}$$
$$V_p = \text{volume of particles (no air, no water)}$$

This sludge is composed of 98 per cent water $+ 2$ per cent sludge particles. A mass balance gives

$$M_{\text{water}} + M_{\text{sludge}} = M_{\text{total}}$$
$$\rho_w V_w + \rho_p V_p = \rho_T V_{\text{total}}$$
$$1000 \times \frac{98}{100} + 1700 \times \frac{2}{100} = \rho_T \frac{100}{100}$$

Therefore

$$\rho_T = 1014 \text{ kg/m}^3$$

The sludge (water + particles) density is not much more than the density of water. Therefore, the assumption in part (a) of taking the sludge density to be the same as water density is safe enough. Even at dry solids as high as 20 per cent, the density of the sludge is still only 1140 kg/m^3.

12.7 SECONDARY TREATMENT

Municipal wastewater pollutants are summarily described by the following parameters:

- Total solids—suspended (40 per cent), colloidal (10 per cent) or dissolved (50 per cent); organic (50 per cent), inorganic (50 per cent)
- Biochemical oxygen demand
- Chemical oxygen demand
- Nutrients—nitrogen and phosphorus

In Sec. 12.6, it was seen that primary settling removed about 60 per cent of the suspended solids and about 30 per cent of the BOD_5. About 65 per cent of the solids removed (settleable) are organic and the remainder inorganic. Figure 12.17 shows a typical parameter set for a municipal wastewater prior to secondary treatment.

From Fig. 12.17, it is seen that the main purpose of secondary treatment is to reduce the BOD_5 value which does not benefit as much as SS from primary settling. In other words, secondary treatment should be a process that is capable of biodegrading the organic matter into non-polluting end products, e.g. H_2O, CO_2 and biomass (sludge). The end product liquid effluent should be well stabilized or well oxidized so that it does not provide a food source for aerobic bacteria in the receiving water body. Then the discharge to a water body should lead to little or no removal of dissolved oxygen by bacterial action. To produce a well-oxidized liquid effluent a vast array of biological processes exist, some general and some proprietary, that are capable of removing the organic matter from the wastewater.

12.7.1 Principles of Biological Oxidation

The mechanisms of removal of organic matter include:

- Biodegradation
- Air stripping
- Adsorption

Adsorption of non-degradable organics onto biological solids is not significant, but it does occur for particular organics including the pesticides, e.g. lindane. Heavy metals will adsorb on to biomass and bioaccumulate, resulting in end product sludges containing heavy metals.

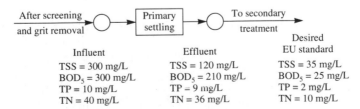

Figure 12.17 Typical municipal wastewater parameters before and after primary settling.

Air stripping of volatile organic carbon (VOC) occurs in aerobic systems. The breakdown of carbonaceous material by aerobic degradation emits CO_2 and other VOCs into the atmosphere.

Biodegradation is the dominant mechanism of organics removal for municipal and most industrial wastewaters. Most treatment plants now use the activated sludge system for this purpose. Figure 12.18 is a schematic of the activated sludge system. It comprises of two 'box' structures, the aeration tank and the secondary clarification tank. The aeration tank, while having many possible configurations, basically retains the influent wastewater for a number of hours (or days) in a well-mixed/aerated environment, before forwarding the effluent for further settling to the secondary clarification tank. The end products of the clarification tank are clarified liquid effluent, ready for discharge to open water bodies, and a liquid–solid sludge. A fraction (about 20 per cent) of the sludge is returned to the aeration tank and is called returned activated sludge (RAS). The sludge contains a high density of live microbial biomass and, in returning part of it, an active population of microbes are always maintained in the aeration tank. The influent wastewater is the food source for the resident microbes in the aeration tank. The microbes biodegrade this feedstock into new microbial cells in the presence of aerated water. Other end products include CO_2, NO_3^-, and SO_4^-. In the clarifier, the excess biomass settles out as sludge and about 80 per cent of this is removed for further treatment and subsequent disposal.

The mixed 'liquor' suspension in the aeration tank contains wastewater, living and dead micro-organisms and inert biodegradable and non-biodegradable suspended and colloidal matter. The particulate fraction of the mixed liquor is called the 'mixed liquor suspended solids' (MLSS). This value is generally of the order of 2000 to 4000 mg/L for a healthy microbial suspension in the aeration tank. The MLSS of the RAS is of the order of 10 000 to 20 000 mg/L. This is a measure of the microbial population and essential to keep the MLSS in the aeration base > 2000 mg/L; otherwise the microbial population is not large enough to biodegrade the incoming organics.

The understanding of the activated sludge process, since it was first introduced in Manchester by Arden and Lockett (1914), has made significant advances. The process is governed by the micro-organism

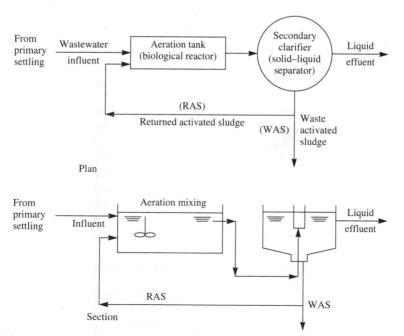

Figure 12.18 Typical layout of activated sludge system.

characteristics and the physical configuration of the aeration basin (plug flow, completely mixed, etc., explained later). As such, the biological kinetics and the process kinetics are inseparable, as the process is dynamic. Whether a secondary treatment system is of the suspended growth type (e.g. activated sludge) or fixed film type (using a medium, e.g. stone, plastic, etc., on which to grow organisms) will depend on the same biological principles; i.e. a microbial population is used to biodegrade carbonaceous matter.

12.7.2 Bacterial Growth in Pure Cultures

The bacterial population (e.g. in an aeration tank) is expressed as either the number of cells (N) per unit volume or the mass of cells (M) per unit volume. They are not directly translatable unless it is known where they reside on the growth curve of Fig. 12.19. At different points on this curve the cells may have a different size and a different mass of sorbed substrata. Figure 12.19 is a schematic of the growth curve for a typical bacteria cell culture.

Bacteria reproduce by binary fission and a significant parameter is called the regeneration rate or doubling time. This varies depending on the position on the growth curve. When a bacterial innoculum is placed in a closed vessel with ideal conditions and excess food, the growth and decay of the bacterial cell numbers follows the curve of Fig. 12.19. During the acclimation period, there is no growth as the organisms acclimate to their new environment. This acclimation period may be up to six weeks for some complex industrial wastes. This phase is followed by an exponential growth, when the cells reproduce at their optimum regeneration times. The doubling time is typically 20 to 60 min, although it may take a few days in exceptional environments. In the stationary phase the population remains constant with the rate of new cell synthesis equal to the death rate of old cells. Around this time, the substrate (food) is becoming limited or exhausted or there is a nutrient deficiency. The following phase, called the endogenous phase, is where the bacteria are surviving off their own stored energy and consuming the dead cells. The number of cells begins to decline and there is then an accelerated death phase when there is no biodegradable substrate. This schematic is atypical of the real microbial population of activated sludge systems, where the population tends to be complex, interrelated mixed populations with each species having its own growth curve. Most organisms in activated sludge are single-cell bacteria with soil origins. The objective

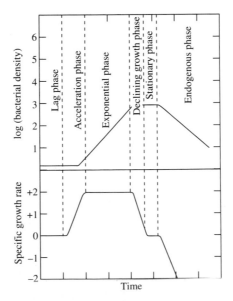

Figure 12.19 Characteristic growth curves of cultured micro-organisms (adapted from Monod, 1949).

in operating the activated sludge system is to try to retain the mixed population within a desirable range on the growth curve for optimum performance. Outside the range there may be an excess of organisms or a deficiency, and both lead to operational problems.

12.7.3 Kinetics of Bacterial Growth

When a biodegradable organic food source is supplied to a heterotrophic (utilizing organic materials for energy, as distinct from autotrophic which use CO_2 as the carbon source) micro-organism population in a well-aerated environment, the response is as follows, adapted from Ekama and Marais (1984):

1. The readily soluble biodegradable COD goes through the cell wall and is metabolized quickly.
2. The slowly biodegradable particulate COD is adsorbed on to the organisms and stored. This quick reaction removes all the particulate and colloidal COD. Over time the stored COD is broken down by extracellular enzymes, transferred through the cell wall and metabolized. The rate limiting step in the overall synthesis is the enzymatic breakdown at rates about 10 per cent of the readily biodegradable COD rate.
3. Some of the COD metabolized is converted to new cells, while the remainder is lost as heat in the energy process required for the new cell synthesis. Oxygen externally supplied is used up in this energy process, such that the amount of oxygen utilized is proportional to the COD lost.
4. At the same time, there is a net loss of live biomass, termed endogenous mass loss, where some of the organisms utilize as food their own stored food materials and dead cells. This endogenous degradation is continuous and relatively constant at about 10 to 20 per cent per day.

The biochemical equation for bacterial cell respiration and synthesis in using organic matter as substrate in the activated sludge process is:

$$\text{Organic matter} + O_2 + \text{nutrients} \xrightarrow{\text{bacteria}} CO_2 + NH_3 + \text{new biomass} + \text{other end products} \tag{12.7}$$

For mixed cultures, as in activated sludge, biomass is the parameter of interest, rather than the number of organisms. In the log growth phase of Fig. 12.20, the rate of biomass increase is proportional to the initial biomass concentration and is represented by the first-order equation:

$$\frac{dX}{dt} = \mu X \tag{12.8}$$

where
$$\frac{dX}{dt} = \text{growth rate of biomass, mg/L/day}$$

$$X = \text{concentration of biomass, mg/L}$$

$$\mu = \text{specific growth rate constant, } d^{-1}$$
$$\text{(mass of cells produced/mass of cells present per unit time)}$$

Letting X_0 represent the biomass at time $t = 0$, integration of Eq. (12.8) is

$$\ln X = \ln X_0 + \mu t$$

$$\ln \left(\frac{X}{X_0} \right) = \mu t$$

$$X = X_0 \cdot \exp (\mu t) \tag{12.9}$$

The growth rate that follows Eq. (12.9) is called the exponential growth rate. It is assumed that Eq. (12.9) holds as long as there is no change in the biomass composition or environmental conditions. However,

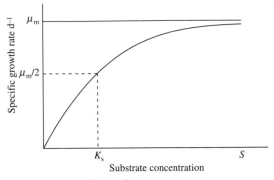

Figure 12.20 Substrate concentration versus specific growth rate.

this cannot always be the case. For instance, a batch culture will incur environmental change during its life time.

In mixed cultures, it is not yet possible to measure μ. Typically, there are essential requirements for microbial growth including availability of substrate or nutrients. If an essential requirement is limited, growth slows or ceases. Monod (1949), noting that the growth rate (dX/dt) was a function of organism concentration (X) and also of a limiting nutrient concentration, showed experimentally that

$$\mu = \mu_m \frac{S}{K_s + S} \tag{12.10}$$

where
$S =$ concentration of limiting substrate, mg/L

$\mu_m =$ maximum growth rate, d^{-1}

$K_s =$ half-saturation constant, i.e. the concentration
of S when $\mu = \mu_m/2$ (mg/L) as shown in Fig. 12.20

Equation (12.9) assumes only growth of micro-organisms. However, there is growth of one set of organisms and die-off of another set, occurring simultaneously. To take account of die-off, an endogenous decay (rate $= k_d$) is used, so Eq. (12.8) becomes

$$\frac{dX}{dt} = \mu X - K_d X$$

Biomass production :
$$\frac{dX}{dt} = \left(\frac{\mu_m S}{K_s + S}\right) X - k_d X \tag{12.11}$$

If all substrate (S) could be converted to biomass (X) then the rate of substrate utilization is

Ideal:
$$-\frac{dS}{dt} = \frac{dX}{dt} \tag{12.12}$$

(Similar to the energy analogy, dissipation = production)

However, such an idealization cannot occur due to inefficiencies in the conversion process and a yield coefficient ($Y < 1$) is introduced such that the rate of substrate utilization is in excess of the rate of biomass generated:

Real:
$$-\frac{dS}{dt} = \frac{1}{Y}\frac{dX}{dt} \tag{12.13}$$

where $\qquad\qquad\qquad$ $Y =$ fraction of substrate converted to biomass,
$\qquad\qquad\qquad\qquad$ mg/L of biomass/mg/L of substrate
$\qquad\qquad\qquad\qquad$ typically 0.4 to 0.8 in aerobic systems and
$\qquad\qquad\qquad\qquad$ 0.08 to 0.2 for anaerobic systems

$$= \frac{\mathrm{d}X}{\mathrm{d}S}$$

Substitution into Eq. (12.12) becomes

Substrate utilization : $\qquad\qquad\qquad -\frac{\mathrm{d}S}{\mathrm{d}t} = \frac{1}{Y}\left(\frac{\mu_\mathrm{m}\,SX}{K_\mathrm{s}+S}\right)$ $\qquad\qquad$ (12.14)

Equation (12.11) for biomass production and Eq. (12.14) for substrate utilization are the fundamental biological process design equations for different reaction configurations. They are used later in the mass balance equations for complete mix and plug flow activated sludge systems. They are also the principal equations in anaerobic digestion (Chapter 13) (see Moletta *et al.*, 1986).

12.7.4 Secondary Treatment Systems

Secondary treatment systems are broadly categorized as:

- Suspended growth
- Attached growth
- Dual biological suspended and attached growth

Suspended growth systems are defined as those aerobic processes that achieve a high micro-organism concentration through the recycle of biological solids. The bacterial organisms convert biodegradable organic wastewaters and certain inorganic fractions into new biomass and other (non-polluting) end products (e.g. water and carbon dioxide). The biomass is removed as sludge and the liquid after settling is removed as clarified effluent. The gases are air stripped. Suspended growth systems and, in particular, the conventional plug flow activated sludge system are the most common processes for treating both municipal and industrial wastewaters.

$\qquad$ Attached growth systems or fixed film reactors allow a microbial layer to grow on the surface of the media (stone, plastic) while exposed to the atmosphere from where it draws its oxygen. The microbial layer is sprayed (from above) with the wastewater. In so doing, the microbial layer converts the biodegradable organic wastewaters to biomass and by-products. The trickling or percolating filter is common for treating municipal wastewaters which already have been through primary settling. Rotating biological contactors and biofilters are in operation for higher strength wastes than municipal. They are sometimes used as roughing filters with subsequent activated sludge or stone media trickling filters for further treatment.

$\qquad$ Dual process systems utilize two stage arrangements of fixed film and suspended growth processes with the objective of achieving very high quality effluent standards. Figure 12.21 itemizes some of the different secondary treatment processes.

12.7.5 The F/M ratio

The F/M (food to microbes) ratio is the most useful design and operational parameter of activated sludge systems. The activated sludge system is a continuous process with growth and decay of micro-organisms. A system achieves equilibrium when the food substrate and the micro-organisms consuming it are in balance. Out of balance can mean too much substrate, too little substrate, too many organisms, too little organisms, etc. The equilibrium parameter is known as the F/M ratio or the food to microbes ratio. This

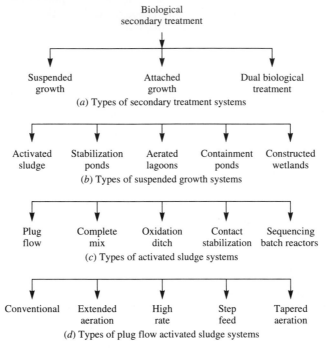

Figure 12.21 Secondary treatment systems.

ratio controls the rate of biological oxidation and the mass of organisms, by maintaining microbial growth in either the log, declining or endogenous phase (Gray, 1990).

The type of activated sludge system can be defined by its F/M ratio (see Fig. 12.22)

- Extended aeration $0.03 < F/M < 0.8$
- Conventional $0.8 < F/M < 2$
- High rate $F/M > 2$

The F/M ratio is defined as

$$F/M = \frac{\text{BOD of sewage (kg/m}^3) \times \text{Influent flow (m}^3/\text{d})}{\text{Reactor solids (kg/m}^3) \times \text{reactor volume (m}^3)}$$

$$= \frac{S_o Q_o}{XV}$$

Therefore

$$F/M = \frac{S_o}{\phi X} \tag{12.15}$$

where

S_o = Concentration of influent BOD, (kg/m^3)

Q_o = Influent flow rate, (m^3/day)

X = concentration of reactor solids i.e. MLSS, (kg/m^3)

V = reactor volume, (m^3)

ϕ = hydraulic retention time, (days)

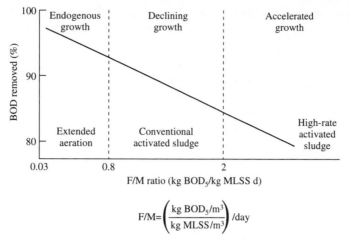

Figure 12.22 Schematic of F/M ratio (not to scale).

In the log or accelerated growth phase there is an excess of substrate, characterizing a high f/m ratio (> 1.0). In the endogenous phase the f/m ratio is low at values generally less than 0.4 and ideally at around 0.2 for plug flow and 0.1 for complete mix systems. Removal rates of BOD are then highest, and this is conventionally called extended aeration (see Fig. 12.22).

12.7.6 Sludge Settlement Parameters

In the process design of activated sludge, a principal feature is the use of recycle sludge. This is where a fraction of the sludge that has settled in the secondary clarifier is used as biomass feedstock for the aeration basin. Note that the secondary clarified sludge is low in organic content and is not a food substrate. It is high in biomass content and is therefore suitable for recycle. If the aeration basin did not receive this recycle sludge, it would ultimately become deficient in micro-organism population. One parameter used in the control of the amount of recycle sludge is the sludge volume index (SVI), determined in a laboratory test, using a 1 L inverted conical flask:

$$SVI = \frac{SV}{MLSS} \times 1000 \qquad (12.16)$$

where
$$SVI = \text{sludge volume index, mL/g}$$
$$SV = \text{volume of settled solids in 1 litre cylinder}$$
$$\text{after 30 min, ml/L}$$
$$MLSS = \text{mixed liquor suspended solids, mg/L}$$

SV is taken from a sample of mixed liquor taken near the exit weir of the aeration basin. MLSS is from a similar sample and is dried out and weighed. A sludge has good settling characteristics if the SVI range is 80 to 120 for an MLSS range of 2000 to 3500 mg/L. As the MLSS increase, the solids loading is higher, producing a lower SVI. An alternative to a lower SVI is to increase the aeration basin size.

12.7.7 Nitrification and Denitrification

The previous sections discussed the stabilization of carbonaceous matter, traditionally the singular objective of wastewater treatment. The nutrients, nitrogen and phosphorus, contribute to eutrophication of lakes and slow moving waters. Therefore, most countries legislate for the removal of nitrogen and phosphorus. As discussed in Chapter 10, the nitrogen cycle in simple terms is

$$N_2 \rightarrow \text{organic N} \rightarrow \underbrace{\overset{\text{ammonia N}}{(NH_3^+ - N)} \xrightarrow{} \overset{\text{nitrite}}{(NO_2^- - N)}}_{\text{Nitrification}} \rightarrow \underbrace{\overset{\text{N}}{(NO_3^- - N)} \rightarrow N_2}_{\text{Denitrification}} \qquad (12.17)$$

The objective of the wastewater processes of nitrification/denitrification is to stabilize the organic N and ammonia N in wastewater, first to nitrate N and secondly to nitrogen gas. Nitrogen in wastewater is generally in the forms of organic N and ammonia N in both soluble and particulate forms. Wastewater contains insignificant amounts of nitrite N and nitrate N. Organic N and ammonia N are undesirable in wastewater effluents since they both have a nitrogenous oxygen demand and ammonia N is toxic to fish life (see Chapters 3 and 7).

Nitrification is the biological process, using nitrifying bacteria (nitrosomonas and nitrobacter) to convert ammonia N to nitrate N in two process steps as follows:

$$NH_4^+ + \tfrac{3}{2}O_2 \xrightarrow[\text{bacteria}]{\text{nitrosomonas}} NO_2^- + 2H^+ + H_2O \qquad (12.18)$$

$$NO_2^- + \tfrac{1}{2}O_2 \xrightarrow[\text{bacteria}]{\text{nitrobacter}} NO_3^- \qquad (12.19)$$

Generally, many environmental authorities are satisfied if wastewater is treated to the level of nitrification. However, the end product of nitrification, nitrate N, still has potential negative impact on receiving water quality. Nitrate N leads to the stimulation of algae growth and is also associated with the infant disease of methanoglobinaemia. As such, many water authorities now require a denitrification process to reduce nitrate N to the inoffensive nitrogen gas (N_2). This process is called denitrification and like nitrification can be attained by biological means, although with some difficulty.

The operation of biological nitrification/denitrification is dependent on maintaining a healthy population of nitrifying/denitrifying bacteria. Each of these bacteria requires different 'climatic' conditions. Conventional activated sludge processes did little for nitrification and usually nothing for denitrification. Extended aeration, however, because of its long hydraulic retention time (HRT) and low oxygen levels (near to anoxic state) towards the downstream end of the oxidation ditch achieves a level of nitrification. As indicated in Sec. 12.10, nutrient removal, not only of nitrogen but also of phosphorus, is achievable purely by biological processes.

In fixed film systems (e.g. percolating filters, see Sec. 12.9) it is also possible to design the process for biological nitrification, but biological denitrification and phosphorus removal have not yet been economically achievable for trickling filters.

12.8 ACTIVATED SLUDGE SYSTEMS

The more common activated sludge systems are:

- Complete mix
- Plug flow
- Oxidation ditch
- Contact stabilization
- Sequencing batch reactors

12.8.1 Complete Mix Reactors

The complete mix reactors have uniform characteristics throughout the entire reactor. They tend to be circular or square in shape and rarely rectangular. Aeration can be provided by surface aerators with adjustable outflow weirs used to vary the depth of submergence of the aerator or by submerged bubble diffuser aeration systems. The effluent quality is the same as the wastewater quality in the basin. As such only a low level of food is available to the microbes. The operating food substrate to microbes ratio (F/M) ranges from 0.04 to 0.07. The volumetric loading is typically less than 1 kg BOD_5/day/m³. Dissolved oxygen (DO) levels are maintained throughout at not less than 2 mg/L. The latter is difficult to maintain, particularly in large-diameter tanks remote from the surface aerator. Generally the returned activated sludge (RAS) from the clarifier is fed directly to the aeration basin where it is completely mixed with the existing contents. A less desirable route for the RAS is to join with the influent stream and mix in the influent pipe before entry to the aeration basin. Advantages of the complete mix system include the ability to withstand shock loads, due to the low F/M ratio, and good flexibility in being able to utilize a wide range of loads. It is particularly suitable for high organic industrial wastewater. Figure 12.23 is a schematic of the complete mix system.

Using the nomenclature of the AWWA (1992), the system of Fig. 12.23 consists of the following:

- Influent stream
- Aeration basin
- Secondary clarifier
- Clarifier effluent
- Sludge waste
- Sludge return to aeration basin

where
Q = flow rate, m³/d
S = substrate food concentration, kg/m³ or mg/L
X = biomass concentration (biological solids), mg/L
V = aeration tank volume, m³

For Fig. 12.23, mass balance equations for biomass production (Eq. (12.11)) and for substrate utilization (Eq. (12.14)) are as follows:

1. *Mass balance of biomass production*   This uses the control volume of the aeration tank and clarifier (shown by dashed lines in Fig. 12.23):

$$\underset{Q_oX_o}{\underset{\text{biomass}}{\text{Influent}}} + \underset{V\frac{dx}{dt}}{\underset{\text{production}}{\text{biomass}}} = \underset{(Q_o - Q_w)X_e}{\underset{\text{biomass}}{\text{effluent}}} + \underset{Q_wX_w}{\underset{\text{biomass}}{\text{sludge wasted}}} \tag{12.20}$$

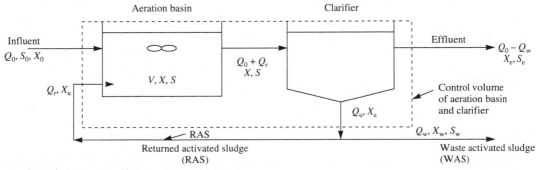

Figure 12.23 Typical complete mix activated sludge system.

Substitute Eq. (12.11):

$$Q_oX_o + V\left(\frac{\mu_mSX}{K_s + S} - k_dX\right) = (Q_o - Q_w)X_e + Q_wX_w \tag{12.21}$$

Assume that the biomass in influent and effluent levels is negligible, i.e. $X_o = X_e = 0$:

$$\frac{\mu_mS}{K_s + S} = \frac{Q_wX_w}{VX} + k_d \tag{12.22}$$

2. *Mass balance of food substrate*   This uses the control volume of the aeration tank and clarifier:

$$
\begin{array}{ccccc}
\text{Influent} & + & \text{substrate} & = & \text{effluent} & + & \text{sludge wasted} \\
\text{substrate} & & \text{consumed} & & \text{substrate} & & \text{substrate} \\
Q_oS_o & & V\dfrac{dS}{dt} & & (Q_o - Q_w)S_e & & Q_wS_w
\end{array}
\tag{12.23}
$$

Substitute Eq. (12.14):

$$Q_oS_o - V\left[\frac{1}{Y}\left(\frac{\mu_mXS}{K_s + S}\right)\right] = (Q_o - Q_w)S_e + Q_wS_w \tag{12.24}$$

Assume that all reactions take place in the aeration basin so that the substrate in the aeration basin is of the same concentration as the substrate in the clarifier and in the effluent (i.e. $S_e = S_w = S$). Rearranging Eq. (12.24) we get

$$\frac{\mu_mS}{K_s + S} = \frac{Q_oY}{VX}(S_o - S) \tag{12.25}$$

Combining Eqs. (12.22) and (12.25):

$$\frac{Q_wX_w}{VX} + k_d = \frac{Q_oY}{VX}(S_o - S) \tag{12.26}$$

Define the hydraulic retention times (HRT) of influent in the aeration basin:

$$\phi = \frac{V}{Q_o}, \text{ units of time} \tag{12.27}$$

Define the mean cell residence time (MCRT) of micro-organisms in the system, also called the solids retention time (SRT) or sludge age, as ϕ_c:

$$\phi_c = \text{total biomass in basin/biomass wasted per unit time}$$

$$\phi_c = \frac{VX}{Q_wX_w}, \qquad \text{units of time} \tag{12.28}$$

As a fraction of the sludge from the clarifier is returned to the aeration, basin, $\phi_c > \phi$. Equation (12.26) becomes

$$\frac{1}{\phi_c} + k_d = \frac{1}{\phi}\frac{Y}{X}(S_o - S) \tag{12.29}$$

Rearranging:

$$X = \frac{\phi_c}{\phi}Y\left(\frac{S_o - S}{1 + k_d\phi_c}\right) \tag{12.30}$$

Equation (12.30) is the concentration of biomass solids in the aeration basin or the mixed liquor suspended solids (MLSS).

A term commonly used is the F/M ratio, or the ratio of food to micro-organisms. Low F/M ratios (introduced in Sec. 12.7.5) as occur in complete mix systems are resistant to shock loads. It is determined as follows:

From Eq. (12.15),

$$F/M = \frac{S_o}{\phi X} \qquad \text{mg BOD applied/mg MLSS/day}$$

$$F/M = \frac{S_o}{(V/Q_o)X} = \frac{Q_o S_o}{VX} \tag{12.31}$$

Example 12.8   In a completely mixed activated sludge system determine:

(a) The aeration basin volume V
(b) The hydraulic retention time (ϕ)
(c) The sludge volume wasted daily Q_w
(d) The mass of sludge wasted daily $Q_w X_w$
(e) The fraction of sludge recycled Q_r/Q_o
(f) The F/M ratio

Given:

- Population equivalent 50 000 (11 250 m³/day)
- Influent BOD_5 (S_o) = 200 mg/L
- Required effluent $BOD_5 \not> 10$ mg/L
- Yield coefficient $Y = 0.6$
- Decay rate $k_d = 0.06$ d⁻¹

Assume:

- MLSS in aeration basin (X) = 3500 mg/L (3.5 kg/m³)
- MLSS in clarifier sludge (X_w) = 15 000 mg/L (15 kg/m³)
- Mean cell residence time (ϕ_c) = 10 days

Solution

(a) Aeration basin volume

Equation (12.30):
$$X = \frac{\phi_c}{\phi} Y \left(\frac{S_o - S}{1 + k_d \phi_c} \right)$$

Equation (12.27):
$$\phi = \frac{V}{Q_o}$$

Therefore

$$X = \frac{\phi_c}{V} Q_o Y \left(\frac{S_o - S}{1 + k_d \phi_c} \right)$$

and
$$V = \frac{\phi_c Q_o Y}{X} \left(\frac{S_o - S}{1 + k_d \phi_c} \right)$$

Consistent units of m³, kg, day:

$$V = \frac{10 \times 11\,250 \times 0.6}{3.5} \left(\frac{0.2 - 0.01}{1 + 0.06 \times 10} \right)$$

$$= 2290 \text{ m}^3$$

(b) Hydraulic retention time

$$\phi = \frac{V}{Q_o} = \frac{2290}{11\,250}$$

Therefore

$$\phi = 0.2 \text{ days} = 4.9 \text{ hours}$$

(c) Volume of sludge wasted daily (Q_w)

Equation (12.28):

$$\phi_c = \frac{VX}{Q_w X_w}$$

Therefore

$$\begin{aligned} Q_w &= \frac{VX}{\phi_c X_w} \\ &= \frac{2290 \times 3.5}{10 \times 15} \\ &= 53.4 \text{ m}^3/\text{day} \end{aligned}$$

(d) Mass of sludge wasted daily ($Q_w X_w$)

$$Q_w X_w = 53.4 \times 15 = 801 \text{ kg/d}$$

(e) The fraction of sludge recycled (Q_r/Q_o)

A biomass mass balance around the clarifier (only) gives

$$\begin{aligned} (Q_o + Q_r)X &= (Q_o - Q_w)X_e + Q_u X_w \\ &= (Q_o - Q_w)X_e + (Q_r + Q_w)X_w \end{aligned}$$

Assume that the biomass concentration in the effluent is zero ($X_e = 0$) and rearranging that

$$\begin{aligned} Q_r &= \frac{Q_o X - Q_w X_w}{X_w - X} \\ &= \frac{11\,250 \times 3.5 - 801}{15 - 3.5} = 3354 \text{ m}^3/\text{d} \\ \frac{Q_r}{Q_o} &= \frac{3354}{11\,250} = 29.8\% \end{aligned}$$

(f) The F/M ratio

Equation (12.31):

$$\begin{aligned} F/M &= \frac{Q_o S_o}{VX} \\ &= \frac{11\,250 \times 0.2}{2290 \times 3.5} \\ &= 0.28 \end{aligned}$$

12.8.2 Plug Flow Reactors

Plug flow means that a 'plug' of substrate influent to an aeration basin is moved forward, without too much interaction with the 'plug' that went before or after it. This means that satisfactory mixing occurs in the lateral direction, but none in the longitudinal direction. Figure 12.24 is a schematic of a plug flow reactor. There is a (BOD) concentration gradient from entry to exit.

The plug flow system is characterized by a high organic loading at the influent end of the basin, reducing as the outfall weir is approached. The aeration basin is typically rectangular or elongated with oval returns (see Fig. 12.26). At the influent end there is an excess of food substrate corresponding to the log growth phase and a high F/M ratio. At the downstream end, there is a shortage of food substrate and the micro-organisms are in the endogenous phase. Through the aeration basin the food substrate decreases while the micro-organism concentration increases. This makes the analysis of kinetics more complex than the complete mix system and in the past, simplifying assumptions were introduced. These include the assumption that the concentration of micro-organisms in the influent to the basin is equal to that in the effluent from the basin. Advantages of plug flow include the ability to treat fully all influent and allow no 'plugs' to go untreated. This is so because the influent typically stays in the aeration basin for much longer periods than in the case of the complete mix system. Dimensions of units tend to be long rectangular with length to width ratios of about 10 to 1 and depths about 2 to 4 m. Figure 12.25 is a layout of a plug flow activated sludge system. As with the complete mix system, the nomenclature is that of the AWWA (1992).

As with the complete mix system of the previous section, it is possible using mass balances of the biomass and the substrate to determine the design and operating parameters. However, this is only possible with the introduction of the simplification that the biomass concentration of influent and effluent of aeration basin is the same, $\overline{X}$. The resulting parameters for plug flow with recycle from Lawrence and McCarty (1970) are:

$$\overline{X} = \frac{\phi_c}{\phi} \; Y\left(\frac{S_o - S}{1 + k_d \phi_c}\right) \tag{12.32}$$

$$\frac{1}{\phi_c} = \frac{\mu m(S_o - S)}{(S_o - S) + (1 - \alpha)\,[K_s \, \ln(S_i/S)]} \tag{12.33}$$

where
$\qquad \alpha =$ recycle ratio

$$S_i = \frac{S_o + \alpha S}{1 + \alpha} \tag{12.34}$$

influent BOD after dilution with recycle flow

Further details are in Schroeder (1977) and Benefield and Randall (1980).

Oxidation ditch   A typical plan configuration of an oxidation ditch is shown in Fig. 12.26. The aeration basin is usually a 'racetrack' configuration with cage or brush aerators at one or more locations. The

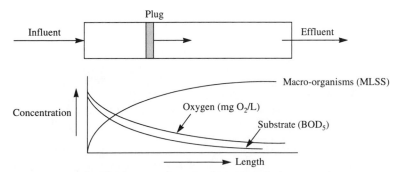

Figure 12.24 Schematic of plug flow system.

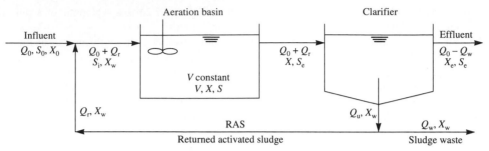

Figure 12.25 Typical plug flow activated sludge system.

influent enters the basin just upstream of the aerator and moves forward as a 'plug', from a location of high oxygen levels to a zone of low oxygen levels. The effluent leaves the basin upstream of the influent. Sludge is recycled, as is typical with activated sludge. Oxidation ditches are characterized by long HRTs ($\sim$24 h) and long SRTs or MCRTs (20 to 30 days). High-quality liquid effluents are also characteristic of oxidation ditches, but a disadvantage is a thin sludge which is difficult to thicken.

Extended aeration, which can be adopted with complete mix systems, is more common with oxidation ditch systems. Extended aeration means long HRTs (16 to 24 h), enabling variable flow rates and flow strengths to be handled. The endogenous phase is the operational phase of the micro-organisms and while nitrification is common, it is also possible to achieve denitrification.

Extended aeration in complete mix reactors is only used for small plants due to the increased size of aeration basin. Primary clarification is omitted as the key objective is to minimize the amount of sludge generated. Ideally, all substrate removal is converted to energy and then oxidized so that no excess biomass is generated and so no sludge handling is required. In practice, some sludge is generated. Based on zero growth then:

$$\left(\frac{dX}{dt}\right)_g = 0 \qquad (12.35)$$

where the subscript 'g' is for generation. Benefield and Randall (1980) derive the equation for the extended aeration tank volume as

$$(V)_{EA} = \frac{YQ_o(S_o - S)}{Xk_d} \qquad (12.36)$$

Example 12.9 Determine the size of an extended aeration basin to treat 30 000 m³/day if the influent BOD is 300 mg/L. The effluent BOD concentration is expected to be 5 mg/L. Given $X = 4$ kg/m³, $Y = 0.4$, $k_d = 0.03$ day^{-1}.

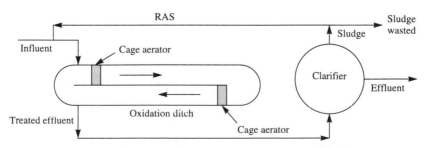

Figure 12.26 Typical layout of an oxidation ditch system.

Solution

$$V_{\text{EA}} = \frac{0.4 \times 30\,000\,(300 - 5)}{4000 \times 0.03}$$
$$= 29\,500 \text{ m}^3$$

Contact stabilization   This system (Fig. 12.27) is a form of activated sludge where aeration is carried out in two phases in two different tanks. In the contact tank the suspended organic matter is adsorbed by the microbial mass and the dissolved organic matter is absorbed by the biomass. The retention time is 30 to 60 min. The second tank, called the stabilization tank, with a 2 to 3 h retention period, is where the solids have been removed in a settlement zone and are further stabilized by re-aeration before being combined with the influent wastewater. The MLSS tends to be about 2000 mg/L in the contact tank and up to 20 000 mg/L in the aeration basin.

The aeration volume requirements are typically 50 per cent lower than conventional plug flow, the model on which it is based. The BOD removal efficiency is in the range 80 to 90 per cent. This system is used for expansion of existing systems (where space is a limitation) and also in package plants (Metcalf and Eddy, 1991).

Sequencing batch reactors (SBRs)   A sequencing batch reactor is a complete mix activated sludge system without a secondary clarifier. Within the single aeration basin, five different sequences are followed, as shown in Fig. 12.28. Aeration and clarification are carried out in one tank. The sequences are as follows:

- Fill   The basin fills with influent
- React   The basin is aerated when 100 per cent full
- Settle   The basin is allowed to settle where sedimentation and clarification occurs
- Draw   The effluent is withdrawn from the top of the tank
- Sludge waste The sludge is wasted from the bottom of the tank

The tank volume, cycle times, cycle purposes and cycles of aeration are also shown in Fig. 12.28. The cycle length varies from 4 to 48 h with SRTs from 15 to 80 days (AWWA, 1992). The F/M ratio varies with cycle length and may range from 0.03 to 0.18. An advantage of the SBR system is that there is no need for a sludge recycle period. Details of SBRs are found in USEPA (1986).

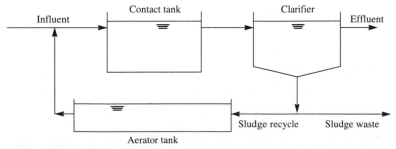

Figure 12.27 Layout of contact stabilization system.

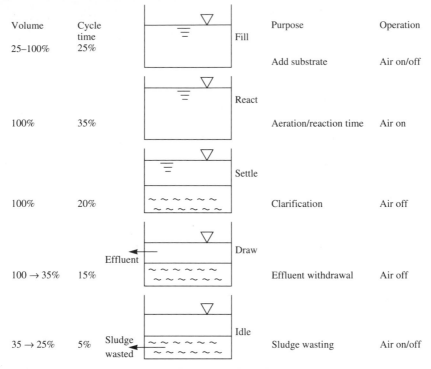

Figure 12.28 Typical configurations for an SBR (adapted from USEPA, 1986).

12.9 ATTACHED GROWTH SYSTEMS

When organic wastewater is 'sprayed' over stones or plastic, a microbial slime layer develops on the surface. This layer has the effect of reducing the BOD_5 of effluents. Traditionally the stones were grouped in a shallow, open-topped cylinder about 1 m deep. The stones were of the order of 25 to 100 mm in size. In more recent times, plastic media have been used instead of stones to encourage the growth of the microbial layer on media with very high surface area to volume ratios. This type of aerobic process is called the attached growth or fixed film system. The earliest version which is still widely used is the percolating or trickling filter. Other systems are the rotating biological contactor, which is used for both low strength and high-strength wastes. In Ireland, effluents from the dairy milk producing industry ($BOD_5 \sim 2000$ mg/L) predominantly use biofilters with plastic media. The latter are often called high-rate biofilters, and are used as roughing filters to reduce the BOD_5 from about 2000 to 300 mg/L, after which it is treated in a conventional activated sludge system. Some of the biofilters are made up of a very lightweight synthetic media called flocor, with media heights up to 6 m. Other variations include the more recent adoption of the aerated biofilter.

12.9.1 Percolating Filters

Percolating or trickling filters were traditionally cylindrical or rectangular boxes of concrete or steel, containing stone media. The media tended to be angular rather than rounded and, as such, limestone was a popular choice. Tank dimensions ranged from 1 to 2.5 m deep and 5 to 50 m in diameter. The floor of the tank has an underdrain system for collecting the treated wastewater. Near the floor of the perimeter walls are openings, to allow an updraught of air for aeration. A schematic is shown in Fig. 12.29. The treated effluent passes to a secondary clarifier. In principle, since the biomass layer is fixed to the stone media, there is no need for recycle sludge as in activated sludge. At the stone media/biomass/air interfaces, aerobic degradation of the substrate takes place. At deeper levels, there may also be biodegradation due to anaerobic processes. As the organic load continues to be sprayed on the media, a buildup of biomass occurs, in excess of the most efficient levels. At this point, some biomass gets washed forward with the effluent. This is known as biomass 'sloughing' and is identified in high BOD_5 readings in the effluent. Recycling of effluent through the filters improves the end quality effluent and, in the case of low-rate units, achieves nitrification.

Maintaining a healthy community of micro-organisms in the trickling filter is essential as it is activated sludge. The significant micro-organisms are essentially facultative bacteria, but fungi, protozoa, algae, worms, insects and snails also reside in the habitat. The bacteria residing at the surface of the filter process the BOD reduction. The nitrifying bacteria resident at lower depths process nitrification.

The factors affecting operation of the process and design are identified by the AWWA (1992) as:

- Wastewater composition and treatability
- Media type and depth
- Hydraulic and organic loading
- Recirculation ratio and arrangement
- Temperature
- Distributor operation

Trickling filters are versatile systems, being able to treat low-strength wastes to advanced standards or to act as a roughing filter to high strength wastes. However, if there is a high recirculation ratio, it is theoretically possible to treat high-strength wastes also to advanced standards. As retention times in trickling filters are short by comparison with conventional activated sludge, it also happens that slowly degradable wastes are partly treatable by trickling filters. This can be the case of wastes with non-flocculent type suspended solids. However, the system is capable of treating soluble organics efficiently and as such is suited to many industrial wastewaters, i.e. milk wastes. The media type varies from dense stone media with shallow depths of about 1 m to lightweight plastic media with depths up to 6 m. The stone media are typically used for low-strength wastes such as municipal and the plastic media typically for industrial wastes, although stone media were used in the past. The hydraulic loading and organic loading vary, depending on the type of filter. Some loading rates and other characteristics are shown in Table 12.13.

Recirculation will normally be the case for all trickling filters except the low rate and the roughing filter. Recirculation ratios may vary from 0.4 to 4, depending on the quality of treatment required. Recirculation may be directly from the effluent of the trickling filter or from the secondary clarifier. Recirculation may re-enter the influent pipe before the filter or directly on to the filter. Which system is chosen will depend on the waste characteristics, loadings, effluent standards and economics.

Table 12.13 Characteristics of trickling filters

Design characteristics	Low rate— conventional	Intermediate rate	High rate	Super rate	Roughing filter
Media	Stone	Stone	Stone	Plastic	Plastic/Stone
Hydraulic loading ($m^3/d/m^2$)	10 000– 40 000	40 000– 100 000	100 000– 400 000	150 000– 900 000	600 000– 1 800 000
Organic loading (kg $BOD_5/d/m^3$)	1–3	3–6	6–12	< 30	> 20
% BOD removal	80–85	50–70	40–80	65–85	40–85
Nitrification	Yes	Some	No	Little	No

Adapted from AWWA, 1992

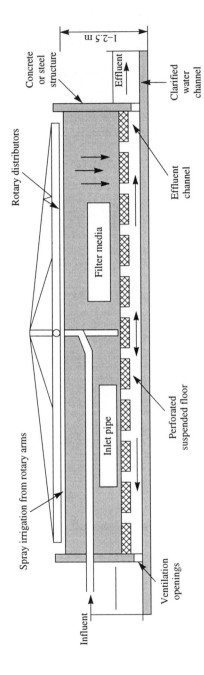

Figure 12.29 Section schematic of percolating stone filter.

Temperature, as with other secondary treatment processes, affects performance. At low winter temperatures, reduced efficiencies occur. At higher summer temperatures efficiency increases due to increased bacterial activity. Filter efficiency and temperature are expressed in the empirical equation

$$E_t = E_{20}\, a^{t-20} \qquad (12.37)$$

where
$$E_t = \text{filter efficiency at temperature } t$$
$$E_{20} = \text{filter efficiency at } 20\,^\circ\text{C}$$
$$a = \text{constant of } 1.035$$

Traditionally, the operation of the distribution arms were at rates of about 1 rev/min with spraying about every 30 s. It has since been established that more efficient operation occurs at spraying frequencies of > 30 min with much slower distributor rotations. In the latter case, problems with flies on the surface and microbial 'sloughing' are reduced.

Process modelling of trickling filters is not well advanced, particularly for stone media, because of the irregularity of the media. More precise process modelling exists for synthetic media. In practice one or other of the many available empirical models is used when designing trickling filter systems. Two particular empirical models are:

1. The National Research Council (US) Model (1948):

$$E = \frac{100}{1 + 0.448\sqrt{W/VF}} \quad \% \qquad (12.38)$$

where
$$E = \text{efficiency of BOD removal}$$
$$W = \text{influent BOD, kg/d}$$
$$V = \text{filter removal, m}^3$$
$$F = \text{recirculation factor} = (1 + R)/(1 + 0.1\,R)^2$$
$$R = Q_r/Q = \text{recirculation flow/wastewater flow}$$

2. The British Manual of Practices Model (1988):

$$\frac{L_\text{I}}{L_\text{o}} = \frac{1}{[1 + K\alpha^{t-15}(As^m/Q^n)]} \qquad (12.39)$$

where
$$L_\text{I} = \text{influent BOD, mg/L}$$
$$L_\text{o} = \text{settled effluent BOD, mg/L}$$
$$K = \text{rate coefficient (0.02 for stone, 0.4 for plastic)}$$
$$As^m = \text{media surface area and coefficient, m}^2/\text{m}^3$$
$$(m = 1.41 \text{ for stone, 0.73 for plastic})$$
$$Q^n = \text{volumetric loading rate and coefficient m}^2/\text{m}^3/\text{d}$$
$$(n = 1.25 \text{ for stone, 1.4 for plastic})$$
$$\alpha = \text{temperature coefficient (1.111)}$$
$$t = \text{temperature of wastewater, }^\circ\text{C}$$

Example 12.10   A municipal wastewater has the following characteristics:

$$\text{Influent BOD}_5 = 360\,\text{mg/L}$$
$$\text{Required effluent standard} = 25\,\text{mg/L}$$
$$\text{Population equivalent} = 20\,000 \text{ at } 225\,\text{L/h/d}$$
$$\text{Wastewater temperature} = 20\,^\circ\text{C}$$

Determine the volume of a single-stage stone trickling filter if the recirculation ratio is 1:1 and 2:1. Use the NRC equation.

Solution

$$\text{Efficiency of BOD removal required} = E$$
$$= \frac{360 - 25}{360} = 93\%$$

Using the NRC equation:

For $R = 1$:   Recirculation factor $F = \dfrac{1+R}{(1+0.1R)^2} = \dfrac{1+1}{(1+0.1)^2} = 1.65$

$$\text{BOD}_5 \text{ influent loading} = 20\,000 \times 225 \times 360 \times 10^{-6} = 1620 \text{ kg/d}$$

$$\text{NRC equation } E = \frac{100}{1 + 0.448\sqrt{W/VF}}$$

$$93 = \frac{100}{1 + 0.448\sqrt{1620/V(1.65)}}$$

$$93 = \frac{100}{1 + 14.04\sqrt{1/V}}$$

$$V = 34\,793 \text{ m}^3$$

If efficiency was 80% then $V = 43\,491 \text{ m}^2$

For $R = 2$:   Recirculation factor $F = \dfrac{1+2}{(1+0.2)^2} = 2.083$

$$\text{NRC equation } 93 = \frac{100}{1 + 0.448\sqrt{1620/V\,(2.083)}}$$

$$V = 27\,550 \text{ m}^3$$

If efficiency was 80 per cent then $V = 31\,212 \text{ m}^2$ or say 8 units of depth 2.5 m and diameter 40 m.

12.9.2 Biotowers

Biotowers use plastic media and achieve a superior hydraulic and organic loading rate to stone trickling filters. They are used primarily for high-strength industrial wastes and are in common use in the dairy milk industry. The design can be based on the following equation (Eckenfelder and Barnhart, 1963):

$$\frac{S_e}{S_o} = \exp^{KD/Q^n} \tag{12.40}$$

where
S_e = effluent substrate concentration, BOD_5, mg/L
S_o = influent substrate concentration, BOD_5, mg/L
D = depth of medium
Q = hydraulic loading rate, $\text{m}^3/\text{m}^2/\text{s}$
k = treatability constant
n = medium characteristic constant

12.9.3 Rotating Biological Contactors

A cylinder with its axis horizontal rotates into and out of the semicircular wastewater holding tank, as shown in Fig. 12.30. The rotating cylinder or reactor is made of high-density plastic. As the cylinder rotates, it builds up on its surface a film of biomass. The wastewater comes into contact with the biomass and with the air. The wastewater trickles down the surface of the contactor and absorbs oxygen from the air. The micro-organisms in the biomass remove organic matter from the wastewater film. The process gained favour in the 1960s and 1970s but has recently fallen out of favour due to problems with excess biomass buildup and structural problems of shafts and media.

12.10 NUTRIENT REMOVAL

In discussing nutrient removal from wastewater, nitrogen and phosphorus are the key nutrients as they cause 'pollution' of the receiving water body. As nitrogen travels through the N cycle it is a water pollutant in four of its oxidation states, which are biochemically interconnectable:

$$\text{Organic N} \rightarrow \text{ammonia N} \rightarrow \text{nitrite N} \rightarrow \text{nitrate N}$$

It is desirable (and regulated) to reduce the 'total nitrogen' (in municipal wastewaters from ~ 40 to $\sim 10\,mg/L$) as defined by

$$\text{Total nitrogen} = \text{organic N} + \text{ammonia N} + \text{nitrite N} + \text{nitrate N}$$

and
$$\text{Kjeldahl N} = \text{organic N} + \text{ammonia N}$$

In wastewaters, there is almost no nitrite and no nitrate, so the total N can be approximated by the Kjeldahl N. However, the process of wastewater treatment, activated sludge, etc., does oxidize the organic N and ammonia N to the higher oxidation states of nitrite and nitrate. Nitrite N is rather unstable and oxidizes easily to nitrate N. It is desirable to further oxidize the nitrate N to nitrogen gas (N_2), which is generally inoffensive. The process of conversion of ammonia N to nitrate N is called nitrification, and takes place in secondary treatment wastewater systems in the presence of nitrifying bacteria (nitrosomonas and nitrobacter). The process of conversion of nitrate N to nitrogen gas (N_2) is called denitrification and takes place in the presence of denitrifying bacteria. Nitrification and denitrification will only work in ideal conditions, specific to each one. Both require different (oxygen level and bacterial) environments, though possibly in the same tank but at different times. The ideal conditions are characterized by:

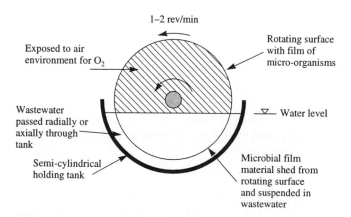

Figure 12.30 Schematic of a rotating biological contactor.

- Dissolved oxygen levels
- Organic loading availability

Like nitrogen, the nutrient phosphorus also has a cycle, as discussed in Chapter 10. Phosphorus occurs in solution, in particles or detritus or in the bodies of micro-organisms, as:

- Ortho-phosphates (PO_4^{3-}, HPO_4^{2-}, $H_2PO_4^{-}$, $H_3PO_4^{-}$)
- Polyphosphate (P_2O_7)   ~ 70 per cent of wastewater phosphorus
- Organically bound P

The sources of phosphorus include:

- Detergents, domestic and commercial
- Land fertilizers
- Potable water treatment additives
- Human excreta in sewage
- Food residues in sewage

Like nitrogen, phosphorus is a water pollutant if it exists in a growth-limiting water environment. It can stimulate the growth of photosynthetic algae, thereby causing eutrophication of lakes or slow-moving rivers. As such, it is desirable to reduce the 'total phosphorus' in municipal wastewaters from ~ 10 to < 2 mg/LP. Traditionally, phosphorus was removed from wastewaters by chemical precipitation resulting from the addition of metal salts, i.e. ferric chloride. Currently phosphorus removal by biological processes is more desirable.

Biological treatment, such as conventional activated sludge converts most P to the *ortho*-phosphate forms ($H_2PO_4^{-}$, HPO_4^{2-} and PO_4^{3-}). These forms are then removed by chemical precipitation, using calcium, aluminium or ferric, such as

$$3HPO_4^{2-} + 5Ca^{2+} + 4OH^- \rightarrow Ca_5(OH)(PO_4)_3 + 3H_2O$$
$$HPO_4^{2-} + Al^{3+} \rightarrow Al\,PO_4 + H^+$$
$$HPO_4^{2-} + Fe^{3+} \rightarrow Fe\,PO_4 + H^+$$

When lime is used as the calcium source, precipitation normally follows biological treatment. With alum or iron as the precipitant, treatment may be effected in the activated sludge operation itself or even in the primary clarifier (Bailey and Ollis, 1986).

12.10.1 Biological Nutrient Removal

Biological nitrogen removal is a two-step process:

- Nitrification in an aerobic environment
- Denitrification in an anoxic environment

Biological phosphorus removal is a one-step process in alternating anaerobic and aerobic environments. Chemical phosphorus removal is a one-step process.

The more common systems include biological N removal with chemical P removal. While biological N with biological P removal systems exist, they tend to be assisted in part by chemical P removal. Many of the bio-N/P removal systems are patented.

12.10.2 Nitrification

Nitrification or ammonia oxidation depends on the population of nitrifying bacteria, which depend on temperature, ammonia concentration, organic substrate, pH, DO concentration, system mean cell residence time and the presence of nitrifying inhibitors. From AWWA (1992), for each kg of ammonia oxidized to nitrate the following occurs:

- 4.18 kg of O_2 are consumed
- 14.1 kg of alkalinity as $CaCO_3$ are destroyed
- 0.15 kg of new cells are produced (extra sludge)
- 0.09 kg of inorganic carbon are consumed

This shows that more oxygen, more alkalinity and more carbon are consumed along with the production of more micro-organism cells. The process design equations for nitrification are detailed in AWWA (1991).

12.10.3 Denitrification

Denitrification follows, on completion of nitrification. It is achieved biologically under anoxic conditions, i.e. in the absence of oxygen. The energy required for denitrification comes from the organic content of the wastewater. As such, the siting of the anoxic zone within the process should be such that adequate internal carbon sources still exist. In the event of carbon being low, external sources such as methanol are added. This is rarely required for municipal plants but may be required for some industrial wastewaters. The denitrifying organisms are primarily facultative heterotrophs (use organic carbon for all cell tissue) that reduce nitrate (NO_3^-) in the absence of molecular oxygen ($DO \sim 0.0$ mg/L). The overall denitrification process is summarized from AWWA (1992) as:

- $NO_3^- \text{-N} \rightarrow N_2$
- Oxygen recovery is 2.86 kg O_2/kg NO_3^--N reduced
- Alkalinity recovery is 3.0 kg $CaCO_3$/kg NO_3^--N
- Biomass production is ~ 0.4 kg VSS/kg COD removed

It is therefore seen that denitrification recovers oxygen and alkalinity and produces more VSS. Therefore, to proceed from nitrification on to denitrification is an economic step regarding oxygen consumption. Typically about 25 per cent less oxygen is used in fully nitrified/denitrified plants. Because denitrification requires a degradable carbon source, this is mostly supplied by recycling a large amount of nitrified effluent to the anoxic basin at the head of the process (see Fig. 12.31).

12.10.4 Combined Nitrification–Denitrification System

Rather than having to use an external carbon source for denitrification and to exploit the oxygen-saving advantages of denitrification, several nitrification–denitrification systems have evolved. Figure 12.31 is the simplest arrangement.

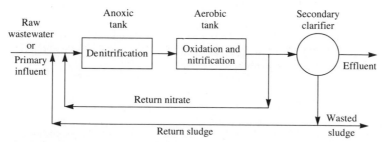

Figure 12.31 Two-stage biological nitrogen removal.

In Fig. 12.31, the raw wastewater or primary clarified effluent is first passed through an anoxic tank where there is no aeration (no mixing). Here no carbon oxidation takes place but denitrification does take place. In the first tank the characteristic of the influent is high BOD, high ammonia and high nitrate levels. Without oxygen, the bacteria for carbon oxidation and ammonia reduction to nitrate do not proliferate. The denitrifying bacteria proliferate in an environment rich in carbon and low in oxygen. In the anoxic tank there will be a minimum of mixing to obviate septicity and to maintain a forward flow. The effluent from the anoxic tank is fed to the aerobic tank where levels of oxygen are maintained in excess of 2 mg/L throughout. Here the environment is suitable for carbon oxidation with a plentiful substrate supply, a high oxygen level and dense biomass of heterotrophic bacteria. With an adequate hydraulic retention time, carbon oxidation is completed in the aerobic tank. With a slightly longer hydraulic retention time, the same aerobic environment is suitable for nitrifying bacteria where nitrification of ammonia to nitrate is completed. The well-nitrified, well-oxidized effluent from the aerobic tank is fed to the secondary clarifier for final settlement. An essential return cycle of nitrified effluent is returned to the anoxic tank for denitrification. Figure 12.32 is a schematic of the oxidation ditch for biological nitrogen removal.

Downstream of the aerators, an aerobic zone (DO > 2 mg/L) is maintained. This 'reduces' to an anoxic zone just upstream of the aerator (DO ~ 0.0 mg/L). The influent wastewater is fed to the anoxic zone where the extensive carbon source of the waste is used as the energy source for the denitrifying bacteria. The effluent is removed at the end of the aerobic zone. Nitrogen removals can be improved by the addition of a second anoxic zone, as in the Bardenpho process (see Fig. 12.35). Process design equations are referenced in Metcalf and Eddy (1991).

12.10.5 Biological Phosphorus Removal

Biological phosphorus removal is based on the idea of forcing the micro-organism to accumulate more phosphorus than is required for cell growth. Typically, wastewater biomass has bacteria cells composed of mainly carbon ($\sim$ 50 per cent), oxygen ($\sim$ 20 per cent), nitrogen ($\sim$ 15 per cent), hydrogen ($\sim$ 10 per cent) and phosphorus (3 per cent). In other words, there appears to be a limit to the amount of phosphorus that can be taken up by the bacteria. Conventional secondary treatment processes remove about 20 per cent of phosphorus influent levels and primary treatment removes about 10 per cent. Raw influent total phosphorus levels are about 10 mg/L for municipal wastewater and regulated standards for effluent total phosphorus are at < 1 mg/L. Conventional treatment processes are inadequate to meet the standards and so chemical precipitation of phosphorus was common until the development of biological P removal systems. Bio-P removal was developed when it was identified that enhanced phosphorus storage by bacteria was possible when they were exposed to alternating anaerobic and aerobic environments. A typical phosphorus content of microbial solids is about 1.5 to 2 per cent on a dry weight basis. A system

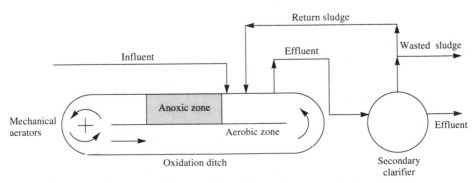

Figure 12.32 Schematic of nitrification–denitrification in an oxidation ditch.

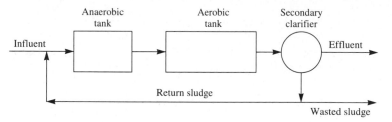

Figure 12.33 Two-stage biological phosphorus removal (A/O process).

with an anaerobic zone followed by an aerobic zone results in the selection of a population rich in organisms capable of taking up phosphorus at levels above the stoichiometric requirements for growth. In this environment the biomass accumulates phosphorus to levels of 4 to 12 per cent of the microbial solids. When these solids are wasted, 2.5 to 4 times more phosphorus removal occurs than for conventional systems. The mechanism of phosphorus removal is via the bacteria *Acinobacter* sp. in the anaerobic environment in the absence of nitrates and dissolved oxygen. The influent to secondary treatment has phosphorus levels reduced by 10 per cent by primary settlement. Biological phosphorus removal systems can further reduce the levels by 70 to 80 per cent. This means that P levels can be reduced from 10 mg/L to 2–3 mg/L by biological P methods. Further reductions often required for sensitive waters are achievable by chemical precipitation. Levels can be reduced to 0.6 to 1 mg/L by the addition of 3 to 6 mg iron/L of wastewater. This compares with 25 mg iron/L for conventional chemical precipitation. Figure 12.33 is a schematic of a two-stage biological phosphorus removal system with hydraulic retention times of 1 to 3 h in the tanks, with the aerobic tank being somewhat longer.

The system shown in Fig. 12.33, called the A/O process, is for combined oxidation and phosphorus removal. Nitrification can be accommodated by an adequate detention time in the aeration basin. In the anaerobic tank, the phosphorus in the wastewater is released as soluble phosphates. The phosphorus is then taken up by the biomass in the aerobic zone and the performance of the system depends on the BOD_5/P ratio of the wastewater. If the ratio is > 10, effluent P levels of < 1 mg/L can be achieved. If the ratio < 10, metal salts may need to be added to achieve low effluent P values (Metcalf and Eddy, 1991).

An alternative bio-P removal system is shown in Fig. 12.34. This system, called the PhoStrip process, removes the phosphorus not in sludge form but in a phosphorus-rich supernatant which can be later chemically precipitated. In the PhoStrip process, some of the waste activated sludge from the secondary clarifier is sent through an anaerobic phosphorus stripper with a retention time of 8 to 10 h. The phosphorus is released in the supernatant while the phosphorus-poor sludge is returned to the aeration tank. The supernatant is chemically 'post' precipitated with lime to produce a sludge and a supernatant which is returned to the aeration basin.

12.10.6 Combined Biological Nitrogen and Phosphorus Removal

Nutrient removal by biological methods is one of the exciting areas of wastewater research in the past two decades. As such, many proprietary processes have developed throughout the world, with leading designs from South Africa, the United States and Europe. Figure 12.35 is a schematic of the five-stage modified Bardenpho process. Details of other processes and references are listed in Randall *et al.* (1992), AWWA (1992) and Metcalf and Eddy (1991).

AWWA (1992) discusses the design information of seven such plants in the United States. Effluent levels for TKN range from 3 to 7 mg/L and for TP range from 0.5 to 3 mg/L. In the modified Bardenpho process, the split of the five tank volumes varies, but those with the best performance are arranged as shown in Fig. 12.35. As with conventional systems, there is also a recycle (liquid) from the end of the first

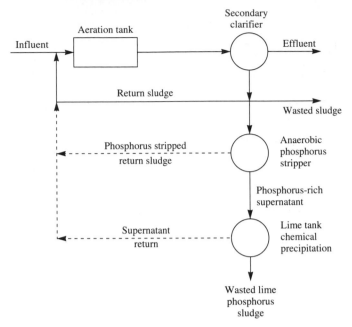

Figure 12.34 PhoStrip bio-P removal process.

aerobic tank to the beginning of the anoxic tank. Process design equations for the many systems are given in Randall *et al.* (1992).

12.11 SECONDARY CLARIFICATION

Retention time in primary clarifiers is typically about 2 hours. Longer than this may produce septicity. Shorter than this can result in inefficient treatment, particularly of suspended solids. The retention time in secondary clarifiers is similar. The function of secondary clarifiers is: clarification and thickening. It is important that no solids should 'escape' in the clarified effluent. This is because the solids are biological and can exert an oxygen demand on the receiving water body. The design parameter of interest for clarification is the surface overflow rate (SOR):

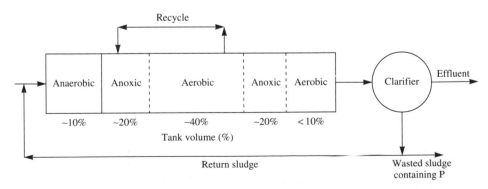

Figure 12.35 Modified Bardenpho process for nitrogen and phosphorus removal.

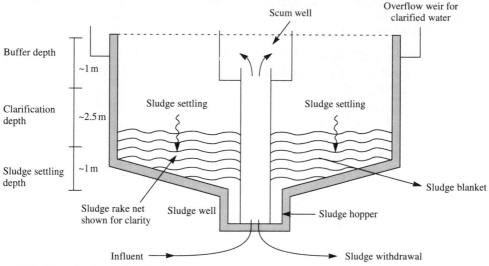

Figure 12.36 Schematic of secondary clarifier.

$$\mathrm{SOR} = \frac{Q}{A} \sim 30\text{--}40 \, \mathrm{m^3/m^2/day} \qquad (12.41)$$

'Sludge thickening' is required so that the returned activated sludge (RAS) to the aeration basin is high enough in concentration to maintain a sufficiently high MLSS concentration in the aeration basin. The design parameter of interest for sludge thickening is the solids loading rate (SLR):

$$\mathrm{SLR} = \mathrm{SOR} \times \frac{\mathrm{MLSS}}{1000} \times \left(1 + \frac{\mathrm{RAS}}{Q}\right) \qquad (12.42)$$

where

$$\mathrm{SLR} = \text{solids loading rate, kg/m}^2 \text{ day}$$
$$\mathrm{SOR} = \text{surface overflow rate, m}^3/\text{m}^2/\text{day}$$
$$\mathrm{MLSS} = \text{mixed liquor suspended solids, mg/L}$$
$$\mathrm{RAS} = \text{returned activated sludge, m}^3/\text{day}$$
$$Q = \text{influent flow rate, m}^3/\text{day}$$

The SLR will depend on the SVI. As the SVI increases, the SLR decreases.

Figure 12.36 is a schematic of a secondary clarifier (similar in construction to primary clarifiers) and Table 12.14 shows the values of some important parameters.

Table 12.14 Maximum typical parameters for design of secondary clarifiers

Parameter	Value
Sludge loading rate	50–300 kg/m^2/d
Surface overflow rate	20–40 m^3/m^2/d
Depth	3–5 m
Detention time	1.5–2.0 h
Shape	Mostly circular

The two functions of clarification (of liquid effluent) and sludge thickening are interdependent. The sludge thickening is provided by a 'deep' tank with adequate settling time. Clarification is produced by ensuring steady state conditions. The design procedure as identified by AWWA (1991) is as follows:

1. Based on F/M, SRT and effluent quality, select MLSS (1000 to 4000 mg/L).
2. Identify the range of MLSS SVI and use the value not exceeded 98 per cent of the time—typically, 100 to 250 mg/L or preferably 150 mg/L.
3. Provide a returned activated sludge (RAS) pumping capacity of 20 to 100 per cent of Q or $> Q$ for an oxidation ditch.
4. With SVI and MLSS, determine the maximum solids loading rate (SLR) from a mass balance around the aeration tank and clarifier.
5. Based on influent flow rates, select a surface overflow rate (SOR) to give the desired clarification (TSS) standard.
6. Select tank geometry with a depth of about 4 to 5 m, to include 0.7 to 1.0 m for thickening, 1 m for buffering and 2.4 m for clarification as per Fig. 12.36.

12.12 ADVANCED TREATMENT PROCESSES

The EU Directive on urban wastewater states that 'more stringent requirements than ... 25 mg/L BOD, 35 mg/L SS and 2 mg/LP ... shall be applied where required to ensure that receiving waters satisfy any other directive'. It also states that 'sensitive areas be subject to more stringent treatment ...' This means that engineers are often required to design plants for standards better than conventional secondary treatment. Lower effluent concentrations of BOD_5/SS may be required to satisfy local water quality standards. Sometimes, trace pollutants such as heavy metals or refractory organics may need to be reduced because of toxicity to aquatic life or interference with downstream potable water supplies. Advanced wastewater treatment addresses:

- Effluent polishing (improving 25/35 to 10/10 for BOD/SS)
- Toxics removal

Nutrient removal is discussed in Sec. 12.10. For effluent polishing and toxics removal the more readily available processes are:

- Granular media filtration
- Adsorption
- Chemical treatment
- Air stripping
- Chlorination

Secondary treatment achieves effluent standards with BOD_5 values around 20 mg/L and corresponding SS concentrations around 30 mg/L. Sometimes long retention time processes, as in oxidation ditches, achieve better standards, but this can also be dependent on low loaded systems. Standards can deteriorate under high loading rates. Effluent polishing is the term used to improve effluent quality from 20/30 to, say, 10/10. Table 12.15 shows achievable standards for some advanced wastewater treatment processes including filtration and/or carbon adsorption.

It is seen from Table 12.15 that typical activated sludge effluents in BOD_5/SS can be reduced from 20/30 to 5/5 by the addition of filtration. Turbidity values also improve greatly. Bio-N–bio-P activated sludge with filtration produces an effluent of 5/5 for BOD_5/SS and better than 2/1 for N/P, with greatly improved turbidity. It is important to have turbidity improvement prior to disinfection processes as turbid waters may mask organisms (pathogenic).

Table 12.15 Advanced wastewater treatment levels

	Effluent parameter					
Treatment process	SS (mg/L)	BOD$_5$ (mg/L)	TKN (mg/L)	NH$_3$-N (mg/L)	PO$_4$-P (mg/L)	Turbidity (NTU)
Activated sludge	20	30	15–35	15–25	4–10	5–15
Activated sludge + filtration	4–6	< 5–10	15–35	15–25	4–10	0.3–5
Activated sludge + filtration + carbon adsorption	< 3	< 1	15–30	15–25	4–10	0.3–3
Activated sludge: nitrification–denitrification + filtration	< 5–10	< 5–10	3–5	1–2	< 1	0.3–3
Bio N–bio P activated sludge + filtration	< 10	< 5	< 5	< 2	< 1	0.3–3

Adapted from Metcalf and Eddy, 1991. Reprinted with permission of McGraw-Hill, Inc.

12.12.1 Granular Media Filtration

Filtration is applied where the effluent SS must be < 10 mg/L. Filtration units are not unlike filtration processes used in potable water treatment described in Chapter 11. They may be downflow (most common) or upflow. They may be dual or multimedia. They may be of natural or synthetic media and they may be pressure or gravity. Figure 12.37 shows a typical layout of a system including filtration. Filtration follows secondary clarification, which in turn may be followed by disinfection.

As SS influent to filters is low, it may be required to assist filtration by chemical coagulants such as polyelectrolytes. These may be added in doses of 0.5 to 1.5 mg/L to the secondary clarifier influent or in doses of 0.05 to 0.15 mg/L to the filter influent. The media of dual type tends to be anthracite on top of sand, with total depths of 1 to 2 m, with filter rates of 5 to 25 m/h. For multimedia filters of anthracite, sand and garnet, the filter depths are 1 to 2 m with filter rates of 5 to 25 m/h. Multimedia filters produce a higher quality filtrate. See AWWA (1991) and Metcalf and Eddy (1991) for further details.

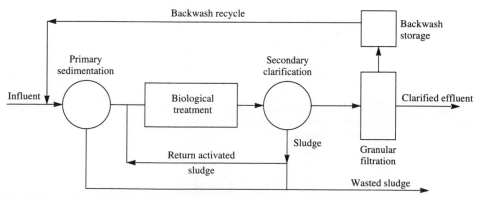

Figure 12.37 Typical layout of a wastewater system including filtration.

12.12.2 Activated Carbon Adsorption

Organic compounds in wastewater may be removed by either powdered activated carbon (PAC) or granulated activated carbon (GAC) as described in Chapter 11. Organics are not usually a problem in municipal wastewaters only but may be an issue if municipal wastewaters are contaminated with industrial organics. Soluble organics (mostly removed by secondary treatment) cause depletion of dissolved oxygen in receiving waters. Toxic organics include: benzenes, chlorinated benzenes, chlorinated ethanes, chlorinated ethers and phenols, dichlorobenzenes and PAHs. In the United States they are classified as priority pollutants and some are listed in Chapter 11; and a full list is given in Eckenfelder (1989).

Activated carbon when contacted with water containing organic matter will selectively remove these compounds by the following processes:

- Adsorption of the less polar molecules
- Filtration of the larger particles
- Partial deposition of colloids on to external surfaces of activated carbon

Carbon adsorption types include:

- Upflow columns
- Downflow columns
- Fixed and expanded beds

The design of a carbon adsorption facility takes account of contact time, hydraulic loading rate, depth of carbon and the number of contactors. Typical values to these parameters are shown in Table 12.16. Typically two or more units are used, sometimes in series to remove the spent carbon. The carbon bed is cleaned by backwashing.

12.12.3 Chemical Treatment

Chemicals may be used for many different problems in wastewater, including:

- Enhancement of sedimentation by metal salts or polyelectrolytes
- Reducing or masking odour
- Chemical precipitation of phosphorus
- Reducing activated sludge problems such as sludge bulking
- pH correction
- Disinfection

Some of the above problems are covered in earlier sections of this chapter. The reader is referred to AWWA (1991), Metcalf and Eddy (1991) and Hosketh and Cross (1989) for further details. Chemical precipitation of phosphorus was briefly dealt with in Sec. 12.10.

Table 12.16 Carbon adsorption parameters

Parameter	Value
Contact time	15–20 min for effluent COD of 10–20 30 + min for effluent COD of 5–15
Hydraulic loading rate	2–7 $L/m^2/s$—upflow 2–7 $L/m^2/s$—upflow
Depth of column	3–12 m Height/diameter > 2

12.12.4 Ammonia Removal by Air Stripping

Many food and pharmaceutical industries produce effluents high in ammonia. Effluent standards are often set for ammonia limits as well as BOD_5/SS limits. If ammonia levels are high, it may not be possible to reduce them by conventional biological nitrogen removal processes. The ammonia stripping process involves raising the pH to a range of 10.8 to 11.5 where the ammonia nitrogen is converted to ammonia gas and released to the atmosphere.

Ammonia nitrogen may also be removed from wastewaters by the addition of chlorine, but only in an effort to polish the effluent and not to remove large amounts of ammonia. When chlorine is added to an ammonia nitrogen wastewater, it produces chloramines in a reaction with hypochlorous acid. With the further addition of chlorine past the 'breakpoint', the chloramines are converted to nitrogen gas.

12.13 WASTEWATER DISINFECTION

Wastewater disinfection is uncommon in most European practice, particularly where treated effluent is discharged to freshwaters, whereas disinfection of potable water supplies is an obligatory requirement. However it is widespread in the United States and to a lesser extent in France. There may be a need to disinfect wastewater if the effluent discharges to:

- Waters used for bathing
- Waters used for shellfish farming
- Waters used for potable water abstraction

These three 'areas' are covered in the EC by the Directives:

- Bathing Water Directive 76/160/EEC
- Shellfish Directive 79/923/EEC
- Surface Water Directive 75/440/EEC

As discussed in Chapters 3 and 11, the objective of disinfection is to eliminate pathogenic organisms. While specific organisms are not routinely monitored, the indicator organisms of faecal and total coliforms are used. The presence of faecal coliforms in a receiving water is indicative of recent pollution by human (sewage) or animal wastes. If non-faecal (non-*E. coli*) coliforms are present, the contamination is from the soil. Untreated domestic wastewater has a total coliform count in the range 10^6 to 10^9 MPN/100 ML. Table 12.17 is a summary of the regulated maximum admissible concentrations of faecal coliforms in various waters, by comparison with MAC (maximum admissible concentration) of zero for drinking water.

Table 12.17 Faecal coliforms standards in the EU and California

EC Directive	EU MAC† total coliform MPN/100 ml	California total coliform MPN/100 ml
Surface water		
A1 water	1 000	
A2 water	5 000	
A3 water	40 000	
Bathing water	$\leq 1\,000$	2300
Shellfish	≤ 300	700
Drinking water	0	

† Flanagan (1990)

In California a principal concern of wastewater discharge to coastal areas is the fouling of beaches. Since the 1940s, the standard for total coliforms at 2300/100 mL was initiated. A total coliform count of 2300 is about equal to a faecal coliform count of 400. By comparison, the California standards that have been in existence since 1945 are more stringent than the EU Bathing Water Directive requirement. In the EU and in California, 80 per cent of the samples are not to exceed the given standard. This means that a limit of 400 sets to 230 as the median. In Europe, very little disinfection of wastewater is provided. By comparison many plants in the United States use disinfection procedures, most commonly chlorination.

The disinfection procedures for wastewater include:

- Chlorine
- Ozone
- Chlorine dioxide
- Ultraviolet radiation

The methodologies of disinfection for wastewater are similar to those described in Chapter 11 for potable water. As potable water is free of BOD_5 and SS, disinfection by chlorine or other oxidant is usually successful, because there is little oxidizing required prior to effecting a residual. In the case of wastewater, because of BOD_5/SS values of, say, 20/30 after secondary treatment, the disinfectant is also an oxidant and will first oxidize BOD_5 before being effective as a disinfectant. As such, to disinfect wastewater effluent, it is usual to first carry out advanced treatment of the wastewater effluent. Advanced treatment methods for BOD_5/SS reduction are described in Sec. 12.12.

Dosages of chlorine for disinfection purposes vary widely depending on the type of wastewater. Typically, domestic wastewater requires about 10 to 12 mg/L. If the wastewater is septic, Cl_2 demand may be 30 to 40 mg/L. If there is industrial wastewater from tanneries or wineries, Cl_2 demand may be 40 to 50 mg/L. Typically, chlorine requirements are about 10 times more than those for potable water requirements. Detailed information on wastewater disinfection is found in White (1992).

12.14 DIFFUSERS FOR WASTEWATER

When wastewater is finally treated to the required standards it is discharged to either a freshwater or seawater environment. Where discharge volumes are low, the discharge may be simply a pipe discharging from the bank of a river. However, for adequate dilution it may be relevant to design and locate a properly engineered outfall. Outfalls may be either of a round jet design or a continuous slit in the pipe or a variation on these two, and may be positioned a distance from the land surface. A jet discharging into an ambient fluid of equal density is shown in Fig. 12.38.

A round jet has a diameter d and an initial exit velocity V_o. At any distance, z, from the exit point, the maximum velocity is v_m. The velocity at a distance z from the exit and a lateral distance x from the centre-line is (from Roberson et al. (1988))

For $z/d > 10$:
$$v_m = 6.2 \, v_o \, (D/z) \tag{12.43}$$

$$v = v_m \exp\left[-87\left(\frac{x}{z}\right)^2\right] \tag{12.44}$$

and
$$\text{Mean dilution } \bar{D} = 0.282 \left(\frac{z}{d}\right) \tag{12.45}$$

The plane jet has a width at the outfall of b.

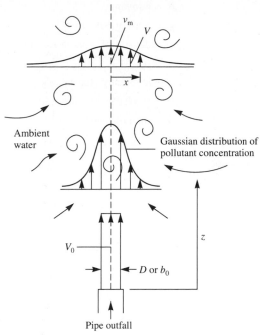

Figure 12.38 Definition sketch for wastewater effluent jet discharging into an ambient fluid (freshwater or estuarine) of non-equal density (jet may be horizontal or vertical).

As for the round jet, the corresponding equations for the velocity are

$$v_\mathrm{m} = 2.41\, v_0 \left(\frac{b}{z}\right)^{1/2} \tag{12.46}$$

$$v = v_\mathrm{m} \exp\left[-74\left(\frac{x}{z}\right)^2\right] \tag{12.47}$$

and the mean dilution

$$\bar{D} = 0.5\left(z/b_0\right)^{1/2} \tag{12.48}$$

The student is referred to Fischer *et al.* (1978) and Roberson *et al.* (1988) for further details.

12.14.1 Diffuser Plume Flow

A plume is flow that is mainly driven upwards by buoyancy, either in the atmosphere as a smoke plume or in a river/estuarine as a liquid pollutant plume. Buoyancy is the density difference between the exit fluid (wastewater) and the ambient fluid (freshwater or seawater). Two actions occur when a plume is discharged. One, the initial momentum on exit, causes mixing. Secondly, the buoyancy thereafter causes mixing:

$$\text{For the round plume } B = g\left(\frac{\Delta\rho}{\rho_\mathrm{ww}}\right)Q = g'Q \tag{12.49}$$

$$\text{For the plane plume } B = g\left(\frac{\Delta\rho}{\rho_\mathrm{ww}}\right)q = g'q \tag{12.50}$$

where

$$B = \text{buoyancy}$$
$$g = \text{gravity}, 9.81\,\text{m/s}^2$$
$$\rho_{\text{ww}} = \text{density of wastewater}, \text{kg/m}^3$$
$$\Delta\rho = \rho_{\text{ww}} - \rho_{\text{w}}$$
$$\rho_{\text{w}} = \text{ambient water density}, \text{kg/m}^3$$
$$q = \text{discharge per unit length}, \text{m}^3/\text{s/m}$$
$$Q = \text{pipe discharge}, \text{m}^3/\text{s}$$

For round plumes, the equations, including buoyancy, from Roberson (1988) are

$$v_m = 4.7 \left(\frac{B}{z}\right)^{1/3} \tag{12.51}$$

$$v = v_m \exp\left[-100\left(\frac{x}{z}\right)^2\right] \tag{12.52}$$

$$\bar{D} = \frac{0.15\,B^{1/3}\,z^{1/3}}{Q} \tag{12.53}$$

For the plane plume, we have

$$v_m = 1.66\,B^{1/3} \tag{12.54}$$

$$v = v_m \exp\left[-74\left(\frac{x}{z}\right)^2\right] \tag{12.55}$$

$$\bar{D} = \frac{0.38\,B^{1/3}z}{q} \tag{12.56}$$

12.14.2 Diffuser Design

The task in the engineering design of a diffuser is to provide adequate dilution and to ensure that the discharged effluent moves on downstream in the case of freshwater or out to sea in the case of seawater. Also the diffuser should be set at an adequate depth to optimize dilution by the time the effluent reaches the surface. A round diffuser typically is a pipeline sitting on the bed with a number of ports over a length of the pipeline. There should be an adequate number of properly spaced ports along the top of the pipeline to ensure a uniform dilution. The sizing of the diffuser pipeline and the ports should be such as to maintain an adequate velocity (not too high, not too low) so that solids do not separate in the pipeline. The pipeline should also be long enough that the wastewater is not returned to shore. The diffuser should also be placed closer to midstream in a river situation so that no wastewater concentrates along the river bank.

Traditionally, there have been many problems associated with diffusers. This was in the time when wastewater could legitimately be discharged to seawaters without treatment (typically, only screening). Since environmental regulations internationally now require that all wastewaters be treated to at least secondary level, the requirements for diffusers and their dilution ability may be less stringent. Problems associated with diffusers in the past have been with the untreated sewage returning to land, pipe and ports clogging, solids separation in the pipeline, differential settlement of pipelines on the bed of coasts, damage during earthquakes, etc.

Example 12.11   A wastewater treated to secondary level is being discharged to the coastal zone. Legislation requires that a dilution of 100 be achieved. The population serviced is 150 000 people equivalent. The temperature of the effluent is 20 °C with a density of 0.998 kg/L, and the density of the seawater is 1.022 kg/L. At what depth (z) should the diffuser be installed if it is a plane diffuser of length 50 m?

Solution   Assume 225 L/person/day.

Flow rate
$$Q = 150\,000 \times 0.225 = 33\,750\,\text{m}^3/\text{day}$$
$$= 0.39\,\text{m}^3/\text{s}$$
$$q = \frac{0.39}{50} = 0.0078\,\text{m}^3/\text{s/m}$$

For a plane diffuser, the dilution, from Eq. (12.56) is
$$\bar{D} = \frac{0.38\,B^{1/3}\,z}{q}$$

Rearranging,
$$z = \frac{\bar{D}q}{0.38\,B^{1/3}}$$

From Eq. (12.50), the buoyancy for a plane diffuser is
$$B = g\left(\frac{\Delta\rho}{\rho_{\text{ww}}}\right)q$$
$$= 9.81\left(\frac{1.022 - 0.998}{0.998}\right)0.0078$$
$$= 1.84 \times 10^{-3}$$

Therefore
$$z = \frac{100 \times 0.0078}{0.38(1.84 \times 10^{-3})^{1/3}} = 16.8\,\text{m}$$
$$= 17\,\text{m}$$

12.15 PROBLEMS

12.1 Identify a geographic area in your country or state and classify the freshwater and saline water bodies into 'sensitive' and 'non-sensitive' areas. Explain briefly your reasons for categorization.

12.2 For a municipal wastewater treatment plant in your area collect the data from management and prepare a graph of hourly rates for flow and waste load (similar to Fig. 12.1). Try to find a plant that treats municipal waste only.

12.3 For the plant you examined in Problem 12.2, determine the watershed area contributing to the plant. Subdivide the watershed into sewage contributing areas. Identify the population statistics. From this survey, using 22 L/head/day, determine the dry weather flow (average daily). How does this compare to your answer to Problem 12.2?

12.4 From your results for Problem 12.3, compute a synthetic hourly flow and waste load rates curve.

12.5 If the BOD$_5$ of a municipal wastewater is 310 mg/L and the BOD$_u$ is 340 mg/L, determine the BOD reaction rate constant (refer to Chapter 7).

Day number	Flow (m^3/day)	BOD$_5$ (mg/L)	SS (mg/L)
1	26 000	260	200
10	36 000	360	310
20	48 000	480	440
30	45 000	450	410
40	42 000	420	375
50	55 000	550	560
60	63 000	630	650
70	58 000	580	590
80	47 000	470	470
90	42 000	420	400
100	31 000	310	250

12.6   What are the mean and standard deviations of flow, BOD_5 and SS for Problem 12.5?

12.7   Design a primary settling tank for an industrial wastewater with a BOD_5 of 600 mg/L and an SS of 400 mg/L to achieve a 60 per cent SS reduction. The flow rate is 35 000 m^3/day.

12.8   Design an equalization basin (for flow) if the flow rates to a municipal wastewater treatment plant are as follows:

12.9   Review the paper, 'Chemically enhanced wastewater treatment—an appropriate technology for the 1990s' by Harleman (1991).

Time (h)	Q_{av} (m^3/s)
0–2	0.32
2–4	0.18
4–6	0.16
6–8	1.2
8–10	0.8
10–12	1.4
12–14	1.2
14–16	0.6
16–18	1.8
18–20	2.2
20–22	1.6
22–24	0.6

12.10 Explain with the aid of sketches the two principal, activated sludge methodologies of complete mix reactors and plug flow reactors.

12.11 In the complete mix activated sludge process, from the starting point of the equations of Sec. 12.8.1, derive Eq. (12.27) for the effluent organic solids concentration.

12.12 For a complete mix activated sludge system, determine the aeration tank volume, the hydraulic retention time, the sludge wasted daily, the mass of sludge wasted daily, the fraction of sludge recycled and the F/M ratio for the following conditions: population equivalent 15 000, influent $BOD_5 = 300$ mg/L, effluent $BOD^5 = 20$ mg/L, yield coefficient $= 0.55$ and a decay rate $= 0.05$ d^{-1}. Assume also that the MLSS in the aeration basin is 4 kg/m^3, in the clarifier sludge is 18 kg/m^3 and the mean cell residence time is 12 days.

12.13 Explain briefly the phenomenon of washout with the aid of schematic graphs (suggest refer to Schroeder, 1977).

12.14 Review the paper 'Unified basis for biological treatment, design and operation', by Lawrence and McCarthy (1970).

12.15 Compute the size of an extended aeration basin to treat wastewater from a population equivalent of 150 000, with a BOD influent of 250 mg/L and a BOD effluent of 10 mg/L, if the MLSS in the aeration basin is 4000 mg/L, the yield coefficient is 0.03 and the decay rate is 0.02 day^{-1}. Explain why extended aeration systems do not need primary clarification.

12.16 Determine the volume of a stone media trickling filter to treat municipal wastewater with a flow rate of 30 000 m^3/day with an influent BOD of 3000 mg/L and an effluent BOD of 20 mg/L if the recirculation ratio (Q_r/Q_0) is 0.1 and 2. The water temperature is assumed to be 20 °C.

12.17 Why is there typically almost no nitrate and nitrite in influent wastewaters to treatment plants? Explain clearly with sketches one method for biological nutrient removal.

12.18 For a wastewater treatment plant treating 30 000 m^3/day compute the total sludge production if a conventional complete mix activated sludge system is used. The influent SS is 350 mg/L and the effluent SS is 40 mg/L.

12.19 For Problem 12.18, if chemically assisted primary sedimentation is the only treatment, using ferrous sulphate in loads of 10 kg per 1000 m^3 determine the total sludge generated.

12.20 For Example 12.2, what are the 90, 50 and 10 per cent of BOD_5 values if a ranking system like that of Chapter 4, Examples 4.3 and 4.9, is used?

12.21 For Example 12.3, repeat the exercise if the storm runoff is included. Use the rational formula $(Q = 0.278cIA)$ to compute the storm flows, when $c \cong 0.8$, $I = $ rain intensity of $20\,mm/h$ over 1 hour and A is the tributary area in km^2 (refer to Chapter 4).

12.22 Compute the density of four sludges if it is made up of the following four dry solids fractions: $DS = 10, 20, 30, 40$ per cent. Assume that the particle density is $2000\,kg/m^3$ (refer to Example 12.6).

12.23 Determine the depth at which a sea outfall diffuser must be set if it is to obtain a dilution of 50. The flow rate q is $0.02\,m^3/S/m$ for a round pipe. Assume $\Delta p = 25\,kg/m^3$.

REFERENCES AND FURTHER READING

American Society of Civil Engineers (ASCE) and Water Environment Federation (WEF) (1991) *Design of Municipal Wastewater Treatment Plants*, Vols. I and II, Alexandria, Virginia, USA.

American Water Works Association. (AWWA) (1992) *Standard Methods for the Examination of Water and Wastewater*, 18th edn.

Apogee Research Inc. (1987) 'Report on wastewater management', National Council on Public Works, May 1987, USA.

Arden, E. and W. T. Lockett (1914) 'Experiments on the oxidation of sewage without the aid of filters.' *J. Soc. Chem. Ind.* **33**, pp. 523, 1122.

Bailey, J. E. and D. F. Ollis (1986) *Biochemical Engineering Fundamentals*, 2nd edn, McGraw-Hill, New York.

Barnes, D. and F. Wilson (1978) *The Design and Operation of Small Sewage Works*, E&FN Spon, London.

Barnes, D., P. J. Bliss, B. W. Gould and H. R. Valentine (1981) *Water and Wastewater Engineering Systems*, Longman Scientific and Technical Publishing, London.

Barnes, D., C. F. Forster and S. E. Hrudey (1987) *Surveys in Industrial Wastewater Treatment*. Vol. 1: *Food and Allied Industries*, Vol. 2: *Petroleum and Organic Chemical Industries*, Vol. 3: *Manufacturing and Chemical Industry*, Longman Scientific and Technical Publishing, London.

Benefield, L. D. and C. W. Randall (1980) *Biological Process Design for Wastewater Treatment*, Prentice-Hall, Englewood Cliffs, New Jersey.

Brassil, L. (1981) *Foul and Surface Water Sewer Design*. Lecture to Cork–Kerry region of Institution of Engineers of Ireland, Cork.

Curi, K. and W. Eckenfelder (1980) *Theory and Practice of Biological Wastewater Treatment*, Sijthoff and Noorohoff, The Netherlands.

Danish Hydraulics Institute (1990) *MOUSE*.

Department of Health (1992) *Public Health Guidelines for the Safe Use of Sewage Effluent and Sewage Sludge on Land*, Department of Health, New Zealand.

Eckenfelder, W. W. (1989) *Industrial Water Pollution Control*. McGraw-Hill, New York.

Eckenfelder, W. W. and W. Barnhart (1963) 'Performance of a high rate trickling filter using selected media', *J. Water Pollution Control Federation*. **35**, 1535.

Eckenfelder, W. W., A. L. Downing, C. J. Appleyard, P. W. Langford, J. L. Musterman and T. Bilstad (1992) *Industrial Water Pollution*, Control Technology, Technomic Publishers. London.

Eikelboom, D. H. and H. J. J. Van Buijsen (1983) *Microscopic Sludge Investigation Manual*, TNO Research, The Netherlands, March.

Ekama, G. A. and G. R. Marais (1984) 'Biological nitrogen removal', in *Theory, Design and Operation of Nutrient Removal Activated Sludge Processes*, Water Resources Commission, Pretoria, South Africa.

Fischer, H. B., E. J. List, R. C. Y. Koh, J. Imberger and N. M. Brooks (1979), *Mixing Inland and Coastal Waters*, Academic Press, San Diego, California, USA.

Flanagan, P. (1992) *Parameters of Water Quality*. Environmental Research Unit, Dublin.

Gray, N. (1990) *Activated Sludge Theory and Practice*, Oxford Science Publishers.

Harleman, D. R. F. (1991) 'Chemically enhanced wastewater treatment—an appropriate technology for the 1990s', in *Water Pollution, Modelling Measuring and Prediction*, Elsevier, Amsterdam.

Haug, T. R. (1980) *Compost Engineering. Principles and Practice*, Ann Arbor Science, Michigan, Illinois.

Heinke, G. and J. A. Tay (1980) 'Effects of chemical addition on the performance of settling tanks', *J. Water Pollution Control Federation*, **52** (12).

Hing, C. L., D. R. Zenz and R. Kuchenrither (1992) *Municipal Sewage Sludge Management: Processing, Utilization and Disposal*, Technomic Publishers, London.

Horan, N. J. (1990) *Biological Wastewater Treatment Systems Theory and Operation*, John Wiley, New York.

Hosketh, H. E. and F. L. Cross (1989) *Odour Control—Including Hazardous and Toxic Odours*, Technomic Publishers, London.

Hydraulics Research (1985) *Design and Analysis of Urban Storm Drainage*, The Wallingford Procedure V1 and V2. Wallingford, UK.

Institution of Water and Environmental Management (1988) *Unit Processes Biological Filtration—Manuals of British Practice in Water Pollution Control*, London, UK.

Lawrence A. W. and P. L. McCarthy (1970) 'Unified basis for biological treatment design and operation', *J. Sanitary Engng. ASCE*, **96** (SA3), 757.

Levine A. D., G. Tchobanoglous and T. Asano (1985) 'Characterisation of the size distribution of contaminants: treatment and reuse applications', *J. Water Pollution Control Federation*, **57** (7), July.

McGhee T. (1991) *Water Supply and Sewage*, McGraw-Hill, New York.

Metcalf and Eddy, Inc. (1991) *Wastewater Engineering: Treatment, Disposal and Reuse*, T. Tchobanoglous and F. Burton (eds), McGraw-Hill, New York.

Moletta R., D. Verrier and G. Albagna (1986) 'Dynamic modelling of anaerobic digestion', *Water Research*, **20** (4), 427–434.

Monod J. (1949) 'The growth of bacterial cultures', *Annual Review of Microbiology*, **3**.

Mudrack K. and S. Kunst (1986) *Biology of Sewage Treatment and Water Pollution Control*, Ellis Horwood, Chichester.

National Research Council (1946) 'Trickling filters in sewage treatment at military installations, *Sewage Works J*, **18** (5).

O'Sullivan G. (1991) 'Wastewater treatment and disposal', in Proceedings of 1st Irish Environmental Engineering Conference, University College Cork, Ireland, May.

O'Sullivan G. (1992) Lecture Notes to Postgraduate Diploma in Environmental Engineering, University College Cork, Ireland.

Pankratz T. M. (1988) *Screening Equipment Handbook for Industrial and Municipal Water and Wastewater Treatment*, Technomic Publishers, London.

Randall C. W., J. L. Barnard and H. D. Stensel (1992) *Design and Retrofit of Wastewater Treatment Plants for Biological Nutrient Removal*, Technomic Publishers, London.

Reynolds T. (1982) *Unit Operations and Processes in Environmental Engineering*, PWS–Kent, Boston, Mass.

Roberson, J. A., J. J. Cassidy and M. H. Chaudhry (1988) *Hydraulic Engineering*, Haughton Mifflin Co., Boston, MA, USA.

Schroeder E. D. (1977) *Water and Wastewater Treatment*, McGraw-Hill, New York.

Sterritt R. M. and J. N. Lester (1988) *Microbiology for Environmental and Public Health Engineers*. E&FN Spon, London.

Thomas G. and R. King (1990) *Advances in Water Treatment and Environmental Management*, Elsevier, Amsterdam.

US Code of Federal Regulations, 40 CFR, Part 33, 1991.

USEPA (1986) *Sequencing Batch Reactors*, EPA/625/8-86/011, Cincinatti, Ohio, October.

USEPA (1988) Barnwell, T. Stormwater Management Model, USEPA, Athens, Georgia, USA.

Water Pollution Control Federation (WPCF) (1990) *Operation of Wastewater Treatment Plants, Manuals of Practice*, Vols. I, II and III, WPCF, Alexandria, Virginia, USA.

Water Pollution Control Federation (WPCF) (1991) *Wastewater Biology: The Microlife*, WPCF.

White, G. C. (1992) *Handbook on Chlorination and Alternative Disinfectants* 3rd edn, Van Nostrand Reinhold, New York, USA.

Wilson F. (1981) *Design Calculations in Wastewater Treatment*, E&FN Spon, London.

THIRTEEN

ANAEROBIC DIGESTION AND SLUDGE TREATMENT

13.1 INTRODUCTION TO ANAEROBIC DIGESTION

The unit treatment process of anaerobic digestion is used world-wide for the treatment of industrial, agricultural and municipal wastewaters and sludges. In recent years it is also being applied to the treatment of municipal solid wastes. By definition, anaerobic digestion (A/D) is: 'the use of microbial organisms, in the absence of oxygen, for the stabilization of organic materials by conversion to methane and inorganic products including carbon dioxide':

$$\text{Organic matter} + H_2O \xrightarrow{\text{anaerobes}} CH_4 + CO_2 + \text{New biomass} + NH_3 + H_2S + \text{heat} \tag{13.1}$$

The process is often used for a first-stage treatment of high-strength organic wastes. The objective is to use A/D to reduce the high organic loads to magnitudes of COD that can be accommodated in conventional aerobic processes, most typically activated sludge. As such A/D is not a complete processor of wastewaters on its own. It is an addendum to existing conventional aerobic processes. The wide range of industrial wastes treated by A/D include:

• Breweries
• Dairy industries
• Food processing
• Chemical industries
• Pharmaceuticals
• Wineries, etc.

Agricultural wastes include those from

• Pigs
• Chickens
• Cattle
• Farmyards

- Waste products—crop residues, etc.
- Offal

Municipal wastes treated include:

- Sludges—principally
  - Raw sewage
  - The organic fraction of municipal solid wastes (OF MSW)

Several locations and studies have combined different feedstocks, e.g. primary sewage sludge + industrial yeast waste + abattoir waste (Sugrue *et al.*, 1992). Great strides have been made in Europe, particularly, into A/D research, funded by the EC (Wheatley, 1991; Ferranti *et al.*, 1987). Extensive applications exist in agriculture and industry. According to Wellinger *et al.* (1992), there are in excess of 500 plants in Europe operating on agricultural farms. Sometimes the term biogas plant is used instead of anaerobic digestion plant and is used interchangeably in this chapter. Typical solids contents for low solids digesters are 3 to 10 per cent, and typically 10 to 30 per cent for high solids digesters. The COD reduction ranges from 75 to 90 per cent. As such, an agricultural waste with, say, an initial COD of 20 000 mg/L will still have a COD content of ≈ 3000 mg/L after anaerobic digestion. It therefore needs further treatment, usually aerobically. The benefits of using A/D include:

- Reduction of pollution potential of waste
- Elimination of pathogens and weed seeds (if mesophilic or thermophilic)
- Improvement in fertilizer/fuel value of waste product
- Production of biogas as an energy source

13.2 MICROBIOLOGY OF ANAEROBIC DIGESTION

An excellent critical review of the kinetics of anaerobic treatment is by Pavlostathis and Giraldo-Gomez (1991), from which some of this section is adopted. Four different trophic microbiological groups (bacteria) are recognized in A/D, and it is the cumulative effect of all of these groups that ensures process continuity and stability. The four metabolic stages required for the production of methane from organic wastes are outlined in Fig. 13.1.

Initially, the complex polymeric materials such as proteins, carbohydrates, lipids, fats and grease are hydrolysed by extracellular enzymes to simpler soluble products of a size small enough to allow their passage across the cell membrane of the micro-organisms. These simple compounds of amino acids, sugars, fatty acids and alcohols are fermented to short-chain fatty acids, alcohols, ammonia, hydrogen and carbon dioxide. These short-chain fatty acids not in acetate form are converted to acetate, hydrogen and carbon dioxide. The final stage is methane production from hydrogen by the hydrogenophilic methanogens and from acetate by the aceticlastic methanogens (Pavlosthatis and Giraldo-Gomez, 1991). Gujer and Zehnder (1983) organized the anaerobic process into seven subprocesses as follows:

- Hydrolyses of complex particulate organic matter
- Fermentation of amino acids and sugars
- Anaerobic oxidation of long-chain fatty acids and alcohols
- Anaerobic oxidation of intermediary products
- Acetate production from CO_2 and H_2
- Conversion of acetate to methane by aceticlastic methanogens
- Methane production by hydrogenophillic methanogens using CO_2 and H_2O.

The biological agents of anaerobic digestion are bacteria but fermentative ciliate and flagellate protozoa and some anaerobic fungi may play minor roles in some systems (Colleran, 1980). The activities of the

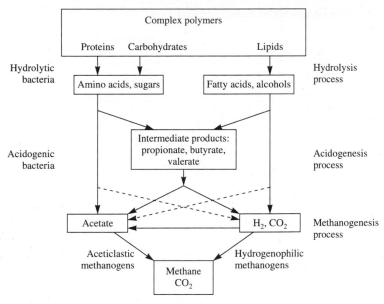

Figure 13.1 Stages in methane production from organic waste.

four trophic groups are briefly described as follows (Colleran, 1980; Wheatley, 1991).

Some typical bacterial species present in the various stages and ranges of respective populations are given in Table 13.1. The huge range of genera and species indicates the complex nature of the microbial population and in each of the stages the population densities (in sewage sludges) range from 10^5 to 10^9 per mL. The bacteria involved in anaerobic digestion have a pH range of 6 to 8 with values close to 7 for optimum activity. Volatile fatty acids depress the pH unless there is sufficient bicarbonate alkalinity present to neutralize the acids. Bicarbonate is formed when CO_2, which is soluble in water, reacts with hydroxide ions to form bicarbonate ions, HCO_3^-. It is important that sufficient alkalinity is available at all times, up to a level of $\sim 3000\,\text{mg/L}$, for sufficient buffering to be maintained.

13.3 REACTOR CONFIGURATIONS

The basic (complete mix—no media) reactor process is shown in Fig. 13.2. The low-rate conventional system shown is made up of several layers. The influent sludge enters the tank close to the top at the location of the supernatant layer (a partially purified liquid layer). Below this is a layer of actively digesting sludge and at the bottom of the tank sits the stabilized sludge, ready for abstraction (withdrawal). Conventional or low-rate digesters are characterized by intermittent mixing, intermittent sludge feeding and intermittent sludge withdrawal (Reynolds, 1982). When mixing is not being carried out the digester contents become stratified, as shown in Fig. 13.2(a).

High-rate digesters are characterized by continuous mixing, except at times of sludge withdrawal. High-rate digesters have HRTs about one-half those of low rate and gas production is as much as twice. The following figures show the many variations of this basic process, including gas recirculation, fixed media or fluidized media, etc.

Reactors can be classified as:

1. First generation type, meaning that the hydraulic retention time is equal to the solids retention time, that is $\phi = \phi_c$ (defined later). As shown in Fig. 13.3, they include:

Table 13.1 Some bacterial species in anaerobic digestion

Stage	Genera/species	Population in Mesophillic sewage sludges
Hydrolytic acidogenic	*Butyrivibrio, Clostridium, Ruminococcus, Acetivibrio, Eubacterium, Peptococcus, Lactabacillius, Streptococcus* etc.	10^8–10^9 per ml
Acetogenic		
Homoacetogenic	*Acetobacterium, Acetogenium, Eubacterium, Pelobacter, Clostridium,* etc.	$\approx 10^5$ per ml
Obiligate proton reducing acetogens	*Methanobacillus omelionskii, Syntrophobacter wolinii, Syntrophomonas wolfei, Syntrophus buswelii,* etc.	
Methanogenic[†]	*Methanobacterium* (many species), *Methanobrevibacter* (many species), Methanococcus (many species), *Methanomicrobium* (many species), *Methanogenium* (many species), *Methanospirillium hungatei,* etc.	$\approx 10^8$ ml

[†]The methanogenic bacteria are usually the limiting population.
Adapted from Wheatley, 1991, and others

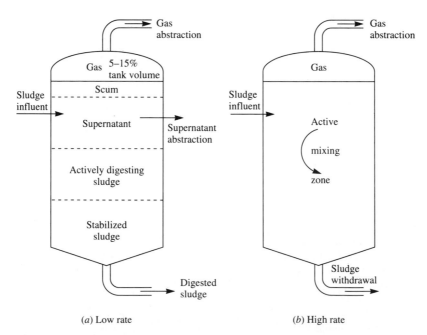

(a) Low rate   (b) High rate

Figure 13.2 Basic anaerobic digestion process.

(a) the batch digester,
(b) the plug flow digester,
(c) the continuously stirred tank reactor (CSTR),
(d) the anaerobic contact reactor.

2. Second generation type, meaning that the solids retention time is greater than the hydraulic retention time, that is $\phi_c > \phi$. As shown in Fig. 13.4, they include:
(a) the upflow–downflow anaerobic filter,
(b) the downflow stationary fixed film reactor,
(c) the fluidized bed reactor,
(d) the upflow anaerobic sludge blanket reactor,
(e) the hybrid anaerobic sludge reactor.

The various configurations have different uses for industrial, agricultural and municipal wastes. For agricultural wastes the batch process is common while in industry the trend is toward fluidized bed or fixed media reactors. Most wastes treated by digesters are liquids of about 2 to 6 per cent solids content. More recently, high solids digesters are being used, particularly for MSW (Kayhanian *et al.*, 1991), where the solids content is 20 to 30 per cent. Solids retention times are typically 10 to 30 days and this is a fundamental design parameter. In those reactors where mixing is provided, it is done by either:

• Mechanical mixers or
• Recirculation of the gas or
• Recycle of the sludge or slurries

Digesters operate at three possible temperatures:

• Psychrophilic (0–$20\,°C$)
• Mesophilic ($\approx 36\,°C$)
• Thermophilic (50–$60\,°C$)

Early digesters operated at ambient temperatures with long retention times. Most digesters now operate at mesophilic temperatures where good stability and gas production results (about $2\ m^3$ biogas/m^3 of digester). Note that biogas $= CH_4 + CO_2$ and is typically 60 per cent CH_4. The optimum temperature to

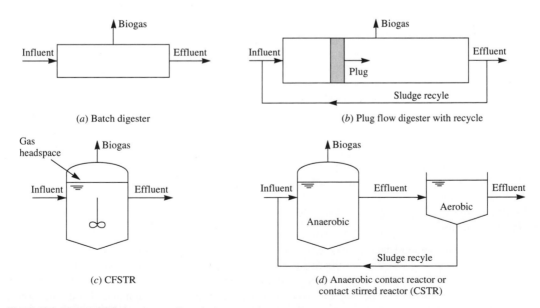

Figure 13.3 First generation reactor configurations.

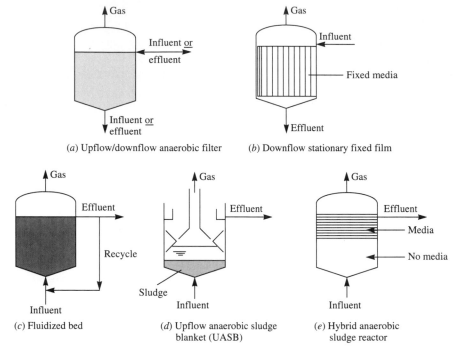

Figure 13.4 Some second generation reactor configurations.

work with is thermophilic where the highest gas production is achieved and pathogen kills are guaranteed. However, at this temperature the process is the least stable and requires more detailed monitoring. Gas production depends on configuration, operating temperature, waste type, etc., but typically ranges from 1 to 5 m^3 biogas per m^3 of digester volume. COD reduction ranges from 70 to 90 per cent. Loading rates defined in kg COD/m^3/day range from 2 to 40. The methane composition of biogas is typically 50 to 70 per cent.

13.3.1 Process Design

The process design involves several parameters including:

- Solids retention time (SRT), days
- Hydraulic retention time (HRT), days
- Volatile solids loading rate, kg VS/m^3/day
- Solids production rate, kg SS/m^3/day
- Gas production, m^3 of CH$_4$ of reactor/day
- Tank configuration
- Mixing systems
- Heating systems

Typical design and operating data for anaerobic digestion in the United States as compiled by the ASCE is shown in Table 13.2. It is notable that the dominant operating temperature is mesophilic (38 °C). This is also true in Europe. Since 1983, anaerobic digestion technology has improved significantly. Many new plants are now operating at the thermophilic conditions for higher gas production and also at high solids content. For digestion systems without recycle, the SRT is the same as the HRT. Adequate sludge

Table 13.2 Design and operating data for anaerobic digesters

Variable	Mean of responses
Sludge feed type:	
Primary (%)	50
Trickling filter (%)	15
WAS (%)[†]	35
Total solids (%)	4.7
Volatile solids (%)	62
Tank diameter (m)	12–38 m
Tank height (m)	5–19
Sludge loading (kg $VSS/m^3/d$)	~5.5
Gas production (m^3/kg VS reduced)	~1
Methane (%)	65
Operating temperature (°C)	38

[†]WAS = waste activated sludge.
Adapted from ASCE, 1983

residence times are provided so as to enable the volatile solids to be fully reduced. SRT and HRT are defined as

$$SRT = \frac{\text{mass of solids in tank, kg}}{\text{rate of solids removed, kg/day}} \tag{13.2}$$

$$HRT = \frac{\text{working volume, L}}{\text{rate of sludge removed, L/d}} \tag{13.3}$$

Typically, municipal wastewater sludge requires an SRT of no less than about 10 days. This is so because the limiting microbiological step is that of methanogenic bacterial growth, which requires about 10 days. Longer periods are often required for mixed sludges, be they municipal, industrial or agricultural. The working range is typically 15 to 30 days.

The volatile solids loading rate is defined as

$$VS \text{ loading rate} = \frac{\text{volatile solids added daily, kg VS/d}}{\text{working volume of the digester, } m^3} \tag{13.4}$$

Typically values are 2 to 3 kg $VS/m^3/d$. Design values may peak at 3.2, because above this ammonia toxicity or metal toxicity may limit operation (WEF, 1991). Gas production rates range from 0.5 to 1.5 m^3 of gas per kg of VS destroyed. The rates will depend on the digester temperature and thermophilic rates are higher. The tank configurations are now primarily cylindrical with diameters of 5 to 50 m and heights of 3 to 25 m. Some digesters have a diameter larger than their height, (e.g. fluidized bed units), while those using gas recirculation as the mixing mechanism may be taller than their diameter. Tank construction is now almost always of steel. The mixing and heating systems may have interdependency. All reactor configurations have thermally insulated containers. Heating mechanisms include external water baths, external jacketed pipes, internal exchangers, steam injection or direct flame. The heat required to raise the temperature of incoming ambient sludge to, say, mesophilic (38°C) and keep it at that temperature is readily computed from WEF (1991) as

$$H = WC\,\Delta T + UA\,\Delta T \tag{13.5}$$

where H = heat required to heat incoming sludge and compensate for heat losses, kg cal/h

       W = mass rate of influent sludge, kg/h

       ΔT = difference between digester temperature and influent sludge temperature

U = heat transfer through tank walls, kg cal/m^2 h°C

A = surface area of digester losing heat

C = mean specific heat of feed sludge, approximately 1 kg cal/kg°C

The mixing of digester contents is essential to maintain uniform sludge quality and prevent blockages of inlet and outlet ports for sludge and gas. The simplest mixing mechanism was reintroducing the gas to the tank. However, this is uneconomic and mechanical systems are now more readily used. The more common method is that of recirculation pumping using an internal draft tube. Further details are given in WEF (1991).

The volume of a batch digester was given in Reynolds (1982) by Fair *et al.* (1968) as

$$V_S = [V_I - \tfrac{2}{3}(V_J - V_f)]t \tag{13.6}$$

where
$$V_S = \text{digester volume}$$
$$V_I = \text{initial sludge volume}$$
$$V_f = \text{final sludge volume}$$
$$t = \text{retention time}$$

13.4 METHANE PRODUCTION

The amount of methane produced will depend on several parameters, but most especially on the waste type, reactor type, temperature and solids content. Bushwell and Mueller (1952) indicated that the stoichiometric equation for methane production was

$$C_nH_aO_b + \left(n - \frac{a}{4} - \frac{b}{2}\right)H_2O \rightarrow \left(\frac{n}{2} - \frac{a}{8} + \frac{b}{6}\right)CO_2 + \left(\frac{n}{2} + \frac{a}{8} - \frac{b}{4}\right)CH_4 \tag{13.7}$$

For example, if pure glucose is used, then

$$\underset{\text{(mol wt = 180)}}{C_6H_{12}O_6} \longrightarrow \underset{\text{(mol wt = 44)}}{3CO_2} + \underset{\text{(mol wt = 16)}}{3CH_4}$$

i.e. 1 kg of glucose produces $3 \times 16/180 = 0.27\,\text{m}^3$ of CH_4. Similarly, 1 kg of carbohydrates produces $\sim 0.35\,\text{m}^3$ of CH_4.

For continuous flow stirred tank reactors (CFSTR) the methane production rate is (Tchobanoglous and Schroeder, 1987)

$$M_{CH_4} = 0.35(nQ\,C_i - 1.42\,r_g\,V) \tag{13.8}$$

where
$$n = \text{fraction of biodegradable COD converted } (\sim 0.85)$$
$$Q = \text{flow rate, m}^3/\text{s}$$
$$C_i = \text{COD loading, kg/L}$$
$$r_g = \text{growth rate, g/m}^3\text{s}$$
$$V = \text{volume, m}^3$$
$$M_{CH_4} = \text{methane production, m}^3/\text{s}$$

As a preliminary estimate of gas production, if growth rate is ignored, then

$$M_{CH_4} \approx 0.3Q\,C_i \tag{13.9}$$

Example 13.1   Determine the volume of a CFSTR required to digest pig slurry from a farm with 2000 pigs. The hydraulic retention time is 15 days.

Solution

$$\text{Assume pig slurry load} \cong 5 \, \text{kg/pig/day (Wheatley, 1991)}$$
$$\text{Load} = 5 \times 2000 = 10\,000 \, \text{kg/day}$$
$$\text{Assume density} \cong 1.05 \, \text{kg/L}$$

Therefore

$$\text{Flow rate } Q \cong \frac{10\,000}{1.05 \times 10^3} = 9.5 \, \text{m}^3/\text{day}$$

$$\text{Required digester volume } V = Q\phi$$
$$= 9.5 \times 15$$
$$= 142.5 \, \text{m}^3$$

Allowing for gas headspace the V design $> 150 \, \text{m}^2$.

Example 13.2 A bench-scale laboratory anaerobic digestion plant of mixed wastes produced the following results for day 65 of a 100 day programme:

$$\text{Reactor volume} = 2 \, \text{L}$$
$$\text{Feedstock COD}_{inf} = 97\,000 \, \text{mg/L}$$
$$\text{COD}_{eff} = 4000 \, \text{mg/L}$$
$$\text{Biogas} = 1.13 \, \text{m}^3/\text{m}^3/\text{day}$$
$$\text{CH}_4 = 59\%$$
$$\text{Daily feed} = 55 \, \text{mL}$$

If growth is ignored in Eq. (13.8), determine the volume of CH_4 per kg of waste digested, i.e. determine a in the following equation (i.e. Eq. (13.9)):

$$M_{CH_4} = anQ \, C_i$$

Solution

$$M_{CH_4} = 0.59 \times 1.13 = 0.67 \, \text{m}^3/\text{m}^3\text{day}$$
$$Q = 55 \, \text{mL/day} = 55 \times 10^{-6} \, \text{m}^3/\text{day}$$
$$nC_i = 97\,000 - 4000 = 93\,000 \, \text{mg/L}$$
$$= 93 \, \text{kg/m}^3$$

$$M_{CH_4} = anQ \, C_i$$
$$M_{CH_4} = a \, \text{m}^3/\text{kg} \times 55 \times 10^{-6} \, \text{m}^3/\text{day} \times 93 \, \text{kg/m}^3$$
$$= a \times 5.12 \times 10^{-3} \, \text{m}^3/\text{day}$$

$$\text{Reactor volume} = 2 \, \text{L}$$

Therefore

$$M_{CH_4} = a \times 5.12 \times 10^{-3} \, \text{m}^3/\text{day} \times 0.5 \times 10^{-3} \, \text{m}^3$$
$$M_{CH_4} = a \times 2.58 \, \text{m}^3/\text{m}^3\text{day}$$
$$0.67 = a \times 2.58$$

Therefore

$$a = \frac{0.67}{2.58} = 0.26$$

that is $0.26\,m^3$ of CH_4 per kg of waste digested (i.e. equal to 75 per cent of that from carbohydrates).

Note This ignores the growth rate, which although likely to be small may not be negligible. If growth were included, a would be larger and therefore closer to 0.35.

Example 13.3 From a municipal wastewater treatment plant calculate the amount of sludge produced daily, the rate of gas production and the anaerobic digester volume if primary sludge is used as a feedstock. The sludge is 2 per cent DS and 70 per cent is volatile solids. Assume 60 per cent volatile solids destruction and a 25 day retention time. The digested solids is 5 per cent DS. The population being served is 50 000.

Solution

(a) From the primary sedimentation tank, assume that 60 per cent of suspended solids is removed. Also assume both the influent SS and BOD are $300\,mg/L$. If the flow per person is assumed as $225\,L/day$, then

$$\text{Total solids per person per day} = \frac{300 \times 225}{10^3} = 67.5\ \text{g/person/day}$$

$$\text{Total solids per day} = 67.5 \times \frac{50\,000}{10^3} = 3375\ \text{kg/day}$$

$$60\%\ \text{removal} = 3375 \times 0.6 = 2025\ \text{kg/day}$$

Therefore

$$\text{Daily sludge} = 2025\ \text{kg dry solids}$$
$$= 101\,250\ \text{kg wet solids (at 2\% DS)} = 101.3\ \text{tonnes/day}$$

(b) Daily gas production. Assume that gas production is $0.15\,m^3/kg$ of VS destroyed.

$$\text{Volatile solids} = 2025 \times 0.7 = 1417\ \text{kg/day}$$
$$\text{Volatile solids destroyed} = 1417 \times 0.6 = 850\ \text{kg/day}$$
$$\text{Gas production} = 0.15 \times 850 = 128\ m^3/\text{day}$$

Assume gas production is 50 per cent CO_2 and 50 per cent CH_4. Therefore

$$\text{Methane production} = 128 \times 0.5 = 64\ m^3/\text{day}$$

(c) Digester volume

$$\text{Volume dimensions} = 64 \times 25 = 1600\ m^3$$

Consider two digesters of size $10\,m$ diameter $\times$ $15\,m$ high (allowing headspace)

Example 13.4 Determine the volume of an anaerobic digester to treat waste activated sludge from a population equivalent of 75 000. The daily (dry) sludge is $0.1\,kg\ DS/day/capita$ with a volatile content of 75 per cent of DS. The specific gravity of wet sludge is 1.016 and the sludge is 4 per cent DS. The A/D process destroys 70 per cent of volatile solids. The final sludge is 7 per cent DS with a specific gravity of 1.028.

Solution Assume mesophilic temperatures of $36\,°C$ and HRT ~ 25 days and that the sludge occupies two-thirds of a tank.

$$\text{Influent sludge solids} = 75\,000 \times 0.1 = 7500\ \text{kg/day}$$
$$\text{Volatile suspended solids (VSS)} = 7500 \times 0.75 = 5625\ \text{kg/day}$$
$$\text{Fixed suspended solids (FSS)} = 7500 \times 0.25 = 1875\ \text{kg/day}$$
$$\text{VSS destroyed} = 5625 \times 0.70 = 3938\ \text{kg/day}$$

Mass balance around digester:

$$Accumulation = inflow - outflow + net\ growth$$

$$Outflow = inflow - loss$$

$$Remaining\ VSS\ in\ sludge = 5625 - 3938 = 1687\ kg/day$$

$$Remaining\ FSS\ in\ sludge = 1875\ kg/day$$

$$Total\ solids\ in\ sludge = 1687 + 1875 = 3562\ kg/day$$

$$Influent\ sludge\ volume = 7500\ kg/day \times \frac{100}{4} \times \frac{1}{1.016} = 184.5\ m^3/day$$

$$Digested\ sludge\ volume = 3562\ kg/day \times \frac{100}{7} \times \frac{1}{1.028} = 49.5\ m^3/day$$

$$Volume\ of\ digested\ sludge = V_S = [V_I - \tfrac{2}{3}(V_I - V_f)]\,t$$

$$= [184.5 - \tfrac{2}{3}(184.5 - 49.5)]\,25$$

$$= 2365\ m^3$$

However, as sludge occupies the bottom two-thirds of tank, then

$$Final\ tank\ size\ V = 3550\ m^3$$

13.5 APPLICATIONS OF ANAEROBIC DIGESTION

Anaerobic digestion applications can be broadly listed as:

- Agricultural
- Industrial
- Municipal

13.5.1 The Treatment of Agricultural Wastes

The use of A/D to treat agricultural waste will probably be the largest single process contributor to the reduction of agricultural pollution. Cattle produce daily excreta wastes at a range of 10 to 40 kg/animal with a solids content range of 10 to 14 per cent. Pig wastes are generated at a rate of 5 to 15 kg/animal with a solids content of 5 to 10 per cent (Wheatley, 1990). These wastes are often diluted with yard or rain runoff. The chemical composition of animal slurries is similar to primary sewage sludge (Wheatley, 1991) with higher values of fibre. Three types of digesters have been used:

- Batch
- Continuous flow stirred tank reactor (CFSTR)
- Second generation digesters
  - Filter and fluidized bed
  - Sludge blanket
  - Contact digesters

The CFSTR is the most common digester in agricultural application when the feedstock is available continuously and there is a corresponding requirement for continuous gas production. In the CFSTR the solids and liquids have the same retention time as they are well mixed and the time varies from 12 to 30 days. Generally, the methanogenic bacteria have a doubling time of 9 days and this is the time constraint. Typically, the cattle and pig slurries have their larger solids removed and sometimes diluted down from 5–15 per cent to 1–2 per cent. The latter is easily digested in a low solids digester. The most common operational temperatures are $\sim 36\,^{\circ}$C, but more recently the thermophilic range is being used. The latter produces a higher biogas output and a more pathogenic free, end product. However, the process is also more unstable and thus requires more control.

13.5.2 The Treatment of Industrial Wastes

Anaerobic digestion is widespread not only in the food industry but also in the chemical, pharmaceutical and paper industries. When an industry utilizes continuous processes A/D may be applicable. However, some industries, particularly some of the chemical industries, operate in batch mode, changing the product chemical and thus the composition and character of the waste stream and A/D may then not be suitable. The continuous stirred tank reactor (CSTR) is popular in industry, but more so in the agroindustry. Usually the waste is pre-treated to remove solids or oils or potential inhibiting agents. Typical operating retention times are 3 to 12 days with organic loading rates of 1 to 5 kg COD/m^3/day.

The anaerobic filter is being used more in recent years where the packing material is a lightweight synthetic (e.g. PVC Flocor). The material has a high surface area and void space. It readily becomes fluidized and so can be cleaned by downflow (backwashing).

13.5.3 The Treatment of Municipal Wastes

Municipal waste falls into three categories with respect to anaerobic digestion:

- Sewage
- Sewage sludges (refer to Sec. 13.7)
- Municipal solid waste (refer to Chapter 14)

As sewage sludges, their treatment and disposal is the subject matter of the remainder of this chapter, a brief discussion now relates to anaerobic digestion of raw sewage. Humans produce about 0.5 kg of faeces and 1.2 kg of urine daily. The excreta is complex and of variable composition, physically, chemically and organically. It is ~ 70 per cent volatile solids, ~ 35 per cent cellulose, ~ 6 per cent hemicellulose, ~ 19 per cent crude protein, ~ 14 per cent lipids, ~ 34 per cent ash and C/N ratio ~ 4 to 5. Sewage can be digested at the three temperature conditions, namely psychrophilic, mesophilic and thermophilic. At psychrophilic temperatures the required retention time is ~ 55 days for pathogenic kill. Anaerobic digestion of raw sewage is commonplace for single family and holiday dwellings (the septic tank) but has not been used for large installations. When used, the UASB or the Hybrid UASB has been the reactor type. Murphy (1992), in a laboratory-scale model, used a Hybrid UASB to treat raw sewage (only) at ambient temperatures. The process was rugged and stable and produced gas at $\sim 1\,\text{m}^3/\text{m}^3$ of reactor per day. The SS and COD reduction was ~ 60 to 80 per cent. The UASB process or Hybrids are in small installations in Brazil where daily ambient temperatures are close to mesophilic.

13.6 INTERNATIONAL REGULATIONS FOR BIOSOLIDS

Biosolids is the semi-solid end product of wastewater purification. Most wastewaters treated are either municipal or industrial. Sludge produced from municipal plants is termed biosolids. Sludges produced from industry are also termed biosolids if they are primarily organic in origin. Chemical sludges do not come within the definition of biosolids. Water treatment plant sludges, even though chemical in nature, are considered to be biosolids. The volume of biosolids is expected to increase significantly (by ~ 50 per cent) as a result of a number of recent environmental developments world-wide, including:

1. The Helsinki Agreement called for banning ocean dumping of sludges by 1987. By 1995 most developed countries had put this ban into practice, e.g. New York and most of the United States by 1992; Australia and New Zealand by 1993; most of the European Union, etc.
2. EU Environmental Directive on Urban Wastewater, requiring better than secondary treatment and in sensitive areas nutrient removal also.
3. In the United States, 'Part 503', the USEPA Standards for the use and disposal of sewage sludge (February 1993).
4. In New Zealand, the 'Public health guidelines for the safe use of sewage effluent and sewage sludge on land' (Department of Health, New Zealand, 1992).

13.6.1 Problems of Biosolids Disposal

Many countries with population centres on coastlines, e.g. Australia, traditionally disposed of much of their sludge at sea. For instance, up to 1993, Sydney disposed 60 per cent to the sea, 10 per cent to agriculture and 30 per cent to compost. Brisbane has had a progressive policy of stockpiling 100 per cent, where all is being utilized for site remediation and landscaping by the City Council. In New York, where the ocean disposal route was used for almost all of its sludge, the 1988 ban forced the city to adopt new policies and stop sea dumping by 1992. It is now operating a three-tier policy including thermal drying, land restoration (in Texas) and engineered mono-landfilling. Los Angeles has also stopped sea dumping from its massive Hyperion wastewater plant ($Q = 1.5 \times 10^6 \, \text{m}^3/\text{day}$). Its current activities include anaerobic digestion, sludge dewatering by centrifuge, sludge dehydration and combustion. The problem of disposal of sludge is world-wide. Even Ireland, which ceased sea dumping in 1990, is operating multioptions for disposal, including land injection, land spreading, silviculture application, mono-landfills, anaerobic digestion. Some countries which embarked on composting technologies found public disquiet with fears of heavy metal, microbiological contamination and malodours.

13.6.2 Biosolids Volumes

The amount of biosolids produced depends not only on the population being served but also on the degree of wastewater treatment.

Example 13.5 Determine the volume of biosolids produced at a plant serving a population equivalent of 400 000 if the suspended solids is 250 mg/L at the influent and 25 mg/L at the effluent. Ignore sludge from screening and biomass production.

Solution

$$\text{Through flow} = 400\,000 \times 225 = 90\,000\,000 \text{ L/day}$$
$$= 90\,000 \, \text{m}^3/\text{day}$$
$$\text{Suspended solids removed} = 250 - 25 = 225 \, \text{mg/L}$$
$$\text{Biosolids per day} = \frac{225}{10^6 \times 10^3} \times 90 \times 10^6 = 20.25 \, \text{t/day}$$

If it is assumed that the biosolids are uniform at 4 per cent DS (dry solids), then

$$\text{Tonnage} = 20.25 \, \text{t/day at 100 per cent DS}$$
$$= 506 \, \text{t/day at 4 per cent DS}$$

Typically, in a multiunit process wastewater treatment plant, biosolids are produced at different dry solids percentages. For instance, the primary settling tank produces biosolids at ~ 5 per cent DS while the secondary clarifier produces at ∼2 per cent DS. The belt presses and centrifuge dewatering technologies (discussed later in this chapter) produce biosolids at 15 to 35 per cent DS. According to the above example, the dry solids production per person is 50.6 g/day. This value can vary significantly, particularly if chemicals are used in any of the treatment processes. This occurs if pH correction is required for proper activated sludge operation. Also, polyelectrolytes are almost always used in the drying presses and other dewatering equipment and this causes an increase in sludge volumes. Additional sludge is generated from biomass production. It is important to note that the 50.6 g/person/day is for dry biosolids. Assuming it is ∼4 per cent DS, the actual wet biosolids is ∼1.26 kg/person/day. Table 13.3 shows typical biosolids quantities for some wastewater unit processes.

The estimates given in Table 13.3 are based on water consumption of 225 L per person per day. This is about typical for the European Union although Denmark and Germany consume less than 200 L. Figures for the United States vary from lows of 300 L to highs of 1000 L. In the United States, the flow rates to wastewater plants are higher than those in the European Union. However, the higher flows create higher dilutions and lower SS values. The latter are typically 150 mg/L by comparison with 300 mg/L in Europe. Those areas in the United States where kitchen grinders are used, however, will have values >150 mg/L. Vesilind *et al.* (1986) quote figures of dry sludge production in Europe ranging from 30 to 124 g/person/day. In Metropolitan Seattle, the amount of solids generated per household is 125 g/day (∼35 g/person/day) (Machno, 1992). In the United Kingdom, raw primary sludge produces 52 g/person/day, co-settled activated sludge produces 74 g/person/day and co-settled activated tertiary sludge produces 76 g/person/day.

The annual biosolids production (in 1990) is shown in Table 13.4. It should be noted that while the population of each country is given, the population served by wastewater facilities is typically about one-half of the total. In countries with dense coastal populations (e.g. Australia) such areas were usually only served with primary treatment.

In computing biosolids volumes, the type of treatment and the form of sewer collection must be evaluated. For instance, in the United Kingdom many industries currently discharge effluent to foul sewers which is subsequently treated at a public wastewater plant. If industrial effluent is treated in this way, there will be an 'apparent' high sludge production rate. Also, by improving water quality by going from primary to secondary treatment, sludge volumes are increased by about a factor of 2. By adding nutrient removal and more especially chemical phosphorus precipitation, sludge volume production is about 3 times that of primary treatment alone.

13.7 BIOSOLIDS CHARACTERISTICS

The characteristics of a sludge depend on whether it is biological or chemical. They also depend on whether it is a primary or secondary sludge or a mixture of both. Many wastewater treatment plants mix

Table 13.3 Typical biosolids quantities before dewatering

Unit process	Dry solids (%)	Specific gravity of sludge	g DS/person/day
Primary clarifiers	∼5	∼1.05	∼30
Primary clarifiers (chemically entranced)	∼5	∼1.05	∼100
Secondary clarifiers (activated sludge)	∼1	∼1.0025	∼20
Secondary clarifiers (percolating filters)	∼2	∼1.01	∼20

Table 13.4 Sludge production and methods of disposal in 1990

Country	Population (total) (million)	Population served (%)	Sludge produced (TDS × 1000/yr)	Disposal method (%) Agriculture	Landfill	Incineration	Other
Austria	7.8	48	320	13	56	31	0
Belgium	9.9	33	75	31	56	9	4
Denmark	5.1	100	130	37	33	28	2
France	56	64	700	50	50	0	0
West Germany	62	90	2500	25	63	12	0
Greece	10		15	3	97	0	0
Ireland	3.5	44	24	28	18	0	54
Italy	57	30	800	34	55	11	0
Luxembourg	0.4	92	15	81	18	0	1
Netherlands	15	90	282 (871[†])	44	53	3	0
Portugal	10.3	47	200	80	13	0	7
Spain	39	47	280	10	15	10	30
Switzerland	6.4	80	215	60	30	20	0
UK	57	84	1075	51	16	5	28
Japan	123	42	2440	24	41	22	13
USA	249		800–1600[‡]	16	43	21	21

[†]The production of sludge in the Netherlands is expected to go from 282 000 TDS/yr in 1990 to 871 000 TDS/yr in 2000, i.e. a threefold increase to 160 g/person/day.
[‡]A range of 800 million to 1.6 billion wet tons in the United States.
Adapted from Garvey *et al.*, 1992 with permission

the primary and secondary for ease of disposal; however, subsequent treatments may thus be limited. The characteristics of sludge can further be described as physical, chemical and biological. In the design of an integrated sludge treatment and management plan, data on sludge sources, characteristics and quantities are required. The integrated plan usually involves sludge conveyance and conditioning systems and also thickening and dewatering facilities. Second-stage treatment may involve composting, incineration, anaerobic digestion, melting, etc., as detailed in Sec. 13.10. Finally, a disposal route is required and with the ban on ocean dumping, disposal on land, in one or more of the many possibilities, is examined in Sec. 13.11.

13.7.1 Primary and Secondary Sewage Sludge

The most common unit process in wastewater treatment is primary sedimentation to remove settleable solids that can be thickened by gravity settling. The sludge consists of organic solids, grit and inorganic fines. Because it contains organic matter, it is suitable for further anaerobic digestion. Downstream thickening is provided by picket fence thickeners (or other), followed by stabilization and dewatering.

Secondary sludge is essentially biological, the result of conversion products from soluble (non-settleable) wastes from the primary effluent. They are produced as wasted sludge from secondary clarifiers, after secondary treatment processes, which include activated sludge, percolating filters, RBCs and variations of these unit processes. The lack of organic matter makes secondary sludge less suitable for anaerobic digestion.

Mixed sludges are those from combined primary and secondary sludges. This is sometimes used so that the easily dewatered properties of primary sludges can be used to assist (in dewatering) the more difficult (to dewater) secondary sludges.

Chemical sludges result when lime, aluminium or ferric salts, etc., are added to improve the suspended solids removal or to chemically precipitate phosphorus. Water treatment sludges are chemical sludges. While some chemicals may be beneficial for dewatering (lime), others inhibit dewatering.

13.7.2 Physical Characteristics of Sewage Sludge

Sewage sludge comprises of lumpy, flaky and colloidal solids interspersed with water. This water consists of (a) water between pores (non-bonded), (b) free capillary water, (c) bound water and (d) cell water. The more capillary and bound water there is, the more difficult it is to dewater. Figure 13.5 shows the volume reduction of sludge as it moves through the different stages of treatment.

Some characteristics of significant physical parameters and typical values are listed in Table 13.5. Other characteristics include:

1. Capillary suction time (CST). This is a parameter determined in a laboratory test of a sludge to identify the dewaterability rate. It is a measure in units of time, usually seconds, for how long it takes a sample of sludge to rid its water on to a blotting paper and travel a distance of 1 cm. Values for unconditioned sludges are typically tens or hundreds of seconds. For conditioned sludge, i.e. treated with polymer, the CST value may be ≤ 10 seconds. Further details are given in Vesilind (1988) and Table 13.5.
2. Specific resistance to filtration (SRF) in m/kg. This, like the CST, is carried out on a laboratory sample using the Buckner funnel test. Sludge is poured into a funnel lined with filter paper and the rate of filtrate is measured in a collecting graduated cylinder. The time/volume (s/ml) is plotted against volume of filtrate (mL) and the slope of the line is noted as b. The SRF is then calculated from

$$SRF = \frac{2PA^2b}{\mu w} \tag{13.10}$$

where
P = vacuum pressure applied
A = area of Buckner funnel
b = slope as above
μ = viscosity
w = sludge solids concentration

SRF is defined as the pressure difference required to produce a unit rate of filtrate flow through a unit weight of sludge cake. A linear relationship between CST and SRF is

$$SRF = 4.235 + 0.125 \, CST \tag{13.11}$$

3. Shear strength of sludge is becoming a more relevant parameter as sludge is more and more being disposed on land. In the case of landfills (specifically mono-landfill) sludge should have a DS >35 per

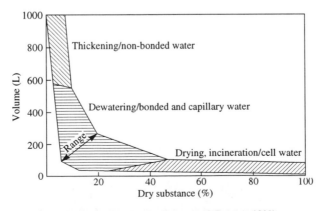

Figure 13.5 Volume reduction by water removal (adapted from Solmaz and Groeger, 1992).

Table 13.5 Physical characteristics of sludge

Parameter	Primary sludge	Secondary sludge	Dewatered sludge
Dry solids	2–6%	0.5–2%	15–35%
Volatile solids	60–80%	50–70%	30–60%
Sludge specific gravity	~1.02	~1.05	~1.1
Sludge solids specific gravity	~1.4	~1.25	~1.2–1.4
Shear strength, (kN/m^2)	<5	<2	<20
Energy content (MJ/kg VS)	10–22	12–20	25–30
Particle size (90%)	<200 μm	<100 μm	<100 μm

Adapted from Andreasen and Nielsen, 1992 with permission

cent and a shear strength of $>30\,kN/m^2$. However, there is some doubt as to the ability of sludge to retain shear strength as some research indicates that sludge loses its strength over time (>2 years).

13.7.3 Chemical Characteristics of Sewage Sludge

The volatile organic substances of the sewage sludge are either solid or liquid. If the water is totally removed the remaining organic volatile matter and inorganic matter (ash) are known as dry solids (DS). The organic volatile matter may be characterized by its net calorific value. Figure 13.6 shows the range of values for sewage sludge. The calorific value has a wide range.

The inorganic content of sewage sludge also varies widely but is typically about:

Waste activated sludge   20–35 per cent DS

Primary sludge   30–45 per cent DS

The inorganic residues of sewage sludges as found from the ash after incineration (Solmaz and Groeger, 1992) are:

S_iO_2	Al_2O_3	CaO	MgO	Fe_2O_3	P_2O_5
40%	14–16%	8–10%	2–3%	7–10%	10–15%

As described in earlier sections, sludge may be either biological or chemical or mixed. However, all will have chemical properties which may include:

- Metals
- Polymers

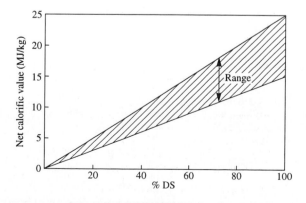

Figure 13.6 Net calorific value of sewage sludge versus per cent DS.

- pH
- Alkalinity
- Nutrients
- PCBs, Dioxins

Comparing US and EU standards Tables 13.6 and 13.7, it is seen that the Netherlands is imposing a very restrictive requirement for land application of biosolids. For instance, most limit values for metals are an order of magnitude less than those accepted for the European Union or the United States. This is due in part to the sensitive nature of soils in Holland and the current levels of soil and groundwater contamination. The latter has been accentuated by the long-term land application of pig slurries from industrial pig farms. However, the severe restrictions on metals (and the banning of ocean dumping) will necessitate innovation and combustion techniques. Table 13.8 gives a range of chemical compositions of sludge from USEPA records.

Table 13.9, again from USEPA, gives the mean concentrations of several elements in different sludge types. Polymers are added to sludge to increase the dewatering rate. Typically they are added in concentrations of 1 to 10 mg/kg of DS. They are long-chain polymers which typically can reduce CST time from, say, 200 to 5 s, thereby decreasing the process turnaround time. However, they are also the dominant operational cost (about 50 per cent) in dewatering.

PCBs and Dioxins are also found in sludge and their quantity depends on the municipality. For instance, dioxins may be an issue in urban areas because of incomplete combustion products from heavily motorized roadways (Steiger, 1992).

13.7.4 Microbiological Characteristics of Sewage Sludge

Like raw wastewater (Chapter 12), sludge may contain bacteria, viruses, protozoa, parasites and other micro-organisms, some beneficial but some perhaps pathogenic. When wastewater is 'purified', the 'purified' liquid effluent will contain almost no pathogens. However, during the treatment process of primary settling, activated sludge, secondary settling, etc., the pathogens become concentrated in the sludge. Therefore raw sludge contains the same infectious agents as raw sewage, but in significantly higher concentrations. Further treatment of sludge (conditioning, thickening, dewatering, etc.) to reduce the water content and so concentrate the solids will also concentrate any pathogens present (Lue-Hing *et*

Table 13.6 Typical values for metals in biosolids and limit values in the United States

			Part 503, 1993			
Parameter	US domestic wastewater sludge (mg/kg)	US domestic + industrial sludge (mg/kg)	USEPA cumulative ceiling concentration (mg/kg)	USEPA cumulative loading rate (kg/ha)	USEPA annual concentration (mg/kg)	USEPA annual loading rate (kg/ha)
As	10–50		75	41	41	2
Cd	10–400	90–240	85	39	39	1.9
Cr	50–200	260–2650	3000	3000	1200	150
Cu	95–700	960–2300	4300	1500	1500	75
Pb	200–500	760–2790	840	300	300	15
Hg	1–11.2	2.6–5	57	17	17	0.85
Mo			75	18	18	0.9
Ni	110–400	200–900	420	420	420	21
Se	10–180		100	100	36	5
Zn	1000–1800	800–460	7500	2800	2800	140

Adapted from Jones, 1981 and Part 503 WEF, 1993b with permission

Table 13.7 Limits for metals in biosolids (EC)†

Parameter	Limit value in soil (mg/kg)	Limit value in sludge (mg/kg)	Annual limit values in sludge for application (kg/ha) based on a 10 yr average	Netherlands post-1995 in sludge (mg/kg)
Cd	1–3	20–40	0.15	1.25
Cu	50–140	1000–1750	12	75
Ni	30–75	300–400	3	30
Pb	50–300	750–1200	15	100
Zn	150–300	2500–4000	30	300
Hg	1–1.5	16–25	0.1	0.75
Cr	—	—	—	75
As				15

†EU Directive 'On the protection of the environment and in particular of the soil, when sewage sludge is used in agriculture', 86/278/EEC.

Table 13.8 Range of typical chemical composition of sludges

Parameter	Primary sludge	Anaerobically digested sludge	Aerobically digested sludge
pH	5–8	6.5–7.5	
Alkalinity (mg/L as $CaCO_2$)	500–1500	2500–3500	
Nitrogen (N% of TS)	1.5–4	1.6–6	0.5–7.6
Phosphorus (P_2O_5% of TS)	0.8–2.8	1.5–4	1.1–5.5
Fats, grease (% of TS)	6–30	5–20	
Protein (% of TS)	20–30	15–20	
Organic acids (mg/L as HAc)	6800–10 000	2700–6800	

Adapted from USEPA, 1977, 1983

Table 13.9 Concentrations of K, Na, Ca, Mg, Ba, Fe and Al in sewage sludge

Parameter	Mean of concentrations (% of dry solids)		
	Anaerobic	Aerobic	All
K	0.52	0.46	0.4
Na	0.7	1.1	0.57
Ca	5.8	3.3	4.9
Mg	0.58	0.52	0.54
Ba	0.08	0.02	0.06
Fe	1.6	1.1	1.3
Al	1.7	0.7	1.2

Adapted from USEPA, 1983

al., 1992). The process of 'sludge stabilization' is therefore required to reduce pathogens and simultaneously eliminate odours. Thus sludge applied to land or water or going into aerosol form may have the potential to impact negatively on human and biotic health. Micro-organism species and density of sludge is related to the community from which it was derived, and particularly its pathogenic content. An important distinction of bacteria and viruses in sludge is that bacteria can increase in numbers in sludge while viruses cannot (the latter require mammalian host cells for replication). Viruses are now known to be prolific within algae, so it is likely that sludges with algal content will contain viruses, e.g. cholera (Colwell and Spira, 1992).

The processes of aerobic and anaerobic digestion, composting, heat treatment, lagooning, etc., all effectively reduce pathogens in sludge. Incineration and liming will totally eliminate pathogens. Disposal of sludge without the above treatments may incur some health risk. The traditional practices (without pre-treatment), including land spreading, landfilling, soil amending and ocean dumping, may pose risks. The treatment processes are discussed later in Sec. 13.9. Table 13.10 contains the levels of indicator and pathogenic organisms in different sludges.

Treated sewage sludge quality requirements The most comprehensive regulations governing quality of the end product sludges is that of WEF (1993b). To satisfy these regulations, the pathogenic requirement of sewage sludges is detailed and explicit. Based on pathogen reduction criteria, sludge is classified as either Class A or Class B.

Class A pathogen reduction This must meet:

- < 1000 MPN faecal coliform per gram of total solids *or*
- < 3 MPN salmonella per four grams of total solids

and one of the following six:

1. Increased treatment time and temperature according to:

Total solids	Temperature	Time	Notes
$\geq 7\%$	$\geq 50\,^{\circ}C$	$\geq 20\,min$	No heating of small particles by warmed gases or liquids
$\geq 7\%$	$\geq 50\,^{\circ}C$	$\geq 15\,s$	Small particles treated
$< 7\%$	$> 50\,^{\circ}C$	$15\,s \leq t < 30\,min$	—
$< 7\%$	$\geq 50\,^{\circ}C$	$\geq 30\,min$	—

2. Alkaline treatment:

$$\text{PH} > 12 \text{ for } 72\,h \text{ with } 12\,h \text{ at temperatures } > 52\,^{\circ}C$$

3. Testing of sludge for virus/helminth ova must satisfy the following, if tested before processing or after processing, for Class A:

$$\text{Enteric virus} < 1 \text{ PFU}/4\,g \text{ TS and}$$
$$\text{Viable helminth ova} < \tfrac{1}{4}\,g \text{ TS}$$

4. To satisfy Class A, if 3 is not used, then testing prior to sale or disposal must satisfy the same requirements of 3.
5. PFRP (process to further reduce pathogens).
6. PFRP equivalent.

Table 13.10 Levels of indicator and pathogenic organisms in sludge bacteria and viruses[†]

Sludge (untreated)[‡]	Total coliforms	Faecal coliforms	Faecal *Streptococci*	*Salmonella* species	*Pseudomonas aeruginosa*	Enteric viruses
Primary	$10^6 – 10^8$	$10^6 – 10^7$	$\sim 10^6$	4×10^2	3×10^3	0.002–0.004 MPN
Secondary	$10^7 – 10^8$	$10^7 – 10^9$	$\sim 10^6$	9×10^2	1×10^4	0.015–0.026 MPN
Mixed	$10^7 – 10^9$	$10^5 – 10^6$	$\sim 10^6$	$\sim 5 \times 10^2$	$\sim 10^3 – 10^5$	—

[†]Note that these populations are 'averages' from various publications cited in Lue-Hing *et al.* (1992). The units are the number of organisms per gram of dry weight (GDW).
[‡]Untreated means not yet stabilized, dewatered, aerobically or anaerobically digested or composted, etc.
Adapted from Lue-Hing *et al.*, 1992

Class B pathogen reduction   This must meet:

- The sewage sludge must be treated to PSRP or equivalent (process to significantly reduce pathogens) *or*
- The mean of seven samples must be: < 2 000 000 MPN or CFU faecal coliforms per gram of total solids

    and the following site restrictions for sludge application must be satisfied:

- For food crops—no harvesting prior to 14 (up to 38) months after application of sludge (depending on crop)
- Feed crops—no harvesting prior to 30 days
- Pasture—no animal grazing prior to 30 days
- Turf—no harvest prior to one year
- Public access—restricted access for 30 days for low exposure areas (up to one year for high exposure)

    Vector attraction reduction requirements are met if any of the alternatives in Table 13.11 are met. *Note*

1. One of 1 to 10 in Table 13.11 must be met when bulk sewage sludge is applied to agricultural land, forest, public lands or reclamation sites.
2. One of 1 to 8 in Table 13.11 must be met when bulk sewage sludge is applied to lawns or home gardens or when sewage sludge is sold or given away for land application.
3. One of 1 to 11 in Table 13.11 must be met when sewage sludge is landfilled.

    No such detailed pathogenic reduction requirements have emerged from the European Union to date. New Zealand has published microbiological guidelines (Department of Health, New Zealand, 1992) and categorized the land application in four zones as shown in Table 13.12. However, these regulations, while

Table 13.11   Part 503 sludge quality requirements for vector attraction reduction

Alternative	Method	Description	Land application	Surface disposal	Septage
1	Anaerobic/ aerobic digestion	38% VS reduction	✓	✓	
2	Anaerobic digestion	If 1 not satisfied + 40 days at 30–37 °C to achieve VS reduction of 17%	✓	✓	
3	Aerobic digestion	If 1 not satisfied + 30 days at 20 °C to achieve VS reduction of 15%	✓	✓	
4	Aerobic digestion	SOUR < 1.5 mg O_2/h gr	✓	✓	
5	Aerobic (composting)	14 days > 40 °C average temperature > 45 °C	✓	✓	
6	Alkaline stabilization	pH > 12 for > 2 h or pH > 11.5 for 22 h	✓	✓	
7	Drying	75% DS	✓	✓	
8	Drying	90% DS	✓	✓	
9	Soil injection	No surface residue after 1 h injected within 8 hours	✓ (no home use)	✓	✓
10	Incorporation	Land applied and incorporated into the soil within 6 h	✓	✓	✓
11	Daily cover	Daily cover to landfilled sludge		✓	✓
12	Septage	pH > 12 for 30 min			✓

Adapted from WEF 1993b

specific to the form of acceptable treatment for sludge (i.e. digestion, composting, lime stabilization), do not spell out numerically the required faecal coliform, or salmonella, or enteric viruses, or viable helminth ova counts as the Rule 503 WEF 1993b does.

13.8 PROCESSING ROUTES FOR BIOSOLIDS

Figure 13.7 is a flow chart of possible routes for biosolids treatment. From a municipal wastewater plant of secondary treatment quality (or better), primary and secondary sludges are produced. As mentioned in Sec. 13.7, primary sludge consists of both organic and inorganic solids that settled, unaided, in a primary sedimentation tank within a retention time of about 1 hour. Secondary sludge is less dense, with smaller particle sizes than primary sludge. Primary sludge is usually thickened by physical processes.

Secondary sludge can also be thickened or pre-dewatered by a similar process to that of primary sludge. However, the selection of the 'thickener' or 'pre-dewatering' unit is likely to be different. This is detailed in Sec. 13.9.2. The next processes may be aerobic digestion or anaerobic digestion. The latter is gaining widespread popularity because of the benefits of methane gas production. Aerobic digestion, unless using covered tanks, is weather dependent and is not suitable for wet climates. Dewatering is the process of increasing the solid content from about 5 per cent DS to anywhere between 15 and 35 per cent DS. It always requires the aid of polymers, a process known as conditioning. Stabilization of sludge is the process of reducing or eliminating the pathogens. If, after dewatering, the sludge is to be landspread or disposed of, it may be necessary to stabilize the sludge before disposal. This traditionally was done using lime, to raise the pH to 12. Other processes for stabilization include anaerobic digestion and several different forms of heat treatment.

13.9 FIRST STAGE TREATMENT OF SLUDGE

We define the following processes as first stage treatment:

- Conditioning
- Thickening

Table 13.12 New Zealand recommended microbiological guidelines for sewage sludge on land

Land application option	Treatment	Controls

Category I
Salad crops, fruit and other
  crops for human consumption
  unpeeled or uncooked
	Heated digestion
	Composting
	Lime stabilization
		Apply to land immediately

Category II Public amenities, sports fields, land reclamation	Heated digestion Composting Lime stabilization	Store for at least one year prior to application
Category III Fodder crops and pasture	Heated digestion Composting Lime stabilization	Sludge may be applied immediately

Category IV
Forest tree lots, bush and scrub
  land
	All sludges except nightsoil and
	  septage—no treatment
		Sludge may be applied
		immediately

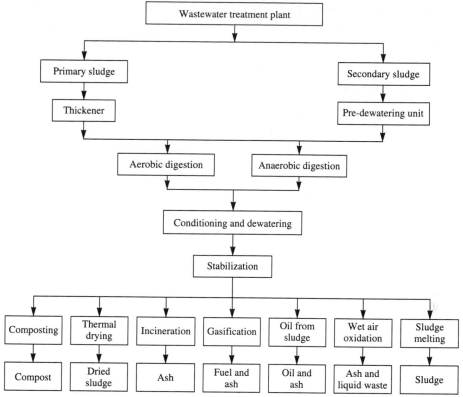

Figure 13.7 Possible processing routes for biosolids (adapted from Barnett, 1993, with permission).

- Dewatering
- Stabilization

The objective of first stage treatment is to reduce the sludge volume by reducing the volume of water. Water in sludge may be of the following types:

- Free capillary water
- Bound water
  - Intercellular (adsorbed)
  - Intracellular (absorbed)

The free water is readily removed from sludges by gravity settling with or without chemicals. The bound water, in intercellular form, is also removable (in part) but requires the addition of polymers. The intercellular water is retained by the sludge by chemical bonding, which may be broken by the addition of polyelectrolytes which cause a change in the electric charges. The intracellular bound water is only possible to remove if the sludge particle walls can be broken by either heating, freezing or electroinduced forces. As such, the free water and intercellular water is removed by mechanical dewatering processes. Dewatering improves the solids content from a range of 2 to 6 per cent to a range of 12 to 35 per cent.

The zeta potential is often used as a measure of the stability of a colloidal particle. It indicates the potential that would be required to penetrate the layer of ions surrounding the particle for destabilization (Fig. 13.8). The major source of stability is the existence of an electrical charge on their surface. This immobile layer of ions which stick tightly to the surface of the colloidal particle is known as the 'Stern layer'. Outside this is a diffused layer of counter ions. The inner shell of charge and outer diffused

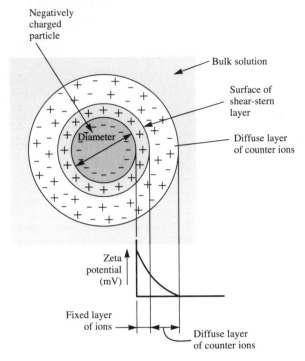

Figure 13.8 Zeta potential (adapted from Eckenfelder, 1980, with permission).

atmosphere is called the 'electrical double layer' (Eckenfelder, 1980). The higher the zeta potential, the more stable the particle and the more tightly the water is bound to the particle. For further details, refer to Eckenfelder (1980) or Schroeder (1977).

13.9.1 Conditioning

Sludge conditioning is the chemical or thermal treatment of sludge to improve the efficiency of thickening and dewatering. The purpose is to reduce the zeta potential by adding specific ions to change the particle charge. Most common is chemical conditioning using either inorganic chemicals or organic polyelectrolytes. The inorganic chemicals used include:

- Ferric chloride
- Lime
- Ferrous sulphate + lime

These are usable principally for conditioning secondary sludge or combined secondary and primary sludge. Their disadvantage is that for each kilogram of inorganic chemical added, an extra kilogram of sludge is produced. The dose range of inorganic chemicals is 100 to 200 kg/tonne of DS. Organic polyelectrolytes or polymers are used in all sludge types and have the advantage of producing less significant increases in sludge volume. The amount of polymer added is in the range of 2 to 100 kg/tonne of DS, but is most typically about 6 kg/tonne of DS.

Organic polymers are long-chain, water-soluble synthetic organic chemicals. They are typically cationic polyacrylamides to destabilize the ionic charge of sludge solids. They are usually supplied in liquid dispersion form and are categorized by:

- Molecular weight (0.5 to 18×10^6)
- Charge density (10 to 100 per cent)
- Active solids levels (2 to 95 per cent)

Many polyelectrolytes are of high molecular weight (about 10^6) and of high charge density. The latter can neutralize the very fine solids in biomass. The high molecular weight provides the floc strength to withstand the shear forces produced by the dewatering equipment, e.g. belt presses and centrifuges. The required dose of polyelectrolyte is critical to the performance of the dewatering equipment and to the cost of the operation. The dose is determined by pilot plant tests or bench tests and may be required to change from time to time as the sludge quality changes. Tests used to estimate dosage are mentioned in WEF (1991) as the:

- Jar test
- Filter leaf test
- Capillary suction time test
- Standard shear test
- Buchner funnel test

The choice of polyelectrolyte depends on the dewatering technology, filter presses, filter belts, centrifuges, etc., and further details are given in WEF (1991) and Metcalf and Eddy (1991).

13.9.2 Thickening

Thickening is the pre-processing of sludge prior to dewatering. Traditionally, the sludge was gently turned in a cylindrical container encouraging the water to rise to the top and to be abstracted as a supernatant. The sludge withdrawn from the bottom would typically be half the volume of the sludge before thickening. Thickeners may be divided into the following types:

- Traditional picket fence thickener
- Continuous gravity thickener
- Gravity belt thickener plus polymers
- Rotary drum thickener plus polymers
- Dissolved air flotation thickener
- Solid bowl centrifugation thickener

Picket fence thickener The sludge is thickened by gravity settling and consolidation of solids. Settling may be:

- Discrete settling
- Hindered or zone settling
- Compression settling

Figure 13.9 shows a schematic of a picket fence thickener. The purpose of the pickets—a collection of vertical bars or plates or angles (angle iron)—is to promote thickening through assisting the separation of liquor from the consolidating sludge and the disengagement of air bubbles (Frost *et al.*, 1992). In some units, ploughs are fixed to the underside of the picket 'gate' to transport the thickened sludge to the centre of the vessel. The ploughs may rotate to suit clockwise or anticlockwise movement. The most common design for the gravity thickener is a circular tank with a side water depth of 3 to 4 m and diameters up to 25 m. Floor slopes to the centre are about 1 in 5 and are steeper than conventional settling tanks. The top 1 m of the tank is the quiescent or clarification zone. Below this for a depth of 1 to 2.5 m is a settlement/consolidation zone. The settling zone sits above the consolidation zone. The bottom 1 m is the transport zone of thickened sludge into the sludge central outlet.

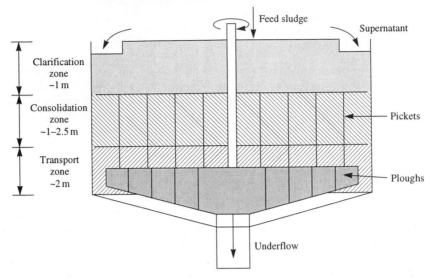

Figure 13.9 Schematic of a picket fence thickener.

The picket fence revolves at a peripheral speed of 2 to 6 m per minute. Higher speeds generate turbulence and prevent consolidation. Also the picket fence is reversed 2 to 4 times per day for 1 to 2 tank revolutions to clean the sludge from the picket or ploughs. The design parameters are:

1. Floor loading of total solids per unit area per unit time ranges from 25 to 250 kg/m^2/day.
2. Supernatant maximum overflow rates in m^3/m^2/day. For primary sludges the range is 15 to 30 m^3/m^2/day. For secondary sludges the rate is halved.

A high overflow rate may produce a carryover of excess solids, so a maximum value is adhered to. On the other hand, a low overflow rate keeps the sludge in the tank for too long and may produce septicity. This is sometimes alleviated by the addition of dilution water or clarified effluent water.

Example 13.6   In the design of a treatment plant for a population equivalent of 400 000, compute the primary sludge, picket fence thickener tank size. Assume that 225 L/person/day is the flow rate and 250 mg/L of suspended solids is the influent solids loading. Also assume that 60 per cent of SS is removed in the primary settling tanks. Assume a solids floor loading of 200 kg/m^2/day. Size the tank for a 1 day retention period and a two per cent DS.

Solution

Influent to WTP $Q = 400\,000 \times 225 = 90 \times 10^6 \text{ L}$

Influent to PFT $Q_{\text{solids}} = 90 \times 10^6 \times (0.6 \times 250) = 13.5 \times 10^9 \text{ mg/day}$

$$= 13.5 \times 10^3 \text{ kg/day}$$

$$Q_{\text{sludge}} = \frac{13.5 \times 10^3}{0.02} = 675 \times 10^3 \text{ kg/day at 2\% DS}$$

$$= 675 \text{ m}^3/\text{day} \qquad (\text{assume SG} = 1.00)$$

Minimum floor area required (based on total solids):

$$A_{\text{floor}} = \frac{Q}{P} = 1.4 \times \frac{13.5 \times 10^3 \text{ kg/day}}{200 \text{ kg/m}^2/\text{day}} \qquad (\text{SG of sludge solids} = 1.4)$$

$$= 94.5 \text{ m}^2$$

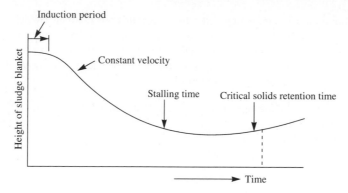

Figure 13.10 Batch consolidation curve.

$$\text{Surface overflow rate} = \frac{675\,\text{m}^3/\text{day}}{94.5} = 7.2\,\text{m}^3/\text{m}^2/\text{day}$$

This surface overflow rate is adequate to prevent septicity. In the event of septicity water is added.

$$\text{Tank size} \cong 11\,\text{m diameter} \times 3 \sim 4\,\text{m high}$$

Continuous gravity thickeners   Traditionally picket fence thickeners operated on a batch process using the fill and draw system. Usually two tanks were provided. More recently, continuous consolidation using picket fence mechanisms is more the standard process. Figure 13.10 is a schematic of the height of the consolidation zone with time. Initially it is tallest and as the water is expunged the height decreases. The process is sensitive and may not operate at optimum heights due to the presence of gas bubbles. The consolidation in continuous thickening is more sensitive than in batch thickening. Retention times in picket fence thickeners is typically:

- 1 day for primary sludge
- 2 days for secondary sludge
- 7 days for digested sludge

Gravity belt thickeners   A recent development to thickening technology is the gravity belt thickener using the principle of gravity dewatering (Sec. 13.9.3). Figure 13.11 is a schematic of this. The filtrate

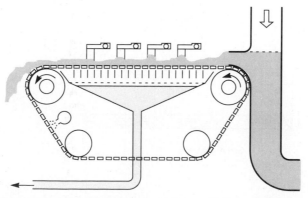

Figure 13.11 Schematic of gravity belt thickener (reprinted by permission of Bellmer GMBH, Germany).

runs off through a continuously circulating belt while the solids are retained. While the sludge is on the belt it is constantly being turned over by collision with streamlined protuberances. Belt speed can be varied to attain the desired percentage of dry solids. Thin sludges (0.2 to 2 per cent) can be thickened more readily by the addition of polyelectrolytes to the value of 5 to 15 per cent. The belt porosity is usually different to that of belts used for dewatering.

Rotary drum thickeners The arrangement for rotary drum thickeners may be cylindrical drums laid horizontally but stacked on each other. Typically the top drum with impermeable walls is to condition and mix polymers to the sludge. The lower level drum or drums are of semi-permeable walls and the sludge is fed through. The filtrate drops through the walls and the thickened sludge is ejected horizontally through one end.

Dissolved air flotation thickeners In Chapter 12, a dissolved air flotation system was described to remove fats and oils from wastewater. In a similar technique air is bubbled through sludge in a container where it is held under pressure. When the pressure is released the bubbles of air attach themselves to solids and now, with a lower specific gravity than water, they rise to the surface. At the surface, the float sludge is skimmed off. The heavier settleable solids are also removed from the floor as in a picket fence type arrangement. The design parameters are:

- Solids loading ratio
- Hydraulic loading ratio
- Air–solids ratio
- SVI of sludge

Further details are given in Metcalf and Eddy (1991) and WEF (1991).

Solid bowl centrifugation thickener The principles of solids–liquids separation are similar to those of gravity thickening except that the applied force is centrifugal and about 2000 times gravity. Only secondary sludges are centrifuged and usually without polymer aids. The outer cylinder is the wall and does not rotate. The inner rotating bowl forces the 'thickened' sludge on to its perimeter wall and the filtrate (now called centrate) is discharged through a port. A helical scroll rotating at a much lower rate pushes the thickened sludge forward towards the tapered end where it is discharged. Figure 13.12 is a schematic of a solid bowl centrifuge. Thickening is achievable to a range of 3 to 6 per cent, but this depends on the bowl type and loading of sludge as well as the sludge characteristics. Centrifugation is more commonly used for dewatering and is described in Sec. 13.9.3.

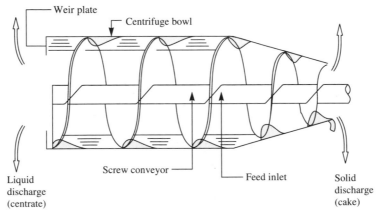

Figure 13.12 Solid bowl centrifuge (adapted from Aqua Enviro, with permission).

13.9.3 Dewatering

Thickening of sludges is defined as taking those thin sludges (0.2 to 5 per cent) and thickening them by physicochemical means to about twice the dry solids content. Dewatering is a similar process but has the objective of taking as much water as possible from the sludge. For instance, a primary sludge of 2 per cent DS can readily be thickened to ~ 4 per cent DS. However, most biological primary sludges can then be further dewatered from 4 per cent DS to 12 to 35 per cent DS. To achieve this level of dewatering the most common technologies are:

- Belt filter press (single or double)
- Plate/membrane filter press
- Centrifuges
  - Solid bowl
  - Press
- Sludge drying beds
- Sludge lagoons
- Sludge freezing beds

Belt filter press This equipment is most common in dewatering municipal sludges, be they primary, waste activated or digested sludges. There are four stages of dewatering according to Bellmer (1992):

1. After conditioning, the sludge is fed across a horizontal belt (Figs 13.13 and 13.14). The filtrate runs off here due to gravity filtration.
2. The two circulating belts form a vertical, wedge-shaped chute with a variable opening. Draining is effected by slowly increasing the pressure from the movement of the belts and the height of the chute.
3. The sludge lying between the two belts is then fed to a dandy roll. Here dewatering is directly to the outside and additionally to the inside through the perforated roller.
4. In the S-shaped press area, the sludge between the belts is further dewatered by press rollers. The resulting kneading action frees enclosed liquid and this further increases the final dry matter content.

In the gravity drainage zone and wedge pressure area, most of the water released is free water. The movement of the sludge through the pressure and shear zone produces high enough shearing action to release some of the bound (intercellular) water. As the sludge moves forward through the press, the decreasing diameters of the rollers progressively increase the pressure on the sludge. Typically belt presses are 1 to 4 m wide and 3 to 8 m long, with heights of 1 to 3 m. The design of a belt press is normally achieved through pilot studies, looking at different polymers to optimize the dewatering and evaluating the capacities of different 'belt' manufactures.

Figure 13.13 Schematic of Bellmer Winkler press (reprinted by permission of Bellmer GMBH, Germany).

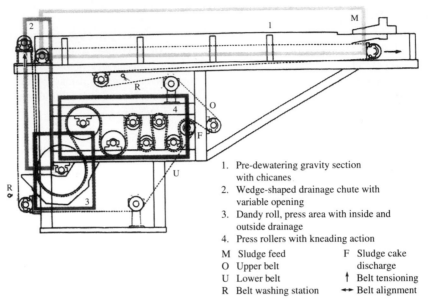

1. Pre-dewatering gravity section
   with chicanes
2. Wedge-shaped drainage chute with
   variable opening
3. Dandy roll, press area with inside and
   outside drainage
4. Press rollers with kneading action

M	Sludge feed	F Sludge cake
O	Upper belt	discharge
U	Lower belt	↑ Belt tensioning
R	Belt washing station	↔ Belt alignment

Figure 13.14 Section through Bellmer Winkler press (reprinted by permission of Bellmer GMBH, Germany).

Example 13.7   Compute the width of a belt press for the following case if the solid loading capacity is 300 kg/m/h (width). The sludge supplied is primary thickened sludge of 4 per cent DS from a wastewater plant of 100 000 p.e.

Solution

$$\text{Primary sludge} = 100\,000 \times 225 \times (0.6 \times 250) \times 10^{-6}$$
$$= 3375\,\text{kg/day solids}$$
$$= 168\,750\,\text{kg/day at 2\% DS}$$
$$\text{Primary thickened} = 84\,375\,\text{kg/day at 4\% DS}$$
$$\text{Width of press} = \frac{84\,375\,\text{kg/day}}{24 \times 300\,\text{kg/m/day}} = 11.7\,\text{m}$$

Therefore use three 4 m wide presses (plus one for standby).

Plate filter presses   While the belt presses can be used for small municipalities, the filter press is more suitable for large population municipalities. The plate filter presses achieve high dry solids end products ($\sim$35 per cent) by forcing the water from the sludge under pressure. Figure 13.15 shows a schematic of a plate press.

The plate press may be either of fixed volume or variable volume. The fixed volume has many plates held tightly in alignment. They are pressed together, usually hydraulically, between a fixed end and a moving end. A filter cloth covers the drainage surface of each plate, providing a filter medium. The sludge is fed through the press and solids collect in the plate chamber until a low feed rate is reached, about one-fifteenth or one-twentieth of initial flow rate (WEF, 1991). The feed is then stopped and the plates are shifted so that the dewatered sludge cake falls down to a collection container. In the variable-volume press, once the plate chamber is filled with cake sludge, a pressure is applied to produce a higher filter cake DS. Filter presses are high capital cost items and require significant amounts of polymers. Operating and maintenance costs are also higher than for belt filter presses.

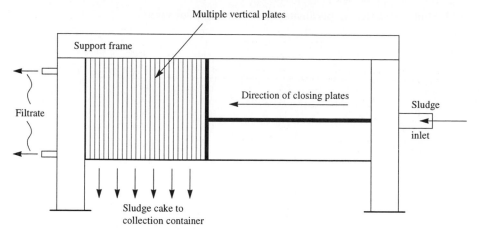

Figure 13.15 Side elevation of plate filter press.

Centrifuges The principal differences between thickening and dewatering centrifuges are the configuration of the conveyor or scroll towards the liquids discharge and the location and configuration of the solids discharge ports. Further details are found in WEF (1991, p. 1187). However, sludges can be dewatered to as high as 30 per cent without too much difficulty. The operational characteristics are shown in Table 13.14.

Sludge drying beds These are outdoor sand bed facilities where the sludge water is dewatered by two mechanisms:

1. The open surface area is permitted to evaporate.
2. By gravity, the filtrate settles through the sand bed and is collected in a series of under-drains, not unlike sand filtration in water treatment.

Figure 13.16 shows a section through a sludge drying bed.

This facility is most successful in dry climates with high evaporation rates. However, it also requires a large land area and as such is not ideal for most urban locations. To determine the requisite areas, evaporation calculations are carried out.

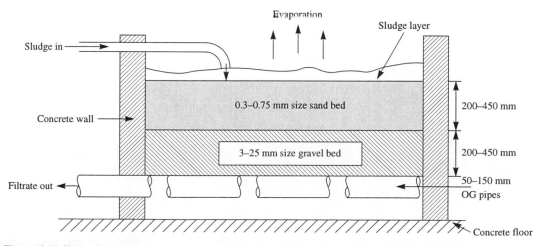

Figure 13.16 Sludge drying bed.

Table 13.13 Sludge dewatering performance of freezing pilot plant

Sludge type	Number of layers applied	Total frozen depth (cm)	Initial (% DS)	Final (% DS)
Anaerobically digested	6	58	6.7	39.3
Anaerobically digested	15	114	5.4	—
Anaerobically digested	12	89	1.1	24.5
Alum sludge	18	99	0.5	82.0

Adapted from Martel, 1992 with permission

Sludge lagoons   Sludge lagoons are more common for industrial plants or very large municipalities with available land space. For instance, food industries, such as the sugar beet process, use large amounts of water in washing. The sludge is allowed to settle and the supernatant is drained or evaporated. Problems with odours prohibit the use of lagoons in developed areas. The cycle time of sludge in lagoons varies from two months to two years, depending on the climate and sludge type. The solids loading of such structures is very low by comparison with the mechanical dewatering systems.

Sludge freezing beds   This is a system that is rarely used but in experimental facilities has been shown to be extremely successful though not necessarily economic. The performance data reported by Martel (1992) is reproduced in Table 13.13. The freezing bed was shown to be most effective for alum sludge and may be a consideration for cold climates. Further details are in the US Cold Regions Research and Engineering Laboratory, CRREL Report 91-6.

Sludge dewatering review   A comparative study of four different types of full scale sludge dewatering units by Andreasen and Nielsen (1992) is reported in Table 13.14. This study concluded that the more economical filter press and solid bowl centrifuge produce dewatered sludge with a lower dry solids content than the more expensive press centrifuge and membrane filter press.

Table 13.14 Comparison of sludge dewatering equipment

Dewatering unit	Sludge 1 Initial (% DS)	Polymer (kg/t DS)	Final (% DS)	Sludge 2 Initial (% DS)	Polymer (kg/t DS)	Final (% DS)	Sludge 3 Initial (% DS)	Polymer (kg/t DS)	Final (% DS)
	Return activated sludge from bio-N and P plant with simultaneous precipitation of P by $FeSO_4$			Gravity thickened activated sludge from bio-N and P plant with simultaneous precipitation of P by $FeSO_4$			Anaerobically digested sludge mixed with gravity thickened activated sludge from bio-N and P plant		
Belt filter press	1	2	21.5	4	3	22	4	5.5	24
Solid bowl centrifuge	1	10	19	4	5.2	18	4	10	23.5
Press centrifuge									
A	1	6.2	22	4	4.5	21	4	13	30.5
B		3.0	24		8.1	20		11	28
Membrane filter press	1	4.6	26	4	5.6	19	4	4.3	33

Adapted from Andreasen and Nielsen, 1992 with permission

13.9.4 Sewage Sludge Stabilization

Stabilization of sludge is any process that renders the sludge or sludge end products pathogen free. Much recent legislative emphasis is placed on successfully stabilizing sludge or compost products. In addition to pathogen reduction, odour reduction and elimination of putrefaction are also achievable by stabilization. The methods used to eliminate the offending micro-organisms include:

- Lime stabilization
- Heat disinfection
- Anaerobic digestion
- Aerobic digestion
- Composting

Anaerobic digestion is the most common sludge stabilization method and with composting and heat drying is covered in Sec. 13.10. The effectiveness of a stabilization process is often measured in the extent of volatile solids destruction for anaerobic and aerobic digestion and for heat drying. However, the most common measurement parameter is that of pathogen indicator organism reduction. Typical reductions are 1 to 4 orders of magnitude, depending on the process.

Lime stabilization   Lime is added to untreated sludge to raise the pH to > 12. Micro-organisms cannot survive in this environment. However, recent research has indicated that micro-organism regrowth may occur. Lime in the form of $Ca(OH)_2$ or CaO or cement kiln dust has been used. The lime may be added before or after dewatering. Typically, smaller rural municipalities use lime directly on the thickened sludge and spread the sludge on land, either as land spread or soil injection (see Sec. 13.11). The use of lime is a low capital cost operation but may be high in operation costs unless the purchase of the lime, i.e. as cement kiln dust, is cheap. Lime stabilization is becoming an outdated technique as the weight of the post-limed sludge has increased by 20 to 40 per cent of its initial weight. This is so because the amount of lime added ranges from 100 to 200 kg/tonne of dry solids.

Heat drying   Heating the sludge to temperatures in excess of 200 °C under pressure is a technique now receiving much attention. This technique not only serves as a stabilizing technique but also as a sludge conditioner, thickener and dewaterer at different temperatures and pressures. Sludge drying is being used and expanded in countries like the Netherlands where serious problems occur with sludge disposal on land. This technique is covered in more detail in Sec. 13.10. Further details on chemical stabilization is found in Metcalf and Eddy (1991).

13.10 SECOND STAGE TREATMENT OF SLUDGE

After conditioning, thickening and dewatering, this text defines second stage sludge treatment as involving technologies for:

- Anaerobic digestion
- Aerobic digestion
- Composting (wind-row and in-vessel)
- Thermal drying
- Incineration
- Pyrolysis
- Gasification
- Wet air oxidation
- Sludge melting

13.10.1 Anaerobic Digestion

This unit process was covered in Secs 13.1 to 13.5. It is now one of those processes that is included in integrated waste management. The anaerobic digestion of biosolids is commonplace where the gas is optimized and used as part of the wastewater treatment plant energy budget. The principal feedstock for municipal wastewater plants is primary sludge as it contains a high amount of organic matter. However, waste activated sludge can also be digested on its own but is often mixed with primary sewage sludge. The attractions of anaerobic digestion of sludge are listed by WEF (1991) as:

- Methane production
- 30 to 50 per cent reduction in sludge volume
- Odour-free sludge end product
- Pathogen-free sludge (more so with thermophilic processes)

13.10.2 Aerobic Digestion

Aerobic digestion is somewhat similar to the activated sludge process. Sludge is fed to a tank where it is mixed aerobically. The main objective of the process is to reduce the solids content for ultimate disposal. The volatile solids are reduced as in anaerobic digestion and thus a stabilized highly fertilizable humus is produced. The principles of activated sludge and more particularly of extended aeration govern the operation of aerobic digestion. Advantages are listed by WEF (1991) as:

- Stabilized end product humus
- Low capital cost
- Easy operation
- Low odour
- Non-explosive gas (CO_2 and NH_3)
- More purified supernatant than in anaerobic digestion

The disadvantages are listed as:

- High operation costs for power and oxygen
- Reduced performance in cold weather
- No methane
- Difficult to dewater sludge

The SRT at 20 °C is typically greater than 15 days. However, practical operations utilize SRTs of 25 to 40 days. Aerobic digestion is more rugged than anaerobic and will operate even at pH values down to 5.5. The system's efficiency is temperature dependent and at higher temperatures the micro-organism activity increases. The volatile solids reduction in anaerobic digestion is typically 60 per cent, while that for aerobic systems is typically 40 per cent. The reduction of biodegradable solids follows a first-order reaction rate expression:

$$\frac{dM}{dt} = -k_d M$$

where

M = concentration of biodegradable volatile solids (BVS) at time t

$\dfrac{dM}{dt}$ = rate of change of BVS

k_d = reaction rate constant, d^{-1}

At 20 °C, $k_d \sim 0.1$ d^{-1} and at 30 °C, $k_d \sim 0.15$ d^{-1}. From Chapter 12, in the discussion of aeration tank sizing in complete mix activated sludge processes, the tank volume (Eqs 12.21 and 12.30) was given as:

$$V = \frac{Q_0 \phi_c Y}{X} \left(\frac{S_0 - S}{1 + k_d \phi_c} \right)$$

This can be adapted for aerobic digestion:

$$V = \frac{Q_0 X_i}{X(k_d + 1/\phi_c)} \tag{13.12}$$

where
V = aerobic digester tank volume, L
Q_0 = influent sludge flow, L/d
X_i = influent BVS, mg/L
X = digester average BVS, mg/L
ϕ_c = SRT, d
k_d = reaction rate, d^{-1}

Example 13.8   Compute the volume of an aerobic digester to treat activated sludge if the following are the conditions:

$Q_0 = 100 \, m^3/day = 100\,000 \, L/day$
$X_i =$ influent biodegradable volatile solids
$\quad = 15\,000 \, mg/L$
$X =$ digester biodegradable volatile solids
$\quad = 10\,000 \, mg/L$
$k_d = 0.1 \, d^{-1}$
$\phi_c = 30 \, days$

Solution

$$V = \frac{Q_0 X_i}{X(k_d + 1/\phi_c)}$$
$$= \frac{100 \times 10^3 \times 15 \times 10^3}{10 \times 10^3 (0.1 + 1/30)}$$

Therefore

$$V = 1\,125\,000 \, L = 1125 \, m^3$$

It is noted from this computation that the tank volume increases as the reaction rate decreases (i.e. as temperature drops). The tank volume reduces as the SRT (ϕ_c) increases. Similarly, the tank increases as the BVS in the tank decreases. The parameters given in the above example can be computed from experience or pilot plant studies.

13.10.3 Composting

Composting is a sludge waste treatment process used in varying degrees of sophistication in all countries. Traditionally, it tended to be a low-cost process with most applications being for single and low-density development. It has not in the past been attractive for large urban wastewater sludge processing. However, much progress has been achieved in recent years in defining sludge and compost characteristics, and, as

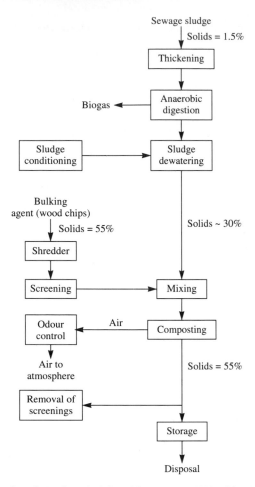

Figure 13.17 Integrated compost flow chart schematic (adapted from Barnett, 1993, with permission).

such, areas in South Africa, Australia, the United States and Europe have invested significantly in full-scale composting facilities.

This section discusses the treatment of dewatered sludge by aerobic decomposition using micro-organisms. However, anaerobic composting is recently under investigation and has been described by Kayhanian and Tchobanoglous (1993) for the biodegradable fraction of municipal solid waste. Composting is a 'dry' process, whereas anaerobic and aerobic digestion are 'wet' processes. Aerobic decomposition occurs naturally in soils and composting artificially creates this 'soil' environment so as to accelerate the organic decomposition. The process of composting, like anaerobic and aerobic digestion, takes about 30 days for complete degradation. If degradation is complete, then the process is irreversible and the end product compost is 'fully' stabilized. However, the reality around the world today is that not all composting facilities produce a 'fully' stabilized product and may not meet standards for pathogen reduction/elimination. Composting with regard to solid waste is discussed in Chapter 14.

Figure 13.17 is a schematic of an integrated compost flow diagram. Sludge is dewatered to about 35 per cent DS and blended with a bulking agent like wood chips. The ratio is approximately 3 part chips to 1 part sludge. The subsequent heat generated by aerobic decomposition evaporates much of the water and kills the pathogens. Mixing and turning is required to ensure that all the mass attains the top temperatures. In colder regions aeration is supplied by under-floor blowers where the air is heated. The chips or bulking agent are recovered by screening after about 20 to 30 days. This is possible because the chips are about 25 mm minimum dimension, whereas the true compost is less than a few millimetres.

The process of composting can be either

- Traditional wind-row
- Aerated static pile
- In-vessel composting

These processes are described in Sec. 14.9. The following parameters are relevant to composting:

- Bulking agent
- Operational temperature
- Moisture content
- Organic and nutrient content

The purpose of the bulking agent is to dry out the blended mix of sludge and agent. Several bulking agents are used, including wood chips, sawdust, leaves, papers, solid waste, etc., but wood chips are the most common. This may be so as it can be screened out of the process at the end and reused several times. Also, chips are unlikely to be contributing to the metal or pathogen levels.

The operational temperature ranges from 40 to 60 °C. If the temperature is too low, the process time increases and pathogens are less likely to be killed. Temperatures above 60 °C may inhibit the composting micro-organisms. Control of temperature is therefore important, particularly in naturally aerated systems. The moisture content of dewatered sludges ranges from 70 to 85 per cent. When blended with chips the moisture content reduces to 40 to 50 per cent. At this level, there is sufficient air space to retain the aerobic environment. It follows that if the moisture levels are too high, a risk exists of producing an anaerobic environment.

As described in Chapter 14, minimum organic and nutrient levels are required for optimum composting. The micro-organisms active in the composting process require carbon for growth and nitrogen for synthesis of protein. The C/N ratio is the parameter of relevance and at C/N levels greater than 30, the composting process is inhibited. Also at C/N levels below 20, incomplete composting occurs. The C/N ratio of dewatered sludge is about 6 and is about 16 for digested sludge. The C/N ratio is increased by the addition of a bulking agent. For instance, the C/N ratio of leaves is about 60. Composting is a more robust process than anaerobic digestion and will operate at a wider pH range of 6.5 to 9.5 (anaerobic digestion has a very narrow pH tolerance range of about 6.8 to 7.3).

The environmental parameters of significance to composting are:

- Odour
- Pathogens
- Metals

Odour is usually not a problem in well-aerated, well-managed systems, although problems with odour have led to the use of in-vessel composting. The major concern for composting is pathogens. The new sludge regulations, Part 503, in the United States (USEPA, 1993) for composting, place severe restrictions on pathogen counts. The New Zealand regulations of 1988 as well as those of USEPA are included in Sec. 13.7.4. Composted sewage sludge products are required to meet (WEF, 1993b) Part 503, Class A, regulations for pathogen reduction. The significant number which the composting industry must achieve is a faecal coliform count of < 1000 MPN/g TS or a *Salmonella* sp. count of < 3 MPN/g TS.

The problem of odour from the composting process is due to ammonia, hydrogen sulphide and organic sulphide compounds. Technologies are now readily available to reduce these. For instance, ammonia gas can be stripped in absorption towers through dilute solutions of sulphuric acid. H_2S and organic sulphide can be removed by their absorption in an alkaline, oxidizing solution which converts these compounds into non-odorous forms. Details of such technologies have been described in full by Leder (1992). Control of ammonia emission is further discussed by Rijs (1992) who points out that

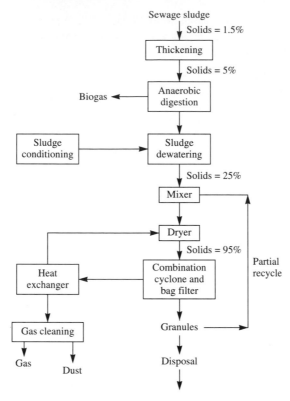

Figure 13.18 Schematic of sludge drying flow chart (adapted from Barnett, 1993, with permission).

ammonia emission is about 10 times less in anaerobic in-vessel composting facilities than in aerobic facilities. It is likely that the future holds good prospects for composting. Traditionally, it was considered a low-technology process. However, for most likely success, high-technology innovative processes must be developed to produce a product that will sell, is not offensive (malodorous) and does not contain pathogens nor is high in heavy metals. A key disadvantage is that aerobic composting is a 'net' importer of energy and anaerobic digestion is a 'net' exporter of energy.

13.10.4 Thermal Drying

Figure 13.18 shows a flow chart schematic of the thermal drying process. The pre-treatment involves thickening, anaerobic digestion, sludge conditioning and dewatering to ~25 per cent DS. The biogas from the anaerobic digester is used as heat supply to the dryer. Sludge drying is a unit operation that involves reducing the water content of the sludge by vaporization. The result is a dried product of ~90 to 95 per cent DS and pathogen free, typically in the form of granules. Conduction, convection and/or radiation are used in heat transfer in thermal drying. Conditions for effective heat transfer to sludge are:

- A large surface area between the sludge and the thermal carrier (condensing steam, hot air, combustion gases, superheated water or thermal oil)
- A high heat content of the thermal carrier (about 250 to 500 °C gases in a rotary dryer)
- A long contact time between the sludge and the thermal carrier (about 30 min in a rotary dryer)

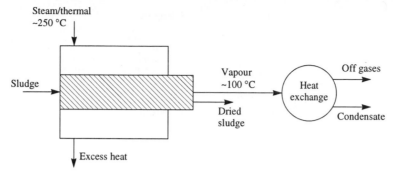

Figure 13.19 Indirect drying.

The heat drying technologies include:

- Flash dryer
- Rotary dryer
- Fluid bed dryer
- Disc dryer
- Multiple effect evaporator (using oil)

Details of the above are found in WEF (1991) and from equipment manufacturers. Sludge drying technology is now finding use internationally. For instance, the Netherlands expects to produce about 870 000 t DS/yr by the year 2000. The process techniques have been decided for 500 000 t DS/yr. Of the latter, 85 000 t DS/yr are to be processed by thermal drying. It is therefore likely that ~ 20 per cent or more of the sewage sludge is to be processed by thermal drying. The sewage sludge from the city of Stuttgart in Germany has been operating disc dryers since 1981. Stuttgart has two sludge streams, one from the primary treatment facilities and the other from the anaerobic digester. Gravity thickening is followed by dewatering with solid bowl centrifuges. From the centrifuges, the 23 per cent TS sludge is pumped to the disc dryers which further dewater the sludge to 43 per cent TS. From the dryers, the sludge goes to the fluid bed reactor incinerator and the waste heat boiler. The sludge at 43 per cent TS burns unaided. The system is considered to be successful. Part of the new sludge management programme for New York is to thermally dry approximately 50 per cent of all sludge (Wagner, E.O. 1992). In the EU there are currently about 35 drying plants (Hall, 1992).

The two methods are principally direct and indirect drying. In direct drying, the sludge is in direct contact with the drying medium, hot air or gas. The water vapour coming from the sludge leaves the dryer with the drying medium. The latter flows in the same direction as the sludge as it not only dries the sludge but also conveys it. Hot air or gas temperatures are in the range 350 to 600°C, which reduces to 80 to 150°C after drying. This method is considered simple, but the drying medium becomes contaminated with malodorous compounds and so has to be gas scrubbed before recirculation, possibly with a modern biofilter. As its name implies, in indirect drying there is no contact between the heating medium and the sludge. The disc dryer is one such case. The party wall between the sludge and heating medium is at a temperature of 150 to 250°C. The heating medium and the sludge water vapour leave the dryer via different routes. Figure 13.19 is a schematic.

13.10.5 Incineration

Incineration of sewage sludge is a technical option used to some extent by most developed countries (see Table 13.4). Incineration with regard to hazardous waste is detailed in Chapter 15. For sewage sludge, fluidized beds, multiple hearth incinerators and powder burning systems are all used. In the Dutch plans

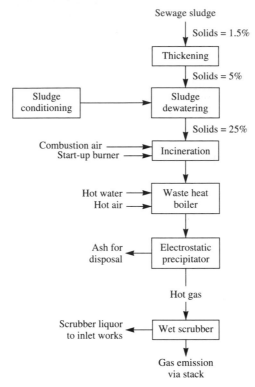

Figure 13.20 Integrated flow chart of incineration (adapted from Barnett, 1993, with permission).

for the year 2000, incineration figures highly with a declared 15 per cent of sludge to be incinerated. A 45 000 t DS/yr plant at Dordrecht and a 70 000 t DS/yr plant at Noord-Brabant are in design/construction. An integrated flow chart is shown in Fig. 13.20. Sewage sludge is first dewatered and can be burned with auxiliary fuel at 20 to 30 per cent DS. At dried cake values of 30 to 50 per cent DS, sludge will burn unaided. Incineration of sludge is the high-temperature combustion ($> 900\,°C$) of the combustible sludge elements—carbon, hydrogen and sulphur in addition to those of grease, carbohydrates and proteins. The end products of the combustion of sludge with excess air are CO_2, SO_2 and water vapour. Ash is also produced. The volume is reduced to about 15 per cent of its original value. The high heating value of sludge varies from 12 to 18 MJ/kg of dry solids. The total heating value can be determined from

$$Hawf = 33.8C + 144.7\,(H - 0.125O) + 9.42S \tag{13.13}$$

where
$$C = \% \text{ carbon}$$
$$H = \% \text{ hydrogen}$$
$$O = \% \text{ oxygen}$$
$$S = \% \text{ sulphur}$$
$$Hawf = \text{high heating value (awf} = \text{ash and water free), MJ}$$

Further details are given in WEF (1991) and Kayhanian and Tchobanoglous (1993). There are examples of sludge incinerators in co-disposal arrangements as follows:

- Incineration of sludge only
- Incineration of sludge + solid waste

- Incineration of sludge + bark waste (Groz, Austria)
- Various other combinations

As described in Chapter 16, significant advances in incineration technology have led to decreased air pollution emissions. Successful technologies for particulates collection and gaseous entrapment are all now possible.

Incineration is an integral part of the treatment policies on sludge of most developed countries. For instance, at Ludwigshafen in Germany, incinerators with the capacity of 120 000 t/yr treat both chemical and municipal sludge. Currently about 3 per cent of sludge in the Netherlands is incinerated with plants at Dordrecht, Amsterdam and Oyen. Of the technologies already specified, incineration will treat approximately 15 per cent of sludge from the Netherlands. Japan, as of 1990, had 220 incinerators of sewage sludge with the favoured methodology being the fluidized bed system.

13.10.6 Pyrolysis

Many organic substances are thermally unstable and upon heating in an oxygen-free atmosphere are split, through a combination of thermal cracking and condensation reactions, into gaseous, liquid and solid fractions. Pyrolysis is the term used to describe this process. In contrast to the combustion process, which is highly exothermic, the pyrolytic process is highly endothermic. Pyrolysis produces:

- A gas stream containing H, CH_4, CO, CO_2 and others, depending on the sludge
- An oil/tar stream
- A char of almost pure carbon

Two forms of pyrolysis under research for sludge are:

- Oil from sludge (OFS) and
- Gasification

In the OFS process, the organics in the sewage (pre-dried to at least 95 per cent DS) are converted to a liquid fuel, not unlike medium fuel oil. The principal piece of technology is the conversion reactor. Firstly, the dried sludge (> 25 per cent DS) is heated to $\sim 450\,°C$ without oxygen. Under these conditions, about half of the sludge is vaporized. These vapours are contacted with the tar residue of the sludge and aliphatic hydrocarbons are produced. The process produces oil and char, non-condensable gas and water. It is possible to combust these by-products at > 800 °C to produce energy for drying. Typical product data from dried raw sludge are 30 per cent oil with a heating value of ~ 40 MJ/kg and 50 per cent char with a heating value of ~ 10 MJ/kg.

Gasification involves taking a dried sludge (> 80 per cent DS) and converting it into a clean combustible gas which may be converted to electricity or hot water for district heating (as is the case in Denmark). The main component of the system is a gasification reactor. The dried sludge (in briquette form) is fed to the reactor and with air is heated to 500 °C in the distillation zone and then to 800 °C in the carbonization zone. As the amount of air is controlled, the solids give off their gases without combustion. The gases then pass through a core with temperatures > 1200 °C, where they are cleaned of tars and oils. The gas then goes to a combined heat and power unit (CHP) to produce electricity and hot water. The two products of gasification are gas and ash.

13.10.7 Wet Air Oxidation

In contrast to most forms of sludge treatment, the process of wet air oxidation does not require the sludge to be dewatered. Pre-treatment is adequate if the sludge is thickened to ~ 5 per cent DS. Wet air oxidation is a process whereby the sludge organic solids are oxidized in an aerobic, high-pressure, high-temperature environment. It is a liquid phase reaction between the organic material in water and oxygen. Temperatures of $\sim 175\,°C$ and pressures of $\sim 10\,MPa$ are used. The subcritical, non-catalytic oxidation of sludge is represented by Rijs (1992) as

$$C_aH_bO_cN_dS_eP_f + (a + \tfrac{1}{4}b - \tfrac{1}{2}c - \tfrac{3}{4}d + \tfrac{3}{2}e + \tfrac{5}{4}f)O_2 \longrightarrow$$

$$aCO_2 + (\tfrac{1}{2}b - \tfrac{3}{2}d - e - \tfrac{3}{2}f)H_2O + dNH_3 + eH_2SO_4 + fH_3PO_4 \qquad (13.14)$$

The objective of wet air oxidation of sludge is to oxidize a large part of the organic sludge components, leaving an almost inorganic sludge which can readily be dewatered. The process can be carried out above ground or below ground. Processes available include: Zimpro, Osaka Gas, Vertech, Kenor, Wetox and Oxidyne. The Vertech system as installed at Apeldorn (Netherlands) is 1225 m deep with an outside shaft diameter of 343 mm. The system consists of a series of concentric pipes as shown in Fig. 13.21. The 5 per cent DS sludge enters the central pipe or 'downcomer'. At a depth of 300 m pure O_2 is added. As the depth increases, the pressure and temperature also increase. At 700 m and a temperature of 175 °C the oxidation process starts and the temperature of the reaction mixture rises to 277 °C as a result of an exothermic reaction. Oxidized sludge leaves the reactor via the 'upcomer'. In the upcomer, pressure gradually reduces and some of the heat generated is transferred to the cold influent downcomer. Outside the two concentric rings is a third and in this annulus a heat exchange medium is transported. The mixture leaving the riser is split into a gas and a liquid stream. The degassed effluent is passed through a lamellar separator, where separation occurs to a suspended solids content in the effluent of $\sim 200\,mg$ TSS/L. The released sludge is further mechanically dewatered to ~ 50 per cent DS. The dry solids is ~ 90 per cent inorganic (Rijs, 1992).

The Zimpro wet air oxidation system in use in Minneapolis–St Paul treats 100 to 150 tons of wet sludge daily (Oslas, 1990). The total process consists of:

1. *Thickening*   Primary sludge is gravity thickened to 6 per cent DS. WAS is thickened by DAF to 3.5 to 4 per cent DS.
2. *Blending and equalization*   The sludge is blended: 72 per cent WAS + 28 per cent primary. Volatile content is ~ 70 per cent.
3. *Wet air oxidation*   Eight oxidation units thermally condition the sludge at 190 °C and 2.2 MPa.
4. *Decanting*   In covered decant tanks, the sludge is thickened to 12 to 18 per cent DS.
5. *Dewatering*   Plate and frame filter presses dewater the almost inorganic sludge to ~ 40 per cent DS.
6. *Combustion*   Dewater cake is burned in six multiple hearth furnaces.
7. *Waste heat recovery*   Heat is recovered to supply energy to the oxidation units and other plant processes.

13.10.8 Sludge Melting

This process is comparable to incineration and is most popular in Japan with 14 plants (1990) treating about 10 per cent of the sludge (Takeshi and Kazuhiro, 1992). Organic sludges are reduced to about one-thirteenth of their original volume by incineration and to about one-thirtieth by the melting process. Melting involves several steps, including:

- Pre-treatment, i.e. dewatering
- Melting
- Heat recovery
- Waste gas purification
- Slag production

In the melting stage, the sludge is heated to $> 1200\,^\circ\text{C}$ to evaporate the water and to thermally decompose and melt the inorganic components. After melting, molten inorganic materials are converted to beneficial end products, e.g. slag. The characteristics of the final slag will depend on the cooling method used. Rapid cooling produces a vitreous low-strength slag, while slow cooling produces a high-strength slag. The types of melting technologies include:

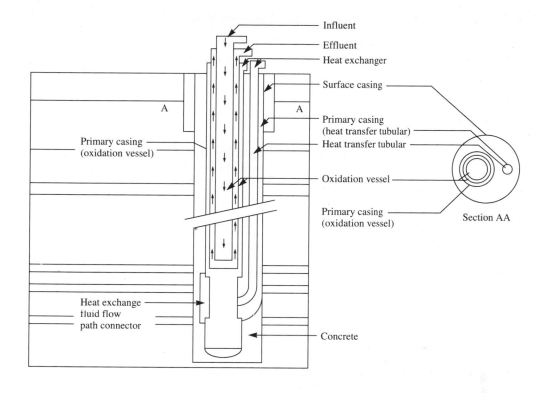

Vertech wet oxidation installation

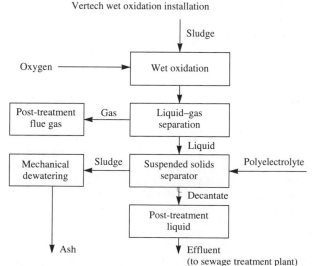

Figure 13.21 Schematic diagram of Vertech wet oxidation process (adapted from Rijs, 1992, with permission).

- Coke bed melting furnace
- Reflector melting furnace
- Cyclone melting furnace

These technologies are described in Takeshi and Kazuhiro (1992).

13.10.9 Summary of Second Stage Sludge Treatment Processes

Table 13.15 summarizes the processes described in the previous sections and suggests that operation costs are about ECU 100 per tonne. Note that capital costs are not included here. For details on these costs refer to Barnett (1993).

13.11 SLUDGE DISPOSAL

Sewage sludges are produced from conventional wastewater plants with dry solids varying from 0.5 to 95 per cent. At the low end, 0.5 to 4 per cent DS, are those sludges produced directly by wastewater treatment plants. These sludges still consist of biodegradable organic fractions, pathogens and lots of water. At the high end, ~ 90 per cent DS, the sludges are typically inorganic, with no biodegradable fractions left, and are likely to be pathogen free. The operational cost of getting from the low solids content to the high solids content is approximately ECU 100 per tonne (excluding dewatering costs) for a plant of population equivalent of 500 000. Dewatering from about 3 to 25 per cent DS has operational costs of the order of ECU 100 per tonne (Andreasen and Nielsen, 1992) for a plant of 50 000 population equivalent. For instance, to treat sludge from a 500 000 population equivalent costs are roughly:

Table 13.15 Sludge treatment processes summary

Process	Input sludge (% DS)	Output	Size for unit cost population equivalent	Cost per tonne (Barnett, 1993)
Anaerobic digestion	6–10	3% DS methane gas	—	—
Aerobic digestion	3–6	4–8% DS reduced biodegradable fraction by $\sim 50\%$	—	—
Composting	25	Compost	500 000	ECU 80
Thermal drying	25	95% DS granules, waste gas	500 000	ECU 85
Incineration	30	Inert ash, waste gas	500 000	ECU 90–150
Pyrolysis OFS	95	Usable oil and char	500 000	ECU 45–120
Gasification	80–85	Usable gas, waste ash	500 000	ECU 110
Wet air oxidation	5	Inorganic sludge, waste gas, purifiable liquid	500 000	ECU 195
Sludge melting	75–90	Recycle gas, usable slag	500 000	ECU 145

Adapted in part from Barnett, 1993 with permission

Assume $\sim$3 per cent DS input sludge:
- Construction costs of dewatering equipment $\sim$ECU 140 per tonne
- Operation costs of dewatering equipment $\sim$ECU 120 per tonne

Output after dewatering is sludge at about 25 per cent DS:
- Thermal drying to 90 per cent DS has a
  - Capital cost of $\sim$ECU 150
  - Operating cost of $\sim$ECU 85

To get a very dry sludge, $\sim$90 per cent DS, the capital and operational costs are therefore of the order of ECU 500 per tonne. There is still sludge to be disposed of. To further treat it by melting and generate a recycle gas and usable slag (with no residual sludge) has

- Capital cost $\sim$ECU 250
- Operating cost $\sim$ECU 145

Therefore, to exploit the best technological means on sludge treatment could cost close to ECU 1000 per tonne. This is a huge difference from the ECU 40 per tonne cost of disposal on land as surface spreading. The latter, of course, was the common traditional disposal route. Technology exists to completely transform sludge to an inert slag but the cost is up to ECU 1000 per tonne, plus the indirect environmental costs of the air emissions, etc. Maybe the technology of treatment and disposal has lost sight of the beneficial uses that can be made of sludge. As such, there are evolving policies internationally to utilize raw sludge as land application or land injection (see Chapters 10 and 17). The disposal routes for sludge include the following:

- Ocean dumping
- Incineration
- Land spreading
- Soil injection
- Land revegetation
- Land reclamation (after mixing, etc.)
- Land reclamation (from the sea)
- Land spreading in forestry
- Landfill—co-disposal
- Landfill—monofill

Table 13.16 outlines the disposal routes of several countries as of 1990. Many changes are happening in the 1990s, especially so as to comply with the elimination of sea dumping. Some countries will increase disposal to land (e.g. United Kingdom, Ireland) while others will reduce disposal to land (e.g. Netherlands, Japan).

13.11.1 Ocean Dumping

This disposal route was most common for island countries and urban areas on the coast; e.g. New York, Los Angeles, London, Tokyo, etc., all disposed of most of their sludges out into the seas. The Helsinki Agreement was signed by most countries to eliminate ocean dumping by 1998. This has been written into the EC Directive on urban wastewater as 31 December, 1998. However, many countries have already stopped. The ban on ocean dumping was signed by the US Congress in 1988 and States were given up to four years to meet the ban. There is therefore no short-term future for ocean dumping.

13.11.2 Incineration

This topic was briefly discussed in Sec. 13.10, where it was considered as a sludge treatment technology rather than a sludge disposal route technology. It is not discussed in any further detail here except to

Table 13.16 Disposal routes for sludge as of 1990
($\times 1000$ tons dry sludge p.a.)

Country	Agriculture	Landfill	Incineration	Sea dumping
Belgium	8	15	6	0
West Germany	698	1286	196	0
Denmark	57	39	35	0
France	234	446	170	0
Greece	30	120	0	0
Ireland	7	4	0	12
Italy	270	440	90	0
Netherlands	150	81	28	22
Portugal	3	7	0	15
Spain	173	28	0	79
UK	524	122	74	225
Sweden	108	72	0	0
Switzerland	113	80	57	0
Austria	57	67	74	0
Japan[†]	200	1400	600	400
USA[†]	48%	33%	14%	5%

[†]Estimated.
Adapted from McIntosh and Matthews, 1992 with permission

reiterate that dewatered sludge of about 25 per cent DS can be incinerated to totally eliminate water, organics and pathogens. What is left is an ash (possibly containing some metals). This slag is best disposed of in a mono-landfill or combined with other materials to produce bricks or road pavements.

13.11.3 Land Spreading and Injection of Sludge

Land application of sewage sludge is here defined as the beneficial application of sludge as a soil amendment to agricultural land. Due to intensive agriculture, some soils became nutrient deficient. Sewage sludges or biosolids are derived from an organic source and so can provide a soil amendment that contains essential plant nutrients in a relatively balanced package. The major nutrient not provided in significant quantities by sludge is potassium (K). Some soils can be deficient in single elements like Zn, which may be amended by sludge. The land spreading of sewage sludge is dependent on many parameters including:

• Climate and rainfall
• Existing soil nutrient status
• Proposed cropping
• Sludge characteristics—chemical
• Potential for crop contamination
• Potential for groundwater (or surface water) contamination

Soil has also been amended with sawdust-stabilized sludge, i.e. one part sludge mixed with two parts sawdust. Some typical US biosolids characteristics are given in Table 13.17.

The level of nutrients in sewage sludge is significant, as seen in Table 13.17. Nitrogen can be as high as 1 per cent. Typical application rates are limited to 10 tons DS/ha or 200 wet tons/ha at 5 per cent DS. The application rate for sludge is typically designed for either the N or P levels of the crop grown on a particular soil. The general approach for determining application rates are:

1. Nutrient crop requirement, taking account of carry-over from previous years.
2. N crop needs, Cd limitation and P crop needs.
3. Sludge application is terminated when cumulative metal limits are reached (see Table 13.6 for limits).

Table 13.17 Some typical US biosolids characteristics

	Parameter (mg/kg)		
	Rock creed (belt press), Portland, Oregon (Lloyd and Wilson, 1992)	Rock creed stockpiled, Portland, Oregon (Lloyd and Wilson, 1992)	Willow Lake WWTP, Salem, Oregon (Cork et al., 1992)
PH	7.8	8.1	—
TS%	14.8	30.5	2.3
TVS%	61	84	—
TKN-N (mg/L)	60 874[†]	25 857[†]	103 075[†]
NH_3-N	9446	4321	42 068
NO_3-N	344	120	—
PO_4-P	47 065	18 869	61 906
K	2209	1149	17 645
As	9	—	—
Cd	4	1	2.8
Cr	57	29	93
Cu	821	296	505
Pb	71	32	181
Hg	2	—	4.5
Ni	22	11	35
Ag	32	—	42
Zn	950	338	817
Fe	—	—	13 000
Ba	—	—	366
Mn	—	—	185
Al	—	—	9950
Ca	—	—	18 500
Mg	—	—	6500
Na	—	—	7320

[†]All including and below this parameter are in concentrations of mg/kg.
Adapted from Wilson, 1992, and Cork et al., 1992

The nitrogen available to plants (N_p) in any particular year is the total of:

1. All NO_3 present in the sludge.
2. All or a fraction of NH_4 in the sludge. Typically, if sludge is land applied 50 per cent of NH_4 is volatilized. Assume 100 per cent of NH_4 is available to plants if the sludge is injected.
3. A fraction of organic N, i.e. N_o in the sludge that is mineralized in the first year after application.

A summation is made of the N_o in the sludge applied during the previous years (if any) which will mineralize during the year being calculated. Thus

$$N_p = 10S(NO_3 + K_vNH_4 + F_{year}N_o) \tag{13.15}$$

where

N_p = plant available N for this year

S = sludge application rate, TDS/ha/yr

NO_3 = % NO_3 in sludge

K_v = volatile factor

NH_4 = % NH_4 in sludge

F_{year} = factor for N_o in sludge in year 0-1, 1-2, etc.

N_o = % N_o (organic nitrogen)

Example 13.9   Determine the amount of N available from sludge in the first year if 5 TDS/ha is added. Sludge characteristics are:

$$N_o \text{ (organic nitrogen)} = 3\%, NO_3 - N = 0\%, NH_4 - N = 1.5\%$$
$$K_v = 0.5, F_{year} = 0.2$$

Solution

$$N_p = 10S(NO_3 + K_v NH_4 + F_{0-1}, N_o)$$
$$= 50[0 + 0.5(1.5) + 0.2(3)]$$
$$= 67.5 \text{ kg } N_p/\text{ha}$$

Example 13.10   For the above example, calculate the N_o available from this year's sludge for the next three years.

Solution

$$\text{Total } N_o = 0.03 \times 5 \times 1000 = 150 \text{ kg/ha}$$

Given the mineralization factors as (USEPA, 1983)

Year	F
0–1	0.2
1–2	0.1
2–3	0.05

$$N_o \text{ mineralized in year } 0\text{–}1 = 0.2 \times 150 = 30 \text{ kg/ha}$$
$$N_o \text{ remaining in year } 1\text{–}2 = 150 - 30 = 120 \text{ kg/ha}$$
$$N_o \text{ mineralized in year } 1\text{–}2 = 0.1 \times 120 = 12 \text{ kg/ha}$$
$$N_o \text{ remaining in year } 2\text{–}3 = 120 - 12 = 108 \text{ kg/ha}$$
$$N_o \text{ mineralized in year } 2\text{–}3 = 0.05 \times 108 = 5.4 \text{ kg/ha}$$

Example 13.11   If the P fertilizer needs of corn is 45 kg/ha, determine the application rate required to satisfy this if the total P in sludge is 2 per cent. Assume that 50 per cent P is plant available.

Solution

$$P_{available} = (\% \ P \text{ in sludge}) \ (50\% \text{ available})$$
$$P_p = 0.02 \times 0.5 \times 1000 = 10 \text{ kg/t}$$
$$\text{Application rate } S_p = \frac{C_p}{P_p}$$
$$= \frac{45}{10} = 4.5 \text{ t/ha}$$

The reader is referred to USEPA (1983) for further and more detailed examples. Locations that use land spreading are many. Traditionally, this was the most common disposal route in many countries for raw sludges and dewatered sludges.

13.11.4 Land Revegetation

Land in many places that has been overgrazed and that has expended its nutrients may be brought back into commission quickly by application of sewage sludge. The conditions for application are in this case more metal limited than nutrient limited which was the case of land spreading. While guidelines for landspreading of sewage sludge on to crop land set a limit of 10 t DS/ha/yr, there is no such limit for land revegetation. The application rates may vary from 3 to 200 t DS/ha/yr. As the application may be a one-off or may be at five year intervals (or so), then more lenient application rates exist. An application of revegetation was described by Wagner E.O., (1992). At that time, some of the excess sludge from POTWs in New York were transported to western Texas to revegetate overgrazed pasture land.

13.11.5 Disturbed Land Reclamation

The application of sewage sludge to revegetate mining sites has been going on for several decades (Sopper, 1993). As mining is an international practice, the opportunities exist for utilizing large volumes of sludge to restore these landscapes in almost all countries. Difficulties have been encountered with revegetation due to low pH of these lands, their low field capacity (unable to hold much water) and the presence of high concentrations of heavy metals. The lands potentially suitable for restoration in this way are:

- Coal mining areas
- Sand and gravel pits
- Other mine areas, e.g. zinc, uranium, etc.
- Landfill cover areas
- Areas blasted for rock, i.e. rock quarries
- Beach retention
- Desert areas
- Slag piles from mining, coal, etc.

The rates of sludge application to revegetate areas is typically one or two orders of magnitude greater than that for sludge applied to cropland. As such, the potential problems are: leaching of nitrogen to groundwater; surface runoff of non-soil-bound metals and pathogen transmission. In the design of a sludge reclamation plan, the same procedures as suggested in USEPA (1983) can be followed. In the latter document, this topic is named 'disturbed land'. As the sludge application is one time application (or maybe at 5-year interludes), a sufficient amount of land should be available for a programme to use up the available sludge.

Example 13.12 (This example is adapted from USEPA, 1983.)   Determine the sludge application rate to disturbed land if the sludge characteristics are:

$$\begin{array}{ll}
\text{TVS \%} = 54 & Pb = 500\,mg/kg\ (0.5\,kg/T) \\
\text{Total N \%} = 1.5 & Zn = 2000\,mg/kg\ (2.0\,kg/T) \\
NH_4\text{--}N\ \% = 0.6 & Cu = 500\,mg/kg\ (0.5\,kg/T) \\
\text{Total P \%} = 0.5 & Ni = 100\,mg/kg\ (0.1\,kg/T) \\
\text{Total K \%} = 0.1 & Cd = 50\,mg/kg\ (0.05\,kg/T)
\end{array}$$

The site characteristics are:

$$\begin{array}{ll}
\text{Area} = 2\,ha & \text{Soil permeability } K = 0.2\ cm/h \\
\text{Soil pH} = 3.9 & \text{Depth to groundwater} = 5\,m \\
\text{Soil CEC} = 13\,meq/100\,g & \text{AAR} = 1200\,mm
\end{array}$$

Assume that the uptake of nitrogen is 300 kg/year.

Solution

(a) *Maximum sludge loading based on metal loadings* From USEPA 40 CFR Part 503, the following are:

$$Pb = 300 \text{ kg/ha} \quad (1120) \quad [150]$$
$$Zn = 2800 \text{ kg/ha} \quad (560) \quad [300]$$
$$Cu = 1500 \text{ kg/ha} \quad (280) \quad [120]$$
$$Ni = 420 \text{ kg/ha} \quad (280) \quad [30]$$
$$Cd = 39 \text{ kg/ha} \quad (10) \quad [1.5]$$

The numbers in round brackets are those that were applicable at the time of USEPA (1983). It is noted that many of the limits have changed. The EC limits as set out in the sludge directive are within the square brackets, based on a 10-year application.

$$T \text{ sludge/ha} = \frac{\text{kg/ha of metal allowed}}{0.001 \times \text{mg/kg of metal in sludge}}$$

$$\text{For Zn} = \frac{2800}{0.001 \times 2000} = 1400 \text{ t/ha}$$

$$Pb = \frac{300}{0.001 \times 500} = 600 \text{ t/ha}$$

$$Cu = \frac{1500}{0.001 \times 500} = 3000 \text{ t/ha}$$

$$Ni = \frac{420}{0.001 \times 100} = 4200 \text{ t/ha}$$

$$Cd = \frac{39}{0.001 \times 50} = 780 \text{ t/ha}$$

The limiting metal is Pb.

(b) *Potential nitrate leaching to groundwater* This calculation is in three steps, i.e.:

(1) Calculate the available nitrogen added by sludge application.
(2) Subtract the estimated nitrogen uptake by vegetation and losses.
(3) Calculate the maximum potential concentration of excess nitrogen percolating to the aquifer.

Assume that 600 t/ha of sludge is applied.

$$\text{NH}_4\text{–N applied} = 600 \times 10^3 \times \frac{0.6}{100} = 3600 \text{ kg/ha}$$

$$\text{Organic N applied} = \frac{1.5 - 0.6}{100} \times 600 \times 10^3 = 5400 \text{ kg/ha}$$

Assume all of the NH_4–N is available in the first year.
Assume the organic mineral rates are:

20%—first year; 10%—second year; 5%—3rd year; following years—3%

For first year:

$$\text{NH}_4\text{–N} = 3600 \text{ kg/ha}$$
$$\text{Organic N} = 5400 \times 0.2 = 1080 \text{ kg/ha}$$
$$\text{Total N} = 4680 \text{ kg/ha}$$

$$\text{Unaccountable losses} \sim 10\% = 468\,\text{kg/ha}$$
$$\text{Vegetation uptake} = 300\,\text{kg/ha}$$
$$\text{Total excess} = 4680 - 468 - 300 = 3912\,\text{kg/ha}$$

For second year:

$$NH_4\text{–}N = 0\,\text{kg/ha}$$
$$\text{Organic N} = 0.1\,(5400 - 1080) = 432\,\text{kg/ha}$$
$$\text{Unaccountable losses} \sim 10\% = 43\,\text{kg/ha}$$
$$\text{Vegetation uptake} = 300\,\text{kg/ha}$$
$$\text{Total excess} = 432 - 343 = 89\,\text{kg/ha}$$

Groundwater contamination potential is only possible in year 1 as the available excess N in year 2 is small and in later years there is a deficit.

$$\text{Gross rainfall} = 1200\,\text{mm}$$
$$\text{Assume evaporation} = 30\% = 360\,\text{mm}$$
$$\text{Net rainfall} = 840\,\text{mm}$$

If all the excess N in the sludge is mobile (assume no N is bonded to sludge/soil particles) then the concentration of nitrate in the percolate is:

$$\text{Total excess N} = 3912\,\text{kg/ha} \times 10^6\,\text{mg/kg}$$
$$= 3912 \times 10^6\,\text{mg/ha}$$
$$\text{N per ha} = \frac{3912 \times 10^6\,\text{mg/ha}}{10\,000\,\text{m}^2/\text{ha}}$$
$$= 3912 \times 10^2\,\text{mg/m}^2$$
$$\text{N per ha per m rain} = \frac{3912 \times 10^2\,\text{mg/m}^2}{0.84\,\text{m}}$$
$$\text{N per m}^3 = 4657 \times 10^2\,\text{mg/m}^3$$
$$= \frac{4657 \times 10^2\,\text{mg/m}^3}{10^3\,\text{L/m}^3}$$
$$\text{N mg/L} = 466\,\text{mg/L}$$

The permissible percolate concentration is here assumed to be 10 mg/L. The proposed EC Landfill Directive on leachate concentrate for inert waste suggests 3 mg/L while for hazardous waste the range is 6 to 30 mg/L. Therefore, the limitation in this example is nitrate leaching. A recalculation based on 10 mg/L suggests that the permissible excess nitrogen in year 1 is

Therefore,

$$\text{Total excess N} = \frac{10}{466} \times 3912 = 84\,\text{kg/ha}$$

$$\text{Maximum allowable excess N} = 84\,\text{kg/ha}$$
$$\text{Vegetation uptake} = 300\,\text{kg/ha}$$
$$\text{Total N} = 384\,\text{kg/ha}$$
$$\text{Assume 10\% losses} = 38\,\text{kg/ha}$$

Then
$$\text{Total} = 422\,\text{kg/ha}$$

and
$$\text{Sludge/year} = \frac{\text{kg/ha of available N} \times 0.1}{\%\ NH_4\text{--}N \times 1.0 + \%\ O - N \times 0.2}$$

$$= \frac{422 \times 0.1}{0.6 \times 1.0 + 0.9 \times 0.2}$$

$$= 54\,\text{t/yr}$$

13.11.6 Land Reclamation from the Sea

It is possible to reclaim sea land with sludges if the outer perimeter boundary of the proposed reclaimed land is first constructed of local dredged material. Like other sludge processes, limits of metals and nutrients are the design criteria. However, in addition to these, there are the structural and geotechnical characteristics.

13.11.7 Land Spreading of Sewage Sludge in Forestry

Land spreading in forest lands is also a beneficial end use. However, forest lands are not always within an economic travel distance of urban wastewater treatment plants. Another difficulty is the mechanics of obtaining an even spread of sludge due to the frequent interruptions by the presence of trees. The methodology of spreading is by spray gun and this can also damage the bark of trees if the force is too strong. Spreading is easier if the access is more frequent, as is sometimes the case with frequent fire-break roads or tracks. Besides the technical difficulties associated with actual spreading and the cost implications of transport, other constraints are related to sludge constituents, including pathogens, and nitrate leaching.

The estimated nitrogen uptake by forest types ranges from about 100 to 300 kg/ha/yr, with older trees having a higher uptake than tree seedlings. The losses due to volatilization of ammonia N are estimated at up to 50 per cent for sludge applied in liquid form (< 4 per cent DS) and close to zero losses for applied dewatered sludge. Accelerated tree growth (200 to 300 per cent) can result from the application of sludge. This of course will change the characteristics of the wood in relation to moisture content, structural properties, etc. In the public forests of the United Kingdom, approximately 0.5 t of fertilizer/ha are added annually. This figure could readily be accomplished by the fertilizer value of sludge. An evaluation has determined that between 6 and 11 per cent of UK sludge proportion could be spread on forest land at economic rates. This could increase if intensive coppice woodland using fast-growing species of willow and poplar are used. This crop is likely to be developed as a bioenergy crop of the future. A success story in relation to forest application of sludge (Leonard *et al.*, 1992) is managed by the Municipality of Metropolitan Seattle. Sludge is transported in dewatered cake, 23 per cent DS, and is wetted on site to 7 to 15 per cent DS. It is then sprayed on to the trees. The application rate is about 19 t DS/ha/yr. The rigorously managed site has had nine years of experience and is planned for at least another eight years. Further details and numerical examples on application rates are found in USEPA (1983).

13.11.8 Landfilling of Sludges

Sludges are landfilled in two methods:

- Mono-landfill
- Co-disposal landfill

The traditional practice has been to co-dispose sludges with municipal solid wastes at municipal landfill sites. The requirements include:

- A dry solids sludge content >35 per cent
- A sludge shear strength $>10\,kN/m^2$
- A mass of municipal waste/sludge ratio of $>10:1$

The methodology most commonly used was:

- Dump the sludge at the tip of the landfill face and spread in a layer $<0.25\,m$ thick
- Spread the municipal waste from the top of the landfill face and follow up by covering the sludge with a layer of municipal waste $>1\,m$ thick
- Use side slopes flatter than $3:1$

The environmental aspects of such a practice included:

- Significantly reduced COD values of leachate
- Reduced VFA values in leachate
- Increased pH values in leachate from about 6.0 to 7.5
- Increased CH_4 production

Some of the methodologies are:

- Sludge-only trenches
- Narrow trench
- Wide trench
- Sludge-only area fill
- Area fill mound
- Area fill layer
- Dyked containment

The sludge and site conditions and design criteria for the above methods as defined by USEPA (1978) are listed in Table 13.18. Note that it is necessary to use clay on synthetic liners for all of these methods.

Example 13.13 Determine the landfill area required to dispose of $50\,m^3/day$, 5 days/week at 25 per cent DS using wide trenches. Assume the life required is 20 years and the trenches are 10 m wide, 3 m deep and 100 m long. Assume 3 m solid ground spacing between trenches and a 20 m buffer to the property boundary.

Solution

$$\text{Required trench volume} = 50 \times 5 \times 52 \times 20 = 260\,000\,m^3$$

$$\text{Number of trenches required} = \frac{260\,000}{10 \times 3 \times 100} = 86.6 \text{ trenches}$$

$$\text{Usable area required} = 87 \times (10+3)(100+3) = 116\,490\,m^2$$

$$= 11.6\,ha$$

$$116\,490\,m^2 = 341\,m \times 341\,m$$

$$\text{Minimum area required} = (341+40)(341+40) = 145\,161\,m^2$$

$$= 14.5\,ha$$

Example 13.14 Determine the trench utilization rate, the sludge application rate, the land utilization rate and the site life for a sludge-only landfill of 50 ha (usable area). The sludge is 14 per cent DS and the wastewater treatment plant produces $25\,t\,DS/day$. The specific gravity of wet sludge is 1.056

Table 13.18 Sludge and site conditions for landfilling of sludges

Parameter	Narrow trench	Wide trench	Sludge-only area fill	Area fill layer	Dyked containment	Area fill mound	Co-disposal	Sludge/soil mixture
Sludge solids	15–20% for 0.6–0.9 m widths	>28% for front end loader, 20–28% for drag line	—	>15%	20–28%	>20%	>3%	>20%
Sludge characteristics	Unstabilized or stabilized	Unstabilized or stabilized	—	Stabilized	Unstabilized or stabilized	Stabilized	Unstabilized or stabilized	Stabilized
Hydrogeology	Deep GW and bedrock	Deep GW and bedrock	Shallow GW or bedrock possible	Shallow GW or bedrock possible	Shallow GW or deep bedrock possible	Shallow GW or bedrock possible	Deep or shallow GW or bedrock	Deep or shallow GW or bedrock
Ground slopes	<20%	<10%	—	—	—	—	<30%	<5%
Trench width	0.6–3.0 m	>3 m	—	—	—	—	—	—
Bulking required	No	No	Yes/soil	Yes/soil 0.25–1 soil: 1 sludge	Not necessary, If yes: 0.25–1 soil: 1 sludge	Yes/soil 0.2–2 soil: 1 sludge	Yes/MSW 4–7 t MSW: t wet sludge	Yes/soil 1:1
Cover soil required	Yes	Yes	—	Yes	Yes	Yes	Yes	No
Cover soil thickness	0.9–1.2 m	0.9–1.5 m	—	0.15–0.3 m interim, 0.6–1.2 m final	0.3–0.9 m interim, 1.2–1.5 final	0.9 m interim, 0.3 m final	0.15–3 m interim, 0.6 final	—
Imported soil required	No	No	—	Yes	Yes	Yes	No	No
Application rate (m³/ha)	2300–10 300	33 200–27 400	—	3800–17 000	9100–28 400	5700–34 600	900–7900	3000

Adapted from USEPA, 1978 with permission

and so the wet sludge is $179\,t/day$ ($= 169.5\,m^3/day$). Assume the trenches are $2.4\,m$ excavated depth with a $1.8\,m$ sludge fill, a $1.2\,m$ cover thickness and a width of $0.6\,m$. The maximum trench length is to be $50\,m$ and trenches are to be spaced at $2.4\,m$ apart.

Solution

$$\text{Trench utilization rate} = \frac{\text{sludge volume/day}}{\text{trench cross-section}}$$

$$= \frac{169.5\,m^3/day}{1.8 \times 0.6} = 157\,m/day$$

$$\text{Sludge application rate} = \frac{\text{cross-sectional area of sludge in trench}}{\text{width of trench} + \text{trench spacing}}$$

$$= \frac{1.8 \times 0.6\,m^2}{0.6 + 2.4\,m}$$

$$= 0.36\,m^2/m$$

$$= 3600\,m^3/10\,000\,m^2$$

$$= 3600\,m^3/ha$$

$$\text{Land utilization rate} = \frac{\text{sludge volume/day}}{\text{sludge application rate}}$$

$$= \frac{169.5\,m^3/day}{3600\,m^3/ha}$$

$$= 0.047\,ha/day$$

$$\text{Site life} = \frac{\text{usable fill area}}{\text{land utilization rate}}$$

$$= \frac{50\,ha}{0.047\,ha/day}$$

$$= 1064\,\text{days} = 2.92\,\text{years}$$

13.12 INTEGRATED SEWAGE SLUDGE MANAGEMENT

Traditionally, municipalities relied on one form of final sludge treatment/disposal. In Ireland, for instance, the municipalities relied on landspreading, but with the environmental awareness, farmers feared metals and pathogens and some inland municipalities lost this disposal route. Many coastal cities, including New York and Los Angeles, lost their sea disposal route which was their main option. Other locations are losing their incineration route. This era in time is one of changing environmental legislation and increased public environmental awareness. It is then pragmatic that municipalities should have a portfolio of final disposal routes, so that if one is closed due to more stringent regulations, then the slack may be taken up by the remaining routes. This chapter has tried to develop the broad picture of sludge treatment and disposal and hopes that the beneficial uses of sludge will be more exploited in the future.

13.13 PROBLEMS

13.1   Visit a municipal wastewater treatment plant in your area. Draw a plan view of the relative location of each of the unit processes. Identify the flow rates, BOD_5, SS loading, sludge production daily and quality of effluent.

13.2   Visit a municipal or industrial wastewater treatment plant in your area that uses anaerobic digestion as well as other wastewater treatment processes. Describe the specific anaerobic digestion process with regard to waste loads, tank type, dimensions, quality of effluent, volume of gas, etc.

13.3   A municipal wastewater treatment plant serves a population equivalent of 200 000. The primary sludge has a 1.5 per cent DS, which is 70 per cent volatile. An anaerobic digestion plant using this primary sludge feedstock has a 60 per cent destruction of volatile solids and a 20-day retention time. The digested solids are 5 per cent DS. Determine:
   (a) the amount of sludge produced daily,
   (b) the rate of gas production daily,
   (c) the digester volume.

13.4   Repeat Problem 13.3 if the retention time is 30 days.

13.5   Review the paper 'Capillary suction time as a measure of sludge dewaterability' by Vesilind (1988).

13.6   Anaerobic digestion feedstock is often computed in terms of equivalent glucose, $C_6H_{12}O_6$. Use the Bushwell and Mueller (1952) equation (13.7) to compute the potential methane production of fructose, sucrose, carbohydrates and glucose.

13.7   A bench-scale laboratory digestion plant of mixed sludges produced the following results for a single day during a long programme:
   Reactor volume $= 5$ L
   Feedstock $COD_{inf} = 3000$ mg/L
   Biogas $= 1.4$ m^3/m^3/day
   % $CH_4 = 53$ per cent
   Daily feed $= 125$ ml

Determine the volume of CH_4 per kg of mixed sludge digested (Eq. (13.8)) if the growth rate is ignored.

13.8   Explain with sketches the ways water is 'tied up' in sludge. Explain the relevance of the 'zeta potential'.

13.9   Explain the following terms as used in sludge processing:
   (a) alkalinity,
   (b) hydraulic detention time,
   (c) solids retention time,
   (d) pH,
   (e) temperature,
   (f) gas production.

13.10   Anaerobic digestion processes sometimes fail due to toxicity of the micro-organisms. Briefly explain the nature of ammonia toxicity, distinguishing between ionized and un-ionized ammonia. At what levels of un-ionized ammonia does toxicity occur? Plot curves relating total ammonia, free ammonia, pH and temperature.

13.11   Review the paper 'Dynamic modelling of anaerobic digestion' by Moletta et al. (1986).

13.12   Review the paper 'A comparative study of full scale sludge dewatering' by Andreasen and Nielsen (1992).

13.13   Determine the amount of plant available nitrogen in sewage sludge if its characteristics are:

Organic $N = 2.5\%$, NO_3–$N = 0.2\%$, NH_4–$N = 1.7\%$

For this sludge, calculate the available plant nitrogen from this year's sludge for the next five years.

13.14   Sludge applied to disturbed land has the following characteristics:

TVS% $= 60$ Pb $= 600$ mg/kg
Total N% $= 1.2$ Zn $= 1500$ mg/kg

$$NH_4-N\% = 0.5 \qquad Cu = 450\,mg/kg$$
$$Total\ P\% = 0.6 \qquad Ni = 250\,mg/kg$$
$$Total\ K\% = 0.2 \qquad Cd = 80\,mg/kg$$

The site characteristics are:

Area $= 25$ ha $\qquad$ Soil permeability $K = 0.05$ cm/h

Soil pH $= 4.2$ $\qquad$ Depth to groundwater $= 8$ m

Soil CEC $= 115$ meq/100 g $\qquad$ AAR $= 800$ mm

Do the calculations for this problem based on Example 13.12.

13.15 Write a brief description of the application of sewage sludge to forest land. Refer to Leonard *et al.* (1992).

13.16 Determine the landfill area to dispose of 250 t/day, 5 days per week, at 15 per cent DS using wide trenches. Assume the minimum life should be 20 years. Also assume trenches to be 8 m wide × 3 m deep × 50 m long. Assume 5 m solid ground spacing between trenches.

13.17 Determine the trend utilization rate, the sludge application rate and the land utilization rate and the site life for a sludge-only landfill of 100 ha usable area. The sludge is 12 per cent DS and the plant produces 55 t DS/day. The specific gravity of wet sludge is 1.05. Assume trenches are 2.5 excavated depth with a 2 m sludge fill, and a 1.0 m cover thickness and a width of 0.6 m. The maximum trench length is to be 50 m and the trenches are to be spaced at 3 m.

13.18 You have just designed a modern municipal wastewater treatment for a population equivalent of 400 000. Prepare a flow chart of the unit processes and where and what characteristics of sludge is produced. Estimate the quantities of sludge at each generation point.

13.19 For Problem 13.18, size an anaerobic digestion plant (maybe several reactors) to treat the sludge.

13.20 For the previous problem, outline the first stage and second stage sludge treatment processes you would recommend.

13.21 For the previous problem, develop a disposal strategy for the sludge to consist of:
   (a) land spreading,
   (b) landfilling,
   (c) composting.
   What fractions do you recommend to go to each stream?

REFERENCES AND FURTHER READING

Albertson, O. E. *et al.* (1991) 'Dewatering municipal wastewater sludges', Pollution Technology Review No. 202, Noyes Data Corporation (NDC), New Jersey.

American Society of Civil Engineers (ASCE) (1983) *A Survey of Anaerobic Digestion Operations*, ASCE, New York, USA.

American Society of Civil Engineers (ASCE) (1987) *Land Application of Wastewater Sludge*, T. M. Younas (ed.), ASCE, New York, USA.

Andreasen, I. and B. Nielsen (1992) 'A comparative study of full-scale sludge dewatering equipment', Proceedings of Conference on *Wastewater Sludge Dewatering*. University of Aalborg, Denmark.

Bailey, J. E. and D. F. Ollis (1986) *Biochemical Engineering Fundamentals*, McGraw-Hill, New York.

Barnett, D. (1993) 'Sewage sludge, advanced treatment and disposal', Unpublished thesis, University College, Cork, Ireland.

Bellmer (1992) Promotional Brochures on Bellmer Winklerpress and Bellmer Turbodrain, Bellmer GmbH, Germany.

Benedict, A. H. and E. Epsteing (1986) *Composting Municipal Sludge: A Technical Evaluation*, USEPA, Cincinnati, USA.

Borchardt, J., W. J. Redman, G. E. Jones and R. T. Sprague (1981) *Sludge and Its Ultimate Disposal*, Ann Arbor Science.

Bruce, A (1984) *Sewage Sludge Stabilization and Disinfection*, Ellis Horwood, Chichester.

Bruce, A. M., F. Colin and P. J. Newman (1989) *Treatment of Sewage Sludge. Thermophilic Aerobic Digestion and Processing Requirements for Landfilling*, Elsevier, Oxford.

Brunner, C. (1980) *Design of Sewage Sludge Incineration Systems*, Noyes Data Corporation (NDC), New Jersey.

Bushwell, A. M. and H. F. Mueller (1952) 'Mechanics of methane fermentation', *J. Industrial Engineering Chemistry*, **44**(3), 550.

Carberry, J. B. and A. Englande (1979) 'Sludge characteristics and behaviour', Proceedings of NATO Advanced Study Institute, Newark, New Jersey.

Ceechi, F., J. Mata-Alvarez and F. G. Pohland (1992) 'Anaerobic digestion of solid waste', Proceedings of International Conference, Vienna, April.

Cheremisinoff, P. N. (1994) *Sludge Management and Disposal*, PTR Prentice-Hall, Englewood Cliffs, New Jersey.

Colin, F., P. J. Newman and Y. J. Pualanne (1991) *Recent Developments in Sewage Sludge Processing*. Elsevier, Oxford.

Colleran, E. (1980) *Microbiology of Anaerobic Digestion*, University College, Galway, Ireland.

Colwell, R. R. and W. M. Spira (1992) 'The Ecology of Vibrio Cholera' Chapter 6 in Barau, D. and W. B. Greenough III, *Cholera*, Plenum Press, New York.

Cork, T., B. Evansen, P. Eckley and G. Hermann (1992) 'Biagro—a 15-year long success story', Proceedings of Conference on *The Future Direction of Municipal Sludge (Biosolids) Management*, Portland, Oregon, Water Environment Federation, Virginia.

Daniel, D. E. (1993) *Geotechnical Practice for Waste Disposal*, Chapman and Hall, London.

Department of Environment (1993) 'UK sewage sludge survey. Final report', Consultants in Environmental Science Limited, February 1993, HMSO, London, UK.

Department of Health, New Zealand (1992) *Public Health Guidelines for the Safe Use of Sewage Effluent and Sewage Sludge on Land*, DOH, Public Health Series, NZ.

Dirkzwager, A. H. and P. L'Hermite (1989) *Sewage Sludge Treatment and Use. New Developments, Technological Aspects and Environmental Effects*, Elsevier, Amsterdam.

EC Directive 86/278/EEC (1986) 'On the protection of the environment and in particular of the soil, when sewage sludge is used in agriculture'.

Eckenfelder, W. W. (1980) *Principles of Water Quality Management*, CBI Publishing Co, Boston, MA, USA.

Ferranti, M. P., G. L. Ferrero and P. L. Hermit (1987) *Anaerobic Digestion: Results of Research and Demonstration Projects*, Elsevier Applied Science, Essex, England.

Frost, R. C., J. Halliday and A. S. Dee (1992) 'Continuous consolidation of sludge in large scale gravity thickeners', Proceedings of Conference on *Wastewater Sludge Dewatering*, University of Aalborg, Denmark.

Garvey, D., R. D. Davis, C. Guarino and C. H. Carlton-Smith (1992) 'Treatment and disposal of sewage sludge—current practice', Proceedings of *Sludge 2000*—An Anglian Water Conference on *Sewage Sludge Use and Disposal—Strategies, Plans and Operational Practice*, Cambridge, September.

Gujer, W., A. J. B. Zehnder (1983) 'Conversion processes in anaerobic digestion', *Water Science Technology*, **15**, 127.

Hall, J. E. (1992) 'Alternative uses for sewage sludge', Proceedings of *Sludge 2000*—An Anglian Water Conference on *Sewage Sludge Use and Disposal—Strategies, Plans and Operational Practice*, Cambridge, September.

Ho, G. and K. Mathew K (1991) 'Appropriate waste management technologies', Proceedings of the International Conference, Perth, Australia, November 1991, Pergamon Press, LAWQ.

Hobson, P. N. and A. D. Wheatley (1993) *Anaerobic Digestion. Modern Theory and Practice*, Elsevier Applied Science, Oxford.

IAWA (1992) Proceedings of International Conference on *Wastewater Sludge Dewatering*, University of Aalborg, Denmark.

IAWA (1992) Proceedings of *Sludge 2000* International Conference, Cambridge, September.

Jones, G. E. (1981) 'Sludge characteristics', In *Sludge and Its Ultimate Disposal*, Ann Arbor Science, MI, USA.

Kayhanian, M. and G. Tchobanoglous (1993) 'Characteristics of humus produced from the anaerobic composting of the biodegradable fraction of municipal solid waste', *Environmental Technology*, **14**, 815–822.

Kayhanian, M., K. Lindenauer, S. Hardy and G. Tchobanoglous (1991) 'Two-stage process combines anaerobic and aerobic methods', *Biocycle*, March.

Leder, C. S. (1992) 'Albuquerque's approach to odour control at a 15 dry ton per day pilot aerated windrow composting facility', Proceedings of Conference on *The Future Direction of Municipal Sludge (Biosolids) Management*, Portland, Oregon, Water Environment Federation, Virginia.

Leonard, P., R. King and M. Lucas (1992) 'Fertilizing forests with biosolids. How to plan, operate and maintain a long term program', Proceedings of Conference on *The Future Direction of Municipal Sludge (Biosolids) Management*, Portland, Oregon, Water Environment Federation, Virginia.

L'Hermite, P. (1991) *Treatment and Use of Sewage Sludge and Liquid Agricultural Wastes*, Elsevier, Amsterdam.

Lloyd, D. and S. Wilson (1992) 'Agricultural land application. A biosolids program that works', Proceedings of Conference on *The Future Direction of Municipal Sludge (Biosolids) Management*, Portland, Oregon, Water Environment Federation, Virginia.

Lowe, T. J. (1990) *Sludge Reduction at the North Charleston Regional Sewer District*, General Environmental Science Ltd, Cleveland, Ohio.

Lue-Hing, C., D. R. Zenz and R. Kuchenrither (1992) 'Municipal sewage sludge management processing, utilization and disposal', Water Quality Management Library, Vol. 4, Chapter 4, Technomic Publishers, London.

McGivern, F. (1992) 'Conditioning, thickening and dewatering of sewage sludge', Thesis for Dip Env Eng, University College Cork, Ireland.

Machno, P. S. (1992) 'Biosolids quantity and quality. A comparison of primary and secondary treatment', In Proceedings of *Sludge 2000*, Cambridge, September.

McIntosh, P. and P. Matthews (1992) 'Sludge management in the U.K. and E.C.', Proceedings of Conference on *The Future Direction of Municipal Sludge (Biosolids) Management*, Portland, Oregon, Water Environment Federation, Virginia.

Malina, J. F. and F. G. Pohland (1992) *Design of Anaerobic Processes for the Treatment of Industrial and Municipal Wastes*, Technomic Publishing, London.

Martel, C. J. (1992) 'Fundamentals of sludge dewatering in freezing conditions', Proceedings of International Conference on *Wastewater Sludge Dewatering*, University of Aalborg, Denmark.

Metcalf and Eddy, Inc. (1991) *Wastewater Engineering Treatment, Disposal and Reuse*, G. Tchobanoglous and F. Burton (eds), McGraw-Hill, New York.

Moletta. R., D. Verrier and A. Albagnac (1986) 'Dynamic modelling of anaerobic digestion', *Water Research*, **20**(4), 427–434.

Montgomery, J. M. (Consulting Engineers) (1985) *Water Treatment, Principles and Design*, John Wiley, Chichester.

Mudrack, K. and S. Kunst (1986) *Biology of Sewage Treatment and Water Pollution Control*, Ellis Horwood, Chichester.

Murphy, J. (1992) Direct Treatment of Urban Wastewater at Ambient Temperatures, unpublished M.Eng.Sc. thesis, University College Cork, Ireland.

Patrick, A. (1992) 'Utilisation of sewage sludge in silviculture', Thesis for Dip Env Eng, University College Cork, Ireland.

Pavlostathis, S. G. and E. Giraldo-Gomez (1991) *Kinetics of Anaerobic Treatment. A Critical Review*, Critical Review in Environmental Control, Vol. 21(5,6), CRC Press, Boca Raton, Florida, pp. 411–490.

Pedersen, D. C. (1981) *Density Levels of Pathogenic Organisms in Municipal Wastewater Sludge. A Literature Review*, Camp Dresser and McKee, Boston.

Reynolds, T. D. (1982) *Unit Operations and Processes in Environmental Engineering*, PWS-Kent Publishing Co., Boston, MA, USA.

Rijs, G. B. J. (1992) 'Advanced techniques for minimizing the volume of sewage sludge', Proceedings of Conference on *The Future Direction of Municipal Sludge (Biosolids) Management*, Portland, Oregon, Water Environment Federation, Virginia.

Schroeder, E. D. (1977) Water and Wastewater Treatment, McGraw-Hill, New York, USA.

Scott, D. and J. Dickey (1992) 'Land application of biosolids. Making it work for the farmer, city and regulator', Proceedings of Conference on *The Future Direction of Municipal Sludge (Biosolids) Management*, Portland, Oregon, Water Environment Federation, Virginia.

Solmaz, S. and G. Groeger (1992) 'Sewage sludge use and disposal', Proceedings of *Sludge 2000* Conference, Cambridge, September.

Sopper, W. E. (1993) *Municipal Sludge Use in Land Reclamation*, Lewis Publishers.

Steger, M. Th. (1992) 'Fate of chlorinated organic compounds during thermal conversion of sewage sludge', Proceedings of *Water Quality International '92*, Washington, D.C., Vol. 5, pp. 2261–2264.

Strauch, D., A. H. Havelaar and P. L'Jermite (1985) *Inactivation of Microorganisms in Sewage Sludge by Stabilization Processes*, Elsevier, Amsterdam.

Sugrue, K., G. Kiely and E. McKeogh (1992) 'A pilot study of a mix of municipal and industrial sludges', Proceedings of IAWQ Conference, Washington, D.C.

Takeshi, K. and T. Kazuhiro (1992) 'Present and future status of sewage sludge treatment and disposal in Japan', Proceedings of *Sludge 2000* Conference, Cambridge, September.

Tchobanoglous, G. and F. Barton (eds) (1991) *Wastewater Engineering: Treatment, Disposal and Reuse*, Metcalf and Eddy Inc., McGraw-Hill, New York.

Tchobanoglous, G. and E. Schroeder (1987) *Water Quality*, Addison Wesley, Reading, Mass.

Torey, S (1979) *Sludge Disposal by Landspreading Techniques*, Noyes Data Corporation (NDC), New Jersey.

USEPA (1977) *Municipal Sludge Management Environmental Factors*, 430.9-77-004, USEPA, Washington, D.C.

USEPA (1978) *Process Design Manual for Municipal Sludge Landfills*, SCS Engineers Revtol, Virginia, for USEPA, October.

USEPA (1983) *Process Design Manual for Land Application of Municipal Sludge*, USEPA Centre for Environmental Research Information, Cincinnati.

USEPA (1985) *Handbook: Estimating Sludge Management Costs*, Technomic Publishers, London.

Vesilind, P. A. (1988) 'Capillary suction time as a fundamental measure of sludge dewaterability', *Journal WPCF*, **60**(2), 215– 220.

Vesilind, P. A. (1979) *Treatment and Disposal of Wastewater Sludges*, Ann Arbor Science Publishers, MI, USA.

Wagner, E. O. (1992) 'Seeking a common vision for sludge', Proceedings of *Sludge 2000* Conference, Cambridge, September.

Wagner, G. (1992) 'The benefits of biosolids, a farmer's perspective', Proceedings of Conference on *The Future Direction of Municipal Sludge (Biosolids) Management*, Portland, Oregon, Water Environment Federation, Virginia.

Water Environment Federation (WEF) (1991) *Sludge Incineration: Thermal Destruction of Residues. Manual of Practice FD-19*, Water Environment Federation, Virginia.

Water Environment Federation (WEF) (1993) *The Future Direction of Municipal Sludge (Biosolids) Management: Where We Are and Where We Are Going*, WEF Proceedings, Vols I and II, Water Environment Federation, Oregon.

Water Environment Federation (WEF) (1993a) *WEF Standards for the Use and Disposal of Sewage Sludge—Final Rule*, Water Environment Federation, Virginia, 19 February 1993.

Water Environment Federation (WEF) (1993b) 'Biosolids and the 503 standards', *Water Environment and Technology*, May 1993. Alexandria, Virginia, USA.

Water Environment Federation (WEF) and American Society of Chemical Engineers (ASCE) (1994) *Design of Municipal Wastewater Treatment Plants*, Vols I and II, Water Environment Federation, Virginia.

Water Quality International (WQI) (1992) *Ludwigshafen Looms Large in Germany's Sludge Burners*, Water Quality International No. 1.

Water Quality International (WQI) (1993) *Anaerobic Composting Saves Wastes from Landfill*, Water Quality International No. 2.

Wellinger, A., K. Wyder and A. E. Metzler (1992) 'Kompogas—a new system for the anaerobic treatment of source separated waste', Proceedings of International Symposium on *Anaerobic Digestion of Solid Waste*, Venice.

Wilson, S., Water Environment Federation (WEF) (1992) *The Future Direction of Biosolids Management*, WEF, Alexandria, Virginia, USA.

Wheatley, A. (1991) *Anaerobic Digestion*: *A Waste Treatment Technology*, SCI and Elsevier Applied Science, Essex, UK.

FOURTEEN

SOLID WASTE TREATMENT

14.1 INTRODUCTION

This chapter examines municipal solid waste, its composition and characteristics—physical, chemical and biological. It identifies the treatment process industry, reduction, recycling, reuse, digestion, incineration, landfill, etc. The student, after studying this chapter, may wish to study further a more detailed text on integrated solid waste management (Tchobanoglous *et al.*, 1993).

An important concern of the developed world is the development of ecologically sound and health promoting ways for the management of the millions of tonnes of urban solid waste that are generated, (1.2 million tonnes per day from OECD countries in 1995). Each household in the western world produces approximately 1 tonne of solid waste per year. This does not include the vast waste products from agriculture, industry, mining and commerce. The 1989 European Charter on the Environment, pledged by 29 countries, stated that '......waste should be managed in such a way to achieve optimal use of natural resources and to cause minimal contamination......'. Traditionally, waste was given a low priority in government policies. This is not the case in the 1990s when governments all over the world are addressing current and past weaknesses in the management of solid waste.

Solid wastes are defined as those wastes from human and animal activities. In the domestic environment the solid wastes include paper, plastics, food wastes, ash, etc. Also included are 'liquid wastes' including paints, old medicines, spent oils, etc. Commercially, paper packagings, timber and plastic containers make up the bulk. Liquid–solid sludges from industry and water/wastewater plants are within this definition. Hazardous wastes requiring special treatment are not included in this definition and are covered in detail in Chapter 15. Wastes accepted by public authorities for ultimate disposal, including hazardous waste, are within this definition.

Improper management of solid wastes has direct adverse effects on health. The uncontrolled fermentation of garbage creates a food source and habitat for bacterial growth. In the same environment, insects, rodents and some bird species (seagulls) proliferate and act as passive vectors in the transmission of some infectious diseases.

The past decade has seen the development of significant items of legislation, protecting humans and their environment from improper treatment and disposal practices of solid waste. In the European Union, the 'Landfill' Directive (1995) specifies the wastes that may be accepted at landfills and the practices of management that must be adhered to in the design, operation and post-closure management of landfills. EU directives exist on waste, waste oils and batteries, with proposed directives on landfill and packaging. In the United States legislation on waste disposal is written into several acts since 1965, beginning with the Solid Waste Act of 1965 up to the California Assembly Bill 939, 1993 (which specifies that 50 per cent of all collected municipal solid waste must be diverted from landfills by the year 2000).

Examples of good treatment and disposal facilities are the exception rather than the rule. The result has been that the principal disposal method of 'landfilling' solid wastes over the past century has left an inheritance of abandoned dump sites, contaminated groundwaters, poisoned lakes and streams, co-disposal sites with toxic soils and methane explosion potential in many locations. Modern legislation addressed to Local Authorities, commerce, industry and the domestic user hopes to halt the negative trend, and is doing so, particularly in the Nordic countries.

Integrated management of solid waste is addressed by the EC Commission and by the US Environmental Protection Agency. Both list the same hierarchy of waste management policies as shown in Table 14.1. It is important to note that the methodology of landfilling, although still the most popular disposal route in most countries, is in both the US and EC policies relegated to the option of last resort. The EU Waste Policy states that 'landfilling is the final resort in waste management'. This means that on the ground there is much change yet to come about in waste management policy implementation.

14.1.1 Public Health Aspects of Solid Waste Management

Solid wastes may contain:

- Human pathogens—diapers, handkerchiefs, contaminated food and surgical dressings
- Animal pathogens—waste from pets
- Soil pathogens—garden waste

Inadequate storage of such wastes provides breeding ground for vermin, flies, cockroaches and birds (seagulls), which may act as passive vectors in disease transmission. The general public, but more particularly the solid waste employees, are at risk. The pathogens that can cause faecal-related diseases are shown in Table 14.2. These pathogens include viruses, bacteria, protozoa and helminths.

For a person to be at risk from solid waste pathogens, suitable conditions must exist and these are:

1. An infectious dose of the pathogen must be present.
2. There must be a transmission route of the pathogens to the person, i.e. aerosol, faecal–oral route, hand to mouth, etc.
3. The person must have no immunity to the pathogen.

Table 14.1 EC and US waste management hierarchy

Hierarchy	EU/US
1	Source reduction
2	Recycling/composting
3	Incineration:
	(a) With energy recovery
	(b) Without energy recovery
4	Landfilling

Table 14.2 Viral, bacterial and protozoal pathogens in faecal contaminated solid wastes

Pathogen	Organism	Disease	Reservoir
Viruses	Poliovirus	Poliovirus	Man
	Hepatitis A	Hepatitis A	Man
	Hepatitis B	Hepatitis B	Man
Bacteria	*Campylobacterfetus sp.*	Diarrhoea	Animals and man
	Pathogenic *E. coli*	Diarrhoea	Man
	Slamonella *S. typhi*	Typhoid fever	Man
	Salmonella *S. paratyphi*	Paratyphoid fever	Man
	Other *Salmonella*	Food poisoning	Animals and man
	Shingella spp.	Bacillary dysentery	Man
	Vibrio cholera	Cholera	Man
	Other *Vibrio*	Diarrhoea	Man
	Yersinia enterocolitica	Diarrhoea	Animals and man
Protozoa	*Balantidium coli*	Diarrhoea, dysentery, colonic ulceration	
			Man, pigs and rats
	Entamoeba histolytica	Colonic ulceration, amoebic dysentery, liver abscess	
	Giardia lamblia	Diarrhoea and malabsorption	Man and animals
Helminths	Flat worms	Digestive disorders	Man and animals
	Round worms		
	Tape worms		
	Trematodes		

Adapted from Feachem *et al.*, 1983

The possible transmission routes for pathogenic transfer are shown in Fig. 14.1.

The three routes of inhalation, percutaneous and ingestion may be negated by good hygiene and dietary habits. However, should the pathogen successfully invade the human host, it is still possible that the host's immunity may prevent the infection. Those involved on a regular basis with solid waste are usually vaccinated for a range of pathogenic diseases. Rodent control is essential as they are carriers of serious illnesses, (Garrett, 1995).

Specific waste storage and handling techniques to reduce the risk of solid waste contamination to employees are discussed in a later section. These techniques include separating at source those wastes most amenable to disease transmission, i.e. food and faecal origin. Less manual handling and more automation on storage, collection and mechanical sorting is desirable for health reasons. It is important

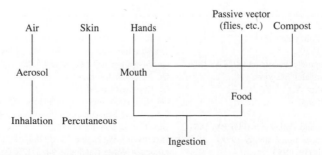

Figure 14.1 Transmission routes for solid waste pathogens.

that employees handling waste are protected against dust and aerosols that might contain endotoxins from bacteria. Gaseous and aerosol emissions from landfills such as benzene and vinyl chloride (both carcinogenic) are serious health concerns (McCall, 1995).

14.1.2 Legal Aspects of Solid Waste

Table 14.3 shows some of the relevant items of legislation covering solid waste in the United States and European Union. While the legal instruments are significant, there has been much procrastination in adherence to the letter of the law, particularly by government agencies and responsible authorities. Areas in the United States and Northern Europe are able to adhere to the legislation by utilizing advanced treatment and disposal technologies, funded by the public at real costs.

In the European Union, significant directives on solid waste are the Directive on Environmental Assessment and the proposed Directive on Landfill. The EIA (Environmental Impact Assessment) Directive lists in its Annexe I that an EIA be carried out for 'waste disposal installations for the incineration, chemical treatment or landfill of toxic and dangerous waste'. It also lists in Annexe II that an EIA be carried out (where the member states consider characteristics so require) for 'installations for the disposal of industrial and domestic waste.' In the United Kingdom, solid waste disposal sites with input exceeding 75 000 tonnes/annum require an EIA. The Landfill Directive sets out parameters for landfill sites and containment principles with systems for landfill gas and leachate collection. An environmental impairment insurance is also a requirement. A landfill aftercare fund based on a levy on all operations may be an alternative to insurance (Mabb, 1993). Requirements on site suitability are defined by geology/hydrogeology/local environment in combination with a bottom liner permeability at values better than 10^{-9} m/s and a thickness > 1 m. Similarly, strict waste acceptance criteria and procedures are set out for:

- Waste acceptance
- Chemical, physical and biological waste properties
- Suitability of wastes
- Waste loading rates
- Water, leachate and gas control
- Water balance and groundwater protection

The objectives of the waste directives are to promote clean technologies and to use landfill only as a last resort. The packaging directive sets the objective of a 90 per cent reduction in overall packaging quantities for landfilling in the first ten years. This, if successful, will significantly reduce the volumes of municipal

Table 14.3 Solid waste legislation in the US and EU

United States	EU Directive
Solid Waste Disposal Act 1965	On Waste 1975
National Environmental Policy Act 1969	On Toxic and Dangerous Waste 1978
Resource Recovery Act 1970	Transfrontier Shipment 1984, 1985, 1986
Resource Conservation and Recovery Act (RCRA) 1976	On Environmental Assessment 1985
Comprehensive Environmental Response	On Access to Information
Compensation and Liability Act	On Beverage Containers
(CERCLA) 1980 (Superfund)	On New MSW Incineration Plants 89/369/EEC
	On Existing MSW Incineration Plants 89/429/EEC
Public Utility Regulation and Policy Act (PURPA) 1981	Directive on Landfill (1995)
RCRA—Response Compensation Liability 1994	Directive on Packaging 93/C/285/01
California Assembly Bill 939, 1993	European Waste Catalogue 94/3/EEC

solid waste as packagings account for approximately 35 per cent of domestic waste and 60 per cent of commercial waste.

In the United States the Solid Waste Disposal Act of 1965 aimed to promote solid waste management and resource recovery and provide guidelines for collection, transport, separation, recovery and disposal. The 1965 law was amended to become the Resources Recovery Act in 1970. This emphasized the shift in objectives from disposal to recycling and reuse of recoverable materials in solid wastes or to the conversion of wastes to energy. The 1976 Act on Resource Conservation and Recovery allowed the EPA to set guidelines for solid waste management to the public. This legislation was the legal basis for implementing the EPA guidelines for solid waste storage, treatment and disposal. The Comprehensive Environmental Response Compensation and Liability Act of 1980, often called CERCLA or, more popularly, Superfund, provided a means for directly responding and funding the activities at uncontrolled hazardous waste sites. Uncontrolled municipal waste sites where known hazardous wastes were dumped are also addressed under CERCLA. The Public Utility Regulation and Policy Act of 1981 directs public and private utilities to purchase power from waste-to-energy facilities. RCRA was updated in 1994. The California Assembly Bill 939 is specific, with a goal that 50 per cent of solid waste collected must be diverted from landfills by the year 2000.

An important aspect of modern legislation on waste is *the duty of care*. This duty, enacted in legislation in the United Kingdom by its EPA Act of 1990, specified those affected by duty of care as being:

- Waste producers (and importers)
- Waste carriers and brokers
- Waste managers

In essence the only exception to the duty of care is for occupiers of domestic property for household waste. Those subject to the duty of care must:

- Prevent any person committing an offence by disposing, treating or storing controlled waste
- Contain any waste within their control
- Ensure that the transport of waste is carried out by authorized or registered carriers and is legally manifested

The duty of care legislation obliges the waste generator to be responsible for that waste from 'cradle to grave'. The implications are severe and it is intended that this legislation will encourage more extensive use of recycling and waste recovery and less dependence on final disposal via traditional landfill.

14.1.3 Environmental Aspects of Solid Waste

Solid waste treatment and disposal methodologies are fraught with problems. Landfill sites, and dump sites in particular, cause soil and groundwater contamination if not properly operated. Additional environmental problems with landfill are odours, litter, scavengers, fires and rat infestation. Waste incineration has had problems with odours and air pollution. Composting has had difficulties with odours, heavy metals and slow compost sales. Transportation problems are more associated with hazardous wastes. Health and hygiene are issues for waste operators. Municipal solid wastes co-disposed with industrial wastes and sludges have been problematic. Liquid–solid sludges applied to landfill have proved very difficult to handle. Landfilling in moist climates like Ireland produces large quantities of leachate which are toxic and of high organic strength and require treatment in wastewater plants. Landfilling in dry climates produces localized air pollution problems. The record shows that landfill, even though still by far the most common disposal route, is undesirable and alternatives must be pursued. These are addressed in the integrated waste management strategies as defined by the hierarchy of Sec. 14.1.4.

14.1.4 Integrated Management of Solid Waste

Integrated management of solid wastes has been promoted in the United States since the 1965 Solid Waste Disposal Act. More recently, the European Union and WHO (World Health Organization) have also been promoting integrated management. The 1993 text on solid waste by Tchobanoglous *et al.* is titled *Integrated Solid Waste Management.* In the latter, ISWM is defined as the selection and application of suitable techniques, technologies and management programmes to achieve specific waste management objectives and goals. The hierarchy in waste management policy for the 1990s is shown in Fig. 14.2.

The priority in a waste management policy is to reduce the amount of waste. This promotes the goal for industry, commerce, agriculture and households not to create waste in the first place. The emphasis is for product manufacture with minimal waste. It promotes good practice of waste monitoring with a view to minimization. Regular 'waste audits' by various producers and consumers can identify areas for improvement. Chapter 8 details waste minimization, policy and technologies with some emphasis on the chemical industry.

Recycling and reuse is a 'green' topic only recently finding acceptance by industry. Regarding solid waste, only 5 per cent of Dublin's waste is recycled out of a possible 45 per cent. The more obvious areas of recycling of paper, glass, metals and plastics are, as yet, untapped in many countries. Industry is making great efforts at product reuse, reprocessing and remanufacture. Recycling and reuse is not simply for municipal solid waste, but now includes the product manufacturers, commerce and agriculture. For instance, liquid–solid sludges from wastewater treatment plants may be more suited to land spreading as soil nutrients rather than landfilling. Waste transformation, be it in the form of incineration, composting or biogas production, all serve to reduce the volume of waste where the final destination is landfilling. Incineration reduces the volume to about 15 per cent of its input and produces energy. Composting transforms organic waste to a nutrient product. Biogas produces energy from organic wastes usually co-digested with industrial or agricultural waste contributions or with municipal sludge.

The avenue of last resort, as defined by the US EPA, WHO, the EC, the UK EPA and many others, is landfilling. Ideally, landfilling is essential only for about 20 per cent of municipal solid wastes. This is the inert (rich in metals) content left after recycling 45 per cent and anaerobically digesting 35 per cent. This inert metal content is also potentially recyclable. After recycling, anaerobic digestion of the organic (mainly food) fraction is most likely to be the next breakthrough in waste transformation technologies. It is possible to transform some of the 20 per cent by way of new products usable in construction such as brick pavings or earthen embankments on roadways. However, it appears that there will always be a need for landfilling, but it is technologically feasible to reduce dramatically the amount of municipal waste going to landfill to about 10 to 20 per cent. The issues in achieving this are not only engineering but also economic and social.

14.2 SOURCES, CLASSIFICATION AND COMPOSITION OF MSW

Municipal solid waste (MSW) is all waste collected by private or public authorities from domestic, commercial and some industrial (non-hazardous) sources. Solid waste is non-standard, and typically, no two wastes are the same. Domestic waste from a single house will vary from week to week and from

Figure 14.2 Hierarchy of integrated solid waste management.

season to season. In some countries, up to 50 per cent of solid waste are ashes in the winter season, with no ashes in summer time. Waste varies from socioeconomic groups within a country and from country to country. Domestic wastes co-disposed with industrial waste can be very different from conventional municipal solid waste.

14.2.1 Sources of Solid Waste

Proper planning, treatment and disposal practices rely on accurate data regarding composition and waste generation rates. Examining industrial waste, composition depends on the industrial classification (e.g. electrical products, wood products). The sources and types of solid wastes according to WHO (1991) are itemized in Table 14.4.

14.2.2 Composition of Solid Waste

In many situations, domestic and commercial wastes are collected and transported by similar authorities and sometimes the fraction/composition is similar in both. Table 14.5 shows possible compositions of domestic and commercial wastes. The most basic arrangement is to identify waste as being organic or non-organic. This may be satisfactory for particular end uses (i.e. landfill the inert and biologically transform the organic). The detail given on composition will reflect the proposed treatment. For instance, if it is proposed to incinerate the waste then it is essential to evaluate the heating value, and the waste would be arranged into combustible and non-combustible. If the plan is to anaerobically digest the organic food fraction, then it may be adequate to detail the food fraction and lump all else into 'others'. The wastes are also itemized with a view to recover the materials and this would identify the glass, metals, plastics and wood in greater detail. Solid waste from industry is typically collected by private hauliers, as is solid waste from the construction industry. Industries tend to identify the composition of their waste according to industry type, e.g. paper products, metal products. WHO (1991) organize industrial waste into three broad categories:

- Non-hazardous industrial waste
- Hazardous waste
- Hospital waste

Table 14.4 Sources and types of solid wastes

Source	Facility	Type of waste
Domestic	Single family dwelling, multifamily dwelling, low, medium and high-rise apartments	Food, paper, packagings, glass, metals, ashes bulky household waste, hazardous household waste
Commercial	Shops, restaurants, markets, office buildings, hotels and motels, institutions	Food, paper, packagings, glass, metals, ashes, bulky household waste, hazardous household waste
Industrial	Fabrication, light and heavy manufacturing, refineries, chemical plants, mining, power generation	Industrial process wastes, metals, lumber, plastics, oils, hazardous wastes
Construction and demolition		Soil, concrete, timber, steel, plastics, glass, vegetation

Table 14.5 Physical composition of solid waste

General composition	Typical composition	Detailed composition
Organic	Food putrescible	Food
		Vegetables
	Paper and cardboard	Paper
		Cardboard
	Plastics	Polyethylene terephthalate (PETE)
		High-density polyethylene (HDPE)
		Polyvinyl chloride (PVC)
		Low-density polyethylene (LDPE)
		Polypropylene (PP)
		Polystyrene (PS)
		Other multilayer plastics
	Clothing/fabric	Textiles
		Carpets
		Rubber
		Leather
	Yard waste	Garden trimmings
	Wood	Wood
	Miscellaneous organics	Bone
Inorganic	Metals	Tin cans
		Ferrous metals
		Aluminium
		Non-ferrous metals
	Glass	Colourless
		Coloured
	Dirt, ash, etc.	Dirt, screenings
		Ashes
		Stone
		Bricks
	Unclassified	Bulky items

In the United States, the Standards Industrial Classification Manual (SIC) 1972 identifies industrial waste by industry and product type. A sample of these is shown in Table 14.6.

The wastes generated within a municipality (excluding industrial and agricultural) will vary widely, depending on the community and its level of commercialism and institutionalism. The data on waste will also depend on the level of sophistication of the waste management operation. It is seen in Table 14.7 that typically 0.1 per cent of waste by weight generated in a community is hazardous. This is similar to Danish figures. However, Ireland records 0 per cent hazardous waste at municipal disposal sites. The actual Irish figures for the hazardous waste fraction at municipal landfill sites are likely to be in excess of the US figures, particularly since Ireland has no facility for hazardous waste treatment. This is even more so as

Table 14.6 Sample industrial waste types

SIC classification	Waste generating process	Wastes
Food and food products	Processing, packaging, shipping	Meats, fats, oils, bones, offal, vegetables, fruit, etc.
Paper and applied products	Paper manufacture, cardboard boxes, etc.	Paper, cardboard, inks, glues, chemicals, etc.
Electrical products	Manufacture of electrical appliances	Scrap metal, plastics, rubber, glass, textiles, etc.
Chemical products	Manufacturing of drugs, etc.	Organic and inorganic chemicals, metals, plastics, solvents, paints, etc.

Ireland practises co-disposal of domestic and industrial waste in the same landfill. Co-disposal is a landfill practice currently being questioned as the wastes from municipal and industrial sources are not necessarily compatible. The practice of disposing of industrial sludges of different origins in municipal landfills has caused problems with operation due to the continuous wet nature of the open cells. Table 14.7, adapted in part from Tchobanoglous *et al.* (1993), lists the breakdown of municipal solid waste in a typical US community.

Wastes delivered to the landfill site at Cork, Ireland (the only disposal facility for the area), is approximately 100 000 tons per annum. Of this, 35 per cent is domestic, 46 per cent is commercial, 11 per cent is industrial and 8 per cent is miscellaneous (builder's rubble, street sweepings, etc.). Table 14.8 is a detailed breakdown of waste composition for Denmark. Table 14.9 shows typical domestic waste composition by country and Table 14.10 shows the results for four sites within Ireland.

From Tables 14.9 and 14.10 the following observations are made:

1. Food wastes constitute about 30 per cent of all domestic waste. The low figure of 9 per cent for the United States is due to the use of kitchen sink grinders. Such utensils increase the organic solid loading to municipal wastewater plants.
2. Paper and cardboard constitute about 35 per cent or greater.
3. Glass constitutes about 8 per cent.
4. Metals constitute about 5 per cent or greater.
5. Plastics constitute about 8 per cent with the figure for Ireland being double that. Plastic is ubiquitous in Ireland, being used as milk and yoghurt containers, soft drink containers, household goods containers, etc.
6. The amount of ash reduces internationally with increasing GDP (gross domestic product), being about 50 per cent in Poland and China, 10 to 20 per cent in Ireland and less than 5 per cent in the United States and Denmark. The use of coal in lower income countries is balanced by very little domestic usage in high income countries. In Ireland, for instance, the ash content is up to 40 per cent in the lower socioeconomic groups during winter, while in the higher income groups (who use oil or gas for heating) the ash content is less than 20 per cent.

Commercial waste will depend on the types of commerce, e.g. restaurants, retail outlets, offices, etc. In Cork, the commercial waste was made up of 20 per cent food waste, 60 per cent paper and cardboard, 6 per cent plastics and 11 per cent glass, with 2 per cent metals.

Table 14.7 Breakdown of MSW excluding industrial and agricultural (US)

Waste source	% by weight	
	Range	Typical
Residential and commercial (not hazardous)	50–75	62
Special waste (bulky, etc.)	3–12	5
Hazardous	0.01–1	0.1
Institutional	3–5	4
Construction and demolition	8–20	14
Municipal services		
Street sweepings	2–5	4
Landscaping	4–9	6
Treatment plant sludges	3–8	5

Adapted from Tchobanoglous *et al.*, 1993. Reprinted by permission of McGraw-Hill, Inc.

Table 14.8 Danish figures on waste composition from difference sources (1985)

Type of waste	Waste fraction (%)							
	Food waste organic	Card-board	Paper	Plastic	Other combustibles	Glass	Metal	Other non-combustibles
Domestic waste								
Household waste	35	5	30	6	8	8	4	4
Bulky waste								
Garden waste	2	5	6	2	44	5	6	30
Waste from office, trade and institutions								
Wholesale	17	40	14	8	10	4	2	5
Retail sale	3	70	9	5	5	5	2	1
Restaurants, hotels	55	11	14	2	—	12	6	—
Office	25	9	58	4	1	2	1	—
Kindergarten, etc.	38	13	28	12	3	—	3	3
Military institutions	30	8	21	6	6	1	15	13
24-h nursing homes	50	12	26	5	2	3	2	—
Hospitals	22	5	51	10	12	—	—	—
Schools	31	13	44	6	—	—	6	—
Industrial waste								
Food industry	20–45			0–15	†	0–15	0–10	10–55
Leather and textile	—		20–30	0–10		—	0–5	40–80
Wood and furniture	—		0–20	—		—	0–5	75–100
Paper and printing	—		50–100	0–5		—	0–5	0–50
Chemical industry	0–5		15–40	10–30		—	0–20	5–60
Stone, ceramic, glass	—		10–15	—		0–30	20–30	35–70
Iron and metal casting	—		10–30	0–5		—	5–75	10–85
Iron and metal, other	—		10–30	0–5		0–30	5–75	10–85
Other industries	5		40			—	15	40
Building industry	—		5	1	40	2		45

†'Other combustible' and 'other non-combustible' united in one group.
Adapted from Danish EPA, 1985

Table 14.9 Typical composition of domestic waste by country

Component	US (Tchobanoglous et al., 1993)	Denmark (Mortensen, 1992)	UK national average (WHO, 1991)	London 1993 (WHO, 1991)	Poland 1991 (Mortensen, 1992)	China 1985 (Mortensen, 1992)	Ireland Dublin (Dennison and Dodd, 1992)
Food wastes	9	35	25	26.7	24	36	34.2
Paper, cardboard	40	35	29	35.5	11	2	18.7
Plastics	7	6	7	5.2	2	1.5	16.1
Glass	8	8	10	10.8	6	1	5.4
Metals	9.5	4	8	6	2	1	2.9
Clothing/textiles	2	8	3	3.4	10	1.5	2.6
Ashes, dust	3	4	14	5	45	57	17.2
Unclassified							
(garden, yard, wood)	21.5		4	7.1			2.9

Table 14.10 Typical composition of domestic waste in Ireland

Component	Dublin City	Cork City	Mallow Co. Cork	Bandon Co. Cork	Ireland national average
Food wastes	34.2	15	42	37	31.5
Paper, cardboard	18.7	44	15		24.5
Plastics	16.1	3	12		14
Glass	2.9	9	7		3
Metals	5.4	8	10		7.5
Clothing/textiles	2.6	2	1		3
Ashes, dust	17.2	4	10	12	12
Unclassified				All other	
(garden, yard, wood)	2.9	15	3	51	

14.2.3 Waste Tonnages

Again the waste tonnages can vary widely, but in the higher income countries the current production is about 1 tonne/household per year. This figure of 1 tonne is an 'overall' one to include the total municipal waste in a community, including all waste delivered to municipal sites, be it domestic, commercial or institutional, but not industrial. Some figures on the tonnages are shown in Table 14.11 for different countries. Table 14.12 is a detailed breakdown of unit waste generation of municipal wastes from Denmark. For instance, the domestic component is 225 kg/head/yr or with a household density of about 2.5 (Denmark) gives 0.56 tonnes/household/yr. In addition, in Denmark there is 100 to 150 kg/head/yr of bulky waste (household furniture, etc.) and 20 to 25 kg/head/yr of garden waste. This total accounts for 0.93 tonnes/household/yr. Then an extra amount must be added to find the total municipal waste/household/yr. However, this will vary, as mentioned, depending on the amount of commercial or industrial activity in the community. Care in computing this figure is required, as sometimes the bulky waste/garden waste may be recorded as 'industrial'.

14.2.4 Some International Data on MSW

Table 14.13 shows the quantities of municipal waste generated internationally and the disposal routes. The developed world shows many similar figures. Japan, interestingly, shows almost no glass or metal waste but much higher paper and cardboard waste than other countries. Table 14.13 also suggests that Ireland

Table 14.11 Municipal solid waste by country (tonnes/head/yr)

Waste Source	North Cork	Cork City	UK national	Denmark	US
Domestic					
tonnes/house/yr	0.81		0.55	0.93	
tonnes/head/yr		0.25		0.37	
Domestic and Commercial					
tonnes/house/yr	1.51				
tonnes/head/yr		0.57			0.62
Total municipal					
tonnes/house/yr	1.58				
tonnes/head/yr		0.68			1.0

Table 14.12 Unit figures on waste generation in Denmark, established for national waste planning in 1985

Class	Type of waste	Unit waste generation	Estimated annual growth (%)
I	Domestic waste		
	Household waste from residential areas	225 kg/head/yr	1
	Household waste from summer houses	75 kg/house/yr	0
	Bulky waste	100–150 kg/head/yr	0
	Garden waste	20–25 kg/head/yr	0
II	Waste from offices, trades and institutions		
	Wholesale	910 kg/head/yr	
	Retail sale	1030 kg/house/yr	
	Restaurant and hotels	1270 kg/head/yr	
	Bank—financing, insurance, public offices, etc.	100 kg/head/yr	
	Kindergartens	30 kg/child/yr	1
	Military institutions	300 kg/head/yr	
	Nursing and rest homes	310 kg/bed/yr	
	Hospitals	550 kg/bed/yr	
	Schools	15 kg/student/yr	
III	Industrial waste		
	Food industry	1400 kg/employee/yr†	
	Leather and textile	750 kg/employee/yr	
	Wood and furniture	1000 kg/employee/yr	
	Paper and printing	1780 kg/employee/yr	
	Chemical industry	1150 kg/employee/yr	0–0.5
	Stone, ceramic and glass	4000 kg/employee/yr	
	Iron and metal casting	500 kg/employee/yr	
	Iron and metal industries	790 kg/employee/yr	
	Other industries	570 kg/employee/yr	
IV	Building industry	2500 kg/head/yr	0.5
V	Waste from energy production		
	Coal fire plants		
	Combustion on grates		
	Slag	127.5 kg/ton coal	
	Fly ash	22.5 kg/ton coal	
	Dust combustion		0
	Slag	15 kg/ton coal	
	Fly ash	135 kg/ton coal	
VI	Waste from sewage treatment		
	Mechanical	12 kg/pe/yr	
	Mechanical, biological	22 kg/pe/yr	0
	Mechanical, biological, chemical	33 kg/pe/yr	

†Per head means per employee.
From Danish guidelines concerning mapping and planning of waste disposal—Danish EPA C.35 1985, 03, 25

has about twice the waste in plastics than other countries. Denmark is the higher waste generator of paper and cardboard.

The variations in Table 14.13 may include variations in consumer traditions, e.g. whether they use bottle return systems for beer, milk, etc., or paper or plastic cartons which go to waste. It may also cover variations in recycling systems, e.g. if waste is collected at drop centres (e.g. bottle banks) it may not be recorded as waste. Finally, it may also include for variations in definitions, e.g. sometimes coated paper (milk cartons) is recorded as paper/packaging and sometimes as other wastes.

Table 14.13 International MSW quantities

Country	Quantity (1000 t)	Disposal method % by weight				Composition % by weight				
		Compost	Incineration	Landfill	Other	Paper, cardboard	Plastic	Glass	Metal	Other
USA	208 760					35	7	9	9	40
Japan	48 283					45	8	1	1	44
Belgium	3 470	11	23	50	16	28	8	8	4	52
Denmark†	2 400	2	50	11	7	30	7	6	3	54
West Germany	19 483	2	28	69		18	5	9	3	65
Greece	3 147			100		20	7	3	4	66
Spain	12 546	16	6	78		20	7	6	4	63
France	17 000	8	36	47	9	27	4	8	7	54
Ireland	1 100			100		25	14	8	3	50
Italy	17 300	6	19	35	34	22	7	6	3	62
Luxembourg	170		95	5		17	6	7	3	67
Netherlands	6 900	4	36	57	4	24	7	7	3	59
Portugal	2 350	16		23	58	19	3	3	3	72
UK	18 000		6	92		30	7	10	7	46
Total EC	104 000	5	19	67	9					
Total all	423 000									

Note that the following countries produce the following waste quantities ($\times 1000$ t):
Canada $= 164\,000$ t, Austria $= 2700$ t, Finland $= 2500$ t, Norway $= 2000$ t, Sweden 2650 t, Switzerland $= 2850$ t, Turkey $= 19\,500$ t.
†Denmark EPA 1994 excludes source separated waste for recycling.

14.3 PROPERTIES OF MSW

Traditionally, (and still in many countries) waste handlers did not need to know much about the physical, chemical and biological properties of solid waste, since all MSW was dumped to landfill. As proper waste management now involves recycling, reuse, transformation and disposal, it is relevant to know the details of the waste with regard to physical, chemical, energy and biological properties of waste.

14.3.1 Physical Properties of MSW

The most relevant physical properties are:

- Density and moisture content (kg/m^3)
- Particle size distribution (range in mm)
- Field capacity (per cent)
- Hydraulic conductivity (m/day)
- Shear strength (kN/m^2)

Density and moisture content   The density of solid waste varies with its composition, its moisture content and its degree of compaction. Table 14.14 shows figures for waste density. Food wastes range from 100 to 500 kg/m^3 with corresponding moisture contents at 50 to 80 per cent. MSW normally compacted in landfill has a density of 200 to 400 kg/m^3 with a moisture content of 15 to 40 per cent. Further densities and moisture contents on other wastes are found in Tchobanoglous *et al.* (1993). Moisture content of wastes are relevant when estimating the calorific value, landfill sizing, reactor sizing, etc.

Table 14.14 Density and moisture content of municipal solid waste

Waste source	Component of waste	Density (kg/m³)	Moisture content (% by weight)
Domestic	Food	290	70
	Paper and cardboard	70	5
	Plastics	60	2
	Glass	200	2
	Metals	200	2
	Clothing/textiles	60	10
	Ashes, dust	500	8
Municipal			
Uncompacted		100	20
In compaction truck		300	20
Normally compacted in landfill		500	25
Well compacted in landfill		600	25

Adapted in part from Tchobanoglous *et al.*, 1993

Particle size distribution   The particle size distribution, like the percentage of combustibles, is relevant to incineration and biological transformation methods. Knowing the particle size is also relevant for recycling and reuse and for equipment sizing for further treatment. For instance, aluminium soft drink containers are typically 0.15 m high by 0.06 m diameter and are categorized by an effective size which might be the largest dimension or $\sqrt{LD} = \sqrt{0.15 \times 0.06} \cong 0.1$m. Waste components are often described by length × breadth × height (L × B × H). Knowing the largest dimension is important for sizing facilities like conveyor belts, grinders, etc., and these dimensions are given in Table 14.15 for domestic wastes. It is seen from Table 14.15 that the average particle size is about 100 mm. Shredders and separators are used to reduce this to desirable sizes for the treatment of composting, etc.

Field capacity   Field capacity (FC) was defined in Chapter 4 for soils as being the maximum percentage of volumetric soil moisture that a soil sample will hold freely against gravity. Above FC, water drains away freely. In Chapter 4 it is shown that FC for soils varied from about 5 per cent for sandy soils to about 30 per cent for dry soils. Similarly, solid waste (in landfills) will have an FC that decreases with the overburden pressure.

Uncompacted municipal solid waste has a field capacity of about 50 to 60 per cent. Water in excess of FC will drain away as leachate. It is therefore important to determine the FC of a particular waste and its landfill disposal methodology so as to limit the amount of leachate generation. An empirical equation from Tchobanoglous *et al.* (1993) for field capacity is

$$FC = 0.6 - 0.55\left(\frac{W}{4500 + W}\right) \tag{14.1}$$

Table 14.15 Typical particle size distribution of MSW

Component	Size range (mm)	Typical (mm)
Food	0–200	100
Paper and cardboard	100–500	350
Plastics	0–400	200
Glass	0–200	100
Metals	0–200	100
Clothing/textiles	0–300	150
Ashes, dust	0–100	25

where $\qquad$ FC = field capacity, % of dry weight of waste

$\qquad\qquad$ W = overburden weight calculated at mid-height of waste in lift, kg

Example 14.1 Determine the field capacity of a landfill site for the following conditions, after 1 year of operation, and also compute the amount of water that can be held in the waste:

$$\text{Density of compacted solid waste} = 600 \text{ kg/m}^3$$
$$\text{Moisture content of waste} = 25\% \text{ by volume}$$
$$\text{Lift after 1 year} = 6 \text{ m}$$
$$\text{Net annual rainfall} = 400 \text{ mm}$$

Solution

$$FC = 0.6 - 0.55 \left(\frac{W}{4500 + W} \right)$$

$$\text{Dry density of solid waste} = 466 \text{ kg/m}^3$$

$$W = \tfrac{1}{2}(6 \times 466) = 1400 \text{ kg/m}^2$$

$$FC = 0.6 - 0.55 \left(\frac{1400}{4500 + 1400} \right)$$

$$FC = 0.47$$

$$\text{Total water storage capacity} = 0.47 \times 480 \times 6 = 1353 \text{ kg/m}^2$$

$$\text{Surplus storage after year 1} = 1353 - 6(600 \times 0.5) = 453 \text{ kg/m}^2$$

Therefore

$$\text{Theoretical leachate generated} = 0$$

Hydraulic conductivity of wastes Sludges in landfills tend to resist the movement of water down through them due to low hydraulic conductivity by virtue of very high moisture content. Instead, rainfall is converted to surface runoff and the sludge material is transported overland to surface streams. Other solid waste material, i.e. paper and packaging, has almost no resistance to rain infiltration. Each site and material is site specific, as regards the hydraulic conductivity of the solid waste material. Later sections of this chapter deal with hydraulic conductivity of liners, be they natural clay or synthetic. The hydraulic conductivity (permeability to soils engineers) is relevant in that it governs the transport rate of leachate and other fluid/microbiological contaminants within the solid waste fill. Dense baled waste has a hydraulic conductivity of 7×10^{-6} m/s while loose samples of solid waste have a K of 15×10^{-5} m/s. Shredded waste has hydraulic conductivities of 10^{-4} to 10^{-6} m/s. Typically, then, the hydraulic conductivity of solid waste is about 10^{-5} m/s, but depends on the density. Solid waste is not necessarily homogeneous and so the hydraulic conductivities are not isotropic.

Shear strength The shear strength of sludges in landfill sites is about zero. When co-disposed with dry refuse (paper and cardboard), their shear strength improves. Solid wastes have a shear strength which is known to be highest shortly after compaction and decreases over time, sometimes to zero after several years in the landfill. Recent discussions in the European Union suggest that sludges, when disposed to landfill, will be required to have shear strengths of $> 15 \text{ kN/m}^2$ with dry solids > 30 per cent. In the United States many states require sludge dry solids > 51 per cent and in some states the requirement is 5 parts solid waste to 1 part sludge by weight. The latter approximates to 15 to 1 parts by volume.

14.3.2 Chemical and Energy Properties of MSW

Traditionally, all MSW was 'dumped' so there was not much need to evaluate the properties, least of all the chemical properties. Because there are several possible recycle, reuse and transformation technologies, one of the first steps in identifying the most suitable treatment technology is to determine its chemical properties. These are:

- Proximate analysis
- Ultimate analysis
- Energy content

Included in proximate analysis are:

- Moisture content by percentage weight
- Volatile matter
- Fixed carbon
- Non-combustible fraction (ash)

Table 14.16 gives proximate analysis and energy content values for typical MSW. Robinson (1986) quoted a comparative proximate analysis and energy content between a US military base MSW and its surrounding county MSW. All figures were very similar with energy content being about 15 per cent higher for the military base. This was due to higher volatile matter and lower moisture content. The energy values as shown in Table 14.16 are indicative for waste as collected, dry and ash-free. The nomenclature for energy content is:

$$H_u = \text{lower heat value, i.e. from waste as collected}$$
$$H_{wf} = \text{normal heat value, i.e. from water-free waste (dry)}$$
$$H_{awf} = \text{higher heat value, i.e. from ash- and water-free waste}$$

In Europe, calculations are normally based on the lower heating value, H_u, computed from

$$H_u = H_{awf} \times B - 2.445 \times W \qquad \text{MJ/kg} \tag{14.2}$$

where
$$B = \text{flammable fraction (\textit{i.e.} volatile matter + fixed carbon)}$$
$$W = \text{moisture content fraction by weight}$$

Typically, for MSW, the higher heating value (HHV or H_{awf}) is about 20 MJ/kg. Table 14.17 is a listing of energy values of MSW components in the United States and Denmark.

Example 14.2   If $H_{awf} = 20$ MJ/kg, compute the lower heating value of MSW if:

$$W, \text{ water content} = 21\%$$
$$B, \text{ flammable} = 59\%$$
$$A, \text{ Ash} = 20\%$$

Solution

$$H_u = H_{awf} \times B - 2.445 \times W$$
$$= 20.0 \times 0.59 - 2.445 \times 0.21$$
$$H_u = 11.29 \text{ MJ/kg}$$

Table 14.16 Typical proximate analysis and energy content in MSW

Waste type	Proximate analysis (% by weight)				Energy content (MJ/kg)		
	Moisture	Volatiles	Fixed carbon	Non-combustible (ash)	As collected lower heat value H_u	Dry normal, water free H_{wf}	Dry ash and water-free higher heat value H_{awf}
Food mixed	70	21	3.6	5.0	4.2	13.9	16.7
Fats	2	95	2.5	0.2	37.4	38.2	39.1
Fruit	79	16	4.0	0.7	4.0	18.6	19.2
Meat	39	56	1.8	3.1	17.6	28.9	30.4
Paper mixed	10.2	76	8.4	5.4	15.7	17.6	18.7
Newspapers	6	81	11.5	1.4	18.5	19.7	20.0
Cardboard	5.2	77	12.3	5.0	26.2	27.1	27.4
Plastics mixed	0.2	96	2	2	32.7	33.4	37.1
Polyethylene	0.2	98	< .1	1.2	43.4	43.4	43.9
Polystyrene	0.2	99	0.7	0.5	38.0	38.1	38.1
Polyurethane	0.2	87	8.3	4.4	26.0	26.0	27.1
PVC	0.2	87	10.8	2.1	22.5	22.5	22.7
Textiles	10	66	17.5	6.5	18.3	20.4	22.7
Yard wastes	60	30	9.5	0.5	6.0	15.1	15.1
Wood mixed	20	68	11.3	0.6	15.4	19.3	19.3
Glass	2			96–99	0.2	0.2	0.15
Metals	2.5			94–99	0.7	0.7	0.7
Domestic MSW	15–40	40–60	4–15	10–30	11.6	14.5	19.3
Commercial MSW	10–30				12.8	15.0	
MSW	10–30				10.7	13.4	

Adapted from Tchobanoglous *et al.*, 1993; Robinson, 1986; Mortensen, 1993

Table 14.17 Energy content of MSW, Denmark and the United States (H_u, lower heating value)

MSW source	Energy content H_u, as collected (MJ/kg)	
	Denmark	US
Household (normal)	8.4–9.2	11.6
Household (extreme)	6.3–10.5	
Commercial	10.5–12.5	12.8
Waste with higher content of wood, paper, plastic	Higher	Higher

Ultimate analysis of MSW   The most important elements in waste energy transformation are:

Carbon   C

Hydrogen   H

Oxygen   O

Nitrogen   N

Sulphur   S

Ash

It is relevant to know the chemical composition or ultimate analysis (see Table 14.18) for the purposes of waste to energy processes either by combustion or by biological transformation. For instance, a waste high in plastics is very suitable for incineration but is totally unsuitable for biological transformation. Similarly, a waste high in wood or yard wastes is good for incineration but not suitable for biological conversion. The latter is so, even though wood is organic, because it has a high lignin content which breaks down very slowly in biological processes. The C/N ratio, as discussed in composting (Sec. 14.9), is a relevant parameter for biological processes. Aerobic biological composting processes operate best at C/N ratios of $30:1$. The energy content of MSW can be determined from the Dulong equation.

$$H_{\text{awf}} = 337C + 1419(H_2 - 0.125O_2) + 93S + 23N \tag{14.3}$$

where C, H, O_2, S and N are the percentage by weight of each element, and H_{awf} is the net calorific value without water and without ash.

Khan *et al.* (1991) have shown that the energy content of MSW can be estimated from the following equation:

$$E = 0.051[F + 3.6(CP)] + 0.352(PLR) \tag{14.4}$$

where

E = energy content, MJ/kg

F = % of food by weight

CP = % of cardboard and paper by weight

PLR = % of plastic and rubber by weight

This equation works well if there is little or no yard or garden waste. It is thus suitable in places like the European Union but less applicable in, say, the west coast of the United States, where garden trimmings account for about 20 per cent of waste by weight.

Table 14.18 Typical ultimate analysis of MSW

Component	% by dry weight					
	Carbon	Hydrogen	Oxygen	Nitrogen	Sulphur	Ash
Food wastes	48	6	38	2.5	0.5	5
Paper and cardboard	43.5	6	44	0.3	0.2	6
Plastics	60	7	23			10
Glass	0.5	0.1	0.4	<0.1		99
Metals	5	0.6	4.3	0.1		90
Clothing/textiles	55	7	30	5	0.2	3
Ashes/dust	26	3	2	0.5	0.2	68

Example 14.3   Compute the lower heat value (H_u) of the domestic MSW shown below, using each waste component and its associated MJ/kg.

Component	% by weight	Component weight (tonnes)	Lower heat value, H_u (MJ/kg or GJ/tonne)	Total energy (GJ)
Food waste	46	5 129	4.2	21 541
Paper and cardboard	11	1 226	16.5	20 229
Plastics	9	1 003	32.7	32 798
Glass	7	780	0.2	156
Metals	5	558	0.7	390
Clothing/textiles	1	111	18.3	2 031
Ashes, dust	19	2 118	6.9	14 614
Unclassified	2	223	—	—
Total	100	11 150		91 759

Solution

$$\text{Total lower heat value } (H_u) = 91\,759 \text{ GJ} = 8.23 \text{ MJ/kg}$$

The above example is MSW from a small Irish urban area. It includes only domestic MSW and not commercial. The energy content of 8.2 MJ/kg is below US or Danish averages of about 11 MJ/kg. This can be explained by the high food waste component. This example was from a survey in early spring. If surveyed in mid-winter, the values of ash would be higher as many households use coal fires for heating.

Example 14.4   Use Khan's equation (14.4) to compute the energy value of the waste in Example 14.3.

Solution

$$E = 0.051[F + 3.6(CP)] + 0.352(PLR)$$
$$F = 46\%$$
$$P = 11\%$$
$$PLR = 9\%$$

Therefore

$$E = 0.051[46 + 3.6(57)] + 0.353(9)$$
$$= 7.53 \text{ MJ/kg}$$

This compares with 8.23 MJ/kg of Example 14.3.

Example 14.5   Compute the lower heat value (H_u) of domestic MSW if the chemical composition is:

$$C_{450} H_{2050} O_{950} N_{12} S$$

Component	Number of atoms per mole	Atomic weight	Weight of each element	%
Carbon	450	12	5 400	23.6
Hydrogen	2050	1	2 050	9.0
Oxygen	950	16	15 200	66.4
Nitrogen	12	14	168	0.7
Sulphur	1	32	32	0.3
Total			22 850	

Solution

$$H_u = 337C + 1419(H_2 - 0.125O_2) + 93S + 23N$$
$$= 337 \times 23.6 + 1419(9.0 - 0.125 \times 66.4) + 93 \times 0.7 + 23 \times 0.3$$
$$= 7950 + 993 + 65 + 6.9$$
$$= 9015 \text{ J/kg} = 9.02 \text{ MJ/kg}$$

Example 14.6   Compute the heat value (H_u) of MSW from Denmark if the total MSW quantity is 2.4×10^6 tonnes (use the compositions shown in Table 14.12).

Component	% by weight	Component weight (per 1000 t)	H_u (MJ/kg)	Total energy (GJ/1000 t)
Food waste	35	350	4.2	1 470
Paper	30	300	15.2	4 560
Cardboard	5	50	26.2	1 310
Plastic	6	60	32.7	1 962
Other combustibles	8	80	18.3	1 464
Glass	8	80	0.2	16
Metal	4	40	0.7	28
Non-combustibles	4	40	0	—
Total	100	1000		10 810

Solution

$$\text{Total energy} = 10\,810 \text{ GJ/1000 t}$$
$$= 10.81 \text{ MJ/kg}$$

Therefore

$$\text{Total energy } H_u = 10.81 \times 2.4 \times 10^6 \times 10^3 \text{ MJ}$$
$$= 25.9 \times 10^6 \text{ GJ}$$

14.3.3 Biological Properties of MSW

Biological properties of MSW are relevant because of the technology of aerobic/anaerobic digestion to transform waste into energy and beneficial end products. Biodegradation can be aerobic or anaerobic. Anaerobic composting is the biological decomposition of 'food wastes' with end products of methane, carbon dioxide and others. Anaerobic digestion of the food fraction of MSW has been used at full scale. Some organic MSW components are undesirable for biological conversion, that is plastic, rubber, leather and wood. The relevant fractions for biological transformation include fats, oils, proteins, lignin, cellulose, hemicellulose, lignocellulose and water-soluble constituents.

The biodegradability of the food fraction of MSW is given by

$$BF = 0.83 - 0.028 \text{ LC} \qquad (14.5)$$

where   BF = biodegradable fraction expressed on a volatile solids (VS) basis

LC = lignin content of VS, % of dry weight

Table 14.19 shows the biodegradability of various MSW components. It is seen that some components,

Table 14.19 Biodegradability of MSW components

Component	VS as % of TVS	LC as % of VS	BF
Food waste	7–15	0.4	0.82
Newsprint	94	21.9	0.22
Office paper	96	0.4	0.82
Cardboard	94	12.9	0.47
Yard wastes	50–90	4.1	0.72

Adapted from Tchobanoglous *et al.*, 1993. Reprinted by permission of McGraw-Hill, Inc.

e.g. newsprint and cardboard, have a high lignin content but a low biodegradability. Those components with very low lignin content are highly biodegradable, e.g. food wastes.

14.4 SEPARATION

Separation can be either at source in the household (or industry) or at the transfer station or at final destination where mechanical separation/sorting is possible. Many urban areas throughout the world practice limited forms of source separation, i.e. the separation of the different fractions (at home) into units collectable by a haulier. If MSW is separated at source, it eliminates the need for expensive and difficult manual and/or mechanical sorting. In Sec. 14.2, the composition of wastes were identified. Overall MSW can be divided into:

• Wastes that are desirably separated at source (for regular collection by public/private haulier)
• All other household waste

The wastes that are desirably separated at source are:

• Food wastes—household source separation
• Paper and cardboard—household source separation
• Plastic—household source separation
• Metals ferrous—household source separation
• Metals non-ferrous—community recycling at drop-off centres
• Glass—community recycling at drop-off centres, household source separation

All other household waste that can be delivered to drop-off centres include:

• Bulky waste (furniture, tyres, etc.)
• Yard waste
• Hazardous household waste (in some countries collected at source)

14.4.1 Requirements for Source Separation

Source separation provides the cleanest and most well-defined fractions of waste suitable for subsequent recycling or reuse (but has the highest collection cost). Mechanical or manual sorting (at destination) tends to provide fractions that may be comprised of more than one group. There are health hazards associated with manual sorting. Mechanical sorting works best if there is a limited number of fractions that have well-defined physical properties (e.g. density). Source separated wastes may be either collected at the doorstep or kerbside or delivered to a drop-off centre. In practice, it is a combination of

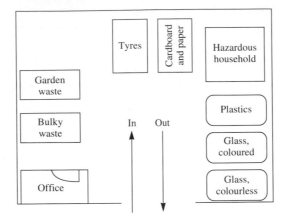

Dimension ~ 50 m × 50 m for population of ~10 000

Figure 14.3 Schematic of public drop-off centre.

collection/delivery services that are used. For source separation to work, the following infrastructure is required:

1. Community drop-off centres for glass and non-ferrous metals.
2. Public drop-off centres, often called civic amenity centres, where bulky, yard and household hazardous waste may be dropped off. In these centres, several different 'skip'-type containers are clearly labelled for the reception of the individual wastes, as shown in Fig. 14.3.
3. Environmental advertising programmes to firstly educate the public as to the required degree of source separation. In the first instance this may mean separating:

    Food
    Paper—newsprint
    　　—magazines
    　　—cardboard
    Plastics—seven types (Table 14.5)
    Metals—tin cans
    　　—others

Figure 14.4 is a schematic of a domestic facilitiy for source separation. The level and type of source separation will depend on people's attitudes and the end use. Ideally, the paper, glass and non-ferrous metals could be recycled. The remainder may only need to be separated into combustibles and non-combustibles if the treatment process is incineration. If the food fraction is to be transformed to biogas and compost, then the food fraction must be separated from the plastics and other non-biodegradable fractions. As commercial MSW is made up of similar components to household MSW (but with much higher paper/cardboard content), separation at source is also desirable.

14.4.2 Manual and Mechanical Sorting

If the waste is not sorted at source and it is intended to make use of different fractions, then sorting can be carried out either manually or mechanically at the final destination. Manual sorting can only be recommended for clean, dry and more or less pre-sorted waste. It is more applicable in separating dry fractions of paper or plastics. There are occupational risks with manual sorting on wet waste. The methodology of manual sorting is 'negative sorting', i.e. the unwanted materials are removed. 'Negative sorting' generates a greater volume than 'positive sorting'.

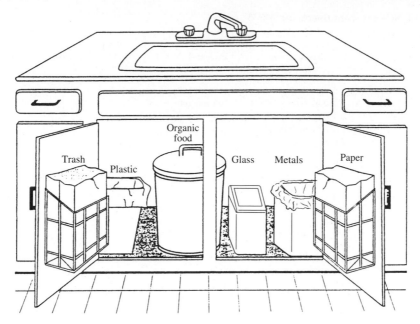

Figure 14.4 Source separation in the kitchen. (Adapted from Cunningham and Saigo, 1992, with permission).

Mechanical sorting is more commonly used. In theory, all fractions can be sorted. However, in practice there are many difficulties, particularly with 'wet' waste. Central sorting, whether it be mechanical or manual, produces fractions that are not as clean as when source separated. Magnetic separators are used to separate the ferrous and non-ferrous components. Vibrating screens or rotating screens are used to separate fractions by particle size. Air screening and ballistic screens are used to separate light from heavy materials.

Volume reduction by compaction is carried out on different waste fractions, particularly paper, cardboard, cans and plastic. Many refuse vehicles compact the collected waste from densities of about 100 to 300 kg/m^3, or a volume reduction of three. The baling of particular recyclable waste fractions is common, e.g. paper/cardboard.

14.5 STORAGE AND TRANSPORT OF MSW

Table 14.20 lists some of the types and sizes of storage containers used for MSW. They range from small plastic or paper bags of 25 L capacity to large containers with capacity up to 40 000 L. The most commonly used for household waste (detached houses) tend to be the smaller sized wheeled bins of 120 to 390 L capacity. For apartment buildings, 600 to 1000 L sizes are used. The type of storage used depends on the collection facility, which may be:

- Doorstep collection
- Regular kerb collection
- Civic amenity drop-off
- Haulier for skip collection of bulky items
- Community recycle bins
- Vacuum trucks

Table 14.20 Storage containers for MSW

Type of container	Container size (m)	Volume (L)	Used for household MSW	Used for commercial MSW
Bags				
Plastic	1 × 0.9	110 and 160	✓	✓
Paper	1 × 0.9	110 and 160	✓	✓
Bags in bin holders	1 × 0.9	110 and 160	✓	✓
Wheeled bins	1 × 0.54 × 0.48	120	✓	✓
	1.1 × 0.61 × 0.55	190	✓	✓
	1.1 × 0.72 × 0.58	240	✓	✓
	1.1 × 0.79 × 0.77	390	✓	✓
Multihousehold wheeled bins	~1.5 × 2 × 1.2	600–1100	✓	✓
Community bins Glass/metal	~2 × 2 × 1.5	~3000	✓	✓
recycle				
Skips				
Mini	1.2 × 1.2 × 1.2	1700	✓	✓
Maxi	1.1 × 3 × 1.8	6000	✓	✓
Jumbo	1.8 × 4.5 × 1.8	14 500	✓	✓
Containers		10 000–25 000		✓
Compacting	2.4 × 3.6 × 6	40 000		
Vacuum truck containers			✓	✓

Figure 14.5 shows the traditional refuse sack (plastic/paper) in a steel sack holder. Problems with handling occur with age and also with hygiene. Typically, the 110 L container weighs 13 kg when empty and 28 kg when full. Figure 14.6 is the modern HDPE wheeled bin, now used all over the world. It is automatically lifted on to the refuse truck. The capacities vary from 120 to 390 L. Empty, the 120 L bin weighs 7 kg and full weighs 27 kg, by comparison with 15 kg and 79 kg for the 390 L bin. Figure 14.7 is an HDPE bin used in multihousehold or apartment locations. It also is automatically lifted on to the refuse truck. The capacity of these units varies from 1100 to 2500 L. The dimensions of these units are typically 1.2 m wide by 2 to 3 m in length. Figure 14.8 is the typical community drop-off centre recycle bin for items such as glass, non-ferrous metals, clothes, paper, etc. Figure 14.9 is a compartmentalized bin for home wastes. Figure 14.10 shows a compartmentalized refuse truck and Fig. 14.11 shows a modern wheeled bin truck loader.

Skips or open top containers are regularly used for builders' and demolition rubble. They are also used for large garden waste and community clean-ups. The size range in capacity is from 1500 to 15 000 L. These skips are loaded on to skip trucks as shown in Fig. 14.12. Containers are used for the largest of bulk transport, generally over longer distances. They are typically used in transporting waste

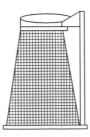

Figure 14.5 Traditional refuse sack in steel bin capacity ~150 L.

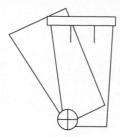

Figure 14.6 Single-household wheeled bin, capacity 120 to 390 L.

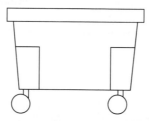

Figure 14.7 Multihousehold wheeled bin, capacity ∼1500 L.

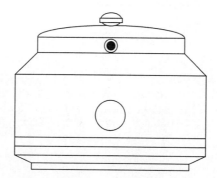

Figure 14.8 Community recycle bin for glass/metal, capacity ∼3000 L.

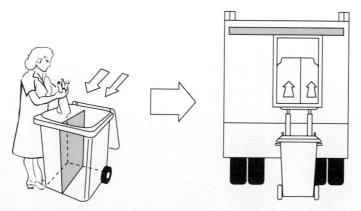

Figure 14.9 Source separation—multicompartment wheeled bin. (Adapted from WHO, 1991. Reprinted by permission of the World Health Organization.)

Figure 14.10 Multicompartment refuse truck. (Adapted from WHO, 1991. Reprinted by permission of the World Health Organization.)

Figure 14.11 Automatic loading of wheeled bin on refuse truck. (Reprinted by permission from Partek-Norba AB, Sweden.)

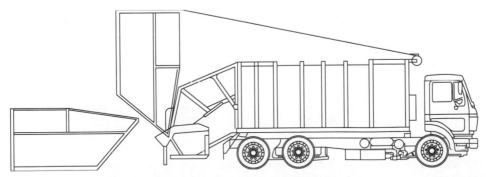

Figure 14.12 Skip loading refuse truck. (Reprinted by permission of Renoflex AS, Denmark.)

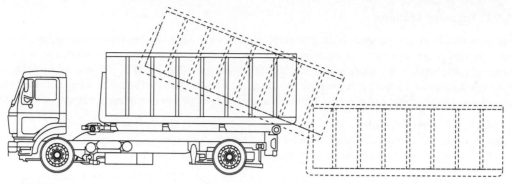

Figure 14.13 Container loading refuse truck. (Reprinted by permission of Renoflex AS, Denmark.)

from transfer stations to the treatment plant site. Containers are used in commercial applications, department stores, warehouses, etc., where in many cases the wastes are compacted with either static or dynamic compactors to further increase their capacity.

A multitude of transport vehicles are used for the transport of MSW including:

- Traditional open top trucks used in low income countries
- Traditional refuse trucks taking primarily bagged waste
- Modern single compartment trucks taking wheeled bins from single unit dwellings (Fig. 14.11)
- Single-compartment trucks taking larger wheeled bins from multiunit apartment buildings and commercial outlets
- Multicompartment trucks taking source separated waste (Fig. 14.10). A multicompartment truck capable of also taking several different recyclables is the most modern and efficient (one operator) system for suburban collection.
- Skip trucks taking skip loads (Fig. 14.12)
- Trucks taking container loads, either closed or open topped (Fig. 14.13)
- Vacuum trucks (Fig. 14.14), used in areas with limited accessibility, with tube lengths up to 100 m

Figure 14.14 Vacuum truck for semi-solid sludges. (Reprinted by permission of Mobilsug AB, Sweden.)

14.5.1 Transfer Stations

Transportation costs can be significant if haulage distances are large. Rather than encouraging all waste vehicles, particularly of small and medium size, to travel the full distance to a treatment site, a transfer station situated between the waste source and its final destination (e.g. landfill) is used. All vehicles with loads less than about 15 tonnes deposit their waste at the transfer station. Here, the waste is loaded on to large closed-top containers of capacity up to 30 tonnes. The objectives in using a transfer station are:

- Reduction in transport costs
- Reduction in trafficking of smaller vehicles at the treatment (disposal) site
- Reduction in refuse crew costs due to waiting time for vehicles in transit

The Greater London area uses transfer methods for 65 per cent of MSW. It uses bulk road transporters, river barges or rail wagons to convey refuse for long distances. WHO (1991) indicate that transfer stations are uneconomical if the haul travel distance (source to treatment) is less than about 20 km. However, it is the travel time rather than travel distance that is the more significant cost factor.

Figure 14.15 is a cross-section through a simple arrangement of a transfer station. The elements of a transfer station include:

1. Level 1 is a concrete paved area where refuse is tipped from refuse vehicles. This area may have the facility for two or more vehicles to unload simultaneously.
2. The waste falls into a hopper where a control breaker (gate) permits the waste to further fall in controlled volumes into a compacting unit. This unit is on level 2, about 5 m below level 1.
3. The compacting unit pushes the compacted waste horizontally into a waiting container (in the docking area).
4. The containers which can be 30 to 40 m^3 capacity are loaded on to waiting multiaxle trucks. The containers are on a 'rail-line' which permits easy movement and storage capacity for a number of containers.

Different designs of transfer stations are in operation, in different countries.

14.5.2 Collection Systems

The collection and transporting of waste, from household to transfer station to materials recovery facility to landfill or wherever, is an exercise in vehicle selection and route optimization. All urban areas have some type of collection system, be it elementary or sophisticated. Urban areas are continuously evolving and collection systems become more efficient with time, except in the supercities (e.g. Mexico) or the developing world, where the collection/treatment system is not keeping pace with population growth. The integration of a route system needs data on the:

- Area to be served
- Types and weights/volumes of waste generated
- Presence or lack of material recycling facility (MRF)
- Presence or lack of transfer stations
- Treatment systems—landfill, anaerobic digestion, composting, incineration, etc.
- Environmental constraints
- Economic constraints
- Vehicular fleet, size and quality

With CAD (computer aided design) and spreadsheets it is now possible to examine many alternative collection routes and identify the optimum, based on the resources available. With modern collection vehicles, it is now possible to operate routes with only one person. The driver doubles as the collector in many US and EU cities. There may be a need for a driver and people who load trucks in downtown areas

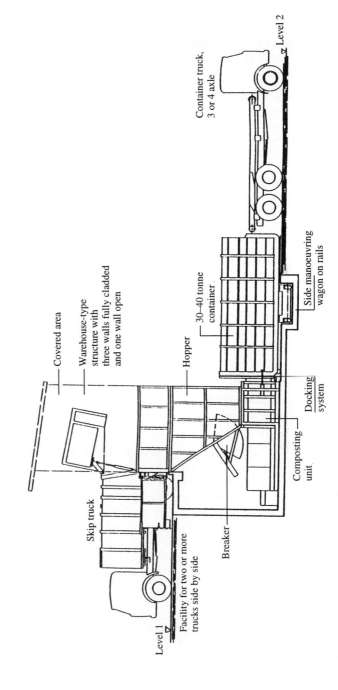

Figure 14.15 Section through a transfer station. (Reprinted by permission of Norba AS, Denmark.)

Level 2

Container truck, 3 or 4 axle

Side manoeuvring wagon on rails

30–40 tonne container

Hopper

Docking system

Composting unit

Breaker

Warehouse-type structure with three walls fully cladded and one wall open

Covered area

Skip truck

Facility for two or more trucks side by side

Level 1

or heavily trafficked areas. In laying out the routes for an urban/suburban complex, ideally the routes travelled by different vehicles on each day should be balanced. Rather than giving an example on collection routeing, Problem 14.7 is set as a student exercise at the end of the chapter.

14.6 MSW MANAGEMENT

The various technologies of treatment and ultimate disposal of MSW include:

- Waste minimization
- Reuse and recycling
- Biological treatment
- Thermal treatment
- Landfilling

The following sections detail each of the above. In an integrated waste management plan, there is room for an aspect of each of the above.

14.7 WASTE MINIMIZATION OF MSW

Waste minimization (regarding products manufacture) is covered in detail in Chapter 18. Waste minimization as regards MSW means reducing the amount that is generated at source. Some of the sources are as follows:

- Production units for food and household products
- Production units for commercial products
- Shopping outlets
- Households
- Offices, commercial properties and institutions

For example, in households, food wastes can be minimized. In the home–office or other institution, paper can be printed (or photocopied) on both sides, consumers can buy goods with least packaging, etc.

14.8 REUSE AND RECYCLING OF MSW FRACTIONS

Legislation is driving innovation and changes in recycling. For instance, in California, regulation AB939 (1993) requires that 50 per cent of the MSW be diverted from landfill by the year 2000. In Europe, the proposed packaging directive (1992) requires that within 10 years of implementation, 90 per cent by weight of packaging must be recovered and at least 60 per cent must be recycled. Currently Denmark has decided not to allow any organic or combustible waste to its urban landfills from 1997. Studies of urban waste streams have indicated that much can be recovered, reused or recycled. For instance, the city of Fullerton, California (Hay *et al.*, 1993), has identified that its MSW can be classified as:

- 40 per cent recyclable
- 29 per cent compostable
- 12 per cent potentially compostable
- 19 per cent other

Items in the waste stream that are currently popular in recycle streams or potentially recyclable are briefly discussed in the following paragraphs.

Aluminium cans Recycling of aluminium cans is at about 85 per cent at Davis, California, and improving. Internationally, it is the most successful of the recyclables and is so because of economics.

Simply, it is more expensive to import raw aluminium (to the United States) than it is to recycle it. Aluminium cans constitute about 1 per cent of the total domestic solid waste stream.

Paper and cardboard   Internationally, paper and cardboard form about 35 per cent of the MSW stream. Recycling has been slow, even in some EU countries. In Denmark, such recycling is becoming widespread. For instance, all corrugated cardboard is now made from recycled paper. Like aluminium, it is a matter of economics. It is cheaper (in many countries) to use raw wood for paper production than it is to use recycled paper. Some paper/cardboard is recycled and reused as fibre board. Parts of the west coast of the United States send their recovered waste paper/cardboard to countries like Taiwan, where wood raw materials are in short supply.

Glass   Traditionally, glass was well recycled. It used to be that milk bottles, beer bottles, soda bottles, etc., were returned and recycled. Again, recycling glass is getting attention. For instance, beer bottles in Denmark are used approximately 35 times. The main problem with glass is the proliferation of colours. However, there are many obvious advantages to glass recycling as it accounts for 5 to 10 per cent of the waste stream. Recycled glass has many uses. Ideally, glass can be reused. If not, it can be crushed and adopted for producing new glass, with energy savings.

Plastics   Plastics are for the most part non-biodegradable and most undesirable in landfills. They do, however, have a high heating value and are desirable for incineration plants. Plastic is typically 5 to 15 per cent of the MSW stream. Plastics are potentially reusable as other forms and this has been applied in several countries, e.g. recycled plastic into park benches, roadside delineators, etc. The potential is for reforming the plastics into another shape. To date, only approximately 5 per cent of plastics are recycled and these consist of PET and HDPE drink containers. In the United States other plastics, including PVC, LDPE, PP and PS, lag far behind in recycling (Truax, 1993). However, in the European Union, LDPE recycling has become more common.

Yard waste   Yard wastes can be shredded and composted and subsequently reused as soil amendments or other agricultural uses. Composting has had a variably successful history, but problems with odours, heavy metals and pathogens are now technologically solvable. Essentially, yard waste should not be landfilled. If not composted, the wood fraction especially has a relatively good heating value for incineration plants.

Organic food fraction   This fraction is about 10 per cent in the United States, about 35 per cent in Europe and Australia and sometimes 50 per cent or more in developing countries. The four options for this stream are:

- Composting
- Anaerobic digestion
- Reuse as animal foodstock
- Incineration

Miscellaneous items of MSW stream   Items like waste oil, car batteries, wood, computers, TVs, refrigerators, clothes, etc., are all recyclables and in the case of waste oil and batteries the trade is successful. Electrical and electronic appliances are now receiving attention by the manufacturers and this is a potential growth area. Table 14.21 shows approximate figures for recyclables arriving at material recovery facilities (MRFs) from curbside recycling. Paper/cardboard and glass make up as much as 93 per cent of total recyclables.

14.8.1 Materials Recovery Facility (MRF)

Materials recovery facilities recover as much reusable waste material as possible, including paper and cardboard, glass, metals, aluminium cans, PET and HDPE plastics, etc. There are many applications of the two formal designs of MRFs:

Table 14.21 Approximate recyclable composition, US

Material	% by weight
Aluminium cans	1
Metal cans	4
Glass bottles	19
Plastic (PET and HDPE)	2
Newspaper	33
Mixed paper	41
Total	100

Adapted from Smoley, 1993 with permission

- MRFs for source separated material
- MRFs for unseparated (commingled) MSW

The key to the variability of MRFs is the percentages of each material recovered. Depending on the extent of contaminants, the recovery factor for paper, for instance, can vary from 30 to 70 per cent. It is therefore important to note that only a fraction of materials arriving at an MRF (source separated or commingled) are recycled. Table 14.22 gives approximate values of recovered materials for source separated and commingled MSW. The percentage recovery for commingled waste is typically less than that of source separated, but it depends on the technology used. Figure 14.16 is a flow chart of a typical MRF for source separated materials.

The details of the 'separation and process' depend on the material being used. For instance, for paper the process is typically:

1. Separate out large cardboard containers.
2. All other paper goes through a 'trommel' to eliminate contaminants like stones, metals, foreign objects, etc. The newspaper goes through its own trommel and the mixed paper goes through a different trommel.
3. Manual sorting is carried out on a conveyor belt, from which sometimes mixed paper is separated out.
4. The final process is baling, which may be of newspaper only or mixed paper.

Other details of other processes for aluminium, tin, plastics, data are found in Cal Recovery and Peer Consultants (1993) and the *Journal of MSW Management.*

Table 14.22 Percentage recovery of source separated and commingled MSW

Material	Source separated MSW	
	Range (%)	Typical (%)
Aluminium cans	80–95	90
Metal cans	70–85	80
Plastic (mixed)	30–70	50
Plastic (PET, HDPE)	80–95	
Glass (mixed)	50–80	65
Glass (mixed/whole)	70–95	
Glass (by colour)	80–95	
Paper (mixed)	40–60	50
Newspaper	60–95	
Cardboard	25–40	30

Adapted from Smoley, 1993 with permission

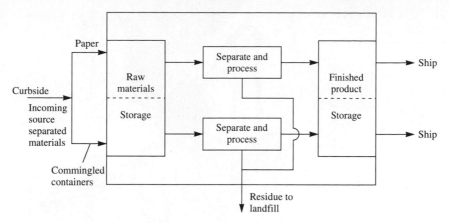

Figure 14.16 Flow chart for source separated materials. (Adapted from Cal Recovery and Peer Consultants, 1993, with permission.)

Figure 14.17 is a three-dimensional schematic of an operational resource recovery and solid waste facility at San Marco, San Diego, California. It is the largest such facility in the world and is a state of the art MRF that recovers recyclable materials from over 2100 tons per day of MSW. The recovered material is:

- Aluminium cans
- Ferrous metals
- Mixed non-ferrous
- HDPE
- PET
- Film plastics
- Newspaper
- Mixed paper
- Corrugated cardboard
- Glass (three colours)

Example 14.7 Determine the composition of recoverable materials for an urban area if each household generates 25 kg of MSW per week. Assume the material distribution is as per column (6) of Table 14.10. The percentage recoverables are taken from Table 14.22.

Solution

Component	% MSW by weight	Percentage recoverable	Weight before recycling (kg)	Weight recovered (kg)	Percentage of recoverables
Food wastes	31.5	—	7.9	—	—
Paper/cardboard	24.5	40	6.1	2.44	40
Plastics	14	50	3.5	1.75	28
Glass	3	65	0.8	0.52	8
Metals	7.5	80	1.9	1.5	24
Clothing/textile	3	—	0.8	—	—
Ashes/dust	12	—	3.0	—	—
Other	4.5	—	1.0	—	—
Total			25	6.2	100

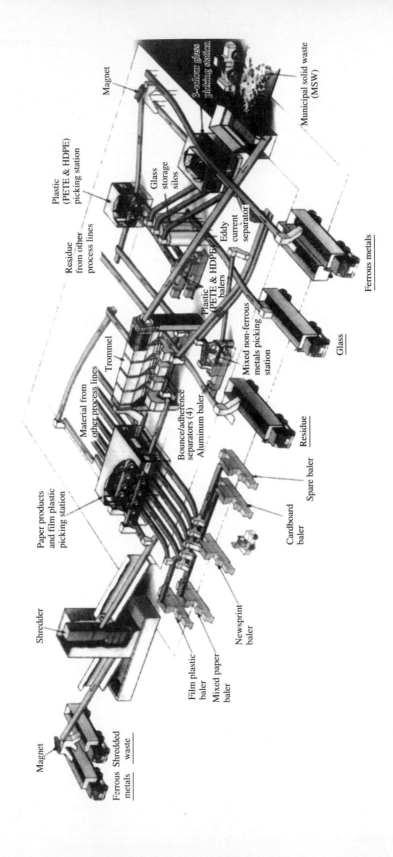

Figure 14.17 Three dimensions of resource recovery and solid waste reduction facility at San Marco, San Diego, California. (Reprinted by permission of National Ecology Company, Babcock & Wilcox, Timonium, Maryland.)

Magnet

Plastic (PETE & HDPE) picking station

Glass storage silos

Residue from other process lines

2-colour glass picking station

Municipal solid waste (MSW)

Eddy current separator

Ferrous metals

Plastic (PETE & HDPE) balers

Mixed non-ferrous metals picking station

Glass

Trommel

Material from other process lines

Bounce/adherence separators (4)

Aluminum baler

Residue

Paper products and film plastic picking station

Spare baler

Cardboard baler

Newsprint baler

Shredder

Film plastic baler

Mixed paper baler

Magnet

Shredded waste

Ferrous metals

It is noted that only 6.2/25 or 24.8 per cent of the full MSW stream, is potentially recoverable based on curbside (kerbside) recycling or source separated waste. This is regarded as a low figure, since internationally the food waste, clothing and textile waste, and ashes are high components in the above waste stream.

14.9 BIOLOGICAL MSW TREATMENT

At least three options exist for biological MSW treatment:

- Aerobic or composting
- Anaerobic or biogas
- Combined anaerobic and aerobic

Traditionally, the practice with composting was to utilize most of the MSW biodegradable material from unsorted collections. Components other than food were being used and as such contaminants such as heavy metals were showing up. These were due to degraded paper, cardboard and other degradable household products. The modern practice with composting is to use only the source separated food fraction of MSW, thereby reducing the occurrence of heavy metal contamination. Biogas production of the food fraction of MSW is raising much interest and the first full-scale MSW biogas plant was built in Helsingore, Denmark, in 1992. Combined aerobic and anaerobic systems using the organic fraction is an operation using the proprietary patented process including Dranco, Valorga and BTS. Research in this area is going on in Denmark, Holland, Italy and the United States.

14.9.1 Composting

Composting is an aerobic process where micro-organisms, in an oxygen environment, decompose the organic food wastes as follows:

$$\text{Organic matter} + O_2 \xrightarrow{\text{aerobic bacteria}} \text{new cells} + CO_2 + H_2O + NH_3 + SO_4 \tag{14.6}$$

The key inorganic nutrients required are nitrogen, phosphorus, sulphur, potassium, magnesium, calcium and sodium. Nutrients are typically present using the correct mix of waste fractions. The final product compost consists of minerals and humus (complex organic material).

Process requirements   The process parameters of relevance are:

- Temperature
- Moisture content
- Oxygen
- C/N ratio
- pH
- Biochemical composition and texture

Temperature   The composting process is exothermic and goes through temperature variations throughout its development:

- Psychrophilic—15 to 20°C
- Mesophilic—25 to 35°C
- Thermophilic—50 to 60°C

Best results are achieved if the thermophilic stage can be reached within the first few days and the compost then maintained at this temperature. Temperatures above the optimum thermophilic range inhibit biological activity but also improve the hygienic condition of the compost. If the compost is exposed to temperatures $> 55\,^{\circ}\mathrm{C}$ for two weeks, pathogenic kill is achieved, as occurs for temperatures $> 70\,^{\circ}\mathrm{C}$ for one hour.

Moisture content   Water is required for microbial activity which in turn produces water. Below 20 per cent moisture, biodegradation ceases. The optimum moisture content is 50 to 60 per cent, below which metabolic activity slows. Above 50 to 60 per cent, water fills the voids between particles, inhibiting oxygen access and causing temperature reduction and anaerobic conditions and malodours to occur. If moisture content decreases, then water must be added.

Oxygen   If oxygen is less than 10 per cent by volume, composting is inhibited. Optimum oxygen levels are 15 to 20 per cent. Oxygen is essential for aerobic decomposition, though decompostion can occur anaerobically with low oxygen levels. This is malodourous and the process of biodegradation is slower. Turning and ventilating compost are meant to keep the oxygen content at a sufficient level.

C/N ratio   The C/N ratio is a measure of the optimum biochemical conditions and occurs at a C/N ratio of 30. This ensures adequate nitrogen for cell synthesis and carbon as the energy source. This ratio of 30 is not so high that N escapes as NH_3. Composting operates down to C/N ratios of 20. Ratios in excess of 30 mean too little nitrogen to form microbial cells, and thus force the micro-organisms to go through additional cycles of carbon consumption, cell synthesis, decay, etc., and so slow down the composting process. Table 14.23 shows the C/N ratios for different MSW fractions.

pH   The optimum pH range is 6 to 8. In the initial days the pH reduces to about 5 as organic acids are formed. Then the pH rises as these acids are consumed in the thermophilic stage. Anaerobiosis sets in at pH below 4.5.

Biochemial composition and texture   The composition of the waste influences the process rate. Some materials are easily degradable, like plants, manure, primary wastewater sludge and food wastes. Straw, wood, leaves, yard wastes and paper with high contents of lignin are slowly biodegradable. The texture influences the process in its variable surface areas as habitats for the micro-organisms and in its ability to retain moisture or oxygen.

Table 14.23 Carbon/nitrogen ratios of some waste materials

Material	% nitrogen (dry)	C/N ratio
Fish scrap	6.5–10	—
Farmyard manure	2.15	14–1
Kitchen waste	2.0	25–1
Seaweed	1.92	19–1
Wheat straw	0.32	128–1
Rotted sawdust	0.25	200–1
Raw sawdust	0.11	510–1
Food wastes	2.0–3.0	15–1
Total refuse	0.5–1.4	30/80–1
Wood	0.07	700–1
Paper	0.2	170–1
Grass clippings	2.2	20–1
Weeds	2.0	19–1
Leaves	0.5–1.0	40/80–1
Fruit wastes	1.5	35–1
Sewage sludge		
Activated	5.6	6–1
Digested	1.9	16–1

Composting systems   Three different composting systems are practised:

- Traditional wind-row
- Aerated static pile
- In-vessel composting

Traditional wind-rows   After the non-biodegradable and/or slowly biodegradable fractions are removed, the MSW is piled up in long rows of almost triangular cross-section on hard surfaces. The height is 1 to 2 m and the width of the base 3 to 4 m. Wind-rows may be outdoors or in a covered shed. Regular turning of the piles is essential to oxygenate the entire material. Full development is usually in excess of 3 months, after which the compost is allowed to cure for up to 12 months.

Aerated static pile   The MSW is piled up in wind-rows about 1 to 2 m high, and 3 to 4 m wide and 20 m long and laid on floors of ventilation piping systems. To reduce odours, a stabilized compost cover is placed on the piles. The piles are aerated by forcing air through the perforated pipes at regular intervals. Temperature probes activate the aeration system. Decomposition occurs in 4 to 6 weeks by comparison with 12 weeks for the traditional wind-rows. This process had the advantage of better control than the traditional wind-row. Figure 14.18 is a schematic of the pH and temperature development for unaerated and aerated composting.

In-vessel composting   This method of composting is carried out in different vessels (reactors):

- Horizontal plug flow reactors
- Vertical continuous flow reactors
- Rotating drums

In the horizontal plug flow (similar to plug flow in activated sludge wastewater systems), a plug of waste is fed through a 'tunnel' where aeration is supplied continuously. In the vertical continuous flow reactor, odour is contained in the vessel. Difficulties are encountered with compaction of the compost at the bottom of the reactor. In the rotating drum, refuse is fed through for a retention time of 4 to 6 hours, after which it is aerated and homogenized. The output of raw compost is ready for aerobic aeration in wind-rows or in the aerated static pile.

Figure 14.19 is a layout plan of a composting plant by Kruger Waste Systems, Denmark, for 24 000 t/yr. The essential plant units are the weighbridge, the shredder, the bulking agent additive (wood chips, straw, etc.), the magnetic separator and the in-vessel compost reactor. Figure 14.20 is a process flow diagram of a wind-row sludge composting operation. Prior to composting a bulking agent of wood chips or straw is added. Other additives like lime may be used to correct the pH, but only at the beginning of the

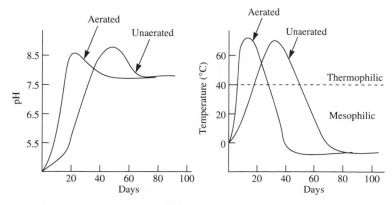

Figure 14.18 Schematic of pH and temperature development in unaerated and aerated composting.

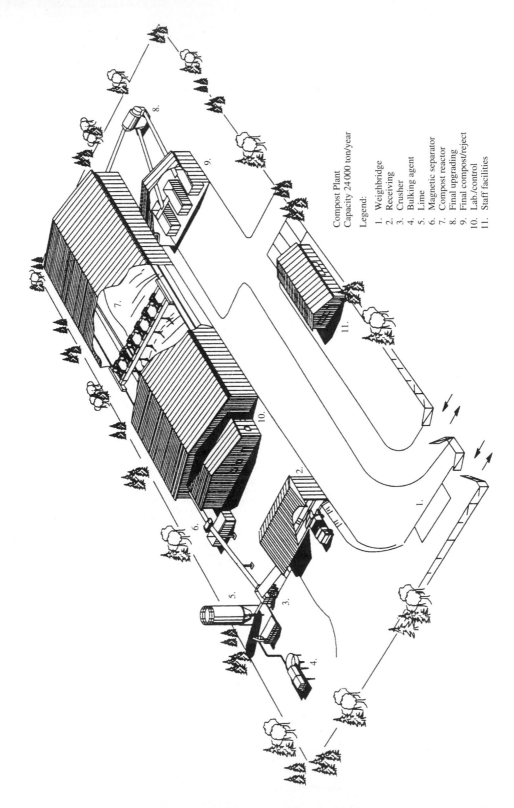

Compost Plant
Capacity 24 000 ton/year

Legend:

1. Weighbridge
2. Receiving
3. Crusher
4. Bulking agent
5. Lime
6. Magnetic separator
7. Compost reactor
8. Final upgrading
9. Final compost/reject
10. Lab./control
11. Staff facilities

Figure 14.19 Layout of a composting plant. (Adapted from Krüger Engineering, Denmark. Reprinted by permission.)

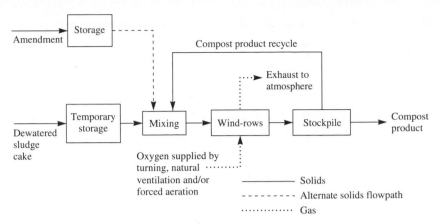

Figure 14.20 Process flow diagram of a wind-row sludge composting operation. (Adapted from Haug, 1980, with permission.)

plant. In practice, lime is rarely used. Caustic soda is more commonly used. Figure 14.21 is a materials balance observed during wind-row composting. What is of main interest here is the use of recycle to supplement the input sludge. Because of the recycle line the amendment mix needed at the beginning is only 1.75/1 and not the more typical 3/1 (in the absence of recycle).

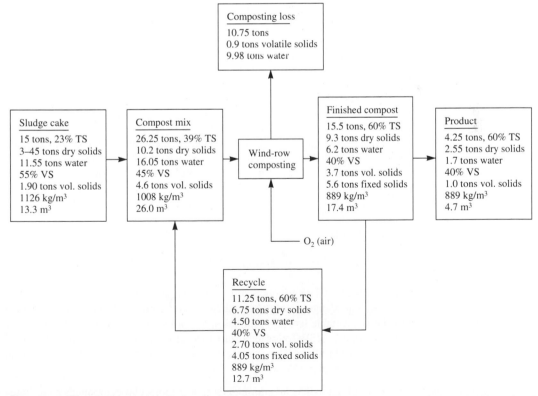

Figure 14.21 Materials balance of wind-row composting of digested sludge cake with recycle. (Adapted from Horath, 1978, with permission.)

Environmental aspects of composting The environmental parameters of relevance in composting are:

- Heavy metals
- Odour
- Sterilization
- Inert contaminants

Heavy metals The presence of heavy metals in compost has caused public concern, leading to reduced acceptance of the compost product. The metals of concern include mercury, cadmium, copper and zinc. These metals are found in general MSW from batteries, paints, plastics, papers, etc. Copper and zinc are the least offensive as they are an essential trace element usually found in excess. Cadmium, chromium and mercury are taken up by plants and can enter the food chain. They are highly toxic and the sources include batteries, leather and plastics. Composting a source separated food fraction produces a compost acceptable in heavy metal concentrations.

Odour Odours are common in composting plants, but more so in unaerated systems. In-vessel composting and aerated static pile systems should be managed to reduce odour levels. Air from ventilating the compost is often treated in biofilters. In outdoor systems, covering the pile or wind-row with a stabilized compost reduces the odour.

Sterilization Sanitization is achieved at thermophilic temperatures (55 to 60 °C) lasting 2 to 3 weeks. It is also achievable at higher temperatures for shorter periods, but this is less complete and pathogenic revival has occurred in compost. It is not possible to verify whether 100 per cent of the compost is sterilized.

Inert contaminants Pre-sorting of wastes is essential if contaminants of glass, metal, rubber and partially decomposed paper are to be insignificant.

14.9.2 Anaerobic Digestion

The WHO document on Urban Solid Waste Management (1991) stated that 'it is clear that energy production from MSW may be optimized by including anaerobic digestion into integrated waste separation and materials recovery systems'. Anaerobic digestion (A/D) of the unseparated MSW is unsuitable due to the presence of plastics, glass, metals, textiles, etc. The source separated food fraction of MSW is approximately 35 per cent by weight. Of this, 70 per cent is moisture. Traditional anaerobic digestion operated with low solids (4 to 10 per cent). Modern technologies in use in France and Holland and in the research stage at the University of California at Davis are using high solids reactors (up to 30 per cent). This is a significant breakthrough in that it now permits high solids MSW to be anaerobically digested. This latter technology is still in development for MSW. Anaerobic digestion itself has grown in use in the past decade, particularly for agricultural waste, but more recently for MSW solids and municipal sludges and sewage (refer to Chapter 15).

Anaerobic digestion is described by

$$\text{Organic matter} + H_2O \xrightarrow{\text{anaerobic bacteria}} \text{new cells} + CO_2 + CH_4 + NH_3 + H_2S \qquad (14.7)$$

The beneficial end product is methane. Other products are sludge water, carbon dioxide and traces of ammonia and hydrogen sulphide. The sludge water produced can be dewatered to produce a supernatant and a filter cake. The filter cake is a suitable fertilizer and has some soil conditioning effects. The filtrate can be used in pre-mixing of the organic MSW to create a slurry feedstock or it can be fed directly to the digester. It can also be used as liquid fertilizer. The process of anaerobic digestion can be described in three phases (for further details, see Chapter 13).

- Hydrolysis, i.e. the breakdown of high molecular compounds to low molecular compounds as in lipids to fatty acids, polysaccharides to monosaccharides, proteins to amino acids, etc.
- Acidogenis, where the lower molecular components of fatty acids, amino acids and monosaccharides

are converted to lower molecular intermediate compounds such as propionate, butyrate, formate, methanol and acetate

- Methanogenesis, where the intermediate compounds are converted to the final products of methane and carbon dioxide

In these three phases, a different group of bacteria are active and are called the hydrolysing bacteria, the non-methanogenic (or acidogens) and methanogens respectively. The latter are strictly anaerobes while the acidogens are facultative and obligate anaerobes.

The maintenance of an environment that keeps the acidogens and methanogens in dynamic equilibrium is required to:

- Be oxygen free
- Not contain inhibiting salts (may be heavy metals or excess ammonia)
- Have a $6.5 < pH < 7.5$
- Be of adequate alkalinity, 1500 to 7500 mg/L
- Have sufficient nutrients, phosphorus and nitrogen
- Be temperature steady at either mesophilic or thermophilic conditions
- Have a constant solids loading rate

Generally, anaerobic digestion has a poor performance at psychrophilic temperatures. Also, temperature steadiness is essential and only a few degrees of variation can upset the dynamic balance, especially at thermophilic temperatures. The feedstock needs to be consistent and variations in alkalinity reduce the biogas production rate. There are two distinct variations on solids concentrations:

- Low solids anaerobic digestion
- High solids anaerobic digestion

Low solids anaerobic digestion   The most conventional system of anaerobic digestion of organic MSW is to use solids concentrations between 4 and 10 per cent. This process is well proven, having been applied in the agricultural and industrial sector throughout the world. Using MSW feedstock of 20 per cent solids or more means that the waste must be diluted to achieve the 4 to 10 per cent solids concentration. This is often achieved by adding sludge to the MSW. Figure 14.22 is a flow diagram of the process.

Typical production values are a gas with a 50 to 70 per cent methane content and production rates in the range of 1.5 to 2.5 m^3/m^3 (biogas) of reactor size or approximately 0.25 to 0.45 m^3/kg of biodegradable volatile solids destroyed. The cycle time in the reactors varies and is typically about 20 days. The cycle time depends on the loading rate and the biodegradability of the feedstock and on the process temperature. Lower time cycles can be achieved but at the cost of reduced loading rates and reduced biogas production.

Figure 14.23 shows the typical retention time for MSW digesters at the different operating temperatures. Clearly, it is desirable to operate at thermophilic temperatures, but due to potential instability problems, this temperature operation is still not proven technology for MSW digestion. Most MSW digesters internationally operate at mesophilic temperatures.

High solids anaerobic digestion   This is an anaerobic process where the solids content in the reactor are 25 to 35 per cent. The Dranco process for organic waste systems in Belgium uses a solids concentration of 30 to 35 per cent. Dranco means dry anaerobic composting. Many processes are in existence internationally. Figure 14.24 is a flow chart of the Dranco process treating organic MSW. The cycle time is 16 to 21 days and biogas production is 5 to 8 m^3/m^3 of reactor. The gas content is 55 per cent CH_4. Energy production is 140 to 200 m^3 biogas per tonne of raw organic waste at 40 to 60 per cent dry solids.

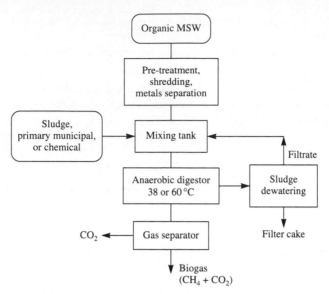

Figure 14.22 Flow diagram of low solids anaerobic digestion.

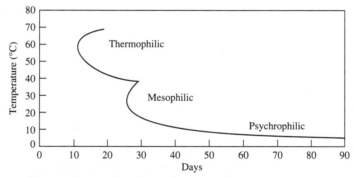

Figure 14.23 General trend for the digestion time of a certain amount of organic matter. The curve might differ according to physical/chemical composition of feedstock. (Adapted from Danish EPA, 1994.)

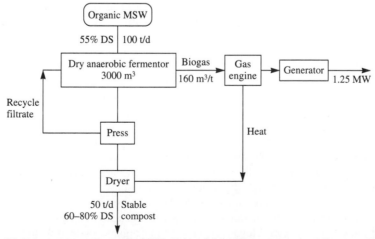

Figure 14.24 High solids Dranco process for organic MSW. (Adapted from Dranco promotional literature, 1992. Reprinted by permission of Organic Waste Systems N.V.)

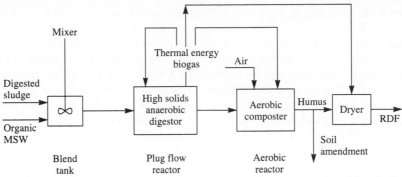

Figure 14.25 Flow diagram for combined treatment of organic MSW and municipal sludge. (Adapted from Kayhanian *et al.*, 1991, with permission.)

14.9.3 Two-Stage Anaerobic and Aerobic Methods

The two-stage process using high solids anaerobic digestion followed by in-vessel aerobic composting is currently an active area of research. The advantages include:

1. No external water is needed for producing a slurry feedstock to the anaerobic digester.
2. No effluent water is produced.
3. There are high rates of biogas production of 5 to $10 \, m^3/m^3$ of reactor.
4. A stabilized humus product is produced.
5. A refuse derived fuel (RDF) is produced.

Figure 14.25 shows the flow process diagram for organic MSW feedstock blended with municipal sludges. In the Kayhanin *et al.* (1991) experiments, biogas production was up to $6 \, m^3/m^3$ of reactor, using an organic MSW feedstock of food waste (38 per cent), office paper (3.8 per cent), newsprint (12 per cent) and yard waste (12 per cent).

14.10 THERMAL TREATMENT—COMBUSTION/INCINERATION

Thermal degradation of organic material can be carried out with or without oxygen. If thermal degradation takes place with a deficit of oxygen (partial combustion), some of the energy stored as chemical energy from the organic material will be released as burnable gases. This process is called *gasification*. If the gasification is dry distillation (heating without access for air, oxygen or steam), the process is called *pyrolysis*.

Gasification has the advantage that the gases are more economical to purify than the flue gases in pyrolysis, due to the large volume of flue gases generated from combustion. Furthermore, the gases can often be utilized directly in gas engines. Gasification of biomass is still at research level. Operational problems are caused by a high content of water or a low density of biomass. Difficulties encountered in crushing and feeding are reasons that the technology has not yet been developed to a commercial level. Because municipal solid waste is more non-homogeneous than other biomass fuels (straw and wood), it is also more difficult to treat by gasification.

Thermal degradation with excess oxygen is called *combustion*. When the fuel is waste, this is called *incineration*. Incineration is chemical oxidation at high temperatures, where organic material is converted into energy (heat), flue gas and slag. Incineration of waste is, in principle, similar to combustion of other solid fuels like coal, wood, etc. In practice, waste differs from other solid fuels in being non-homogeneous and having a higher content of water than coal or wood. This has, in some cases, been solved by turning waste into pellets, named RDF (refuse-derived fuel). RDF pellets can be used in furnaces which burn traditional solid fuels, often in a mixture with coal. They can also be stored just as coal or other solid fuels.

Incineration plants for solid waste are based on technologies developed especially for incineration of waste. In most cases, incineration is more economical than firstly converting the waste into RDF pellets. A number of companies have developed the technology for waste incineration into commercial concepts (e.g. Vølund, Denmark). The layout of the incineration plant varies from company to company. Some use movable grates alone. Others use movable grates in combination with rotary kilns. The design varies in the type of feeding system, grates, furnace, slag removal system and boiler. Except for some experimental incinerators using fluidized beds and mass combustion, generally the same design and layout principles are used for MSW and for coal combustion plants design and layout.

14.10.1 Historical Background

The first waste incineration plant was built in the United Kingdom in the 1870s. Compared to the present-day technology, it was a very simple system with manual stoking, manual slag removal, no regulation of air supply with combustion air, nor any flue gas cleaning system. One reason why waste incineration was introduced was the increasing interest in hygienic aspects, caused by the outbreak of a number of cholera epidemics towards the end of the nineteenth century. At the beginning of the twentieth century, technology improved and plants were built for the generation of electricity by steam production. Energy production based on waste became uneconomic and the technology faded. The renaissance of waste incineration came in the late 1960s. Today's waste incineration is a key element in waste treatment in countries like Japan, Switzerland and Denmark. Table 14.24 outlines the state of MSW incineration in the European Union, the United States and Canada.

14.10.2 Design Principles for Waste Incinerators

The general principles of waste incineration are similar to those for combustion of other solid fuels. However, due to the difference in the fuel, waste incineration plants have modified layout arrangements.

Table 14.24 Incineration data for the United States, the European Union and Canada

Country	Total number of plants	Average quantity of incinerated waste per year ($\times 10^3$ tonnes)	Number of plants without heat recovery	Number of plants with heat recovery		
				Industrial use and heating	Electricity production	Electricity + heating or steam production
Belgium	28	56.0	15	7	4	2
Federal Germany (1986)	47	183.0	3	7	7	30
Denmark (1992)	38	52.5	—	34	—	4
Spain (1987)	8	415.0	5	—	2	1
France (1985)	280	84.5	216	43	2	19
Italy (1987)	94	26.0	78	12	4	—
Luxembourg	1	100.0	—	—	1	—
Netherlands	11	158.2	5	2	3	1
Portugal (1987)	—	—	—	—	—	—
United Kingdom	48	66.7	43	4	1	—
EC total	557	—	367	111	24	53
United States†	83	—	14	42	11	16

†The US figure includes five Canadian mass-burning plants.

Figure 14.26 is a cross-section of a typical municipal waste incinerator at Reno Nord, Aalborg, in Denmark. It has a capacity of 12.5 t/h with power generation delivering surplus heat for district heating. The principal segments of this incineration plant are shown on the figure.

Unloading and storage of waste   When delivered waste has been recorded, normally on a weighbridge at the entrance gate, it is then unloaded into a storage pit. It is useful to have a flat paved area in front of the pit, where it is possible to unload and examine delivered waste if spot inspections are required for quality control purposes with regard to the origin and composition of the waste. With some bulky waste (i.e. furniture), it is desirable to first reduce this in a shredder. This simplifies its passage through the loading hopper and encourages sastisfactory out-burning at the minimum retention time on the grates. The receiving area in front of the storage pit should have the capacity for a suitable number of trucks unloading at the same time. The capacity of the pit should be enough to level out variations in deliveries over the 7-day week. It should also be large enough to accommodate additional back-up loads if the incineration plant breaks down. The layout of the pit must also enable the operator of the crane to mix different kinds of waste so as to reduce variations in the calorific value between, say, household food waste ($H_u \sim 4.2$ MJ/kg) and waste from commercial activities with a high content of paper ($H_u \sim 16.5$ MJ/kg) and plastic ($H_u \sim 32.7$ MJ/kg). The layout and capacity of the pit should also ensure that all areas of the pit can be emptied regularly, so that waste does not stagnate and age in the pit. Septicity may occur with age of the MSW. It is sometimes desirable to establish the unloading area and the pit inside a building, for the purpose of minimizing odour emission.

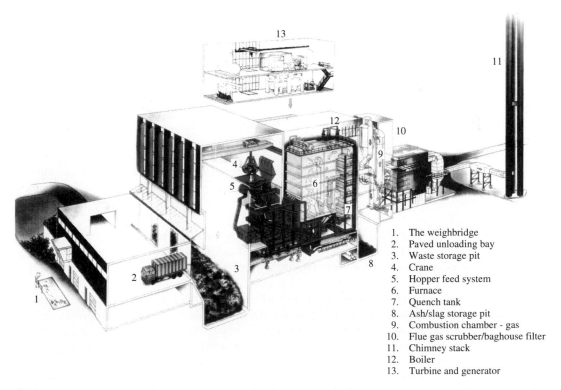

1. The weighbridge
2. Paved unloading bay
3. Waste storage pit
4. Crane
5. Hopper feed system
6. Furnace
7. Quench tank
8. Ash/slag storage pit
9. Combustion chamber - gas
10. Flue gas scrubber/baghouse filter
11. Chimney stack
12. Boiler
13. Turbine and generator

Figure 14.26 Municipal waste incinerator at Reno Nord, Aalborg, Denmark. (Reprinted by permission of Krüger Engineering, Denmark.)

Crane   The crane is a key element in waste incinerators. Besides the capacity and considerations of stand-by requirements, it is important that the crane can remove items (i.e. refrigerators) that are unsuitable for incineration.

Hopper and feeding system   The design of the receiving hopper should enable a buildup of a high column of waste without blocking. The column of waste should prevent the ingress of excess air to the incinerator chambers. At the top of the hopper, below the funnel, there should be a gate flap arrangement that can be closed in case of fire or when starting up or closing down the incinerator. At the bottom of the hopper, there should be a piston moving the waste forward to the grate (see Fig. 14.26).

Grates   Proper design of the grates is essential for satisfactory combustion. The grates move the waste forward and turn it over in a way that ensures an optimal contact between the waste and the burning gases. However, the movement should not be so fast that excess dust is released or that the ignited waste might quench. The grates must be divided into at least three sections with individual air supply systems to enable regulation of the combustion in each phase. It is further important that the design of the grates produces a satisfactory distribution of primary air supply through the grates. The capacity of the grates is normally in the range of 2 to 3 GJ/m^2 h. If the load is too high, there might be problems with fixation of slag. If the load is too low, there might be problems in reaching the desired temperature required for complete combustion.

Furnace   The furnace can be divided into two zones: the primary combustion chamber (PCC) and the secondary combustion chamber (SCC). The primary combustion chamber will normally be designed for a capacity of 0.5 to 0.7 GJ/m^3 h. The secondary combustion chamber is designed for an appropriate gas retention. As an example, the Danish EPA guidelines on waste incineration specify two seconds of retention time and a Reynolds number $> 60\,000$ is required to ensure adequate mixing of waste and air. The purpose of the minimum retention time and turbulence is to ensure destruction of potential dioxins and dibenzofurans (see also Chapter 15).

Air supply   The air for incineration is often taken from the discharge hall (unloading bay). This creates a slight vacuum and reduces the risk of emissions to the surrounding areas. Combustion air is divided into primary and secondary air. Primary air is injected from beneath the grates, with control for each section of the grate. If the incinerator is designed for waste with a high water content, it is normal that air injected at the first grate be pre-heated. Secondary air is injected in the top of the furnace before the secondary combustion chamber. The relationship between primary and secondary air depends on the composition of the waste. A high content of fugitive hydrocarbons, as may occur with plastics, increases the requirement of secondary air.

Boiler system   Energy produced in incinerators can be utilized in boilers positioned after the secondary combustion chamber. The choice depends on what is more feasible for the particular location. It may be steam for generating electricity or hot water for district heating.

Slag and ash removal   The residues at the end of the grates (slag or bottom ash) are removed continuously without permitting the inlet of false air. Another requirement is that the opening for slag removal should be bigger than the opening in the hopper to ensure that large unburnable items that have passed the furnace can also be removed. The solution is often a trap consisting of a chamber with water beneath the end of the grates and a piston at the bottom of the water chamber to remove the slag to a conveyor system outside the furnace. Besides the slag, there is also ash and fine material that pass the grates and need to be removed. There will also be flyash to remove from the bottom of the boiler. Flyash will normally be removed with screw conveyors.

14.10.3 The Combustion Process in Waste Incinerators

Waste incineration is, in principle, similar to combustion of other solid fuels. The combustion process can be divided into three phases:

- Drying
- Gas pyrolysis and gas combustion
- Combustion of carbon residue

It is shown in Fig. 14.27 that drying takes place at the first part of the grate. When the waste is dried and heated, gasification takes place at zone A. Here, easily evaporated hydrocarbons are released and some with low combustion temperatures are burned just above the waste. Others escape to the secondary combustion chamber where the combustion takes place at a higher temperature, 950 to 1050 °C at zone B. The solid residue, mainly consisting of ash and carbon, is burned at the third grate as nearly pure carbon combustion at zone C. Operating with only small rates of excess air (3 to 6 per cent) in zone A reduces the generation of NO_x.

14.10.4 Waste as an Incinerator Fuel

Waste is sometimes compared to coal and wood as a fuel source. However, because of the non-homogeneity of waste, it is nearly impossible to collect a representative sample for a calorimetric test.

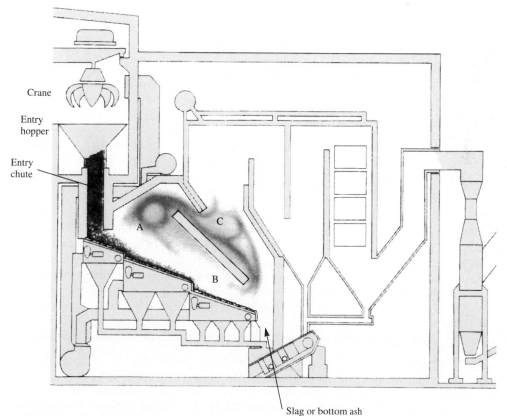

Figure 14.27 Three phase locations of incineration: gasification (zone A), high temperature combustion (zone B) and pure carbon combustion (zone C). (Reprinted by permission of Vølund Ecology Systems, A/S Denmark.)

Furthermore, the composition of waste varies during the year and often even during the week. Fuel characterization of waste can be made by processing a sample of MSW through an existing incinerator plant. This requires an existing plant within a reasonable distance and a representative sample big enough to run a test for 4 to 5 hours. The other alternative is to characterize the waste by content of different materials, estimating calorific value of each fraction and then calculating the average calorific value.

Ash is a fundamental part of all wastes. Ash is inert in the combustion process. The influence in the amount of energy produced from the fuel is only slightly influenced by a small amount of heat lost by heating the ash from the inlet temperature to the outlet temperature for ash and slag. The content of water is also inert. Water leaves the furnace and boiler by evaporation. Besides the energy loss in heating water, there will also be a great energy loss in evaporation of water. The burnable part of the waste is that part of the waste that can release energy in the oxidation process. A simple approximation is given for estimation of the net calorific value in Eq. (14.2):

$$H_{(\text{net})} = H_{\text{u}} = H(\text{awf}) \times B - 2.445 \times W \qquad (14.2)$$

where   B and W = fractions of burnables and water content respectively

$H(\text{net})$ = net calorific value, MJ/kg

$H(\text{awf})$ = net calorific value of the burnable part of waste, MJ/kg
(ash and water free)

Experiments show that, for practical purposes, it is possible to estimate the energy potential based on an average value of $H(\text{awf}) \approx 20$ MJ/kg. For more detailed calculation, the values in Table 14.16 can be used.

14.10.5 Air Emission and Flue Gas Cleaning

The combustion process will break down the complex organic compounds of waste into more simple compounds and subsequently generate new compounds. Many of these compounds will be present in the flue gas. Carbon, hydrogen and sulphur will be converted to CO_2, H_2O and SO_2. Nitrogen will be converted to N_2 and NO_x. Chlorine will become HCl and may also be converted to complex organic chlorides. Inorganic substances will be oxidized and escape as particles in the ash or as vapour in the flue gas. There are three main factors that influence the content of the raw flue gas from MSW incinerators:

1. The composition of waste has an impact on pollutants such as SO_2, HCl, HF, organo-chlorides and metals.
2. Thermal processes have an impact on burned or partially burned gases such as carbon, hydrogens and chlorinated organics. Analyses of these compounds are very complex. Therefore CO is often used as an indicator. These compounds can be limited by careful control of combustion and temperature. Excess air is also an important factor. On the other hand, too much excess air reduces the combustion temperature and increases the heat loss by flue gas. Other pollutants that depend on the thermal process are NO_x. The smallest quantities of NO_x are produced in the temperature range of 800 to 1000 °C. This is the optimum temperature range for the NH_3 reduction of the NO_x. At approximately 1200 °C, the amount of NO_x increases rapidly.
3. Mechanical processes influence the amount of flyash produced in the combustion chambers.

Typical pollutant content levels of raw flue gas from MSW incinerators are given in Table 14.25 corresponding to dry flue gas. Flue gas typically contains 5 to 15 per cent H_2O. Until recently, most attention was paid to the flyash particulate matter. As such, most flue gas cleaning consisted of flyash removal by the use of cyclones, electric filters, bag filters or scrubbers. The most popular was the electrostatic precipitator. The effect of different filters on typical flue gases from incinerators is given in Table 14.26.

Table 14.25 Typical pollutant content levels of dry flue gas before cleaning and EU emission standards

Pollutant	Typical content: mg/N m³ at 9% CO₂			
	Typical untreated flue gas	EC nominal capacity (t/h)		
		< 1 t/h	1–3 t/h	>3 t/h
Dust	1500–8000	200	100	30
SO_2	400		300	300
HCF	500	250	100	50
HF	5		4	2
NO_x	300			—
CO	100			150
Organic vapours	5			20
Hg vapours	0.05–0.5			0.2†
$Ni + As + Pb + Cr + Cu + Mn$	0.05–0.5			0.2†

† Total Hg and Cd.

Table 14.26 The effect of different filters on typical flue gases from incinerators

Filter type	Particle separation down to	Effectiveness (%)	Dust emission (mg/N m³)	Notes
Cyclone	10 µm	90	Seldom less than 500	Low capital cost Big pressure loss (running costs)
Bag filters	0.3 µm	> 99	< 150	Lower capital cost than electric Larger pressure loss than electric Required temperature less than 160 °C
Electrostatic precipitation	0.3 µm	> 99	< 150	Expensive establishment Low pressure loss
Scrubber	1–2 µm	> 95	< 500	Low capital cost Big pressure loss (running costs) Cooling of flue gas

In the last few years, attention has also been directed towards other elements in the flue gas, especially acids (HCl, HF and SO_2), heavy metals, particularly Hg, and dioxins and furans (PCDD and PCDF). The EU-guidelines are shown in Table 14.26. The PCDD and PCDF, together with polyaromates (PAH), can be limited by good design and efficient management. To reduce the emission of acids, new flue gas cleaning equipment has been developed. Two types of absorption can be used:

- Absorption with lime
  - Dry absorption
  - Semi-dry absorption

- Absorption in water
  - Scrubbing with added water
  - Condensation

Dry absorption (see Fig. 14.28) works with injection of dry $Ca(OH)_2$, to absorb HCl and form $CaCl_2$ and water. The water will evaporate due to the heat in the flue gases, and a dry residue will remain. The dry residue, a mixture of $CaCl_2$ and flyash, can then be separated in bag filters. Due to the high content of chloride salts there is a high leakage of chloride and also of heavy metals. Further treatment is therefore

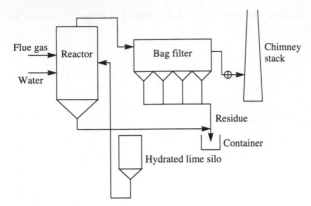

Figure 14.28 Layout for a dry flue gas cleaning process. (From information from contractor on flue gas cleaning system. Reprinted by permission of ABB Fläkt.)

necessary before final disposal. One method is to wash the material with water to encourage settling of heavy metals before discharging. Fixation has also been investigated, but so far without success, due to the high chloride content.

The semi-dry absorption process is similar to dry absorption with one exception (see Fig. 14.29). The Ca(OH$_2$) is injected in a water solution where the amount of water is balanced so that heat excess from the flue gas can evaporate the water. Scrubbing and condensation are normally positioned after a particle filter, often an electrostatic precipitator. An example is shown in Fig. 14.30. When the acid gases (together with other pollutants) have been separated in water, further separation is achieved by treating the water. The water treatment process ends up with treated water of high chloride content to be discharged and a smaller amount of metal contaminated sludge that must be fixated with cement or flyash before disposal in a landfill site.

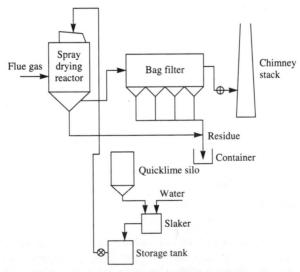

Figure 14.29 Layout for a semi-dry flue gas cleaning process. (From information from contractor on flue gas cleaning system. Reprinted by permission of ABB Fläkt.)

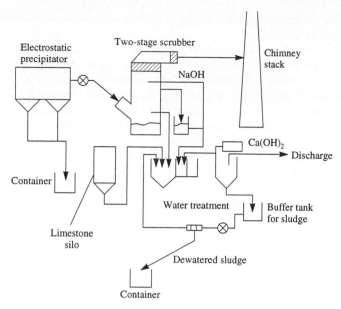

Figure 14.30 Layout for a wet flue gas cleaning process. (From information from contractor on flue gas cleaning system. Reprinted by permission of ABB Fläkt.)

14.10.6 Flyash and Bottom Ash

Before the introduction of modern advanced flue gas cleaning, the only residue to be disposed from incineration was flyash and bottom ash (slag). The major part of the incombustible matter ends in the bottom ash. It is made up of incombustible items like bottles, cans, etc., and it is the ash fraction of burnable material like paper, plastic, wood, etc. A small part of the ash fraction ends as flyash. One tonne of waste in a mixture of MSW from domestic, commercial and industrial activities gives about 200 kg of bottom ash and 30 to 40 kg of flyash. If the waste contains any pollutants like heavy metals that cannot be destroyed by incineration, they will often be found concentrated in the bottom ash and flyash. As an example, lead has a tendency to be concentrated mostly in bottom ash and cadmium in the flyash. The heavy metals in bottom ash are not prone to leaching while they seem to be more leachable in the flyash. Because leachability is low from bottom ash, it has become common in Scandinavia to utilize bottom ash for road construction. Since 1990, nearly all bottom ash in Denmark is beneficially used. This is due to the waste tax of approximately 25 ECU per tonne of waste disposed to landfill. Since flyash has higher leachability than bottom ash, it is only disposed at secure landfills.

14.11 MSW LANDFILL

The most traditional method of disposing of MSW was landfill. Over the past two decades the practice has changed from 'dumping' to well-engineered landfilling. Modern landfill practices include monitoring programmes for incoming waste, for gas, for leachate, etc., to control pollution of the surrounding environments, particularly groundwater, surface water and air.

An MSW landfill is a repository for MSW. In the United States, MSW landfills do not accept hazardous wastes but may accept industrial sludges which pass the US EPA leachate test. There is as yet (1996) no published leachate test for the European Union. Hazardous waste residues from hazardous waste treatment facilities go to 'secure' landfills. However, the design of MSW landfills and secure landfills is similar. The key parameter in the design of either is to attain an impermeability of 10^{-9} m/s, and so prevent leachate breakthrough.

Leachate is the water-type liquid that seeps out of a landfill. It is from rain infiltration through the landfill and also from the moisture fraction of MSW. Because it has a long residence time in a landfill, in a predominantly anaerobic environment, it is contaminated with organics and heavy metals and, as such, is highly toxic. It is essential in landfill practice that the leachate is not allowed to enter surface water or groundwater, but that it should be collected and treated in a conventional municipal wastewater treatment plant.

Landfill gas is produced in landfill sites due to the anaerobic degradation of biodegradable organic wastes. The gas produced is typically about 60 per cent methane and 40 per cent carbon dioxide. Landfill gas, with its high content of methane, is potentially explosive and, as such, needs to be controlled. If some means of controlling (extracting) the gas is not used, the gas can migrate off-site, causing problems to the surrounding environment.

Figure 14.31 is a schematic of landfill operations and processes. In broad terms these are:

- Landfill design
  - Foundation design
  - Liner design
  - Leachate collection and gas collection
  - Drainage design
  - Filling design
  - Runoff collection, etc.
  - Closure design

- Landfill operations
  - Waste inventory, loads, types, etc.
  - Cell layout
  - Cells for hazardous waste
  - Cells for non-hazardous waste

- Biochemical reactions in landfill
  - Biological decay rates
  - Slowly biodegradable
  - Rapidly biodegradable
  - Non-biodegradable

- Leachate management
  - Collection
  - Treatment
  - Monitoring
  - Reuse

- Landfill gas management
  - Monitoring
  - Collection
  - Flaring or using
  - Quantity and quality

- Environmental monitoring
  - Air quality and odour monitoring
  - CH_4, H_2S, VOCs, etc.
  - Groundwater monitoring
  - Pests and litter
  - Traffic

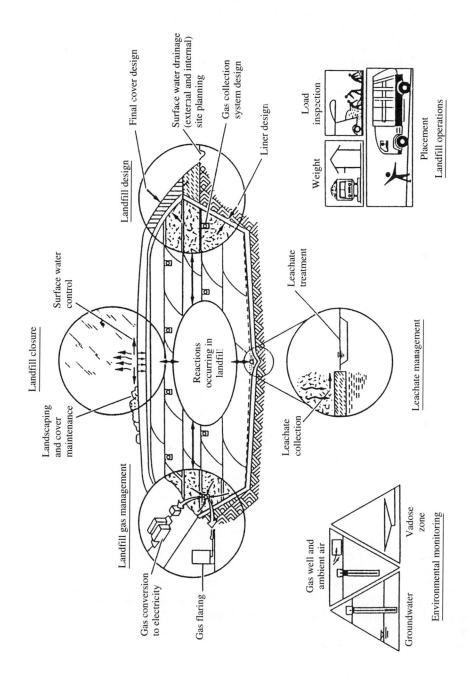

Figure 14.31 Schematic of double liner, leachate collection and landfill operations and processes. (Adapted from Tchobanoglous *et al.*, 1993. Reprinted by permission of McGraw-Hill, Inc..)

675

14.11.1 EU Landfill Legislation

The EU Directive on the Landfill of Waste (1995) sets the following objectives: landfilling, as with all other waste treatment processes, should be controlled and managed in a direct way and so prevent and limit the potential negative impact on environmental and human health.

Landfills are categorized as being:

- Sites for hazardous waste
- Sites for non-hazardous waste
- Sites for inert waste

Wastes unsuitable for landfilling are defined as:

- Liquid wastes incompatible with other wastes
- Liquid wastes on their own
- Oxidizing, explosive or flammable waste
- Infectious wastes
- Mixed hazardous and inert for the purpose of dilution

Annexe I of the proposed EU directive sets out the general requirements to be investigated for all landfills. Among the methodology for controlling leachate from landfills are guidelines for lining construction. This can be a combination of a geological barrier and a sealing with an artificial bottom.

The recommendation for the geological barrier is:

- Sites for hazardous waste $\quad K \leq 1 \times 10^{-9}\,\mathrm{m/s} \quad$ thickness $\geq 5\,\mathrm{m}$
- Sites for non-hazardous waste $\quad K \leq 1 \times 10^{-9}\,\mathrm{m/s} \quad$ thickness $\geq 1\,\mathrm{m}$
- Sites for inert waste $\quad K \leq 1 \times 10^{-9}\,\mathrm{m/s} \quad$ thickness $\geq 1\,\mathrm{m}$

Sealing with an artificial liner is recommended for hazardous and non-hazardous landfill sites. This sealing should be accompanied by a drainage layer of at least 0.5 m. An exception from the recommendation requires documentation in an EIA that verifies the exception at the specific location.

14.11.2 Landfill Types

In engineering terms, landfills are of two types:

- Attentuate and disperse sites
- Containment sites

The attentuate and disperse sites were a traditional form of landfilling. The attentuation mechanisms were dilution and dispersion through pores and microfissures into the underlying saturated zones. It was not possible to monitor or track the leachate pollutants, and in some areas leachate transport was too quick, causing poorly diluted leachate to reach surface waters and groundwaters. This type of site is now unacceptable and does not meet the EU directive requirements. As per the directive, this type of site is still suitable on the basis of K for inert waste, but the environmental monitoring requirements cannot be satisfied.

All new landfill sites are containment sites. This is where the wastes, its leachate and gas are isolated from the surrounding environment. The containment is achieved either by natural clay bottom liners or synthetic liners or a combination of both. Facilities for leachate and gas collection and removal are installed and regular monitoring is possible. The containment site is expected to be 'leak free'.

14.11.3 Landfill Practices

Landfills have been established in at least four different terrain types:

- Excavated cells
- Ground cover
- Depression infill
- Reclaimed land

Excavated cells   Excavated cells are suitable where the water table is well below ground level and ideally more than 1 m below the excavated level. Typically, cells are excavated to depths of 1 to 3 m, with side slope of 2 : 1 or 3 : 1 (3 horizontal). The length and width depends on the loading rates and rainfall intensities. A site in rural Cork with loading rates of 80 t/day uses cells of 100 m long, 2 m deep and 10 m wide, at the base. The side slopes are 4 : 1, giving a top width of 26 m. The topsoil of 0.3 m deep is underlain by 30 m of unweathered red boulder clay of permeability 10^{-9} m/s or better (Aherne, 1991). In excavated cell landfills, the excavated material is used for trench bunding, daily and final cover. The maximum working area for a landfill can be calculated approximately by

$$A_{\text{max}} = \frac{0.1\,W}{R} \tag{14.8}$$

where
A_{max} = maximum working area

R = average annual rainfall, m

W = average annual waste input, tonnes

This assumes an absorption capacity of the waste of $0.1\,\text{m}^3/\text{tonne}$.

   Example 14.8   Compute the maximum working area if W is 20 000 t/yr and R is 1.2 m/yr. Comment on the result.

   Solution

$$A_{\text{max}} = \frac{0.1 \times 20\,000}{1.2} = 1667\,\text{m}^2$$

If the cell length is 100 m, then the average width would be 16.7 m.

Ground level landfills   MSW can be deposited on the surface (after topsoil removal) if K is 10^{-9} m/s or better. This value is achievable either with natural clay liners or synthetic liners. However, it is unlikely that natural liners will exist so close to the surface. Disadvantages exist for this type of landfill in that material for daily and final cover is imported and environmental considerations of litter, noise, etc., are more acute and visible than excavated cells.

Depression infill   Most traditional dumpsites were in depressions, either naturally occurring or man-made. Mining quarries and dry valleys were commonly used. Many problems occur in this type of site, particularly in relation to surface water control and inadequate impermeability.

Reclaimed land   The recent commissioning of the major new landfill site for MSW in Copenhagen deserves attention in that the site is on the coast on reclaimed land. The tidal range is about 2 m. An embankment separates the landfill from the open sea. The site invert is 0.5 m below sea-level and designed and established for leachate collection. However, in the event of failure of the lining system, seawater will leak into the landfill (due to water pressure) and leachate will not leak out.

14.11.4 Landfill Gas

Landfill gas and leachate are the result of biochemical reactions within the landfill site. The parameters relevant to the production of gas and leachate, in terms of quantities and quality, are:

- Organic matter feedstock
- Rain infiltration
- Anaerobic environment within the landfill
- Age of landfill

Landfill gas is produced in an anaerobic environment within the landfill and the composition of the gas is particularly dependent on the composition and age of the wastes. Table 14.27 shows a typical landfill gas composition. These values vary with age and content of landfill. An idealized description of the metabolism of a landfill is shown in Fig. 14.32 and is considered to consist of four sequential phases:

- Initial aerobic phase
- First transition stage
- Second transition stage
- Methane phase

Initial aerobic phase   This phase will only take a few days to a few weeks. During this time, O_2 is allowed to diffuse into the waste or the original content of O_2 is being used up, as some of the more easily degradable organics are being broken down (non-lignin wastes). The result is that CO_2 begins to be generated with a temperature increase, sometimes of 15 to 20 °C. Aerobic conditions continue while O_2 is still available.

First transition stage   This follows the aerobic phase. Here anaerobic conditions begin to develop. This stage can also be called the acid stage as the pH decreases to 4 to 6, due to the development of organic

Table 14.27 Typical landfill gas composition

Component	Typical value (% volume)	Observed maximum
Methane	63.8	77.1
Carbon dioxide	33.6	89.3
Oxygen	0.16	20.93
Nitrogen	2.4	80.3
Hydrogen	0.05	21.1
Carbon monoxide	0.001	
Saturated hydrocarbons	0.005	0.074
Unsaturated hydrocarbons	0.009	0.048
Halogenated compounds	0.000 02	0.032
Hydrogen sulphide	0.000 02	0.0014
Organosulphur compounds	0.000 01	0.028
Alcohols		0.127
Component		
Temperature, °C	38–50	
Specific gravity	1.02–1.06	
Moisture content	Saturated	
HHV (Hawf), kJ/L	15–38†	

† The value of 38 kJ/L is the energy value of pure methane while the low value (15) corresponds to biogas with its many contaminants.
Adapted from Mortensen, 1995

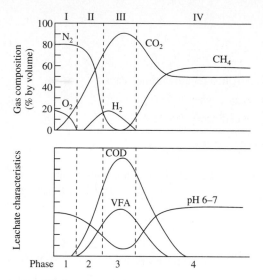

Figure 14.32 Metabolic phases in a landfill site.

acids, firstly as fatty acids, which are later converted to volatile fatty acids (e.g. acetic acid). This stage lasts from a few weeks to a few months.

Second transition stage In the third stage, the methanogenic bacteria are developed and begin to convert simple acids like acetic and formic acids and methanol to methane gas. This stage is unstable until a balance between the acid generation and methane production is established. It is often 3 to 5 years before the balance between acid generation and the methane production is stable. If the pH is sufficiently depressed (in stage 2) it takes longer before the process is stabilized, as the optimum conditions for methanogenesis is pH 6 to 7, which is similar to in-vessel anaerobic digestion.

Methane phase The fourth phase is the methanogenic phase where the process is stable because all the organic acids are being continuously used up by the methanogenic bacteria. This phase is described as:

$$\text{CHONS} + \text{H}_2\text{O} \xrightarrow{\text{anaerobic bacteria}} \text{H}_2\text{O} + \text{CO}_2 + \text{CH}_4 + \text{H}_2 + \text{NH}_4 + \text{HS} \tag{14.9}$$

The products from the stabilization process can be found in the leachate and in the landfill gas. In this phase, methane and carbon dioxide are each about 45 to 65 per cent of gas by volume.

Methane gas, process biochemistry The generalized biochemical reaction for methane fermentation of municipal waste given by Emcon Associates (1980) is

$$C_n H_a O_b N_c + \left(2n + c - b - \frac{9sd}{20} - \frac{de}{4}\right)\text{H}_2\text{O}$$

$$= \left(\frac{de}{8}\right)\text{CH}_4 + \left(n - c - \frac{sd}{5} - \frac{de}{8}\right)\text{CO}_2 + \frac{sd}{20}C_5 H_7 O_2 N$$

$$+ \left(c - \frac{sd}{20}\right)\text{NH}_4{}^+ + \left(c - \frac{sd}{20}\right)\text{HCO}_3{}^- \tag{14.10}$$

where

$$d = 4n + a - 2b - 3c \tag{14.11}$$

s = fraction of waste COD converted to biomass

e = fraction of waste COD converted to CH_4

and

$$s = a_e \left(\frac{1 + 0.2f\, \theta_c}{1 + f\, \theta_c} \right)$$ (14.12)

where
θ_c = solids retention time, days
$a_e = s_{\max}$ when $\theta_c = 0$
f = cell decay rate, d^{-1}

Emcon Associates (1980) assume $e \cong 0.96$.

Example 14.9   Determine the methane production rate of MSW if the ultimate analysis without sulphur but including water is $C_{70} H_{150} O_{70} N$. Assume CHON is 60 per cent of the wet weight of waste.

Solution

$$\text{Vol. (L) of } CH_4/kg \text{ waste} = \frac{(de/8) \text{ volume STP}}{\text{molecular weight of waste/CHON in waste}}$$

$$\text{Molecular weight} = 12 \times 70 + 150 \times 1 + 70 \times 16 + 1 \times 14 = 2124 \text{ g/mole}$$

From Eq. (14.11), $d = 4n + a - 2b - 3c$

$$= 4 \times 70 + 150 - 2 \times 70 - 3 \times 1 = 287$$

Moles of CH_4 per mole of waste, from Eq. (14.10)

$$\frac{de}{8} = \frac{287 \times 0.96}{8} = 34.5$$

Volume of methane per unit weight of waste (L/kg)

$$= \frac{34.5 \times 22\,266}{2124/0.6}$$

$$= 217 \text{ L } CH_4/kg \text{ of wet waste}$$

Values for gas production of MSW range from 150 to 250 L of CH_4 per kg of wet waste. However, in a landfill site much less than the above values are obtained and values in the range of 60 ± 30 are suggested by Emcon Associates (1980).

14.11.5 Leachate in Landfills

Leachate is the 'contaminated' water in landfills, which arrived at the landfill site through external precipitation. The chemical composition of typical leachates is shown in Table 14.28 and it must be noted that there is a range of values for each parameter. Leachate from young sites is much more of a contaminant than that from mature landfills. With time, the pH moves from slightly acid to neutral, and the ratio of BOD/COD decreases. The ratio of SO_4/Cl also decreases with time. Most of the leachate collected today is treated with municipal wastewater or, in some cases, specific wastewater plants treat leachate only. Landfills can be designed with recirculation of leachate from new stages to old stages with full anaerobic conditions. Such recirculation uses the effect of the anaerobic biological reactor of old landfills to reduce the high oxygen demand of 'young' leachates. This is more ideal in drier climates.

Leachate is conventionally treated on site at an adjacent wastewater treatment plant or transported to such a plant. In some cases (the drier climates), leachate is collected and sprayed over the landfill site. This serves to dilute the contamination strength. Leachate is also tested on a regular basis to identify the strength and its composition. Before being allowed to be treated in a conventional wastewater treatment plant, the leachate quality should satisfy the standards given in Table 14.29.

Table 14.28 Typical leachate chemical composition

Parameter Leachate	Units	Leachate 'young sites' Weak	Leachate 'young sites' Strong	Leachate 'general' Weak	Leachate 'general' Strong	Leachate 'old sites' Weak	Leachate 'old sites' Strong
Conductivity	mS/m	500	3000			250	1500
TSS	mg/L			500	2500		
VSS	mg/L	3000	8000			1000	2000
TOC	mg/L	3000	15 000			150	750
COD	mg O_2/L	5000	30 000			1000	5000
BOD_5	mg O_2/L	4000	20 000			200	1000
Cl^-	mg/L			1000	3000		
SO_4^-	mg S/L	50	400			10	30
Total N	mg N/L			500	1500		
Ammonia-N	mg N/L			200	1200		
Total P	mg P/L			5	100	5	10
Na	mg/L			500	2000		
Ca	mg/L	500	1500			80	200
Fe	mg/L	200	1000			20	100
Cd	μg/L			10	100		
Cr	μg/L			20	1000		
Cu	μg/L			10	1000		
Ni	μg/L			50	2000		
Pb	μg/L			20	1000		
Zn	mg/L			0.1	10		
Phenol	mg/L			0.5	5		
Oil/grease	mg/L			2	20		

From Christensen *et al.*, 1982 and Mortensen, 1993

Table 14.29 Proposed EU leachate standards for hazardous and inert waste

Parameter	Hazardous waste range	Inert waste range
pH	4–13	4–13
TOC	40–220 mg/L	< 200 mg/L
Arsenic (II)	0.2–1.0 mg/L	< 0.1 mg/L
Lead	0.4–2.0 mg/L	
Cadmium	0.1–0.5 mg/L	
Chromium (VI)	0.1–0.5 mg/L	
Copper	2–10 mg/L	Total < 5 mg/L
Nickel	0.4–0.2 mg/L	
Mercury	0.02–0.1 mg/L	
Zinc	2–10 mg/L	
Phenols	20–100 mg/L	< 10 mg/L
Fluoride	10–50 mg/L	< 5 mg/L
Ammonium	0.2–1 g N/L	< 50 mg/L
Chloride	1.2–6 g/L	< 0.5 g/L
Cyanide	0.2–1 mg/L	< 0.1 mg/L
Sulphate	0.2–1 g/L	< 1 g/L
Nitrite	6–30 mg/L	< 3 mg/L
AOX†	0.6–0.3 mg/L	< 0.3 mg/L
Solvents‡	0.02–0.10 mg Cl/L	< 10 μg Cl/L
Pesticides‡	1–5 μg/L	< 0.5 μg/L

† Adsorbed organically bound halogens.
‡ Chlorinated.

Leachate collection The drainage system has to be connected to some wells and, perhaps, pump stations. These installations are dangerous to enter when landfilling has started. This is due to methane gases from the landfill and the absence of oxygen. These problems have to be taken into account in the design stage, as well as in the day-to-day operation of the landfill. If the leachate cannot be pumped directly to the wastewater treatment plant, a storage tank should be provided. This should have a capacity for a maximum flow produced when a new landfill stage has been commissioned. A typical vertical section showing the liner composition is shown in Figs 15.23 and 15.24.

14.11.6 Water Balance in Landfills

The amount of leachate produced in a landfill depends on its water balance, which is

$$LC = PR + SRT - SRO - EP - ST \qquad (14.13)$$

where

        LC = leachate

        PR = precipitation

        SRO = surface runoff

        SRT = surface runto (landfills should be designed so that water outside the site should not enter the site and then SRT = 0)

        EP = evapotranspiration

        ST = change in water storage

In catchment hydrology, the water balance equation can be simplified if the time frame is, say, a year, when ST is taken as zero. However, this simplification is not possible in landfills as EP and ST keep changing due to biochemical reactions and ongoing loading of the site. The change in water storage, ST, is not simply a change due to infiltration variations; it also needs to account for water consumed in the landfill in the production of landfill gas and water lost as gas vapour and also account for water introduced to the landfill by 'wet' solids, 'wet' sludges and 'wet' cover material. It is the objective of the design of landfill sites (and the EU directive on landfilling) to minimize the quantities of leachate. In a continuously moist climate like Ireland, this requires detailed landfill design with specific attention to top covers and liners and leachate collection. The elements of water balance are shown in Fig. 14.33. The computer program HELP by Schroeder (1994) is a hydrological simulation package that can be used to determine the hydrologic water balance of landfills and thus investigate the leachate volumes generated.

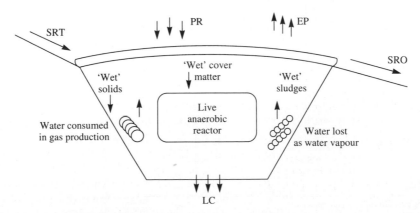

Figure 14.33 Elements of a landfill water balance.

In landfills, the water is absorbed in two ways:

- For anaerobic metabolism
- To decrease the water deficit

The water consumption in anaerobic processes is very small. Total biological degradation requires approximately 65 to 80 litres per m^3 of waste (of density 700 to 900 kg/m^3). If the degradation time is 50 years this requires only 2 mm of water annually.

> **Example 14.10** Compute the moisture deficit in a landfill for each m^3 of waste if the parameters are
>
> $$\text{Density of waste at time of deposit} = 800 \text{ kg/m}^3$$
> $$\text{Field capacity} = 60\% \text{ by weight}$$
> $$\text{Water content of waste being deposited} = 30\% \text{ by weight}$$
>
> *Solution*   At the time of deposit the waste characteristics are
>
> $$\text{Water content} = 30\% \text{ of } 800 \text{ kg} = 240 \text{ kg}$$
> $$\text{Solids content} = 70\% \text{ of } 800 \text{ kg} = 560 \text{ kg}$$

For field capacity:

$$\text{Total solids} = 40\% = 560 \text{ kg}$$
$$\text{Water capacity} = 60\% = 840 \text{ kg}$$
$$\text{Water deficit} = 840 - 240 = 600 \text{ kg}$$

There is a moisture deficit of 600 kg/m^3 .

During the deposit period, the water deficit will decrease as there is little evident evapotranspiration and in well-managed landfills there should be no surface runoff. Because of the difficulties and likely low values, the change in water content of landfills is often neglected in water balance/leachate computations.

14.11.7 Geotechnical Aspects of Landfill Sites

The selection of a landfill site, while depending on many factors already discussed, will be especially dependent on the soil conditions at the invert level of the landfill. The objective is to site a landfill on soil that has a very low hydraulic conductivity, $< 10^{-9}$ m/s, so as to prevent leachate penetration through the unsaturated zone and into the groundwater, thus causing contamination.

Field permeability tests are used to measure the *in situ* hydraulic conductivity. The test is usually conducted in a test boring or a monitoring well. The test is influenced by the position of the water table, type of material, depth of test zone, hydraulic conductivity of the test zone (Oweis and Khera, 1990). The test may be either a constant head test, a variable head test or a pump test, described in conventional soil mechanics textbooks. The hydraulic conductivity can also be determined from laboratory tests and typical values are given in Chapter 4.

If the invert level of the proposed landfill site does not satisfy $K < 10^{-9}$ m/s and a thickness of > 1 m, then additional protection beyond the *in situ* soil material is required. This may mean excavating the existing soil and replacing with better clay. The EU directive requires 0.5 m for this layer. Improvement can also be made by using bentonite on top of existing clay or by using a synthetic liner.

Lining system   To avoid or minimize the seepage of leachate from landfills, it has become general practice to provide a protective lining system. Depending on the type of waste to be disposed of and the sensitivity of the location, the liner can be either a single or multilayer system. Liners can be made of:

- Naturally *in situ* clay
- Imported clay
- Bentonite-improved earth
- Synthetic

***In situ* liners**   As a rule-of-thumb, *in situ* clay liners should be not less than 2 m thick. Furthermore, the top 25 to 30 cm should be kneaded and compacted again to avoid non-homogeneous areas in the surface. The requirement for impermeability varies from country to country. In some countries the requirement is that it be better than 10^{-10} m/s, in others better than 10^{-9} m/s.

Imported clay   Imported clay should be applied for a minimum total depth of 0.5 m thick and the clay has to be placed in layers of not more than 20 cm at a time, with careful compaction between each layer and of course k should be $< 10^{-9}$ m/s.

Bentonite   Bentonite is a clay material with swelling qualities. It can be used to improve the impermeability of different types of sand, clay or soil materials. Bentonite can be spread on the surface alone or together with some other soil materials. Several companies marketing bentonite install it between two layers of cardboard or geotextile. The guidelines for thickness and impermeability of clay or bentonite liners varies from country to country. A usual requirement is 10^{-9} to 10^{-10} m/s. An impermeability at 10^{-10} m/s can, in conjunction with a good drainage system, remove more than 95 per cent of all leachate generated. If leachate is not removed immediately, the thickness of the liner is of importance because it provides resistance to hydraulic pressure.

Synthetic liners   Many synthetic liners are available. Most of them are made of polyethylene (LDPE or HDPE). They are normally available in thicknesses of 0.5 to 2.0 mm. It is very important that all constructions and operations with synthetic liners, including the welding process, should be carried out and controlled very carefully. The liner should be protected underneath with sand or geotextile. The layout and the welding, melting or gluing have to be done very carefully. Finally, the first layer should consist of selected waste. Transport should not be allowed on top of liners before at least 1 m of waste has been deposited.

Drainage and protection system   The surface of the liners should be drained carefully. For that purpose, a combined protection and drainage system has to be established on the top of the liner. For drainage purposes, pipes and/or gravels are used, covered with small stones and finally protected with a layer of rough sand of at least 50 cm thick.

14.11.8 Design of Landfills

Due to potential negative impact and conflicting interests, the location of the landfill is a significant element in the general process of planning, design and construction. As landfills always have some impact on their surroundings, it is not possible to find the perfect site. The task is to balance all interests to find the best compromise location. The pre-planning stage of a landfill should include not only an Environmental Impact Statement (EIS) (see Chapter 19), but also a Health Impact Statement (HIS). The interests to take into account include:

- Ecological and biological conditions
- Geological and hydrogeological conditions
- Existing and potential water supply sources
  - groundwater
  - surface water
  - other water recipients

- Historical and archaelogical
- Recreational
- Other planning, e.g. town and industrial development, agriculture

These interests need to be balanced with infrastructure requirements:

- Access possibilities
- Transport distance
- Treatment possibilities of the leachate, e.g. transport distance

These interests should be determined and evaluated in the planning and location phase. This phase can be divided into elements:

- Pre-investigation and draft design
- Planning
- Selection
- Decision

The planning task is to determine all interests in a certain area. In principle, they can be drawn on a map, each with its own colour or shade. Location with no conflicts or a minimum of conflicts can then be selected and a number of potential sites can be identified for further evaluation and final decision. For selection and final decision, it is useful to use a ranking system to select the most suitable of the sites. In the ranking system, one of the interests may be predominant. When only a few potential sites remain, more detailed information on each site has to be evaluated, and a proposal of ranking has to be made for the final political decision.

After the decision has been made, pre-investigation has to give information for the design and to verify that the assumptions on which the decision was based were correct. At any point in the planning procedure, it may become obvious that the selection has been inadequate, and then it can be necessary to reconsider the situation and to supplement with new potential site locations.

When a suitable site has been located, the landfill has to be designed. This is now a specialized area covering the disciplines of structures, hydraulics and hydrogeology. As regards the surroundings, the site should be landscaped. When the shape has been decided, the useful volume can be estimated. The site can be planned with fences, embankments and planting to surround the site. The access roads, receiving areas, garage and staff facilities can be designed. Furthermore, the deposit area has to be designed with a lining system, leachate collection and storage system, and a gas collection or a ventilation system. Finally, a monitoring system (for water quality of leachate, gas quality of methane, hydrology and hydrological parameters) has to be planned and implemented. For some parameters, the monitoring system has to be implemented before the establishment of the site, for instance monitoring groundwater.

Receiving area A well-designed site should have a receiving area with facilities, including:

- Weighbridge for recording incoming waste according to type and origin
- Office and staff building with recording system for the weighbridge, dining-room, toilets and shower rooms for staff
- Garage for compaction machinery, excavator, etc., with facilities for repair, oil changing, cleaning, etc.
- Containers where waste from individuals can be received so that they do not have to go to the deposit area
- Storage facilities for small amounts of oil and chemical waste.

An outline layout of a receiving area is shown in Fig. 14.34.

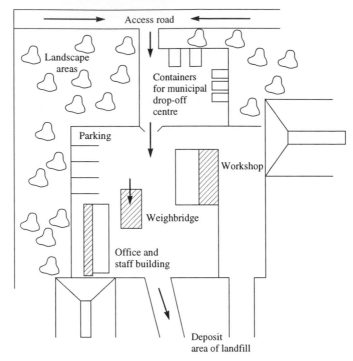

Figure 14.34 Receiving area of a landfill site.

Internal roads

1. The internal roads should be constructed for heavy traffic with a pavement all the way to the deposit area.
2. Temporary roads at the deposit area can normally be made of construction debris.
3. Roads for compactors have to be made with a surface of stone or gravel due to the steel teeth on the wheels.

Deposit area   Normally the design of the deposit area can profitably be divided into several stages. The number, size and order of the stages have to consider practical local conditions, daily amounts of waste and a practical minimum size for handling. The demarcation has to eliminate exchange with surface water and create a boundary for the waste storage. A fence is erected to eliminate wind-blown litter in the form of paper and plastic and to protect the surroundings against noise and an unpleasant appearance.

Control and monitoring   Due to the fact that landfills can have a negative impact on the environment, a control and monitoring system is required. The control system has to ensure that banned categories of waste are not disposed of on the site. These may include hazardous waste properly destined for other disposal routes, e.g. incineration. Simultaneously, records of the amount and type of the waste should be made.
The monitoring system should cover:

- Leachate
- Surface runoff
- Recipients
- Groundwater
- Noise

Operation   Before a landfill site is commissioned, an operating manual should be produced and the staff instructed in the correct operation of the site. The manual should include instructions for:

- Receiving control

- Recording the amount and type of waste received
- Rates and rating system for receiving waste
- Instruction for deposit of the first layer of waste
- Routines for deposit of different types of waste, including daily cover
- Regulation of traffic and behaviour at the site, with regard to occupational safety
- Who is to be admitted to the site and how they should be instructed with regard to occupational safety
- Regulations and routines for work with leachate
- Routines for cleaning the roads and site, including collection of littering paper
- Routines to avoid dust and smoke from the site
- Routines for fire control

An operation manual may seem like an obvious requirement, but experience shows that only the best-managed landfills implement such a practice.

Co-disposal   Landfills with organic biodegradable waste act as an anaerobic digester, and this fact can be exploited in co-disposal. The philosophy of co-disposal is that organic solvents and other organic chemicals can be degraded anaerobically due to their long retention time in the landfill. Similar arguments are used for fixation of heavy metals in, for example, sludge from galvanizing industries. The anaerobic reduction of SO_4^{2-} to S^{2-} can be used for fixation of heavy metals as heavy soluble sulphide. The counter-argument is that nobody can predict with 100 per cent accuracy what new processes will take over when the biological processes end after 100 or more years. It is therefore not possible to predict whether any of the heavy metals will become mobile.

Example 14.11   Determine the area required for a new landfill site with a projected life of 20 years for a population of 150 000 generating 25 kg per household per week. Assume the density of waste is $500\,kg/m^3$. A planning restriction limits the height of the landfill to 10 m.

Solution   Assume 3.5 persons per household.

$$\text{Waste generated} = \frac{150\,000}{3.5} \times \frac{25}{10^3} = 1071 \text{ tonnes/week}$$

$$= 55\,700 \text{ tonnes/year}$$

$$\text{Volume of landfill space required} = \frac{55\,700 \times 10^3}{500} = 111 \times 10^3 m^3/\text{year}$$

If the height is 10 m,

$$\text{Required land area} = \frac{111 \times 10^3}{10} = 11\,100 \text{ m}^2$$

$$= 1.1 \text{ ha}$$

This value will need to increase by about 50 per cent to allow for daily cover, roads, receiving areas, fencing, etc. Therefore,

$$\text{Required area for 20 years} \cong 1.1 \times 20 \times 1.5$$

$$\cong 33 \text{ ha}$$

Example 14.12   Compute the volumetric flow rate through a natural clay liner if the area is 1 ha and the liner thickness is 1 m with a hydraulic conductivity of $10^{-9}\,m/s$. Assume 0.5 m of head of water.

Table 14.30 Options for integrated waste management of MSW

Component	Waste minimization	Recycle	Reuse	Incineration	Compost	Biogas	Landfill
Food and organic	A		✓	✓	C	B	✓
Paper and cardboard	A	B	C	✓		✓	✓
Plastics	A	✓	B	C			✓
Glass	A	B	C				✓
Metals F	A		B				C
Metal NF	A	B	C				✓
Textiles	A	B	C	✓			✓
Others	A						✓

A = most desirable option, B = next most desirable option, C = less desirable, ✓ = possible but undesirable.

Solution   From Darcy's law,

$$Q = KA\frac{dh}{dL} \cong KA\frac{h}{L}$$

$$= 10^{-9}\ \text{m/s} \times 10^4\ \text{m}^2\,\frac{1.5\ \text{m}}{1\ \text{m}}$$

$$= 1.5 \times 10^{-5}\ \text{m}^3/\text{s}$$

Therefore

$$Q = 1.296\ \text{m}^3/\text{day}$$

14.12 INTEGRATED WASTE MANAGEMENT

Integrated waste management utilizes a variety of technologies to treat and dispose of its waste. Table 14.30 indicates some options for the broad listed components of MSW.

It is seen from a qualitative assessment that waste minimization should figure highly. However, it is not currently possible to minimize to zero. In the intervening period, the ideal programme might have the following priorities:

1. Minimize all component waste fractions.
2. Recycle what is possible of paper, cardboard, glass, non-ferrous metals and textiles.
3. Reuse plastics, ferrous metals and glass.
4. Biogas or compost the food fraction of MSW.
5. Incinerate only the remaining plastics or food waste.
6. Landfill only the remaining 20 per cent.

14.13 PROBLEMS

14.1   Is the domestic garbage being adequately quantified in your area? If so, compile:
    (a) the population being served (numbers of persons and numbers of households,
    (b) the total weight (and volume where applicable) being collected.
    What is the waste production per person and per household in tonnes/year?

14.2   For Problem 14.1, compile a waste classification as per: food waste, paper and cardboard, plastics, glass, metals and yard wastes.

14.3   Use the energy content rates of Table 14.15 and Khan's equation (14.4) to compute the energy value of the following waste:

Component	% by weight
Food	35
Paper, cardboard	35
Plastics	8
Glass	5
Metals	4
Yard wastes	10
Other	3

If the total annual weight is 150 000 tonnes, what is the potential energy value of this waste?

14.4   A domestic waste has been analysed and found to be: $C_{420} H_{1950} O_{850} N_{14} S$. Compute the energy (low heat value).

14.5   'Recycling is nothing more than the heavy hand of legislation cramping the innovation and freedom of industry and entrepreneurs who materially improve our lifestyle.' Discuss in support of this view.

14.6   Review the paper, 'MRF design—an uncertain act' by Hess (1993).

14.7   Use a draughting package (AutoCad or the like) to optimize the collection route for the following residential development. Assume the following hold:
(a) frequency of collection once per week at kerbside,
(b) household generation, 25 kg/week at 300 kg/m^3,
(c) a one man crew on a 10m^3 truck.

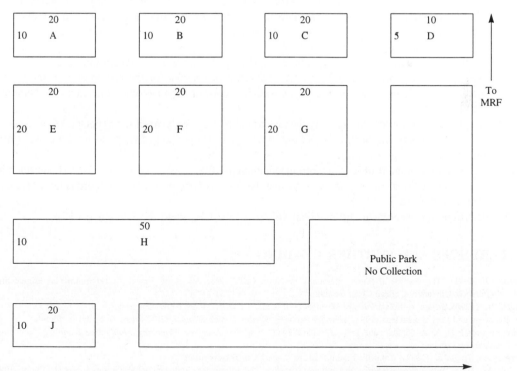

Assume the truck can only ride on one side of the street and the driving position is closest to the kerbside.

14.8 Estimate the composition of the recoverable materials in a kerbside recycling programme if an MRF exists for separating and processing paper, cardboard, aluminium cans, metal tins, glass and mixed plastics. Use the data of Example 14.3.

14.9 Outline in flow chart form the layout of an MRF for source separated MSW. Provide detailed flow charts for paper, aluminium, tin, glass and plastics.

14.10 Review, the paper 'Understanding the total waste stream, Fullerton, California' by Hay *et al.* (1993).

14.11 Discuss the economics of using a transfer station for an urban area with a population of 150 000 if the landfill site (assume all waste goes to landfill) is 40 km from the centre of the collection area, which has a radius of 5 km.

14.12 Design a drop-off centre for Problem 14.11. Estimate the costs of setting up a centre, exclusive of land purchase. Also estimate the operational costs.

14.13 For Problem 14.12, what costs per kg of weight delivered should be levied by the local authority if set-up costs are to be paid off in 6 years? Assume that the land purchase cost is ECU 200 000.

14.14 For the compost exercise of Fig. 14.2, produce a material if there is no recycle. Assume the input compost mix (straw or wood chips) is 15 tons of 23 per cent DS sludge cake plus 45 tons of 60 per cent DS wood chips. The influent mix is 40 per cent VS.

14.15 For the waste composition of Example 14.3, consider the annual waste generated as 111 500 tonnes/year (factor by 10). Estimate how much energy can be produced if all is incinerated. Assume a net calorific value of 8.4 MJ/kg. Estimate the calorific value if:
 (a) all the paper is recycled,
 (b) the food waste is retrieved for composting.

14.16 The flue gas from the incinerator of Problem 14.15 is 7 N m^3 flue gas per kg of waste. If the chloride content is 500 mg/N m^3, estimate the minimum amount of $Ca(OH)_2$ required to reduce the chloride content to 50 mg/N m^3. Also determine the amount of $Ca(Cl)_2$ produced in a dry or semi-dry flue gas cleaning process.

14.17 For the waste composition of Example 14.3, consider an annual generation rate of 111 500 tonnes/year. Develop an integrated waste management strategy for the full amount of the waste. Identify the volumes, etc., going to recycling, composting, anaerobic digestion, incineration, landfill, etc.

14.18 Design a landfill for the full quantity of wastes (111 500 tonnes/year) if the expected life is to be 20 years. By comparison, then design a landfill (size) for that portion of waste of Problem 14.17 that you designated to go to landfill.

14.19 Determine the amount of leachate produced annually if the waste of Example 14.3 is landfilled in total on a new lined site of 5 ha. Assume that the AAR is 1000 mm/year and there is no SRT. State other assumptions.

14.20 Determine the theoretical and practical volume of the CH_4 yield for Problem 14.19.

REFERENCES AND FURTHER READING

Aherne, J. (1991) 'The control of waste operations in North Cork', Proceedings of First Irish Environmental Engineering Conference, University College Cork, Ireland.

Bagchi, A. (1990) *Design, Construction and Monitoring of Sanitary Landfill*, John Wiley, Chichester.

Cal Recovery and Peer Consultants (1993) *Material Recovery Facility Design Manual*, CRC Press, Boca Raton, Florida.

Cheremisinoff, P. N., K. A. Gigliello and T. K. O'Neill (1984) *Groundwater Leachate, Modelling, Monitoring, Sampling*, Technomic Publishers, London.

Christensen, Th., *et al.* (1982) 'Controlled landfill issues (Danish), in *Teknisk forlag*.

Collins, M. (1992) 'Waste minimization programme for a hospital', NUI thesis for Graduate Diploma in Environmental Engineering, University College Cork, Ireland.

Commission of the European Communities (1991) 'Proposal for a Council directive on the landfill of waste', Com. (91), Brussels, May.

Corbitt, R. A. (1990) *Standard Handbook on Environmental Engineering*, McGraw-Hill, New York.

Crawford, J. F. and P. G. Smith (1985) *Landfill Technology*, Butterworths, Oxford.

Cunningham, W. and B. Saigo (1990) *Environmental Science—A Global Concern*, Wm. C. Brown, Dubuque, Iowa.

Cunningham, W. P. and B. W. Saigo (1992) *Environmental Science*, W & B Publishers.

Danish EPA (1994) *Biogas from Municipal Solid Waste—Overview of Systems and Markets for Anaerobic Digestion of MSW.* International Energy Agency Agreement, Herning Municipalities, Herning, Denmark.

Davis, M. L. and D. A. Cornwall (1991) *Introduction to Environmental Engineering*, McGraw-Hill, New York.

Dennison, G. J. and V. A. Dodd (1992) 'Recent experience of kerb side recycling in Dublin', *Proc. Instn Civ. Engrs, Mun. Engr*, **93**, 143–150, September.

Department of the Environment (UK) (1983) 'Wastes from tanning, leather dressing and fellmongering', Waste Management Paper No. 17, Her Majesty's Stationery Office, London.

Department of the Environment (UK) (1986) 'Landfilling and wastes', Waste Management Paper No. 26, Her Majesty's Stationery Office, London.

Duane, B. M. (1991) 'Preliminary feasibility study of a source separation programme and composting plant in North Cork', NUI thesis for Graduate Diploma in Environmental Engineering, University College Cork, Ireland.

EC Directive 89/369/EEC (1989) 'Council directive on the prevention of air pollution from new municipal waste incineration plants'.

EC Directive 89/429/EEC (1989) 'Council directive on the reduction of air pollution from existing municipal waste incineration plants', No. L. 203/50, 15 July.

EC (1992) 'The state of the environment in the Community—overview', Com. (92), 23 final, Vol. III, Brussels, 27 March.

EC 92/C263/01 (1992) 'Proposal for a Council directive on packing and packaging waste', No. C 263/1, 12 October.

EC 93/C129/08 (1993) 'Opinion on the proposal for a Council directive on packing and packaging waste', No. C129/18, 5 May.

EC 91/C190/1. (1995). Proposed Council Directive on landfill of waste. No. C190, 1995. Unpublished proposed directive.

Emcon Associates (1980) *Methane Generation and Recovery from Landfills*, Ann Arbor Science, Michigan.

Environmental Information Service (ENFO) (1990) 'Environmental protection fact and action sheets', The Environmental Information Service, Dublin.

Feacham, R. G., D. J. Bradley, H. Garelick and D. D. Mara (1983) *Sanitation and Disease—Health Aspects of Excreta and Wastewater Management*, John Wiley, New York.

Flynn, J. D. (1992) 'Landfill gas', NUI thesis for Graduate Diploma in Environmental Engineering, University College Cork, Ireland.

Garrett, L. (1995) *The Coming Plague: Newly Emerging Diseases in a World out of Balance*, Penguin Books, New York.

Goldman, L. J., A. S. Damle, C. M. Northerm, L. I. Greenfield, G. L. Kingsburg and R. S. Truesdale (1990) *Clay Liners for Waste Management Facilities*, Pollution Technology Review No. 178, Noyes Data Corporation (NDC), New Jersey.

Hartless, B. (1992) *Methane and Associated Hazards to Construction—A Bibliography*, BRE and CIRIA Special Publication 79. Building Research Establishment (BRE), Garston, Watford, UK.

Haug, R. T. (1980) *Compost Engineering—Principles and Practice*, Ann Arbor Science, Michigan.

Hay, J., A. Wesner, M. McGee and G. Buell (1993) 'Understanding the total waste stream, Fullerton, California', *Biocycle*, July, Emmaus, PA, USA.

Hennessy, A. (1991) 'Solid waste management for Cork City', NUI thesis for Graduate Diploma in Environmental Engineering, University College Cork, Ireland.

Hess, S. (1993) 'MRF design—an uncertain act', *Journal in MSW Management*, October.

Jackson, M. H., G. P. Morris, P. G. Smith and J. F. Crawford (1990) *Environmental Health Reference Book*, Butterworth Heinemann, London.

Kaiser, E. R. (1964) Proceedings of National Incineration Conference, Denmark.

Kayhanian, M., K. Lindernauer, S. Hardy and G. Tchobanoglous (1991), 'The recovery of energy and production of compost from the biodegradeable fraction of MSW using the high solids anaerobic digestion/acrobic biology process'. University of California at Davis Report. Civil Engineering Dept., Davis, CA, USA.

Khan, Z., Ali, and Z. H. Abu-Ghurrah (1991) 'New approaches for estimating energy content in MSW', *ASCE Journal of Environmental Engineering*. **117** (3).

Kharbanda, O. P. and E. A. Stellworthy (1990) *Waste Management—Towards a Sustainable Society*, Gower.

Kosaric, N. and R. Velayudhan (1991) 'Biorecovery processes: fundamental and economic considerations', in *Bioconversion of Waste Materials to Industrial Products*, A. M. Martin (ed.), Elsevier, Amsterdam.

Leach, B. A. and H. K. Goodgar (1991) *Building on Derelict Land*, CIRIA Special Publication 78.

Mabb G. T. (1993) 'Waste—A duty of care', *Proc. Instn Civ. Engrs*, Mun. Engr, **98**, 49–52, March.

Macdonald, C. (1991) *Municipal Solid Waste Conversion to Energy—a Summary of Current Research and Development Activity in Denmark*, Environmental Safety Centre, AEA Technology, Harwell, UK.

Martin, A. M. (1991) *Bioconversion of Waste Materials to Industrial Products*, Elsevier, Oxford.

Mortensen, E. (1990–93) 'Introduction to solid waste', Lecture notes to Graduate Diploma in Environmental Engineering, University College Cork, Ireland.

Noyes, R. (1991) *Handbook of Pollution Control*, Noyes Publications (NP), New Jersey.

O'Brien, M. (1993) 'The role of Cork County Council in the disposal of refuse in North Cork', *Local Authority News*, **9** (2).

Organic Waste Systems N.V. (1990) *Dranco-Dry Anaerobic Composting*, Organic Waste Systems N.V., Ghent, Belgium.

Oweis, J. S. and R. P. Khera (1990) *Geotechnology of Waste Management*, Butterworths, Oxford.

Peavy, H., D. Row and G. Tchobanoglous (1985) *Environmental Engineering*, McGraw-Hill, New York.

Peyton, R. L. and P. R. Schroeder (1990) 'Evaluation of landfill liner designs', *ASCE Journal of Environmental Engineering*, **116** (3).

Robinson, W. D. (1986) *The Solid Waste Handbook—A Practical Guide*, John Wiley, Chichester.

Schroeder, P. R. (1984) *The Hydrologic Evaluation of Landfill Performance (HELP) Model*, Risk Reduction Engineering Laboratory Office of Research and Development, USEPA, Cincinnati, Ohio, USA.

Smoley, C. K. (1993) *Material Recovery Faulty Design Manual*, CRC Press, Boca Raton, Florida.

Suess, M. J. (1985) *Solid Waste Management Selected Topics*, World Health Organization (WHO) Regional Office for Europe, Denmark.

Tchobanoglous, G., H. Theisen and S. Vigil (1993) *Integrated Solid Waste Management—Engineering Principles and Management Issues*, McGraw-Hill, New York.

World Health Organization (WHO) (1991) *Urban Solid Waste Management*, Instituto per i Rapporti Internationali di Sanita (IRIS), Firenze, Italy.

CHAPTER FIFTEEN

HAZARDOUS WASTE TREATMENT

15.1 INTRODUCTION

The USEPA (40 CFR 261.20–261.24) considers a substance hazardous if it exhibits one or more of the following characteristics:

- Ignitable—the substance causes or enhances fires
- Reactive—the substance reacts with others and may explode
- Corrosive—the substance destroys tissues or metals
- Toxic—the substance is a danger to health, water, food and air

This definition suggests the ubiquitous nature of hazardous waste. It is found in our homes, our factories, our workplaces, our farms and in the natural environment. The identification of whether a waste is hazardous or not is a complex process and is examined in Sec. 15.2.

The public perception of hazardous waste is that it is a problem and those that generate hazardous waste must be somehow stopped from this nefarious practice. Contaminated sites are viewed with equal distaste. However, as engineers and trained problem solvers, it is our task to address the treament and safe disposal of hazardous waste and remediation of contaminated sites in the same rigorous way as we would address the problem of delivering quality drinking water to a metropolis. Hazardous waste and contaminated sites are engineering problems that can be solved, or at least diminished, by engineering and scientific principles. The first ethic of engineering is to encourage industry and waste generators to reduce or eliminate the waste streams, by waste minimization techniques discussed in Chapter 18. If hazardous waste is still generated after state-of-the-art technology and waste minimization techniques, then our tasks include:

- Identifying the hazardous waste streams (solid, liquid or gaseous)
- For each stream, quantifying the wastes (e.g. m^3/day of liquid waste in a pipe)
- For each stream, characterizing the waste according to whether its form is
  - Physical
  - Chemical
  - Biological

- Recommending waste minimization measures
- Identifying hazardous waste treatment options (e.g. incineration, landfill, etc.)
- Specifying safe disposal routes for waste residues of treated waste

In addition, we must address the problems associated with the transport and storage of hazardous waste. However, before examining any of the above, we must first define hazardous waste and what items of waste, be they industrial, commercial, household, medical, agricultural, etc., are considered hazardous. The objective of this chapter then is to examine those points identified in the previous paragraphs. This chapter aims to be more qualitative than the previous chapters. This is due to the complexity of the hazardous waste treatment process and our effort here is meant to be introductory. See Wentz, 1995, Perry *et al.*, 1984 and Freeman (1992) or any specific engineering text on hazardous waste for further details.

15.1.1 Love Canal

In history there have been many cases of irresponsible hazardous waste management. Before defining hazardous waste, it is instructive to discuss the story of one of the more serious incidents in the history of hazardous waste. In the following pages, the 'Love Canal' incident is described as an example of what can happen *even* when the waste generator is careful.

The Love Canal is near Niagara Falls in the United States. The place is named after William T. Love who went to Niagara Falls in 1892 where he planned to build a canal to provide inexpensive transport to the local industrial community. Excavation work started in 1894 but was never completed and in 1910 the project was finally abandoned. The control of the site was later taken over by the firm Hooker Electrochemical, which intended to utilize the site for other purposes.

From 1941 onwards this company disposed of 20 000 tonnes of waste from their chlorine alkaline process and other processes, into the unfinished canal. The canal, merely a ditch, was very suitable for waste disposal, its sides being lined with an impermeable quality clay. The ditch was filled with waste and the whole area capped with some 3 ms of clay. The company fulfilled all the regulations (at the time) for the construction of hazardous waste dumping sites with a broad margin of safety. The site, when filled, was no longer used by the company, but it had over the years become an item of valuable real estate and, as such, attracted the attention of the Niagara Falls Board of Education. They proposed to build a school and a park on the site. The Hooker Company sold the site to the Board for 'one dollar' and stated in the deed that:

> The premises ... have been filled with waste products and the Board [of Education] assumes all risk and liability ... no claim, suit, action or demand by the [Board] for injury or death.

This was in 1953. Following development of the site, some 7700 m^3 of the cap was removed, reducing the thickness of the cap to as little as 1 m. The Board had further plans for the site and in 1957 they proceeded in transferring land to private landowners with the purpose of developing the site into a residential area.

The Hooker Company protested against the development plans and stated that:

> ... the land was not suitable for construction where underground facilities would be necessary ... [Hooker] could not prevent the Board from selling the land, but the property should not be divided for the purpose of building homes ... [they] hoped that no one would be injured.

This was not the only warning to the Board of Education not to proceed with the plans. More followed, but in spite of this the City constructed storm sewers (in 1957 and 1960) across the landfill site, thus severing the top cap and the side walls of the landfill.

The first reports of skin irritations and burns started in the late 1950s. The first media reports of chemicals seeping into the basements of some of the homes sited on the edge of the landfill, allegedly causing a multitude of illnesses, was made public in October 1976 in the local press, *The Niagara*

efforts of the authorities to try to identify the problems and rectify them, their action came too late and was not regarded as sufficient, at least by the new residents in Love Canal. In 1978 the nearby school was temporarily closed and pregnant women and children living in the immediate vicinity were evacuated. The authorities were also more or less forced to buy out 250 families, while the remediation work on the canal site started.

In 1979 a number of birth defects and miscarriages were registered in the population in Love Canal. The authorities failed to ascribe the illnesses to the seepage of toxic gases to the basements of private homes or to the ambient air. Despite the repeated warnings from the Hooker Company, the state of New York filed a US$ 635 million lawsuit against Hooker's then parent company Occidental Petroleum. In addition to this, the mass media created such a heated public opinion that the case even became an issue in the American presidential election campaign in 1980. In the end the USEPA went into action with a comprehensive cleanup programme. The Love Canal is only one in many tens of thousands of landfill sites containing hazardous waste situated all over the world—waiting to be identified.

15.1.2 The Seveso Dioxin Accident

On Saturday morning 10 July 1976 in the little town of Seveso, 20 km north of Milan in Northern Italy, a release of approximately 2 kg of the extremely toxic compound 3,4,7,8-tetradibenzoparadioxine occurred in a runaway chemical reaction in a factory producing hexachlorophene and the herbicide 2,4,5-trichlorophenoxy acetic acid. The cloud of dioxin resulted in as many as 37 000 people being exposed to measurable amounts of the chemical. This accident, which did not cause any human fatalities at the time (although many animals died), triggered so much attention internationally that work began shortly afterwards to draw up regulations for all types of facilities handling dangerous compounds, whether storing, manufacturing or treating them. The Seveso accident is described in more detail in Wentz (1995).

The Seveso Directive or the EC Major Accident Directive   On 24 June 1982 the European Community passed the Council Directive on the major accident hazards of certain industrial activities. This Directive was later to be known as the 'Seveso Directive', (Council Directive 82/501/EEC with amendments 87/216/EEC and 88/610/EEC). It was the first EC law requiring hazardous waste information to be provided and exchanged across national frontiers to both the public and governments.

This Directive is meant to be protection for the public and the environment, and also for safety and protection in the workplace. It calls for particular attention to be given to certain industrial activities capable of causing major accidents. Such industrial accidents have already occurred in the European Union and have had serious consequences for workers and, more generally, for the public and the environment, e.g. the Sandoz accident in Switzerland in 1989 caused extensive environmental damage to the ecology of the River Rhine.

This Directive also regulates industrial activity which involves, or may involve, dangerous substances and which, in the event of a major accident, may have serious consequences for man and the environment. The manufacturer of a hazardous compound must take all necessary measures to prevent such accidents and to limit their consequences. In doing so, the training and information of persons working on an industrial site plays an important preventative role.

If a manufacturer implements industrial activities that include compounds that are particularly dangerous in certain quantities, the competent authorities must be informed about the intended activities, including information concerning the substances in question, enabling them to take the necessary steps to reduce the consequences of a possible accident. The manufacturer must also (should an accident occur) immediately inform the competent authorities and communicate the necessary information for assessing the impact of the accident. This means that a contingency plan must be worked out beforehand, allowing the authorities and their acting bodies to respond with the necessary speed and efficiency. In the Seveso Directive the seven annexes listed in Table 15.1 describe in detail the areas covered by the directive.

Table 15.1 Description of EU Seveso Directive annexes

Annexe number	Annexe description
Annexe I	A list of the type of installations covered in the Directive
Annexe II	The list of substances subject to the Directive when they are kept in storage
Annexe III	The list of 178 substances subject to systematic controls
Annexe IV	Indicative criteria for substances that are toxic, very toxic, flammable or explosive
Annexe V	Details of information to be supplied in notification to competent authorities
Annexe VI	Indications of information to be supplied to the European Commission by member states when reporting a major accident
Annexe VII	A statement pledging member states to consult on measures to avert major accidents and to limit the consequences for people and the environment

the authorities and their acting bodies to respond with the necessary speed and efficiency. In the Seveso Directive the seven annexes listed in Table 15.1 describe in detail the areas covered by the directive.

A hazardous waste treatment facility (i.e. an incineration plant) is required to comply with the directive according to Annexe I. According to the Directive, the authorities and the general public liable to be affected should be informed about accidents. This is an important part of the contingency plan for the area in question. All this activity has in turn created a high awareness of the necessity for security within chemical plant management at all levels, forcing the companies to perform risk analysis at regular intervals. Going through such exercises has given plant management a very useful tool in avoiding accidents. This is also true for hazardous waste treatment plants, so that their activities in general can be regarded as safe. The Seveso Directive represents a fundamental step towards the prevention, control and limitation of hazards arising from industrial activities. It provides common standards of responsibility in industry and in the member states of the European Union for the prevention of major accidents and the limitation of their consequences.

15.2 DEFINITION OF HAZARDOUS WASTE

The following two sections discuss the definition of hazardous wastes in the United States and in the European Union. It must be understood that those compounds or wastes or substances that come within the definition change with time as we learn more about their impacts. Additionally, the number of synthetic organic compounds is increasing daily and many of these are considered hazardous.

15.2.1 Definition of Hazardous Waste in the US

In the United States, the definition of hazardous is identified by the USEPA, CERCLA (Comprehensive Environmental Response, Compensation and Liability Act), also known as Superfund (see Chapter 1) and the US DOT (Department of Transport). Their definitions are listed in Table 15.2

15.2.2 Definition of Hazardous Waste Within the EU

Within the European Union, efforts have been made to produce comprehensive legislation dealing with hazardous waste. The directive (91/689/EEC) on hazardous waste and the 1995 directive on the incineration of hazardous waste are therefore relevant. The definition of the properties that render a compound or a waste hazardous is given in Annexe III of the EU Hazardous Waste Directive 91/689/EEC (included in Sec. 15.13). These are listed briefly in Table 15.3. For instance, compounds of

Table 15.2 Definition of 'hazardous'

HAZARDOUS WASTE
USEPA definition (40 CFR 260.10):
A solid waste that may cause or significantly contribute to an increase in mortality or an increase in serious irreversible or
  incapacitating reversible illness; or pose a substantial present or potential hazard to human health or the environment when it is
  improperly treated, stored, transported, disposed of or otherwise managed; and the characteristic can be measured by a
  standardized test or reasonably detected by generators of solid waste through their knowledge of their waste. The characteristics of
  hazardous waste are: ignitability, corrosivity, reactivity, EP toxicity (40 CFR 261.20–261.24). Hazardous wastes are listed in 40
  CFR Subpart D (Parts 261.30-261.33).
The US DOT (Department of Transport) also subscribes to this definition.

HAZARDOUS SUBSTANCE
Definition of CERCLA:
Any substance designated pursuant to Section 102, Section 307(a) and Section 311(b)(2)(A) of the Federal Water Pollution Control
  Act. Any hazardous waste having the characteristics identified under or listed pursuant to Section 3001 of the Solid Waste
  Disposal Act. Any hazardous air pollutants listed under Section 112 of the Clean Air Act. Any imminently hazardous chemical
  substance or mixture with respect to which the Administrator of the USEPA has taken action pursuant to Section 7 of the Toxic
  Substances Control Act.
Definition of the US DOT:
A material and its mixtures or solutions identified in 49 CFR 172.101 when offered for transportation under specific conditions of
  packaging and when the quantity of the materials equals or exceeds the reportable quantity.

HAZARDOUS MATERIAL
US DOT definition:
A designated substance or material that has been determined by the Secretary of Transportation to be capable of posing an
  unreasonable risk to health, safety and property when transported in commerce.

Adapted from Bellandi, 1988

category H1 are 'explosive' substances and preparations which may explode under the effect of flame or which are more sensitive to shocks or friction than dinitrobenzene. Category H7 are 'carcinogenic' substances and preparations which if they are inhaled or ingested, or if they penetrate the skin, may induce cancer or increase its incidence. In Annexe 1.A and 1.B, categories or generic types of hazardous waste are listed according to their nature or the activity that generated them. Waste may be a liquid, a sludge or solid in physical appearance. This listing is complemented with an index of constituents in the wastes given in Annexe I.B which render them hazardous when they have the properties described in Annexe III.

Table 15.3 EU hazard categories of hazardous waste

hazard category	Directive 91/689/EEC, Annexe III
H1	Explosive
H2	Oxidizer
H3A	Highly flammable (extremely flammable incl.)
H3B	Flammable
H4	Irritant
H5	Harmful
H6	Toxic (very toxic incl.)
H7	Carcinogenic
H8	Corrosive
H9	Infectious
H10	Teratogenic
H11	Mutagenic
H12	Water contact liberates toxic gas
H13	Source of hazardous substance
H14	Ecotoxic

The European Waste Catalogue (EWC), Directive 94/904/EC, establishes a list of wastes by many and varied activities and processes.

In order to decide whether a given waste falls within the (EU) hazardous waste classification the decision tree analysis approach is a useful tool, and is shown in Figs 15.1 and 15.2. The decision tree has two key levels. The first level is 'general' and requires no test procedures, only a visual inspection along with assessment of a manifest documentation and relevant legislation. If the assessor has doubt about the 'waste', a specific evaluation is recommended, usually by testing. The evaluation tests may be physical, chemical or leaching. On the result of this evaluation, the waste may be deemed:

• Non-hazardous and suitable for landfill
• Hazardous and suitable for incineration
• Hazardous and suitable for incineration only after pre-treatment
• Hazardous inorganic and treatable at an inorganic plant (described later) without incineration

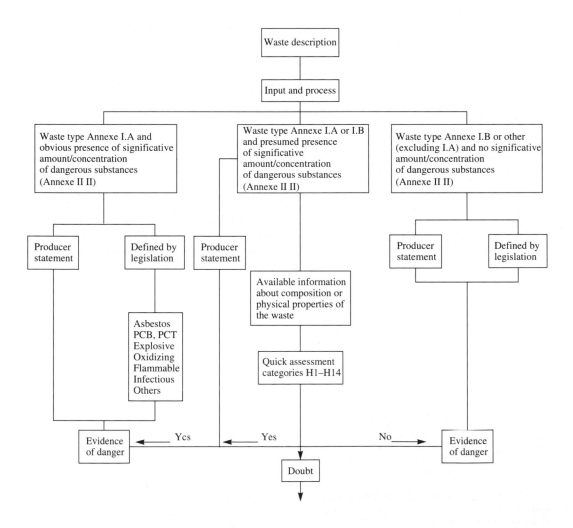

Figure 15.1 Decision tree to define hazardous waste—phase 1.

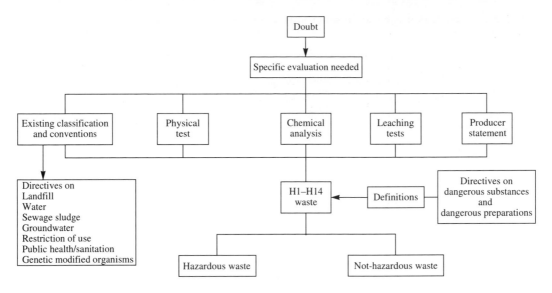

Figure 15.2 Decision tree to define hazardous waste—phase 2.

The definition of a hazardous waste, as per the Annexes of the EU Directive, is particularly useful for statistical purposes, but is not ideal for classification purposes and for day-to-day management of a waste treatment facility. A classification that can be used in the day-to-day handling of waste has been developed based on a much more simplistic system. In Denmark such a simplified system has been in use since 1972 at Kommunekemi, the National Hazardous Waste Treatment Facility. Kommunekemi is centrally located in Denmark, and all known hazardous wastes requiring treatment and/or disposal are sent to this site, where facilities for incineration, etc., exist. In this system, hazardous wastes are classified into one of eight categories, A, B, C, H, T, X, K or Z, as shown in Table 15.4. The groups are also listed in Fig. 15.3 together with a guide for selection of the different groups. For more exact statistics the different main groups are divided into subgroups, but the waste handling is always decided by using the main groups. The methodology displayed in Fig. 15.3 is used as an initial screening of incoming hazardous waste to the treatment facility.

15.3 HAZARDOUS WASTE GENERATION

Hazardous waste may account for approximately 2 to 20 per cent of all wastes in the European Union. The amount of hazardous and toxic waste generated in the European Union is estimated at 22 million tons per year (by comparison with 420 million tons of MSW). A breakdown, by country, is shown in Table 15.5.

It is seen that in the low-income countries of the European Union most of the hazardous waste is directly landfilled, and not necessarily in specially engineered hazardous waste designed landfill. In the more developed countries, incineration is used on about a quarter of the waste. While the figures of Table 15.5 might look definitive, it should be realized that some of these figures, particularly for the lower-income EU countries, are estimates. Also, to balance the figures of Table 15.5, another column called 'other' (e.g. export) is required. Some of the most important categories of waste are:

- Liquid organic chemicals and solvents
- Heavy metals containing wastes

Table 15.4 Hazardous waste classification at Kommunikemi, Denmark

Group	Type	Examples	Treatment process simplified
A	Mineral oil waste	Pumpable wastes containing mineral oil, e.g. lubricating oil, hydraulic oil, heat transmission oil, drilling oil, synthetic oil, oil from intercepting traps	Recovery and Incineration
B	Halogenous solvent waste	Pumpable wastes containing halogenated solvents such as trichlor-ethylene, perchlor-ethylene, tetrachlor-ethylene, chloroform, chlorotene, genklene, freon. Pumpable and non-pumpable halogen or sulphur containing organic chemical wastes	Incineration
C	Solvent waste	Pumpable wastes containing non-halogenous solvents such as gasoline, turpentine, solvent, xylene, ethyl alcohol, propyl alcohol, thinners, octane, MIBK, MEK, ethylacetate, butylacetate	Incineration
H	Organic chemical wastes, halogen and sulphur free	Used paint, paint sludge, distillation residues, organic chemical by-products, tar, deep frying oil, organic acids and their salts, glue waste, used developer, alkaline, cyanide-free washing baths, bitumen, grease, solid fuel oil, soap wastes	Incineration
T	Pesticide containing wastes	Insecticides, fungicides, weed killers, rat poison. Seed grain containing mercury	Special sorting + incineration or special disposal
X	Inorganic chemical wastes	Used packing acid, electroplating baths, metal hydroxide sludge, wastes from regeneration of ion exchange, contaminated sulphuric acid, block metal acid, soda lye, ammonia water, alkaline cyanide containing degreasing baths, hardening salts, salts, caustic soda	Neutralization, detoxification and precipitation
Z	Other wastes	Isocyanate (MDI and TDI)-containing wastes	Pre-treatment + incineration
		Oil polluted soil	Incineration
		Pharmaceutical wastes	Incineration
		Wastes from laboratories in small containers (lab packs)	Incineration after sorting
		Spray cans	Incineration
		Chemical wastes from households	Incineration after sorting
		Used mercury batteries	Special disposal
		Used drums and used small containers packed in drums, sacks or the like	Incineration

Table 15.6 lists some industries producing hazardous wastes. These industries are found in all developed countries. Heavy metals, organics, and hydrocarbons are produced.

Copper (Cu), zinc (Zn), lead (Pb) and mercury (Hg) are widely used. Copper and zinc are of less concern than lead, mercury and cadmium (Cd). Copper is used as an electrical conductor, as a basis in alloys. Zinc is used in alloys, pigments, plastics, batteries, galvanizing and plating. By comparison, lead is highly toxic. It impairs haemoglobin synthesis, particularly in children, and may cause neurological disorders. Lead is found in paints, pipes, batteries and in some petrol types. Mercury is also highly toxic, causing damage to the central nervous system and kidney malfunction. It bioaccumulates and has been responsible for high mortality in birds. It is used in the pharmaceutical and chemical industry. Cadmium is a pollutant in the air, soil and water environment. Cadmium is released into the air from waste incineration, fuel combustion, cadmium plating industries, and from rechargeable batteries. Cadmium bioaccumulates and may lead to kidney malfunction. All of the metals and hazardous substances listed in Table 15.6 have maximum permissible values in the different environments and it is the goal of the EU directives to achieve values below these limits.

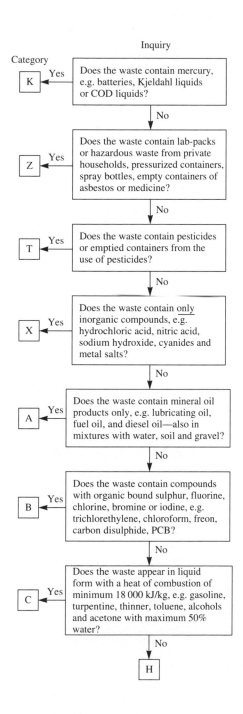

Figure 15.3 Hazardous waste categorization at Kommunikemi, Denmark, for screening of incoming wastes. (From Chemcontrol, Denmark, educational material. Reprinted by permission.)

Table 15.5 EU toxic and hazardous waste, quantities and treatment method

Country	Amount (million tons/year)	Landfilled (%)	Incinerated (%)	Physicochemical treatment (%)
Belgium	0.7			
Denmark	0.5		28	
Germany	4.9	48	8	8
Greece	0.3			
Spain	1.7			
France	4.0	10	25	10
Ireland	0.08	60	14	7
Italy	3.8			
Luxembourg	0.08			
Netherlands	1.0	30	25	15
Portugal	1.0	30	25	15
UK	3.7	80	2	8
Total EU	21.8			

A final column for 'other' disposal routes for the remaining percentage is not included, as the specifics of 'other' are unknown.
Adapted from EC 1993, State of the Environment

15.3.1 Hazardous Waste Inventory

Assessment of waste sources is normally a difficult task as waste generators (industrial, local authorities, agriculture, etc.) do not always retain inventories of their waste streams. Even when they do, they tend to be discreet about it, realizing that hazardous waste is a liability and a cost burden. Two tasks must be undertaken:

1. The waste generators must be identified.
2. The waste streams must be categorized and inventories made.

Past experience with hazardous waste surveys has shown them to be somewhat unreliable. Some surveys have even turned out to be so much off their target that the use of them for any planning purpose would not be advisable. Waste assessment has thus become a highly and somewhat expensive exercise, requiring time, patience, imagination, money, training and the ability to co-ordinate and create a high level of interagency and multidisciplinary co-operation. In countries where no reliable data are available a number of quick methods for the initial waste assessment include:

1. Utilize existing trade registries and membership lists to identify the largest potential waste generators.

Table 15.6 Some industries producing hazardous waste

Industry type	Hazardous substance
Batteries	Cd, Pb, Ag, Zn, NO_2
Chemical manufacturing	Cr, Cu, Pb, Hg, organics, hydrocarbons*
Electrical/electronic	Cu, Co, Pb, Hg, Zn, Se, organics, hydrocarbons*
Printing	As, Cr, Cu, Pb, Se, organics
Electroplating	Co, Cr, Cn, Cu, Zn
Textiles	Cr, Cu, organics
Pharmaceuticals	As, Hg, organics
Paints	Cd, Cr, Cu, Co, Pb, Hg, Se, organics
Plastics	Co, Hg, Zn, organics, hydrocarbons*
Leather	Cr, organics

*Including halogenated organic compounds
Adapted from Nemerow and Dasgupta, 1991

2. In co-operation with the environmental authorities issue a relatively simple questionnaire in order to verify the 'capturable' waste amounts being produced as well as the already existing storage of waste.

3. Verify the amount of waste reported by making plant visits to selected groups of potential waste generators.

4. Compare the results with data from other countries. A number of possibilities exist, e.g. the 'Invent' computer program based on amounts of waste produced by each employee in specific types of industry. The program uses Italian data, but others, e.g. Danish and Australian statistics, can be used. All the different approaches are then evaluated, and the result of this may give a reasonable description of the waste streams in the area.

Once the problem of environmental impact and the cause of it (the waste generation) have been identified, and provided that the environmental legislation is in place, including the enforcement system, considerations can be made for solutions to problems as well. This includes regulations, which may already exist, the interaction with the different environmental agencies, both central and local, and the possibility of adopting existing international regulations as a supplement (e.g. ADR, RID classifications for transportation of hazardous waste) to the country's own legislation. The combined efforts can then be listed in the preparation of a *Hazardous action plan* which should include the following:

- Establishment of waste treatment and disposal facilities
- Legislation to set acceptable standards for waste handling facilities and to require monitoring and reporting of waste operations
- An administration to enforce the legislation and to monitor wastes
- Establishment of adequate infrastructure and technical support services, training institutions, information services, waste data monitoring banks, etc.

It is important that the different elements in the plan are timed correctly as they are dependent on each other.

15.4 MEDICAL HAZARDOUS WASTE

Traditionally, medical waste was all dumped at landfills, and in some underdeveloped countries (UDCs) this is still the practice. Some hospitals had very basic combustion units, operating at temperatures well below the complete combustion temperature. While hospitals had always been particularly aware of potential infectious waste, other non-infectious hazardous waste was not treated with the same rigour and so was routinely disposed at landfill. Much of the hazardous waste generated from medical establishments can only be properly disposed of by incineration, although microwave treatment techniques are now being used for small volume wastes. This type of waste belongs to a special group of waste, adding further dimensions to the term 'hazardous' inasmuch as it can be infectious and cytostatic (cytostatica can promote the development of cancer). Medical hazardous waste consist of the following groups of material:

- Obsolete medicines past the expiry date or medicine that is not, for one reason or another, going to be used any more
- Cytostatica with a special demand for careful handling
- Infectious material, e.g. bed linen, used wound dressings, used transfusion equipment, etc.
- Pathological waste, e.g. waste from operating theatres, etc.
- Sharp and pointed items
- Waste from dental clinics

Waste from the first group may be produced in a number of different areas, e.g.:

- In private households
- In old peoples' homes and in other forms of nursing homes
- In doctors' clinics
- In pharmacies
- At wholesalers' premises
- At hospitals

The very nature of medical waste has in many cases called for specific legislation regulating the many aspects of its collection and handling. The flow sheet shown in Fig. 15.4 indicates the many waste streams appearing in this sector:

1. Special hospital waste types include:
   (a) Biological waste from humans
   (b) Infectious waste from humans
   (c) Biological waste from animals
   (d) Infectious waste from animals
   (e) Microbiological waste

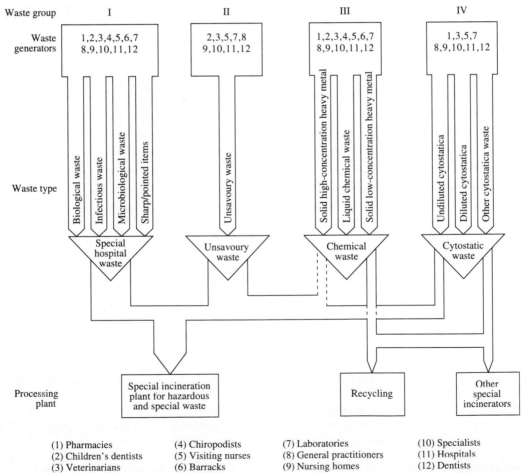

Figure 15.4 Waste streams from medical sources. (From HOSPICONS, hospital waste handling company educational material. Reprinted by permission.)

(f) Diluted cytostatica/antibiotica

(g) Other cytostatic polluted waste

(h) Sharp and pointed items

2. Unsavoury waste:

(i) Classified as daily collection

3. Chemical waste types include:

(j) Solid chemical waste with a high content of heavy metal

(k) Solid chemical waste with a low content of heavy metal

(l) Liquid chemical waste

(m) Undiluted cytostatica

15.5 HOUSEHOLD HAZARDOUS WASTE

Significant amounts of hazardous waste can be identified in households. Table 15.7, adapted from the US Water Pollution Control Federation (1987) itemizes some of the possible wastes and their location in the household. Also included are possible disposal routes for spent items, either to landfill, recycle or those that specifically require a licensed hazardous waste haulier for disposal. It is noted that the domestic 'garage' typically contains most hazardous waste, much of which is required to be disposed of by a licensed haulier. In Chapter 14, it was noted that in general 0.01 to 1.0 per cent of MSW by weight disposed at landfills in the United States, is estimated to be hazardous.

15.6 TRANSPORTATION OF HAZARDOUS WASTE

Hazardous waste covers a very small segment of the large area that is classified as 'dangerous goods". Only in recent years has the term 'hazardous waste' or simply 'waste' been even mentioned in international regulations applied to transport. However, this situation is now changing. In Denmark a system for the regulation of transport of hazardous waste has been under continuous development since 1975. This system is now being adjusted to the inter-European system, progressively introduced by the European Union. The transport of hazardous waste, according to the previous mentioned classifications, is governed by a number of regulations. They are primarily derived for use in the transport of dangerous goods, but can with minor amendments also be used for the transport of hazardous waste.

The regulations relevant in this context are:

- ADR (*A*ccord Européen relatif au transport international des marchandise *D*angereuse par *R*oute)—transport of dangerous goods by road
- RID (*R*eglement *I*nternational concernant le transport des marchandise *D*angereuse par chemin de fer)—transport of dangerous goods by rail
- IMDG (*I*nternational *M*aritime *D*angerous *G*oods code)—transport of dangerous goods by ship
- ICAO (technical instructions)—transport of dangerous goods by air

In this chapter only the two first regulations are described as they are the most commonly used in connection with the transport of hazardous waste (even though hazardous waste does travel by air). The two regulations are fairly similar and work is in progress to amalgamate them.

15.6.1 The ADR/RID Classification System

Both sets of regulations are operating with a nine-classification system, as shown in Table 15.8. For each class shown in Table 15.8, there is a specific warning label as shown in Fig. 15.5. For instance, Class 6.1 are toxic compounds, identified by a white diamond sign with an insert of a skull and cross bones.

Table 15.7 Hazardous household waste

Location in household	Type of waste	Safe to pour down the drain	Safe to landfill	Recyclable	Dispose by licensed hazardous haulier
Kitchen	Aerosol cans		✓		
	Aluminium cleaners	✓			
	Ammonia-based cleaners	✓			
	Bug sprays				✓
	Drain cleaners	✓			
	Floor care products				✓
	Metal polish with solvent				✓
	Window cleaner	✓			
	Oven cleaner		✓		
Bathroom	Alcohol lotions (perfumes, etc.)	✓			
	Bath and toilet cleaners	✓			
	Disinfectants	✓			
	Hair relaxers	✓			
	Medicines (expired)	✓			
	Nail polish (solidified)		✓		
Garage	Antifreeze	✓			
	Batteries (including acids)			✓	✓
	Brake fluid				✓
	Car wax with solvent				✓
	Diesel fuel			✓	✓
	Gasoline			✓	✓
	Kerosene			✓	✓
	Metal polish with solvent				✓
	Motor oils			✓	✓
	Paint brushes			✓	✓
	Paint latex		✓		
	Paint oil				✓
	Paint thinner			✓	✓
	Paint stripper				✓
	Glue		✓		
	Varnish, wood preservative				✓
Garden	Fertilizer	✓			
	Fungicide, herbicide, insecticide				✓
	Rat poison				✓
	Weed killer				✓

Adapted from WPCF, 1994, with permission

Guidelines are also given regarding 'keeping dry' identified by an umbrella, 'handle with care' identified by a drinking glass and 'this side up' by two upright arrows.

15.6.2 Pre-transport and Preparation of Waste

Each waste type must be classified according to the above-mentioned transportation code before transport can take place. In each class there is a listing of compounds for that class. This listing is further divided into three risk groups, a, b and c, with the a group carrying the highest risk. As regards the restricted classes, only compounds specifically mentioned in the listing can be transported in these classes. All other compounds are excluded from transport in the classes. Open classes can be used for transport of the listed compounds and other compounds with similar properties, including hazardous waste.

Table 15.8 ADR/RID Classification

Class	Item description	Open/restricted
1a	Explosives and explosive items	Restricted class
1b	Items loaded with explosives	Restricted class
1c	Detonators, fireworks and similar types of goods	Restricted class
2	Gases, compressed, condensed or dissolved under pressure	Restricted class
3	Flammable liquids	Open class
4.1	Flammable solids	Open class
4.2	Self-ignitable compounds	Open class
4.3	Compounds which develop flammable gases in contact with water	Open class
5.1	Oxidizing compounds	Open class
5.2	Organic peroxides	Open class
6.1	Toxic compounds	Open class
6.2	Infectious or loathsome acting compounds	Open class
7	Radioactive compounds	Restricted class
8	Corrosive compounds	Open class
9	Miscellaneous dangerous goods	Open class

Once the waste has been classified according to the ADR/RID regulations, the waste generator must fulfil a number of other tasks before the waste is ready for transport. Firstly, an appropriate container (drum) must be selected. In the ADR/RID regulation, specifications in selecting the containers (drums) are given. Secondly, a manifest (or declaration) must be filled out. The manifest includes information on:

- The waste generator
- The composition of the waste
- The physical appearance
- The method of packaging, etc.
- The ADR/RID classification
- The simple Danish classification
- The UN number

The waste must then be packed in the specified containers, equipped with warning labels and marked with the manifest number both on the top and on the sides. The transport system and its different steps are shown in Fig. 15.6.

15.6.3 Transport in Bulk

Not all waste is transported in drums and small containers. Some derived from large-scale waste generators are packed in so-called IBCs (intermediate bulk containers) or tank containers with capacities of 20 to 30 m^3. Transport of these containers can be made on road and rail. The truck carrying a tank container with dangerous goods must be equipped with a set of orange warning signs with a black rim, one in the front of the truck and one in the rear. In the case of a normal dangerous goods transport, the sign will show two sets of numbers, one set above a black divider line (a two- or three-digit number) and one set below (always a four-digit number), as shown in Fig. 15.7. The upper set of numbers refers to the properties of the goods, e.g. 3 for flammable (ADR/RID class 3: flammable). The second digit (3) indicates that the liquid is 'very flammable'. Another example is the number 268, which means that the liquid belongs to ADR/RID class 2, compressed gases, it is toxic (6) and finally the compound is corrosive (8).

Below the divider line there is a four-digit number giving the identity of the transported compound. The numbers are the so-called 'UN' numbers. They are all shown in a list issued by the United Nations. A specimen of the list is given in Table 15.9. From the table it can be seen that, for example, 1005 is

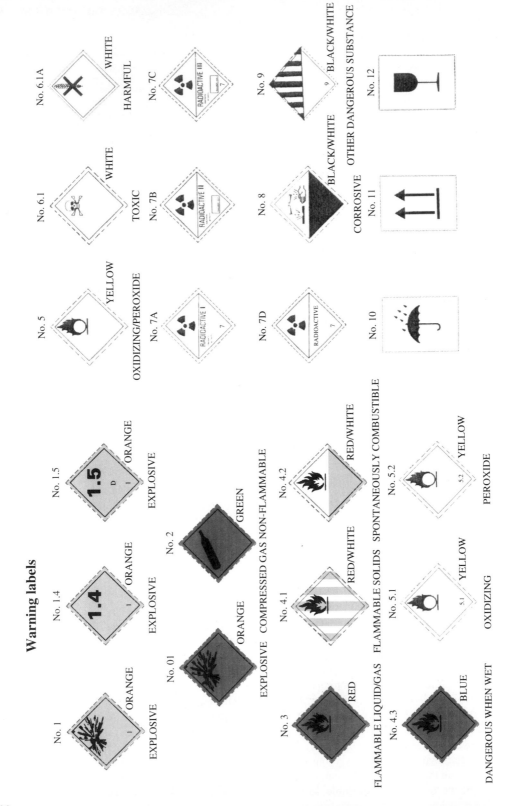

Figure 15.5 Warning labels for transport of dangerous goods. (ADR/RID classification system.)

708

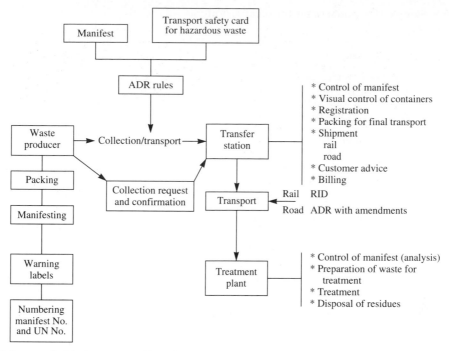

Figure 15.6 Steps required in the transport of hazardous waste in Denmark.

ammonia (upper number 268). For hazardous waste transport, special numbers are given. Examples commonly used are:

- 1759   Corrosive solid compounds
- 1760   Corrosive liquids
- 1906   Waste acids
- 1992   Flammable liquids, toxic
- 1993   Flammable liquids,
- 2920   Corrosive liquids, flammable
- 2922   Corrosive liquids, toxic

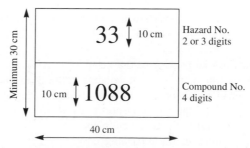

Figure 15.7 Sample of warning numbers of bulk transport of dangerous goods.

Table 15.9 Sample UN numbers for the transport of hazardous goods

UN number	Hazardous item
1001	Acetylene (ethylene), dissolved
1002	Compressed air
1003	Liquefied air, refrigerated
1005	Ammonia, dry, liquefied or soluted in water
1008	Boron trifluoride
1009	Trifluorbromomethane (Freon 13B1)
1010	Butadiene, stabilized
1011	Butane or mixtures with butane
1012	Butylene
1013	Carbon dioxide
1014	Carbon dioxide in mixtures with oxygen
1015	Carbon dioxide in mixtures with dinitrogenoxide

15.6.4 Collection Systems for Hazardous Waste

In an integrated waste treatment system, the collection plays an important role. If a collection service is not offered or if the collection procedure is too complicated there is a risk of waste being disposed of in an inappropriate manner. Taking into account the multitude of existing waste sources, an efficient collection system must be able to operate at different levels suiting the many types of waste generators, such as:

- Large industrial enterprises
- Small to medium companies
- Small companies
- Private households
- Health sector—e.g. doctors' surgeries, care centres, clinics and hospitals

To make the collection system as cost efficient as possible, different models can be used, but only the Danish system is described here. It operates with a set of 18 transfer stations strategically sited throughout Denmark in the areas to be served by the collection system and they form the basic collection network. A subset of this network is a system of collection centres having at least one centre situated in each municipality, the larger municipalities having more than one. In general the transfer stations are used by industry while the collection centres are used mainly by the public. Figure 15.8 shows the waste collection routes.

15.6.5 Transfer Station

A transfer station serves a multitude of tasks, including:

1. Organizing a 'pick-up' service. The service will, on request, use a certified transporter and collect the agreed waste type and amount. A condition for collection is that the waste is packed and marked according to the ADR transport rules.
2. Make a visual assessment of the packing of the waste.
3. Inspect the documentation (the manifest). All columns in the form must be completed.
4. Reload the waste onto either trucks or railroad wagons for final transport to the treatment plant.
5. Fill out the necessary papers for the final transport.
6. Transmit data from the manifest to the treatment plant.
7. Sell suitable containers (drums) for waste packing.
8. Organize the distribution of multiple journey containers (pallet tanks, etc.).

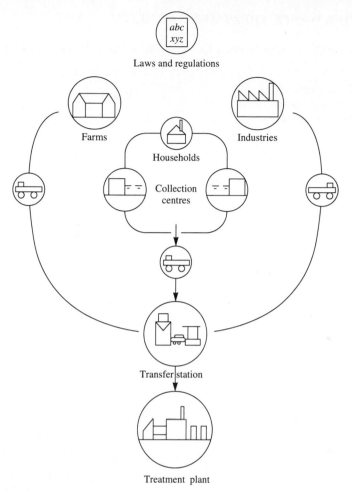

Figure 15.8 Hazardous waste collection routes.

9. Give advice to waste generators about classification, handling, packing and transport of hazardous waste.

A transfer station for MSW is shown in Chapter 14 and is similar to that for hazardous waste.

15.6.6 Collection Centre

A collection centre is a minor installation meant for receiving waste from the general public, particularly households. The waste, mostly in small packs, is sorted into different categories and packed in drums for shipment to a transfer station. The private waste generator does not complete a manifest. The centre will fill out the necessary manifest for shipment to the transfer station, the drums will be marked as 'hazardous waste from households' and in the manifest the collection centre is the sender. Supplementing the activities of the collection centre, a number of other schemes are used, e.g. collection trucks calling on all households or calling on special places at regular intervals. Also pharmacies and chemists are used to act as receivers of waste (e.g. drugs past their use-by date, paints).

15.7 HAZARDOUS WASTE TREATMENT FACILITY

The available commercial treatment facilities for hazardous waste (Palmark, 1986) include:

- Thermal treatment
  - Rotary kiln incinerators
  - Liquid injection incinerators
  - Plasma arc incineration
  - Wet air oxidation
  - Fluidized bed combustion
- Chemical treatment
  - Neutralization
  - Detoxification
  - Precipitation
  - Ion exchange
- Physical treatment
  (separation)
  - Filtration
  - Flocculation
  - Sedimentation
  - Centrifugation
- Disposal
  - Direct to landfill
  - Pre-treatment and then to landfill
  - Discharge of wastewater
  - Discharge to the air

A facility for the treatment of hazardous waste will, in addition to an incineration plant, have other plants for the treatment of waste oils and inorganics. Figure 15.9 shows the different plant facilities in Kommunekemi and the distribution of incoming wastes. The essential plants and the percentage of waste treated by each are:

- The incineration $\sim$ 53 per cent
- The inorganic plant $\sim$ 14 per cent
- The waste oil plant $\sim$ 32 per cent $\rightarrow$ to incinerator

It is noted that 53 per cent of the incoming waste goes direct to incineration. After the oil is separated from the water and sludge, this 32 per cent component goes to the incinerator in different streams.

The wastes received at Kommunekemi for 1994 are shown in Table 15.10. Waste oils make up approximately one-third, while organics (halogen and sulphur free) make up another third. The inorganic fraction is 11 per cent. The total is 87 243 tonnes for the year.

15.7.1 Reception of Waste at the Kommunekemi Treatment Plant

At the reception, the waste is put on hold while being sampled and analysed. There are in principle two different approaches to the sampling procedure:

- Pre-sampling
- Random sampling

The pre-sampling method is based on a procedure where a representative sample is taken by the waste generator and sent to the treatment plant for 'acceptance analysis'. At the plant the sample is analysed for

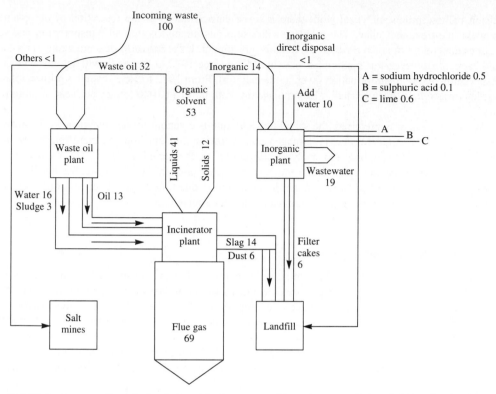

Figure 15.9 Waste streams at Kommunekemi. Quantities in per cent of incoming waste. (Adapted from Palmark, 1986, with permission of Chemcontrol A/S, Denmark.)

Table 15.10 1994 wastes to Kommunekemi

Waste group KK class	Waste type	Quantity (tonnes)	% by weight
A	Mineral oil	18 787	21.5
B	Halogeneous solvents	2 597	3
C	Solvents	6 403	7.5
H	Organics, halogen and sulphur free	42 056	48.2
T	Pesticides	717	0.8
K	Mercury containing waste	453	0.5
X	Inorganic	10 956	12.5
Z	Other	5 274	6.0
Total		87 243	

a number of key parameters. If accepted, the waste generator is notified and the waste consignment can proceed for shipment. If rejected, further discussions are held between the treatment plant and the waste generator. The basic problem with this method is to perform the difficult task of taking a representative sample and practice has shown that it is a demanding operation. On arrival at the treatment plant further samples are taken to ensure that the consignment is in accordance with the pre-sample. This last procedure is known as 'finger printing'.

By random sampling the waste is first analysed on arrival at the plant. The number of samples to be taken for analysis depends on the waste generator's 'track record' and the amount of waste supplied. The decision of how much analytical work there is to be carried out is at the discretion of the evaluation chemist at the treatment plant. The problem with using this method is that all transportation and safety measures depend on the data given in the manifest document. If the manifest data are wrong, an accident could happen. However, this method has now been used since 1972 in Denmark without any accidents in transportation due to misinformation given in the manifest. From 1972 to 1993, over 1.3 million tonnes of hazardous waste have been treated at Kommunekemi, with up to 110 000 tonnes per year in more recent years.

Whatever system is preferred, the sampling technique is essential to the analytical work carried out afterwards, as waste is far from being homogeneous. Knowing the data from the manifest, the evaluation chemist writes out a sampling and analyses instruction to the sampling team. The results from the laboratory investigations are entered into the customer file under the manifest number heading. At the same time the evaluation chemist will be notified and will in due time issue a treatment order. Only then can the waste be moved from the holding area at the waste reception.

15.7.2 Analytical Aspects

Dealing with waste means that a large number of compounds can and might be involved. Trying to analyse for every one of the probable waste constituents would neither be practical nor possible. It is, of course, very interesting to know what the exact composition of a waste is, but seen from a waste treater's perspective only few parameters are of more than academic interest. The properties of a waste which are of interest are those having an influence on handling, mixing and treatment, as well as those that influence the composition on effluent water and flue gas. The list of key waste parameters shown below is not to be regarded as comprehensive but it gives an idea of the parameters of interest for the waste treater:

- Chemical reactivity
- Radioactivity
- BTU value or calorific content
- Halogens (F, Cl, Br, I)
- Sulphur
- Phosphorus
- Arsenic
- Heavy metal compounds including mercury
- For inorganic waste, complexing compounds (e.g. NH_4^+), cyanides and nitrates.

15.7.3 Pre-treatment Prior to Incineration

On arrival at the treatment plant the waste in general is not in a condition where it can be presented directly to the treatment plants. There will nearly always be a number of operations to be carried out before efficient and safe treatment can be performed. In preparing waste for incineration, the operations performed are shown in Fig. 15.10. They will for a greater part consist of a variety of opening and emptying operations of drums and other forms of containers—homogenizing, neutralization and mixing. These operations traditionally involved relatively large amounts of manual handling, but in more modern plants most of the work is now carried out automatically, thus avoiding the exposure of the staff to the waste. This has been made possible through the introduction of automatic drum handling equipment in conjunction with a much better knowledge of the composition of the different waste types.

Automatic drum emptying Drums containing pumpable waste are introduced in a nitrogen blanketed sluice, as shown in Fig. 15.11. The sluice is followed by a double set of shredders (also nitrogen blanketed to avoid ignition of the organics due to friction in the shredder). This disintegrates the drum. The

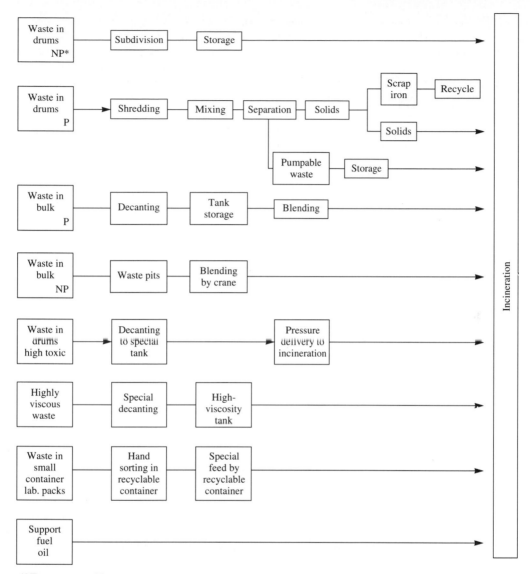

Figure 15.10 Variety of feedstock flow to incineration.

*NP = non-pumpable

contained waste together with the pieces of the drum will then enter a set of mixers. The mixing action will blend the different waste types from the different drums and also help to clean the pieces of the shredded drum by rubbing the metal and plastic chunks against each other. The mix will then enter a vibration sieve which will separate the mix into two streams:

- An organic liquid/sludge ready for storage and subsequent incineration and
- A mixture of scrap steel, plastic and some remaining waste

The scrap metal, plastic, etc., will then be washed in an alkaline washing bath. The cleaned steel and plastic is separated by a magnetic separator. The steel fraction goes to a scrap iron works and the plastic fraction goes to the incinerator.

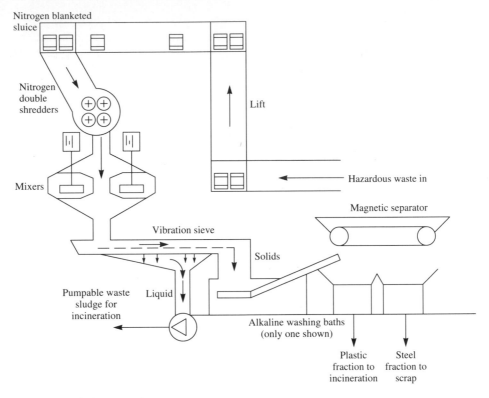

Figure 15.11 Automatic drum emptying system.

Some of the waste types are dealt with manually. In this category are found highly reactive waste types, e.g. isocyanates and highly toxic compounds (e.g. off-spec pesticides). Provisions must also be made for the handling of acidic organic waste which must be neutralized (with sodium hydroxide solution) before entering the storage tanks.

15.8 PLANNING A HAZARDOUS WASTE INCINERATOR

Figure 15.12 shows some of the main waste data necessary for planning a waste incinerator. Data for the heat value of the waste categories are necessary for estimating the kiln and secondary combustion chamber (SCC) capacity or the proportion of wastes that can be loaded on to the kiln and secondary burner. Also, the heat value will determine the requirement of support fuel. Information on the contents of halogens, sulphur and heavy metals are used for designing the flue gas cleaning system.

15.9 PLANNING AN INORGANIC WASTE TREATMENT PLANT

Figure 15.13 shows the parameters needed in designing an inorganic treatment plant for plating and other inorganic wastes. The design is based on:

- Detoxification of cyanide and chromate
- Precipitation of the metals as hydroxides

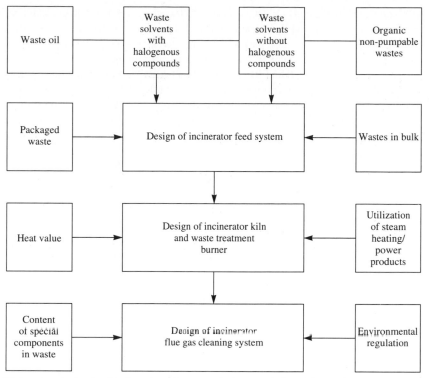

Figure 15.12 Waste categorization in planning a hazardous waste incinerator. (Adapted from Palmark, 1986, with permission of Chemcontrol A/S, Denmark.)

The parameters in Fig. 15.13 are a prerequisite for designing the size of the holding tanks, reaction vessels, filters, centrifuges, pumps, etc.

15.9.1 Inorganic Plant

In the plant for inorganic chemical waste at Kommunekemi, shown in Fig. 15.14, spent pickling bath waste from the electroplating industry as well as other waste of an inorganic nature is treated. The treatment of the waste takes place in stages, starting with a detoxification of the cyanide-bearing waste

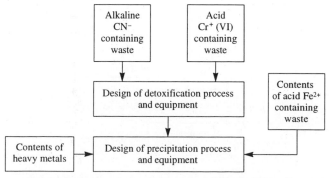

Figure 15.13 Parameters used in inorganic plant design. (Adapted from Palmark, 1986, with permission of Chemcontrol A/S, Denmark.)

(oxidation with sodium hypochlorite) (NaOCl). This is followed by a redox process reducing hexavalent chromium to trivalent chromium, while bivalent iron is oxidized to trivalent iron. The pH level is then adjusted to pH 9.8 to 10.0. Here the major part of the heavy metals is precipitated as heavily soluble metal hydroxides. The hydroxides, which mainly consist of iron, chromium, nickel, copper, aluminium and zinc, are filtered off in a filter press. The filter water is then neutralized (with sulphuric acid, H_2SO_4) and analysed. If the results are satisfactory (corresponding to the outlet criteria set in the operational permit), the water is drained off into the municipal sewage works. The filter cakes are disposed of at a secured landfill (for hazardous wastes). The cakes are covered with a plastic membrane to prevent precipitation exposure. The filter cakes may potentially in the future form the raw materials for a recovery process extracting the heavy metals.

15.9.2 Waste Oil Plant

In the waste oil plant, ingoing waste oil mixtures are separated into three fractions by thermal treatment into three fractions:

- Oil
- Oil-contaminated water
- Sludge

The contaminated water and the sludge are transferred directly to the incineration plant, whereas the oil phase is further treated (filtering, reheating), ending up as a fuel oil suitable for use as supplementary fuel for the incineration plant. The processes are shown in Figure 15.15.

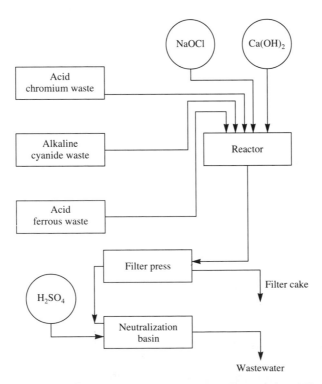

Figure 15.14 Processes in an inorganic plant. (Adapted from Palmark, 1986, with permission of Chemcontrol A/S, Denmark.)

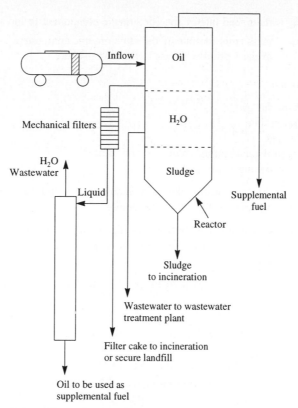

Figure 15.15 Waste oil plant. (Adapted from Palmark, 1986, with permission of Chemcontrol A/S, Denmark.)

15.10 TREATMENT SYSTEMS FOR HAZARDOUS WASTE

The three incineration treatment systems examined are:

- Incineration
- Wet oxidation
- Fluidized bed combustion

15.10.1 Incineration

Second to landfilling, the most universal way of treating hazardous waste is by incineration. In the incinerator, the waste is oxidized in an oxygen-rich enviroment at elevated temperatures. An incineration plant means any technical equipment used for the incineration by oxidation of hazardous wastes including pre-treatment as well as pyrolysis or other treatment process, e.g. plasma process, insofar as their products are subsequently incinerated with or without heat recovery. This includes plants burning such wastes as a regular or additional fuel for any industrial process. The most important criteria for hazardous waste incinerators is the complete destruction of the major hazardous compounds while confining the level of air emissions. Typically, a destruction and removal efficiency (DRE) of 99.999 per cent is required. The DRE for an incineration is calculated by

$$DRE = \frac{1 - W_{out}}{W_{in}} \times 100\% \qquad (15.1)$$

where $\qquad W_{in}$ = mass feed rate of specific organic component to incinerator

$\qquad\qquad\qquad W_{out}$ = mass emission rate of the same organic component in the exhaust prior to release to the atomosphere

Satisfactory DRE requires:

- Adequate residence time for gas
- Adequate solids retention time
- High enough temperatures
- Enough O_2 (or hydrogen if incinerating chlorine or bromine)
- Sufficient turbulence for mixing waste and oxygen

To evaluate the stoichiometric combustion air flow requirements and to predict the gas flow and composition, the following analyses of the waste are required:

- Proximate analysis—moisture content, volatile solids, fixed carbon, non-combustibles
- Ultimate analysis—carbon, hydrogen, oxygen, nitrogen, sulphur, halogens, phosphorus
- Energy content—net heating value

This is similar to that discussed in solid waste treatment, Chapter 14. The combustion efficiency (CE) is calculated from

$$CE = \frac{C_{CO_2} - C_{CO}}{C_{CO_2}} \times 100 \text{ in per cent} \qquad (15.2)$$

where $\qquad\qquad\qquad C_{CO_2}$ = concentration of CO_2

$\qquad\qquad\qquad\qquad C_{CO}$ = concentration of CO

Costs of incineration of hazardous waste   The estimated capital cost of a new plant to treat 50 000 tonnes/annum is given in Table 15.11. This assumes one line of incineration and 1996 costs. Operational costs are shown in Table 15.12.

The chemistry of incineration   Incineration can be defined as the controlled high-temperature oxidation of primarily organic compounds to produce carbon dioxide and water:

$$\text{Organic waste} \xrightarrow{\text{incineration}} CO_2 + H_2O + \text{inoffensive by-products} \qquad (15.3)$$

The mechanisms of heat transfer, conduction, convection and radiation occur with solids, liquids, liquid solids and gases at high temperatures. The fluid environment may be laminar or turbulent but generally the latter. This suggests that incineration is a very complex process, involving chemistry, physics, thermodynamics and fluid mechanics. When ethane is incinerated the reaction is:

$$\underset{\text{ethane}}{2C_2H_6} + \underset{\text{oxygen}}{7O_2} \xrightarrow{\text{incineration}} \underset{\substack{\text{carbon} \\ \text{dioxide}}}{4CO_2} + \underset{\text{water}}{6H_2O} \qquad (15.4)$$

When aromatic hydrocarbons are incinerated the reaction is:

$$\underset{\text{aromatic hydrocarbon}}{CH_3-C_6H_5} + 9O_2 \xrightarrow{\text{incineration}} 7CO_2 + 4H_2O \qquad (15.5)$$

During incineration, the chemical bonds between the various elements of the reacting mixtures are broken

Table 15.11 Capital cost of a 55 000 tons/year incineration plant†

Item	Costs (million ECU)
Receiving facilities including laboratory	1.0
Organic waste shredder	2.6
Stabilization/solidification	1.4
Organic liquid drum emptying	1.6
Oil recovery	2.0
Common facilities	0.7
Incinerator train with WHB	16.0
Steam boiler	5.0
Total estimate	30.0
Contingencies (10%)	3.0
Interest, 10% over 3 years	5.0
Total	38
Investment per tonne/year 1: 350 ECU	

† Costs of setting up the incinerator line with the adjacent pre-preparation units for the waste. 1996 rates.

Table 15.12 Operational and maintenance costs per ton of incineration of hazardous waste for 50 000 tpa†

Item	Units
Depreciation period	10 years
Interest rate	10%
Instalment rate	16%/year
Personnel, 40 000 ECU/yr each	120 personnel
Annual load factor	80%
Operating cost	120 ECU/t
Personnel	60 ECU/t
Capital cost (Table 15.11)	350 ECU/t
Total	530 ECU/t

†1996 rates

and free radicals are formed, producing the complete combustion products of CO_2 and H_2O. Some of the stoichiometric reactions that occur in an incinerator are shown in Table 15.13.

N_2 will form different forms of oxides of N, i.e. NO_2, NO and N_2O, depending on the circumstances (temperature, etc.). NO_2 is reduced in the flue gas by introduction of NH_3 with the following reaction:

$$6NO_2 + 8NH_3 \rightarrow 12H_2O + 7N_2$$

The reaction will demand a catalyst and temperatures $>300\,°C$.

Thermodynamics of incineration   The first law of thermodynamics states that: 'While energy has many forms, and interchanges from one to another, the total amount of energy is constant, i.e. the law of conservation of energy.' The second law of thermodynamics states that: 'Heat is transferred from a region of higher temperature to a region of lower temperature.'

Table 15.13 Some stoichiometric reactions in an incinerator

Elements $+ O_2$ or H_2O $\longrightarrow$		Compound
$C + O_2$	$\longrightarrow$	CO_2
$H_2 + 0.5O_2$	$\longrightarrow$	H_2O
$S + O_2$	$\longrightarrow$	SO_2
$2P + 2.5O_2$	$\longrightarrow$	P_2O_5
$N_2 + 3H_2O$	$\longrightarrow$	$2NH_3 + 0.5O_2$
$Cl_2 + H_2O$	$\longrightarrow$	$2HCl + 0.5O_2$
$F_2 + H_2O$	$\longrightarrow$	$2HF + 0.5O_2$
$Br_2 + H_2O$	$\longrightarrow$	$2HBr + 0.5O_2$

All incineration processes conform to these two laws of thermodynamics. Heat is transferred between mediums as a result of a temperature gradient. A hazardous waste mixture has a heating value that is released during incineration. Each hazardous waste component has a particular heating value (assuming complete heat release). The total heating value of a waste mixture is the summation of the heating value of the individual waste components. The temperature required for incineration can be determined from the net heating value (NHV) of the waste mixture and the following assumptions:

- Adiabatic conditions in the incinerator
- The heat capacity of excess air (EA) of 1.25 kJ/kg °C
- Air requirement of 0.268 L/kJ at standard temperature and pressure (STP)

The net heating value (NHV) for the waste mixture is

$$NHV = 1.25(T - 15)[1 + 0.268(NHV + EA)] \tag{15.6}$$

$$T = 15 + \frac{NHV}{1.25[1 + 0.268(NHV + EA)]} \tag{15.7}$$

where T = the required temperature for incineration, $°C$

Heat transfer of incineration   During incineration it is desirable to optimize the heat transfer which occurs by conduction, convection and radiation. At low temperatures, conduction and convection control the mechanism, while at high temperatures, radiation controls. Convection is the least significant mechanism.

Heat transfer by conduction is represented by

$$Q = -kA\frac{dT}{dx} \tag{15.8}$$

Heat transfer by convection is represented by

$$Q = hA(T_2 - T_1) \tag{15.9}$$

Heat transfer by radiation is represented by:

$$Q = cAE\left[\left(\frac{T_4}{100}\right)^4 - \left(\frac{T_3}{100}\right)^4\right] \tag{15.10}$$

where

Q = rate of heat transfer

k = thermal conductivity

A = cross-sectional area

dT/dx = temperature gradient

h = a film coefficient of heat transfer

c = coefficient of radiation

T_1 = ambient temperature

T_2 = interface temperature

T_3 = absolute temperature of lower temperature element

T_4 = absolute temperature of higher temperature element

E = emissivity, varying with colour and texture of particles

The functional basis for using incineration as a waste treatment process is encompassed in the three Ts:

- Temperature
- Time
- Turbulence

Simply put, proper decineration and total destruction requires a sufficiently high temperature ($>1000\,°C$), an adequate residence time at this temperature and finally a turbulent environment in the incineration chamber. The heat balance in an incineration as given by the WPCF (1990) is

Heat in = heat out

where   Heat in = fuel value of waste

+ heat from auxiliary fuel

+ air pre-heat (air rate × heat capacity × T_{air})

where   pre-heat = primary and secondary combustion air

and   Heat out = heat out from stack

+ heat out in hot ash

+ heat loss (through kiln walls and SCC, boiler and flue gas cleaning treatment)

Heat loss is given as approximately 5 per cent of total heat. Heat from stack = actual gas flow × heat capacity × T_{air}.

An incinerator must be big enough, have adequate volume and have a high enough temperature to allow the waste to undergo complete combustion, before the waste gases and particulates go to a gas cleaning system prior to emission to the atmosphere. Temperatures for the various types of incinerators are described in the following sections. However, temperatures for complete combustion are usually in excess of $1000\,°C$.

The gas residence time (at the elevated temperatures), while typically about two seconds, depends on the incinerator volume in the combustion chamber and on the proximate analysis and energy content of the wastes. The parameter of solids retention time is relevant to rotary kiln incinerators and depends on feed rate, kiln length, diameter, slope and rotation speed.

Oxygen is an essential requirement for combustion and the amount required depends on the reaction stoichiometry, the amount of oxygen in the waste feed and the excess air. Excess air is that supplied in excess of what is needed and is usually expressed as %EA (excess air). Often we express the oxygen content by measuring O_2 in the stack. The O_2 content should then be 8 to 10 per cent.

Turbulence is essential in providing adequate mixing of the wastes with oxygen and to ensure that all the waste achieves the required temperature. The level of turbulence is affected by:

- Rotation speed
- Type of incinerator
- Liquid atomization

15.10.2 Types of Incinerators

The types of incinerators vary widely, depending on the age of the structure and available economics. The five principal types of incinerator technology are:

- Rotary kiln
- Liquid injection
- Plasma arc—a special case of very high temperature treatment
- Wet air oxidation
- Fluidized bed.

Rotary kiln incineration   The rotary kiln is the most common technology for treating hazardous waste for multivariable waste streams from different sources. The processing of raw materials in the solid, semi-solid and liquid states at elevated temperatures using rotary kilns has been used in the industry for many years. It is routinely used in the cement, lime, clay, phosphate, iron ore and coal industries. Traditional cement kilns are now also opening their chambers to be used as hazardous waste incinerators. Rotary kilns provide a number of functions necessary for incineration. They provide for:

- Conveyance of solids
- Mixing of solids
- Containment of heat for heat exchange
- Chemical reactions

They provide the means of ducting the flue gases to the secondary combustion chamber and further on to the flue gas cleaning system. Rotary kilns are equally capable of dealing with bulk solids, sludges, liquids and containerized waste. A typical rotary kiln is shown in Fig. 15.16.

Temperature, time and turbulence   The kiln used for hazardous waste destruction is relatively short at 10 to 12 m with a diameter of about 3.5 to 4.8 m. The temperature in the kiln run, in the slagging mode, is normally 1100 to 1300 °C. The regulatory demands of an incineration system are, however, normally not focused on the temperatures in the kiln itself but on the conditions measured in the secondary combustion chamber (SCC). The temperature prescribed here is 1200 °C at standard conditions. When special waste types are treated, e.g. chlorinated aromatic hydrocarbons, temperatures in the range of 1250 to 1350 °C are essential.

The retention time for the incineration gases in the different parts of the incineration process plays a very important role. In the 'combustion' end of the system, regulations dictate in most cases that the retention time must be at least 2 s at the prescribed temperature in the SCC. At higher temperatures than 1200 °C the time can be reduced. The solids residence time in a rotary kiln, from the WPCF (1988) is given by

$$t_s = \frac{0.19(L/D)}{SR} \qquad\qquad (15.11)$$

where
$$t_s = \text{solids retention time, h}$$
$$L/D = \text{length/diameter ratio} \sim 1:5$$
$$S = \text{slope} \sim 3\text{–}9\%$$
$$R = \text{rotation speed} \sim 0.03\text{–}0.3 \text{ rev/min}$$

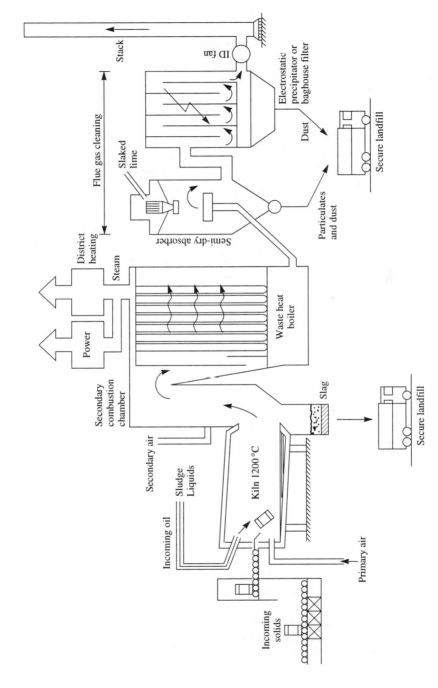

Figure 15.16 Rotary kiln incineration plant. (Adapted from Chemcontrol, Denmark, educational material. Reprinted by permission.)

Typically, the solids retention time is a few hours.

The gas retention time (t_g) for 99.99 per cent destruction of a specific compound is given by:

$$\ln t_g = \ln \frac{9.21}{A} + \frac{E}{RT} \qquad (15.12)$$

where
A = Arrhenius pre-exponent frequency, s^{-1}
E = energy of activation, J/kg mol
R = universal gas constant = 8314 J/kg mol
T = absolute temperature, K

A and E are usually known for a particular compound, so t or T can be calculated. It is of no value to fulfil the requirements of temperature and time if the vaporized waste is never exposed to the right degree of mixing. To satisfy a condition of an adequate turbulent environment the Reynolds number of the fluid flow in the incineration chamber should be greater than 60 000.

Secondary combustion chamber (SCC)   Downstream of the rotary kiln is the secondary combustion chamber (SCC). To ensure that all organic material is completely burned up, additional fuel and secondary combustion air is introduced. Regulations demand that the SCC has normally a 2 second retention time at 1200 °C. The turbulence in the SCC in modern designs is secured by a circular cross-section, where the burners are arranged in a tangential position.

Waste heat recovery boiler   In Europe the normal practice is to install a waste heat boiler (WHB) in the incinerator train. In the WHB the temperature drops from 1050 to 280 °C, utilizing the steam generated for power production or for district heating. The relatively slow cooling of the flue gases while passing through the WHB increases the risk of dioxin reformation. Recent American designs use a shock cooling quench system instead of a WHB. The disadvantage of quenches is that there is no energy recovery and the quench water must be treated before recycling or disposal. However, by using the most modern techniques the formation of dioxins can be reduced to a very low level even without utilizing quenches. No rotary kiln system is, however, better than its flue gas cleaning system, see WPCF (1988) for details.

Liquid injection incineration   In many process industries liquid injection incineration has gained widespread use in dealing with the great variety of liquid waste streams containing minor concentrations of organic contaminants. When treating liquid organic waste types with a high calorific content (even used as support fuel) and aqueous waste containing inorganic salts (ash) and organics, the purpose of the operation is to oxidize the organics and still be able to handle the inorganic part of the waste components in an operationally safe way. In doing so all the normal procedures mentioned for the rotary kiln section applicable. Conditions must therefore be established to secure the necessary waste preparation, mixing and atomization and in the reactor to secure the time, temperature and turbulence exposure, in order to obtain the required destruction and removal efficiency (DRE). Also the downstream flue gas cleaning facilities are in many cases identical with the ones used in the rotary kiln type, since the emission level standards are identical to other high-temperature treatment processes.

In the field of liquid injection incinerators it is very often seen that the waste does not contain the necessary heat value to sustain its own combustion. Waste with a calorific value < 2500 kcal/kg does not burn by itself. In this case, an auxiliary fuel in the form of high calorific value liquid waste, e.g. organic waste solvents or waste oil, or the supply of oil or natural gas is required. The introduction can either be made by mixing the low calorific waste with compounds of a higher calorific value in the feed tanks prior to injection. Another method is the mixing of a high and low calorific 'fuel' in a co-firing burner. The concept of two separate injectors can also be used, one the 'burner' and the other acting only as a 'spraying' nozzle. The conceptual design of a liquid injection incineration system can be made in many ways and a multitude of commercial systems are in operation in industry. Figure 15.17 shows one of the types of liquid injection incinerators.

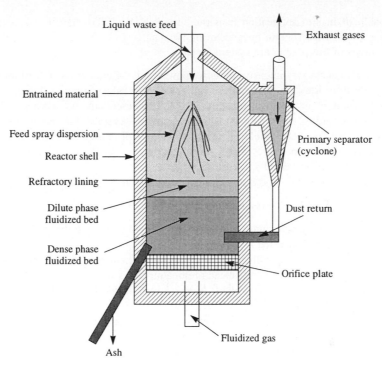

Figure 15.17 Liquid injection incineration system. (Adapted from Chemcontrol, Denmark, educational material. Reprinted by permission.)

Plasma arc destruction incineration   The use of an electrical arc for waste treatment is not a new application. The method has for many years been used within the chemical and metallurgical industry as a tool providing extremely high temperatures, as high as 28 000 °C. In recent times serious attempts have been made to use this method for the treatment of hazardous waste, with the advantage that the waste materials exposed to the very high temperatures are broken down to their elementary components. The most widespread method of plasma generation is the electrical discharge through and along a gas. The type of gas used is relatively unimportant but the choice of gas will of course have some influence on the composition of the compounds created by the process. The systems atmosphere could be oxidizing, reducing or inert, thereby changing its function. For instance, an oxidizing atmosphere is usually needed to destroy organic hazardous waste, while a reducing atmosphere is used to extract metals from ore or to recover metals from dust from electric arc furnaces (EAFs). In a plasma arc reactor, a co-linear electrode arrangement creates an electric arc which is stabilized by a field coil medium through which an electrical current is passed. Passing through the gas, the electrical energy is converted to thermal energy. It is absorbed by the gas molecules, driving them into an ionized atomic state, thereby losing electrons. Ultraviolet radiation is then emitted when the excited molecules or atoms return to their normal stable energy level. The result of this process is an outward electrically neutral plasma consisting of charged and neutral particles reaching the above-mentioned very high temperatures. This very hot plasma will then act as a heat transfer medium to the waste exposed to it. Exposed to the heat, the waste will momentarily be pyrolysed (incinerated without oxygen), atomized and broken down into its elementary components. In theory the waste would be broken down to its most basic elements or at least into very simple molecules, e.g. hydrogen, carbon monoxide, carbon and hydrochloric acid, etc. As with all other high temperature processes the end result, given as the composition of the flue gases, will be highly dependent on the flue gas treatment downstream from the reactor. Rapid development in the applications of the plasma arc is occurring and in recent years a number of commercial concepts are now available. Additionally, this process is capable of in situ application in waste sites (e.g. landfill) where it can atomize the waste, thereby reducing the landfill volume by a factor of 10.

The advantages of the plasma arc system include:

1. The plasma system has a very intensive radiative power and is therefore capable of transferring its heat to the waste (be it in liquid or solid form) much faster than other forms of thermal treatment.
2. The plasma system is a pyrolysis process. It requires practically no oxygen. Consequently the downstream gas cleaning and treatment systems can be designed for much smaller gas streams compared to systems used by standard incinerators. This will reduce the construction costs.
3. Due to the very high temperature used in the process even chlorinated hydrocarbons, which are difficult to destroy, can be treated successfully.
4. In general a DRE (destruction and removal efficiency) of more than 99.999 can be achieved.

The disadvantages of the plasma arc system include:

1. Operating at such high temperatures creates a very high stress on the construction material. A substantial amount of maintenance is required.
2. An arc system is very sensitive to fluctuations in the material balance. This means that great emphasis must be put on the development of advanced pre-treatment systems, which will be able to present the waste to the plasma system in an acceptable homogeneous form. This is of special importance when the waste is from different sources.
3. The creation of the arc consumes a considerable amount of energy. It is therefore of importance to the system that relatively cheap electricity is available. Some form of heat recovery reduces the cost of power, i.e. district heating systems.

A typical plasma arc design is shown in Fig. 15.18.

Wet air oxidation Wet oxidation is an aqueous phase oxidation where organic and inorganic materials suspended, emulsified or dissolved in water are exposed to a gaseous source of oxygen, often air but also pure oxygen. The temperature of operation ranges from 150 to 325 °C with an operational pressure between 2000 kPa and 20 000 kPa, regulating the speed of reaction and controlling the evaporation.

The process shown in Fig. 15.19 may be carried out in a batch reactor, but usually a continuous reactor is used. This reactor can be designed in many ways, but recent concepts use either a horizontal reactor tube which could be as long as 1400 m or a vertical reactor formed as a drilled, cased well going down to as much as 1600 m below ground level. The latter type was briefly disucssed in Chapter 14. The advantage of the deep well concept is that the necessary pressure for the oxidation to take place can be obtained without utilizing high-cost, high-pressure pumps. A disadvantage, however, is that it is not easy to perform maintenance and repairs on the reactor since it is a long way underground. Environmental agencies might be somewhat reluctant to give permits for this type of construction due to fear of undetected leakages deep underground.

Due to the high pressure and the relatively high temperature, the construction material of the reactor must be extremely corrosion resistant and close control must be exercised over the composition of the waste and the inlet rate of the oxygen in order to prevent runaway reactions. An external water-based cooling system is normally also provided. If the waste is in suspension, emulsion or solution with an oxygen demand (OD) in the range of about 20 to 200 g/L, it can be pumped directly into the reactor. More concentrated waste streams must be diluted before treatment. During treatment in a horizontal reactor, oxygen (air or air/oxygen mixtures) are injected at certain positions into the reactor. A continuous oxidation takes place and the majority of the organic compounds are stoichiometrically oxidized according to Table 15.13.

The most resistant compounds are the halogenated aromatics. A typical decomposition rate is only 70 per cent where other organics show a rate of 99 per cent or more under identical conditions. In practice the final oxidation products are thus not entirely CO_2 and H_2O. When a destruction rate of approximately

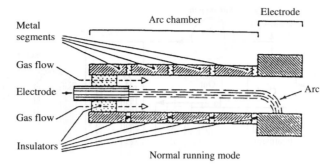

Normal running mode

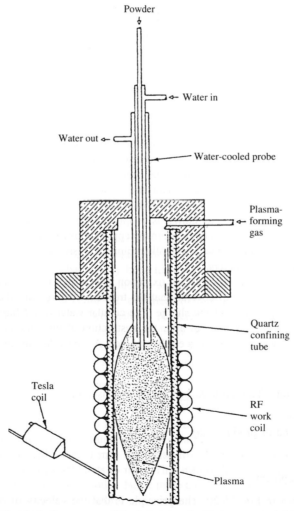

The single-flow, vortex-stabilized plasma torch with the water-cooled
probe used for powder injection

Figure 15.18 Plasma arc incineration system. (Adapted from Chemcontrol, Denmark, educational material. Reprinted by permission.)

15.10.3 Flue Gas Cleaning Systems for Incinerators

A very significant pollutant potential of the incineration process is the emission gases to the atmosphere. Traditionally, incinerators had a poor record in this area. Nowadays, the technology does exist to come close to 'zero' emissions. This is achievable by sophisticated flue gas cleaning systems. The cleaning procedures will normally utilize both mechanical and chemical processes.

Mechanical methods   These include:

- The electrostatic filter
- The baghouse filter
- Activated carbon

Electrostatic filter   The gases pass a number of ducts formed by two vertically placed collecting plates with a set of electrodes placed between them. The high negative voltage applied to the electrodes creates a strong electrical field between the plates and the electrodes, resulting in an electrical breakdown of the gases near to the electrodes. This phenomenon is called the 'corona'. The corona produces a large number of gas ions. The positive ions will immediately be attracted to the electrodes. The dust particles suspended in the gas attract themselves to the negative ions, which migrate towards the collection plates. Here they are unloaded and build up a layer of dust which is dislodged by 'rapping'. The filters are used to remove particles from 0.05 to 200 μm. The pressure drop across the filter is quite small and it can operate with relatively high gas temperatures.

Baghouse filter   The use of baghouse filters has a long history. A baghouse filter generally is any porous structure composed of fibrous material which tends to retain suspended particulates as the carrier gas passes through the voids of the filter. Cleaning of the filters normally takes place by mechanical vibration or shaking, pulsating jets or reverse air flow. The filter material can be made of many different materials, e.g. wool, cotton, nylon, polyesters, aromatic polyamides, teflon, glass fibres or metal web, depending on the actual application, temperature and the desired efficiency. Fabric filters can remove particles down to 0.01 μm with reasonable operating pressure drops and power requirements. One of the main advantages is the high efficiency over a broad range of particle sizes.

Other mechanical means   Cyclones have been used, but the efficiency is very low compared to baghouse or electrostatic filters. See Chapter 16 for further details.

Activated carbon   Activated carbon and special activated carbon can be used as an adsorber material for a range of non-polar substances, i.e. NH_3, SO_2, NO_x, H_xC_y and some trace elements, e.g. Hg. Activated carbon has a very porous structure and thus a very large surface area. During operation activated carbon is polluted and after a certain period it requires regeneration or it must be disposed of either by combustion or to landfill. The efficiency of the filter depends on the gas velocity and the temperature. The higher the temperature the less efficiency. In recent projects activated carbon filters are used as the last unit in the flue gas cleaning train, acting as a polishing filter capturing dioxins.

Chemical methods: Chemical flue gas cleaning may, broadly, be divided into four main processes:

- Dry gas cleaning
- Semi-dry cleaning
- NO_x reduction
- Wet cleaning

Dry gas cleaning   The flue gas after leaving the SCC (secondary combustion chamber) enters a reactor in which lime hydrate powder, $Ca(OH)_2$, will be injected. The acidic compounds HCl, HF and SO_2 react with the lime forming $CaCl_2$, CaF_2, $CaSO_3$ and $CaSO_4$, as shown in the following stoichiometric equations:

$$Ca(OH)_2 + 2HCl \rightarrow CaCl_2 + 2H_2O \tag{15.13}$$

$$Ca(OH)_2 + 2HF \rightarrow CaF_2 + 2H_2O \tag{15.14}$$

$$Ca(OH)_2 + SO_2 \rightarrow CaSO_3 + H_2O \tag{15.15}$$

$$Ca(OH)_2 + SO_3 \rightarrow CaSO_4 + H_2O \tag{15.16}$$

The reaction temperature is typically 140 to 220 °C. The optimum reaction speed is obtained in the presence of water, which is also used to control the temperature in the reactor. Normally the stochiometric proportion is 2:1 between the lime and the acidic gases. The lime is utilized most efficiently when a baghouse filter is used downstream of the reactor. The surplus lime acts as a filter coating and this secures an extra contact between the flue gas and the lime. Also the content of Hg compounds, being in the particulate or gaseous state, is reduced due to the extended contact time. The most efficient temperature is below 150 °C. A part of the waste product from the baghouse filter may be recirculated to the reactor for better utilization of the lime.

Semi-dry cleaning   The chemical reactions are identical to the ones mentioned for dry cleaning, but instead of injecting lime as a powder, lime slurry is injected as an aerosol formed by a fast rotating atomizer. The degree of atomization is set by the peripheral speed of the atomizer head. The lime/water slurry is prepared in a slaker house. The waste content of the slurry must fit exactly to the amount evaporating during the reaction between the acidic compounds in the flue gas and the lime to ensure a dry end product extracted from the bottom of the reactor.

NO_x reduction   Removal of NO_x (NO_2, NO, N_2O) may also take place in a selective catalytic reactor (SCR) in which NO_x in the presence of a catalyst reaction with NH_3 forms N_2 and H_2O as follows.

$$NO + 2/3 \ NH_3 \rightarrow 5/6 \ N_2 + H_2O \tag{15.17}$$

$$NO_2 + 4/3 \ NH_3 \rightarrow 7/6 \ N_2 + 2H_2O \tag{15.18}$$

$$2N_2O + 4/3 \ NH_3 \rightarrow 8/3 \ N_2 + 2H_2O \tag{15.19}$$

Experience shows that the stochiometric proportion between NH_3 and NO_x is about 1:1. The catalytic reaction requires a temperature between 300 and 400 °C. The catalyst may be inactivated by being poisoned by sulphur and the SCR process will therefore take place after the scrubber chamber. Research work is ongoing in order to try to reduce the reaction temperature. Further, it has been shown that the catalyst used for this process is also active in reducing the level of dioxins.

Wet scrubbing   Before the wet scrubbing process the flue gases pass a filter, often an electrostatic filter, and the gases during the process are cooled to about 60 to 100 °C. The scrubbing process will typically take place in several stages. The first stage is a water rinsing, often called a quench, where the flue gases are partly washed and partly cooled down. The fast cooling of the gases will, among other things, reduce the time in which the gases dwell within the critical zone for a reformation of chlorinated dioxins (range 800 to 250 °C). The strong acids, HCl and HF, formed in the combustion zone of the incinerator are absorbed in the water and the pH in the rinsing water adjusts to 0.5 to 1.0. During this stage SO_2 is not absorbed because of the low pH level. The second stage works with water, with a pH level > 7. This level is reached by adding a mixture of $CaCO_3$, $Ca(OH)_2$ and NaOH. The trend is to use NaOH instead of lime, even if NaOH is more expensive than $Ca(OH)_2$, as lime is likely to give undesirable precipitations. In addition to the reactions mentioned under the 'dry cleaning' heading, the following reaction also takes place:

$$2NaOH + SO_2 \rightarrow Na_2SO_3 + H_2O \tag{15.20}$$

The wet cleaning processes produce acidic and neutral/alkaline wastewater in which considerable amounts of trace elements, especially mercury (Hg), are retained and the water must therefore be treated before being discharged to the sewer system. The normal treatment consists of a neutralization and heavy metal precipitation by adding lime, $Ca(OH)_2$. In some cases Na_2S in solution can be used as a precipitation compound. After the precipitation the water must be filtered. The residue from the filtration is landfilled after stabilization. The stoichiometric proportion is 1:1 between the addition of lime, limestone, sodium hydroxide and the acid gases.

Wet scrubber chambers are normally used with one or more scrubber tower(s), either packed or empty, in which water or $NaOH/Ca(OH)_2$ slurry is injected, or with a venturi scrubber with throat injection. Often the venturi scrubber acts as the quenching section followed by a scrubbing tower. In the case of wet scrubbing the cleaned flue gases are reheated before passing the stack. The wet cleaning process may be modified by using a semi-dry absorber section as the last stage, resulting in a dry end product so that the wastewater treatment may be avoided.

Venturi scrubbers Venturi scrubbers are normally used for the removal of particles from flue gas streams, but in connection with an incinerator process it can be used for two other purposes:

- As a quench, cooling the flue gas cools down quickly through the so-called 'temperature risk zone' where the formation of halogenated dioxins and dibenzofuranes are likely to be formed (at 800 to 200 °C)
- Removal of acid compounds such as SO_2 and HCl from the flue gas

It is important for the efficiency of the scrubber system to have good atomization of the spray water to obtain the maximum contact with the gas. Despite the fact that the scrubber operates in a concurrent mode, the sought-after efficiency can only be obtained at the expense of a relatively large gas-side pressure drop and a consequent larger power consumption when compared with a spray absorber. There are many designs of a venturi scrubber. The one shown in Fig. 15.21 is an ejector venturi design. The washwater must be supplied in sufficient quantities partly to make up for water loss due to evaporation

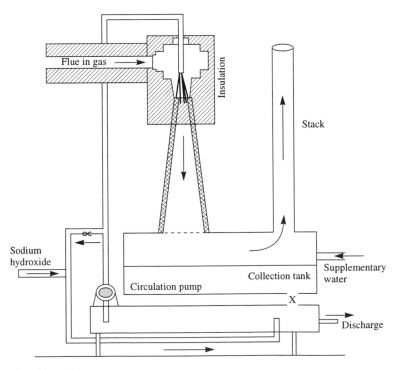

Figure 15.21 Venturi scrubber system.

and partly to absorb the acid compounds of the flue gas. For some minor part the scrubber is acting also as a gas mover, although it is not very efficient for that purpose. In an incineration line the ID fan (induced draft fan) will do the major part of the gas transfer. The function as a quench results in the formation of relatively large amounts of water vapour. The volume of this vapour will add to the cost of further downstream equipment, as it must be able to handle this increased volume. The scrubber effluent must be treated before discharge in a wastewater treatment plant as heavy metals, halogens and sulphur dioxide are dissolved in the water.

15.10.4 Emission Standards for Incinerators

The emission polluting substances from an incineration plant of principle concern are:

- Organic compounds
- HCl—chlorine compounds
- HF—fluorine compounds
- Heavy metals
- Halogenated dioxins and halogenated furans

Other emission substances include:

- Dust
- Carbon monoxide
- Sulphur dioxide
- Nitrogen oxides (NO_x)

The objectives of setting emission standards in the EU Directive on Incineration of Hazardous Waste (1994) are:

- To minimize environmental damage caused by incineration of hazardous waste
- To impede the flow of waste to lower cost incinerators
- To promote and ensure a reduction in waste movement

As incineration of hazardous waste is a rather new area, (with not more than two decades of experience), most countries do not have fixed levels for maximum permissible emissions. If regulatory levels do exist, they are often given in connection with incineration in general, with the opportunity of issuing special regulations with stricter emission levels for the individual hazardous waste incinerator. Table 15.14 presents the emission standards for several countries but it is very difficult to compare the figures directly, as their terms of reference may not be the same. Average levels are not necessarily expressed for the same time period. Some of the figures are given for incineration of household waste only. Others are specified for hazardous waste incineration only. Some of the figures are regulated by legislation, while others are given for a specific plant. The European Union has set stringent emission standard levels for incinerators.

15.10.5 Monitoring of Emissions

As per the EU Directive (94/67/EC) (1994) on the incineration of hazardous waste, the parameters to be monitored are listed in Table 15.15. A comparison of some emission standards in EU member states are shown in Table 15.16.

Table 15.14 Emission standards after flue gas cleaning in different countries

Country (a)	Dust (mg/m³)	CO (mg/m³)	SO₂ (mg/m³)	NOₓ (mg/m³)	HCl (mg/m³)	HF (mg/m³)	TOC (mg/m³)	Dioxin (b) (mg/m³)	Metals (mg/m³)
EEC 1991 (1)	5/10 (2/3)	50/150 (2/3)	25/50 (2/3)		5/10 (2/3)	1/2 (2/3)	5/10 (2/3)	0.1 (5)	$Hg < 0.05$, $Cd + Ti < 0.05$ (4); $Sb + As + Pb + Cr + Co + Cu + Mn + Ni + V + Sn < 0.5$ (4)
EEC-househ 1989	30	100	300		50	2	20		$Hg + Cd + Ti < 0.2$, $Ni + As < 1$, $Pb + Cr + Cu + Mn < 5$
DK-KKIV 1989 (7)	30 (8)	100 (9)	300 (9)	200 (9)	100 (8)	2 (9)	20 (9)		$Pb < 1.4$, $Hg + Cd + Ti + As + Cr + Ni < 0.2$ (9)
DK-KKIII 1982 (10)	100	150	750	300	300	5	300		
DK-KKI 1975 (10)	150	19000	2200	300	600	5			
NL 1990 (1,6)	5	50	40	70	10	1	10	0.1	$Hg < 0.05$, $Cd < 0.05$, total heavy metals < 1.0
D-BimSchV 1990 (1)	10/30 (2/3)		50/200 (2/3)	200/400 (2/3)	10/60 (2/3)	1/4 (2/3)	10/20 (2/3)	0.1 (2)	$Hg < 0.05$, $Cd + Ti < 0.05$ (2); $Sb + As + Pb + Cr + Co + Cu + Mn + Ni + V + Sn < 0.5$ (2)
D-TA 1986 (1)	30	100	100		50	2	20	0.5	$Hg + Cd + Ti < 0.2$, $As + Co + Ni + Se + Te < 1$
S-SAKAB 1989 (7)	10		100		50	3		1.0	$Sb + Pb + Cr + CN + F + Cu + Mn + Pt + Pd + Ph + V + Sn < 5$
SF-Ecochem 1989 (6,11)	10	70	200	200	20	2	20	0.1	$Hg < 0.03$, $Cd < 0.05$
A-midle 1989 (1,2)	20	50	100	300	15	0.7	20	0.1	$Hg < 0.1$, $Cd < 0.05$, $As + Co + Ni < 1$, $Pb + Zn + Cr < 3.0$
CH 1985 (1)	50	*	*	*	*	*	*		$Hg < 0.1$, $Cd < 0.1$, $Pb + Zn < 5$; $Hg + Cd + Tl + Sb < 3$
Hong Kong 1989 (11)	75	150	750	500	38	7.5	35	0.1	$Pb + Cu + As + Ni + Cr < 5$, total heavy metals < 10
I < 3 t/h 1990	30	100	300		50	2	20	4	$Hg + Cd + Ti < 0.2$, $Se + Te < 1$; $Sb + Cd + Cr + Mn + Pd + Pb + Pt + SiO2 + Cu + Rh + Sn + V < 5$

Table 15.14 Emission standards after flue gas cleaning in different countries (*Continued*)

Country (a)	Dust (mg/m³)	CO (mg/m³)	SO2 (mg/m³)	NOₓ (mg/m³)	HCl (mg/m³)	HF (mg/m³)	TOC (mg/m³)	Dioxin (b) (ng/m³)	Metals (mg/m³)
I > 3 t/h 1990	100	100	300	100	100	4	20	4	$Hg + Cd + Tl < 0.2$, $Se + Te < 1$ $Sb + Cd + Cr + Mn + Pd + Pb + Pt + SiO2 + Cu + Rh + Sn + V < 5$ $Hg < 10$, $Cd < 15$, $Pb + Sb + As < 25$, $Zn + Cu < 100$
Malaysia 1978	400		200	2000†	400	100			
British Columbia no information to date	20	55	180	380	50	4	32		$Tl + Cd + Hg < 0.15$, $As + Cr + Co + Ni + Se + Te < 0.7$ $Pb + Sb + Cu + Mn + V + Zn < 3.6$

(a)

EEC	The new regulations in the EEC
EEC-househ	ECC rules for household incinerators
DK-KKIV	Denmark, regulatory levels for Kommunekemi incinerator FIV
DK-KKIII	Denmark, regulatory levels for Kommunekemi incinerator III
DK-KKI	Denmark, regulatory levels for Kommunekemi incinerator I
NL	Regulatory levels, the Netherlands
D-BimSchV	Regulatory levels in Germany
D-TA	Regulatory levels according to TALuft, Germany
S-SAKAB	Regulatory levels, SAKAB, Sweden
SF-Ecochem	Regulatory levels, Ecochem, Finland
A-midle	Regulatory levels, ESB, Vienna, Austria
CH	Regulatory levels, Switzerland
Hong Kong	Regulatory levels, Hong Kong
I < 3 t/h	Regulatory levels, Italy
I > 3 t/h	Regulatory levels, Italy
Malaysia	Regulatory levels, Malaysia (to be updated)
British Columbia	Regulatory levels, British Columbia, Canada

(b) Expressed in toxic equivalents 2,3,7,8 TCDD.
  (1) 11% O_2, 273 K, 1 atm, dry gas.
  (2) Daily average values.
  (3) Half-hourly average values.
  (4) All average values over the sample period of a minimum of 1 hour and a maximum of 4 hours.
  (5) All average values over the sample period of a minimum of 6 hours and a maximum of 16 hours.
  (6) Hourly average values.
  (7) 10% O_2, 273 K, 1 atm, dry gas.
  (8) Monthly average values.
  (9) Yearly average values.
  (10) 7% CO_2, 273 K, 1 atm, dry gas.
  (11) 12% O_2 273 K, 1 atm, dry gas.
  (12) 10 minutes average values.
  * Values not specified separate.
  † 2000 mg/N m³ of SO_3.

Table 15.15 Monitoring parameters for incinerator plant

Flue gas temperature in the ...	Concentrations in flue gas of ...	Combustion efficiency and emission parameters	Particulate emissions, metals and acid gases, parameters
Kiln room	O_2	NO_x	Particulate matter
SCC	CO	SO_2	Hg, Cd
Boiler	HCl	Suspended matter	As, Ni, Pb, Cr, Cu, V
Before the filter	Suspended matter	Dust	Cr, Cu, V
After the filter	Dust	TOC	HCl, HF, HBr
			SO_2, SO_3
			TOC
			PCB, PCT
			Halogens and H_2S
			Dioxins and dibenzofurans
			P compounds
			Odorous compounds

Table 15.16 Some emission standards for hazardous waste incinerators

Pollutant	Range* (mg/m³)	EU Directive* (mg/m³)	Germany* (mg/m³)	Netherlands* (mg/m³)
Total dust particles	5–30	10	10	5
Total organic carbon	5–20	10	10	10
Inorganic chlorine compounds		10	10	10
Inorganic fluorine compounds		1	1	1
Sulphur oxides		15	50	40
Carbon monoxide		50	50	50
Cadmium ⎫		0.05	0.05	
Thallium ⎭			0.05	0.05
Mercury		0.05	0.05	0.05
Other heavy metals Total		0.5	0.5	1
Nitrogen oxides	~200			
Dioxins and furans**	0.1 ng/m³	0.1 ng/m³	0.1 ng/m³	0.1 ng/m³

15.11 HANDLING OF TREATMENT PLANT RESIDUES

From Figs. 15.9 and 15.22, it is seen that from Kommunekemi the final residues are:

From the incinerator:

- Slag from the rotary kiln (RK)
- Dust from the waste heat boiler (WHB)
- Reaction products from the chemical flue gas cleaning
- Dust from the baghouse/textile filters (BHF)

From the inorganic plant:

- Filter cakes
- Wastewater

The waste oil plant produces no waste *per se* as all of it is forwarded to the incinerator. Figure 15.22 shows an estimated materials balance for Kommunekemi.

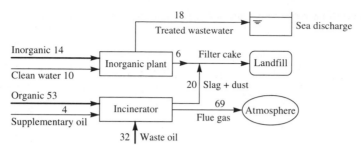

Figure 15.22 Material balance of residues from Kommunekemi, Denmark. (Adapted from Palmark, 1986, with permission of Chemcontrol A/S, Denmark.)

The input to the plant is 100 units (about 100 000 tons/yr) divided as 32 per cent waste oil, 53 per cent as organic waste and 14 per cent as inorganic waste. There is < 2 per cent of special waste that goes to a salt mine for immediate disposal. The heavy lines in Fig. 15.22 are the input waste streams. Additionally, about 10 units of clean water are added to the inorganic wastewater treatment plant and about 4 units of supplementary oil are added to the incinerators. It is seen from Fig. 15.22 that most of the effluent from the plant is in the form of flue gas while smaller fractions are of wastewater and slag/dust and filter cake. The wastewater is treated in a conventional industrial-type wastewater plant. The flue gas is cleaned before emission to the atmosphere. The solid/sludge waste goes to landfill.

Before the solid waste can go to landfill, it must pass the USEPA leachate test. This is described in USEPA CFR 40 Appendix 11, part 261. Only products satisfying this test can go to landfill. If they fail the test, they are required to be further treated. Some of the contaminants and their regulating loads are shown in Table 15.17. Experience at Kommunekemi indicates that operating the rotary kilns in the slagging mode, produces a slag that passes the leachate test. The dust from the WHB and reaction products from the semi-dry absorber (SDA) and from the baghouse/textile filter are collected in one system and require further treatment to pass the leachate test. The filter cakes from the inorganic plant also require further treatment.

15.11.1 Secure Landfill

Secure landfills (residual repositories) are the disposal sites for the filter cakes, slag and dust from a full treatment facility. The design of a secure landfill is similar to that of an MSW landfill (described in Chapter 14). The principal phases in the overall design for a secure landfill are:

- Preliminary assessment
- Site location
- Design of landfill
- Construction of landfill
- Operation
- Re-establishment of original landscape
- Closure
- Post-closure controls

As there is almost no organic material in hazardous waste landfills, there is likely to be little gas generation. However, leachate generation does occur and it may be contaminating. The practice in Denmark was to construct the landfill close to the coast, where a leachate breakthrough would not damage groundwater/freshwater. Figure 15.23 shows a schematic of a secure landfill with a double liner system. Figure 15.24 shows in more detail the type of construction.

Table 15.17 Toxicity characteristics contaminants and regulatory levels—USEPA leachate test requirements

EPA hazardous waste number	Contaminant	Chronic toxicity reference level (mg/L)	Basis†	Regulatory level (mg/L)‡
D004	Arsenic	0.05	MCL	5.0
D005	Barium	1.0	MCL	100.0
D018	Benzene	0.005	MCL	0.5
D006	Cadmium	0.01	MCL	1
D019	Carbon tetrachloride	0.005	MCL	0.5
D020	Chlordane	0.0003	RSD	0.03
D021	Chlorobenzene	1	RfD	100.0
D022	Chloroform	0.06	RSD	6.0
D007	Chromium	0.05	MCL	5.0
D023	o-Cresol	2	RfD	200.0*
D024	m-Cresol	2	RfD	200.0*
D025	p-Cresol	2	RfD	200.0*
D026	Cresol	2	RfD	200.0*
D016	2,4-D	0.1	MCL	10.0
D027	1,4-Dichlorobenzene	0.075	MCL	7.5
D028	1,2-Dichloroethane	0.005	MCL	0.5
D029	1,1-Dichloroethylene	0.007	MCL	0.7
D030	2,4-Dinitrotoluene	0.000 5	RSD	0.13*
D013	Endrin	0.000 2	MCL	0.02
D031	Heptachlor (o)	0.000 08	RSD	0.008
D032	Hexachlorobenzene	0.000 2	RSD	0.13*
D033	Hexachloro-1,3-butadiene	0.005	RSD	0.5
D034	Hexachlorethane	0.03	RSD	3.0
D008	Lead	0.05	MCL	5.0
D013	Lindane	0.004	MCL	0.4
D009	Mercury	0.002	MCL	0.2
D014	Methoxychlor	0.1	MCL	10.0
D035	Methyl ethyl ketone	2	RfD	200.0
D036	Nitrobenzene	0.02	RfD	2.0
D037	Pentachlorophenol	1	RfD	100.0
D038	Pyridine	0.04	RfD	5.0*
D010	Selenium	0.01	MCL	1.0
D011	Silver	0.05	MCL	5.0
D039	Tetrachloroethylene	0.007	RSD	0.7
D015	Toxaphene	0.005	MCL	0.5
D040	Trichlorethylene	0.005	MCL	0.5
D041	2,4,5-Trichlorophenol	4	RfD	400.0
D042	2,4,6-Trichlorophenol	0.02	RSD	2.0
D017	2,4,5-TP (Silvex)	0.01	MCL	1.0
D043	Vinyl chloride	0.002	MCL	0.2

† MCL = maximum contaminant level or National Interim Primary Drinking Water Standard; RSD = risk-specific dose; RfD = reference dose.
‡ The regulatory level equals the chronic toxicity level times a dilution/attenuation factor (DAF) of 100, unless otherwise noted (shown by *).

15.11.2 Stabilization (Solidification)

Should a waste ready for landfilling not conform to the leachate test (TCLP) further treatment is necessary. Waste stabilization procedure is one possible option. A large number of stabilization systems, (many proprietary products), are based on some form of silicate chemistry. A careful selection of the offered systems should be performed before selection. Some of the selection criteria could be:

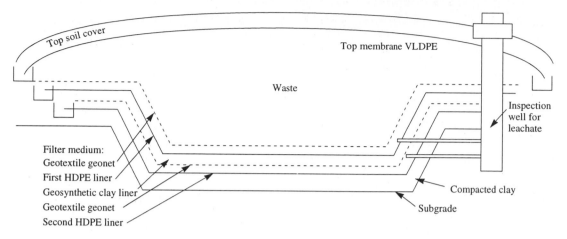

Top soil cover

Top membrane VLDPE

Waste

Inspection well for leachate

Filter medium:
Geotextile geonet
First HDPE liner
Geosynthetic clay liner
Geotextile geonet
Second HDPE liner

Compacted clay

Subgrade

Figure 15.23 Double liner system for secure landfill.

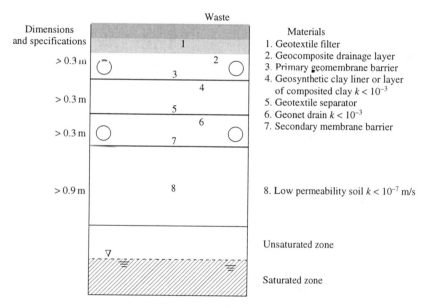

Waste

Dimensions and specifications

> 0.3 m

> 0.3 m

> 0.3 m

> 0.9 m

Materials

1. Geotextile filter
2. Geocomposite drainage layer
3. Primary geomembrane barrier
4. Geosynthetic clay liner or layer of composited clay $k < 10^{-3}$
5. Geotextile separator
6. Geonet drain $k < 10^{-3}$
7. Secondary membrane barrier

8. Low permeability soil $k < 10^{-7}$ m/s

Unsaturated zone

Saturated zone

Figure 15.24 Schematic of components of double liner system for secure landfill.

1. There should be ready availability of the stabilization materials in sufficient quantities.
2. The stabilization materials should be comparatively cheap.
3. The quantity of stabilization material necessary to perform the desired reaction should be minimal so as not to increase the total volume of material to be disposed of.
4. Sensitivity to certain types of 'contamination' should be minimal, i.e. copper, organic material, etc.

The stabilization procedure is then a fairly simple process where all the ingredients are mixed in a continuous mixer. The final grout is transferred to a prepared landfill cell where the grout is allowed to solidify into a rock-like structure. Samples are taken over time to ensure that the process is running according to design and that the final result will conform with the regulatory procedures. A flow diagram for a stabilization procedure is shown in Fig. 15.25.

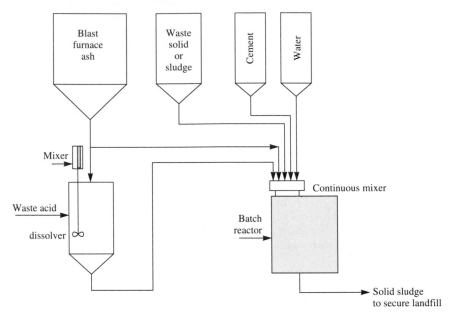

Figure 15.25 Component flow chart of solidification plant.

15.12 CONTAMINATED SITES

A site is designated contaminated only after a very thorough investigation. From this investigation, the type and extent of contamination is determined (size, nature of contaminants, quantities, etc.). The difficulty in making this evaluation is to define what would be the normally acceptable level of contamination left after remediation, especially regarding heavy metal contamination. This normal level will in most cases depend very much on the future use of the site and also on the risk for further contamination of groundwater. Site contamination can be divided into the following classes:

- Soil contaminated with mineral oil or mineral oil substances
- Soil contaminated with halogenated solvents
- Soil contaminated with PCB, PCT, dioxins, etc.
- Soil contaminated with gas cleaning compounds from old city gas works
- Soil contaminated with heavy metals (e.g. Cd, Pb, Hg, Cr, Ni, As, Cu)

Many sites show a mixture of these classes and this list is not comprehensive.
    The process involved in a feasibility study of a contaminated site includes:

- Site survey to establish historical uses of site
- Establishing the *in situ* soil type, classification, particle size analysis, porosity, bulk density, level of concentration, etc.
- Delineation of extent of soil contamination
- Delineation of groundwater/soil water contamination by boreholes
- Assess contaminants from soil samples
- Determine magnitude of each contaminant in soil samples
- Assess contaminants in groundwater samples
- Determine magnitude of contaminants in groundwater
- Assess contaminants in the local air environment due to the contaminated site

- Determine the level of air contaminants
- Determine the extent of contaminant movement by boreholes 'downstream' of the contaminated site

With the above information the following may be established:

- Level of contamination
- Extent of contamination
- The groundwater contaminant plume, its current status and future tracking, which may be determined by using computer modelling

Knowing the current status of contamination and the possible future contamination, remediation methods and costs can be addressed and may include:

- Abandoning the site and closing off
- Removing soil and transporting to an incineration plant for decontamination or other thermal treatment option
- *In situ* bioremediation, using microbes to decontaminate the soil (used for more than just hydrocarbon cleanups)
- Extraction of contaminated groundwater and replacing with clean water

The level of decontamination and cleanup will, of course, depend on the acceptable standards, costs, ecology and future use. The reader is referred to Bellandi (1988) for further details.

For the remedial action of sites a multitude of different processes are available but two methods are most common:

- Thermal treatment
- Bioremediation

15.12.1 Thermal Treatment

This is a non *in-situ* method and is suitable to be used with organic contamination (including oil contamination). The process utilizes the principle of heating the contaminant up to its boiling point, where it will transfer into gas form. This gas can then be incinerated in the conventional way. The end temperature in the reactor is normally not exceeding 800 to 900 °C, thus leaving the physical structure of the soil undisturbed so that the treated soil can be used as normal soil. The thermal treatment can either be performed in a rotary kiln or in special reactors. One such special reactor is shown in Fig. 15.26, where the process is divided into two steps:

Step 1: heating up to approximately 320 °C
Step 2: final heating to 700 to 800 °C and as far as 1100 °C for particular contaminants

15.12.2 Bioremediation

All organic materials are prone to biodegradation at varying rates. This fact is used to decompose organic contaminants. In principle there are two basic processes:

- Aerobic degradation
- Anaerobic degradation

Both principles utilize the natural bacterial population in soils or by adding selected bacterial strains tailor-made for the specific contaminant in question. Optimal environmental conditions are achieved for the bacteria to grow, breaking down the contaminant in doing so. The right temperature, humidity, level of available oxygen and basic nutrients, should be present. Regarding nutrients, the level should be only so high to force the bacteria to utilize the contaminant as a foodstuff. The biological processes are slow,

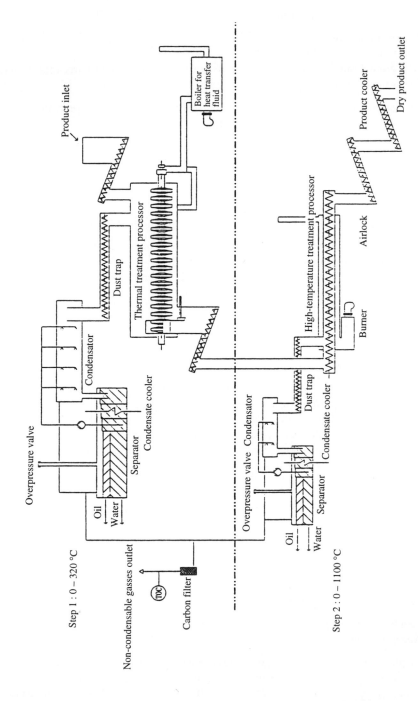

Product inlet

Boiler for
heat transfer
fluid

Overpressure valve

Condensator

Dust trap

Thermal treatment processor

High-temperature treatment processor

Separator

Condensate cooler

Product cooler

Airlock

Dry product outlet

Oil

Water

Burner

Step 1 : 0 – 320 °C

Non-condensable gasses outlet

Overpressure valve   Condensator

Dust trap

Condensate cooler

Separator

TOC

Carbon filter

Oil

Water

Step 2 : 0 – 1100 °C

Figure 15.26 Soil recovery high-temperature treatment plant. (Adapted from Chemcontrol, Denmark, educational material. Reprinted by permission.)

compared to thermal processes, but the investment in equipment is low, so if it is possible to wait for the process to finish it is more economical than thermal treatment, see Wentz (1995) for details.

15.13 EU HAZARDOUS WASTE DIRECTIVE (91/689/EEC) ANNEXES I, II, III

15.13.1 Annexe I

Categories or generic types of hazardous waste listed according to their nature or the activity which generated them (waste may be liquid, sludge or solid in form)

Annexe I.A Wastes displaying any of the properties listed in Annexe III and which consist of:

1. Anatomical substances; hospital and other clinical wastes
2. Pharmaceuticals, medicines and veterinary compounds
3. Wood preservatives
4. Biocides and phyto-pharmaceutical substances
5. Residue from substances employed as solvents
6. Halogenated organic substances not employed as solvents excluding inert polymerized materials
7. Tempering salts containing cyanides
8. Mineral oils and oily substances (e.g. cutting sludges, etc.)
9. Oil/water, hydrocarbon/water mixtures, emulsions
10. Substances containing PCBs and/or PCTs (e.g. dielectrics, etc.)
11. Tarry materials arising from refining, distillation and any pyrolytic treatment (e.g. still bottoms, etc.)
12. Inks, dyes, pigments, paints, lacquers, varnishes
13. Resins, latex, plasticizers, glues/adhesives
14. Chemical substances arising from research and development or teaching activities which are not identified and/or are new and whose effects on man and/or the environment are not known (e.g. laboratory residues, etc.)
15. Polytechnics and other explosive materials
16. Photographic chemicals and processing materials
17. Any material contaminated with any congener of polychlorinated dibenzo-furan
18. Any material contaminated with any congener of polychlorinated dibenzo-p-dioxin

Annexe I.B.
19. Animal or vegetable soaps, fats, waxes
20. Non-halogenated organic substances not employed as solvent
21. Inorganic substances without metals or metal compounds
22. Ashes and/or cinders
23. Soil, sand, clay including dredging spoils
24. Non-cyanidic tempering salts
25. Metallic dust, powder
26. Spent catalyst materials
27. Liquids or sludges containing metals or metal compounds
28. Residue from pollution control operations (e.g. baghouse dusts, etc.) except 29, 30 and 33
29. Scrubber sludges
30. Sludges from water purification plants
31. Decarbonization residue
32. Ion-exchange column residue
33. Sewage sludges, untreated or unsuitable for use in agriculture
34. Residue from cleaning of tanks and/or equipment

35. Contaminated equipment
36. Contaminated containers (e.g. packaging, gas cylinders, etc.) whose contents included one or more of the constituents listed in Annexe II
37. Batteries and other electrical cells
38. Vegetable oils
39. Materials resulting from selective waste collections from households and which exhibit any of the characteristics listed in Annexe III
40. Any other wastes which contain any of the constituents listed in Annexe II and any of the properties listed in Annexe III

15.13.2 Annexe II

Constituents of the wastes in Annexe I.B which render them hazardous when they have the properties described in Annexe III

Wastes having as constituents:

C1 Beryllium; beryllium compounds
C2 Vanadium compounds
C3 Chromium (VI) compounds
C4 Cobalt compounds
C5 Nickel compounds
C6 Copper compounds
C7 Zinc compounds
C8 Arsenic; arsenic compounds
C9 Selenium; selenium compounds
C10 Silver compounds
C11 Cadmium; cadmium compounds
C12 Tin compounds
C13 Antimony; antimony compounds
C14 Tellurium; tellurium compounds
C15 Barium compounds, excluding barium sulphate
C16 Mercury; mercury compounds
C17 Thallium; thallium compounds
C18 Lead; lead compounds
C19 Inorganic sulphides
C20 Inorganic fluorine compounds, excluding calcium fluoride
C21 Inorganic cyanides
C22 The following alkaline or alkaline earth metals: sodium, potassium, calcium, magnesium in uncombined form
C23 Acidic solutions or acids in solid form
C24 Basic solutions or bases in solid form
C25 Asbestos (dust and fibres)
C26 Phosphorus; phosphorus compounds, excluding mineral phosphates
C27 Metal carbonyls
C28 Peroxides
C29 Chlorates
C30 Perchlorates
C31 Azides
C32 PCBs and/or PCTs

C33 Pharmaceutical or veterinary compounds

C34 Biocides and phyto-pharmaceutical substances (e.g. pesticides, etc.)

C35 Infectious substances

C36 Creosotes

C37 Isocyanates; thiocyanates

C38 Organic cyanides (e.g. nitriles, etc.)

C39 Phenols, phenol compounds

C40 Halogenated solvents

C41 Organic solvents, excluding halogenated solvents

C42 Organohalogen compounds, excluding inert polymerized materials and other substances referred to in this Annexe

C43 Aromatic compounds; polycyclic and heterocyclic organic compounds

C44 Aliphatic amines

C45 Aromatic amines

C46 Ethers

C47 Substances of an explosive character, excluding those listed in this Annexe

C48 Sulphur organic compounds

C49 Any congener of polychlorinated dibenzo-furan

C50 Any congener of polychlorinated dibenzo-p-dioxin

C51 Hydrocarbons and their oxygen, nitrogen and/or sulphur compounds not otherwise taken into account in this Annexe.

15.13.3 Annexe III

Properties of wastes which render them hazardous

Il1 'Explosive': substances and preparations which may explode under the effect of flame or which are more sensitive to shocks or friction than dinitrobenzene

H2 'Oxidizing': substances and preparations which exhibit highly exothermic reactions when in contact with other substances, particularly flammable substances

H3-A 'Highly flammable':

- Liquid substances and preparations having a flash point below 21 °C (including extremely flammable liquids) or
- Substances and preparations which may become hot and finally catch fire in contact with air at ambient temperature without any application of energy or
- Solid subtances and preparations which may readily catch fire after brief contact with a source of ignition and which continue to burn or to be consumed after removal of the source of ignition or
- Gaseous substances and preparations which are flammable in air at normal pressure or
- Substances and preparations which, in contact with water or damp air, evolve highly flammable gases in dangerous quantities

H3-B 'Flammable': liquid substances and preparations having a flash point equal to or greater than 21 °C and less than or equal to 55 °C

H4 'Irritant': non-corrosive substances and preparations which, through immediate, prolonged or repeated contact with skin or mucous membrane, can cause inflammation

H5 'Harmful': substances and preparations which, if they are inhaled or ingested or if they penetrate the skin, may involve limited health risks

H6 'Toxic': substances and preparations (including very toxic substances and preparations) which, if they are inhaled or ingested or if they penetrate the skin, may involve serious, acute or chronic health risks and even death

H7   'Carcinogenic': substances and preparations which, if they are inhaled or ingested or if they penetrate the skin, may induce cancer or increase its incidence

H8   'Corrosive': substances and preparations which may destroy living tissue on contact.

H9   'Infectious': substances and preparations containing viable micro-organisms or their toxins which are known or reliably believed to cause disease in man or other living organisms

H10 'Teratogenic': substances and preparations which, if they are inhaled or ingested or if they penetrate the skin, may induce non-hereditary congenital malformations or increase their incidence

H11 'Mutagenic': substances and preparations which, if they are inhaled or ingested or if they penetrate the skin, may induce hereditary genetic defects or increase their incidence

H12 Substances and preparations which release toxic or very toxic gases in contact with water, air or an acid

H13 Substances and preparations capable of any means, after disposal, of yielding another substance, e.g. leachate, which posseses any of the characteristics listed above

H14 'Ecotoxic': substances and preparations which present or may present immediate or delayed risks for one or more sectors of the environment

15.14 PROBLEMS

15.1 An automobile garage is producing many different types of waste. List at least six different types of waste and classify them according to the Kommunekemi classification code.

15.2 You are the environmental manager at a bicycle factory, where all the components for making a bicycle are manufactured except:
(a) steel tubes,
(b) ball bearings,
(c) tyres.
List the possible types of hazardous waste and give suggestions for their proper treatment.

15.3 A groundwater supply has been found to contain hydrocarbons. Outline the procedures to:
(a) assess the magnitude of the problem,
(b) remediate the problem.

15.4 Review the paper 'Diffusive contaminant transport in natural clay. A field example and implications for clay lined waste disposal sites' by Johnson *et al.* (1989).

15.5 Review the paper, 'Evaluation of landfill liner designs' by Schroeder and Peyton (1990).

15.6 Review the paper, '*In situ* biorestoration of organic contaminants in the subsurface' by Thomas and Ward (1989).

15.7 For a hospital in your area, try to obtain (or prepare) an inventory of all of its waste streams, by quantity, material and classification. Also identify the treatment processes and ultimate fate of all the waste streams.

15.8 For Problem 15.7, once you have prepared the waste inventory, assume you have a free hand to treat and dispose of all streams with state-of-the-art technology. Then identify your new integrated waste management policy/strategy for the hospital's waste.

15.9 Discuss the suitability of the 'Kommunekemi model' for adoption in your jurisdictions.

15.10 Briefly describe three different forms of incineration plant, using sketches.

15.11 You are the project control engineer of a municipal landfill site. You know that up to 2 per cent of the wastes incoming are hazardous. Design a programme to identify, characterize, quantify and redirect such wastes.

15.12 Review the current literature on *in-situ* bioremediation of oil contaminated sites and write a three-page summary of the state of the art.

REFERENCES AND FURTHER READING

Bellandi, R. (1988) *Hazardous Waste Site Remediation. The Engineer's Perspective*, O'Brien and Gere Engineers Inc. Van Nostrand Rheinhold, New York.

Cairney, T. (1993) *Contaminated Land. Problems and Solutions*, Blackie Academic and Professional, Lewis Publishers U.K.

Daniel, D. E. (1993) *Geotechnial Practice for Waste Disposal*, Chapman and Hall, London.

Domenico, P. A. and F. W. Schwartz (1990) *Physical and Chemical Hydrogeology*, John Wiley, New York.

EC (1994) Directive on incineration of hazardous waste, EC official publication, Brussels.

EC (1993) *The State of the Environment in the European Community*, EC official publication, Brussels.

Freeman, H. M. (1992) *Hazardous Waste Minimization*, McGraw-Hill, New York.

Freeman, H. M. (1985) *Standard Handbook of Hazardous Waste Treatment and Disposal*, McGraw-Hill, New York.

Johnson, R. L., J. A. Chery and J. F. Pankow (1989) 'Diffusive contaminant transport in natural clay', *Environmental Science and Technology*, **23**, 340–349.

Kharbanda, O. P. and E. A. Stallworthy (1988) *Safety in the Chemical Industry*, Heinemann Professional Publishing, London.

Kolaczkowski, S. T. and B. D. Crittenden (eds.) (1987) *Management of Hazardous and Toxic Wastes in the Process Industries*, Elsevier Applied Science, Oxford.

Lehman, J. P. (1983) *Hazardous Waste Disposal*, Plenum Press, New York.

McCoy Associates (1985) *Index to Hazardous Waste Regulations*, McCoy and Associates, Lakewood, Colorado, p.147.

Major, D. W. and J. Fitchko (1992) *Hazardous Waste Treatment—On Site and In Situ*, Butterworth-Heinemann, London.

Martin, A. M. (1991) *Bioconversion of Waste Materials to Industrial Products*, Elsevier Applied Science, Oxford.

Nemerow, N. L. and A. Dasgupta (1991) *Industrial and Hazardous Waste Treatment*, Van Nostrand Reinhold, New York.

Palmark M. (1980) Chemical Waste Association of Danish Engineers.

Palmark, M. (1986) *Types of Chemical Waste in Denmark*. Chemcontrol A/S, Company Literature, Nyborg, Denmark.

Perry, R H, D. W. Green and J. O. Maloney (1984) *Chemical Engineering Handbook*, 6th edn. McGraw-Hill, New York.

Pojasek, R. B. (1979) *Toxic and Hazardous Waste Disposal*, Vol. 1, Ann Arbor Science Publishers Inc., Michigan.

Schroeder, R. R. and R. L. Peyton (1990) 'Evaluation of landfill liner designs', *J. Environmental Engineering Division of ASCE*, **116** (3), 421–437.

Thomas, J. M. and C. H. Ward (1989) '*In situ* biorestoration of organic contaminants in the subsurface', *Environmental Science and Technology*, **23**, 760–766.

US Water Pollution Control Federation (WPCF) (1987) *Household Hazardous Waste—Chart*, WPCF, Virginia.

US Water Pollution Control Federation (WPCF) (1988) *Hazardous Waste Site Remediation*, WPCF, Virginia.

US Water Pollution Control Federation (WPCF) (1990) *Hazardous Waste Treatment Processes Including Environmental Audits and Waste Reduction*, WPCF, Virginia.

Wentz, C. A. (1995) *Hazardous Waste Management*, McGraw-Hill, New York.

Williams M. J. and M. E. Redman (1990) 'Unmixing Mixed Waste', *Civil Engineering*, ASCE, April 1990.

CHAPTER SIXTEEN

INDUSTRIAL AIR EMISSIONS CONTROL

16.1 INTRODUCTION

Industrial air emissions are regulated, so as to keep the air quality acceptable to national and international standards. Industry achieves these standards by utilizing a variety of air emission abatement technologies. The waste stream is often 'air' which may contain a variety of gaseous and particulate contaminants of different densities, particle sizes and volatility, etc. For instance, coal-fired power plants generate a waste gas stream containing gases such as oxides of nitrogen and sulphur and particulates such as fly ash from the burnt coal.

The design of air emissions abatement equipment is a difficult and complex task. It requires not only a knowledge of the physical and chemical properties of the stream but also an appreciation of the vast array of equipment available and how this equipment operates. The equipment itself must, in operation, consistently comply with very tight performance criteria.

It is important for the engineer who designs such equipment to realize that environmental legislation is now setting the emission standards which are required of such equipment. The TA Luft (1986) regulations in Germany have been used as a basis for much legislation throughout Europe. TA Luft specifies the maximum allowable concentrations to be emitted from a whole range of industries. The US air emission standards required by industry are discussed in Chapters 1 and 8. Also, increasing emphasis is now placed on a systems management approach rather than an end of pipe control approach. This is becoming evident in documents such as the British Standard BS 7750 (1992), which is an attempt to do to environmental control systems what ISO 9000 (international quality control standard) did to the manufacturing and service industries. Such legislative requirements implicitly demand that the equipment design basics is clear and that the equipment performs accordingly. This chapter aims to introduce the reader to the basics of engineering design with regard to industrial air emission control.

Ideally, the engineer should not have to design such equipment at all. The offending constituents of the stream should be removed at source. Waste minimization is covered in Chapter 18. However, quite often it is only when a stream is fully characterized that the possibility of changing its composition is realized; this should always be borne in mind in the design process, especially at the preliminary stages.

There are three general options in air emissions abatement:

- Waste minimization (of raw materials, a product or a by-product)
- Recovery and recycling
- Destruction or disposal

Each of these cases is shown in Fig. 16.1.

It is clear from the Fig. 16.1 that the only benefit option 3 offers is regulatory compliance. The principal type of equipment used to destroy streams is the incinerator. Aspects of incineration are covered in Chapter 15, for both municipal and industrial waste streams.

16.2 CHARACTERIZING THE AIR STREAM

Wastewater treatment relies on the microbiological treatability or chemical composition of the liquid waste stream to effect purification. Water purification manipulates the chemistry of a waste stream to purify it. However, the purification of a gaseous stream involves manipulating the physical, chemical and, sometimes, the biological properties of the stream. The reason for this is because a vast choice of abatement equipment exists today, each manipulating a different property of the gas stream. For example, condensers utilize the relationship between a liquid and its vapour. It is well known that if an air stream, saturated with steam at a temperature of, say, 60°C, is cooled down, the steam will then condense. The principle of operation of an industrial condenser is the same. Cyclones, on the other hand, separate particles from a gas by utilizing the fact that their densities are different. To choose and design the most cost effective equipment the engineer must therefore characterize the stream fully. The most important measurable properties of a stream include:

- The stream composition
- The stream flowrate
- The stream temperature
- The stream pressure

Based on this information the choice of abatement equipment can be short listed. Further information which is required to make the final choice is:

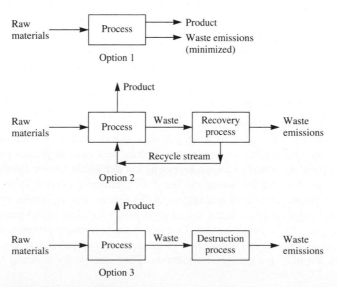

Figure 16.1 Flow sheet options in emissions control.

1. The variability in the composition, flowrate, temperature and pressure of the stream (i.e. due to startup, shutdown).
2. The explosivity of the stream. This is especially important where VOCs are concerned, e.g. petrol. Properties such as the flash point, auto-ignition temperature and the concentration at which it forms an explosive mixture with air, known as the lower explosive limit (LEL), should be known.
3. The corrosiveness of the stream in both liquids and gases; e.g. if SO_2 combines with water vapour and condenses, it forms corrosive sulphuric acid (H_2SO_4).

With regard to each constituent of the stream, some, or all, of the following information may be required:

- Molecular formula and weight
- Freezing and boiling point
- Solubility
- Adsorptive and absorptive properties
- Chemical behaviour/reactivity
- Heats of condensation, adsorption and solution
- Particle size distribution and densities of any solids
- Odour threshold
- Health effects
- pH
- Vapour pressure curve

16.3 EQUIPMENT SELECTION

Equipment selection first considers the type of compounds which are required to be removed. These may be considered under three broad categories:

- VOCs, defined as 'Any organic substance or mixture which can release vapour to the atmosphere and with the potential to cause environmental effects at low atmospheric levels', (Chemical Industries Association (CIA) 1992)
- Inorganic compounds
- Particulate matter

The stream contents will determine the type of equipment which may be used in any application. The major types of equipment in use today are:

- Incinerators
- Adsorbers
- Condensers
- Filters
- Scrubbers
- Absorbers
- Various particle collection devices

Their applications are listed in Table 16.1.

Adsorbers utilize a granular porous solid, usually a special form of carbon for VOC removal. The carbon, because of its porosity, has a large internal surface area. Gas molecules penetrate the pores and adhere to the walls by weak adhesion forces, known as van der Waals forces. Another type of adsorption is chemisorption. This is where the compound becomes chemically bonded to the adsorbent and is not easily removed, e.g. odour control by adsorption. In the former, the molecules can be removed by the application of heat. In absorption, a liquid is contacted with a gas. The liquid preferentially 'absorbs' or removes one component from the gas stream. Decreasing the temperature or increasing the pressure of a gas stream will cause vapours to condense. It is the former method that is most commonly used in condensers. Filters are used for gas–solid separation. They operate by utilizing a porous solid barrier

Table 16.1 Air emissions control technologies

Method	Organic vapours	Inorganic vapours	Particulate matter
Incineration	✓		
Adsorption	✓		
Condensation	✓		
Absorption	✓	✓	
Filtration			✓
ESP			✓
Scrubbers		✓	✓

which prevents solid particles passing through it. Most other gas solid separation devices utilize the density difference between a gas and a solid to effect separation. The two major exceptions are scrubbers and electrostatic precipitators (ESP). The former use a liquid spray to remove the particles and the latter use the fact that particles become charged upon passing through an electric field. Table 16.1 lists technologies suitable for particular stream types.

Once the most appropriate abatement technologies have been determined, one, or a combination of them, must be selected to accomplish the task. The same degree of stream purification cannot be achieved with all devices. Table 16.2 shows the removal efficiencies achievable with different technologies for different stream inlet concentrations (McInnes *et al.*, 1990). These figures are typical efficiencies. Some technologies, such as condensation, are highly dependent on the compounds involved and the operating temperatures to determine the efficiency of the equipment. Similar data for gas–solid separation is given in Table 16.3.

Table 16.2 VOC removal efficiencies

Technology	Inlet concentration (ppm)	Efficiency (%)
Condensation	> 5000	95 +
	> 2500	90 +
	> 500	50 +
Absorption	> 5000	99 +
	> 500	95 +
	> 200	90 +
Adsorption	> 5000	99 +
	> 1000	95 +
	> 200	90 +
Thermal incineration	> 100	99 +
	> 20	95 +
Catalytic incineration	> 100	95 +
	> 50	90 +

Table 16.3 Technologies of gas–solid

Technology	Applicable particle size (μm)	Efficiency
Gravity settling chamber	> 150	95% on particles > 300 μm
Cyclone	> 10	80% on particles < 20 μm
Spray tower	> 3	98% on particles > 5 μm
Filter	> 0.5	95–99% on particles < 5 μm
ESP	> 0.001	80–99%

16.4 EQUIPMENT DESIGN

16.4.1 Condensation

Condensers transfer heat from a vapour stream to a cooling stream. This removal of heat cools the gases present and condenses some, or all, of the vapour. They can be of the direct or indirect contact type. In the former, the cooling stream is contacted directly with the vapour stream while in the latter, contact is effected through a solid barrier. It is the latter type that is most commonly used today. They are traditionally of the shell and tube variety but other designs such as spiral condensers are becoming very popular. A schematic of a shell and tube exchanger is shown in Fig. 16.2. The schematic shows a horizontal condenser with the coolant on the tube side.

Operating principles of a condenser Condensation involves removing the heat from a stream to reduce the temperature and thereby cause the pollutant to condense from its vapour to its liquid form. The basic equation of heat transfer is

$$Q = UA\Delta T \tag{16.1}$$

where
Q = heat transferred, kW

U = heat transfer coefficient, kW/m^2 °C

A = heat transfer area, m^2

ΔT = temperature difference between hot and cold fluid, °C

The heat transferred is dependent on the temperatures and compositions of both the inlet and outlet streams, and therefore can be determined by performing an energy balance around the system. Luyben and Wenzel (1988) detail mass and energy balances. An energy balance determines the value of Q. The heat transfer coefficient is a measure of the ease with which heat is transferred through the gas, the condensing vapours, the heat exchanger wall and into the cold fluid. Traditionally each of these parameters is considered to have its own resistance to heat transfer. This is calculated for each of them in turn; they are summed and the result is the U value in Eq. (16.1). Other terms also contribute to the U value and are included in any detailed sizing. The calculation of the U value is one of the most important aspects of condenser design. For condensers in abatement applications it is difficult to determine U because there is usually a large amount of non-condensable gas present; this is discussed further in the next section. The design engineer wishes to determine the area required to achieve a specified outlet

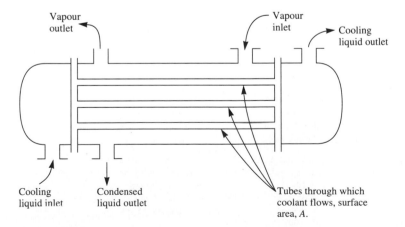

Figure 16.2 Schematic of a condenser.

concentration. For that reason, A in Eq. (16.1) is usually unknown. The average temperature difference is determined by the inlet and outlet conditions. It is calculated as follows for counter-current flow, i.e. the cold fluid enters from the opposite end to the hot fluid:

$$\Delta T = \frac{(T_3 - T_2) - (T_4 - T_1)}{\ln\left[(T_3 - T_2)/(T_4 - T_1)\right]} \tag{16.2}$$

where

T_1 = the cooling fluid inlet temperature

T_2 = the cooling fluid outlet temperature

T_3 = the gas stream inlet temperature

T_4 = the gas stream outlet temperature

T_3 is determined by the process generating the gas stream and normally cannot be changed. T_4 is determined by the composition of the outlet stream and cannot be changed, except to say that the lower the outlet temperature, the better for condensing material. T_1 is determined by the services available. This may be anything from a municipal water supply to liquid nitrogen. T_2 is an economic choice. The important point is that doubling the ΔT halves the required heat exchanger area but doubles the cooling liquid flow requirement. Therefore, the only real unknown at this stage is the heat transfer coefficient value (U).

Calculation of the overall heat transfer coefficient As stated previously, the overall heat transfer coefficient is the sum of a number of individual resistances. For a typical shell and tube heat exchanger, the heat exchanger coefficient is given by the following equation (Coulson and Richardson, 1991):

$$\frac{1}{U} = \frac{1}{h_o} + \frac{1}{h_{od}} + d_o \ln \frac{d_o/d_i}{2k_w} + \frac{d_o d_i}{h_{id}} + \frac{d_o d_i}{h_i} \tag{16.3}$$

where

U = overall heat transfer coefficient, W/m^2 °C

h_o = outside fluid film coefficient, W/m^2 °C

h_{od} = outside fouling factor, W/m^2 °C

d_o = tube outside diameter, m

d_i = tube inside diameter, m

k_w = thermal conductivity of the tube wall, W/m °C

h_i = inside fluid film coefficient, W/m^2 °C

h_{id} = inside fouling factor, W/m^2 °C

Typical values of the outside fluid film coefficient are given in the literature (Coulson and Richardson, 1991). If not, it can be calculated from the physical properties of the fluid using one of the standard equations.

The inside and outside fouling factors are very much a function of the operating conditions and the fluids. They cannot be calculated from first principles and are time dependent. Regular cleaning is the key to minimizing them. Normally, cooling water would have cleaning agents added to it if it is recirculated. The thermal conductivity of the tube wall is a function of the materials of construction and is detailed in any physical properties handbook. For commonly used materials in heat exchangers thermal conductivities can be found in a number of sources (e.g. Simonson, 1984). This leaves only the inside fluid heat transfer coefficient to be determined. This is a complex function of the rate of heat, mass and momentum transfer on the condensing surface. It is complicated by the fact that, as condensation progresses, the flow regime over the surface may change as the liquid phase increases in size. Add to this the fact that it is influenced by whether the condensation is inside or outside the tubes, and also the orientation of the condenser. A proper calculation of the heat transfer rate involves boundary layer calculations. In the presence of a non-condensable gas, the inside film heat transfer coefficient is normally the limiting resistance. One method which may be used to calculate the vapour/liquid film heat transfer coefficient is that of Silver. This method was further developed by Bell and Ghaly and is explained by Whalley (1987). The method can be developed into a computer or spreadsheet application. This method is only applicable, however, to single-component condensation. Other calculation methods also exist. Lydersen (1979) details a semi-empirical equation developed by Renker to calculate the condensing film heat transfer coefficient. This is more cumbersome than the above method, but does take into account the diffusivity of the gas–vapour mixture. However, before applying any of the above equations, care should be taken to ensure that they are applicable to the case in question. This is because the differences between vertical and horizontal condensation, inside and outside tube condensation, etc., can be significant. When more than one component is present, i.e. water and a VOC being a typical case, the water will condense first; however, as it condenses it may absorb some of the VOC from the gas stream, unpredictably improving its performance. Coulson and Richardson (1991) give a good introduction to heat exchange and exchanger design. For more detailed information on the heat exchanger design numerous textbooks exist. The American TEMA standards (Tubular Exchanger Manufacturers Association) are applied internationally in design and there are numerous journals including the *International Journal of Heat and Mass Transfer*.

Calculation of the outlet stream composition from its temperature and pressure   The partial pressure of a compound in its vaporous or gaseous state is proportional to the number of moles of it present. If the temperature, total pressure and compounds present in a stream are known, then the composition of the stream can be calculated by first calculating the vapour pressure of each compound using the Antoine equation:

$$\ln (VP) = A - B/(T + C) \tag{16.4}$$

where
$$A, B, C = \text{constants}$$
$$VP = \text{vapour pressure}$$
$$T = \text{temperature, } ^\circ K$$

Dividing each of these by the total pressure then gives the volumetric fraction of each component present which can be converted into mass concentration units using the molar mass. The units used should be consistent with the constants. This method gives the maximum quantity in the stream at the specified temperature, because it uses the saturated vapour pressure.

Example 16.1   A gaseous stream is at 25 °C and is known to contain only acetone and air. The total pressure is 760 mm Hg absolute. Determine the following:

(a) the mole fraction of acetone present,

(b) the ppm of acetone present.

Solution   Dalton's law of partial pressures states that the sum of the pressures exerted by the individual gases is equal to the total pressure. Therefore,

$$P_t = P_{\text{acetone}} + P_{\text{air}}$$

where
$$P_t = \text{total pressure}$$
$$P_{acetone} = \text{partial pressure of acetone}$$
$$P_{air} = \text{partial pressure of air}$$

Assuming that the air is saturated with acetone, the partial pressure of acetone may be calculated using the Antoine equation and inserting the constants appropriate to acetone. From Reid *et al.* (1977), the constants for acetone are

$$A = 16.6513$$
$$B = 2940.46$$
$$C = -35.93$$

The Antoine equation is:

$$\ln(\text{VP}) = A - B/(T + C)$$

Where T is temperature in kelvin and the units of vapour pressure are mm Hg. Thus

$$\ln(\text{VP}) = 16.6513 - 2940.46/(273 + 25 - 35.93)$$
$$\text{VP} = 228 \text{ mm Hg}$$

The mole fraction of acetone present is given by $228/760 = 0.3$ and the ppm of acetone present is $0.3 \times 1\,000\,000 = 300\,000$ ppm.

16.4.2 Absorption

In an absorber, the gas, rich in the compound which it is required to remove, enters the bottom of a tall narrow column down through which liquid is poured. The liquid selectively absorbs the compound, thereby producing a purer gas stream from the top of the column, while liquid rich in the substance removed from the gas is removed from the bottom of the column. A schematic of a typical column is shown in the Fig. 16.3. This is known as simple absorption and its main disadvantage is that it generates a liquid waste stream. An example of this is flue gas desulphurization (FGD) which is discussed in Sec. 16.5.

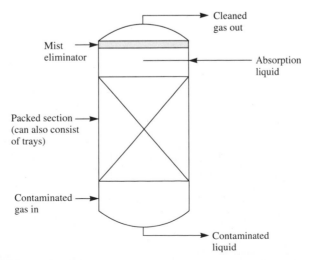

Figure 16.3 Simple absorption system.

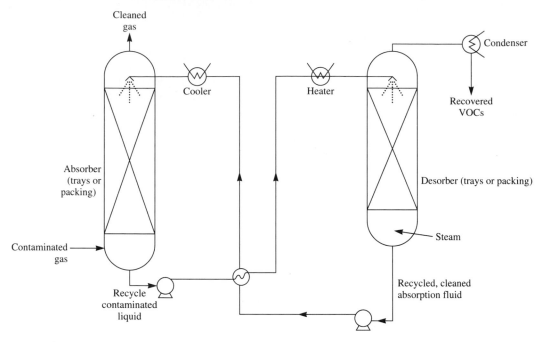

Figure 16.4 Complex absorption system.

A more environmentally acceptable alternative is complex absorption, where the compound absorbed in the absorber is removed from the liquid in a second column called a desorber. This is achieved by the application of heat, in the form of steam and sometimes under vacuum. It is not applicable in cases such as FGD where a chemical reaction takes place. However, it is widely used in the pharmaceutical industry for recovering VOCs. The liquid used is usually an oil. A complex absorption system is shown in Fig. 16.4.

Consider a section of the absorption column. Performing a mass balance around, say, the bottom of the column, as shown in Fig. 16.5, then

$$GY_0 + LX_{n+1} = GY_n + LX_1 \qquad (16.5)$$

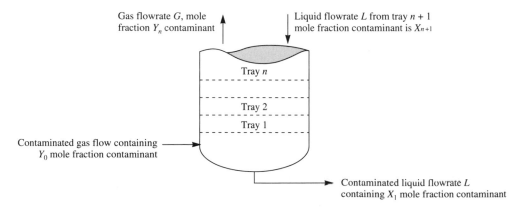

Figure 16.5 Mass balance diagram of absorption column.

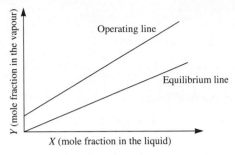

Figure 16.6 The equilibrium and the operating line of an absorption column.

Therefore,

$$Y_n = (L/G)X_{n+1} + (Y_0 - LX_1/G) \tag{16.6}$$

This is the equation of a straight line, known as the system operating line. The equilibrium relationship expressing the relationship between the composition of a liquid and its vapour phase can be written as

$$Y = mX \tag{16.7}$$

where
X = the fraction in the liquid phase

Y = the fraction in the vapour phase

m = the equilibrium constant

Plotting the above two equations on a graph of Y versus X gives Fig. 16.6.

For more details on the design of absorption columns, desorption columns, determining whether to use packing or trays, and design data, Perry (1984), Rousseau, (1987) or Coulson and Richardson (1991) should be consulted.

Example 16.2   The vapour liquid equilibrium relationship for ammonia (NH_3) at 30°C is given by

$$Y = 1.32X$$

A gas stream flowing at 50 kg mol/h contains 5 per cent v/v ammonia. It is desired to reduce the concentration to 0.5 per cent v/v. For the design of an absorber, determine:

(a) the minimum liquid flowrate required;
(b) the liquid flowrate required at 1.5 times the minimum liquid to gas ratio, L/G_{min};
(c) the number of equilibrium stages required at 1.5 times L/G_{min};
(d) how many trays would be required in such a column;
(e) if packing is used instead of trays calculate the height of packing required at $1.5L/G_{min}$ given that the height of packing equivalent to a theoretical stage (HETS) is given by

$$HETS = 0.75 \text{ m}$$

Solution

(a) Calculation of the minimum liquid flowrate required. Assumptions:

(1) L and G are constant.
(2) $X_{n+1} = 0$.

At the minimum value of L, the operating line will touch the equilibrium line, minimizing its slope. The applicable equation is

$$Y_n = (L/G)X_{n+1} + [Y_0 - (L/G)X1]$$

Given:

$$X_{n+1} = 0$$
$$Y_n = 0.005$$
$$Y_0 = 0.05$$

X_n can be calculated from the equilibrium relationship:

$$Y = 1.32X$$
$$X1 = Y_0/1.32 \text{ at } L/G_{min}$$

Therefore,

$$X1 = 0.05/1.32$$
$$= 0.038$$

Inserting these values into the Eq. (16.6),

$$0.005 = (L/G)_{min} \times 0 + [0.05 - (L/G) \times 0.038]$$
$$0.005 = 0 + 0.05 - 0.038(L/G)$$
$$0.045 = 0.038(L/G)$$
$$L/G = 1.184$$

Therefore, the minimum liquid flowrate required is

$$L = 1.184 \times G$$
$$= 1.184 \times 50$$
$$= 59.2 \text{ kg mol/h}$$

(b) Calculation of the liquid flowrate required at 1.5 times L/G_{min}:

$$L = 1.5 \times 59.2$$
$$= 88.8 \text{ kg mol/h}$$

(c) Calculation of the number of equilibrium stages required at 1.5 times L/G_{min}.
The equation of the operating line is:

$$Y_n = 1.776X_{n+1} + (0.05 - 1.776X1)$$

To use this equation $X1$ must be known. $X1$ can be determined by a mass balance using Eq. (16.5):

$$GY_0 + LX_{n+1} = GY_n + LX1$$

Inserting the values calculated above:

$$50 \times 0.05 + 0 = 50 \times 0.005 + 88.8 \times X1$$

Therefore,

$$X1 = (50 \times 0.05 - 50 \times 0.005)/88.8$$
$$= 0.0253$$

The equation of the operating line is therefore

$$Y_n = 1.776 \times X_{n+1} + 0.005$$

This equation and the equilibrium equation given in the problem are plotted on a graph of Y versus X shown in Fig. 16.7. This graph is then stepped off, as shown, to determine the number of equilibrium stages required to effect the separation.

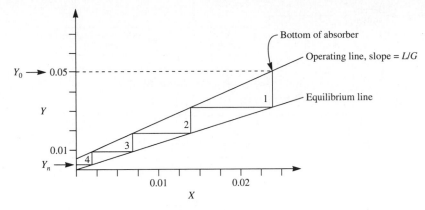

Figure 16.7 Determining the number of stages required in absorption.

(d) The number of trays required. If perfect equilibrium is achieved then four trays are required.
(e) The height of packing required:

$$\text{Height of packing} = 4 \times 0.75$$
$$= 3 \text{ m}$$

16.4.3 Adsorption

Adsorption is most commonly used to recover compounds from very dilute streams. It can be used in a number of different ways including fluidizing the adsorbent, rotating it through the vapour or passing the gas stream through a fixed bed of carbon. The latter is by far the most common method employed and is examined here. A typical fixed bed adsorber is shown in Fig. 16.8. It consists of two units, through one of which the gas stream is passed while the other is being regenerated.

Fixed bed adsorption   Consider activated carbon, the most commonly used adsorbent for the recovery of volatile organic carbons from gas streams. Perry (1984) states that activated carbon will adsorb 8 kg of acetone, a VOC commonly used in the electronics and the pharmaceutical industries, per 100 kg of fresh carbon. Therefore, if an air stream of $100 \text{ m}^3/\text{h}$ containing 1 kg of acetone per 100 m^3 of air was passed through a bed of 100 kg of fresh carbon, the bed should adsorb no more acetone after 8 hours. If the concentration of acetone in the outlet stream was measured, the concentration profile should look like Fig. 16.9, curve A.
In practice, various factors combine to change the shape of the above curve such that, in reality, it would be shaped like that shown in Fig. 16.9, curve B. This discrepancy occurs because the VOC molecules are in equilibrium with the gas stream. The engineer uses the activated carbon to change this equilibrium and the molecules react against this. This is termed mass transfer resistance. If there was no resistance to mass transfer, the outlet concentration curve would be exactly as shown in Fig. 16.9, curve A. The greater the resistance, the more the curve moves from that of Fig. 16.9, curve A, to that of curve B. Furthermore, adsorption occurs in a narrow band within the bed of carbon; this is known as the adsorption zone. The width of this zone is proportional to the shape of the outlet curve. Also, because the flow through an adsorber is normally plug flow, VOCs should not be detected in the outlet stream until the adsorption zone reaches it. For example, the bed in Fig. 16.9, curve A, would have no adsorption zone, whereas in Fig. 16.9, curve B, it would be large. A number of factors can now be identified, including:

1. It is theoretically possible to recover 100 per cent of the VOC.
2. The longer the bed, the more it can adsorb with 100 per cent efficiency.

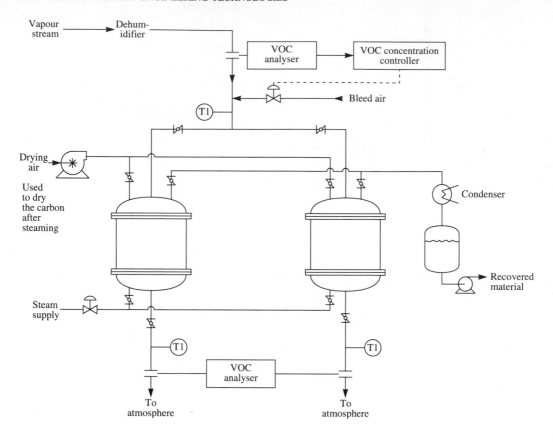

Figure 16.8 A fixed bed adsorption unit.

3. The rate of mass transfer from the gas phase to the solid phase should be maximized, thereby minimizing the size of the adsorption zone.

Mathematical analysis of fixed bed adsorption   A typical outlet curve is given in Fig. 16.10. The velocity with which the adsorption wave moves through the bed is given by

$$u = \frac{L}{t_b} \tag{16.8}$$

where u = the adsorption wave velocity, m/s

L = the length of the bed, m

t_b = time at which breakthrough occurs

The length of the mass transfer zone (MTZ) is given by

$$L_{MTZ} = u(t_0 - t_b) \tag{16.9}$$

If the outlet concentration was as in Fig. 16.11, the bed would be utilized to its maximum. However, because of the actual outlet concentration profile, a certain amount of underutilization of the carbon occurs. This is frequently expressed in terms of unused bed length (LUB). Mathematically, this can be shown to be given by

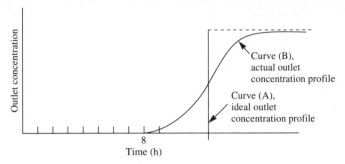

Figure 16.9 Ideal and actual outlet concentration profiles of adsorption.

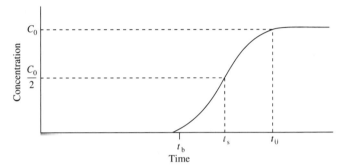

Figure 16.10 Mathematical analysis of fixed bed adsorption.

$$\text{LUB} = L\left(1 - \frac{t_b}{t_s}\right) \tag{16.10}$$

where t_s = time at which the outlet concentration equals half the inlet concentration

Of course, the most important parameter is the time at which breakthrough occurs, t_b. This can be calculated as follows, given the length of the bed:

$$t_b = \frac{L - L_{\text{MTZ}}}{u} \tag{16.11}$$

Example 16.3   Evaluation of an adsorber. A carbon bed of 1 m height, containing 400 kg of carbon is adsorbing acetone. Breakthrough occurs after 70 minutes (t_b) and the outlet concentration equals half the inlet concentration after 80 minutes (t_s). Calculate (a) the length (height) of the unused bed, (b) the length (height) of the mass transfer zone and (c) the adsorption wave velocity.

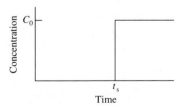

Figure 16.11 Ideal outlet concentration profile of fixed bed adsorption.

Solution

(a) Length of unused bed, LUB:

$$\text{LUB} = L\left(1 - \frac{t_b}{t_s}\right)$$
$$= 1\left(1 - \frac{70}{80}\right)$$
$$= 0.125 \text{ m}$$

(b) Length of the mass transfer zone:

$$L_{MTZ} = u(t_0 - t_b)$$

However,
$$u = \frac{L}{t_b}$$

Therefore,
$$L_{MTZ} = \frac{L}{t_b}(t_0 - t_b)$$
$$= \frac{1}{70}(90 - 70)$$
$$= 0.286 \text{ m}$$

(c) Adsorption wave velocity, u:

$$u = \frac{L}{t_b}$$
$$= \frac{1}{70}$$
$$= 0.0143 \text{ m/min}$$
$$= 0.86 \text{ m/h}$$

16.4.4 Filtration

A filter comprises a solid porous media which allows only gas and very small particles to pass through it. Gas filters normally consist of either bags or cartridges through which the gas passes. The construction of a typical gas filter is shown in Fig. 16.12. The performance of a filter is characterized by two parameters:

- Particle collection efficiency
- Pressure drop

Both are a function of the filter media, though the latter is also a function of the filtration velocity.

It is important to know the particle size distribution of the dust to allow the most appropriate filter media to be determined. Once the filter is in operation, a thin layer of dust will penetrate the filter media and form a thin film on it, thereby improving its performance. However, such a film also increases the pressure drop through a filter, reportedly sometimes by a factor of ten (Perry *et al.*, 1984). The pressure drop through a filter may be determined from the following equation:

$$\Delta P = k_1 V_f + k_2 W V_f \tag{16.12}$$

where
$$\Delta P = \text{pressure drop}$$
$$k_1 = \text{filter media constant}$$
$$k_2 = \text{solids constant}$$

W = solids per unit area

V_f = filtration velocity

The above parameters should be in consistent units. Perry *et al.* (1984) gives typical values for the constants for some dusts. However, it is universally accepted that unless a safety factor is used in the sizing of a filter, then on-site pilot trials should first be carried out. Lydersen (1979) suggests using the fact that because the pores are so small, flow will be laminar and the pressure drop can be easily estimated. However, this still requires a knowledge of the drag coefficient of the particles.

Filter media selection is based on the operating conditions. The main parameters that should be considered are temperature, acid, alkali and moisture resistance, strength and efficiency. There are numerous options in both media and the methods of construction. A review of the available media is given in the Institution of Chemical Engineers (UK) (IChemE) *Guide to Dust and Fume Control* (1985). Filters use a combination of mechanisms. Particles larger than the pore size of the filter are sieved out of the gas stream. Particles smaller than the pore size are removed by interception with the filter media, by impaction or by gravitational settling and, for small particles, diffusion and electrostatic attraction are important. In addition, a pre-coat layer can be used to improve collection efficiency. The pore size of a filter media can vary substantially with time because of the buildup of a dust layer on it. The filter media commonly used include:

- Woven fabric
- Needle felt
- Paper
- Plastic
- Ceramic filters (for high-temperature applications)

Operation of a gas filter Unlike liquid–solid filtration, gas–particulate filtration is carried out in two principal types of equipment:

- The bag filter
- The cartridge filter

The design of both is quite similar. They typically consist of a number of filter bags or cartridges suspended vertically. The gas stream may pass through them from either side, depending on the method of solids removal. Solids removal can be achieved in a number of ways. Two units can be employed with one cleaned while the other is on line. If filter bags are being used, they can be shaken by mechanical action

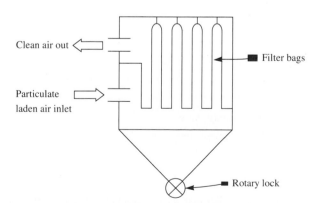

Figure 16.12 Construction of a typical bag filter unit.

(rapping), without interrupting the filtration. Acoustical action can also be used. The air flow through the filter can be reversed temporarily or a periodic or continuous jet of air can be used to dislodge the solids.

Which method to choose depends on the operating conditions and the material concerned. For example, for difficult dust cake or for particle penetration of the filter media, a pulsed jet of air at a high pressure, typically up to 700 kPa, is most effective. For a continuous jet the air pressure does not need to be as high. Because the cake is the major contributor to the pressure drop through the unit, and therefore determines the size of the fan, the frequency of cleaning is very important; i.e. if it can be cleaned frequently, the pressure drop and therefore the fan size will not be excessive.

Design of a gas filter   The cost of a gas filter is primarily a function of the filtration area required. The area required is the volumetric flowrate divided by the velocity through the filter. The velocity through the filter is limited by both the pressure drop through the unit and the 'can' velocity. This is the velocity of the gas in the passages between the filter units in the filter house. It is a frequently overlooked design parameter. There are two ways to size a gas filter. One is to calculate the velocity through the filter, often called the air to cloth ratio, by deciding on a cleaning frequency and then calculating the area required based on the maximum allowable pressure drop, using Eq. (16.12). The other is to pick a filtration velocity from a table of typical values and install a cleaning unit whose cleaning frequency can be adjusted to suit the performance of the filter when operational. The latter method is the most commonly used for small installations and is done in conjunction with the equipment vendor. Tables of typical filtration velocities are given both in the IChemE Guide (1981) and Croom (1993). Typically values for filtration velocities are between 0.5 and 1 m/min for all types of cleaning mechanisms except pulsed jet, which can have a filtering velocity of up to 5 m/min. The 'can' velocity should not normally exceed about 75 m/min (Croom, 1993).

16.4.5 Impingement Separators

Because of the density difference between solids and gases, in laminar flow their stream lines are different if the direction of flow is changed. This fact is frequently exploited to separate a solid particle from a gas stream, usually by suddenly changing the direction of flow of the gas stream.

The settling chamber   A settling chamber reduces the velocity of the gas stream so that the particles drop out by gravity. It is a large device not often used as a final control mechanism. If it is assumed that Stokes' law applies (see Chapter 11), then the particle size which will be removed with 100 per cent efficiency is given by the following equation:

$$d_p = \sqrt{\frac{18\mu_g uh}{\rho_p gL}} \tag{16.13}$$

where

d_p = particle diameter, m

μ_g = gas viscosity, kg/m s

u = gas velocity through the chamber, m/s

h = chamber height, m

g = gravity, m/s^2

L = chamber length, m

ρ_p = particle density, kg/m^3

The efficiency with which particles smaller than d_p are removed is the ratio of their settling velocity to that of d_p, as calculated by Stokes' law (see Chapter 11 on sedimentation).

Cyclone separators The use of gravity alone to separate particles from a gas stream has the twin disadvantages of not being very efficient and requiring a large area. Utilizing a centrifugal instead of a gravitational force to effect the separation requires a lower area and ensures a greater efficiency. A cyclone separator is used to achieve this. The (centrifugal) force exerted on a particle in a cyclone is given by the following equation:

$$F = \frac{mv^2}{R}$$

(16.14)

where

F = centrifugal force, N

m = particle mass, kg

v = particle velocity, m/s

R = cyclone radius, m

The construction of a typical reverse flow cyclone is shown in Fig. 16.13.

Design of a cyclone 'Cyclone specification and design is based on empirical studies' (Heumann, 1991). The equations outlined here are based on work carried out in the 1950s. More detailed design methods are available in Coulson and Richardson (1983) and Flagan and Seinfeld (1988), which review in the various cyclone design methods. A commonly used equation predicting the size of particle that can be removed with 50 per cent efficiency was developed by Lapple (1951):

$$d_{50} = \sqrt{\frac{9\mu_g h}{2\pi N_e V_i (\rho_p - \rho_g)}}$$

(16.15)

where

d_{50} = diameter of the particle removed with 50 per cent efficiency, m

μ_g = gas viscosity, kg/ms

h = width of cyclone inlet, m

N_e = number of effective turns within the cyclone

V_i = inlet gas velocity, m/s

ρ_p = gas density, kg/m^3

ρ_p = particle density, kg/m^3

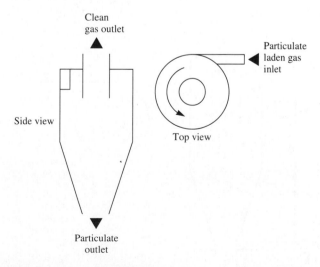

Figure 16.13 The construction of a reverse flow cyclone.

The efficiency of a cyclone in removing particles of other sizes is given in Peavy *et al.* (1986). The number of effective turns which the gas makes within the cyclone is given by the following equation:

$$N_e = \frac{1}{c}(2d + e) \tag{16.16}$$

The dimensions of a reverse flow cyclone are given in Fig. 16.14. The inlet velocity is normally between 15 and 20 m/s, however, Perry *et al.* (1984) suggests that in agreement with the above equation, substantially higher velocities can be used.

Example 16.4   Design of a cyclone separator. An air stream is flowing at the rate of $1000 \, \text{m}^3/\text{h}$ at a temperature of $50°C$. It contains particles with a density of $1200 \, \text{kg/m}^3$. Determine the diameter of the particles that will be removed with 50 per cent efficiency if the inlet air velocity cannot exceed $10 \, \text{m/s}$.

Solution

$$\text{Air at } 50 \,°C, \text{ density } \rho = 1.25 \text{ kg/m}^3$$
$$\text{Viscosity } \mu = 1.8 \times 10^{-5} \text{ N-s/m}^2$$
$$\text{Inlet velocity} = 10 \text{ m/s}$$

Therefore,

$$\text{Inlet area} = \frac{1000}{3600} \times \frac{1}{10} = 0.027\,77$$
$$= c \times h$$

However,
$$c = b = 0.5a = 2h$$

Therefore,

$$0.0278 = 2h \times h$$
$$= 2h^2$$
$$h = \sqrt{0.0278/2}$$
$$= 0.118 \text{ m}$$

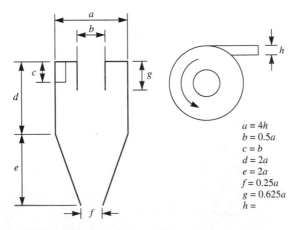

$u = 4h$
$b = 0.5a$
$c = b$
$d = 2a$
$e = 2a$
$f = 0.25a$
$g = 0.625a$
$h =$

Figure 16.14 The relative dimensions of a reverse flow cyclone.

and

$$a = 4 \times 0.118$$
$$= 0.4714 \text{ m}$$
$$b = 0.5 \times 0.4714$$
$$= 0.236 \text{ m}$$
$$c = 0.236$$
$$d = 2 \times 0.4714$$
$$= 0.9428$$

also

$$e = 0.9428$$

Therefore, from Eq. (16.16)

$$N_e = \frac{1}{0.236} \times (3 \times 0.9428)$$
$$= 11.98$$
$$= 12 \text{ turns}$$

and from Eq. (16.15)

$$d_{50} = \sqrt{\frac{9 \times 1.8 \times 10^{-5} \times 0.118}{2\pi \times 12 \times 10(1200 - 1.25)}}$$
$$= \sqrt{2.115 \times 10^{-11}}$$
$$= 4.6 \times 10^{-6} \text{ m} \simeq 5 \ \mu\text{m}$$

So this design is capable of removing 5 μm particles with a 50 per cent removal efficiency.

16.4.6 Scrubbers

Scrubbers, also known as wet collectors, employ a liquid to remove particles from a gas stream. In use it is either the liquid droplets which collect the particles or liquid that is poured continuously on to a porous packing and the particles collected by both the liquid droplets and adhesion to the liquid on the packing. Figure 16.15 shows the construction of a basic scrubber. The use of packing allows a smaller tower to be used but the pressure drop is higher (thereby increasing inefficiency). The pressure drop through a spray

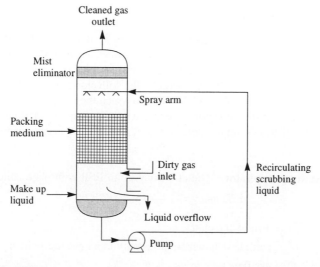

Figure 16.15 A typical spray tower.

tower is typically between 0.25 and 0.5 kPa. For a packed bed it is between 0.25 and 2.0 kPa. The liquid to gas ratio in a spray tower is typically between 1.3 and 2.7 L/m^3. In a packed tower it is normally between 0.1 and 0.5 L/m^3 (Corbitt, 1989).

Basically, the contaminated gas (e.g. biogas as CH_4 and CO_2 contaminated with excess NH_3) enters the tower at a low level and rises due to its buoyancy. The scrubbing liquid enters the top of the tower and sprays down on top of the vertically rising dirty gas. At interception the contaminants adsorb to the falling liquid and the purified gas continues to rise and emits from the top of the tower. Packing (e.g. random plastic pieces) improve the adsorption efficiency.

The nozzles on the spray arms atomize the liquid which then falls through the tower at the droplet terminal settling velocity. The efficiency of a spray tower is given by the following equation (Flagan and Seinfeld, 1988):

$$\text{Efficiency} = 1 - \exp[-1.5 \times n \times (V_t/v) \times W \times L/(V_g \times A_c \times D_s)] \qquad (16.17)$$

where
$\qquad V_t =$ droplet terminal velocity

$\qquad v = V_t - V_g$

$\qquad V_g =$ gas velocity

$\qquad A_c =$ cross-sectional area

$\qquad L =$ height of the tower

$\qquad D_s =$ droplet diameter

$\qquad n =$ the collection efficiency of particles on a droplet

The most difficult variable to quantify here is the collection efficiency of particles on a droplet. It must either be developed from pilot plant studies or calculated from an experimental correlation, such as that of Slinn, as detailed by Flagan and Seinfeld (1988). The droplet diameter is a function of the liquid being used, the spray arm nozzles and the flow through them. Lydersen (1979) states that the most effective drop size for a spray tower is 400 to 1000 μm. He also gives performance details for a Pease–Anthony venturi scrubber in various applications. Spray towers can be used on biogas from anaerobic digesters to remove the ammonia from the biogas, as ammonia is a toxin to methanogenic bacteria in the reactors.

16.4.7 Electrostatic Precipitators

These are used to remove very small liquid and solid particles from a gas stream, and are used mainly in the utility industries. They operate by generating a corona between a high-voltage electrode, usually a fine wire, and a passive earthed electrode, such as a plate or pipe. Particles passing through such an electric field are ionized by ions migrating from the discharge to the collector electrode, with whom they collide. These particles then drift towards the collector electrode to which they are held by electrostatic attraction. The particles are removed from the collector by either a water spray or 'rapping' it periodically. The collectors can either be flat plates or tubular units. Usually, a number of discharge electrodes will hang as shown in Fig. 16.16.

Design of an electrostatic precipitator The efficiency of an ESP with plate collectors is given by

$$\text{Efficiency} = 1 - \exp(-AW/Q) \qquad (16.18)$$

where
$\qquad A =$ area of the plates, m^2

$\qquad W =$ particle sedimentation velocity in an electric field, m/s

$\qquad Q =$ gas flow rate, m^3/s

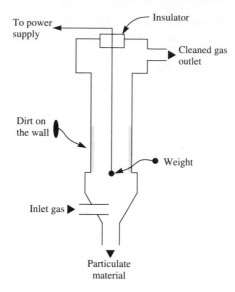

Figure 16.16 An electrostatic precipitator.

Equation (16.18) is known as the Anderson–Deutsch equation. In reality, such parameters as the particle sedimentation velocity are not constant. For further information, Oglesby *et al.* (1978) should be consulted.

Example 16.5   Design of an electrostatic precipitator. A quantity of $50\,\text{m}^3/\text{s}$ of air flows from a cement manufacturing facility. It contains cement particles whose settling velocity is $0.12\,\text{m/s}$. If 99 per cent removal efficiency is required, calculate the surface area of the electrostatic precipitator.

Solution

$$\text{Efficiency} = 1 - \exp(-AW/Q)$$
$$0.99 = 1 - \exp(-0.12\,A/50)$$
$$-A\frac{0.12}{50} = \ln(1 - 0.99)$$
$$-A = \frac{50}{0.12}\,\ln(0.01)$$
$$A = -416.67\,\ln(0.01)$$
$$= 1919\,\text{m}^2$$

16.4.8 Odour Abatement

Odours are not easily characterized or quantified and therefore represent a particular design problem. Control of odours is best achieved at source. This involves identifying the cause of the odour instead of the odour itself and then changing the operating conditions, methods, design or raw materials to eliminate or minimize the odour. Failing this, a number of options exist, such as:

- Adsorption, usually onto activated carbon
- Incineration
- Absorption/scrubbing
- Use of masking agents (all too common)
- Biofiltration (soil, peat beds, biological scrubbers)

Incineration and absorption are discussed elsewhere and therefore will not be discussed in detail here. The design of absorption and incineration equipment is the same as for vapour recovery or destruction. For more detailed information on odour control, Hesketh and Cross (1989) or Martin *et al.* (1992) should be consulted.

Adsorption   Adsorption for odour control differs from the adsorption discussed earlier in that concentrations are much lower. A single bed, usually a carbon filter, is used and the unit is normally non-regenerative because the odorous compounds are chemically bonded to the carbon. The adsorption wave would normally be quite narrow and its velocity would be low because of the low loading. An equation that may be used to approximate the filter life is Turk's equation (WPCF, 1979):

$$t = \frac{WS}{ERC} \tag{16.19}$$

where
$t = $ filter life, days

$W = $ mass of adsorbent, g

$S = $ the fraction of filter weight that can be adsorbed, usually 0.165 to 0.5

$E = $ efficiency of the filter

$R = $ gas flow rate, m^3/min

$C = $ concentration of odorous material in the gas, ppm

The use of masking agents   The use of masking agents involves the addition of specially formulated compounds to the waste stream. Addition is at source, i.e. addition to an activated sludge aeration tank, or by dilution with water and application via a spray tower. These compounds act in a number of ways, depending on the odour that they must remove. They can dissolve the offending gas by reacting with it so that it decomposes or, as in the case of an organic acid, they can convert it to a harmless organic salt. On-site or pilot trials are recommended to determine the effectiveness of any such compound.

For example, one masking agent Epoleon® acts as follows to control the odour associated with hydrogen sulphide:

$$CH_2COONa + H_2S \longrightarrow NaHS + CH_2COOH$$

For ammonia:

$$CH_2COOH + NH_3 \longrightarrow CH_2COONH_4$$

For methyl mercaptan:

$$CH_2COONa + CH_3SH \longrightarrow CH_3SNe + CH_2COOH$$

Epoleon chemically converts the offending compounds into harmless inoffensive compounds. (These examples are taken from Epoleon® Corporation promotional literature.)

Soilbeds   Commonly used in parts of the United States, soilbeds have not found widespread use in Europe to date because of both the cost of land and the possibility of frost reducing their efficiency. Typically about 1 metre deep, the gas flow rate is about $10 \, m^3/hm^2$. Moist loam soil is normally used. They are reasonably resilient to sudden stream composition changes, adapting themselves quickly to changes in loading circumstances. It is the microbial population within the soil that removes the odorous compounds. It is usually aided by the planting of shallow rooted plants to keep the soil loose. They can also have the secondary effect of keeping sulphate concentrations in the soil under control.

Peat bed filters   These consist of a coarse-fibred, dust-free peat from the top layer of a bog. Typically a *Spagnum* or *Trichophorum* peat is used. The peat bed itself is usually not more than 1 metre deep to avoid

Table 16.4 Typical efficiencies of peat bed filter for odour removal

Odorous compound	Inlet concentration (ppm)	Reduction across filter (%)
Ethanol	10– 70	> 99
Methanol	50– 300	> 99
Acetone	70– 400	> 97
Ethyl acetate	500– 5000	> 75

compaction. The bed temperature is maintained between 10 and 45 °C. The pH should always be kept within the range 4 to 7.5. For highly odorous streams the air flow rate should not be more than 110 to 130 m^3/hm^2 of bed. For less intensive odours, the flow rate can be increased as high as 200 m^3/hm^2. Table 16.4 gives the performance of a biofilter for a range of odorous compounds.

The odour intensity is measured using a trained panel of observers and a device called an olfactometer. Peat beds are mainly used in Europe where they have been in use, in agriculture and fish processing. The construction of a typical peat bed filter is shown in Fig. 16.17.

16.5 SPECIAL TOPICS

16.5.1 Flue Gas Desulphurization

There are two main types of flue gas desulphurization (FGD) systems. One generates a residue which must be disposed. The other converts the sulphur dioxide (and sulphur trioxide) to a marketable product. Approximately 95 to 97 per cent of the world's FGD systems are of the former type, i.e. non-regenerable. The limestone–gypsum process is the more economical to run and accounts for 40 per cent of the installed systems. In the longer term it is most likely that regenerable and catalytic processes will emerge as the desirable method of desulphurization. This is for two reasons. Firstly, the quantities of sludge that are required to be landfilled are growing with the increasing numbers of FGD plants. This is combined with the fact that the cost of landfilling is likely to increase substantially with increased monitoring and control requirements. Secondly, the public are unlikely to accept that a non-regenerable waste is the best

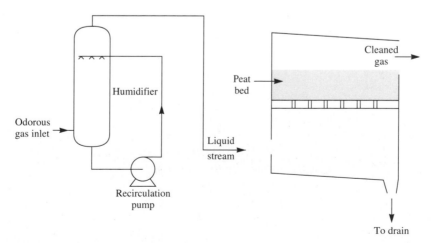

Figure 16.17 Schematic of a typical peat bed filter.

practicable environmental option when a number of regenerable stream options also exist. Most flue gas desulphurization processes centre around an absorption tower in which the sulphur dioxides are chemically absorbed into an alkaline liquid stream. They are further classified as either wet or dry, depending on the phase in which the reaction occurs. Most FGD systems in use today are wet non-regenerable processes.

The chemistry of flue gas desulphurization Non-regenerative systems are based around the reaction of sulphur dioxide with one or more of the following compounds: lime (CaO), caustic soda (NaOH), soda ash (Na_2CO_3), ammonia (NH_3), slaked lime ($Ca(OH)_2$), limestone ($CaCO_3$) and dolomite ($MgCO_3 + CaCO_3$, $MgCO_3$ mainly). The reaction mechanisms are not well defined and influenced by both the source and the method of processing of the raw material (Davis and Cornwell, 1991). Some of the reactions are as follows:

$$SO_2 + CaO \longrightarrow CaSO_3$$
$$2CaSO_3 + O_2 \longrightarrow 2CaSO_4$$
$$CaCO_3 + SO_2 \longrightarrow CaSO_3 + CO_2$$

The above equations show how the two most commonly used compounds in FGD operate. Lime reacts to produce calcium sulphite ($CaSO_3$), part of which is converted to calcium sulphate ($CaSO_4$) by both reacting with the excess oxygen in the flue gas and by aeration of the sludge afterwards. The sulphite is a thixotropic gel, difficult to landfill. The sulphate is a stable solid; hence settling ponds are often aerated to separate the sulphate from the gel. Also, the sulphate (known as gypsum) is the prime ingredient in plasterboard. One power station currently being built in the United Kingdom will produce enough gypsum to supply about a third of the British plasterboard market. It is, however, to be landfilled. Carbon dioxide is generated in limestone FGD systems, and up to 5 per cent of the total power produced by a power plant is required to operate an FGD system.

Example 16.6 Consider a hypothetical coal (1 per cent sulphur, 8 per cent ash and a calorific value of 30 MJ/kg)-fired power plant of 915 MW capacity with a load factor of 72.5 per cent and an efficiency of 40 per cent. An FGD scrubber is installed prior to the ESP. If the scrubber removes 60 per cent of fly ash and 90 per cent of SO_2, determine the amount of sludge generated by the scrubber.

Solution

$$
\begin{aligned}
\text{Coal requirement} &= 915 \text{ MW} \\
&= 915 \text{ MJ/s} \\
&= 3294 \times 10^3 \text{ MJ/h} \\
&= \frac{3294 \times 10^3 \times 0.725}{0.4 \times 30} \text{ kg/h} \\
&= 199\,012.5 \text{ kg/h} \\
&= 199 \text{ t/h} \\
\text{Sulphur content} &= 0.01 \times 199 \text{ t/h} \\
&= 1.99 \text{ t/h} \\
SO_2 &= 3.98 \text{ t/h} \\
\text{Fly ash} &= 0.08 \times 199 \text{ t/h} \\
&= 15.92 \text{ t/h}
\end{aligned}
$$

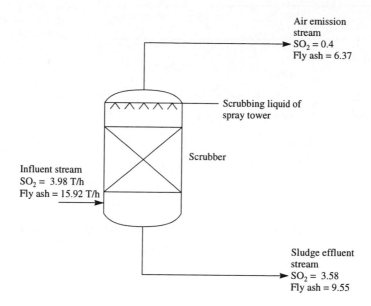

Mass balance for sulphur dioxide removal. Oxygen is required for SO_2 removal. The reactions are:

$$SO_2 + CaCO_3 \longrightarrow CaSO_3 + CO_2$$
$$2CaSO_3 + O_2 \longrightarrow 2CaSO_4$$

The overall reaction can therefore be shown as follows:

$$O_2 + 2SO_2 + 2CaCO_3 \longrightarrow 2CaSO_4 + 2CO_2$$

The molecular weights are

$$32 + 2 \times 64 + 2 \times 100 \longrightarrow 2 \times 136 + 2 \times 44$$

Thus 32 kg of oxygen is required for every 128 kg of SO_2 treated. Therefore

$$\text{Amount of oxygen required} = \frac{32}{128} \times (3.98 - 0.4)$$
$$= 0.895 \text{ T/h}$$
$$\text{Amount of calcium carbonate required} = \frac{200}{128} \times (3.98 - 0.4)$$
$$= 5.594 \text{ T/h}$$
$$\text{Amount of sludge (calcium sulphate) generated} = \frac{272}{128} \times (3.98 - 0.4)$$
$$= 7.608 \text{ T/h}$$
$$\text{Amount of } CO_2 \text{ generated} = \frac{88}{128} \times (3.98 - 0.4)$$
$$= 2.461 \text{ T/h}$$

Mass balance table for SO₂ removal (all units T/h)

	Inlet gas (1)	Outlet gas (2)	Inlet liquid (3)	Outlet sludge (4)
SO_2	3.98	0.4	—	—
O_2	0.895	—	—	—
CO_2	0	2.461	—	—
$CaCO_3$	—	—	5.594	—
$CaSO_4$	—	—	—	7.608
Total	4.875	2.861	5.594	7.608

$$\text{Mass balance}: \ (1) + (3) = (2) + (4)$$
$$4.875 + 5.594 = 2.861 + 7.608$$
$$10.469 = 10.469$$

Therefore, the amount of sludge generated is 7.608 T/h from SO_2 removal and 9.55 T/h of fly ash.

$$\text{Total sludge} = 7.608 + 9.55$$
$$= 17.158 \ \text{T/h}$$
$$= 150\,304 \ \text{T/yr}$$

16.5.2 NO$_x$ Removal

The term NO$_x$ implies two major oxides, nitrogen oxide (NO) and nitrogen dioxide (NO_2). In combustion, NO is the dominant of the two, NO_2 mainly a downstream derivative of NO. There are three main mechanisms of NO$_x$ production from combustion processes:

- From the reaction of N_2 in the fuel air with oxygen at the high temperatures of a burner chamber
- From nitrogen existing in the fuel
- From reactions of fuel-derived radicals with N_2 ultimately leading to NO

To control NO$_x$ emissions effectively, the dominant formation mechanism must be known. There are a number of ways of then controlling the NO$_x$ emissions.

The chemistry of NO$_x$ formation   The simplest, and most widely used, model of NO$_x$ formation is the Zeldovich mechanism, which may be expressed chemically as follows:

$$N_2 + O = NO + N$$
$$O_2 + N = NO + O$$
$$OH + N = NO + H$$

These reactions may be considered to be independent of the main fuel burning reaction which takes place in boilers and furnaces. These reactions occur because of the high temperature that exists in the flame itself. Indeed, it has been observed that NO$_x$ formation does not occur to a significant extent below 1800 K and that the NO$_x$ present is dependent on temperature (Lefebvre, 1983).

Minimizing NO$_x$ formation   Therefore, burner design needs to balance the combustion process requirements with NO$_x$ formation. This is done in a number of ways. Firstly, the chemical reactions outlined above can be manipulated. This is done, for example, by minimizing the quantity of oxygen present to minimize NO$_x$ formation. In practical terms this translates into using low excess air.

Recirculating the flue gas is also common. This acts by reducing the peak flame temperature and the quantity of oxygen present. By injecting water into the combustion chamber, the temperature is reduced due to the energy being taken up by the latent heat of water. Finally, staged combustion can be used to reduce the peak temperature. Up to three stages can be used, but this requires very tight control over both the fuel and air flow rates to each stage. In the 1970s, it was found that injecting ammonia into the furnace flame reduced the NO_x formation. This idea was developed further by the EPRI in the United States who found that injecting urea worked even better than ammonia. Today ammonia is used in conjunction with a catalyst in a technique known as selective catalytic reduction (SCR). Ammonia is injected into the flue gas upstream of the catalyst bed. The NO_x combine with the ammonia on the catalyst surface where they decompose to harmless nitrogen and water. Such systems give good NO_x removal rates but are expensive. Selective non-catalytic reduction (SNCR) is a post-combustion method of eliminating NO_x emissions. It involves the injection of ammonia or urea into the combustion chamber or an area of high temperature downstream. It requires a high temperature to achieve the reaction activation energy and hence is not applicable to certain types of equipment.

Selecting NO_x abatement equipment   Table 16.5 shows typical efficiencies and the applicability of the various methods of NO_x reduction.

16.5.3 Fugitive Emissions

Fugitive emissions are industrial emissions from both point and non-point sources. These sources may be considered the equipment and methods associated with the transferring, conveying, loading, unloading, storage, packaging and processing of materials. Depending on the process involved, fugitive emissions may be negligible or, as in, say, open quarrying, they may be the main emissions source.

Control of fugitive emissions   The control of fugitive emissions depends on the compounds involved, the quantities involved and the equipment being used. In industry, commonly used control methods are equipment modification and local ventilation/extraction to an emissions control unit from where the cleaned gas is emitted as a point source.

Fugitive dust sources and their control   The main sources of fugitive dust are open mining/quarrying, unpaved roads, construction activity and the open or semi-open storage or movement of solid material. Control of such sources can involve anything from the repeated application of water to chemical stabilization, surface cleaning, windbreaks or simply good management.

Fugitive VOC sources and their control   This was identified in the United States as having a significant impact on ambient air quality standards when 32 per cent of the releases of organic and halogenated organic compounds to the atmosphere in 1987 were defined as fugitive. Industrial valves, pumps and

Table 16.5 Methods of NO_x reduction

Method	Applicability	NO_x reduction (%)
Flue gas recirculation	Thermal NO_x reduction	70–80
Low NO_x burners	Fuel and thermal NO_x reduction	10–25
Staged burners	Fuel and thermal NO_x reduction	40–70
Selective catalytic reduction	Fuel and thermal NO_x reduction	80–90
Selective non-catalytic reduction	Fuel and thermal NO_x reduction (not applicable to small boilers or gas turbines)	60–80

flanges have been identified as the main contributors to fugitive emissions from industrial facilities and legislation is in place in the United States to minimize such emissions. Similar legislation is not yet in place in Europe. Legislation in the United States has developed from industries reporting, as annual totals, the fugitive emissions from a facility to a more comprehensive, legislative-driven emissions minimization programme. A rule is now in place where industries whose fugitive emissions increase must inspect more frequently and replace troublesome equipment instead of simply paying a fine. The legislation even includes a quality improvement programme (QIP). This is for industries with severe difficulties controlling fugitive emissions. Such a programme requires a company to identify the problem, investigate alternative solutions and apply the most appropriate one. This must all be completed within a given time-scale. Such legislation provides a compelling incentive for companies to do better and is an excellent example of well-developed legislation. Emissions from many diverse sources are now not only coming under public scrutiny but under legislative control, again, especially in the United States, where VOC abatement technologies are being adapted for control of petroleum fume emissions at marine loading/unloading stations and filling station forecourts.

16.6 PROBLEMS

Stream characterization and equipment selection
16.1 What is the difference between an organic and an inorganic compound?
16.2 Classify the following compounds as either organic, inorganic or particulate:
   (a) dry cleaning agents,
   (b) petrol,
   (c) ammonia,
   (d) boiler flue gases.
16.3 Under what circumstances should incineration be considered as a method of air emissions control?
16.4 Figure 16.1 outlines the principal flow sheeting alternatives for air emissions control. List all the control methods you know of and classify each of them under the options outlined in Fig. 16.1.

Equipment design
16.5 An air stream of 500 N m^3/h at a temperature of 50 °C and at atmospheric pressure, contains only acetone. Determine the maximum mass flow rate of acetone. If cooling water is available at a temperature of 25 °C and a temperature rise of 5 °C is permissible, determine the amount of acetone which could be condensed and the amount of heat that must be removed to do so. Calculate also the cooling water flow requirement.
16.6 Discuss simple versus complex absorption.
16.7 How would the choice of an absorbant for absorption be influenced by the use of simple or complex absorption?
16.8 What are the important parameters in determining the optimum absorption column size?
16.9 The vapour liquid equilibrium relationship for ammonia at 10 °C is given by

$$Y = 0.0225X$$

Repeat Example 16.2 using this relationship. Comment on the results.
16.10 Figure 16.8 shows the principal valves on a twin bed, steam desorbed adsorption unit for the recovery of VOCs. List the various stages in the adsorption–desorption cycle, label the valves in the diagram and for each stage, state what valves should be opened and what valves should be closed.
16.11 An electrostatic precipitator has a 5000 m^2 plate area. It is 99.0 per cent efficient at treating 100 m^3/s of flue gas from a power plant. How large would the plate area have to be to increase its efficiency to 99.9 per cent?

16.12   A scrubber followed by two electrostatic precipitators is used to clean the flue gas from a power plant. Downstream of the precipitators there is a bed of catalyst for NO_x reduction. The precipitators are in parallel and have an efficiency of 97 per cent. Every 24 hours both precipitators are bypassed for 10 minutes to clean.

Sketch the system. What, do you think, is the function of the absorber? Is this the best arrangement for the equipment? Estimate the reduction in the amount of dust escaping through the ESPs if they are cleaned at different times.

If a third precipitator is installed to increase the overall efficiency, determine whether it should be in parallel or series with the other two and when it should be cleaned.

16.13   A gas stream has a flow rate of 1500 N m^3/h, a density of 1.15 kg/m^3 and a viscosity of 0.018 cP. It contains particles with a density of 1200 kg/m^3 and a diameter of 6 μm. If the pressure drop is limited to 150 mm HG and the pressure drop through a cyclone is given by

$$\Delta P = 4\rho(V^2)$$

where
$$\rho = \text{density of the gas, kg/m}^3$$
$$V = \text{inlet gas velocity, m/s}$$

determine what fraction of particles will be removed:

(a) by one cyclone,
(b) by four cyclones in series,
(c) by five cyclones in parallel,
(d) by a gravity settling chamber.

Note: this question requires the student to investigate the relationship between d_{50} and other particle sizes for a cyclone.

NO_x control

16.14   Why would urea be preferable to ammonia as a method of controlling NO_x emissions?

16.15   From the point of view of generating the minimum amount of waste, which is the most suitable NO_x control method?

Odour control

16.16   An activated sludge effluent treatment plant is experiencing complaints from the surrounding area. The plant is a secondary treatment plant. Discuss the possible sources of odours and the ways to control them.

16.17   A municipal landfill site wishes to install an odour abatement plant. If the area of it is 2 km^2, discuss the options for odour abatement.

REFERENCES AND FURTHER READING

BS 7750 (1992) *Specification for Environmental Managements Systems*, British Standards Institution, London.
Chemical Industries Association (CIA) (1992) *Guidance on the Management of VOC emissions*, Chemical Industries Association, London.
Corbitt, R. A. (1990) *Standard Handbook of Environmental Engineering*, 1st edn, McGraw-Hill, New York.
Coulson, J. M. and J. F. Richardson (1983) *Chemical Engineering*, Vol. 6, 1st edn, Pergamon Press, Oxford.
Coulson, J. M. and J. F. Richardson (1991) *Chemical Engineering*, Vol. 2, 4th edn, Pergamon Press, Oxford.
Croom, M. L. (1993) 'Effective selection of filter dust collectors', in *Chemical Engineering*, Vol. 100, No. 7, July 1993, McGraw-Hill, New York.
Davis, M. L. and D. A. Cornwell (1991) *Introduction to Environmental Engineering*, 2nd edn, McGraw-Hill, New York.
Flagan, R. C. and J. H. Seinfeld (1988) *Fundamentals of Air Pollution Control*, 2nd edn, Prentice-Hall, Englewood Cliffs, New Jersey.

Hanley, J. and J. Petchonka (1993) 'Equipment selection for solid–gas separation', in *Chemical Engineering*, July, McGraw-Hill, New York.

Hesketh, H. E. and F. L. Cross (1989) *Odor Control Including Hazardous/Toxic Odors*, 1st edn, Technomic Publishing Company, Pennsylvania.

Heumann, W. L. (1991) 'Cyclone separators: a family affair', in *Chemical Engineering*, June, McGraw-Hill, New York, pp. 118–123.

Institution of Chemical Engineers (IChemE) (1985) *A User Guide to Dust and Fume Control*, 2nd edn, Institution of Chemical Engineers, London.

Lefebvre, A. H. (1983) *Gas Turbine Emissions*, 1st edn, McGraw-Hill, New York, p. 481.

Lapple, C. E. (1951) 'Processes use many collection types', *Chem. Engng.* **58**(5), 144, May.

Luft, T. A. (1986) 'Technical instructions on air quality control', German Federal Emission Control Law, 27 February.

Luyben, W. L. and L. A. Wenzel (1988) *Chemical Process Analysis*, 1st edn, Prentice-Hall, Englewood Cliffs, New Jersey.

Lydersen, A. L. (1979) *Fluid Flow And Heat Transfer*, 1st edn, John Wiley, Chichester.

McInnes, R., S. Jelinek and V. Putsche (1990) 'Cutting toxic organics', in *Chemical Engineering*, September, McGraw-Hill, New York.

Martin, A. M., S. L. Nolen, P. S. Gess and T. A. Baesen (1992) 'Controlling odors from CPI facilities', *Chemical Engineering Progress*, **88**(12), 53–61, December.

Oglesby, jr. S. and G. B. Nichols (1978) *Electrostatic Precipitation*, Marcel-Dekker Inc., New-York.

Peavy, H., D. Row and G. Tchobanoglous (1985) *Environmental Engineering*, McGraw-Hill, New York.

Perry, R. H, D. Green and J. O. Maloney (1984) *Perry's Chemical Engineers' Handbook*, 6th edn, McGraw-Hill, New York.

Reid, R. C., T. K. Sherwood and J. M. Prausnitz (1977) *The Properties of Gases and Liquids*, 3rd edn, McGraw-Hill, New York.

Rousseau, R. W. (1987) *Handbook of Separation Process Technology*, 1st edn, John Wiley and Sons, New York.

Shanahan, I. (1990) *Biofilters for Odour Control*, Bord na Móna, Ireland.

Simonson, J. R. (1988) *Engineering Heat Transfer*, 2nd edn., MacMillan, London.

Treybal, R. E. (1981) *Mass-Transfer Operations*, 3rd edn, McGraw- Hill, New York.

Whalley, P. B. (1987) *Boiling, Condensation and Gas–Liquid Flow*, 1st edn, Oxford University Press.

WPCF (1979) *Odour Control For Wastewater Facilities*, Manual of Practice No. 22, Water Pollution Control Federation, Washington D.C.

AGRICULTURAL POLLUTION CONTROL

17.1 INTRODUCTION

The primary objective of agricultural enterprise is the optimization of profit. In the past, maximizing output has been the most expedient way of achieving this objective; farmers have accomplished this by modifying the agro-ecosystem. Modifications over recent decades include increased use of inorganic fertilizers and greater use of chemicals for controlling weeds and other pests, both resulting in improved plant yields and quality. Other changes include larger animal herds, higher stocking rates, increased use of feed concentrates and improved animal performance from breeding programmes. Modified animal housing designs and the use of high-density, confined animal production practices are other changes in agricultural practice. The results of these changes have been higher concentrations of nutrients, organic matter and chemicals on modern farms compared to those operating a few decades ago.

Consequently, the pollution potential of modern farms is significantly greater than that of the more extensive farming systems of previous decades. In addition, agricultural land is now used as a receptor of non-farm organic wastes, such as food processing wastes and municipal wastewater sludges. The need to control pollution from agriculture is more obvious, now that efforts begun in the 1970s to reduce pollution from industries and municipalities have taken effect.

This chapter introduces concepts for developing and evaluating effective agricultural pollution control strategies. The primary emphasis is on controlling water pollution.

17.2 OBSTACLES TO AGRICULTURAL POLLUTION CONTROL

Several unique characteristics of the agricultural industry make pollution control difficult. Agricultural production occurs under circumstances that are quite different from those common to most industries required to control pollution. Firstly, the basic medium for almost all agricultural production (i.e. the soil) is a non-homogeneous biological system, with physical, chemical and biological characteristics that may vary widely, even within a few metres. Other industries take elaborate quality control precautions to minimize variability in their production base (e.g. machinery, raw products and facilities). Secondly, unlike manufacturing industries that are based in *confined* areas under controlled conditions (inside factories, for example), land-based agricultural production systems involve *large land areas*. Land-based agricultural production systems are open to the effects of uncontrollable and largely unpredictable climatic events, which add variability to production conditions. In contrast, other production industries strive for closely monitored and regulated production environments (e.g. ventilation, relative humidity and lighting levels) and predictable, if not controllable, supplies of raw inputs. Finally, but importantly, production of basic (not processed) agricultural products involves small profit margins. Increased production costs associated with pollution control are very difficult to transfer to consumers, partly because of farm price support mechanisms often operated by national governments and international trading blocks.

From an environmental management perspective, the intensification of agricultural production and its conduct in the open and variable environment present many difficulties. Production inputs, which may become pollutants, such as nitrogen (N) and phosphorus (P), cannot be collected and removed once they are added to the production system. Instead, these inputs become integrated into the production system and follow through natural cycles (Chapter 10). This situation is dramatically different from that posed by, for example, a manufacturing facility where there is greater control over the flow of inputs if production (or pollution) problems arise. From a waste treatment perspective, the agricultural scenario is also quite different from that encountered at manufacturing plants and most urbanized areas. Wastes can be collected from these sources and diverted to a common location (wastewater treatment plant) where highly trained individuals can manage the unit treatment operations under controlled conditions to yield an effluent of a given quality. The 'collection and treatment' approach of pollution control is neither economically nor technologically feasible for land-based agricultural systems because collection often is not practical (e.g. runoff from farm fields) and their occurrence often is unpredictable (i.e. weather dependent). Finally, although farmers are highly trained and experienced in agricultural production, they are not experts in pollution control.

17.3 AGRICULTURAL WATER POLLUTION CONTROL PRINCIPLES

Controlling agricultural pollution requires an interdisciplinary approach that combines the expertise of engineers, agronomists, soil scientists and, in some situations, biologists. As discussed in Sec. 17.1, pollutants of primary concern from agriculture include nutrients (N and P), organic matter (BOD_5), pathogens (bacteria), synthetic organic chemicals (pesticides) and, in arable areas, eroded soil (TSS). Agricultural pollution derives both from *point* (i.e. well-defined) sources around the farmyard, such as slurry tanks, and from *non-point* (i.e. ill-defined or diffuse) sources, such as fields or portions of fields. The same physical, chemical and biological principles used to control industrial and municipal sources of pollution are applicable to agricultural sources, but must be applied in the context of an open, uncontrolled and variable environment instead of a controlled treatment system. Further, the techniques used to minimize pollution risk from the two types of agricultural pollution sources are different, although a total quality management (TQM) approach or systems approach is a basic requirement for the success of all methods.

17.3.1 Point Sources

Key point sources of pollution in agricultural systems are the farmyard itself (uncovered exercise or feeding areas, soiled water storage tanks), facilities used to store animal wastes (slurry pits and tanks, dungsteads), facilities for collecting and storing silage effluent (pads and tanks) and facilities used for storing and handling pesticides (storage sheds, filling and rinseate collection areas). These sources pose threats to the environment because they concentrate large amounts of potential pollutants in a relatively compact volume and area.

Minimizing pollution risks from point sources depends on properly designing, constructing and managing the facilities. From an engineering perspective, 'properly designed and constructed' facilities are those that satisfy their intended purpose at minimum cost consistent with accepted factors of safety. The intended purposes of controls for most agricultural point sources of pollution are to contain pollutants and prevent their uncontrolled release to the environment. The basic requirement for such facilities is adequate waste storage capacities, structural integrity and careful site locations. Many countries have research-based design criteria for these facilities, which are specified by appropriate government agencies. Accepted construction practices must be followed to guarantee that the facilities will perform as designed. Just as important, however, operators (farmers) must utilize, or manage, these facilities properly to achieve design objectives, i.e. follow a TQM approach to pollution control.

17.3.2 Non-point Sources

Non-point sources of agricultural pollution are the land areas on which agricultural production is accomplished. Although these 'fields' can be identified by physical dimensions, the precise origin of pollutants from these areas cannot be clearly identified, especially within the context of an entire catchment. The diffuse nature of this type of pollution gives rise to its definition as 'non-point source' (NPS) pollution.

Precipitation excess and snowmelt (creating surface runoff and interflow) are the transport agents of non-point source pollutants to surface waters; drainage water through the soil profile is the transporting agent for pollutants to groundwater. Wind can also transport non-point source pollutants to surface waters, but, except in special circumstances, the relative importance is small compared to water. In general, it is not feasible to collect runoff, interflow or drainage water so that entrained pollutants can be removed.

Thus, controlling agricultural non-point source pollution depends largely on preventing pollutants from leaving the production system (i.e. the soil). The edges of fields and bottom of the root zone are convenient demarcations for land-based agricultural systems. Since most agricultural pollutants are initially production inputs (e.g. nutrients), keeping these constituents within the confines of the root zone and field edge is economically, as well as environmentally, sound.

Management practices are the most important agricultural non-point source pollution control techniques, supplemented in some cases by structural or vegetative controls. Managerial practices must be practicable (technologically and economically achievable) and adaptable to currently used farming techniques to be effective on either a voluntary or regulatory basis. Structural practices, such as ponds, terraces and diversions, must be designed following engineering principles; vegetative controls must incorporate both physical design and agronomic principles.

17.4 POINT SOURCE CONTROLS

17.4.1 Site Selection

The proximity of any source of pollution to receiving waters is a major determinant in the relative pollution risk posed by the source. Site selection is thus the primary step in designing facilities that will contain agricultural pollutants. Codes of good agricultural practice typically give guidance for siting point source pollution controls. As a rule, these facilities should be located as far as practicable from surface waters and down gradient from nearby (50 to 75 m) groundwater wells. In addition, facilities should be placed where surface and subsurface soil conditions are suitable (low organic matter, low shrink-swell potential, good compactability and adequate bearing capacity, deep water table, etc.), as determined by soil borings or test pits. The number and depth to which borings are made should be consistent with providing enough information on which to base a safe structural design. Where animal waste lagoons are proposed, a site investigation should confirm that subsurface soil conditions will prevent excessive leakage of the lagoon contents or that unsuitable conditions can be ameliorated by site modification (such as incorporation of low-permeability soils) or by the use of synthetic liners. Sites with creviced bedrock, karst limestone, shallow or gravelly soils, and soils with high water tables pose special challenges for a safe design of any facility that is to contain agricultural pollutants.

Local siting regulations may exist and must be followed when designing facilities to control agricultural pollutants. Dairies are often subject to special health regulations that may restrict the siting of animal waste storage facilities. Siting of point source pollution control facilities must respect the health and safety of both humans and animals, making the direction of prevailing winds and the location of existing structures important considerations. For both safety and utility, facilities should be located to provide convenient all-weather access.

17.4.2 Sizing of Structural Facilities

The underlying purpose of agricultural point source pollution control facilities is to contain potential pollutants (e.g. dirty or soiled water, animal wastes, pesticide rinseate). Achieving this objective is dependent on providing structurally sound facilities with adequate capacities to hold the pollutants (and rainwater in the case of uncovered structures) until they can be further managed, typically by application to land. Design capacities are determined by the rate at which the pollutants are generated and the required storage period. The durations of required storage periods are often specified by local regulations and typically are related to the availability of suitable periods during which wastes can be applied to land. If the facilities are to afford some degree of pollutant treatment, such as anaerobic lagoons, sizing of the facilities also depends on biochemical principles (Chapter 12). While design principles are similar to those used for wastewater treatment facilities, the characteristics of agricultural wastes dictate that a somewhat different application of these design procedures be used (Barth, 1985; Merkel, 1981). Figure 17.1 is a generic representation of the relative volumes that must be accommodated in an uncovered waste storage or treatment structure.

Excreta rates for various types of farm animals are given in Table 17.1. Management practices, animal diets, animal ages and productivity levels of the animals affect waste excreta rates and composition. Consequently, the values given in Table 17.1 are only *approximations* of excreta rates found in a specific situation. Values in the table also disregard the volumes of water, bedding or litter, runoff from yard areas or other wastes (such as dairy parlour washings or silage effluent) that are routinely mixed with excreted wastes. These additions must be considered when sizing facilities to contain animal wastes.

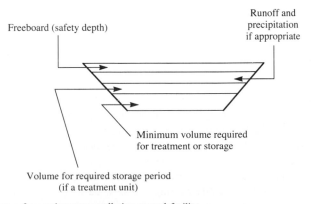

Figure 17.1 Relative volumes for a point source pollution control facility.

Table 17.1 Approximate daily livestock excreta production

Type of livestock	Body weight (kg)	Excreta (litres/day approximately) Range	typical	Excreta (approximate %, dry matter)
Cattle				
Calf—up to 2 months fed concentrated liquid feed	73	4.0–6.1	5.0	12–14
Calf—up to 6 months	160	6.3–7.8	7.5	12–14
Heifer—up to 12 months	270		15.0	12–14
Heifer—up to 18 months	380		20.0	12–14
Beef store—up to 12 months	400	10–34	27.0	12–14
Dairy cow	500	32–54	41.0	13
Horse	680		30.0	9
Mink			0.2	13
Pigs				
Piglet—up to 3 weeks	5		1.0	10
Weaner	12	1.5–2.5	2.0	10
Fattener fed dry meal	50	2.0–5.5	4.0	10
Fattener fed water; meal Fattener fed 2:1	50	2.0–5.5	4.0	10
Fattener fed 4:1	50	4.0–9.0	7.0	6
Fattener fed swill	50	Very variable	15.0	3–5
Fattener fed whey	50	14.0–17.0	14.0	2
Boar			5.0	10
Dry sow			4.5	10
Sow and litter to 3 weeks			15.0	10
Poultry				
Broiler (including shavings litter)	2		0.04	60
Duck	2		0.03	12
Goose	3.5		0.55	25
Laying hen	2	0.10–0.14	0.114	25
Rabbit	2.7	0.39		
Turkey	7		0.17	23

Adapted from Grundy, 1980, with permission

The contributions of direct precipitation must also be considered for uncovered facilities exposed to the atmosphere. The amount of precipitation requiring containment typically is specified in local regulations by defining the return period and duration of a 'design storm', e.g. precipitation over 24 hours that occurs, on average, once every 10 years (i.e. a 10-year, 24-hour event). Such a specification also allows, if necessary, calculation of the amount of runoff that must be accommodated in the facility. Precipitation is extremely location dependent: for example, the 10-yr, 24-hr rainfall for Dublin (eastern Ireland) is 54 mm, whereas for Malin Head (Northern Ireland) it is 42 mm. A prudent design incorporates on-farm observations of excreta rates and additions of extraneous matter (i.e. water, bedding), as well as local climatic data.

Example 17.1 A dairy farmer milking 60 cows wants to add a below-ground slurry pit to contain animal wastes and runoff from a 15 m × 45 m exercise yard for a period of 4 months, during which time the long-term average precipitation totals 560 mm. What size of tank should be constructed?

Solution

(a) Animal waste generation:

$$60 \text{ cows} \times 41 \text{ L/day animal} \times 120 \text{ days} = 295.2 \times 10^3 \text{ L} = 295.2 \text{ m}^3$$

(b) Runoff diverted into tank from exercise yard:

$$15 \text{ m} \times 45 \text{ m} \times 0.56 \text{ m} = 378 \text{ m}^3$$

(c) Minimum required volume $= 295.2 \text{ m}^3 + 378 \text{ m}^3 = 673.2 \text{ m}^3$.

(*Note*: Approximately 0.5 m should be added to the resulting tank depth for freeboard as an additional measure of safety.)

Effluent resulting from silage-making operations is a serious pollutant requiring collection and storage until it can be utilized on agricultural land. Table 17.2 contains estimates of the volumes of silage effluent produced when grasses of various moisture contents are ensiled. These data provide estimates of the capacity required for dedicated effluent containment facilities or the additional volume required if the effluent is diverted into other facilities (such as a slurry tank).

Water is the predominant component of pollutants held in point source pollution control facilities. Thus, minimizing the size of these facilities is accomplished by excluding as much rainwater as possible. For most facilities this can be accomplished by:

- Diverting clean runoff or roof drainage away from areas that are contaminated with wastes, such as exercise yards
- Minimizing the area of uncovered 'dirty' farmyard areas
- Using low-volume parlour washing procedures and
- Using 'demand' type drinkers and correcting leaks in drinkers, sprinklers and supply pipes

17.4.3 Design and Construction

Most agricultural point source pollution control facilities are reinforced concrete or steel structures that must be designed according to accepted engineering standards and procedures. The corrosive nature of many agricultural pollutants (such as animal wastes and silage effluent) requires that special precautions be taken to select resistant construction materials and follow accepted construction practices. Construction of most facilities should be supervised by a construction engineer or other competent inspector. Care must be exercised to ensure that foundations are sufficient to prevent differential settling of facilities that would weaken structural integrity or cause leakage. Design guidelines for most types of agricultural point source

Table 17.2 Approximate effluent volume from grass ensiling operations

Grass dry matter content at filling (%)	Litres of effluent per tonne of silage
16	220
18	170
20	130
22	90
26	40
30	0

From Teagasc, 1989, with permission

pollution control facilities are widely available (Midwest Plan Service, 1985; ASAE, 1990; Department of Agriculture, 1985; MAFF, 1991). Design and construction specifications typically are available from local or national government agricultural agencies for facilities constructed using government financial assistance. Information about climatic factors (predominant wind speed and direction, air and soil temperatures, precipitation amounts and patterns) can be obtained from national or local meteorological offices and agricultural research/advisory agencies.

17.5 NON-POINT SOURCE (NPS) CONTROLS

Agricultural non-point source pollution controls are managerial, structural or vegetative practices. Regardless of type, the objective of all techniques is to prevent or reduce the availability, release or transport of agricultural pollutants to receiving waters. Agricultural NPS pollution control practices can be viewed as on-site controls. The term 'best management practices' (BMPs) is often used collectively for these pollution control techniques. 'Best' implies that an individual practice or combination of practices is the most effective and practicable control technique for a particular combination of farm characteristics and pollution problems.

Practicability is an absolutely essential component of BMPs to control agricultural non-point source pollution. Agricultural production practices at any given geographic location are the result of evolution and adaptation over centuries to the interactive influences of climate, soils and topography.

NPS pollution control techniques must be suited to these same influences and be economically viable as well. Designing BMPs for a given farm requires a field-by-field examination of soils, slope lengths, steepness and proximity to groundwater or surface waters, as well as a whole-farm evaluation of the type of enterprise, financial resources available and level of managerial expertise. The techniques for making such evaluations are given elsewhere (Hudson, 1981; Novotny and Chesters, 1981; Schwab, *et al.*, 1993).

Tables 17.3 and 17.4 contain a list of frequently used best management practices that can be effective in controlling losses of N, P and eroded soil from arable agricultural systems in the northern hemisphere. Not all practices will be appropriate for every farm situation. Integrated pest management (IPM) practices are also applicable for arable systems in which pesticides are used. As its name implies, IPM is a comprehensive pesticide management strategy that enhances the effective use of pesticides, thereby reducing pollution risks by combining non-chemical and chemical controls, field examinations of pest infestations and delaying pesticide applications until a predefined economic threshhold for pest damage has been reached.

For grassland systems, the nutrient management practices in Table 17.3 are entirely appropriate to guide the environmentally responsible, agronomically effective use of both animal wastes and inorganic fertilizers. In addition, livestock and pasture or grazing areas should be managed to minimize animal damage to soils and grass quality, and to prevent uncontrolled access of animals to surface waters. These objectives are achieved by managing stocking densities, the timing and duration of grazing, and providing access to shade and water away from receiving streams.

17.6 LAND APPLICATION OF WASTES

Land application of wastes is the most economical, practical and environmentally sustainable method for managing agricultural wastes, especially animal wastes. Application of agricultural wastes to the land recycles valuable nutrients and organic matter into the system from which they originated. Land application can also be an effective component of management strategies for other organic wastes, such as wastewater treatment sludges and food processing wastes.

Table 17.3 Effective best management practices for controlling nitrogen and phosphorus losses from cropland†

Practice	Description
1. Correct nutrient application rates	Match crop needs to nutrient application rates. Establish optimum economic yield goals consistent with soil capability and managerial expertise. Apply nutrients according to soil test programme recommendations. Account for all sources of nutrients applied to the crop.
2. Appropriate timing of nutrient application	Applications should correspond as closely as possible to plant needs. Avoid late autumn applications of nitrogen for spring-planted arable crops. Avoid application of nutrients on frozen soils. Where consistent with other recommended agronomic practices, use split applications of nitrogen to increase N uptake efficiency.
3. Appropriate method of nutrient application	Use methods that promote efficient nutrient use. Band or incorporate fertilizers or slurries where consistent with other crop management practices. Use only calibrated application machinery.
4. Reduced tillage practices	For arable crops, select tillage practices that are consistent with soil properties, topography, climate and the overall farming system. Minimize the number of tillage operations to reduce erosion risks.
5. Crop rotations	Incorporate legumes where possible to assist in reducing nitrogen additions for following crops. Most arable crop rotations that include a sod crop reduce erosion risks, reduce leaching and improve soil structure.
6. Cover crops	Where climate permits, plant cover crops to use residual nutrients, particularly nitrate N, remaining after main crop harvest, to reduce erosion and to decrease leaching. Legume cover crops may reduce the need for nitrogen additions to the following crop (e.g. grassland).
7. 'Critical area' seeding	Planting highly erodible areas in permanent cover reduces transport of eroded soil and any attached pollutants.
8. Pond	In certain situations, a permanent water impoundment can trap sediment and attached phosphorus and reduce nitrogen losses to receiving waters by promoting volatilization and denitrification.

† Practices 1 to 3 are also applicable to grassland systems.

Table 17.4 Effective best management practices for reducing losses of phosphorus and eroded soil for arable crops†

Practice	Description
1. Contour cultivation	Field operations are performed on the contour of the land, reducing runoff, nutrient and soil losses.
2. Strip cropping	Alternating strips of row crops and close growing crops, such as cereals, are planted on the contour of the land. Runoff is filtered and infiltration is increased.
3. Grass filter (buffer) strips	A permanent sod strip at the base of the slope of a field. Where runoff occurs as sheet flow, filter strips remove particulates from runoff and increase infiltration.
4. Natural vegetation filter area	Permanent indigenous vegetation at the base of a catchment, which filters particulates from runoff and increases the potential for denitrification.
5. Terrace	A constructed channel perpendicular to field slope and wide enough to be cropped, which reduces slope length and runoff velocity. Runoff is reduced and infiltration is enhanced.
6. Diversion	A narrow channel constructed perpendicular to the field slope and planted to grass, which diverts excess runoff to areas where it can be managed to reduce erosion potential.
7. Grassed waterway	A sod channel that transports runoff at non-erosive velocities to a stable outlet.
8. Windbreaks	Strips or belts of trees or shrubs established within or next to a field, farmstead or feedlot to decrease wind velocities and reduce erosion potential.

† Most of these practices reduce runoff and increase infiltration: consequently, care must be exercised in their application on freely draining soils that overlie sensitive groundwater resources.

When designed and managed 'properly', systems for the land application of wastes do not pose undue threats to environmental quality. One aspect of proper design and management involves applying wastes at the correct rates, at the correct time, using the correct methodology. Hydrologic and agronomic principles govern the design of land application systems.

17.6.1 Rate of Application—Organic

From a waste treatment perspective, the soil system can be viewed as a fixed film biological reactor (recall the percolating filter in Chapter 12), which, due to an immense microbial population, has a large—although finite—waste assimilative capacity. For the soil system to function effectively, wastes must be applied to land at rates that do not exceed either the instantaneous or long-term assimilative capacity of the soil system.

The instantaneous assimilative capacity of a soil system is related most closely to an ability to degrade organic matter aerobically, thereby avoiding problems resulting from septic soil conditions (odours and plant damage). This in turn is dependent on the aeration status of the soil (a function of soil texture and structure and moisture content), temperature and the organic strength of the waste. Organic strength of wastes is measured by BOD_5, BOD_{ult} and/or COD, as appropriate for a specific waste.

Typically, BOD is associated with the solid fraction of organic wastes, which in land application systems remain on the soil/plant surface or within the upper few millimetres of the soil (assuming the wastes are surface applied without incorporation). Here, oxygen transfer rates from the atmosphere are suitable for aerobic waste degradation. For this reason, problems resulting from organic overloading are rarely encountered when animal wastes are surface applied at agronomically acceptable rates.

Conversely, COD is generally associated with dissolved pollutants and is often the most appropriate measure of waste organic strength when land application systems are being designed. While a portion of these pollutants may be retained by physical (i.e. filtering) or chemical processes (e.g. ionic adsorption) near the soil surface (where aeration rates are high), the liquid fraction of wastes applied to land will move deeper into the soil profile than the solid (sludge) fraction. Because oxygen transfer rates decrease steadily with depth in the soil profile, acceptable organic loading rates also decrease with depth below the soil surface.

Application rates for wastes and wastewaters should be based on the characteristics of the most restrictive soil horizon. Oxygen diffusion rates of at least $0.2\,\mu g/cm^2/min$ are needed for most agricultural crops. Because diffusion rates through water are limited, achieving adequate aeration in a soil depends on controlling soil moisture. A minimum air-filled volume of 10 per cent in the root zone is recommended for most agronomic crops (Wesseling *et al.*, 1957). On well-aerated soils (e.g. light textured, free draining sands and loams) total oxygen demand loadings of 56 to 112 kg/ha day have been used successfully (EPA, 1975).

An appropriate method of assessing proposed organic loading rates is to compare them with oxygen utilization rates in productive agricultural soils, which reach 3.4 to 6.8 kg/ha hr. The actual organic loading rates that are acceptable for a given site depend on waste characteristics (degradability), soil characteristics (oxygen movement), temperature (bacterial activity) and precipitation (degree of saturation of soil pore space). Methods by which to ascertain oxygen diffusion rates in soil may be found (Hillel, 1980).

Example 17.2   Assume that a beverage company wishes to pay a nearby farmer to accept processing wastewater for application to the farmer's land. The soil is a sandy loam. The wastewater has a COD of 15 000 mg/L and will be generated at an average volume of $37.85\,m^3$ per week. Assuming soil and weather conditions permit year-round application, how much area is required to achieve an acceptable organic loading rate of 110 kg/ha day? (Neglect long-term assimilation capacity.)

Solution

(a) Mass of COD produced:

$$(15\,000\,\text{mg/L} \times 37\,850\,\text{L/wk}) \times 10^{-6}\,\text{kg/mg} = 568\,\text{kg/wk}$$

(*Note:* assuming two equal applications per week, the COD load per application is 284 kg.)

(b) Area required:

$$(284\,\text{kg})/(110\,\text{kg/ha}) = 2.6\,\text{ha}$$

(*Note:* the hydraulic loading rate required to achieve this organic loading rate should also be checked for acceptability.)

17.6.2 Rates of Application—Other Parameters

Hydraulic   Hydraulic loading influences the instantaneous assimilative capacity of soil when liquid or semi-solid wastes are applied. Application rates that exceed the infiltration rate of the soil will result in surface ponding, surface runoff and the consequent transport of pollutants. Likewise, application rates that exceed the storage capacity of the soil profile will result in leaching and, potentially, the transport of dissolved pollutants downwards and out of the root zone.

Infiltration rates decrease with time, approaching as a limit the saturated hydraulic conductivity of the soil (see Chapter 4). Characteristics of the applied liquid, the tendency of the soil surface to crust or seal, soil moisture content and the presence and type of vegetation all influence the infiltration rate. Infiltration capacity, sometimes referred to as the final infiltration rate, is the rate at which liquid will enter the soil surface and is limited only by soil factors (soil structure and pore size distribution). Table 17.5 contains 'typical' infiltration capacities for clear water for several soil types on negligible (0 to 3 per cent) slopes. As with all soil characteristics, actual values for infiltration capacities can vary widely from those given in Table 17.5, making on-site soil investigations imperative for a complete design. In addition, wastewaters usually behave differently to clear water, resulting in infiltration capacities that may in some cases be as little as 10 per cent of clear water values.

While infiltration capacity is one factor controlling instantaneous hydraulic loading rates, the amount of waste application must also be a hydraulic consideration. Applying liquid wastes in quantities greater than the soil retention capacity results in leaching from the root zone and increased potential for groundwater contamination. The moisture retention or storage capacity of a soil is defined in terms of the amount of *plant available* water that can be retained and is usually defined as *available water capacity* (see Chapter 4).

One 'end point' for available water capacity is the amount of moisture that is retained in the soil against the force of gravity. A combination of forces in the soil matrix create what is equivalent to a negative pressure, or 'suction', that controls the movement of soil moisture. Depending on the soil, the negative pressure at which moisture is retained against gravity is -10 to -33 kPa (-0.1 to -0.3 bar); in this condition the soil is said to be at *field capacity*. The other end point for available water capacity is the *permanent wilting point*, which is the amount of water retained at the maximum negative pressure

Table 17.5 Typical clear water† infiltration capacities for various soils

Soil type	Infiltration capacity (mm/h)
Coarse textured loamy sands, loamy fine sands	18–38
Medium textured very fine sandy loams, loams and silt loams	8–18
Fine textured sand clays, silty clays and clays	3–6

† Infiltration capacities for wastewaters can be much smaller than clear water values

Table 17.6 Typical available water capacities for various soil types

Soil type	Available water capacity (mm/m)
Coarse textured loamy sands, loamy fine sands	83
Medium texture very fine sandy loams, loams and silt loams	166
Fine textured sandy clays, silty clays and clays	192

($-1500\,$kPa or $-15\,$bar) that plants can exert. Table 17.6 contains typical values for available water capacities for several soil types in terms of depth of moisture retained per unit depth of soil. Values in Table 17.6 are only guides, actual values vary widely for a particular soil.

More detailed techniques for determining acceptable hydraulic loading rates are available elsewhere (Halley and Soffe, 1988; Hillel, 1980; McCuen, 1989; Shaw, 1988).

Nutrients   Whereas the short-term or instantaneous assimilative capacity of the soil system is most closely related to organic and hydraulic loadings, the long-term assimilative capacity of a waste application site is more aligned with nutrient application rates. (For wastewater treatment sludges, heavy metals may determine long-term loading rates.) For environmental sustainability, wastes should be applied at rates that supply the nutrient needs of the crop produced (and, where appropriate, the buildup of soil fertility). Unfortunately, organic wastes rarely contain the major plant nutrients (N, P and K) in the relative proportions that are required by plants. In addition, the N content of organic wastes (especially animal wastes and sewage sludges) tends to be 'unreliable' because it must be mineralized from organic form to be useful to plants. As discussed in Chapter 10, mineralization rates are variable and dependent on uncontrollable factors such as weather. For these reasons, inorganic fertilizers are often used in combination with organic wastes to meet the nutrient needs of crops. Analysis of both the soil (to determine fertility status) and the organic wastes (to determine nutrient contents) is imperative to minimize pollution potential and the need for inorganic fertilizers.

Table 17.7 contains approximate 'crop uptake' values for a variety of crops and are indications of nutrient application rates that should be used in designing a land application system for wastes. Actual fertilizer recommendations will deviate from values in Table 17.7 when soil fertility levels are considered. More definitive nutrient application guidance should be obtained from university or government agricultural research and advisory services. While data in Table 17.7 are normalized on the basis of yield, plant uptake at these rates nevertheless is predicated on achieving good agronomic crop yields. Because crop production occurs under uncontrollable weather conditions, achieving these yields is not always straightforward, and requires careful attention to both the nutrient and water requirements of the crop. Nutrient uptake rates in Table 17.7 should be viewed as approximations to maximum nutrient application

Table 17.7 Approximate nutrient uptake by various crops, kg/t (fresh weight basis)

Crop	Nutrient uptake (kg/t)[†]			% dry matter
	N	P	K	
Cereals				
Grain	17.0	3.4	4.7	85
Straw	6.0	0.7	6.8	85
Sugar beet				
Roots	1.7	0.3	1.8	22
Tops	3.2	0.5	4.8	16
Grass				
Silage	6.4	0.6	4.0	20
Hay	14.0	2.6	15.0	85

[†]kg/t of fresh weight.
Adapted from Archer, 1988, with permission

rates, except where soil fertility levels are deficient and require correction (such as by increasing soil P levels).

Timing of waste applications   The timing of waste applications is dependent on a combination of crop, site and weather factors. In general, wastes should be applied at times that will supply nutrients when the crop requires them. Practical considerations associated with agricultural production systems and weather act to constrain precisely when wastes are land applied.

Crops incorporate nutrients into their growing biomass; ideally nutrients should be supplied (by release from the soil and by application of wastes and/or fertilizers) at the rate at which they are required by the crop. As a practical matter, it is rarely possible to 'spoon feed' nutrients to either tillage crops or grassland. Instead, good agricultural practice consists of applying nutrients (wastes or fertilizers) as early in the growing season as possible and as frequently as economically and practically achievable. Frequency of application is governed by fuel, labour and machinery costs as well as by soil compaction risks. The timing of nutrient applications for efficient crop production is typically recommended by agricultural research and advisory services.

Recommended waste application times must be adjusted to accommodate weather and soil conditions. For example, operation of waste spreading equipment when soils are saturated damages soil structure and creates a pollution potential from runoff. Applying wastes when heavy precipitation (in excess of infiltration capacity) is imminent also increases pollution potential. Pollution potential from runoff at waste application sites decreases exponentially as the time between waste application and the occurrence of runoff increases. Adequate storage capacity and attention to management is essential to being able to apply wastes at the proper times.

Example 17.3   The wastewater described in Example 17.2 is to be applied to grassland for silage. The wastewater has a total N content of 45 mg/L (as $NH_4^+ - N$) and a total P content of 30 mg/L. The expected yield of silage is 12 t/ha (dry matter). At what rate should the wastes be applied? Are the 2.6 ha computed in Example 17.2 adequate?

Solution

(a) Calculate annual N and P production:

$$N: 45\,mg/L \times 37\,850\,L/wk \times 10^{-6}\,kg/mg \times 52\,wk/yr = 89\,kg/yr$$

$$P: 30\,mg/L \times 37\,850\,L/wk \times 10^{-6}\,kg/mg \times 52\,wk/yr = 59\,kg/yr$$

(b) Calculate application rate. (From a nutrient management perspective, year-round wastewater application would not be recommended. Assume wastewater will be applied once per week during the months of April, May, June, July and August when crop uptake of water and nutrients will be active. Once-per-week applications would be expected to allow ample resting of the application site.)

Total wastewater generation:

$$37\,850\,L/wk \times 52\,wk/yr \times 10^{-3}\,m^3/L = 1.968 \times 10^3\,m^3/yr.$$

Weekly application amount (20 applications):

$$(1.968 \times 10^3\,m^3)/20\,applications = 98.4\,m^3/application.$$

(c) Check application rate against infiltration capacity:

$$(98.4\,m^3/application)/(2.6\,ha) \times 10^{-4}\,ha/m^2 \times 10^3\,mm/m = 3.78\,mm/application.$$

Even if wastewater were applied within only 1 hour (3.8 mm/h), this hydraulic application rate would be acceptable (see Table 17.5).

(d) Re-check organic loading rate:

$$15\,000\,\text{mg/L} \times 98.4\,\text{m}^3/\text{application} \times 10^3\,\text{L/m}^3 \times 10^{-6}\,\text{kg/mg} = 1476\,\text{kg/application}$$

$$(1476\,\text{kg/application} \times 1\,\text{application/day})/2.6\,\text{ha} = 568\,\text{kg/ha/day}$$

This is unacceptably high; to reduce this to 110 kg/ha/day (see Example 17.2), approximately 13.5 ha of application area is required:

$$(1476\,\text{kg/application} \times 1\,\text{application/day})/110\,\text{kg/ha/day} = 13.4\,\text{ha}$$

(e) Check nutrient application rate with revised application area of 13.5 ha:

Crop uptake of $N = ?$

$$\text{Expected yield} = 12\,\text{t DM/ha} = 60\,\text{t/ha @ 20\%DM}$$

$$\text{from Table 17.7, N uptake rate} = 6.4\,\text{kg/t}$$

$$\text{Therefore uptake of } N = 60 \times 6.4 \times 13.5 = 5184\,\text{kg}$$

(Since the annual N supply from the wastewater is only 89 kg, the crop nutrient needs will have to be supplemented by commercial fertilizer or other nutrient source.)

In addition to matching nutrient needs by nutrient supply, wastewaters are typically applied at frequencies dictated by the *irrigation requirement* of the crop produced at the application site. Irrigation requirement is the difference in volume between effective precipitation and evapotranspiration and is, therefore, both crop and weather dependent. In addition, the crop's sensitivity to both saturated and excessively dry soil moisture conditions must be considered. In humid regions, irrigation is often utilized when 50 per cent of the available water capacity of the soil has been depleted. This guideline also determines the volume of water to be supplied by irrigation to meet crop needs. It also provides a 'rest period' for the application site, giving time for the soil to re-aerate and to recover infiltration capacity.

Limiting factor   Successful land application of wastes depends on many factors as described above. One *limiting factor* will, however, ultimately govern the utilization of wastes on land. Designs may be limited by hydraulic loading, organic loading, nutrient loading or, in the case of wastewater sludges, metals loadings. If nutrient loading is the limiting design criterion, local circumstances will dictate whether the design must be based on nitrogen or phosphorus application rates. Generally, where surface waters must be protected, nutrient loadings should be based on phosphorus, especially if catchment soils are prone to runoff. Conversely, if groundwater aquifers are to be protected, nutrient loadings should be based on achieving nitrogen balances.

Application methodology   The techniques by which wastes are applied to land are dictated by the waste characteristics, the production system and sometimes by regulatory considerations. Wastes that are 15 per cent or more in dry matter content are considered solid wastes and are land applied by flail-type application machinery. Wastes with between 4 and 15 per cent dry matter are considered to be slurries and can be land applied as liquids using specialized application equipment. Wastes with less than 4 per cent dry matter are dilute liquids and can be satisfactorily land applied using irrigation equipment. Whatever the application technique, care must be exercised to assure that the application equipment is calibrated and operated to give the target application rate.

Example 17.4   How much swine slurry is required to supply the nutrient needs (N and P) of first-cut silage from grassland? The first-cut yield is 5 tDM/ha. Assume 20 per cent DM (i.e. 5 t fresh weight per tDM).

Solution
Crop uptake of N (from Table 17.7):

$$(6.4\,\text{kg/t} \times 5\,\text{t fresh weight/tDM} \times 5\,\text{tDM/ha} = 160\,\text{kg/ha}$$

Crop uptake of P (from Table 17.7):

$$0.6\,\text{kg/t} \times 5\,\text{t fresh weight/tDM} \times 5\,\text{tDM/ha} = 15\,\text{kg/ha}$$

N supply in swine slurry (from Table 10.2): 4.6 kg/t
P supply in swine slurry (from Table 10.2): 0.9 kg/t

Slurry required for N needs: $(160\,\text{kg})/(\text{ha}/4.6\,\text{kg/t}) = 35\,\text{t/ha}$
Slurry required for P needs: $(15\,\text{kg/ha})/(0.9\,\text{kg/t}) = 17\,\text{t/ha}$

Therefore, P is the limiting factor. Slurry application will be restricted to 17 t/ha to satisfy P needs; commercial fertilizer or other suitable nutrient source must be used to supplement slurry to meet N needs.

Depending on the waste and characteristics of the land application site, regulatory requirements may dictate the waste application methodology to be used. Certain wastewater sludges are required to be injected into the soil to minimize the risk of disease dispersal. Injection is also sometimes required (or recommended) as a means to minimize gaseous emissions (of either odours or ammonia) from land-applied wastes. When injection of wastes is practised, shallow injection (into the root zone) would be preferred to deep injection (below the root zone) to achieve an environmentally sustainable land application system based on nutrient uptake.

17.7 CODES OF PRACTICE FOR LAND APPLICATION OF ANIMAL AND OTHER WASTES

Animal wastes have been applied to the land as a source of nutrients and organic matter for all of recorded history. Not surprisingly, codes of practice have been developed to guide the utilization of these wastes to meet agronomic and environmental goals. These codes also offer appropriate guidance for the land application of other organic wastes. A concise summary of a typical code of practice is given in Table 17.8. Other codes of practice are available elsewhere (e.g. MAFF, 1991; MAFF, 1992).

Table 17.8 Typical code of good practice for slurry spreading

1. Apply slurry at rates that take account of crop nutrient requirements and soil fertility levels.
2. Use a regular programme of soil and slurry testing to determine nutrient needs and supplies.
3. Apply slurry earlier rather than later in any crop growing season.
4. Avoid applying slurry on wet or waterlogged soils, on frozen or snow-covered soils, and on areas near surface waters and groundwater wells.
5. Always check weather forecasts before applying slurry; avoid spreading if precipitation predicted likely to produce runoff within 48 hours.
6. Use calibrated application equipment and operate it according to specification for achieving desired waste application rates.
7. Avoid direct contamination of surface waters and groundwater by maintaining sufficient safety margin (buffer zones or unsaturated soil respectively) between these resources and the slurry application site.
8. Where possible, avoid leaving bare soil over winter.
9. Take all reasonable steps to minimize odour emissions (incorporate wastes immediately, if possible; do not apply slurry when prevailing winds are in the direction of nearby residences; use low-trajectory instead of high-trajectory splash plates).

17.8 AGRICULTURAL AIR POLLUTION CONTROL

As discussed in Chapter 10, the major air pollutants from agriculture are odorous emissions related to the storage and handling of animal wastes. There are three main areas on farms with which odours are generally associated: the housing/yard area, the waste treatment/holding area and the waste utilization area. Odours in the housing and yard areas originate from animals themselves (body odours), from feed materials, and from decomposition of wastes excreted by the animals. Odours from the waste treatment/storage area are compounds derived from the incomplete anaerobic, microbial breakdown of organic matter in the wastes (carbohydrates, proteins and fats). At the waste utilization area, odorous compounds that developed during waste storage/treatment and handling escape to the atmosphere as the wastes are applied to the land. Most odour complaints from the public are associated with the land application portion of the waste management process.

17.8.1 Controlling Odours from Housing/Yard Areas

Practices associated with good animal husbandry facilitate odour control from housing and yard areas. For facilities designed for continuous waste removal, maintenance of the mechanical apparatus to assure efficient waste removal is essential. Providing adequate ventilation for humidity and temperature control in buildings that house animals should disperse odours at non-objectionable levels in the outside atmosphere. Keeping yards clean of animal wastes and spilled feeds will not only minimize odour generation but will also control flies and other pests, as well as contribute to the appearance of an efficiently operated enterprise.

17.8.2 Controlling Odours from Waste Treatment/Holding Areas

For aerobic lagoons, maintaining sufficient oxygen in the lagoons to assure aerobic conditions at all times is essential for odour control, as well as effective waste treatment. Likewise, maintaining anaerobic conditions in anaerobic lagoons is equally important, as is achieving environmental conditions in the lagoon that facilitate acid-forming and methane-forming bacteria to work in tandem. Assuming lagoons of either type are designed (i.e. sized) correctly (Barth, 1985; Merkel, 1981), loading the lagoons with wastes at rates that avoid 'shocks' to the bacterial populations and provide sufficient energy supplies to the bacteria results in near-odour-free operation.

    For structures that function strictly as waste storage devices, providing a covering for the waste can reduce odour emissions. Crusting that forms naturally when dairy cattle wastes are stored serve this purpose. Chopped straw can be used to provide an 'artificial' cover for other wastes.

17.8.3 Controlling Odours from Waste Utilization Areas

Reductions in odour emissions associated with land spreading are achieved by reducing the concentrations of odorous compounds in the air following landspreading. This can be accomplished readily by changing the method of waste application. Only approximately 1 per cent of the odorous emissions emanate from the actual spreading of wastes; 99 per cent of the odours are emitted after the wastes are applied. Consequently, incorporating wastes immediately after application normally controls odours to non-objectionable levels. For grassland systems where waste incorporation is not feasible, odour reductions can be achieved by using bandspreading or shallow injection application equipment. In addition, codes of good practice should be followed when applying wastes to land. Recommended practices include avoiding spreading at times when the risk of causing odour nuisance to the public is high (e.g. on weekends and holidays), avoiding spreading when prevailing winds are in the direction of neighbouring housing and population centres and avoiding the use of application techniques that tend to atomize wastes or otherwise disperse them into the atmosphere.

17.9 PROBLEMS

17.1 Animal wastes are added to a mineral soil in the spring of year 1 at a rate of approximately 100 kg N/ha. How much nitrogen could be expected to be available for plant uptake immediately, 1 year later and 5 years later?

17.2 A dairy farmer wants to grow maize, a cereal, for grain on a mineral soil. The target yield is 9.5 t/ha. Assuming that only 50 per cent of the total nitrogen in cattle slurry will become available to the crop, how much slurry should be applied to meet the nutrient needs of the maize? Specify how much, if any, supplemental fertilizer will be required.

17.3 Determine the size of a circular, above-ground storage tank to serve an 80-cow dairy herd. Wastes must be contained for a minimum of 4 months, during which time the total precipitation is 250 mm. The tank is also to store silage effluent also. The farmer has 75 ha of grassland from which an average yield is expected of 12 t/ha, at 20 per cent dry matter.

17.4 Acme Industrial Limited, wish to dispose of process water on agricultural land. A soils analysis revealed that most of the soils at the intended application site contain 10 per cent sand, 75 per cent silt and 15 per cent clay. What frequency and maximum application rate would be recommended for the land application system?

17.5 Consider the following hydrologic data:

Month	Jan	Feb	Mar	Apr	May	Jun	Jul	Aug	Sep	Oct	Nov	Dec
Precipitation, mm	100	80	70	60	50	50	50	70	90	100	110	120
Evapotranspiration, mm	20	20	25	40	50	80	90	70	50	30	20	15

From a hydrologic standpoint determine the most appropriate times of the year to apply wastes to land to avoid pollution of groundwaters.

REFERENCES AND FURTHER READING

Alexander, M. (1977) *Introduction to Soil Microbiology*, 2nd edn, John Wiley, New York.

American Society of Agricultural Engineers (ASAE) (1990) *Standards, Engineering Practices and Data*, 37th edn, American Society of Agricultural Engineers, St Joseph, Michigan.

Archer, J. (1988) *Crop Nutrition and Fertiliser Use*, 2nd edn, Farming Press Limited, Ipswich.

Barth, C. L. (1985) 'The rational design standard for anaerobic livestock lagoons'. in *Agricultural Waste Utilization and Management: Proceedings of the 5th International Symposium on Agricultural Wastes*, American Society of Agricultural Engineers, St Joseph, Michigan.

Department of Agriculture (1985) *Guidelines and Recommendations on Control of Pollution from Farmyard Wastes* (revised), Department of Agriculture and Food, Dublin, Ireland.

Environmental Protection Agency (EPA) (1975) *Land Treatment of Municipal Wastewater Effluents: Design Factors II*, US Environmental Protection Agency, Washington, D.C.

Grundy, K. (1980) *Tackling Farm Waste*, Farming Press Limited, Ipswich.

Halley, R. J. and R. J. Soffe (1988) *The Agricultural Notebook*, 18th edn, Blackwell Scientific Publications, Oxford.

Hillel, D. (1980) *Fundamentals of Soil Physics*, Academic Press, New York.

Hudson, N. (1981) *Soil Conservation*, 2nd edn, Cornell University Press, Ithaca.

Lee, J. and B. Coulter (1990) 'A macro view of animal manure production in the European Community and implications for environment', in *Manure and Environment Seminar—VIV Europe*, Utrecht, The Netherlands, 14 November.

McCuen, R. H. (1989) *Hydrologic Analysis and Design*, Prentice-Hall, Englewood Cliffs, New Jersey.

MAFF (1991) *Code of Good Agricultural Practice for the Protection of Water*, Ministry of Agriculture, Fisheries and Food, London.

MAFF (1992) *Code of Good Agricultural Practice for the Protection of Air*, Ministry of Agriculture, Fisheries and Food, London.

Merkel, J. A. (1981) *Managing Livestock Wastes*, AVI Publishing, Westport, Connecticut.

Midwest Plan Service. *Livestock Waste Facilities Handbook*, 2nd edn (MWPS-18), Midwest Plan Service, Iowa State University, Ames, Iowa.

Novotny, V. and G. Chesters (1981) *Handbook of Nonpoint Pollution*, Van Nostrand Reinhold Co., New York.

O'Bric, C., O. T. Carton, P. O'Toole and A. Cuddihy (1992) 'Nutrient values of cattle and pig slurries on Irish farms and the implications for slurry application rates', *Irish J. Agric. Res.*, **31**(1) 89–90.

Schwab, G. O., D. D. Fangmeier, W. J. Elliot, and R. K. Frevert (1993) *Soil and Water Conservation Engineering*, 4th edn, J. Wiley, Somerset, New Jersey.

Shaw, E. M. (1988) *Hydrology in Practice*, 2nd edn, Chapman and Hall, London.

Teagasc (1989) *Farmyard Wastes and Pollution*, Agriculture and Food Development Authority, Dublin, Ireland.

Teagasc (1992) Miscellaneous data (unpublished). Johnstown Castle Research and Development Centre, Wexford, Ireland.

Tunney, H. and S. M. Molloy (1975) 'Variations between forms of N, P, K, Mg and dry matter composition of cattle, pig and poultry manures', *Irish J. Agric. Res.*, **14** 71–79.

U.S. Department of Agriculture (1975) *Agricultural Waste Management Field Manual*, Soil Conservation Service, US Department of Agriculture, Washington, D.C.

Wesseling, J., W. R. van Wijk, M. Fireman, B. D. van't Woudt and R. M. Hagan (1957) 'Land drainage in relation to soils and crops', in *Drainage of Agricultural Lands*, J. N. Luthin (ed.). American Society of Agronomy, Madison, Wisconsin.

FOUR

ENVIRONMENTAL MANAGEMENT

EIGHTEEN
WASTE MINIMIZATION

18.1 INTRODUCTION

The limitations in the recuperative capacity of the ecosystem have been recognized since about 1960. The alleviation measures taken by industry to pollution were add-on installations (end-of-pipe treatments) and these succeeded in bringing about a relatively quick improvement. However, since the mid-1970s the limitations of these methods have also been recognized, and a waste minimization approach has been seen as the only sustainable means of dealing with the waste problem. Within any waste management system, the primary concern should be to reduce the quantities of waste material produced. This avoids the necessity to treat and dispose of such materials. This chapter aims to introduce the reader to the qualitative aspects of waste minimization, with particular case study references to the chemical industry.

Several reasons may be put forward in favour of waste minimization (Grüjer, 1991). These are:

1. The generation of large volumes of waste correlates with the depletion of mostly non-renewable resources.
2. The energy requirement for the transformation and upgrading of wastes is in proportion to the quantities treated and rises exponentially with increasing dilution of the waste.
3. The increasing total costs for collection, segregation, intermediate storage, transport, treatment and final storage make waste minimization economically attractive.
4. Increased public and legislative pressures seem likely to be mitigated only by waste reduction/minimization.
5. Since waste equals inefficiency, reducing waste increases efficiency and hence profitability.

Many organizations, such as the International Chambers of Commerce and the Chemical Manufacturers Association, have endorsed the concepts of waste minimization and sustainability (Willums and Golüke, 1992; CMA, 1991).

In many cases good operating practices, housekeeping, etc., can lead to a substantial reduction in industrial and other wastes. However, there is a need for innovation in the design and operation of plant and equipment, in order to fully achieve the waste prevention goals which are being and will be set. These innovations are the so-called clean technologies.

18.1.1 Definition of Clean Technology

In recent years the drive towards improved environmental performance has seen the emergence of a new approach to the solution of waste problems. This has variously been termed clean or cleaner technology, cleaner production, waste minimization, waste reduction, pollution prevention and so on. There are many definitions of 'Clean Technology'. However, all incorporate the following same two ideas, namely:

1. The emphasis is on the generation of less waste and on the consumption of fewer raw materials and less energy. Thus a simple but satisfactory definition of clean technology is 'any technology or process which uses fewer raw materials and/or less energy, and/or generates less waste than an existing technology or process'.
2. The avoidance of 'end-of-pipe' emission reduction is also emphasized. End-of-pipe methods are those that attempt to reduce the environmental impact of a waste, after that waste has been produced.

Although the concept of zero emission processes has been espoused, such a target is thermodynamically impossible for a manufacturing process, if such a process is regarded as an open system (a system that exchanges both material and energy with its surroundings). Manipulating the system boundary in an attempt to produce a closed system (one that exchanges only energy and not materials with its surroundings) is analogous to the end-of-pipe solution to material problems, which merely transfers matter from one medium to another. Enlarging the systems boundary to incorporate the energy supply facility reveals that the enlarged system is, in fact, open and depositing material into the surroundings (biosphere).

It should also be understood that traditional treatment methods merely alter the form of waste or transfer it from one medium to another. Thus, the total quantity of waste is not reduced.

As a consequence of such an analysis, the following tenets arise:

1. There cannot be zero waste from any manufacturing process.
2. Once created, waste cannot be destroyed.

It is important to note that there is no thermodynamic restriction regarding elimination of a particular waste, nor on the transformation of a waste material to a more innocuous one, provided that the conservation laws are observed. The only reasonable outcome from thermodynamic considerations is, therefore, that waste emissions from a manufacturing process can be minimized in terms of both quantity and toxicity, so that they fall within the assimilative capacity of the biosphere. This outcome leads to the principle of waste minimization, and ultimately provides the goal for setting environmental impact standards for manufacturing industry.

18.2 LIFE CYCLE ASSESSMENT

Before examining waste minimization techniques and strategies, let us first examine what is now called the life cycle assessment of a product. The objective of this section is to show that a systems approach is required when examining pollution sources in process operations. Initially the life cycle of the product or service should be assessed. This will indicate the relative contribution of the life cycle stages to environmental impact. Failure to do this may allow attention to be focused on a stage which is most topical but has least significance. For example, the usage of water, detergent and electricity during the operation of washing machines is much more important than the emissions during their manufacture or disposal. The emissions during the manufacture of pharmaceuticals may be more significant than the emissions during their distribution or after their excretion from the body. The general structure of processing systems and their component operations are discussed. Likely emission sources are identified. Following this, the need for an integrated pollution prevention and control approach is emphasized. The reader is referred to the reading list for further information. This brief section can only serve as an introduction.

Life cycle assessment is a developing environmental management technique which has been applied to a greater or lesser extent for two decades, but has been the focus of intense interest since 1990. It is an attempt to attribute all the environmental impacts in the life cycle of a marketable product. It recognizes that raw materials production and eventual disposal may be as significant environmentally as the product manufacture. Another name, disliked by the purists, is a 'cradle-to-grave' study. The first reported life cycle assessment was carried out for Coca-Cola in 1969, on the choice of beverage containers. In the 1970s, energy supply systems were the main concern. This illustrates that the technique may also be applied to services and activities, not just artefacts. However, the then limits to the theory and available data meant interest waned. Through the 1980s, the increasing problem of limited landfill capacity drew attention to the disposal problems associated with the 'throw-away society' in general and packaging in particular. Life cycle studies were awakened and we have tangible consequences in the German and EU packaging restrictions.

Since 1990, there has been tremendous interest in the technique. There is considerable work under way in research institutes, governments and large corporations seeking to develop strategic policy and small flexible enterprises trying to anticipate markets. In Europe, the emergence of the Eco-label and the environmental marketing concerns of major retail outlets demand that every enterprise is aware of this technique and the possible implications for its survival (Willums and Golüke, 1992).

The use of life cycle assessment is an attempt to apply rational criteria to deciding where to invest resources in an environmentally responsible manner. At the level of the consumer, where is she or he going to best spend their money? For an enterprise, how may it promote its products to this newly educated consumer? The enterprise must also examine where it should concentrate its limited resources (money, staff) to minimize its impact on the environment. It may need to improve its existing products or develop new ones to ensure its own survival in the market. Governments must decide how to deal with the large-scale problems of waste disposal and emissions from energy consumption. This may result in product bans, e.g. non-returnable beverage containers in Denmark, taxes, e.g. packaging in Germany, or incentives, e.g. pollution prevention grants in the United States.

18.2.1 What is Life Cycle Assessment (LCA)?

It may be formally defined as 'a systematic inventory and comprehensive assessment of the environmental effects of two or more alternative activities involving a defined product in a defined space and time including all steps and co-products in its life cycle' (Pedersen, 1993). Other terms used in Europe are 'ecoprofile', 'ecobalance' and 'product life assessment'. If the assessment stage is omitted, we can use the term 'life cycle analysis'. Any product may have the following stages in its life cycle, as shown in Fig. 18.1

- Raw materials acquisition
- Bulk material processing
- Engineered and speciality materials production
- Manufacturing and assembly
- Use and service
- Retirement
- Disposal

The term 'life cycle' in this context is not the same as the business life cycle. Instead, the physical life is considered. Raw materials acquisition includes mining non-renewable material, e.g. coal, and harvesting biomass, e.g. forestry. These bulk materials are processed into base materials by separation and purification, e.g. converting bauxite to aluminium. Some base materials are converted into engineered and speciality materials, e.g. ethylene polymerization to polyethylene pellets.

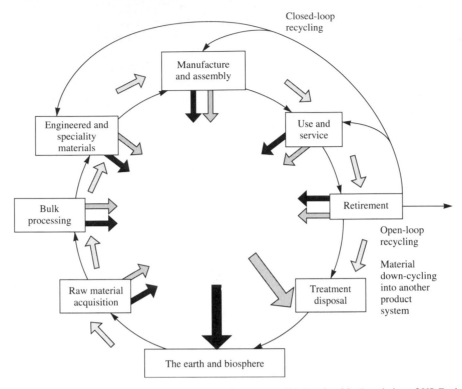

Figure 18.1 The product life cycle system. (From Koeleian and Menerey, 1993. Reprinted by permission of US Environmental Protection Agency.)

At every stage there can be material, energy and labour inputs. At each stage there may be waste that is treated or untreated before disposal to air, water or land. Within the cycle, wastes and retired materials may be re-used, re-manufactured or further processed (recycled) to minimize the net output. However, all of these consume energy and generally suffer some degradation of material. Eventually the material may be 'downcycled' from its original product cycle to another product cycle with less demanding requirements, e.g. white notepaper to computer printout paper to cardboard (see Fig. 18.2).

18.2.2 An Outline of the Steps Necessary to Conduct a Life Cycle Assessment

An LCA has the following phases:

- Planning
- Screening
- Data collection (inventory)
- Data treatment (aggregation/classification)
- Evaluation

Planning The planning phase is critical. At this stage, decisions are taken which determine the complexity of the study. They also determine the range of results and what may be gained from these results. Goal definition is important. One must determine who is going to use the results and for what purpose. Different users have different demands. An enterprise seeking to promote a product has a different outlook from a Government determining policy. If the results are to be applied to a particular

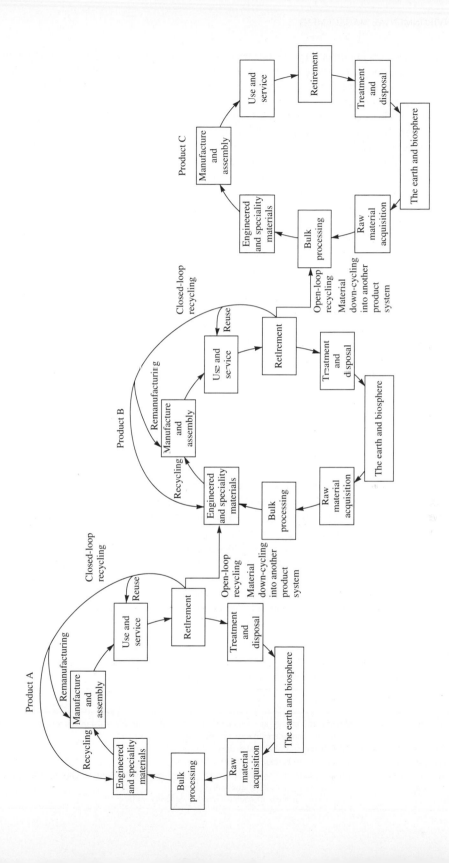

Figure 18.2 Material downcycling. (From Koeleian and Menerey, 1993. Reprinted by permission of US Environmental Protection Agency.)

region, e.g. Europe, it is important that the data used are relevant. For example, one should not presume an 80 per cent recycle rate of retired products, e.g. glass bottles, if the actual rate is 20 per cent. The time horizon is also important. Is one trying to satisfy current or anticipated legislative limits, or plan for longer term sustainable development?

The product or service being compared must be precisely defined. When examining washing machines, we are really concerned with washing equal amounts of clothes to a specified standard of cleanliness. When examining milk containers, we want to deliver a specified quantity of milk to a particular location with a specified shelf life. We cannot directly compare pint and litre containers, for example. If we compare disposable nappies with cloth nappies, do we presume that the cloth nappies will be washed domestically or will a 'nappy service' be considered?

Each alternative can be represented graphically in a cycle or 'process tree'. Refer to Fig. 18.3 which could represent a detergent product system. Such a tree can be complex, but rather than start with a simplified tree, it may be preferable to construct the complicated tree and explicitly simplify. Unfortunately, this may contradict the idea of screening (described later).

Data must be locally relevant. When considering the disposal of products, we must assess the future consequences. Unfortunately, if we are examining the eventual pathways of landfilled material we may be forecasting centuries into the future. Incineration may have a more immediate effect. Sometimes the primary environmental burden associated with a product is actually a by-product. This must not be omitted from the examination. Choosing the environmental parameters is a contentious area. Factors that may be considered are listed in Table 18.1.

Having recognized the wide spectrum of parameters to be considered, it is obvious that some simplification is required. Effects must be aggregated to avoid a morass of detail. This leads to the difficulty of comparing the ozone-depleting potential of one substance with the aquatic toxicity of another.

Screening   Having identified the parameters of interest and the alternatives of concern, one must decide where and how to gather the significant data. At this stage, decisions should be taken to limit the study to consequential factors only, i.e. those that will distinguish one product from another. A preliminary LCA should be carried out after the plan has been formulated. This should be a coarse and simplified study. Its purpose is to validate the plan and correct errors before too much effort has been spent. This may provide enough information to satisfy the initial objectives (Lindfors, 1992).

Data collection   This phase is often known as 'inventory'. It may be defined as 'an objective data-based process of quantifying energy and raw material requirements, air emissions, waterborne effluents, solid waste and other environmental releases throughout the life cycle of a product, process or activity'. Data may be collected from relevant literature studies, databases and reports (e.g. BUWAL, 1991). On-site measurements, records and personnel estimates may be available. Theoretical calculations and finally informed judgement may provide the necessary information.

Table 18.1 Environmental factors to be considered in LCA

Parameter	Comment
Resource depletion	Material and energy are consumed and one must consider whether the sources are renewable. Scarcity and alternative or potential uses are also issues
Wastes	Solid, liquid, airborne and energy
Ecological degradation	Scarcity, diversity and sensitivity of ecosystems
Human health and safety	The population at risk, workers, users or general community must be identified
Economic effects	Financial costs, including alternative uses
Social effects	Degradation of areas of social interest or amenity value

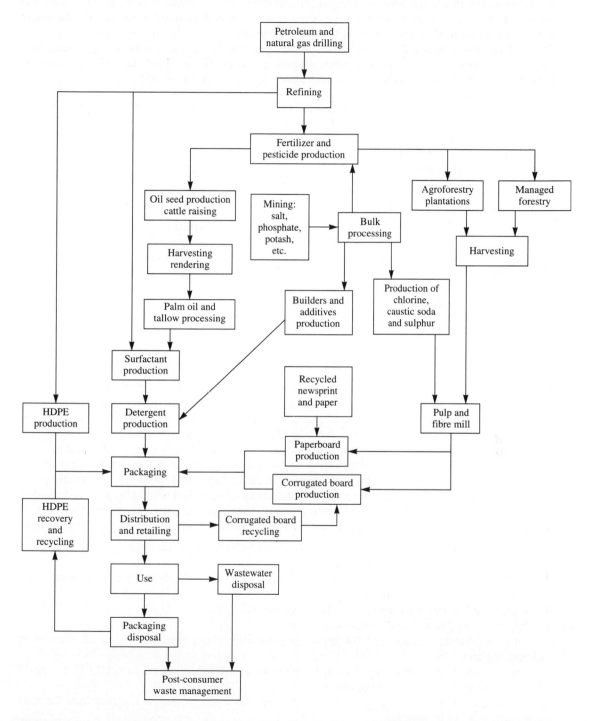

Figure 18.3 Life cycle of a detergent product system. (From Koeleian and Menerey, 1993. Reprinted by permission of US Environmental Protection Agency.)

Data treatment   This is often described as 'classification' or 'aggregation'. Quality control must be exercised on the data. Extensive data are available, but dispersed. Allowance must be made for time and regional differences. Different sources must be compared, and measured and reported values assessed against theoretical predictions. Data validity generally must be verified. Having now accumulated all the data and calculated their effects through the process tree, the many results must be aggregated to a smaller, manageable number. There is often a desire to reduce all the results to a single parameter, particularly financial. While useful, care must be taken that this is not simplistic.

Evaluation   Finally, conclusions must be drawn from the exercise. A sensitivity analysis of the data will assess the significance of differences. This may also indicate that certain areas require even more consideration before a conclusion may be drawn. Impacts and improvements may be analysed. Depending on the objectives, scope for process improvement will be identified or guidance provided in selecting alternative products or processes.

18.2.3 Critique

Life cycle assessment is a useful, though underdeveloped, environmental management tool. It must be recognized that LCA is a developing technique. Contradictory studies have been published on the same topic. This can easily arise through the application of inadequate, inconsistent or inappropriate data. Different activities may be included or excluded from the cycle. The sheer scale of any problem necessitates simplification. This can distort the assessment. Selection of appropriate environmental performance indicators and criteria is difficult. At present, it is difficult to separate objective and subjective elements.

Finally as the methodology and available data are developed, it is worth revisiting previous life cycle assessments. As knowledge improves, it should be possible to place more confidence in our detailed assessments and also facilitate rapid, simpler, but nevertheless valid shortcut assessments.

Example 18.1   Recycling of glass. This example is based on Swiss data (BUWAL, 1991). It refers primarily to the inventory phase of an LCA and is intended to illustrate the effects of recycling. Comparison is not made between glass and alternative containers. The benefits from recycling glass are that finite raw materials are conserved and that only 75 per cent of the energy required to make fresh glass is required when handling glass splinters. Glass is manufactured from glass sand, soda, limestone, dolomite and feldspar. A small quantity of special components are added to impart particular colours and properties. Water is contained in the raw materials. Glass is produced as white (clear), brown and green. The data refers to an average mixed glass based on the normal colour distribution. This glass is the molten 'glass drop' prior to container formation. In 1989, the overall Swiss recycling rate was 56.2 per cent for glass, but this particular factory achieved 74.8 per cent. Examination of the data reveals the following:

1. Recycling is not 'free'—there is an energy requirement for transporting and processing the scrap glass. This is offset by the energy saved in melting.
2. There is an increase in lead emissions with recycling. The lead arises from wine bottle caps. Segregation of the scrap is necessary.
3. There are substantial reductions in landfill requirements.
4. The spectrum of wastes to be considered is wide.
5. The environmental impacts and the energy requirements of raw material mining are reduced as recycling increases.
6. Even with the measure of energy, there is a need to use an equivalencing measure, UCPTE (Union of the Connection of Electricity Production).
7. A small quantity of solid waste, e.g. aluminium, iron, paper and plastics, is removed from the glass splinters prior to processing.

The ecobalance of glass is illustrated in Table 18.2.

Table 18.2 The ecobalance of glass

Balance per kg mixed glass

		Raw materials recycling rate (%)		
	Unit	56.2	74.8	100.00
Sand	g	281.1	173.9	0.0
Soda	g	79.8	48.2	6.3
Limestone	g	48.2	29.3	0.0
Dolomite	g	49/2	30.3	0.0
Feldspar	g	25/0	15.6	0.0
Miscellaneous	g	4.7	1.7	1.5
Splinters	g	562.0	748.0	1010.0
Total	g	1050.0	1047.0	1017.8

Energy carrier requirement for 1 kg mixed glass excluding transportation and electricity production

Natural gas	dm^3	19.3	12.6	3.4
Propane	g	4.1	4.1	4.1
Fuel oil extra-light	g	0.9	0.5	0.0
Fuel oil heavy	g	138.1	130.6	120.5

Energy requirement

Material input	MJ		0.000 0		0.000 0		0.000 0
Mining of raw materials	kWh	0.010 1		0.006 2		0.000 3	
	MJ		0.659 7		0.400 1		0.041 1
Transportation of raw materials	kWh	0.004 7		0.002 9		0.000 0	
	MJ		0.000 0		0.000 0		0.000 0
Glass production	kWh	0.035 5		0.035 5		0.0355	
	MJ		5.937 6		5.631 2		5.216 2
Use of waste heat	MJ		−0.274 0		−0.274 0		−0.274 0
Pre-combustion	MJ		0.628 6		0.574 5		0.500 6
Transport scrap glass	kWh	0.002 8		0.003 8		0.005 1	
	MJ		0.011 9		0.015 8		0.021 4
Reprocessing scrap glass	kWh	0.001 4		0.001 9		0.002 6	
Disposal							
Credit WIP	MJ		0.000 0		0.000 0		0.000 0
Transplantation	MJ		0.018 5		0.010 7		0.000 0
Total	kWh	0.054 5		0.050 2		0.043 5	
	MJ		6.982 2		6.358 4		5.505 3
Energy equivalent as per UCPTE 88	MJ		7.5		6.8		5.9

(continued)

18.3 ELEMENTS OF A WASTE MINIMIZATION STRATEGY

18.3.1 The Waste Management Hierarchy

It is clearly seen from the foregoing that the prevention of waste generation is preferable to attempting to 'clean up' the waste after its production. Indeed, this concept leads directly to the hierarchy of preferred options, in Fig. 18.4, which is the hallmark of the waste minimization philosophy. The various elements of this hierarchy are explained below, together with some illustrative examples.

Reduction at source As the name implies, the reduction or elimination of wastes at source, usually within a process, is an established waste minimization technique. It is the most effective means of waste minimization, and the one that should always be considered first. Source reduction measures include:

Table 18.2 The ecobalance of glass *(Continued)*

	Unit	Atmospheric emissions recycling rate (%)		
		56.2	74.8	100.00
Dust	g	8.029	5.060	0.428
CO	g	0.078	0.069	0.057
HC	g	1.631	1.413	1.113
NO_x	g	1.866	1.759	1.586
N_2O	g	0.053	0.036	0.012
SO_2	g	3.079	2.899	2.652
HCl	g	0.036	0.023	0.006
HF	g	0.014	0.018	0.024
Pb	g	0.009	0.012	0.016
Aldehydes	g	0.006	0.005	0.005
Other organic compounds	g	0.009	0.008	0.007
NH_3	g	0.003	0.003	0.002

	Unit	Water consumption pollution (dm^3)		
		0.1	0.1	0.0
Dissolved solids	g	1.782	1.681	1.542
Suspended solids	g	0.001	0.001	0.001
BOD	g	0.001	0.001	0.001
COD	g	0.003	0.002	0.002
Oils	g	0.024	0.022	0.020
Phenols	g	0.000	0.000	0.000
Ammonia	g	0.000	0.000	0.000

Solid wastes	Unit			
Mining of raw materials	g	0.0	0.0	0.0
Glass production	g	0.0	0.0	0.0
Recycling	g	15.2	20.2	27.0
Pre-combustion	g	0.3	0.3	0.2
Electricity production	g	2.7	2.5	2.1
Total production	g	18.2	22.9	29.3
Disposal				
WIP		350.4	201.6	0.0
Landfill		87.6	50.4	0.0
Total production + disposal	g	456.2	274.9	29.3
Landfill volume	cm^3	237.9	147.2	24.2

From BUWAL, 1991. Reprinted by permission of Bundesant für Umwelt, Wald und handschaft Bibliothek

- Process modifications
- Feedstock purity improvements
- Housekeeping and management practice changes
- Increases in efficiency of equipment
- Recycling within a process

Examples   Improved purchasing procedures can prevent materials becoming waste by being out-of-date. Less packaging will result in reduced packaging waste. Altering reactor conditions to improve the yield reduces the quantities of by-products and unreacted starting materials which must be dealt with. Substitution by a more benign solvent eliminates the potential for environmental release of a hazardous one. All of these examples will reduce or eliminate the quantities of waste being generated.

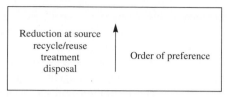

Figure 18.4 Waste minimization hierarchy.

Recycling/reuse   This is the use or reuse of a waste as an effective substitute for a commercial product or as an ingredient or feedstock in an industrial process. It includes:

- Reclamation of useful constituent fractions within a waste material
- Removal of contaminants from waste to allow reuse

The distinction between recycling and reuse is not always easy to make.

Examples

Recycling   Materials are often recycled within a process; e.g. solvents can be separated from products and used again. Recycling outside a process also offers opportunities for waste reduction. Perhaps the three best-known examples of this are glass, metal and paper recycling. Recycling can reduce the depletion of resources, since fresh material requirements are lowered. There is a need to be careful, however, that the reprocessing procedure does not generate additional waste. The recycling process will consume energy and may use additional chemicals, for example bleaches for the de-inking of paper.

Reuse   Often it is possible to find a use for a waste material. An excellent example of this is the reuse of whey. Whey, a waste product from cheese manufacture, is converted to ethanol by fermentation using the yeast *Kluyveromyces fragilis* and eventually produces a liqueur.

Treatment   Waste treatment incorporates any method, technique or process that changes the physical, chemical or biological character of a waste. The objective of waste treatment may be to accomplish one of the following:

- Neutralize the waste
- Recover energy or material resources from the waste
- Render such waste
  - Non-hazardous
  - Less hazardous
  - Safer to manage
  - Amenable for recovery
  - Amenable for storage
  - Reduced in volume

Examples   This category incorporates the so-called end-of-pipe methods and includes traditional technologies such as biological wastewater treatment, gas scrubbing, adsorption, etc. The end-of-pipe methods may result in transfer of a waste from one medium to another, resulting in a continuing problem. For example, scrubbing may result in an uncontaminated gas stream—but at the expense of a contaminated liquid stream. Biological treatment produces solid sludge. Judicious choice of a treatment system can result in effective reuse of a waste material. The anaerobic treatment of high BOD aqueous waste is a successful example of this. The process produces methane, which can then be used as a boiler fuel or converted to electricity. Thus, energy is obtained from waste.

Disposal   Disposal is the discharge, deposit, injection, dumping, spilling, leaking or placing of waste into or on any land, or water or into the air.

Examples It is plain that concentration on the source reduction, recycle and reuse aspects of a waste problem will reduce the need for treatment and disposal. However, since there will always be some waste which is not amenable to the previous methods, an effective and environmentally safe means of ultimate disposal is necessary. Traditionally this has seen the widespread use of landfill and incineration. However, with landfill sites becoming more scarce and increasingly stringent legislation governing both landfill and incineration, it is necessary to carefully manage such ultimate disposal facilities. Nevertheless, properly designed, controlled and managed landfill and incineration can result in low hazards.

Example 18.2 Metal parts are degreased by spraying with solvent. To protect workers' health, an air extraction system is installed over the workbenches. This leads to a problem with atmospheric emissions of VOCs.

(a) Outline some possible methods of overcoming this problem.
(b) In each case, identify the type of solution (e.g. source reduction, recycling, etc.).
(c) Give the preferred order of priority of the solutions.

Solution
Install a fume incinerator This is a disposal technique. It is an end-of-pipe method and results at best in carbon dioxide and water being emitted to atmosphere. In theory it is possible to recover the heat generated, but this is unlikely to be efficient in this case because of the low loads.
Install a scrubber to remove VOCs from air This is a treatment or end-of-pipe solution. As with all such solutions, it results in a contaminated stream—in this case solvent laden scrubber liquor.
Install a condenser to recover the solvent for reuse This is an acceptable waste minimization approach, but it is unlikely that condensation will be efficient unless it is cryogenic, well designed and the ratio of solvent to air is high.
Use a mechanical or aqueous cleaning method, if such an approach is possible This is a source reduction technique. Bear in mind, however, that the result may be an aqueous stream contaminated with grease and detergents. This will increase the load on the wastewater treatment facility. The preferred hierarchy of solutions is:
mechanical cleaning > aqueous cleaning > condensation > scrubbing > incineration.

18.4 BENEFITS OF WASTE MINIMIZATION

Companies often claim waste reduction credit for activities such as incineration that follow generation of a waste, rather than avoiding waste creation. Many companies use a definition of waste reduction that gives them credit for improved waste management and pollution control. Often the benefits of waste minimization are not seen. The apparent slowness in adopting methods which are clearly beneficial prompts the following questions:

1. Why do companies opt for end-of-pipe solutions, rather than clean technology?
2. Why should clean technologies be adopted?
3. What are the incentives and disincentives for adopting clean technologies?

The answers to these questions are complex. Several possible answers suggest themselves.

18.4.1 Why are Clean Technologies Not Widely Used?

In many cases the principles of waste minimization are poorly understood. The reduction of emissions to the atmosphere is seen as minimization. However, as has already been explained, waste once created cannot be destroyed. Thus, a reduction in emissions to atmosphere must necessarily lead to either an increase in emissions to land or water, or emissions of a different type to the atmosphere. While such emissions may be preferable to the original ones, their generation as a replacement cannot be regarded as waste minimization. In addition, companies are under legislative and public pressure to reduce waste. Hence, treatment systems that reduce the quantities of a certain type of waste emitted to a particular environment are claimed as waste reduction. Companies must invest in end-of-pipe technology to comply with emissions regulations. This diverts funding from waste reduction programmes. Some disincentives to the adoption of clean technologies are listed below:

- Lack of appreciation of economic benefits due to accounting systems that do not allocate total environmental costs of production profit centres
- Competing production priorities
- Belief that legally required pollution control is good enough
- Incomplete data on exact sources and amounts of wastes
- Difficulty of simultaneously spending resources on regulatory compliance and waste reduction

18.4.2 Why Should Clean Technologies be Adopted?

Enlightened self-interest   Globally, many enterprises have seen that it is in their interest to minimize waste. The International Chamber of Commerce has prepared a Business Charter for Sustainable Development (Willums and Golüke, 1992). These 16 principles for environmental management are intended to assist enterprises in fulfilling their commitment to environmental stewardship in a comprehensive fashion. Principle 8 states: 'Facilities and Operations: to develop, design and operate facilities and conduct activities taking into consideration the efficient use of energy and materials, the sustainable use of renewable resources, the minimization of adverse environmental impact and waste generation, and the safe and responsible disposal of residual wastes'. Companies such as Aer Lingus, 3M, Dow, IBM, Apple and DuPont have subscribed to these principles. Internationally, within the chemical process industry adherents include, Henkel, Johnson and Johnson, Pfizer, SmithKline Beecham and Syntex Corporation.

In the United States, the Chemical Manufacturers Association, as part of its Responsible Care programme, has produced a Pollution Prevention Code. Members 'have committed to the goal of minimising wastes, reducing releases to the environment, and managing generated wastes in a manner that is environmentally acceptable and protects human health'. According to the Code: 'the goal is to establish a long-term, substantial downward trend in the amount of wastes generated and contaminants and pollutants released'.

In Europe, the organization PREPARE provides a network for countries with waste minimization programmes. PREPARE is an initiative of the Dutch government, but now encompasses 14 countries. PREPARE organizes expert workshops in various industrial sectors, with a view to determining relevant clean technologies. The factors that prompt an enterprise to take waste minimization initiatives are many and complex, but it is worth examining three in particular: economics, legislation and community response.

Economics   Waste, by definition, equals inefficiency. Inefficiency costs money. Reducing waste must therefore present opportunities for improving profitability. It also eases the burden of regulatory compliance. Treatment and disposal costs are reduced and potential liabilities lessened. Product quality may also improve through better operation. It is interesting to examine the position of two companies to adopt clean technology. 3M describes its programme as Pollution Prevention Pays (3Ps) and Dow calls its Waste Reduction Always Pays (WRAP). 3M claims to have saved about $300 million through eliminating wastes.

Legislation The European Union has expressed its preference for waste minimization and some member countries, e.g. Denmark, with its cleaner technologies programme, has been putting this into practice. The sharpest expression of this preference comes in the US Pollution Prevention Act of 1990. This policy has been actively promoted by the US Environmental Protection Agency and is seen as a landmark in US American environmental legislation. The rising costs of treatment and disposal, and the continued tightening of environmental legislation means that a reduction in the quantity of waste generated may offer the only viable and economic strategy.

Community response Adoption of waste minimization programmes is a way of demonstrating corporate and employee commitment to the environment and the community. Customer behaviour may be guided by a company's environmental performance. While this may be most immediate in affecting the producer of consumer goods for the public, the idea of audits in quality is a familiar one. Similar environmental audits will lead to the primary producers. Product life cycle assessment may currently be of concern for the manufacturer of car components or packaging but, as this technique develops, its extension to chemical products can be anticipated. Finally, it must be recognized that pressure groups are demanding that the waste problem be addressed.

18.4.3 What are the Incentives for Adopting Clean Technologies?

Some incentives to the adoption of clean technologies are listed below:

- Often improved process economics
- Reduced treatment costs
- Reduced disposal costs
- Reduced liability
- Reduced risk of fines for breaches
- Increased public satisfaction

Example 18.3 A chemical company uses 2 kg reactants and 5 kg of solvent per kg of finished product manufactured. Examine the waste minimization alternatives if the relevent production and pricing data are as follows:

Solvent purchase price:	ECU5/kg
Incineration costs for solvent:	ECU0.5/kg
Reactant costs:	ECU20/kg
Product purification costs:	ECU10/kg
Solvent losses during processing:	5%
Product sales price:	ECU100/kg
Production rate:	10 000 kg/annum

The economic implications of some waste minimization options can be considered.

Solution:
Current economics

Reactants:	20 000 kg at ECU20/kg	= ECU400 000
Solvents:	50 000 kg at ECU2/kg	= ECU100 000
Incineration:	0.95(50 000) kg at ECU0.5/kg	= ECU23 500
Separation:	10 000 kg at ECU10/kg	= ECU100 000
Total		ECU623 500
Product:	10 000 kg at ECU100/kg	= ECU1 000 000
Profit		= ECU376 500

Waste minimization options:

- Increase conversion
- Recover solvents
- Reduce separation costs (increase separation efficiency)
- Eliminate losses

Increase conversion For a 10 per cent increase 2 kg reactants yields 1.1 kg of product. Thus,

Sales revenue = 11 000 kg at ECU100/kg = ECU1 100 000 (versus ECU1 000 000)
New profit = ECU476 500

and pro rata for other conversion increases. There will, of course, be other savings, such as decreased separation costs.

Recover solvents (assume 90 per cent efficiency)
Quantity recovered = 0.9(0.95 × 50 000) = 42 750 kg/annum
Savings in solvent purchase = 42 750 × ECU2 = ECU85 500
Savings in incineration costs = 42 750 × ECU0.5 = ECU21 375
Total savings = ECU106 875 (minus recovery cost)

Increase separation efficiency (assume 20 per cent improvement)

Savings = 10 000 × 0.2 at ECU10 = ECU20 000

Eliminate losses

50 000 × 0.05 = 2500 kg at ECU2/kg = ECU5000

Comparison of alternatives

Method	Savings
10% conversion increase	ECU100 000
90% solvent recovery	ECU106 875
20% separation improvement	ECU20 000
Eliminate losses	ECU5000

It should be possible to combine all four techniques.

18.5 ELEMENTS OF A WASTE MINIMIZATION PROGRAMME

In accordance with the waste management hierarchy, all strategies aimed at waste minimization should follow the priority sequence as illustrated in Fig. 18.5. To ensure an optimum waste minimization

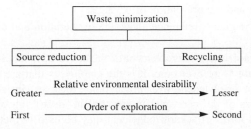

Figure 18.5 Waste minimization priorities.

programme a wide-ranging strategy must be employed. This is, in many ways, similar to the safety strategy employed by many large companies such as DuPont.

The basic parts of the waste minimization strategy may be summarized as follows:

- Management involvement
- Setting of goals
- Selection of targets
- Technical and economic evaluation
- Implementation of programmes
- Follow-up assessment and monitoring

18.5.1 Management Involvement

It is essential that top management be committed to the idea of waste minimization if the programme is to be successful. The management of a company will support a waste prevention programme if it is convinced that the benefits of such a programme can reduce its costs and improve its environmental performance. The potential benefits have already been outlined. The potential cost includes both the direct and indirect costs which ensue from the various investments. The management, once committed to the objective of waste minimization, should issue a policy statement, together with environmental guidelines. In addition, it is essential to involve the whole organization if the waste minimization programme is to maximize its chances of success.

Policy statement   A formal environmental policy document or manual or guideline is the best way to communicate the objectives of a waste elimination programme. This is akin to a safety statement and should emphasize the company's commitment to clean technology.

An example of such a policy statement by PREPARE might be (De Hoo, 1991):

> (A chemicals company)... Undertakes the obligation to supervise and channel protection of the environment. Environmental protection is one of the primary responsibilities of management as well as the responsibility of all personnel.
>
> As we intend to adhere to this policy it is our aim, as a company, to limit the generation of waste materials and emissions and to ensure that, through practising environmental management, the adverse effects on the air, the soil and water will be kept to a minimum.

Environmental guidelines   The environmental guidelines include:

1. Protection of the environment is a line responsibility and an important criterion for measuring the achievements of employees. Also, each employee carries just as much responsibility for environmental protection as he or she carries with regard to safety and other company objectives.
2. The prevention of waste and emissions is, and will continue to be, a major consideration in research, the development of production methods and the running of the company; the management sets this on a par with safety, profits and the prevention of damage.
3. Reuse of materials will be given preference above the incineration and disposal of waste.

Communications: line organization   Involvement of the entire company is essential if conflicts are to be resolved and obstacles are to be overcome. It is the employees that are the key to the programme's success through their direct involvement with production processes, installations, waste streams and emissions. The inventiveness of the personnel is essential in identifying opportunities for prevention. Bonuses, rewards and other forms of acknowledgement may be used to motivate employees. Achieving prevention goals may be used as a measure for assessing the performance of managers and workers.

18.5.2 Setting of Goals

Corporate level   Apart from qualitative objectives such as those in the waste minimization policy statement, quantitative goals should be set. Examples of this are:

- 35 per cent reduction in waste by 1992 when compared to 1982 values (DuPont)
- 60 per cent waste reduction from 1990 to 1994 (Chevron)

A simple statement like 5 per cent reduction per annum would be a realistic goal.

Site-specific level   These can be more specific than the corporate goals, e.g.

- 20 per cent reduction in air emissions from 1988 to 1990 (Michigan Division of DOW, USA)

18.5.3 Selection of Targets

There are usually many opportunities for minimizing waste. Notwithstanding any corporate policies, it should be the responsibility of individual facilities to select targets and implement waste reduction. This may be carried out on a 'freelance' basis or it may be as part of a more structured company ethos.

For example, the DOW Chemical Company requires each facility to develop an action plan to:

- Inventory all process losses to air, water and land
- Identify sources, establish priorities, quantify losses and ratio to production
- Evaluate environmental impact and risk
- Set action priorities
- Determine cost-effective actions
- Set reduction goals
- Determine resources necessary to accomplish goals
- Track and communicate performance and plan for future reductions

In any event it is necessary for each facility to select target areas for waste reduction. These should bear in mind the size and nature of the waste stream, its source, the cost savings brought about by reduction, the technical feasibility of any solution, the required investment and the feasibility of monitoring the effects of the proposed initiative. To facilitate the selection of target areas or candidate projects a waste minimization audit or assessment is necessary. A waste minimization audit is *not* an environmental audit. Its purpose is not to measure compliance with regulations but to identify areas where waste reduction may be achieved. Several approaches to carrying out a waste minimization assessment have been documented. These differ in detail only and the general consensus is that a two-tier approach is most beneficial. The components are variously termed as follows:

- Pre-assessment phase or first-tier investigations
- Assessment phase or second-tier investigations

First-tier investigations   This phase is a screening operation which is used to identify the viable areas of priority for preventing waste. It is a means of obtaining an overall picture of the waste streams and company activities with limited means and within a short period. A pre-assessment can help substantially increase motivation in the company to launch a full-scale assessment.
The steps in the pre-assessment phase may be as follows:

1. Establish a waste minimization assessment team.
2. Make an inventory of the waste streams and omissions generated on site.
3. Make an inventory of prevention opportunities and bottlenecks.

Second-tier investigations   This is a natural follow-on from the pre-assessment phase and involves further investigations on the prioritized options already identified. The assessment team may have to be strengthened, with additional personnel from affected areas of the plant becoming involved. The assessment procedure is similar, but more detailed than in pre-assessment. Process flow diagrams and material balances are useful, but more information is needed, and visits to plants together with interviewing plant personnel are seen as essential.

The type of information to be gathered should include:

Design information

- Process flow diagrams
- Material application diagrams
- Piping and instrumentation diagrams
- Equipment lists, specifications, drawings, data sheets, operating and maintenance manuals
- Plot plans, arrangement drawings and work flow diagrams

Environmental information

- Hazardous waste manifests
- Emission inventories and waste assays
- Biennial hazardous waste reports
- Environmental (compliance) audit reports
- Permits and permit applications
- Spill/release prevention and countermeasure plans

Raw material/production information

- Composition sheets
- Material safety data sheets
- Batch sheets
- Product and raw material inventory records
- Production schedules
- Operator data logs

Economic information

- Waste treatment and disposal costs
- Product, utility and raw material costs
- Operating and maintenance labour costs

Other information

- Company environmental policy statements
- Company and department standard operating procedures
- Organization charts

For very small installations, the use of a two-tier approach may not be strictly necessary, since it may be possible to identify prospective projects quickly. Even in such cases, however, it is often advantageous to carry out a brief screening procedure.

18.5.4 Technical and Economic Evaluation

Technical evaluation   The final product of the assessment phase is a list of waste minimization options postulated for the assessed area. The assessment will have screened out the impractical or unattractive options. The next step is to determine whether the remaining options are technically and economically feasible.

The technical evaluation determines whether a proposed waste minimization option will work in a specific application. Typical technical evaluation criteria include the following (Hanlon and Fromm, 1990):

1. Is the system safe for workers?
2. Will product quality be maintained?
3. Is the new equipment, material or procedure compatible with production operating procedures, work flow and production rates?
4. Is additional labour required?
5. Are utilities available? Or must they be installed, thereby raising capital costs?
6. How long will production be stopped in order to install the system?
7. Is special expertise required to operate or maintain the new system?
8. Does the system create other environmental problems?
9. Does the vendor provide acceptable service?

If, after the technical evaluation, the project appears unfeasible or impractical, it should be dropped.

Economic evaluation   All proposed projects in industry are subject to economic scrutiny. Waste minimization projects are no different and the usual decision criteria should be applied. However, it is now being increasingly recognized that knowledge about the true costs associated with generating hazardous waste is limited. The availability of such information would heighten the awareness of management to the need for different waste management programmes and it would support many capital projects aimed at reducing waste. The four basic costs incurred when a company generates waste arise from:

- Underutilizing the value of the raw materials
- General management costs associated with moving the waste around the site, storage, keeping track of waste records and shipping
- Disposing of the waste
- The associated costs with third-party liabilities if the waste is improperly disposed of

Attempting to determine the true management costs associated with generating wastes at a particular company, either at a plant site or from a product line, will give managers the opportunity to truly realize waste management costs and waste minimization will become much more attractive.

18.5.5 Implementation of Programmes

Waste minimization options that involve operational, procedural or materials changes (without additions or modifications to equipment) should be implemented as soon as the potential cost savings have been determined. For projects involving equipment modifications or new equipment, the installation of a waste minimization project is essentially no different from any other physical plant improvement project.

18.5.6 Follow-up Assessment and Monitoring

A useful measure of the effectiveness of a waste minimization project is its payback. The project should pay for itself through reduced waste management costs and reduced raw material costs. In very many cases payback is two years or less. Thus, it can be expected that waste minimization projects increase profitability.

However, it is also important to measure the actual reduction of waste brought about by the waste minimization project. The easiest way to measure waste reduction is by recording the quantities of waste generated before and after a waste minimization project has been implemented. Since waste production is normally a function of plant throughput it is important that waste measurements reflect this. Thus the quantity of a waste generated per unit product or per unit raw material used is felt to be an acceptable measure.

18.5.7 Continuing the Policy

A waste minimization programme is not a once-off programme, but rather an on-going one. After having implemented projects and reduced waste in priority areas, other, lower, priority areas should be assessed. The overall objective is to reduce waste generation to the greatest possible degree. It is important to realize that reduction of hazardous waste is only a first step. All industrial discharges should be reduced, since their generation implies inefficiency and lost profits. In many companies a safety or quality culture has been established. A waste minimization culture should similarly be developed. Thus, waste minimization must be an integral part of a company's operations, and be on a par with safety, quality and production.

18.6 WASTE REDUCTION TECHNIQUES

Section 18.5 outlined the ethos behind implementing a waste minimization policy. The question arises as how best to implement the elements of such a policy. For example, what techniques are available for source reduction, recycling, etc. The answer to this question is not straightforward and depends on the application. The so-called clean technologies do not exist in their own right. If a clean technology is regarded as one that reduces waste generation, and/or material and energy usage, then it can be seen that what may be regarded as a cleaner technology in one industrial sector or application is not necessarily so regarded in another sector. For example, distillation is widely used throughout the chemical industry to recover materials for reuse. Consequently, it would not normally be regarded as a clean technology in this context. Research is ongoing to attain better and more energy efficient separation techniques. On the other hand, in a small paint spraying operation which produces pigment-contaminated solvents, the installation of a small distillation column for solvent recovery would indeed be a cleaner technology.

It can thus readily be seen that there is no panacea for reducing wastes. Nevertheless, several general guidelines have been established, which can be supplemented by examples gleaned from experience. The techniques for bringing about waste reduction can be broken down into four major categories as follows (Hunt, 1990):

- Inventory management
  - Inventory control
  - Materials control
- Production process modification
  - Operational and maintenance procedures
  - Materials change
  - Process equipment modification
- Volume reduction
  - Source segregation
  - Concentration
- Recovery
  - On-site
  - Off-site

These techniques are discussed in more detail in the following sections. Figure 18.6 is often used as an alternative means of conveniently summarizing possible waste minimization techniques. It should be noted that waste reduction techniques are generally used in combination so as to achieve maximum effect at the lowest cost. It is important to realize that the impact of a waste reduction technique on all waste

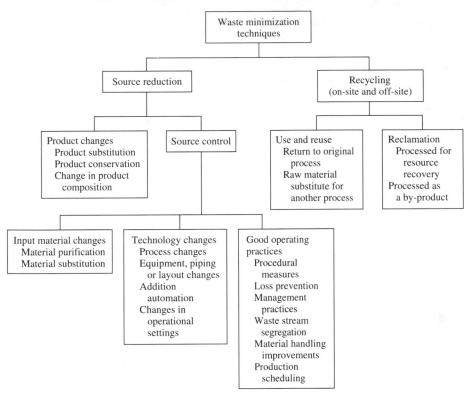

Figure 18.6 Hazardous waste minimization techniques. (Adapted from Freeman, 1990. Reprinted by permission of McGraw-Hill, Inc.)

streams must be considered. For example, switching from solvent-based to water-based methods may result in lower organic emissions to atmosphere, but can lead to an increased burden on wastewater treatment.

18.6.1 Inventory Management

Proper inventory control over raw materials, intermediate products, final products and the associated waste streams is an important waste reduction technique. In many cases waste is just out-of-date, off-specification, contaminated, or unnecessary raw materials, spill residues or damaged final products. The cost of disposing of these materials not only includes the actual disposal costs but also the cost of the lost raw materials or product. Inventory management incorporates both inventory and material control.

Inventory control   Inventory control involves techniques to reduce inventory size and hazardous chemical use while increasing inventory turnover. Proper inventory control can help reduce wastes occurring as a result of the following: excess, out-of-date and no-longer-used raw materials. Methods that can be used are purchasing in small quantities, purchasing in appropriate container sizes and just-in-time purchasing.

Material control   It is essential to have proper control over the storage of raw materials, products and process waste and the transfer of these items within the process and around the facility. This will minimize losses through spills, leaks or contamination. It will also ensure that the material is efficiently handled and used in the production process and does not become waste. A list of typical sources of losses is given in Table 18.3.

18.6.2 Production Process Modification

Waste can be significantly reduced by improved process efficiency. Such improvements can range from simple, inexpensive changes in production procedures to the installation of state-of-the-art equipment. Three techniques for production process modification have been identified. These are improved operation and maintenance, material change and equipment modifications.

Operational and maintenance procedures   Improvements in operation and maintenance are usually relatively simple and cost effective, and may lead to significant waste reduction. The techniques used are not new, but have not generally been applied to waste reduction problems. A list of examples of operational changes is given in Table 18.4. A strict maintenance programme which stresses corrective and preventive maintenance can reduce waste generation caused by equipment failure. Such a programme can help spot potential sources of release and correct a problem before any material is lost.

Material change   The replacement of materials, used in either a product formulation or in a production process, can either result in the elimination of a hazardous waste or facilitate recovery of a material. For example, CFCs are gradually being replaced by more ozone-friendly products. This is an example of eliminating a hazardous waste source. Replacement of a solvent by one with a different vapour pressure may allow separation and recovery by distillation, condensation, etc. Table 18.5 gives some examples of waste reduction through material change. As previously emphasized, care must be taken to examine the impact of changes on the total wastes from a process. This is particularly important in the case of material changes. The effect on aqueous wastes of changing from organic solvents to water is a case in point.

Table 18.3 Potential sources of process material loss

Area	Source
Loading	Leaking fill hose or fill line connections
	Draining of fill lines between filling
	Punctured, leaking or rusting containers
	Leaking valves, piping and pumps
Storage	Overfilling of tanks
	Improper or malfunctioning overflow alarms
	Punctured, leaking or rusted containers
	Leaking transfer pumps, valves and pipes
	Inadequate diking or open drain valve
	Improper material transfer procedures
	Lack of regular inspection
	Lack of training programme
Process	Leaking process tanks
	Improperly operated and maintained process equipment
	Leaking valves, pipes and pumps
	Overflow of process tanks; improper overflow controls
	Leaks and spills during material transfer
	Inadequate diking
	Open drains
	Equipment and tank cleaning
	Off-specification raw materials

Adapted from Hunt, 1990. Reprinted by permission of McGraw-Hill, Inc.

Table 18.4 Examples of operational changes to reduce waste generation

Reduce raw material and product loss due to leaks, spills, drag-out, and off-specification process solution

Schedule production to reduce equipment cleaning, e.g. formulate light to dark paint so the vats do not have to be cleaned out between batches

Inspect parts before they are processed to reduce number of rejects

Consolidate types of equipment or chemicals to reduce quantity and variety of waste

Improve cleaning procedures to reduce the generation of dilute mixed waste with methods such as using dry cleanup techniques, using mechanical wall wipers or squeegees and using compressed gas to clean pipes and increasing drain time

Segregate wastes to increase recoverability

Optimize operational parameters (such as temperature, pressure, reaction time, concentration and chemicals) to reduce by-product or waste generation

Develop employee training procedures on waste reduction

Evaluate the need for each operational step and eliminate steps that are unnecessary

Collect spilled or leaked material for reuse

Adapted from Hunt, 1990. Reprinted by permission of McGraw-Hill, Inc.

Table 18.5 Examples of waste reduction through material change

Industry	Technique
Household appliances	Eliminate cleaning step by selecting lubricant compatible with next process step
Printing	Substitute water-based ink for solvent-based ink
Textiles	Reduce phosphorus in wastewater by reducing use of phosphate-containing chemicals
	Use ultraviolet light instead of biocides in cooling towers
Air conditioners	Replace solvent-containing adhesives with water-based products
Electronic components	Replace water-based film-developing system with a dry system
Aerospace	Replace cyanide cadmium-plating bath with a non-cyanide bath
Ink manufacture	Remove cadmium from product
Plumbing fixtures	Replace hexavalent chrome-plating bath with a low-concentration trivalent chrome-plating bath
Pharmaceuticals	Replace solvent-based tablet-coating process with a water-based process

Adapted from Hunt, 1990. Reprinted by permission of McGraw-Hill Inc.

Process equipment modification The installation of more efficient equipment or the modification of equipment can reduce the generation of waste. Modifications may be simple, such as improved mixing, or better pump seals. On the other hand, installation of completely new equipment may be involved. In most cases the investment will pay for itself in a relatively short space of time, due to increased production rates, reduced waste handling costs, etc. Some examples of process modifications are given in Table 18.6

18.6.3 Volume reduction

While reduction in volume does not, of itself, constitute waste reduction, it frequently facilitates separation and recovery. The reduction in volume may involve complex concentration technologies or may simply consist of source segregation. Some examples of waste reduction through volume reduction are given in Table 18.7.

Source segregation This very simple operation is a basic tenet of good waste minimization. The segregation of wastes allows them to be more readily removed or recovered. For example, the mixing of small quantities of organic solvents with aqueous streams both militates against effective recovery and puts an increased load on biological treatment systems. Keeping the organic streams separate may allow recovery, incineration, etc. A commonly used waste segregation technique is to collect and store washwater or solvents used to clean process equipment (such as tanks, pipes, pumps or printing presses)

Table 18.6 Examples of production process modifications for waste reduction

Process step	Technique
Chemical reaction	Optimize reaction variables and improve process controls
	Optimize reactant-addition method
	Eliminate use of toxic catalysts
	Improve reactor design
Filtration and washing	Eliminate or reduce use of filter aids and disposable filters
	Drain filter before opening
	Use counter-current washing
	Recycle spent washwater
	Maximize sludge dewatering
Parts cleaning	Enclose all solvent cleaning units
	Use refrigerated freeboard on vapour degreaser units
	Improve parts draining before and after cleaning
	Use mechanical cleaning devices
	Use plastic-bead blasting
Surface finishing	Prolong process bath life by removing contaminants
	Redesign part racks to reduce drag-out
	Reuse rinse water
	Install spray or fog nozzle rinse systems
	Properly design and operate all rinse tanks
	Install drag-out recovery tanks
	Install rinse water flow control valves
	Install drip racks and drain boards
Surface coating	Use airless air-assisted spray guns
	Use electrostatic spray-coating system
	Control coating viscosity with heat units
	Use high-solids coatings
	Use powder coating systems
Equipment cleaning	Use high-pressure rinse system
	Use mechanical wipers
	Use counter-current rinse sequence
	Reuse spent rinse water
	Use 'pigs' to clean lines
	Use compressed gas to blow out lines
Spills and leaks	Use bellow-sealed valves
	Install spill basins or dikes
	Use seal-less pumps.
	Maximize use of welded pipe joints.
	Install splash guards and drip boards
	Install overflow control devices

Adapted from Hunt, 1990. Reprinted by permission of McGraw-Hill, Inc.

Table 18.7 Examples of waste reduction through volume reduction

Industry	Technique
X-ray film	Segregate polyester film scrap from other production waste and recycle
Resins	Collect waste resin and reuse in next batch
Printed circuit boards	Use filter press to dewater sludge to 60 per cent solids and sell sludge for metal recovery
Pesticide formulation	Use separate bag houses at each process line and recycle collected dust into product
Research laboratory	Segregate chlorinated and non-chlorinated solvents to allow off-site recovery
Aircraft components	Use ultrafiltration to remove recoverable oil from spent coolants
Paint formulation	Segregate and reuse tank-cleaning solvents in paint formulations
Furniture	Segregate and reuse solvents used to flush spray-coating lines and pumps as coating thinner

Adapted from Hunt, 1990. Reprinted by permission of McGraw-Hill, Inc.

for reuse in the production process. This technique is used by paint, ink and chemical formulators, as well as by printers and metal fabricators.

Concentration The concentration of a waste involves the removal of some portion of it, e.g. water. There are many techniques available to enable wastes to be concentrated. These include physical separation of solids and liquids (filtration, centrifugation, gravity settling, etc.), membrane separations such as ultrafiltration, reverse osmosis, etc., as well as energy intensive operations such as evaporation.

18.6.4 Recovery

Waste recovery should only be considered after all other waste reduction options have been instituted. The recovery of waste costs money in terms of energy and/or material input. Reduction of waste generation at source is more cost effective, since it represents a reduction in lost raw materials, intermediates, products, etc. Nevertheless, since there will always be some waste generation, recovery represents a viable and cost effective waste management alternative. Effective recovery is enhanced by segregation of materials, as already outlined. Recovery may be carried out on-site or off-site. On-site recovery is preferable, where possible, since it reduces possible handling losses and allows the management of the waste to remain within the compass of the producer. On-site recovery is particularly appropriate where the recovered material can be reused as a raw material. Table 18.8 gives some examples of waste recovery techniques.

Most on-site recovery systems will generate some type of residue (i.e. contaminants removed from the recovered material). This residue can either be processed for further recovery or properly disposed of. The economic evaluations of any recovery technique must include the management of these residues.

In the event that on-site recovery is not feasible, for economic or other reasons, off-site recovery should be considered. In some situations a waste may be transferred to another company for use as a raw material in the other company's manufacturing process.

Example 18.4 A production process uses a raw material that contains a small amount of sulphur as an impurity. After processing the sulphur appears as sulphate in an aqueous waste stream. This waste is anaerobically biodegraded to produce a biogas, which is used as a boiler fuel. In the anaerobic digestor,

Table 18.8 Examples of waste reduction through recovery and reuse

Industry	Technique
Printing	Use a vapour-recovery system to recover solvents
Photographic processing	Recover silver, fixer and bleach solutions
Metal fabrication	Recover synthetic cutting fluid using a centrifuge system
Mirror manufacturing	Recover spent xylene using a batch-distillation system
Printed circuit boards	Use an electrolytic recovery system to recover copper and tin/lead from process wastewater
Tape measures	Recover a nickel-plating solution using an ion-exchange unit
Medical instruments	Use a reverse-osmosis system to recover a nickel-plating solution
Power tools	Recover alkaline degreasing baths using an ultrafiltration system
Textiles	Use an ultrafiltration system to recover dyestuffs from wastewater
Hosiery	Reconstitute and reuse spent dye baths
Food processing	Send all solids off-site for by-product recovery
Wastewater treatment	Reuse waste caustic solids to treat acid waste stream
Pickles	Transfer waste brine pickle solution to a textile plant as a replacement for virgin acetic acid
Chemicals	Use spent electrolyte from one division as raw material in another; purify hydrochloric acid in waste stream and sell as a product
Industrial and consumer products	Segregate and sell office paper, corrugated cardboard, paper trimming and rejected paper products
Aluminium die-caster	Sell waste fumed amorphous silica for use in concrete

Adapted from Hunt, 1990. Reprinted by permission of McGraw-Hill, Inc.

the sulphate is converted to hydrogen sulphide, which results in sulphur dioxide emissions in the boiler flue gases. Outline various methods of reducing the SO_2 emissions and identify them as specific waste reduction techniques.

Solution

(a) Purchase alternative sulphur-free raw material. This is material substitution.
(b) Divert waste from this production process away from the anaerobic digestor. This is source segregation
(c) Recover hydrogen sulphide from the gas stream as pure sulphur for sale. This is material reuse

18.7 CONCLUSION

The basic elements of a waste minimization have been outlined. Many techniques and technologies exist to reduce the generation of waste and to recover wastes once generated. However, a waste minimization programme should not rely on technology alone. Senior management commitment, a rigorous waste management programme and a continuing emphasis on reduction at source are prerequisites for success. The reward is increased competitiveness, reduced waste treatment and disposal problems, legislative compliance and an improved public image. Some waste will always exist, even after rigorous implementation of a waste minimization programme. Such waste must be treated and disposed of in an environmentally safe manner. It is the producer's responsibility to ensure the safe handling and disposal of such wastes. Finally, no matter how successful a waste minimization programme or project, it should never be considered completed. It is not a once-off programme, but rather a continuing one. Constant improvements must be made and new methods of reducing waste sought.

18.8 CASE STUDY—PAINT INDUSTRY (USEPA 1990)

The paint and allied industries are engaged in the manufacture of paint, varnish, lacquer, enamels, cleaners, putties and allied products. In the United States, 60 per cent of paint is solvent based, 35 per cent water based and the rest allied products. Raw materials used in the industry include resins, solvents (aromatics, alcohols, aliphatics), pigments (titanium dioxide, inorganics), extenders (calcium carbonate, talc, clay) and other compounds such as plasticizers, and drying oils.

In the production of solvent-based paints some of these materials are blended in a high-speed mixer. Solvents and plasticizers are also added. Frequently, this operation is followed by grinding and mixing in a mill. Afterwards, the paint base is transferred to an agitated vessel where tints and thinners are added along with the remaining resins and solvents. After reaching the proper consistency, the mixture is filtered and transferred to a filling hopper. The water-based paint process is very similar except that water is used predominantly and the sequencing of material additions is different. Waste from a typical paint process can be generated at any one of several stages, as shown in Fig. 18.7. The main wastes can arise from raw material storage containers, off-specification products, dust, spills and equipment cleaning wastes.

18.8.1 Plant Waste Streams

Equipment cleaning wastes Equipment is routinely cleaned to prevent product contamination and maintain efficiency. Depending on the process unit and product type, cleaning is accomplished by flushing with water, or solvent, alkaline cleaning solution, pressure jet, or mechanically or manually scraping.

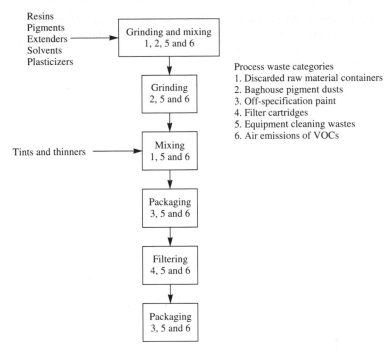

Figure 18.7 Paint process waste categories. (Adapted from USEPA, 1990.)

Off-specification and obsolete stock Off-specification stock is produced by errors in batch formulation, quality control failure and other process errors. This may lead to products being returned by customers, leading to disposal problems.

Spills Spillages are inadvertent discharges that occur at various locations in the plant. Water-based paints are washed to the waste treatment plant while solvent paints are recovered and stored.

Filters, bags, packages Filter bags and cartridges are used as the dust control equipment for both solvent and water-based processes. Spent bag filters are washed and disposed of as non-hazardous waste. Hazardous waste materials (lead and chromium packaging) need to be disposed of in a controlled fashion.

18.8.2 Source Reduction Measures

1. A caustic cleaning solution was replaced by a proprietary cleaning solution, which reduced by one half the previous waste volume.
2. In the water-based process, low-pressure water was used to clean vessels. A portable high-pressure system was purchased, contributing to a 25 per cent reduction in wastewater volume.
3. Dedication of vessels to specific products required that cleaning be carried out after several batches as opposed to after each batch. The residual solids were removed manually.
4. Certain batches are sequenced in order of light to dark paint manufacture. This, in some cases, eliminated the requirement of intermediate cleaning.
5. Lead- and chromium-based pigments are being phased out as these pigments are prohibited in consumer products.
6. The raw material inventory is controlled and monitored carefully. Only raw materials that satisfy stringent quality control standards are purchased, are used as quickly as possible and are returned to the supplier if it becomes obsolete.

7. Obsolete finished material can be reduced again by proper planning and inventory control.
8. Off-specification products are reworked on-site to produce marketable products. Reduction of off-specification material can be achieved by proper process automation and quality control.
9. Spills are recovered by manual scooping or by vacuum and then reworked into useful products.
10. Most of the pigments used are in slurry or paste format, keeping packaging material to a minimum.
11. Cleaning solvents are reused several times for rinsing tanks. When the material is considered too dirty for cleaning purposes, it is distilled on-site. The distilled materials are recycled to the cleaning operation, while the still bottoms are used to produce a primer product.
12. In the past, the water-based cleaning agent produced a sludge residue which was partially dewatered and then landfilled. The company decided to eliminate the need to landfill this waste totally. Currently, the waste stream is blended with additives after flocculation, to produce a beige-coloured product now sold as a general purpose paint.

18.9 PROBLEMS

18.1 Motor fuel is produced by refining crude-oil.
   (a) Identify the stages in the life cycle of this product.
   (b) Define a system on which a life cycle assessment of such a product might be conducted.
   (c) What environmental parameters should be chosen?
   (d) What data would you need to gather?
   (e) In your opinion, what stage of the life cycle is most important, and why?
18.2 A car (automobile) manufacturer is designing a new bumper (fender). The material choice lies between plastic or steel. Applying the principles of LCA, which do you think is preferable?
18.3 Liquid milk may be sold in any of four containers, each of one litre capacity:
   (a) waxed paper cartons,
   (b) disposable plastic bottles,
   (c) re-usable plastic bottles,
   (d) re-usable glass bottles.
   Qualitatively assess which is least detrimental to the environment. Explain what factors must be considered.
18.4 Consider the Swiss glass bottle data in the text. Is it environmentally favourable to recycle as much glass as possible? Examine the sensitivity of your conclusion to variations in the data.
18.5 Five methods for reduction of waste at source are given in the text. Prioritize these in terms of preferred order of implementation. Explain why you have made such a decision.
18.6 A simple block diagram of a manufacturing facility is given below:

   (a) Draw an envelope into which all feeds enter and out of which all feeds leave.
   (b) Redraw the diagram for the situation where all waste is recycled/reused (the so-called zero waste process). Is this a true representation of the emissions due to the manufacturing facility?
   (c) Include the energy production facility and show all input and output streams.

18.7 Outline some methods for minimizing waste in the domestic kitchen/bathroom. Identify the generic type of each method (i.e. source reduction, etc.)

18.8 What do you think are the main reasons that people do not readily adopt the methods for waste miminization that you have identified in Problem 18.3?

18.9 Draw up a plan for a preliminary assessment of waste minimization opportunities in a small automobile repair shop. Identify, in advance, possible target areas for waste reduction.

18.10 Conversion may be defined as the percentage of feed material that is transformed to product. A process uses a feed that is 90 per cent pure. The conversion is 35 per cent to desired product and 10 per cent to waste by-product. Based on 100 kg feed, determine all input and output flow rates. Outline some methods that might be used to minimize waste.

18.11 A spray paint shop produces items in 46 different colours. Identify possible waste streams and outline methods of reducing waste. Prioritize your reduction methods, assigning generic types to them (e.g. recycling) and explaining your reasons for choosing them. If the paint is currently organic solvent based, what are the advantages and disadvantages of converting to water-based paints?

18.12 Consider a wastewater treatment plant serving a population equivalent of 150 000. Identify the key unit processes and materials quantities in a conventional treatment plant. Now, suppose you have been given the task of designing a waste minimization programme. Identify what you would do in both a qualitative and a quantitative way. What volume reduction can you achieve and what amount of energy can you save?

18.13 A block diagram of a chemical manufacturing facility is shown below. Process step efficiencies are reacion, 60 per cent; solvent recovery, 80 per cent; separation, 90 per cent.

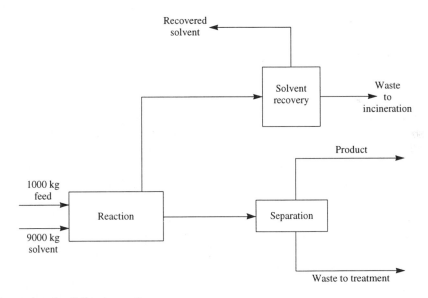

(a) Determine the following ratios:
   (i) waste to product,
   (ii) product to feed.
(b) Determine the effect on waste of a 5 per cent increase in efficiency for each operation.

REFERENCES AND FURTHER READING

BUWAL (1991) *Ecobalance of Packaging Materials State of 1990*, BUWAL, Switzerland.

CEC (1990) 'Technical note on best available technologies not entailing excessive cost for nitric acid production', EUR 13004 EN, CEC, Brussels.

CEC (1992) *European Community Environmental Legislation*, Vols 1–7, DGXI.

CEC (1992) 'Proposal for a directive on the limitation of emissions of sulphur dioxide from large combustion plants', COM(92) 563 final.

CEC (1992) 'An EC programme of policy and action in relation to the environment and sustainable development', COM(92) 23 final.

CEC (1993) 'Proposal for a council directive on integrated pollution prevention and control', COM(93) 423 final.

Chemical Manufacturers Association (CMA) (1991) *Pollution Prevention Resource Manual*, Chemical Manufacturers Association. Washington D.C., USA.

Coulson, J. M., J. F. Richardson and R. K. Sinnott (1983) *Chemical Engineering*, Vol. 6, *Design*, Pergamon, Oxford.

Cunningham, J. D. and J. Moriarty (1993) *Clean Technologies in the Fine Chemicals/Pharmaceutical Industry*, Clean Technology Centre, Ireland.

Cunningham, J. D. (1993) *Estimation of Volatile Organic Compound Emissions from Non-point Sources in the Chemical Process Industries*, Clean Technology Centre, Ireland.

De Hoo, S., H. Dielman, R. Von Berkel, F. Reigenge, H. Brezet, J. Cramer and J. Schot (eds) (1991) *PREPARE*, Dutch Ministry of Economic Affairs.

Freeman, H. (1990) *Hazardous Waste Minimisation*, McGraw-Hill, New York.

Grüjer, U. (1991) *Waste Minimisation: A Major Concern for the Chemical Industry*, *Water Sci. Technology*, **24**(12), 43–56.

Hanlon, D. and C. Fromm (1990) 'Waste minimization assessments', in *Hazardous Waste Minimisation*, H. Freeman, (ed.), McGraw-Hill, New York, pp. 71–126.

Hunt, G. E. (1990) 'Waste reduction techniques and technologies', in *Hazardous Waste Minimisation*, H. Freeman (ed.), McGraw-Hill, New York, pp. 25–54.

Johannson, A. (1992) *Clean Technology*, Lewis Publishers.

Keoleian, G. A. and D. Menerey (1993) *Life Cycle Design Guidance Manual*, EPA600/R-92/226, USEPA.

King, C. J. (1980) *Separation Processes*, 2nd edn, McGraw-Hill, New York.

Kirk, R. E., D. F. Othmer, *et al.* (1984) *Kirk–Othmer Encyclopaedia of Chemical Technology*, Wiley–Interscience, New York.

Levenspiel, O. (1972) *Chemical Reaction Engineering*, 2nd edn, John Wiley, New York.

Lindfors, L.-G. (ed.) (1992) *Product Life Cycle Assessment—Principles and Methodology*, Nordic Council of Ministers.

McCabe, W. L., J. C. Smith and P. Harriott (1993) *Unit Operations of Chemical Engineering*, 5th edn, McGraw-Hill.

McKetta, J. J. and W. A. Cunningham (eds) (1990) *Encyclopaedia of Chemical Processing and Design*, Marcel Dekker, New York.

Pedersen, B. (ed.) (1993) *Environmental Assessment of Products—A Course on Life Cycle Assessment*, 2nd edn, UETP-EEE, Helsinki.

Perry, R. H. and D. W. Green (1984) *Perry's Chemical Engineers' Handbook*, 6th edn, McGraw-Hill, New York.

Peters, M. S. and K. D. Timmerhaus (1991) *Plant Design and Economics for Chemical Engineers*, 4th edn, McGraw-Hill, New York.

Theodore, L. and Y. C. McGuinn (1992) *Pollution Prevention*, Van Nostrand Rheinhold, New York.

USEPA (1990) *Guides to Pollution Prevention: The Paint Manufacturing Industry*, EPA/625/7-90/005, USEPA. Washington, DC, USA.

Willums, J. O., and U. Golüke (1992) *From Ideas to Action: Business from Sustainable Development*, ICC. International Chamber of Commerce, Paris, France.

ENVIRONMENTAL IMPACT ASSESSMENT

19.1 INTRODUCTION

Environmental impact assessment (EIA) is a process that requires consideration of the environment and public participation in the decision-making process of project development. An environmental impact statement (EIS) is a review document prepared for assessment in the EIA process. In some countries EIA is a direct legal requirement, while in others it is enforced indirectly under general planning, health or pollution control powers. Consequently, it is appropriate to review the origin of EIA, alternate procedures, project screening, EIS scoping, preparation and review. We also examine multidisciplinary project management as engineering practice plays a key role in all stages of project development. This chapter aims to introduce the reader to the EIA process and investigate the depth of study required in the production of an EIA.

The stages of EIA include the following:

1. *Screening*, to decide which projects should be subject to environmental assessment. Criteria used include threshold, size of project and sensitivity of the environment.

2. *Scoping* is the process which defines the key issues that should be included in the environmental assessment. Many early EIAs (pre-1985) were criticized because they were encyclopaedic and included irrelevant information.

3. *EIS preparation* is the scientific and objective analysis of the scale, significance and importance of impacts identified. Various methods have been developed to assist this task.

4. *Review.* As environmental assessments are normally produced by the project proponent, it is usual for a review to be undertaken by a government agency or an independent review panel. The review panel guides the study and then advises the decision makers.

These elements in addition to multidisciplinary project management are considered in detail below. Examples of EIS checklists for road, waste and mining projects are also provided for reference.

19.2 ORIGINS OF EIA

All ecosystems, including human communities, have thresholds of tolerance for pollution and disturbance beyond which the system may suffer anything from temporary upsets to complete destruction. In this regard the industrial and agricultural practices after the Second World War began to cause environmental damage which crossed these thresholds.

Wide-scale environmental interest and concern was awakened by Rachel Carson's book *Silent Spring* (1962). This book set out to show the American people how their land and lives were affected by the large-scale spraying of crops with insecticides and herbicides. She showed that agricultural pesticides were being found in environments outside their target areas. She succeeded in making the public aware of the ecological consequences of introducing toxic chemicals into the natural food chains and the effects of cumulative dosage with apparent small quantities of agricultural chemicals. From this beginning arose public concern for the environment (biologists and ecologists had long been aware of the dangers) and eventually pressure by the public and environmentalists forced state and federal authorities to exert some control over the release of toxic chemicals into the environment. The National Environmental Policy Act (NEPA) was introduced in the United States in 1969 and required environmental statements to be prepared for federally funded or supported projects that were likely to have impacts on the environment. The US Council for Environmental Quality (CEQ) was given the task of developing standard procedures for environmental statements which were published in August 1973.

Environmental assessment was accepted in principle at the United Nations Conference on the human environment in 1972 at Stockholm when the framework of modern environmental international and national policy was laid down. The conference generated a concern for the environment which resulted in the 1980 publication of a world conservation strategy by the International Union for the Conservation of Nature (IUCN), the United Nations Environment Programme (UNEP) and the World Wildlife Fund (WWF), and subsequent launching of a series of national policies on environmental conservation and control, and the slow, controversial but definite progress of the European Community towards formal European legislation.

The European Community initiated environmental action programmes in 1973, 1977, 1983, 1987 and 1992, the principle being that the best environmental policy consists of preventing the creation of pollution or nuisances at source, leading to the need to consider developments before construction and the consequent need to create legislation to enforce such consideration.

19.3 EIA PROCEDURE

There are three options to consider in establishing EIA procedures. They include:

- Legislative option—formal legal approach in which environmental assessment procedures become law and are enforced by the courts
- Middle ground option—adoption of environmental assessment principles within accepted planning procedures
- Policy option—systems are developed and incorporated within the administrative machinery of government

The legislative option is the more formal legal approach in which the EIA procedures become law and are enforced by the courts.

The advantages are:

- Mandatory procedures
- Regulations developed to direct and control activities
- Enforceable requirements

The disadvantages are:

- The cost of the bureaucratic machinery required for administration
- The time lost when the law is challenged in the courts
- The loss of flexibility in dealing with unique types of projects and or environments

The policy option as a basis for EIA procedures means that systems are developed and incorporated within the administrative machinery of government. Under this option the rules and regulations are not enforceable in a legal sense.

The advantages are:

- Greater direct control over the process
- More opportunity to alter procedures in the light of experience
- Less administrative costs
- Avoidance of long delays on projects due to legal arguments

The disadvantages are:

- The entire system is more vulnerable to political whims.
- It may be difficult to force agencies to take responsibilities seriously and the concerned public may not be able to directly challenge a final decision with which they disagree.

The middle ground option clearly takes elements of both the legislative and policy options as outlined above.

19.3.1 The Legislative Option (NEPA)

The concept of EIA was first put into practice in the United States with the passage of the National Environmental Policy Act (NEPA) in 1969. The Act requires all federal agencies to consider the environmental consequences of their proposed developments and, where significant, an EIS needs to be prepared, reviewed by other interested agencies and then submitted to the Council of Environmental Quality (CEQ), the agency established in 1970 to oversee the implementation of NEPA. Since the Act specifies what is to be included in the EIS, the responsible agency can be challenged in the courts for failure to comply with the legal requirements. In 1978, the CEQ published regulations under the Act which were directed towards streamlining the process, making the output more useful to planners and decision makers, encouraging public involvement and ensuring that the agencies follow up on their responsibilities after the approval of the EIS.

NEPA is very much a reflection of the importance placed on 'the legislative or democratic approach' under the US system of government, under which the general public, operating through the courts, act as the watchdog of government. Such an approach requires the presence of well-organized, well-funded and influential environmental lobby groups since concerned individuals seldom have the resources to undertake legal proceedings against governments or large industries.

Under the legislative option for EIA, as practised in the United States, 'enforcement' is achieved by public pressure operating through the judiciary who are the final interpreters of the law.

19.3.2 The Policy Option (EARP)

In 1973, the Canadian government implemented the Environmental Assessment Review Process (EARP) which was based on a decision of the Federal Cabinet. Under this policy all agencies of the federal government are to take account of potential environmental and social consequences of development early in project planning. The Federal Environmental Assessment Review Office (FEARO) was established to oversee the implementation of EARP.

The Environmental Assessment Review Process has a number of characteristics:

- It is based on the concept of self-assessment, i.e. the agencies themselves are responsible for determining whether their activities pose significant environmental problems. Their decisions must be defensible and they can be challenged by the Federal Environmental Assessment Review Office (FEARO).
- The policy is based on a hierarchical approach in which only the potentially most environmentally damaging projects are subjected to a comprehensive public review with resulting recommendations to the ministerial level of government. Projects considered to have less potential impact are reviewed through more normal administrative procedures.
- It operates on the principle of 'the polluter pays', i.e. the industrial proponent pays for the cost of the EIS (where it is required) and any resulting project changes, delays or mitigation measures.

Under the review process, federal agencies conduct their own screening of projects for environmental problems. If the problems are considered to be minor, they are dealt with through regular interdepartmental consultative mechanisms which ensure that all existing environmental standards and regulations are applied. If the potential environmental or directly related socioeconomic impacts are considered to be significant, then the project is referred to the FEARO which establishes a review panel.

The American and Canadian approaches to EIA represent opposite ends of a spectrum but there are signs that in practice they are converging on a 'middle ground'. Both NEPA and EARP are trying to promote the adoption of EIA principles within accepted planning procedures.

19.3.3 The Middle Ground Option (EC Directive)

The EU Environmental Assessment Directive, which was agreed in July 1985 and came into force three years later, is arguably the most significant of all the existing pieces of Community environmental legislation and is an example of a middle ground option. The requirements of the Directive reach deeply into the decision-making processes of Member States. It should, however, be stressed from the start that the Directive is based on what is essentially an underlying faith as it is focused on procedural rather than substantive obligation. In itself it does not ensure or require that Member States refuse to approve projects that are damaging to the environment.

Where the Directive does apply, it simply provides that the information on the environmental effects of the project gathered in accordance with the Directive must be 'taken into consideration in the development consent procedure'. The policy assumption is that if the procedures are followed, then eventual decisions are likely to improve from an environmental perspective. Nevertheless, the final decision-making on individual projects and the complex weighing up of environmental, economic and other interests that are inevitably involved are in principle explicitly delegated to Member States and their competent authorities. As a result, the whole issue of compliance with the Directive is heavily infused with questions of discretion and the principles on which the Commission or courts should review questionable decisions by national administrations.

The Directive consciously addresses an all-encompassing range of potential environmental assets, from nature protection to cultural heritage. The complexities were well illustrated by the radically different methodologies that emerged during the development of the Directive, particularly those that concerned one of the central issues—the selection of projects to be subject to assessment.

One of the original drafts argued strongly in favour of the Directive specifying general selection criteria only and against the idea of listing project types; later drafts specified a number of mandatory project types, but left others to be identified by Member States according to criteria specified in the Directive; the final structure, which retains the mandatory list, includes a lengthy discretionary list of projects, but largely leaves it to Member States to determine themselves the criteria necessary to identify which particular projects within the latter should be subject to assessment.

The most critical step in understanding the Directive is to appreciate that 'assessment' as it appears in the Directive does not refer to a single document, however lengthy or comprehensive, but implies a process of what may well be iterative steps towards decision making. Once a project has been determined to fall within the scope of the Directive, the assessment takes the form of five obligatory steps:

- The supply of information on environmental effects by the developer
- Consultation with other environmental agencies
- Consultation with the public
- Consultation with other Member States where the project has transnational implications
- The requirement on the part of the competent authority to take into consideration information gathered during these previous steps before making its decision to authorize the project.

No single step in itself is an 'environmental assessment' for the purposes of the Directive—only in cumulation do they amount to assessment.

19.4 PROJECT SCREENING FOR EIA

19.4.1 Introduction

The success of EIA is only as effective as the coverage of projects to which it is applied. Little attention has been paid to the selection of those projects that should be subject to EIA. This has often led to the development of case law which indicates to the agencies developments for which an EIA will be required. In countries without such a procedure, guidelines rarely exist, the decision being dependent upon the scale of the proposal, its environmental setting, uniqueness of the project and the likely degree of public opposition. Because countries have adopted different approaches large disparities exist as to the number of EIAs prepared.

There are a number of means by which project selection for EIA can be undertaken. Specified development types accompanied, in some instances, by thresholds of size, cost or power requirements may automatically require EIA. In some countries, EIAs are mandatory for certain classes of development which exceed a specific financial threshold. This approach unfortunately neglects the importance of the environmental setting of a project. The extent and significance of a particular impact depends not only on the causative agent, for example the amount of a pollutant, but also on the sensitivity of the receiving environment. Alternatively, environmentally sensitive areas can be designated in which an EIA is required for specific developments. This approach contains the implicit assumption that only specific development types are detrimental to certain environmental features. The identification of environmentally sensitive areas, (e.g. wetlands) is an ongoing process in most countries.

19.4.2 Methods

Outlined below are a number of methods commonly used to select projects for EIA:

- The use of positive and negative lists
- The use of project criteria.
- Sensitive area criteria
- Matrices
- Initial environmental evaluations

Positive and negative lists These are lists of project types which are considered to be candidates for EIA (positive) or projects for which an EIA will not be required (negative). This is further refined by having two (positive) lists: (a) projects for which EIA will be required in every circumstance, (b) projects for which

EIA is required only in certain circumstances. See Appendix 19.1 at the end of this chapter for an example of the list approach adopted by EC Directive 85/337.

Project thresholds   Thresholds can assist in project screening, where projects exceeding predetermined thresholds are considered to be candidates for EIA. Thresholds can relate to size of project, production or area of land required.

Sensitive area criteria   Two approaches can be discussed:

1. The first refers to the assimilative capacity of the environment and implies a predetermined set of values which describe the resilience of the environment to known levels of disturbance.
2. The second relies on the determination of environmentally sensitive areas in terms of features of interest which in combination warrant the designation of an EIA. Any project within this area would become an automatic candidate for EIA.

Matrices   The matrix compares project activities listed under the main headings of site investigation and preparation, construction, operation and maintenance, future and related activities, with the environmental consequences which may be physical/chemical, ecological, aesthetic, social. Matrix interactions may have no significant effect, be mitigated through an appropriate design measure, have unknown effects, have potentially adverse effects, and not have significant effects. On completion of this stage a panel of experts is convened to assess the results and decide whether an EIA should be required.

Initial environmental evaluation (IEE)   IEE is a 'mini EIA'. It requires a description of the environment and the development and the identification (not assessment or interpretation) of environmental impacts that are anticipated.

19.4.3 Practice

In the US, the NEPA applies to all types of activities undertaken by all federal agencies. The only 'screen' is the interpretation of the word 'significant' with respect to potential impacts. Since this is poorly defined in the Act and Regulations, agencies have tended to consider questionable impacts as significant to avoid being in contempt of legislation. This has resulted in the production of a significant number of EISs for relatively minor projects.

In Canada, under EARP the initiating departments are given the responsibility for determining which projects are insignificant. If there is not enough information to arrive at a decision a more detailed analysis will be carried out. In practice, it is only the large projects that potentially involve a wide variety and intensity of impacts which are referred to the FEARO for a comprehensive public review. In this manner, the agencies tend to incorporate the environmental considerations for smaller and more conventional projects within regular planning procedures.

The EU Directive applies to proposals for projects falling within various classes specified in the Annexees to the Directive (see end of this Chapter). Annexe I contains nine classes of projects and subject to special exemption procedures. Annexe II contains a much more extensive list of 83 types of project divided into 12 classes. These projects are subject to assessment 'where Member States consider that their characteristics so require'. This apparent discretion has proved to be one of the major areas of difficulty of interpretation, with conflicts of approach between the EU Commission and some Member States. Initially, a number of Member States took the wording of the Directive at its face value and decided that they had the power to exclude whole categories of Annexe II projects from the range of their national legislation on the grounds that they did not consider their potential for environmental impact warranted the assessment procedures (see Appendix 19.1 for project listings).

The EU Commission eventually took a much stricter interpretation. Although it was accepted that Member States have some discretion to determine which particular projects falling within the classes of Annexe II should be subject to assessment, this discretion has be set against what was considered to be the fundamental requirement of the Directive:

> Member States shall adopt all measures necessary to ensure that before consent is given, projects likely to have significant effects on the environment by virtue *inter alia* of their nature, size or location are made subject to an assessment with regard to their effects.

According to this interpretation, the wording of Article 2 is a quasi-objective statement of the goals of the Directive, and Member States therefore do not have an unlimited discretion to determine which projects in Annexe II shall be subject to assessment or not. At the very least, Member States must incorporate within their legislation the potential for including any Annexe II project within the assessment procedures, together with the methodology adopted to decide which particular projects are likely to have significant environmental impacts.

The state of implementation with regard to the most basic framework for Annexe II projects remains unsatisfactory in a number of Member States: Spain, for example, has national assessment legislation in relation to Annexe I projects only but has left it to regional authorities to decide how to proceed with Annexe II projects; Italy is in the same position; the framework decree in Portugal was only issued on 6 June 1991; Greece has no legislation implemented with respect to Annexe II projects; German legislation has excluded a number of whole categories of Annexe II projects.

As to methodology to decide which particular Annexe II projects should be subject to assessment, the Directive gives Member States a wide choice: Article 4(2) provides that:

> Member States may *inter alia* specify certain types of project as being subject to an assessment or may establish the criteria and/or thresholds necessary

Some of the countries that have developed legislation in relation to Annexe II have adopted radically different approaches. The Dutch legislation, for example, contains detailed and precise thresholds for the various categories of project. The British legislation has avoided the use of legal thresholds, but relies upon a case-by-case sifting, set against central government advisory criteria for some of the project types. In Belgium there are both examples, with the legislation in the Walloon and Flemish regions adopting opposite methods. The absolute legal threshold and the case-by-case approach are at extreme ends of a spectrum of methodologies, and both have pitfalls from the point of view of the Directive.

The legal threshold approach has the advantage of certainty and transparency, but a project falling below thresholds may as a result be excluded from the need for an assessment even though in particular and unpredictable circumstances (such as location of a small-size project in a sensitive area) significant environmental effects are likely. The case-by-case approach for Annexe II projects is probably more truly in accordance with the aims of the Directive and has the advantage of site sensitivity. However, from the Union perspective it is vulnerable to local abuse (assuming individual determinations of significance are made at local level) and makes the task of reviewing and comparing the application of the Directive in different Member States complex. Under the British Regulations, no reasons need be given where a local planning authority determines that a particular proposal is not likely to have significant environmental effects and third parties have no right of appeal against the decision. Nor is the Department kept regularly informed of negative decisions. However, as a matter of European Union law, and in response to complaint proceedings, it may be that more detailed justification would have to be shown to establish that the Directive was being properly implemented in practice—in particular, both the EU Commission and other Member States are likely to be concerned at local decisions on significant environmental impacts which are contrary to published government guidelines on the subject.

19.5 SCOPE STUDIES FOR EIS

19.5.1 Introduction

Scoping is the procedure for establishing the terms of reference for the Environmental Impact Statement (EIS). In general the objectives would be to identify the concerns and issues that warrant attention, to provide for public involvement and to prepare a detailed brief for the investigation of specific issues associated with the development.

The US Council on Environmental Quality (CEQ) introduced, in 1978, regulations establishing scoping as a formal requirement of the EIA process. Essentially, its role is to 'single out', at the outset of any assessment, the most important environmental issues relevant to the study. Very often, these are dictated by the concerns of the local community, although other issues such as the importance of a locality for nature conservation may emerge as a key issue. In this way, the EIA process is better focused and more streamlined as a result, and specific issues that are not directly material to the authorizing decision fall outside the remit of the EIA.

Institutional structures for the EIA vary and, with them, the responsibility for implementing different parts of the EIA process. In Europe, with the exception of Holland, scoping is not an explicit part of the EIA process and it is implied that the responsibility for determining the terms of reference for the EIA falls to the project proponent (although limited guidance as to the information required to be presented in the environmental report is given by Annexe III of the EU Directive). It should be noted that some guidance is often given by the competent authority. The panel produces a guideline or terms of reference as a result of this exercise, which set out a detailed brief to the developer describing the scope of the EIA. In the United States, scoping is a part of the EIA and is the responsibility of the project proponent, although this is done in close consultation with the federal agency to which the EIS will be submitted, as well as other agencies with a contribution to make to the assessment.

Ideally, scoping should be a process of open dialogue and consultation. While it is appropriate that the financial burden of the activity should be met by the proponent, local authorities, public and environmental agencies have a considerable contribution to make to establishing the terms of reference of a study. This requires the willing participation of such agencies and a need to enter into discussions with a project proponent.

Definitions of the term 'significance' are various and often controversial. Mostly, they are a function of the perception of certain groups in society who possess certain predetermined values against which the significance of certain activities is judged. The determination of significant issues is similarly, therefore, a phrase that is open to misinterpretation and for this reason is often substituted by the term 'key issues'.

Most scoping studies have either implicitly or explicitly defined key issues to be those that are identified by the organizations, bodies and local groups who have an interest in a particular development proposal. In some cases, many of the concerns of different organizations may, in fact, be similar. For example, residents' groups may be concerned about noise, dust, amenity and conservation. These concerns may mirror those of business interests such as hoteliers and those involved in the tourist trade who see a high-quality environment as vital to the continuation of their business. Similarly, noise, dust and amenity may be the concerns of agriculturalists who perceive disturbance to their activities from pollution. For this reason, scoping has focused on facilitating the involvement of interested parties and the community to identify the legitimate concerns of these groups, thereby establishing the terms of reference for any subsequent EIA.

19.6 PREPARATION OF AN EIS

One of the important elements of any project is the preparation of documentation to communicate the findings and conclusions of the study. No amount of data collection can compensate for a poorly conceived or written report. Furthermore, the value of a project is sharply diminished if its findings do not reach its intended audience.

A number of considerations need to be addressed:

- Planning
- Purpose
- Audience
- Structure

19.6.1 Planning

There are considerable benefits to be gained in pre-planning the preparation of a document before drafting. Three issues should be considered: the purpose of the report, its intended audience and the report's structure.

Purpose   The purpose of a project report is to communicate its findings. In the case of environmental study projects, it is to set out in a clear and logical manner the likely environmental consequences of pursuing certain courses of action with the intention that eventual decisions will be better informed and appraised of the possible implications of that action.

Audience   Having stated that the purpose is to communicate findings, an understanding and appreciation of the intended audience must be gained. In many cases, technical reports have a large non-technical readership. Environmental assessment reports, for example, are aimed at planning officials and decision makers who cannot be expected to have an all-encompassing appreciation of technical matters. The general public represents a second audience that needs to be accommodated within the overall style of the document. Very often the problems posed by non-technical readerships can be overcome by the incorporation into the document of a non-technical summary which highlights the salient features of a project and reports the key issues and implications associated with it. Technical information pertinent to the study can be included in appendices to the main report.

Structure   Documents should have a logical structure and avoid repetition. A common failing with environmental study documents is that, due to the interrelated and multidisciplinary nature of many environmental issues, documents tend to be very repetitious. This is due to the fact that each contributor (and there are usually several different expert contributions) is required to address the development in terms of:

- Baseline conditions
- Development features
- Anticipated effects
- Mitigation
- Residual impact

Early impact document layouts reflected the presentation needs in terms of a verbatim reading of the EU Directive. This approach considered, for example, each baseline element such as air quality, water quality, noise and traffic, etc., in a baseline section, while impacts and mitigation measures were considered separately in an effects section. However, a much more efficient cohesive and less repetitive form of presentation is to review each feature of a single study element in one section. For instance a section on water quality would consider anticipated effects mitigation with residual impacts as a discrete entity. Consequently, a typical EIS document would now comprise the following sections:

- Planning framework
- Development proposal
- Archaeology

- Ecology
- Landscape
- Water quality
- Air quality (if applicable)
- Waste
- Traffic
- Social effects

Non-technical summary   As stated earlier, a non-technical summary is a desirable element of a report, especially in the case of technical assessments which require to be communicated to a wide audience. A common failing of many non-technical summaries, however, is that they tend to be little more than a précis of the complete document, rather than a summary of the key issues and salient features of a proposal.
In the case of an environmental assessment, the summary should highlight the main characteristics of the project and set out in a clear and concise form the environmental impacts associated with it. It should not summarize the whole document. On the other hand, it should explain the rationale behind its conclusions. In essence, the non-technical summary should restrict its attention to the issues most relevant to the eventual decision.

19.6.2 Public Participation

Like the EIA itself, the nature of public involvement varies with different institutional arrangements. In some countries, formal procedures exist to involve the public in scoping, while in others public involvement is undertaken at the discretion of the party responsible for the process. In the United Kingdom, neither formal nor informal procedures for the involvement of the public exist at any stage. However, a large number of EIAs undertaken in the United Kingdom to date have sought the involvement of the public by some means. Mostly, this is through the circulation of information to public bodies, outlining the scope of an EIA as envisaged by the project proponent and inviting comment. This may be followed up by meetings with certain agencies and groups.

In Canada, public involvement is more formalized. A panel of experts is assembled, responsible for the development of guidelines for the EIA and reviewing the environmental report submitted by the project proponent. The scoping activity is conducted by the panel as a part of their determination of EIA guidelines—the 'brief' for the EIA. To aid in this task, a panel will frequently elicit views of the public, communicated via written comment, workshops or public meetings.

Public meetings achieve a number of ends:

1. They provide the panel with questions and concerns expressed by the local people and other agencies and groups with an interest in the project.
2. They allow the project proponent to explain the details of a project in a non-adversarial environment.
3. They allow contentious issues to be discussed in an open forum.

The strength of such meetings lies in their informality. In the absence of an adversarial or legalistic atmosphere, issues and concerns can be discussed openly and the details of proposed project actions explained. However, care is needed to ensure that meetings do not provide opportunities for conflict to develop. Skilful management and control of the meetings is essential if their quality is to remain uncompromised.

While it would be appropriate to establish the *bona fide* concerns of the public, the value of conducting pro and anti ballot polls early in the EIA process must be questioned. The EIS must insofar as possible identify and address the *issues of concern* of third parties. However, attempting to measure the *level of support or opposition* is extremely difficult for a number of reasons.

Finally, it should be said that during the planning phase a development proposal can change radically in scale and content. In addition, awareness of the project reaches an ever-increasing audience which changes its concept of key concerns as a project develops. Consequently, opinion poll data can only be related to the level of project information available at a particular point in the pre-planning phase. These poll data should not be presented as representative of the support or opposition for a project at any other time, as this could lead to an erroneous interpretation.

19.7 REVIEW OF EIS

Because most EIA systems state that an assessment must be produced by the project proponent there is usually a need for an impartial, scientific and independent review. This is not to imply that all EIAs are biased and play down adverse impacts and emphasize positive ones, although there is some evidence to suggest that this occurs. The public need to be confident of impartiality and for this reason there is a need for some form of independent review.

The review authority is often likely to be the authority from which authorization for the development is requested. Questions relating to impartiality may arise when the authorizing agency has been responsible for the EIA. An independent review agency may remove any suspicion of bias in those cases where the authorizing agency is an advocate for the development or holds unreasonable views against the development.

The functions of the review authority may include:

- The 'scope' of the assessment, i.e. which projects should be subjected to a full or partial EIS
- General or specific guidelines and advice on methods of EIS.
- Formulate the terms of reference and initiate a detailed EIS
- Ensure that the EIS had been adequately completed within the terms of reference

It is essential that the EIA is not regarded as a procedure that is only to be utilized at the decision-making stages. The EIA ought to be regarded as an adaptive process which continues after the decision. It would ensure that the project conforms to the standards detailed in the relevant permissions and it would provide a database for any subsequent impact study as well as allowing the monitoring and control programme to adapt to changing circumstances or increased knowledge.

19.8 MULTIDISCIPLINARY TEAM MANAGEMENT

19.8.1 Introduction

The operation of collecting environmental information, carrying out an environmental impact study and preparing the environmental statement takes time. The skill of the specialist in environmental information collection lies in knowing exactly what data are relevant and as far as scientifically feasible just how each estimated effect is likely to affect the environment. Once the main data have been collected and examined it is very often the case that only a small number of effects appear to be really decisive in determining whether or not the project should be given planning permission (consent). It is these significant effects that should be the subject of closer and deeper analysis by the environmental assessment team rather than a wider and shallower coverage of all the projected effects.

Project management as a means of controlling the timetable, budget and employment of consultants is essential. When the environmental study is carried out by a consortium of consultant firms, each professional naturally has his or her own method of work and unless the project manager can keep overall control of the programme, the progress of the assessment may not be as rapid or efficient as it should be. Whether the control is exercised by an independent manager appointed for the environmental study or by the client's own project manager or by one of the assessment study team is not important. What does matter is the efficiency and tact with which the job is done as personality may make all the difference to the quality of the environmental study and the conviction carried by the document.

For very large projects there is some merit in appointing two project managers, one to carry out the purely administrative tasks of budget and programme control, and supplies and general office services, and a professional consultant as a technical manager who is responsible for the quality and quantity of work produced by a team and who has the authority to override individual consultant's decisions. Such a person may well be responsible for the final preparation of the text of the environmental study, but to be effective will need to carry the confidence of the whole assessment team.

19.8.2 Project Management Framework

Certain tasks must be undertaken to implement a project within an environmental study framework. Project implementation can be divided into four main tasks:

- Selection of study team and leader
- Study approach.
- Study management
- Report preparation and review

19.8.3 Selection of Study Team

Four questions should be addressed before the study team and team leader(s) are selected:

1. What are the key experience requirements?
2. Should the team be inter- or multidisciplinary?
3. Are any special studies required?
4. What tasks are required of the team leader?

Key experience requirement   Key experience requirements are detailed by the characteristics of both the project and environment. Extractive projects are likely to demand experience in the geological or hydrogeological sciences whereas road schemes necessitate engineering, ecological and landscape assessment skills. In determining experience requirements, it is also important to establish the role of each individual in the environmental assessment.

Inter- or multidisciplinary   Again, this decision is influenced largely by the characteristics of the project and the environment. Typically, the larger the size and scope of a project, the more likely the involvement of a range of disciplines. A new airport, for example, may require the skills of transportation engineers and planners, ecologists, noise specialists, economists, etc. An estuarine barrage scheme on the other hand may only require a small number of disciplines (e.g. hydraulics engineers, surveyors), although it is likely that a range of expertise may be appropriate, e.g. marine and terrestrial ecologists, ornithologists, etc.

Special studies   The likelihood of specialist studies is a major determinant of the skill requirement of an assessment study team. Industrial projects may warrant specialists studies on the dispersion and impacts of certain emissions. Large-scale projects may necessitate specialist investigations on the availability of labour and the economic implications of labour visibility.

Team leader key tasks   Establishing the tasks of the team leader helps to establish the most appropriate team member for this task. Very often the selection of the team leader will be made on the basis of relevant experience in performing similar tasks, although the anticipated workload inputs of team members are an important criteria. For example, it may not be sensible to select as leader a member of the team with a major input into the study, for the simple reason of avoiding overcommitment. The following are typical functions that a team leader would fulfil:

- The development of specific work schedules and project targets
- Co-ordination with the project client
- Expenditure monitoring and budget management
- The review of progress and revision of schedules
- Planning and initiation of team meetings (frequency, agenda, short-term/long-term targets, etc).

19.8.4 Study Approach

Any study, whatever its nature, can be divided into four basic steps: the acquisition of information, the analysis of information, the communication of conclusions and the selection of appropriate action. These four components are relevant to environmental assessment projects as they form the building blocks of an approach that is both logical and structured. Following this 'checklist' of tasks helps to promote both a comprehensive and organized study and also assists in developing an approach that is consistent, irrespective of the nature of the project or environmental assessment.

Description of need   What is the justification for a particular course of action? Can end objectives be met by other means?

Description of project   What are the physical and process characteristics of the project? What land use requirements, emission quantities and types are expected? What alternatives (technology/location) have been examined?

Relevant institutional information   What local, regional, national, or international laws, regulations, standards, etc., are relevant?

Identification of issues   What bodies and organizations (statutory/non-statutory) are likely to have an interest in the project? What issues or concerns are expressed?

Description of affected environment   What components of the environment will be affected or are most relevant to investigate (on the basis of the previous task)? What baseline data requirements exist? Are physical or chemical data relevant? What sources of data should be used? What literature is available?

Impact prediction   What components of the project and environment will interact? Detailed information of the nature of the project, e.g. information on the nature and composition of fuel types, waste streams, employee requirements, etc., may be relevant. Research data may be available to predict the nature of likely impacts. Where possible, the prediction should be made in quantitative terms, e.g. waste gas concentrations, change in ambient noise level, area of habitat affected, duration of impact, probability of occurrence, etc. However, many cases will arise where only a qualitative assessment will be possible, e.g. manner of landscape intrusion, psychological effect. In such cases, the assumptions relevant to each prediction should be made explicit.

Impact assessment   What is the importance/significance of an impact, i.e. how significant is the change in waste gas concentration or noise level? How important is the loss of habitat? Certain criteria against which the importance of impacts can be assessed may take the form of institutional information, e.g. air quality standards or consent limits, or international guidelines on standards to protect health. Other criteria include professional judgement, comparisons with previous projects and concerns expressed by the community.

Impact mitigation   What steps will be taken to minimize or avoid adverse environmental impacts resulting from the proposed project? How will these be implemented? Who will be responsible for implementation? How will their appropriateness and effectiveness be ensured? Certain negative impacts can be mitigated in the course of construction, through the adoption of environmental management plans.

19.8.5 Study Management

Two management tasks need to be addressed; technical and financial management. Technical management serves to guide the project through the collection and interpretation of information, so that sensible conclusions can be reached at the end of the study. Responsibility for technical management lies most appropriately with the study team leader. A number of tasks can be identified:

1. *Team meetings and discussions*   During team meetings, progress in the acquisition or analysis of information can be discussed. Interaction between team members is encouraged, and problems and ideas can be discussed and explored. Deadlines can be re-evaluated and if necessary, the project programme revised.

2. *Meetings with client*   In cases where a project is being undertaken on behalf of a client or organization, it is invariable that regular progress meetings will be held. These provide an opportunity to report adherence to schedules, to highlight problem areas that may emerge and to indicate preliminary observations. The latter is especially important where environmental assessment projects are undertaken as a component of larger development projects. Highlighting possible environmental problems early in the project cycle provides an opportunity to modify potentially impacting features of a proposal in order to reduce or avoid environmental effects.

3. *Peer review*   Peer review is another component of the technical management of a project. It is especially relevant to the prediction and assessment phases of the study, where conclusions and interpretation may be modified in the light of other professional opinions. Peer review is also helpful during the preparation of reports. Advice from colleagues not directly involved in the project can be invaluable in providing a critical appraisal of the style and expression of documentation produced.

Financial management is the second important management function. In some cases, responsibility may lie with the study team leader; in others it may lie more appropriately with financial management personnel. Financial management ensures that project budgets are not exceeded. Employee time, materials, travel and subsistence expenditures, reproduction costs and in some cases certain overheads should be recorded and monitored. Complete financial records must be kept so that the project budget can be audited and all expenditures accounted for.

19.9 EXAMPLES OF PROJECT EIS

19.9.1 Introduction

Every environmental impact statement is unique. While professional guidelines exist, site location, nature of project and interest group concerns combine to produce a matrix of issues that are unique to each project. The elements of an EIS framework, however, can be described with general comment on project type variation. Included here for example and reference are three project types, i.e. mining, motorways and wastewater treatment plants. Each project type is considered in terms of the checklist required for an EIS. The checklists are summarized under project description, environmental effects, possible and mitigation options. A case study of a proposed combined road/rail bridge between Denmark and Sweden is described in Sec. 19.10.

19.9.2 Mining Projects

- Surface industrial installations, principally the preliminary processing of the ores, which usually involves the crushing and sorting of ore from the parent minerals. These processes use reagents that can be highly toxic in minute quantities.
- Refined ores can become sources of potential pollution if spilled or blown from equipment, storage or transportation.
- Other impacts are similar to chemical industries.
- Mineral extraction (other than quarrying).
- Surface industrial installations. This activity has considerable potential to cause significant environmental impacts. Some of these impacts can be permanent and virtually irreversible.
- Spoil heaps and tailings ponds are the surface manifestations, these can be very large. There are other sources of problems because of physical instability, difficulty of re-vegetation, air and water pollution, and problems of finding low maintenance after uses.
- Underground workings produce contamination of ground and surface waters, subsidence and hazards.

This form of development is distinguished by the extent to which it relies on calculation, interpretation and remote sensing to predict impacts.

Project description   Impact assessment is critically dependent on the availability of accurate, relevant data on mineralogy and hydrogeology. BAT (best available technology) is usually justifiable for survey, analysis and prediction, given the permanence and irreversibility of many of the potential impacts.

Checklist of items to be described:

- Construction
  - Survey and exploration activities
  - Environmental measures during decline development
  - Typical issues, noise, dust, vibration, destruction of habitats, dewatering, hazard and traffic
  - Initial dewatering and monitoring for associated settlement
- Operation (including available alternatives)
  - Times of blasting
  - Times of shift changes
  - Monitoring techniques
  - Dewatering
  - Mining methods
  - Subsidence control
  - Ore pressure
  - Reagents used
  - Process control systems
  - Tailings/spoil heap
  - Control of mine drainage
  - Transportation and handling
  - Ancillary facilities
  - Potential for in-mine pollution
  - Potential for tailings/spoil pollution
- Decommissioning (if applicable)
  - Phasing of rehabilitation
  - Interim and eventual land uses
  - Strategies for re-vegetation
  - Monitoring proposals
  - Provision for aftercare

- Financial bonds
- Growth
  - Potential for additional mineral reserves
  - Capacity for additional spoil heaps/tailing impoundments
- Associated developments
  - Transportation infrastructure
  - Monitoring facilities
  - Post-decommissioning land uses
  - Water supply schemes
  - Ore processing
  - Power supply

Environmental effects   Typical significant impacts likely to affect:

- Human beings
  - Health and safety
  - Employment (new)
  - Employment (existing)
- Flora
  - Obliteration by surface installations
  - Changes in surface water
  - Changes in groundwater
  - Dust deposition and photosynthesis
  - Phytotoxic metals and reagents
  - Yields of crops and pasture
  - New habitats after decommissioning
- Fauna
  - Aquatic fauna threatened by water pollution
  - Aquatic habitats threatened by water control measures
  - Aquatic habitats threatened by post-closure leachates
  - Agriculture stock affected by dust (especially metals)
  - Agriculture stock affected by vibration and blasting
- Soils
  - Dust deposition and contamination
  - Contaminated post-mining surfaces
  - Re-vegetation of tailing surfaces
  - Capping of tailings ponds
- Surface water
  - Changes in character (temperature, dissolved gas, etc.) due to dewatering related surface discharges
  - Contamination with in-mine pollutants discharged to surface
  - Contamination with process water discharges
  - Contamination with runoff from stockpiles, spoil heaps, tailings and settling ponds
  - Contamination with runoff from site structures made of mine rock
  - Contamination from accidental spills, reagents, fuels, ores, process water, etc.
- Groundwater
  - Contamination from reaction with mine workings, backfill, oxidized in-mine mineralogy
  - Contaminations from in-mine operations, spills, sewage, etc.
  - Movement of contaminated minewaters after mine closure
  - Widespread lowering of water levels due to dewatering

- Air
  - Air quality from mine vents (fumes from blasting and engines)
  - Dust from transportation, handling and stockpiles
  - Dust from 'beaches' of tailings and settling ponds
  - Dust from site roads
  - Fumes from process vents
- Noise
  - Transportation, storage and handling equipment
  - Klaxons, warnings and PA systems
  - Vent raises
  - Maintenance areas
  - Exploration drilling
- Vibration
  - From mine development
  - From mine operations
- Climate
  - Combinations of climatic conditions (i.e. wind after frost) can cause dust blows from tailings ponds
- The landscape
  - Visual impacts of overground structures
  - Visual impacts of spoil/tailings depositories
  - Rehabilitation after decommissioning
  - Land use/landscape changes due to mining effects
- The interaction of the foregoing
  - Poor rehabilitation/re-vegetation leads to air pollution by dust blows
  - Poor rehabilitation leads to water pollution with impacts for farming, fishing, tourism and residential amenity
- Cultural heritage
  - Obliteration of sites by surface installations
  - Damage to old structures by vibration
  - Changes in soils moisture can affect presentation of buried features
- Material assets
  - Agriculture can be affected by many mining impacts
  - Surface structures can be affected by vibration and settlement

Possible mitigation options

- Dewatering methods
- Process reagents
- Mining methods
- Disposal of tailings
- Mine closure methods
- Rehabilitation objectives

19.9.3 Motorway Projects

- Other roads
- Bridges
- Permanent racing and test tracks for cars and motorcycles

Developments in this category tend to have widespread impacts because of their linear nature. Impacts are typically numerous and varied. Principal concerns would normally include noise, vibration, air quality, material assets (roads), landscape issues, safety (to humans and fauna) and cultural heritage.

Project description   Checklist of items to be described:

- Construction
  - Site preparation
  - Materials sourcing
  - Transportation
  - Drainage works
  - Watercourse diversion/coffer dams
  - Employment
  - Time of year and duration
  - Hours of operation
  - Lighting
  - Noise
  - Vibration
  - Dust
- Operation (including available alternatives)
  - Noise
  - Vibration
  - Maintenance
  - Lifespan
  - Vehicle fumes
  - Dust generation
  - Traffic loading
  - Lighting
  - Safety
- Decommissioning (if applicable)
- Growth
  - Is development big enough to cater for predicted future loadings?
  - Will it need to be widened/strengthened?
- Associated developments
- Developments that require road transport availability (including housing and industry)
- Catering/fuel, etc., services

Environmental effects   Typical significant impacts likely to affect:

- Human beings
  - Noise impacts
  - Vibration impacts
  - Air quality impacts
  - Quality of life
  - Travel times
  - Safety
- Flora
  - Clearance of existing cover
  - Colonization of new roadside habitats
  - Effects of dust and fumes
  - Habitat changes due to gritting/salting during operation

- Fauna
  - Disturbance
  - Danger
  - Migratory obstacle
  - Value of new habitat
- Water
  - Interference with drainage patterns
  - Runoff pollutants
  - Construction impacts to watercourse
  - Erosion
- Air
  - Noise
  - Dust
  - Air
- Climate
  - Ozone depletion, local heating effects, smog
- The landscape
  - Surfacing finish
  - Landscaping proposal
  - Roadside furniture (including signage, safety, rails, lighting)
  - Bridge design/aesthetics
  - Viewing stands/other facilities (test/racing tracks)
  - Car parking (racing tracks)
- The interaction of the foregoing
- Material assets
  - Infrastructural upgrading
  - Power (lighting)
- Cultural heritage
  - Disturbance to items of cultural value.

Possible mitigation options

- Routing/siting alternatives
- Design alternatives including
  - Width of carriageway
  - Surface alternatives
  - Road markings/signage
  - Lighting
  - Railings
  - Landscape
  - Drainage
- Hours of operation (race/test tracks)
- Frequency of usage (race/test tracks)
- 'Do nothing' option
- Underpasses/bridges for humans/wildlife/agricultural livestock

19.9.4 Wastewater Treatment Projects

- Installations for the disposal of industrial and domestic wastewater
- Sludge deposition sites

The problem is to try to balance positive impacts of wastewater treatment with potentially negative impacts such as below:

- Health hazards through biotic factors
- Visual and landscape impacts
- Nuisance of odour, vermin, traffic

Project description Checklist of items to be described:

- Construction
  - Access
  - Traffic
  - Site preparation
  - Landscaping
- Operation (including available alternatives)
  - Hours of operation
  - Capacity of facility
  - Safety and hazard control
  - Pest and odour control
  - Perimeter security
  - Monitoring facilities
  - Quality of waste
  - Management procedures
  - Transportation of sludge
  - Removal of paper and plastic in initial stages
  - Personnel
- Decommissioning (if applicable)
- Growth
  - Phases of expansion
- Associated developments

Environmental effects Typical significant impacts likely to affect:

- Human beings
  - Health and safety
  - Disposal of sludges
  - Handling of sludges
  - Transportation of sludges
  - Nuisance
  - Residential amenity
  - Overall benefit of treatment of wastewater—higher quality of water entering outfall area, with little risk of polluted waters
- Flora
- Fauna
  - Attraction of pests—insects, rodents, starlings. These in turn act as vectors for disease

- Water
  - Improvement in quality of water discharged into outfall area
  - Contamination by uncontrolled surface runoff (water pollution if not properly treated and disposed of)
  - Pathogens released with water
- Air
  - Odours
  - Noise of machinery (e.g. agitators) and transportation trucks
- Climate
  - Odour dispersal/concentration
- The landscape
  - Perimeter fences
  - Access roads, entrances
  - Exposed waste
  - Site structures
- The interaction of the foregoing
  - Climatic effects can concentrate or disperse airborne impacts and nuisances
- Material assets
  - Diminution of amenities for residential and leisure land uses
- Cultural heritage

Possible mitigation options

- Site alternatives
- Site layout to minimize proximity to sensitive receptors
- Landscaping
- Monitoring
- High standards of site management including control of waste acceptance

19.10 CASE STUDY

19.10.1 Swedish Environmental Control Programme for a proposed combined road/rail bridge across the Straits of Öresund between Denmark and Sweden

In this case study (see Fig. 19.1) an environmental control programme has been prepared by the proponent and monitoring started in parallel to the final project planning and EIA. It is an example of what was considered to be the main issues and content of a monitoring programme designed and started before final project approval.

A preliminary programme for control of the environmental impacts has been proposed and is expected to be approved under the Swedish Environmental Protection Act. In order to obtain data in the pre-project situation, monitoring was started by the consortium who were responsible for the project. The consortium started construction work before the end of 1993.

The key issues, where control of environmental impacts is necessary, are considered to be:

- Impacts on marine life
- Impacts on marine environment
- Impacts on groundwater

A hydrological control programme, covering about ten wells, has been ongoing since September 1992. It provided basic data concerning:

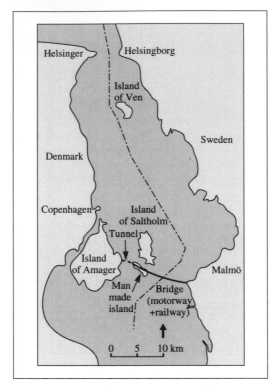

Figure 19.1 Situation of the proposed road/rail connection across the Straits of Öresund (Copenhagen/Denmark-Malmö/Sweden). (From OECD, 1994, *Environmental Impact Assessment of Roads.* Reproduced by permission of the OECD, Paris.)

- Natural variations in the position of the salt front
- Natural variations in the chemical composition of the groundwater (with regard to its interaction with the waters of the Öresund)
- Natural variations in the groundwater level

An analysis of substances that could be future pollutants from the construction and maintenance activities (e.g. petrol, hydrocarbons and heavy metals) will be carried out.

   The programme will provide basic data which may be needed in case of any possible future dispute about groundwater quality in the area. Any possible problems will be identified early on and mitigation measures instigated (e.g. artificial filtration of the freshwater).

   Monitoring of the marine environment has also started. It includes:

- Pelagical investigations (biota, chemical situation)
- Benthical fauna and sediment
- Benthical flora
- Coastal birds
- Seals
- Environmentally poisonous substances
- Fish health

An oceanographical programme has also begun in accordance with the EIA statement. A fishery control programme has been started by the Fishery Authorities in order to collect statistical data.

19.11 PROBLEMS

19.1 Discuss the origin of the EIA and the key elements of the process.

19.2 Distinguish between policy and legislative implementation of the EIA and provide examples.

19.3 Discuss the screening and scoping stages of the process. Provide examples of screening methods.

19.4 Prepare a scope review for a proposed motorway project.

19.5 Provide an assessment of the multidiscipline project management requirements of EIS.

19.6 Discuss mechanisms by which the EIA process could be applied to government and federal policies and programmes.

19.7 Identify any weaknesses in the EU, EIA directive.

19.8 Review the current positive and negative environmental impacts of the Danish/Swedish road/rail bridge project. Use literature or the Worldwide Web to research your review.

REFERENCES AND FURTHER READING

An Foras Forbartha (1986) *EEC Directive on Environmental Assessment*, Dublin.

Bartlett R. V. (1989) *Policy Through Impact Assessment. Institutionalised Analysis as a Policy Strategy*, Greenwood Press, Westport, Connecticut, USA.

Carson, R. (1962) *Silent Spring*, Houghton-Mifflin, Boston, Mass., USA.

Clarke, M. and G. Herington. (1989) *The Role of Environmental Impact Assessment in the Planning Process*, Mansell Publishing, London, UK.

Dixon J. A., R. A. Carpenter, L. A. Fallon, P. B. Sherman and S. Manipomoke (1988) *Economic Analysis of the Environmental Impacts of Development Projects*. Earthscan Publications, London, UK.

Doctor Institute for Environmental Studies (1991) *European Environmental Yearbook*, Doctor International.

EC (1986) *Directive on Impact Assessment 86/337/EEC*, European Commission.

Forsund, F. R. and S. Strom (1988) *Environmental Economics and Management: Pollution and Natural Resources*, Croom Helm Publishers.

Fortlage, C. A. (1990) *Environmental Assessment—A Practical Guide*, Gower Technical Publishers.

OECD (1994) *Environmental Impact Assessment of Roads*, Road transport research, Organisation for Economic Co-operation and Development, Paris.

O'Sullivan, M. (1990) *Environmental Impact Assessment. A Handbook*, REMU, University College Cork, Ireland.

Wathern, P. (1988) *Environmental Impact Assessment. Theory and Practice*, Unwin Hyman Publishers, London, UK.

APPENDIX 19.1: PROJECT SCREENING

Example of project lists (adapted from EC Directive 85/337)

Projects for which EIA is required in every circumstance (Annexe I)

1. Crude oil refineries (excluding undertakings manufacturing only lubricants from crude oil) and installations for the gasification and liquefaction of 500 tonnes or more of coal or bituminous shale per day.
2. Thermal power stations and other combustion installations with a heat output of 300 megawatts or more and nuclear power stations and other nuclear reactors (except research installations for the production and conversion of fissionable and fertile materials, whose maximum power does not exceed 1 kilowatt of continuous thermal load).
3. Installations solely designed for the permanent storage or final disposal of radioactive waste.
4. Integrated works for the initial melting of cast iron and steel.
5. Installations for the extraction of asbestos and for the processing and transformation of asbestos and products containing asbestos: for asbestos-cement products, with an annual production of more than 20 000 tonnes of finished products, for friction material, with an annual production of more than 50 tonnes of finished products, and for other uses of asbestos, utilization of more than 200 tonnes per year.
6. Integrated chemical installations.
7. Construction of motorways, express roads and lines for long-distance railway traffic and of airports with a basic runway length of 2100 m or more.
8. Trading ports and also inland waterways and ports for inland waterway traffic which permit the passage of vessels of over 1350 tonnes.
9. Waste-disposal installations for the incineration, chemical treatment or landfill of toxic and dangerous wastes.

Projects for which EIA is required only in certain circumstances (Annexe II)

1. Agriculture
   (a) Projects for the restructuring of rural land holdings
   (b) Projects for the use of uncultivated land or semi-natural areas for intensive agricultural purposes
   (c) Water-management projects for agriculture
   (d) Initial afforestation where this may lead to adverse ecological changes and land reclamation for the purposes of conversion to another type of land use
   (e) Poultry-rearing installations
   (f) Pig-rearing installations
   (g) Salmon breeding
   (h) Reclamation of land from the sea

2. Extractive industry
   (a) Extraction of peat
   (b) Deep drillings with the exception of drillings for investigating the stability of the soil and in particular:
      (i) Geothermal drilling
      (ii) Drilling for the storage of nuclear waste material
      (iii) Drilling for water supplies
   (c) Extraction of minerals other than metalliferous and energy-producing minerals, such as marble, sand, gravel, shale, salt, phosphates and potash
   (d) Extraction of coal and lignite by underground mining
   (e) Extraction of coal and lignite by open-cast mining

    (f) Extraction of petroleum

    (g) Extraction of natural gas

    (h) Extraction of ores

    (i) Extraction of bituminous shale

    (j) Extraction of minerals other than metalliferous and energy-producing minerals by open-cast mining

    (k) Surface industrial installations for the extraction of coal, petroleum, natural gas and ores, as well as bituminous shale

    (l) Coke ovens (dry coal distillation)

    (m) Installations for the manufacture of cement

3. Energy industry

    (a) Industrial installations for the production of electricity, steam and hot water

    (b) Industrial installations for carrying gas, steam and hot water; transmission of electrical energy by overhead cables

    (c) Surface storage of natural gas

    (d) Underground storage of combustible gases

    (e) Surface storage of fossil fuels

    (f) Industrial briquetting of coal and lignite

    (g) Installations for the production or enrichment of nuclear fuels

    (h) Installations for the reprocessing of irradiated nuclear fuels

    (i) Installations for the collection and processing of radioactive waste

    (j) Installations for hydroelectric energy production

4. Processing of metals

    (a) Iron and steelworks, including foundries, forges, drawing plants and rolling mills

    (b) Installations for the production, including smelting, refining, drawing and rolling, of non-ferrous metals excluding precious metals

    (c) Pressing, drawing and stamping of large castings

    (d) Surface treatment and coating of metals

    (e) Boilermaking, manufacture of reservoirs, tanks and other sheet-metal containers

    (f) Manufacture and assembly of motor vehicles and manufacture of motor vehicle engines

    (g) Shipyards

    (h) Installations for the construction and repair of aircraft

    (i) Manufacture of railway equipment

    (j) Swaging by explosives

    (k) Installations for the roasting and sintering of metallic ores

5. Manufacture of glass

6. Chemical industry

    (a) Treatment of intermediate products and production of chemicals

    (b) Production of pesticides and pharmaceutical products, paint and varnishes, elastomers and peroxides

    (c) Storage facilities for petroleum, petrochemical and chemical products

7. Food industry

    (a) Manufacture of vegetable and animal oils and fats

    (b) Packing and canning of animal and vegetable products

    (c) Manufacture of dairy products

    (d) Brewing and malting

    (e) Confectionery and syrup manufacture

(f) Installations for the slaughter of animals

(g) Industrial starch manufacturing installations

(h) Fish-meal and fish-oil factories

(i) Sugar factories

8. Textile, leather, wood and paper industries

   (a) Wool scouring, degreasing and bleaching factories

   (b) Manufacture of fibre board, particle board and plywood

   (c) Manufacture of pulp, paper and board

   (d) Fibre-dyeing factories

   (e) Cellulose-processing and production installations

   (f) Tannery and leather-dressing factories

9. Rubber industry

   Manufacture and treatment of elastomer-based products

10. Infrastructure projects

    (a) Industrial-estate development projects

    (b) Urban-development projects

    (c) Ski-lifts and cable-cars

    (d) Construction of roads, harbours, including fishing harbours, and airfields

    (e) Canalization and flood-relief works

    (f) Dams and other installations designed to hold water or store it on a long-term basis

    (g) Tramways, elevated and underground railways, suspended lines or similar lines of a particular type, used exclusively or mainly for passenger transport

    (h) Oil and gas pipeline installations

    (i) Installation of long-distance aqueducts

    (j) Yacht marinas

11. Other projects

    (a) Holiday villages, hotel complexes

    (b) Permanent racing and test tracks for cars and motor cycles

    (c) Installations for the disposal of industrial and domestic waste

    (d) Wastewater treatment plants

    (e) Sludge-deposition sites

    (f) Storage of scrap iron

    (g) Test benches for engines, turbines or reactors

    (h) Manufacture of artificial mineral fibres

    (i) Manufacture, packing, loading or placing in cartridges of gunpowder and explosives

    (j) Knackers' yards

ENVIRONMENTAL IMPACT OF TRANSPORTATION

20.1 INTRODUCTION

To assess the environmental impact of a proposed development, be it a housing project, a new industry, a new highway, etc., the transportation impacts must always be examined and quantified. We sometimes think of transportation impacts as singularly traffic air pollution, but the impacts are much wider. Transportation impacts differ from many other environmental impacts since they may be more severe at a considerable distance from the development location (e.g. traffic congestion, noise, etc.). These impacts, which are mainly adverse, can be divided into the following four categories.

Vehicular impacts   Noise, vibration, air pollution, litter and anxiety resulting from the traffic generated by a proposed development, a road improvement or a traffic management scheme.

Safety and operational impacts   Impact on the existing roadway system including additional delays to all road users.

Roadway impacts   Visual intrusion, severance, disturbance of archaeological, historical or amenity areas, effects on the aquatic ecosystem, demolition of property, impacts on the urban fabric, on employment, etc. These impacts are usually associated with the construction of major new roads, railways or airports or what is generally termed infrastructural developments.

Impacts during construction   These include the impacts of construction traffic and other temporary disturbances such as those resulting from the temporary diversion of streams, construction noise, etc. In some countries the safety and capacity impacts of road construction and improvement proposals are considered separately from the project environmental assessment since these impacts can be incorporated into an economic evaluation using standard monetary costs for accidents, for travel time and for vehicle operating costs. However, the total impact of a development on the receiving environment should be included in an environmental assessment. This is necessary in order to satisfy environmental directives internationally.

This chapter addresses roads and traffic impacts on the environment but the material presented may also be relevant to other forms of transport such as air and rail. The following material examines the issues, more in a qualitative sense rather than being rigorously quantitative. However, some areas requiring rigour are addressed in technical journals.

20.2 TRANSPORTATION AND DEVELOPMENT

The motor vehicle is a profligate user of energy and land and a large source of pollution in the world today. However, we have made transport vital to both our economic and social well-being since the production and distribution of goods and services are dependent on it. Economic development has always followed the available lines of communication. In European Union countries about 10 per cent of the GDP and 9 per cent of employment are generated by the transport industry. Over 90 per cent of motorized person trips and more than 80 per cent of all freight movements are by road. The OECD (1986) estimates that the socioeconomic cost of road transport—pollution, congestion and accidents—is as high as 5 per cent of GDP (not including any contribution towards global warming). In the most developed countries (MDCs) people believe that they are entitled to mobility and the absence of adequate transport is considered a handicap. It is worth noting that transportation and telecommunications are partly substitutable (as evidenced by the growth of telecommuting) but the evidence to date suggests that increased use of advanced telecommunications media will not significantly reduce the total amount of travel. However, it is expected that the distribution of travel over space and time will be altered by telecommunications and telecommuting. Consequently, the volume of car traffic generated by a specific development could be reduced by the appropriate use of telecommunications.

Telecommuting is the concept that the people living in suburbia, instead of driving (commuting) one or two hours to the office in the city, work from home using the telecommunications network or work from offices in suburbia which have telecommunication links to head offices in the city. Of course, this is not an option for all work, but is an option for many office-type businesses. As such, the employee may only visit the headquarters once a week or so. This frees motorway space with all its advantages of lesser pollution, etc. A surge in telecommuting was experienced in Los Angeles in 1994, when the Northridge earthquake inhibited motorway traffic.

20.3 TRANSPORTATION PLANNING

Traffic congestion and its associated environmental effects are among the most obvious problems in urban areas, while even in the most remote rural area it is difficult to entirely escape from traffic noise and vehicular-borne litter, although the latter is a diminishing problem due to our increased anti-litter mentality. Large increases in car ownership and use are predicted for all European countries. Typical car ownership growth rates of about 3 per cent per annum are expected (this implies a doubling in the number of cars over 25 years) while average trip lengths are also increasing. There is general agreement that car traffic must be restrained in urban areas but none of the existing traffic policies has been effective in reducing car use except to a limited extent in the centres of large cities. It appears that only severe pricing policies (which are politically very unpopular) are effective. Commercial vehicle use also continues to grow and the widespread implementation of 'just-in-time' manufacturing and retailing is expected to lead to increases in the numbers of commercial vehicles on our roads and motorways. In the longer term, issues such as global warming may necessitate reductions in car use, but such issues are outside the scope of this chapter which concentrates on the impacts of traffic on the local and regional environment. It is anticipated that 10 per cent of road vehicles in California will be electric by the year 2000.

20.4 MATRIX OF ENVIRONMENTAL IMPACT AND TRANSPORTATION SYSTEM STAGES

In evaluating the potential environmental impacts of any proposed change in the transportation system or the transportation implications of a proposed development, the following questions should be asked for each of the six environmental subsystems shown in Table 20.1:

1. Is there a potential impact on the environment, and is it positive or negative?
2. How serious is such a change?
3. What is its expected magnitude of change?
4. How certain is the effect to occur?
5. What further assessment or research is required?

Environmental effects, including health effects, might occur in any of the four stages of the transportation system:

- Production of vehicles and fuels
- Construction of the transportation infrastructure
- Operation of the transportation system
- Decommissioning and disposal of vehicles and infrastructure

The impacts of the first stage, which range from runoff and groundwater pollution from metal mining to fuel spillages during transport from plant to sales point, are not considered here. Table 20.1 summarizes typical environmental concerns resulting from the second and third stages (infrastructure construction and transportation operations), while disposal of engine lubricating oil on land, disposal of metals such as nickel, cadmium and chromium, burning of tyres and batteries, car junk yards, derelict vehicles, etc., are the principal concerns in the fourth stage of the transportation system. A more comprehensive matrix of

Table 20.1 Typical environmental impacts resulting from transportation operations and infrastructure construction

Environmental subsystem	Transportation operations	Infrastructure construction
Ecosystem	Runoff from fuelling facilities Ingestion of lead	Disruption of streams Blocking of animal movements
Resources	Consistent fuel supply Maintenance skills for electric vehicles and other advanced technologies	Parking spaces Capital, particularly for urban mass transit system construction
Physical environment	Vehicle exhaust emissions Deposition in water or rain of exhaust emissions Noise Aesthetic impacts	Dust Runoff Noise Aesthetic impacts
Health	Spills of toxic fluids HC, CO, NO_x, SO_2 Particulates Photochemical smog	Injestion of dust Noise HC
Safety	Spills of hydrocarbons Traffic accidents	Construction accidents
Socioeconomic environment	Maintenance costs Enforcement problems Impact on local employment	Resistance to new construction/fuel storage

Adapted from OECD, 1986

the areas of environmental impact and each of the above transportation stages should be developed for actual projects.

20.5 THE ENVIRONMENTAL EFFECTS OF ROADS AND TRAFFIC

As previously noted, the environmental impacts resulting from changes in transportation are usually perceived as being adverse, however transportation also has positive impacts such as improving the residential environment by permitting the development of homes in areas with new roads or by the removal of through traffic following the construction of new urban roads. It is interesting to recall that transportation played a major role in the elimination of cholera from American cities by replacing horses with cars. For instance, at the turn of the century there were 3.5 million horses in American cities, used for transportation and haulage. Every city had large horse populations: 83 000 in Chicago, 12 000 each in Detroit and Milwaukee, 120 000 in New York. Experts at the time reckoned that horse excrement was from 7 to 15 kg per horse per day and this caused immense problems in air contamination, noxious odours and noise—all potentially health damaging (Petulla, 1987). It is also interesting that the environmental problems that we now associate with cars are the same as those we associated with the horse in 1910. The populace then supported the introduction of the 'horseless carriage'.

The principal adverse environmental impacts, are summarized in Table 20.2. These cover a wide spectrum of both physical and perceived impacts but only a limited number normally result from a specific development. Generally the physical impacts such as noise can be quantified, but the perceived impacts (e.g. anxiety) are difficult to evaluate except in qualitative terms. It is also difficult to determine acceptable values of many of the physical impacts; (e.g. what levels of noise or air pollution are acceptable?). Few legal thresholds or standards exist for the impacts shown in Table 20.2 since they are seldom directly harmful to people or to the local ecology in the short term, but are primarily concerned with the quality of life and as such are subjective. Also our perceptions of many of these impacts change over time.

Table 20.2 The environmental effects of roads and traffic

Vehicular impacts	Noise
	Vibration
	Air pollution
	Litter
	Physical damage
	Anxiety
Safety and capacity impacts	Accidents
	Effects on the operation of roads and intersections
Roadway impacts	Visual intrusion and aesthetics
	Severance
	Land consumption and loss of property
	Changes in land access and land values
	Planning blight
	Effects on wildlife, plants and the aquatic ecosystem
	Impacts on historic and cultural resources
	Impacts on utilities and drainage
	Employment/business impacts
Construction impacts	Damage to local roads
	Disturbance to roadside residents and other road users
	Effects on ecosystem and drainage
	Impact at source of materials
	Litter, mud, odours, etc.

20.6 VEHICULAR IMPACTS

These are impacts on the receiving environment caused by the traffic generated by a proposed development. The effects of this traffic on the operation and safety of the roadway system are considered separately.

20.6.1 Noise

Traffic noise can interfere with speech communication, can disturb sleep and relaxation and interfere with the ability to perform complex tasks (as discussed in Chapter 9). Surveys in many countries have shown that traffic noise is one of the principal environmental nuisances in urban areas. The number of complaints received by local authorities about traffic noise is usually less than those concerning industrial noise or even noise from neighbours, probably because no one person or organization is seen as responsible for traffic noise and also because it does not usually result in any long-term damage. Since noise is easily measured and analysed, it is frequently used in the assessment of road proposals and traffic management schemes as a proxy for other less easily quantified environmental effects of traffic such as air pollution. Although the latter can readily be measured, some authorities do not have the technical competence to do so. However, there is little justification for this practice except in the case of airborne vibrations resulting from traffic. A major study in the United Kingdom (Watts, 1990) concluded that the noise level was the best predictor of annoyance from such vibrations, which is not surprising since both noise and vibration result from the same physical phenomena. Chapter 9 considers noise pollution in general and reference should be made to this chapter for the explanation of noise terms, measurement techniques and instrumentation, etc.

Noise criteria   There is general agreement that traffic noise (and industrial noise) can be appropriately measured in A-weighted (dBA) units. An increase of 2 or 3 dBA is just noticeable while an increase of 10 dBA approximates to a doubling of loudness. However, traffic noise levels are seldom constant and specific criteria are required to quantify varying levels. The severity of a noise problem is then determined by the amount the noise level exceeds a threshold or standard value of a selected noise criterion. Either the L_{Aeq} or the L_{A10} averaged over a specific time period is normally used (see Chapter 9). A 12 hour period (08.00 to 20.00 hours) is often used with the L_{Aeq} while an 18 hour period (06.00 to 24.00 hours) is usually associated with the L_{A10}. The latter criterion is specified in the British regulations governing the provision of insulation against traffic noise arising from new roads (DOT, 1988). Consequently, the L_{A10} (18 hour) is mostly used in the United Kingdom for road design. However, the L_{Aeq} is more universally accepted and is recommended by International Standard ISO 1996 (1971). There is usually only a relatively small difference between the values of the L_{A10} and the L_{Aeq} at a particular site (the L_{A10} is typically 2 or 3 dBA higher) and the validity of either criterion for quantifying annoyance from traffic noise is not well established. Consequently, either the L_{A10} or the L_{Aeq} are appropriate for quantifying traffic noise levels unless national regulations specify otherwise. Although a criterion that is more closely correlated with annoyance is desirable, the development of such a criterion is difficult due to subjective differences between peoples' reactions to noise.

Noise thresholds   The severity of a noise problem is usually indicated by the extent the noise level exceeds a threshold or standard value of a specified noise criterion (e.g. L_{A10} of 18 hour period). A range of thresholds may be used either for different locations (city centre, suburban, rural) or for different time periods (e.g. day and night). There are no universally accepted threshold levels but it is generally accepted that outdoor facade noise levels above 70 dBA constitute a problem. Noise insulation treatment is provided in the United Kingdom where the L_{A10} (18 hour) from a proposed new road is predicted to exceed 68 dBA in the fifteenth year after the road is opened (subject to certain other criteria). A typical facade insulation of

20 dBA was provided by the houses included in the surveys which preceded the UK Land Compensation Act: Noise Insulation Regulations. This implies an internal L_{A10} level of 48 dBA. Recent UK assessment studies of urban transport proposals have assumed that L_{A10} (18 hour) outside levels up to 65 dBA 'do not constitute a problem' while values between 66 and 70 dBA constitute a 'slight problem'. 'Severe problems' are assumed to occur either above 70 dBA or above 75 dBA. In other countries L_{Aeq} outside levels of 60 to 65 dBA are frequently used. The cost of remedial treatment and the number of people exposed are often taken into account in the selection of threshold values. Studies carried out by O'Cinnéide (1991) suggest that the threshold range is appropriate but that a substantial proportion of people (40 per cent at one location) are still annoyed at 65 dBA.

It appears that the above thresholds are set in relation to existing noise levels in urban areas rather than on desirable values with a low level of annoyance (15 per cent annoyed). A WHO task group has recommended daytime noise limits of about 55 L_{Aeq} as a general health goal for outdoor noise in residential areas (OECD, 1986). At night, an outdoor level of about 45 L_{Aeq} is required to meet sleep criteria. However, the maximum noise level, rather than the average, has more influence on the risk of wakening individuals. Also the range of noise thresholds used in assessment studies (65 to 75 dBA) is equivalent to a doubling in subjective loudness and would involve a tenfold increase in traffic volume. This again suggests a lack of sensitivity in the noise criteria. Consequently, a detailed site survey involving the adjacent residents may be necessary to correctly determine traffic noise impacts.

Major factors affecting traffic noise   The noise generated by a stream of traffic depends on the following factors:

- Traffic volume and speed
- Traffic composition (percentage of heavy commercial vehicles)
- Road gradient
- Traffic flow conditions (free flowing or stop and go)
- Road surface type and irregularities

For free-flowing traffic with at least 5 per cent heavy vehicles, the traffic noise level drops to a minimum at an average speed of 30 to 40 km/h irrespective of the traffic volume (DOT, 1988).

The noise level at the reception point is influenced by:

- The distance from the road to the reception point
- The height of the reception point above the road
- The intervening ground surface conditions
- The presence of obstructions (including noise barriers) between the road and the reception point
- The presence of nearby buildings, walls or ground surfaces which reflect noise

The wind speed and direction will also influence the noise level, but this is frequently omitted from consideration.

Noise level measurement   Although high indoor noise levels cause most annoyance, traffic noise levels are normally measured outside buildings because of differences in the noise insulation provided by buildings. Sound level meters are used to record the required noise criterion (e.g. L_{Aeq}) over the required time period (see Chapter 9).

The recommended position for traffic noise measurement in the United Kingdom (DOT, 1988) is at a distance of 1.0 m from the front of a residence and at a height of 1.2 m above the ground or at the most exposed window. The road surface should be dry, the average wind speed midway between the road and the reception point should be less than 2 m/s in the direction from the road to the reception point. The wind speed at the microphone in any direction should not exceed 10 m/s and the peaks of wind noise at the microphone should be 10 dBA or more below the measured value of L_{A10}. Usually measurements need only be taken on one weekday for the required time period since substantial changes in traffic volumes are required to significantly change noise criterion levels.

Traffic noise prediction methods Accurate methods for the prediction of the L_{A10} and the L_{Aeq} for both free-flowing and non-free-flowing traffic conditions have been developed (DOT, 1988; Ministere de l'Environnement *et al.*, 1980; Nordic Council of Ministers, 1980; Transportation Research Board, 1976) and convenient computer programmes are available (Staunton, 1991). Accuracy of prediction depends on the geometry of the situation being analysed. Prediction is often better than ± 1 dBA where propagation is unobstructed and ± 2 dBA where obstruction is significant (Transportation Research Board, 1976). Accuracy decreases where there is significant screening combined with reflection, scattering and diffraction between the façades of buildings, but these are situations not often encountered in practice. Because of the insensitivity of the noise criteria to annoyance and the accuracy of these prediction methods, it is seldom necessary to take actual traffic noise level measurements in practice unless required for legal or public relations reasons. However, baseline noise measurements are usually required for environmental impact assessment purposes.

Amelioration techniques The maximum permitted noise levels emitted by new vehicle engines, particularly heavy vehicle engines, have been progressively reduced by EU regulations. However, the amount of reduction that is technically possible is limited by tyre noise which is the dominant noise source above 80 km/h. Also, the growth in vehicle numbers is likely to negate the benefits of reductions in the noise levels of individual vehicles and other amelioration methods are often necessary, particularly in noise sensitive areas such as near schools or hospitals. One such method is the use of special road surfacings (porous asphalt) which can significantly reduce tyre noise. Table 20.3 lists a variety of methods that can be used to reduce traffic noise levels. It is expected that, by the year 2000, electric vehicles which are practically noiseless, will make up 2–3 per cent of urban vehicles.

Frequently the only options available to reduce excessive traffic noise are the construction of barriers or the provision of building insulation. The maximum barrier reduction attainable seldom exceeds 10 dBA (equivalent to half the loudness) and this usually requires very long and high barriers. The installation of double-glazed windows can reduce the external sound level by 30 to 35 dBA. Higher levels of sound insulation (up to about 40 dBA) can be achieved by comprehensive building insulation (see Chapter 9). However, the cost of such sound insulation is high. BRE/CIRIA (1993) in their document 'sound control for homes' have developed many useful quantitative aids for noise calculating.

Table 20.3 Methods for reducing the impact of traffic noise

Road planning	Avoidance of noise-sensitive areas
	The creation of special environmental areas
Road design	Attention to the horizontal and vertical alignment
	Siting roads in cuttings and tunnels
	Construction of noise barriers
	Use of 'quiet' road surfaces
Traffic management	Concentration of traffic on the principal road network
	Restrictions on heavy vehicles
	Smoothing of traffic flow and minimization of stops
Building design and layout	Shielding of noise-sensitive buildings (by intervening buildings)
	Single aspect housing
	Dwelling insulation
Vehicle design change	Use of electric vehicles

Example 20.1 Traffic noise level calculation

Brief. Using *Calculation of Road Traffic Noise* (DOT, 1988) predict the noise level at a house located 60 m from the edge of a proposed new main road. The ground between the house and the road is level and grassed. The only screening is provided by a thin hedge.

Data required. The location of the most exposed window, which is on the first floor, is 4.5 m above ground, and the following is the traffic and road data:

- Traffic volume, 20 000 vehicles per 18 hour day
- Mean traffic speed, 74 km/h
- Proportion of heavy vehicles, 20 per cent
- Road gradient, zero
- Noise source height 0.5 m (standard value)
- Height of receiver point above source, 4.0 m
- Average height of propagation, 2.5 m (0.5 + 4/2)

Use BRE/CIRIA (1993) to compute the noise corrections due to the individual structural elements and ultimately the noise level of the house.

Solution

	dBA
Basic noise level (Fig. 20.1(a))	72.0
Correction for mean traffic speed and heavy vehicles (Fig. (20.1(b))	+3.6
Impervious road surface	−1.0
Distance correction, hard ground (Fig. (20.1(c))	−6.7
Soft ground correction (Fig. 20.1(d))	−3.5
Reflection effect of building façade	+2.5
	———
Predicted noise level of L_{A10} (18 hour)	66.9

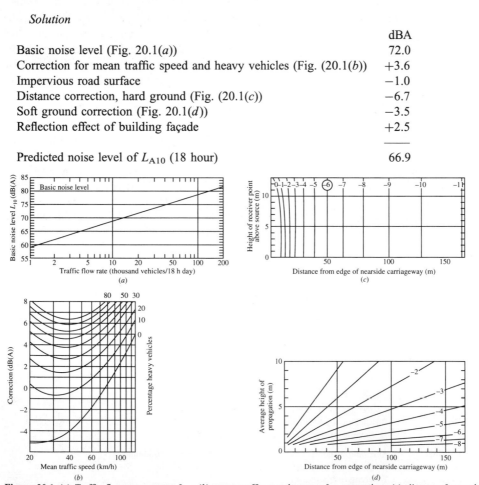

Figure 20.1 (*a*) Traffic flow rate versus L_{10}; (*b*) mean traffic speed versus L_{10} correction; (*c*) distance from edge of nearside carriageway versus height of receiver; (*d*) distance from edge of nearside carriageway versus average height of propagation. (Adapted from BRE/CIRIA, 1993, and DOT, 1988. Reproduced from Building Research Establishment and CIRIA report *Sound Control for Homes* by permission of the Controller of HMSO.)

It should be noted that additional corrections to the above noise level would be required if the angle of view of the road was restricted by adjacent buildings or if there were intervening noise barriers (the hedge is not a barrier). The reader is referred to BRE/CIRIA (1993) for more detailed examples. Figures 20.1 (*a*), (*b*), (*c*) and (*d*) are adaptations of BRE/CIRIA (1993) and DOT (1988).

20.6.2 Traffic-induced Vibrations in Buildings

Vibrations from passing traffic are a common source of environmental nuisance, particularly for those living beside main roads. They may be the most frequent environmental complaint made by residents to local authorities, probably because of the fear that the vibrations could damage buildings. Traffic-induced vibrations may be airborne or groundborne and are almost entirely associated with heavy vehicles.

Airborne vibrations   Airborne vibrations are caused by low-frequency sound (50 to 100 Hz) produced by large vehicle engines and exhausts. The resonant frequencies of rooms may be excited by acoustic coupling through windows and doors. This produces annoying rattling of doors, windows and small objects in the front rooms of buildings. At the most exposed locations acoustically induced floor vibrations can become perceptible (Watts, 1990). The presence of such airborne vibrations can be detected by noting whether front room windows and doors rattle when a bus or another heavy vehicle passes. Low-frequency noise can be perceived directly and can sometimes lead to annoying muffled sensations in the ears and perceptible chest vibrations (Watts, 1990).

Groundborne vibrations   Groundborne vibrations are caused by varying forces between the tyres of heavy vehicles and road surfaces which result from irregularities in the road surface. They can become perceptible in buildings located within a few metres of the carriageway when heavy vehicles pass over irregularities of the order of 20 mm in the road surface. Groundborne vibrations are of lower frequency than airborne vibrations (8 to 20 Hz) and enter buildings through the foundations. Both compression and shear waves are produced in the ground, which can result in structural damage to poorly maintained buildings. Consequently groundborne vibrations are potentially more serious than airborne vibrations. Their presence can be felt as short duration impulsive vibrations, particularly in the middle of upper floors of buildings. However, a major investigation (including extensive site studies) carried out by the Transport and Road Research Laboratory (TRRL, UK) concluded that there is no evidence to support the assertion that traffic vibrations can cause significant damage to buildings, although severe nuisance to occupants can occur (OECD, 1986).

Vibration measurement and thresholds   Vibration measurements involve the use of accelerometers, velocity pick-up and laser systems. They are usually expressed in terms of the peak particle velocity (p.p.v.). The following threshold values are often used:

$$\text{Perception} = 0.3 \text{ mm/s (p.p.v.)}$$
$$\text{Annoyance} = 1 \text{ mm/s (p.p.v.)}$$
$$\text{Structural damage} = 10 \text{ mm/s (p.p.v.)}$$

A threshold value of 5 mm/s for structural response is often associated with road traffic or construction vibration.

Vibration prediction   The Transport and Road Research Laboratory study (Watts, 1990) concluded that the best predictor of airborne vibrations was either the L_{Aeq} (18 hour) or the L_{A10} (18 hour) level as used in noise measurement. British surveys showed that between 23 and 44 per cent of roadside residents were bothered by vibrations at an L_{A10} level of 68 dBA (houses with double-glazed windows had lower numbers of people bothered). At higher L_{A10} levels the percentage bothered increased rapidly. Broadly similar results have been reported elsewhere (O'Cinnéide, 1991).

The presence and magnitude of groundborne vibrations depend on the characteristics of the passing heavy vehicles (axle load, suspension design, speed), the road surface and the intervening ground. The median vibration nuisance prediction found in the TRRL study, (Watts, 1990) was proportional to the 18 hour heavy vehicle flow and to the distance from the carriageway.

Vibration amelioration   The action required to ameliorate traffic-induced vibrations depends on the type of vibration. The installation of better fitting windows appears to be the best remedy for reducing airborne vibrations since it is seldom practical to ban passing heavy vehicles. Double glazing or double windows are likely to be particularly effective. Groundborne vibrations do not normally arise if road surfaces and buildings are maintained to a reasonable standard—the removal of existing road surface irregularities may eliminate groundborne vibrations. Speed and weight restrictions on heavy vehicles can also be useful. The isolation of buildings by resilient mounts or filled trenches is expensive and would normally only be considered for the preservation of historic buildings in poor structural condition.

20.6.3 Air Pollution from Vehicles

Air pollution is considered in general in Chapter 8 while industrial air pollution control and abatement is considered in Chapter 16. Pollution from vehicles is only summarized here. The major sources of atmospheric pollution caused by motor vehicles are from exhaust gases, evaporative losses from the fuel tank and carburettor, crank case losses and dust from tyres (rubber), brake linings and clutch plates (asbestos). Typically vehicular air pollution is less than that from industry and homes but, with increasing vehicle numbers, is a serious concern in urban areas. High proportions of the total amount of carbon monoxide, hydrocarbons and lead in the atmosphere result from vehicle emissions. Figure 20.2 shows the constituents of vehicle exhaust gas divided into noxious and innocuous compounds. Although carbon dioxide is not an innocuous compound (vis-à-vis health), it is undesirable as it is a contributor to global warming.

Petrol engines are considered less obnoxious than diesel engines although diesel engines burn fuel more efficiently resulting in lower pollution emissions (e.g. diesel engines emit only one-fifth the quantity of carbon monoxide). However, since diesel engine pollution is more visible and emits odours, it causes more public disquiet and the relatively high emissions of inhalable particulate matter and associated hydrocarbons raises health and environmental concerns.

Since lead is taken up in the biological chain and has been shown to cause behavioural problems in children, the EU and USEPA have pursued a policy of reducing the amount of lead in petrol engines (there is no lead in diesel fuel). However, the reduction in lead levels may be outweighed by the continuing

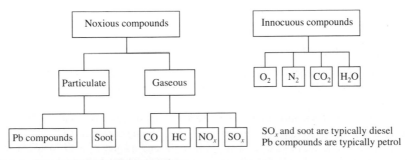

Figure 20.2 The constituents of vehicle exhaust emissions.

increase in car numbers. Readers are also referred to Chapter 8 for discussions on lead and the amount still in automotive gasoline in European countries.

Nitric oxide and sulphur oxides contribute to acid rain and the sulphur content of diesel fuel is being reduced by EU regulations. Under conditions of bright sunlight and still air, hydrocarbons and nitrogen oxides can, in the presence of sunlight, combine to form secondary products such as ozone or photochemical smog. The latter causes eye irritation, is injurious to plants and is almost entirely caused by motor vehicles. This is discussed in greater detail in Chapter 8.

A large number of different hydrocarbons are emitted by petrol vehicles. Benzene, which comprises about 5 per cent of hydrocarbon emissions from petrol engines, is particularly hazardous. To combat vehicle pollution all new petrol engine cars sold in the European Union must be fitted with catalytic converters which reduce the emissions of hydrocarbons, nitrogen oxides and carbon monoxide. However, catalytic converters are only effective when the engine is warm and they increase the volume of carbon dioxide emissions (contributing to global warming). Diesel engines produce up to 25 per cent less carbon dioxide than petrol engines. Further pollution reduction measures may be necessary in future because of the predicted increase in car numbers and the prevailing high levels of urban air pollution. Additionally, the soiling of buildings with diesel particulate emissions is common.

In the United States in 1986, there were 186 million registered vehicles, 135 million of which were passenger cars. They consumed 1.3 billion gallons of fuel. They emitted 58 per cent of the American total of CO, 38 per cent of lead, 34 per cent of NO, 27 per cent of VOCs and 16 per cent of the particulates (Masters, 1991). The Californian highway emission standards (more stringent than those of the Federal Clean Air Act) for passenger cars are 0.25 g/km for HC, 2.1 g/km for CO and 0.25 g/km for NO_x. These standards are achieved by means of the compulsory smog test, which tests every car on a regular basis for pollutant emission rates. The three-way catalytic converter installed in the exhaust system is now the norm in new vehicles and achieves the above standards where:

- Hydrocarbons are oxidized to CO_2
- Carbon monoxide is oxidized to CO_2
- NO_x are reduced to N_2

The reader is referred to Powell and Brennan (1988) and Masters (1991) for further details.

Typical levels of air highway pollution are shown in Table 20.4 (from Ball *et al.* 1991). It is emphasized that pollutant levels vary in space and time. A coefficient of variation of 0.3 is common and 90 per cent of recorded values might be expected to be within a factor of 2 of the typical values. For trace constituents, greater variation occurs and pollutant levels in dust may vary by a factor of 10 (Ball *et al.*, 1991).

The emission rates of vehicle pollutants from highway sources are listed in Table 20.5 for vehicles with no control technology fitted. Control technology, now obligatory in the United States and even more specifically in California, is having excellent success with reducing emissions.

Air pollution measurement and thresholds   Ambient air pollution recommended threshold levels are considered in Chapter 8. Typical threshold values for vehicular-related air pollutants are presented in Table 20.6.

Vehicular air pollution prediction   Air pollution modelling is considered in Chapter 21. A number of computer-based models are available for the prediction of vehicular air pollution, particularly in the public domain from the USEPA. The input parameters are typically the traffic volume, the mean speed, the ground topography, the road layout and the prevailing meteorological conditions (wind direction and velocity, temperature, amount of sunlight and stability conditions). Gaussian models that predict the distribution from a point source have been found to be fairly accurate for typical meteorological conditions. However, when the source is close to the ground surface, these models have been found to underestimate concentrations of pollutants, particularly for crosswind conditions. For this condition a more sophisticated

Table 20.4 Typical pollutant levels in highway environments

Pollutant	Typical level
Particulate matter	
Total airborne suspended particulates	$100\ \mu g/m^3$
Atmospheric dark smoke	$30\ \mu g/m^3$
Total suspended solids in runoff	$100\ \mu g/m^3$
Surface dust	$50\ \mu g/m^3$
Airborne particulate elemental carbon	$3\ \mu g/m^3$
Gaseous air pollutants	
CO	$2\ mg/m^3$ (1.7 ppm)
SO_2	$60\ mg/m^3$ (22 ppb)
NO_2	$50\ mg/m^3$ (25 ppb)
NO_x	$150\ mg/m^3$ (100 ppb)
Total HC	$1500\ mg/m^3$
VOC	$500\ mg/m^3$
Heavy metals	
Pb in air	$0.3\ \mu g/m^3$
Pb in dust	$350\ \mu g/g$
Pb in stormwater	$1000\ mg/m^3$
Zn in air	$0.1\ \mu g/m^3$
Zn in dust	$350\ \mu g/g$
Zn in stormwater	$100\ mg/m^3$
Cd in air	$0.001\ \mu g/m^3$
Cd in dust	$1\ \mu g/g$
Cd in stormwater	$2\ mg/m^3$
Cu in air	$0.01\ \mu g/m^3$
Cu in dust	$60\ \mu g/g$
Cu in stormwater	$50\ mg/m^3$

Adapted from Ball *et al.*, 1991. Reprinted by permission of Elsevier Science

Table 20.5 Vehicular emission rates of pollutants

Pollutant	Emission rate (g/km/vehicle)		
	Gasoline	Diesel	Californian standard (with control technology)
	(no control technology)		
CO	10	1	2.1
Total HC	1	0.3	0.25
NO_x	3	6	0.25
SO_2	0.03	0.2	
Smoke	0.1	0.4	
Dark smoke	0.04	1.2	
Pb (assuming 0.15 g Pb/L)	0.01	0.0	
Zn	0.003	0.003	
Cd	10^{-8}	10^{-8}	
Cu	5×10^{-5}	5×10^{-5}	

Adapted from Ball *et al.* 1991. Reprinted by permission of Elsevier Science

Table 20.6 Threshold limit value of air pollutants

Pollutant	Health protection criteria ($\mu g/m^3$)		Source
Carbon monoxide	1700		WHO
Hydrocarbons	*		
Nitrogen dioxide	200	Hourly maximum	EEC/85/203
Sulphur dioxide	350	Daily maximum†	EEC/80/779
Smoke	213	Daily maximum‡	EEC/80/779
Lead	2	Annual mean	EEC/80/882

* Separate limits for each HC in future
† 98th percentile of daily values; varies with smoke level.
‡ 98th percentile of daily values for year.

model such as the Stanford Research Institute Street Canyon Model would be appropriate. In using any of these models, we must be aware that they require many parameters and an inexperienced user may obtain meaningless results. However, using these models is certainly a great learning tool and allows us to look at the impacts of different proposed road layouts.

In predicting the likely air pollution from a proposed new road, a worst case scenario, with the wind blowing from the road during the one hour period of maximum daily traffic flow, is frequently assumed. Predictions are made for the likely levels of each traffic-related pollutant. The predicted vehicular pollution is then added to the existing annual maximum air pollution level for each pollutant.

Amelioration of vehicular air pollution The reduction of vehicular air pollution depends primarily on reducing the pollution from individual vehicles at source. Methods available, apart from the installation of catalytic converters, include the re-design of carburettors, the use of more efficient lean burn engines and the use of electric vehicles. These methods all impose additional costs, but the predicted increase in the total number of vehicles may require their implementation. Other methods for reducing vehicular air pollution include the improvement of vehicular flow conditions (since vehicular emissions are almost totally associated with decelerating, idling and accelerating vehicles), better engine maintenance (estimated to reduce pollution from current petrol engines by one-third), the planting of trees and shrubs and the restriction of traffic from highly polluted areas such as city centres. Indeed, air quality requirements may eventually require controls of car use throughout our urban areas.

Example 20.2 Air pollution calculation

Brief. Estimate the total hydrocarbon concentration during the peak hour traffic at a point that is 300 m from a proposed new road.

Data required. Peak hour traffic volume and speed is 6000 vehicles per hour at 65 km/h. Find the emission of total hydrocarbons from an average vehicle at the mean peak hour speed, assuming a value of 2×10^{-2} g/s of total hydrocarbons per vehicle at a speed of 65 km/h. The meteorological conditions assume an overcast day with the wind blowing directly from the road at a speed of 4 m/s.

Solution The road is considered as a continuous infinite line source. The source strength, q, is first calculated. This is the number of vehicles per metre (the concentration or density) multiplied by the emission per vehicle:

$$\text{Concentration} = 6000/65$$
$$= 9.23 \times 10^{-2} \text{ vehicles/m}$$
$$\text{Source strength } q = (9.23 \times 10^{-2})(2 \times 10^{-2})$$
$$= 1.846 \times 10^{-3} \text{ g/m s}$$

Under overcast conditions with a wind speed of 4 m/s stability class D applies (see Chapter 8) for class D at $x = 300$ m, $\sigma_z = 12$ m (see Fig. 8.24).

The concentration equation from an infinite line source is a modification of those in Chapter 8 (Turner, 1964):

$$C(x, y, 0 : H) = \left(\frac{2q}{\sqrt{2\pi}\sigma_z u} \right) \exp\left[-\frac{1}{2} \left(\frac{H}{\sigma_z} \right)^2 \right]$$

$$C(300, 0, 0 : 0) = \frac{2q}{\sqrt{2\pi}\sigma_z u}$$

$$= \frac{2(1.846 \times 10^{-3})}{2.507 \times 12 \times 4} = 3.1 \times 10^{-5} \text{ g/m}^3 \text{ of HC}$$

Therefore

$$C = 31 \ \mu g/m^3$$

20.6.4 Vehicular-borne Litter

Litter is reported as a major environmental nuisance in public opinion surveys and certain areas, such as the outskirts of towns, can be spoiled by car-borne litter. Abandoned cars and illegal dumping may also constitute problems. The provision of convenient litter receptacles appears necessary in such circumstances since exortation or enforcement alone are often insufficient. Many countries, in northern Europe and northern America particularly, have succeeded in reducing such litter.

20.6.5 Physical Damage

Heavy vehicles cause physical damage not only to the road pavement itself but also to roadside objects such as signs, footpaths and gateposts, particularly on narrow roads. Cars and heavy vehicles splash pedestrians in wet weather. Adequate road maintenance and design, heavy vehicle entry restrictions and enforcement appear the only available remedies.

20.6.6 Anxiety

Anxiety or the feeling of not being safe (unsafety) are strongly associated with heavy vehicles and with high speeds. Older people are particularly affected as are parents who worry about their childrens' safety on the roads; parents of young children who cycle are especially worried. An allied environmental impact is the disturbance to social life by not being able to talk on the footpath because of passing traffic. The implementation of traffic calming methods such as speed control and heavy vehicle restrictions can reduce these vehicular impacts. Generally a detailed local study is necessary in order to determine the necessity for traffic calming and to identify appropriate remedial measures.

20.7 SAFETY AND CAPACITY IMPACTS

These include the impacts of a proposed development or transportation scheme on the existing roadway system. The transport of hazardous waste is considered in Chapter 15.

20.7.1 Accidents

The traffic generated by a proposed development (including pedestrians and cyclists) may affect the safety of the existing road network, particularly if there is a high accident rate at present. Accident records and rates on the existing network should be examined for a period of at least the previous three years. Comparisons with national accident rates for similar road types should be made. A safety audit into the impact of the proposed development can then be carried out. The potential safety impact on vehicle users, pedestrians and cyclists should be determined separately.

20.7.2 Effects on the Operation on Roads and Intersections

These include the effects on traffic congestion and delay and on parking. The traffic volumes (annual average daily and peak volumes) on the existing road system may be available either directly from the local authority or from published traffic count data. If not, traffic counts will probably be necessary. Short-term traffic counts (even one day) can be extrapolated (with caution) to approximate AADT (annual average daily traffic) values by the use of the appropriate expansion factors. The capacities of the existing roads and key intersections can be estimated from the appropriate road design standards manuals. Where congestion exists, travel time studies may be required. The impact of the additional traffic should be quantified for vehicle users, cyclists and pedestrians in terms of delays or time savings (total estimated hours/day). Changes in vehicle operating costs and the effects on emergency services and parking should also be determined. In evaluating the impact of a proposed new road, the additional traffic generated by the road must be included.

20.8 ROADWAY IMPACTS

As previously noted, these impacts are mainly associated with the construction of new transportation schemes (especially motorways) rather than with the traffic generated by other types of developments. However, the latter may require the construction of new road schemes.

20.8.1 Visual Intrusion and Aesthetics

The most significant visual impacts include the following:

- Scenic incompatibility, which may occur either in an urban or in a rural area
- Obscuring of existing views
- The creation of gaps in the urban fabric
- Loss of sunlight
- Loss of privacy

Depending on the viewpoint of the observer, these impacts may vary. Consequently three different viewpoints must be considered: that of the adjacent property occupiers, that of the community or users of the area and that of drivers. The above effects are usually assessed separately for each group. The impact on the adjacent property occupiers is usually considered the most significant.

A visual envelope map outlining the limits of visibility of the proposed development may be drawn using contoured maps and field surveys. The quality of the landscape, the activities affected and the extent of obstruction to the field of vision can then be examined by detailed surveys. Visual obstruction (blocking of a view) may be quantified in terms of the solid angle subtended by the new structure from a given observation point in units of steradians or described in qualitative terms. A steradian is the angle subtended at the centre of a sphere of unit radius by unit area of the surface. Alternatively, visual obstruction, loss of sunlight and loss of privacy may be presented in qualitative terms (e.g. the number of properties with severe or moderate visual obstruction) since the relationship between steradians and the perceived environmental impact is not clear. Large articulated vehicles passing in front of residences and overnight lorry parking are common forms of visual intrusion, particularly in urban areas. Visual intrusion and the other aesthetic impacts listed above depend on the existing landscape or townscape. The likely intrusive effect of the new development must ultimately rely on a qualitative assessment. Physical models, photomontages and computer-generated graphics can be used to aid this assessment. The removal of unique buildings, vistas, rows of trees, etc., are other visual impacts associated with new road construction. Each such input should be quantified where possible (e.g. the number of mature trees removed) as part of the environmental assessment. Proposed remedial measures should also be included.

20.8.2 Severance

Severance occurs when a new development forms either a physical or a psychological boundary between different areas. It is primarily associated with large transportation schemes such as motorways but can also result from increased speeds or traffic volumes on existing roads. Apart from those whose lands or properties are directly effected, severance can have a major impact on communities, for example the separation of residents from facilities and services, from their place of employment or from friends and relatives. It is most directly felt by pedestrians in built-up areas. However, severance can occasionally be beneficial, for instance by providing a boundary between areas of different land use. In urban areas particular attention is focused on the facilities used by vulnerable groups in society. Research by Clark *et al.* (1991) concluded that 'severance is a complex effect, in which a bundle of negative aspects of trafficked roads are linked together, including pedestrian delay, trip diversion and trip suppression, noise pollution, perceived danger and general unpleasantness'. Detailed studies have shown that the overall impact of this bundle of aspects is substantial and that the public is significantly concerned by these impacts. Also the above research suggests that changes in severance effects should be assessed throughout the affected network and that this should be done by considering the effects on specific groups of people rather than by merely identifying particular locations and functions that may be affected. The factors giving rise to and determining the potential for severance from road proposals were found to be:

- Facilities to which access is inhibited
- Catchment areas for those facilities
- Numbers and types of persons affected
- Levels of traffic causing the severance
- Extent of mitigation of road crossing difficulties

There is as yet no agreement on how to combine these effects into a measure of severance or on the relative importance of severance in comparison with other factors such as monetary costs or time savings. Subjective judgement appears to be the only practical approach to the evaluation of severance impacts. Consequently, community participation is essential. This requires meetings with those affected and social surveys as well as measurements of traffic volumes and speeds, pedestrian delays, etc. Checklists such as those contained in the *Manual of Environmental Appraisal* (DOT, 1983) or in the more recent TRRL Contractor Report 135 (Clark, 1991) are useful. Land use and ownership maps are also of value in assessments, particularly in rural areas.

20.8.3 Land Consumption and Loss of Property

The number of properties demolished is usually an important measure of the impact of a proposed transportation development, especially where there are employment implications. However, other impacts such as the displacement and relocation of people, the loss of recreational areas or wetlands, even the loss of parking, may also be significant and should be quantified. Frequently new road construction through urban areas results in awkwardly shaped underdeveloped sites adjoining the road.

20.8.4 Changes in Land Access and Land Values

Transportation improvements may result in either increased or decreased access to adjacent properties. Normally increased access occurs which may stimulate development and lead to increased land values. Local businesses and retailers usually benefit from increased access. However, increased access can sometimes lead to undesirable developments and decreased land values (e.g. where a new road is constructed through an established housing area). Also some businesses may suffer from increased levels of passing traffic due to parking difficulties or traffic congestion. Decreased access can cause problems for emergency and service vehicles and may lead to other detrimental effects, particularly for commercial properties. When a new road bypasses an area, local businesses can suffer trade losses. The severity of these impacts should be examined by observing the routeways used. Planning restrictions on lands adjacent to a proposed development (e.g. access restrictions) may reduce property values.

20.8.5 Planning Blight

The announcement of a proposed new transport development can lead to the gradual rundown of the directly affected properties since little or no maintenance of these properties is likely to be undertaken. Consequently, unless the development takes place within a relatively short time, planning blight may follow. A short-term planning strategy for the affected area is essential.

20.8.6 Effects on Wildlife, Plants and the Aquatic Ecosystem; Impacts on Historic and Cultural Resources; Impacts on Utilities and Drainage

These are largely similar to those described in Chapter 19 for proposed developments in general, except that they can occur over a wider area. Pollution of water courses adjacent to roads can result from the concentrated runoff during periods of heavy rain. Even at low concentrations there can be long-term impacts on the surrounding environment. However, spillages of toxic materials in accidents are usually much more significant. Regular monitoring of watercourses adjacent to roads is desirable in order to limit consequential damage.

20.8.7 Employment and Business Impacts

The construction of a new development such as a retail centre may have severe impacts on existing retail businesses over a wide area but only transportation-related impacts are considered here. Changes in employment and business activities are usually consequences of previously considered roadway impacts such as severance, changes in land access or demolition of property, but should be considered separately because of their importance. Transportation developments can stimulate employment and business activity but the contrary can also occur; e.g. the construction of a new motorway can lead to a reduction in distribution sector jobs because a larger area can be served in the same time compared with the pre-motorway situation. The severity of the likely impact may be estimated by observing the existing routes used by business customers, suppliers, etc.

20.9 CONSTRUCTION IMPACTS

The traffic generated by the construction of a new development can cause major impacts, particularly for those living along the routes used by this traffic. Since construction is labour intensive, large numbers of cars access construction sites at the start and end of the working day. However, heavy vehicles carrying material to and from the site and the movement of large earthmoving equipment usually cause the principal problems. The noise and vibration from these vehicles disturb roadside residents, traffic delays occur, mud is deposited on the roads and the road pavements and verges may be physically damaged. Pedestrians and cyclists may require special facilities. A safety assessment should always be carried out to identify potential hazards. The effects of the proposed construction on surrounding areas (traffic queues, diversions, accesses and exits) must also be determined. Although construction impacts are of temporary duration they can be severe and can last for a considerable period. Consequently mitigation measures must be considered to reduce these impacts. Preferred routes for construction traffic, avoiding heavily populated areas, are often specified. The construction of new roads may also involve temporary disturbance to drainage, to the ecosystem and to roadside utilities. Mitigation measures must again be considered. The safety of construction workers must be ensured under the EU Framework Directive (89/392/EEC) on health and safety at work.

20.10 THE TRAFFIC GENERATED BY PROPOSED DEVELOPMENTS

In order to assess the environmental impacts of a proposed development, the traffic generated during the construction phase and afterwards must be quantified. The amount of car traffic and freight traffic should be separately estimated.

20.10.1 Car Traffic

The volume of car traffic attracted to a proposed development depends on the type and size of the development, its location and on the number of employees and their car ownership level. For a development located in a rural area not served by public transport, it can be usually assumed that most employees and visitors (including customers) will arrive by car. By assuming a car occupancy rate for both visitors and employees (e.g. one car for each visitor and 1.5 employees per car), the volume and distribution of car traffic throughout the day can be predicted from the estimated number and arrival/departure times of employees and visitors. Peak period traffic is of special concern and should be predicted separately. For developments located in urban areas, the proportions of employees and visitors arriving by public transport, on foot or cycling must be estimated before the volume and distribution of car traffic can be estimated.

20.10.2 Light Goods Vehicles

The number of delivery vehicles and other light goods vehicles and their distribution throughout the day will depend on the activities being carried out at the proposed development. This traffic is often combined with the estimated car traffic since the environmental impacts are similar.

20.10.3 Heavy Goods and Public Service Vehicles

Heavy vehicles usually cause the principal environmental impacts resulting from transportation. The number of heavy vehicles (including delivery vehicles and waste removal vehicles) required to service the

Table 20.7 Generic checklist of impacts

Impact group	Impacts
Road users	Time savings/delays Vehicle operating costs Accident costs Driver comfort and convenience View from road
Physical environment	Landscape Infrastructure Air quality Nature conservation
Social environment	Community severance Employment Aesthetics (intrusion) Culture and heritage
Occupiers of property	Demolition Severance (land) Noise Visual obstruction

activities to be carried out at a proposed development and their routeing must be estimated by an examination of the projected level of these activities and of the likely origins and destinations of deliveries. Specific attention must be paid to vehicles which would carry hazardous or odorous materials (see Chapter 15). A number of trip rate databases for specific retail, residential and commercial uses are available. These are particularly useful where similar developments are being evaluated.

Table 20.8 Typical assessment techniques for environmental impacts

Impact areas	Effects	Unit of measurement
Road users	Time savings Vehicle operating costs Accident costs Comfort and convenience View from the road	Monetary Monetary Monetary Scaled (low, moderate or high driver stress) Scaled (no view, restricted, intermittent, open view)
Physical environment	Landscape Infrastructure Air quality Nature conservation	Descriptive Descriptive Ceiling levels Descriptive (initial and final appraisal)
Social environment	Community severance Employment Aesthetic (intrusion) Culture, etc.	Scaled (none, slight, moderate, severe) Descriptive Scaled (slight, moderate, severe) Descriptive
Occupiers of property	Noise Visual obstruction Severance of land Demolition	Scaled (increase of 3–5, 5–10, 10–15 dBA, etc.) Scaled: >150 steradian–high 50–150 steradian—moderate 25–50 steradian—light Number of properties severed, plus description of quality of land Number of properties demolished

Adapted from Rogers, 1992, with permission of ICE, UK.

20.11 THE ENVIRONMENTAL IMPACT ASSESSMENT OF PROPOSED ROAD DEVELOPMENTS

Environmental impact assessment (EIA), which is considered in Chapter 19, consists of a systematic investigation into the effects of a proposed development on the environment. Such assessments are now required as an integral part of the evaluation process for all large transportation proposals. Traditionally, the evaluation of proposed road schemes was largely based on cost benefit analysis which consisted of estimating the likely construction costs and assigning monetary values to the predicted benefits such as savings in travel time and accidents. Environmental impacts tended to be omitted since it is difficult to determine their monetary values. This frequently resulted in public opposition to proposed road schemes and resulted in the inclusion of environmental impacts in the evaluation procedures. Under the 1985 EU Directive on the Environment at Impact Assessment, large transportation projects must be subjected to an EIA. The EU legislation does not specify detailed assessment methods and, in general, the various national regulations incorporating the EU Directive allow considerable latitude in the methods used. This is probably because of the subjective nature of many environmental impacts and disagreement on how they should be quantified.

20.11.1 Contents of a Road Environmental Impact Assessment

A typical assessment for a proposed road would include the following:

- A summary of the proposed road development and of the principal environmental impacts
- General project description and a description of the alternatives considered
- A baseline survey of the existing environment
- Assessment of the environmental impacts
- The implications for the land use and development plans for the affected area
- The financial implications
- Mitigation measures proposed to reduce the negative impacts
- A synoptic table summarizing the individual impacts and costs for each of the alternatives considered
- Conclusions

See Chapter 18 for further details on EIA.

20.11.2 Impact Identification

Checklists are commonly used to ensure that all relevant impacts are identified. Table 20.7 shows an example of a generic checklist.

20.11.3 Impact Prediction/Assessment

As previously discussed, it is difficult to assess many of the impacts shown in Table 20.7 in quantitative terms. Table 20.8 gives an indication of the level of assessment typically used for each individual impact. An overview of the use of EIA techniques and methodologies for road projects is given by Rogers (1992).

20.12 PROBLEMS

20.1 List the major factors that affect the noise generated by a stream of traffic at the façade of a building and briefly discuss each factor.

20.2 Describe three alternative methods for the reduction of traffic noise in buildings.

20.3 Explain the causes of airborne and groundborne traffic-induced vibrations in buildings. How would you detect the presence of each type of vibration? What amelioration techniques are available?

20.4 State the principal air pollutants emitted by (a) petrol engines and (b) diesel engines. Which pollutants are considered the most hazardous? Discuss how vehicular air pollution can be reduced in city centres.

20.5 Discuss the safety and road system impacts you would expect from the construction of a multistorey residential building in a densely populated area. Explain how you would quantify each of these impacts.

20.6 Describe the visual impacts that usually result from the construction of an elevated urban motorway. Discuss how each impact may be assessed.

20.7 List and explain the factors that determine the extent of the severance impacts of major urban roads.

20.8 Briefly describe the contents of an environmental impact assessment of a proposed rural motorway.

20.9 Repeat Example 20.1 where the house is located 20, 40 and 80 m from the roadside edge. Is the noise trend linear or non-linear as the road approaches the house?

REFERENCES AND FURTHER READING

Ball, D. J., R. S. Hamilton and R. M. Harrison (1991) *The Influence of Highway Related Pollutants on Environmental Quality in Highway Pollution*, R. S. Hamilton and R. M. Harrison (eds), Elsevier Science, Amsterdam).

Building Research Establishment (BRE)/Construction Industry Research and Information Association (CIRIA) (1993) *Sound Control for Homes*, Building Research Establishment, Watford, UK, and Construction Industry Research and Information Association, Building Research Establishment (BRE). London, UK.

Clark J. M., *et al.* (1991) *The Appraisal of Community Severance*, CR135, Transport and Road Research Laboratory (TRRL), Crowthorne, Berkshire.

Department of Transport (DOT) (1983) *Manual of Environmental Appraisal*, Department of Transport, Assessments and Policy Division, HMSO, London.

Department of Transport (DOT) (1988) *Calculation of Road Traffic Noise*, Technical Memorandum, HMSO, London.

ISO 1966 (1971). *Assessment of Noise with respect to Community Response*. International Standards Organisation, Geneva, Switzerland.

Masters, G. (1991) *Introduction of Environmental Engineering and Science*, Prentice-Hall, Englewood Cliffs, New Jersey.

Ministere de l'Environnement *et al.* (1980) *Prevision des Niveaux Sonores*, CETUR, Paris.

Nordic Council of Ministers (1980) *The Computing Model for Road Traffic Noise*, Liber Distribution, Stockholm.

O'Cinnéide, D. (1991) 'The selection of thresholds in noise evaluation', in *Proceedings of First Irish Environmental Engineering Conference, University College, Cork*, G. Kiely, P. O'Kane and E. McKeog (eds). Allied Print, Cork, Ireland.

OECD (1986) *Environmental Effects of Automotive Transport*, The OECD Compass Project, OECD, Paris.

Petulla, G. (1987) *Environmental Protection in the United States*. San Francisco Study Center, Market Street, San Francisco, CA, USA.

Powell, J. D. and R. D. Brennan (1988) *The Automobile, Technology and Society*, Prentice-Hall, Englewood Cliffs, New Jersey.

Rogers, M (1992) 'The use of EIA techniques and methodology within the road planning system in the Republic of Ireland', *Proc. Instn Civ. Engrs, Mun. Engr*, **93**, 39–50, March 1992.

Staunton, M. M. (1991) *Soundtrack*, RT 389, Environmental Research Unit, Dublin.

Turner, D. B. (1964), 'A diffusion model for an urban area'. *Journal of Applied Meteorology*, February.

Transportation Research Board (1976) 'Highway noise—a design guide for prediction and control'. Kugler, G. (ed), NCHRP Report 174, Transportation Research Board, Washington, DC.

Watts, G. R. (1990) *Traffic Induced Vibrations in Buildings*, Research Report 246, Transport and Road Research Laboratory, Crowthorne, Berkshire.

TWENTY-ONE

ENVIRONMENTAL MODELLING

21.1 INTRODUCTION

The mathematical modelling of any environmental engineering process is made up of at least the following four steps:

1. After studying the physical, chemical or biological process, identify the essential and dominant mechanisms, e.g. bacterial population dependent on the initial population and a growth rate.
2. Develop a mathematical expression for the system. Typically, this may be as simple as a single algebraic or differential equation or as complex as a set of differential equations, e.g.

$$\frac{dP}{dt} = kP$$

where
P = population (perhaps concentration, mg/L)

k = growth rate, day^{-1}

(see Example 21.1)

3. Analytically solve the equation(s) if possible. If not, solve the equation(s) numerically:

$$P(t) = P_0 e^{kt}$$

where
P_0 = initial population (concentration)

4. Check to see whether the solution to the 'model' satisfies known data. If not, start the process again and repeat until the model solution is acceptable.

Modelling of environmental processes and contaminant transport is essentially second nature now to the modern engineer and scientist. The scope of engineering problems that can be addressed by modelling is truly infinite. There are 'off the shelf' models or package programmes for almost every conceivable task, be it wastewater treatment, optimizing the routes for municipal solid waste haulage trucks, identifying the optimum location for a sewage outfall (to maximize mixing and dilution) or, of course, air quality modelling. A good source of low cost package public domain software is the US EPA. Questions to your favourite internet environmental engineering forum will also show up locations of software availability. Traditional physical models of riverine systems have almost been replaced by computer models, primarily because of the lower cost involved and the ease with which problems can be run and rerun with different data at quick speed. Modelling may be generally simplified as shown in Fig. 21.1.

Let us suppose our task is riverine modelling, the input data may be rainfall, ground topography, river bed levels, estuarine bathymetry, etc., while the expected output might be water levels or water quality in the river system. To determine the output or to forecast what might occur if a set of inputs were supplied is the task of modelling. The level of sophistication varies hugely from what are known as simple box models (used still in the air environment) to very sophisticated three-dimensional modelling. The sophistication of the model depends on the level of understanding of the system. For instance, to model the hydrodynamics of river flow, the most commonly used models are one-dimensional, i.e. they give a single value of velocity at each cross-section. At the most sophisticated end is the full numerical solution of the three-dimensional St Venant equations (see any fluid mechanics book, e.g. Liggett, 1994) which fully describe the total flow. Sometimes, the system or problem definition is not readily understood. For instance, the activated sludge process in wastewater treatment is a very complicated biochemical process, not yet fully understood, so the simulation system knowledge is still incomplete with many 'fudge' factors. This is the situation for many real world problems. For instance, the movement of a contaminant in groundwater or in the vadoze zone is inhibited by incomplete knowledge of field parameters, like three-dimensional hydraulic conductivity. Many engineering 'textbook' problems are solved analytically for steady state conditions (i.e. the problem conditions do not vary with time). In reality, many problems do have conditions changing with time and such models, where $d/dt \neq 0$, require numerical solution techniques. The steps outlined by Tanji (1994) in the development and application of system simulation models are shown in Table 21.1.

This chapter is meant to introduce the student to the general concepts of modelling in environmental systems. It is not meant to be an introduction to numerical analysis which is the solution technique to

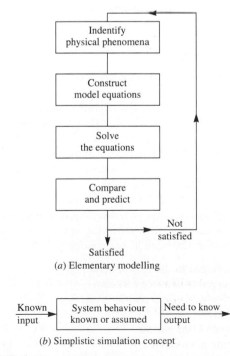

(a) Elementary modelling

(b) Simplistic simulation concept

Figure 21.1 Fundamentals of modelling.

Table 21.1 Steps in development and application of systems simulation models

---

Problem situation and study objectives
  Recognition of a particular problem situation
  Identification of modelling objectives

System analysis
  Isolate the system of interest from system environment
  Evaluation of existing data
  Conceptualization of model
  Mathematical model formulation

System synthesis
  Computer modelling, programming
  Verification of numeric scheme
  Calibration of model
  Sensitivity of analysis
  Validation of model using other data sets
  Model runs

Simulation analysis
  Evaluation and interpretation of simulated results
  Re-examination of systems analyses and synthesis

Adapted from Tanji, 1994, with permission

solve the model equations derived from a description of the problem. The student is referred to, for example, the text *Numerical Models for Engineers* by Chapra and Canale (1988) for a rigorous introduction to this subject. The basic equations to many physical, chemical and biochemical problems are introduced in this chapter, but not always solved. A discussion of their solution is included with the purpose of defining the breadth of scope for modelling uses in environmental engineering.

21.2 MECHANISM OF POLLUTANT FATE IN THE ENVIRONMENT

The fate of a slug (once off) of pollutant entering a riverine system is defined as the origin to destiny, distribution and concentration of that pollutant. For instance, a sewage discharge of 30 mg/L BOD concentration may have a fate of 1 mg/L, 5 km downstream (from its outfall), where it has diffused into the riverine background concentration of BOD. It may be of interest to track the change in concentration of that pollutant from the discharge point to 5 km downstream. It may also be of interest to determine the lateral distribution (across the river) of the pollutant for this river reach of 5 km and interest may also be in the vertical distribution through the river depth for several sections in the river reach (length). In other words, we may be interested in the three dimensional distribution of the pollutant at different times along its track. This is a rather mammoth task. The following are examples of pollutants, whose fate is commonly required to be known in environmental engineering:

- Chimney stack pollutants discharged to the atmosphere
- Conservative pollutants discharged to a riverine system
- Non-conservative pollutants discharged to a riverine system
- Raw sewage into a biological reactor
- Noise source emitted to an urban environment
- Agricultural pollutant dispersing in an aquifer
- Industrial wastewater discharged through an ocean outfall

The three basic physical mechanisms that are responsible for the transport of pollutants in fluid bodies are:

1. *Advection* is the transport due to the bulk movement of fluid (a dynamic property).
2. *Diffusion* is the non-advective transport due to
   (a) molecular or Brownian motion (at the microscopic scale) and
   (b) eddy or turbulent motion (at all scales).
3. *Buoyancy* is the transport due to a vertical temperature gradient.

In riverine systems, advection is typically the dominant transport mechanism with (molecular) diffusion playing an insignificant role. Buoyancy is relevant in systems where temperature gradients exist in almost static water bodies, i.e. lakes, estuaries or oceans, or sometimes in an atmospheric environment. Diffusion is most significant in quiescent systems if the concentration gradient of the pollutant is significant. Other physical/chemical/biological mechanisms, although non-transport in essence, contribute to the fate of a pollutant. These include:

- Volatilization
- Absorption
- Adsorption
- Evaporation
- Bacterial decay
- Sedimentation

For instance, the fate of a groundwater contaminant may also depend on absorption by the soil particles, adsorption on to underground rocks or bacterial decay due to biological reactions of soil microfauna. The fate of pollutants in the different environments with some of their specific processes are shown in Fig. 21.2. It is also relevant to realize that there is much interaction between the atmosphere, surface water (both freshwater and oceans) and groundwater environments.

21.2.1 Mathematics of Motion: Continuity and Momentum Equations

There are fundamental equations to understanding the movement, transport and diffusion of pollutants, be they in the air, water or soil environments. These are:

- Continuity or conservation of mass
- Momentum
- Advective diffusion.

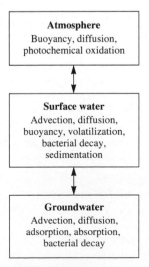

Figure 21.2 Some of the pollution dilution mechanisms in atmosphere, surface water and groundwater.

Continuity   The continuity or mass conservation equations are derived with reference to the volume element of Fig. 21.3. Consider the thought experiment, where water is flowing through a length of river, of variable cross-section A. There is a streamwise flow of fluid with mean velocity u and also an inflow laterally from the adjacent fields. The lateral inflow may be imagined as, say, overland flow entering a stream along the length, dx:

$$\text{The mass inflow through the upstream boundary} = \rho u A$$

where

$$\rho = \text{density of water, kg/m}^3$$
$$u = \text{velocity in } x \text{ direction, m/s}$$
$$A = \text{cross-sectional area, m}^2$$

$$\text{the mass inflow through the downstream boundary} = \rho u A + \rho \frac{\partial}{\partial x}(uA)\,\mathrm{d}x$$

where

$$\rho \frac{\partial}{\partial x}(uA)\,\mathrm{d}x = \text{the change in discharge over d}x$$

$$\text{The lateral mass inflow} = \rho q_s\,\mathrm{d}x$$

where

$$q_s = \text{lateral inflow, m}^3/\text{m s}$$

The time rate of change of mass within the volume element $= \rho \dfrac{\partial A}{\partial t}\,\mathrm{d}x$

We are dealing with the partial derivative since $A = f(x, t)$. The mass balance for the volume element is then

$$\text{Inflow} \quad - \quad \text{outflow} \quad = \text{change in storage}$$

$$(\rho u A + \rho q_s\,\mathrm{d}x) - \left[\rho u A + \frac{\partial}{\partial x}(uA)\,\mathrm{d}x \right] = \rho \frac{\partial A}{\partial t}\,\mathrm{d}x \qquad (21.1)$$

For a constant density, this reduces to

$$\frac{\partial A}{\partial t} + \frac{\partial}{\partial x}(Au) - q_s = 0 \qquad (21.2)$$

For, say, a rectangular channel of constant width W and depth h $[h = h(x, t)]$, then

$$W\frac{\partial h}{\partial t} + W\frac{\partial(uh)}{\partial x} - q_s = 0 \qquad (21.3)$$

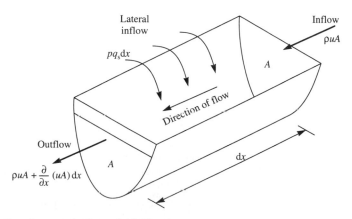

Figure 21.3 Flows through a control volume of a fluid.

$$\frac{\partial h}{\partial t} + u\frac{\partial h}{\partial x} + h\frac{\partial u}{\partial x} - \frac{q_s}{W} = 0 \tag{21.4}$$

For no lateral inflow and one-dimensional flow, i.e. in the stream, there is only streamwise velocity u, in the direction x:

$$\frac{\partial h}{\partial t} + \frac{\partial(uh)}{\partial x} = 0 \tag{21.5}$$

For two-dimensional flow, there is a velocity u in the x direction and also a velocity v in the y direction (assume y is orthogonal to x),

$$\frac{\partial h}{\partial t} + \frac{\partial(uh)}{\partial x} + \frac{\partial(vh)}{\partial y} = 0 \tag{21.6}$$

For three-dimensional flow, that is there are u, v and w velocities in the fluid flow,

$$\frac{\partial h}{\partial t} + \frac{\partial(uh)}{\partial x} + \frac{\partial(vh)}{\partial y} + \frac{\partial(wh)}{\partial z} = 0 \tag{21.7}$$

where u, v, w are the mean velocities in the x, y and z direction respectively. For steady flow,

$$\frac{\partial}{\partial t} \to 0$$

Equations (21.5), (21.6) and (21.7) are known as the continuity or mass conservation equations.

Momentum   The conservation of the momentum equation can be expressed in a linear or angular momentum form, but the former is only of interest here. Momentum is the product of mass by velocity. This derivation can be found in most texts on fluid mechanics. However, it is derived here in the interests of completeness.

Consider the volume element or control volume shown in Fig. 21.4 and inspect the forces acting on it. The forces acting on the fluid element of length Δx shown in Fig. 21.4, between sections 1 and 2 are:

$$\text{Gravity (downstream component) } F_g = \rho g A \Delta x S_0$$

$$\text{Friction } F_f = \rho g A \Delta x S_f$$

$$\text{Hydrostatic } F_1 - F_2 = \frac{1}{2}\rho g \frac{\partial}{\partial x}(y^{-2}A)$$

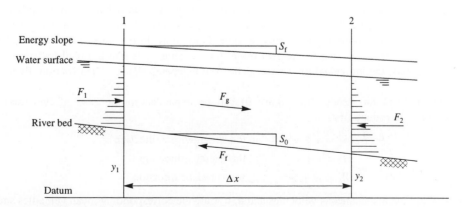

Figure 21.4 Force balance on a fluid element.

The driving force is the streamwise component of gravity and the resistance force is that due to bed friction, where

$$\Delta x = \text{length of fluid element, (m)}$$
$$A = \text{average cross-section area, (m}^2)$$
$$S_f = \text{slope of the energy line, (m/m)}$$
$$S_0 = \text{bed slope, (m/m)}$$
$$\bar{y} = \text{centroid height of the hydrostatic force, (m)}$$

The energy line represents the total head line at each section, i.e.

$$\text{Total head} = \text{datum} + \text{water depth} + u^2/2g$$

where
$$u^2/2g = \text{the 'velocity' head}$$

The energy slope line, the water surface slope and the bed slope are sometimes simplified and assumed to be the same. For this derivation, the energy line has a slope of S_f and the bed level has a slope of S_0. The conservation of momentum equation is thus:

The rate of change of momentum for the volume element = the resultant of the forces acting on this volume element

$$\frac{d}{dt}(v) = F_g + (F_1 - F_2) - F_f \tag{21.8}$$

$$\frac{d}{dt}(v) = \rho g A \Delta x S_0 + \frac{1}{2}\rho g \frac{\partial}{\partial x}(\bar{y}^2 A) - \rho g A \Delta x S_f \tag{21.9}$$

$$\rho A \Delta x \left(\frac{\partial u}{\partial t} + u\frac{\partial u}{\partial x}\right) = \rho g A \Delta x (S_0 - S_f) + \frac{1}{2}\rho g \frac{\partial}{\partial x}(\bar{y}^2 A) \tag{21.10}$$

For a constant cross-section and one-dimensional flow, Eq. (21.10) reduces to

$$\frac{\partial u}{\partial t} + u\frac{\partial u}{\partial x} = g(S_0 - S_f) \tag{21.11}$$

It was mentioned earlier that S_f is the slope of the energy line. If there were no friction losses or any other losses between section 1 and 2, then $S_f \rightarrow 0$. Therefore, S_f takes account of the losses incurred by the fluid in getting from section 1 to section 2. These losses are potentially of many sources. Typically, the most significant of these are bed friction losses. In addition, there are losses due to the internal fluid friction, as each horizontal layer of fluid shears over the next layer, producing a shear loss. Also, in turbulent flow, and real rivers always experience turbulent flow, there are the losses due to the turbulence generated, i.e. due to the fluctuating components of velocity in three dimensions. In the case of flow around bends in rivers there are further losses, due to the centrifugal forces. Thus S_f is like a global loss parameter. In sophisticated models, the individual losses are spelled out independently, e.g. bed friction, fluid friction, etc.

Equation (21.11) for steady flow ($\partial/\partial t = 0$) can be expanded to the case of two (and three)-dimensional flow. Introduce first,

$$U = u + u' \quad \text{in the streamwise } x \text{ direction}$$
$$V = v + v' \quad \text{in the lateral } y \text{ direction}$$
$$W = w + w' \quad \text{in the vertical } z \text{ direction}$$

where U, V, W are the instantaneous velocities and u, v, w are the corresponding mean velocities and u', v', w' the fluctuating velocity components. The momentum equation (for mean flow) for turbulent steady flow in the streamwise direction is

$$u\frac{\partial u}{\partial x} + v\frac{\partial u}{\partial y} + w\frac{\partial u}{\partial z} = gS_0 + \frac{\partial}{\partial x}(\overline{-u'^2}) + \frac{\partial}{\partial y}(\overline{-u'v'}) + \frac{\partial}{\partial z}(\overline{-u'w'}) + v\nabla^2 u \qquad (21.12)$$

where g = gravity, S_0 = bed slope = $\sin\theta$, where θ is the angle of bed slope; v is the kinematic viscosity (orders of magnitude less than the eddy viscosity). This equation can be found in almost any fluids book (e.g. Liggett, 1994). If we assume the flow to be one dimensional, that is the mean quantities $u \neq 0$, $v = w = 0$. However, as the flow is turbulent, the fluctuating quantities are non-zero, that is $u' \neq v' \neq w' \neq 0$. We also assume uniform flow, that is $\partial/\partial x = 0$. Equation (21.12) then reduces to

$$gS_0 + \frac{\partial}{\partial y}(\overline{-u'v'}) + \frac{\partial}{\partial z}(\overline{-u'w'}) = 0 \qquad (21.13)$$

The first term in the above equation is the driving force while the other two are resistances to flow. We can integrate Eq. (21.13) over the depth of the flow, let us say in a rectangular open channel, of depth H. Equation (21.13) becomes, on integration,

$$gHS_0 + \int_0^H \frac{\partial}{\partial y}(\overline{-u'/v'})\,dz + \frac{\partial}{\partial z}(\overline{-u'w'})\,dz = 0 \qquad (21.14)$$

If we first examine the third term of the above equation, the $\overline{-u'w'}$ term, it is thought of as a shear stress and so

$$\int_0^H \frac{\partial}{\partial z}(\overline{-u'w'})\,dz = -\frac{1}{\rho}\tau_b = -\frac{f}{8}u^2$$

where τ_b is the bed shear stress and f is a friction factor, from the Chezy equation. Thus Eq. (21.14) then becomes

$$gHS_0 + \int_0^H \frac{\partial}{\partial y}(\overline{-u'v'})\,dz - \frac{\tau_b}{\rho} = 0 \qquad (21.15)$$

Equation (21.15) can be further simplified if we can introduce a 'model' for the second term. This term is due to the Reynolds stress in the fluid motion or the internal fluid friction (ξ_{yx}). The most simple model for this term is known as the eddy viscosity, where the turbulent 'mixing' coefficient or eddy viscosity is introduced along the lines of the Boussinesque approximation (constant ρ). Therefore, the second term can be simplified to

$$\int_0^H \frac{\partial}{\partial y}(\overline{-u'v'})\,dz = H\frac{\partial}{\partial y}\xi_{yx}$$

where
$$\xi_{yx} = \rho v_t \frac{\partial u}{\partial y} \qquad (21.16)$$

where v_t is the eddy viscosity, which is here assumed to be constant across the section, though to be precise it is not constant. Finally, Eq. (21.15) becomes

$$gS_0 + v_t\frac{\partial^2 u}{\partial y^2} - \frac{fu^2}{8h} = 0 \qquad (21.17)$$

This is known as the DELV equation, or the depth average equation for the lateral distribution of streamwise velocity.

This second order, partial differential equation can be solved analytically and numerically to give a distribution for $u = u(y)$, where y is lateral and u is the depth integrated or depth averaged streamwise velocity. The solution of the model of Eq. (21.17) is by no means easy, certainly for the undergraduate

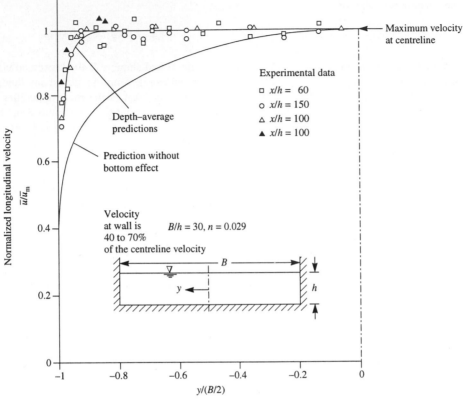

Figure 21.5 Horizontal distribution or depth-averaged velocity for fully developed flows in a rectangular open channel. (Adapted from Nezu and Nagawawa, 1993, with permission of IAHR.)

student. The interested reader is referred to the IAHR monograph, *Turbulence in Open Channel Flows* by Nezu and Nakagawa (1993).

Figure 21.5 shows some experimental results (from Nezu and Nakagawa, 1993) for a rectangular open channel. The vertical axis is u/u_m, where u_m is the maximum depth average velocity at the centre-line of the channel. It is seen that the velocity is not the same across the full width. As one approaches the wall away from the centre-line, the effect of wall friction comes into effect and reduced velocities occur. This figure is for a very wide channel, with a roughness coefficient n (Mannings n of 0.029, refer to any hydraulics text for the Mannings equation). The solution of Eq. (21.17) is dependent on the boundary conditions and a model for eddy viscosity. Some understanding of the hydrodynamics is essential before attempting to model water quality in river systems.

21.3 MATHEMATICS OF MASS TRANSPORT: DIFFUSION–ADVECTION

The transport of constituents by advection (moving water) and diffusion is dependent on the hydrologic and hydrodynamic characteristics of the particular environment, i.e. surface water. Advective transport dominates river flow that results from surface runoff and groundwater inflow. Diffusion dominates in estuaries or rivers experiencing tidal action. A conservative pollutant or constituent is one that does not undergo biological, chemical or physical reactions, i.e. salinity. Many constituents are assumed to be conservative for the purposes of mathematical simplification. The simulation of one- or two-dimensional depth averaged hydrodynamic systems with constant water density is carried out prior to any mass transport-diffusion simulation. The velocity fields produced from the hydrodynamics are used as input to the transport-diffusion (water quality) simulation. Mass transport is the flux or movement of a foreign constituent (e.g. a pollutant discharged to a river) from one site to another. This material transport is brought about by the mechanisms of advection and diffusion. The concentration of material A, C_A, in moles/L is defined as

$$C_A = C_A(x, y, z, t) \tag{21.18}$$

i.e. the concentration is a function of position and time. In one dimension, it is

$$C_A = C_A(x, t)$$

We can discuss mass transport or flux with reference to Fig. 21.6. A flux of a material goes from a zone of higher concentration to a zone of lower concentration. The flux is introduced as q_x. With reference to Fig. 21.6 we have

q_x is the flux in the x direction
$q_{x/x}$ is the unit flux in the x direction at $x = x_1$.
$q_{x/x+\Delta x}$ is the unit flux in the x direction at $x = x_2 = x + \Delta x$.

The concentration gradient of A is defined as $\partial C_A / \partial x$. Fick's law of molecular diffusion (see Fig. 21.6) states that 'the rate of mass transport of a material A, through a unit cross-sectional area of fluid by molecular diffusion is proportional to the concentration gradient of the material in the fluid':

$$\text{Fick's law } q = -D_M \frac{\partial C_A}{\partial x} = -D_M \left(\frac{C_1 - C_2}{X_2 - X_1} \right) \tag{21.19}$$

where

$q =$ flux of the material, as the mass of material being transferred per unit time, per unit cross-sectional area, in the x direction, kg/m^2 s $(q = C_A U_A)$

$D_M =$ molecular diffusion coefficient, with the negative sign indicating that transport (or flux) is from the area of high concentration to low concentration, m^{2}s

$U_A =$ molecular diffusion velocity $\ll u$ (the advective velocity)

Molecular diffusion is a function of the concentration gradient and occurs even in static waters. Sometimes the word dispersion is used as a 'catch-all' to account for the molecular and eddy diffusion and dispersion due to advection (i.e. moving waters). As such, when material transport is being discussed for a fluid in motion, the term dispersion is more commonly used.

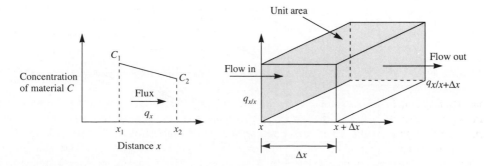

Figure 21.6 Molecular diffusion.

21.3.1 Molecular Diffusion

A slug of a pollutant is discharged to a fluid at rest. Assume molecular diffusion only applies (see Fig. 21.7). For simplicity, assume diffusion only in the x direction. (Of course, we know that molecular diffusion is isotropic, i.e. it has the same magnitude in all directions.) The mass balance for molecular diffusion through the static volume element shown in Fig. 21.7 is:

Rate of change of material A in the volume element = flux in − flux out + reactions

$$\text{The time rate of change in mass of material A} = \frac{\partial \rho_A}{\partial t} \, dx \, dy \, dz \tag{21.20}$$

$$\text{The flux into the element in } x \text{ direction} = \rho_A U_A \, dy \, dz \tag{21.21}$$

$$\text{The flux out of the element in } x \text{ direction} = \left[\rho_A U_A + \frac{\partial}{\partial x}(\rho_A U_A) \, dx \right] dy \, dz \tag{21.22}$$

$$\text{The rate of chemical or biological production} = r_A \, dx \, dy \, dz \tag{21.23}$$

where r_A is a production rate. In the x direction this mass balance reduces to

$$\frac{\partial \rho_A}{\partial t} = -\frac{\partial(\rho_A U_A)}{\partial x} + r_A \tag{21.24}$$

Neglecting production (or any internal reactions) and dividing through by the molecular weight M_A to get the molar concentration, i.e. $C_A = \rho_A / M_A$, Eq. (21.24) becomes

$$\frac{\partial C_A}{\partial t} = -\frac{\partial(C_A U_A)}{\partial x} \tag{21.25}$$

With no production, in three dimensions Eq. (21.25) becomes

$$\frac{\partial C_A}{\partial t} = -\left[\frac{\partial(C_A U_A)}{\partial x} + \frac{\partial(C_A V_A)}{\partial y} + \frac{\partial(C_A W_A)}{\partial z} \right] \tag{21.26}$$

This is the continuity equation for molecular diffusion in a static fluid in molar concentration form with units of moles/L. From Fick's law:

$$q = C_A U_A = -D_M \frac{\partial C_A}{\partial x} \tag{21.27}$$

Substituting into Eq. (21.25), we get

$$\frac{\partial C_A}{\partial t} = D_M \frac{\partial^2 C_A}{\partial x^2} \tag{21.28}$$

In three-dimensions we get

$$\frac{\partial C_A}{\partial t} = D_M \left(\frac{\partial^2 C_A}{\partial x^2} + D_M \frac{\partial^2 C_A}{\partial y^2} + D_M \frac{\partial^2 C_A}{\partial z^2} \right) \tag{21.29}$$

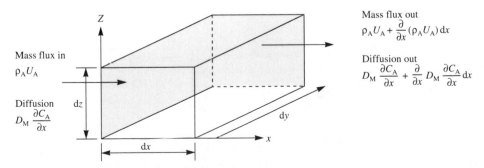

Figure 21.7 Continuity and diffusion in an elemental volume.

However,

$$\nabla^2 = \frac{\partial^2}{\partial x^2} + \frac{\partial^2}{\partial y^2} + \frac{\partial^2}{\partial z^2}$$

and therefore

$$\frac{\partial C_A}{\partial t} = D_M \nabla^2 C \qquad (21.30)$$

Equation (21.30) is known as the diffusion equation for a pollutant diffusion into a static fluid. It takes no cognisance of the fluid in motion (advection) and considers molecular diffusion only. It can be developed also for a fluid in motion, as described later.

The analytic solution of the I-D form of Eq. (21.30) for a slug of a pollutant of mass M is

$$C(x, t) = \frac{M}{\sqrt{4\pi Dt}} \exp\left(\frac{-x^2}{4Dt}\right) \qquad (21.31)$$

where $C(x, t)$ is the pollutant concentration at x downstream of the discharge point, at time t. Now we have replaced C_A with C and D_M with D, for clarity. At $t = 0$, Eq. (21.31) reduces to

$$C(x, 0) = 0$$

At $x = 0$, Eq. (21.31) reduces to

$$C(0, t) = \frac{M}{\sqrt{4\pi Dt}}$$

This solution is, of course, of Gaussian shape which we have already encountered (e.g. air pollution). Figure 21.8 is a distribution plot of the concentration, for the case where $M = 1$ and $D = 1/4$.

Along the centre-line of the 'plume', i.e. at $x = 0$ and time $t = 0$, we get maximum concentration. As time progresses, molecular diffusion sets in and at $t = \pi/4$, spreading is significant, the distribution reduces to a spike. For further details refer to Fischer *et al.* (1979). Figure 21.8 also shows what happens as we move out from the centre-line of the plume, i.e. $x > 0$.

It is seen from the previous discussion that molecular diffusion is a slow process. Realistically in riverine systems, we always have advection (moving waters, and, as such, the molecular diffusion contribution is insignificant). However, lakes may be a case where molecular diffusion is more important than advection.

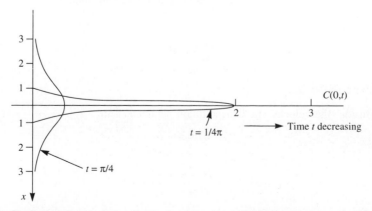

Figure 21.8 Gaussian distribution of concentration for $M = 1$ and $D = 1/4$. (Adapted from Fischer *et al.*, 1979, with permission.)

21.3.2 Advective Diffusion

Consider the case of a moving fluid, into which a slug of a pollutant is discharged. Diffusion will be by molecular plus advective diffusion. Advective diffusion may be much more significant than molecular diffusion. The flux is then:

$$\text{Fickian diffusion } q = C_A U_A = -D\frac{\partial C_A}{\partial x}$$

$$\text{Fickian plus advective diffusion } q = C_A U_A + C_A U \tag{21.32}$$

where C_A is the concentration of pollutant A. From now on, we will replace C_A with C. U_A is the diffusive (molecular only) velocity, while U is the instantaneous advective velocity. Similarly, D is the molecular diffusion (replacing D_M) and is initially assumed to be the same in all directions. Thus

$$q = \underset{\substack{\text{diffusive}\\\text{flux}}}{-D\frac{\partial C}{\partial x}} + \underset{\substack{\text{advective}\\\text{flux}}}{UC} \tag{21.33}$$

The streamwise point velocity is here assumed constant.

Recall the continuity Eq. (21.25) for diffusion only:

$$\frac{\partial C}{\partial t} = -\frac{\partial CU}{\partial x} \tag{21.25}$$

This can be written to be independent of molecular or advective diffusion, but simply called diffusion, as

$$\frac{\partial C}{\partial t} = -\frac{\partial q}{\partial x} \tag{21.34}$$

Combining Eq. (21.34) and (21.33), for a constant D, we get

$$\frac{\partial C}{\partial t} = -\frac{\partial}{\partial x}\left(-D\frac{\partial C}{\partial x} + UC\right)$$

$$= D\frac{\partial^2 C}{\partial x^2} - \frac{\partial(UC)}{\partial x} \tag{21.35}$$

In three-dimensions, we get

$$\boxed{\frac{\partial C}{\partial t} = D\nabla^2 C - \nabla UC} \tag{21.36}$$

Equation (21.36) is the advection–diffusion equation and has an additional term to the straight diffusion equation, which was

$$\boxed{\frac{\partial C}{\partial t} = D\nabla^2 C} \tag{21.30}$$

21.3.3 Turbulent Diffusion

In turbulent diffusion, the velocity and concentration parameters are

$$\begin{aligned} U &= u + u' \\ V &= v + v' \\ W &= w + w' \\ C &= c + c' \end{aligned} \tag{21.37}$$

where U, V, W and C are the time-varying point velocity and point concentrations. The mean values are represented by u, v, w and c. The fluctuations are represented by u', v', w' and c'. Consider Eq. (21.36):

$$\frac{\partial C}{\partial t} = D\nabla^2 C - \nabla UC \tag{21.36}$$

Neglecting the molecular diffusion component, we get

$$\frac{\partial C}{\partial t} = -\nabla UC \tag{21.38}$$

Expanding Eq. (21.38), in three-dimensions,

$$\frac{\partial C}{\partial t} + \frac{\partial (UC)}{\partial x} + \frac{\partial (VC)}{\partial y} + \frac{\partial (WC)}{\partial z} = 0 \tag{21.39}$$

Substituting Eq. (21.37) for the point velocities and concentrations and introducing a time average, long enough so that the average of the turbulent fluctuating velocities and fluctuations of concentrations are zero, Eq. (21.39) in one-dimension becomes

$$\frac{\partial \bar{C}}{\partial t} + \frac{\partial}{\partial x}\overline{(u+u')(c+c')} = 0 \tag{21.40}$$

where the overbar means the long time average. Expanding, we get

$$\frac{\partial \bar{C}}{\partial t} + \frac{\partial}{\partial x}(\overline{uc}) + \frac{\partial}{\partial x}(\overline{uc'}) + \frac{\partial}{\partial x}(\overline{u'c}) + \frac{\partial}{\partial x}(\overline{u'c'}) = 0$$

Since the time average of c' and u' are assumed zero, then we have

$$\frac{\partial c}{\partial t} + \frac{\partial}{\partial x}(\overline{uc}) + \frac{\partial}{\partial x}(\overline{u'c'}) = 0 \tag{21.41}$$

Expanding Eq. (21.41) we get

$$\frac{\partial c}{\partial t} + \bar{u}\frac{\partial \bar{c}}{\partial x} + \bar{c}\frac{\partial \bar{u}}{\partial x} + \frac{\partial}{\partial x}(\overline{u'c'}) = 0 \tag{21.42}$$

Removing the overbars on individual terms, we get

$$\frac{\partial c}{\partial t} + u\frac{\partial c}{\partial x} + c\frac{\partial u}{\partial x} + \frac{\partial}{\partial x}(\overline{u'c'}) = 0$$

Reintroducing three-dimensions, we get

$$\frac{\partial c}{\partial t} + u\frac{\partial c}{\partial x} + v\frac{\partial c}{\partial y} + w\frac{\partial c}{\partial z} + c\left(\frac{\partial u}{\partial x} + \frac{\partial v}{\partial y} + \frac{\partial w}{\partial z}\right) + \frac{\partial}{\partial x}(\overline{u'c'}) + \frac{\partial}{\partial y}(\overline{v'c'}) + \frac{\partial}{\partial z}(\overline{w'c'}) = 0 \tag{21.43}$$

From continuity, we have

$$\frac{\partial u}{\partial x} + \frac{\partial v}{\partial y} + \frac{\partial w}{\partial z} = 0$$

Therefore Eq. (21.43) in three-dimensions reduces to

$$\frac{\partial c}{\partial t} + u\frac{\partial c}{\partial x} + v\frac{\partial c}{\partial y} + w\frac{\partial c}{\partial z} + \frac{\partial}{\partial x}(\overline{u'c'}) + \frac{\partial}{\partial y}(\overline{v'c'}) + \frac{\partial}{\partial z}(\overline{w'c'}) = 0 \tag{21.44}$$

By similarity with the Fickian molecular diffusion model ($q = -D\,\partial c/\partial x$), we introduce the turbulent equivalents, or eddy diffusion, ε_x, ε_y, ε_z. These are the constants in the relationship of flux being proportional to the concentration gradient. As such, ε_x, ε_y, ε_z are referred to as the 'Fickian turbulent diffusion coefficients'. Molecular diffusion involves small-scale motions. For the latter reason, the coefficients are called the eddy diffusivities (Fischer *et al.*, 1979), and are defined as

$$\overline{c'u'} = -\varepsilon_x \frac{\partial c}{\partial x}$$

$$\overline{c'v'} = -\varepsilon_y \frac{\partial c}{\partial y} \qquad (21.45)$$

$$\overline{c'w'} = -\varepsilon_z \frac{\partial c}{\partial z}$$

Substitute Eq. (21.45) into (21.44) and replace $\bar{c}$ with c; we get

$$\frac{\partial c}{\partial t} + u\frac{\partial c}{\partial x} + v\frac{\partial c}{\partial y} + w\frac{\partial c}{\partial z} - \frac{\partial}{\partial x}\left(\varepsilon_x \frac{\partial c}{\partial x}\right) - \frac{\partial}{\partial y}\left(\varepsilon_y \frac{\partial c}{\partial y}\right) - \frac{\partial}{\partial z}\left(\varepsilon_z \frac{\partial c}{\partial z}\right) = 0 \qquad (21.46)$$

If we introduce C for c (the mean value of concentration), Eq. (21.46) becomes

$$\boxed{\frac{\partial C}{\partial t} + u \cdot \nabla C = \frac{\partial}{\partial x}\left(\varepsilon_x \frac{\partial C}{\partial x}\right) + \frac{\partial}{\partial y}\left(\varepsilon_y \frac{\partial C}{\partial y}\right) + \frac{\partial}{\partial z}\left(\varepsilon_z \frac{\partial C}{\partial z}\right) = 0} \qquad (21.47)$$

where
$$u \cdot \nabla C = u\frac{\partial C}{\partial x} + v\frac{\partial C}{\partial y} + w\frac{\partial C}{\partial z}$$

or, in one dimension, Eq. (21.47) becomes

$$\frac{\partial C}{\partial t} = D_x \frac{\partial^2 C}{\partial x^2} - U\frac{\partial C}{\partial x}$$

where D_x is the one-dimensional turbulent mixing coefficient. Equation (21.47) is known as the turbulent diffusion equation in three-dimensions.

The basic one-dimensional transport/turbulent diffusion equation for mass transport (our interest being mass pollutants) as solved by QUAL2 (US EPA, 1987), the US EPA water quality computer programme for freshwater and seawater, described in Section 21.6.3 is

$$\frac{\partial M}{\partial t} = \underbrace{\frac{\partial}{\partial x}\left(A_x D_x \frac{\partial C}{\partial x}\right)dx}_{\text{dispersion}} - \underbrace{\frac{\partial}{\partial x}(A_x \bar{u}C)\,dx}_{\text{advection}} + \underbrace{A_x\,dx\frac{dC}{dt}}_{\substack{\text{constituent}\\\text{reactions/}\\\text{interactions}}} + \underbrace{S}_{\substack{\text{sources/}\\\text{sinks}}} \qquad (21.48)$$

where
M = mass of pollutant
x = streamwise distance
t = time
C = concentration
A_x = cross-sectional area of river
D_x = turbulent mixing or dispersion coefficient, m^2/s
$\bar{u}$ = mean longitudinal velocity
S = external sources/sinks

This can be derived from Eq. (21.47). Note that the dispersion coefficient, $D_x \gg D_M$ (the molecular diffusion coefficient). For further details refer to Fischer *et al.* (1979). This is examined in further detail in Sec. 21.6.

21.4 POPULATION MODELS AND MODELS OF PHYSICAL SYSTEMS

Population growth models have been introduced in Chapters 3 and 12 with regard to growth of micro-organisms in wastewater. We can write

$$\frac{dP}{dt} = kP$$

where P is the actual population at any time (t). We used X instead of P in Chapter 12 to represent microbial population, i.e.

$$\frac{dX}{dt} = \mu X$$

where μ is a growth rate or proportionality factor. The units of μ are day^{-1}. The solution for the above population model is

$$X = X_0 e^{\mu t} \tag{12.9}$$

where X_0 is the microbial population at time zero. In Chapter 12, we had a model, called the saturation growth rate model for the growth rate, as

$$\mu = \mu_{max} \frac{S}{K_s + S} \tag{12.10}$$

where μ_{max} is the maximum growth rate of microbial population, S is the substrate or food concentration and K_s is the substrate concentration at a value of $\mu = \mu_{max}/2$. Frequently, K_s is called the half saturation constant. The constants μ_{max} and K_s are empirical values determined from experimentation. In the case of biomass and wastewater, this is typically a pilot plant or laboratory experiment. For instance, it is common to model at small scales a proposed treatment process, so as to understand the physics (flows), the chemistry (reactions) and the microbiology (biomass) before building a full-scale plant. The following example is that of a biochemical model.

Example 21.1 Consider a laboratory experiment with the following data. Determine the growth rate constant μ_{max}, and the half-saturation constant K_s.

S(mg/L)	μ(day^{-1})
14	0.14
18	0.19
30	0.24
50	0.32
80	0.40
150	0.48
200	0.49
300	0.53

Solution, from Eq. (12.10)

$$\mu = \mu_{max} \frac{S}{K_s + S} \tag{12.10}$$

Inverting we get

$$\frac{1}{\mu} = \frac{K_s + S}{\mu_{max} S} = \frac{K_s}{S} \frac{1}{\mu_{max}} + \frac{1}{\mu_{max}}$$

By this manipulation, Eq. (12.10) has been transformed into a linear form; i.e. $1/\mu_{max}$ is a linear function of $1/S$ with a slope of K_s/μ_{max} and an intercept of $1/\mu_{max}$. We can therefore produce the following table:

$1/S$(L/mg)	$1/\mu_{max}$ (day)
0.0714	7.14
0.0555	5.26
0.0333	4.17
0.0200	3.13
0.0125	2.50
0.0066	2.08
0.0050	2.04
0.0033	1.89

From the least squares fit in Fig. 21.9(a), it is seen that $\mu_{max} = 0.66$ and $K_s = 64$ mg/L. This results in the biochemical model, with parameters μ_{max} and K_s, such that

$$\frac{dX}{dt} = 0.66 \frac{S}{64 + S} X$$

The above equation or model can be solved for $X(t)$ analytically or numerically when the initial condition, $S = S(t)$ is known. In biological reactors, we will typically know what the feed rate is (i.e. $S(t)$), so the task is to determine the population $X(t)$. In activated sludge aeration basins, $X(t)$ is typically the mixed liquor suspended solids, or MLSS, with concentration in mg/L.

The technique used here to determine the parameters μ_{max} and K_s was the linearization of Eq. (12.10). Another approach is to use non-linear regression on Eq. (12.10). This technique requires initial guessing of μ_{max} and K_s. Many software graphics packages contain such facilities of linear/non-linear regression so the student may not have to resort to the mathematics of differential equations or numerical analysis.

Two types of hydraulic models can form the basis of many natural fluid systems, especially in water and wastewater treatment processes, but also in riverine systems. These are:

- The continuous flow stirred tank reactor model (CFSTR) and
- The plug flow reactor model (PFR)

21.4.1 The CFSTR Model

Consider a reach of river/lake as a CFSTR process. Allow the discharge of a non-reactive tracer (or salt solution) into this reach continuously. The continuous input flow rate is Q, while the influent tracer concentration is C_i. The assumption of a CFSTR model is that the volume of fluid in the river reach is well mixed. This means there is no concentration gradient within the reach. The outflow flow rate is Q and the outflow concentration is C. It is required to know the status of C with time (i.e. $C = C(t)$). At the beginning, the concentration of tracer in the river is zero. As time goes on, the tracer concentration in the river and in the outflow is increasing. The CFSTR system is represented in Fig. 21.10. The outflow concentration is assumed equal to the concentration in the river (assuming well mixed).

A mass balance is:

$$\text{Inflow} - \text{outflow} - \text{production} = \text{accumulation}$$

$$QC_i \quad - \quad QC \quad + \quad rV \quad = \quad \frac{dC}{dt} V$$

(21.49)

where

$$Q = \text{flow rate, m}^3/\text{s}$$
$$C_i = \text{influent tracer concentration, mg/L or g/m}^3$$
$$C = \text{effluent concentration, mg/L or g/m}^3$$
$$V = \text{volume of river reach or lake or tank}$$

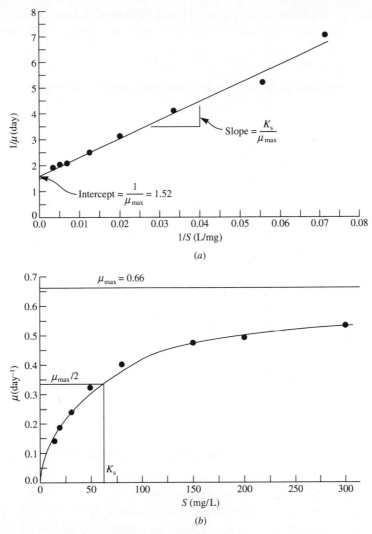

Figure 21.9 (*a*) Linearized version of the saturation growth rate model. The line is a least square fit to give the model coefficients $\mu_{max} = 0.66$ and $K_s = 64$ mg/L. (*b*) Fit of the saturation growth rate model.

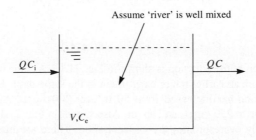

Figure 21.10 CFSTR model.

Assuming the influent is a non-reactive tracer, then $r = 0$, and for outflow and tank concentration, retain C as a variable. Our interest is in the changing of the value of the concentration C with time:

$$Q(C_i - C) = \frac{dC}{dt} V$$

Introduce the mean hydraulic retention time $\phi = V/Q$ (assumed constant). Rearranging,

$$\frac{dC}{C_i - C} = \frac{1}{\phi} dt$$

The initial condition is $C = C_0$ at $t = 0$ so,

$$\int_{C_0}^{C} \frac{dC}{C - C_i} = -\frac{1}{\phi} \int_0^t dt$$

$$\ln |C - C_i|_{C_0}^{C} = -\frac{1}{\phi} | t |_0^t$$

$$\ln \left(\frac{C - C_i}{C_0 - C_i} \right) = -\frac{1}{\phi} t$$

$$\frac{C - C_i}{C_0 - C_i} = e^{-t/\phi}$$

$$C - C_i = (C_0 - C_i)e^{-t/\phi}$$

$$C = C_i + (C_0 - C_i)e^{-t/\phi}$$

$$= C_i(1 - e^{-t/\phi}) + C_0 e^{-t/\phi} \tag{21.50}$$

i.e. the concentration in the river at any time t is a function of C_i, C_0, V/Q and, of course, t. If C_0 is zero, as is often the case, then Eq. (21.50) reduces to

$$C = C_i(1 - e^{-t/\phi})$$

Example 21.2 Determine the concentration, with time, of a non-reaction tracer discharged into a river block with the following conditions:

$$C_i = 300 \text{ mg/L}, \quad \text{continuous discharge}$$
$$Q = 700 \text{ m}^3/\text{day} \quad \text{flow rate}$$
$$C_0 = 50 \text{ mg/L}, \quad \text{the pre-existing concentration in the river}$$
$$V = 200 \text{ m}^3, \quad \text{the volume of river reach of interest}$$

Solution

$$\phi = V/Q = 200/700 = \frac{1}{3.5} \text{ day} = 6.7 \text{ h}$$

Eq. (21.50) can be written as

$$C = 300(1 - e^{-3.5t}) + 50 \, e^{-3.5t}$$

The exact analytic solution of the equation is shown in Fig. 21.11. At time 0, the tracer concentration is 4.0 mg/L (assume this is from an earlier tracer experiment in the same day). At one hydraulic retention time, $t = 6.7$ h, the concentration has increased from 50 to over 200 mg/L. At two hydraulic retention times, the concentration is close to 250 mg/L and so on. Also shown in Fig. 21.11 is a numerical solution of the equation, which slightly overestimates the concentration. There are many choices of numerical techniques but the one shown is the Euler method, with large time steps.

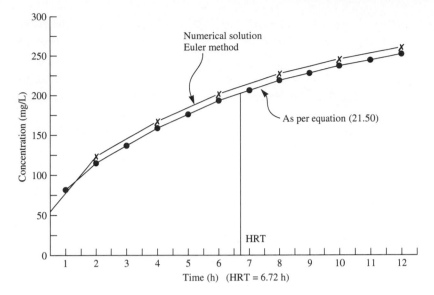

Figure 21.11 CFSTR model of concentration versus time.

When bacteria are involved, as is the case of the CFSTR of an activated sludge wastewater treatment plant, then the above Eq. (21.49) and its solution, Eq. (21.50), are only part of the story. In such a case, two mass balances are required, one for the biomass and another mass balance for the food substrate supply. As discussed in Chapter 12, the biomass mass balance is written in terms of concentration X. The substrate is written in terms of concentration S. They are coupled equations. This is discussed in more detail in Sec. 21.11.

Application of the CFSTR model can be utilized in evaluating water or wastewater treatment plants where many of the tanks are completely mixed and can be modelled using a series of CFSTRs. Similarly, preliminary analysis can be obtained for constituents discharged to riverine systems if the system is assumed to be made up of a discrete number of well-mixed reaches. Alternatively, if an effluent is discharged to a well-mixed lake, so too can a preliminary estimate be made of the concentrations of the lake pollution level, by using the above analysis. A more complicated analysis is involved when there is a reaction and if there are variations on the rate of reaction. The response of what happens in a length of river can be modelled in an elementary way by considering the river as a series of well-mixed CFSTRs, as shown in Example 21.3.

Example 21.3 Model a reach of river as three CFSTRs in series, as shown in Fig. 21.12. Consider a non-reactant dye injected upstream of reach 1. Determine the steady state concentration at points 300, 600 and 900 m downstream if the tracer concentration is 100 mg/L (100 g/m^3) at the upstream end.

Solution Assume that the tracer concentration decays as a first-order reaction, i.e. $r = -kC$ and assume that $k \cong 0.3 \text{ days}^{-1}$. The mass balance for the first reach is as follows:

$$\text{inflow} - \text{outflow} + \text{production} = \text{accumulation}$$

$$QC_{\text{in}} - QC_1 + rV_1 = \frac{dC}{dt}V_1$$

The steady state case is:

$$QC_{\text{in}} - QC_1 - kC_1V_1 = 0$$

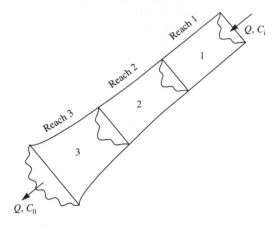

Figure 21.12 Modelling of a river as a series of CFSTRs.

The reach volumes are
$$V_1 = 1.8 \times 10^4 \, \text{m}^3$$
$$V_2 = 2.7 \times 10^4 \, \text{m}^3$$
$$V_3 = 3.6 \times 10^4 \, \text{m}^3$$
$$Q = 12 \, \text{m}^3/\text{s}$$

The mean hydraulic retention times are
$$\emptyset = V/Q$$
$$\emptyset_1 = V_1/Q = 0.018 \, \text{days}$$
$$\emptyset_2 = V_2/Q = 0.027 \, \text{days}$$
$$\emptyset_3 = V_3/Q = 0.035 \, \text{days}$$

Therefore, the mass balances for the three reaches are

$$QC_{in} - QC_1 - kC_1V_1 = 0$$
$$QC_1 - QC_2 - kC_2V_2 = 0$$
$$QC_2 - QC_3 - kC_3V_3 = 0$$

where C_1, C_2, C_3 are the tracer concentrations at the exit of each of reaches 1, 2 and 3 respectively. Introduce the mean hydraulic retention time for each reach as $\phi_1 = V_1/Q$, etc. Then rearranging the above equations and solving for C_1, C_2 and C_3:

$$C_1 = \frac{C_{in}}{1 + k\phi_1} = \frac{100 \, \text{g/m}^3}{1 + 0.3 \times 0.018} = 99.5 \, \text{g/m}^3$$

$$C_2 = \frac{C_{in}}{(1 + k\phi_1)(1 + k\phi_2)} = 98.7 \, \text{g/m}^3$$

$$C_3 = \frac{C_{in}}{(1 + k\phi_1)(1 + k\phi_2)(1 + k\phi_3)} = 97.6 \, \text{g/m}^3$$

Clearly, there is almost no decay of this tracer over the first kilometre. It is seen from the above that the decay is dependent on the product of the reaction rate and the mean hydraulic retention time. If the reach lengths were an order of magnitude larger then the concentrations would be

$$C_1 = 94.8 \, \text{g/m}^3$$
$$C_2 = 87.6 \, \text{g/m}^3$$
$$C_3 = 79.3 \, \text{g/m}^3$$

for reach lengths of 3, 6 and 9 km respectively. This simple modelling approach can be applied to a river system and lends itself to computerization very easily. Also, for more exact results the reaches can be as short as one wishes.

21.4.2 The Plug Flow Reactor (PFR) Model

A characteristic of the PFR model is the existence of a longitudinal concentration gradient. It is assumed that there is no concentration gradient laterally across the reactor. As used in the activated sludge biological wastewater treatment system, a continuous input of a constituent is put through at the upstream end. As the 'plug' of wastewater moves through the longitudinal reactor, it becomes 'diluted' such that when it leaves the downstream end it is 'purified'. This is sketched in Fig. 21.13. The plug flow system is the most common arrangement of activated sludge model, but as for the CFSTR model, it can be used as an elementary water quality modelling technique in rivers.

Consider a river reach, again, where a tracer is input upstream. Assume a plug flow model, and then determine the concentration of the tracer in the outflow. A mass balance on the elemental volume $\Delta V = A \Delta x$, as sketched in Fig. 21.13 is

$$\text{Inflow} - \text{outflow} + \text{production} = \text{accumulation}$$

$$QC(x) - QC(x + \Delta x) + r \, \Delta V = \frac{\partial C(x)}{\partial t} \Delta V \qquad (21.51)$$

$$QC(x) - Q\left[C(x) + \frac{\partial C}{\partial x}\Delta x\right] + r \, A \, \Delta x = \frac{\partial C(x)}{\partial t} A \, \Delta x$$

where

$$C(x) = \text{concentration at point } x, \text{g/m}^3$$

$$C(x + \Delta x) = \text{concentration at point } x + \Delta x, \text{g/m}^3$$

$$Q = \text{flow rate, m}^3/\text{s}$$

$$r = \text{reaction rate, g/m}^3 \text{ s of tracer}$$

Dividing across by $A \, \Delta x$ and taking the limit:

$$\frac{\partial C}{\partial t} = -\frac{Q}{A}\frac{\partial C}{\partial x} + r$$

For the steady state case,

$$\frac{Q}{A}\frac{dC}{dx} = r$$

Consider the rate of reaction in the case of a PFR to be first order:

$$r = -kC$$

where
$$k = \text{a reaction rate constant, day}^{-1}$$

Therefore,

$$\frac{Q}{A}\frac{dC}{dx} = -kC$$

$$\frac{dC}{C} = -\frac{A}{Q}k \, dx$$

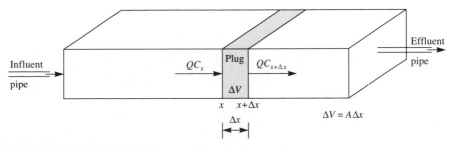

Figure 21.13 Plug flow reactor (PFR) model.

Introduce the mean hydraulic retention time

$$\frac{V}{Q} = \phi$$

where $\qquad\qquad\qquad\qquad V$ = reactor volume, m^3

and

$$\frac{A}{Q}dx = d\phi$$

Integrating between limits $C_i \rightarrow C$ and between $\phi = 0 \rightarrow \phi$,

where $\qquad\qquad\qquad C_i$ = tracer concentration entering the river reach, g/m^3

and $\qquad\qquad\qquad\quad C$ = tracer concentration leaving the river reach, g/m^3

then

$$\int_{C_i}^{C} \frac{dC}{C} = -\int_{0}^{\phi} k \, d\phi$$

$$\ln\frac{C}{C_i} = -k\phi$$

$$\frac{C}{C_i} = e^{-k\phi}$$

$$C = C_i e^{-k\phi} \qquad\qquad\qquad (21.52)$$

It is seen from Eq. (21.52) that the tracer concentration in the outflow is simply a function of the inflow concentration. It is merely transformed in time, as shown in Fig. 21.14. For $k \sim 0.25$ day^{-1} and $\phi = V/Q = 200/700$, $C = 0.93C_i$, so if C_i is a constant, so is C.

As with the CFSTR, the PFR model may be used for constituent dilution in river analysis or other environmental engineering applications, if the assumptions are appropriate to PFR behaviour.

We will now consider the more complex plug flow scenario by examining what happens within the river/reactor as we proceed downstream along x. Rewriting Eq. (21.51) we get

$$\underbrace{QC(x)}_{\text{flow in}} - \underbrace{Q\left[C(x) + \frac{\partial C(x)}{\partial x}\Delta x\right]}_{\text{flow out}} - \underbrace{DA\frac{\partial C(x)}{\partial x}}_{\text{dispersion in}} + \underbrace{DA\left[\frac{\partial C(x)}{\partial x} + \frac{\partial}{\partial x}\frac{\partial C(x)}{\partial x}\Delta x\right]}_{\text{dispersion out}} - \underbrace{kC(x)V}_{\text{decay rate}} = V\frac{\Delta C(x)}{\Delta t} \qquad (21.53)$$

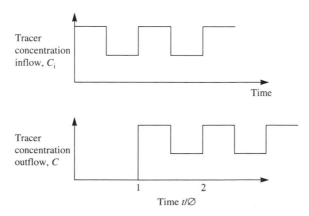

Figure 21.14 Ideal response of a plug flow reactor to a stepped function tracer input.

where D is a dispersion coefficient (m^2/day) and k is a first-order decay coefficient (day^{-1}) and A is the cross-section area. From Eq. (21.19), we have

$$q = -D \frac{\partial C}{\partial x}, \qquad \text{Fick's law}$$

In this example, D represents the turbulent mixing coefficient which tends to move the 'plug' of water from regions of high to regions of low concentration (i.e. from left to right in Fig. 21.13). In the limit as Δx and $\Delta t \to 0$, we have Eq. (21.53) reducing to

$$\frac{\partial C}{\partial t} = D \frac{\partial^2 C}{\partial x^2} - U \frac{\partial C}{\partial x} - kC \qquad (21.54)$$

where $U = Q/A$, the streamwise velocity of water flow along the river/reactor. This equation is the advection–diffusion equation (21.47). Equation (21.54) is a second-order partial differential equation. For the steady state case, it reduces to a second-order ordinary differential equation (ODE):

$$O = D \frac{d^2 C}{dx^2} - U \frac{dC}{dx} - kC \qquad (21.55)$$

At time $t < 0$, the tank is full of water, but the tracer has not been injected. At $t = 0$, the tracer is injected at the upstream end, and so we have the following boundary condition:

$$QC_{\text{in}} - QC_0 = -DA \frac{dC_0}{dx}$$

where C_{in} is the tracer concentration at the pipe entry to the reactor and C_0 is the tracer concentration at the most upstream node of the reactor, i.e. at $x = 0$.

The solution of Eq. (21.55) can be computed by a number of numerical techniques. Here we use what is called the centred finite difference method and rewrite Eq. (21.55) as

$$O = D \frac{C_{i+1} - 2C_i + C_{i-1}}{\Delta x^2} - U \frac{C_{i+1} - C_{i-1}}{2\Delta x} - kC_i \qquad (21.56)$$

where C_i is the concentration at node i, C_{i+1} at the adjacent downstream node $i+1$ and C_{i-1} at the adjacent upstream node. Multiply across by $\Delta x / U$ and collecting terms we get

$$-\left(\frac{D}{U\Delta x} + \frac{1}{2} \right) C_{i-1} + \left(\frac{2D}{U\Delta x} + \frac{k\,\Delta x}{U} \right) C_i - \left(\frac{D}{U\Delta x} - \frac{1}{2} \right) C_{i+1} = 0 \qquad (21.57)$$

This equation can be written for each node in the longitudinal direction x. At the upstream end, this becomes

$$-\left(\frac{D}{U\Delta x} + \frac{1}{2} \right) C_{-1} + \left(\frac{2D}{U\Delta x} + \frac{k\,\Delta x}{U} \right) C_0 - \left(\frac{D}{U\Delta x} - \frac{1}{2} \right) C_1 = 0 \qquad (21.58)$$

C_{-1} is replaced because of the first boundary condition, as follows. At the inlet we have

$$QC_{\text{in}} = QC_0 - DA \frac{dC_0}{dx} \qquad (21.59)$$

where C_0 is the concentration at $x = 0$. This says that the concentration of reactant carried into the reactor by advection is the same as that carried away from the inlet pipe by both advection and turbulent diffusion in the tank. We can write Eq. (21.59) numerically as

$$QC_{\text{in}} = QC_0 - DA \frac{C_1 - C_{-1}}{2\,\Delta x} \qquad (21.60)$$

Equation (21.60) is solved for C_{-1}:

$$C_1 - C_{-1} = -\frac{2\Delta x}{DA}(QC_{\text{in}} - QC_0)$$

However, $Q = UA$. Therefore,

$$C_1 - C_{-1} = -\frac{2\Delta x}{D}(UC_{in} - UC_0)$$

$$C_{-1} = C_1 + \frac{2\Delta x}{D}UC_{in} - \frac{2\Delta x}{D}UC_0 \qquad (21.61)$$

Substituting Eq. (21.61) into (21.58) and collecting terms, we get

$$\left(\frac{2D}{U\Delta x} + \frac{k\,\Delta x}{U} + 2 + \frac{\Delta xU}{D}\right)C_0 - \left(\frac{2D}{U\Delta x}\right)C_1 = \left(2 + \frac{\Delta xU}{D}\right)C_{in} \qquad (21.62)$$

This is the equation at the upstream end. At the downstream end we have the form of Eq. (21.57), which is

$$\left(\frac{-D}{U\Delta x} + \frac{1}{2}\right)C_{n-1} + \left(\frac{2D}{U\Delta x} + \frac{k\,\Delta x}{U}\right)C_n - \left(\frac{D}{U\Delta x} - \frac{1}{2}\right)C_{n+1} = 0 \qquad (21.63)$$

where n is the nth node at the downstream end. The downstream boundary condition is

$$QC_n - DA\frac{dC_n}{dx} = QC_n \qquad (21.64)$$

Differencing reduces Eq. (21.64) to

$$QC_n - DA\frac{C_{n+1} - C_{n-1}}{2\Delta x} = QC_n \qquad (21.65)$$

Equation (21.65) implies that $C_{n+1} = C_{n-1}$, i.e. the gradient at the outlet is zero. Substituting C_{n-1} for C_{n+1} in Eq. (21.63), we get

$$\left(\frac{2D}{U\Delta x}\right)C_{n-1} + \left(\frac{2D}{U\Delta x} + \frac{k\,\Delta x}{U}\right)C_n = 0 \qquad (21.66)$$

The equations requiring solution are now (21.58), (21.62) and (21.66). They form a system of n tridiagonal equations with n unknowns and can be solved if we have system parameters (e.g. D) and boundary conditions. Example 21.4 is such an exercise.

Example 21.4   Consider $D = 1.5\,\text{m}^2/\text{s}$, $U = 0.5\,\text{m/s}$, $\Delta x = 2.0\,\text{m}$, $k = 0.25\,\text{day}^{-1}$, with $C_{in} = 300\,\text{mg/L}$. Determine the response of the PFR reactor in the longitudinal direction, x.

Solution   Infill the tridiagonal matrix as described in the above equations, as follows:

$$\begin{bmatrix} +4.66 & -3.0 & - & - & - \\ -1.5 & +3.33 & -1.0 & & \\ & -1.5 & +3.33 & -1.0 & \\ & & -1.5 & +3.33 & -1.0 \\ & & & -3.0 & +3.33 \end{bmatrix} \begin{Bmatrix} C_0 \\ C_1 \\ C_2 \\ C_3 \\ C_4 \end{Bmatrix} = \begin{Bmatrix} 300 \\ 0 \\ 0 \\ 0 \\ 0 \end{Bmatrix}$$

This matrix is solved to give $C_0 = 263.2$, $C'_1 = 142.2$, $C_2 = 78.6$, $C_3 = 48.6$, $C_4 = 43.7$. The solution of this set of equations is fairly straightforward and numerical methods such as Gauss elimination work well. The student is advised to refer to Chapra and Canale (1988) for excellent documentation on suitable numerical methods and more detailed CSTR and PFR case studies.

These results are plotted in Fig. 21.15, for two different values of the turbulent diffusion coefficient, D, i.e. $D = 1.5$ and $D = 6\,\text{m}^2/\text{day}$. It is seen from Fig. 21.15 that the concentration decreases as the tracer moves through the tank, due to the decay reaction. As D goes from 1.5 to 6, it is seen that the curve flattens significantly. If D is further decreased, it is noted that the mixing becomes less important relative

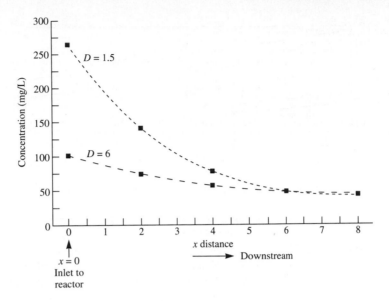

Figure 21.15 Plug flow reactor—concentration versus distance.

to advection and decay. The above development and example can be used to determine more detailed analysis with regard to PFR systems. The number of Δxs can be increased and so increase the detail of the analysis.

21.5 HYDRODYNAMIC MODELLING OF RIVERS

Before water quality in riverine systems can be evaluated, the hydrodynamics of the system must first be determined. This means determining the hydraulic flow characteristics of depth, slope, velocity, discharge. These parameters may be a function of time if the flow is considered unsteady. In riverine systems the flow is usually non-uniform with a gradually varied depth. However, the water quality characteristics are dependent on the hydrodynamics of flow and so the equations for diffusion and mass transport are coupled to those of the hydrodynamics. Typically modelling involves simplification processes and means of uncoupling these equations. The most simple problem in hydrodynamics is when the flow problem can be defined as steady and the water levels are dependent only on downstream conditions such as is the case of backwater analysis. The water levels upstream are dependent on the flow rate and on the initial water level of, say, a bridge or weir downstream. The levels are computed from the downstream level, proceeding to calculate the water levels at each cross-section as we move upstream. The opposite is the case when flood routeing is involved. This is an unsteady case and the flow rates change over time. The conditions are the input hydrograph from upstream, which starts with a flow rate and water level and the hydrograph is routed in a direction downstream by one of the many techniques, e.g. kinematic wave routeing.

21.5.1 Introduction to Hydrodynamic Models

HEC-2 The US Army Corps of Engineers make available a backwater analysis programme which is in widespread use. It is capable of producing longitudinal water surface profiles along riverine systems subject to steady flow, if given a starting water level and flow at the downstream end (assuming subcritical

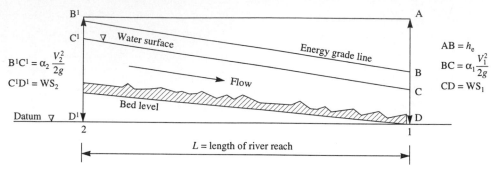

Figure 21.16 Flow between two river cross-sections (2 and 1).

flow). The computational procedure is based on the solution of the one-dimensional energy equation using the standard step iterations method, HEC-2 (US Army Corps of Engineers, 1990) to calculate the unknown water surface elevation at a cross-section (see Fig. 21.16). We are, of course, talking about the steady state flow, i.e. $d/dt = 0$. We are talking about subcritical flow, i.e. the Froude number is < 1. This means that there are no 'jumps', e.g. 'hydraulic jumps', in the longitudinal water surface profile.

The total energy, measured as 'head of water' at any channel section is

$$H = \text{WS} + \alpha \frac{V^2}{2g} \tag{21.67}$$

where

$\text{WS} = $ depth to water surface above a datum (potential energy)

$\alpha \dfrac{V^2}{2g} = $ kinetic energy term

Equating the total energy at section 2 to that of section 1 gives

$$\text{WS}_2 + \alpha_2 \frac{V_2^2}{2g} = \text{WS}_1 + \alpha_1 \frac{V_1^2}{2g} + h_e \tag{21.68}$$

where

$V_2, V_1 = $ mean velocities $(Q/A_1, Q/A_2)$

$\alpha_2, \alpha_1 = $ velocity coefficients

$h_e = $ energy head loss between sections 2 and 1, i.e. the reduction in water surface level between sections 1 and 2 due to overcoming the bed frictional and other retardation forces between 1 and 2

Also,

$$h_e = L\bar{S}_f + C\left(\alpha_2 \frac{V_2^2}{2g} - \alpha_1 \frac{V_1^2}{2g}\right) \tag{21.69}$$

$\bar{S}_f = $ representative friction slope $ = (\text{WS}_2 - \text{WS}_1)/L$

$C = $ expansion or contraction loss coefficient

$L = $ length of river reach

As the HEC-2 program is a competent number-cruncher, it has the facilities to accommodate complex cross-sections, including main channels and floodplains, bridges, culverts, etc., and data can be input to describe even minor changes in the lateral bed profile.

The unknown water surface elevation at a cross-section is determined by an iterative solution of Eqs (21.68) and (21.69). The computational procedure as listed in HEC-2 (US Army Corps of Engineers, 1990) is:

1. Assume a water surface elevation at the upstream cross-section (WS$_2$) for the subcritical flow case.
2. Determine the downstream and upstream section conveyance and velocity heads. Conveyance is defined as

$$K_1 = \frac{1}{n_1} A_1 R_1^{2/3} \qquad (21.70)$$

Eq. (21.70) is a form of the Manning equation, i.e.

$$Q = \frac{1}{n} A R^{2/3} S^{1/2}$$

where

n_1 = Manning's roughness coefficient
(0.01 for smooth concrete, 0.03 for typical rivers); tables for n are given in most hydraulics textbooks (e.g. Chow, 1959)

A_1 = cross-sectional area at section 1

R_1 = hydraulic radius of section 1

$\quad = \dfrac{A_1}{P_1}$, where P_1 = wetted perimeter

S = gradient of the energy (or water surface) line

3. Compute the representative friction slope $\bar{S}_f$:

$$\bar{S}_f = \frac{Q_1 + Q_2}{K_1 + K_2} \qquad (21.71)$$

and solve Eq. (21.69) for h_e, after identifying values for α_1, α_2 and C. For the simple typical case of steady flow, $Q_1 = Q_2$ (and no lateral inflow). Other methods exist to compute $\bar{S}_f$ (US Army Corps of Engineers, 1990).
4. Solve Eq. (21.68) for WS$_2$.
5. Compare the computed value of WS$_2$ to the initial estimate of WS$_2$. If the difference is less than, say, 0.01 m, proceed upstream to the next cross-section. If not, try another WS$_2$ and repeat the steps again.

An application in the widespread use for HEC-2 is the determination of longitudinal water surface profiles, for floods of different magnitudes (different return periods). This is carried out in the initial event for, say, a river. If a flood study is being carried out, compound channel modifications are possibly entered in the revised cross-section data and a new set of water surface profiles are obtained. The benefits of channel modification can then be assessed.

An excellent description of HEC-2 is found in Bedient and Huber (1988). The reader is referred to this text and the examples and exercises in it. Information on HEC 2 software is available from the US Army Corps of Engineers Headquarters in Davis, California or from suppliers of public domain software.

HEC-1   This is a computer program from the US Army Corps of Engineers (1990) that simulates the surface runoff response of a river basin to precipitation by representing the basin as an interconnected system of hydrologic and hydraulic components (Fig. 21.17). Each component models an aspect of the precipitation–runoff process within a sub-basin. A component may represent a surface runoff entity, a stream channel or a reservoir. The output is in the form of streamflow hydrographs at sections along the river reach. It can be used for both large rural catchments or more urbanized catchments (US Army Corps of Engineers, 1992). The kinematic wave routeing method is used for the urbanized catchments while the user can choose either the kinematic wave method or the Muskingum–Cunge flood routeing methods for rural catchments. These two methods are described in most hydraulic textbooks (see Bedient and Huber, 1988).

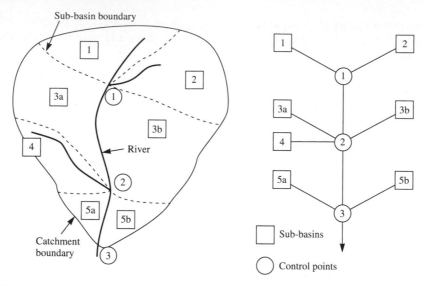

Figure 21.17 River basin and schematic river basin. (Adapted from HEC-1, 1990, with permission.)

Water is transported either by surface runoff or channel runoff to the stream network. The rainfall hyetograph produces surface runoff and/or channel runoff. The rainfall hyetograph is the excess rainfall (over time) found by subtracting infiltration and detention losses from precipitation. The rainfall and infiltration are assumed uniform over the sub-basin. The resulting rainfall excesses are then routed by the unit hydrograph method or kinematic wave method to produce the sub-basin hydrograph, which serves as partial input for the next downstream sub-basin. This suggests that the sub-basin should be small in area and constant in topography and hydrogeology. As rainfall excess is assumed uniform over a sub-basin, it is best to utilize small areas as sub-basins.

Details of the rainfall transformation to rainfall excess and rainfall hyetograph are given in Section 3 of HEC-1 (US Army Corps of Engineers, 1990). The user may adopt a synthetic storm based on the standard projected storm (details for the United Kingdom/Ireland are available in the NERC *Flood Studies Report*, 1975). An empirical exponential loss rate function may be used to compute rainfall excess. The Soil Conservation Service (SCS) method, Holtan method or Green-Ampt (see Dingman, 1994) method may also be used.

In kinematic wave routeing (of the hydrograph and lateral inflow along the channel) the one dimensional governing equations of flow are the St Venant equations:

Continuity :
$$\frac{\partial Q}{\partial t} + \frac{\partial Q}{\partial x} = q_L + (i - f) \tag{21.72}$$

$$\underset{\substack{\text{rate of} \\ \text{rise term}}}{} \quad \underset{\substack{\text{storage} \\ \text{term}}}{} \quad \underset{\substack{\text{lateral} \\ \text{inflow}}}{} \quad \underset{\substack{\text{excess} \\ \text{rainfall}}}{}$$

This equation is written with Q replaced by q for the overland flow case:

Momentum :
$$\frac{\partial u}{\partial t} + u\frac{\partial u}{\partial x} + g\frac{\partial y}{\partial x} = g(S_0 - s_f) - q_L\frac{u - v}{y} \tag{21.73}$$

$$\underset{\substack{\text{acceleration} \\ \text{or dynamic} \\ \text{term}}}{} \quad \underset{\substack{\text{velocity} \\ \text{head or} \\ \text{convective} \\ \text{interia}}}{} \quad \underset{\substack{\text{depth} \\ \text{slope}}}{} \quad \underset{\substack{\text{bed slope} \\ \text{and} \\ \text{slope}}}{} \quad \underset{\substack{\text{lateral} \\ \text{inertia}}}{}$$

where $\qquad$ q = discharge per unit width of cross-sections; q_L = lateral inflow

y = mean depth of channel

x = distance downstream

u = x component of velocity

v = x component of velocity of lateral inflow

f = infiltration rate

i = rainfall rate

S_o = bed slope

S_f = friction slope

The above equations are simplified when the following assumptions are taken into account:

1. There is no water surface superelevation and no secondary circulations.
2. Flows are gradually varied with hydrostatic pressure prevailing throughout, such that vertical accelerations are neglected.
3. Boundaries are fixed and non-erodable.
4. Bed slopes are flatter than $\sim 1{:}10$.
5. Empirical equations are used to evaluate flow resistance (Manning or Chezy equation).

Kinematic waves are long shallow waves where surface perturbations are assumed not to exist. Kinematic waves occur when the dynamic terms in the momentum equation(s) tend to zero. Therefore Eq. (21.73) reduces to

$$S_o = S_f \qquad (21.74)$$

i.e. the bed slope is approximately equal to the friction slope. Therefore, under the condition of negligible backwater effects, the discharge can be described in stage/discharge terms as

$$Q_c = \alpha_c A_c^{m_c} \qquad \text{or} \qquad q_o = \alpha_0 y_0^{m_0} \qquad (21.75)$$

where α and m are parameters related to flow geometry and roughness. The subscript c refers to the main channel and the subscript o refers to the overland flow. The flow at any point can then be computed from the Manning equation:

$$Q = \frac{1}{n} A R^{2/3} S^{1/2} \qquad (21.76)$$

for a wide rectangular channel of width W and depth y, where $W \gg y$. Recall the hydraulic radius $R = A/P$, where P is the wetted perimeter:

$$Q = \frac{1}{n} A \left(\frac{A}{P}\right)^{2/3} S^{1/2}$$

$$= \frac{1}{n} A \frac{A^{2/3}}{W^{2/3}} S^{1/2}$$

$$= A^{5/3} \frac{1}{n} S^{1/2} W^{-2/3}$$

$$Q = A^m \alpha$$

Therefore

$$m = \frac{5}{3} \qquad \text{and} \qquad \alpha = \frac{1}{n} S^{1/2} W^{-2/3} \qquad (21.77)$$

As the momentum equation was reduced to an empirical relationship between area and discharge, the movement of the flood wave is fully described by the continuity equation:

$$\frac{\partial A}{\partial t} + \frac{\partial Q}{\partial x} = q \tag{21.78}$$

By comparison with backwater analysis (HEC-2), which has its initial condition set at the downstream boundary, the kinematic wave routeing model has its initial condition set at the upstream boundary. The initial and boundary conditions for the kinematic wave are determined from the upstream boundary hydrograph. The governing equation (21.78) is modified by combining with Eq. (21.75) to become

$$\frac{\partial A}{\partial t} + \frac{\partial(\alpha A^m)}{\partial x} = q$$

However, since Q is a function of A, i.e. $Q = \alpha A$, i.e. $Q = \alpha A^m$, then

$$\frac{\partial Q}{\partial x} = \frac{\partial Q}{\partial A}\frac{\partial A}{\partial x} = \alpha m A^{(m-1)}\frac{\partial A}{\partial x}$$

and

$$\frac{\partial A}{\partial t} + \alpha m A^{(m-1)}\frac{\partial A}{\partial x} = q \tag{21.79}$$

since α and m are constants for a particular geometric hydraulic shape and roughness. Therefore, the only dependent variable is A, the cross-sectional area. Equation (21.79) is written for both the overland flow and main channel routeing as follows:

Overland flow :
$$\frac{\partial y_o}{\partial t} + \alpha_o m_o y_o^{(m_o-1)}\frac{\partial y_o}{\partial x} = i - f \tag{21.80}$$

Main channel flow :
$$\frac{\partial A_c}{\partial t} + \alpha_c m_c A_c^{(m_c-1)}\frac{\partial A_c}{\partial t} = q_o \tag{21.81}$$

where the subscript c relates to main channel and o refers to overland flow. These equations are solved (independently) by a method of finite difference to produce a relationship for y_o in terms of x, t and $i - f$ and for A_c in terms of x, t and q_o. These values of y_o (or A_c) are then substituted into Eq. (21.75) to solve for Q_c (or q_o). This procedure allows the calculation of Q_c (or q_o) as a function of segment (reach) length (L_c) and time (t). Therefore, the hydrograph for each reach can be determined. The procedure followed in HEC-1 is first to calculate the overland flows, then the flows through the collector channel system and finally the discharges in the main stream. The reader is referred to the manuals of HEC-1 and to Bedient and Huber (1988) where excellent details, examples and case studies using HEC-1 are found. Schematics of hydrograph shapes are shown in Fig. 21.18 where the resultant inflow hydrograph catchment was determined from knowledge of the rainstorm and characteristics of the catchment and HEC-1 was used to route this hydrograph through a lake and so determine the downstream outflow hydrograph. The use of

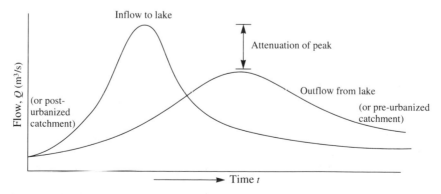

Figure 21.18 Schematic of inflow/outflow hydrographs to a lake.

models such as HEC-1, DAMBRK, etc., can determine the attenuating effects of lakes, floodplains, river storage, etc. Very often in flood studies or sizing of upstream flood reservoirs the determination of a specific attenuation is required and hence the design size of a retention storage reservoir.

As these models can be used to determine the attenuation effects of lakes or flow over floodplains, so also can they be used to determine the impact of proposed urbanization on a rural catchment (Fig. 21.18) (US Army Corps of Engineers, 1992). Another excellent program that is capable of flood routeing is that of DAMBRK by Fread of the US National Weather Service. There are many other excellent hydrodynamic programs available, some in the public domain and others proprietary. The reader is recommended to obtain a copy of any of the above to gain practice in the use of flood routeing as a forerunner to river water quality modelling.

21.6 WATER QUALITY MODELLING IN RIVERINE SYSTEMS

In analysing and predicting water quality, specific constituents may need to be examined. Among the non-conservative (those that change or react) constituents of interest in riverine water quality are:

- Dissolved oxygen (DO)
- Biochemical oxygen demand (BOD)
- Temperature
- Algae as chlorophyll-a
- Organic nitrogen (N_o)
- Ammonia nitrogen ($NH_3 - N$)
- Nitrite nitrogen ($NO_2 - N$)
- Nitrate nitrogen ($NO_3 - N$)
- Organic phosphorus
- Dissolved phosphorus
- Coliforms

Conservative constituents may also be of interest and these include:

- Sediment
- Dissolved solids or salts
- Metals (Pb, Cu, Hg, Cr, Cd)
- Non-reactive dye tracers for experimental purposes

21.6.1 Modelling of Conservative Constituents

The following example of determining the concentration of dissolved solids in riverine systems is adapted from O'Connor (1976). Conservative dissolved solids are defined as those that enter river systems primarily from natural geologic backgrounds, for instance calcium salts. Other constituents derived from the rock strata include iron, magnesium, sodium, potassium, aluminium, lead, mercury, etc. 'Conservative' means that the compounds are not chemically reactive. These constituents (assuming that there are no anthropogenic sources) are contributed to the river system via groundwater. If it is assumed that the groundwater contribution to river flow does not change hugely (even in flood times) then these dissolved salts will be somewhat constant throughout the year. Consider a riverine control volume with streamwise flow Q with a concentration of a conservative constituent of C. Additionally there is a lateral inflow of surface runoff ΔQ_s with concentration C_s. There is also a groundwater contribution of flow ΔQ_g and concentration C_g. This is shown in Fig. 21.19. We are interested in determining the flow and water quality impacts of the surface runoff and groundwater on the main river flow.

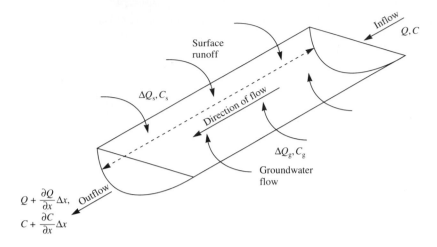

Figure 21.19 Flows and concentrations through a riverine control volume.

Consider a mass balance around the control volume:

$$\text{Inflow} - \text{outflow} + \text{generation} = \text{accumulation}$$

Assume no generation, i.e. conservative constituents:

$$QC + \Delta Q_s C_s + \Delta Q_g C_g - \left(Q + \frac{\partial Q}{\partial x}\Delta x\right)\left(C + \frac{\partial C}{\partial x}\Delta x\right) = \frac{\Delta VC}{\Delta t} \tag{21.82}$$

where
$$Q = \text{flow, m}^3/\text{s}$$
$$C = \text{concentration, g/m}^3 \text{(mg/L)}$$
$$\Delta x = \text{length of river, km}$$

Recall that

$$\left(C\frac{\partial Q}{\partial x} + Q\frac{\partial C}{\partial x}\right)\Delta x = \frac{\partial(CQ)}{\partial x}\Delta x$$

Neglecting second-order terms and dividing through by Δx:

$$C_s\frac{\Delta Q_s}{\Delta x} + C_g\frac{\Delta Q_g}{\Delta x} - \frac{\partial(CQ)}{\partial x} = \frac{A\Delta xC}{\Delta t\Delta x} \tag{21.83}$$

where
$$Q_g = \text{groundwater flow}$$
$$C_g = \text{groundwater flow concentration}$$
$$Q_s = \text{surface overland flow}$$
$$C_s = \text{concentration of surface flow}$$

Taking the limit, the general expression of mass flow is

$$C_s\frac{\partial Q_s}{\partial x} + C_g\frac{\partial Q_g}{\partial x} - \frac{\partial(CQ)}{\partial x} = \frac{\partial(AC)}{\partial t} \tag{21.84}$$

The steady state condition reduces to

$$C_s\frac{\partial Q_s}{\partial x} + C_g\frac{\partial Q_g}{\partial x} - \frac{\partial(CQ)}{\partial x} = 0 \tag{21.85}$$

Integrating over the river length of Δx we get

$$C_s Q_s + C_g Q_g - CQ = 0$$

Therefore

$$CQ = C_s Q_s + C_g Q_g \tag{21.86}$$

That is, the product of flow and concentration of the total is the sum of its surface and groundwater contributions. Introduce

$$r = \frac{Q_g}{Q} \quad \text{and} \quad (1 - r) = \frac{Q_s}{Q}$$

Then Eq. (21.86) becomes

$$C = rC_g + (1 - r)C_s \tag{21.87}$$

Introduce $Q_0 =$ base flow. For

$$Q < Q_0 \quad \text{then let} \quad Q = Q_g \text{ and so } C = C_g \tag{21.88}$$

Thus, for flows less than base flows, the concentrations would be essentially those from groundwater; i.e. in summer times we expect high salts concentration (if they exist in the groundwater). Usually surface water flows are much lower in salts concentration than groundwater. Rearranging Eq. (21.87):

$$C = r(C_g - C_s) + C_s$$

For river flows higher than base flows, then

For $Q > Q_0$ introduce $Q_g = \beta Q^n$

Then

$$r = \frac{Q_g}{Q} = \frac{\beta Q^n}{Q} = \beta Q^{n-1}$$

and

$$C = C_s + \frac{\beta}{Q^{1-n}}(C_g - C_s) \tag{21.89}$$

During the dry periods of the year, when there is no surface water contribution, the concentration in the river is due to groundwater and is independent of flow magnitude, Eq. (21.88). At other times, the concentration is a combination of groundwater and surface water contributions. Equation (21.89) suggests that, at those times, the concentration is dependent on the river flow magnitude. O'Connor (1976) has shown that Eq. (21.89) holds for a number of rivers in the United States over a range of flow magnitudes. The reader is referred to this classic paper for further details.

Fig. 21.20(a), adapted from O'Connor (1976), shows the relationship of river flow with groundwater. It is seen that for $Q < Q_0$, all the flow is from Q_g. Above Q_0, there can be different contributions of Q_s to the total flow. Fig. 21.20(b) shows the relationship of the ratio of groundwater and river flow with river flow and Fig. 21.20(c) shows the relationship of concentrations.

21.6.2 Modelling of Non-conservative Constituents

Non-conservative constituents are those that are chemically/biochemically reactive. Oxygen in rivers or the lack of it was first modelled by Streeter and Phelps (1925). The oxygen concentration in a riverine system changes with time and space and is dependent on a host of other parameters and other constituents. The dissolved oxygen saturation concentration is ~ 14.5 mg/L at $0\,^\circ C$, ~ 9.0 mg/L at $20\,^\circ C$ and ~ 6.0 mg/L at $40\,^\circ C$. If water lies quiescent in a shallow lake, the top surfaces will increase in temperature (during the warm seasons) and thereby the DO saturation value falls. Should there be an excess nutrient supply of phosphorus, then algae may grow and exert a demand on the oxygen, thereby

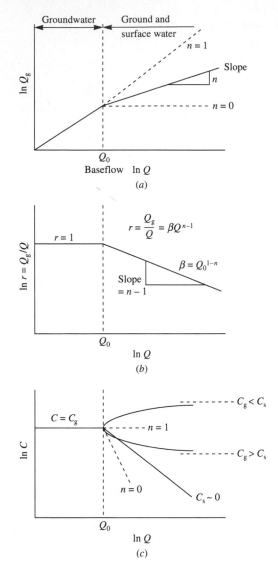

Figure 21.20 The relationship between the river flow and (a) groundwater, (b) ratio of river flow to groundwater and (c) the concentrations of dissolved solids. (Adapted from O'Connor, 1976. Copyright by the American Geophysical Union.)

reducing the oxygen level further. Further growth leads to further reduction in oxygen, ultimately leading to a eutrophic lake. BOD wastes discharged into riverine systems exert demands on oxygen levels and may draw the oxygen down to levels too low to sustain aquatic life (about $< 3 \, \mathrm{mg/L}$). The concentration of DO is represented as the resultant of two principal competing processes:

- Deoxygenation and
- Aeration

Deoxygenation This occurs where the rate of uptake of oxygen is dependent on its remaining BOD concentration:

$$r_{01} = \frac{\mathrm{d}L}{\mathrm{d}t} = -K_1 L \qquad (21.90)$$

where $\quad\quad\quad\quad r_{01}$ = rate of deoxygenation, mg/m^3 d

$\quad\quad\quad\quad\quad\quad\quad\quad L$ = BOD remaining, mg/m^3

$\quad\quad\quad\quad\quad\quad\quad\quad K_1$ = first-order rate coefficient (temperature dependent), d^{-1}

The integrated form of Eq. (21.90) is

$$L = L_0 e^{-K_1 t} \tag{21.91}$$

where $\quad\quad\quad\quad L_0$ = concentration of ultimate BOD

Therefore

$$r_{01} = \frac{\mathrm{d}L}{\mathrm{d}t} = -K_1 L_0\, e^{-K_1 t} \tag{21.92}$$

A more detailed rate equation allows for further deoxygenation due to settling (benthic deposits) such that

$$r_{01} = \frac{\mathrm{d}L}{\mathrm{d}t} = -K_1 L - K_3 L$$

where $\quad\quad\quad\quad K_3$ = rate of deoxygenation due to settling (temperature dependent)

Reoxygenation or reaeration   This occurs where the rate of replenishment of oxygen is dependent on the DO deficit and a reaeration rate coefficient K_2:

$$r_{02} = K_2(C_s - C) = K_2 D \tag{21.93}$$

where $\quad\quad\quad\quad r_{02}$ = rate of reaeration, mg/m^3 d

$\quad\quad\quad\quad\quad\quad\quad\quad K_2$ = reaeration rate coefficient, d^{-1}

$\quad\quad\quad\quad\quad\quad\quad\quad C_s$ = dissolved oxygen (DO) saturation level, mg/L

$\quad\quad\quad\quad\quad\quad\quad\quad C$ = actual DO level, mg/L

$\quad\quad\quad\quad\quad\quad\quad\quad D$ = DO deficit, mg/L

Streeter and Phelps (1925) (see Sec. 7.3) expressed the time rate of charge of dissolved oxygen deficit as

$$\frac{\mathrm{d}D}{\mathrm{d}t} = K_1 L - K_2 D \tag{21.94}$$

Consider a riverine control volume with inflows and outflows as shown in Fig. 21.21. Consider a mass balance around the flow volume:.

$$\text{Inflow} - \text{outflow} + \text{deoxygenation} + \text{reaeration} = \text{accumulation}$$

$$QC|_x - QC|_{x+\Delta x} + r_{01}\Delta V + r_{02}\Delta V = \frac{\partial C}{\partial t}\Delta V \tag{21.95}$$

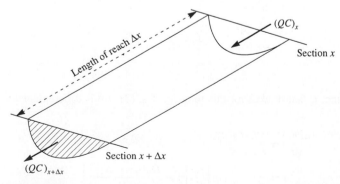

Figure 21.21 Riverine control volume.

where $\qquad C =$ actual concentration of oxygen, mg/L

$\Delta V =$ incremental volume, over the incremental length Δx

Taking the limit as $\Delta x \to 0$ and for the steady state case $(\partial C / \partial t = 0)$,

$$-Q\frac{\partial C}{\partial V} - K_1 L + K_2 D = 0 \tag{21.96}$$

However, $t = V/Q$ so Eq. (21.96) becomes the differential equation for the simple oxygen sag model:

$$\frac{dC}{dt} = -K_1 L + K_2 D \tag{21.97}$$

However $\qquad D = C_s - C$ and C_s is a constant

Therefore,

$$\frac{dD}{dt} = -\frac{dC}{dt}$$

and, from Eq. (21.94) $\qquad \frac{dD}{dt} = K_1 L - K_2 D$

and, from Eq. (21.91) $\qquad K_1 L = K_1 L_0 e^{-K_1 t}$

Therefore

$$\frac{dD}{dt} + K_2 D = K_1 L_0 e^{-K_1 t} \tag{21.98}$$

This ordinary first-order, linear, non-homogeneous differential equation (refer to Waite, 1977, p. 86) is readily solvable to

$$D(t) = \frac{K_1 L_0}{K_2 - K_1}(e^{-K_1 t} - e^{-K_2 t}) + D_0 e^{-K_2 t} \tag{21.99}$$

where $\qquad D(t) =$ oxygen deficit at any time $t = C_s - C(t)$

$L_0 =$ ultimate BOD at the discharge site

$D_0 =$ initial oxygen deficit at the BOD discharge site

Equation (21.99) is the Streeter–Phelps oxygen sag equation, and is represented in Fig. 21.22. Typically there is a need to know the minimum DO in the river and to see whether this becomes anoxic or worse. Also, it is sometimes relevant to know how far downstream it is expected before this worst condition occurs.

The maximum oxygen deficit D_c occurring at a distance x_c downstream of the point of discharge may be computed by setting $dD/dt = 0$, i.e.

$$\frac{dD}{dt} = K_1 L_0 e^{-K_1 t} - K_2 D_c = 0$$

$$D_c = \frac{K_1}{K_2} L_0 e^{-K_1 t} \tag{21.100}$$

To determine the time, t, that it takes for this to occur, Eq. (21.99) is differentiated with respect to t and this solves for t:

$$\frac{dD(t)}{dt} = \frac{K_1 L_0}{K_2 - K_1}(K_2 e^{-K_1 t} - K_1 e^{-K_2 t}) - K_2 D_i e^{-k_2 t} = 0$$

$$t = \frac{1}{K_2 - K_1}\ln\left\{\frac{K_2}{K_1}\left[1 - \frac{D_0(K_2 - K_1)}{K_1 L_0}\right]\right\} \tag{21.101}$$

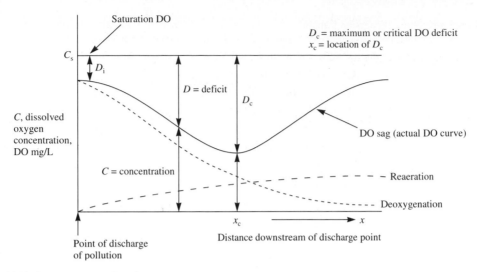

Figure 21.22 Oxygen sag curve in a river.

Further details and examples are included in Chapter 7 on water quality.

In the form presented above, the oxygen sag model is incomplete, but serves to get initial values on behaviour. Other parameters that influence the BOD/DO relationship include:

- Sedimentation
- Photosynthesis
- Re-suspension
- Advection
- Dispersion

A more sophisticated version of Eq. (21.99) to include some of the above parameters is

$$\frac{\partial L}{\partial t} = -U\frac{\partial L}{\partial x} + E\frac{\partial^2 L}{\partial x^2} - (K_1 + K_3)L + La \tag{21.102}$$

or

$$\frac{\partial C}{\partial t} = \underbrace{-U\frac{\partial C}{\partial x}}_{\substack{\text{advection}\\\text{term}}} + \underbrace{E\frac{\partial^2 C}{\partial x^2}}_{\substack{\text{dispersion}\\\text{term}}} - \underbrace{K_1 L}_{\text{deoxygenation}} + \underbrace{K_2(C_s - C)}_{\text{reaeration}} - \underbrace{B}_{\substack{\text{benthic}\\\text{rate}\\\text{of } O_2\\\text{uptake}}} \tag{21.103}$$

where L is the BOD and C is the DO concentration and La is the rate of addition of BOD along the river reach. The second-order PDE, Eq. (21.102) can be solved analytically (one dependent variable L) for the simplified steady state case $(\partial L/\partial t) = 0)$, but for the unsteady state case, it is common to use a numerical scheme—either finite difference or finite element. However, since Eq. (21.103) has two dependent variables, C and L, it cannot be solved analytically, even for the steady state. Equation (21.103) is said to be coupled to the solution of Eq. (21.102) and so Eq. (21.102) is first solved for L and then Eq. (21.103) is solved for C. The reader is referred to James (1993) for further details on the numerical solution of such coupled equations as Eq. (21.102) and Eq. (21.103).

21.6.3 QUAL2—The US EPA, Enhanced Water Quality Model

This model is briefly described in the following pages as it is in widespread use, being available in the public domain from the US EPA, Athens, Georgia. The basic equation solved by QUAL2 is the one-dimensional advection–dispersion transport equation, which is numerically integrated over space and time for each water quality constituent. This equation includes the effects of advection, dispersion, dilution, constituent reactions and interactions, and sources and sinks. For any constituent, this equation is

$$\frac{\partial M}{\partial t} = \frac{\partial(AD_L \partial C/\partial x)}{\partial x} \, dx - \frac{\partial(A\bar{u}C)}{\partial x} \, dx + S \qquad (21.48)$$

where

M = mass of constituent, mg

C = concentration, mg/L

V = volume, m^3 or L

$\bar{u}$ = mean velocity of river flow, m/s

S = external source/sink, mg/s

D_L = longitudinal dispersion (eddy) coefficient

Since

$$M = VC$$

$$\frac{\partial M}{\partial t} = \frac{\partial}{\partial t}(VC) = V\frac{\partial C}{\partial t} + C\frac{\partial V}{\partial t}$$

Assuming steady flow ($\partial Q/\partial t = \partial V/\partial t = 0$), then

$$\frac{\partial M}{\partial t} = V\frac{\partial C}{\partial t}$$

Noting that $V = A_x \, dx$, then Eq. (21.48) becomes

$$\frac{\partial C}{\partial t} = \underbrace{\frac{1}{A_x} \frac{\partial(AD_L \partial C/\partial x)}{\partial x}}_{\text{dispersion}} - \underbrace{\frac{1}{A_x} \frac{\partial(A\bar{u}C)}{\partial x}}_{\text{advection}} + \underbrace{\frac{S}{V}}_{\substack{\text{sources/} \\ \text{sinks}}} \qquad (21.104)$$

Solving Eq. (21.104) involves first writing it in finite difference form, in two steps. First the advection and diffusion terms are differentiated once with respect to x, giving

$$\frac{\partial C_i}{\partial t} = \frac{(AD_L \partial C/\partial x)_i - (AD_L \, \partial C/\partial x)_{i-1}}{V_i} - \frac{(A\bar{u}C)_i - (A\bar{u}C)_{i-1}}{V_i} + \frac{S_i}{V_i} \qquad (21.105)$$

The nodes are identified as $i-1$, i, $i+1$, with i the centre node.

Secondly, the spatial derivatives of the diffusion terms and then the time derivatives of C are expressed in finite difference terms:

$$\frac{C_i^{n+1} - C_i^n}{\Delta t} = \frac{(AD_L)_i C_{i+1}^{n+1} - (AD_L)_i C_i^{n+1}}{V_i \Delta x_i}$$

$$- \frac{(AD_L)_{i-1} C_i^{n+1} - (AD_L)_{i-1} C_{i+1}^{n+1}}{V_i \Delta x_i}$$

$$- \frac{Q_i C_i^{n+1} - Q_{i-1} C_{i+1}^{n+1}}{V_i} + \frac{S_i}{V_i} \qquad (21.106)$$

Equation (21.106) is grouped into the form

$$a_i C_{i-1}^{n+1} + b_i C_i^{n+1} + c_i C_{i+1}^{n+1} = Z_i \qquad (21.107)$$

where the coefficients a_i, b_i, c_i and z_i are all defined in the QUAL2 manual.

The governing equation (21.107), represents a set of simultaneous equations whose solution provides the value of C_i^{n+1} (the constituent concentrations) for all i's. These are accomplished in QUAL2 where the boundary conditions to be set up are

- Upstream and
- Downstream

The upstream boundary condition is typically the concentration at a point at the beginning of a river or estuarine reach. The downstream boundary condition is specified as either:

- A zero concentration gradient or
- A fixed downstream constituent concentration (usually for each constituent) of, say, 10 per cent of the upstream discharge concentration

This is not unlike the boundary conditions for the plug flow reactor of Sec. 21.4.2.

In the case of QUAL2, the hydraulic regime is assumed steady, i.e. $\partial Q/\partial t = 0$, and so

$$\left(\frac{\partial Q}{\partial x}\right)_i = (Q_x)_i = \text{sum of external inflows}$$

The longitudinal dispersion used is

$$D_L \propto Kn\,\bar{u}\,d^{5/6}$$

where
$$D_L = \text{longitudinal dispersion, m}^2/\text{s}$$
$$K = \text{dispersion coefficient (dimensionless)}$$
$$n = \text{Manning's roughness } n \text{ (dimensionless)}$$
$$\bar{u} = \text{mean velocity, m/s}$$
$$d = \text{mean depth, m}$$

Many empirical equations like the dispersion equation are used in QUAL2 and reference should be made to the manual QUAL2 (US EPA, 1987). However, it is of interest to note that much detail is included in the empirical equations for particular entities. For instance, the dissolved oxygen differential equation included is

$$\underset{\text{reaeration}}{\frac{dC}{dt} = K_2(C^* - C)} + \underset{\substack{\text{algal} \\ \text{demand}}}{(\alpha_3\mu - \alpha_3\rho)A} - \underset{\substack{\text{BOD} \\ \text{deoxygenation}}}{K_1L} - \underset{\substack{\text{sediment} \\ \text{demand}}}{\frac{K_4}{d}} - \underset{\substack{\text{ammonia} \\ \text{demand}}}{\alpha_5\beta_1N_1} - \underset{\substack{\text{nitrate} \\ \text{demand}}}{\alpha_6\beta_2N_2} \qquad (21.108)$$

All of these terms are explained in detail in the manual to QUAL2.

The reader is recommended to study the detail of this program and use it if possible. Many programs are available for water quality studies in freshwater. For instance, the US Army Corps of Engineers at HEC have a program HEC-5Q (US Army Corps of Engineers, 1986) which can examine water quality in reservoirs and lakes. The water quality capabilities include two options:

Option A: 1. Water temperature
2. Up to three conservative constituents
3. Up to three non-conservative constituents
4. Dissolved oxygen

Option B: 1. Water temperature
2. Total dissolved solids
3. Nitrate nitrogen
4. Phosphate phosphorus
5. Phytoplankton
6. Carbonaceous BOD
7. Ammonia nitrogen
8. Dissolved oxygen

21.7 WATERSHED MODELLING

Before briefly discussing some of the available software packages for watershed modelling, it is instructive to introduce some watershed modelling definitions.

21.7.1 Classification of Watershed Models

The following is adapted in part from Internet literature produced by Hydrocomp Inc., Stanford, California, (Linsley, 1976). Hydrologic simulation models use mathematical equations to calculate results like runoff volume or peak flow. These models can be classified as either theoretical or empirical models.

A theoretical model includes a set of general laws or theoretical principles. If all the governing physical laws were well known and could be described by equations of mathematical physics, the model would be physically based. However, all existing theoretical models simplify the physical system and often include obviously empirical components (e.g. the conservation of momentum equation used to describe surface flow includes an empirical hydraulic resistance term and the Darcy equation used in subsurface problems is an empirical equation), so they are considered conceptual models. An empirical model omits the general laws and is in reality a representation of data. A simplified spectrum of mathematical watershed models is shown in Fig. 21.23.

Depending on the character of the results obtained, models are classified as stochastic or deterministic. If one or more of the variables in a mathematical model are regarded as random variables having distributions in probability, then the model is stochastic. If all the variables are considered to be free from random variation, the model is deterministic (even though some 'deterministic models' may include stochastic processes to add the dimension of spatial and temporal variability to some of the subprocesses, such as infiltration).

Event versus continuous models   An *event* model is one that represents a single runoff event occurring over a period of time, ranging from about an hour to several days. The initial conditions in the watershed for each event must be assumed or determined by other means and supplied as input data. The accuracy of the model output may depend on the reliability of these initial conditions.

A *continuous watershed* model is one that operates over an extended period of time, determining flow rates and conditions during both runoff periods and periods of no surface runoff. Thus the model keeps a continuous account of the basin moisture condition and therefore determines the initial conditions applicable to runoff events. At the beginning of the run, the initial conditions must be known or assumed. However, the effect of the selection of those initial conditions decreases rapidly as the simulation advances. Most continuous watershed models utilize three runoff components: direct or surface runoff, shallow surface flow (interflow) and groundwater flow, while an event model may omit one or both of the subsurface components and also evapotranspiration.

Increasing physical information and increasing complexity			
Empirical model	Linear systems transfer functions	Explicit moisture accounting model (conceptual)	Physical process model
Example:	*Example:*	*Example:*	*Example:*
Regression equations FSR (UK)	Unit Hydrograph— SCS methods	HSPF SHE WATFLOOD	None available

Figure 21.23 Mathematical watershed models. (Adapted from Linsley, 1976, with permission.)

Scope of the model: complete versus partial models   This classification relates to what parts of the hydrologic cycle are included in the model. *Complete or comprehensive* watershed models are models for which the primary input is precipitation and other meteorologic data and the output is the watershed hydrograph. The model represents in more or less detail all hydrologic processes significantly affecting runoff and it maintains the water balance by solving the continuity equation of precipitation, evapotranspiration and runoff (i.e. the hydrologic cycle):

$$\text{Precipitation} - \text{actual evapotranspiration} = \text{runoff} \pm \text{change in storage}$$

Solving the water balance equation increases the accuracy of the model and therefore it constitutes one of the most important advantages of complete models over partial models.

A *partial* model represents only a part of the overall runoff process. For example, a water yield model gives runoff volumes but no peak discharges.

Calibrated parameter versus measured parameter models   A *calibrated parameter* model is one that has one or more parameters that can be evaluated only by fitting computed hydrographs to the observed hydrographs. Calibrated parameters are usually necessary if the watershed component has any conceptual component models, which is true for most presently used watershed models. Thus, with a calibrated parameter model, a period of recorded flow is needed, usually several years, for determining the parameter values for a particular watershed.

A *measured parameter* model, on the other hand, is one for which all parameters can be determined satisfactorily from known watershed characteristics, either by measurement or by estimation. For example, watershed area and channel length can be determined from existing maps, channel cross-sections can be measured in the field and soil characteristics can be determined in the laboratory (though not easily). Characteristics like channel roughness are often estimated. A measured parameter model can be applied to totally ungauged watersheds and is therefore highly desirable (e.g the catchment characteristics model from the UK, (SERC, 1975) see Chapter 4). However, the development of such a model that is continuous, acceptably accurate, and generally applicable is a goal that has not yet been attained.

Lumped versus distributed models   *Lumped* models do not explicitly take into account the spatial variability of inputs, outputs or parameters. They are usually structured to utilize average values of the watershed characteristics affecting runoff volume. Averaging a certain parameter also implicitly averages the process being represented. Because of non-linearity and threshold values, this can lead to significant error.

Distributed models include spatial variations in inputs, outputs and parameters. In general, the watershed area is divided into a number of elements and runoff volumes are first calculated separately for each element.

It is necessary to be careful when classifying a model according to this category. Models have often been mistakenly classified as 'lumped', even though they can represent spatial variability by subdividing the basin into segments with representative 'lumped' parameters for each segment (e.g. HSPF, on Hydrologic Simulation Program Fortran from Hydrocomp Inc., California).

General models versus special purpose models   A *general* model is one that is acceptable, without modifications, to watersheds of various types and sizes. The model has parameters, either measured or calibrated, that adequately represent the effects of a wide variety of watershed characteristics. In order to achieve this, it is generally necessary to use conceptual models which have parameters that require calibration.

A *special purpose* model is one that is applicable to a particular type of watershed in terms of topography, geology or land use, e.g. an urban runoff model. Usually, such models can be applied to watersheds of different sizes, as long as the characteristics of the watersheds are the same. This is often the case for estuarine models.

Model selection Physically based, continuous soil moisture accounting models are the most accurate models currently available and the HSPF is a most comprehensive model of this type. However, this does not mean that the HSPF is the best model to use in all circumstances. The decision to use a model and which model to use is an important part of water resource plan formulation. Even though there are no clear rules on how to select the right model to use, a few simple guidelines can be stated

1. The first step is to define the problem and determine what information is needed and what questions need to be answered.
2. Use the simplest method that can provide the answer to your questions.
3. Use the simplest model that will yield adequate accuracy.
4. Do not try to fit the problem to a model, but try to select a model that fits the problem.
5. Question whether increased accuracy is worth the increased effort. (With advances in computer technology, computational cost is hardly an issue any more.)
6. Do not forget the assumptions underlying the model used and do not read more significance into the simulation results than is actually there.

21.7.2 Generalities of Watershed Models

Catchment or watershed modelling is developed to the extent that not only are there existing software packages available in the public domain (HSPF, WATFLOOD) but there are also many available commercially (SHE and MIKE-SHE). Many were developed for a specific watershed and then expanded to be 'general'. Catchment models may include all or some of the following capabilities:

- Model the hydrodynamics and water quality of land surface runoff
- Model the groundwater
- Model the freshwater hydrodynamics and water quality

Typically, a three-dimensional topographic, grid type, model of the land surface is set up. Depending on the detail of the soil in the unsaturated and saturated zones a similar three-dimensional subsurface model is set up. The rainfall runoff relationships are established and typically time series data on rainfall or stream runoff are required for model verification. In addition to the hydrodynamics, the water quality in the various steps may be included in the model, for instance rainfall being transformed to surface runoff or groundwater flow and thence into streamflow. All steps will have different water quality depending on its travel history. For instance, rain falling on farmland, high in phosphorus, may re-suspend the phosphorus in the surface runoff and carry it to the stream. Similarly, rain infiltrating through the unsaturated zone may transport nitrates with it to the groundwater and ultimately be discharged to a stream or lake. Hence the soil chemical status is required and the chemical kinetics of rain plus excess land fertilizers and subterranean leaching is also required. Such detailed hydrodynamics and water quality modelling is analogous to the photochemical models for air pollution. While the topography/geometry of catchment models might be set up in three dimensions, the surface flow, chemical/sediment transport and streamflow is most likely to be two-dimensional and many times one-dimensional. For instance, the HSPF is essentially a one-dimensional program. An attraction with it is its continuous simulation feature, i.e. it is an unsteady state model.

In the HSPF, the various hydrologic processes are represented mathematically as flows and storages. In general, each flow is an outflow from a storage, usually expressed as a function of the current storage amount and the physical characteristics of the subsystem. Thus the overall model is physically based, although many of the flows and storages are represented in a simplified or conceptual manner. Although this requires the use of calibrated parameters, it has the advantage of avoiding the need for giving the physical dimensions and characteristics of the flow system. This reduces input requirements and gives the model its generality.

For simulation with the HSPF, the basin has to be represented in terms of land segments and reaches/reservoirs. A land segment is a subdivision of the simulated watershed. The boundaries are established according to the user's needs, but generally a segment is defined as an area with similar hydrologic characteristics. For modelling purposes, water, sediment and water quality constituents leaving the watershed move laterally to a downslope segment or to a reach/reservoir. A segment of land that has the capacity to allow enough infiltration to influence the water budget is considered pervious. Otherwise it is considered impervious. The two groups of land segments are simulated independently.

In pervious land segments the HSPF models the movement of water along three paths: overland flow, interflow and groundwater flow. Each of these three paths experiences differences in time delay and differences in interaction between water and its various dissolved constituents. A variety of storage zones are used to represent the storage processes that occur on the land surface and in the soil horizons. Snow accumulation and melt are also included, so that the complete range of physical processes affecting the generation of water and associated water quality constituents can be approximated.

Processes that occur in an impervious land segment are also simulated. Even though there is no infiltration, precipitation, overland flow and evaporation occur and water quality constituents accumulate and are removed.

The hydraulic and water quality processes that occur in the river channel network are simulated by reaches. The outflow from a reach or completely mixed lake may be distributed across several tagets to represent normal outflow, diversions and multiple gates on a lake or reservoir. Evaporation, precipitation and other fluxes that take place in the surface are also represented. Routeing is done using a modified version of the kinematic wave equation.

The HSPF is a popular model (in the United States) that can simulate the continuous, dynamic event or the steady state behaviour of both hydrologic/hydraulic and water quality processes in a watershed. The model is unusual in its ability to represent the hydrologic regimes of a wide variety of streams and rivers with reasonable accuracy. Thus, the potential applications and uses of the model are comparatively large and include flood mapping, urban drainage studies, river basin planning, studies of sedimentation and water erosion problems and in-stream water quality planning.

The SHE model from the Danish Hydraulic Institute is a deterministic, distributed and physically based modelling system for simulation of all the major hydrological processes of the land phase of the hydrological cycle. A three-dimensional groundwater component, transport modelling through the complete subterranean land phase, includes chemical reactions, irrigation effects and soil erosion. SHE is applicable to a wide range of water resources problems related to surface and groundwater management, point and non-point contamination and soil erosion. It is capable of either long-term or short-term regional studies. The spatial variation in the meteorological input and the catchment characteristics are represented in the horizontal plane by a network of square grids. Within each grid square, the soil profile is further divided into a series of horizontal layers. The river network is assumed to run along the boundaries of square grids. The model describes the processes of interception, snowmelt, infiltration, subsurface flow in the unsaturated and saturated zones, evapotranspiration and surface runoff or land and stream channels.

HEC-1 can simulate the hydrologic processes during flood events. While primarily for looking at unsteady flow in riverine systems, the preceding hydrologic studies also include it as a program capable of watershed modelling. Other programs commonly used for this purpose include: MITCAT, SWMM, ILLUDAD, STORM, USGS, DWOPER, etc., for which information is available from NTIS or the US EPA. NTIS is the US National Technical Information Services, with offices world-wide. Enquiries on the Internet will give details of the watershed modelling software available to satisfy most students' requirements.

21.8 MODELLING WATER QUALITY IN ESTUARIES

Modelling of estuaries tends to incur more approximations than are normally adopted for freshwater river flow. This is due to the added complexity caused by:

- More complex geometries
  - Changes in width
  - Changes in depth
  - Variation in cross-section bed profile
- Two-directional flow
  - Downstream freshwater flow
  - Upstream saline flow
- Wetting and drying of parts of the cross-sectional floor
- Density differences between freshwater and saline water
- More suspended sediment than freshwater systems
- More alluvial beds where sediment can become re-suspended
- Wave action generating currents at the surface

As such, not only is evaluation of the hydrodynamics more complex but the water quality, being dependent on hydrodynamics, is extremely complex due to the interaction of:

- Freshwater quality processes
- Reservoir quality processes
- Tidal/oceanic quality processes (salinity)

Fischer (1979) considered turbulent mixing in estuaries to be composed of:

- Vertical mixing with a coefficient (eddy velocity)

$$\varepsilon_v \cong 0.07 \ du* \tag{21.109}$$

- Transverse mixing with a coefficient

$$\varepsilon_t \cong 0.15 \ du* \tag{21.110}$$

- Longitudinal dispersion, which is dependent on the freshwater input flow and its momentum, the tidal interaction with the bathymetry and wind and local effects
- Dispersion by gravitational circulation, i.e. internal circulation driven by density variation and explained by an 'Estuarine Richardson number',

$$R = \frac{\Delta \rho}{\rho} \frac{g Q_f}{b u^3}$$

and that the transition from strongly stratified to well-mixed estuaries is $0.08 < R < 0.8$
- Dispersion by tide and wind
- Dispersion by shear within the vertical velocity profile

where
$$d = \text{depth}$$
$$u* = \text{root mean square (rms) tidal velocity}$$
$$\Delta \rho = \text{density difference}$$
$$Q_f = \text{freshwater flow}$$
$$b = \text{estuary average width}$$
$$u = \text{mean velocity}$$

The simplest model for longitudinal dispersion in an estuary is obtained by a mass balance of a conservative tracer:

$$A \frac{\partial C}{\partial t} + Q_f \frac{\partial C}{\partial x} = \frac{\partial}{\partial x} \left(KA \frac{\partial C}{\partial x} \right) + \text{source or sink terms} \tag{21.111}$$

where
Q_f = freshwater flow rate

C = the tidal cycle average concentration

A = the cross-sectional area

K = the mixing coefficient within the tidal cycle

The above equation is written in a pseudo-steady state for time averaged over the tidal cycle. Equation (21.111) is solved analytically for the steady state case as

$$U_f \frac{\partial C}{\partial x} = K \frac{\partial^2 C}{\partial x^2} - kC \tag{21.112}$$

where U_f is the freshwater flow rate and k is a decay rate coefficient ($dC/dt = kC$). Consider a tracer of concentration C_i discharged at $x = 0$ into an estuary. The solution for the tracer concentration from O'Connor (1960, 1965) is

At $x < 0$:

$$\text{Upstream of peak} \quad \frac{C}{C_i} = \exp\left[\frac{U_f}{2D}\left(1 + \sqrt{\frac{U_f^2 + 4kD}{U_f^2}}\right)x \right] \tag{21.113}$$

At $x > 0$:

$$\text{Downstream of peak} \quad \frac{C}{C_i} - \exp\left[\frac{U_f}{2D}\left(1 - \sqrt{\frac{U_f^2 + 4kD}{U_f^2}}\right)x \right] \tag{21.114}$$

where U_f = net downstream velocity from freshwater flow at $x = 0$, where the tracer concentration is $C = C_i$.

At $x = 0$:

$$C_i = \frac{M}{Q\sqrt{(U_f^2 + 4kD)/U_f^2}} \tag{21.115}$$

where
M = mass of non-conservative tracer/contaminant entering per time, kg/s

Q = freshwater flow rate, m^3/s

Values for the dispersion coefficient vary from 5 to $1000\,m^2/s$ and Harleman (1964) suggested:

$$D_L = 63n\, U_T R_H^{5/6} \quad m^2/s \tag{21.116}$$

where
n = Manning's roughness coefficient

U_T = maximum tidal velocity, m/s

R_H = hydraulic radius, m

Figures 21.24 and 21.25 depict the relative concentrations of tracer upstream and downstream of the discharge point in an estuarine regime. These figures can be computed from the previous set of equations.

From Fig. 21.24 it is seen that for line 1, with the higher net downstream velocities, the upstream concentration is less but the downstream concentration is higher. This means that the higher velocities help to dilute the upstream concentrations and that they only transport but weakly dilute the downstream concentrations. From Fig. 21.25, examining the effect of two different dispersion rates, it is seen that the higher dispersion rate causes increased concentration upstream, but has little effect downstream of the discharge point.

When wastes (with BOD strengths = L) are discharged to estuaries, they negatively impact on the dissolved oxygen levels. As such, the profile of DO will depend on the BOD loading. From O'Connor

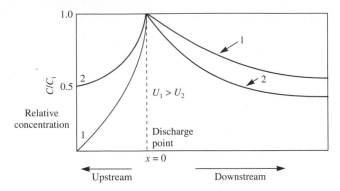

Figure 21.24 Schematic of distribution of reactant with respect to initial concentration (C_i) for two different velocities and similar k and D.

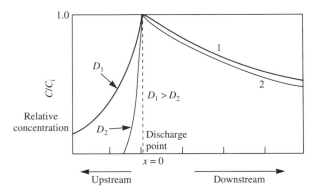

Figure 21.25 Schematic of distribution of reactant with respect to initial concentration (C_i) for two different D values and similar k and V.

(1965), the one-dimensional, differential equations for steady flow are:

For BOD :
$$D_L \frac{d^2L}{dx^2} - U_f \frac{dL}{dx} - kL = 0 \tag{21.117}$$

For DO :
$$D_L \frac{d^2D}{dx^2} - U_f \frac{dD}{dx} - K_2 D + kL_x = 0 \tag{21.118}$$

where
$$D = \text{dissolved oxygen deficit, mg/L}$$
$$L = \text{BOD remaining, mg/L}$$
$$k = \text{reaction coefficient, d}^{-1}$$
$$K_2 = \text{reaeration coefficient, d}^{-1}$$
$$L_x = \text{BOD concentration at point } x, \text{mg/L}$$
$$D_L = \text{mixing dispersion coefficient}$$

O'Connor's solutions are:

$$L = \frac{M}{Q m_1} \exp\left[\frac{U_f}{2 D_L} (1 \pm m_1) x \right] \tag{21.119}$$

$$D = \frac{kM}{(K_2 - k)Q} \left\{ \frac{1}{m_1} \exp\left[\frac{U_f}{2 D_L} (1 \pm m_1) x \right] - \frac{1}{m_2} \exp\left[\frac{U_f}{2 D_L} (1 \pm m_1) \right] \right\} \tag{21.120}$$

where
$$m_1 = \left(\frac{U_f^2 + 4kD_L}{U_f^2}\right)^{1/2} \tag{21.121}$$

and
$$m_2 = \left(\frac{U_f^2 + 4k_2D_L}{U_f^2}\right)^{1/2} \tag{21.122}$$

and
$$M = \text{mass of BOD pollutant per time, kg/s}$$

The positive sign in the exponent of Eqs. (21.119) and (21.120) refers to $x < 0$ and the negative sign refers to the case of $x > 0$ (downstream).

More complex models involving two- and three-dimensional and non-conservative contaminants are available, but the solutions are numerical. The reader is referred to Tchobanoglous and Schroeder (1987), James (1993), Fischer *et al.* (1979) and Orlob (1981) for more detailed treatment of estuarine water quality. Many programs exist for computer modelling of flow in estuaries including King (1990), Leenderste (1970) and Falconer (1991).

21.9 MODELLING WATER QUALITY IN LAKES AND RESERVOIRS

Water quality in lakes was introduced in Chapter 7. This section concerns itself with the modelling of the dynamics of long residence times in water bodies, natural or man-made. The key driving force of lake dynamics is usually temperature. Its vertical distribution defines whether a lake is stratified or not. The mathematical modelling of lakes is based on the following equations:

- Conservation of mass
- Conservation of momentum
- Transport of contaminants
- Chemical and biological process kinetics
- Conservation of heat

The first two equations are the foundations of all hydrodynamic problems, while the third is used in the water environment and other environments (air, soil, biota). The fifth, the conservation of heat, has not been used so far in this text and while it may be applicable to the air environment, it is usually not of relevance to rivers. Usually lakes require to be modelled when there are problems with water quality and this most often occurs in both shallow and deep lakes when they are stratified. Stratification (as discussed in Chapter 7) is when there is a significant temperature difference between the epilimnion (upper waters) and the hypolimnion (lower waters). The narrow band between both with a steep temperature gradient is the thermocline. If the densiometric Froude number (defined in Chapter 7) is < 0.01, then the water body is stratified. The physical processes involved in the heat budget of a lake are depicted in Fig. 21.26.

One technique to model stratified lakes is to divide the lake into a number of horizontal layers as shown in Fig. 21.27. The more simplistic model will have three 'layers': the epilimnion, the thermocline and the hypolimnion. More capable models will have several layers. On each layer, a mass balance is computed, allowing for the transfer of flow and constituents from one layer (vertically) to another. For instance, the mass flow balance of the jth (horizontal) layer is

$$\frac{\partial V_j}{\partial t} = Q_{j\text{in}} + (Q_{j-1} + Q_{j+1}) + P \qquad \text{Inflow}$$
$$- E - Q_{j\text{out}} - Q_j \qquad \text{Outflow}$$
$$\pm Q_g \qquad \text{Groundwater} \tag{21.123}$$

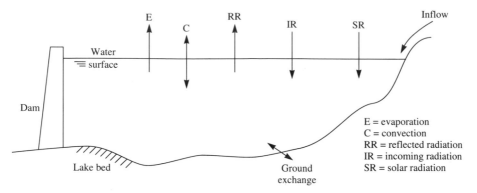

Figure 21.26 Conceptual model of heat budget of a lake.

where

V_j = volume of the jth layer

$Q_{j\,in}$, $Q_{j\,out}$ = horizontal surface inflow and outflow (abstraction) from the jth layer

Q_j = outflow vertically from the jth layer

Q_{j-1}, Q_{j+1} = inflow vertically to the jth layer from the $j-1$ and $j+1$ layers

Several of the above terms require information on the heat budget of the lake as depicted in Fig. 21.25. Assuming that P, E and Q_g are included in the vertical fluxes (Q_{j-1} and Q_{j+1}) then the mass conservation for a typical layer of thickness Δz is

$$\frac{\partial V_j}{\partial t} = Q_{j\,in} \pm (Q_{j-1} + Q_{j+1})_z - Q_{j\,out} \qquad (21.124)$$

where

$Q_{j\,in}$, $Q_{j\,out}$ = horizontal flow advection

Q_{j-1}, Q_{j+1} = vertical flow (into or out of the element)

The conservation of heat equation stored in a horizontal control volume layer is described by Orlob (1981). Solution of the mass balance Eq. (21.124) requires information on the heat energy balance. The end result of modelling of reservoirs is the production of vertical profiles of temperature to determine the lake status for different inputs of solar heat and freshwater flows.

Detailed descriptions of the above and solution techniques are to be found in Orlob (1981), Tchobanoglous and Schroeder (1987), US Army Corps of Engineers (1986), James (1993) and Fischer *et al.* (1979). A computer program DYRESM detailed by Fischer *et al.* (1979) and Imberger (University of West Australia) is commercially available to solve lake hydrodynamic and lake water quality problems.

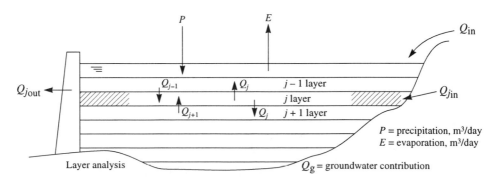

Figure 21.27 Conceptual model of a stratified lake.

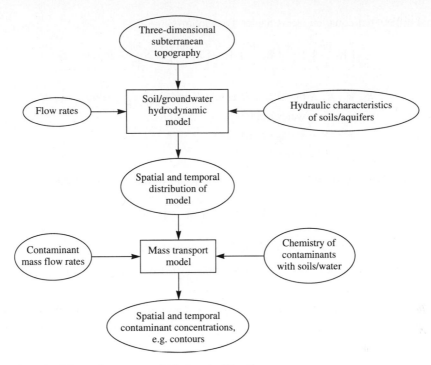

Figure 21.28 Schematic of flow and transport model in the subsoil/aquifer.

These models are used to predict temperature profiles in stratified lakes and are ideal for lake water quality management in identifying incipient seasonal eutrophication. Complex models may extend to two and three dimensions and/or incorporate reaction chemistry and ecological models. Recent work by Casamitjana and Schladow (1993) has added particle movement analysis to the existing hydrochemical lake models, DYRESM.

21.10 GROUNDWATER MODELLING

Groundwater flow is reasonably well defined for aquifer (rock) structures. However, water flow is extremely complex in the unsaturated soil layers above the rocks. This zone of subsoil is sometimes called the unsaturated zone as it is subject to wetting and drying, depending on many parameters not least of all rainfall and infiltration. Correspondingly, then, the flow of solutes (pollutants) in soil water/groundwater is definable for aquifers but is complex in the unsaturated zone. This section will briefly examine modelling with respect to:

- Flow–hydrodynamics and
- Water quality—mass transport.

Both are represented in the schematic of Fig. 21.28.

21.10.1 Flow Modelling in Groundwater

For flow modelling, the Laplace equation combines the continuity equation and Darcy's law into a second-order partial differential equation (see Chapter 4) as follows:

$$\frac{\partial}{\partial x}\left(K_x \frac{\partial h}{\partial x}\right) + \frac{\partial}{\partial y}\left(K_y \frac{\partial h}{\partial y}\right) + \frac{\partial}{\partial z}\left(K_z \frac{\partial h}{\partial z}\right) = S\frac{\partial h}{\partial t} - R(x, y, z, t) \tag{21.125}$$

where K_x, K_y, K_z = the hydraulic conductivity in the x, y, z directions

h = water head

S = storage coefficient

R = recharge

For the two-dimensional case with steady flow and no recharge this equation reduces to

$$\frac{\partial}{\partial x}\left(K_x \frac{\partial h}{\partial x}\right) + \frac{\partial}{\partial z}\left(K_x \frac{\partial h}{\partial z}\right) = 0 \tag{21.126}$$

Typically $K_y \cong K_x$ in the horizontal plane. If $K_x = K_z$ then this reduces to

$$\frac{\partial^2 h}{\partial x^2} + \frac{\partial^2 h}{\partial z^2} = 0 \tag{21.127}$$

Finite difference models commonly used in the past have been those developed by Prickett and Lonnquist (1971) and Trescott *et al.* (1976). Both of these models solved a form of the unsteady flow equation which allowed for heterogeneous and anisotropic soils/aquifers. The equation solved by them was

$$\frac{\partial}{\partial x}\left(K_x \frac{\partial h}{\partial x}\right) + \frac{\partial}{\partial z}\left(K_z \frac{\partial h}{\partial z}\right) = S\frac{\partial h}{\partial t} - R(x, z, t) \tag{21.128}$$

McDonald and Harbaugh (1984) extended this model to that of three dimensions. The reader is referred to Wang and Anderson (1982), Bedient and Huber (1988) and McDonald and Harbaugh (1984) for further details and model applications.

21.10.2 Contaminant Transport

Extensive modelling effort has been applied to the mass transport of contaminants in soil/groundwater over the past two decades. The mechanism of pollutant transport depends on hydraulic conductivity of the soil/aquifer. If the hydraulic conductivity is very low, as in some aquifers and clays, then the transport mechanism may be primarily by diffusion. For high conductivity, advection is the dominant transport mechanism. The problem of transport becomes more complex if the contaminant is a reactive chemical. In such a situation, chemical reaction rates and even microbial rates may need to be considered. In the case of bioremediation of contaminated soils, using specific microbes the reaction rates for microbes are significant and are considered in the BIOPLUME II model by Bedient and Rifai (1993).

The one-dimensional advection–dispersion equation derived in this chapter applies also to the movement of a contaminant in the subsurface environment:

$$\frac{\partial C}{\partial t} = D\frac{\partial^2 C}{\partial x^2} - u\frac{\partial C}{\partial x} \tag{21.129}$$

where C = concentration of non-reactive contaminants, g/m^3

D = hydrodynamic dispersion

u = average fluid velocity

and
$$D\frac{\partial^2 C}{\partial x^2} = \text{Fick's second law for diffusive/dispersive flow}$$

and
$$u\frac{\partial C}{\partial x} = \text{the mass flow or convective advective flow}$$

D has many definitions, particularly in the context of solute movement in the ground:

$$D = D_0\tau + \alpha v \qquad (21.130)$$

where
$$D_0 = \text{free solution diffusion coefficient}$$
$$\tau = \text{tortuosity factor}$$
$$\alpha = \text{dispersive parameter}$$
$$v = \text{flow velocity}$$

Figure 21.29 indicates the variation of longitudinal and molecular dispersion with respect to flow, where D_L is the longitudinal dispersion coefficient, D_0 is the molecular diffusion coefficient and d is the average grain diameter. It is seen that at small velocity the molecular diffusion dominates, while at higher velocity rates advection dominates; there is a transition zone between the two while this is to be expected, it is seen from Fig. 21.29 that greater than one or two orders of magnitude can be the difference in relative dispersion rates.

There are analytical solutions for some specific problems in groundwater flows. Consider the case with the following boundary conditions, where a non-reactive contaminant is applied to a soil column (in a laboratory) as a step function C_0. The boundary conditions $C(x, t)$ are

$$C(x, 0) = 0 \text{ for } x \geq 0$$
$$C(0, t) = C_0 \text{ for } t \geq 0$$
$$C(\infty, t) = 0 \text{ for } t \geq 0$$

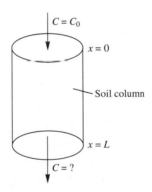

The task is to solve the advective diffusion, Eq. (21.129) for the above boundary condition.

The Ogata and Banks (1961) solution for this is

$$\frac{C}{C_0} = \frac{1}{2}\left[\operatorname{erfc}\left(\frac{x - \bar{v}t}{\sqrt{4Dt}}\right) + \exp\left(\frac{\bar{v}x}{D}\right)\operatorname{erfc}\left(\frac{x - \bar{v}t}{\sqrt{4Dt}}\right)\right] \qquad (21.131)$$

where
$$C_0 = \text{the initial contaminant concentration, g/m}^3$$
$$C = \text{the concentration at any distance } x > 0$$
$$v = \text{average velocity, m/s}$$
$$D = \text{dispersion coefficient}$$
$$\operatorname{erfc} = \text{complementary error function}$$

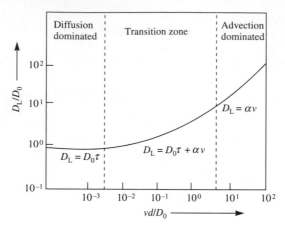

Figure 21.29 Longitudinal and molecular diffusion versus flow in subsoil/aquifer.

Sometimes the second term in the above equation is dropped if x is large or if t is great, so the reduced form is

$$\frac{C}{C_0} = \frac{1}{2}\left[\operatorname{erfc}\left(\frac{x - \bar{v}t}{\sqrt{4DT}}\right)\right] \tag{21.132}$$

Example 21.5 Compute the relative concentration (C/C_0) if $\bar{v} = 5\,\text{cm/h}$, $D = 2\,\text{cm}^2/\text{h}$ for $x = 10\,\text{cm}$ and $t = 1.6\,\text{h}$.

Solution

$$\operatorname{erfc}\left(\frac{x - \bar{v}t}{\sqrt{4DT}}\right) = \operatorname{erfc}\left(\frac{10 - 5 \times 1.6}{\sqrt{4 \times 2 \times 1.6}}\right) = 0.4002$$

$$\exp\left(\frac{\bar{v}x}{D}\right) = \exp\left(\frac{5 \times 10}{2}\right) = 7.2 \times 10^{10}$$

$$\operatorname{erfc}\left(\frac{x + \bar{v}t}{\sqrt{4DT}}\right) = \operatorname{erfc}\left(\frac{10 + 5 \times 1.6}{\sqrt{4 \times 2 \times 1.6}}\right) = 0$$

$$\frac{C}{C_0} = \frac{1}{2}\left[\operatorname{erfc}\left(\frac{x - \bar{v}t}{\sqrt{4DT}}\right)\right], \qquad \text{i.e. reduced form}$$

$$= 0.2001$$

Fig. 21.30 shows the breakthrough curves of the relationship of C/C_0 with respect to advection, molecular diffusion and dispersion. If there is only advection, the contaminated wetting front will advance in a 'square' form. For the above example, at $\bar{v} = 4\,\text{cm/h}$ and $t = 1.6\,\text{h}$ the front has travelled a distance of 3.13 cm to line (A). If there is also molecular diffusion, curve (B) is found. If in addition there is hydrodynamic dispersion then the front advances to curve (C).

In addition to advection–diffusion transport, the following mechanisms may also occur:

- Contaminant reaction with water
- Transport with adsorption to soil particles or water particles
- Transport with volatilization
- Transport with precipitation

Further details on these processes may be found in Crank (1956), Ghadiri and Rose (1992) and Richter (1990).

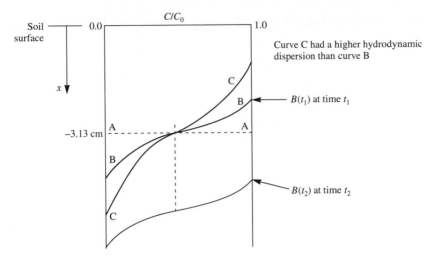

Figure 21.30 Breakthrough curves for contaminant flow through soil/aquifer.

21.10.3 Modelling Water Balance in Landfills

A computer program available through the US EPA, for landfill design and water balance of landfills, is called HELP (hydrologic evaluation landfill program). The input to the program is data on rainfall (time series), landfill properties (porosity, soil moisture, hydraulic conductivity, etc.). The output is a time series distribution of the rainfall distributed amongst in to infiltration, evapotranspiration and runoff, if it exists.

The input data required are:

- Climate
  - Daily precipitation
  - Mean monthly solar radiation
  - Mean monthly temperature

- Soil
  - k_{sat} (saturated hydraulic conductivity)
  - Porosity
  - Evaporation coefficient
  - Field capacity
  - Wilting point
  - Maximum infiltration rate
  - Curve number (this is a US term indicative of the percentage of runoff from rainfall)
- Design
  - Number of layers
  - Layer thickness
  - Layer slope
  - Lateral flow distance
  - Surface area of landfill
  - Leakage fraction
  - Runoff fraction from waste

The program as set up for the United States has a database of precipitations for 184 cities (in the US and Canada) allows up to 12 layers of soil and up to 4 soil liners. This program is simple to use and can be used both at the design stage of landfill or the operational stage to determine the water balance characteristics.

21.11 MODELLING OF WASTEWATER TREATMENT: ACTIVATED SLUDGE

There are a multitude of wastewater treatment processes, some described in Chapter 12 and all amenable to computer modelling. However, this chapter only considers the most common biological process—activated sludge. Several computer programs are available, including UCTOLD from the University of Capetown in South Africa, IAWPRC from the International Association of Water Quality and SSSP from Clemson University. Figure 21.31 is a schematic of the activated sludge process, where Q is the flow rate in m^3/d, S is the substrate (food) concentration in mg/L and X is the biomass concentration in mg/L.

From Chapter 12 the equations for mass balances with recycle are:

$$V\frac{dX}{dt} + (Q_0 + Q_r)X - V(\mu - kd)X - Q_r X_u = 0 \tag{21.133}$$

$$V\frac{dS}{dt} + (Q_0 + Q_r)S - \frac{V\mu X}{Y} - Q_r S - Q_0 S_0 = 0 \tag{21.134}$$

These are the two basic differential equations which can be solved as simultaneous equations. The more simple equations were solved in Chapter 12 for the steady state case. Here, the initial conditions can be obtained by first solving the steady state case. Further complexities to account for the nitrification step can also be added. The reader is referred to James (1984) for further details. The programs UCTOLD and IAWPRC can be used to predict the response of single or in series completely mixed reactors with or without recycle. The programs predict the response of the following compounds:

- COD
- Oxygen
- VSS (volatile suspended solids)
- Nitrogen
- Alkalinity

The models simulate the biological system behaviour, accounting for a large number of processes. The models quantify, for each process, both the kinetics (rate–concentration dependence) and the stoichiometry (effects on the masses of compounds involved). The representation of the process model is in matrix format, explained in Dold *et al.* (1991).

Both programs are available and can be used in the design stage for a plant or in the operational stage to model the efficiencies. Readers are recommended to practice on one or both of the programs.

21.12 FUGACITY MODELLING

With the many thousands of chemicals available today, some find their way into the different environments (air, water, soil). As such we are interested in quantifying, if say a chemical or pesticide is accidentally released or applied in agriculture, where the remains of that chemical go. We know that some of it stays in the air, some in the water, some in the soil and some in the biota. The fate of chemicals in the

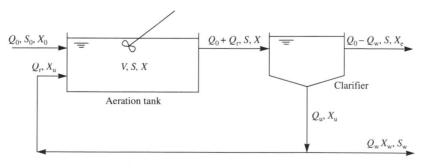

Figure 21.31 Schematic of conventional activated sludge system.

environment may to some extent be predicted using mathematical models. Fugacity models describe the chemical's partitioning, transport and transformation processes. The concept of *fugacity* is the tendency of a given molecular species to escape from one phase to another. So the higher the fugacity, the lower the chemical concentration. Its units are pressure. Fugacity is related to the chemical concentration by

$$f = \frac{C}{Z} \tag{21.135}$$

where
f = fugacity, Pa

C = concentration of the chemical, mol/m^3

Z = fugacity capacity, representing the solubility
of the chemical in the responsive state, mol/m^3 Pa

For example, if two phases are in equilibrium, i.e. a gas and a liquid phase, then

$$f_1 = f_2$$

Therefore

$$\frac{C_1}{Z_1} = \frac{C_2}{Z_2}$$

Define partition coefficient as

$$K_{12} = \frac{C_1}{C_2} = \frac{Z_1}{Z_2}$$

A chemical is in equilibrium with several phases, if

$$f_{\text{air}} = f_{\text{water}} = f_{\text{soil}} = f_{\text{sediment}} = f_{\text{biota}}$$

where f_{air} = fugacity of the chemical in the air environment.

The fugacity capacities of a chemical in the different environments are:

In air:
$$Z_{\text{air}} = \frac{1}{RT} \tag{21.136}$$

In water:
$$Z_{\text{water}} = \frac{1}{H} = \frac{C^{\text{s}}}{P_{\text{s}}} \tag{21.137}$$

In solid (soil, etc.):
$$Z_{\text{soil}} = \frac{K_{\text{sw}}\rho_{\text{s}}}{H} = K_{\text{sw}}\rho_{\text{s}}(Z_{\text{water}}) = K_{\text{oc}}Z_{\text{w}} \tag{21.138}$$

In biota:
$$Z_{\text{biota}} = \frac{K_{\text{fw}}\rho_{\text{f}}}{H} - K_{\text{fw}}\rho_{\text{f}}(Z_{\text{water}}) \tag{21.139}$$

where
$R = 8.314$ Pa m^3/mol K (gas constant)

T = temperature, °K

C^{s} = aqueous solubility, mol/m^3

P^{s} = vapour pressure, Pa

H = Henry's law constant, Pa m^3/mol

K_{sw} = partition coefficient—soil to water, L/kg

ρ_{s} = density of soil, kg/L

K_{fw} = bioconcentration factor—biota to water, L/kg

ρ_{f} = density of biota, kg/L

K_{oc} = organic carbon partition coefficient, L/kg

Example 21.6   Consider the hypothetical chemical with the following properties:

$$\text{Molecular weight } (M) = 150 \, \text{g/mol}$$
$$\text{Aqueous solubility } (C^s) = 50 \, \text{g/m}^3 = 0.333 \, \text{mol/m}^3$$
$$\text{Vapour pressure}(P^s) = 1\text{Pa}$$
$$\text{Temperature} = 25 \, ^\circ\text{C} = 298 \, \text{K}$$
$$\text{Linear sorption coefficient } (K_p) = 75.9 \, \text{L/kg}$$
$$\text{Bioconcentration factor } (K_B) = 188.4 \, \text{L/kg}$$
$$\text{Soil density } (\rho_s) = 1.5 \, \text{kg/L}$$
$$\text{Biota density } (\rho_B) = 1.0 \, \text{kg/L}$$

Determine the fugacity capacities in the various environments and then the concentration in the different environments.

Solution

$$Z_{\text{air}} = \frac{1}{RT} = \frac{1}{8.32 \times 298} = 404 \times 10^{-4} \, \text{mol/m}^3 \, \text{Pa}$$

$$Z_{\text{water}} = \frac{1}{H} = \frac{C^s}{P^s} = \frac{0.33}{1} = 0.333 \, \text{mol/m}^3 \, \text{Pa}$$

$$Z_{\text{soil}} = \frac{K_p \rho_s}{H} = K_p \rho_s Z_w$$

$$= 75.9 \times 1.5 \times 0.333 = 37.9 \, \text{mol/m}^3 \, \text{Pa}$$

$$Z_{\text{biota}} = \frac{K_B \rho_B}{H} = K_B \rho_B Z_w$$

$$= 188.4 \times 1.0 \times 0.333 = 62.8 \, \text{mol/m}^3 \, \text{Pa}$$

For this chemical, to determine the chemical concentrations in the four environments, assuming each environment was of equal volume, then the highest concentrations align with the lowest fugacities, as shown below (but volumetrically all environments are very different):

$$\rightarrow \text{ increasing fugacity capacities}$$
$$\text{Air} \rightarrow \text{ water} \rightarrow \text{ soil} \rightarrow \text{ biota}$$
$$\leftarrow \text{ increasing chemical concentrations}$$

A simplistic model reproduced here, developed by McCall *et al.* (1982), is of chemical partitioning in model ecosystems. The ecosystem data are reproduced in Fig. 21.32. Table 21.2, reproduced from Tanji (1994), illustrates the fate of the hypothetical chemical of Example 21.6 in the unit world environment of Fig. 21.32, where

$$f = \frac{M}{\Sigma V_i Z_i} = \frac{100 \, \text{mol}}{7.32 \times 10^6 \, \text{mol/Pa}} = 1.363 \times 10^{-5} \, \text{Pa}$$

$$m_i = f \, V_i Z_i$$

and
$$C_i = m_i / V_i$$

It is noted that the greatest 'dilution' is in the air and water environments and least in the biota and soil/sediments. Therefore the highest concentrations occur in the biota at 860 ppm.

There have been models developed from fugacity fundamentals. Such equilibrium models include those developed by the US EPA such as: EXAMS, HydroQual, TOX-SCREEN, CREAMS and ENPART.

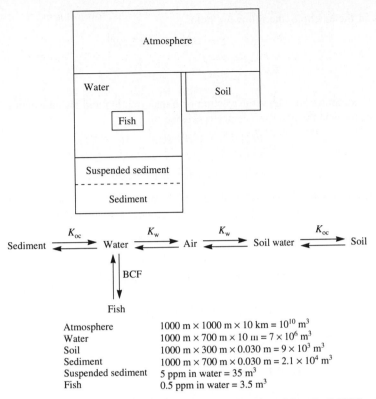

$$
\begin{array}{ll}
\text{Atmosphere} & 1000\ m \times 1000\ m \times 10\ km = 10^{10}\ m^3 \\
\text{Water} & 1000\ m \times 700\ m \times 10\ m = 7 \times 10^6\ m^3 \\
\text{Soil} & 1000\ m \times 300\ m \times 0.030\ m = 9 \times 10^3\ m^3 \\
\text{Sediment} & 1000\ m \times 700\ m \times 0.030\ m = 2.1 \times 10^4\ m^3 \\
\text{Suspended sediment} & 5\ ppm\ in\ water = 35\ m^3 \\
\text{Fish} & 0.5\ ppm\ in\ water = 3.5\ m^3
\end{array}
$$

Figure 21.32 Model ecosystem of a similar type to that of McCall *et al.* (1982), adapted from Tanji (1994), with permission.

There are various levels of sophistication in fugacity modelling, defined by Paterson (1985) as levels I, II, III and IV.

The level I computations are the simplest and the assumption (as in the previous example) is that each compartment of the unit world is well mixed and there are no reactions or advection in or out of the system. At equilibrium:

$$f_1 = f_2 = f_3 = \cdots = f_i$$

where $1, 2, \ldots, i$, are the compartments of the unit world and a total amount of chemical M(mol) is

Table 21.2 Fate of hypothetical chemical in a model environment

Environment compartments	Environment volume V_i (m³)	Fugacity capacity Z_i (mol/m³ Pa)	$V_i Z_i$ (mol/Pa)	Mass m_i (mol)	Concentration m_i/V_i (mol/m³)
Air	10^{-10}	4.03×10^{-4}	4.03×10^6	55.0	5.5×10^{-9}
Water	7×10^6	0.333	2.33×10^6	31.8	4.5×10^{-6}
Sediments	2.1×10^4	37.9	0.79×10^6	10.9	5.2×10^{-4}
Biota	3.5	62.8	220	0.003	8.6×10^{-4}
Suspended sediment	35	37.9	1330	0.018	5.2×10^{-4}
Soil	9×10^3	1.9×10^1	1.7×10^5	2.33	2.6×10^{-4}
Total			7.3×10^6	100	

Adapted from Tanji, 1994, with permission

introduced. As in the example, the following hold:

$$M = \Sigma m_i = \Sigma C_i V_i = f\Sigma V_i Z_i$$
$$m_i = f V_i Z_i$$
$$C_i = f Z_i$$

The level II computations allow for reaction or transformation and also advection and so residence times are significant parameters. The average residence time is

$$\tau_R = \frac{\text{total amount in system}}{\text{total reaction rate}} = \frac{M}{E} = \frac{\Sigma V_i C_i}{\Sigma V_i C_i k_i}$$

where k_i = reaction rates for each compartment

The level III computations consider a steady state, non-equilibrium model. Recall that the previous two cases were equilibrium models. Therefore, for the non-equilibrium case:

$$f_i \neq f_2 \neq f_3 \neq \ldots \neq f_i$$
$$N = D_{ij}(f_i - f_j)$$

where D_{ij} is the transport or diffusion rate coefficient in mol/h Pa between compartments and N is the flux in mol/h.

The level IV fugacity model is a dynamic version of level III where emissions and thus concentrations vary with time.

21.13 AIR QUALITY MODELLING

Air quality modelling is used to predict air quality and assist with policy and planning decisions with respect to industrial and infrastructural development and management. Air quality as a systems analysis is represented in Fig. 21.33.

Air quality modelling was briefly introduced in Chapter 8. The science of modelling of air quality and specific constituents is extremely sophisticated, particularly where adequate input data on meteorological, topographical and chemistry constituents is available. Many urban areas, like the Los Angeles authorities, have detailed air modelling capacities, traditionally because of such serious air pollution problems. The classes of models include:

- Simple deterministic
- Statistical
- Local plume and puff
- Box and multibox
- Finite difference and grid
- Particle

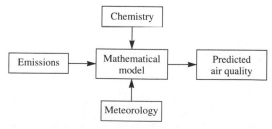

Figure 21.33 Air quality system analysis.

- Physical—wind tunnels
- Regional

as categorized by Szepesi (1989).

21.13.1 Simple Deterministic Models

These are based on empirical data and formulated in terms of algebraic relationships. These include:

- Modelling of air pollution indices
- Area source models

Air pollution index modelling is typically based on a function α, where α is ascribed a number indicating good quality, satisfactory quality, unhealthy quality and hazardous. This index is used sometimes in the United States at a level understandable by the public and is called the PSI, or Pollution Standards Index. The function α may be related to a specific parameter like CO, smog or SO_2 or any parameter that is listed in the Air Quality Standards (see Chapter 8) or to a group of parameters. This modelling method takes the weighted values of individual pollutant parameters measured at spatial points and then compares this to the single number in the Air Quality Standards. The attraction of this model is that the number α is a non-dimensional number (not ppm or mg/m^3) and a value of, say, 400 for CO or SO_2 may be deemed hazardous for both, even though the corresponding concentrations may be, say, 400 ppm (CO) and 1 ppm (SO_2). The non-dimensional α (an interpolation equation) can be obtained from

$$\alpha = \alpha_i + \frac{\alpha_{i+1} - \alpha_1}{C_{i+1} - C_i}(C - C_i) \tag{21.140}$$

where

α = Pollutant Standards Index

C = corresponding pollutant concentration

α_1 = breakpoint PSI from one quality to another,
say 100 between satisfactory and unhealthy quality

Example 21.7   Determine the Pollutant Standard Index for SO_2 if the concentration is 0.9 ppm. The breakpoints are:

PSI	Description	Concentration (ppm)
< 50	Good quality	< 0.07
50	Satisfactory	0.14
100	Unhealthy	0.3
200	Hazardous—alert	0.3
300	Hazardous–warning	0.6
400	Hazardous—emergency	0.8
500	Hazardous—serious harm	1.0

Solution

$$\alpha = \alpha_1 + \frac{\alpha_{i+1} - \alpha_1}{C_{i+1} - C_i}(C - C_i)$$

$$= 400 + \frac{500 - 400}{1 - 0.8}(0.9 - 0.8) = 450$$

$$PSI = 450$$

In the same way, a concentration of 45 ppm CO converts to a PSI of 450, numbers readily understood by the public. For further details, refer to Szepesi (1989).

The simple area source model of Gifford and Hanna (1974) is

$$C = \frac{C_i Q}{U} \tag{21.141}$$

where
C = time averaged pollutant concentration, mg/L

Q = source strength per unit area, kg/m^2

U = annual average wind speed, m/s

C_i = parameter proportional to the city size and specific pollutant

Example 21.8 Determine the annual average concentration of SO$_2$ due to the hypothetical coal-fired power station of Example 8.11 if $C_i \cong 1$, $Q \sim 1.1$ kg/s and $U \sim 5$ m/s. Assume this affects an area of 100 km^2.

Solution

$$Q = 1.1 \text{ kg/s}/100 \text{ km}^2$$
$$= 0.011 \text{ kg/s/km}^2$$
$$= 11\,000 \text{ mg/s/km}^2$$
$$= 0.011 \text{ mg/s/m}^2$$
$$C = \frac{C_i Q}{U}$$

Therefore,

$$C = \frac{1 \times 0.011}{5}$$
$$= 0.0022 \text{ mg/m}^3$$
$$= 2.2 \ \mu\text{g/m}^3$$

If this affected an area of 10 km^2 then the concentration goes up to 22 μg/m^3. The annual limit standard for SO$_2$ (from Chapter 8) is ~ 60 μg/m^3.

21.13.2 Box Models

This is a common simple model used to get an initial estimate of concentration values. It is based on the mass conservation of a pollutant in a box. The reference frame is Eulerian, i.e. fixed frame, rather than Lagrangrian, i.e. reference frame moving with the velocity of the pollutant. The box or volume may represent a city or region as defined in Fig. 21.34. The plan area over a city is represented by $\Delta x \, \Delta y$ and Δz is the vertical dimension of the airshed.

Consider the wind entering the airshed with a velocity U and a concentration C_{in}. Assume no pollutants leave the side walls of the box and full mixing occurs within the box. The pollutants for simplicity are assumed to be conservative (i.e. the generation/decay rate $\to 0$):

$$\begin{array}{c} \text{Rate of change} \\ \text{of pollution} \\ \text{in box} \end{array} = \begin{array}{c} \text{rate of pollution} \\ \text{entering} \end{array} - \begin{array}{c} \text{rate of pollution} \\ \text{leaving} \end{array} + \begin{array}{c} \text{rate of pollution} \\ \text{generation or decay} \end{array}$$

$$\Delta x \, \Delta y \, \Delta z \frac{\partial C}{\partial t} = \Delta x \, \Delta y q_s + \Delta y \, \Delta z U (C_{in} - C_{out}) + \Delta x \, \Delta y \, \Delta z \, r_g \tag{21.142}$$

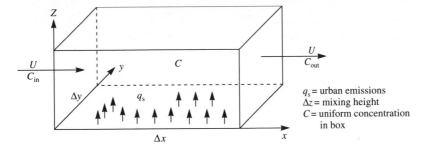

Figure 21.34 Schematic of a box model in the atmosphere.

For no generation within the box, $r_g = 0$,

$$\frac{\partial C}{\partial t} = \frac{q_s}{\Delta z} + \frac{U}{\Delta x}(C_{in} - C_{out}) \tag{21.143}$$

Assuming constant concentration throughout the box, then $C_{out} = C$ (similar to a CFSTR tank), and taking the limit,

$$\frac{dC}{dt} = \frac{q_s}{H} + \frac{U}{L}(C_{in} - C) \tag{21.144}$$

where

$$\Delta x = L = \text{length of box}$$
$$\Delta z = H = \text{mixing height}$$

The steady state case reduces to

$$C = \frac{q_s L}{UH} + C_{in} \tag{21.145}$$

The unsteady case becomes

$$\frac{dC}{dt} + \left(\frac{U}{L}\right)C = \frac{q_s}{H} + \frac{U}{L}C_{in} \tag{21.146}$$

$$\frac{L}{U}\frac{dC}{dt} + C = \frac{q_s L}{UH} + C_{in}$$

and the analytic solution is:

$$C(t) = \left(\frac{q_s L}{UH} + C_{in}\right)(1 - e^{(-Ut)/L}) + C(0)e^{(-Ut)/L} \tag{21.147}$$

If there is only pollution entering the box from city emissions and the initial concentration in the box is zero, then Eq. (21.129) reduces to

$$C(t) = \frac{q_s L}{UH}(1 - e^{(-Ut)/L}) \tag{21.148}$$

Example 21.9 Consider an urban area with a population of 150 000, and 50 000 vehicles, trafficking within a 100 km^2 area with an average travel distance of 10 km from 8 to 10 a.m. daily. Assume each vehicle emits 4.0 g/km of CO. Determine the CO concentration.

Solution

$$C(t) = \frac{q_s L}{UH}(1 - e^{(-Ut)/L})$$

$$\text{Time } t = 2\,\text{h} \ (7200\text{ s})$$
$$L = 10\,\text{km (assume a square box)}$$

Assume
$$H = 30\,\text{m}$$
and wind speed,
$$U = 2\,\text{m/s}$$
$$q_s = \text{emission rate per m}^2$$
$$= \frac{50\,000 \times 10 \times 4}{100 \times 10^6 \times 7200} = 2.6 \times 10^{-6}\ \text{g/m}^2\,\text{s}$$

Therefore,

$$C(t) = \frac{2.6 \times 10^{-6} \times 10 \times 10^3}{2 \times 30}\ (1 - e^{-1.42})$$
$$= 57.4\ \text{mg/m}^3$$

which is $> 30\,\text{mg/m}^3$, the WHO standard for 1 h.

21.13.3 Multibox Models

As its name implies, multibox models are extensions of the box model. The air and pollutants are assumed to be well mixed in each box, and each box is linked. Within a box, reaction and removal processes are allowed. Multibox models give better resolution in time and space than a single box. At box junctions, the interactions between boxes are defined as in boundary-type conditions.

21.13.4 Gaussian Modelling

The reader is referred to Bibbero and Young (1974) for an introduction to this area. The Gaussian plume model for single and multiple sources are the most common air pollutant models. Equation (21.149) is the equation that describes the three-dimensional concentration field generated by point source(s) under stationary meteorological and emission conditions:

$$C(x, y, z) = \frac{Q}{2\pi\sigma_y\sigma_z u}\exp\left[-\frac{1}{2}\left(\frac{y}{\sigma_y}\right)^2\right]\left\{\exp\left[-\frac{1}{2}\left(\frac{z-H}{\sigma_z}\right)^2\right] + \exp\left[-\frac{1}{2}\left(\frac{z+H}{\sigma_z}\right)^2\right]\right\} \qquad (21.149)$$

where $C(x, y, z) = $ pollutant concentration, kg/m^3 or m^3/m^3 or ppb, etc.

$Q = $ stack emission rate, g/s

$\sigma_y, \sigma_z = $ standard deviation of the plume concentration distribution along y or z

This equation, its parts and its use are explained in Chapter 8. It is straightforward to model this equation and to link it to a graphical package to output contours of concentrations at distances from the source(s). In many cases, most interest is in concentrations at ground level and so Eq. (21.137) reduces to

$$C(x, y, z = 0) = \frac{Q}{2\pi\sigma_y\sigma_z u}\exp\left[-\frac{1}{2}\left(\frac{y}{\sigma_y}\right)^2\right]\exp\left[-\frac{1}{2}\left(\frac{H}{\sigma_z}\right)^2\right] \qquad (21.150)$$

where $H = $ height of the emission

If the emission is at ground source, as in fires, then for concentration Eq. (21.150) reduces to

$$C(x, y = 0, z = 0) = \frac{Q}{\pi\sigma_y\sigma_z u} \qquad (21.151)$$

The values of σ_y, σ_z are the Pasquill–Gifford parameters and depend on the atmospheric stability class and the distance from the source, as given in Chapter 8. The modelling of these equations, with facilities for multiple sources and complex topography, is a feature of many proprietary computer models. However, as the model is based on stationary meteorological and emission conditions, it is not usable in pollution/air quality forecasting. Dry deposition, wet deposition and chemical transformation processes can be accounted for in the Gaussian models (Zannetti, 1990) by multiplying the basic Eq. (21.149) by exponential terms such as

$$\exp\left(-\frac{t}{T}\right)$$

where
$$t = \frac{x}{u} = \text{travel time}$$
$$T = \text{time-scale}$$

According to Zanneti (1990), the time-scales for dry deposition can be expressed as

$$T_{\mathrm{d}} = \frac{\Delta h}{V_{\mathrm{d}}} \tag{21.152}$$

where
$$\Delta h = \text{vertical plume thickness}$$
$$V_{\mathrm{d}} = \text{deposition velocity}$$

Dry deposition does not occur within the immediate downwind distance of the emission and Zanneti (1990) suggests that Eq. (21.149) be applied further downwind of x_d where

$$x_d = \frac{H_{\mathrm{e}}}{2\sigma_z}$$

where
$$H_{\mathrm{e}} = \text{plume effective length}$$

Similarly, the time-scale for wet disposition is expressed as

$$T_{\mathrm{w}} = \frac{3.6 \times 10^6 P_{\mathrm{L}}}{S_{\mathrm{r}} P_{\mathrm{R}}} \tag{21.153}$$

where
$$P_{\mathrm{L}} = \text{thickness of precipitation layer } (\sim 4000 \text{ m})$$
$$S_{\mathrm{r}} = \text{scavenging ratio } (\sim 4.2 \times 10^5)$$
$$P_{\mathrm{R}} = \text{precipitation rate, (mm/h)}$$

The time-scale for chemical transformation depends on a list of parameters: whether the pollutant is organic or inorganic, the temperature, the gas solubility, etc. General air pollution modelling tends to operate on the basis of conservative pollutants, but most air pollutants are reactive. The time-scale for transformation of SO_2 is ~ 24 h.

Gaussian models can be spatially integrated to simulate the effects of line (traffic on a highway), area (industrial building complex) and volume sources, though this integration may be numerical rather than analytical. Extra diffusion caused by building wakes is included in many programmes by altering the σ_y, σ_z terms to

$$\sigma_{yw} = \left(\sigma_y^2 + \frac{k_1 A}{\pi}\right)^{1/2}$$

$$\sigma_{zw} = \left(\sigma_z^2 + \frac{k_1 A}{\pi}\right)^{1/2}$$

The Gaussian model can be modified to allow for the fanning, fumigation, looping, coning and lofting profiles as described in Chapter 8. Fanning is a slow vertical diffusion during stable conditions and is therefore represented by Eq. (21.149) with the receptor at ground level. Here $\sigma_y \gg \sigma_z$. Fumigation is the accelerated downward mixing of a plume trapped aloft by a stable inversion at height h_i. Equation (21.149) is integrated over the height Z and the plume is evenly distributed over this height h_i, (Bibbero and Young 1974).

$$C = \frac{Q}{(2\pi)^{1/2} u\, h_i \sigma_y} \exp\left[-\frac{1}{2}\left(\frac{y}{\sigma_y}\right)^2 \right] \tag{21.154}$$

21.13.5 Photochemical Models

Photochemical air pollution is formed as a result of complex interactions between sunlight, meteorology and primary emissions of nitrogen oxides and reactive hydrocarbons. Photochemical models have been used in urban areas burdened with smog. These models must account for the chemical reactions including chemical depletion and deposition. This increased complexity is addressed by examining the atmospheric diffusion equation with facilities for reactions, sources and sinks. The box models and Gaussian models, while capable of accounting for a wide range of source geometries and boundary conditions, are limited in use with respect to photochemicals and non-conservative pollutants. The atmospheric diffusion, as written by Bibbero and Young (1974), is

$$\frac{\partial C_i}{\partial t} + \nabla \cdot (\bar{u} C_i) = -\nabla \cdot \bar{q}_i + R_i + S_i \tag{21.155}$$

where
C_i = the time averaged concentration of the ith chemical species

$\bar{u}$ = the vector wind velocity

R_i = the rate of chemical production or depletion of species i

S_i = the emission rate of i species from sources

$\bar{q}$ = the mass flux of i species due to eddy diffusion

where
$i = 1, 2, \ldots, p$, the number of independent chemical species

and
$\nabla = \dfrac{\partial}{\partial x} + \dfrac{\partial}{\partial y} + \dfrac{\partial}{\partial z}$

Equation (21.155) expanded is

$$\frac{\partial C_i}{\partial t} + u\frac{\partial C_i}{\partial x} + v\frac{\partial C_i}{\partial y} + w\frac{\partial C_i}{\partial z} = \frac{\partial}{\partial x}\left(K_x \frac{\partial C_i}{\partial x}\right) + \frac{\partial}{\partial y}\left(K_y \frac{\partial C_i}{\partial y}\right) + \frac{\partial}{\partial y}\left(K_z \frac{\partial C_i}{\partial z}\right) + R(C_1, C_2, C_3, \ldots, C_i) + S_i \tag{21.156}$$

McRae *et al.* (1982) at Caltech developed a mathematical model of photochemical air pollution using Eq. (21.156) as the starting point. Their work involved the numerical solution of Eq. (21.156). To do this the 'airshed' is first divided into an array of grid cells with each cell having horizontal dimensions of Δx, Δy of a few kilometres and vertical dimensions Δz of several tens of metres. They applied their model to the Los Angeles area (400 km × 100 km grid) with subsets at 10 km × 10 km. Their model was based on the following:

- Species continuity equation incorporating:
  - Advective transport
  - Turbulent diffusion
  - Chemical reactions
  - Source emissions
  - Surface removal processes
- Three dimensions
- Lagrangian trajectory
- Vertically integrated single-cell model

Their model adequately predicts levels for ozone and nitrogen dioxide for a period over the Southern Californian region. The chemical reactions in photochemical type models are described by kinetic mechanisms. For instance, in a typical urban atmosphere there are hundreds of different hydrocarbons (McRae *et al.*, 1982). The detailed kinetic model may need to take into account that n single phase species, R_i, $i = 1, 2, \ldots, n$, simultaneously participate in m elementary reaction steps. This level of complexity is discussed by McRae *et al.* (1982).

21.13.6 The US EPAs Preferred Air Quality Models

Among the list of preferred air quality models, many as submodels within the EPA-UNAMAP system include:

1. ISC—Industrial Source Complex Model. The ISC is a steady Gaussian plume model to assess the concentrations of conservative pollutants from a wide variety of sources associated with an industrial source complex. It can account for settling and dry deposition of particulates; downwash effects; area, line and volume sources; separate point sources; and varying ground topography. It can model long-term and short-term averages. It is considered suitable for distances of less than ~ 50 kms.
2. Caline—Highway Pollutant Model. Caline is a model to determine the concentrations of conservative pollutants from highway traffic. It is a steady state Gaussian model to be used with uncomplicated topography. Any wind direction, highway orientation or receptor location is possible. Particulate concentrations can also be modelled with time averages of 1 to 24 h.
3. UAM—Urban Airshed Model. This is a complex, urban type, three-dimensional, grid type numerical solution of the diffusion equations. It can model the photochemical kinetics of reactive pollutants including O_3, NO_x and VOC. It is appropriate for single urban areas and one hour averaging times.
4. Further details of these and a multitude of other programs are listed in Zanetti (1990) and are available from NTIS, US Department of Commerce, Springfield, Virginia.

21.14 PROBLEMS

21.1   With regard to the completely stirred tank reactor (CFSTR) of Sec. 21.4.1, rewrite the mass balance equation for a reactive (non-conservative) constituent. If the reaction of material A to material X is a first-order reaction represented by $r_A = - kC_A$, show that the concentration ratio is: $C_0/C_i = 1/(1 + k\phi)$, where ϕ is the hydraulic retention time.

21.2   For Problem 21.1, write a short FORTRAN program to compute C_0/C_i and, using a plotting routeing, plot the concentration ratio C_0/C_i versus the time ratio t/ϕ for values of $k = 5, 3, 1, 0.5$ and 0.1.

21.3   Use a spreadsheet to do Problem 21.2.

21.4   A river reach is 20 km long and is divided into five subreaches each of 4 km long. The flow rate is $25 \, m^3/s$ and a pollutant is discharged at the head of the reach with a concentration of $500 \, mg/L$. Assume a first-order reaction rate with $k = 0.1 \, day^{-1}$. The cross-sectional areas of the respective reaches are $A_1 = 28 \, m^2$, $A_2 = 30 \, m^2$, $A_3 = 32 \, m^2$, $A_4 = 34 \, m^2$, $A_5 = 36 \, m^2$ and $A_6 = 38 \, m^2$. Determine the pollutant concentrations for each reach, assuming a series of CFSTRs.

21.5   Write a FORTRAN program to solve Problem 21.4. Plot the longitudinal concentration profiles for values of $k = 0.1, 0.2, 0.5, 1$ and $5 \, day^{-1}$.

21.6   Apply the theory of a PFR to Problem 21.4. What is the predicted concentration at 20 km downstream of the outfall? (Do not divide into subreaches.) Compare your answer with that of Problem 21.4.

21.7 Use the US Army Corps of Engineers Program HEC-2 to compute the backwater, water surface profile for a river flow of $200 \, m^3/s$, with only main channel flow and an assumed Manning roughness coefficient of 0.4. The starting water level downstream is unknown, so assume critical flow. The cross-sectional data are as follows:

Section number	Section chainage (km)	Bed level (m)	Section width (m)	Bank height (m)
0	0.0	200	70	3.0
1	1.5	205	68	4.0
2	3.5	211	66	4.1
3	4.8	218	64	3.7
4	6.1	220	62	3.0
5	7.1	225	60	4.1
6	9.8	228	58	4.0

21.8 Based on the equations in Sec. 21.6.1 from O'Connor (1976), prepare a plot of the relationship between river flow and the groundwater and river flow and dissolved salts concentration, using a spreadsheet representation of the equations.

21.9 Using the theory on estuarine flow in Sec. 21.8, determine the distribution of a reactant around its outfall if k is 0.1 day^{-1} and the freshwater flow velocity is 0.3 m/s and the dispersion coefficient is $10 \, m^2/s$. Plot the profile distribution of C/C_i versus distance for dispersions of 10 and $100 \, m^2/s$.

21.10 In the estuarine problem of 21.9, for $k = 0.1$ day^{-1} and dispersion coefficient $D = 200 \, m^2/s$, compute and plot the distribution of C/C_i for the freshwater velocities of 0.1 m/s and 1.0 m/s.

21.11 Use a spreadsheet to solve Problem 21.10.

21.12 For a contaminant flow in soils/aquifers, use the Ogata and Banks solution of Sec. 21.10.2 to determine the breakthrough curves if the dispersion is $4 \, cm^2/h$, $x = 20$ cm and the mean velocity is 4 cm/h. Determine the curves for a time of 1, 2 and 3 h.

21.13 For the activated sludge wastewater treatment process use either (a) SSSP (Clemson) (b) UCTOLD (South Africa) or (c) IAWPRC computer model to size an activated sludge aeration tank and clarifier to treat $90\,000 \, m^3/day$ with a BOD/SS influent of 300/350 to a standard of 20/30. State your assumptions and reaction coefficients selected.

21.14 Develop a flow chart of the equations to be solved for the unsteady case of activated sludge process. Refer to IAWPRC, UCTOLD and SSSP.

21.15 Compute the average annual concentration of SO_2 if the coal-fired power station of Example 8.11 emits $Q = 1.1$ kg/s and $C_i \cong 1$ for an average wind speed of 2 m/s. Assume the concentration is effected over an area of $1000 \, km^2$.

21.16 Write a short FORTRAN program to predict the PSI (Pollution Standard Index) if the SO_2 concentration for weeks 1 to 10 follows the relationship $C = 2 - 0.1 w_i$ where w_i is the week number. Input as data the standards as shown in Example 21.7.

21.17 For Example 21.9, if in addition to vehicular pollution a wind speed U of 4 m/s introduces a concentration of CO of $C_{in} = 20 \, mg/m^3$ and the initial concentration $C_{(a)}$ in the 'box' prior to 8 a.m. is $10 \, mg/m^3$, determine the concentration after 2 h of traffic.

21.18 If for Problem 21.17 traffic lasted only 1 h, what is the concentration after that 1 h?

21.19 Write a FORTRAN program to use the Gaussian plume equation (see Chapter 6) to determine air pollutant concentrations using the Pasquill–Gifford indices. Input the indices as data. Now use your model to determine the concentration of SO_2 emitted from the coal-fired power station of Example 8.11.

21.20 Simplify your program of Problem 21.19 to be able to compute concentrations of pollutants emitted at ground level. Use Example 8.10.

21.21 Repeat Example 21.3 if the reaches and volumes are the same but the flow rate, Q, is $2 \, m^3/s$ and the reaction rate $k = 0.5$ day^{-1}.

REFERENCES AND FURTHER READING

Air and Waste Management Association (1992) *Air Modelling*, Proceedings of 85th Annual Meeting, Kansas, 21–26 June, Air and Waste Management Association, Pittsburg, Pennsylvania.

Bear, J and A. Verruijt (1987) *Modelling Groundwater Flow and Pollution. Theory and Applications of Transport in Porous Media*, D. Reidel Publishing Co. (member of Kluwer Academic Publishing), Dordrecht, The Netherlands.

Beck, M. B. (1991) 'Principles of modelling', *Water Sci. Technol.*, **24**(6), 1–6.

Bedient, P. B. and W. C. Huber (1988) *Hydrology and Flood Plain Analysis*, Addison-Wesley Publishing, Reading, Massachusetts.

Bedient, P. B. and H. S. Rifai (1993) 'Modelling *in situ* bioremediation', in *in situ bioremediation*, National Research Council, National Academic Press.

Benarie, M.M. (1980) *Urban Air Pollution Modelling*, MIT Press, Cambridge, Massachusetts.

Bibbero, R. and J. S. Young (1974) *Systems Approach to Air Pollution Modelling*, John Wiley, New York.

Bidstrup, S. M. and L. P. Grady (1967) *Manual for SSSP. Simulation of Single Sludge Processes for Carbon Oxidation, Nitrification and Dentrification*, Clemson University, South Carolina.

Biswos, A. K. (1981) *Models for Water Quality Management*, McGraw-Hill, New York.

Bordon, S. I. (1985) *Computer Models in Environmental Planning*, Van Nostrand Reinhold, New York.

Casamitjana, X. and G. Schladow (1993) 'Vertical distribution of particles in stratified lake', *ASCE J. of Environ. Engng*, **119**(3).

Chapra, S. C. and R. P. Canale (1988) *Numerical Methods for Engineers*, McGraw-Hill, New York.

Chow, V. T. (1959) *Open Channel Hydraulics*, McGraw-Hill, New York.

Crank, J. (1956) *The Mathematics of Diffusion*, Oxford University Press.

Dingman, L. (1994) *Physical Hydrology*, Macmillan, London.

Dold, P. L., M. C. Wentzel, A. E. Billing, G. A. Ekama and GvR Marais (1991) *Activated Sludge System Simulation Program*, Water Research Commission, Pretoria, South Africa.

Falconer, R. A. (1991) 'Review of modelling flow and pollutant transport processes in hydraulic basins', *Proceedings of Water Pollution: Modelling Measuring and Prediction*, L. C. Wroebel and C. A. Brebbia, Southampton.

Fischer, H. B. (1976) 'Mixing and dispersion in estuaries', *Annual Rev. of Fluid Mechanics*, **8**.

Fischer, H. B., E. J. List, R. C. Y. Koh, J. Imberger and N. H. Brooks (1979) *Mixing in Inland and Coastal Waters*, Academic Press, New York.

Geankoplis, C. J. (1993) *Transport Processes and Unit Operations*, Prentice-Hall, Englewood Cliffs, New Jersey.

Ghadiri, H. and C. A. Rose (1992) *Modelling Chemical Transport in Soils. Natural and Applied Contaminants*, Lewis Publishers, Ann Arbor, Michigan.

Gifford, E. A. and S. R. Hanna (1974) 'Modelling urban air pollution', *Atmospheric Environment*, **8**, 870–871.

Grady, W. G. (1986) *Physics Based Modelling of Lakes, Reservoirs and Impoundments*, ASCE, New York.

Hamming, R. W. (1973) *Numerical Methods for Scientists and Engineers*, Dover Publications, London.

Harleman, D. R. F. (1964) 'The Significance of Longitudinal Dispersion in the Analysis of Pollution of Estuaries', *Proc. 2nd Int. Conf. on Water Pollution Res., Tokyo*. Pergamon Press, New York.

Havis, R. N. and D. Ostendorf (1989) 'Approximate dynamic lake phosphorus budget models'. *ASCE J. Environ. Engng.*, **115**(4), August.

Horvath, I. (1984) *Modelling in the Technology of Wastewater Treatment*, Pergamon Press, Oxford.

Hromadka, T. V., R. H. McCuen, J. J. De Vries and T. J. Durbin (1993) *Computer Methods in Environmental and Water Resource Engineering*, Lighthouse Publications, Mission Viejo, California.

IAWPRC Task Group—M. Henze, C. P. L. Grady Jr, W. Gujer, GvR. Marais and T. Matsuo (1987) *Activated Sludge Model No. 1*, IAWPRC Scientific and Technical Report 1, Pergamon Press, London.

James, A. *An Introduction to Water Quality Modelling*, John Wiley, New York.

James, A. *An Introduciton to Water Quality Modelling*, John Wiley, New York.

James, D. J. G. and J. J. McDonald (1981) *Case Studies in Mathematical Modelling*, John Wiley, New York.

Jousma, G., J. Bear, Y. Y. Haimes and F. Walter (1987) *Groundwater Contamination: Use of Models in Decision Making*. Kluwer Academic Publishing, Dordrecht, The Netherlands.

Kapur, J. N. (1988) *Mathematical Modelling*, Wiley Eastern Limited. New Delhi, India.

King, I. (1990) *RMA4—A Two Dimensional Finite Element Water Quality Model*, Department of Civil Engineering, University of California at Davis, March.

Leenderste, J. J. (1970–1978) *A Water Quality Simulation Model for Well Mixed Estuaries and Coastal Seas*, Vols 1 to 9, Rand Corporation, Santa Monica, California.

Liggett, J. A. (1994) *Fluid Mechanics*, McGraw-Hill, New York.

Linsley, K. (1976). 'Why Simulation, Hydrocomp Simulation' *Network Newsletter*, Vol. 8, No. 5. Sept. 1976. Hydrocomp Inc., Stanford CA, USA.

McCall, P. J., R. L. Swan and D. A. Laskowski (1982) *Partition Models for Equilibrium Distribution of Chemicals in Environmental Departments*, ACS Symposium Series 225, Kansas, September.

McDonald, M. G. and A. W. Harbaugh (1984) *A Modular Three Dimensional Finite Difference Ground Water Flow Model*, Open File Report 83–875, US Department of the Interior, USGS National Center, Reston, Virginia.

McRae, G. J., W. R. Goodin and J. H. Seinfeld (1982) *Mathematical Modelling of Photochemical Air Pollution*, EQL Report No. 18, California Institute of Technology, Pasadena, California.

Maki, D. P. and M. Thompson (1973) *Mathematical Models and Applications*, Prentice-Hall, Englewood Cliffs, New Jersey.

Maskimovic, C. and M. Radojkovic (1986) *Urban Drainage Modelling*, Pergamon Press, Oxford.

Marchuk, G. I. (1986) *Mathematical Models in Environmental Problems*, North-Holland, Amsterdam, The Netherlands.

Meyer, W. J. (1984) *Concepts of Mathematical Modelling*, McGraw-Hill, New York.

Morley, D. A. (1979) *Mathematical Modelling in Water and Wastewater Treatment*, Applied Science Publishers, London.

Natural Environmental Research Council (NERC) (1975) *Flood Studies Report*, Natural Environmental Research Council, London.

Nezu, I. and H. Nakagawa (1993) *Turbulence in Open Channel Flows*, IAHR Minograph, Balkema, The Netherlands.

O'Connor, D. J. (1965) 'Estuarine distribution of non-conservative substances', *J. Sanitary Engng. Div. ASCE*, **91-SAI**, 23.

O'Connor, D. J. (1976) 'The concentration of dissolved salts and river flow', *Water Resources. Res.*, **12**(2), 279–294.

O'Connor, D. J. and W. E. Dobbins (1958). 'Mechanisms of reaeration in natural streams' *Transactions of ASCE*. Vol. 123, pp. 641–666.

O'Connor, D. J. and W. E. Dobbins (1956). 'Mechanism of reaeration in natural streams' *J. Sanitory Engng. Div. ASCE*, Vol. 82, SA6, p. 1115.

O'Connor, D. J. (1960). 'Oxygen balance of an estuary'. *J. Sanitary Engng. Div. ASCE*, Vol. 86, SA3, p. 35.

Ogata, A. and R. B. Banks (1961) *A Solution of the Differential Equation of Longitudinal Dispersion in Porous Media*, Professional Paper 411–9, US Geological Survey, Washington, D.C.

Orlob, G. T. (1981) 'Models for stratified impoundments' in *Models for Water Quality Management*, A. K. Biswas (ed.), McGraw-Hill, New York.

Paterson, S. (1985) 'Equilibrium models for the initial integration of physical and chemical properties', Chapter 9 in *Environmental Exposure from Chemicals*, Vol. 1, W. B. Neely and G. E. Blau (eds), CRC Press, Boca Raton, Florida, pp. 217–229.

Patry, G. G. and D. Chapman (1989) *Dynamic Modelling and Export Systems in Wastewater Engineering*, Lewis Publishers.

Peyton, R. L. and P. R. Schroeder (1988) 'Field verification of HELP model for landfills', *ASCE J. Environ. Engng.*, **114**(2), April.

Pratt, J. W. (1974) *Statistical and Mathematical Aspects of Pollution Problems*, Marcel Dekker, New York.

Prickett, T. A. and C. G. Lonnquist (1971) *Selected Digital Computer Techniques for Groundwater Resource Evaluation*, Illinois State Water Survey Bulletin 55.

Richter, J. (1990) *Models for Processes in Soils: Programs and Exercises*, Catena Verlag, Berlin.

St. John Buckley, M. (1993) 'Computational water quality and hydrodynamic models of Cork Harbour, Unpublished BE thesis, University College Cork, Ireland.

Schoellhamer, D. H. (1988) 'Lagrangian transport modelling with QUAL II kinetics', *ASCE J. Environ. Engng.*, **114**(2), April.

Stephenson, D. and M. E. Meadows (1986) *Kinematic Hydrology and Modelling*, Elsevier, New York.

Streeter, A. W. and E. B. Phelps (1923) *A Study of the Pollution and Natural Purification of the Ohio River*, US Public Health Bulletin 146.

Szepesi, D. J. (1989) *Compendium of Regulatory Air Simulation Models*, Akademiai Kiadó, Budapest.

Tanji, K. K. (1994) 'Hydrochemical modelling', Class Notes, Land Air and Water Resources, University of California at Davis.

Tchobanoglous, G. and E. D. Schroeder (1987) *Water Quality*, Addison Wesley, Reading, Massachusetts.

Trescott, P. C., G. F. Pinder and S. P. Larsen (1976) *Finite-Difference Model for Aquifer Simulation in Two Dimensions with Results of Numerical Experiments*, USGS Technique of Water Resources Investigations, Book 7, US Geological Survey, Washington, D.C.

US Army Corps of Engineers (1986) *HEC-5Q Simulation of Flood Control and Conservation Systems—Appendix on Water Quality Analysis*, CPD-5Q, US Army Corps of Engineers, HEC Center, Davis, California, September.

US Army Corps of Engineers (1990) *HEC-1, Food Hydrograph Package. Users' Manual*, US Army Corps of Engineers, HEC Center, Davis, California.

US Army Corps of Engineers (1992) *Introduction and Application of Kinematic Wave Routing Techniques Using HEC-1*, TD-10, US Army Corps of Engineers, HEC Center, Davis, California.

US Army Corps of Engineers (1990). *HEC-Z. Water Surface Profiles, Users' Manual*, Hydrological Engineering Center, Davis, CA, USA.

US EPA (1978) *Guidelines on Air Quality Models*, US EPA EPA-450/2-78-027R, US EPA, Athens, Georgia, July.

US EPA (1987) *QUAL2—The Enhanced Stream Water Quality Models*, EPA/600/3-87/007, US EPA, Athens, Georgia.

Waite, T. D. and N. J. Freeman (1977) *Mathematics of Environment Processes*, Lexington Books, London.

Wang, H. F. and M. P. Anderson (1982) *Introduction to Groundwater Modelling. Finite Difference and Finite Element Methods*, W. H. Freeman and Company, San Francisco, California.

Zanetti, P. (1990) *Air Pollution Modelling. Theories, Computational Methods and Available Software*, Van Nostrand Reinhold, New York.

Acidification A form of pollution of surface water due to acid rain.

Acidogenesis The conversion of the lower molecular components of fatty acids, amino acids and monosaccharides to lower molecular intermediate compounds.

Acidity The capacity to neutralize a base.

Acid rain Rain with a pH less than 5.7, caused chiefly by the dissolution of sulphur dioxide and nitrogen oxides.

Actinomycetes A group of bacteria existing in the soil whose cells are arranged in fine filaments and who assist in the decomposition of organic matter.

Activated sludge A population of bacteria, protozoa and other micro-organisms, in a suspended floc. Their function aeration of the sewage thereby replacing the dissolved oxygen as rapidly as it is taken up by oxidation of the organic content of the sewage.

Adsorption A physical and chemical process by which a sorbed substance adheres to the surface of a sorbing solid.

Advection Heat transfer by a horizontal flow of air or liquid.

Aerated lagoon A lagoon, operating on the principle of the activated sludge process without sludge return, in which wastewater undergoes biological treatment.

Aeration zone The uppermost layers of soil where a high percentage of air exists within the pores.

Aerobic Requiring oxygen.

Aerosol propellant Halocarbons and chlorofluorocarbons found in aerosol cans.

Airborne vibrations Vibrations caused by sounds in the frequency range 50-100 Hz, produced by passing heavy vehicles.

Airshed A control volume of air, typically over an urban district, extending from street level to the atmospheric boundary layer.

Air screening A process whereby separation of the light and heavier components of MSW is accomplished, based on the weight difference of the components in an air stream.

Air stripping A process to remove ammonia from wastewater by passing it downwards through a packed tower, countercurrent to an induced air flow passing upwards.

Aliphatic compound A compound whose molecules contain chains of carbon atoms.

Alkalinity The capacity to neutralize acid.

Allochthonous A term given to describe micro-organisms which have been introduced into an alien environment.

947

Anthropogenic   Relating to the science of man.

Amino acids   Compounds, containing within their structure at least one amino group (NH_2) and one carboxylic group (-COOH)). Sub-units of amino acids make up proteins.

Amplitude   The magnitude of the maximum displacement of an oscillating sound wave.

Anaerobic digestion   Digestion of organic matter by anaerobic microbial action, resulting in the production of methane gas.

Anaerobiosis   The presence of life in an anaerobic environment.

Anion   A negatively charged ion.

Anion exchange capacity   The ability to exchange positively charged particles of two or more compounds, measured in milliequivalents per 100 grams.

Aquaclude   Rocks and soils which transmit water with difficulty, e.g. clay, shale and unfractured granite.

Aquifer   Rocks and soils which transmit water with ease through their pores and fractures, e.g. limestone, sandstone and fractured granite.

Aromatic compounds   Compounds which contain a six-membered ring of carbon atoms, known as a benzene ring, e.g. vitamin B.

Atmospheric boundary layer   A region near the earth in which the relative velocity increases from zero with elevation, due to the motion of the atmosphere relative to the earth's surface.

Atmospheric inversion   A term which describes the rising of warm air above cold air when two air masses of different temperatures, humidity and pressure meet.

Attached growth   Fixed microbial growth on the media surface in a trickling filter.

Attentuate and disperse landfill sites   The traditional type of landfill site from which the leachate produced seeps through soil fissures and pores into the underlying saturated zone, where it is diluted.

Attrition   A gradual abrasion.

Auto-ignition temperature   The temperature to which a reactive mixture must be raised so that, under certain conditions of pressure and after a specific period of time, that mixture will spontaneously ignite.

Autotrophic   A term applied to organisms which produce their own organic constituents from inorganic compounds utilizing energy from sunlight or oxidation processes.

Available water content   The water available in the soil for plant use, i.e. the difference between the permanent wilting point and the field capacity.

Avogadro number   The number of atoms of carbon in exactly twelve grams of the carbon-12 isotope, i.e. $6.023*10^{23}$.

Bag filter   A row of fabric bags through which a gas stream is passed for the removal of particulate matter.

Baling   A compaction process in which the volume of waste requiring to be stored or transported is reduced.

Bandspreading   The spreading of fertilizers in thick bands, 300 mm apart.

Baseflow   Water which enters streams from persistent, slowly varying sources and maintains streamflow between water-input events.

Basidia   The reproduction cell of the fungal group, *Basidiomycetes*, which contains the mushroom, puffballs and rust.

Bearing capacity   A measure of the load per unit area that a material can withstand before failure.

Benthic   Of the bottom.

Bentonite   A natural clay whose particles form a skin of very low permeability on an excavated soil face. The skin supports the soil hydrostatic pressure, thereby promoting stability.

Benzene ring   The basic structure of benzene: six carbon atoms arranged in a ring, each with a hydrogen atom attached.

Binary fission   A form of reproduction of micro-organisms in which the cell mass is passed on as two new individuals to the succeeding generation and the biomass is retained within the population.

Biochemical oxygen demand (BOD)   A measure of the amount of oxygen used by bacteria in the degradation of organic matter.

Biocide   A chemical toxic or lethal to living organisms.

Biodegradable   Capable of decomposition by living matter.

Biodiversity   The infinite range of living organisms found within an ecosystem.

Biome   A major regional ecological community, characterized by distinct life forms and principal plant or animal species.

Biosolids   The semi-solid end product of wastewater treatment.

Biosphere   The part of the earth and the atmosphere in which life can occur.

Biotope   The smallest geographical unit of the biosphere or of a habitat, characterized by its biota, that can be defined by convenient boundaries.

Bluff body   A body which is of angular, rather than aerodynamic, shape.

Bound water   A thin film of water held by adhesion to the surface of soil particles.

Buoyancy   The upward force that acts on a body which is totally immersed in a fluid and is equal to the weight of the fluid displaced by the body.

Bubbling bed   Expansion and fluidization of the sand of a fluid-bed incinerator caused by high rate passage of air.

Budding   A type of asexual reproduction in which new cells are formed as outgrowths of a parent cell.

Buffer   A solution which undergoes only a slight change in pH when H^+ or OH^- ions are added to it.

Buffer stripping   The cultivation of narrow strips of land across the slope of the land rather than parallel to it, with the aim of reducing soil erosion.

Bulking agent   A low density material, usually domestic refuse, straw or woodchips which is mixed with compost to permit air circulation while the compost is digesting..

Buttress zone   Zone of protection.

Can velocity   The velocity of the gas in the passages between the filter units in the filter house of a gas filter.

Capilliarity   The rise of water in tubes of small bore due to the adhesion between the water molecules and the surface of the vessel wall.

Capilliary suction time (CST)   A laboratory-determinable parameter defining the dewaterability rate of a wastewater sludge.

Capilliary water   Water held in soil micropores by weak capilliary forces.

Capsid   A protein coat surrounding the nucleic acid of a simple virus.

Capsule   A layer of well-organized materials lying outside and adhering to the bacterial cell wall.

Carotenes   See carotenoids.

Carotenoids   A group of plant pigments of an orange, yellow or red colour which assist in photosynthesis, absorb light in the violet–blue range but whose presence is usually masked by chlorophyll. They contain the groups carotenes and xanthophylls.

Catalyst   A substance which alters the rate of a chemical reaction but which is not used up and is unchanged chemically at the end of the reaction.

Catchment   A natural drainage basin which channels rainfall into a single outflow.

Cation   A positively charged ion.

Cation exchange capacity   The ability to exchange negatively charged particles of two or more compounds, measured in milliequivalents per 100 grams.

Cavity zone   A region within which there is little mixing of air.

Cellular storm   A rainfall event consisting of a number of discreet rainfall-bearing cells (clouds).

Cell wall   The outer supporting layer of a plant cell made by the protoplast and consisting largely of cellulose.

Chemical oxygen demand (COD)   A quick chemical test to measure the oxygen equivalent of the organic matter content of wastewater that is susceptible to oxidation by a strong chemical.

Chemisorption   Adsorption involving very strong bonding forces.

Chemotrophic A term applied to organisms which produce their own organic constituents from inorganic compounds utilizing the energy obtained from the oxidation of hydrogen sulphide.

Chloracne A widespread acneform eruption due to exposure to compounds such as dibenzofurans, dibenzodioxins and chlorodiphenyls.

Chloramine A compound composed of chlorine and ammonia.

Chlorination A disinfection technique used in water treatment, involving the addition of Cl_2 gas, chlorine dioxide, sodium hypochlorite or calcium hypochlorite.

Chlorofluorocarbons (CFCs) Compounds containing chlorine, fluorine or bromine, used as aerosol propellants, refrigerants, foaming agents and solvents and which, on decomposition by sunlight, produce oxides of chlorine responsible for the removal of ozone from the stratosphere.

Chlorophyll A photosynthetic plant pigment which absorbs red and blue light but reflects green light. The chlorophyll molecule has a square head, magnesium at the centre and a long tail.

Chlorophyll a The most important of the pigments in chlorophyll, found in all photosynthetic plants except bacteria.

Chlorophyll b One of the constituent pigments of chlorophyll, found in higher plants and green algae.

Chloroplast A chlorophyll-containing, cytoplasmic body of plant cells where photosynthesis occurs.

Chromatophore A plastic containing coloured pigment.

Cilia Whiplike structures of 5–20 μm length which allows bacterial mobility by beating with a swimming action.

Circulating bed Recovery of solids from the gas phase of a fluidized bed combustion reactor, followed by reinjection into the sand bed.

Closed loop recycling The remanufacture of a new product from a retired product of the same type.

Coagulation The water/wastewater treatment process of destabilizing colloidal particles to facilitate particle growth during flocculation by either double-layer compression, charge neturalization, interparticle bridging or precipitate enmeshment.

Coarse fish Fish, e.g. mullet, which are able to tolerate low oxygen levels.

Coepod species A phylum containing the Crustacea, i.e. small freshwater and marine animals, of which some plankton is composed.

Conductivity A measure of the ability of a solution to conduct an electrical current and is proportional to the concentration of ions in the solution.

Coliforms Non-pathogenic bacteria present in the intestines of warm-blooded animals, water and wastewater, whose numbers indicate contamination.

Colloids Very small particles in suspension, e.g. clays.

Combustion A high temperature process involving the decomposition of organics in an excess of air.

Completely mixed reactor An aeration tank in which, on entering, the influent wastewater is dispersed immediately throughout the reactor volume.

Composting The biological stabilization of wastes of biological origin under controlled conditions.

Compound A substance, the molecules of which consist of two or more different kinds of atoms.

Compression settling Particles are present in such a high concentration that they touch each other and settling can occur only by compression of the particle mass.

Condensation point That level above the earth's surface to which a parcel of unsaturated air must ascend before becoming saturated.

Constructed wetland A biological wastewater treatment system which utilizes plants for the degradation of organic waste.

Contact stabilization A wastewater treatment plant in which there are two tanks, one for the adsorption of organic matter onto the suspended solids and another for oxidation of the adsorbed materials.

Containment landfill sites The modern landfill site, in which the leachate generated is contained by bottom liners, collected and treated.

Contaminated site A landfill into which hazardous polluting waste has been dumped.

Contour ploughing Ploughing across the slope of the land rather than with it, to prevent soil erosion.

Convection   Transport of heat by vertical movement of a heated body.

Convective precipitation   When a parcel of air which is less dense than the air surrounding it, rises, it cools and loses moisture which falls to the earth as rain.

Coriolis force   A transverse force, caused by the movement of the earth about the sun, which causes a build-up in the level of water to the right of a tidal current in the northern hemisphere and to the left in the southern hemisphere.

Corona   The upper portion of a body part.

Criteria pollutant   Emissions to the urban air traditionally seen as polluting, e.g. carbon monoxide (CO_2), sulphur dioxide (SO_4).

Cryogenic   Producing very low temperatures.

Cyanide   A highly poisonous salt of hydrocyanic acid, used frequently in the extraction of gold and silver.

Cyclone separator   A means of purifying an air stream by using both gravitational and centrifugal forces.

Cytotoxic   Damaging to cell structure and cell division.

Daphnid species   The phylum containing the Branchiopodia, i.e. marine solitary, benthic animals with a shell of two valves.

Denitrification   The chemical reduction of nitrate and nitrite to gaseous forms: nitric oxide, nitrous oxide and dinitrogen: $NO_3^- \rightarrow NO_2^- \rightarrow NO \rightarrow N_2O \rightarrow N_2$

Deoxyribose nucleic acid (DNA)   A large organic molecule found in the cell nucleus, containing a phosphate group, five-carbon sugars (deoxyribose) and four different nitrogenous bases in a repetitive structure.

Detritivores   Organisms which feed on fragmented particulate organic matter.

Dewatering of sludge   A mechanical unit operation which increases the dry solids concentration of the sludge from 3.9 per cent after digestion to 25–30 per cent thereby ensuring that the sludge effectively behaves as a solid for handling purposes.

Diffusion   The process by which gases and liquids spread themselves throughout any space into which they are put.

Dilute-phase bed   The stage in fluidized bed combustion at which the bubbling of the reactor bed becomes so great that the boundary between the bed and the gas above it becomes indistinct.

Dimiclic   A term to describe a lake whose thermocline is disrupted due to two periods of free circulation or overturn per year in the lake.

Dioxin   Tetrachlorodibenzoparadioxin (TCDD), a highly toxic and environmentally persistent product of the manufacture of the pesticide 2,4,5-T.

Direct contact condenser   The vapour stream is in direct contact with the drying medium, hot air or gas. The drying medium (hot air or flue gas) leaves the drier with the water vapour coming from the sludge. The drying temperature is 80–150 °C.

Directivity index   The difference between the measured sound power level and the value based on the assumption of uniform radiation in all directions.

Discrete settling   Particles settle as independent units, without interaction of flocs.

Disinfection   The removal or inactivation of pathogenic organisms.

Dissolved oxygen   A measure of the amount of oxygen dissolved in water, expressed as either:
  (i) mg/l—which is the absolute amount of oxygen dissolved in the water mass
  (ii) as percentage saturation of the water with O_2 (% sat)

Dissolved solids   The total colloidal and suspended solids in a liquid. Any particle passing a 1.2 μm filter is defined as dissolved.

Dominant group   The highest ranking group in a social order of dominance sustained by aggressive or other behavioural patterns.

Downflow column   e.g. Sand filtration where water flows through the filter by gravity. Also used in anaerobic digestion, where the wastewater enters at the upper levels and flows down through a packed medium. Opposite to upflow column.

Downwash   The drawdown of a plume after emission due to a low pressure area downwind of the stack.

Dry absorption   A method of controlling acids in flue gas emissions, by injection of dry calcium hydroxide into the gases leaving the furnace of an incinerator.

Dry weather flow   The combination of wastewater and dry weather infiltration flowing in a sanitary sewer during times of low precipitation.

Ecology   That branch of science dealing with living organisms and their surroundings.

Ecosystem   A community of interdependent organisms together with the environment which they inhabit and with which they interact, e.g. a pond.

Ecotron   A controlled, in-house, ecological experiment to recreate a particular ecosystem.

ECU   The EU unit of monetary currency.

Effluent   The outflow from a sewage treatment plant.

Electrical double layer   A name given to the combination of the Stern layer and the diffused layer of both negatively and positively charged ions which surround it.

Electron   Negatively charged particle contained within an atom, the weight of which is about two thousand times less than that of the hydrogen atom.

Electrostatic precipitation   A means of purifying an air stream by attraction and adhesion of ionized particles to an electrode.

Element   A substance, the molecules of which have all the same atoms.

Endotoxin   An environmental toxin which attacks the endocrine glands, i.e. kidney, liver, etc.

Environmental impact assessment (EIA)   A review to which all commencing projects must be subjected with regard to their impact on the environment.

Enzyme   A substance produced by living cells which acts like a catalyst in promoting reactions within the organism.

Epilimnion   The zone in a stratified lake just below the near-surface water in which temperature decreases rapidly with depth.

Epilithic   Relating to organisms growing on rocks or on other hard, inorganic substances.

Equilibrium concentration   The concentration of the dissociated ions when the rates of both backward and forward reactions are equal.

Equilization basin   A holding tank within which variations in sewage inflow rate and liquid nutrient concentrations are averaged.

Equivalence   The number of protons donated in an acid-base reaction or the total change in valence in an oxidation-reduction reaction.

Eucaryotic cell   A cell whose nucleus is enclosed by a membrane, e.g. algae, higher plants and animals.

Euphotic zone   The surface zone of large lakes through which sufficient light penetrates for photosynthesis to occur.

European Communities (EC)   A precursor to the EU, created by the merger of the European Coal and Steel Community, the European Economic Community and the European Atomic Energy Community.

European Economic Community (EEC)   An organization established in 1957 under the Treaty of Rome to co-ordinate the activities of its member countries in the coal and steel industry, the establishment of a common market and the pooling of atomic energy resources.

European Union (EU)   A supranational organization which replaced the EEC in 1993, with the objective of peace and prosperity for its members by achieving complete economic and political union.

Eurytopic   A term describing an organism which is tolerant of a wide range of habitats.

Eurotrophic   A term describing freshwater bodies which are rich in plant nutrients and therefore highly productive.

Eutrophication   An increase in the concentration of nutrients in an aquatic ecosystem, causing:
    (i) the increased productivity of autotrophic green plants, leading to the blocking out of sunlight
    (ii) elevated temperatures within the water body
    (iii) depletion of the water's oxygen resources
    (iv) increased algal growth

(v) reduction in the level of and variety of fish and animal life

Evaporation   The changing of liquid water from rivers, lakes, bare soil and vegetative surfaces into water vapour.

Evapotranspiration   A collective term for all the processes by which water in the liquid or solid phase at or near the earth's land surfaces becomes atmospheric water vapour.

Exothermic reaction   A chemical reaction during which heat is liberated.

Extended aeration   Involves an aeration period of more than 24 hours and a high rate of return sludge to allow cell decay during the endogenous respiration phase of the growth curve.

Facultative aerobes/anaerobes   Having the ability to live either with or without oxygen.

Fickian diffusion   Molecular diffusion, governed by Fick's law, which says that the rate of flow of molecules across a unit area of a certain plane is directly proportional to the concentration gradient.

Field capacity   The amount of water which can be held in the soil against the force of gravity, i.e. water which will not drain freely out of the soil.

Filtration   A process whereby suspended and colloidal matter is removed from water and wastewater by passage through a granular medium.

Five-day biochemical oxygen demand (BOD_5)   A measure of the amount of oxygen used by bacteria to degrade organic matter in a sample of wastewater over a 5 day period at 20 °C, expressed in mg 1^{-1}.

Fixed bed   A bed of dry carbon which recovers volatile organic carbons from an air stream.

Flagellae   Whiplike structures of 100–200 μm length which allow bacterial mobility by undulating in planar or helical waves.

Flash point   The lowest temperature at which a flammable vapour/air mixture exists at the surface of a combustible liquid.

Flocculation   The water treatment process in which particle collisions are induced in order to encourage the growth of larger particles.

Flotation   A process by which suspended matter is lifted to the surface of a liquid to facilitate its removal. Frequently done by the bubbling of air through the liquid.

Flow duration curve   A means of summarizing temporal variability by averaging precipitation over a selected time period.

Flowing well   When the groundwater is flowing in a confined aquifer, it is under hydrostatic pressure. Should a standpipe be inserted into the aquifer, the water will rise in the standpipe.

Flue gases   Gas by-products of the incineration process whose temperature is a measure of incinerator efficiency and whose constituents may be polluting.

Flue gas scrubber   Equipment used for the removal of suspended particulates and acid gases from flue gas emissions.

Fluidized bed combustion   An incineration technique in which waste is destroyed by combustion on a bubbling bed.

Fluoridation   The addition of fluoride to drinking water within the limits 0.7–1.2 mg/l^{-1} to help prevent the occurrence of tooth decay.

Foaming agent   Anti-foaming chemicals added to wastewater in the aeration tank to disperse the contaminating foam caused by the action of the surface aerators and the presence of detergents in the wastewater.

Food/micro-organism ratio (F/M)   A measure of the organic loading rate of a wastewater treatment system, i.e. the ratio between the daily BOD load and the quantity of activated sludge in the system (microbes).

Fugitive emissions   Emissions from non-point sources, e.g. loading/unloading, transferring, transporting, storing and processing of materials.

Fumigating   A term describing a plume from an emission stack which is trapped by a stable inversion above the stack mouth, thereby hitting the ground level very close to the stack.

Functional group   A group of atoms on which the characteristic properties of a particular homologous series depend, e.g. the alkanes, alcohols and esters.

Furans   Compounds causing chloracne, liver damage and liver cancer. Strictly C_4H_4O, but more commonly one of a range of polychlorinated dibenzofurans that are produced as contaminants from the incomplete incineration of chlorinated hydrocarbons.

Gamete   A mature cell, involved in reproduction.

Gas chromatography   A process whereby compounds become separated by being physically carried by a gas over a liquid of a high molecular weight.

Gas flaring   The burning of recovered landfill gas from a stack under controlled conditions to help eliminate the discharge of harmful constituents to the atmosphere.

Gasification   A high temperature process involving the decomposition of organics in the absence of oxygen. Some of the energy stored as chemical energy from the organic material will be released as burnable gas.

Genotype   The genes which an organism possesses or the genetic make-up of an organism.

Groundborne vibrations   Vibrations caused by the reaction of tyres of heavy vehicles with irregularities in the road surface.

Groundwater   Water under a pressure greater than atmospheric pressure which is present in the saturated zone of the soil.

Haematins   A group of coloured plant pigments, including the red pigment, haematochrome.

Haloform   A basic organic unit of the halogen group.

Halogen   The reactive members of Group 7 of the Periodic Table, including chlorine, bromine, fluorine and iodine.

Hardness in water   The sum of the calcium and magnesium ion concentrations. A hard water will leave a scale on the inside of kettles and will form a scum rather than a lather with soap.

Hazardous waste   A substance which exhibits ignitability, reactivity, corrosivity, and/or toxicity.

Heat of adsorption   Adsorption is the process of retaining a gas molecule by either physical or chemical means onto an adsorbent (a solid, e.g. activated carbon). The heat change taking place during this process (loss of heat of gas, increase in temperature of adsorbent) is the heat of adsorption.

Heat of condensation   The quantity of heat required to bring about a phase change from a gas to a liquid.

Heat of solution   The heat change which takes place when one mole of a substance is dissolved in excess solvent.

Heavy metal   Inorganic species of large atomic weight. Usually chromium (Cr^{3+}), lead (Pb^{2+}), mercury (Hg^{2+}), zinc (Zn^{2+}), cadmium (Cd^{2+}) and barium (Ba^{2+}).

Herbivores   Animals which feed on plant material only, e.g. rabbits.

Heterotrophic   A term applied to organisms which need ready-made food materials from which to produce their own constituents and to obtain all their energy.

High rate aeration   An increased rate of aeration of MLSS in an activated sludge system requiring less activated sludge and shorter aeration periods.

Homogenous   Consisting of only one phase.

Humus   The vegetative upper layers of the soil.

Hydraulic conductivity   See permeability.

Hydrograph   A graph of stream discharge versus time.

Hydrolysis   The breakdown of high molecular compounds to low molecular compounds.

Hydrophilic   Displaying an affinity for water.

Hydrophobic   Displaying an aversion for water.

Hydrothermal vent   An opening in the earth through which heated or superheated water is ejected.

Hydraulic jump   An area of turbulence and of loss of energy associated with the transmission from shooting to tranquil flow.

Hydrological cycle   The endless recirculatory transport process of the earth's water resources, linking the atmosphere, the land and the oceans.

Hyetograph   A graph of water input to a catchment versus time.

Hypha   A tubular filament which is the basic unit structure of most fungi and some bacteria.

Hypolimnion   The lower layer of water in stratified lakes which retains the winter temperature.

Ion   Atoms or groups of atoms which have either lost or gained electrons and so have become either positively or negatively charged.

Ion exchange   Ion exchange can be illustrated by the following reaction: $Ca^{2+} + Na_2Z \rightarrow CaZ + 2Na^+$.

Incineration   Chemical oxidation at high temperatures where organic material is converted into heat energy, flue gas and slag.

Inclusion body   Organic or inorganic bodies containing glycogen, protein or lipids, present in the cytoplasm of a bacterium.

Indirect contact condenser   A condenser in which there is no direct contact between the heating medium and the vapour stream, but a partition divides the two. The water vapour is removed separately from the heating medium. The drying temperature is $100–250\,°C$.

Invertebrate   Animal without cranium and spinal column.

Irrigation requirement   The difference in volume between effective precipitation and evapotranspiration.

Isohyet   A line on a map connecting areas of equal precipitation.

Isocyanate   Derivatives of nitrogen-substituted carbamic acids, containing carbon, oxygen and an organic amine group.

Isotropic   A substance whose physical properties are the same in all directions.

Jute   Fibre from the bark of some plants, used mostly for sacking.

Karst   Landforms of chemically weathered limestone, characterized by underground channels and caverns, swallow holes and open joints.

Ketone   An organic compound containing three carbon atoms, one of which is double-bonded to an oxygen atom, the other two each attached to three hydrogen atoms e.g. acetone.

Landfill   A repository in the ground for unwanted waste.

Landfill gas   This is produced principally from the anaerobic decomposition of biodegradable organic waste and includes ammonia, carbon dioxide, carbon monoxide, hydrogen, hydrogen sulphide, methane, nitrogen and oxygen.

Landfill liner   Used to limit the movement of leachate and landfill gases from the landfill site. Can be made of natural clay material or composite geomembrane and clay materials.

Lapse rate   The rate of temperature change with height for a parcel of dry air rising adiabatically.

Latent heat of evaporation   The quantity of heat required to bring about a phase change from a liquid to a vapour.

Leachate   Liquid, composed of external rainfall, groundwater, etc. which has percolated through solid waste and has extracted both biological and chemical, dissolved or suspended materials.

Legumes   Legumes are specific plants, e.g. clover, soybeans and lupins, which carry nodules on their roots and, together with bacteria of the genus *Rhizobium* are responsible for the biological fixation of nitrogen in the soil.

Life cycle assessment (LCA)   The assessment of the steps in a product life cycle, including: raw materials acquisition, bulk material processing, materials production, manufacture, assembly, use, retirement and disposal.

Ligand   Molecules of a complexing agent in a complex ion, i.e. an aggregate formed when a metal ion bonds to several other ions or molecules which cluster around it. In the reaction $AgCl + 2NH_3 \rightarrow Ag(NH_3)_2^+ + Cl^-$, NH_3 is the ligand.

Light compensation point   The depth in a sea or lake below which, because of low light intensities, plants use up more organic matter in respiration than they make during photosynthesis.

Liquid injection incineration   A method of incineration of liquid waste by high-rate injection into a combustion chamber.

Lithotrophic   A term describing organisms which use inorganic compounds as electron donors in their energetic processes.

Littoral zone   The shore of a lake to a depth of about 10 metres.

Lofting A term describing a plume from an emission stack which remains aloft due to a stable inversion below the mouth of the stack.

Lower explosive limit (LEL) The concentration at which a gas forms an explosive mixture with air.

Lysis The rupture of cells.

Magnetic separation A process which utilizes the magnetic properties of ferrous metals to extract them from the waste stream.

Masking agent A substance which will remove an offending odour from an air stream by decomposition or conversion to an organic salt.

Materials recovery facility (MRF) Depots where reusable waste material is recovered.

Mean cell residence time (MCRT) The average time a single microbe will remain in an activated sludge system and is calculated by

$$\frac{\text{Total mass of cells}}{\text{Rate of cell wastage}}$$

Meiosis A type of cell nuclear division in which the daughter nuclei receive only half the original number of chromosomes in the parent nucleus.

Membrane process The removal of dissolved solids from water by passage through a membrane of minute pore diameter ($3*10^{-10}$ m).

Mesophilic temperatures Those temperatures in the range 10–45 °C.

Mesotrophic A term to describe waters having intermediate levels of the minerals required by green plants.

Methanogenesis Intermediate compounds are converted to the final products of methane and carbon dioxide.

Methanogenic bacteria Obligate anaerobes and methanobacteria (e.g. methanosarcina, methanobacilli) which produce methane gas from the decomposition of acids and alcohols:

$$CH_3COOH \rightarrow CH_4 + CO_2$$
$$CO_2 + H_2O + NH_3 \rightarrow NH_4HCO_3$$

Micro-organisms Neither plant nor animal, these are small, simple organisms which are either unicellular or multicellular, consisting of protozoa, algae, fungi, ricettsiae, viruses and bacteria.

Mineralization The process by which organic N is reconverted to mineral form by a wide variety of heterotrophic organisms—bacteria, fungi and actinomycetes.

Mixed liquor suspended solids (MLSS) The microbial suspension in the aeration tank containing living and dead micro-organisms and inert biodegradable matter, the operating concentration of which may vary in the range 1500 to 4000 mg/1^{-1}.

Mole A mole of any substance is that amount of it which contains the Avogadro Constant number of particles. A mole of any substance is equal to its molecular mass or atomic mass expressed in grams.

Molecular diffusion The drifting of molecules under random kinetic motion from a low concentration region to a high concentration region.

Monomiclic A term to describe a lake having a single period of free circulation or overturn per year.

Morphology The study of the form of animals and plants.

Mouse system Software for the hydrodynamic and hydrochemical design of a wastewater collection system from the Danish Hydraulic Institute.

Mutagenic Causing alteration of the genetic material of an organism, leading to inherited differences.

Mycelium A mat of branching hyphae found particularly in actinomycetes.

Negative project A commencing project for which an Environmental Impact Assessment will not be required.

Negative sorting Manual sorting of waste to remove the unwanted fractions. Is recommended only for dry waste.

Neutrality An ion or ion group which has an equal number of electrons and protons, i.e. neither a positive nor a negative overall charge.

Niche   The ecological role of a species in a community.

Nitrification   The conversion of the ammonium ion, NH_4^+, into the nitrite ion, NO_3^+. It occurs in two steps:

  (i) $2NH_4^+ + 3O_2 = 2NO_2^- + 2H_2O + 4H^+$ by the bacteria genus *Nitrosomonas*

  (ii) $2NO_2^- + O_2 = 2NO_3^-$ by the bacteria genus *Nitrobacter*.

Non-point source pollution   Pollution from diffuse and not easily identifiable sources, e.g. a field.

Normality   A concentration unit which is defined as:

$$\frac{\text{number of equivalents of solute}}{\text{number of litres of solution.}}$$

Nucleic acid   See Deoxyribose nucleic acid and Ribonucleic acid.

Nucleoid   An irregularly shaped region in the procaryotic cell containing the genetic material.

Nucleus   The well-defined region surrounded by the cell wall which contains the chromosomes, i.e. the materials of inheritance of the cell.

Nutrient removal   Tertiary treatment introduced to remove some of the trace compounds and elements contained in most domestic wastewaters, e.g. inorganic ammonia, nitrates, phosphates and sulphates, which are little affected by conventional treatment processes.

Octave band   The interval between a given frequency and twice that frequency within the audible frequency range.

Odour threshold   The minimum level or value of an odour necessary to elicit a public response.

Off-specification stock   Stock which is produced by errors in processing, leading frequently to customer dissatisfaction.

Olfactometer   An instrument used to measure relative odour levels.

Oligotrophic   A term describing freshwater bodies which are poor in plant nutrients and are therefore unproductive.

Open loop recycling   The manufacture of a new and simpler product from a retired, downcycled, more complex product.

Organochlorine   A highly persistent and carcinogenic compound containing one or more chlorine atoms found in pesticides such as DDT.

Organoleptic parameters   These are properties which can be detected by the human senses—eyes, nose and mouth.

Organotrophic   A term describing organisms which use organic compounds as electron donors in their energy-producing processes.

Overflow rate   The rate at which water is drawn off from the surface of primary and secondary clarification tanks. It is an important tank design parameter and is derived from the analysis of settling particles.

Overland flow   The lateral movement of water over the ground surface due to gravitational forces.

Oxidizing waste   A waste which loses electrons in an oxidation-reduction reaction, thereby becoming reduced itself.

Oxidation   A process in which there is loss of electrons from an element or ion.

Oxidation ditch   A ring-shaped channel, 1–1.5 m deep, around which wastewater circulates at 0.3–0.6 m/s^{-1}, is aerated by mechanical rotors and undergoes biological treatment by the resident microbes.

Oxidation number   The charge which an atom of an element has, or appears to have, in a compound, e.g. chlorine has an oxidation number of -1.

Oxygen sag curve   The longitudinal profile of oxygen concentration in a river.

Ozone   A triatomic gas, particularly prevalent in the stratosphere, formed by the reaction:

$$O + O_2 + \text{energy} \rightarrow O_3 + \text{energy.}$$

Ozone-depleting gases   Gases (e.g. oxides of nitrates, chlorine nitrate, halocarbons and water vapour) which cause destruction of the ozone layer, thereby allowing increased amounts of sunlight to reach the earth.

PAH   Polycyclic aromatic hydrocarbons.

PAN   Peroxyacetyl-nitrate.

Parasite   An organism which lives on or in another living organism of a different species (the host), from which it obtains food and protection, e.g. tapeworms, greenflies.

Parr   Fry, i.e. young fish.

Partial pressure   The pressure a single gas within a contained mixture of gases would exert if it were the only gas in the container.

Peat bed filter   Odorous compounds are removed from an air stream by passage through a bed of uncompacted peat from the upper layer of a bog.

Percentage exceedance   The inverse of the return period, i.e. a frequency (e.g. 90 per cent) of occurrence of a flood event or rainfall whose associated depth is exceeded that percent (90 per cent) of the time.

Periplasmic space   A space between the plasma membrane and the outer membrane of a biological cell, sometimes filled with a loose network of peptidoglycan.

Permanent wilting point   The water content of the soil beyond which plants cannot exert sufficient suction to extract moisture.

Permeability   The rate at which a fluid flows through a porous medium under the hydraulic head operating within the medium. Usually, the greater the porosity, the greater the permeability.

Pesticide   A material used for the mitigation, control or elimination of plants or animals detrimental to human health or economy.

pH   A measure of the acidity or basicity of a solution i.e. the negative of the logarithm of the hydrogen ion concentration.

Phagocytose   Ingestion of solid particulate matter by a cell.

Phenol   An organic compound with a hydroxyl (OH) group bonded directly to a benzene ring.

Phenotic compound   Compounds containing a phenol group, i.e. those containing hydrogen, six carbon atoms joined by alternating single and double bonds, and a hydroxyl group attached to the first carbon atom.

Photo-oxidation   Oxidation initiated by sunlight.

Photolytic process   A process in which radiant energy causes chemical decomposition.

Phototrophic   An orientation response to light.

Physico-chemical parameters   Instrumental methods of analysis such as turbidimetry, colorimetry, polarography, adsorption spectrometry, spectroscopy and nuclear radiation.

Physiology   The science of functioning of living organisms.

Phytoplankton   Plankton consisting of photosynthesising plants, such as algae.

Plasma arc destruction   A method of incineration in which very hot plasma, heated by the conversion of electrical to thermal energy, pyrolyses and atomizes waste.

Plasma membrane   A membrane of 5–10 nm containing proteins and lipids, surrounding the cytoplasm of all cells.

Plastid   A membrane-bounded body found in the cytoplasm of most plant cells. See chloroplast.

Point-source pollution   Pollution from sources which are easily identified, e.g. slurry tank.

Pollination   The transfer of pollen, usually by insects or wind, from the anther of a stamen (male part of the flower) to the stigma of a carpel (female part of the flower).

Polyampholite   A type of polymer.

Polyaromatic compounds (PAHs)   Long chain compounds, very persistent in nature, containing the hydroxyl group in a cyclic structure.

Polychlorinated biphenyls (PCBs)   A generic term covering a family of chlorinated isomers of biphenyl found in sewage outfalls and industrial and municipal solid wastes.

Polyelectrolytes Long-chain molecules used in the conditioning of sludge which, by neutralizing surface charges, cause bridging across fine particles or flocs to form larger particles.

Polymer Giant molecules built up from thousands of smaller molecules, combined together to form a repetitive structure.

Porosity The proportion of void spaces in the soil. The porosity of fine soils, e.g. clay, is low, whereas that of coarser gravelly soils is higher.

Positive project A commencing project for which an Environmental Impact Assessment is considered essential.

Precipitation The depth of rainfall plus the water equivalent of snow, sleet and hail falling during a given measurement period.

Precipitation reaction A physical or chemical reaction which results in the precipitation of one of the products formed.

Predator An organism which lives by killing and consuming other living things, e.g. sparrows, rabbits.

Primary consumer Organisms which feed directly on the primary producers. These include herbivores, detritus feeders, scavengers and decomposers (animals which feed on dead plant remains).

Primary pollutant Air pollutants which are emitted from an identifiable source, e.g. carbon monoxide from the car engine.

Primary producer Organisms which are capable of using solar energy to make food by the process of photosynthesis, e.g. plants.

Primary succession An ecological succession commencing in a habitat or on a substrate that has never previously been inhabited.

Procaryotic cell A cell which lacks a distinct nucleus, e.g. bacteria.

Profundal zone The zone of a lake lying below that depth at which the light compensation point occurs.

Project thresholds Pre-determined levels relating to project size, production or site required which, if exceeded by a commencing project, define the necessity for an Environmental Impact Assessment.

Protista The microbial kingdom to which unicellular or cell groups of eucaryotic organisms which lack true tissues, e.g. protozoa, belong.

Protozoa Aquatic, free-living and parasitic organisms, these are the most basic of all animals, with only one single cell and measuring no more than 5–1000 μm in size.

Proteins Substances containing the elements carbon, hydrogen, oxygen, nitrogen and occasionally sulphur, whose main function is cell growth and repair.

Proton Positively charged particle contained within the nucleus of an atom.

Protoplasm The living matter of a cell, comprising both the nucleus and the cytoplasm.

Pseudopodium A temporary protrusion of cytoplasm from the surface of a cell which serves for both cell motion and ingestion.

Psychrophilic temperatures Those in the range 0–10 °C.

Pyrolysis A high temperature process involving the decomposition of organics without oxygen, air or steam. Burnable gas is released as a by-product.

Quench tank A tank containing water to cool ashes and unburned materials which fall from the grates into a residue hopper during combustion.

Radiation The emission of rays and particles characteristic of radioactive substances.

Radical An element or atom, or group of these, normally forming part of a compound and remaining unaltered during that compound's ordinary chemical changes.

Radioactive cloud An artificially generated atmospheric cloud containing radioactive compounds.

Rapid gravity filter A filter used in water treatment which removes suspended solids from water by passing it through a sand bed, where the solids collect as a surface mat and in the sand interstices. The water should previously have been treated by coagulation, flocculation and sedimentation.

Recharge The process of renewing underground water by infiltration.

Redox potential The oxidizing or reducing power of a reactant.

Reduced groundwater Groundwater which contains no oxygen.

Reduction   A process in which an atom or ion gains electrons.

Regeneration rate   The rate of reproduction of bacteria, the method of which is usually by binary fission.

Regolith   All loose earth material above the underlying soil rock.

Retention time   The length of time a wastewater remains in a clarification tank, an important design parameter in the optimization of settling of suspended solids.

Return activated sludge   Settled activated sludge from the clarifier which is returned to the aeration tank to ensure an active population of microbes will be mixed with the incoming wastewater.

Return period   The long-term average of the intervals between successive exceedances of a flood magnitude.

Reverse osmosis   A membrane process in which solutions of two different concentrations are separated by a semi-permeable membrane. An applied pressure gradient greater than the osmotic pressure ensures flow from the more concentrated to the less concentrated solution.

Reynold's number   The ratio of inertial to viscous forces in a fluid, the value of which will determine whether the fluid flow is turbulent or viscous.

Ribonucleic acid (RNA)   A large organic molecule found in the cell cytoplasm containing a phosphate, five-carbon sugars (ribose) and four nitrogenous bases in a repetitive structure.

Ribosomes   Tiny bodies containing RNA responsible for protein synthesis and found in the bacterial cell cytoplasm.

Riffle area   An area of the river bed covered by grains of too large a size and weight to be carried by the water and so were dispersed.

Root zone   The soil layer from which plant roots can extract water during transpiration.

Rotary screen   See trommel.

Rotary kiln incineration   A process by which waste enters an inclined, rotating kiln, is mixed with air and combusted.

Rotating biological contactor   A form of biological treatment in which fixed media is grown on circular discs mounted on a horizontal axle. These discs are partially submerged in wastewater while the axle rotates, allowing bio-oxidation of the wastewater, using oxygen from the air.

Roughing filter   A high-rate trickling filter of depth 1–2 m, hydraulic loading 10–40 $m^3/m^2/d$ and organic loading 0.32–1.0 kgBOD/m^3/d through which wastewater may be passed prior to an activated sludge treatment.

Salmonoids   Belonging to the salmon family, Salmonidae.

Saprophytic   An organism which obtains food by absorbing dissolved organic materials resulting from organic breakdown and decay.

Schmutzdeck   The surface mat of suspended particles which forms on the surface of a slow sand filter.

Scoping   The second stage of an Environmental Impact Assessment which decides the key issues for review within the EIA.

Screening   (i) The final sorting stage necessary for high-quality compost, during which uncomposted particles such as wood, glass or plastic are removed by passing through a fine mesh. (ii) The first stage of an Environmental Impact Assessment (EIA) in which the projects to be subjected to an EIA are chosen.

Scrubbing   A process by which suspended particles and acid gases are removed from a flue gas stream, the former by absorption onto liquid droplets and the latter by diffusion into the liquid phase.

Scum well   A box used to store the scum which forms on the surface of a wastewater in a clarification tank. Scum is usually drawn off by a horizontal, slotted pipe that can be rotated by a lever or a screw.

Secondary consumers   Organisms which feed on the herbivores or other primary consumers, e.g. foxes, lions.

Secondary pollutant   Air pollutants which are formed in the atmosphere by chemical reactions, e.g. ozone.

Secondary production   The assimilation of organic matter by a primary consumer.

Secondary succession   An ecological succession that takes place in an area where a natural community existed and was removed.

Selective catalytic reduction (SCR)   A pre-combustion method of decomposition of NO_x in an air stream to nitrogen and water by injection of ammonia into the catalytic bed of a combustion chamber.

Selective non-catalytic reduction (NSCR)   A post-combustion method of decomposition of NO_x in an air stream to nitrogen and water by injection of ammonia downstream of a combustion chamber.

Semi-dry absorption process   A method of controlling acids in flue gas emissions, by injection of a calcium hydroxide and water solution into the gases leaving the furnace of an incinerator.

Sensible heat   That portion of the heat radiated by the sun which is required to heat the earth.

Sensitive area   A water body which may intermittently suffer eutrophication.

Sequencing batch reactor   A time-stepped batch process for the biological treatment of liquid hazardous waste.

Sere   A term used to describe a succession of communities, each following one after the other and finally reaching a stable state.

Settling chamber   The purification of an air stream by reducing the velocity of the gas so that the particles drop out by gravity.

Settling tank   A rectangular or circular tank in which particle velocities within the liquid are sufficiently reduced to allow the suspended material to be removed from the liquid by gravity settling.

Settling velocity   This is the velocity at which a particle will fall to the bottom of a settling tank and is equal to the surface overflow rate for a rectangular tank.

Severance   The physical or psychological division of an existing community or property due to traffic development.

Sewage   Wastewater and other refuse such as faeces, carried away in sewers.

Sewerage   System of pipes and treatment plants which collect and dispose of sewage in a town.

Sheath   A hollow, tubelike structure found in most bacteria surrounding a chain of cells.

Sisal   Strong, durable white fibre of agave used in the making of ropes.

Slag   The fused bottom ash produced by the incineration process containing incombustibles, the ash fraction of combustibles and any undestroyed pollutants.

Sloughing   A term which describes the falling off of the slime layer of micro-organisms on the media of a trickling filter due to the development of anaerobic conditions and lack of food caused by an increase in slime thickness.

Slow sand filter   A filter which removes suspended solids from raw water by passing it through a sand bed, where the solids collect as a surface mat and in the sand interstices. Filtration rates are in the order of $2–5 \, l/m^2/min$.

Sludge   The accumulation of solids resulting from chemical coagulation, flocculation and sedimentation after water or wastewater treatment.

Sludge bulking   A phenomenon caused when a large number of filamentous micro-organisms present in the mixed liquor interferes with the compaction of the floc and produces a sludge with a poor settling rate.

Sludge conditioning   Addition of chemicals, polyelectrolytes or heat treatment to improve the rate of dewatering.

Sludge dewatering   The mechanical unit operation used to reduce the moisture content of sludge to 70–75 per cent and thus ensure that the remaining sludge residue effectively behaves as a solid for handling purposes.

Sludge stabilization   The process of destroying or inactivating pathogens.

Sludge volume index (SVI)   A measure of the ability of sludge to settle, coalesce and compact on settlement.

Smog   Dense, smoky fog, the formation of which is promoted by reactions between unsaturated hydrocarbons and oxides of nitrogen in the presence of sunlight and under stable meteorological conditions.

Soilbed A large tract of land, the microbes within which remove odorous compounds from an air stream.

Soil horizons The soil layers seen in a vertical soil profile, characteristic of soil-forming processes over time.

Soil profile A vertical cut through the soil revealing a sequence of horizons.

Soil suction Water pressure within a soil which is less than atmospheric pressure.

Solid waste All the wastes arising from human and animal activities which are normally solid and are discarded as useless or unwanted.

Solubility product The equilibrium constant for a reaction involving a precipitate and its constituent ions, e.g. for magnesium sulphate $MgSO_4 = Mg^{+2} + SO_4^{-2}$, the solubility product $= [Mg^{+2}][SO_4^{-2}]$

Solute A substance dissolved in a fluid.

Solution The conversion of a solid or gas into liquid form by mixing with a solvent.

Solvent A liquid capable of or used for dissolving something.

Sound exposure level (SEL) Used to express the energy of isolated noise events, the SEL is that constant level in decibels lasting for one second which has the same amount of acoustic energy as a transient noise.

Sound intensity The average sound power per unit area normal to the direction of propagation of a sound wave.

Sound power The rate, measured in watts, at which energy is transmitted by oscillating sound waves.

Spates A river in flood.

Specific flux A measure of rate of flow per unit area.

Specific resistance to filtration (SRF) A laboratory-determinable wastewater sludge parameter.

Spectrophotometry An instrumental method of measuring the intensity of light in various parts of the spectrum.

Spore A unicellular or multicellular microscopic body involved in plant, bacteria and protozoan reproduction.

Stable inversion On moving downwards through the atmosphere, a cool parcel of air becomes heated and less dense than the surrounding air, thereby being pushed back up. It finds itself in a stable position—wanting neither to move up nor down.

Stabilization pond A quiescent, diked pond in which wastewater undergoes biological treatment under microbial action.

Stenotopic A term describing an organism which is tolerant of a narrow range of habit.

Step feed aeration An aeration system in which a portion of the sewage load is added at each of several inlets, thus spreading out the oxygen demand over the length of the tank so that oxygen utilization is more efficient.

Stern layer The innermost ion layer tightly attached to the surface of a colloidal particle.

Stratosphere The temperature-constant region of the atmosphere above the troposphere which contains oxygen and ozone.

Supernatant The partially purified water, high in suspended solids and ammoniacal nitrogen, which is released during the digestion process and whose quality and amount is dependent on the type and settling quality of the waste and on the digester system efficiency.

Surface tension The minimization of the surface of a free body of liquid due to the unbalanced attractions exerted by the liquid and the air on the liquid surface molecules.

Surge channel A channel or basin designed to take excess flow.

Suspended growth The free-moving, aerobic, microbial culture used in the biological treatment of wastewater by the activated sludge process.

Suspended solids Solids in suspension in a water or wastewater which can be removed by filtration.

Suspension A substance consisting of particles suspended in a medium.

Sustainable development Projects undertaken with care to preserve and manage resources, use genetic engineering with responsibility, search for technical alternatives to existing energy sources and control land, water and air pollution.

Synoptic storm A storm covering several hundred miles, associated with frontal activity and/or intense low pressure centres.

Synthetic organics Man-made, organic compounds, some of which are carcinogenic, including surfactants pesticides, cleaning solvents and trihalomethanes.

Tapered aeration An aeration system which equalizes the quantity of air supplied to the demand for air exerted by the micro-organisms as the liquor flows through the aeration tank.

Temperate climate A climate not exhibiting extremes of either heat or cold, e.g. the Irish climate.

Tempering A process which brings metals to the proper hardness and elasticity by heating after quenching.

Teratogenic Causing developmental malformations.

Tertiary consumer Organisms which feed on secondary consumers, e.g. man.

Thermal drying An operation which involves reducing the water content of sludge by vaporization of water to air, resulting in a granular dried product of 92–95 per cent dry solids concentration.

Thermal plume Heated effluent from an outfall, usually less dense than the receiving water, causing increased growth rates and species changes due to local warming.

Thermocline A horizontal temperature discontinuity layer in a lake in which the temperature falls by at least $1\,°C$ per metre depth.

Thermophilic temperatures Those in the range 45–75 °C.

Thickening of sludge A process which facilitates disposal of sludge by increasing the solids content to approximately 4 per cent.

Thiocyanates Pseudohalide ions, formed from the oxidation of a CN^- group, containing an SCN^- group.

Threshold of hearing $10^{-12}\,W/m^{-2}$, i.e. the lowest sound intensity to which the human ear can respond.

Toxin A specific poison of biological organic origin.

Transfer station A location to accomplish transfer of solid wastes from collection and other small vehicles to larger transport equipment, with the aim of economizing on waste transportation.

Transmissivity A measure of the rate of flow of water through a water-bearing rock.

Trapping A term describing a plume from an emission stack which is trapped by a stable inversion above the stack mouth, but due to mixing below the mouth level, hits the ground downwind rather than beside the stack.

Trickling filter A biological reactor in which micro-organisms, growing as a slime on the surface of fixed media, oxidize the colloidal and dissolved organic matter in wastewater using atmospheric oxygen which diffuses into the thin film of liquid as the wastewater is trickled over the slimed surfaces at regular intervals.

Transpiration The loss of water vapour from the surface of the plant due to evaporation.

Trommel A rotary drum screen used to separate out the various size fractions of municipal solid waste.

Trophic levels One of the hierarchical strata of the food web characterized by organisms which are the same number of steps removed from the primary producers.

Tropopause The interface between the troposphere and stratosphere.

Troposphere The layer of atmosphere extending from the earth's surface to the stratosphere.

Tubificid worm *(Potamothrix hammonensis)* A benthic worm, tolerant of low oxygen conditions, belonging to the genus *Tubifex*.

Tundra A vast, level, treeless region with an arctic climate and vegetation.

Turbidity The clarity of water, i.e. a measure of the accumulation of colloidal particles, determined by light transmission through the water.

Turbulent mixing   When a flow of liquid or air becomes large, the streamlines become irregular and parcels of the flowing substance begin to move in a highly irregular path while maintaining a nett downstream velocity.

Ultrafiltration   Filtration technique used in water treatment to separate out bacteria larger than 10^{-6}–10^{-4} mm.

Upflow column   Where the water/wastewater flows upward under pressure through a column or tank, instead of downward by gravity. When used in water filtration, it is akin to the backwashing process in rapid gravity filters. Also used in aerobic digestion of industrial wastewaters where the column is packed with aggregate or synthetic material.

Vadose zone   The entire zone of negative water pressures above the water table, the lowest portion of which is permanently saturated by capillary rise.

Valency   The number of electrons which an atom of an element must either lose or gain to achieve a noble gas structure.

Van der Waals forces   The forces which exist between the molecules in a crystal.

Vibrating screen   Used to remove undersized components of municipal solid waste.

Virion   A mature virus.

Volatile acid   A fatty acid with, at most, six carbon atoms which are water soluble.

Volatile solid   Solids, frequently organic, which volatilize at a temperature of 550 °C.

Vortex shedding   Turbulent eddies which are shed from the downstream corners of buildings in a wind/water environment.

Waste minimization   The general trend in developed countries to reduce the quantities of waste material produced.

Watershed   Line between the headstreams of river systems, dividing one catchment from another.

Water table   The level of water within the soil at which the pore water pressure is equal to the atmospheric pressure.

Wavelength   The horizontal distance between two successive wave crests or between two wave troughs or between any two corresponding points on the wave surface.

Waveperiod   The time taken for two successive wave crests or two wave troughs or any two corresponding points on successive waves to pass a fixed point in space.

Wet oxidation   A method whereby waste, either dissolved in water or emulsified, is oxidized at very high temperatures and pressures.

Windrow   A form of composting in which pretreated refuse is laid out in heaps with a triangular cross section of 2–3 m width at the base and a height of 2 m and turned at regular intervals.

Zeta potential   A measure of the charge on a colloidal particle.

Zone settling   Particles are so close together that interparticle forces hinder the settling of neighbouring particles, causing all the particles to remain in a fixed position relative to each other and to settle at a constant velocity.

ATOMIC NUMBERS AND ATOMIC MASSES

Actinium	Ac	89	227.027 8	Mercury	Hg	80	200.59
Aluminium	Al	13	26.981 54	Molybdenum	Mo	42	95.94
Americium	Am	95	(243)	Neodymium	Nd	60	144.24
Antimony	Sb	51	121.75	Neon	Ne	10	20.179
Argon	Ar	18	39.948	Neptunium	Np	93	237.048 2
Arsenic	As	33	74.921 6	Nickel	Ni	28	58.70
Astatine	At	85	(210)	Niobium	Nb	41	92.906 4
Barium	Ba	56	137.33	Nitrogen	N	7	14.006 7
Berkelium	Bk	97	(247)	Nobelium	No	102	(259)
Beryllium	Be	4	9.012 18	Osmium	Os	76	190.2
Bismuth	Bi	83	208.980 4	Oxygen	O	8	15.999 4
Boron	B	5	10.81	Palladium	Pd	46	106.4
Bromine	Br	35	79.904	Phorphorus	P	15	30.973 76
Cadmium	Cd	48	112.41	Platinum	Pt	78	195.09
Calcium	Ca	20	40.08	Plutonium	Pu	94	(244)
Californium	Cf	98	(251)	Polonium	Po	84	(209)
Carbon	C	6	12.011	Potassium	K	19	39.098 3
Cerium	Ce	58	140.12	Praseodymium	Pr	59	140.907 7
Cesium	Cs	55	132.905 4	Promethium	Pm	61	(145)
Chlorine	Cl	17	35.453	Protactinium	Pa	91	231.038 9
Chromium	Cr	24	51.996	Radium	Ra	88	226.025 4
Cobalt	Co	27	58.933 2	Radon	Rn	86	(222)
Copper	Cu	29	63.546	Rhenium	Re	75	186.207
Curium	Cm	96	(247)	Rhodium	Rh	45	102.905 5
Dysprosium	Dy	66	162.50	Rubidium	Rb	37	85.467 8
Einsteinium	Es	99	(254)	Ruthenium	Ru	44	101.07
Erbium	Er	68	167.26	Samarium	Sm	62	150.4
Europium	Eu	63	151.96	Scandium	Sc	21	44.955 9
Fermium	Fm	100	(257)	Selenium	Se	34	78.96
Fluorine	F	9	18.998 40	Silicon	Si	14	28.085 5
Francium	Fr	87	(223)	Silver	Ag	47	107.868
Gadolinium	Gd	64	157.25	Sodium	Na	11	22.989 77
Gallium	Ga	31	69.72	Strontium	Sr	38	87.62
Germanium	Ge	32	72.59	Sulphur	S	16	32.06
Gold	Au	79	196.966 5	Tantalum	Ta	73	180.947 9
Hafnium	Hf	72	178.49	Technetium	Tc	43	(97)
Helium	He	2	4.002 60	Tellurium	Te	52	127.60
Holmium	Ho	67	164.930 4	Terbium	Tb	65	158.925 4
Hydrogen	H	1	1.007 9	Thallium	Tl	81	204.37
Indium	In	49	114.82	Thorium	Th	90	232.038 1
Iodine	I	53	126.904 5	Thulium	Tm	69	168.934 2
Iridium	Ir	77	192.22	Tin	Sn	50	118.69
Iron	Fe	26	55.847	Titanium	Ti	22	47.90
Krypton	Kr	36	83.80	Tungsten	W	74	183.85
Lanthanum	La	57	138.905 5	Uranium	U	92	238.029
Lawrencium	Lr	103	(260)	Vanadium	V	23	50.941 4
Lead	Pb	82	207.2	Xenon	Xe	54	131.30
Lithium	Li	3	6.941	Ytterbium	Yb	70	173.04
Lutetium	Lu	71	174.97	Yttrium	Y	39	88.905 9
Magnesium	Mg	12	24.305	Zinc	Zn	30	65.38
Manganese	Mn	25	54.938 0	Zirconium	Zr	40	91.22
Mendelevium	Md	101	(258)				

*From *Pure Appl. Chem.*, vol. 47, p. 75 (1976). A value in parentheses is the mass number of the longest lived isotope of the element.

MECHANICAL PROPERTIES OF AIR AT STANDARD ATMOSPHERIC PRESSURE

Temperature	Density, ρ kg/m^3	Specific weight, γ N/m^3	Dynamic viscosity, μ N·s/m^2	Kinematic viscosity, ν m^2/s
$-20\,°C$	1.40	13.7	1.61×10^{-5}	1.16×10^{-5}
$-10\,°C$	1.34	13.2	1.67×10^{-5}	1.24×10^{-5}
$0\,°C$	1.29	12.7	1.72×10^{-5}	1.33×10^{-5}
$10\,°C$	1.25	12.2	1.76×10^{-5}	1.41×10^{-5}
$20\,°C$	1.20	11.8	1.81×10^{-5}	1.51×10^{-5}
$30\,°C$	1.17	11.4	1.86×10^{-5}	1.60×10^{-5}
$40\,°C$	1.13	11.1	1.91×10^{-5}	1.69×10^{-5}
$50\,°C$	1.09	10.7	1.95×10^{-5}	1.79×10^{-5}
$60\,°C$	1.06	10.4	2.00×10^{-5}	1.89×10^{-5}
$70\,°C$	1.03	10.1	2.04×10^{-5}	1.99×10^{-5}
$80\,°C$	1.00	9.81	2.09×10^{-5}	2.09×10^{-5}
$90\,°C$	0.97	9.54	2.13×10^{-5}	2.19×10^{-5}
$100\,°C$	0.95	9.28	2.17×10^{-5}	2.29×10^{-5}
$120\,°C$	0.90	8.82	2.26×10^{-5}	2.51×10^{-5}
$140\,°C$	0.85	8.38	2.34×10^{-5}	2.74×10^{-5}
$160\,°C$	0.81	7.99	2.42×10^{-5}	2.97×10^{-5}
$180\,°C$	0.78	7.65	2.50×10^{-5}	3.20×10^{-5}
$200\,°C$	0.75	7.32	2.57×10^{-5}	3.44×10^{-5}

APPROXIMATE PHYSICAL PROPERTIES OF WATER AT ATMOSPHERIC PRESSURE

Temperature	Density, ρ kg/m^3	Specific weight, γ N/m^3	Dynamic viscosity, μ N·s/m^2	Kinematic viscosity, ν m^2/s	Vapour pressure N/m^2
0 °C	1000	9810	1.79×10^{-3}	1.79×10^{-6}	611
5 °C	1000	9810	1.51×10^{-3}	1.51×10^{-6}	872
10 °C	1000	9810	1.31×10^{-3}	1.31×10^{-6}	1 230
15 °C	999	9800	1.14×10^{-3}	1.14×10^{-6}	1 700
20 °C	998	9790	1.00×10^{-3}	1.00×10^{-6}	2 340
25 °C	997	9781	8.91×10^{-4}	8.94×10^{-7}	3 170
30 °C	996	9771	7.96×10^{-4}	7.99×10^{-7}	4 250
35 °C	994	9751	7.20×10^{-4}	7.24×10^{-7}	5 630
40 °C	992	9732	6.53×10^{-4}	6.58×10^{-7}	7 380
50 °C	988	9693	5.47×10^{-4}	5.54×10^{-7}	12 300
60 °C	983	9643	4.66×10^{-4}	4.74×10^{-7}	20 000
70 °C	978	9594	4.04×10^{-4}	4.13×10^{-7}	31 200
80 °C	972	9535	3.54×10^{-4}	3.64×10^{-7}	47 400
90 °C	965	9467	3.15×10^{-4}	3.26×10^{-7}	70 100
100 °C	958	9398	2.82×10^{-4}	2.94×10^{-7}	101 300

CARBONATE EQUILIBRIUM CONSTANTS AS A FUNCTION OF TEMPERATURE

$T,$ °C	K_m	$K_1,$ mol/L	$K_2,$ mol/L	$K_{sp},$* mol^2/L^2
5		3.02×10^{-7}	2.75×10^{-11}	8.13×10^{-9}
10		3.46×10^{-7}	3.24×10^{-11}	7.08×10^{-9}
15		3.80×10^{-7}	3.72×10^{-11}	6.03×10^{-9}
20		4.17×10^{-7}	4.17×10^{-11}	5.25×10^{-9}
25	1.58×10^{-3}	4.47×10^{-7}	4.68×10^{-11}	4.57×10^{-9}
40		5.07×10^{-7}	6.03×10^{-11}	3.09×10^{-9}
60		5.07×10^{-7}	7.24×10^{-11}	1.82×10^{-9}

* Solubility product constant for $CaCO_3$.

INDEX

Absorption processes, 671–2
Absorption system, 757–61
Acid-base reactions, 86, 92
Acid deposition, 358
Acid neutralizing capacity (ANC), 66–70
Acid rain, 91, 100
Acidification, 259
  biological effects, 286
  freshwater, 284–5
  lakes, 283
  surface waters, 283–8
Acidity of water, 66–70
Acoustic far field, 411
Acoustic intensity, 410
Acoustic near field, 411
Activated carbon adsorption, 554
Activated sludge systems, 525–6, 532–9
  domestic effluent, 140
  F/M ratio, 529–31
  industrial effluent, 140–1
  modelling, 932
Adaptation, 45–6
Adiabatic lapse rate, 367–8
ADR/RID classification system, 705
Adsorption processes, 485, 761–4
  for odour control, 772
Advection, 886
Advective diffusion, 890
Aerobes, 108
Aerobic degradation, 743
Aerobic digestion, 595–7
  two-stage process, 665
Agrichemicals, 74
Agricultural pollution, 420–34
  legislation, 432–3
  potential pollutants, 431–2
  waste production, 427–8
  (*see also* Farm waste production)
Agricultural pollution control, 781–97
  design and construction of facilities, 786–7
  land application of wastes, 787–94
  non-point sources, 783, 787

obstacles to, 781–2
odorous emissions, 795
point sources, 782–7
site selection, 783–4
sizing of structural facilities, 784–6
water, 782–3
Agricultural systems:
  industrialization of, 7
  nutrient cycles in, 421–4
  potential pathways from land-based, 430
Agricultural wastes, anaerobic digestion, 563–4, 573
Air emissions:
  abatement equipment, 750–1
    design, 754–73
    selection, 752–3
  control
    flow sheet options, 751
    technologies, 753
  industrial, 750–80
Air environment:
  EU Directives, 13
  history, 4–7
  quality standards
    EU, 15–17
    US, 24
Air pollutants:
  Ambient Air Quality Standards, 338
  criteria pollutants, 338, 340–57
  inorganic substances, 364
  non-criteria pollutants, 362
  primary pollutants, 337
  secondary pollutants, 338
  units of concentration, 338–40
Air pollution, 4-7, 334–89
  episodes, 6
  EU standards, 16, 335
  index modelling, 937
  meteorology, 366–74
  modelling, deterministic models, 937–8
  prediction, vehicular, 867–8

source, 335
trace species concentrations, 336
vehicles, 866–70
Air Pollution Act 1987, 335
Air Pollution Control Act 1955, 6, 10, 334, 335
Air quality legislation, US, 21
Air quality modelling, 936–43
  box models, 938–40
  Gaussian modelling, 940–2
  multibox models, 940
  photochemical models, 942–3
  US EPAs preferred models, 943
Air quality standards, 335, 937
Air stream:
  characterization, 751–2
  measurable properties, 751
Air stripping, ammonia removal by, 555
Air toxics, 10
Airborne vibrations, 865
Airport noise, land use guidance (LUG) zones for, 415–16
Alanine deaminase (ALDA), 258
Algae, 110, 279–80, 283, 292, 314
  classification, 110
Alkalinity, 259
  of water, 66–70
Aluminium, 259
Aluminium sulphate, 458–9
Ambient lapse rate, 367–8, 371–3
Ammonia:
  in groundwater, 218
  removal by air stripping, 555
Amoebae, 111
Anabaena, 280
Anaerobes, 108
Anaerobic degradation, 743
Anaerobic digestion, agricultural wastes, 563–4, 573
Anaerobic digestion (A/D), 563–74, 595, 596, 662–5
  agricultural applications, 563–4
  agricultural wastes, 573

applications, 573–4
bacterial species, 566
basic process, 566
benefits of using, 564
biological agents, 564–5
definition, 563
design and operating data, 569
high solids, 663
hydraulic retention time (HRT),
    568–9
industrial applications, 563
industrial wastes, 574
low solids, 663
methane production, 570–3
microbiology, 564–5
municipal wastes, 574
process design, 568–70
reactor configurations, 565–8
solids retention time (SRT), 568–9
subprocesses, 564
two-stage process, 665
Animals (worms) and water quality, 108
Anion-cation balance, 61, 90
Anion exchange capacity (AEC), 426
Antoine equation, 756, 757
Anxiety caused by vehicles, 870
Aquatic pollutants, 276
Aquifers, 200–1
    confined, 212
    flow, 203
    unconfined, 213
    unconfined flow, 205
    unconfined island with recharge, 207
    water yielding capacity, 202
Areal reduction factor (ARF) for
    precipitation, 158
Aspergillus flavus, 109
Aspidisca, 111
Asteriorella, 279
Aswan High Dam, 243–5
Atlantic Ocean, 248, 249
Atmospheric boundary layer (ABL), 96
Atmospheric chemistry, 95–102
Atmospheric dispersion, 374
    and lapse rates, 371–2
    Gaussian modelling, 374–6
    terrain effects on, 373–4
Atmospheric stability, 368–9
    discontinuities, 372
Attached growth systems, 529, 540–5
Attenuation:
    atmospheric, 412
    by trees, 413
    due to distance, 411
    due to meteorological conditions, 412
    effect of ground topography, 413
    ground surface effects, 412
    reflecting surfaces and noise barriers,
        413
Audiogram, 399
Autotrophic organisms, 108
Auxins, 297
Average Chandler Score, 271
Avogadro number, 54

Bacillus, 118

Bacillus megaterium, 116
Bacteria, 115–23
    commonly encountered, 116
    growth curve, 121
    indicator, 122
    metabolic groups, 118
    morphological characteristics, 116
    of special interest to environmental
        engineering, 122–3
    pathogenic, 122–3
    physical parameters affecting, 118–19
    physiological characteristics, 116
Bacterial cells:
    basic elements, 117–18
    composition and characterization,
        118
    growth and death, 119–22
    respiration and synthesis, 527
Bacterial growth:
    in pure cultures, 526–7
    kinetics, 527–9
Bacterial rate processes, 127–9
Bantry Bay, 295
Barophil, 119
Base neutralizing capacity (BNC), 66
Batch culture, 121
Batch process, 137–8
Batch reactors, 136
Bathing waters, 15
    EU Directive, 495
Beaches, cleaning, 294
Bellmer Winkler press, 591
Belt filter press, 591–2
Benthic zone, oxygen diffusion into, 315
Best management practices (BMPs), 787
Bhopal disaster, 6
Bilham's equation, 161
Binary fission, 119
Bioaccumulation, 256–7
Biochemical models, 893–4
Biochemical oxygen demand (see BOD)
Biochemical reactions, 124–9
Bioconcentration, 256
Biodegradation:
    organic carbon, 494
    organics, 525
Biodiversity, 30–1, 33
Biogeochemical cycles, 41–3
Biological nitrogen and phosphorus
        combined removal, 549–60
Biological nutrient removal, 546
Biological oxidation, principles of, 524–6
Biological phosphorus removal, 548–9
Biological toxins, 281
Biomagnification, 256–7
Biomes, 36
BIOPLUME II model, 928
Bioremediation, contaminated sites,
        743–5
Biosolids:
    characteristics, 576–84
    disposal problems, 575
    international regulations, 574–6
    metals in, 580, 581
    processing routes, 584–606
    volumes produced, 575–6

(see also Sludge; Sewage sludge;
        Solid waste treatment)
Biosphere, 36
Biotic component, 34–5
Biotic index for rivers, 270
Biotowers, 544
Blue-green algae, 279, 281, 282
Blue Nile, 243, 244
BOD (biochemical oxygen demand), 3,
        76, 291, 292, 303–9, 493–4,
        552, 789
    and dissolved oxygen in streams,
        309–11
Boreholes, 209
Bowen ratio, 179
Briggs equation, 381–3
British Standard BS 7750, 750
Buffers, 91–2
Buoyancy, 557–8
Business impacts, 873

Cadmium, 297, 352, 700
Calcium carbonate, 86
    mass concentrations as, 55
Calibrated parameter model, 919
California Assembly Bill 939, 627
Canadian Clean Air Act 1971, 335
Capitella, 292
Carbohydrates, 74
Carbon biogeochemical cycle, 43
Carbon dioxide, 99, 360
Carbon monoxide, 99, 340–1
Carbonaceous biochemical oxygen
        demand (CBOD), 307
Carbonaceous organic matter, biological
        oxidation, 312–13
Carbonate system, 86–92
    major equilibria, 88
Carbonic acid, 91
Carboxylic acids and esters, 74
Carchesium, 111
Carcinogenicity, 363, 478
Catchment modelling, 187–9
    relationship of water to soil, 189
Catchments:
    gauged, 193–4
    ungauged, 192–3
Cation/anion exchange, 104–6
Cation exchange capacity (CEC), 105–6,
        426
Centrifuges, 593
Chandler Score biotic index scheme, 270,
        275
Chemical compounds, lethal and
        sublethal effects, 258
Chemical energy, 38
Chemical industry, 6
Chemical oxidation, 485–6
Chemical oxygen demand (COD), 76–7,
        84, 124, 127, 309, 494, 527,
        789
Chemical precipitation, water softening
        by, 482–4
Chemical reactions, 124–9
Chemistry, basic principles, 52–8
Chemoautotrophs, 108

Chemoheterotrophs, 108
Chemotrophs, 108
Chernobyl nuclear accident, 6, 8, 42
Chezy equation, 885
Chick's law, 471
Chilodonella, 111
Chimney stack:
    emissions, 371–3
    plume behaviour, source effects,
        386–7
    plume characteristics, 375
    plume dispersion, 374
    plume rise, 380–5
Chloramines, 473
Chloride in groundwater, 219
Chlorinated aromatic compounds, 75
Chlorination, 474–8
Chlorine, 556
Chlorine dioxide, 472, 556
Chlorofluorocarbons (CFCs), 8, 11,
    360–1
Chlorophyll-a, 279, 326
Chloroplasts, 110
Chromatophores, 110
Ciliophora, 111
Citrobacter fruendii, 123
Cladophora, 280
Clarification, secondary, 550–2
Clean Air Act 1956, 6, 334, 335
Clean technology, 812–15
    definition, 802
Clean Water Act, 433
Climatic changes, 44, 48, 358–62
Clostridium, 118
Clostridium perfringens, 122
Coal:
    for energy and heating, 4
    SO_2 from, 6
Coal combustion, fly ash composition,
    101
Coastal Zone Management Act (1972),
    26
Coliform count, *E. coli*, 292, 555–6
Colloids, 59
Combined heat and power (CHP),
    603
Combustion, 665
Commercial products, 32
Comminutors, 513
Communities, 35
    disturbance resistance, 232
    dynamically fragile, 232
    overall stability, 232
Complete mix reactors, 533–6
Complete or comprehensive watershed
    models, 919
Completely stirred tank reactor (CSTR)
    process, 138
Composting, 597–600, 657–62
Comprehensive Environmental
        Response, Compensation and
        Liability Act 1980
        (CERCLA), 10, 25, 627
Concentration, methods of expressing,
    53–5
Condensation, 147, 754–7

heat transfer coefficient, 755–6
operating principles, 754–5
outlet stream composition, 756–7
Conductivity factors, 72
Conductivity of water, 71–3
Conservation of momentum equation,
    883–6
Conservation quality standards, US, 26
Conservative constituents, modelling,
    909–11
Construction impacts, 874
Contact stabilization, 539
Contaminants:
    in water, 74
    transport modelling, 928–30
Contaminated sites, 11, 742–5
    bioremediation, 743–5
    thermal treatment, 743
Continuity or mass conservation
    equations, 882–3, 927
Continuous flow stirred tank reactors
    (CFSTR), 570, 573
    model, 894–8
Continuous gravity thickeners, 589
Continuous stirred tank reactor (CSTR),
    136, 574
Continuous watershed model, 918
Cooling waters, 298, 755
Copper, 700
Coriolis effect, 248
Cost-benefit analysis, 34
Crop protection chemicals, 432
Cryptosporidium, 111
Cryptosporidium cysts, 466
Cultural eutrophication, 278–80
Cultured micro-organisms, charactcristic
        growth curves, 526
CWA, wastewater standard, 24
Cyanobacteria, 280
Cyanophyta, 281
Cyclone separators, 767–9

Dams, 243–5
Darcy's law, 170, 203, 204, 206, 212, 927
DDD, 257
DDT, 257, 259
Decibels, 394, 398
Decomposer food chain, 40–1
Decomposition, 40–3
DELV equation, 885
Denitrification, 531–2, 545, 547
Denmark, environmental legislation, 10
Deoxygenation, 912–13
Diatoms, 279
Dibenzofuran (PCDF), 364
Diffusers:
    design, 558–9
    for wastewater treatment, 556–9
    plume flow, 557–8
Diffusion, 886
Diffusion equation, 377–8
Dinoflagellates, 110
Dioxins, 6, 75, 256, 364, 431, 580, 671,
    695–6
Disinfectants, 472
Disinfection, 471–8

wastewater, 555
Dissolved air flotation thickeners, 590
Dissolved oxygen (DO), 303–4
    and biochemical oxygen demand
        (BOD) in streams, 309–11
Distributed models, 919
Diversity indices, 271
Drinking water, 4, 23–4
    EU Directive, 432, 439, 442
    pathogenic bacteria, 122
    quality guidelines, 267
    standards, 443, 444
    US primary standards, 487
Dry adiabatic lapse rate (DALR), 368
Dual biological suspended and attached
        growth systems, 529
Dunne surface runoff mechanism, 182–3
Dupuit equation, 206
Dupuit parabola, 206
DYRESM computer program, 926

Ear:
    and sound perception, 398
    damage mechanisms, 399–400
Earth's atmosphere:
    average composition, 97
    chemical composition, 98
    contamination gases, 98–102
    primary pollutants, 99
    secondary pollutants, 99
    structures of, 96–8
    variable constituents, 98
    vertical temperature profile, 97
Ecological concepts, 29–51
Ecological perspective, 29–30
Ecological systems, 34–5
    disturbances and pollution, 231–62
Ecosystem processes, 36–43
Ecosystems:
    disruption, 34
    disturbance resistance, 232
    dynamics, 35
    low range, 37–8
    middle range, 38
    normal range, 37
    transportation effects, 873
    very high range, 38
EEC treaty 1958, 9
Effluent National Pollutant Discharge
        Elimination System
        (NPDES), 495
Effluent treatment, 139–42
Eichornia cassipes, 280
Ekman spiral, 248
Electrodialysis, 486, 487
Electrostatic precipitators (ESP), 753,
    770–1
El Nino Southern Oscillation (ENSO),
    251
Emission standards:
    incineration, 735
    industrial sources, 362–4
    petrol vehicles, 335
    waste incinerators, 363–4
Employment impacts, 873
Endangered Species Act 1973, 26, 44

Energy balance, 152–3
Energy budget, 152–3
  evaporation, 177, 179–80
Energy flow, 36–9, 41
  hydraulic analogy, 39
Energy pyramid, 39
Energy sources, 36
Engineering, ethics and the environment, 7–8
Entamoeba histolytica, 111
Enterobacter aerogenes, 123
Enterobacteria, 119
Enterococcus, 116
Enteromorpha, 292
Environmental Assessment Review Process (EARP), 833–4
Environmental audit, 30–3
Environmental changes:
  normal responses, 46–8
  responses over time-scales, 48–9
Environmental components, functions and attributes, 31
Environmental engineering practice, 7
Environmental ethics, 7–8, 34
Environmental gradients, 44–8
Environmental impact assessment (EIA), 831–56
  case study, 851–2
  EU Directives, 19, 834–5, 836–7
  initial environmental evaluation (IEE), 836
  legislative option (NEPA), 833
  middle ground option (EC Directive), 834–5
  noise section, 416–17
  origins, 832
  policy option (EARP), 833–4
  practice, 836–7
  procedure, 832–5
  project screening, 835–7
  road developments, 876
  stages, 831
Environmental impact of transportation, 857–77
Environmental impact statement (EIS):
  examples of, 844–7
  landfills, 684–5
  mining projects, 845–7
  motorway projects, 847–9
  multidisciplinary team management, 841–4
  noise section, 416–17
  planning, 839–40
  preparation, 838–41
  project management, 842
  public participation, 840–1
  review, 841
  scope studies, 838
  selection of study team, 842–3
  study approach, 843–4
  wastewater treatment projects, 850–1
Environmental modelling, 878–946
  fundamentals, 879
  mechanism of pollutant fate in environment, 880–6
Environmental Research Unit (ERU)

biotic index scheme, 270
EPA Act 1990, 627
Equalization, 516–18
  facilities, 517
Equipment modification to reduce waste generation, 823
Equivalent continuous level ($L_{Aeq'}$), 402–3
ERU biotic index scheme, 273
Escherichia coli, 111, 117, 119, 122, 123, 217, 292, 441
Estuaries:
  modelling water quality in, 921–5
  origin of pollutants, 289–91
  pollution, 288–91
  water quality, 288–91
EU Directives, air environment, 13
Euglena, 111
European Union (EU):
  air environment quality standards, 15–17
    comparison with US, 26
  air pollution standards, 335
  environmental legislation, 8–19
  habitat quality standards, 18–19
  noise quality standards, 18
  pollutant standards in air environment, 16, 335
  waste quality standards, 17–18
  water environment quality standards, 13–15
European Union (EU) Directives:
  bathing water, 495
  drinking water, 432, 439, 442
  environmental, 12–14
  environmental impact assessment (EIA), 19, 834–7
  hazardous waste, 745–8
  landfills, 624, 626, 676
  major accidents, 695–6
  noise pollution, 400
  noise standards, 404
  soil, 13
  wastewater treatment, 15, 493, 495, 552
  water quality standards, 13, 15, 439
Eutrophication, 110, 239–40, 278–83
  effects on man, 280–2
  reduction and control of, 282–3
Evaporation, 147, 149, 174–81
  energy budget, 177, 179–80
  factors causing, 175, 176
  from open water surface, 177
  mass transfer method for, 178–9
  water balance method for, 180
Evapotranspiration (ET), 173–81
  importance of, 178
  water balance method for, 180
Event model, 918
Exfiltration flow rates, 496–7
Extended aeration, 538

Facultative anaerobes, 108
Faecal bacteria in groundwater, 217
Faecal coliforms standards, 555–6
Faecal contamination of seawater, 292

Farm waste production, 427–33
  pollution potential, 428–9
Federal Insecticide, Fungicide and Rodenticide Act 1972 (FIFRA), 25
Federal Pollution Control Act (FPCA), 10
Federal Water Pollution Acts, 11
Federal Water Pollution Control Act (FWPCA), 10
Ferric chloride, 458
Ferrous sulphate, 458
Fickian diffusion, 890
Fickian molecular diffusion model, 891
Fickian plus advective diffusion, 890
Fick's law, 901
Field capacity, 169
Filtration, 465–70
  classification systems, 466
  gas, 764–6
  granular media, 553
  mechanisms and performance, 467
  percolating filters, 540–4
  rapid gravity, 468–70
  slow sand, 466–8
  trickling filters, 540–4
First-order reactions, 125–6
Fish habitats, 15
Fixed bed adsorption, 761–3
Flagellates, 111
Flood flows, 191–5
  determination for gauged and ungauged catchments, 192–5
  versus return periods, 195, 196
Flood peak data series, 194
Flooding:
  engineering implications, 236
  rivers and streams, 234–6
Flotation systems, 515–16
Flow duration curve, 196, 197
Flow equalization, 516
Flow modelling in groundwater, 927–8
Flue gas cleaning, 670–3, 732–5, 757
Flue gas desulphurization (FGD), 773–6
  chemistry, 774
Fluidized bed combustion, 730–1
Fluoridation, 478–9
Fly ash composition of coal combustion, 101
F/M (food to microbes) ratio, 529–31
Food, 31–2
Food chains, 38
Food source, 38
Food webs, 38
Forest and Rangeland Renewable Resources Planning Act 1974, 26
Forestry, land spreading in, 614
Fragillaria, 280
Freshwater systems, 233–45
  acidification, 284–5
  catastrophic disturbances, 235–6
  chemistry, 237–8
  consumer processes of resources, 263
  current factor, 233–7
  current, substrate and longitudinal changes, 236–7

euphotic zone, 238
eutrophic, 239
flooding disturbance, 234–6
hydrogen ion toxicity in, 259
lentic, 233
light penetration, 238
littoral zone, 238
lotic, 233
oligotrophic, 239
organic pollution in, 277–8
pollution, 276–88
profundal or deeper zone, 238
quality issues, 266
Froude number, 325, 925
Fugacity modelling, 932–6
Fugitive emissions, 777–8
  legislation, 778
Fungi, 109
Furans, 671

Gas-solid separation devices, 753
Gas-solid technologies, 753
Gas solubility in water, 81
Gasification, 665
Gaussian distribution of concentration,
  889
Gaussian modelling, 374–6, 940–2
General model, 919
Generation time, 119
Germany, environmental legislation, 10
Giardia cysts, 466
Giardia lamblia, 111
Granulated activated carbon (GAC), 554
Gravity belt thickeners, 589–90
Great Barrier Reef, 250
Great Bitter Lakes, 251
Great Plains, 254
Great Smoky Mountains, Tennessee, 47
Greenhouse effect, 43
Greenhouse gases, 8, 334, 358–62
Gringorton equation, 195
Grit channels, 513–18
Gross primary production, 37
Ground level emissions, 379–80
Groundborne vibrations, 865
Groundwater, 15, 23–4, 166, 200–13
  chemistry, 214–21
  contaminant transport modelling,
    928–30
  contamination, 328–9
  contamination indicators, 217–19
  flow, 203–8
  flow in saturated medium, 203–5
  flow modelling, 927–8
  hardness, 214
  hydrogeological pollution factors,
    220
  investigations, 208–9
  iron in, 480
  manganese in, 480
  modelling, 927–32
  pollution, 215–17
  quality, 214–15, 328–9, 446
  ridging, 184
  unconfined flow, 205
  vulnerability to pollution, 219–21

assessment, 221
mapping, 221

Habitats, 35
  quality standards, EU, 18–19
Haloforms, 75
Halogenated compounds, 74
Halogenated hydrocarbons, 75
Halophiles, 119
Hazardous air pollutants (HAPs), 84
Hazardous chemicals, 6
Hazardous substance, definition, 693
Hazardous waste:
  ADR/RID classification system, 705
  categorization, 701
  classification, 700
  collection centre, 711
  collection systems, 710
  decision tree, 698–9
  definition, 696–9
  EU Directive, 745–8
  EU hazard categories, 697
  generation, 699–700
  household, 705
  industries producing, 700, 702
  inventory, 702–3
  legislaiton, US, 22
  medical, 703–5
  preparation, 706–7
  pre-transport, 706–7
  public perception, 693
  transfer station, 710–11
  transport in bulk, 707–10
  transportation, 705–11
  ubiquitous nature of, 693
Hazardous waste treatment, 693–749
  analytical aspects, 714
  facilities for, 712–16
  handling of residues, 738–41
  incineration, 714–16, 719–38
  inorganic wastes, 716–18
  oil, 718
  pre-treatment prior to incineration,
    714–16
  reception, 712–14
  sampling procedure, 712–14
  systems for, 719–35
  tasks involved, 693–4
Health Impact Statement (HIS), landfills,
  684–5
Hearing damage, 399–400
Hearing process, 398
Heat effects, 298–9
  (see also Thermal treatment)
Heat exchanger, 755
Heat transfer coefficient, condensation,
  755–6
Heavy metal pollution, 271
Heavy metals, 256, 274, 291, 295–7, 352,
  671, 700
HEC-1 program, 905–9, 921
HEC-2 program, 903–5
HEC-5Q program, 917
HELP computer program, 931
Henry's law, 86
Henry's law constant, 82, 86

Herbicides, 297–8, 364
Heterotrophic organisms, 108, 109
Heterotrophic slimes, 109
Holland, environmental legislation, 10
Holland equation, 382
Holland formula, 380
'Honour' system, 8
Horton surface runoff mechanism, 182–3
HSPF model, 187, 920–1
Human dimension, 43–4
Humanity, influences on terrestrial
  ecosystems, 253
Hydraulic conductivity, 170, 202, 203
Hydraulic gradient, 202, 203, 205
Hydraulic models, 894
Hydraulic retention time (HRT), 138
Hydrocarbons, 74, 102, 348–50
  alicyclic, 102
  aliphatic or acyclic, 102
  aromatic, 102
Hydrodynamic models:
  introduction to, 903–9
  of rivers, 903–9
Hydrofluosilicic acid, 479
Hydrogen ion toxicity in freshwater
  ecosystems, 259
Hydrogen ions, 65
Hydrogen sulphide in groundwater, 215
Hydrogeology, 200
Hydrograph, 184–5
Hydrological cycle, 147–50
  components of, 148
  material balance, 148
Hydrology, 146–227
  instrumentation, 189–90
  low flows, 195 7
  remote sensing, 190
  urban, 198–200
Hydroxide ions, 65
Hydroxyl ions, 65
Hypochloric acid, 475

IAWPRC program, 932
Imhoff cone, 62
Impingement separators, 766–9
Incineration, 601–3, 607–8, 665–73
  air emission, 670–3
  bottom ash, 673
  chemistry, 720
  combustion process, 669
  costs, 720
  design principles, 666–8
  emission standards, 363–4, 735
  flue gas cleaning, 670–3, 732–5
  flyash, 673
  hazardous waste treatment, 714–16,
    719–38
  heat balance, 723
  heat transfer, 722–4
  historical background, 666
  monitoring of emissions, 735
  stoichiometric reactions, 722
  thermodynamics, 721
  three phase locations, 669
  types of incinerators, 724–8
  waste as fuel, 669–70

Index of Community Sensitivity, 271
Indian Ocean, 251
Individual organisms, 35
Industrial air emissions (*see* Air emissions)
Industrial emissions, 362–4
Industrial products, 32
Industrial Revolution, 4
   and public health, 5
Industrial solvents in groundwater, 216
Industrial wastes, 291
   anaerobic digestion (A/D), 574
Infiltration, 163–74
   ϕ index method, 173
   cumulative, 173
   excess overland flow model, 184
   flow rates, 496–7
   Horton's equation, 172–3
   potential, 172
   simple models, 172–4
Inorganic plant nutrients, 278
Inorganic waste treatment plant, 716–18
Integrated Solid Waste Management, 628
Integrated waste management, 688
Interception, 149
International environmental agreements, 11
Intolerance zone, 45
Inventory control, 821
Inventory management, 821
Inverse square law, 410
Ion exchange processes, 484–5
Iron:
   in groundwater, 215, 219, 480
   in surface water, 480
   in water, 480
   removal procedures
      for high concentrations, 481
      for low concentrations, 481
      for poorly buffered waters, 481
   undesirable properties, 480–1
ISO 9000, 750

Kinetics, 124–9
Kluyveromyces fragilis, 811
Kommunekemi treatment plant, 712–14

Lactococcus, 116
Lake Clear, California, 257
Lake Kariba, 244
Lake Nasser, 243
Lake Washington, 280
Lakes:
   acidification, 283
   classification, 239, 240
   diffusion coefficients, 327
   eutrophic, 240
   modelling water quality in, 925–7
   monomictic, 239
   numerical concepts, water quality in, 324
   oligotrophic, 240
   oxygen-depth profile, 242, 243
   phosphorus in, 327
   temperature-depth profile, 242, 243
   thermal stratification, 241

trophic quality, 326
   water quality, 303–33
   zonation, 238–9
Land access, changes in, 873
Land application of wastes, 787–94
   application rates, 789–94
   codes of practice, 794
   nutrients, 791–2
Land consumption, 872–3
Land reclamation, 611–14
   from the sea, 614
   landfills, 677
Land revegetation, 610–11
Land spreading, 608–10
   in forestry, 614
Land use guidance (LUG) zones for airport noise, 415–16
Land values, changes in, 873
Landfills, 614–17, 673–88
   co-disposal, 687
   control and monitoring, 686
   deposit area, 686
   depression infill, 677
   design, 684–7
   EU Directive, 624, 626, 676
   excavated cells, 677
   gas, 678–80
   geotechnical site aspects, 683–4
   ground level, 677
   internal roads, 686
   land reclamation, 677
   leachate in, 680–2
   leachate test, 740–1
   legislations, 676
   lining system, 683–4
   methane, 679–80
   operations and processes, 674, 687
   practices, 677
   receiving area, 685
   secure, 739
   stabilization (solidification), 740–1
   types, 676
   water balance, 682–3
      modelling, 931
Laplace equation, 205, 927
Lapse rates, 367–8
   and atmospheric dispersion, 371–2
'Law of the Sea', 11
Leachate in landfills, 680–2
Leachate test, landfills, 740–1
Lead, 297, 352, 700
Legislation, 7–9
   agricultural pollution, 432–3
   air quality, US, 21
   Denmark, 10
   environmental impact assessment (EIA), 833
   EU, 9–19
   fugitive emissions, 778
   Germany, 10
   Holland, 10
   landfills, 676
   levels of, 9
   solid waste, 21, 626
   US, 10–11, 19–26
   waste minimization, 814

water quality, 20
Less developing nations, 8
Lethal effects, 258
Life cycle assessment (LCA), 802–8
   critique, 808
   data collection, 806
   data treatment, 808
   definition, 803–4
   evaluation, 808
   planning, 804–6
   screening, 806
Light, 252–3
Lime stabilization, 595
Limiting factors, 46
Lipids, 74
Liquid injection incineration, 726
Litter, vehicular-borne, 870
Los Angeles Basin, 334
Loss of property, 872–3
Loudness levels, 398–9
Lough Neagh, 282
Love Canal, 10, 694–5
Low flows, 195–7
   frequency, 197
   parameters of, 196
Lumped models, 188, 919

Macropore flow, 184
Maintenance procedures, 822
Major Accident Directive, EU, 695–6
Major ionic species, in natural waters, 61
Manganese:
   in groundwater, 215, 219, 480
   in surface water, 480
   in water, 480
   removal procedures, 482
   undesirable properties, 480–1
Manning equation, 905
Marine pollution, 292–9
Marine systems, 245–51
Masking agents, 772
Mass concentrations as $CaCO_3$, 55
Mass/mass method, 53
Mass transfer method for evaporation, 178–9
Mass transport, mathematics of, 886–92
Mass/volume method, 53
Mastigophora, 111
Material balances, 130–6
   hydrological cycle, 148
   methodology, 133
Material change, 822
Material control, 821
Material loss, 822
Materials recovery facilities (MRF), 653–7
Mathematical modelling, 878
Measured parameter model, 919
Medical waste, 703–5
Medicines, 32
Mediterranean Sea, 251
Membrane filter method (MF), 123
Membrane processes, 486–7
Mercury, 259, 296, 352, 671, 700
Metals, in biosolids, 580, 581
Methane, 361

landfills, 679–80
production, 570–3
Microbial classification, 107–8
Microbial nomenclature, 108
Microbiology, 107–23
Microcystis, 281
Microfiltration, 486
Micro-organisms, 107
  and temperature, 119
    generation times, 120
Mining projects, EIS, 845
Minor ionic species in natural waters, 61
Modelling (*see* Environmental
        modelling; Mathematical
        modelling)
Molality, 54
Molarity, 54
Mole, 54
Mole fraction, 54
Molecular diffusion, 888–9
Monod model, 128, 129
Montreal Protocol, 8, 11
Most probable number method (MPN),
        123
Motor Vehicle Act 1960, 10
Motorway projects, environmental impact
        assessment (EIS), 847–9
Moulds, 109
Multiple-Use Sustained-Yield Act 1960,
        26
Municipal wastes:
  anaerobic digestion, 574
  solid waste (*see* Solid waste)
  wastewater (*see* Wastewater
        treatment)
Mushrooms, 109
Mutagenicity, 478
Mycoplasma, 117

Naegleria fowleri, 111
Naphthalenes, 364
National Ambient Air Quality Standards
        (NAAQS), 334
National Emission Standards for
        Hazardous Air Pollutants
        (NESHAP), 10
National Environmental Policy Act
        (NEPA) 1969, 25, 833
Natural environment, major components
        and subcomponents, 29
Natural parks, 7
Natural resources, 29–51
Natural waters:
  major ionic species in, 61
  minor ionic species in, 61
Net primary production, 37
Nitrate in groundwater, 217–18
Nitrate leaching, 429–30
Nitrate-nitrogen concentration in rivers,
        264
Nitric acid, 100
Nitric oxide, 100
Nitrification, 531–2, 545–7
Nitrification-denitrification combined
        system, 547–8
Nitrobacter, 119

Nitrogen:
  in water, 63–4
  removal from wastewater, 545–50
Nitrogen cycle, 421–3
Nitrogen dioxide, 100
Nitrogen oxides (NO$_x$), 100, 341–3
  abatement equipment, 777
  formation, 776–7
  methods of reduction, 777
  removal, 776–7
Nitrogenous organic matter, biological
        oxidation, 312–13
Nitrosomonas, 119
Nitrous oxide, 361
Noise, vehicular (*see* Traffic noise)
Noise barriers, 413
Noise contours, 415–16
Noise control, 417–18
Noise criteria, 401–3, 861
Noise emissions:
  environmental impact assessment
        (EIA) and statement (EIS),
        416
  vehicles, 392
Noise exposure:
  patterns, 401
  permissible levels, 404
Noise impact, statement of, 417
Noise levels, 391–2
  prediction
    outdoor, 413–15
    traffic, 863
Noise measurement, 404–8, 862–3
  case study, 407–8
  physical conditions for, 406
  recommended procedure, 406
  report, 406–8
Noise pollution, 390–419
  annoyance, 400
  EC Directive, 400
  official complaints, 391
  remedial measures, 417
  sources, 390–1
  speech interference, 400
  work interference, 400
Noise quality standards, EU, 18
Noise standards, 404
  EU Directive on, 404
Noise survey, EIS, 416
Noise thresholds, 861–2
Non-conservative constituents,
        modelling, 911–15
Non-renewable resources, 30
Non-volatile fraction, 60
Normality, 54
Nuclear power, 6
Numerical Models for Engineers, 880
Nutrient balancing, 516
Nutrient cycles in agricultural systems,
        421–4
Nutrient losses, 429–30
Nutrient recycling, 40–3
Nutrient removal:
  biological, 546
  from wastewater, 545–50
Nutrients, 252–3

in water, 63
land application, 791–2
sources of, 279

Oak Ridge formula, 380
Occult deposition, 284
Ocean dumping, 8, 607
Oceans:
  anthropogenic disturbances, 251
  circulation, 247–9
  coverage, 245
  mixing properties, 247
  natural disturbances, 250–1
  oxygen in, 247
  physicochemical properties, 246
  salinity, 246
  stratification and productivity, 246–7
  surface currents, 248–9
  temperature, 246
  waves, 249
Octave band analysis, 397
Odour abatement, 771–3
Odour control, agricultural, 795
Oil pollution, 293–5
  and public health, 295
  commercial damage, 295
  ecological impact, 294
Oil waste treatment, 718
Operational changes to reduce waste
        generation, 822
Opercularia, 111
Organic content and water quality, 304–9
Organic equalization, 516
Organic poisons, 297
Organic pollution in freshwater systems,
        277–8
Organic waste, 291
Organochlorine compounds, 274, 291,
        297
Organophosphorus compounds, 297
O'Shaughnessy Reservoir, 7
Owens Valley, 7
Oxidation agents, 93
Oxidation ditch, 537–9
Oxidation number, 92–3
Oxidation-reduction, 92–5
Oxygen:
  diffusion into benthic zone, 315
  in freshwater systems, 233
  in oceans, 247
  sag curve, 277
  transfer in water bodies, 319–22
Ozone, 351–2
  depletion, 8
  disinfection, 473, 556
Ozone-depleting gases, 11

Panama Canal, 251
Paramecium, 111
Partial contributing area model, 184
Partial models, 919
Particle size:
  classification, 61
  distribution, 636
Particulate matter, 6, 101–2, 345–8
Pasquill-Gifford curves, 378–9

Pasquill-Gifford stability classes, 368, 369
Pasquill stability classes, 378–9
PCBs (*see* Polychlorinated biphenyls)
Peat bed filters, 772–3
Percolating filters, 540–4
Periphyton, 109
Permeability in groundwater pollution, 220
Pesticides, 6, 74, 291, 297, 432
pH:
  balancing, 516
  seawater, 247
  soil, 427
  water, 65–6, 259, 285–8
pH scale, 67
Pharmaceutical industry, 6
Phosphorus:
  in lakes, 327
  in water, 65–6
  removal from wastewater, 545–50
  runoff, 431
Phosphorus cycle, 423–4
Photoautotrophic organisms, 108
Photochemical models, 942–3
Photoheterotrophic organisms, 108
Photosynthesis, 37, 252, 314–15
Phototrophic organisms, 108, 109
Physiological stress, 45
Phytoplankton, 314–15
Picket fence thickener, 587–9
Plant pigments, 74
Plants, 109
Plasma arc destruction incineration, 727–8
Plasmodium, 111
Plastids, 110
Plate filter presses, 592
Plug flow reactors (PFR), 136–7, 536–9
  model, 898–903
Pollutants:
  classification, 255
  definition, 255
  effect on physical environment, 255
  EU standards in air environment, 16
  in natural waters, 268
  mechanism of fate in environment, 880–6
  mixtures of compounds, 257–8
  movement, transport and diffusion equations, 881–6
  persistence, 256
  toxic, 256
'Polluter pays principle', 8
Polyaromates (PAHs), 671
Polychlorinated biphenyls (PCBs), 251, 256, 274, 291, 297, 298, 364, 431, 580
Polycyclic aromatic hydrocarbons (PAHs), 295
Polyelectrolytes, 459
Polynuclear aromatic hydrocarbons (PAHs), 75
Populations, 35
  models, 893–903
Porosity values by material, 202

Positive lapse rate, 96
Potable water, 14
  bacterial counts, 123
  coliform standards, 124
  (*see also* Drinking water)
Potassium in groundwater, 218
Potential evapotranspiration (PE), 180–1
Powdered activated carbon (PAC), 554
Precipitation, 79, 147–50, 154–63
  analysis, 156–63
  areal, 156–7
  areal reduction factor (ARF), 158
  depth-area-duration analysis, 157–9
  forms, 154
  frequency, 159
  intensity-duration-frequency (IDF) analysis, 160–3
  measurement, 154–6
Precipitation-evapotranspiration from continents, 174
Primary production, 37–8
Production process modification, 822
Proteins, 74
Protoporphyrin IX, 258
Protozoa, 110–11
Pseudopodia, 111
Public health and Industrial Revolution, 5
Public lands quality standards, US, 26
Public Utility Regulation and Policy Act 1981, 627
Publicly owned treatment works (POTWs), 495
Pyrolysis, 603, 665

QUAL2 model, 915–17
Quality improvement programme (QIP), 778

Radioactive ions, 256
Radioactivity, 298
Rainfall (*see* Precipitation; Runoff)
Rainfall-runoff relationships, 181–9
Raingauges, 154–5
Raw water sources from freshwaters including groundwater, 15
Reactions of variable order, 139–42
Reactor analysis, 137
Reactor configurations, 136–42
Reaeration, 913–15
  water bodies, 313
Recreation, 32
Recurrence interval, 191
Recycling/reuse, 652–7, 811, 825–6
Red Sea, 251
Redox equation, 92
Redox reactions, 93–5
Reduction (*see* Oxidation-reduction)
Refined mineral oils in groundwater, 216–17
Refuse-derived fuel (RDF), 665–6
Renewable resources, 30
Reoxygenation, 913–15
Reservoirs, water quality in, 324–8, 925–7
Resistivity tests, 208
Resource Conservation and Recovery Act

(RCRA) 1976, 25, 627
Resource Conservation and Recovery Act (RCRA) 1994, 10
Resource Recovery Act 1970, 25
Respiration, 314–15
Return period, 191
  versus flood flows, 196
Reverse flow cyclone, 767
Reverse osmosis, 486–7
Rio Declaration, 11
River Pollution Act 1876, 3
River pollution surveys, 274
Rivers:
  acidification, 283
  biotic index for, 270
  classification, 239
  hydrodynamic models of, 903–9
  nitrate-nitrogen concentration in, 264
  sewage discharge to, 3
  turbulent mixing in, 322–4
  water quality, 271, 303–33
Roads:
  developments, environmental impact assessment (EIA), 876
  environmental effects, 860
Roadway impacts of transportation schemes, 871–3
Roseires Dam, 244
Rotary drum thickeners, 590
Rotary kiln incineration, 724–6
Rotating biological contactors, 545
Runoff, 181–4
  computation, rational method, 187
  Dunne mechanism, 182–3
  flow rates, 497–8
  generation mechanisms, 185
  Horton mechanism, 182–3
  urban, 198–200
    FSR assessment, 198–200
    modelling, 199–200
    rational method, 199
Runoff coefficient for different surfaces, 199

Safe Drinking Water Acts (SDWA), 11, 19, 23
Safe minimum standards (SMS) criterion, 34
Salinity, 246
Sarcodina, 111
Saturated wedge flow, 184
Saturation excess overland flow, 184
Scrubbers, 753, 769–70
Seawater, 246
  density, 246
  faecal contamination of, 292
  pH, 247
Secondary production, 38–9
Second-order reactions, 126–7
Sediment oxygen demand (SOD), 304, 315
Seismic surveys, 208–9
Sequencing batch reactors (SBRs), 539
Settling chamber, 766
Seveso dioxin accident, 6, 695–6
Sewage, 291–3, 493

Sewage discharge to rivers, 3
Sewage discharge water, 4
Sewage sludge:
  chemical characteristics, 579
  composition, 581
  microbiological characteristics, 580–4
  microbiological guidelines, 584
  physical characteristics, 578–9
  primary and secondary, 577
  quality requirements, 582–4
  (see also Sludge)
Sewage treatment/disposal, 3–4
Sewer networks, 495
Shannon-Wiener diversity index, 271
SHE models, 187
Silica in water, 63
Single European Act, 9
Sludge:
  blanket clarifiers, 463–5
  chemical composition, 581
  conditioning, 586–7
  dewatering, 591–4
  dewatering equipment review, 594
  disposal, 606–17
  drying beds, 593
  first stage treatment, 584–95
  freezing beds, 594
  heat drying, 595
  injection, 608–10
  lagoons, 594
  melting, 604–6
  production and disposal, 574, 577
    (see also Biosolids)
  quantities from primary settling, 522–4
  second stage treatment, 595–606
  settlement parameters, 531
  stabilization, 581, 595
  thickening, 551, 587–9
Slurries, mineral/nutrient composition, 428
Slurry spreading, code of good practice, 794
Smog, 6
Smoke pollution, 6
Sodium chloride in groundwater, 215
Sodium fluoride, 479
Sodium silicofluoride, 479
Soil water content, 103, 167–9
  phases, 168–9
  zones, 168
Soilbeds, 772
Soils, 252–3
  air, 103
  bulk density, 164–5
  chemical composition, 103–4
  chemical properties, 424, 426–7
  chemistry, 102–6
  degree of saturation, 166
  elemental properties, 164–6
  EU directives, 13
  exchangeable ions, 104–6
  horizons, 166–7
  minerals, 105
  moisture characteristics and soil

  texture, 169
  moisture deficit (SMD), 171–2
  particle density, 165
  pH, 427
  physical properties, 424–6
  porosity, 166
  reaction, 427
  salinity, 106
  structure, 426
  suction, 170
  system, use of term, 424
  tension, 170
  texture, 425
    and soil moisture characteristics, 169
  use of term, 424
  volumetric water content, 166
  water movement, 169–70
Solar radiation, 176
Solid bowl centrifugation thickener, 590
Solid waste:
  composition, 629–31
  definition, 623
  environmental aspects, 627
  faecal contaminated, 625
  health hazard, 623
  industrial types, 630
  integrated management, 624, 628
  legislation, 21, 626
  municipal, 10, 623, 628–34
    biodegradability, 643
    biological properties, 642–3
    biological treatment, 657–65
    chemical properties, 638–42
    collection systems, 650–2
    energy content, 638–42
    international data, 633–4
    physical properties, 635–7
    separation, 643–5
    sorting, 644–5
    storage, 645–52
    transfer stations, 650
    transport, 645–52
  pathogens, 625
  physical composition, 630
  public health aspects, 624–6
  quality standards, US, 24–5
  sources, 629
  tonnages, 633
  treatment, 623–92
  types, 629
Solid Waste Act 1965, 25
Solid Waste Disposal Act 1965, 10, 627
Solids:
  settling (sedimentation), 314
  solubility in water, 79–80
Solubility of water, 78–86
Solvents in groundwater, 216
Sound:
  classification, 398
  frequencies, 396–7
  insulation, 417–18
  intensity, 393–4
  perception, 398
  physical properties, 392–8
Sound level exceeded for N% of the time

  in dBA (L_{AN}), 403
Sound level meters, 404–6
  calibration, 406
Sound power, 393–4, 410
Sound pressure levels, 409, 410
  attenuation, 411–13
  combining, 395–6
Sound pressures, 394
Sound propagation:
  directivity, 410–11
  geometrical spreading, 409–10
  outdoor, 409–15
Sound sources:
  directivity, 410–11
  non-point, 411
Sound waves, 392–3
South Coast Air Quality Management District, 334
Spaerotilus natans, 118
Special purpose model, 919
Species:
  depletion, 33
  extinction, 33–4
Species niche, 35
Specific yield, 202
Sporozoa, 111
Stack emissions, 371–3
Staphylococcus, 116
Stockholm Declaration on the Human Environment, 8, 9, 11
Stoichiometry, 80
  examples, 55–7
Stokes' Law, 766
Storage coefficient, 202
Storm events, 181–2
Straits of Öresund, case study, 851–2
Stratosphere, 96
Streeter-Phelps oxygen sag equation, 914
Streeter-Phelps oxygen sag model, 315–18
Streptococcus, 116
Sublethal effects, 258
Subsoils in groundwater pollution, 220
Subsurface streamflow, 184
Suctoria, 111
Suez Canal, 251
Sulphate in groundwater, 215
Sulphur dioxide, 98, 99
  emissions, 10
  from coal, 6
Sulphur oxides (SO_x), 99, 344–5
Sulphur trioxide, 99
Sulphuric acid, 91, 99, 100
Superfund Amendments and Reauthorization Act (SARA), 11, 25
Surface active agents, 75
Surface overflow rate (SOR), 550
Surface runoff (see Runoff)
Surface water:
  iron in, 480
  manganese in, 480
  quality parameters, 446
Suspended growth systems, 529
Suspended solids fraction, 59–60
Sustainability, 34

Synthetic organic chemicals (SOC), 291, 459
System simulation models, 879

TA Luft (1986) regulations, 750
2,3,7,8-TCDD, 364
Tennessee Valley, 7
Terrestrial ecosystems, 252–5
  influence of humanity, 253
  moisture, 252
  natural changes in vegetation and disturbance, 253–5
  temperature, 252
Thermal drying process, 600–1
Thermal treatment:
  contaminated sites, 743
  organic material, 665–73
Thermal waste, 291
Titration curves, 91
Tolerance, 44–6
Total coliform count, 555–6
Total organic carbon (TOC), 77
Total quality management (TQM), 782, 783
Total solids, 59
Tourism, 32
Toxic pollutants, 256
Toxic Substance Control Act (TSCA), 25
Toxic substances in water, 441
Toxicity, environmental factors affecting, 258–60
Traffic, environmental effects, 860
Traffic accidents, 871
Traffic congestion, 871
Traffic generated by development proposals, 874–5
Traffic-induced vibrations, 865
Traffic noise, 861, 862
  amelioration techniques, 863
  methods for reducing impact, 863
  prediction, 863
Transformation processes in water bodies, 311–18
Transmissivity, 202, 210–11
  feedback, 184
Transpiration, 149
Transport processes in water bodies, 318–19
Transportation:
  and development, 858
  environmental impact of, 857–77
  matrix of environmental impact, 859–60
  planning, 858
Transportation schemes:
  roadway impacts of, 871–3
  safety and capacity impacts, 870–1
Trent Biotic Index, 269, 272
Trent River, 269
Trickling filter, 540–4
  domestic waste, 141–2
  industrial effluent, 142
Trihalomethanes (THMs), 75
Trophic levels, 38, 39
Tropopause, 96
Troposphere, 96

Turbulent diffusion, 890–2
Turbulent mixing in rivers, 322–4

UCTOLD program, 932
Ultrafiltration, 486
Ultraviolet radiation, 474, 556
Ulva, 292
Unit hydrograph, 185–6, 188
United Nations Environment Programme (UNEP), 9
United States:
  air environment quality standards, 24
  air quality legislation, 21
  Clean Air Acts, 10
  conservation quality standards, 26
  environmental legislation, 8–11, 19–26
    comparison with EU, 26
  hazardous waste legislation, 22
  public lands quality standards, 26
  solid waste quality standards, 24–5
  solid wastes legislation, 21
  water environment quality standards, 23–4
  water quality legislation, 20
Unsaturated/saturated zone, 167
USEPA, 19, 20

Vaporization/volatilization process, 84–6
Vegetation:
  natural changes in, 253–5
  secondary succession, 253
  semi-natural, 253
Vehicles:
  air pollution, 866–70
    amelioration, 869
    prediction, 867–8
  anxiety caused by, 870
  exhaust emissions, 866
  physical damage caused by, 870
Vehicular-borne litter, 870
Vehicular environmental impacts, 861–71
Vibration:
  airborne, 865
  amelioration, 866
  groundborne, 865
  measurement, 865
  prediction, 865
  threshold values, 865
  traffic-induced, 865
Vibrio cholera, 118
Vibrios, 116
Viruses, 113–15
  classification, 114
  diseases caused by, 115
  in groundwater, 217
  life cycle, 114
  structural arrangement of capsid, 113–14
Volatile fraction, 60
Volatile organic carbon (VOC), air stripping, 525
Volatile organic compounds (VOCs), 84–6, 304, 348, 756, 761
  fugitive sources, 777
  removal efficiencies, 753

Volatilization, 84–6
Volume reduction to reduce waste generation, 823
Vorticella, 111

Waste concentration, 825
Waste disposal, 811
Waste elimination:
  communications: line organization, 816
  environmental guidelines, 816
  policy statement, 816
Waste management, integrated, 688
Waste minimization, 652, 801–30
  benefits of, 812–15
  case study, 826–8
  community response, 814
  elements of, 809–12
  follow-up assessment, 819–20
  implementation of programmes, 819
  legislation, 814
  management involvement, 816
  monitoring, 819–20
  policy continuation, 820
  priorities, 815
  programme elements, 815–20
  selection of targets, 817–18
  setting of goals, 817
  source segregation, 823
  technical evaluation, 818–19
  techniques, 693
Waste quality standards, EU, 17–18
Waste recovery, 825–6
Waste reduction techniques, 820–6
Waste treatment, 811
Wastes, land application of, 787–94
Wastewater:
  CWA standard, 24
  disinfection, 555
  environment, history, 3
  EU Directive on, 15, 493, 495, 552
  nutrient removal from, 545–50
  US quality standards, 24
Wastewater flow rates:
  domestic, 495–7
  homes and commercial premises, 496
  industrial, 501–5
Wastewater treatment, 493–562
  advanced processes, 552–5
  chemical pre-treatment of industrial effluents, 512
  chemical treatment, 554
  chemically enhanced primary sedimentation, 521–2
  clarification, 519
  classification of wastewater prarameters, 499
  diffusers for, 556–9
  domestic wastewater, 493
  domestic wastewater characteristics, 498–501
  equalization, 516–18
  EU Directives, 15, 493, 495, 552
  flow chart, 510
  industrial, 493

industrial wastewater characteristics, 501–2
key effluent parameters, 493
mass balance for industrial wastewater, 502–5
modelling, 932
municipal wastewater, 291, 493, 495, 508
municipal wastewaters organics, 501
objectives, 494–5
pollutant characteristics of raw domestic wastewater, 500
pre-treatment, 511–13
primary treatment, 519–24
processes, 508–10
projects, environmental impact statement (EIS), 850–1
screenings, 512–13
secondary clarification, 550–2
secondary treatment, 524–32
sedimentation, 519
settling, 519
urban, 493, 494
(see also Anaerobic digestion (A/D); Biosolids)
Water, 57–95
  acidity of, 66–70
  alkalinity of, 66–70
  chemistry, 237–8
  colour, 58
  conductivity of, 71–3
  contaminants in, 74
  density, 239–41
  EU directives, 13, 439
  gross chemical properties, inorganic, 65–73
  hardness/softness, 70–1, 259, 441, 482
  impurities, 443
  inorganic chemical properties, 60–73
  iron in, 480
  manganese in, 480
  nitrogen in, 63–4
  nutrients in, 63
  odour, 58–9
  organic chemical properties, 73–8
  organic compounds in, 74
  organic content, 75–8
  pH of, 65–6, 259, 285–8
  phosphorus in, 65–6
  physical properties, 58–60
  raw water analysis, 445–6
  regulation, 241–5
  silica in, 63
  softened, 441
  solids content, 59–60
  solubility of, 78–86
  synthetic substances in, 74
  taste, 59
  temperature, 59
  thermal stratification, 239–41
  toxic substances in, 441

turbidity, 58
Water balance, 150–1
  and land use, 151
  global annual, 149
  method for evaporation, 180
  method for evapotranspiration, 180
  of continents, 152
  world, 149
Water bodies:
  oxygen transfer in, 319–22
  reaeration of, 313
  transformation processes in, 311–18
  transport processes in, 318–19
Water consumption:
  average daily per capita (ADPC), 438
  examples, 438
  projected, 439
Water environment, history, 3
Water environment quality standards:
  European Union, 13–15
  US, 23–4
Water environments, E_h/pH, 95
Water pollution, ecological perspectives, 263–302
Water pollution control, agricultural, 782–3
Water Pollution Control Act Amendments 1972, 433
Water quality, 264–76
  and organic content, 304–9
  and water sources, 443–6
  assessment, 267–76
  biological assessment techniques, 268
  chemical assessment techniques, 268
  description of, 267
  ecological assessment methods, 268
  effects of acidification, 285–8
  estuarine, 288–91
  faecal contamination, 441
  in estuaries, modelling, 921–5
  in lakes
    modelling, 925–7
    numerical concepts, 324
  in reservoirs, 324–8
    modelling, 925–7
  in rivers and lakes, 303–33
  indicator organisms, 441
  legislation, 20
  microbiological parameters, 441
  modelling in riverine systems, 909–17
  monitoring, 267, 441
  odour, 440
  organoleptic parameters, 440
  physicochemical parameters, 440
  raw water, 445
  rivers, 271
  standards and parameters, 265–7
  substances undesirable in excessive amounts, 440
  taste, 440
  toxic substances parameters, 441

turbidity, 440
Water quality parameters, 439
  testing, 439
Water quality standards, 439–43
  EU directives, 439
Water resources, 263–4
Water softening by chemical precipitation, 482–4
Water sources, and water quality, 443–6
Water supplies, requirements, 438–9
Water treatment, 437–92
  advanced processes, 479–87
  aeration, 450
  chemical pre-treatment, 450–1
  classes, 446–8
  coagulants, 458–60
  coagulation, 457–63
  flocculation, 460–1
  objectives, 437–8
  pre-treatment, 449–51
  processes, 446–8
  recommended treatment for specific impurities, 448
  screening, 449
  sedimentation, 451–7
    discrete particles type I, 451–6
    flocculent particles type II, 456–7
    of flocculent particles, 463
  selection of processes, 448
  storage: equalization and neutralization, 449
  turbidity, 459
  (see also Disinfection; Chlorination; Filtration; Fluoridation)
Water vapour (H_2O), 361–2
Watershed models:
  classification, 918–20
  generalities, 920–5
Waves, 249
  (see also Sound waves)
Weibull formula, 195
Weight/weight method, 53
Well tests, 209–11
Wells, steady state hydraulics, 212–13
Wet air oxidation, 603–4, 728–30
Wilting point, 169
Wind direction, variation with altitude, 370–1
Wind speed, variation with altitude, 369–70
Wind systems and surface currents, 248
Wind tunnels, 374
World Conservation Strategy, 34
World Health Organization (WHO), 9
Worms and water quality, 108

Yeasts, 109

Zero-order reactions, 125
Zinc, 700